An Introduction to Genetic Analysis

A Note from the Publisher

Since the publication of the first edition, *An Introduction to Genetic Analysis* has set the standard for genetics textbooks with its analytical approach and balanced coverage of classical and modern genetics.

This vision of how students best learn genetics was established by David Suzuki and Anthony Griffiths in 1976. Just as the field of genetics has evolved, so have the text and the contributions of the authors. Richard Lewontin and Jeffrey Miller joined the author team in subsequent editions, offering their expertise in population genetics and molecular genetics, respectively. In this edition we welcome William Gelbart, a leading developmental geneticist from Harvard University.

Our expanded author team reflects the positive changes of the fifth edition and the ongoing evolution of the authors' roles, as indicated by the new order of names on the cover.

An Introduction to Genetic Analysis

Fifth Edition

Anthony J. F. Griffiths
University of British Columbia

Jeffrey H. Miller
University of California, Los Angeles

David T. Suzuki
University of British Columbia

Richard C. Lewontin
Harvard University

William M. Gelbart
Harvard University

W. H. Freeman and Company / New York

The Cover

One of the most exciting observations in modern genetics is that quite divergent animals use the same types of master regulatory genes to develop body plans that are very different in structure. The Hox (homeobox) gene cluster in vertebrates and the HOM-C (homeotic complex) cluster in insects are evolutionarily conserved sets of genes that control the individual identities of the body segments from head to tail. By classical genetic and recombinant DNA–based "gene knockout" experiments, it has been possible to demonstrate that the Hox and HOM-C genes function in parallel ways: when either a Hox or HOM-C gene is inactivated, a posteriorly located segment is converted to a more anterior identity. Normally, the fruit fly *Drosophila* has only one pair of wings, coming from the second thoracic segment. When one of the HOM-C genes is inactivated, the third thoracic segment also generates a pair of wings, producing a four-winged fly. In the mouse, the lumbar vertebrae do not have ribs. When one of the Hox genes is inactivated, the first lumbar vertebra (circled in the skeleton) is transformed into a thoracic vertebra bearing ribs. (See Chapter 22 for details.) Cover illustration by Neil Brennan, copyright 1993.

Library of Congress Cataloging-in-Publication Data

An Introduction to genetic analysis. — 5th ed. / Anthony J.F. Griffiths . . . [et al.]
p. cm.
Includes bibliographical references and index.
ISBN 0-7167-2285-2
1. Genetics. 2. Genetics—Methodology. I. Griffiths, Anthony J. F.
[DNLM: 1. Genetics. QH 430 I608]
QH430.I62 1993
575.1—dc20
DNLM/DLC
for Library of Congress 92-48977
CIP

Printed in the United States of America

2 3 4 5 6 7 8 9 0 RRD 9 9 8 7 6 5 4 3

Contents in Brief

www.whfreeman.com/genetics.

Contents

Preface

Our Traditional Strengths

Genetics has become an indispensable component of almost all research in modern biology and medicine. This position of prominence has been achieved through the powerful merger of classical and molecular approaches. Each analytical approach has its unique strengths: classical genetics is unparalleled in its ability to explore uncharted biological terrain; molecular genetics is equally unparalleled in its ability to unravel cellular mechanisms. It would be unthinkable to teach one without the other, and each is given due prominence in this book. Armed with both approaches, students are able to form an integrated view of genetic principles.

Figure 2-13

A Balanced Approach

The partnership of classical and molecular genetics has always presented a teaching dilemma: which of the two partners should the student be introduced to first, the classical or the molecular? We believe that students begin much as biologists did at the turn of the century, asking general questions about the laws governing heredity. Therefore the first half of the book introduces the intellectual framework of classical eukaryotic genetics in more or less historical sequence. Although molecular information is provided where appropriate, it is not emphasized in this half. Having acquired the classical framework, the student then proceeds to the second half of the book, which hangs molecular genetics onto this framework. The coverage of genetic mutation is a case in point. In Chapter 7 the student is treated to the classical principles of gene mutation, while Chapter 18 expands this knowledge to include molecular aspects. Because progression from general to specific is a natural one, this approach makes sense not only in research, but in teaching about research.

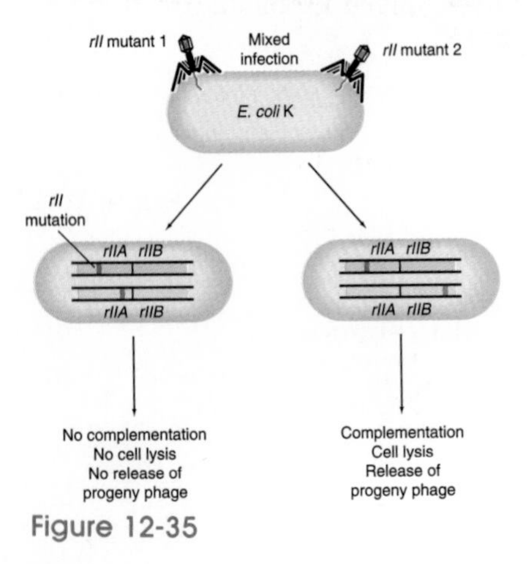

Figure 12-35

Focus on Genetic Analysis

True to its title, the theme of this book is genetic analysis. This theme emphasizes our belief that the best way to understand genetics is by learning how genetic inference is made. On almost every page we recreate the landmark experiments in genetics and have the students analyze the data and draw conclusions as if they had done the research themselves. This proactive process teaches students how to think like scientists. The modes of inference and the techniques of analysis are the keys to future exploration.

Similarly, quantitative analysis is central to the book because many of the new ideas in genetics, from the original conception of the gene to modern techniques such as RFLP mapping, are based on quantitative analysis. The problems at the ends of the chapters provide students with the opportunity to test their understanding in quantitative analyses that effectively simulate the act of doing genetics.

Study Aids

Strengths of the previous editions have been retained and reinforced. The **Key Concepts** at the chapter openings give an overview of the main principles to be covered in the chapter, stated in simple prose without genetic terminology. These provide a strong pedagogic direction for the reader. Throughout the chapters, boxed **Messages** provide convenient milestones at which the reader can pause and contemplate the material just presented. **Chapter Summaries** provide a short distillation of the chapter material and an immediate reinforcement of the concepts. All these items are useful in text review, especially for exam study.

The problems at the end of each chapter are prefaced by **Solved Problems** that illustrate the ways that geneticists apply principles to experimental data. Research in science education has shown that this application of principles is a process that professionals find second nature, whereas students find it a major stumbling block. The Solved Problems demonstrate this process and prepare the students for solving problems on their own. Finally, the **Problems** themselves continue to be one of the strengths of the book. The problems are generally arranged to start from simple and proceed to the more difficult. Particularly challenging problems are marked with an asterisk. All problems have been classroom tested. Answers to selected problems are found at the back of the book, and the full set of solutions is in the *Student Companion,* all prepared by Diane Lavett (Emory University).

New Features of the Fifth Edition

REVISED AND UPDATED

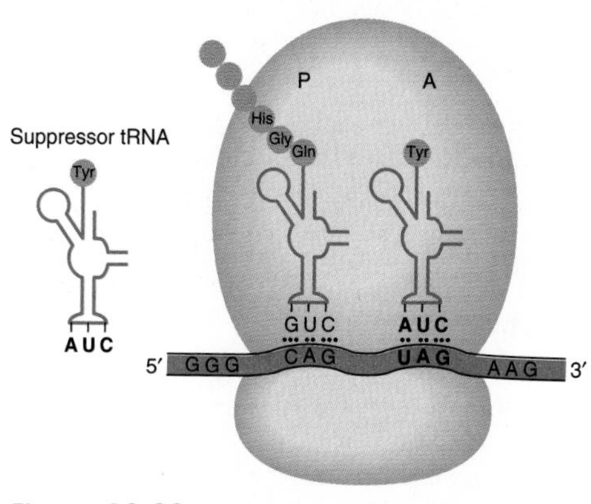

Figure 13-29

Chapter 1 now contains a summary of basic cell biological concepts that relate to genetics. This summary reinforces material from previous courses and provides a framework for the new material of this course. Chapter 1 also discusses genetics and human affairs. Chapters 11 and 13 now contain greatly expanded and updated versions of DNA replication and protein synthesis. The chapter on DNA manipulation in the fourth edition has now been expanded and split into two more manageable chapters, 14 and 15. These chapters contain new material on reverse genetics, the polymerase chain reaction, and cosmid vectors. Chapter 16, "The Structure and Function of Eukaryotic Chromosomes," has been completely reorganized

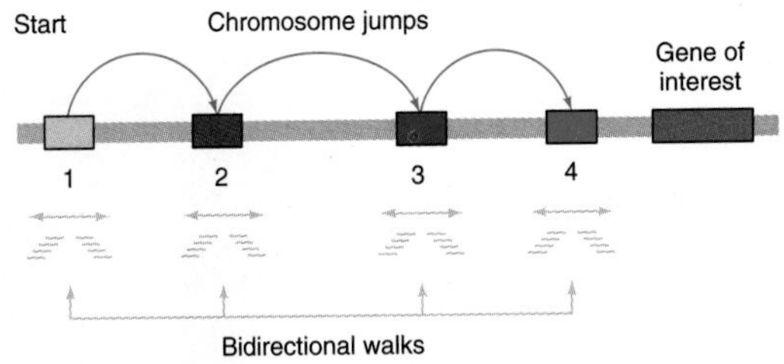

Start
Chromosome jumps
Gene of interest
1
2
3
4
Bidirectional walks

Figure 15-28

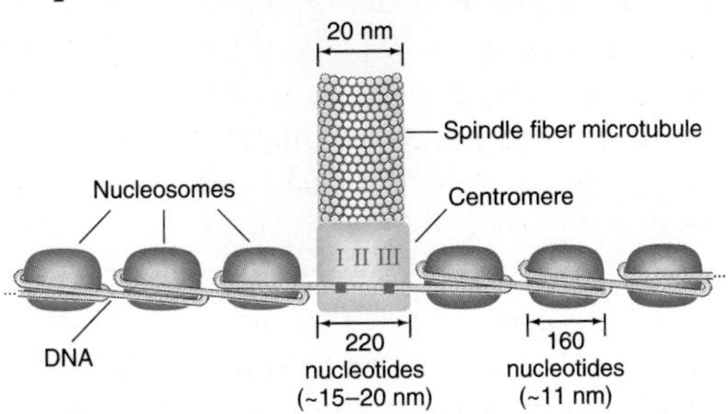

Figure 16-15

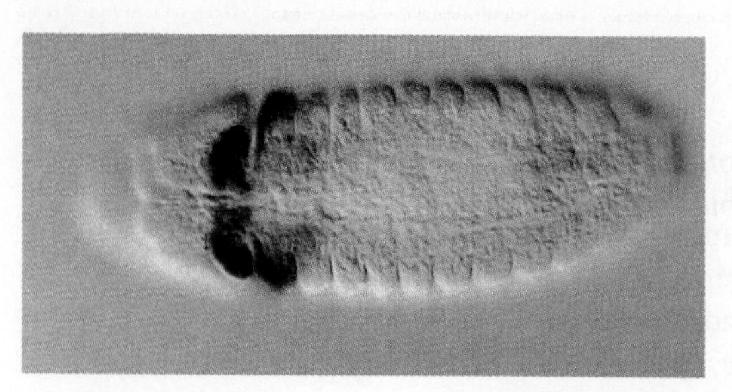

Figure 22-28

along new principles that better reflect research progress. The new version shows what researchers have learned about the organization of chromatin and the modulation of chromatin condensation. A new section describes the genetic organization of the DNA of chromosomes, with details on various kinds of repetitive DNA, the recent sequencing of the entire yeast chromosome 3, and on DNA fingerprinting.

The two chapters on developmental genetics (22 and 23) have been completely rewritten. Chapter 22 focuses on the establishment of cell fate, or how a cell acquires its unique identity, and how this identity is integrated into the requirements to construct the entire body plan. In Chapter 23 we examine differential gene activity in normal development and how aberrations in genes controlling differentiation can result in cancer.

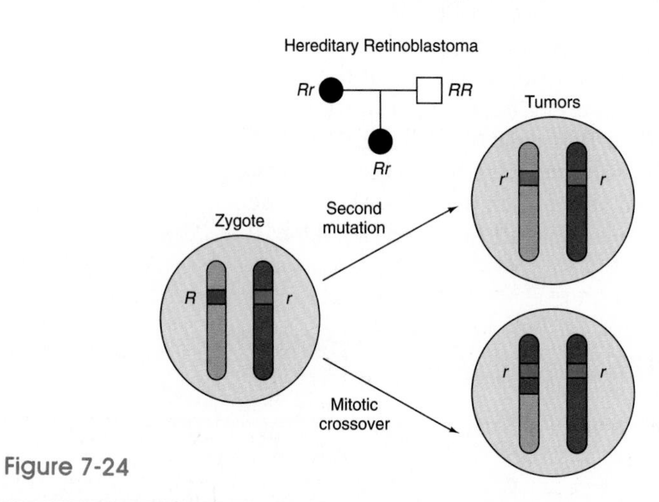

Figure 7-24

■ MORE HUMAN EXAMPLES

In response to comments from users and reviewers, we have added many more examples and problems from human genetics. The additions are most conspicuous in Chapters 2, 3, 7, 8, 9, and 21. The genetic aspects of human cancer have also been emphasized in several different chapters, including 7, 8, 9, 18, and 23. However, in all chapters, a broad representation of other organisms is retained.

■ MORE INVITING PRESENTATION

Thorough re-editing of the entire text has made this edition clearer, more accessible, and easier to read.

Figure 4-11

■ NEW ILLUSTRATIONS

The most prominent new feature is the use of full-color illustrations. Every attempt has been made to use colors pedagogically in the diagrams, and the photographs show readers the colorful array of organisms and processes that are the source of a great deal of inspiration for genetic research. The color photographs have been fully integrated as part of the linear text sequence, thus enriching the learning experience.

CHAPTER INTEGRATION PROBLEMS

Each Solved Problem section now includes a new type of solved problem that emphasizes concept integration. These chapter integration problems help to show how one set of learned skills builds on previous ones. They also enable students to develop a holistic perspective as they begin to organize diverse concepts into a coherent body of knowledge.

NEW PROBLEMS

We have added more than 100 new problems, including easier drill problems. To make room for these, we have selectively retained only the best of the problems from previous editions.

CONCEPT MAPS

Another new end-of-chapter feature is the problems requiring concept mapping. Concept maps grew out of the constructivist movement in education, which asserts that student learning is most effective when new information is brought into direct conflict with previous understanding. The concept map provides a useful structure for visualizing and resolving such conflicts. Suggested graphic solutions are included for instructors in the *Test Bank.*

Concept Map

A selection of terms is given, and the challenge is to draw lines or arrows between the ones that you think are related, with a description on the arrow showing just what the relationship is. Draw as many relationships as possible.

Here is the concept map for this chapter:

genotype / phenotype / norm of reaction / developmental noise / environment / development / organism

NEW AUTHOR

Last but not least, we are delighted to welcome our new coauthor, William Gelbart. He brings to us an expertise in developmental genetics and a great enthusiasm for teaching.

Course Syllabi

For a two-semester course, the entire text provides an appropriate course structure and syllabus that reflects the range of modern genetics. A syllabus for a one-semester course can be designed around selected chapters. For students with prior exposure to DNA structure and function from introductory biology or cell biology courses, a possible selection of chapters for a one-semester course is Chapters 2, 3, 4, 5, 7, 9, 10, 12, 14, 16, 22, and 25. A one-semester course in molecular genetics could be based on Chapters 10 through 23.

Acknowledgments

Thanks are due to the following people at W. H. Freeman and Company for their considerable support throughout the preparation of this edition: Linda Chaput, former President; Mary Shuford, Director of Development; Randi Rossignol, Senior Developmental Editor; Philip McCaffrey, Managing Editor; Penelope Hull, Associate Managing Editor; Janet Tannenbaum, Project Editor; Nancy Singer, Designer; John Hatzakis, Production Artist; Christine McAuliffe, Illustration Coordinator; Travis Amos, Senior Photo Editor; Larry Marcus, Assistant Photo Editor; and Ellen Cash, Production Manager. We also thank the copy editors, William O'Neal, Gloria Hamilton, and Cathy Lundmark; the indexer, Ellen Murray; and the proofreader, Chrishanthi Squires.

We appreciate the contribution of several other individuals; in particular Tony Griffiths thanks Barbara Moon for her useful ideas on pedagogy, Dick Lewontin thanks Rachel Nasca for her work in preparing the manuscript, and Jeffrey Miller and Bill Gelbart thank Kim Anh Miller and Janet and Marnie Gelbart for their constant support.

We thank Diane Lavett for critiquing the problem sets and for the solutions to the Chapter Integration Problems for Chapters 11, 15, and 19 and the Solved Problems in Chapters 14 and 15. Finally, we extend our thanks and gratitude to our colleagues who reviewed this edition and whose insights and advice were most helpful:

Michael Abruzzo	California State University at Chico
M. T. Andrews	North Carolina State University at Raleigh
Frank Baker	Indiana University of Pennsylvania
Ramesh N. Bhambhani	University of Alberta
C. W. Birky, Jr.	Ohio State University
F. M. Butterworth	Oakland University
Kathy Dunn	Boston College
Victoria Finnerty	Emory University
William Fixsen	Harvard University
Robert Fowler	San Jose State University
Peter Gergen	State University of New York at Stony Brook
Stephen Goldman	University of Toledo
Thomas Gray	University of Kentucky
Tulle Hazelrigg	Columbia University
R. Hodgetts	University of Alberta
Margaret Hollingsworth	State University of New York at Buffalo
Andrew Hoyt	Johns Hopkins University
David R. Hyde	University of Notre Dame
James Jacobson	University of Houston
Mitrick A. Johns	Northern Illinois University
James T. Kadonaga	University of California at San Diego
Muriel Nesbitt	University of California at San Diego
Mark Sanders	University of California at Davis
Trudi Schupbach	Princeton University
David Sheppard	University of Delaware
David J. Stanton	Eastern Michigan University
Laurie Tompkins	Temple University
Monte Turner	University of Akron

We believe this edition to be a true celebration of genetics. As authors, we hope that our love of the subject comes through and that the book will stimulate the reader to do some first-hand genetics, whether as professional scientist, student, amateur breeder, or naturalist. Failing this, we hope to impart some lasting impression of the incisiveness, elegance, and power of genetic analysis.

1

Genetics and the Organism

Genetic variation in the color of corn kernels. Each kernel represents a separate individual with a distinct genetic makeup. The photograph symbolizes the history of humanity's interest in heredity. Humans were breeding corn thousands of years before the rise of the modern discipline of genetics. Extending this heritage, corn today is one of the main research organisms in classical and molecular genetics. (William Sheridan, University of North Dakota; photo by Travis Amos)

KEY CONCEPTS

Genetics unifies the biological sciences.

▪

Genetics is of direct relevance to human affairs.

▪

Genetics may be defined as the study of genes through their variation.

▪

Genetic variation contributes to variation in nature.

▪

As an organism develops, its unique set of genes interacts with its unique environment in determining its characteristics.

Why study genetics? There are two basic reasons. First, genetics has come to occupy a pivotal position in the entire subject of biology; for any serious student of plant, animal, or microbial life, an understanding of genetics is thus essential. Second, genetics, like no other scientific discipline, has become central to numerous aspects of human affairs. It touches our humanity in many different ways. Indeed, genetic issues seem to surface daily in our lives, and no thinking person can afford to be ignorant of its discoveries. In this chapter we take an overview of the science of genetics, showing how it has come to occupy its crucial position. In addition we provide a perspective from which to view the subsequent chapters.

First we need to define what genetics is. Some define it as the study of heredity, but hereditary phenomena probably have been interesting to humans since before the dawn of civilization. Long before biology or genetics existed as the scientific disciplines we know today, ancient peoples were improving plant crops and domesticated animals by selecting desirable individuals for breeding. They also must have puzzled about the inheritance of traits in humans, and asked such questions as Why do children resemble their parents? and How can various diseases run in families? But these people could not be called geneticists. Genetics as a set of principles and analytical procedures did not begin until the 1860s when an Augustinian monk named Gregor Mendel (Figure 1-1) performed a set of experiments that pointed to the existence of biological elements called **genes.** The word "genetics" comes from "genes," and genes provide the focus for the subject. Whether geneticists study at the molecular, cellular, organismal, family, population, or evolutionary level, genes are always central in their studies. Simply stated, genetics is about genes.

Figure 1-1 Gregor Mendel. (Moravian Museum, Bruno)

What is a gene? A gene is a section of a threadlike molecule called **deoxyribonucleic acid,** abbreviated as **DNA.** DNA, the hereditary material that passes from one generation to the next, dictates the inherent properties of a species. Each cell in an organism has one or two sets of the basic DNA complement, called a **genome.** The genome itself is made up of one or more extremely long molecules of DNA that are called **chromosomes.** The genes are the functional regions of the DNA and are simply active segments ranged along the chromosomes. In complex organisms chromosomes generally number in the order of tens, but genes number in the order of tens of thousands. Armed with these definitions of genetics and genes, let us go on to see how understanding these subjects has become so important. We will start with the impact of genetics on ourselves.

Genetics and Human Affairs

Genetics seems to hold a special place in human affairs. Not only is it relevant in the same sense that other scientific disciplines are, but it also has much to tell us about the nature of our humanity, and in this sense it holds a special place among the biological sciences.

Modern society depends on genetics. Take a look at the clothes you are wearing. The cotton of your shirt and jeans came from cotton plants that differ from their wild ancestors because they have been through intensive breeding programs involving the methodical application of standard genetic principles. The same could be said for the sheep that produced the wool for your sweater and coat. Think also about your most recent meal. You can be sure that the rice, the wheat, the chicken, the beef, the pork, and all the other major organisms that feed the planet have been specially engineered with the application of standard genetic procedures (Figure 1-2). The so-called Green Revolution, which dramatically increased crop productivity on a global scale, is a genetic success story about the breeding of highly productive strains of certain major crop species (Figure 1-3).

To maximize crop production, a farmer may plant a vast area with seeds of a single genetic constitution, a practice called monoculture. Because pathogenic and

Figure 1-2 A geneticist making controlled crosses in corn. Paper bags are used to prevent random cross pollinations, and to collect pollen to make the required specific crosses. (Chuck O'Rear/Woodfin Camp)

predatory organisms attack plant crops, special resistance genes are bred into crop lines to protect them. But the protection is only temporary. Random genetic changes occur constantly in pathogen populations, and such changes sometimes confer new pathogenic ability. This leaves the entire monoculture at risk. For this reason, plant geneticists always have to be one step ahead of the pathogens to prevent widespread epidemics of plant disease or destruction that might have devastating effects on the food supply (Table 1-1 and Figure 1-4).

Fungi and bacteria have also been specially bred for human needs. Yeast is an obvious example; it forms the basis of multibillion dollar industries producing baked goods, alcoholic beverages, and fuel alcohol. Fungi provide the antibiotic penicillin, the immunosuppressant drug cyclosporin that prevents rejection of organ transplants, and a whole range of important industrial compounds such as citric acid and amylase. Bacteria provide such antibiotic chemicals as streptomycin to medical science. Most of the industries using fungi and bacteria have benefited from the application of classic genetic principles, but now we are entering a new age where molecular genetic techniques enable scientists to synthesize entirely novel strains of microbes that are created in a test tube and tailored exclusively to human needs. For example, from such syntheses we now have bacterial strains producing mammalian substances such as insulin for diabetes treatment and growth hormone for treatment of pituitary dwarfism. The engineered insulin product is the genuine human type in contrast to the previous product, which was from cattle and pigs. Some genetically engineered substances are extremely difficult to obtain from other sources; for example one year's treatment of growth hormone for one child previously had to be extracted from the pituitary glands of approximately 75 human cadavers. In addition, special derivatives of fungal and bacterial strains are being genetically engineered to produce larger amounts of their natural beneficial products.

Molecular genetic engineering was first applied to microbes, but now the same techniques are being applied to plants and animals, resulting in engineered types that could never have been produced with classical genetics. The application of genetic engineering is broadly

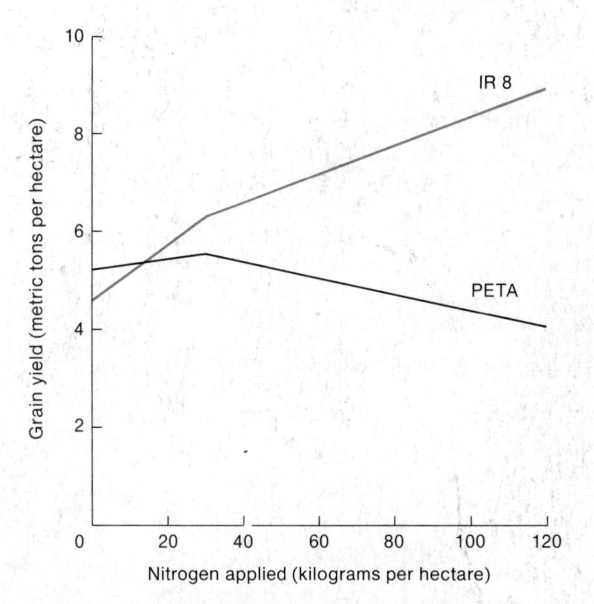

Figure 1-3 The specially bred strain of dwarf rice called IR 8 owes part of its success to its remarkable response to the application of fertilizer. PETA, an older, nondwarf strain, shows a more typical response. (From Peter R. Jennings, "The Amplification of Agricultural Production." 1976. *Scientific American*)

Table 1-1. Pest-resistant Strains of Rice

Strain	Year Developed	Diseases					Insects		
		Blast Fungus	Bacterial Blight	Leaf-streak Virus	Grassy Stunt Virus	Tungro Virus	Green Leafhopper	Brown Hopper	Stem Borer
IR 8	1966	MR	S	S	S	S	R	S	MS
IR 5	1967	S	S	MS	S	S	R	S	S
IR 20	1969	MR	R	MR	S	R	R	S	MS
IR 22	1969	S	R	MS	S	S	S	S	S
IR 24	1971	S	S	MR	S	MR	R	S	S
IR 26	1973	MR	R	MR	MR	R	R	R	MR

NOTE: The entries describe each strain's susceptibility or resistance to each pest as follows: S = susceptible; MS = moderately susceptible; MR = moderately resistant; R = resistant. (From *Research Highlights,* International Rice Research Institution, 1973, p. 11.)

called biotechnology. It is estimated that the biotechnology industry, which has genetics as its basis, will become one of the top money-making industries in the coming decades. Whether this is true or not, the biotechnology industry joins the genetic cornucopia that supports mankind on this planet at its present high population size and living standards.

Genetics is a crucial component of medicine. A large proportion of human ill health has a genetic basis. For example, it has been estimated that at least 30 percent of pediatric hospital admissions have a genetic component. However, current research is revealing more and more genetic predispositions to serious conditions as well as milder ailments, so this figure is almost certainly an underestimate. Genetic ill health can be divided into three major types. The first type is **inherited genetic diseases** such as cystic fibrosis, phenylketonuria, and muscular dystrophy, which are caused by abnormal genes passed on from one generation to the next. The second type is **somatic genetic disease,** which is caused by the sudden appearance of an abnormal form of a gene in one part of the body. Cancer is the most significant example of this kind of disease. It is little appreciated that cancer, which touches all our lives in some way, is a genetic disease. Although somatic changes are not passed on to the next generation, various predispositions to cancer are inherited as abnormal genes. The third type is **chromosomal aberrations,** such as Down syndrome and cri du chat syndrome. These are caused by inherited abnormalities of chromosomal structure or number.

Figure 1-4 A specially bred strain of dwarf wheat *(right)* resists crop damage far better than the normal strain *(left)*. (The Rockefeller Foundation)

Genes are at the root of a large number of diseases, but genetics will provide relief from suffering in many cases. Already molecular genetic probes are being used to detect defective genes in prospective parents. Furthermore, the defective genes themselves are being isolated and characterized by molecular genetic techniques. For example, the DNA of the gene that causes cystic fibrosis was recently isolated and analyzed. From the gene's molecular structure, as revealed by this work, scientists can now investigate the physiological defect that causes the disease. Once the disease is understood, better therapeutic approaches can be designed. Ultimately, there is the hope that direct gene therapy will relieve many genetic diseases. Gene therapy inserts a copy of a normal gene into cells carrying the defective counterpart, so that the normal function of the inserted gene can compensate for the abnormal function of the defective gene.

Geneticists are also at work studying the human immunodeficiency virus (HIV) that causes acquired immune deficiency syndrome (AIDS). As a natural part of their reproduction, viruses such as HIV insert copies of their genetic material into the chromosomes of the individuals they infect. Therefore, in a sense this too is a genetic disease, and an understanding of how such viral genes integrate and function is an important step in overcoming the diseases caused by these viruses.

Human genetics obviously holds an important position in human affairs, and a measure of this is the amount of money that society is spending on human genetic research. Society rarely allots billions of dollars for single biological projects, but $3 billion has been committed to finance the complete sequencing of the human genome. This will be an international collaborative effort by many laboratories, each working on specific chromosomal regions. The potential benefits for medicine are immense. But perhaps most exciting is the opportunity finally to see a glimmer of the grand design in the blueprints of what is arguably the most complex structure in the known universe, *Homo sapiens* (Figure 1-5).

Genetics affects one's world view. We each acquire our individual view of the universe and of our own position in that universe gradually, from the beginning of our consciousness. This viewpoint represents our identity as individuals. It drives our attitudes and our actions, and as such determines the kind of people we are and ultimately the kind of society we live in. Any new knowledge either has to be accommodated into this world view or the world view has to be changed to make it fit. Ignorance or rejection of new knowledge leads to closed-mindedness and bigotry. Genetics has provided some powerful new concepts that have radically changed humanity's view of itself and its relation to the rest of the universe.

Probably the best example of how genetics changes a person's world view comes from genetic, cytogenetic, and molecular studies that show we are related not only to apes and other mammals, but, more surprisingly, to *all* the other living things on the planet, including plants, fungi, and bacteria. Living things share a common system for storing and expressing information and show homology in many structures, even down to the genes themselves. That there is a continuous spectrum of relatedness within the living world is a powerful intellectual notion that unifies us with other living organisms. This notion radically affects one's world view. It suggests a view of humanity not as the pinnacle or the center of creation, but as one form equal to other life forms. Admittedly, this brings us into the domain of philosophy and religion, but that is the point: genetics forces us to consider issues that question how we see ourselves.

Some of the world's biggest and most pressing social issues have an indirect genetic component. For example, some major problems of prejudice and social suffering center on behavioral differences between races and between the sexes. Genetics provides a way of analyzing and thinking about these complex and unresolved issues.

One of the biggest global concerns of biologists is the alarming rate at which we are destroying natural habitats, especially in the tropics, which hold vast reservoirs of plant and animal life. Here again the problem has a crucial genetic component because the issue is one of conserving genetic diversity and genetic resources. An understanding of the full impact of destroying natural habitats requires an understanding of genetics.

Another genetic issue with potential global impact is the genetic health of our populations. Many geneticists are concerned because our human genomes are under assault by an ever increasing array of environmental agents, mainly radiation and chemicals, that are capable of causing random changes in genes. The vast majority of such changes would inevitably be deleterious. In the short run they might not significantly increase the frequency of inherited disease, but over the long run the changes could accumulate, eventually surfacing as a "genetic time bomb."

The above examples show that many of the issues that confront us daily as individuals and as societies intimately involve genetics. Only by understanding the genetic component of these issues can we as global citizens hope to make wise decisions for the future of an ever more complex and unstable world.

Message Genetic insight helps us understand humanity and human society.

Figure 1-5 Two molecular geneticists analyzing DNA sequences. The dark bands are photographic images of radioactive DNA segments. (David Parker/Science Source/Photo Researchers)

Genetics and Biology

Biology is a huge subject. The planet Earth contains a staggering array of life forms. Already we know about the existence, for example, of 286,000 species of flowering plants, 500,000 species of fungi, and 750,000 species of insects. In addition, many more species are still undiscovered. Fifty years ago, the science of biology was divided into separate disciplines, each analyzing life at a different level. There was morphology, physiology, biochemistry, taxonomy, ecology, genetics, and so on, all working largely in separate compartments. However, discoveries in genetics have provided unifying themes for the whole of biology, so that now conceptual threads link the disciplines.

The major thematic linking thread is in reality a thread, the molecule DNA. We now know that DNA is the basis for all of the processes and structures of life. The DNA molecule has a structure that accounts for two of the key properties of life, replication and generation of form. We will learn that DNA is a double helical structure that has the inherent property of being able to make copies of itself, and it is this property that enables replicas of cells and organisms to be made and persist through time. Furthermore, written into the linear sequence of the building blocks of a DNA molecule is a code that contains the instructions for building an organism; we can view this as information, or "that which is necessary to give form." The unique features of a species, whether structures or processes, are seen to be under the influence of DNA. So underlying the structures studied by morphologists, the reactions studied by physiologists, the homologies studied by evolutionists, and so on, we see the unifying thread of the DNA molecule.

DNA works in virtually the same way in all organisms. This in itself provides another unifying theme, but in addition it means that what is learned in one organism can often be applied in principle to others. For this reason, genetics has made extensive use of model organisms, many of which will appear in the pages of this book. In fact, the advances made in human genetics over the recent decades have been possible in large part because of the advances made with such lowly model organisms as bacteria and fungi.

Genetics has also provided some of the most incisive analytical approaches now being used across the spectrum of the biological disciplines. Foremost is the technique of **genetic dissection.** In this experimental approach, any structure or process can be picked apart, or "dissected," by discovering which genes influence it. For example, in the study of development each abnormal gene that produces a developmental abnormality identifies a component in the normal process of development. Then the larger picture can be assembled by interrelating all the genetically controlled components. Another successful technique is the use of specific genes as **markers.** Just as you might use brightly colored tags to mark animals or plants in some biological study, geneticists use variant genes to keep track of specific chromosomes, cells, or individuals. This technique has found application across the breadth of the biological disciplines, from cell biology to evolution and ecology. For example, genes for human diseases are now being isolated by virtue of their chromosomal proximity to unrelated marker sequences.

We have seen that molecular genetic engineering has opened up new vistas in applied biotechnology, but the same techniques are just as useful in basic research. With the ability to move genes from organism to organism, scientists have produced plants that glow because they express the phosphorescence genes of fireflies, plants that acquire cold hardiness by expressing antifreeze genes from fish, and giant mice that are expressing the growth hormone genes of rats. Scientists have manipulated genes in yeast to produce totally artificial, functional chromosomes. The ability to isolate a gene in a test tube, to modify its structure in specific ways, and then to reinsert it back into the organism has provided the sharpest of scalpels for genetic dissection. Altogether, genetic engineering has revolutionized the biological sciences, and no biologist today can afford to be ignorant of this powerful analytical tool.

Message Genetic analysis is used not only in the study of heredity, but in all other areas of biology.

But perhaps the biggest biological success story of all is the elucidation of precisely how the genes do their job, in other words, how information becomes form. It is a marvellous story that has developed with amazing rapidity within the span of the careers of scientists now still in their fifties, scientists who never in their wildest dreams imagined that by the 1990s geneticists would be sequencing entire genomes. But today the coding and flow of genetic information in cells are foundations of modern biological thought and baselines from which experimental explorations start off. It is worth summarizing the highlights of how genetic information flows — or, equivalently, how genes act — which has been called the new paradigm of biology. Figure 1-6 shows diagrammatically the essentials of gene action in a generalized cell of a **eukaryote.** Eukaryotes are those organisms whose cells have a membrane-bound nucleus. Animals, plants, and fungi are all eukaryotes. Inside the nucleus are the chromosomes and outside the nucleus are a complex array of membranous structures, including the endoplasmic reticulum and Golgi apparatus, and organelles such as the mitochondria and chloroplasts.

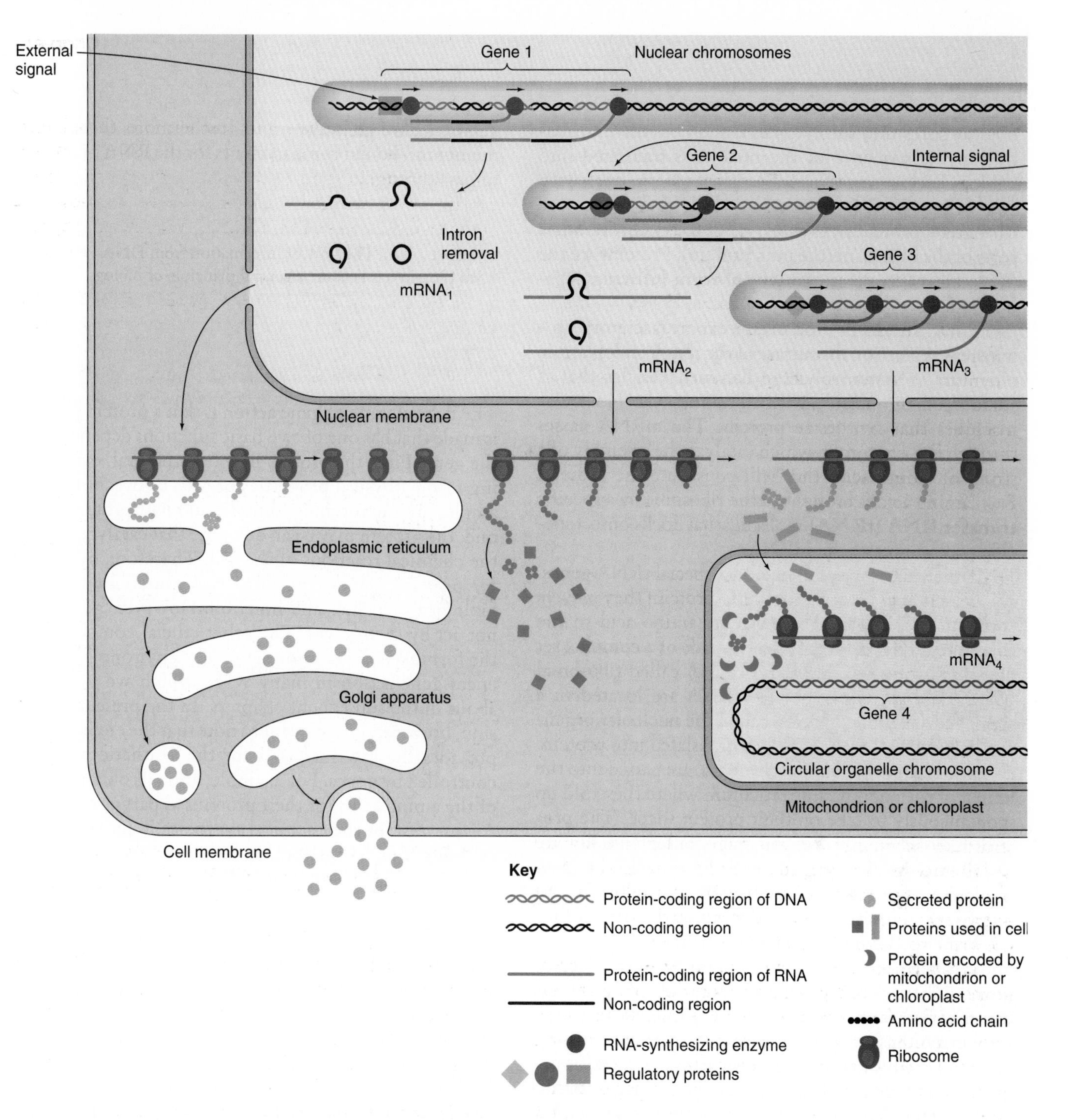

Inside the eukaryotic nucleus, some genes are active more or less constantly, but others have to be turned on and off to suit the needs of the cell or the organism. The signal to activate a gene may come from outside the cell, from, for example, a substance such as a steroid hormone. Or the signal may come from within the cell, for example from a special regulatory gene whose job it is to turn other genes on and off. The regulatory substances bind to a special region of the gene and initiate the synthesis of copies of the gene's DNA. These copies are in the form of a molecule called RNA. Noncoding regions of the gene called **introns** are cut out at this RNA stage, and the remaining RNA sequence is called **messenger RNA,** or **mRNA.** The mRNA molecules pass out through the

Figure 1-6 Simplified view of gene action in a eukaryotic cell. The basic flow of genetic information is from DNA to RNA to protein. Four types of genes are shown. Gene 1 responds to external regulatory signals and makes a protein for export; gene 2 responds to internal signals and makes a protein for use in the cytoplasm; gene 3 makes a protein to be transported into an organelle; gene 4 is part of the organelle DNA and makes a protein for use inside its own organelle. Most eukaryotic genes contain introns, regions (generally noncoding) that are cut out in the preparation of functional messenger RNA. Note that many organelle genes have introns and that an RNA-synthesizing enzyme is needed for organelle mRNA synthesis. These details have been omitted from the diagram of the organelle for clarity. (Introns will be explained in detail in subsequent chapters.)

nuclear pores into the cytoplasm and here the information in the sequence of the mRNA is translated into protein. Each gene codes for a separate protein, each with specific functions either within the cell (for example the dark green protein in Figure 1-6) or for export to other parts of the organism (the pink protein). Proteins are the most important manifestations of form in living organisms.

The synthesis of proteins for export (secretory proteins) takes place on the surface of the rough endoplasmic reticulum, a system of large flattened vesicles that is studded on the outside with ribosomes, the molecular machines that synthesize protein. The mRNA passes through the ribosomes, which catalyze the assembly of a string of amino acids that will constitute the protein. Each amino acid is brought to the ribosome by a specific **transfer RNA (tRNA)** molecule that docks onto a specific coding unit of the mRNA.

The tRNAs are synthesized off special tRNA genes. No tRNAs are ever translated into protein; they recycle constantly, delivering their specific amino acid to the ribosomes. The ribosome itself is made of a complex set of proteins plus several kinds of RNA called ribosomal RNA (rRNA). The genes for rRNA are located in a special chromosomal region called the nucleolar organizer. Like tRNA, rRNA is never translated into protein.

The completed amino acid chains are passed into the lumen of the endoplasmic reticulum where they fold up spontaneously to take on their protein shape. The proteins may be modified at this stage, but eventually are passed into the chambers of the Golgi apparatus, and on into secretory vessels that eventually fuse with the cell membrane and release their contents to the outside.

Proteins destined to function in the cytosol or in mitochondria and chloroplasts are synthesized on ribosomes unbound to membranes. For example, proteins that function as enzymes in glycolysis follow this route. Protein synthesis occurs by the same mechanism using the same kinds of tRNAs. The proteins destined for organelles are specially tagged to target their insertion into the organelle. Mitochondria and chloroplasts have their own small circular DNA chromosomes. Synthesis of proteins encoded by genes on mitochondrial or chloroplast chromosomes takes place on ribosomes inside the organelles themselves. Therefore the proteins in the mitochondria and chloroplasts are of two different origins, either nucleus coded and imported into the organelle, or organelle coded and synthesized within the organelle compartment.

Prokaryotes are organisms such as bacteria whose cells have a simpler structure; there is no nucleus or other membrane-bound structures within these cells. Protein synthesis in prokaryotes is generally similar, using mRNA, tRNA, ribosomes, and the genetic code, but there are some important differences. For example, prokaryotes have no introns, and, furthermore, there are no membrane-bound compartments for the RNA or protein to pass through.

Message The flow of information from DNA to RNA to protein has become a central principle of biology.

Genes and Environment

The net outcome of gene action is that a protein product is made that has one of two basic functions depending on the gene. First, the protein may be structural, contributing to the physical properties of cells or organisms. Examples are microtubule, muscle, and hair proteins. Second, the protein may be an **enzyme** that catalyzes one of the chemical reactions of the cell. Therefore, by coding for proteins, genes determine two important facets of biological structure and function. However, genes cannot act by themselves. The other crucial component in the formula is the environment. The environment influences gene action in many ways, which we will learn about in the subsequent chapters. In the present discussion, however, it is relevant to note that the environment provides the raw materials for the synthetic processes controlled by genes. For example, animals obtain several of the amino acids for their proteins as part of their diet. Again, most of the chemical syntheses in plant cells use carbon atoms taken from the air as carbon dioxide. Finally, bacteria and fungi absorb from their surroundings many substances that are simply treated as carbon and nitrogen skeletons, and their enzymes convert these into the compounds that constitute the living cell. Thus through genes an organism builds the orderly process that we call life out of disorderly environmental materials.

Genetic Determination

From our brief look at gene action, we can see that living organisms mobilize the components of the world around themselves and convert these components into their own living material. An acorn becomes an oak tree, using in the process only water, oxygen, carbon dioxide, some inorganic materials from the soil, and light energy.

An acorn develops into an oak, while the spore of a moss develops into a moss, although both are growing side by side in the same forest (Figure 1-7). The two plants that result from these developmental processes resemble their parents and differ from each other, even though they have access to the same narrow range of inorganic materials from the environment. The parents pass to their offspring the specifications for building living cells from environmental materials. These specifica-

Figure 1-7 The genes of a moss direct environmental components to be shaped into a moss, whereas the genes of a tree cause a tree to be constructed from the same components. (André Bärtschi)

tions are in the form of genes in the fertilized egg. Because of the information in the genes, the acorn develops into an oak and the moss spore becomes a moss.

Just as genes maintain differences between species such as the oak and moss, they also maintain differences within species. Consider plants of the species *Plectritis congesta,* the sea blush. Two forms of this species are found wherever the plants grow in nature: one form has wingless fruits, and the other has winged fruits (Figure 1-8). These plants will self-pollinate, and we can observe the offspring that result from such "selfs" when these are grown in a greenhouse under uniform conditions: we commonly observe that the selfed progeny of a winged-fruited plant are all winged-fruited and that the selfed progeny from a wingless-fruited plant all have wingless fruits. Because all the progeny were grown in the same environment, we can rule out the possibility that environmental differences cause some plants to bear wingless fruits and others, winged. We can safely conclude that the fruit-shape difference between the original plants, which each passed on to its selfed progeny, results from the different genes they carry.

Plectritis has two inherited forms that are both perfectly normal. The determinative power of genes is probably more often demonstrated by differences in which one form is normal and the other abnormal. The human inherited disease sickle-cell anemia provides a good example. The underlying cause of the disease is a variant of hemoglobin, the oxygen-transporting protein molecule found in red blood cells. Normal people have a type of hemoglobin called hemoglobin A, the information for which is encoded in a gene. A minute chemical change at the molecular level in the DNA of this gene results in the production of a slightly changed hemoglobin, termed hemoglobin S. In people possessing only hemoglobin S, the ultimate effect of this small change is severe ill health and usually death. The gene works its effect on the organism through a complex "cascade effect," as summarized in Figure 1-9.

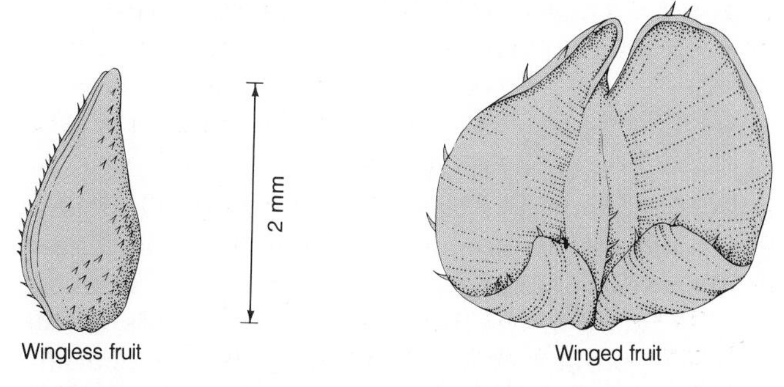

Figure 1-8 The fruits of two different forms of *Plectritis congesta,* the sea blush. Any one plant has either all wingless or all winged fruits. In every other way the plants are identical. A simple genetic difference determines the difference in the fruits.

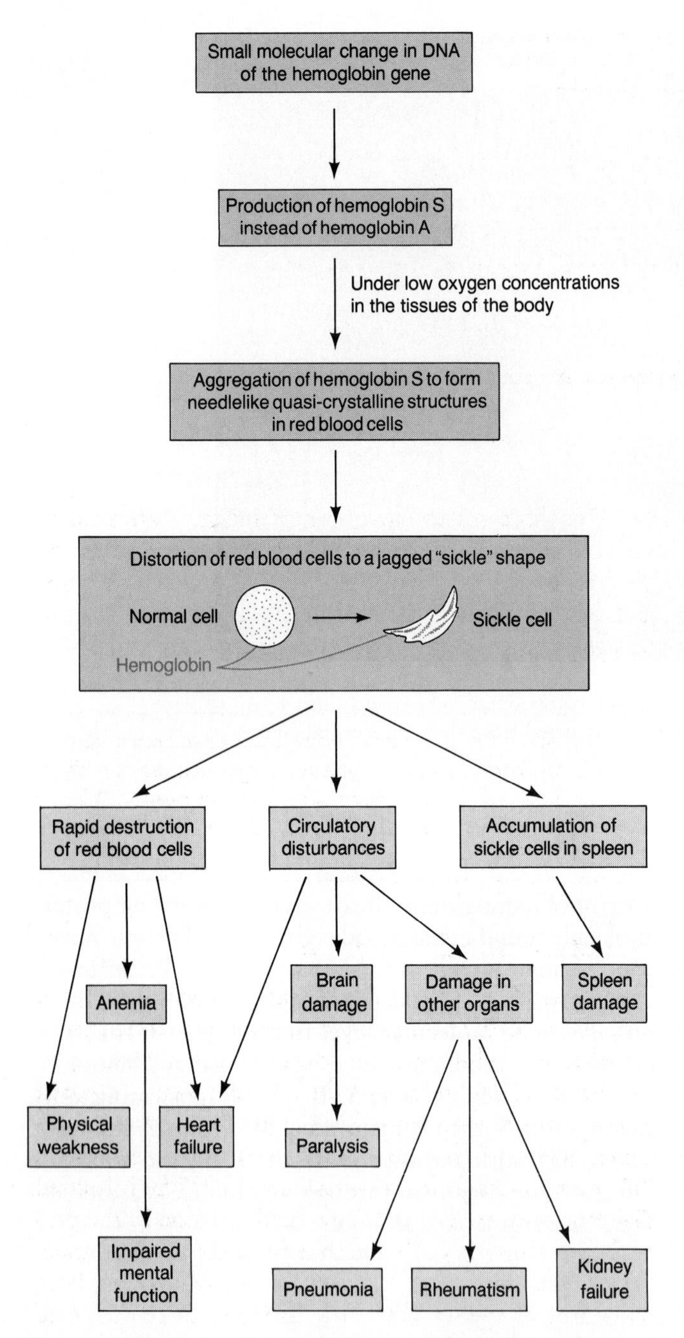

Figure 1-9 Chain of events in human sickle-cell anemia.

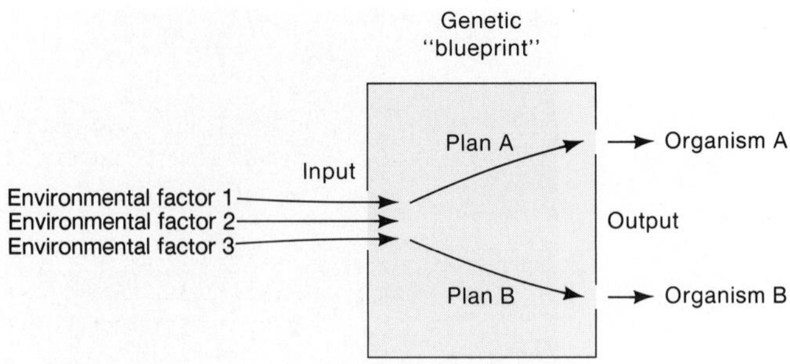

Figure 1-10 A model of determination that emphasizes the role of genes.

Observations like these lead to the model of how genes and the environment interact shown in Figure 1-10. In this view, the genes act as a set of instructions for turning more-or-less undifferentiated environmental materials into a specific organism, much as blueprints specify what form of house is to be built from basic materials. The same bricks, mortar, wood, and nails can be made into an A-frame or a flat-roofed house, according to different plans. Such a model implies that the genes are really the dominant elements in the determination of organisms; the environment simply supplies the undifferentiated raw materials.

Environmental Determination

But now consider two monozygotic ("identical") twins, the products of a single fertilized egg that divided and produced two complete individuals with identical genes. Suppose that the twins are born in England but are separated at birth and taken to different countries. If one is raised in China by Chinese-speaking foster parents, she will speak Chinese, while her sister raised in Budapest will speak Hungarian. Each will absorb the cultural values and customs of her environment. Although the twins begin life with identical genetic properties, the different cultural environments in which they live will produce differences between the sisters (and differences from their parents). Obviously, the differences in this case are due to the environment, and genetic effects are of little importance in determining the differences.

This example suggests the model of Figure 1-11, which is the converse of that shown in Figure 1-10. In the model in Figure 1-11, the genes impinge on the system, giving certain general signals for development, but the environment determines the actual course of action. Imagine a set of specifications for a house that simply calls for "a floor that will support 300 pounds per square foot" or "walls with an insulation factor of 15"; the actual appearance and other characteristics of the structure would be determined by the available building materials.

Our different types of examples — of purely genetic effect versus that of the environment — lead to two very different models. First, consider the seed example: given a pair of seeds and a uniform growth environment, we would be unable to predict future growth patterns solely from a knowledge of the environment. In any environment we can imagine, if the organisms develop, the acorn becomes an oak and the spore becomes a moss. The model of Figure 1-10 applies here. Second, consider the

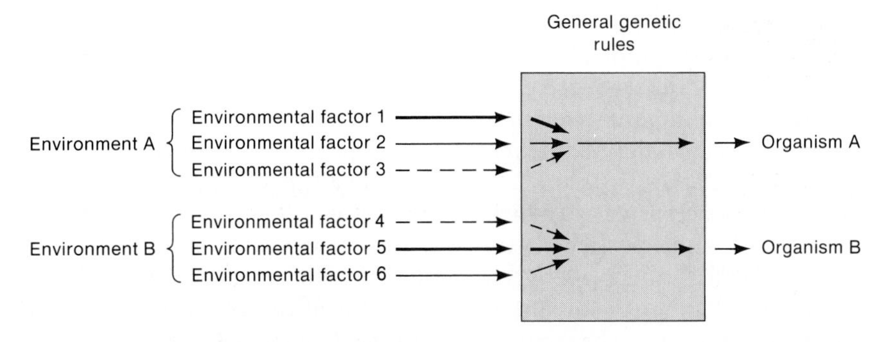

Figure 1-11 A model of determination that emphasizes the role of the environment.

twins: no information about the set of genes they inherit could possibly enable us to predict their ultimate languages and cultures. Two individuals that are *genetically different* may develop differently in the *same environment,* but two *genetically identical* individuals may develop differently in *different environments.* The model of Figure 1-11 applies here.

In general, of course, we deal with organisms that differ in both genes and environment. If we wish to predict how a living organism will develop, we must first know the genetic constitution that it inherits from its parents. Then we must know the *historical sequence* of environments to which the developing organism is exposed. Every organism has a developmental history from birth to death. What an organism will become in the next moment depends critically both on the environment it encounters during that moment and on its present state. It makes a difference to an organism not only what environments it encounters but in what sequence it encounters them. A fruit fly *(Drosophila)* develops normally at 20°C. If the temperature is briefly raised to 37°C early in its pupal stage of development, the adult fly will be missing part of the normal vein pattern on its wings. However, if this "temperature shock" is administered just 24 hours later, the vein pattern develops normally.

Our discussion of how genes and the environment interact has brought us to a point where we can appreciate a more general model (Figure 1-12) in which genes and the environment jointly determine (by some rules of development) the actual characteristics of the organism.

Message As an organism transforms developmentally from one stage of its life to another, its genes interact with its environment at each moment of its life history. The interaction of genes *and* environment determines what organisms are.

Genotype and Phenotype

In studying how genes and the environment interact to produce an organism, geneticists have developed some useful terms, which are introduced in this section.

A typical organism resembles its parents more than it resembles unrelated individuals. Thus, we often speak as if the individual characteristics themselves are inherited: "He gets his brains from his mother," or "She inherited diabetes from her father." Yet our discussion in the preceding section shows that such statements are inaccurate. "His brains" and "her diabetes" develop through long sequences of events in the life histories of the affected

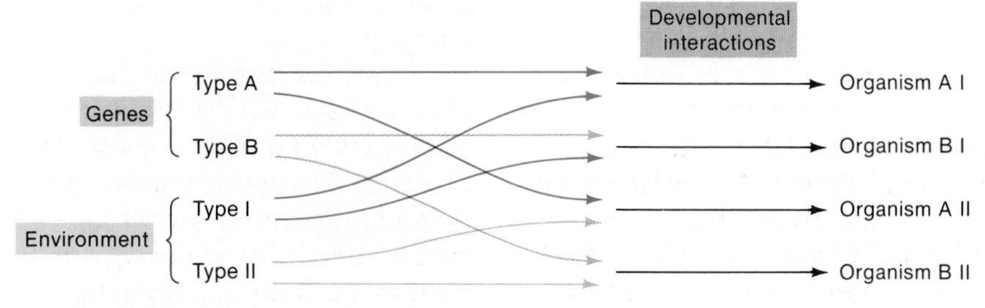

Figure 1-12 A model of determination that emphasizes the interaction of genes and environment.

persons, and both genes and environment play roles in those sequences. In the biological sense, individuals inherit only the molecular structures of the fertilized eggs from which they develop. Individuals inherit their *genes,* not the end products of their individual developmental histories.

To prevent such confusion between genes (which are inherited) and developmental outcomes (which are not), geneticists make a fundamental distinction between the genotype and the phenotype of an organism. Organisms share the same **genotype** if they have the same set of genes. Organisms share the same **phenotype** if they look or function alike.

Strictly speaking, the genotype describes the complete set of genes inherited by an individual and the phenotype describes *all* aspects of the individual's morphology, physiology, behavior, and ecological relationships. In this sense, no two individuals ever belong to the same phenotype, because there is always some difference (however slight) between them in morphology or physiology. Also, except for individuals produced from another organism by asexual reproduction, any two organisms differ at least a little in genotype. In practice, we use the terms genotype and phenotype in a more restricted sense. We deal with some partial phenotypic description (say, eye color) and with some subset of the genotype (say, the genes that influence eye pigmentation).

Message When we use the terms phenotype and genotype, we generally mean "partial phenotype" and "partial genotype" and we specify one or a few traits and genes that are the subsets of interest.

Note one very important difference between genotype and phenotype: the genotype is essentially a fixed character of an individual organism; the genotype remains constant throughout life and is essentially unchanged by environmental effects. Most phenotypes change continually throughout the life of an organism as its genes interact with a sequence of environments. Fixity of genotype does not imply fixity of phenotype.

The Norm of Reaction

How can we quantify the relation between the genotype, the environment, and the phenotype? For a particular genotype, we could prepare a table showing the phenotype that would result from the development of that genotype in each possible environment. Such a set of environment-phenotype relationships for a given genotype is called the **norm of reaction** of the genotype. In practice, of course, we can make such a tabulation only for a partial genotype, a partial phenotype, and some particular aspects of the environment. For example, we might specify the eye sizes that fruit flies would have after developing at various constant temperatures; we could do this for several different eye-size genotypes to get the norms of reaction of the species.

Figure 1-13 represents just such norms of reaction for three eye-size genotypes in the fruit fly *Drosophila melanogaster.* The graph is a convenient summary of more extensive tabulated data. The size of the fly eye is measured by counting its individual facets, or cells. The vertical axis of the graph shows the number of facets (on a logarithmic scale); the horizontal axis shows the constant temperature at which the flies develop.

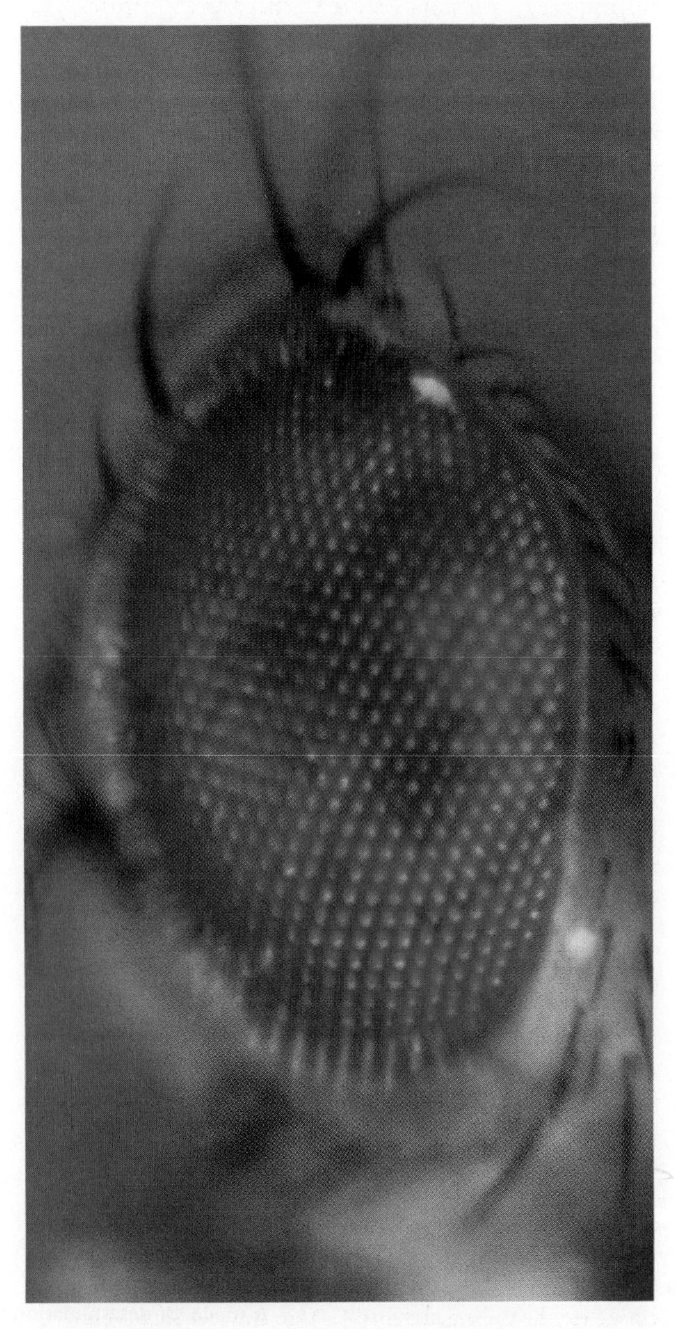

(a)

Three norms of reaction are shown on the graph. When flies of the *wild-type* genotype that is characteristic of flies in natural populations are reared at higher temperatures, they develop eyes that are somewhat smaller than those of wild-type flies reared at cooler temperatures. The graph shows that wild-type phenotypes range from more than 700 to 1000 facets—the wild-type norm of reaction. A fly that has the *ultrabar* genotype has smaller eyes than wild-type flies regardless of temperature during development. Temperatures have a stronger effect on development of *ultrabar* genotypes than *wild-type* genotypes, as we see by noticing that the ultrabar norm of reaction slopes more steeply than the wild-type norm of reaction. Any fly of the *infrabar* genotype also has smaller eyes than any wild-type fly, but temperatures have the opposite effect on flies of this genotype; infrabar flies raised at higher temperatures tend to have larger eyes than those raised at lower temperatures. These norms of reaction indicate that the relationship between genotype and phenotype is complex rather than simple.

Message A single genotype may produce different phenotypes, depending on the environment in which organisms develop. The same phenotype may be produced by different genotypes, depending on the environment.

If we know that a fruit fly has the *wild-type* genotype, this information alone does not tell us whether its eye has 800 or 1000 facets. On the other hand, the knowledge that a fruit fly's eye has 170 facets does not tell us whether its genotype is *ultrabar* or *infrabar*. We cannot even make a general statement about the effect of temperature on eye size in *Drosophila,* because the effect is opposite in two different genotypes. We see from Figure 1-13 that some genotypes do differ unambiguously in phenotype, no matter what the environment: any wild-type fly has larger eyes than any ultrabar or infrabar fly. But other genotypes overlap in phenotypic expression: the eyes of an ultrabar fly may be larger or smaller than those of an infrabar fly, depending on the temperatures at which the individuals developed.

To obtain a norm of reaction like the norms of reaction in Figure 1-13, we must allow different individuals of identical genotype to develop in many different environments. To carry out such an experiment, we must be able to obtain or produce many fertilized eggs with identical genotypes. For example, to test a human genotype in 10 environments, we would have to obtain genetically identical sibs and raise each individual in a different milieu. Obviously, that is possible neither biologically nor socially. At the present time, we do not know the norm of reaction of any human genotype for any character in any set of environments. Nor is it clear how we can ever acquire such information without the unacceptable manipulation of human individuals.

For a few experimental organisms, special genetic methods make it possible to replicate genotypes and thus to determine norms of reaction. Such studies are particularly easy in plants that can be propagated vegetatively (that is, by cuttings). The pieces cut from a single plant all have the same genotype, so all offspring produced in this way have identical genotypes. Such a study has been done on the yarrow plant, *Achillea millefolium* (Figure 1-14a). The experimental results are shown in Figure 1-14b. Many plants were collected, and three cuttings were

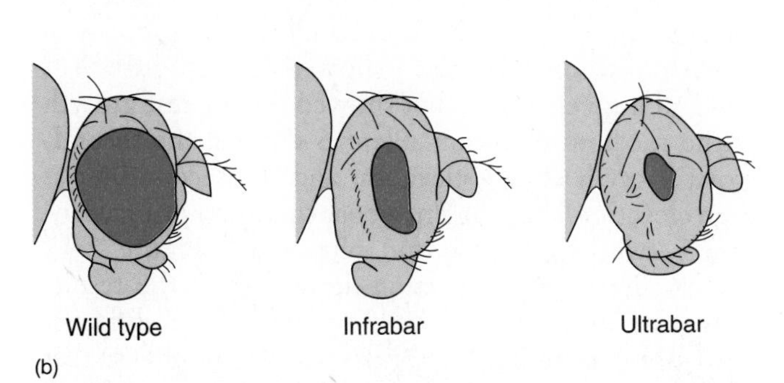

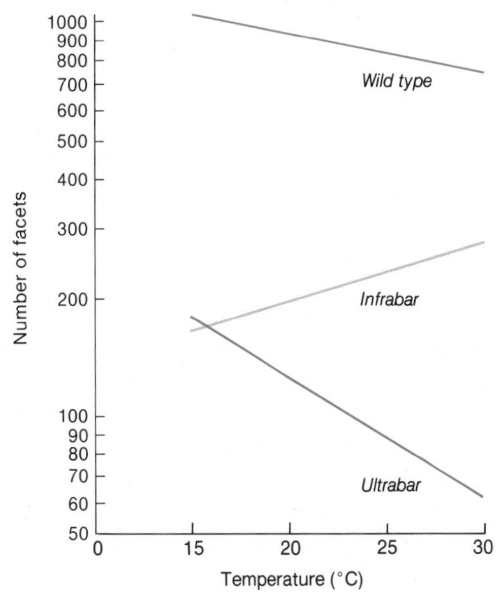

Figure 1-13 Norms of reaction to temperature for three different eye-size genotypes in *Drosophila.* (a, *facing page*) Closeup showing how the normal eye comprises hundreds of units called facets. The number of facets determines eye size. (b) Relative eye sizes of wild-type, infrabar, and ultrabar flies reared at the higher end of the temperature range. (c) Norms of reaction curves for the three genotypes. (a, Don Rio and Sima Misra, University of California, Berkeley)

Figure 1-14 (a) *Achillea millefolium.* (Harper Horticultural Slide Library) (b) Norms of reaction to elevation for seven different *Achillea* plants (seven different genotypes). A cutting from each plant was grown at low, medium, and high elevations. (Carnegie Institution of Washington)

(a)

High elevation

50 cm 0 50 0

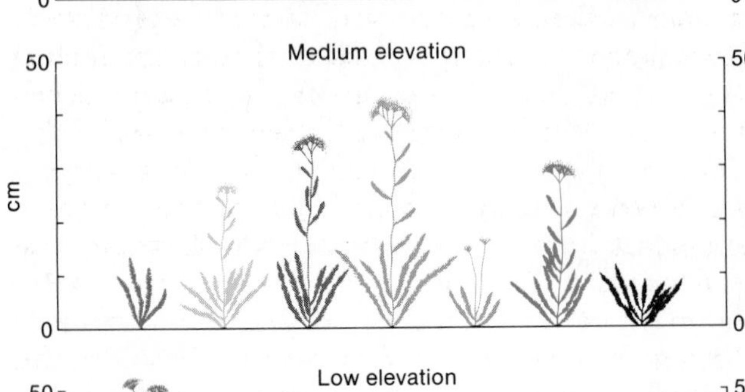

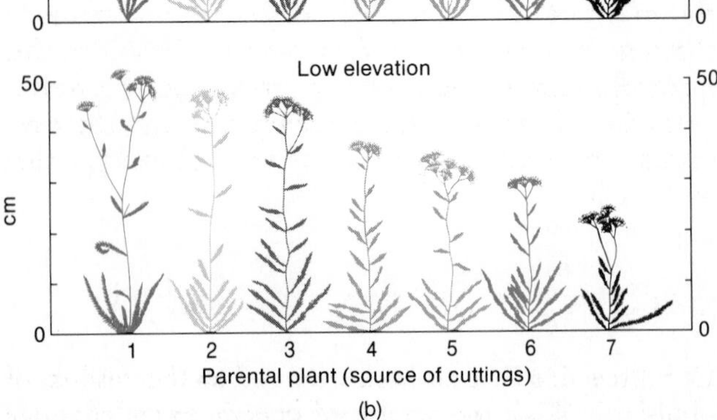

(b)

taken from each plant. One cutting from each plant was planted at low elevation (30 meters above sea level), one at medium elevation (1400 meters), and one at high elevation (3050 meters). Figure 1-14b shows the mature individuals that developed from the cuttings of seven plants; each set of three plants of identical genotype is aligned vertically in the figure for comparison.

First, we note an average effect of environment: in general, the plants grew poorly at the medium elevation. This is not true for every genotype, however; the cutting of plant 4 grew best at the medium elevation. Second, we note that no genotype is unconditionally superior in growth to all others. Plant 1 showed the best growth at low and high elevations but showed the poorest growth at the medium elevation. Plant 6 showed the second-worst growth at low elevation and the second-best at high elevation. Once again, we see the complex relationship between genotype and phenotype. Figure 1-15 graphs the norms of reaction derived from the results shown in Figure 1-14b. Each genotype has a different norm of reaction, and the norms cross one another so that we cannot identify either a "best" genotype or a "best" environment for *Achillea* growth.

We have seen two different patterns of reaction norms. The difference between the *wild-type* and the other eye-size genotypes in *Drosophila* is such that the corresponding phenotypes show a consistent difference, regardless of the environment. Any fly of *wild-type* genotype has larger eyes than any fly of the other genotypes,

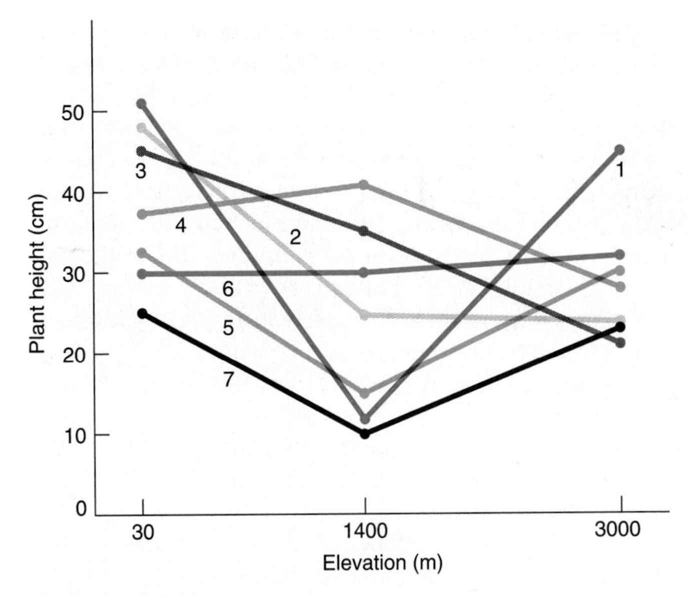

Figure 1-15 Graphic representation of the complete set of results of the type shown in Figure 1-14. Each line represents the norm of reaction of one plant.

so we could (imprecisely) speak of "large-eye" and "small-eye" genotypes. In this case, the differences in phenotype between genotypes are much greater than the variation within a genotype caused by development in different environments. But the variation for a single *Achillea* genotype in different environments is so great that the norms of reaction cross one another and form no consistent pattern. In this case, it makes no sense to identify a genotype with a particular phenotype except in terms of response to particular environments.

Developmental Noise

Thus far, we have assumed that a phenotype is uniquely determined by the interaction of a specific genotype and a specific environment. But a closer look shows some further unexplained variation. According to Figure 1-13, a *Drosophila* of *wild-type* genotype raised at 16°C has 1000 facets in each eye. In fact, this is only an average value; one fly raised at 16°C may have 980 facets and another may have 1020. Perhaps these variations are due to slight fluctuations in the local environment or slight differences in genotypes. However, a typical count may show that a fly has, say, 1017 facets in the left eye and 982 in the right eye. In another fly, the left eye has slightly fewer facets than the right eye. Yet the left and right eyes of the same fly are genetically identical. Furthermore, under typical experimental conditions, the fly develops as a larva (a few millimeters long) burrowing in homogeneous artificial food in a laboratory bottle and then completes its development as a pupa (also a few millimeters long) glued vertically to the inside of the glass high above the food surface. Surely the environment does not differ significantly from one side of the fly to the other! But if the two eyes experience the same sequence of environments and are identical genetically, then why is there any phenotypic difference between the left and right eyes?

Differences in shape and size are partly dependent on the process of cell division that turns the zygote into a multicellular organism. Cell division, in turn, is sensitive to molecular events within the cell, and these may have a relatively large random component. For example, the vitamin biotin is essential for *Drosophila* growth, but its *average* concentration is only one molecule per cell! Obviously, the rate of any process that depends on the presence of this molecule will fluctuate as the concentration of biotin varies. But cells can divide to produce differentiated eye cells only within the relatively short developmental period during which the eye is being formed. Thus, we would expect random variation in such phenotypic characters as the number of eye cells, the number of hairs, the exact shape of small features, and the variations of neurons in a very complex central nervous system—even when the genotype and the environment are precisely fixed. Even such structures as the very simple nervous systems of nematodes vary at random for this reason. Random events in development lead to variation in phenotype; this variation is called **developmental noise.**

> Message In some characteristics, such as eye cells in *Drosophila,* developmental noise is a major source of the observed variations in phenotype.

Like noise in a verbal communication, developmental noise adds small random variations to the predictable development governed by norms of reaction. Adding developmental noise to our model of phenotypic development, we obtain something like Figure 1-16. With a given genotype and environment, there is a range of possible outcomes for each developmental step. The developmental process does contain feedback systems that tend to hold the deviations within certain bounds, so that the range of deviation does not increase indefinitely through the many steps of development. However, this feedback is not perfect. For any given genotype, developing in any given sequence of environments, there remains some uncertainty as to the exact phenotype that will result.

Techniques of Genetic Analysis

Our discussion thus far has been based on the wisdom of hindsight. With the wealth of genetic knowledge we now share, we can make generalizations about DNA,

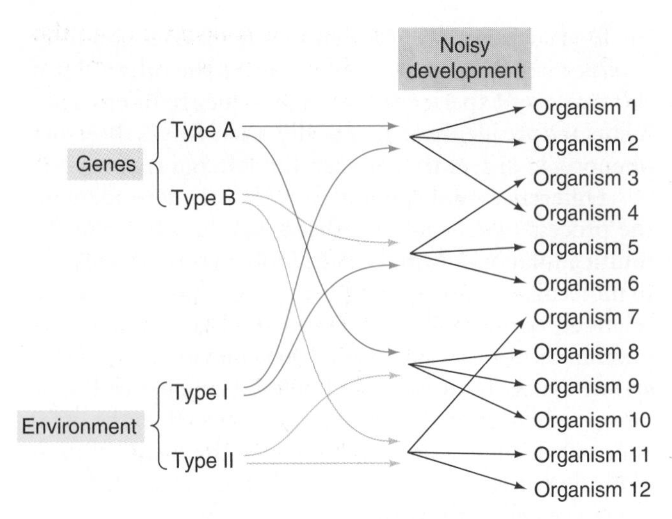

Figure 1-16 A model of phenotypic determination that shows how genes, environment, and developmental noise interact to produce a phenotype.

genes, phenotypes, and genotypes as though these concepts were self-evident. But obviously this was not always the case; the current wisdom was acquired only after extensive genetic research over the years. Mendel, for example, almost certainly was completely without a conceptual basis for his research at the beginning of his work, but he was able to piece together his genetic principles from the results of his many experiments. This is true of genetic research in general: we start in the unknown, and then ideas and facts emerge out of experimentation. But how does everyday genetics work? We shall explore the answer to this question in much of the rest of this book, but we begin here with an overview of the principles of genetic research.

We have seen that the process of identifying the specific hereditary components of a biological system is called **genetic dissection.** In the same way that an anatomist probes biological structure and function with a scalpel, the geneticist probes biological structure and function armed with genetic variants, usually abnormal ones. If the geneticist is interested in a biological process X, he or she embarks on a search for genetic variants that affect X. Each variant identifies a separate component of X. In much the same way that a novice auto mechanic can learn a lot about how an internal combustion engine works by pulling out a spark-plug lead, for example, the geneticist "tinkers" with a living system. This approach is tremendously effective in charting the unknown — the invisibility and magnitude of which are often not appreciated by those who have not attempted some kind of research — and represents a truly powerful tool.

Message Genetic dissection is a powerful way of discovering the components of any biological process.

Geneticists use and analyze hereditary variants from the molecular level, probing cellular and organismal processes, all the way up to the population level, where the variants reveal evolutionary processes. The spectrum of variants that has been obtained and studied effectively by geneticists is staggering; as we have seen, there are variants affecting shape, number, biochemical function, and so on. In fact, the general finding has been that genetic variants can be obtained for virtually any biological structure or process of interest to an investigator.

Recall that the genotypes underlying a set of variants of a given biological character fall into either of two broad categories: (1) those whose norms of reaction do not overlap, and (2) those whose norms of reaction do overlap. The techniques of genetic analysis are different for the two categories.

Geneticists have elucidated the cellular and molecular processes of organisms mostly by analyzing genotypes with nonoverlapping norms of reaction. The appeal of such systems from the experimenter's viewpoint is that the environment can virtually be ignored, because each genotype produces a discrete, identifiable phenotype. In fact, geneticists have deliberately sought and selected just such variants. Experiments become much easier — actually, they only become possible — when the observed phenotype is an unambiguous indication of a particular genotype. After all, we cannot observe genotypes; we can visibly distinguish only phenotypes.

The availability of such variants made Mendel's experiments possible. He used genotypes that were established horticultural varieties. For example, in one set of variants, one genotype always produces tall pea plants and another always produces dwarf plants. Had Mendel chosen variants with genotypes that have overlapping norms of reaction, he would have obtained complex and patternless results like those in the *Achillea* experiment (Figure 1-14b) and could never have identified the simple relationships that became the foundation of genetic understanding. Much of modern molecular genetics is based on research with variants of bacteria that similarly show distinctive characters with genotypes whose norms of reaction do not overlap. For instance, modern DNA manipulation technology relies heavily on selection systems based on bacterial genes for drug resistance; the expression of such genes is very clear-cut and reliable.

The variants in the nonoverlapping category are often strikingly different. Many such variants could never survive in nature, simply because they are so developmentally extreme. In other cases, such as the winged and wingless fruits of *Plectritis,* the strikingly different forms do appear regularly in natural populations.

Reflecting the importance of discrete genetic variants, most of the rest of this book is devoted to the analysis of this kind of variation. The story begins with Mendel's research and theories, proceeds through classical genetics, and ends with the startling discoveries of molecular biology. It must be emphasized that the entire development of this knowledge critically depended on the availability of phenotypes that have simple relationships to genotypes. Because our study of genetic analysis necessarily puts so much emphasis on such simple trait differences, you may get the impression that they represent the most common relationship between gene and trait. They do not. Geneticists have to pick and choose to find sets of variants whose genotypes correspond simply to phenotypes. For example, although hundreds of *Drosophila* mutations have been well studied, only about one-quarter are ideally suited for the purposes of careful genetic analysis. Environment and developmental age can be quite important to the expression of even the most useful variants. The widely used mutation *purple* produces an eye color like the normal ruby color in very young flies, which darkens to a distinguishable difference only with age. The mutation *Curly*, one of the most important tools in genetic analysis in *Drosophila*, results in curled wings at 25°C (Figure 1-17) but in normally straight wings at 19°C.

Simple one-to-one relationships of genotype to phenotype dominate the world of experimental genetics, but in the natural world such relationships are rare. When obtained for organisms from natural populations, norms of reaction for size, shape, color, metabolic rate, reproductive rate, and behavior almost always turn out to be like those of *Achillea*. The relationships of genotype to phenotype in nature are almost always one-to-many rather than one-to-one. This explains the rarity of discrete phenotypic classes in natural populations.

Obviously, the analysis of these one-to-many relationships is far more complex. The researcher is confronted with a bewildering range of phenotypes. Special statistical techniques must be used to disentangle the genetic, environmental, and noise components. Chapter 24 deals with such techniques.

Figure 1-17 The curled wing variant of *Drosophila melanogaster*, caused by a simple genetic change in one gene. In wild-type flies, the wings are flat. (Peter Bryant/BPS)

There is another problem that is distinct from the purely analytic difficulties. The major discoveries of genetics came out of studies on phenotypes with simple relationships with genotypes. Although these discoveries have revolutionized pure and applied biology, great care must be taken in extrapolating these ideas to the type of variation found in natural populations, because the phenotypes underlying this variation nearly always show complex relationships with genotypes. For example, finding that the difference between yellow-bodied and gray-bodied lab stocks of *Drosophila* is due to a simple gene difference does not tell us much about the variation of pigment intensity in natural populations of *Drosophila*, and even less about skin pigment variation in human populations.

Message In general, the relationship between genotype and phenotype cannot be extrapolated from one species to another or even between phenotypic traits that seem superficially similar within a species. A genetic analysis must be carried out for each particular case.

Summary

Genetics is the study of genes at all levels from molecules to populations. As a modern discipline, began in the 1860s with the work of Gregor Mendel, who first formulated the idea that genes exist. We now know that a gene is a functional region of the long DNA molecule that constitutes the fundamental structure of a chromosome.

Genetics has had a profound impact on human affairs. Much of our food and clothing comes from genetically improved organisms, and molecular genetic engineering is expanding the scope of applied genetics. In addition, a large proportion of human ill health has a genetic component. Genetic research has also provided important insights on the way we see ourselves in relation to the organic world and to the rest of the universe.

Genetics has revolutionized biology by showing that most organisms on the planet work on a common information storage and expression system, centered on DNA —a system in which information flows from DNA to RNA to protein. Genetic dissection, moreover, is an incisive analytical tool used in all of the biological sciences.

Genes do not act in a vacuum, but interact with the environment at many levels in producing a phenotype. The relationship of genotype to phenotype across an environmental range is called the norm of reaction. Most of the major advances in genetics have come from studying laboratory systems having a simple, one-to-one correspondence of genotype to phenotype. However, in natural populations phenotypic variation generally shows a more complex relationship to genotype and not a one-to-one correspondence.

Concept Maps

Each chapter contains an exercise on drawing concept maps. As you draw these maps, you will be organizing knowledge you have just acquired. As we learn we must place new knowledge into the overall conceptual framework of the discipline. An essential part of this is to discern the interrelationships between new knowledge and previously acquired knowledge. It is not as easy as it sounds and the mind can sometimes trick us into believing that we have the overall picture. Concept maps are a good way of proving to ourselves that we really do know how the various structures, processes, and ideas of genetics interrelate. The maps can also help in pinpointing gaps in our understanding.

A selection of terms is given, and the challenge is to draw lines or arrows between the ones that you think are related, with a description on the arrow showing just what the relationship is. Draw as many relationships as possible. Force yourself to make connections, no matter how remote they might first seem. The arrow description must be clear enough that another person reading the map can understand what you are thinking. Sometimes the maps reveal a simple series of consecutive steps, sometimes a complex network of interactions. There is no one correct answer; there are many correct variations. But of course, you may make some incorrect connections, and finding these misunderstandings is part of the purpose of the exercise.

Here is an example from bioenergetics:
Draw a concept map interrelating the terms

ATP / breathing / cell respiration / cow / grass / sunlight / photosynthesis

One possible map is shown in Figure 1-18.

Here is the concept map for this chapter:
Draw a concept map interrelating as many of the following terms as possible. Note that the terms are listed in no particular order.

genotype / phenotype / norm of reaction / developmental noise / environment / development / organism

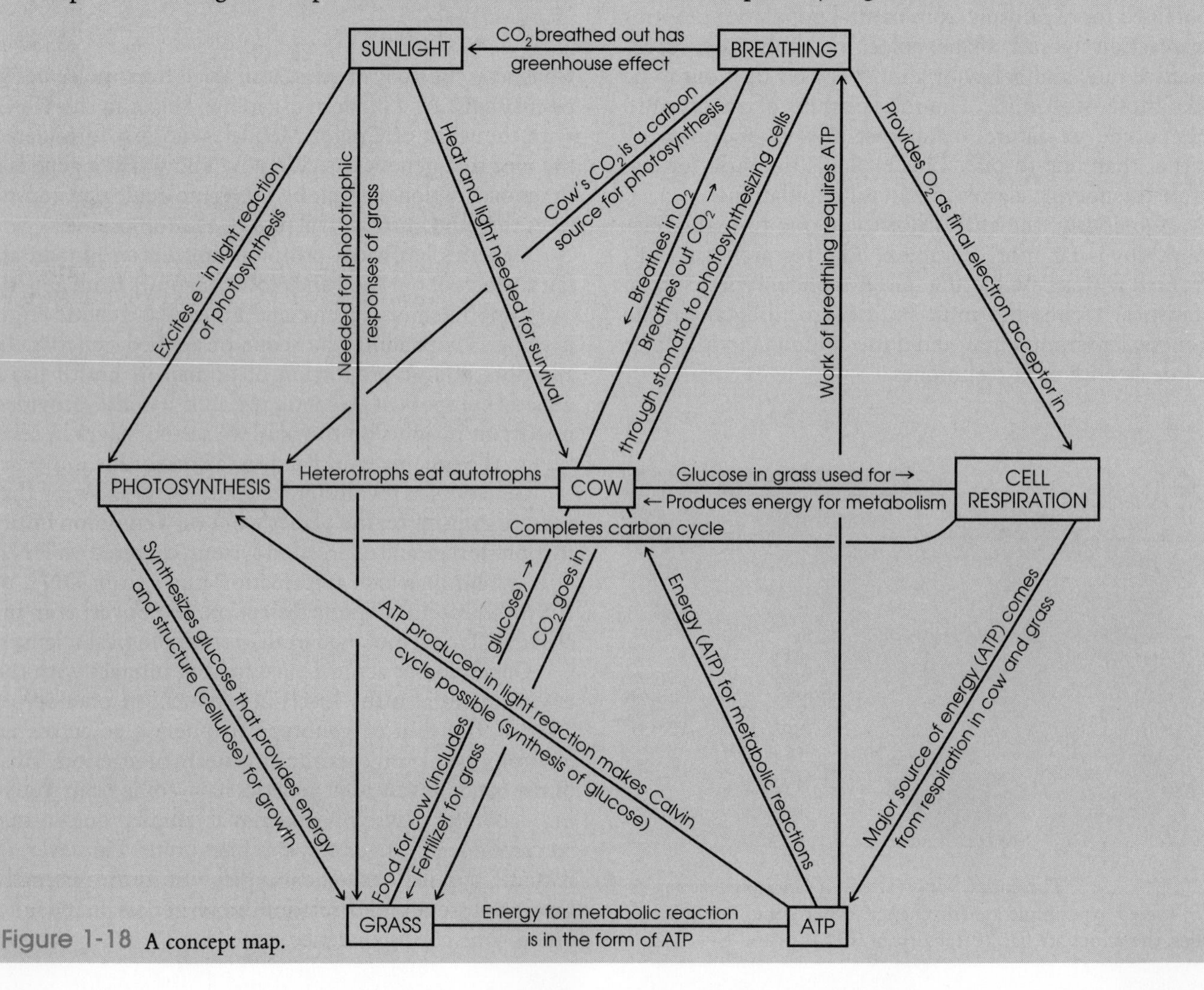

Figure 1-18 A concept map.

2
Mendelian Analysis

Flowers of *Pisum sativum* (garden pea), the experimental organism used by Gregor Mendel, founder of the science of genetics. (Jeremy Burgess/Science Photo Library/Photo Researchers)

KEY CONCEPTS

The existence of genes can be inferred by observing certain progeny ratios in crosses between hereditary variants.

A discrete character difference is often determined by a difference in a single gene.

In higher organisms, each gene is represented twice in each cell.

During sex-cell formation, each member of a gene pair separates into one-half of the sex cells.

During sex-cell formation, different genes are often observed to behave independently of one another.

The gene, the basic functional unit of heredity, is the focal point of the discipline of modern genetics. In all lines of genetic research, the gene provides the common unifying thread to a great diversity of experimentation. Geneticists are concerned with the transmission of genes from generation to generation, with the physical structure of genes, with the variation in genes, and with the ways in which genes dictate the features of a species.

In this chapter we trace how the concept of the gene arose. We shall see that genetics is, in one sense, an abstract science: most of its entities began as hypothetical constructs in the minds of geneticists and were later identified in physical form.

The concept of the gene (but not the word) was first proposed in 1865 by Gregor Mendel. Until then, little progress had been made in understanding heredity. The prevailing notion was that the spermatozoon and egg contained a sampling of essences from the various parts of the parental body; at conception, these essences somehow blended to form the pattern for the new individual. This idea of **blending inheritance** evolved to account for the fact that offspring typically show some characteristics that are similar to those of both parents. However, there are some obvious problems associated with this idea, one of which is that offspring are not always an intermediate blend of their parents' characteristics. Attempts to expand and improve this theory led to no better understanding of heredity.

As a result of his research with pea plants, Mendel proposed instead a theory of **particulate inheritance.** According to Mendel's theory, characters are determined by discrete units that are inherited intact down through the generations. This model explained many observations that could not be explained by the idea of blending inheritance. It also served well as a framework for the later, more detailed understanding of the mechanism of heredity.

The importance of Mendel's ideas was not recognized until about 1900 (after his death). His written work was then rediscovered by three scientists, after each had independently obtained the same kind of results. Mendel's work constitutes the prototype for genetic analysis. He laid down an experimental and logical approach to heredity that is still used today.

Mendel's Experiments

Mendel's studies provide an outstanding example of good scientific technique. He chose research material well suited to the study of the problem at hand, designed his experiments carefully, collected large amounts of data, and used mathematical analysis to show that the results were consistent with his explanatory hypothesis. The predictions of the hypothesis were then tested in a new round of experimentation.

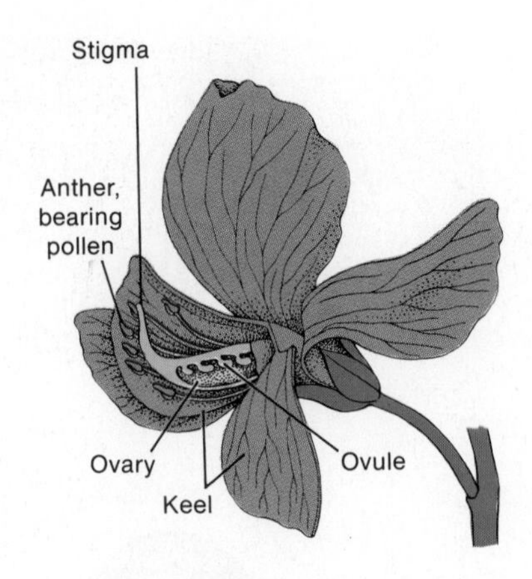

Figure 2-1 A pea flower with the keel cut and opened to expose the reproductive parts. The ovary is also shown in a cutaway view. (After J. B. Hill, H. W. Popp, and A. R. Grove, Jr., *Botany*. Copyright 1967 by McGraw-Hill.)

Mendel studied the garden pea *(Pisum sativum)* for two main reasons. First, peas were available from seed merchants in a wide array of distinct shapes and colors that could be easily identified and analyzed. Second, peas can either **self** (self-pollinate) or be cross-pollinated. The peas self because the male parts (anthers) and female parts (ovaries) of the flower — which produce the pollen containing the sperm and the ovules containing eggs, respectively — are enclosed by two petals fused to form a compartment called a keel (Figure 2-1). The gardener or experimenter can **cross** (cross-pollinate) any two pea plants at will. The anthers from one plant are removed before they have opened to shed their pollen, an operation called emasculation that is done to prevent selfing. Pollen from the other plant is then transferred to the receptive stigma with a paintbrush or on anthers themselves (Figure 2-2). Thus, the experimenter can readily choose to self or to cross the pea plants.

Other practical reasons for Mendel's choice of peas were that they are cheap and easy to obtain, take up little space, have a relatively short generation time, and produce many offspring. Such considerations enter into the choice of organism for any piece of genetic research. The choice of organism is a crucial decision and is often based on not only scientific criteria but also a good measure of expediency.

Plants Differing in One Character

Mendel chose several *characters* to study. Here, the word **character** means a specific property of an organism; geneticists use this term as a synonym for characteristic or trait.

For each of the characters he chose, Mendel obtained lines of plants, which he grew for two years to make sure they were pure. A **pure line** is a population that breeds

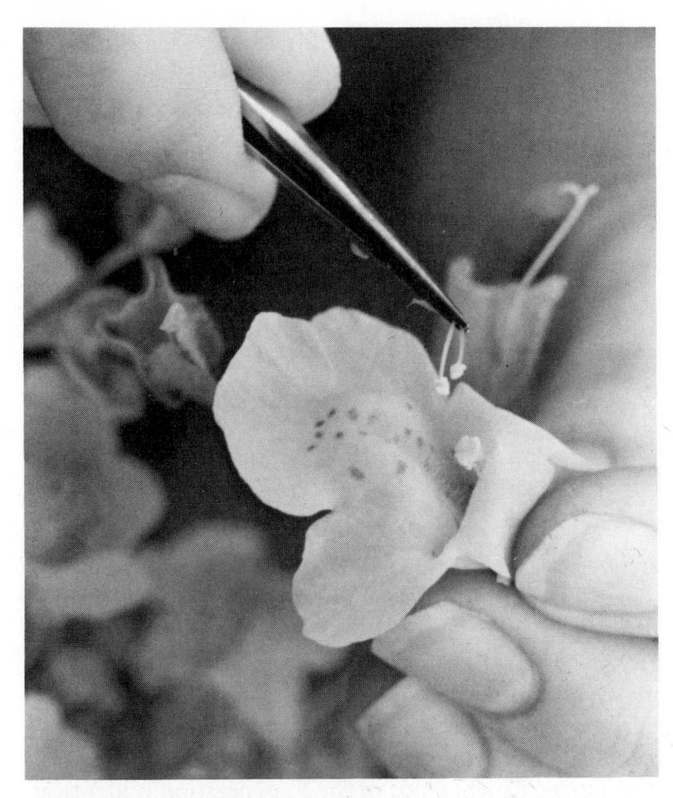

Figure 2-2 One technique of artificial cross-pollination, demonstrated with *Mimulus guttatus,* the yellow monkey flower. To transfer pollen, the experimenter touches anthers from the male parent to the stigma of an emasculated flower, which acts as the female parent. (Anthony Griffiths)

true for, or shows no variation in, the particular character being studied; that is, all offspring produced by selfing or crossing within the population show the same form of this character. By making sure his lines bred true, Mendel had made a clever beginning: he had established a fixed baseline for his future studies so that any changes observed following deliberate manipulation in his research would be scientifically meaningful; in effect, he had set up a control experiment.

Two of the pea lines Mendel grew proved to breed true for the character of flower color. One line bred true for purple flowers; the other, for white flowers. Any plant in the purple-flowered line — when selfed or when crossed with others from the same line — produced seeds that all grew into plants with purple flowers. When these plants in turn were selfed or crossed within the line, their progeny also had purple flowers, and so on. The white-flowered line similarly produced only white flowers through all generations. Mendel obtained seven pairs of pure lines for seven characters, with each pair differing in only one character (Figure 2-3).

Each pair of Mendel's plant lines can be said to show a **character difference** — a contrasting difference between two lines of organisms (or between two organisms) in one particular character. The differing lines (or individuals) represent different forms that the character may take: they can be called character forms, character variants, or **phenotypes.** The term phenotype (derived from Greek) literally means "the form that is shown"; it

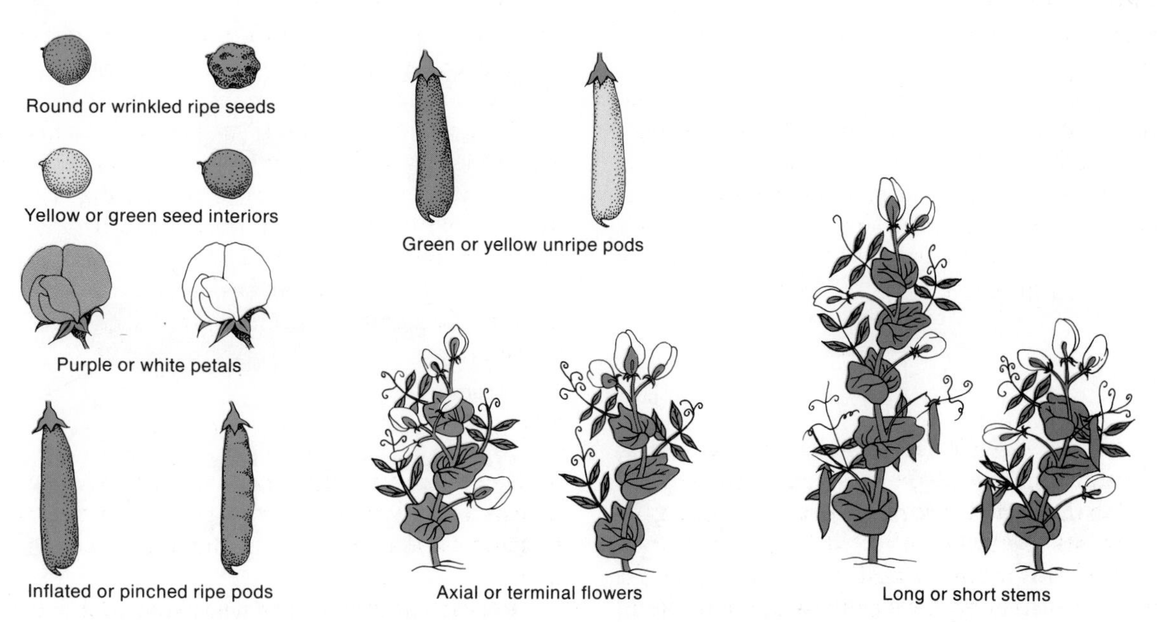

Figure 2-3 The seven character differences studied by Mendel (After S. Singer and H. Hilgard, *The Biology of People.* Copyright 1978 by W. H. Freeman and Company.)

is the term used by geneticists today. Even though such words as gene and phenotype were not coined or used by Mendel, we shall use them in describing Mendel's results and hypotheses.

Figure 2-3 shows seven pea characters, each represented by two contrasting phenotypes. Contrasting phenotypes for a particular character are the starting point for any genetic analysis. This illustrates the point made in Chapter 1 that variation is the raw material for any genetic analysis. Of course, the delineation of characters is somewhat arbitrary; an organism may be "split up" into characters in many different ways. For example, we can state one character difference of the pea plants in at least three ways:

Character	Phenotypes
flower color	purple versus white
flower purpleness	presence versus absence
flower whiteness	absence versus presence

In many cases, the description chosen is a matter of convenience (or chance). Fortunately, the choice does not alter the final conclusions of the analysis, except in the words used.

We turn now to some of Mendel's experiments with the lines breeding true for flower color. In one of his early experiments, Mendel pollinated a purple-flowered plant with pollen from a white-flowered plant. We call the plants from the pure lines the **parental generation** (P). All the plants resulting from this cross had purple flowers (Figure 2-4). This progeny generation is called the **first filial generation** (F_1). (The subsequent generations produced by selfing are symbolized F_2, F_3, and so on.)

Mendel made **reciprocal crosses**. In most plants, any cross can be made in two ways, depending on which phenotype is used as male (♂) or female (♀). For example, the two crosses

phenotype A ♀ × phenotype B ♂

phenotype B ♀ × phenotype A ♂

are reciprocal crosses. Mendel's reciprocal cross in which he pollinated a white flower with pollen from a purple-flowered plant produced the same result (all purple flowers) in the F_1 (Figure 2-5). Mendel concluded that it makes no difference which way the cross is made. If one pure-breeding parent is purple-flowered and the other is white-flowered, all plants in the F_1 have purple flowers. The purple flower color in the F_1 generation is identical to that in the purple-flowered parental plants. In this case, the inheritance obviously is not a simple blending of

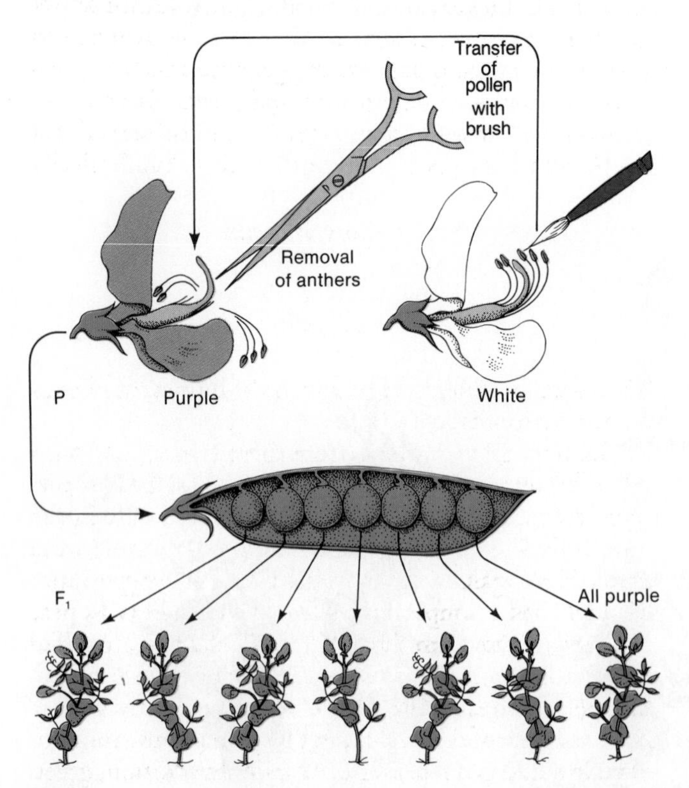

Figure 2-4 Mendel's cross of purple-flowered ♀ × white-flowered ♂.

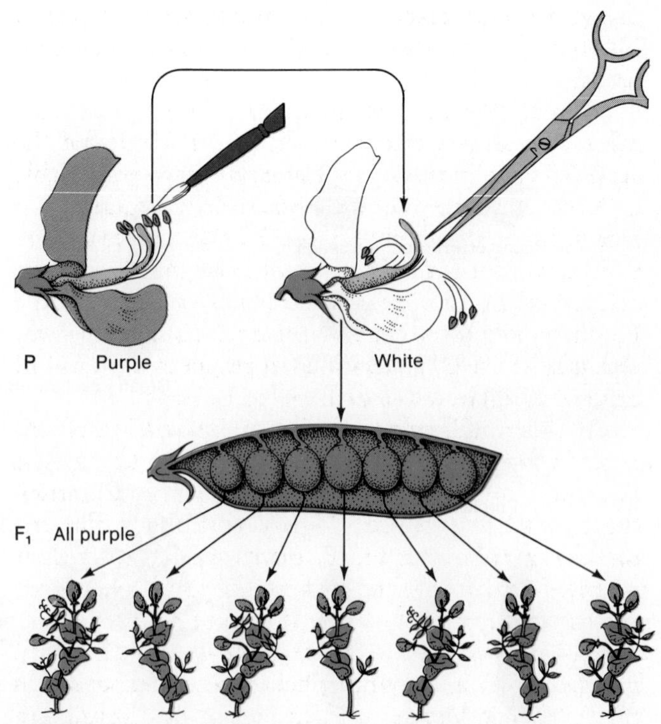

Figure 2-5 Mendel's cross of white-flowered ♀ × purple-flowered ♂.

Table 2-1. Results of All Mendel's Crosses in Which Parents Differed for One Character

Parental phenotypes	F_1	F_2	F_2 Ratio
1. Round × wrinkled seeds	All round	5474 round; 1850 wrinkled	2.96:1
2. Yellow × green seeds	All yellow	6022 yellow; 2001 green	3.01:1
3. Purple × white petals	All purple	705 purple; 224 white	3.15:1
4. Inflated × pinched pods	All inflated	882 inflated; 299 pinched	2.95:1
5. Green × yellow pods	All green	428 green; 152 yellow	2.82:1
6. Axial × terminal flowers	All axial	651 axial; 207 terminal	3.14:1
7. Long × short stems	All long	787 long; 277 short	2.84:1

purple and white colors to produce some intermediate color. To maintain a theory of blending inheritance, we would have to assume that the purple color is somehow "stronger" than the white color and completely overwhelms any trace of the white phenotype in the blend.

Next, Mendel selfed the F_1 plants, allowing the pollen of each flower to fall on its own stigma. He obtained 929 pea seeds from this selfing (the F_2 individuals) and planted them. Interestingly, some of the resulting plants were white flowered; the white phenotype had reappeared. Mendel then did something that, more than anything else, marks the birth of modern genetics: he *counted* the numbers of plants with each phenotype. This procedure had seldom, if ever, been used in genetic studies before Mendel's work. Indeed, others had obtained remarkably similar results in breeding studies but had failed to count the numbers in each class. Mendel counted 705 purple-flowered plants and 224 white-flowered plants. He noted that the ratio of 705:224 is almost a 3:1 ratio (in fact, it is 3.1:1).

Mendel repeated the crossing procedures for the six other pairs of pea character differences. He found the same 3:1 ratio in the F_2 generation for each pair (Table 2-1). By this time, he was undoubtedly beginning to believe in the significance of the 3:1 ratio and to seek an explanation for it. In all cases, one parental phenotype disappeared in the F_1 and reappeared in one-fourth of the F_2. The white phenotype, for example, was completely absent from the F_1 generation but reappeared (in its full original form) in one-fourth of the F_2 plants.

It is very difficult to devise an explanation of this result in terms of blending inheritance. Even though the F_1 flowers were purple, the plants evidently still carried the *potential* to produce progeny with white flowers. Mendel inferred that the F_1 plants receive from their parents the ability to produce both the purple phenotype and the white phenotype and that these abilities are retained and passed on to future generations rather than blended. Why is the white phenotype not expressed in the F_1 plants? Mendel used the terms **dominant** and **recessive** to describe this phenomenon without explaining the mechanism. In modern terms, the purple phenotype is dominant to the white phenotype and the white phenotype is recessive to purple. Thus the operational definition of dominance is provided by the phenotype of an F_1 established by intercrossing two pure lines. The parental phenotype that is expressed in such F_1 individuals is by definition the dominant phenotype.

Mendel went on to show that in the class of F_2 individuals showing the dominant phenotype there were in fact two genetically distinct subclasses. In this case he was working with seed color. In peas the color of the seed is determined by the genetic constitution of the seed itself, and not by the maternal parent as in some plant species. This is convenient because the investigator can treat each pea as an individual and can observe its phenotype directly without having to grow up a plant from it as must be done for flower color. This also means much larger numbers can be examined, and it is more convenient to extend studies into subsequent generations. The seed colors Mendel used were yellow and green. He crossed a pure yellow line with a pure green line, and observed that the F_1 peas that appeared were all yellow. Symbolically

P yellow × green
↓
F_1 all yellow

Therefore, by definition, yellow is the dominant phenotype and green is recessive.

Mendel grew F_1 plants from these F_1 peas, and then selfed the plants. The peas that developed on the F_1 plants represented the F_2 generation. He observed that in the pods of the F_1 plants $\frac{3}{4}$ of the peas were yellow and $\frac{1}{4}$ were green. Here again in the F_2 we see a 3:1 phenotypic ratio. Mendel took a sample consisting of 519 yellow F_2 peas and grew plants from them. These F_2 plants were selfed individually and the peas that developed were noted. Mendel found that 166 of the plants bore only yellow peas, and each of the remaining 353 plants bore a mixture of yellow and green peas in a 3:1 ratio. In addition, green F_2 peas were grown up and selfed, and were found to bear only green peas. In summary, all of the F_2 greens were evidently pure breeding like the green parental line, but of the F_2 yellows $\frac{2}{3}$ were like the F_1 yellows (producing yellow and green seeds in a 3:1 ratio) and $\frac{1}{3}$ were like the pure-breeding yellow parent. Thus, the study of the next generation (the F_3) revealed that underlying the 3:1 phenotypic ratio in the F_2 generation there was a more fundamental 1:2:1 ratio:

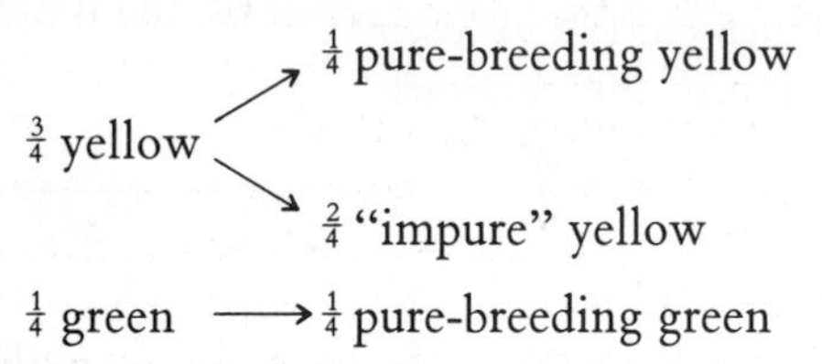

Further studies showed that such 1:2:1 ratios underlay all of the phenotypic ratios that Mendel had observed. Thus, the problem really was to explain the 1:2:1 ratio. Mendel's explanation was a classic example of a creative model or hypothesis derived from observation and well suited for testing by further experimentation. He deduced the following explanation of the 1:2:1 ratio.

1. There are hereditary determinants of a particulate nature. (Mendel saw no blending of phenotypes, so he was forced to draw this conclusion.) We now call these determinants genes.
2. Each adult pea plant has two genes—a **gene pair**—in each cell for each character studied. Mendel's reasoning here was obvious: the F_1 plants, for example, must have had one gene that was responsible for the dominant phenotype and another gene that was responsible for the recessive phenotype, which showed up only in later generations.
3. The members of the gene pairs segregate (separate) equally into the gametes, or eggs and sperm.
4. Consequently, each gamete carries only one member of each gene pair.
5. The union of one gamete from each parent to form the first cell (or zygote) of a new progeny individual is random—that is, gametes combine without regard to which member of a gene pair is carried.

These points can be illustrated diagrammatically for a general case, using A to represent the gene that determines the dominant phenotype and a to represent the gene for the recessive (as Mendel did). This is similar to the way a mathematician uses symbols to represent abstract entities of various kinds. In Figure 2-6, these symbols are used to illustrate how the above five points explain the 1:2:1 ratio.

The whole model made logical sense of the data. However, many beautiful models have been knocked down under test. Mendel's next job was to test his model. He did this by taking (for example) an F_1 plant that grew from a yellow seed and crossing it with a plant grown from a green seed. A 1:1 ratio of yellow to green seeds could be predicted in the next generation. If we let Y stand for the gene that determines the dominant phenotype (yellow seeds) and y stand for the gene that determines the recessive phenotype (green seeds), we can diagram Mendel's predictions, as shown in Figure 2-7. In this experiment, he obtained 58 yellow (Yy) and 52 green (yy), a very close approximation to the predicted 1:1 ratio and confirmation of the equal segregation of Y and y in the F_1 individual. This concept of **equal segregation** has been given formal recognition as Mendel's first law.

Mendel's First Law The two members of a gene pair segregate from each other into the gametes, so that one-half of the gametes carry one member of the pair and the other one-half of the gametes carry the other member of the pair.

Now we need to introduce some more terms. The individuals represented by Aa are called **heterozygotes,** or sometimes **hybrids,** whereas the individuals in pure lines are called **homozygotes.** In words such as these, "hetero-" means different and "homo-" means identical. Thus, an AA plant is said to be **homozygous dominant;** an aa plant is homozygous for the recessive gene, or **homozygous recessive.** As we saw in Chapter 1, the designated genetic constitution of the character or characters under study is called the **genotype.** Thus, YY and Yy, for example, are different genotypes even though the seeds of both types are of the same phenotype (that is, yellow). In such a situation the phenotype can be thought of simply as the outward manifestation of the underlying genotype. Note that underlying the 3:1 phenotypic ratio in the F_2 there is a 1:2:1 genotypic ratio of YY:Yy:yy.

Note that strictly speaking the expressions "dominant" and "recessive" are properties of the phenotype. The dominant phenotype is established in analysis by the

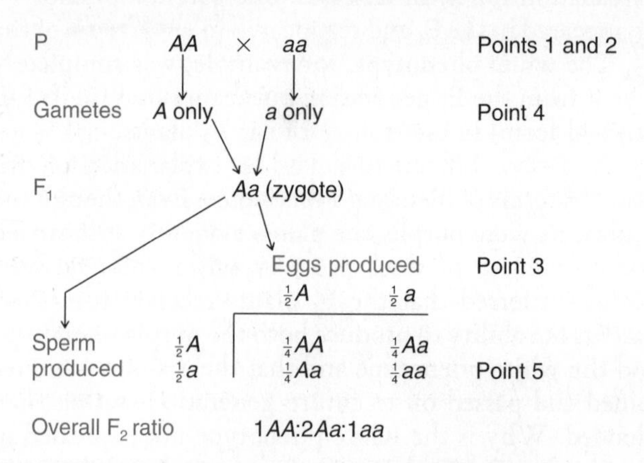

Figure 2-6 Mendel's model of the hereditary determinants of a character difference in the P, F_1, and F_2. The five points are those listed in the text.

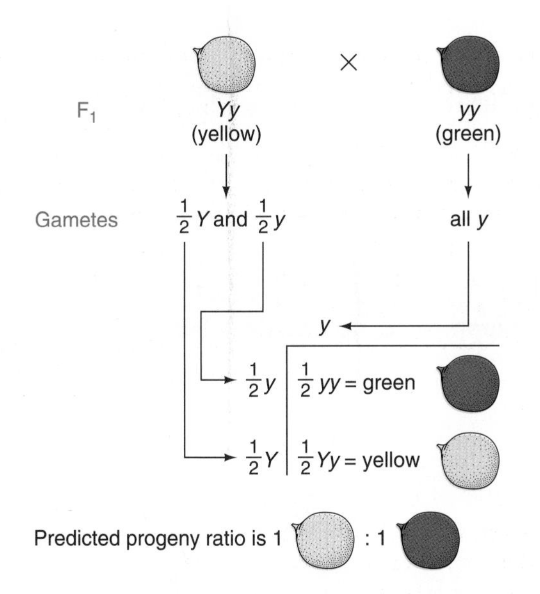

Figure 2-7 Prediction of a 1:1 phenotypic ratio in a cross of an F_1 yellow with a green in peas.

appearance of the F_1. Obviously, however, a phenotype (which is merely a description) cannot really exert dominance. Mendel showed that the dominance of one phenotype over another is in fact due to the dominance of one member of a gene pair over the other.

Let's pause to let the significance of this work sink in. What Mendel did was to develop an analytic scheme for the identification of genes regulating any biological character or function. Let's take petal color as an example. Starting with two different phenotypes (purple and white) of one character (petal color), Mendel was able to show that the difference was caused by one gene pair. Modern geneticists would say that Mendel's analysis had identified a gene for petal color. What does this mean? It means that in these organisms there is a gene that has a profound effect on the color of the petals. This gene can exist in different forms: the dominant form of the gene (represented by *C*) causes purple petals, and the recessive form of the gene (represented by *c*) causes white petals. The forms *C* and *c* are called **alleles** (or alternative forms) of that gene for petal color. They are given the same letter symbol to show that they are forms of one gene. We could express this another way by saying that there is a kind of gene, called phonetically a "see" gene, with alleles *C* and *c*. Any individual pea plant will always have two "see" genes, forming a gene pair, and the actual members of the gene pair can be *CC*, *Cc*, or *cc*. Notice that although the members of a gene pair can produce different effects, they obviously both affect the same character.

The related terms gene and allele are potentially confusing, but to clarify the concepts let us jump ahead to think about what they signify at the level of DNA. When alleles like *A* and *a* are examined at the DNA level using modern technology, it is generally found that they are identical for most of their sequence, and differ only at one or a few nucleotides out of the thousands that make up the gene. Therefore, we see that the alleles are truly different versions of the same basic gene. Looked at another way, gene is the generic term and allele is specific.

We have seen that although the terms dominant and recessive are defined at the level of phenotype, the phenotypes clearly reflect the different actions of various alleles. Therefore we can legitimately use the phrases dominant allele and recessive allele as the determinants of dominant and recessive phenotypes. (Note that this usage is more correct than the terms dominant gene and recessive gene, although these terms are sometimes used.)

How does the allele terminology relate to the gene pair concept? A gene pair can consist of identical alleles (in homozygotes) or different alleles (in heterozygotes). When a gene pair segregates as described by Mendel's first law, then it can be identical alleles that segregate, as in homozygotes, or different alleles that segregate, as in heterozygotes. Of course it will only be the heterozygous segregation that can lead to a phenotypic segregation (that is, to two separate phenotypes) in the progeny.

The basic route of Mendelian analysis for a single character is summarized in Table 2-2.

Message The existence of genes was originally inferred (and is still inferred today) by observing precise mathematical ratios in the descendants of two genetically different parental individuals.

Plants Differing in Two Characters

In the experiments described so far we have been concerned with what is sometimes called a **monohybrid cross,** the crossing or selfing of heterozygotes produced by a mating between individuals from two pure lines that differ in a single gene that controls a character difference. Now we can ask what happens in a **dihybrid cross,** in which the pure parental lines differ in two genes that control two separate character differences. We can use the same symbolism that Mendel used to indicate the genotypes of seed color (*Y* and *y*) and seed shape (*R* and *r*). *R* gives round seeds, and *r* gives wrinkled. In a monohybrid cross, a ratio of $\frac{3}{4}$ round and $\frac{1}{4}$ wrinkled is observed in the F_2 (Figure 2-8).

A pure line of *RR yy* plants, on selfing, produces seeds that are round and green. Another pure line is *rr YY;* on selfing, this line produces wrinkled yellow seeds. When Mendel crossed plants from these two lines, he obtained round, yellow F_1 seeds, as expected. The results in the F_2 are summarized in Figure 2-9. Mendel per-

Table 2-2. Summary of the Modus Operandi for Establishing Simple Mendelian Inheritance

Experimental procedure:
1. Choose pure lines showing a character difference (purple versus white flowers).
2. Cross the lines.
3. Self the F_1 individuals.

Results: F_1 is all purple; F_2 is $\frac{3}{4}$ purple and $\frac{1}{4}$ white.

Inferences:
1. The character difference is controlled by a major gene for flower color.
2. The dominant allele of this gene causes purple petals; the recessive allele causes white petals.

Symbolic interpretation:

Character	Phenotype	Genotype	Allele	Gene
Flower color	Purple (dominant)	*CC* (homozygous dominant) *Cc* (heterozygous)	*C* (dominant) *c* (recessive)	Flower-color gene
	White (recessive)	*cc* (homozygous recessive)		

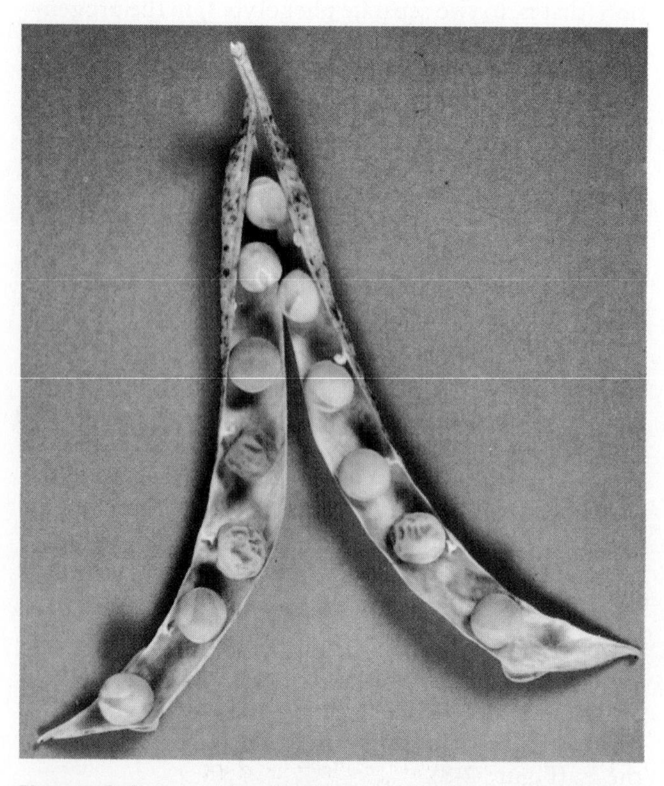

Figure 2-8 Round ($R-$) and wrinkled (rr) peas in a pod of a selfed heterozygous plant (Rr). The phenotypic ratio in this pod happens to be precisely the 3:1 ratio expected on average in the progeny of this selfing. (Recent molecular studies have shown that the wrinkled allele used by Mendel is produced by insertion into the gene of a segment of mobile DNA of the type to be discussed in Chapter 20.) (Madan K. Bhattacharyya)

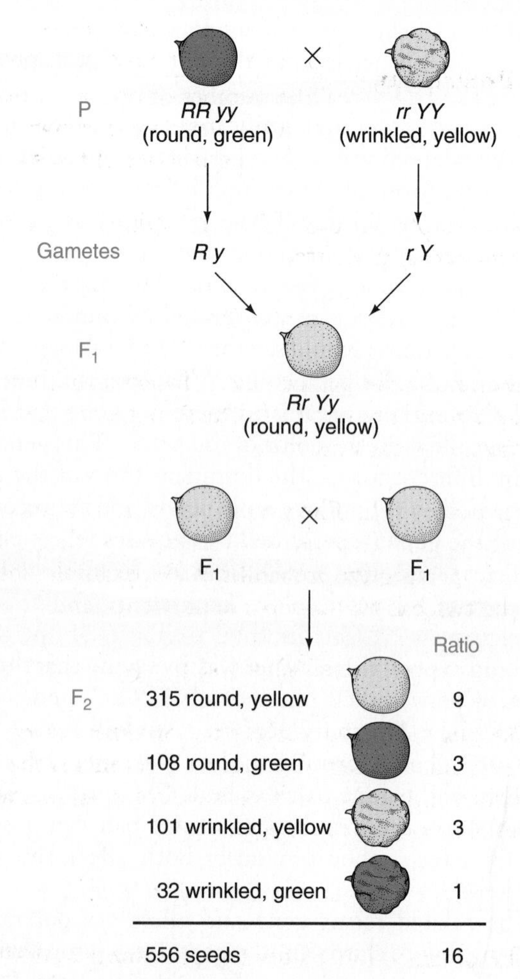

Figure 2-9 The F_2 generation resulting from a dihybrid cross.

formed similar experiments using other pairs of characters in many other dihybrid crosses; in each case, he obtained 9:3:3:1 ratios. So, he had another phenomenon to explain, and some more numbers to turn into an idea.

Mendel first checked to see whether the ratio for each gene in the dihybrid cross was the same as that for a monohybrid cross. If you look at only the round and wrinkled phenotypes and add up all the seeds falling into these two classes in Figure 2-9, the totals are $315 + 108 = 423$ round and $101 + 32 = 133$ wrinkled. Hence, the monohybrid 3:1 ratio still prevails. Similarly, the ratio of yellow seeds to green seeds is $(315 + 101):(108 + 32) = 416:140$, again very close to 3:1. From this clue, Mendel concluded that the two hereditary systems are independent. He was mathematically astute enough to realize that the 9:3:3:1 ratio is nothing more than a random combination of two independent 3:1 ratios.

This is a convenient point at which to introduce some elementary rules of probability that we will use often throughout this book.

1. *Definition of probability.*

$$\textbf{Probability} = \frac{\text{The number of times an event is expected to happen}}{\text{The number of opportunities for an event to happen (or the number of trials)}}$$

For example, the probability of rolling a 4 on a die in a single trial is written

$$p(\text{of a 4}) = \tfrac{1}{6}$$

because the die has six sides. If each side is equally likely to turn up, then on the average one 4 should turn up for each six rolls.

2. *The product rule.* The probability that two independent events will occur simultaneously is the product of their respective probabilities. For example, rolling a die twice is two independent events, and

$$p(\text{of two 4s}) = \tfrac{1}{6} \times \tfrac{1}{6} = \tfrac{1}{36}$$

3. *The sum rule.* The probability of *either one* of two independent (mutually exclusive) events is the sum of their individual probabilities. For example, when two dice are rolled together,

$$p(\text{of two 4s } or \text{ two 5s}) = \tfrac{1}{36} + \tfrac{1}{36} = \tfrac{1}{18}$$

The composition of the F_2 from the pea dihybrid cross can be predicted if the mechanism for putting R or r into a gamete is *independent* of the mechanism for putting Y or y into the gamete. The frequency of gamete types can be calculated by determining their probabilities according to the rules just given. Thus, if you pick a gamete at random, the *probability* of picking a certain type of gamete is the same as the frequency of that type of gamete.

We know from Mendel's first law that a heterozygote produces gametes in the proportions

$$Y \text{ gametes} = y \text{ gametes} = \tfrac{1}{2}$$

$$R \text{ gametes} = r \text{ gametes} = \tfrac{1}{2}$$

An $Rr\,Yy$ plant forms four types of gametes. The probability that a gamete carries R and Y is written $p(R\,Y)$. Similarly, $p(R\,y)$ denotes the probability that a gamete carries R and y. Under the assumption that the segregation of R or r into a gamete is independent of the segregation of Y or y into that same gamete, we can use the product rule to calculate the probability of each gametic combination:

$$p(R\,Y) = \tfrac{1}{2} \times \tfrac{1}{2} = \tfrac{1}{4}$$

$$p(R\,y) = \tfrac{1}{2} \times \tfrac{1}{2} = \tfrac{1}{4}$$

$$p(r\,y) = \tfrac{1}{2} \times \tfrac{1}{2} = \tfrac{1}{4}$$

$$p(r\,Y) = \tfrac{1}{2} \times \tfrac{1}{2} = \tfrac{1}{4}$$

Thus, we can represent the F_2 generation by a grid named (after its inventor) a Punnett square, as shown in Figure 2-10. The columns of the square tabulate the contributions of the male parents to the F_2 and the rows tabulate those of the female parents.

The probability of $\tfrac{1}{16}$ shown for each box in the square is also obtained using the product rule. The constitution of a zygote is the result of two independent events — the event that formed the male gamete and the event that formed the female. Thus, for example, the probability (or frequency) of $RR\,YY$ zygotes (combining an $R\,Y$ male gamete with an $R\,Y$ female gamete) is $\tfrac{1}{4} \times \tfrac{1}{4} = \tfrac{1}{16}$. Grouping all the types that look the same from Figure 2-9, we find the 9:3:3:1 ratio (now not so mysterious) in all its beauty:

round, yellow	$\tfrac{9}{16}$ or 9
round, green	$\tfrac{3}{16}$ or 3
wrinkled, yellow	$\tfrac{3}{16}$ or 3
wrinkled, green	$\tfrac{1}{16}$ or 1

The independence of the allelic segregations for two different genes is an important concept. It is called **independent assortment,** and its general statement is now known as Mendel's second law.

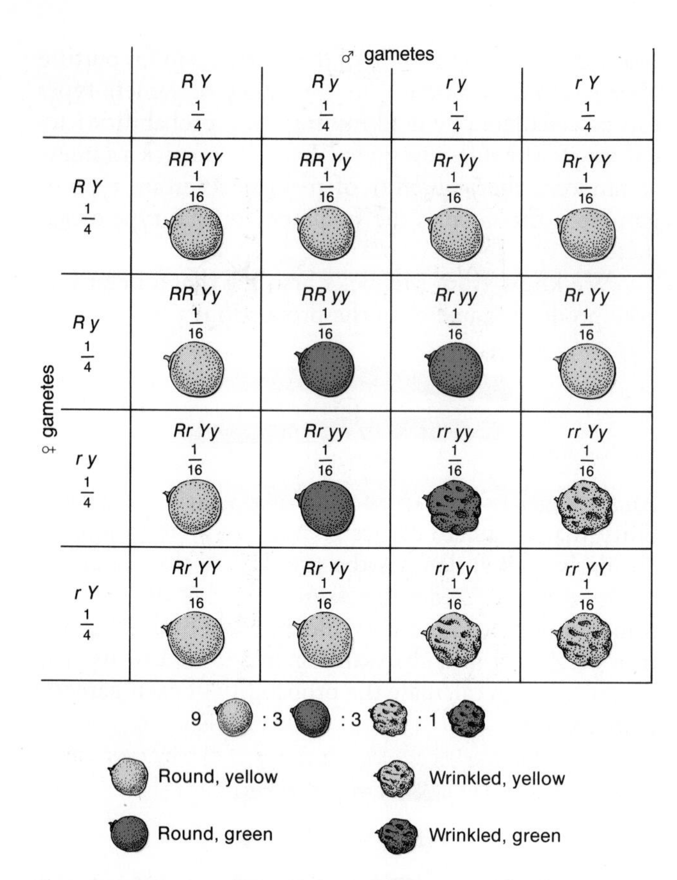

Figure 2-10 Punnett square showing predicted genetic and phenotypic constitution of the F_2 generation from the dihybrid cross shown in Figure 2-9.

Mendel's Second Law During gamete formation the segregation of alleles of one gene is independent of the segregation of alleles of another gene.

A note of warning: we shall see later that a phenomenon called gene linkage results in an important exception to Mendel's second law.

Note how Mendel's counting led to the discovery of such unexpected regularities as the 9:3:3:1 ratio and how a few simple assumptions (such as equal segregation and independent assortment) can explain this ratio that initially seems so baffling. Although it was unappreciated at the time, Mendel's quantitative approach provided the key to an understanding of genetic mechanisms.

Of course, Mendel went on to test this second law too. For example, he crossed an F_1 dihybrid *Rr Yy* with a doubly homozygous recessive strain *rr yy*. A cross to a homozygous recessive is now known as a **testcross.** Testcrosses allow the experimenter to focus on the genotype underlying a dominant phenotype in one individual because the homozygous recessive parent contributes only recessive alleles to the progeny. We shall see this kind of cross many times in this book. For his testcross, Mendel predicted that the dihybrid *Rr Yy* would produce the gametic types *R Y, R y, r y,* and *r Y* in equal frequency—that is, as shown along one edge of the Punnett square in Figure 2-10, in the frequencies $\frac{1}{4}$, $\frac{1}{4}$, $\frac{1}{4}$, and $\frac{1}{4}$. On the other hand, because it is homozygous, the *rr yy* plant produces only one gamete type (*r y*), regardless of equal segregation or independent assortment. Thus, the progeny phenotypes should reflect directly the gametic types from the *Rr Yy* parent (because the recessive *r y* contribution from the *rr yy* parent does not alter the phenotype indicated by the other gamete). Hence, Mendel predicted a 1:1:1:1 ratio of *Rr Yy, Rr yy, rr yy,* and *rr Yy* progeny from this testcross, and his prediction was confirmed. He tested the concept of independent assortment intensively on four different combinations of his characters and found that it applied to every combination.

Of course, the deduction of equal segregation and independent assortment as abstract concepts that explain the observed facts leads immediately to the question of what structures or forces can account for such behavior of genes. The idea of equal segregation seems to indicate that both alleles of a gene actually exist in some kind of orderly, paired configuration from which they can separate cleanly during gamete formation (Figure 2-11). If alleles of another gene behave independently in the same way, then we have independent assortment (Figure 2-12). But this is all speculation at the present stage of our discussion, as it was after the rediscovery of Mendel's work. The actual mechanisms are now known, and in Chapter 3 we shall see that it is the chromosomal location of genes that is responsible for their equal segregation and independent assortment.

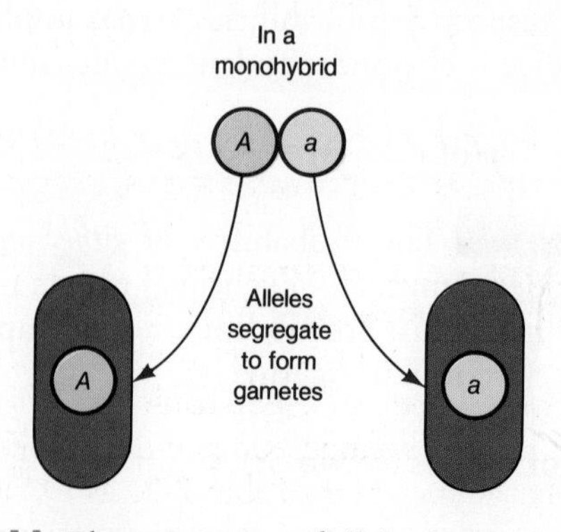

Figure 2-11 The segregation of alleles into gametes.

Message Mendel established the ground rules for genetic analysis. His work made it possible to infer the existence and nature of hereditary units and mechanisms without ever seeing them. The analysis of phenotypic frequencies in the progenies of controlled crosses is still an integral part of the experimental approach used in much of modern genetics.

Methods for Calculating Genetic Ratios

We pause here for a few words on calculating phenotypic and genotypic ratios. The Punnett square is graphic and reliable, but it is unwieldy; it is suited only for illustration, not for efficient calculation.

A branch diagram is useful for solving some problems. For example, the 9:3:3:1 phenotypic ratio can be derived by drawing a branch diagram and applying the product rule to determine frequencies. (Note the use of the convention that $R-$ represents both RR and Rr; that is, either allele can occupy the space indicated by the dash.)

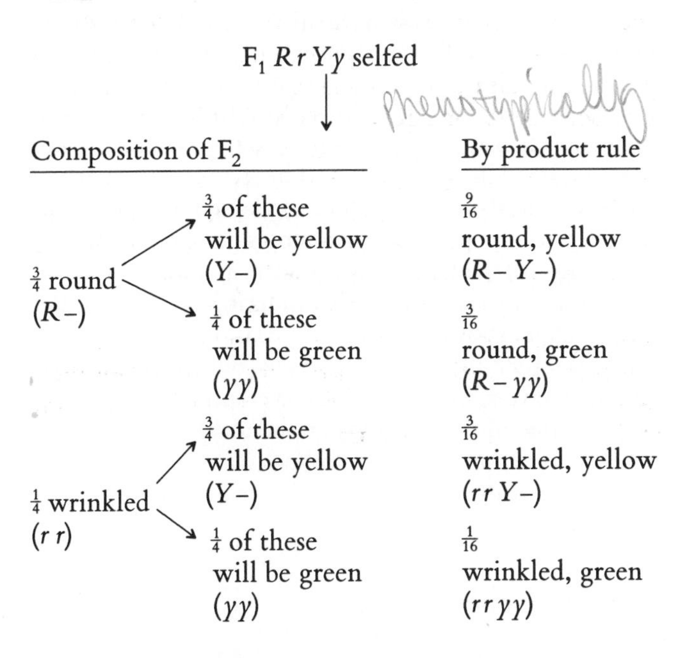

The branch diagram is, of course, a graphic expression of the product rule. It can be used for phenotypic or genotypic ratios.

The diagram can be extended to a trihybrid ratio (such as $AaBbCc \times AaBbCc$) by drawing another set of branches on the end. However, as the number of genes increases, the number of identifiable phenotypes rises startlingly and the number of genotypes climbs even more steeply, as shown in Table 2-3. With such large numbers, even the branch method becomes unwieldy.

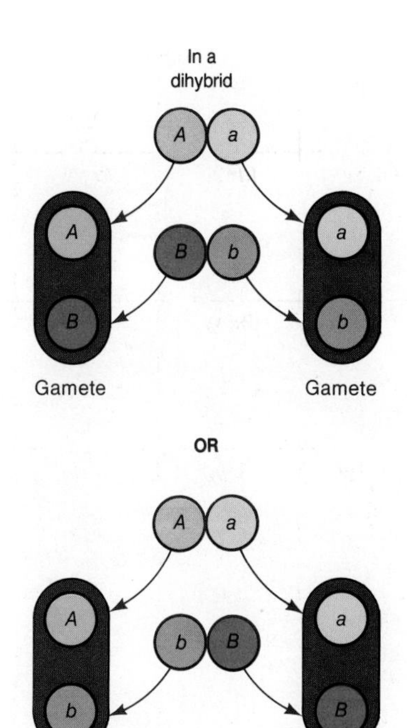

Figure 2-12 The segregation of alleles of two independent genes into gametes.

Table 2-3. Rise in Number of Phenotypes and Genotypes as the Power of the Number of Segregating Gene Pairs

Number of segregating gene Pairs	Number of phenotypes	Number of genotypes
1	2	3
2	4	9
3	8	27
4	16	81
.	.	.
.	.	.
.	.	.
n	2^n	3^n

In such cases, we must resort to devices based directly on the product and sum rules. For example, what proportion of progeny from the cross

$$AaBbCcDdEeFf \times AaBbCcDdEeFf$$

will be $AAbbCcDDeeFf$? The answer is easily obtained if the pairs of alleles all assort independently, thereby allowing use of the product rule. Thus, $\frac{1}{4}$ of the progeny will be AA, $\frac{1}{4}$ will be bb, $\frac{1}{2}$ will be Cc, $\frac{1}{4}$ will be

Figure 2-13 A 9:3:3:1 ratio in the phenotypes of kernels of corn. Each kernel represents a progeny individual. The progeny result from a self on an individual of genotype *Aa Bb*, where A = purple, a = yellow, B = smooth, and b = wrinkled. (Anthony Griffiths)

DD, $\frac{1}{4}$ will be *ee*, and $\frac{1}{2}$ will be *Ff*, so we obtain the answer by multiplying these frequencies:

$$p(AA\,bb\,Cc\,DD\,ee\,Ff) = \tfrac{1}{4} \times \tfrac{1}{4} \times \tfrac{1}{2} \times \tfrac{1}{4} \times \tfrac{1}{4} \times \tfrac{1}{2} = \tfrac{1}{1024}$$

Let us return now to Mendel's work. When Mendel's results were rediscovered in 1900, his principles were tested in a wide spectrum of eukaryotic organisms (organisms with cells that contain nuclei). The results of these tests showed that Mendelian principles were generally applicable. Mendelian ratios (such as 3:1, 1:1, 9:3:3:1, and 1:1:1:1) were extensively reported (Figure 2-13), suggesting that equal segregation and independent assortment are fundamental hereditary processes found throughout nature. Mendel's laws are not merely laws about peas but laws about the genetics of eukaryotic organisms in general. The experimental approach used by Mendel can be extensively applied in plants. However, in some plants and in most animals, the technique of selfing is impossible. This problem can be circumvented by crossing identical genotypes. For example, an F_1 animal resulting from the mating of parents from differing pure lines can be mated to its F_1 siblings (brothers or sisters) to produce an F_2. The F_1 individuals are identical for the genes in question, so the F_1 cross is equivalent to a selfing.

Simple Mendelian Genetics in Humans

Other systems present some special problems in the application of Mendelian methodology. One of the most difficult, yet most interesting, is the human species. Obviously, controlled crosses cannot be made, so human geneticists must resort to scrutinizing records in the hope that informative matings have been made by chance. Such a scrutiny of records of matings is called **pedigree analysis.** A member of a family who first comes to the attention of a geneticist is called the **propositus.** Usually the phenotype of the propositus is exceptional in some way (for example, the propositus might be a dwarf). The investigator then traces the history of the character in the propositus back through the history of the family and draws up a family tree or pedigree, using certain standard symbols given in Figure 2-14. (The terms autosomal and sex-linked in the figure will be explained later; they are included to make the list of symbols complete.)

Many pairs of contrasting human phenotypes are determined by pairs of alleles inherited in exactly the same manner shown by Mendel's peas. Pedigree analysis can reveal such inheritance patterns, but the clues in the pedigree have to be interpreted differently depending on whether one of the contrasting phenotypes is a rare disorder or whether both phenotypes of a pair are part of normal variation. These are considered separately in the following sections.

Medical Genetics

Medicine is concerned with the disorders of human beings, and many of these disorders are inherited as dominant or recessive phenotypes in a simple Mendelian manner. Generally, the propositus, or index case, is an individual who comes to the attention of a physician as a patient. If there is a family history of the disorder, a pedigree is constructed as far as available information allows, and then the pedigree is analyzed. Let us consider the inheritance of recessives first.

The unusual condition of a recessive disorder is determined by a recessive allele, and the corresponding

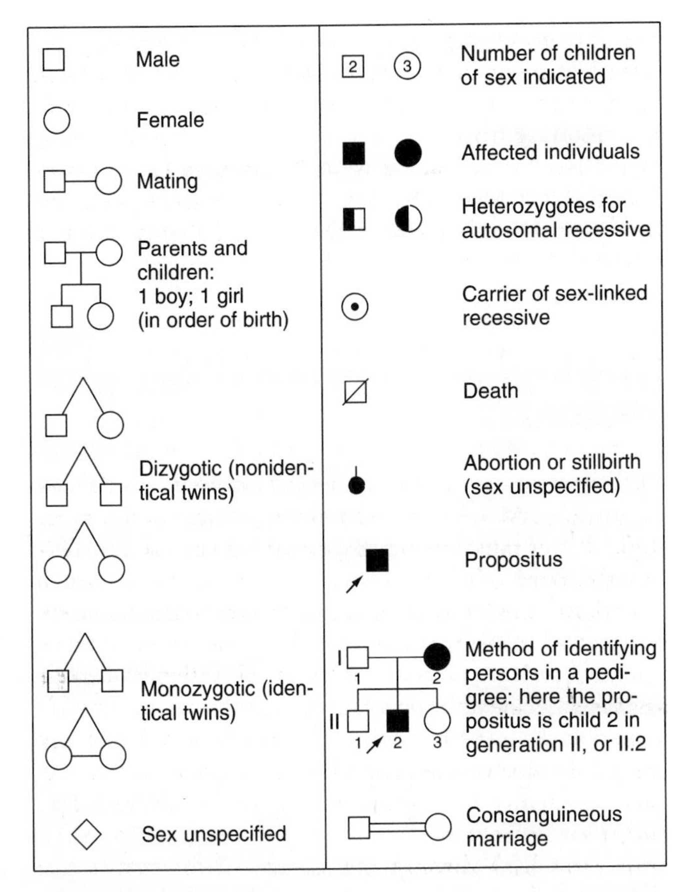

Figure 2-14 Symbols used in human pedigree analysis. (After W. F. Bodmer and L. L. Cavalli-Sforza, *Genetics, Evolution, and Man*. Copyright 1976 by W. H. Freeman and Company.)

unaffected phenotype is determined by a dominant allele. For example, the human disease phenylketonuria (PKU) is inherited in a simple Mendelian manner as a recessive phenotype, with PKU determined by an allele p and the normal condition by P. Therefore, sufferers from this disease are of genotype pp, and people who do not have the disease are either PP or Pp. What patterns in a pedigree would reveal such an inheritance? The two key points are that generally the disease appears in the progeny of unaffected parents and that the affected progeny includes both males and females. When we know that both male and female progeny are affected, we can assume that we are dealing with simple Mendelian inheritance, and not a special type of sex-linked inheritance that we will discuss in Chapter 3. The following typical pedigree illustrates the key point that affected children are born to unaffected parents:

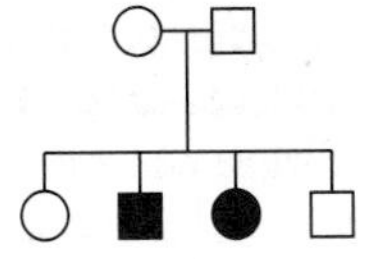

From this pattern we can immediately deduce simple Mendelian inheritance of the recessive allele responsible for the exceptional phenotype (indicated by shading). Furthermore, we can deduce that the parents are both heterozygotes, say $A\,a$: both must have an a allele because each contributed an a allele to each affected child, and both must have an A allele because they are phenotypically normal. We can identify the genotypes of the children (in the order shown) as $A\,-$, $a\,a$, $a\,a$, and $A\,-$. Hence, the pedigree can be rewritten

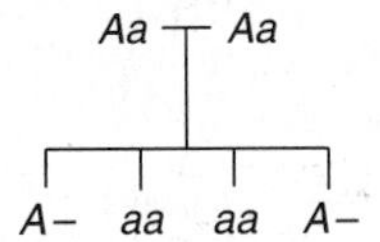

Notice another interesting feature of pedigree analysis: even though Mendelian rules are at work, Mendelian ratios are rarely observed in families because the sample size is too small. In the above example, we see a 1:1 phenotypic ratio in the progeny of a monohybrid cross. If the couple were to have, say, 20 children, the ratio would be something like 15 unaffected children and 5 with PKU (a 3:1 ratio), but in a sample of four any ratio is possible, and all ratios are commonly found.

The pedigrees of recessive disorders tend to look rather bare, with few shaded symbols. A recessive condition shows up in groups of affected siblings, and the people in earlier and later generations tend not to be affected. To understand why this is so, it is important to have some feel for the genetic structure of populations underlying such rare conditions. By definition, if the condition is rare, most people do not carry the abnormal allele. Furthermore, most of those people who do have the abnormal allele are heterozygous for it rather than homozygous. The basic reason that heterozygotes are much more common than recessive homozygotes is that to be a recessive homozygote, both of your parents must have had the a gene, but to be a heterozygote all you need is one parent with the gene.

Geneticists have a quantitative way to connect the rareness of an allele with the commonness or rarity of heterozygotes and homozygotes in a population. They obtain the relative frequencies of genotypes in a population by assuming that the population is in Hardy-Weinberg equilibrium, to be fully discussed in Chapter 25. Under this simplifying assumption, if the relative proportions of two alleles A and a in a population are p and q respectively, then the frequencies of the three possible genotypes are given by p^2 for $A\,A$, $2pq$ for $A\,a$, and q^2 for $a\,a$. A numerical example illustrates this. If we assume that the frequency, q, of a recessive, disease-causing allele is $\frac{1}{50}$, then p, is $\frac{49}{50}$, and the frequency of homozygotes with the disease is $q^2 = (\frac{1}{50})^2 = \frac{1}{2500}$, and the frequency of het-

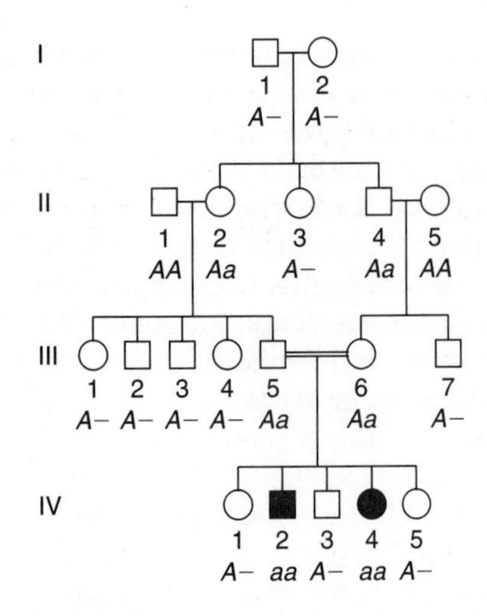

Figure 2-15 Pedigree of a rare recessive phenotype determined by a recessive allele *a*. Gene symbols normally are not included in pedigree charts, but genotypes are inserted here for reference. Note that individuals II-1 and II-5 marry into the family; they are assumed to be normal because the heritable condition under scrutiny is rare. Note also that it is not possible to be certain of the genotype in some individuals with normal phenotype; such individuals are indicated by *A*–.

erozygotes is $2pq = 2 \times \frac{49}{50} \times \frac{1}{50} =$ approximately $\frac{1}{25}$. Hence, for this example, we see that heterozygotes are 100 times more frequent than disease sufferers, and this ratio increases the rarer the allele.

The formation of an affected individual usually depends on the chance union of unrelated heterozygotes. However, inbreeding (mating between relatives) increases the chance that a mating will be between two heterozygotes. An example of a cousin marriage is shown in Figure 2-15. You can see from the figure that an ancestor who is a heterozygote may produce many descendants who are also heterozygotes. Matings between relatives thus run a higher risk of producing abnormal phenotypes caused by homozygosity for recessive alleles than do matings between non-relatives. It is for this reason that first cousin marriages contribute a large proportion of the sufferers of recessive diseases in the population.

What are some examples of recessive disorders in humans? We have already used PKU as an example of pedigree analysis, but what kind of phenotype is it? PKU is a disease in which the body cannot properly process the amino acid phenylalanine, a component of all proteins in the food we eat. This substance builds up in the body and is converted to phenylpyruvic acid, which interferes with the development of the nervous system, leading to mental retardation. Babies are now routinely tested for this processing deficiency upon birth. If the deficiency is detected, phenylalanine can be withheld by use of a special diet, and the development of the disease can be arrested.

Cystic fibrosis is another disease inherited according to Mendelian rules as a recessive phenotype. This disease has received much recent attention because the allele that causes it was isolated in 1989 and the sequence of its DNA determined. This has led to an understanding of gene function in affected and unaffected individuals, giving hope for more effective treatment. Cystic fibrosis is a disease whose most important symptom is the secretion of large amounts of mucus into the lungs, resulting in death from a combination of effects, but usually precipitated by upper respiratory infection. The mucus can be dislodged by mechanical chest thumpers, and pulmonary infection can be prevented by antibiotics, so with treatment, cystic fibrosis patients can live to adulthood.

Albinism (Figure 2-16) is a rare condition that is inherited in a Mendelian manner as a recessive phenotype in many animals, including humans. The striking "white" phenotype is caused by the inability of the body to make melanin, the pigment that is responsible for most of the black and brown coloration of animals. In humans, such coloration is most evident in hair, skin and retina, and its absence in albinos (who have the homozygous recessive genotype *aa*) leads to white hair, white skin, and eye pupils that are pink because of the unmasking of the red hemoglobin pigment in blood vessels in the retina.

Figure 2-16 Albinism in an Indian man. The phenotype is caused by homozygosity for a recessive allele, say *aa*. The dominant allele *A* determines one step in the chemical synthesis of the dark pigment melanin in the cells of skin, hair, and eye retinas. In *aa* individuals this step is nonfunctional, and the synthesis of melanin is blocked. (Joe McDonald/Visuals Unlimited)

Message In pedigrees, simple Mendelian inheritance of a recessive disorder is revealed by the appearance of the phenotype in the male and female progeny of unaffected individuals.

What about disorders inherited as dominants? Here the normal allele is recessive, and the abnormal allele is dominant. It might seem paradoxical that a rare disorder can be dominant, but remember that dominance and recessiveness are simply properties of how alleles act, and are not defined in terms of predominance in the population. A good example of a rare dominant phenotype with Mendelian inheritance is achondroplasia, a type of dwarfism (see Figure 2-17). Here, people with normal stature are genotypically *dd*, and the dwarf phenotype in principle could be *Dd* or *DD*. However, it is believed that the two "doses" of the *D* allele in *DD* individuals produce such a severe effect that this is a lethal genotype. If true, all achondroplastics are heterozygotes.

In pedigree analysis, the major clues for identifying a dominant disorder with Mendelian inheritance are that the phenotype tends to appear in every generation of the pedigree and that affected fathers and mothers transmit the phenotype to both sons and daughters. Again, the representation of both sexes among the affected offspring rules out the sex-linked inheritance mentioned in our discussion of recessive disorders. The phenotype appears in every generation because generally the abnormal allele carried by an individual must have come from a parent in the previous generation. Abnormal alleles can arise de novo by the process of genetic change called mutation. This is relatively rare, but must be kept in mind as a possibility. A typical pedigree for a dominant disorder is shown in Figure 2-18. Once again, notice that Mendelian ratios are not necessarily observed in families. As with recessive disorders, individuals bearing one copy of the rare allele (*A a*) are much more common than those bearing two copies (*A A*), so most affected people are heterozygotes, and virtually all matings involving dominant disorders are *A a* × *a a*. Therefore, when the progeny of such matings are totaled, a 1:1 ratio is expected of unaffected (*a a*) to affected individuals (*A a*).

Figure 2-17 The human achondroplasia phenotype, illustrated by a family of five sisters and two brothers. The phenotype is determined by a dominant allele, which we can call *D*, that interferes with bone growth during development. Most members of the human population can be represented as *dd* in regard to this gene. This photograph was taken upon the arrival of the family in Israel after the end of the Second World War. (UPI/Bettmann News Photos)

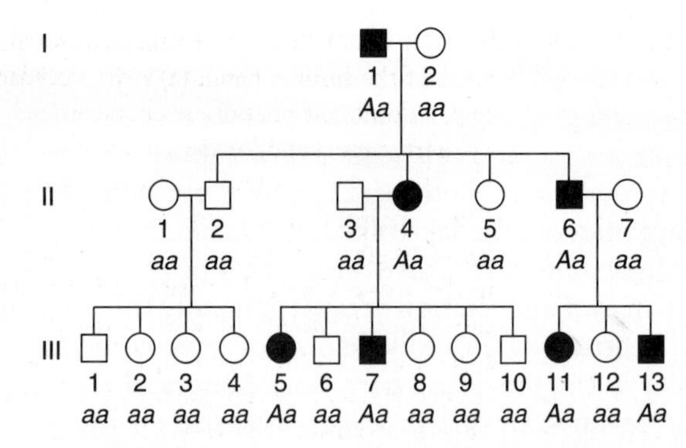

Figure 2-18 Pedigree of a dominant phenotype determined by a dominant allele A. In this pedigree, all of the genotypes have been deduced.

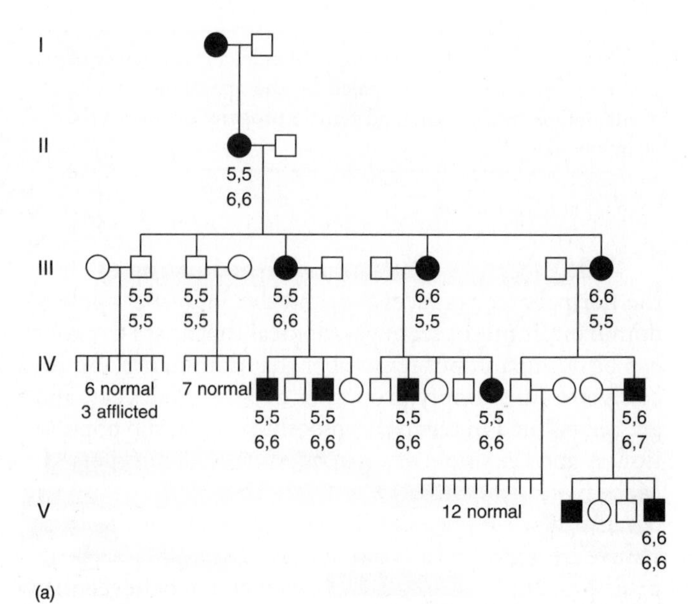

Figure 2-20a *(see legend on facing page.)*

Huntington's disease is an example of a disease inherited in a Mendelian manner as a dominant phenotype. The phenotype is one of neural degeneration, leading to convulsions and premature death. However, it is a late-onset disease, the symptoms generally not appearing until after the individual has begun to have children (see Figure 2-19). Each child of a carrier of the abnormal allele stands a 50 percent chance of inheriting the allele and the associated disease. This tragic pattern has led to a great effort to find ways of identifying individuals who carry the abnormal allele before they experience the onset of the disease. The application of molecular techniques has resulted in a promising screening procedure.

Some other rare dominant conditions are polydactyly (extra digits) and brachydactyly (short digits), shown in Figure 2-20, and piebald spotting, shown in Figure 2-21.

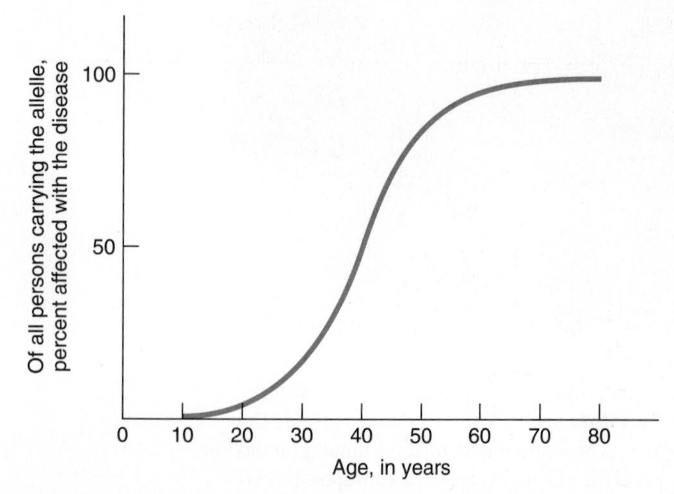

Figure 2-19 The age of onset of Huntington's disease. The graph shows that persons carrying the allele generally do not express the disease until after child-bearing age.

> **Message** Pedigrees of Mendelian dominant disorders show affected males and females in each generation and also show affected men and women transmitting the condition to equal proportions of their sons and daughters.

Normal Phenotypic Variants

All populations of plants and animals show variation, and human populations are no exception. Because isolated populations diverge genetically, some human variation corresponds to ethnic differences. We all know, however, that even within ethnic groups there is a vast amount of variation, the type of variation that allows us to distinguish each other. Normal variation is of two types, continuous and discontinuous.

Continuous variation is that shown by measurable characters such as height or weight, in which the phenotypes are arranged along a continuous spectrum. Continuous variation requires special techniques of analysis that we cover in Chapter 24. Discontinuously varying characters, on the other hand, can be classified into distinct phenotypes, such as those analyzed by Mendel. In humans there are many examples of discontinuous characters, such as brown versus blue eyes, dark versus blonde hair, chin dimples versus none, widow's peak versus none, attached versus free earlobes, and so on. Before we look at a sample pedigree, let's consider a useful new term. A set of two or more common, alternative, normal phenotypes is called a **polymorphism,** a word derived from the Greek for "many forms." The alternative phenotypes are called morphs. Primarily we will be concerned with sets of two contrasting morphs, the simplest type of polymorphism, called dimorphism. So using one

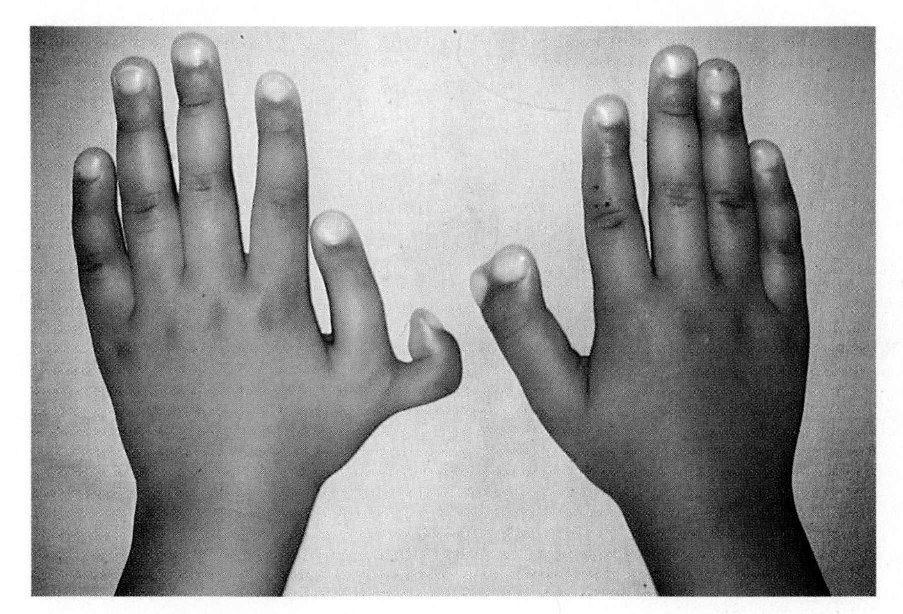

Figure 2-20 Some rare dominant phenotypes of the human hand. (a) *(left)* Polydactyly, a dominant phenotype characterized by extra fingers and/or toes, determined by an allele *P*. The numbers in the accompanying pedigree *(facing page)* show the number of fingers in the upper lines, and toes in the lower. (Note the variation in expression of the *P* allele, a topic we will cover specifically in Chapter 4.) (b) *(below)* Brachydactyly, a dominant phenotype of short fingers, determined by an allele *B*. Note the very short terminal bones in the fingers compared with those in a normal hand. The pedigree for a family with brachydactyly shows a typical inheritance pattern for a rare dominant condition. All affected individuals are *Bb* and unaffected individuals are *bb*. (a, photo: Lester Bergman & Assoc.; b, based on C. Stern, *Principles of Human Genetics,* 3d. ed., copyright 1973 by W. H. Freeman and Company.)

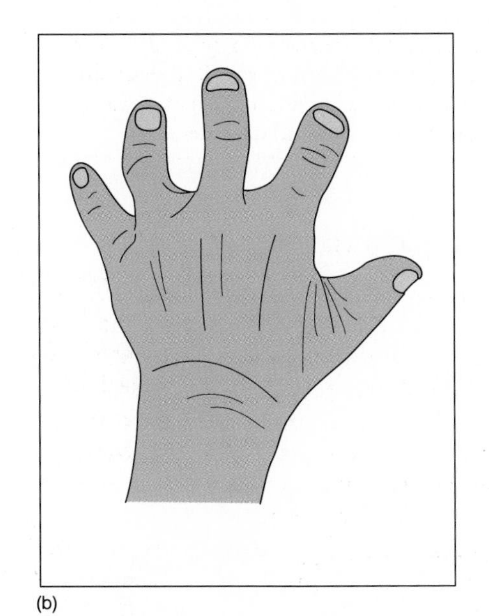

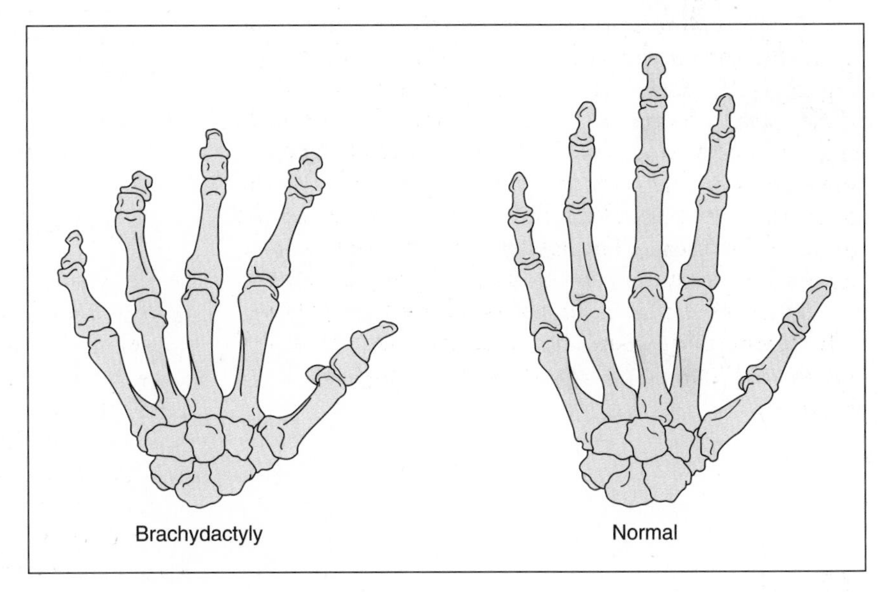

(b)

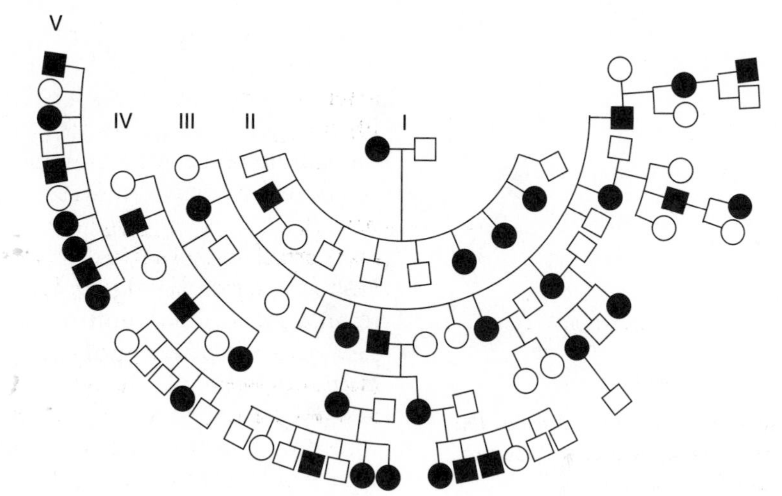

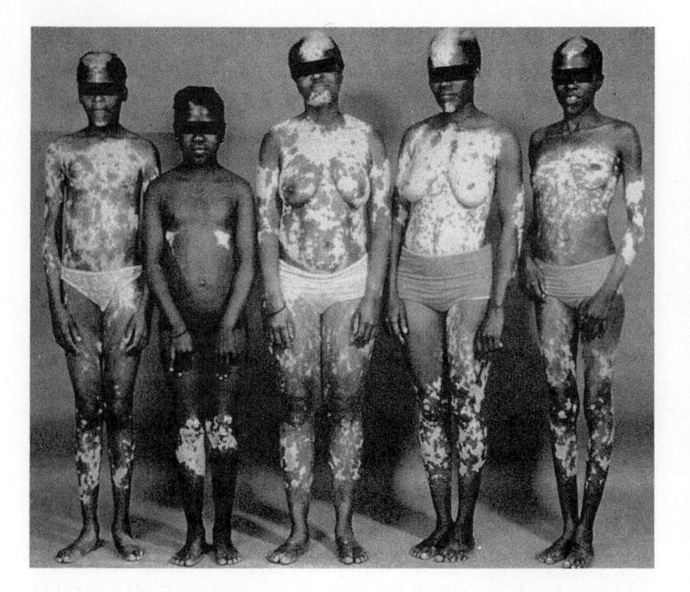

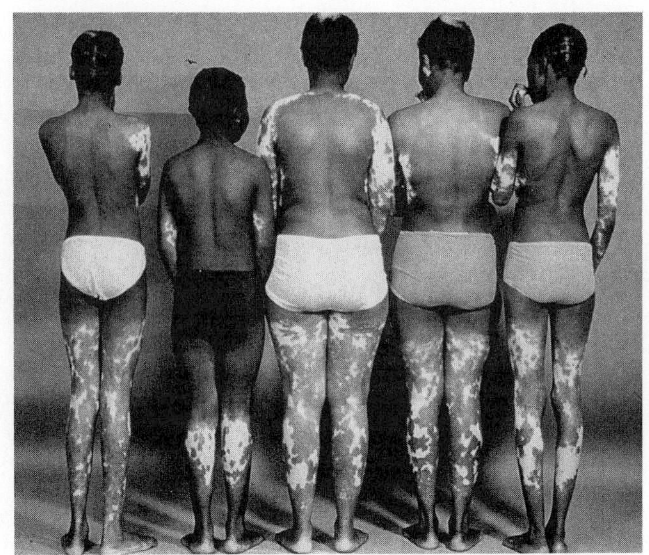

(a)

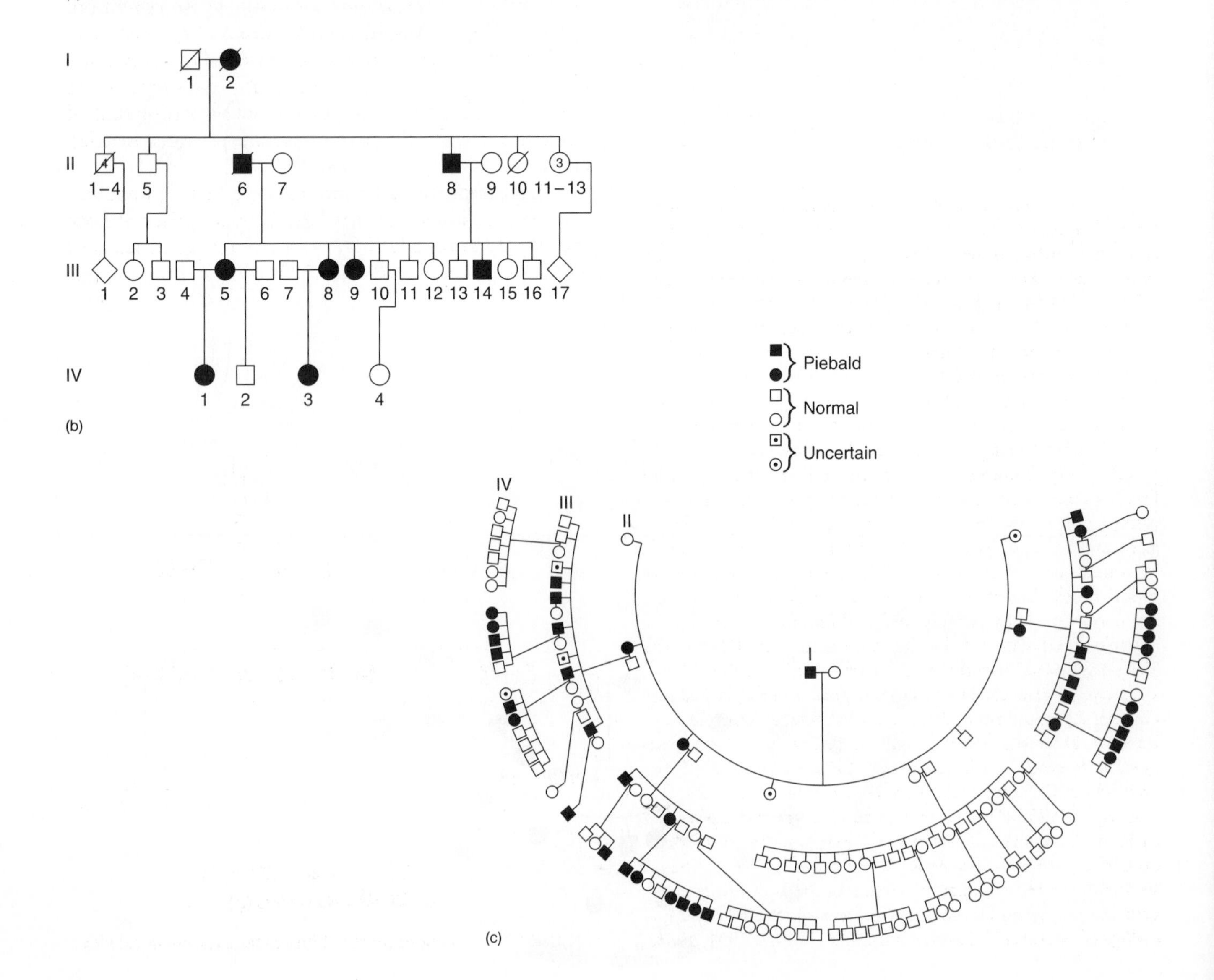
I
1 2
II
1–4 5 6 7 8 9 10 11–13
III
1 2 3 4 5 6 7 8 9 10 11 12 13 14 15 16 17
IV
1 2 3 4
Piebald
Normal
Uncertain
IV
III
II
I

(b)

(c)

Figure 2-21 Piebald spotting, a rare dominant human phenotype. Although the phenotype is encountered sporadically in all races, the patterns show up best in those with dark skin. (a) The photographs show front and back views of affected individuals IV-1, IV-3, III-5, III-8, and III-9 from the family pedigree shown in (b). Notice the variation between family members in expression of the piebald gene. A larger pedigree of a Norwegian family is shown in (c). It is believed that the patterns are caused by the dominant allele interfering with the migration of melanocytes (melanin-producing cells) from the dorsal to the ventral surface during development. The white forehead blaze is particularly characteristic and is often accompanied by a white forelock in the hair. The same basic condition is known in mice, and again the melanocytes fail to cover the top of the head and the ventral surface (d). Piebaldism is not a form of albinism; the cells in the light patches have the genetic potential to make melanin, but since they are not melanocytes they are not developmentally programmed to do so. In true albinism, the cells lack the potential to make melanin. (The DNA of the piebald allele has recently been characterized as an allele of c-*kit*, a type of gene called a proto-oncogene, to be discussed in Chapter 7.) (a, b from I. Winship, K. Young, R. Martell, R. Ramesar, D. Curtis, and P. Beighton, "Piebaldism: An Autonomous Autosomal Dominant Entity," *Clinical Genetics* 39, 1991, 330; c from C. Stern, *Principles of Human Genetics,* 3d ed., Copyright 1973 by W. H. Freeman and Company; d provided by R. A. Fleischman, University of Texas, Southwestern Medical Center, Dallas—also see R. A. Fleischman, D. L. Saltman, V. Stastny, and S. Zneimer, "Deletion of the c-*kit* Protooncogene in the Human Developmental Defect Piebald Trait," *Proceedings of the National Academy of Sciences USA* 88, 1991, 10885.)

of the above examples we would say that earlobes are dimorphic, with attached and free as the two major morphs. The morphs of a polymorphism are often determined by the alleles of one gene, inherited in the simple Mendelian manner described in this chapter.

The interpretation of pedigrees for polymorphisms is somewhat different, because by definition the morphs are common. Let's look at a pedigree for an interesting human dimorphism. Most human populations are dimorphic for the ability to taste the chemical phenylthiocarbamide (PTC). That is, people can either detect it as a foul, bitter taste, or—to the great surprise and disbelief of tasters—cannot taste it at all. From the pedigree in Figure 2-22, we can see that two tasters sometimes produce nontaster children. This makes it clear that the allele that confers the ability to taste is dominant and that the allele for nontasting is recessive. Notice that almost all persons who marry into this family carry the recessive allele either in heterozygous or homozygous condition. Such a pedigree thus differs from those of rare recessive abnormalities for which it is conventional to assume that all who marry into a family are homozygous normal. As both PTC alleles are common, it is not surprising that all but one of the family members in this pedigree married carriers of the recessive allele.

Polymorphism is an interesting genetic phenomenon. Population geneticists have been surprised at how much polymorphism there is in natural populations of plants and animals generally. Furthermore, even though

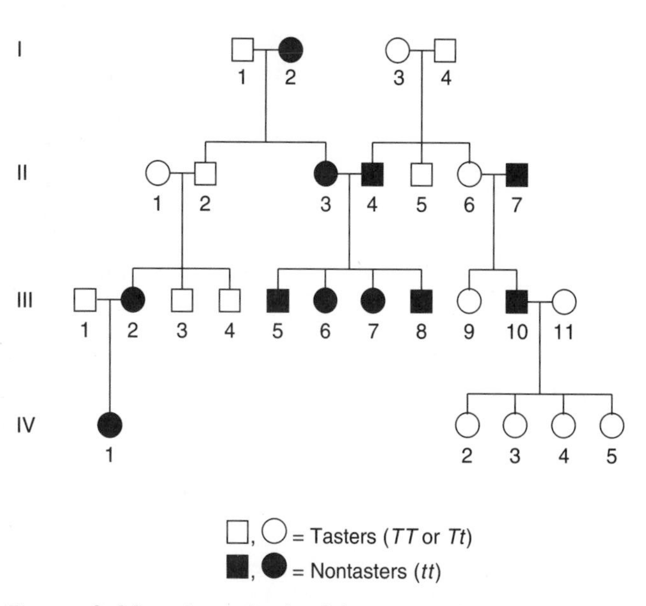

Figure 2-22 Pedigree for the ability to taste the chemical PTC.

Generation	*AA*	*Aa*	*aa*
0	0	100	0
1	25	50	25
2	37.5	25	37.5
3	43.75	12.5	43.75
4	46.875	6.25	46.875
∞	50	0	50

(a)

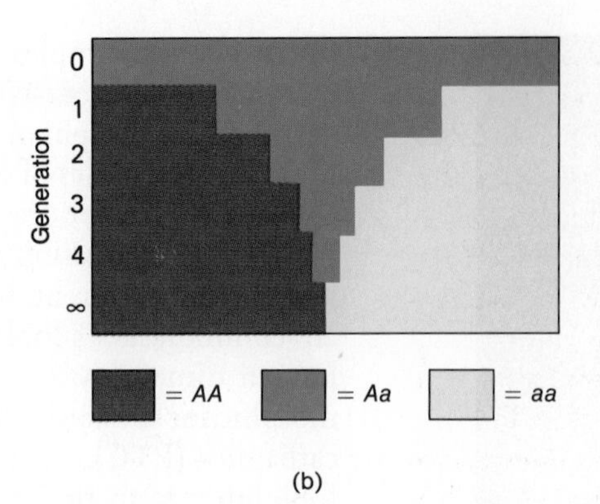

Figure 2-23 Repeated generations of selfing increase the proportion of homozygotes. (a) The percentages of the three genotypes are shown through several generations, assuming that all individuals in generation 0 are *Aa* and that all individuals reproduce at the same rate. At each generation, *AA* and *aa* breed true, but *Aa* individuals produce *Aa*, *AA*, and *aa* progeny in a 2:1:1 ratio. (b) A graphic depiction of how the proportions of genotypes change.

the genetics of polymorphisms is straightforward, there are very few polymorphisms for which there is satisfactory explanation for the coexistence of the morphs. But polymorphism is rampant at every level of genetic analysis, even down to the DNA level, and indeed polymorphisms observed at the DNA level have been invaluable as landmarks to help geneticists find their way around the chromosomes of complex organisms.

Message Populations of plants and animals (including humans) are highly polymorphic. Contrasting morphs are generally determined by alleles inherited in a simple Mendelian manner.

Simple Mendelian Genetics in Agriculture

Plant breeding methods used by Neolithic farmers were probably the same as those used until the discovery of Mendelian genetics. Basically, the approach was to select superior phenotypes that often appear as rare variants in natural populations and then propagate these over the years. Particularly desirable were pure lines of these favorable phenotypes, because such lines produced constant results over generations of planting. Without the knowledge of Mendelian genetics, how is it possible to develop pure lines? It so happens that self-pollinating plants, such as many crop plants, naturally tend to be homozygous, because selfing reduces heterozygosity in the population as is apparent from considering the genotypes of parents and the genotypes possible among their progeny:

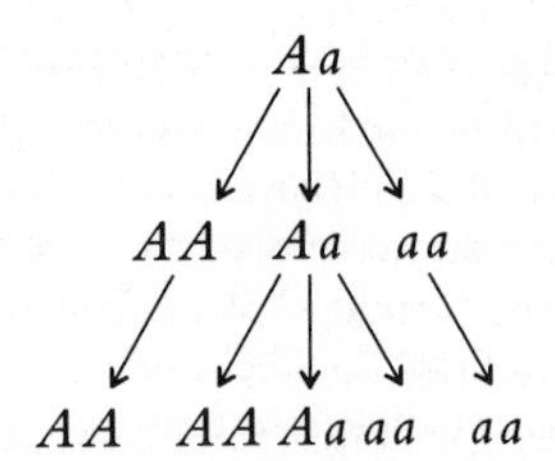

This is illustrated in more detail in Figure 2-23. Thus, pure lines have developed automatically over the years.

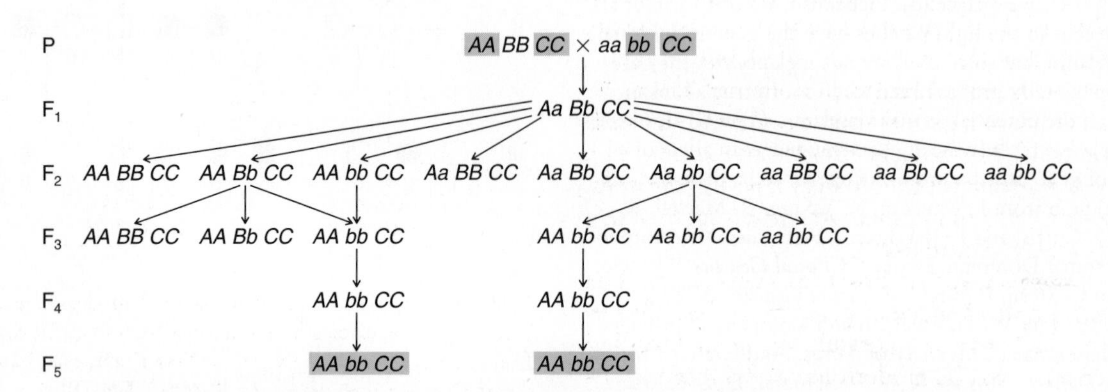

Figure 2-24 The basic technique of plant breeding. A hybrid is generated, and breeding stock is selected in subsequent generations to obtain improved pure lines. For simplicity, we assume here that the parental lines differ in only two genes (*A* and *B*). In this example, the *AA bb CC* genotype is the one desired; that is, tests on the F_2 generation show this to be the most desirable phenotype.

However, these unsophisticated breeding practices suffered from a major problem: the breeder was forced to rely on favorable natural combinations of genes. With the advent of Mendelian genetics, it became evident that favorable qualities from different lines could be combined through hybridization and subsequent gene reassortment. This procedure forms the basis of modern plant breeding.

The breeding of plants to produce new and improved genotypes works in basically the following way. For naturally self-pollinating plants, such as rice or wheat, two pure lines (each of different favorable genotype) are hybridized by manual cross-pollination and an F_1 is developed. The F_1 is then allowed to self, and its heterozygous pairs of alleles assort to produce many different genotypes, some of which represent desirable new combinations of the parental genes. A small proportion of these new genotypes will be pure-breeding already; in those that are not, several generations of selfing will produce homozygosity of the relevant genes. Figure 2-24 summarizes this method. An example of the importance of such methodology is in the breeding of highly productive, dwarf lines of rice, one of the world's staple foods. The breeding program is shown in Figure 2-25.

An example of genetic improvement in a species more familiar to most is the tomato. Anyone who has read a recent seed catalog will be familiar with the abbreviations V, F, and N next to a listed tomato variety. These represent, respectively, resistance to the pathogens *Verticillium, Fusarium,* and nematodes—resistance that has been bred into domestic tomatoes, typically from wild relatives. A useful tomato phenotype is determinate growth. The tips of the branches of most tomato species can keep growing (indeterminate growth), but in the determinate lines most branches end in a growth-arresting inflorescence. Determinate plants are bushier and more compact, and they do not need as much staking (Figure 2-26). Determinate growth is caused by a recessive allele *sp* (self-pruning), which has been crossed into modern varieties. The symbol *sp* represents a single gene; many gene symbols are more than one letter. A useful allele of another gene is *u* (uniform ripening); this allele eliminates the green patch or shoulder around the stem on the ripe fruit. Tomato geneticists have produced a large array of different morphological types of tomatoes, many of which have found their way into supermarkets and garden stores (Figure 2-27).

Such examples could be listed for many pages. The point is that simple Mendelian genetics, as described in this chapter, has provided agricultural plant breeding with its rationale and its modern methods. Entire complex genotypes may be constructed from an array of ancestral lines, each showing some desirable feature. When we think of genetic engineering, we think of the molecular techniques of the 1980s and 1990s, but genetic engineering for plant improvement began long ago.

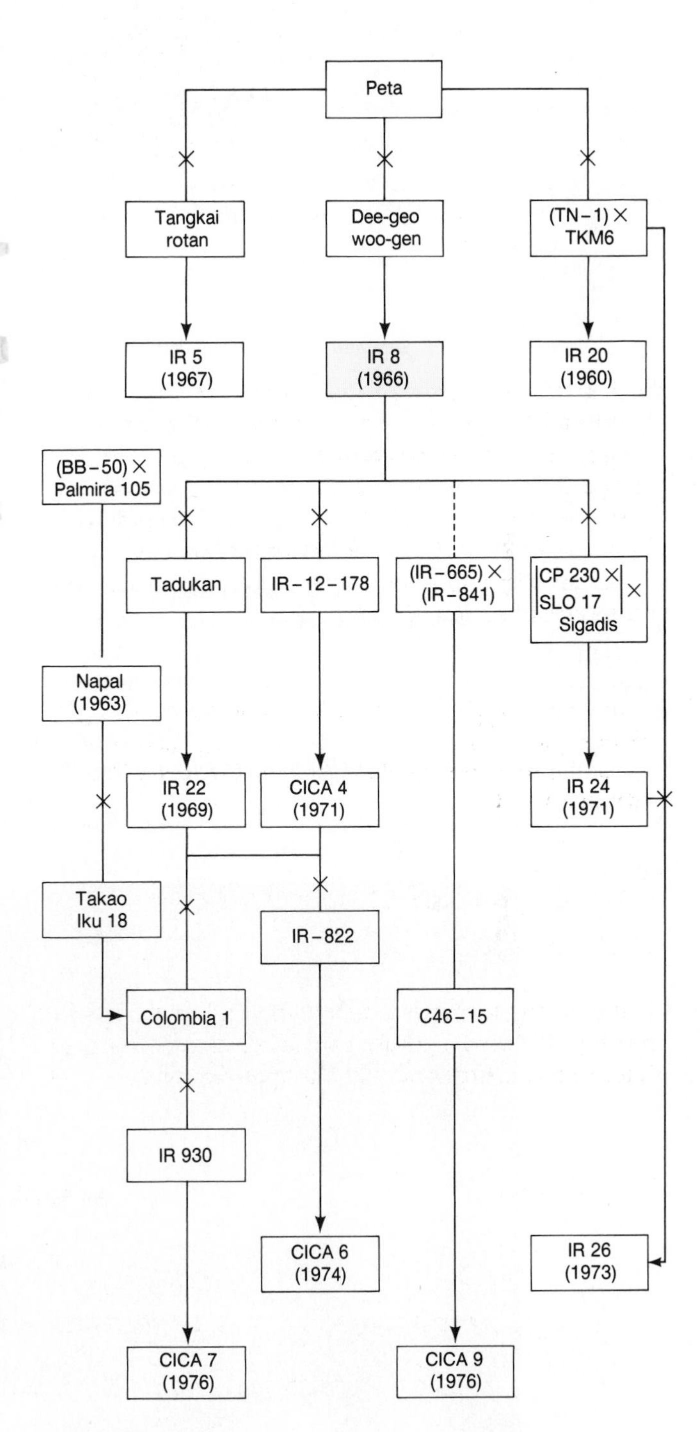

Figure 2-25 The complex pedigree of modern rice varieties. The progenitor of most modern dwarf types was IR 8, selected from a cross between the vigorous Peta (Indonesian) and the dwarf Dee-geo woo-gen (from Taiwan). Most of the other crosses represent progressive improvement of IR 8. The diagram illustrates the extensive plant breeding that produced modern crop varieties. (From Peter R. Jennings, "The Amplification of Agricultural Production." Copyright 1976 by Scientific American, Inc. All rights reserved.)

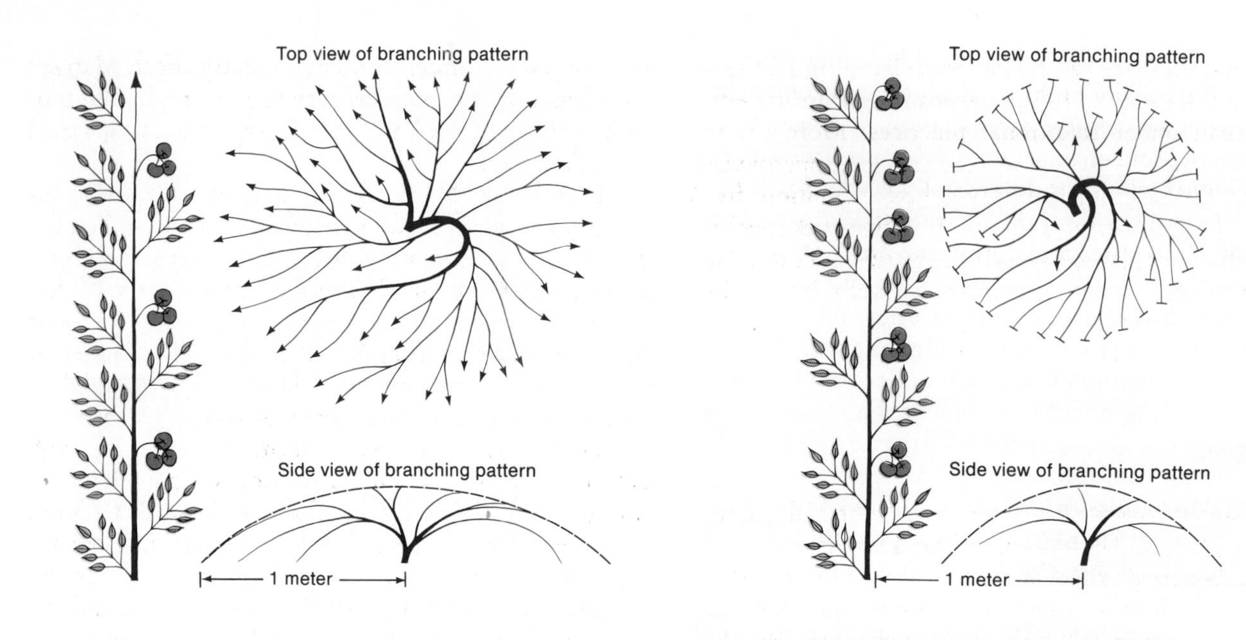

Figure 2-26 Growth characteristics of indeterminate (*Sp* –) and determinate (*spsp*) tomatoes. Arrows show growth; tomatoes indicate locations of inflorescences. Determinate has progressively fewer leaves between inflorescences, and the determinate stems end with a size-arresting inflorescence. Individual stem branches are shown, plus typical branching patterns for mature plants in top and side views. (In the top view, bars indicate inflorescences that terminate branch growth.) Note the size difference. (From Charles M. Rick, "The Tomato." Copyright 1978 by Scientific American, Inc. All rights reserved.)

Figure 2-27 Some of the many tomato phenotypes produced by breeders. (Ray Clark)

Mendelian genetics also has provided a formal theoretical basis for animal breeding, enabling a greater efficiency than under traditional practices. More recent techniques, such as the use of frozen semen, superovulation, artificial insemination, test-tube fertilization, frozen embryos, and surrogate mothers in livestock species, have enabled breeders to amplify the number of offspring of a specific genotype — a number normally limited by the life span of the animal.

Variants and Genetic Dissection

Genetic analysis, as we have seen, must start with parental differences. Without variants, no genetic analysis is possible. Where do these variants — these raw materials for genetic analysis — come from? This is a question that can be answered in full only in later chapters. Briefly, most of the variants like the ones used by Mendel (and by ancient and modern breeders of plants and animals) arise spontaneously in nature or in the breeders' populations without the deliberate action of geneticists.

Let us emphasize again that variants can range from rare to common. Some rare variants are abnormal. Undoubtedly in a natural setting many of them would be weeded out by natural selection, but they can be kept alive by nurture so that the alleles responsible can be studied. On the other hand, for many genes there are two or more common alleles in a population, resulting in genetic polymorphism — the coexistence of genetically determined variant phenotypes in a population. Although the reason for the existence of polymorphisms is usually not easy to discover, for the geneticist they are a useful source of variant alleles for study.

We have seen that the genetic analysis of variants can identify a particular gene that is important for a biological process. This central aspect of modern genetics is called **genetic dissection.** Mendel was the first genetic surgeon. Using genetic analysis, he was able to identify and distinguish among the several components of the hereditary process in a way as convincing as if he had microdissected those components. The fact that the genes he was using were for pea shape, pea color, and so on, was largely irrelevant. Those genes were being used simply as **genetic markers,** which enabled Mendel to trace the hereditary processes of segregation and assortment. A genetic marker is a variant allele that is used to label a biological structure or process throughout the course of an experiment. It is almost as though Mendel were able to "paint" two alleles two different colors and send them through a cross to see how they behaved! Genetic markers are now routinely used in genetics and in all of biology to study all sorts of processes that the marker genes themselves do not directly affect. Markers are often morphological, as in the pea example. But molecular variants (DNA and proteins) are increasingly used as markers too.

Probably without realizing it, Mendel had also invented another aspect of genetic dissection in which the precise genes used *were* important. In this type of analysis, genetic variants are used in such a way that by studying the variant gene function, we can make inferences about the "normal" operation of the gene and the process it controls. Every time a gene is identified by Mendelian analysis, it identifies a component of a biological process.

Once a gene has been identified as affecting, say, petal color in peas, we call it a major gene, but what does this really mean? It is major in that it is obviously having a profound effect on the color of the petals. But can we conclude that it is *the* single most important step in the determination of petal color? The answer is no, and the reason may be seen in an analogy. If we were trying to discover how a car engine works, we might pull out various parts and observe the effect on the running of the engine. If a battery cable were disconnected, the engine would stop; this might lead us erroneously to conclude that this cable is *the* most important part of the running of the engine. Other parts are equally necessary, and their removal could also stop or seriously cripple the engine. In a similar way it can be shown (as we will see in Chapter 4) that several genes can be identified, all of which have a major and similar effect on petal coloration.

Mendel's work has withstood the test of time and has provided us with the basic groundwork for all modern genetic study. Yet his work went unrecognized and neglected for 35 years following its publication. Why? There are many possible reasons, but here we shall consider just one. Perhaps it was because biological science at that time could not provide evidence for any real physical units within cells that might correspond to Mendel's genetic particles. Chromosomes had certainly not yet been studied, meiosis had not yet been described, and even the full details of plant life cycles had not been worked out. Without this basic knowledge, it may have seemed that Mendel's ideas were mere numerology.

Message Mendel's work was the prototypical genetic analysis. As such, it is significant for the following reasons.

1. It showed how it is possible to study biological processes by using genetic markers.
2. It showed how the functions of genes themselves can be elucidated from the study of variant alleles.
3. It had far-reaching ramifications in agriculture and medicine.

In the next chapter, we focus on the physical locations of genes in cells and on the consequences of these locations.

Summary

Modern genetics is based on the concept of the gene, the fundamental unit of heredity. In his experiments with the garden pea, Mendel was the first to recognize the existence of genes. For example, by crossing a pure line of purple-flowered pea plants with a pure line of white-flowered pea plants and then selfing the F_1 generation, which was entirely purple, Mendel produced an F_2 generation of purple plants and white plants in a 3:1 ratio. In crosses such as those of pea plants bearing yellow seeds and pea plants bearing green seeds, he discovered that a 1:2:1 ratio underlies all 3:1 ratios. From these precise mathematical ratios Mendel concluded that there are hereditary determinants of a particulate nature (now known as genes). In higher plant and animal cells, genes exist in pairs. Variant forms of a gene are called alleles. Individual alleles can be either dominant or recessive.

In a cross of heterozygous yellow (*Yy*) plants with homozygous green (*yy*) plants, a 1:1 ratio of yellow to green plants was produced. From this ratio Mendel confirmed his so-called first law, which states that two members of a gene pair segregate from each other during gamete formation into equal numbers of gametes. Thus, each gamete carries only one member of each gene pair. The union of gametes to form a zygote is random as regards which allele the gametes carry.

The foregoing conclusions came from Mendel's work with monohybrid crosses. In dihybrid crosses, Mendel found 9:3:3:1 ratios in the F_2, which are really two 3:1 ratios combined at random. From these ratios Mendel inferred that alleles of the two genes in a dihybrid cross behave independently. This concept is Mendel's second law.

Although controlled crosses cannot be made in human beings, Mendelian genetics has great significance for humans. Many diseases and other exceptional conditions in humans are determined by recessive alleles inherited in a Mendelian manner; other exceptional conditions are caused by dominant alleles. In addition, Mendelian genetics is widely used in modern agriculture. By combining favorable qualities from different lines through hybridization and subsequent gene reassortment, plant and animal geneticists are able to produce new lines of superior phenotypes.

Finally, Mendel was responsible for the basic techniques of genetic dissection still in use today. One such technique is the use of genes as genetic markers to trace the hereditary processes of segregation and assortment. The other is the study of abnormal variants to discover how genes operate normally.

Concept Map

Draw a concept map interrelating as many of the following terms as possible. Note that the terms are listed in no particular order.

independent assortment / genotype / phenotype / dihybrid / testcross / 9:3:3:1 ratio / self / gametes / fertilization / 1:1:1:1 ratio

Chapter Integration Problem

Each chapter has one solved problem in which we stress the integration of concepts from different chapters. Learning in any discipline is a linear process, moving from topic to topic in some kind of appropriate sequence. But of course the discipline itself is not linear, but a set of integrated parts that the professional sees as one whole. We hope that focusing on integration will clarify the overall structure of genetics and that the reader will not see the contents of each chapter in isolation. As we pass from chapter to chapter, the levels of understanding build on the previous ones, and the subject is assembled like the layers of an onion.

Crosses were made between two pure lines of rabbits that we can call A and B. A male from line A was mated with a female from line B, and the F_1 rabbits were subsequently intercrossed to produce an F_2. It was discovered that $\frac{3}{4}$ of the F_2 animals had white subcutaneous fat, and $\frac{1}{4}$ had yellow subcutaneous fat. Later, the F_1 was examined and was found to have white fat. Several years later, an attempt was made to repeat the experiment using the same male from line A and the same female from line B. This time, the F_1 and all the F_2 (22 animals) had white fat. The only difference between the original experiment and the repeat that seemed relevant was that in the original all the animals were fed on fresh vegetables, and in the repeat they were fed on commercial rabbit chow. Provide an explanation for the difference and a test of your idea.

Solution

The first time the experiment was done, the breeders would have been perfectly justified in proposing that a pair of alleles determine white versus yellow body fat. This is because the data clearly resemble Mendel's results in peas. White must be dominant, so we can represent the white allele as *W*, and the yellow allele as *w*. The results can then be expressed

P $WW \times ww$

F_1 Ww

F_2 $\frac{1}{4}\ WW$
$\frac{1}{2}\ Ww$
$\frac{1}{4}\ ww$

No doubt if the parental rabbits had been sacrificed it would have been predicted that one (we cannot tell which) would have had white fat, and the other yellow. Luckily, this was not done, and the same animals were bred again, leading to a very interesting, different result. Often in science, an unexpected observation can lead to a novel principle, and rather than moving on to something else it is useful to try to explain the inconsistency. So why did the 3:1 ratio disappear? Here are some possible explanations.

First, perhaps the genotypes of the parental animals had changed. This type of spontaneous change affecting the whole animal, or at least its gonads, is very unlikely, because even common experience tells us that organisms tend to be stable to their type. Nevertheless, this is a reasonable idea.

Second, in the repeat, the sample of 22 F_2 animals did not contain any yellow simply by chance ("bad luck"). This again seems unlikely, because the sample was quite large, but it is a definite possibility.

A third explanation draws upon the principle covered in Chapter 1 that genes do not act in a vacuum; they depend upon the environment for their effect. Hence, the useful catchphrase arises, "genotype plus environment equals phenotype." A corollary of this catchphrase is, of course, that genes can act differently in different environments, so

genotype 1 plus environment 1 equals phenotype 1

and

genotype 1 plus environment 2 equals phenotype 2

In the present question, the different diets constituted different environments, so a possible explanation of the results is that the recessive allele *w* produces yellow fat only when the diet contains fresh vegetables. This explanation is testable. One way is to repeat the experiment again using vegetables in the food, but the parents might be dead by this time. A more convincing way is to interbreed several of the white-fatted F_2 rabbits from the second experiment. According to the original interpretation, about $\frac{3}{4}$ would bear at least one recessive *w* allele for yellow fat, and if their progeny are reared on vegetables, yellow should appear in Mendelian proportions. For example one might choose two rabbits that turn out to be *Ww* and *ww*; if so, the progeny would be $\frac{1}{2}$ white and $\frac{1}{2}$ yellow.

If this did not happen, and no yellows appeared in any of the F_2 matings, one would be forced back onto explanations one or two. Explanation two can be tested by using larger numbers, and if this explanation doesn't work, we are left with number one, which is difficult to test directly.

As you might have guessed, in reality the diet was the culprit. The specific details illustrate environmental effects beautifully. Fresh vegetables contain yellow substances called xanthophylls, and the dominant allele *W* gives rabbits the ability to break down these substances to a colorless ("white") form. However, *ww* animals lack this ability, and the xanthophylls are deposited in the fat, making it yellow. When no xanthophylls have been ingested, both *W* – and *ww* animals end up with white fat.

Solved Problems

This section in each chapter contains a few solved problems that show how to approach the problem sets that follow. The purpose of the problem sets is to challenge your understanding of the genetic principles learned in the chapter. The best way to demonstrate an understanding of a subject is to be able to use that knowledge in a real or simulated situation. Be forewarned that there is no machinelike way of solving these problems. The three main resources at your disposal are the genetic principles just learned, common sense, and trial and error.

Here is some general advice before beginning. First, it is absolutely essential to read and understand all of the question. Find out exactly what facts are provided, what assumptions have to be made, what clues are given in the question, and what inferences can be made from the available information. Second, be methodical. Staring at the question rarely helps. Restate the information in the question in your own way, preferably using a diagrammatic representation or flow chart to help you think out the problem. Good luck.

1. Consider three yellow, round peas, labeled A, B, and C. Each was grown into a plant and crossed to a plant grown from a green, wrinkled pea. Exactly 100 peas issuing from each cross were sorted into phenotypic classes as follows:

A: 51 yellow, round
49 green, round

B: 100 yellow, round

C: 24 yellow, round
26 yellow, wrinkled
25 green, round
25 green, wrinkled

What were the genotypes of A, B, and C? (Use gene symbols of your own choosing; be sure to define each one.)

Solution

Notice that each of the crosses is

yellow, round × green, wrinkled
↓
progeny

Because A, B, and C were all crossed to the same plant, all the differences among the three progeny populations must be attributable to differences in the underlying genotypes of A, B, and C.

You might remember a lot about these analyses from the chapter. This is fine, but let's see how much we can deduce from the data. What about dominance? The key cross for deducing dominance is B. Here, the inheritance pattern is

yellow, round × green, wrinkled
↓
all yellow, round

So yellow and round must be dominant phenotypes, because dominance is literally defined in terms of the phenotype of a hybrid. Now we know that the green, wrinkled parent used in each cross must be fully recessive; we have a very convenient situation because it means that each cross is a testcross, which is generally the most informative type of cross.

Turning to the progeny of A, we see a 1:1 ratio for yellow to green. This is a demonstration of Mendel's first law (equal segregation) and shows that for the character of color, the cross must have been heterozygote × homozygous recessive. Letting Y = yellow and y = green, we have

$$Yy \times yy$$
$$\downarrow$$
$$\tfrac{1}{2}\,Yy \text{ (yellow)}$$
$$\tfrac{1}{2}\,yy \text{ (green)}$$

For the character of shape, because all the progeny are round, the cross must have been homozygous dominant × homozygous recessive. Letting R = round and r = wrinkled, we have

$$RR \times rr$$
$$\downarrow$$
$$Rr \text{ (round)}$$

Combining the two characters, we have

$$Yy\,RR \times yy\,rr$$
$$\downarrow$$
$$\tfrac{1}{2}\,Yy\,Rr$$
$$\tfrac{1}{2}\,yy\,Rr$$

Now, cross B becomes crystal clear and must have been

$$YY\,RR \times yy\,rr$$
$$\downarrow$$
$$Yy\,Rr$$

because any heterozygosity in pea B would have given rise to several progeny phenotypes, not just one.

What about C? Here, we see a ratio of 50 yellow : 50 green (1 : 1) and a ratio of 49 round:51 wrinkled (also 1:1). So both genes in pea C must have been heterozygous, and cross C was

$$Yy\,Rr \times yy\,rr$$
$$\downarrow$$
$$\tfrac{1}{2}\,Yy \begin{cases} \tfrac{1}{2}\,Rr \longrightarrow \tfrac{1}{4}\,Yy\,Rr \\ \tfrac{1}{2}\,rr \longrightarrow \tfrac{1}{4}\,Yy\,rr \end{cases}$$
$$\tfrac{1}{2}\,yy \begin{cases} \tfrac{1}{2}\,Rr \longrightarrow \tfrac{1}{4}\,yy\,Rr \\ \tfrac{1}{2}\,rr \longrightarrow \tfrac{1}{4}\,yy\,rr \end{cases}$$

which is a good demonstration of Mendel's second law (independent behavior of different genes).

How would a geneticist have analyzed these crosses? Basically, the same way we just did but with fewer intervening steps. Possibly something like this: "yellow and round dominant; single-gene segregation in A; B homozygous dominant; independent two-gene segregation in C."

2. Phenylketonuria (PKU) is a human hereditary disease that prevents the body from processing the chemical phenylalanine, which is contained in the protein we eat. PKU is manifested in early infancy and, if it remains untreated, generally leads to mental retardation. PKU is caused by a recessive allele with simple Mendelian inheritance.

A couple intends to have children but consults a genetic counselor because the man has a sister with PKU and the woman has a brother with PKU. There are no other known cases in their families. They ask the genetic counselor to determine the probability that their first child will have PKU. What is this probability?

Solution

What can we deduce? If we let the allele causing the PKU phenotype be p and the respective normal allele be P, then the sister and brother of the man and woman, respectively, must have been pp. In order to produce these affected individuals, all four grandparents must have been heterozygous normal. The pedigree can be summarized as follows:

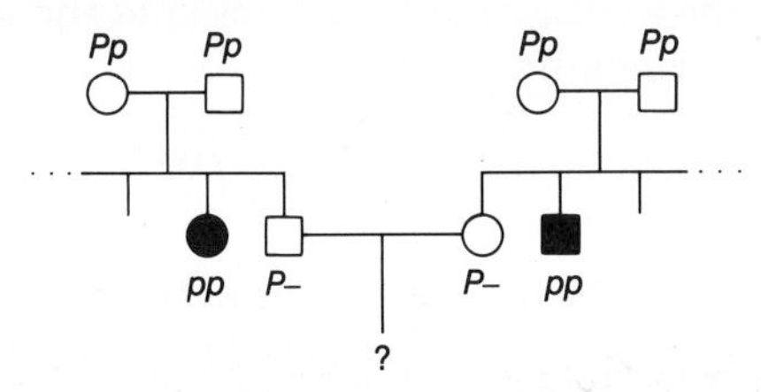

Once these inferences have been made, the problem is reduced to an application of the product rule. The only way the man and woman can have a PKU child is if both of them are heterozygotes (it is obvious that they themselves do not have the disease). Both the grandparental matings are simple Mendelian monohybrid crosses expected to produce progeny in the following proportions:

$$\left.\begin{array}{l}\frac{1}{4}\,PP\\ \frac{1}{2}\,Pp\end{array}\right\}\ \text{Normal}\ \frac{3}{4}$$

$$\frac{1}{4}\,pp\qquad \text{PKU}\ \frac{1}{4}$$

We know that the man and the woman are normal, so the probability of either being a heterozygote is $\frac{2}{3}$, because within the $P-$ class, $\frac{2}{3}$ are Pp and $\frac{1}{3}$ are PP.

The probability of *both* the man and the woman being heterozygotes is $\frac{2}{3} \times \frac{2}{3} = \frac{4}{9}$. If they are both heterozygous, then one-quarter of their children would have PKU, so the probability that their first child will have PKU is $\frac{1}{4}$, and the total probability of their being heterozygous *and* of their first children having PKU is $\frac{4}{9} \times \frac{1}{4} = \frac{4}{36} = \frac{1}{9}$, which is the answer.

Problems

Generally the straightforward problems are at the beginning of a set. Particularly challenging problems are marked with an asterisk.

1. What are Mendel's laws?

2. If you had a fruit fly *(Drosophila melanogaster)* that was of phenotype A, what test would you make to determine if it was AA or Aa?

3. Two black guinea pigs were mated and over several years produced 29 black and 9 white offspring. Explain these results, giving the genotypes of parents and progeny.

4. A woman had a sister who died of Tay-Sachs disease (a rare recessive disease). The woman is now worried that her unborn child will also be affected. How would you counsel her?

5. You have three dice: one red (R), one green (G), and one blue (B). When all three dice are rolled at the same time, calculate the probability of the following outcomes:
   - **a.** 6(R) 6(G) 6(B)
   - **b.** 6(R) 5(G) 6(B)
   - **c.** 6(R) 5(G) 4(B)
   - **d.** no sixes at all
   - **e.** two sixes and one five on any dice
   - **f.** three sixes or three fives
   - **g.** the same number on all dice
   - **h.** a different number on all dice

6. **a.** You have three jars containing marbles, as follows:

jar 1	600 red	and	400 white
jar 2	900 blue	and	100 white
jar 3	10 green	and	990 white

   If you blindly select one marble from each jar, calculate the probability of obtaining
   - **(1)** a red, a blue, and a green
   - **(2)** three whites
   - **(3)** a red, a green, and a white
   - **(4)** a red and two whites
   - **(5)** a color and two whites
   - **(6)** at least one white

   ***b.** In a certain plant, R = red and r = white. You self a red Rr heterozygote with the express purpose of obtaining a white plant for an experiment. What minimum number of seeds do you have to grow to be at least 95 percent certain of obtaining at least one white individual? (HINT: consider your answer to Question 6a(6).

7. Holstein cattle normally are black and white. A superb black and white bull, Charlie, was purchased by a farmer for $100,000. The progeny sired by Charlie were all normal in appearance. However, certain pairs of his progeny, when interbred, produced red and white progeny at a fre-

quency of about 25 percent. Charlie was soon removed from the stud lists of the Holstein breeders. Explain precisely why, using symbols.

8. Maple syrup urine disease is a rare inborn error of metabolism. It derives its name from the odor of the urine of affected individuals. If untreated, affected children die soon after birth. The disease tends to recur in the same family, but the parents of the affected individuals are always normal. What does this information suggest about the transmission of the disease: is it dominant or recessive? Explain.

9. In humans, the disease galactosemia is inherited as a recessive trait in a simple Mendelian manner. A woman whose father had galactosemia intends to marry a man whose grandfather was galactosemic. They are worried about having a galactosemic child. What is the probability of this outcome?

10. Suppose that a husband and wife are both heterozygous for a recessive gene for albinism. If they have dizygotic (two-egg) twins, what is the probability that both of the twins will have the same phenotype for pigmentation?

11. The plant blue-eyed Mary grows on Vancouver Island and on the lower mainland of British Columbia. The populations are dimorphic for purple blotches on the leaves—some plants have blotches, and others don't. Near Nanaimo, one plant in nature had blotched leaves. This plant, which had not yet flowered, was dug up and taken to a laboratory, where it was allowed to self. Seeds were collected and grown into progeny. One randomly selected (but typical) leaf from each of the progeny is shown in the figure (left column).

a. Formulate a concise genetic hypothesis to explain these results. Explain all symbols and show all genotypic classes (and the genotype of the original plant).

b. How would you test your hypothesis? Be specific.

12. Can it ever be proved that an animal is *not* a carrier of a recessive allele (that is, not a heterozygote for a given gene)? Explain.

13. In nature, the plant *Plectritis congesta* is dimorphic for fruit shape; that is, individual plants bear either wingless or winged fruits, as shown in the figure.

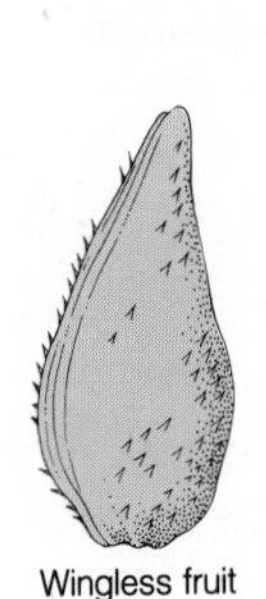

Wingless fruit

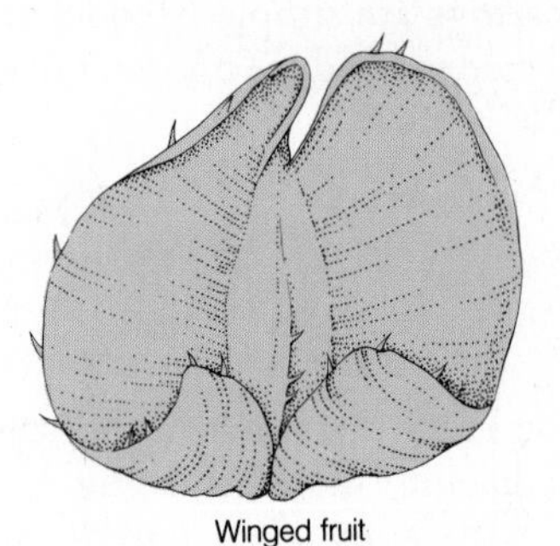

Winged fruit

Plants were collected from nature before flowering and were crossed or selfed with the following results:

Pollination	Number of progeny	
	Winged	Wingless
Winged (selfed)	91	1*
Winged (selfed)	90	30
Wingless (selfed)	4*	80
Winged × wingless	161	0
Winged × wingless	29	31
Winged × wingless	46	0
Winged × winged	44	0
Winged × winged	24	0

Interpret these results, and derive the mode of inheritance of these fruit-shape phenotypes. Use symbols. (NOTE: the phenotypes of progeny marked by asterisks probably have a nongenetic explanation. What do you think it is?)

14. Here are four human pedigrees. The black symbols represent an abnormal phenotype inherited in a simple Mendelian manner.

1

2

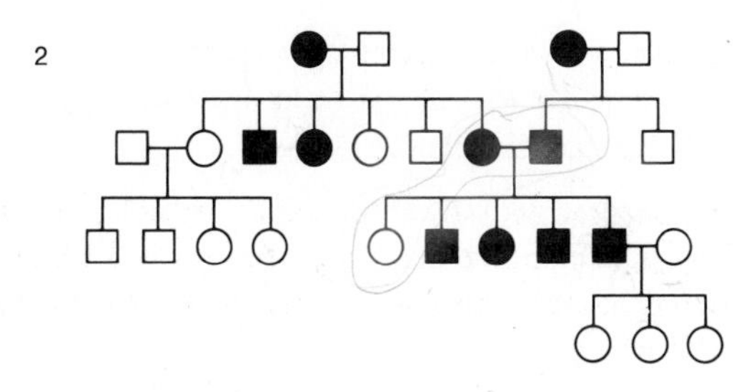

3

4

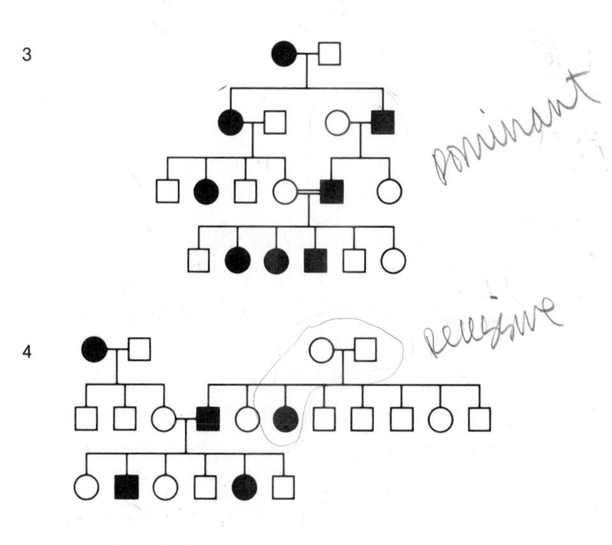

a. For each pedigree, state whether the abnormal condition is dominant or recessive. Try to state the logic behind your answer.

b. In each pedigree, describe the genotypes of as many individuals as possible.

15. Tay-Sachs disease ("infantile amaurotic idiocy") is a rare human disease in which toxic substances accumulate in nerve cells. The recessive allele responsible for the disease is inherited in a simple Mendelian manner. For unknown reasons, the allele is more common in populations of Ashkenazi Jews of Eastern Europe. In the following pedigree, the great-great-grandmother (*) was known to be heterozygous for the Tay-Sachs allele. What is the probability that the child of the consanguineous mating will have Tay-Sachs disease?

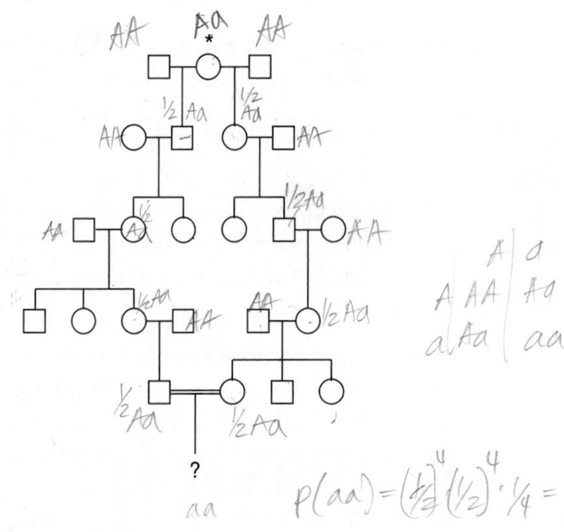

16. The following pedigree was obtained for a rare kidney disease:

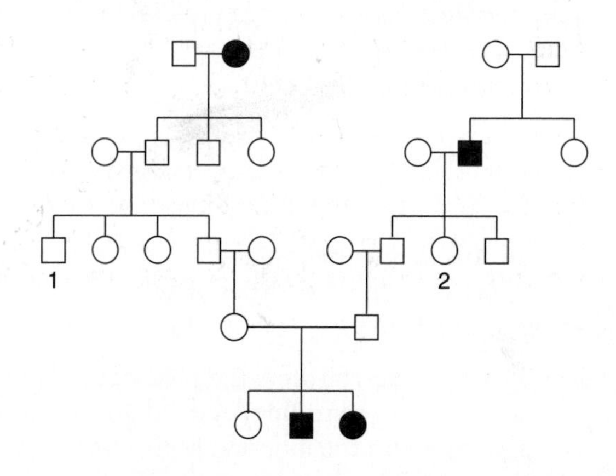

a. Deduce the inheritance of this condition, stating your reasons.

b. If individuals 1 and 2 marry, what is the probability that their first child will have the kidney disease?

17. a. A curious polymorphism in human populations has to do with the ability to curl up the sides of the tongue to make a trough ("tongue rolling"). Some people can do this trick, and others simply cannot. Hence it is an example of a dimorphism. Its significance is a complete mystery. In one family, a boy was unable to roll his tongue, but to his great chagrin his sister could. Furthermore, both his parents were rollers, and so were both grandfathers and one paternal uncle and one paternal aunt. One paternal aunt, one paternal uncle, and one maternal uncle could not. Draw the pedigree

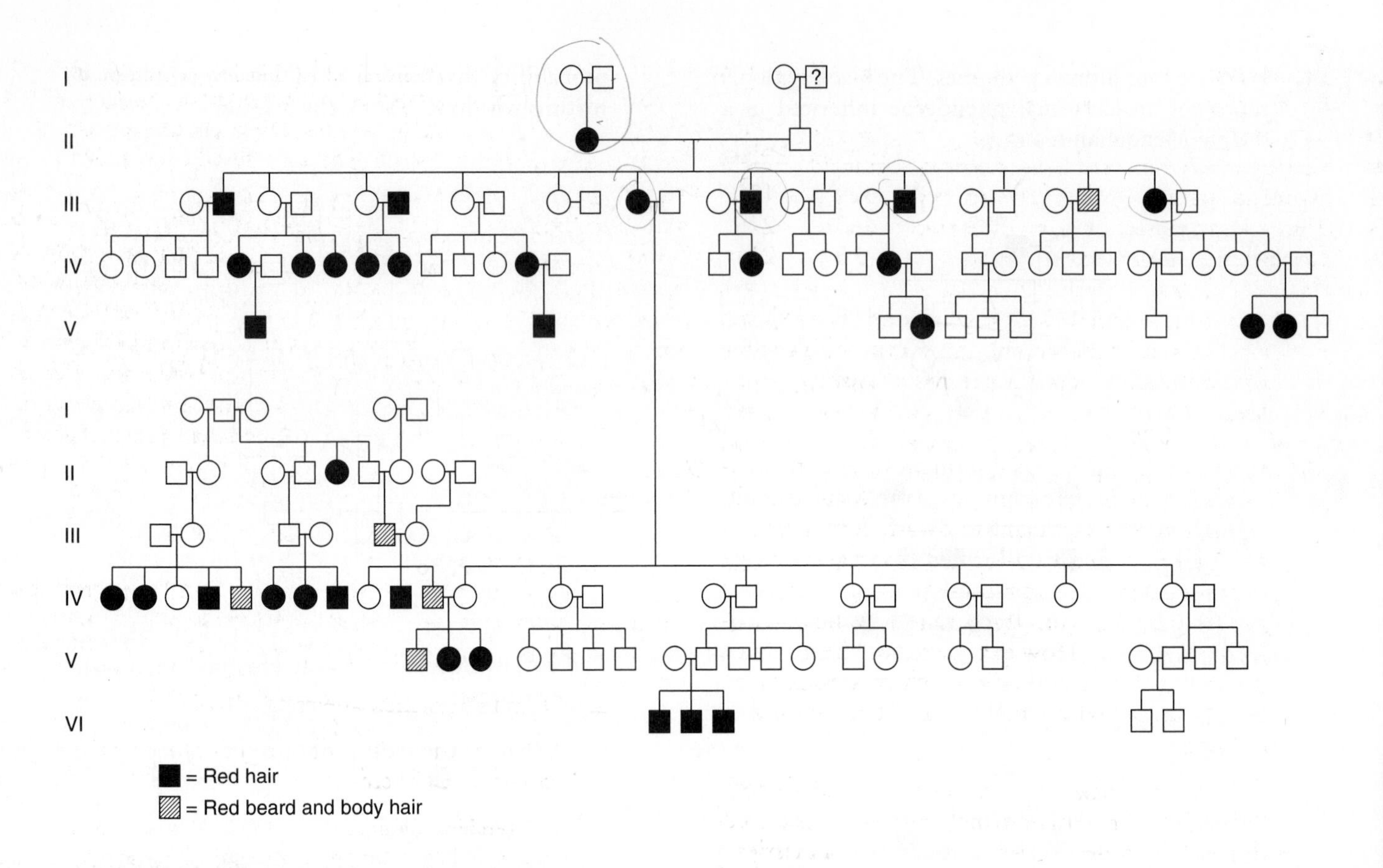

for this family, defining your symbols clearly, and deduce the genotypes of as many individuals as possible.

b. The pedigree you drew is typical of the inheritance of tongue rolling, and led geneticists to come up with the inheritance mechanism that no doubt you came up with. However, in a study of 33 pairs of identical twins, it was found that both members of 18 pairs could roll, neither member of 8 pairs could roll, and one of the twins in 7 pairs could roll but the other couldn't. Because identical twins are derived from the splitting of one fertilized egg into two embryos, the members of a pair must be genetically identical. How can the existence of the seven discordant pairs be reconciled with your genetic explanation of the pedigree?

18. A rare, recessive allele inherited in a Mendelian manner causes the disease cystic fibrosis. A phenotypically normal man whose father had cystic fibrosis marries a phenotypically normal woman from outside the family, and the couple consider having a child.

a. Draw the pedigree as far as described above.

b. If the frequency in the population of heterozygotes for cystic fibrosis is 1 in 50, what is the chance that the couple's first child will have cystic fibrosis?

c. If the first child does have cystic fibrosis, what is the probability that the second child will be normal?

19. In human hair, the black and brown colors are produced by various amounts and combinations of chemicals called melanins. However, red hair is produced by a different type of chemical substance about which little is known. Red hair runs in families, and the figure above shows a large pedigree for red hair. Does the inheritance pattern in this pedigree suggest that red hair could be caused by a dominant or a recessive allele of a gene that is inherited in a simple Mendelian manner?

(Pedigree from W. R. Singleton and B. Ellis, *Journal of Heredity* 55, 261, 1964.)

20. When many families were tested for the ability to taste the chemical PTC, the matings were grouped into three types, and the progeny totaled, with the results shown below.

Parents	Number of families	Children: Tasters	Children: Nontasters
Taster × taster	425	929	130
Taster × nontaster	289	483	278
Nontaster × nontaster	86	5	218

Assuming that PTC tasting is dominant (*P*) and nontaster is recessive (*p*), how can the progeny ratios in each of the three types of mating be accounted for?

21. In tomatoes, red fruit is dominant to yellow, two-loculed fruit is dominant to many-loculed fruit, and tall vine is dominant to dwarf. A breeder has two pure lines: red, two-loculed, dwarf and yellow, many-loculed, tall. From these he wants to produce a new pure line for trade that is yellow, two-loculed, and tall. How exactly should he go about doing this? Show not only which crosses to make, but also how many progeny should be sampled in each case.

22. In humans, achondroplastic dwarfism and neurofibromatosis are both extremely rare dominant conditions. If a woman with achondroplasia marries a man with neurofibromatosis, what phenotypes could be produced in their children, and in what proportions? (Be sure to define any symbols you use.)

23. In dogs, dark coat color is dominant over albino and short hair is dominant over long hair. Assume that these effects are caused by two independently assorting genes and write the genotypes of the parents in each of the crosses shown below, where D and A stand for the dark and albino phenotypes, respectively, and S and L stand for the short-hair and long-hair phenotypes.

Parental phenotypes	Number of progeny: D, S	D, L	A, S	A, L
a. D, S × D, S	89	31	29	11
b. D, S × D, L	18	19	0	0
c. D, S × A, S	20	0	21	0
d. A, S × A, S	0	0	28	9
e. D, L × D, L	0	32	0	10
f. D, S × D, S	46	16	0	0
g. D, S × D, L	30	31	9	11

Use the symbols *C* and *c* for the dark and albino coat-color alleles and the symbols *S* and *s* for the short-hair and long-hair alleles, respectively. Assume homozygosity unless there is evidence otherwise.

24. In tomatoes, two alleles of one gene determine the character difference of purple (P) versus green (G) stems and two alleles of a separate, independent gene determine the character difference of "cut" (C) versus "potato" (Po) leaves. The results for five matings of tomato plant phenotypes are shown below.

Mating	Parental phenotypes	Number of progeny: P, C	P, Po	G, C	G, Po
1	P, C × G, C	321	101	310	107
2	P, C × P, Po	219	207	64	71
3	P, C × G, C	722	231	0	0
4	P, C × G, Po	404	0	387	0
5	P, Po × G, C	70	91	86	77

a. Determine which alleles are dominant.

b. What are the most probable genotypes for the parents in each cross?

25. We have dealt mainly with only two genes, but the same principles hold for more than two genes. Consider the cross

$$AaBbCcDdEe \times aaBbccDdee$$

a. What proportion of progeny will *phenotypically* resemble (1) the first parent, (2) the second parent, (3) either parent, and (4) neither parent?

b. What proportion of progeny will be *genotypically* the same as (1) the first parent, (2) the second parent, (3) either parent, and (4) neither parent?

Assume independent assortment.

26. Most *Drosophila melanogaster* have brown bodies but members of the species that are homozygous for the recessive allele *y* have yellow bodies. However, if larvae of pure *YY* lines are reared on food containing silver salts, the resulting adults are yellow. These are called **phenocopies,** phenotypic copies of specific genotypes. If you were presented with a single yellow fly, how would you determine if it was a yellow genotype or a yellow phenocopy? (Can you think of examples of phenocopies in humans?)

3

Chromosome Theory of Inheritance

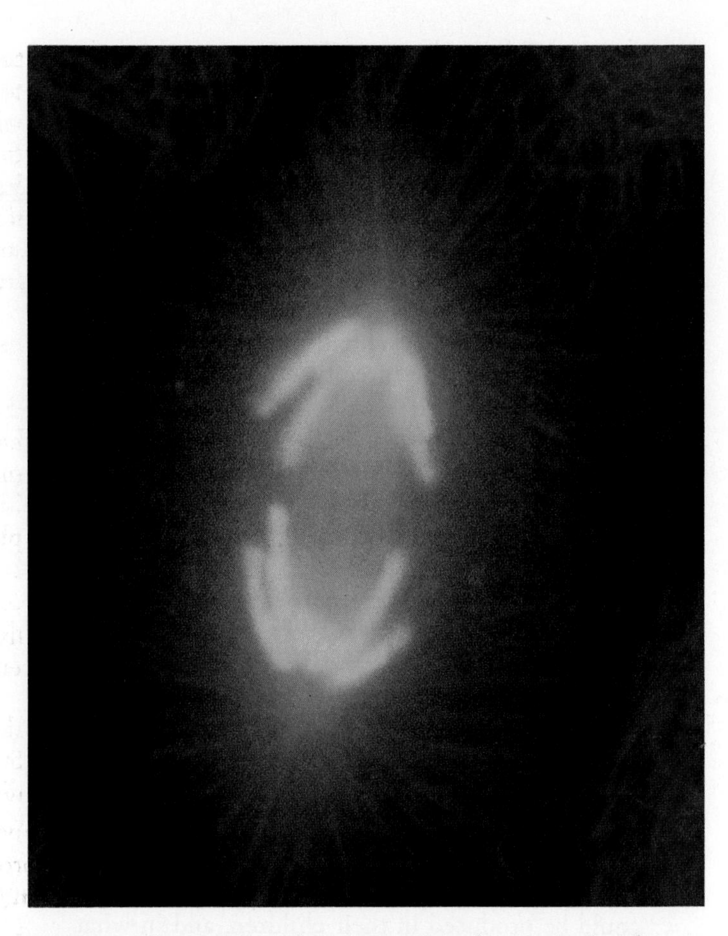

A human cell in mitosis. Chromosomes have been stained with a dye that detects DNA, and microtubules have been stained with a fluorescent antibody. (J. M. Murray)

KEY CONCEPTS

Genes are parts of chromosomes.

Mitosis is the nuclear division that results in two daughter nuclei each with genetic material identical to that of the original nucleus.

Meiosis is the nuclear division by which a reproductive cell with two equivalent chromosome sets divides into four meiotic products, each of which has only one set of chromosomes.

Mendel's laws of equal segregation and independent assortment are based on the separation during meiosis of members of each chromosome pair and on the independent meiotic behavior of different chromosome pairs.

Special phenotypic progeny ratios, differing for males and females, mirror the inheritance patterns of genes on the sex chromosomes.

The beauty of Mendel's analysis is that we need not know what genes are or how they control phenotypes to analyze the results of crosses and to predict the outcomes of future crosses from the laws of equal segregation and independent assortment. We can do all this simply by representing abstract, hypothetical factors of inheritance (genes) by symbols—without any concern about their molecular structures or their locations in a cell. Nevertheless, our interest naturally turns to the next obvious question: what structures within cells correspond to these hypothetical genes?

The development of genetics took a major step forward by accepting the notion that the genes, as characterized by Mendel, are parts of specific cellular structures, the chromosomes. This simple concept has become known as the **chromosome theory** of heredity. Although simple, the idea has had profound implications, inextricably uniting the disciplines of genetics and cytology and providing a means of correlating the results of breeding experiments with the behavior of structures that can be actually seen under the microscope. This fusion is still an essential part of genetic analysis today and has important applications in medical genetics, agricultural genetics, and evolutionary genetics.

Mitosis and Meiosis

How did the chromosome theory take shape? Evidence gradually accumulated from a variety of sources. One of the first lines of evidence came from observations of how chromosomes behave during the division of a cell's nucleus. The observations leading up to the discovery of the two different types of nuclear division, termed *mitosis* and *meiosis,* were as follows. In the interval between Mendel's research and its rediscovery, many biologists were interested in heredity even though they were unaware of Mendel's results, and they approached the problem in a completely different way. These investigators wanted to locate the hereditary material in the cell. An obvious place to look was in the gametes, because they are the only connecting link between generations. Egg and sperm were believed to contribute equally to the genetic endowment of offspring even though they differ greatly in size. Because an egg has a great volume of cytoplasm but a sperm has very little, the cytoplasm of gametes seemed an unlikely seat of the hereditary structures. The nuclei of egg and sperm however, were known to be approximately equal in size, so the nuclei were considered good candidates for harboring hereditary structures.

What was known about the contents of cell nuclei? It became clear that the most prominent components were the chromosomes, which proved to possess unique properties that set them apart from all other cellular structures. A property that especially intrigued biologists was the constancy of the number of chromosomes from cell to cell within an organism, from organism to organism within any one species, and from generation to generation within that species. The question therefore arose: how is the chromosome number maintained? The question was answered by observing the behavior of chromosomes under the microscope during mitosis and meiosis; from those observations developed the postulation of the chromosome theory—that chromosomes are the structures that contain genes.

Mitosis is the nuclear division associated with the division of somatic cells, that is, cells of the eukaryotic body that are not destined to become sex cells. This kind of nuclear division produces a number of genetically identical cells from a single progenitor cell, as for example, in the division of a fertilized human egg cell to become a multicellular organism composed of trillions of cells. Each single mitosis is associated with a single cell division that produces two genetically identical daughter cells.

Meiosis is the name given to the nuclear divisions in the special cells that are destined to produce gametes. Such a cell is called a **meiocyte.** There are two cell divisions of each meiocyte and two associated meiotic divisions of the nucleus. Hence, each meiocyte generally produces four cells, which we shall call **products of meiosis.** In humans and other animals, meiosis takes place in the gonads, and the products of meiosis are the gametes—sperm (more properly, spermatozoa) and eggs. In flowering plants, meiosis takes place in the anthers and ovaries, and the products of meiosis are **meiospores,** which eventually give rise to gametes. We now turn to the details of these two basic kinds of nuclear division. The following descriptions of the various stages are as general as possible and are applicable to mitosis and meiosis in most organisms in which such divisions take place. Note, however, that the photographic illustrations are all of one organism (a flowering plant, *Lilium regale*) and therefore cannot be completely general for all details of mitosis and meiosis. Hence, a parallel series of idealized drawings is also included.

Mitosis

The cell cycle is the series of events from any stage in a cell to the equivalent stage in a daughter cell. It can be divided into several periods: **M, S, G1,** and **G2** (Figure 3-1). Mitosis (M) is usually the shortest period of the cycle, lasting for approximately 5 to 10 percent of the cycle. DNA synthesis takes place during the S period. G1 and G2 are gaps between S and M. Together, G1, S, and G2 constitute **interphase,** the time between mitoses. (Interphase used to be called "resting period"; however, cells actively function in many ways during interphase, not the least of which, of course, is in synthesizing

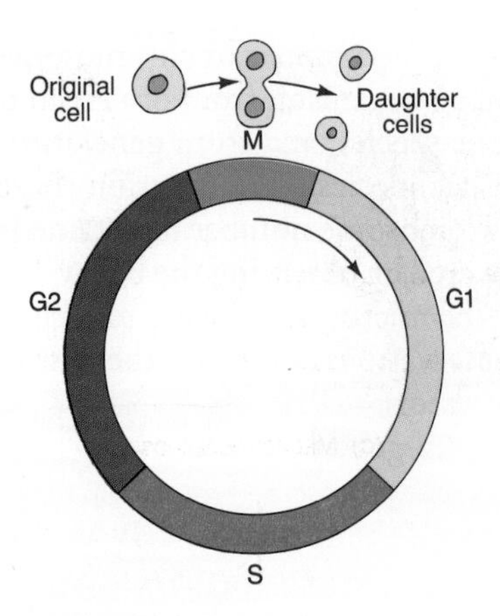

Figure 3-1 Stages of the cell cycle. M = mitosis, S = DNA synthesis, G = gap.

DNA.) The chromosomes cannot be seen during interphase (Figure 3-2a), mainly because they are in an extended state and become intertwined with each other like a tangle of yarn.

The net achievement of mitosis is that each chromosome in the nucleus duplicates longitudinally, and then this double structure splits to become two daughter chromosomes, each going to a different daughter nucleus. Mitosis produces two daughter nuclei identical to each other and to the nucleus from which they were derived. For the sake of study, scientists divide mitosis into four stages called **prophase, metaphase, anaphase,** and **telophase.** It must be stressed, however, that any nuclear division is a dynamic process on which we impose such arbitrary stages only for our own convenience.

Prophase. The onset of mitosis is heralded by the chromosomes becoming distinct for the first time (Figure 3-2b). They get progressively shorter through a process of contraction, or condensation, whereby the chromosomes contract into a series of spirals or coils; the coiling produces structures that are more easily moved around (for the same reason, cotton fibers are packaged commercially on spools). As the chromosomes become visible, they appear double-stranded, each chromosome being composed of two longitudinal halves called **chromatids** (Figure 3-2c). These "sister" chromatids are joined together at a region called the **centromere.** The **nucleoli** —large intranuclear spherical structures—disappear at this stage. The nuclear membrane begins to break down, and the nucleoplasm and cytoplasm become one.

Metaphase. At this stage, the **nuclear spindle** becomes prominent. This is a birdcage-like structure that forms in the nuclear area; it consists of a series of parallel fibers that point to each of two cell poles. The chromosomes move to the equatorial plane of the cell and the centromere of each becomes attached to spindle fibers. (Figure 3-2d).

Anaphase. This stage begins when the pairs of sister chromatids separate, one of a pair moving to each pole (Figure 3-2e). The centromeres, which now appear to have divided, separate first. As each chromatid moves, its two arms appear to trail its centromere; a set of V-shaped structures results, with the points of the V's directed at the poles.

Telophase. Now a nuclear membrane re-forms around each daughter nucleus, the chromosomes uncoil, and the nucleoli reappear—all of which effectively re-form interphase nuclei (Figure 3-2f). By the end of telophase, the spindle has dispersed and the cytoplasm has been divided into two by a new cell membrane.

In each of the resultant daughter cells, the chromosome complement is identical to that of the original cell. Of course, what were referred to as chromatids now take on the role of full-fledged chromosomes in their own right.

Message Mitosis produces two daughter nuclei that have a chromosomal constitution identical to that of the original nucleus.

The role of chromatin in directing development was hotly debated in the late nineteenth and early twentieth centuries. Wilhelm Roux's position was that if chromosomes played a role in development, they would have to be differentially partitioned to the various cell lineages. When his ideas were disproved, the importance of chromatin in development was discounted for a time. The fact that the chromosomal material is precisely maintained in development was not proved until later, when nuclear transplantation experiments were successful.

It was evident that mitosis maintains the chromosome number in each nucleus, but this made early investigators puzzle over the joining of two gametes in the fertilization event. They knew that during this process, two nuclei fuse but that the chromosome number nevertheless remains constant. What prevented the doubling of the chromosome number at each generation? This puzzle was resolved by the prediction of a special kind of nuclear division that *halved* the chromosome number. This special division, which was eventually discovered in the gamete-producing tissues of plants and animals, is called meiosis.

Meiosis

Meiosis is preceded by a premeiotic S phase in the meiocyte. Most of the DNA for meiosis is synthesized during this S phase, but some is synthesized during the first

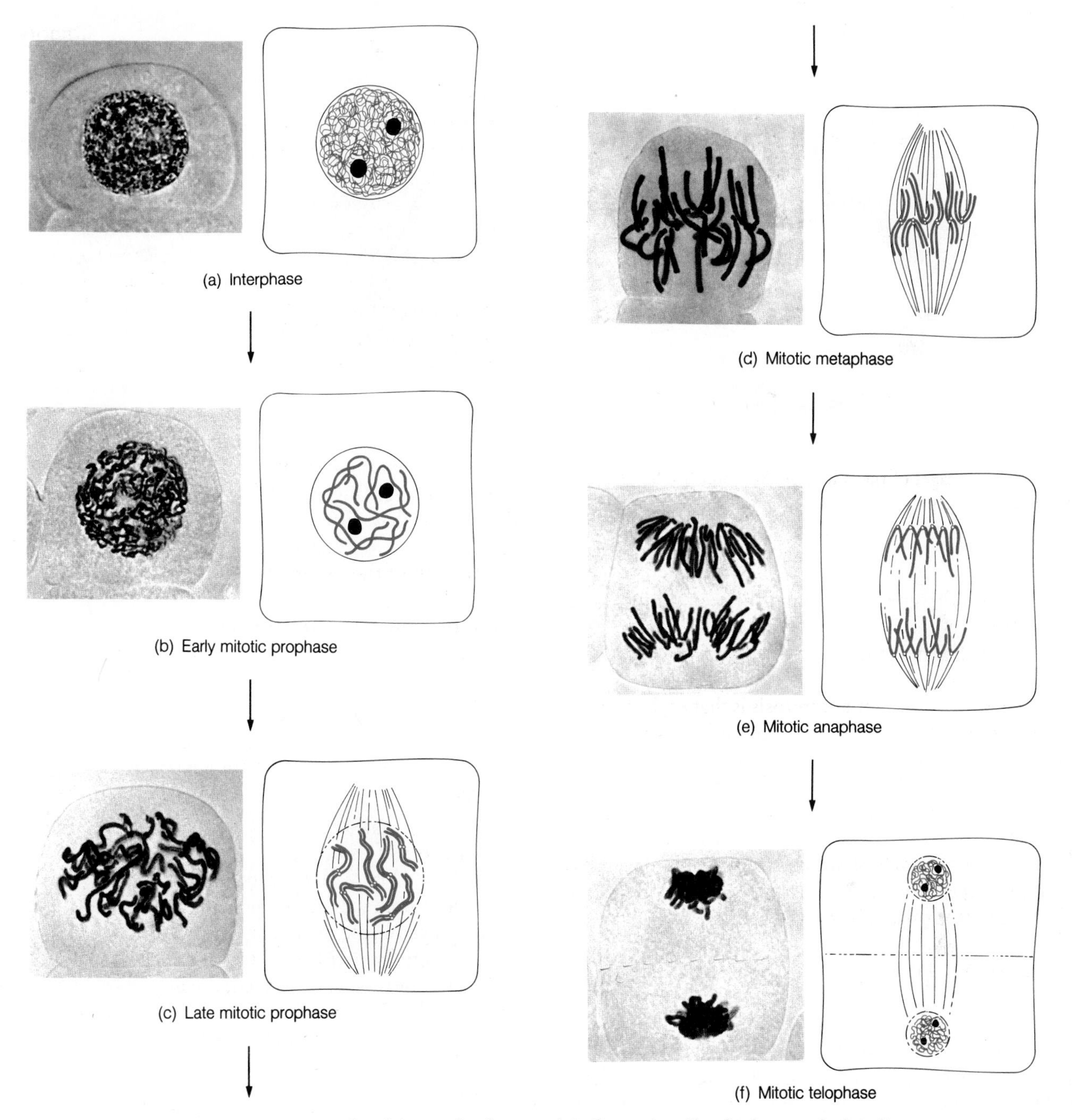

Figure 3-2 Mitosis. The photographs show nuclei of root tip cells of *Lilium regale.* (Modified from J. McLeish and B. Snoad, *Looking at Chromosomes.* Copyright 1958, St. Martin's, Macmillan.)

prophase of meiosis. Meiosis consists of two cell divisions distinguished as meiosis I and meiosis II. The events of meiosis I are quite different from those of meiosis II, and the events of both differ from those of mitosis. Each meiotic division is formally divided into prophase, metaphase, anaphase, and telophase. Of these stages, the most complex and lengthy is prophase I, which has its own subdivisions: **leptotene, zygotene, pachytene, diplotene,** and **diakinesis.** Once again, try to imagine those processes as merging dynamically into each other with no clear borders.

Prophase I. LEPTOTENE. The chromosomes become visible at this stage as long, thin, single threads (Figure 3-3a). The process of chromosome contraction continues during leptotene and throughout the entire prophase. During leptotene small areas of thickening called **chromomeres** develop along each chromosome, which give it the appearance of a bead necklace.

ZYGOTENE. This is a time of active pairing of the threads (Figure 3-3b), making it apparent that the chromosome complement of the meiocyte is in fact two complete

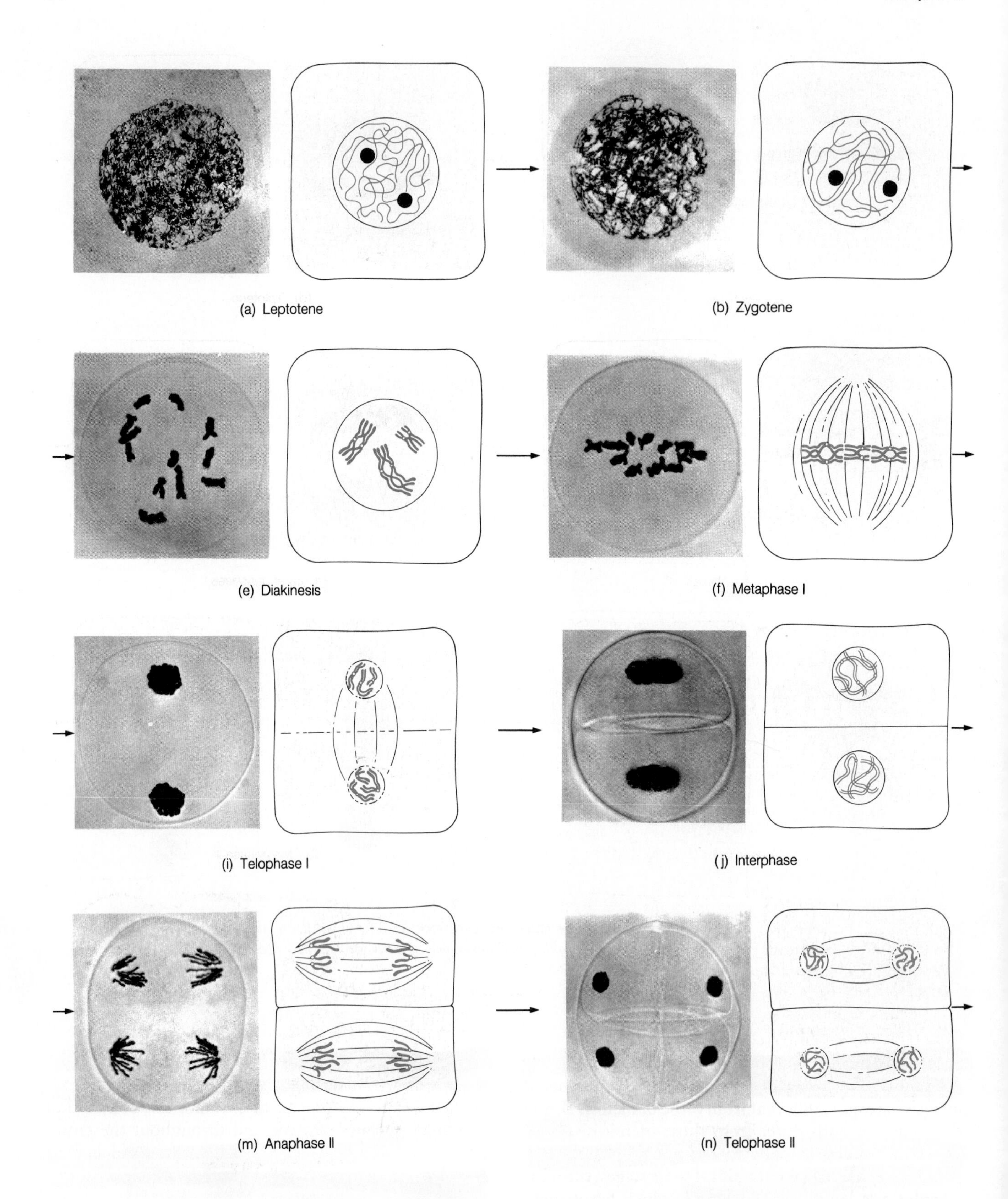

Figure 3-3 Meiosis and pollen formation. The photographs are of *Lilium regale.* Note: For simplicity, multiple chiasmata are drawn as involving only two chromatids; in reality, all four chromatids can be involved. (Modified from J. McLeish and B. Snoad, *Looking at Chromosomes.* Copyright 1958, St. Martin's, Macmillan.)

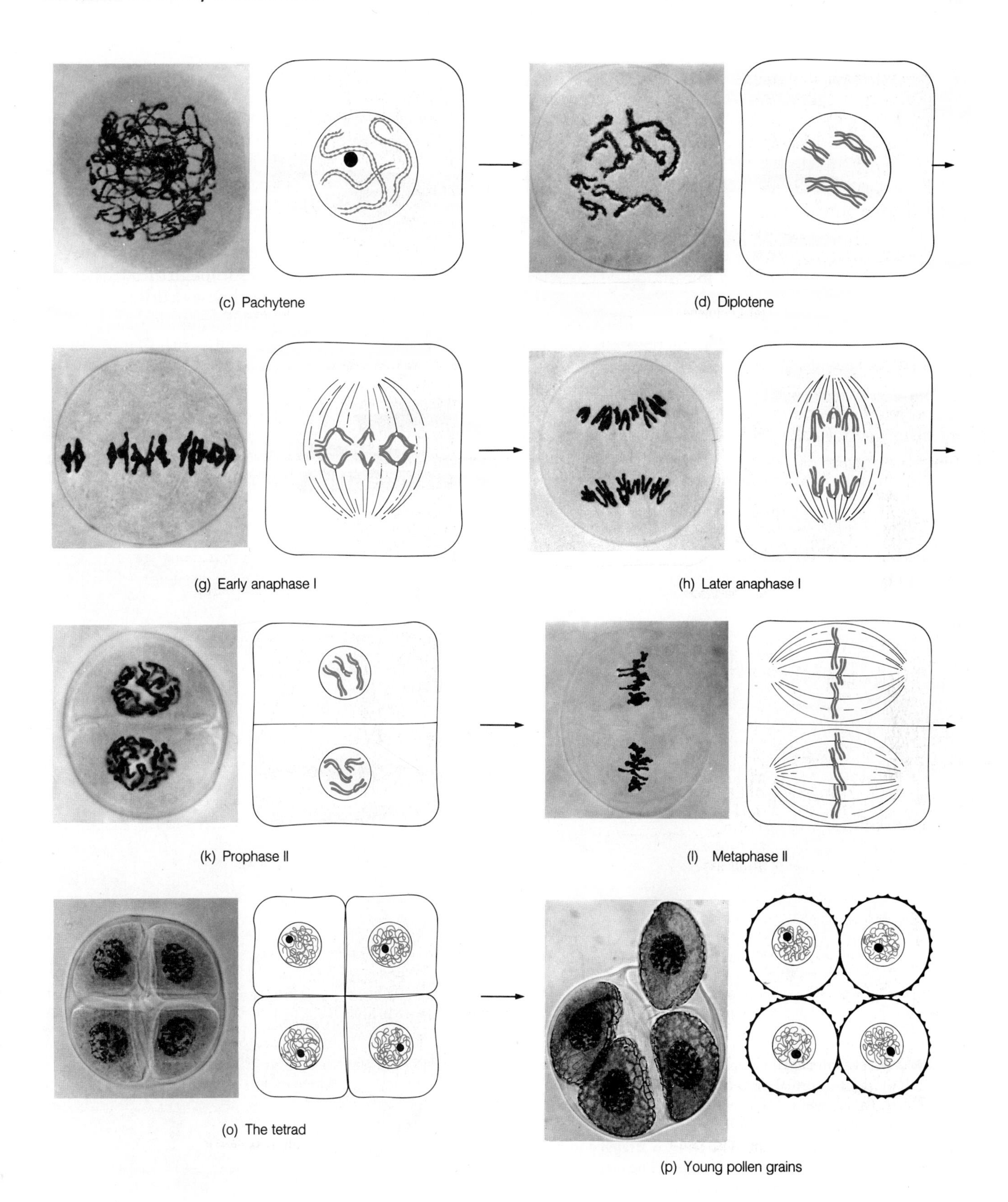

(c) Pachytene

(d) Diplotene

(g) Early anaphase I

(h) Later anaphase I

(k) Prophase II

(l) Metaphase II

(o) The tetrad

(p) Young pollen grains

chromosome sets. Thus, each chromosome has a pairing partner, and the two become progressively paired, or **synapsed,** side by side as if by a zipper.

At this point we need to introduce some terminology. Each chromosome pair is called a **homologous pair,** and the two members of a pair are called **homologs.** It should be noted that the pairing is a striking difference from mitosis, in which there is no such process. Furthermore, whereas cells with any number of chromosome sets may undergo mitosis, only cells with two chromosome sets (two **genomes**) undergo meiosis. Cells with two chromosome sets are called **diploid** and are represented symbolically as $2n$, where n is the number of chromosomes in a set. In contrast, cells with only one set (n) are called **haploid.** In most complex organisms such as mammals and flowering plants, the cells of the organism are normally diploid and the meiocytes are simply a subpopulation of cells that are set aside to undergo meiosis. In haploid organisms, as we shall see later, a diploid meiocyte is constructed as part of the normal reproductive cycle.

How do two homologs find each other during zygotene? The probable answer to this is that the ends of the chromosomes, the **telomeres,** are anchored in the nuclear membrane and it is likely that homologous telomeres are close, so that the zippering-up process can begin with them. How does the zippering-up work? What is the mechanism whereby two homologs can pair so precisely along their length? Although the mechanism is not precisely understood, we know that it requires an elaborate structure composed of protein and DNA, called a **synaptonemal complex** (Figure 3-4), that is always found sandwiched between homologs during synapsis.

PACHYTENE. This stage is characterized by thick, fully synapsed threads (Figure 3-3c). Thus, the number of homologous pairs of chromosomes in the nucleus is equal to the number n. Nucleoli are often pronounced

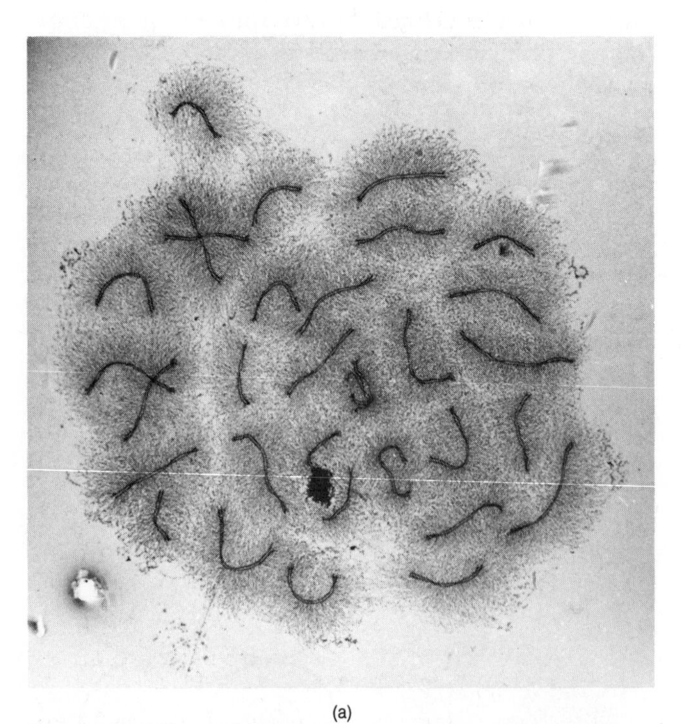

(a)

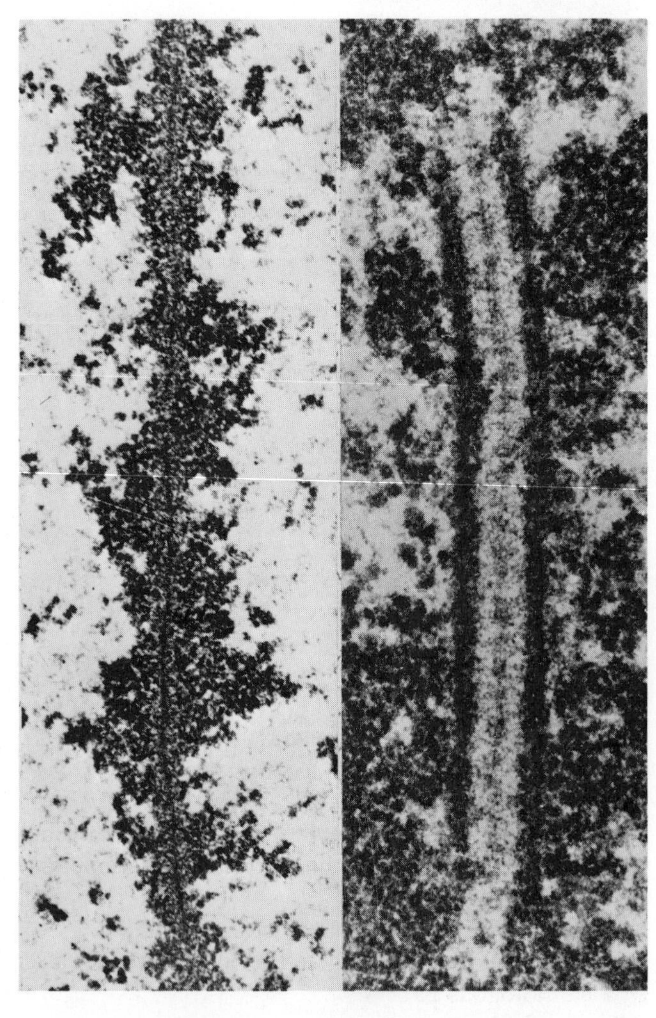

(b)

Figure 3-4 Synaptonemal complexes. (a) In *Hyalophora cecropia,* a silk moth, the normal male chromosome number is 62, giving 31 synaptonemal complexes. In the individual shown here, one chromosome *(center)* is represented three times; such a chromosome is termed trivalent. The DNA is arranged in regular loops around the synaptonemal complex. The black, dense structure is the nucleolus. (b) Regular synaptonemal complex in *Lilium tyrinum.* Note *(right)* the two lateral elements of the synaptonemal complex and also *(left)* an unpaired chromosome, showing a central core corresponding to one of the lateral elements. (Parts a and b courtesy of Peter Moens.)

during pachytene. The beadlike chromomeres align precisely in the paired homologs, producing a distinctive pattern for each pair.

Diplotene. Although each homolog appeared to be a single thread during leptotene, the DNA had, in fact, already replicated during the premeiotic S phase. This becomes manifest during diplotene as a longitudinal doubleness of each paired homolog (Figure 3-3d). As in mitosis, the divided subunits are called chromatids. Hence, since each member of a homologous pair produces two sister chromatids, the synapsed structure now consists of a bundle of four homologous chromatids. At diplotene, the pairing between homologs becomes less tight; in fact, they appear to repel each other, and as they separate slightly, cross-shaped structures called **chiasmata** (singular, chiasma) appear between nonsister chromatids. Each chromosome pair generally has one or more chiasmata. Chiasmata are the visible manifestations of events called **crossovers** that occurred earlier, probably during zygotene or pachytene. Crossovers represent one major way in which meiosis differs from mitosis—crossovers during mitosis are rare. A crossover is a precise breakage, swapping, and reunion between two nonsister chromatids. Studies performed on abnormal lines of organisms that undergo crossing-over very inefficiently, or not at all, show severe disruption of the orderly events that partition chromosomes into daughter cells at meiosis. Thus, crossing-over obviously helps determine how paired homologs behave, and the presence of at least one crossover per pair is usually essential for proper segregation. Crossovers have another interesting role, which, as we shall see in Chapter 5, is to make new gene combinations, an important source of genetic variation in populations.

Diakinesis. This stage (Figure 3-3e) does not differ appreciably from diplotene, except for further chromosome contraction. By the end of diakinesis, the long, filamentous chromosome threads of interphase have been replaced by compact units that are far more maneuverable in the movements of the meiotic division.

Metaphase I. The nuclear membrane and nucleoli have disappeared by metaphase I, and each pair of homologs takes up a position in the equatorial plane (Figure 3-3f). At this stage of meiosis, the centromeres do not divide; this lack of division represents a major difference from mitosis. The two centromeres of a homologous chromosome pair attach to spindle fibers from opposite poles.

Anaphase I. As in mitosis, anaphase begins when chromosomes move directionally to the poles. The members of a homologous pair move to opposite poles (Figures 3-3g and 3-3h).

Telophase I. This telophase (Figure 3-3i) and the ensuing "interphase," called **interkinesis** (Figure 3-3j), are not universal. In many organisms, these stages do not exist, no nuclear membrane re-forms, and the cells proceed directly to meiosis II. In other organisms, telophase I and the interkinesis are brief in duration; the chromosomes elongate and become diffuse, and the nuclear membrane re-forms. In any case, there is never DNA synthesis at this time, and the genetic state of the chromosomes does not change.

By convention, the two nuclei that result from meiosis I are considered to be haploid. This might seem perverse, because each chromosome is composed of a pair of sister chromatids. Nevertheless, since the sister chromatids remain attached at the centromere, they are counted as one chromosome. Viewed this way, haploidy is best demonstrated by counting centromeres. The key point is that the total number of chromosomes in each cell has been reduced by one-half. For this reason, meiosis I is called a **reduction division.** In contrast, meiosis II is similar to a mitotic division in that the number of chromosomes (as defined above) is the same in the original and product cells. This type of division is called **equational division.**

Prophase II. The presence of the haploid number of chromosomes in the contracted state characterizes prophase II (Figure 3-3k).

Metaphase II. The chromosomes arrange themselves on the equatorial plane during metaphase II (Figure 3-3l). Here the chromatids often partly dissociate from each other instead of being closely appressed as they are in mitosis.

Anaphase II. Centromeres split and sister chromatids are pulled to opposite poles by the spindle fibers during anaphase II (Figure 3-3m).

Telophase II. The nuclei re-form around the chromosomes at the poles (Figure 3-3n).

The four products of meiosis are shown in Figure 3-3o. In the anthers of a flower, each of the four products of meiosis develops into pollen grains; these are shown in Figure 3-3p. In other organisms, differentiation produces other kinds of structures from the products of meiosis, such as sperm cells in animals.

In summary, meiosis requires one doubling of genetic material (the premeiotic S phase) and two cell divisions. Inevitably, this must result in products of meiosis that each contain one-half the genetic material of the original meiocyte.

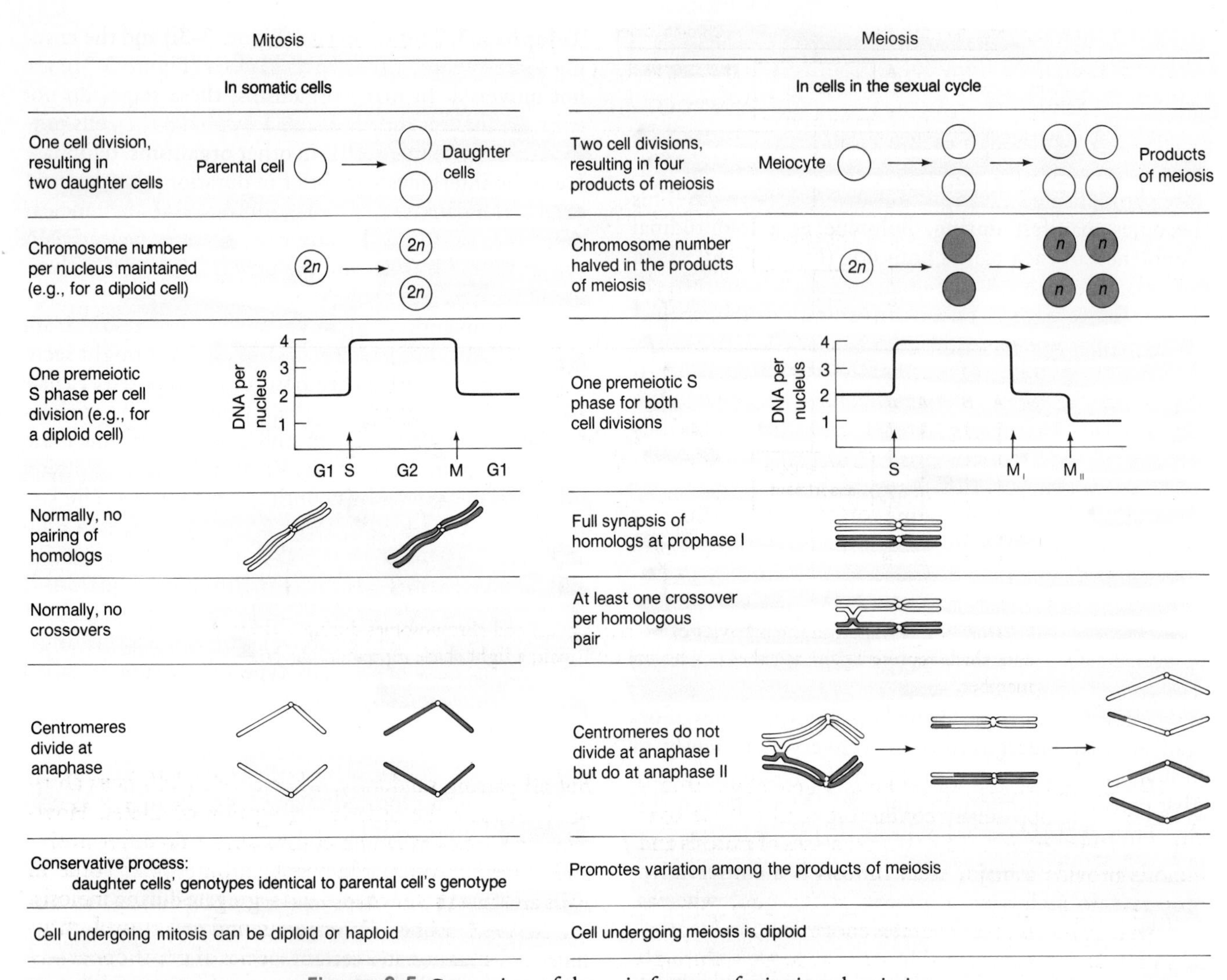

Figure 3-5 Comparison of the main features of mitosis and meiosis.

Message Meiosis generally takes place in a diploid meiocyte and yields four haploid products of meiosis.

A summary of the net events of mitosis and meiosis is shown in Figure 3-5.

The Chromosome Theory of Heredity

Credit for the chromosome theory of heredity—the concept that genes are parts of chromosomes—is usually given to both Walter Sutton (an American who at the time was a graduate student) and Theodor Boveri (a German biologist). In 1902, these investigators recognized independently that the behavior of Mendel's particles during the production of gametes in peas precisely parallels the behavior of chromosomes at meiosis: genes are in pairs (so are chromosomes); the alleles of a gene segregate equally into gametes (so do the members of a pair of homologous chromosomes); different genes act independently (so do different chromosome pairs). After recognizing this parallel behavior (which is summarized in Figure 3-6), both investigators reached the same conclusion.

Message The parallel behavior of genes and chromosomes led to the suggestion that genes are located on chromosomes.

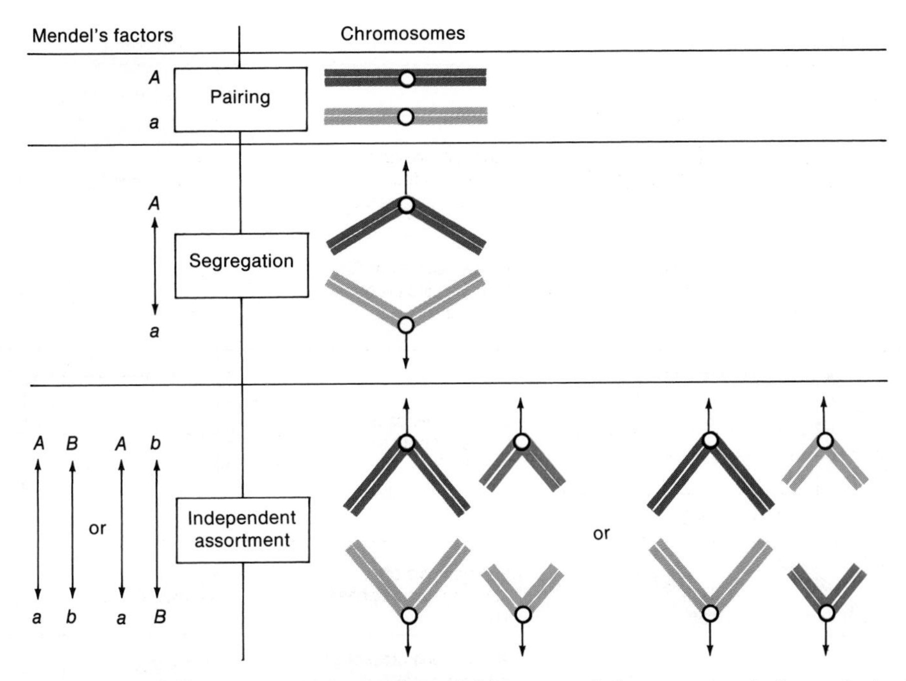

Figure 3-6 Parallels in the behavior of Mendel's genes and chromosomes during meiosis. A dark shade represents one member of a homologous pair; a light shade represents the other member.

We saw in Chapter 1 that a major goal of genetics is to explain two apparently conflicting forces of biology: heredity and variation. The two processes of mitosis and meiosis provide a major clue: mitosis is a conservative process that maintains a genetic status quo, whereas meiosis is a process that generates enormous combinatorial variation, rather like shuffling the gene pack, through independent assortment and (as we shall see later) through crossing-over.

To modern biology students, the chromosome theory may not seem very earthshaking. However, early in the twentieth century, Sutton's and Boveri's hypothesis (which potentially united cytology and the infant field of genetics) was a bombshell. Of course, the first response to publication of the hypothesis was to try to pick holes in it. For years after, there was a raging controversy over the validity of what became known as the Sutton-Boveri chromosome theory of heredity.

It is worth considering some of the objections raised to the Sutton-Boveri theory. For example, at the time, chromosomes could not be detected during interphase (between cell divisions). Boveri had to make some very detailed studies of chromosome position before and after interphase before he could argue persuasively that chromosomes retain their physical integrity through interphase, even though they are cytologically invisible at that time. It was also pointed out that in some organisms several pairs of chromosomes all look alike, making it impossible to say from visual observation that they are not all pairing randomly, whereas Mendel's laws absolutely require the orderly segregation of alleles. However, in species in which chromosomes do differ in size and shape, it was verified that chromosomes come in pairs and that these synapse and segregate during meiosis.

In 1913, Elinor Carothers found an unusual chromosomal situation in a certain species of grasshopper—a situation that permitted a direct test of whether different chromosome pairs do indeed segregate independently. Studying grasshopper testes, she found one chromosome pair that had nonidentical members; this is called a **heteromorphic pair,** and the chromosomes presumably show only partial homology. Furthermore, she found that another chromosome, unrelated to the heteromorphic pair, had no pairing partner at all. Carothers was able to use these unusual chromosomes as visible cytological markers of the behavior of chromosomes during assortment. By looking at anaphase nuclei, she could count the number of times that each dissimilar chromosome of the heteromorphic pair migrated to the same pole as the chromosome with no pairing partner (Figure 3-7). She observed the two patterns of chromosome behavior with equal frequency. Although these unusual chromosomes obviously are not typical, the results do suggest that nonhomologous chromosomes assort independently.

Other investigators argued that because all chromosomes appear as stringy structures, qualitative differences between them are of no significance. It was suggested

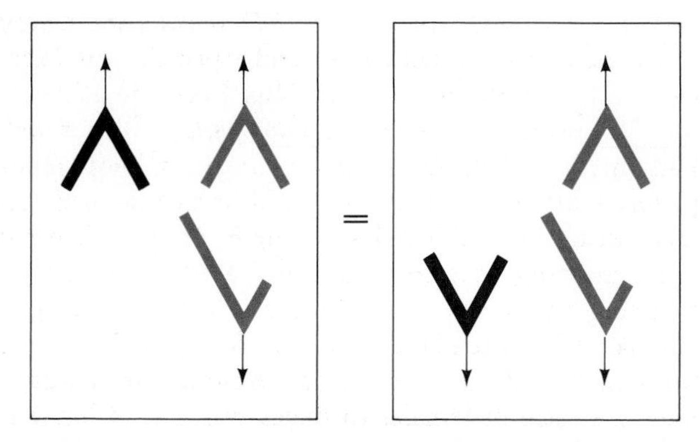

Figure 3-7 Two equally frequent patterns by which a heteromorphic pair and an unpaired chromosome move into gametes, as observed by Carothers.

Figure 3-8 Fruits from *Datura* plants, each having one different extra chromosome. Their characteristic appearances suggest that each chromosome produces a unique effect. (From E. W. Sinnott, L. C. Dunn, and T. Dobzhansky, *Principles of Genetics* 5th ed. McGraw-Hill Book Co.)

that perhaps all chromosomes were just more or less made of the same stuff. It is worth introducing a study out of historical sequence that effectively counters this objection. In 1922, Alfred Blakeslee performed a study on the chromosomes of jimsonweed *(Datura stramonium)*, which has 12 chromosome pairs. He obtained 12 different strains, each of which had the normal 12 chromosome pairs plus an extra representative of one pair. Blakeslee showed that each strain was phenotypically distinct from the others (Figure 3-8). This result would not be expected if there were no genetic significance to the differences among the extra chromosomes.

All these results indicated that the behavior of chromosomes closely parallels that of genes. This of course made the Sutton-Boveri theory attractive, but there was as yet no real proof that genes are located on chromosomes. Further observations, however, did provide such proof, and these began with the discovery of sex linkage.

The Discovery of Sex Linkage

In the crosses discussed thus far, it does not matter which sex of parent is from which strain. Reciprocal crosses (such as strain A ♀ × strain B ♂ and strain A ♂ × strain B ♀) yield similar progeny. The first exception to this pattern was discovered in 1906 by L. Doncaster and G. H. Raynor. They were studying wing color in the magpie moth *(Abraxas)*, using two different lines: one with light wings; the other with dark wings. If light-winged females are crossed with dark-winged males, all the progeny have dark wings, showing that the allele for light wings is recessive. However, in the reciprocal cross (dark female × light male), all the female progeny have light wings and all the male progeny have dark wings. Thus, this pair of reciprocal crosses does not give similar results, and the wing phenotypes in the second cross are associated with the sex of the moths. Note that the female progeny of this second cross are phenotypically similar to their fathers, as the males are to their mothers. How can we explain these results? Before attempting an explanation, let's consider another example.

William Bateson had been studying the inheritance of feather pattern in chickens. One line had feathers with alternating stripes of dark and light coloring, a phenotype called barred. Another line, nonbarred, had feathers of uniform coloring. In the cross barred male × nonbarred female, all the progeny were barred, showing that the allele for nonbarred is recessive. However, the reciprocal cross (barred female × nonbarred male) gave barred males and nonbarred females. Again, the result is that female progeny have their father's phenotype and male progeny have their mother's. Can we find an expla-

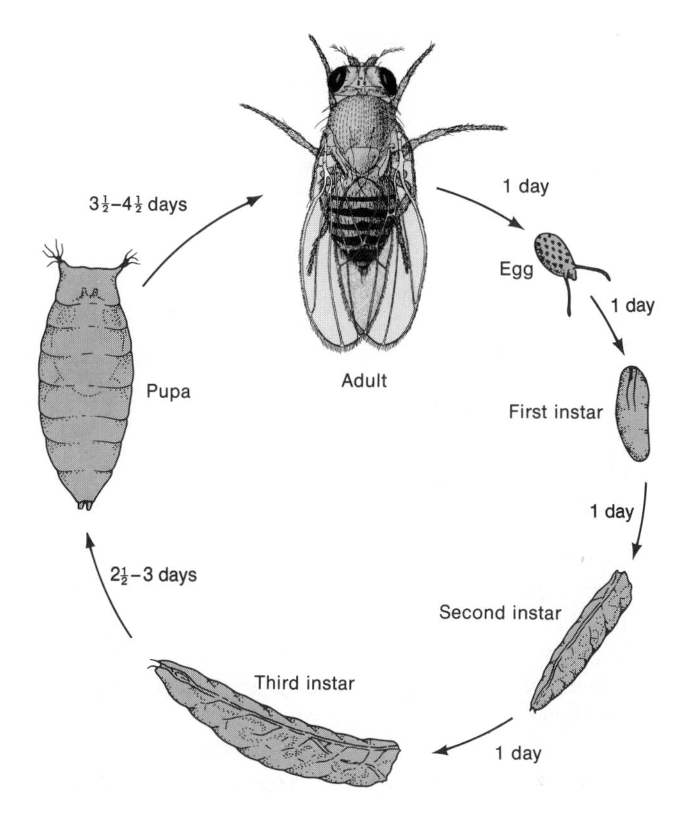

Figure 3-9 Life cycle of *Drosophila melanogaster,* the common fruit fly.

nation for these similar results with moths and with chickens?

An explanation came from the laboratory of Thomas Hunt Morgan, who in 1909 began studying inheritance in a fruit fly *(Drosophila melanogaster)*. Because this organism has played a key role in the study of inheritance, a brief digression about the creature is worthwhile.

The life cycle of *Drosophila* is typical of the life cycles of many insects (Figure 3-9). The flies grow vigorously in the laboratory. In the egg, the early embryonic events lead to the production of a larval stage called the first instar. Growing rapidly, the larva molts twice, and the third-instar larva then pupates. In the pupa, the larval carcass is replaced by adult structures, and an imago, or adult, emerges from the pupal case, ready to mate within 12 to 14 hours. The adult fly is about 2 millimeters long, so it takes up very little space; thousands of flies can be maintained in vials kept on a single laboratory shelf. The life cycle is very short (12 days at room temperature) in comparison with that of a human, a mouse, or a corn plant; thus, many generations can be reared in a year. Moreover, the flies are extremely prolific: a single female is capable of laying several hundred eggs. Perhaps the beauty of the insect when observed through a microscope added to its early allure as a research organism. In any case, as we shall see, the choice of *Drosophila* was a very fortunate one for geneticists—and especially for Morgan, whose work earned him a Nobel prize in 1934.

The normal eye color of *Drosophila* is bright red. Early in his studies, Morgan discovered a male with completely white eyes (Figure 3-10). When he crossed this male with red-eyed females, all the F_1 progeny had red eyes, showing that the allele for white is recessive. Crossing the red-eyed F_1 males and females, Morgan obtained a 3:1 ratio of red-eyed to white-eyed flies, but all the white-eyed flies were males. Among the red-eyed flies, the ratio of females to males was 2:1. What was going on?

Morgan gathered more data. When he crossed white-eyed males with red-eyed female progeny of the cross of white males and red females, he obtained red males, red females, white males, and white females in equal numbers. Finally, in a cross of white females and red males (which is the reciprocal of the cross of the original white male with a normal female), all the females were red and all the males were white. This is very similar to outcomes we still need to explain in the examples of chickens and moths, but note a difference: in chickens and moths the progeny are like the parent of opposite sex when the parental males carry the recessive alleles; in the *Drosophila* cross, this outcome is seen when the female parents carry the recessive alleles.

Before turning to Morgan's explanation of the *Drosophila* results, we should look at some of the cytological information he was able to use in his interpretations. In 1891, working with males of a species of Hemiptera (the true bugs), H. Henking observed that meiotic nuclei contained 11 pairs of chromosomes and an unpaired element that moved to one of the poles during the first meiotic division. Henking called this unpaired element an "X body;" he interpreted it as a nucleolus, but later studies showed it to be a chromosome. Similar unpaired

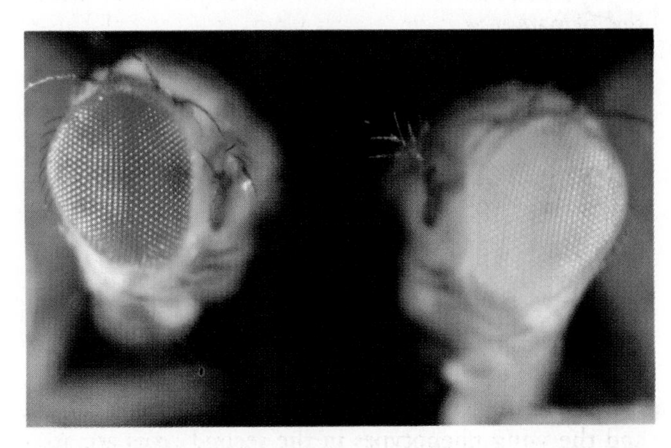

Figure 3-10 Red-eyed and white-eyed *Drosophila.* (Carolina Biological Supply)

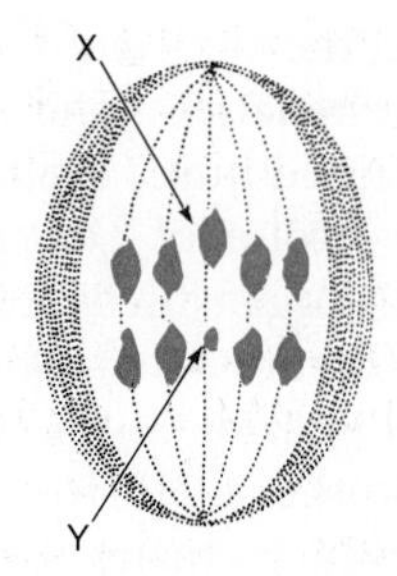

Figure 3-11 Segregation of the heteromorphic chromosome pair (X and Y) during meiosis in a *Tenebrio* male. The X and Y chromosomes are being pulled to opposite poles during anaphase I. (From A. M. Srb, R. D. Owen, and R. S. Edgar, *General Genetics,* 2d ed. Copyright 1965 by W. H. Freeman and Company.)

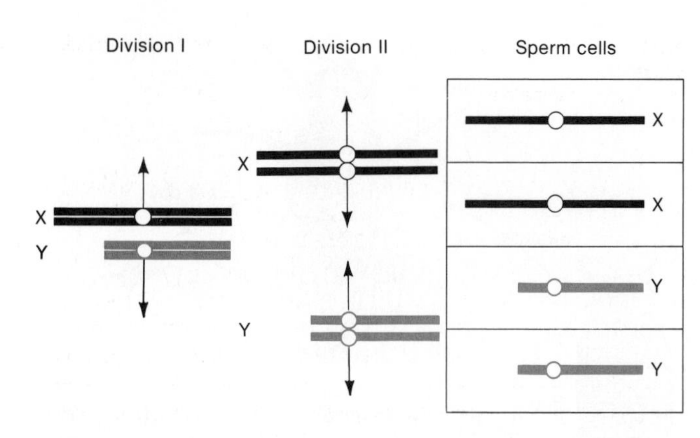

Figure 3-13 Meiotic pairing and the segregation of the X and Y chromosomes into equal numbers of sperm.

elements were later found in other species. In 1905, Edmond Wilson noted that females of *Protenor* (another Hemipteran) have seven pairs of chromosomes, whereas males have six pairs and an unpaired chromosome, which Wilson called (by analogy) the X chromosome. The females, in fact, have a pair of X chromosomes.

Also in 1905, Nettie Stevens found that males and females of the beetle *Tenebrio* have the same number of chromosomes, but one of the chromosome pairs in males is heteromorphic (that is, the two members differ in size). One member of the heteromorphic pair appears identical to the members of a pair in the female; Stevens called this the X chromosome. The other member of the heteromorphic pair is never found in females; Stevens called this the Y chromosome (Figure 3-11). She found a similar situation in *Drosophila melanogaster,* which has four pairs of chromosomes, with one of the pairs being heteromorphic in males. Figure 3-12 summarizes these two basic situations. (You may be wondering about the male grasshoppers studied by Carothers that had both a heteromorphic chromosome pair and an unpaired chromosome. This situation is very unusual, and we needn't worry about it at this point.)

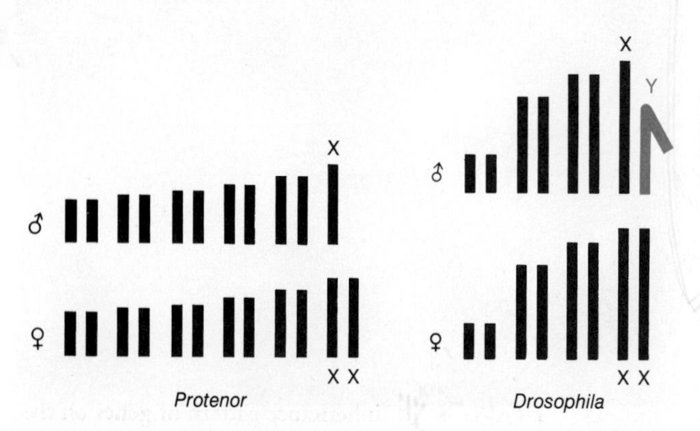

Figure 3-12 The chromosomal constitutions of males and females in two insect species.

With this background information, Morgan constructed an interpretation of his genetic data. First, it appears that the X and Y chromosomes determine the sex of the fly. *Drosophila* females have four chromosome pairs, whereas males have three matching pairs plus a heteromorphic pair. Thus, meiosis in the female produces eggs that each bear one X chromosome. Although the X and Y chromosomes in males are heteromorphic, they seem to synapse and segregate like homologs (Figure 3-13). Thus, meiosis in the male produces two types of sperm, one type bearing an X chromosome and the other bearing a Y chromosome. According to this explanation, union of an egg with an X-bearing sperm produces an XX (female) zygote and union with a Y-bearing sperm produces an XY (male) zygote. Furthermore, approximately equal numbers of males and females are expected due to the equal segregation of X and Y.

Morgan next turned to the problem of eye color. Assume that the alleles for red or white eye color are present on the X chromosome, with no counterpart on the Y chromosome. Thus, females would have two alleles for this gene, whereas males would have only one. This highly unexpected situation proves to fit the data. In the original cross of the white-eyed male with red-eyed females, all F_1 progeny had red eyes, showing that the allele for red eyes is dominant. Therefore, we can represent the two alleles as W (red) and w (white). If we designate the X chromosomes as X^W and X^w to indicate the alleles supposedly carried by them, then we can diagram the two reciprocal crosses as shown in Figure 3-14.

As we can see from the figure, the genetic results of the two reciprocal crosses are completely consistent with the known meiotic behavior of the X and Y chromosomes. This experiment strongly supports the notion that genes are located on chromosomes. However, it is

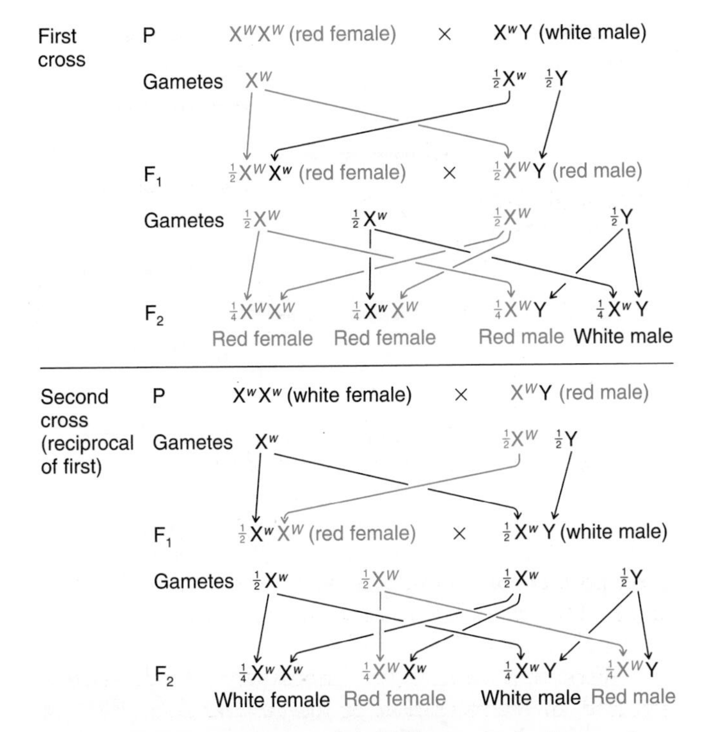

Figure 3-14 Explanation of the different results from reciprocal crosses between red-eyed *(red)* and white-eyed *(white)* *Drosophila.*

only a correlation; it does not constitute a definitive proof of the Sutton-Boveri theory.

Can the same XX and XY chromosome theory be applied to the results of the earlier crosses made with chickens and moths? You will find that it cannot. However, Richard Goldschmidt recognized immediately that these results can be explained with a similar hypothesis, making the simple assumption that the *males* have pairs of identical chromosomes whereas the *females* have a heteromorphic pair. To distinguish this situation from the X-Y situation in *Drosophila,* Morgan suggested that the heteromorphic chromosomes in chickens and moths be called W-Z, with males being ZZ and females being WZ. Thus, if the genes in the chicken and moth crosses are on the Z chromosome, the crosses can be diagrammed as shown in Figure 3-15.

The interpretation is consistent with the genetic data. In this case, cytological data provided a confirmation of the genetic hypothesis. In 1914, J. Seiler verified that both chromosomes are identical in all pairs in male moths whereas females have one heteromorphic pair.

Message The special inheritance pattern of some genes makes it extremely likely that they are borne on the sex chromosomes, which show a parallel pattern of inheritance.

An Aside on Genetic Symbols

In *Drosophila,* a special symbolism for allele designation was introduced to define variant alleles in relation to a "normal" allele. This system is now used by many geneticists and is especially useful in genetic dissection. For a given *Drosophila* character, the allele that is found most frequently in natural populations (or, alternatively, the allele that is found in standard laboratory stocks) is designated as the standard, or **wild type.** All other alleles are then non-wild-type alleles. The symbol for a gene comes from the first non-wild-type allele found. In Morgan's *Drosophila* experiment, this was the allele for white eyes, symbolized by *w.* The wild-type counterpart allele is conventionally represented by adding a + superscript, so the normal red-eye allele is written w^+.

In a polymorphism, several alleles might be common in nature and all might be regarded as wild type. In this case, geneticists use appropriate superscripts to distinguish alleles. For example, two alleles of the alcohol dehydrogenase gene in *Drosophila* are designated Adh^F and Adh^S. (This gene controls the alcohol-metabolizing enzyme alcohol dehydrogenase, and F and S stand for fast and slow movements, respectively, of the enzyme in an electrophoretic gel.)

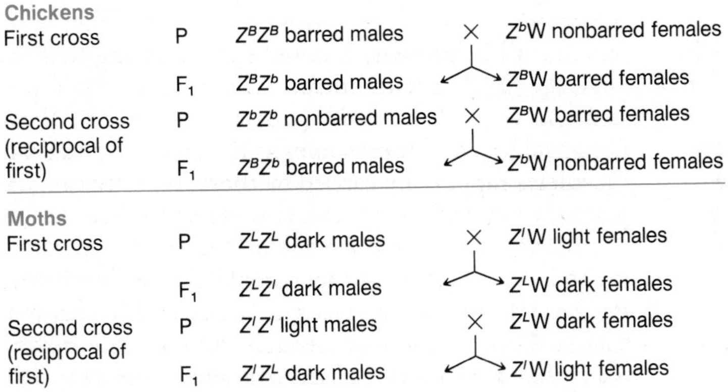

Chickens First cross	P	$Z^B Z^B$ barred males	×	Z^bW nonbarred females
	F_1	$Z^B Z^b$ barred males		Z^BW barred females
Second cross (reciprocal of first)	P	$Z^b Z^b$ nonbarred males	×	Z^BW barred females
	F_1	$Z^B Z^b$ barred males		Z^bW nonbarred females
Moths First cross	P	$Z^L Z^L$ dark males	×	Z^lW light females
	F_1	$Z^L Z^l$ dark males		Z^LW dark females
Second cross (reciprocal of first)	P	$Z^l Z^l$ light males	×	Z^LW dark females
	F_1	$Z^l Z^L$ dark males		Z^lW light females

Figure 3-15 Inheritance pattern of genes on the sex chromosomes of two species having the ZW mechanism of sex determinations.

The wild-type allele can be dominant or recessive to a non-wild-type allele. For the two alleles w and w^+, the use of the lower-case letter indicates that the wild-type allele is dominant over the one for white eyes (that is, w is recessive to w^+). As another example, the wild-type phenotype of a fly's wing is straight and flat. A non-wild-type allele causes the wing to be curled. Because this allele is dominant over the wild-type allele, it is written *Cy* (short for *Curly*), whereas the wild-type allele is written Cy^+. Here note that the capital letter indicates that *Cy* is dominant over Cy^+. (Also note from these examples that the symbol for a single gene may consist of more than one letter.)

The symbolism specifying wild type is useful because it helps geneticists focus on the procedure of genetic dissection. In many situations, the wild-type allele is defined as being responsible for the normal function and non-wild-type alleles (whether recessive or dominant) can be regarded as abnormal. The abnormal alleles then become experimental probes. Geneticists learn how normal functions work by using the abnormal alleles to study the ways in which the normal mechanism can go wrong. Note that Mendel's symbols (*A* and *a*, or *B* and *b*) do not define or emphasize normality. In a pea flower, for example, is purple or white normal? However, the Mendelian symbolism is useful for some purposes; it is used extensively in plant and animal breeding.

Proof of the Chromosome Theory

The correlations between the behavior of genes and the behavior of chromosomes made it very likely that genes are parts of chromosomes. But this was not a proof of the chromosome theory, and debate continued. The critical proof of the Sutton-Boveri theory came from one of Morgan's students, Calvin Bridges. Bridges's work began with a fruit fly cross we have discussed before, now represented in our new symbolism as X^wX^w (white-eyed ♀) × $X^{w^+}Y$ (red-eyed ♂). We know that the progeny are $X^{w^+}X^w$ (red-eyed ♀♀) and X^wY (white-eyed ♂♂). When Bridges replicated the cross on a large scale, he observed a few exceptions among the progeny. About one out of every 2000 F_1 progeny was a white-eyed female or a red-eyed male. Collectively, these individuals were called primary exceptional progeny. All the primary exceptional males proved to be sterile. However, when Bridges crossed the primary exceptional white-eyed females with normal red-eyed males, approximately 4 percent of the progeny were white-eyed females and red-eyed males that were fertile. Thus, exceptional offspring were again recovered, but at a higher frequency, and the males were fertile. These exceptional progeny of primary exceptional mothers were called secondary exceptional offspring (Figure 3-16). How do we explain the exceptional progeny?

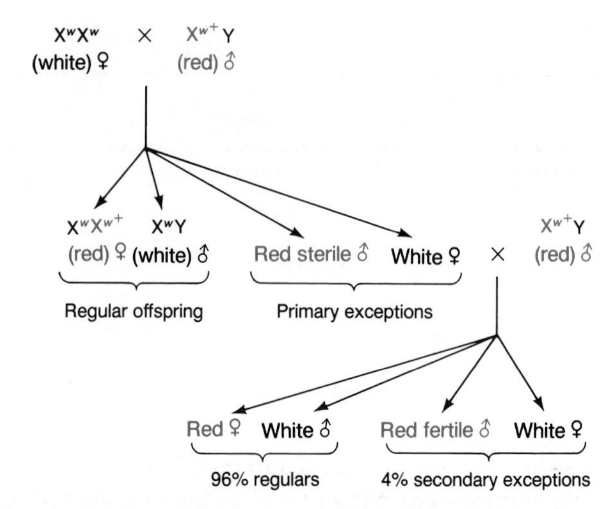

Figure 3-16 *Drosophila* crosses from which primary and secondary exceptional progeny were originally obtained. *Red,* red-eyed, and *white,* white-eyed, *Drosophila.*

It is obvious that the exceptional females—which, like all females, have two X chromosomes—must get both of these chromosomes from their mothers because they are homozygous for w. Similarly, exceptional males must get their X chromosomes from their fathers because they carry w^+. Bridges hypothesized rare mishaps during meiosis in the female whereby the paired X chromosomes fail to separate during either the first or second division. This would result in meiotic nuclei containing either two X chromosomes or no X at all. Such a failure to separate is called nondisjunction; it produces an XX nucleus and a nullo-X nucleus (containing no X). Fertilization of eggs having these types of nuclei by sperm from a wild-type male produces four zygotic classes (Figure 3-17). It is important to note that the line represent-

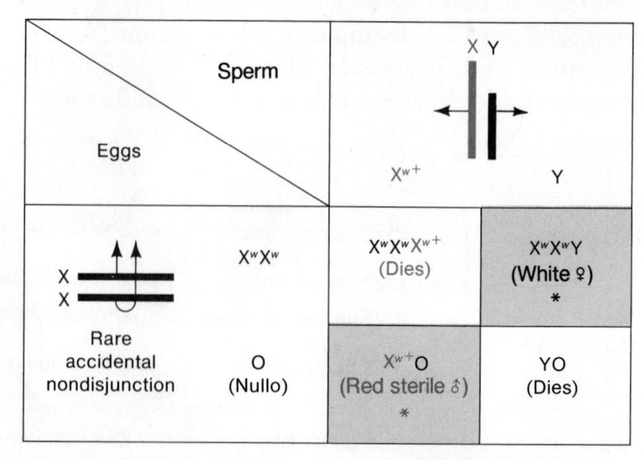

Figure 3-17 Proposed explanation of primary exceptional progeny through nondisjunction of the X chromosomes in the maternal parent. *Red,* red-eyed, and *white,* white-eyed, *Drosophila.*

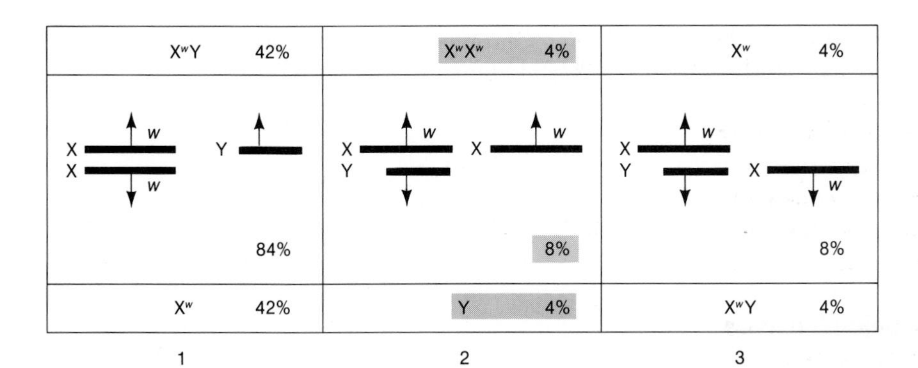

Figure 3-18 Three different segregation patterns in an XXY female fruit fly.

ing a chromosome in these diagrams is in each case not a single chromosome but a pair of daughter chromatids.

If we assume that the XXX and YO zygotes die before development is complete, then the two types of exceptional progeny can be expected to be X^wX^wY (♀) and $X^{w^+}O$ (♂). Notice that it is implicit in Bridges's model that the sex of *Drosophila* is determined not by the presence or absence of the Y chromosome but by the number of X chromosomes with two X chromosomes producing a female and one X chromosome producing a male.

What about the sterility of the primary exceptional males? This makes sense if we assume that a male must have a Y chromosome to be fertile.

How can we explain the secondary exceptional offspring? During meiosis in the XXY females, if the two X chromosomes paired and disjoined all the time while half of the eggs received the unpaired Y chromosome and half did not, there would be equal numbers of X-bearing and XY-bearing eggs. However, we know that the X and Y chromosomes can pair and segregate, because normal males produce equal numbers of X-bearing and Y-bearing sperm. How did this help explain the results Bridges observed? To explain the observed results, he assumed that during meiosis in the XXY females, 84 percent of the pairings were between the two X chromosomes while the Y chromosome successfully paired with an X chromosome in 16 percent of the pairings, leaving the other X chromosome free to move to either pole. One-half (8 percent) of these pairings will result in X^w and X^wY eggs, and the other one-half (8 percent) will result in X^wX^w and Y eggs (Figure 3-18).

We can now look at the results of fertilization by equal numbers of X^{w^+} and Y sperm (Figure 3-19). We find that one-half of the fertilized X^wX^w and Y eggs will produce $X^wX^wX^{w^+}$ and YY zygotes, which presumably die. The other one-half of these fertilized eggs produce the secondary exceptions, X^wX^wY and $X^{w^+}Y$. Now we

Eggs		Sperm: X^{w+} (50%)	Sperm: Y (50%)
X–Y pairing (16%)	X^wX^w (4%)	$X^wX^wX^{w+}$ (dies) (2%)	X^wX^wY (white ♀) (2%)
	Y (4%)	$X^{w+}Y$ (red fertile ♂) (2%)	YY (dies) (2%)
	X^w (4%)	$X^{w+}X^w$ (red ♀) (2%)	X^wY (white ♂) (2%)
	X^wY (4%)	$X^wX^{w+}Y$ (red ♀) (2%)	X^wYY (white ♂) (2%)
X–X pairing (84%)	X^wY (42%)	$X^wX^{w+}Y$ (red ♀) (21%)	X^wYY (white ♂) (21%)
	X^w (42%)	X^wX^{w+} (red ♀) (21%)	X^wY (white ♂) (21%)

Secondary exceptional progeny phenotype (4%) (4% die)

"Regular" (expected) progeny phenotypes (92%)

Figure 3-19 The proposed origin of secondary exceptional progeny as a result of specific gamete types produced by the XXY parent. *Red*, red-eyed, and *white*, white-eyed, *Drosophila*.

see why the secondary exceptional males are fertile: each of them receives a Y chromosome from the XXY mother.

So far, this is all a model—an intellectual edifice. Bridges assumed that w and w^+ are located on X chromosomes, and hypothesized nondisjunction to explain the exceptional progeny. However, if this model is correct, he could make testable predictions.

1. Cytological study of the primary exceptional progeny (which were identified through the genetic study) should show that the females are XXY and the males are XO. Bridges confirmed this prediction.
2. Cytological study of the secondary exceptional progeny (which were identified genetically) should show that the females are XXY and the males are XY. Bridges confirmed this prediction.
3. One-half of the red-eyed daughters of exceptional white-eyed females should be XXY, and one-half should be XX. Bridges confirmed this prediction.
4. One-half of the white-eyed sons of exceptional white-eyed females should themselves give exceptional progeny, and all of those that do should be XYY. Bridges confirmed this prediction.

Thus, Bridges verified all the testable predictions arising from the assumptions that w and w^+ are indeed on the X chromosome and that as a rare meiotic event chromosomes fail to disjoin. These confirmations provide unequivocal evidence that genes are associated with chromosomes and are probably parts of the chromosomes themselves.

Message When Bridges used the chromosome theory to predict successfully the outcome of certain genetic analyses, the chromosome location of genes was established beyond reasonable doubt.

Sex Chromosomes and Sex Linkage

Humans and all mammals show an X-Y sex-determining mechanism, with males being XY and females XX. The mechanism in humans, however, differs from that in *Drosophila:* it is the presence of the Y that determines maleness in humans. This difference is demonstrated by the sexual phenotypes of the abnormal chromosome types XXY and XO (Table 3-1). However, we postpone a full discussion until a later chapter.

Table 3-1. Chromosomal Determination of Sex in *Drosophila* and Humans

	Sex Chromosomes			
Species	XX	XY	XXY	XO
Drosophila	♀	♂	♀	♂
Humans	♀	♂	♂	♀

Vascular plants show a variety of sexual arrangements. Some species have both male and female sex organs on the same plant, either combined into the same flower—making it a **hermaphroditic** species (the rose is an example)—or in separate flowers on the same plant—making it a **monoecious** species (corn is an example). **Dioecious** species, however, have the sexes separate, with female plants bearing flowers containing only ovaries and male plants bearing flowers containing only anthers (Figure 3-20). Some, but not all, dioecious plants have a heteromorphic pair of chromosomes associated with (and almost certainly determining) the sex of the plant. Of the species with heteromorphic sex chromosomes, a large proportion have an X-Y system. Critical experiments in a few species suggest a *Drosophila*-like system. Other dioecious plants have no visibly heteromorphic pair of chromosomes; they may still have sex chromosomes, but not visibly distinguishable types.

After studying meiosis in males, cytogeneticists have divided the X and Y chromosomes of some species into regions that pair and regions that do not, which are called differential regions (Figure 3-21). The pairing regions of the X and Y chromosomes are thought to be homologous. In contrast, the differential region of each chromosome appears to hold genes that have no counterparts on the other sex chromosome. Genes in the differential regions are said to be **hemizygous** ("half-zygous") in males. Genes in the differential region of the X show an inheritance pattern called **X linkage;** those in the differential region of the Y show **Y linkage.** Genes in the pairing region show what might be called **X-and-Y linkage.** In general, genes on the sex chromosomes show **sex linkage.**

We can introduce some other common terminology here. The sex having only one kind of sex chromosome (XX♀ or ZZ♂) is called the **homogametic** sex; the other (XY♂ or ZW♀) is called the **heterogametic** sex. Thus, human and *Drosophila* females (and male birds and moths) are homogametic; that is, each individual produces only one chromosomal type of gamete. The nonsex chromosomes (that we might call the "regular" chromosomes) are called **autosomes.** Humans have 46 chromosomes per cell: 44 autosomes plus two sex chromosomes. The plant *Melandrium album* has 22 chromosomes per cell: 20 autosomes plus two sex chromosomes.

Genes on the autosomes show the kind of inheritance pattern discovered and studied by Mendel. The

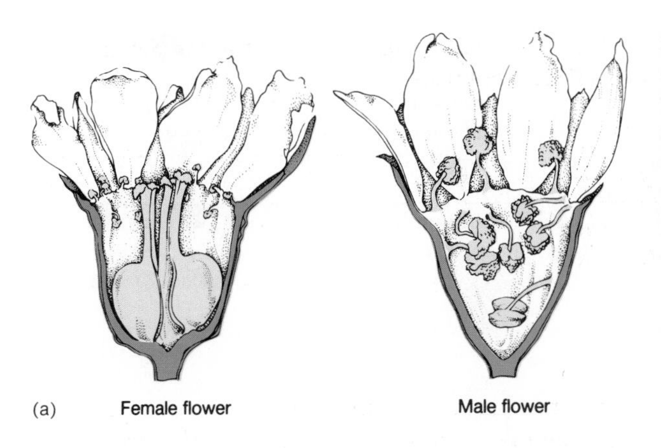

Figure 3-20 Two dioecious plant species. (a) *Osmaronia dioica.* (b) *Aruncus dioicus;* female at left, male at right. (b, Anthony Griffiths)

genes on the differential regions of the sex chromosomes show their own typical patterns of inheritance, as follows.

X-Linked Inheritance

We have already seen an example of X-linked inheritance in *Drosophila:* the inheritance pattern of the white eye phenotype and its wild-type counterpart, red eye. Of course, eye color is not concerned with sex determination, so we see that genes on the sex chromosomes are not necessarily involved with sexual function. The same is true in humans, where pedigree analysis has revealed many X-linked genes, of which few could be construed as being connected to sexual function. Just as we earlier listed rules for detecting autosomal inheritance patterns

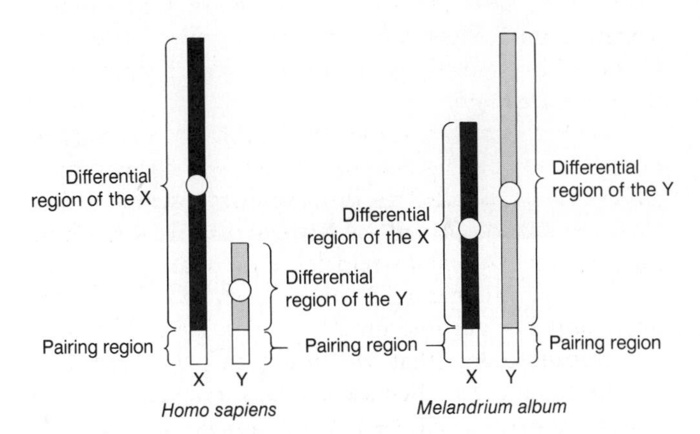

Figure 3-21 Differential and pairing regions of sex chromosomes of humans and of the plant *Melandrium album.* The regions were located by observing where the chromosomes synapsed during meiosis and where they did not.

in humans, we can list clues for detecting X-linked genes in human pedigrees.

X-linked recessive inheritance can be deduced from human pedigrees through the following clues.

1. Typically, many more males than females show the recessive phenotype. This is because an affected female can be produced only from a mating in which both the mother and father bear the allele (for example, $X^AX^a \times X^aY$), whereas an affected male can be produced when only the mother carries the allele. If the recessive allele is very rare, almost all affected persons are males.

The next two clues refer to X-linked recessive phenotypes that are rare in the population. Recall from Chapter 2 our discussion of human phenotypes inherited in a simple Mendelian manner. In the terminology of the present chapter, such inheritance is referred to as **autosomal inheritance,** because the genes concerned are all on the autosomes. Recall, too, that many such phenotypes are rare disorders and because of this certain assumptions could be made about the incidence of specific genotypes in a population. For example, in the case of autosomal recessive phenotypes it is assumed that people marrying into an affected family do not themselves bring the allele in question with them. The same assumption is made in the case of X-linked recessive inheritance. This assumption is particularly relevant to females that enter an affected family, and this point is implicit in the next two pedigree clues.

2. For a rare phenotype, none of the offspring of an affected male are affected, but all his daughters carry the allele, masked in the heterozygous condition; one-half of the sons borne by these daughters are

Figure 3-22 Pedigree showing that X-linked recessives are expressed in males and then carried unexpressed by females in the next generation, to be expressed in their sons. Note that III.3 and III.4 cannot be distinguished phenotypically.

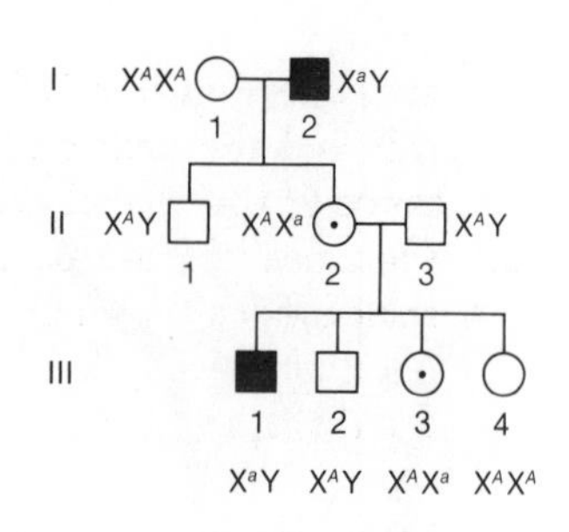

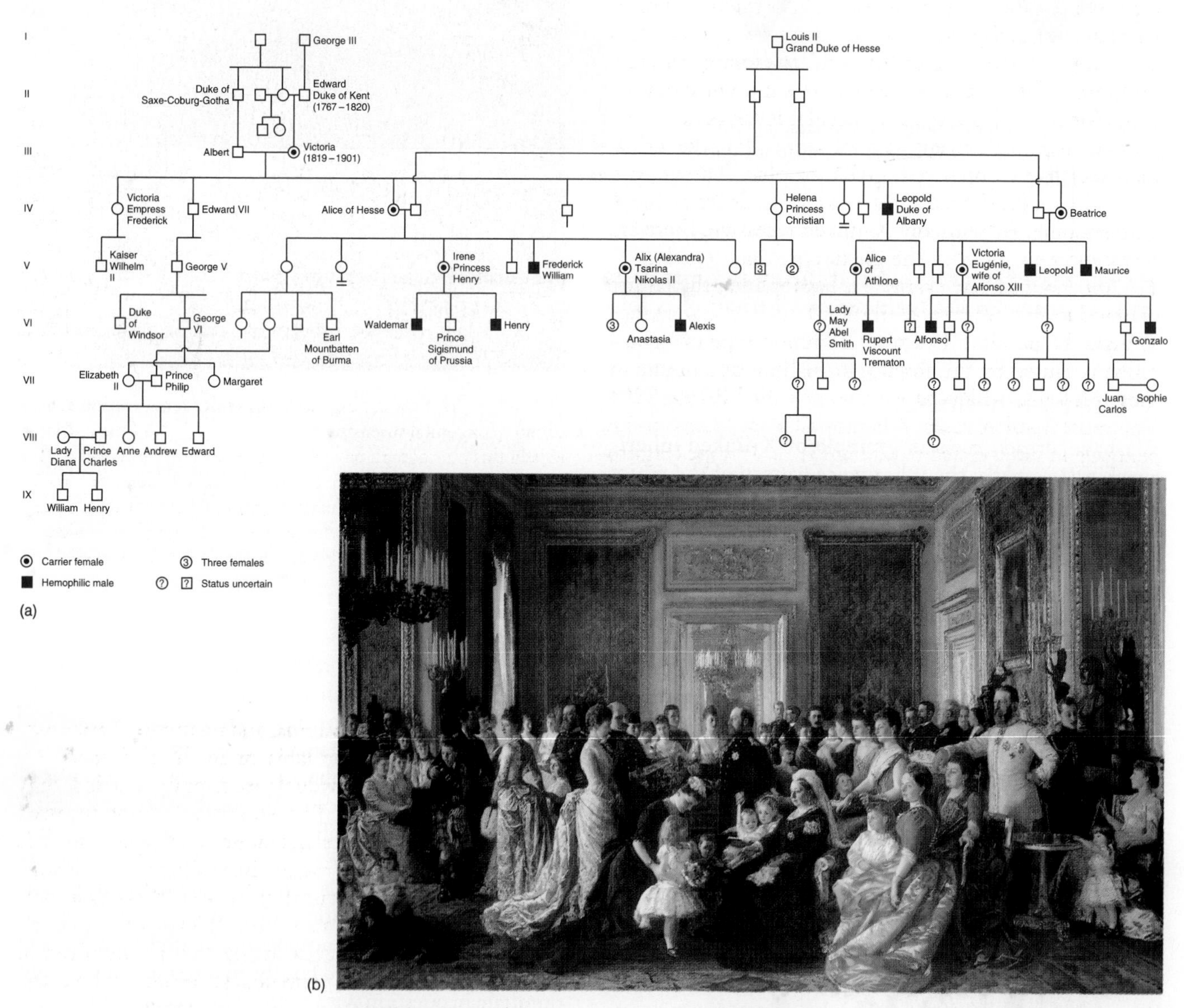

Figure 3-23 The inheritance of the X-linked recessive condition hemophilia in the royal families of Europe. A recessive allele causing hemophilia (failure of blood clotting) arose in the reproductive cells of Queen Victoria, or one of her parents, through mutation. This hemophilia allele spread into other royal families by intermarriage. (a) The partial pedigree shows affected males and carrier females (heterozygotes). Most spouses marrying into the families have been omitted from the pedigree for simplicity. Can you deduce the likelihood of the present British royal family harboring the recessive allele? (b) The painting shows Queen Victoria surrounded by her numerous descendants. (a, modified from C. Stern, *Principles of Human Genetics,* 3d. ed. Copyright 1973 by W. H. Freeman. b, Royal Collection, St. James's Palace. Copyright Her Majesty Queen Elizabeth II.)

affected (Figure 3-22). (If a phenotype is common, this and the next clue may be obscured by inheritance of the recessive allele from the mother.)

3. For a rare phenotype, none of the sons of an affected male inherit the allele, so not only are they free of the phenotype, but also they will not pass the allele to their offspring.

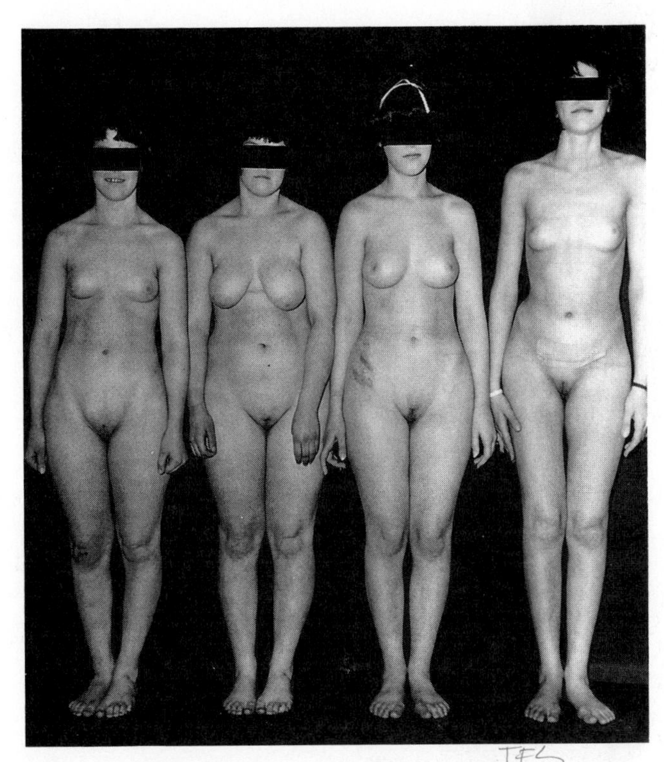

Figure 3-24 Four siblings with testicular feminization syndrome (congenital insensitivity to androgens). All four individuals in this photograph have 44 autosomes plus an X and a Y, but they have inherited the recessive X-linked allele conferring insensitivity to androgens (male hormones). One of their sisters (not shown) was a carrier who bore a child who also showed testicular feminization syndrome. (Leonard Pinski, McGill University)

Let us consider some examples of X-linked, rare, recessive conditions in humans. Perhaps the most familiar example is red-green colorblindness. People with this condition are unable to distinguish red from green, and see them as the same. Quite a lot is now known about the genes for color vision at the molecular level; color vision is based on three different kinds of cone cells in the retina, each sensitive to only red, green, or blue. The genetic determinants for the red and green cone cells are on the X chromosome. As with any X-linked recessive, there are many more males with the phenotype than females.

Another familiar example is hemophilia, the failure of blood to clot. Many proteins must interact in sequence to make blood clot. The most common type of hemophilia is caused by the absence or malfunction of one of these proteins, which happens to be called Factor XIII. The most familiar cases of hemophilia are found in the pedigree of interrelated royal families in Europe (Figure 3-23). The original hemophilia allele in the pedigree arose spontaneously (as a mutation) either in the reproductive cells of Queen Victoria's parents or of Queen Victoria herself. The son of the last Czar of Russia, Alexis, inherited the allele ultimately from Queen Victoria, who was the grandmother of his mother Alexandra. Nowadays, hemophilia can be treated medically, but it was formerly a potentially fatal condition. It is interesting to note that in the Jewish Talmud there are rules about exemptions to male circumcision that show clearly that the mode of transmission of the disease through unaffected carrier females was well understood in ancient times. For example, one exemption was for the sons of women whose sisters' sons had bled profusely when they were circumcised.

Duchenne's muscular dystrophy is a fatal X-linked recessive disease. The phenotype is a wasting and atrophy of muscles. Generally the onset is before the age of six, with confinement to a wheelchair by 12, and death by 20. The gene for Duchenne's muscular dystrophy has now been isolated, holding out hope for a better understanding of the physiology of this condition.

A rare X-linked recessive phenotype that is interesting from the point of view of sex determination is a condition called testicular feminization syndrome, which has a frequency of about one in 65,000 male births. Persons afflicted with this syndrome are chromosomally males, having 44 autosomes plus an X and a Y, but they develop as females (Figure 3-24). They have female external genitalia, a blind vagina, and no uterus. Testes may be present either in the labia or in the abdomen. Although many such individuals are happily married, they are, of course, sterile. The condition is not reversed by treatment with male hormone (androgen), so it is sometimes called androgen insensitivity syndrome. The reason for the insensitivity is that there is a malfunction in the androgen receptor, so male hormone can have no effect on the target organs that are involved in maleness. In humans, femaleness results when the male-determining system is not functional.

X-linked dominant inheritance can be detected in human pedigrees through the following clues.

1. The most important clue here is that affected males pass the condition on to all of their daughters but to none of their sons (Figure 3-25).
2. Females married to unaffected males pass the condition on to one-half of their sons and daughters (Figure 3-26).

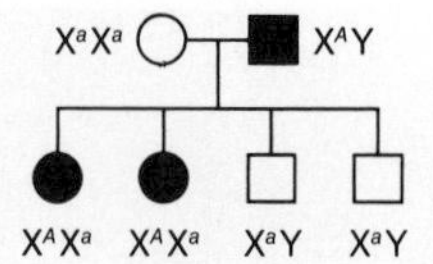

Figure 3-25 Pedigree showing that all daughters of affected males express X-linked dominant phenotypes.

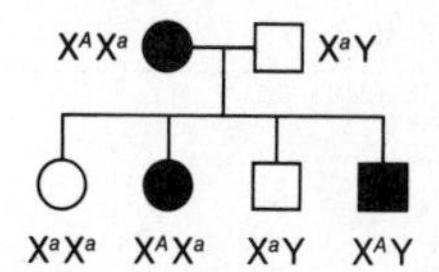

Figure 3-26 Pedigree showing that females affected by an X-linked dominant condition usually are heterozygous and pass the condition to one-half of their sons and daughters.

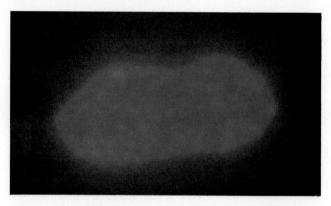

Figure 3-27 A Barr body, condensed inactivated X chromosome, in the nucleus of a cell of a normal woman. Men have no Barr bodies. The number of Barr bodies in a cell is always equal to the total number of X chromosomes minus one. (Karen Dyer, Vivigen)

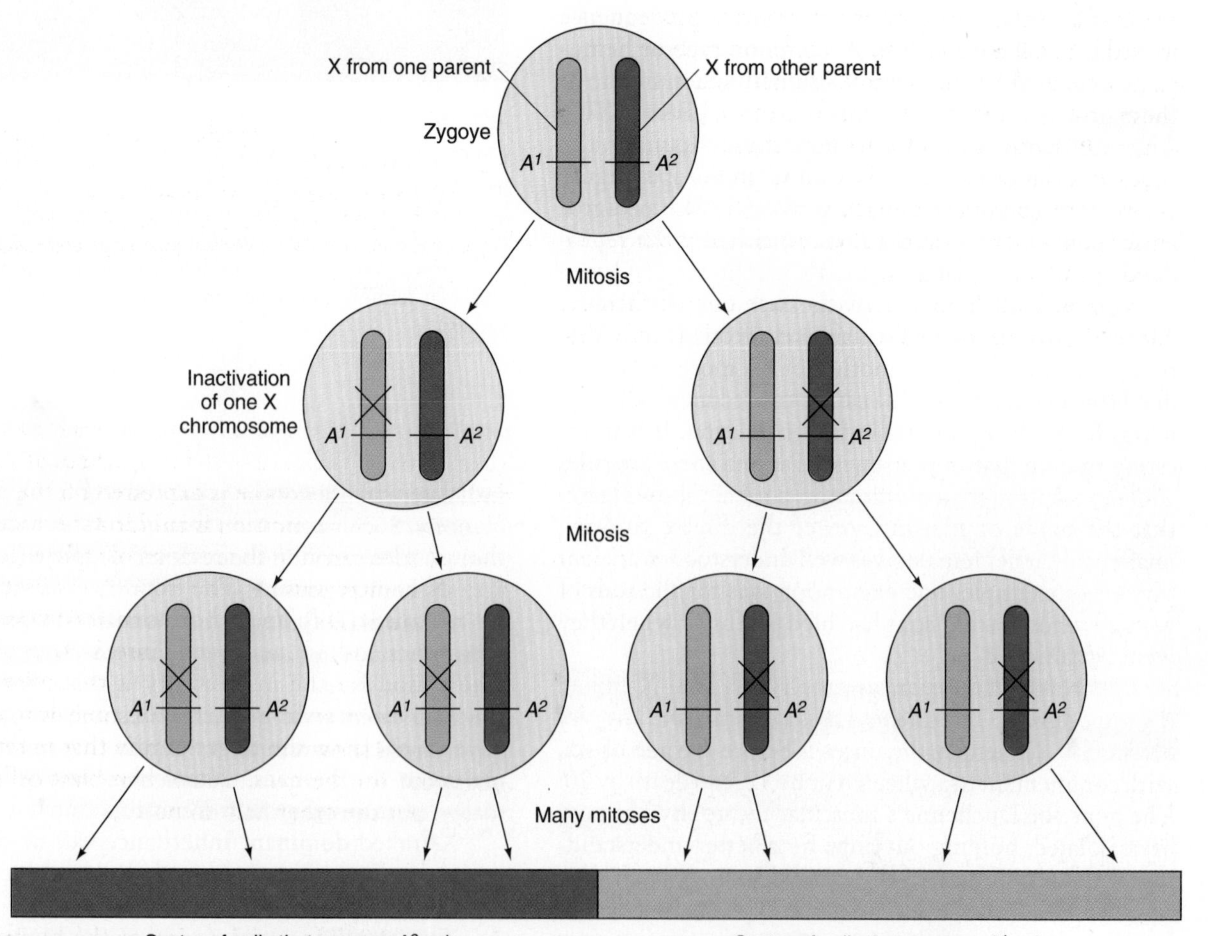

Figure 3-28 X chromosome inactivation in mammals. The zygote of a female mammal heterozygous for an X-linked gene becomes a mosaic adult composed of two cell lines expressing one or the other of the alleles of the heterozygous gene because one or the other X chromosome is inactivated in all cell lines. For simplicity, inactivation is shown at the two-cell stage, but it can take place at other low cell numbers too.

There are few examples of X-linked dominant phenotypes in humans. One example is hypophosphatemia (vitamin D-resistant rickets).

X Chromosome Inactivation

Early in the development of female mammals, one of the X chromosomes in each cell becomes inactivated. The inactivated X chromosome is highly condensed and becomes visible as a densely staining spot called a **Barr body** (Figure 3-27). Surprisingly, this inactivation persists through all the subsequent mitotic divisions that produce the mature body of the animal. The inactivation process is random, affecting either of the X chromosomes. As a result, the adult female body is a mixture, or **mosaic,** of cells having two genotypes corresponding to the inactivation alternatives (Figure 3-28). During the growth of some tissues, the mitotic descendants of a progenitor cell stay next to each other, so if a female is heterozygous for an X-linked gene that has its effect in that tissue, the two alleles of the heterozygote are expressed in patches, or sectors. A mosaic phenotype familiar to you is the coats of tortoiseshell and calico cats (Figure 3-29). Such cats are females heterozygous for the alleles *O* (which causes fur to be orange) and *o* (which causes it to be black). Inactivation of the *O*-bearing X chromosome produces a black patch expressing *o*, and inactivation of the *o*-bearing X chromosome produces an orange patch expressing *O*.

Although all human females have one of their X chromosomes inactivated in every cell, this is only detectable when a female is heterozygous for an X-linked gene. This is particularly striking when, as in tortoiseshell cats, the phenotype is expressed on the exterior of the body. Such a condition is anhidrotic ectodermal dysplasia. Males carrying the responsible allele (let us call it *d*) in its hemizygous condition have no sweat glands. A heterozygous (*Dd*) female has a mosaic of *D* and *d* sectors across her body, as shown in Figure 3-30. Interestingly, the hypothesis that the gene causing testicular feminization is located on the X chromosome was confirmed when it was shown microscopically that in females heterozygous for the gene half of fibroblast cells bind androgen but the other half do not.

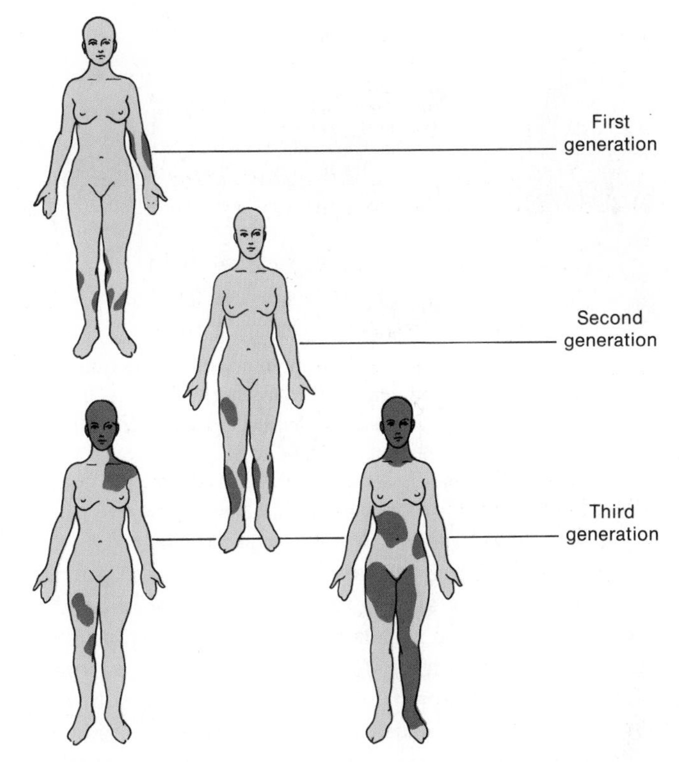

Figure 3-30 Somatic mosaicism in three generations of females heterozygous for sex-linked anhidrotic ectodermal dysplasia, absence of sweat glands. Areas without sweat glands are shown in blue. The extent and location of the different tissues is determined by chance, but each female exhibits the characteristic mosaic pattern.

Figure 3-29 A calico cat. Both calico and tortoiseshell cats are females heterozygous for two alleles of an X-linked coat color gene, *O* (orange) and *o* (black). The orange and black sectors are caused by X chromosome inactivation. The white areas are caused by a separate genetic determinant present in calicos, but not in tortoiseshell cats.

Y-Linked Inheritance

Genes on the differential region of the human Y chromosome are inherited only by males, with fathers transmitting the region to their sons. However, other than maleness itself, no human phenotype has been conclusively proved to be Y linked. Hairy ear rims (Figure 3-31) has been proposed as a possibility. The phenotype is ex-

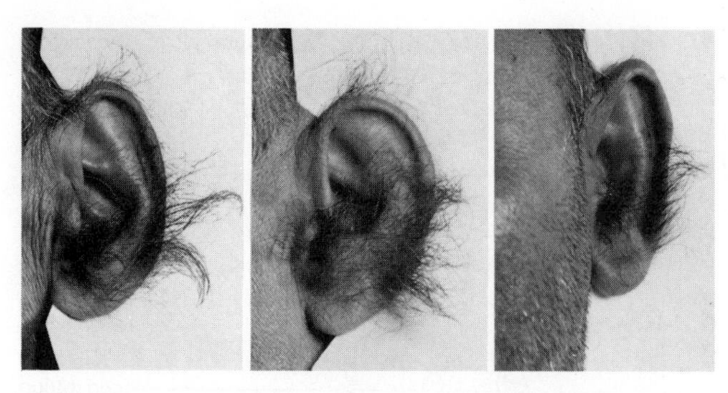

Figure 3-31 Hairy ear rims. This phenotype may be caused by an allele of a Y-linked gene. (From C. Stern, W. R. Centerwall, and S. S. Sarkar, *The American Journal of Human Genetics* 16, 1964; 467. By permission of Grune & Stratton, Inc.)

tremely rare among the populations of most countries but more common among the populations of India. An Indian geneticist, K. Dronamraju, studied the trait in his own family. Every male in the family descended from a certain male ancestor showed the trait. In other Indian families, however, males seem to transmit the trait to only some of their sons, which is part of the reason that the evidence for Y-linked inheritance is considered to be inconclusive.

The Y chromosome of the fish *Lebistes* carries a gene that determines a phenotype consisting of a pigmented spot at the base of the dorsal fin. This phenotype passes from father to son, and females never carry or express the gene, so it seems to be a clear example of Y linkage.

X-and-Y-Linked Inheritance

Are there hereditary patterns that identify genes on the pairing regions of the X and Y chromosomes? There is a gene in *Drosophila* that is inherited in a way that indicates an X-and-Y location. Curt Stern found that a certain non-wild-type recessive allele in *Drosophila* causes a phenotype of shorter and more slender bristles; he called it bobbed *(b)*. If a bobbed X^bX^b female is crossed with a wild-type X^+Y^+ male, all the F_1 progeny are wild-type: X^+X^b ♀♀ and X^bY^+ ♂♂. The same result, of course, would be expected from an autosomal gene. However, the X-and-Y linkage is revealed in the F_2, which shows the following clear, sex-associated pattern.

Sex	Phenotype	Inferred genotype
Males	Wild type	X^+Y^+ and X^bY^+
Females	$\frac{1}{2}$ Bobbed	X^bX^b
	$\frac{1}{2}$ Wild type	X^+X^b

It has recently been shown by a combination of genetic and molecular studies that there is a homologous region of the human X and Y chromosomes that pair and cross over. In fact, it is known that there is an obligatory crossover in this region in every meiosis, so alleles in this vicinity are effectively uncoupled from the unique regions of the X and Y chromosomes and show what is called **pseudoautosomal inheritance.**

Message Inheritance patterns with an unequal representation of phenotypes in males and females can locate the genes concerned to one or both of the sex chromosomes.

The Parallel Behavior of Autosomal Genes and Chromosomes

The patterns of inheritance arising from the normal and the nondisjunctional behavior of the sex chromosomes provide satisfying confirmation of the chromosome theory of inheritance, which was originally suggested by the parallel behavior of Mendelian genes and autosomal chromosomes. Now we should pause and review the situation for the autosomal genes, because these are the genes most commonly studied. Such a summary is best achieved in a diagram: Figure 3-32 illustrates the passage of a hypothetical cell through meiosis. Two genes are shown on two chromosome pairs. The cell has four chromosomes: a pair of homologous long chromosomes and a pair of homologous short ones. (Such size differences between pairs are common.) The genotype of the cell is *Aa Bb*.

Parts 4 and 4′ of Figure 3-32 show that two equally frequent spindle attachments result in two allelic segregation patterns. Meiosis then produces four cells of the genotypes shown from each of these segregation patterns. Because segregation patterns 4 and 4′ are equally common, the meiotic product cells of genotypes *A B, a b, A b,* and *a B* are produced in equal frequencies. In other words, the frequency of each of the four genotypes is $\frac{1}{4}$. This, of course, is the distribution postulated in Mendel's model and is the one we noted along one edge of the Punnett square (see Figure 2-10). We can now understand exactly how chromosomal behaviors produce the Mendelian ratios.

Notice that Mendel's first law (equal segregation) expresses what happens to a pair of alleles when the pair of homologs that carry them separate into opposite cells at the first meiotic division. Notice also that Mendel's second law (independent assortment) results from the independent behavior of separate pairs of homologous chromosomes.

The chromosome theory is important in many ways. At this point in our discussion, we can stress the impor-

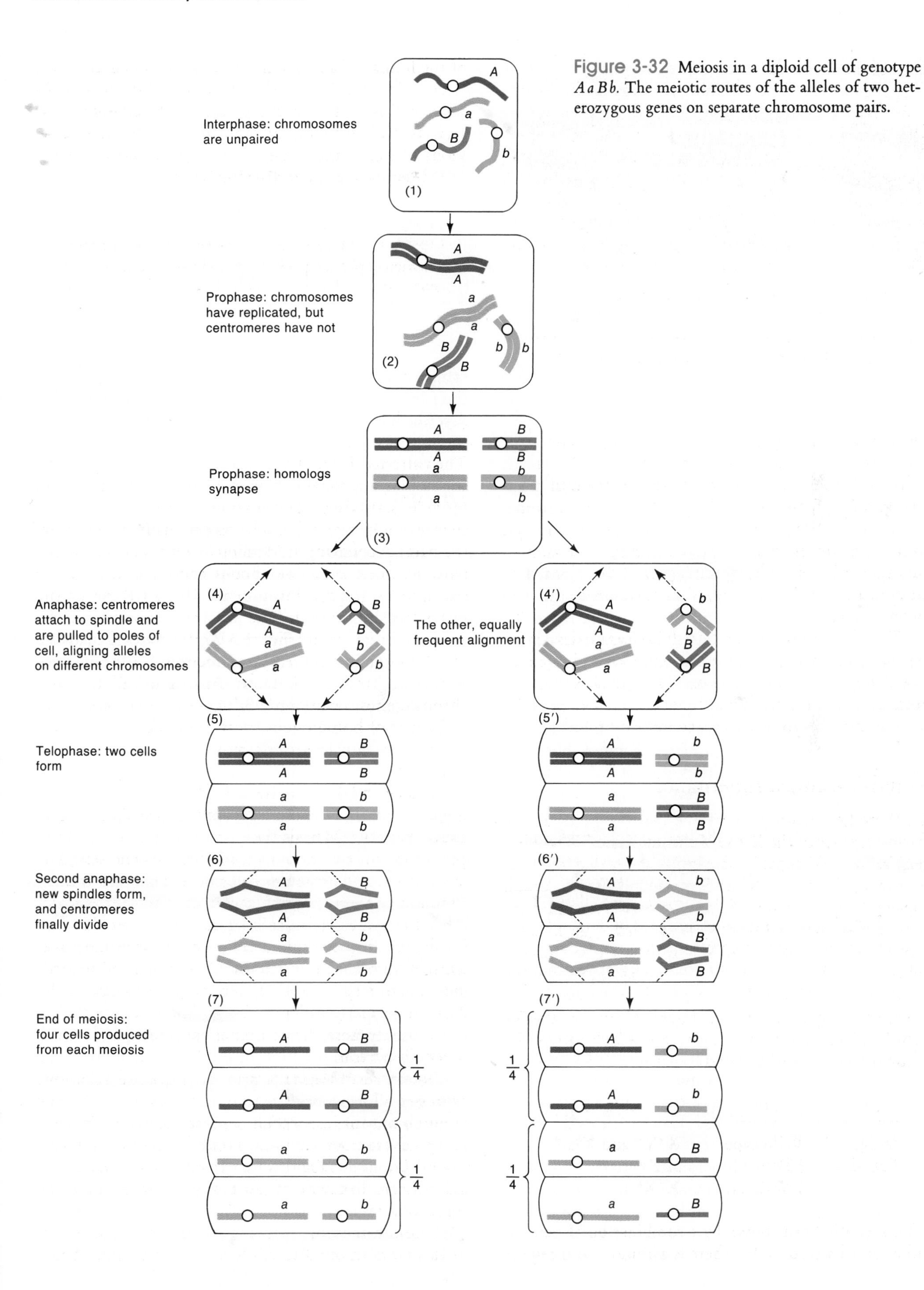
Interphase: chromosomes are unpaired
(1)
A
a
B
b
Prophase: chromosomes have replicated, but centromeres have not
(2)
Prophase: homologs synapse
(3)
Anaphase: centromeres attach to spindle and are pulled to poles of cell, aligning alleles on different chromosomes
(4)
The other, equally frequent alignment
(4′)
Telophase: two cells form
(5)
(5′)
Second anaphase: new spindles form, and centromeres finally divide
(6)
(6′)
End of meiosis: four cells produced from each meiosis
(7)
(7′)
1/4

Figure 3-32 Meiosis in a diploid cell of genotype $AaBb$. The meiotic routes of the alleles of two heterozygous genes on separate chromosome pairs.

tance of the theory in centering attention on the role of the cell genotype in determining the organism's phenotype. The phenotype of an organism is determined by the phenotypes of all its individual cells. The cell phenotype, in turn, is determined by the alleles present on the chromosomes of the cell. When we say that an organism is, for example, *AA*, we really mean that each cell of the organism is *AA*. The cell genotype determines how the cell functions, thereby controlling the phenotype of the cell and hence the phenotype of the organism.

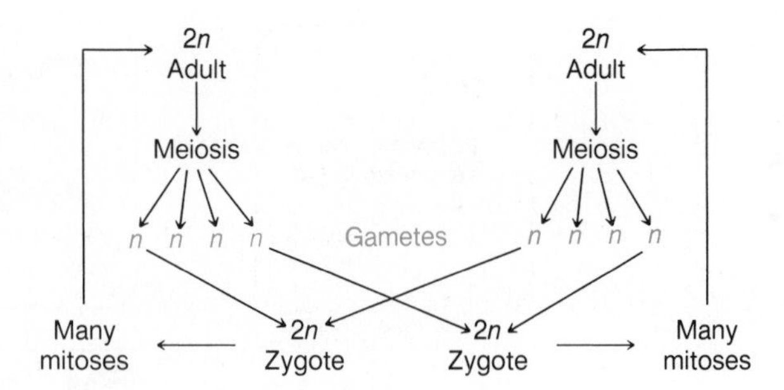

Figure 3-33 The diploid life cycle.

Message The phenotype of an organism is determined by gene action at the cellular level.

Mendelian Genetics and Life Cycles

So far, we have been discussing mainly diploid organisms—organisms with two homologous chromosome sets in each cell. As we have seen, diploid is designated $2n$, where n stands for one chromosome set (for example, the pea cell contains two sets of seven chromosomes, so $2n = 14$). The flowering plants and the animals (including humans) we encounter in our daily existence are diploid in most of their tissues. Nevertheless, a large proportion of the biomass on the earth comprises organisms that spend most of their life cycles in a haploid condition, in which each cell has only one set of chromosomes. Important examples are the fungi and the algae. Bacteria could be considered haploid, but they form a special case because they do not have chromosomes of the type we have been discussing. (Bacterial cycles are discussed in Chapter 13.) Also important are organisms that spend part of their life cycles as haploid organisms and another part as diploid. Such organisms are said to show **alternation of generations,** which refers to the alternation of $2n$ and n stages. All plants, in fact, show alternation of generations. The haploid stage in flowering plants and conifers, however, is inconspicuous and dependent on the diploid plant, appearing as a specialized structure. Other plants, such as mosses and ferns, have independent haploid stages.

Do all these life cycles show Mendelian genetics? The answer is that Mendelian inheritance patterns characterize any species that has meiosis as part of its life cycle, because Mendelian laws are based on the process of meiosis. All the groups of organisms mentioned, except bacteria, utilize meiosis as part of their cycles. In the next sections we consider the inheritance patterns shown by less familiar eukaryotes and compare them to more familiar cycles. This is important because it demonstrates the universality of Mendelian genetics. Furthermore, some of the less well-known organisms have been the subject of extensive genetic research, and a knowledge of their life cycles is essential to their genetic analysis. We describe here three major types of cycle, beginning with the more familiar diploid types.

Diploids

Figure 3-33 summarizes the diploid cycle. This is the cycle of most animals (including humans). Meiosis takes place in specialized diploid cells, the meiocytes, which are set aside for the purpose but which are part of the diploid adult organism. The products of meiosis are the gametes (eggs or sperm). Fusion of haploid gametes forms a diploid zygote, which (through mitosis) produces a multicellular organism. Mitosis in a diploid proceeds as outlined in Figure 3-34.

Haploids

Figure 3-35 shows the basic haploid cycle. Here, the organism itself is haploid. How can meiosis possibly occur in a haploid organism? After all, meiosis requires the pairing of two homologous chromosome sets! The answer is that all haploid organisms that undergo meiosis create a temporary diploid stage that provides the meiocytes. In some cases, unicellular, haploid, adult individuals fuse to form a diploid cell, which then undergoes meiosis. In other cases, haploid cells from different parents fuse to form diploid cells for meiosis. (These fusing cells are properly called the gametes, so we see that in these cases gametes arise from mitosis.) Meiosis, as usual, produces haploid products of meiosis, which are called **sexual spores.** The sexual spores in some species become new unicellular adults; in other species, they develop through mitosis into a multicellular haploid individual. Notice that a cross between two adult haploid organisms involves only one meiosis, whereas a cross between two diploid organisms involves a meiosis in each organism. As we shall see, this simplification makes haploids very attractive for genetic analysis. In haploids, mitosis proceeds as shown in Figure 3-36.

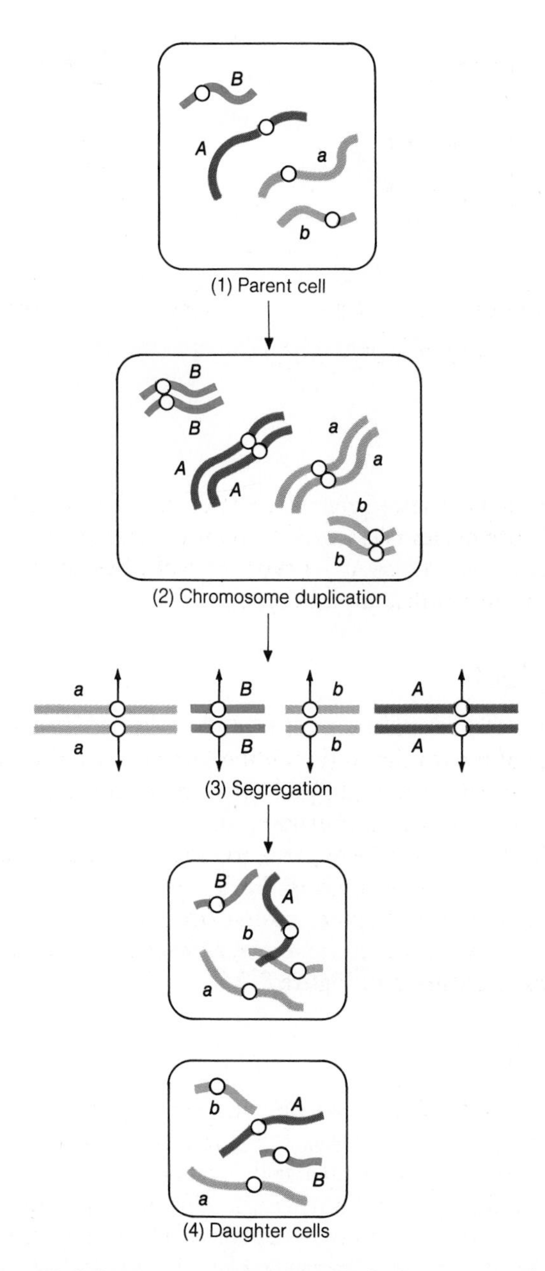

Figure 3-34 Mitosis in a diploid cell of genotype *A a B b*. The heterozygous genes are on separate chromosome pairs.

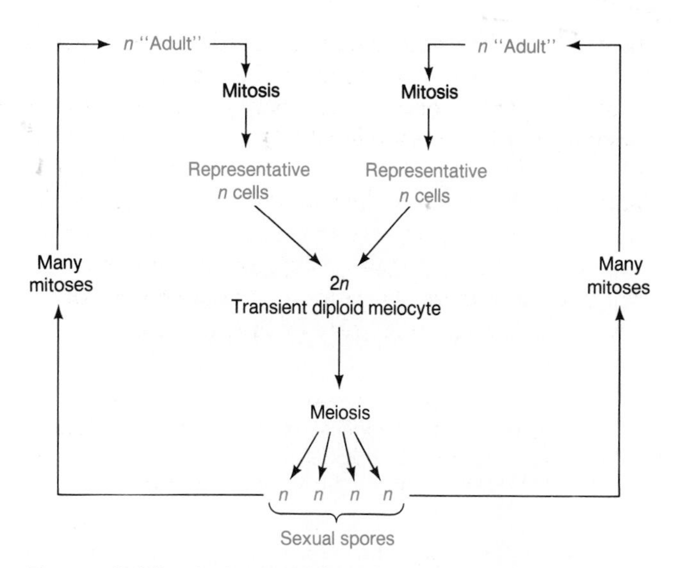

Figure 3-35 The haploid life cycle.

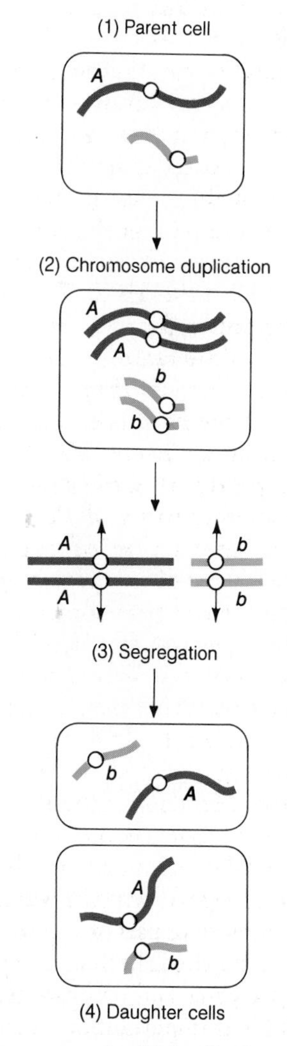

Figure 3-36 Mitosis in a haploid cell of genotype *A b*. The genes are on separate chromosomes.

Let's consider a cross in a specific haploid. A convenient organism is the pink bread mold *Neurospora*. This fungus is a multicellular haploid in which the cells are joined end to end to form **hyphae,** or threads of cells. The hyphae grow through the substrate and also send up aerial branches that bear cells known as **asexual spores.** These can detach and disperse to form new colonies, or alternatively, they can act as male gametes (Figure 3-37). Female gametes develop inside a specialized knot of hyphae.

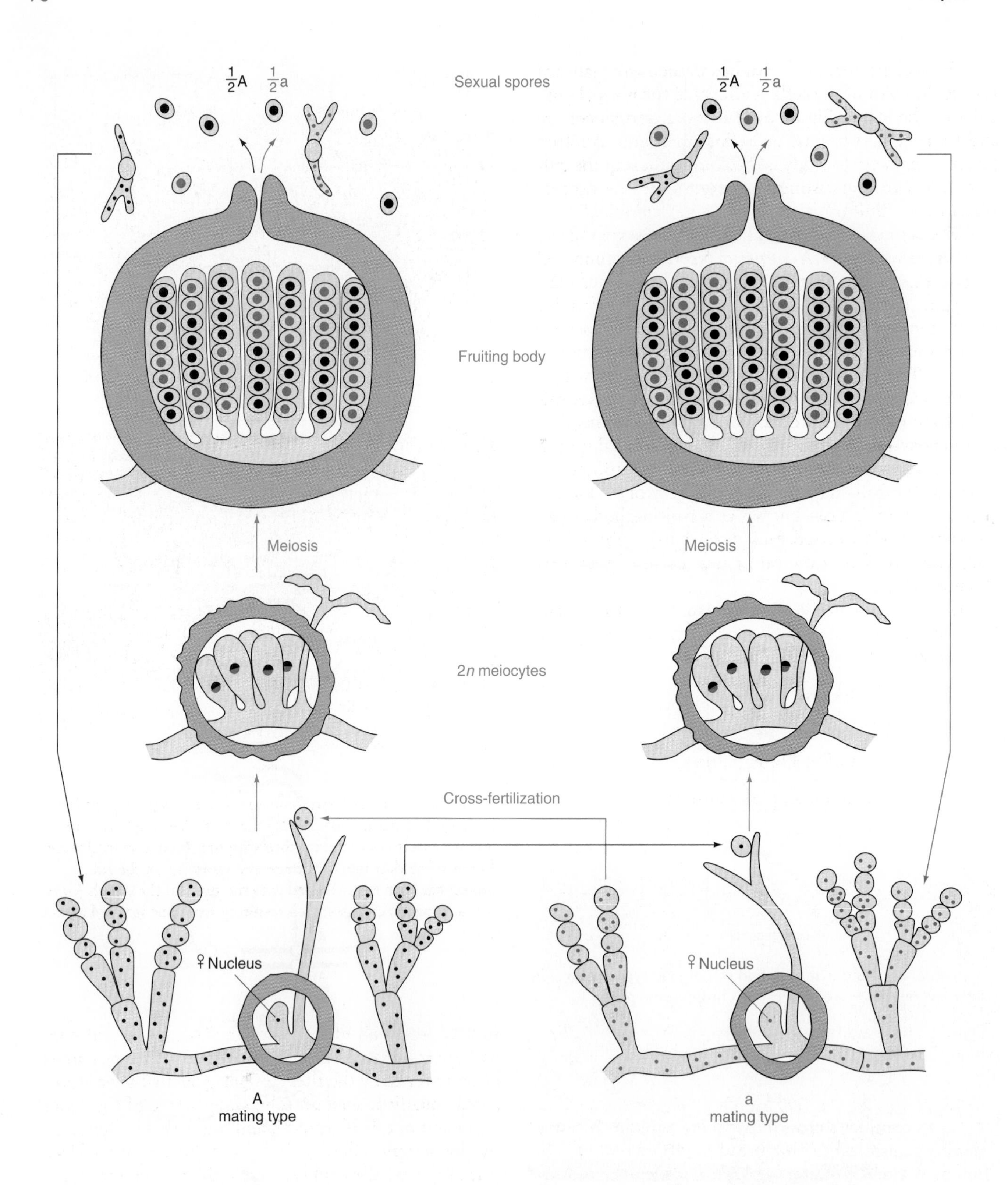

Figure 3-37 The life cycle of *Neurospora crassa,* the pink bread mold. Self-fertilization is not possible in this species: there are two mating types, determined by the alleles *A* and *a* of one gene. A cross will succeed only if it is *A* × *a*. An asexual spore from the opposite mating type fuses with a receptive hair, and a nucleus travels down the hair to pair with a nucleus in the knot of cells. The *A* and *a* pair, then undergo synchronous mitoses, finally fusing to form diploid meiocytes.

What characters can be studied in such an organism? One is the color of the cells. Variants of the normal pink color can be found. Figure 3-38 shows a normal culture and some color variants, including an albino. Another possible character to study is the compactness of the culture, and two contrasting phenotypes are the normal appearance, "fluffy," and a dense type, "colonial."

We can make a cross by allowing the asexual spores to act as male gametes. A culture of *Neurospora* cannot self because this fungus has two genetically determined mating types, *A* and *a*. Fertile crosses can occur only if strains of different mating types are paired $A \times a$. Crosses are carried out by adding asexual spores of one culture to another. The nucleus of an asexual spore pairs with a female nucleus. This pair undergoes synchronous mitotic division, and the products finally fuse, forming diploid meiocytes. Then meiosis occurs, and sexual spores, called **ascospores,** are formed. These ascospores are black and football-shaped; they are shot out of the knot of hyphae, which is now known as a fruiting body. The ascospores can be isolated, each into a culture tube, where each ascospore will grow into a new culture by mitosis (Figure 3-39).

If we cross a fluffy, pink culture with a colonial, albino culture and then isolate and grow 100 ascospores, on average we find

25 fluffy, pink cultures
25 colonial, albino cultures
25 fluffy, albino cultures
25 colonial, pink cultures

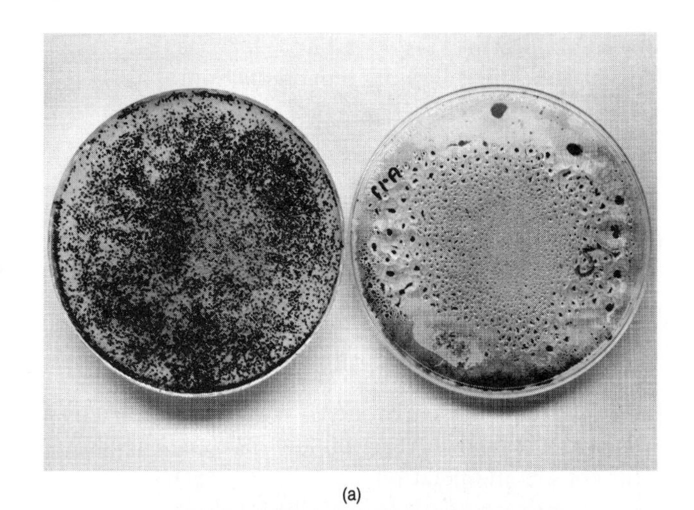

(a)

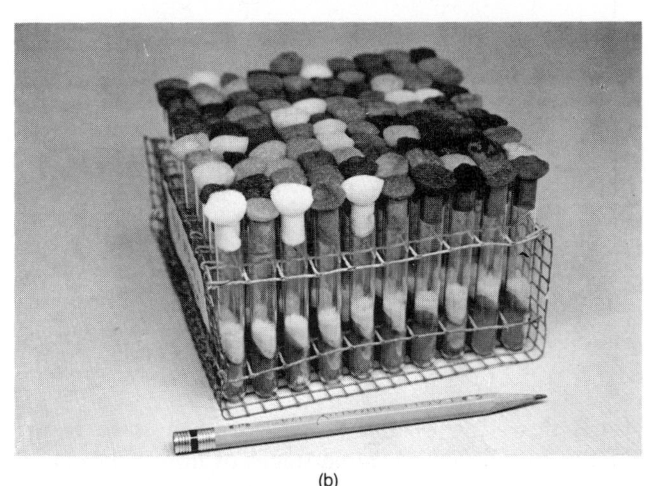

(b)

Figure 3-39 (a) A *Neurospora* cross made in a petri plate (on the left). The many small black spheres are fruiting bodies in which meiosis has occurred; the ascospores (sexual spores) were shot as a fine dust into the condensed moisture on the lid (which has been removed and is to the right of the plate). (b) A rack of progeny cultures, each resulting from one isolated ascospore.

Figure 3-38 Genetically determined color variants of the fungus *Neurospora crassa.* The orange wild-type color is shown on the right, together with albino, yellow, and brown variants. Their genotypes are wild type, $al^+ \; ylo^+ \; ad^+$; albino, $al^- \; ylo^+ \; ad^+$; yellow, $al^+ \; ylo^- \; ad^+$; brown, $al^+ \; ylo^+ \; ad^-$.

In total, one-half of the "progeny" are fluffy and one-half are colonial. Thus, this phenotypic difference must be determined by the alleles of one gene that have segregated equally at meiosis. The same is true of the other character: one-half are pink and one-half are albino, so the phenotypic difference in color is also determined by a pair of alleles. We could represent the four culture types as

$col^+ \; al^+$ (fluffy, pink)
$col \; al$ (colonial, albino)
$col^+ \; al$ (fluffy, albino)
$col \; al^+$ (colonial, pink)

The 1:1:1:1 ratio is a result of independent assortment, as illustrated in the following branch diagram:

$$\frac{1}{2}\,col^+ \begin{cases} \frac{1}{2}\,al^+ \longrightarrow \frac{1}{4}\,col^+\,al^+ \\ \frac{1}{2}\,al \longrightarrow \frac{1}{4}\,col^+\,al \end{cases}$$

$$\frac{1}{2}\,col \begin{cases} \frac{1}{2}\,al \longrightarrow \frac{1}{4}\,col\,al \\ \frac{1}{2}\,al^+ \longrightarrow \frac{1}{4}\,col\,al^+ \end{cases}$$

So we see that even in such a lowly organism, Mendel's laws of equal segregation and independent assortment are still in operation.

Alternating Haploid–Diploid

In an organism with alternation of generations, there are two stages to the life cycle: one diploid and one haploid. One stage is usually more prominent than the other. For example, what we all recognize as a fern plant is the diploid stage, but the organism does have a small, independent, photosynthetic, haploid stage that is usually much more difficult to spot on the forest floor. The green moss plant is the haploid stage and the brownish stalk that grows up out of this plant is a dependent diploid stage that is effectively parasitic on it.

In flowering plants, the main green stage is, of course, diploid. The haploid stages of flowering plants are extremely reduced and dependent on the diploid. These haploids are found in the flower. In the anther and the ovary, meiocytes undergo meiosis; the haploid products of meiosis are called **spores.** The spores undergo a few mitotic divisions to produce a small, multicellular haploid stage. The diploid stage of an organism with alternation of generations is called the **sporophyte,** which means sexual-spore-producing plant, and the haploid stage is called **gametophyte,** which means gamete-producing plant. The male gametophyte of seed plants is known as a pollen grain. Figure 3-40 shows that in flowering plants, cells of the gametophytes act as eggs or sperm in fertilization. The generalized cycle of alternation of generations is shown in Figure 3-41.

In mosses and ferns, the sperm cells are highly motile and have to travel from one gametophyte to another in a film of water to effect fertilization. Let us consider a cross we might make in a moss. The character to be studied, of course, can pertain to the gametophyte or the sporophyte. Assume that we have a gene whose alleles affect the "leaves" of the gametophyte, with w causing wavy edges and w^+ causing smooth edges. Also assume that a separate gene affects the color of the sporophyte, with r causing reddish coloration and r^+ causing the normal brown coloration. We fertilize a smooth-leaved gametophyte (that bears the unexpressed allele r) by transferring onto it male gametes from a wrinkly leaved gametophyte carrying r^+ (Figure 3-42).

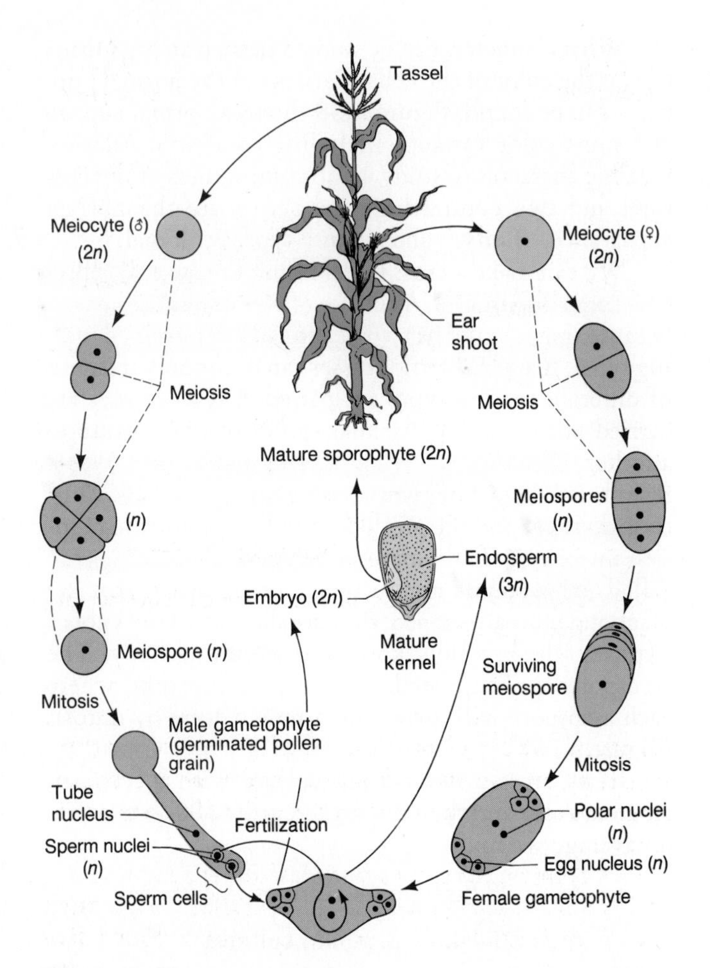

Figure 3-40 Alternation of generations in corn. The male gametophyte arises from a meiocyte in the tassel. The female gametophyte arises from a meiocyte in the ear shoot. One sperm cell from the male gametophyte fuses with the egg nucleus of the female gametophyte, and the diploid zygote thus formed develops into the embryo. The other sperm cell fuses with two nuclei in the center of the female gametophyte, forming a triploid ($3n$) cell that generates the endosperm tissue surrounding the embryo. The endosperm provides nutrition to the embryo during seed germination. Which parts of the diagram represent the haploid stage? Which parts represent the diploid stage?

A diploid sporophyte develops, actually on the gametophyte, and it is brown (because reddish is recessive). This sporophyte produces sexual spores in the proportions

$$\frac{1}{4}\,w^+\,r^+$$
$$\frac{1}{4}\,w^+\,r$$
$$\frac{1}{4}\,w\,r^+$$
$$\frac{1}{4}\,w\,r$$

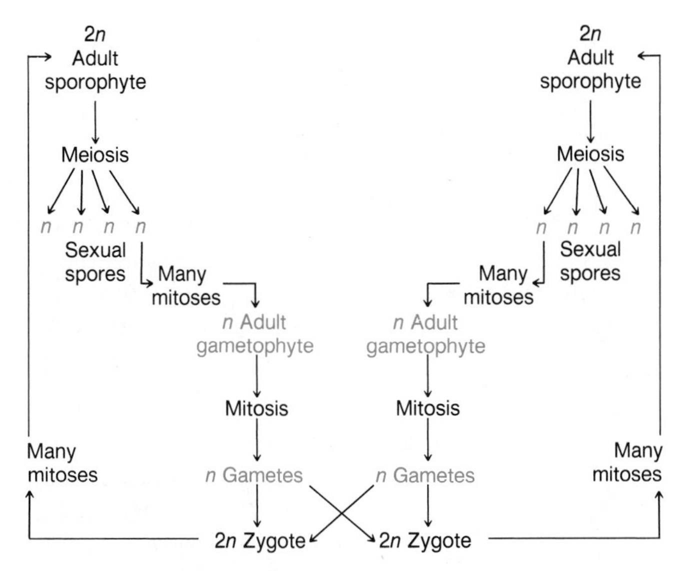

Figure 3-41 The alternation of diploid and haploid stages in the life cycle of plants.

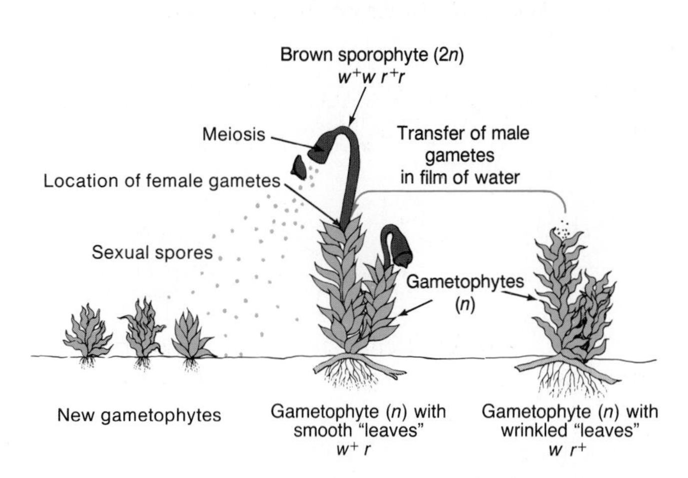

Figure 3-42 Mendelian genetics in a hypothetical cross in a moss. Only the haploid gametophyte express the w^+ and w, and only the diploid sporophyte expresses the r^+ or r alternatives.

Of course, we can identify only the leaf character in these gametophytes; we would have to make the appropriate intercrosses to determine whether each individual is r^+ or is r.

Once again, Mendel's laws describe the inheritance. It is simply a matter of keeping track of the ploidy in each part of the cycle, and observing the simple Mendelian ratios.

Message Mendelian laws apply to the products of meiosis in any organism and may be generally stated as follows:

1. At meiosis, the alleles of a gene segregate equally into the haploid products of meiosis.
2. At meiosis, the alleles of one gene pair segregate independently of the alleles of genes on other chromosome pairs.

Thus, the theory that genes are located on chromosomes perfectly explains inheritance patterns. The chromosome theory of inheritance is no longer in doubt; it forms one of the cornerstones of modern biological theory.

Summary

After the rediscovery of Mendelian principles in 1900, scientists set out to discover what structures within cells correspond to Mendel's hypothetical units of heredity, which we now call genes. Recognizing that the behavior of chromosomes during meiosis parallels the behavior of genes, Walter Sutton and Theodor Boveri suggested that genes were located in or on the chromosomes.

In her experiments with a certain species of grasshopper, Elinor Carothers discovered that nonhomologous chromosomes assort independently. This finding provided further evidence that the behavior of chromosomes closely parallels that of genes. Additional evidence for the validity of the chromosome theory of heredity came from the discovery of sex-linked inheritance and the existence of sex chromosomes. In his studies of *Drosophila,* Thomas Hunt Morgan showed that the inheritance of red or white eye color is completely consistent with the meiotic behavior of X and Y chromosomes. Finally, Calvin Bridges's postulation of nondisjunction (the failure of paired chromosomes to separate) during meiosis enabled him to make testable predictions based on the assumption that the gene for eye color is on the X chromosome. The confirmation of these predictions provided unequivocal evidence that genes are located on chromosomes.

We now know which chromosome behaviors produce Mendelian ratios. Mendel's first law (equal segregation) results from the separation of a pair of homologous chromosomes into opposite cells at the first division. Mendel's second law (independent assortment) results from independent behavior of separate pairs of homologous chromosomes.

Because Mendelian laws are based on meiosis, Mendelian inheritance characterizes any organism with a meiotic stage in its life cycle, including diploid organisms, haploid organisms, and organisms with alternating haploid and diploid generations.

Concept Map

Draw a concept map interrelating as many of the following terms as possible. Note that the terms are listed in no particular order.

chromosomes / meiosis / equal segregation / alleles / diploid / independent assortment / haploid / centromere / products of meiosis / Mendelian ratio / spindle fiber attachment

Chapter Integration Problem

Two *Drosophila* flies were mated that had normal (transparent, long) wings. In the progeny, two new phenotypes appeared, dusky wings (having a semi-opaque appearance) and clipped wings (with squared ends). The progeny were as follows:

Females	179 transparent, long
	58 transparent, clipped
Males	92 transparent, long
	89 dusky, long
	28 transparent, clipped
	31 dusky, clipped

a. Provide a genetic explanation for these results, showing genotypes of parents and of all progeny classes under your model.

b. Design a test for your model.

Solution

a. The first step is to state any interesting features of the data. The first striking feature is, of course, the appearance of two new phenotypes. We encountered the phenomenon in Chapter 2, and explained it there in terms of recessive alleles masked by their dominant counterparts. So first we might suppose that one or both parental flies have recessive alleles of two different genes. This inference is strengthened by the observation some progeny express only one of the new phenotypes. If the new phenotypes always appeared together, we might suppose that the same recessive allele determines both.

However, the other striking aspect of the data, which we cannot explain using the Mendelian principles from Chapter 2, is the obvious difference between the sexes; although there are approximately equal numbers of males and females, the males fall into four phenotypic classes but the females comprise only two. This should immediately suggest some kind of sex-linked inheritance. When we study the data, we see that the long and clipped phenotypes are segregating in both males and females but only males have the dusky phenotype. This suggests that the inheritance of wing transparency differs from the inheritance of wing shape. First, long and clipped are found in a 3:1 ratio in both males and females. This can be explained if the parents were both heterozygous for an autosomal gene; we can represent them as $L\,l$, where L stands for long and l stands for clipped.

Having done this partial analysis, we see that it is only the wing transparency inheritance that is associated with sex. The most obvious possibility is that the alleles for transparent *(D)* and dusky *(d)* are on the X chromosome, because we have seen in this chapter that gene location on this chromosome gives inheritance patterns correlated with sex. If this suggestion is true, then the parental female must be the one sheltering the *d* allele, because if the male had the *d* he would have been dusky whereas we were told that he had transparent wings. Therefore, the female parent would be Dd, and the male D. Let's see if this suggestion works: if it is true, all female progeny would inherit the D allele from their father, so all would be transparent winged. This was observed. Half of the sons would be D (transparent) and half d (dusky), which was also observed.

So overall we can represent the female parent as $Dd\,Ll$ and the male parent as $D\,Ll$. Then the progeny would be

Females

$\frac{1}{2}\,DD$ → $\frac{3}{4}\,L-$ → $\frac{3}{8}\,DD\,L-$
$\frac{1}{2}\,DD$ → $\frac{1}{4}\,ll$ → $\frac{1}{8}\,DD\,ll$
$\frac{1}{2}\,Dd$ → $\frac{3}{4}\,L-$ → $\frac{3}{8}\,DD\,L-$
$\frac{1}{2}\,Dd$ → $\frac{1}{4}\,ll$ → $\frac{1}{8}\,DD\,ll$

→ $\frac{3}{4}$ transparent, long
→ $\frac{1}{4}$ transparent, clipped

Males

$\frac{1}{2}\,D$	$\frac{3}{4}\,L-$	$\frac{3}{8}\,D\,L-$	transparent, long
	$\frac{1}{4}\,ll$	$\frac{1}{8}\,D\,ll$	transparent, clipped
$\frac{1}{2}\,d$	$\frac{3}{4}\,L-$	$\frac{3}{8}\,d\,L-$	dusky, long
	$\frac{1}{4}\,ll$	$\frac{1}{8}\,d\,ll$	dusky, clipped

b. Generally a good way to test such a model is to make a cross and predict the outcome. But which cross? We have to predict some kind of ratio in the progeny, so it is important to make a cross from which a unique phenotypic ratio can be expected. Notice that using one of the female progeny as a parent would not serve our needs: we cannot say from observing the phenotype of any one of these females what her genotype is. A female with transparent wings could be DD or Dd, and one with long wings could be LL or Ll. It would be good to cross the parental female of the original cross with a dusky, clipped son, because the full genotypes of both are specified under the model we have created; according to our model, this cross is

$$Dd\,Ll \times d\,ll$$

From this we predict:

Females

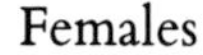

$\frac{1}{2}\,Dd$ → $\frac{1}{2}\,Ll$ → $\frac{1}{4}\,Dd\,Ll$
$\frac{1}{2}\,Dd$ → $\frac{1}{2}\,ll$ → $\frac{1}{4}\,Dd\,ll$
$\frac{1}{2}\,dd$ → $\frac{1}{2}\,Ll$ → $\frac{1}{4}\,dd\,Ll$
$\frac{1}{2}\,dd$ → $\frac{1}{2}\,ll$ → $\frac{1}{4}\,dd\,ll$

Males

$\frac{1}{2}\,D$ → $\frac{1}{2}\,Ll$ → $\frac{1}{4}\,D\,Ll$
$\frac{1}{2}\,D$ → $\frac{1}{2}\,ll$ → $\frac{1}{4}\,D\,ll$
$\frac{1}{2}\,d$ → $\frac{1}{2}\,Ll$ → $\frac{1}{4}\,d\,Ll$
$\frac{1}{2}\,d$ → $\frac{1}{2}\,ll$ → $\frac{1}{4}\,d\,ll$

Solved Problems

1. A rare human disease afflicted a family as shown in the accompanying pedigree.

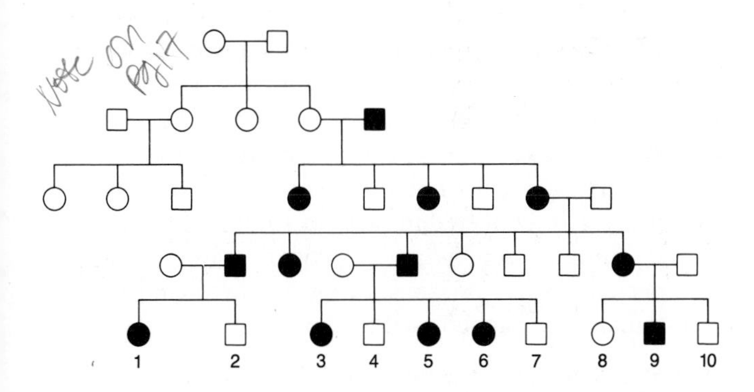

a. Deduce the most likely mode of inheritance.

b. What would be the outcomes of the cousin marriages 1×9, 1×4, 2×3 and 2×8?

Solution

a. The most likely mode of inheritance is X-linked dominant. We assume that the disease phenotype is dominant because once it is introduced into the pedigree by the male in generation II, it appears in every generation. We assume that the phenotype is X-linked because fathers do not transmit it to their sons. If it were autosomal dominant, father-to-son transmission would be common.

In theory, autosomal recessive could work but it is improbable. In particular, note the marriages between affected members of the family and unaffected outsiders. If the condition were autosomal recessive, the only way these marriages could have affected offspring is if each person marrying into the family were a heterozygote; then the matings would be aa (affected) $\times$ Aa (unaffected). However, we are told that the disease is rare; in such a case, it is highly unlikely that heterozygotes would be so common. X-linked recessive inheritance is impossible because a mating of an affected woman with a normal man could not produce affected daughters. So we can let A represent the disease-causing allele and a represent the normal allele.

b. 1×9: Number 1 must be heterozygous Aa because she must have obtained a from her normal mother. Number 9 must be AY. Hence, the cross is Aa ♀ $\times$ AY ♂.

Female gametes	Male gametes	Progeny
$\frac{1}{2}\,A$	$\frac{1}{2}\,A$	$\frac{1}{4}\,AA$ ♀
	$\frac{1}{2}$ Y	$\frac{1}{4}\,A$ Y ♂
$\frac{1}{2}\,a$	$\frac{1}{2}\,A$	$\frac{1}{4}\,Aa$ ♀
	$\frac{1}{2}$ Y	$\frac{1}{4}\,a$ Y ♂

1×4: Must be Aa ♀ $\times$ aY ♂.

Female gametes	Male gametes	Progeny
$\frac{1}{2}\,A$	$\frac{1}{2}\,a$	$\frac{1}{4}\,Aa$ ♀
	$\frac{1}{2}$ Y	$\frac{1}{4}\,A$ Y ♂
$\frac{1}{2}\,a$	$\frac{1}{2}\,a$	$\frac{1}{4}\,aa$ ♀
	$\frac{1}{2}$ Y	$\frac{1}{4}\,a$ Y ♂

2×3: Must be aY ♂ $\times$ Aa ♀ (same as 1×4).
2×8: Must be a Y♂ $\times$ aa ♀ (all progeny normal).

2. Two corn plants are studied; one is Aa, and the other is aa. These two plants are intercrossed in two ways: using Aa as female and aa as male; using aa as female and Aa as male. Recall from Figure 3-38 that the endosperm is $3n$ and is formed by the union of a sperm cell with the two polar nuclei of the female gametophyte.

a. What endosperm genotypes does each cross produce? In what proportions?

b. In an experiment to study the effects of "doses" of alleles, you wish to establish endosperms with genotypes *aaa, Aaa, AAa,* and *AAA* (carrying 0, 1, 2, and 3 "doses" of *A*, respectively). What crosses would you make to obtain these endosperm genotypes?

Solution

a. In such a question, we have to think about meiosis and mitosis at the same time. The meiospores are produced by meiosis; the nuclei of the male and female gametophytes in higher plants are produced by the mitotic division of the meiospore nucleus. We also need to study the corn life cycle to know what nuclei fuse to form the endosperm.

First cross: Aa ♀ × aa ♂

Here, the female meiosis will result in spores of which one-half will be *A* and one-half will be *a*. Therefore, similar proportions of haploid female gametophytes will be produced. Their nuclei will be either all *A* or all *a*, because mitosis reproduces genetically identical genotypes. Likewise, all nuclei in every male gametophyte will be *a*. In the corn life cycle, the endosperm is formed from two female nuclei plus one male nucleus, so two endosperm types will be formed as follows.

♀ spore	♀ polar nuclei	♂ sperm	$3n$ endosperm
$\frac{1}{2}$ *A*	*A* and *A*	*a*	$\frac{1}{2}$ *AAa*
$\frac{1}{2}$ *a*	*a* and *a*	*a*	$\frac{1}{2}$ *aaa*

Second cross: aa ♀ × Aa ♂

♀ spore	♀ polar nuclei	♂ sperm	$3n$ endosperm
all *a*	all *a* and *a*	$\frac{1}{2}$ *A*	$\frac{1}{2}$ *Aaa*
		$\frac{1}{2}$ *a*	$\frac{1}{2}$ *aaa*

Notice that the phenotypic ratio of endosperm characters would still be Mendelian, even though the underlying endosperm genotypes are slightly different. (Of course, none of these problems arise in embryo characters because embryos are diploid.)

b. This kind of experiment has been very useful in studying plant genetics and molecular biology. In answering the question, all we need to realize is that the two polar nuclei contributing to the endosperm are genetically identical. To obtain endosperms, all of which will be *aaa*, any *aa* × *aa* cross will work. To obtain endosperms, all of which will be *Aaa*, the cross must be *aa* ♀ × *AA* ♂. To obtain embryos, all of which will be *AAa*, the cross must be *AA* ♀ × *aa* ♂. For *AAA* obviously any *AA* × *AA* cross will work. Notice that these endosperm genotypes can be obtained in other crosses, but only in combination with other endosperm genotypes.

Problems

1. What are the major differences between mitosis and meiosis?

2. When a cell of genotype *Aa Bb Cc* having all the genes on separate chromosome pairs divides mitotically, what are the genotypes of the daughter cells?

3. In working with a haploid yeast, you cross a purple (ad^-) strain of mating type *a* and a white (ad^+) strain of mating type α. If ad^- and ad^+ are alleles of one gene and *a* and α are alleles of an independently inherited gene on a separate chromosome pair, what progeny do you expect to obtain? In what proportions?

4. The recessive allele *s* causes *Drosophila* to have small wings and the s^+ allele causes normal wings. This gene is known to be X-linked. If a small-winged male is crossed with a homozygous wild-type female, what ratio of normal to small-winged flies can be expected in each sex in the F_1? If F_1 flies are intercrossed, what F_2 progeny ratios are expected? What progeny ratios are predicted if F_1 females are backcrossed to their father?

5. State where cells divide mitotically and where they divide meiotically in a fern, a moss, a flowering plant, a pine tree, a mushroom, a frog, a butterfly, and a snail.

6. Human cells normally have 46 chromosomes. For each of the following stages, state the number of chromosomes present in a human cell: **(a)** metaphase of mitosis, **(b)** metaphase I of meiosis, **(c)** telophase of mitosis, **(d)** telophase I of meiosis, **(e)** telophase II of meiosis. (In your answers, count chromatids as chromosomes.)

7. Four of the following events are part of both meiosis and mitosis, but one is only meiotic. Which one?

(a) Chromatid formation, **(b)** spindle formation, **(c)** chromosome condensation, **(d)** chromosome movement to poles, **(e)** chromosome pairing.

8. Suppose that you discover two interesting *rare* cytological abnormalities in the karyotype of a human male. (A karyotype is the total visible chromosome complement.) There is an extra piece (or satellite) on *one* of the chromosomes of pair 4, and there is an abnormal pattern of staining on one of the chromosomes of pair 7. Assuming that all the gametes of this male are equally viable, what proportion of his children will have the same karyotype he has?

9. Suppose that meiosis occurs in the transient diploid stage of the cycle of a haploid organism of chromosome number n. What is the probability that an individual haploid cell resulting from the meiotic division will have a complete parental set of centromeres (that is, a set all from one parent or all from the other parent)?

10. Assuming the sex chromosomes to be identical, name the proportion of all alleles you have in common with **(a)** your mother; **(b)** your brother.

11. A wild-type female schmoo who is graceful (G) is mated to a non-wild-type male who is gruesome (g). Their progeny consist solely of graceful males and gruesome females. Interpret these results and give genotypes.

(Problem 11 from E. H. Simon and H. Grossfield, *The Challenge of Genetics*. Copyright 1971 by Addison-Wesley.)

12. A man with a certain disease marries a normal woman. They have eight children (four boys and four girls); all of the girls have their father's disease, but none of the boys do. What inheritance is suggested? **(a)** Autosomal recessive, **(b)** autosomal dominant, **(c)** Y-linked, **(d)** X-linked dominant, **(e)** X-linked recessive.

13. An X-linked dominant allele causes hypophosphatemia in humans. A man with hypophosphatemia marries a normal woman. What proportion of their sons will have hypophosphatemia? **(a)** $\frac{1}{2}$, **(b)** $\frac{1}{4}$, **(c)** $\frac{1}{3}$, **(d)** 1, **(e)** 0.

14. A condition known as icthyosis hystrix gravior appeared in a boy in the early eighteenth century. His skin became very thick and formed loose spines that were sloughed off at intervals. When he grew up, this "porcupine man" married and had six sons, all of whom had this condition, and several daughters, all of whom were normal. For four generations, this condition was passed from father to son. From this evidence, what can you postulate about the location of the gene?

15. Duchenne's muscular dystrophy is sex-linked and usually affects only males. Victims of the disease become progressively weaker, starting early in life.

a. What is the probability that a woman whose brother has Duchenne's disease will have an affected child?

b. If your mother's brother (your uncle) had Duchenne's disease, what is the probability that you have received the allele?

c. If your father's brother had the disease, what is the probability that you have received the allele?

16. The following pedigree is concerned with an inherited dental abnormality, amelogenesis imperfecta.

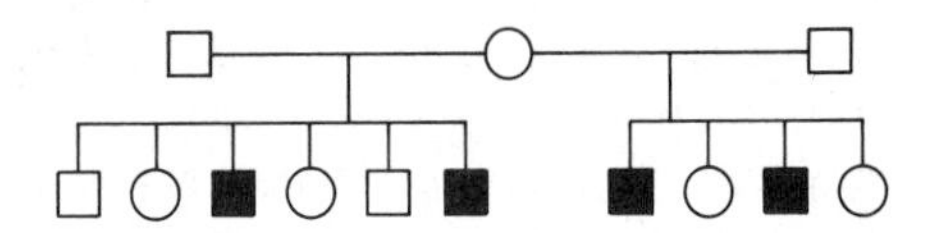

a. What mode of inheritance *best* accounts for the transmission of this trait?

b. Write the genotypes of all family members according to your hypothesis.

17. A sex-linked recessive allele c produces a red-green colorblindness in humans. A normal woman whose father was colorblind marries a colorblind man.

a. What genotypes are possible for the mother of the colorblind man?

b. What are the chances that the first child from this marriage will be a colorblind boy?

c. Of the girls produced by these parents, what proportion can be expected to be colorblind?

d. Of all the children (sex unspecified) of these parents, what proportion can be expected to have normal color vision?

18. Male house cats are either black or orange; females are black, orange, or calico.

a. If these coat-color phenotypes are governed by a sex-linked gene, how can these observations be explained?

b. Using appropriate symbols, determine the phenotypes expected in the progeny of a cross between an orange female and a black male.

c. Repeat part b for the reciprocal of the cross described there.

d. One-half of the females produced by a certain kind of mating are calico, and one-half are black; one-half of the males are orange, and one-half are black. What colors are the parental males and females in this kind of mating?

e. Another kind of mating produces progeny in the following proportions: $\frac{1}{4}$ orange males, $\frac{1}{4}$ orange females, $\frac{1}{4}$ black males, and $\frac{1}{4}$ calico females. What colors are the parental males and females in this kind of mating?

p31 notes

19. A man is heterozygous *Bb* for one autosomal gene, and he carries a recessive X-linked allele *d*. What proportion of his sperm will be *b d*? **(a)** 0, **(b)** $\frac{1}{2}$, **(c)** $\frac{1}{8}$, **(d)** $\frac{1}{16}$, **(e)** $\frac{1}{4}$.

20. The accompanying pedigree concerns a certain rare disease that is incapacitating but not fatal.

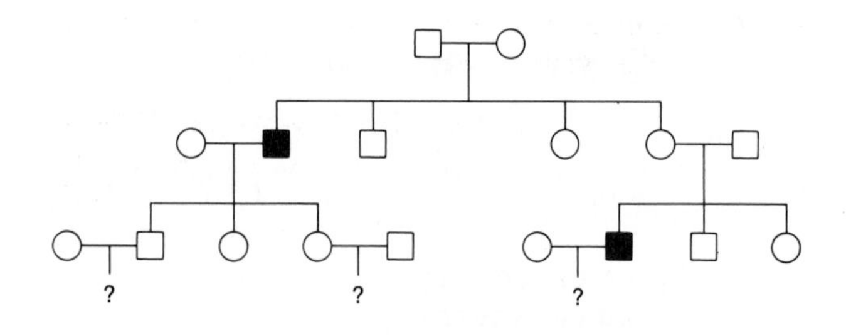

a. Determine the mode of inheritance of this disease.

b. Write the genotype of each individual according to your proposed mode of inheritance.

c. If you were this family's doctor, how would you advise the three couples in the third generation about the likelihood of having an affected child?

P. 30 Notes

21. Assume that this pedigree is straightforward, with no complications such as illegitimacy.

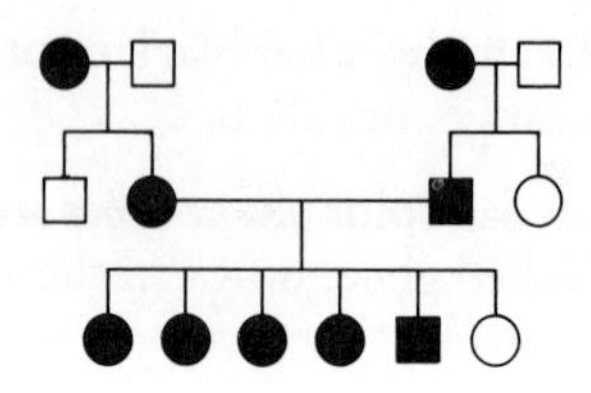

Phenotype W, found in the individuals represented by the shaded symbols, is rare in the general population. Which of the following patterns of transmission for W are consistent with this pedigree? Which are excluded? **(a)** Autosomal recessive, **(b)** autosomal dominant, **(c)** X-linked recessive, **(d)** X-linked dominant, **(e)** Y-linked.

(Problem 21 from A. M. Srb, R. D. Owen, and R. S. Edgar, *General Genetics*, 2d ed. Copyright 1965 by W. H. Freeman and Company.)

22. A mutant allele in mice causes a bent tail. Six pairs of mice were crossed. Their phenotypes and those of their progeny are given below. N is normal phenotype; B is bent phenotype. Deduce the mode of inheritance of this phenotype.

	Parents		Progeny	
Cross	♀	♂	♀	♂
1	N	B	All B	All N
2	B	N	$\frac{1}{2}$ B, $\frac{1}{2}$ N	$\frac{1}{2}$ B, $\frac{1}{2}$ N
3	B	N	All B	All B
4	N	N	All N	All N
5	B	B	All B	All B
6	B	B	All B	$\frac{1}{2}$ B, $\frac{1}{2}$ N

a. Is it recessive or dominant?

b. Is it autosomal or sex-linked?

c. What are the genotypes of all parents and progeny?

pg 18 Notes

23. The normal eye color of *Drosophila* is red but strains in which all flies have brown eyes are available. Similarly, wings are normally long, but there are strains with short wings. A female from a pure line (homo) with brown eyes and short wings is crossed with a male from a normal pure line. The F_1 consists of normal females and short-winged males. An F_2 is then produced by intercrossing the F_1. *Both* sexes of F_2 flies show phenotypes as follows:

$\frac{3}{8}$ red eyes, long wings

$\frac{3}{8}$ red eyes, short wings

$\frac{1}{8}$ brown eyes, long wings

$\frac{1}{8}$ brown eyes, short wings

Deduce the inheritance of these phenotypes, using clearly defined genetic symbols of your own invention. State the genotypes of all three generations and the genotypic proportions of the F_1 and F_2.

24. The wild-type (W) *Abraxas* moth has large spots on its wings, but the lacticolor (L) form of this species has very small spots. Crosses were made between strains differing in this character, with the following results:

	Parents		Progeny	
Cross	♀	♂	F_1	F_2
1	L	W	♀ W	♀ $\frac{1}{2}$ L, $\frac{1}{2}$ W
			♂ W	♂ W
2	W	L	♀ L	♀ $\frac{1}{2}$ W, $\frac{1}{2}$ L
			♂ W	♂ $\frac{1}{2}$ W, $\frac{1}{2}$ L

Provide a clear genetic explanation of the results in these two crosses, showing the genotypes of all individuals.

25. A certain gene that governs the activity of the enzyme glucose-6-phosphate dehydrogenase (G6PD), has two common alleles in Mediterranean and African populations. One allele stands for normal G6PD activity, and the other allele, which stands for reduced G6PD activity, confers resistance to malaria.

When the red blood cells of a certain African woman were examined under the microscope, precisely half the cells were found to contain the malarial parasite, whereas the other half appeared normal. Provide a genetic explanation for this finding.

26. Medical literature records the interesting case of a woman whose right breast was larger than the left, who had no pubic hair to the right of the midline, and who suffered from menstrual irregularities. Upon investigation of her family, it was discovered that a brother, a son, and a grandson showed testicular feminization syndrome. One of her daughters had three normal sons. Draw the pedigree from this information, determine if it fits the inheritance mode described in this chapter, and speculate on the cause of the symptoms in the propositus.

27. The pedigree shown below is for a rare human disease called spastic paraplegia, a nervous disorder in which there is an inability to coordinate voluntary movements.

a. What mode of inheritance is suggested by this pedigree?

b. Which individuals must be heterozygous under your model?

(Pedigree from V. A. McKusick, *On the Chromosomes of Man*. Copyright 1964, American Institute of Biological Science, Washington, D.C.)

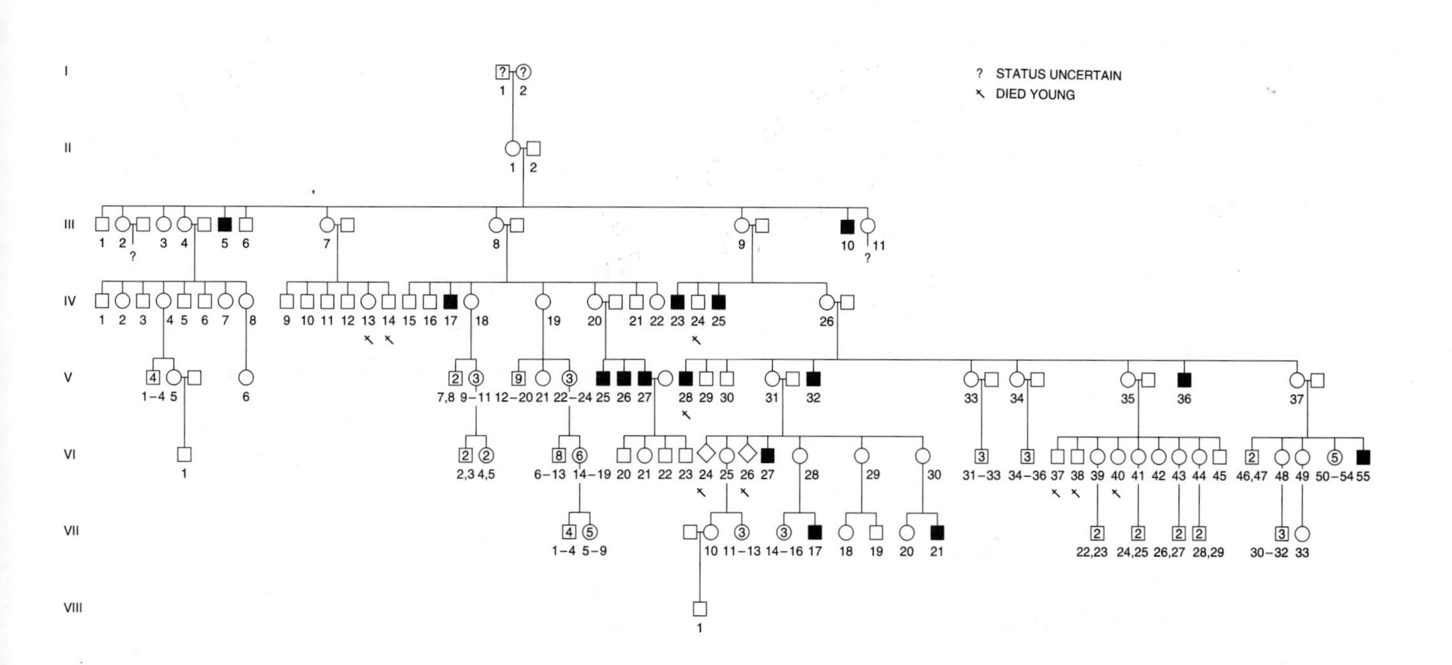

28. The accompanying pedigree shows the inheritance of a rare human disease. Is the pattern best explained as being caused by an X-linked recessive allele or by an autosomal dominant allele with expression limited to males?

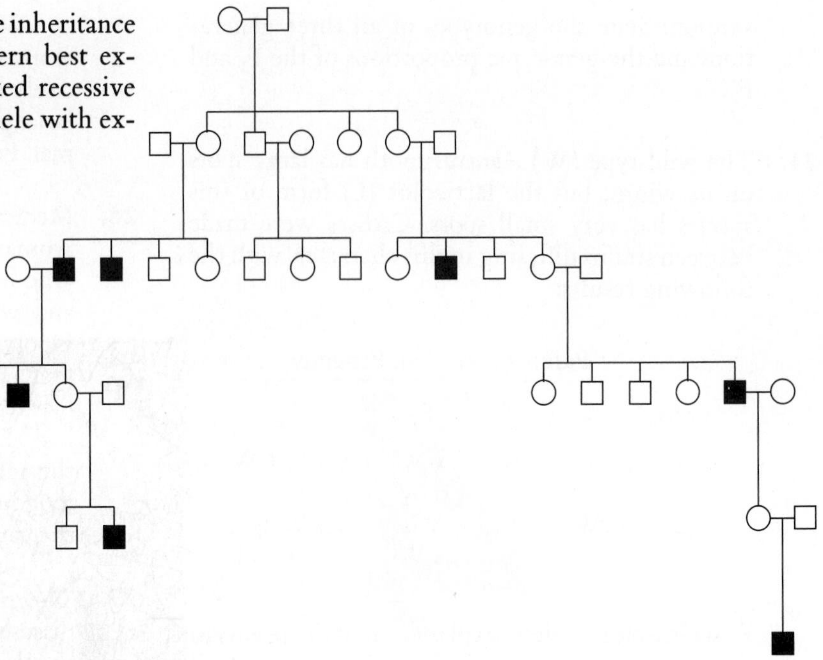

(Pedigree modified from J. F. Crow, *Genetics Notes,* 6th ed. Copyright 1967, Burgers Publishing Company, Minneapolis.)

4

Extensions of Mendelian Analysis

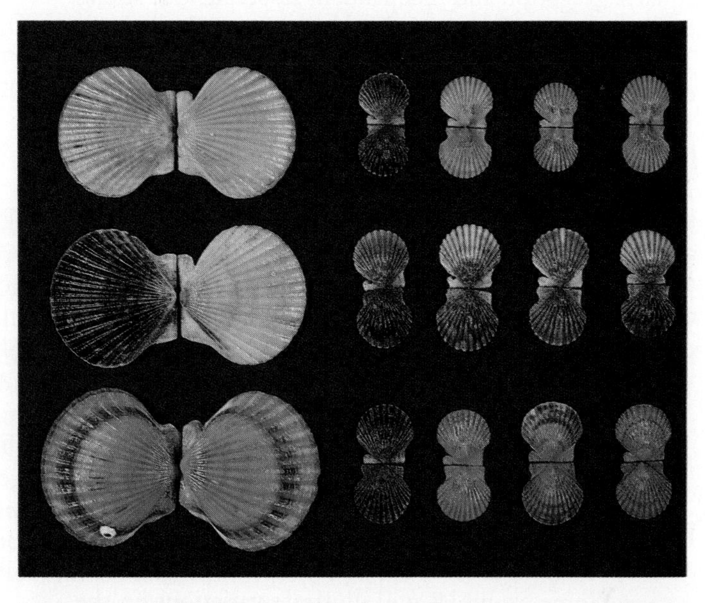

Variation in shell coloration of the bay scallop *(Argopecten irradians)* caused by three alleles of one gene. Yellow, black, and orange are determined by the alleles p^y, p^b, and p^o, respectively. The groups of small shells represent the proportions obtained upon selfing individuals of genotype $p^y p^b$ (top row), $p^b p^b$ (middle row) and $p^o p^b$ (bottom row). These results demonstrate the allelic relationship, and also show that p^y and p^o are both dominant to p^b. (From L. Adamkewicz and M. Castagna, *Journal of Heredity* 79, 1988, 15/BPS.)

KEY CONCEPTS

A gene can have more than two alleles.

Phenotypes of some heterozygotes reveal types of dominance other than full dominance.

Many genes have alleles that can kill the organism.

Most characters are determined by sets of genes that interact with each other and with the environment.

Modified Mendelian ratios reveal gene interactions.

We have seen that Mendel's laws of equal segregation and independent assortment seem to hold across the entire spectrum of eukaryotic organisms. These laws form a base for predicting the outcome of simple crosses. However, it is only a base; the real world of genes and chromosomes is more complex than Mendel's laws suggest, and exceptions and extensions abound. These situations do not invalidate Mendel's laws. Rather, they show that more explanatory elements must be added to the base of equal segregation and independent assortment of alleles to fit these situations into the fabric of genetic analysis. This is the challenge we now must meet. Of course, one extension—sex linkage—has already been explained. This chapter presents an assortment of other extensions, based mainly on the complexities of gene expression. The two following chapters discuss two more major extensions. We shall see that rather than creating a hopeless and bewildering situation, taking these complexities into account reveals a precise and unifying set of principles for the genetic analyst. These principles interlock and support each other in a highly satisfying way that has provided great insight into the mechanisms of inheritance.

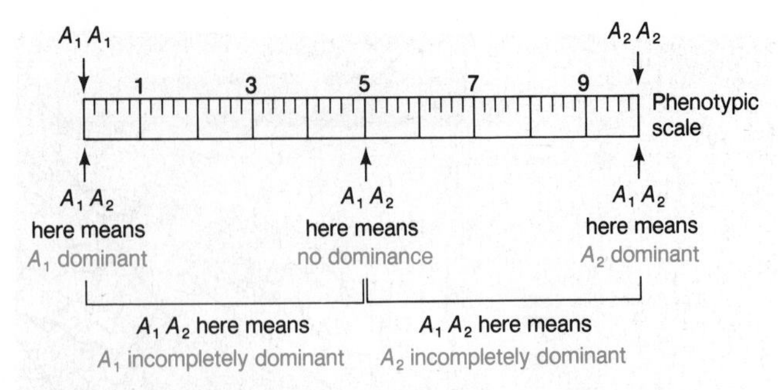

Figure 4-1 Summary of dominance relationships. The ruler represents some sort of phenotypic measurement such as amount of pigment.

Variations on Dominance

Dominance is a good place to start. Mendel reported full dominance (and recessiveness) for all seven genes he studied. Some examples will illustrate variations on the theme of dominance.

Four-o'clocks are plants native to tropical America whose flowers open in the late afternoon. When a pure four-o'clock line with red petals is crossed with a pure line with white petals, the F_1 has pink petals. If an F_2 is produced by intercrossing the F_1, the result is

$\frac{1}{4}$ of the plants have red petals

$\frac{1}{2}$ of the plants have pink petals

$\frac{1}{4}$ of the plants have white petals

Because of the 1:2:1 ratio in the F_2, we can deduce an inheritance pattern based on two alleles of a single gene. However the heterozygotes (the F_1 and one-half of the F_2) are intermediate in phenotype, suggesting **incomplete dominance.** Inventing allele symbols that have no dominance connotation, we can list the genotypes of the four-o'clocks in this experiment as C_1C_1 (red), C_2C_2 (white), and C_1C_2 (pink). (Subscripts and superscripts are often used to denote alleles.) Incomplete dominance describes the general situation in which the phenotype of a heterozygote is intermediate between the two homozygotes on a phenotypic "scale" of measurement. Figure 4-1 gives terms for all the theoretical positions on the scale, but in practice it is difficult to determine exactly where on such a scale the heterozygote is located.

The phenotype of the heterozygote is also the key in the phenomenon of **codominance,** in which the heterozygote shows the phenotypes of *both* the homozygotes. The three MN blood groups found in human populations provide an example. The three blood groups, M, N, and MN correspond to the genotypes L^ML^M, L^NL^N, and L^ML^N, respectively. Blood groups are actually determined by the presence of an immunological antigen on the surface of the red blood cells. As specified by their genotype, people have either antigen M (from L^ML^M) or antigen N (from L^NL^N), or they have both (from L^ML^N). Because the heterozygote has both phenotypes, the two alleles are said to be codominant.

The human disease sickle-cell anemia gives interesting insight into dominance. The gene concerned affects the molecule hemoglobin, which transports oxygen and is the major constituent of red blood cells. The three genotypes have different phenotypes, as follows:

Hb^AHb^A: Normal. Red blood cells never sickle.

Hb^SHb^S: Severe, often fatal anemia. Abnormal hemoglobin causes red blood cells to have sickle shape.

HB^AHb^S: No anemia. Red blood cells sickle only under low oxygen concentrations.

Figure 4-2 shows sickle cells. In regard to anemia, the Hb^A allele is obviously dominant. In regard to blood cell shape, however, there is incomplete dominance. Finally, as we shall now see, in regard to hemoglobin itself there is codominance! The alleles Hb^A and Hb^S actually code for two slightly different forms of hemoglobin and both these forms are present in the heterozygote, showing that the alleles are codominant. The different hemoglobin forms can be visualized using electrophoresis, a technique that separates molecules with different charge and size (Figure 4-3). It so happens that the A and S forms of hemoglobin have different charges, so they can be separated by electrophoresis, (Figure 4-4).

Figure 4-2 Electron micrograph of red blood cells from an individual with sickle-cell anemia. A few rounded cells appear almost normal. (Patricia N. Farnsworth.)

We see that homozygous normal people have one type of hemoglobin (A) and anemics have type S, which moves slower in the electric field. The heterozygotes have both types, A and S. In other words, there is codominance at the molecular level.

Sickle-cell anemia illustrates that the terms incomplete dominance and codominance are somewhat arbitrary. The type of dominance depends on the phenotypic

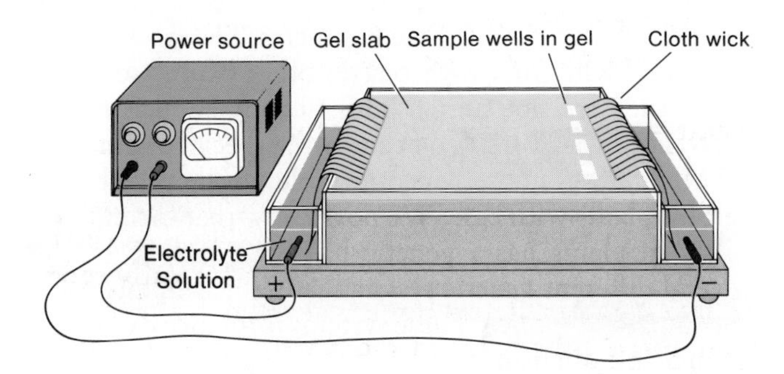

Figure 4-3 Set-up for electrophoresis. Each sample is placed in a well in the gel. The molecules in the samples migrate different distances on the gel due to their different electric charges. Several samples are tested at the same time (one in each well). The positions to which the molecules have migrated are later revealed by staining.

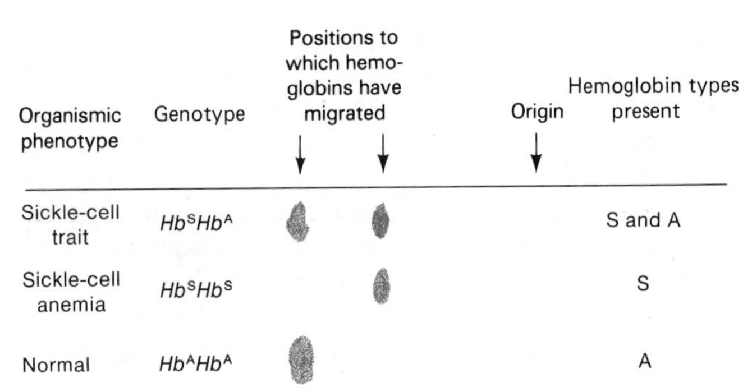

Organismic phenotype
Genotype
Positions to which hemoglobins have migrated
Origin
Hemoglobin types present
Sickle-cell trait
Hb^SHb^A
S and A
Sickle-cell anemia
Hb^SHb^S
S
Normal
Hb^AHb^A
A

Figure 4-4 Electrophoresis of hemoglobin from an individual with sickle-cell anemia, a heterozygote (called sickle-cell trait), and a normal individual. The smudges show the positions to which the hemoglobins migrate on the starch gel.

level at which the observations are being made—organismal, cellular, or molecular.

> Message Complete dominance and recessiveness are not essential aspects of Mendel's laws; those laws govern the inheritance patterns of genes, not the functions of the genes.

Multiple Alleles

Early in the history of genetics, it became clear that it is possible to have more than two forms of a gene. Although a diploid organism can have only two alleles of a gene (and a haploid organism can have only one), in a population the total number of different alleles for a single gene is often quite large. This situation is called **multiple allelism,** and the set of alleles itself is called an **allelic series.** The concept of allelism is a crucial one in genetics, so we consider several examples. The examples themselves serve also to introduce important areas of genetic investigation.

ABO Blood Group in Humans

The human ABO blood-group alleles show multiple allelism. There are four blood types (or phenotypes) in the ABO system, as shown in Table 4-1. The allelic series

Table 4-1. ABO Blood Groups in Humans

Blood phenotype	Genotype
O	ii
A	I^AI^A or I^Ai
B	I^BI^B or I^Bi
AB	I^AI^B

includes three major alleles—i, I^A, and I^B—but, of course, any person has only two of the three alleles (or two copies of one of them). In this allelic series, the alleles I^A and I^B each determine a unique antigen; allele i confers inability to produce an antigen. In the genotypes $I^A i$ and $I^B i$, the alleles I^A and I^B are fully dominant, but they are codominant in the genotype $I^A I^B$.

C Gene in Rabbits

A larger allelic series concerns coat color in rabbits. The alleles in this series are C (full color), c^{ch} (chinchilla, a light grayish color), c^h (Himalayan, albino with black extremities), and c (albino). Here you can see the value of denoting alleles by superscripts: more than the two symbols C and c are needed to describe these multiple alleles. In this series, each allele is dominant to the alleles listed after it in the order C, c^{ch}, c^h, c. Verify this by studying Table 4-2.

Operational Test for Allelism

Now that we have seen two examples of allelic series, it is a good time to pause and consider a question. How do we know that a set of contrasting phenotypes is determined by alleles of one gene? In other words, what is the operational test for allelism? For now, the answer is simply the observation of Mendelian monohybrid F_2 ratios from crosses of all pairwise combinations of pure-breeding lines. For example, consider three pure-line phenotypes in a hypothetical plant species. Line 1 has round spots on the petals; line 2 has oval spots on the petals; and line 3 has no spots on the petals. Suppose that crosses of the three lines yield the following results:

Cross	F_1	F_2
1 × 2	all round-spotted	$\frac{3}{4}$ round: $\frac{1}{4}$ oval
1 × 3	all round-spotted	$\frac{3}{4}$ round: $\frac{1}{4}$ unspotted
2 × 3	all oval-spotted	$\frac{3}{4}$ oval: $\frac{1}{4}$ unspotted

Table 4-2. *C* Gene in Rabbits

Coat-color phenotype	Genotype
Full color	CC or Cc^{ch} or Cc^h or Cc
Chinchilla	$c^{ch}c^{ch}$ or $c^{ch}c^h$ or $c^{ch}c$
Himalayan	c^hc^h or c^hc
Albino	cc

These results tell us that we are dealing with three alleles of a single gene that affects petal spotting because each cross gives a monohybrid F_2 ratio. We can choose any symbols we wish. Because we don't know which phenotype is the wild type, we could follow the rabbit system and use S for the round-spotted allele, s^o for the oval-spotted allele, and s for the unspotted allele. Alternatively, we could use S^r for round, S^o for oval, and s for unspotted. There are no firm rules about whether to use capital or small letters, particularly for the alleles in the middle of the series that are dominant to some of the alleles but recessive to others.

What if crosses between pure-breeding lines differing for one character do not produce monohybrid F_2 Mendelian ratios? This outcome is covered later in the chapter.

Clover Chevrons

Clover is the common name for plants of the genus *Trifolium.* There are many species. Some are native to North America; some grow here as introduced weeds. Much genetic research has been done with white clover, which shows considerable variation among individuals in the curious V, or chevron, pattern on the leaves. Figure 4-5 shows that in this species an allelic series determines the different chevron forms (and the absence of chevrons). Study the photographs to determine the type of dominance, if any, of each allele. List the alleles in a way that expresses how they relate to one another in dominance. Are there uncertainties? Does the photographic evidence permit us to say anything about the dominance or recessiveness of allele v?

Incompatibility Alleles in Plants

It has been known for millennia that some plants just will not self-fertilize. A single plant may produce both male and female gametes, but no seeds will ever be produced from the fusion of two of these gametes. The same plants, however, will cross with certain other plants, so obviously they are not sterile. This phenomenon is called **self-incompatibility.** Of course, the species of pea used by Mendel was not self-incompatible, for he was able to self his plants with ease. We now know that incompatibility in plants has a genetic basis and that there are several different genetic systems acting in different self-incompatible species. These systems form nice examples of multiple allelism.

One of the most common incompatibility systems is that found in sweet cherries, tobacco, petunias, and evening primroses. In each of these species, one gene, S, determines compatibility-incompatibility relations; in any one species, different plants bear different pairs from

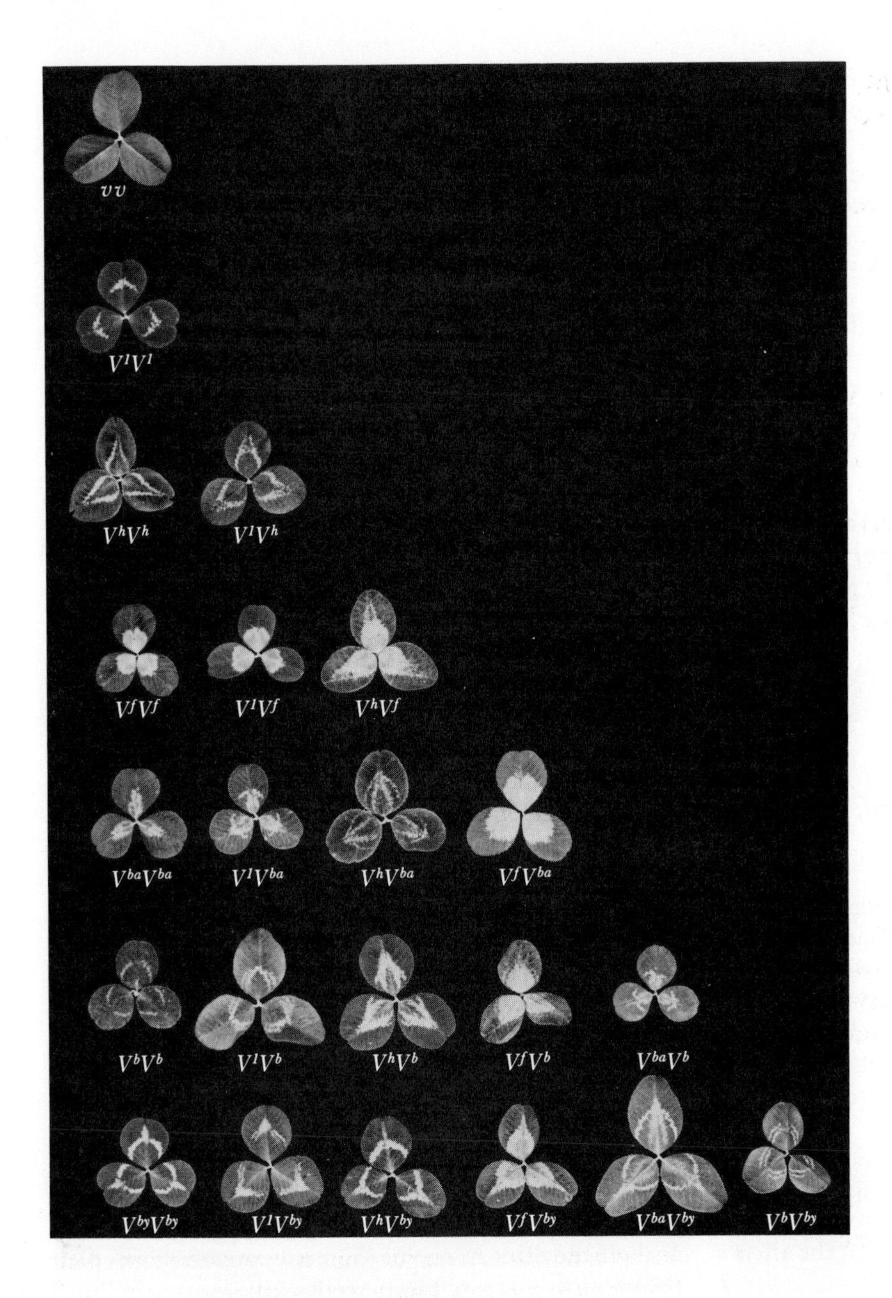

Figure 4-5 Multiple alleles determine the chevron pattern on the leaves of white clover. The genotype of each plant is shown below it. (W. Ellis Davies)

among many different alleles of the gene. Figure 4-6 shows how this system produces a fully incompatible reaction, a semicompatible reaction, and a fully compatible reaction. If a pollen grain bears an *S* allele that is also present in the maternal parent, then it will not grow. However, if that allele is not in the maternal tissue, then the pollen grain produces a pollen tube containing the male nucleus, and this tube effects fertilization. The number of *S* alleles in a series in one species can be very large (it is more than 50 in evening primroses and clover), and some species have more than 100 alleles. It is generally assumed that this type of system has evolved because it promotes outbreeding.

Message A gene can have several different states or forms — a situation called multiple allelism. The alleles are said to constitute an allelic series, and the members of a series can show any type of dominance to one another.

Lethal Alleles

Normal wild-type mice have coats with a rather dark overall pigmentation. In 1904, Lucien Cuenot studied mice having a lighter coat color called yellow. After mating a yellow mouse to a normal mouse from a pure

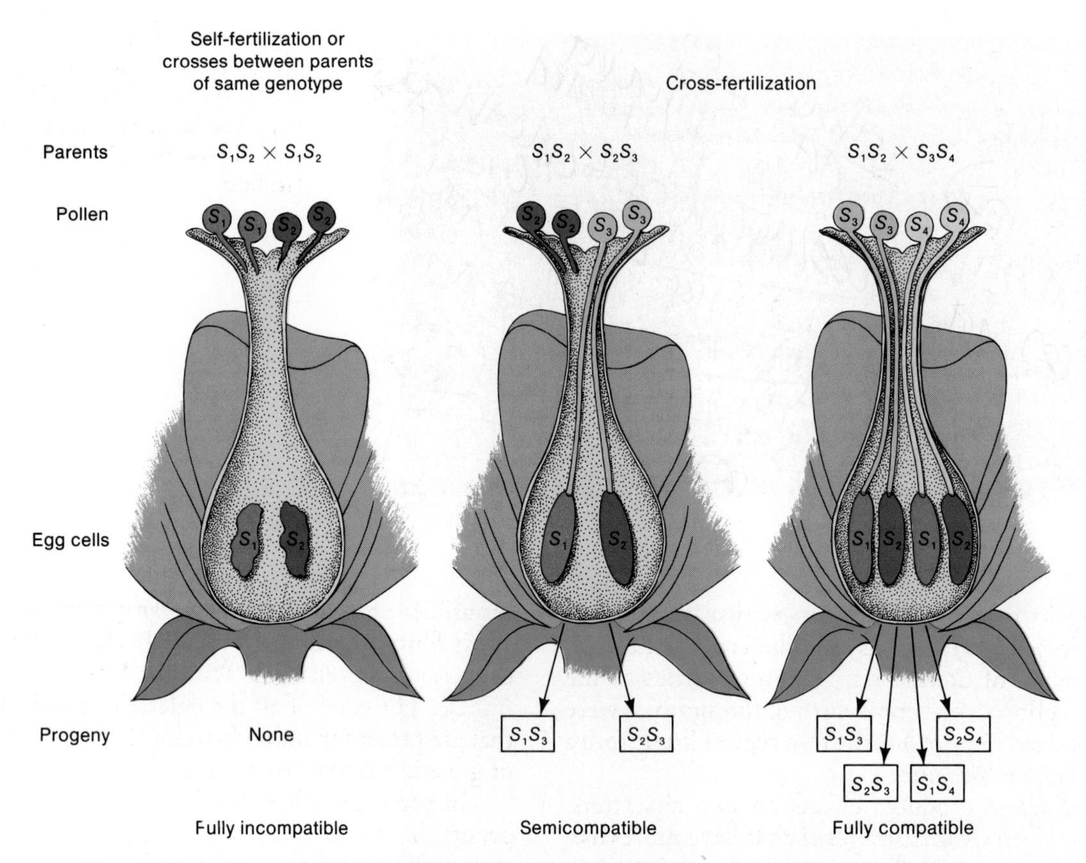

Figure 4-6 How multiple alleles control self incompatibility in certain plants. A pollen tube will not grow if the S allele that it contains is present in the female parent. This diagram shows only four multiple alleles, but many plant incompatibility systems use far larger numbers of alleles. Such systems promote the exchange of genes between plants by making selfing impossible and crosses between near relations very unlikely. (From A. M. Srb, R. D. Owen, and R. S. Edgar, *General Genetics,* 2d. ed. W. H. Freeman and Company, 1965.)

line, Cuenot observed a 1:1 ratio of yellow to normal mice in the progeny. This observation suggests that a single gene determines these phenotypes, that the yellow mouse was heterozygous for this gene, and that the allele for yellow is dominant to an allele for normal color. However, the situation became more confusing when Cuenot crossed yellow mice with one another. The result was always the same, no matter which individual yellow mice were used:

$$\text{yellow} \times \text{yellow} \begin{cases} \frac{2}{3}\ \text{yellow} \\ \frac{1}{3}\ \text{normal color} \end{cases}$$

Note two interesting features in these results. First, the 2:1 ratio is a departure from Mendelian expectations. Second, because no cross of yellow × yellow ever produced all yellow progeny, as there would be if either parent were a homozygote, it appeared that there were no homozygous yellow mice.

Cuenot suggested the following explanation for these results. A cross between two heterozygotes would be expected to yield a Mendelian genotypic ratio of 1:2:1. If all of the mice in one of the homozygous classes died before birth, the live births would then show a 2:1 ratio of heterozygotes to the surviving homozygotes. Specifically, the allele A^Y for yellow might be dominant to the normal allele A with respect to its effect on color, but it might act as a recessive **lethal** allele with respect to a character we would call viability. Thus, a mouse with the homozygous genotype A^YA^Y dies before birth and is not observed among the progeny. All surviving yellow mice must be heterozygous A^YA, so a cross between yellow mice will always yield the following results:

$$A^YA \times A^YA \rightarrow \begin{cases} \frac{1}{4}\ AA & \text{normal color} \\ \frac{2}{4}\ A^YA & \text{yellow} \\ \frac{1}{4}\ A^YA^Y & \text{die before birth} \end{cases}$$

The expected Mendelian ratio of 1:2:1 would be observed among the zygotes, but it is altered to a 2:1 ratio of

Figure 4-7 A mouse litter from two parents heterozygous for the yellow coat-color allele, which is lethal in a double dose. The larger mice are the parents. Not all progeny are visible. (Anthony Griffiths)

viable progeny because the progeny with a lethal A^YA^Y genotype do not survive. This hypothesis was confirmed by the removal of uteri from pregnant females of the yellow × yellow cross; one-fourth of the embryos were found to be dead. Figure 4-7 shows a typical litter from a cross between yellow mice.

The A^Y allele produces effects on two characters: coat color and survival. Such genes that have more than one distinct phenotypic effect are said to be **pleiotropic** genes. It is entirely possible, however, that both effects of the A^Y pleiotropic allele result from the same basic cause, which promotes yellowness of coat in a single dose and death in a double dose.

The tailless Manx phenotype in cats (Figure 4-8) is also produced by an allele that is lethal in the homozygous state. A single dose of the Manx allele, M^L, severely interferes with normal spinal development, resulting in the absence of a tail in the M^LM heterozygote. In M^LM^L homozygotes, the double dose of the gene produces such an extreme developmental abnormality that the embryo does not survive.

There are indeed many different types of lethal alleles. Some lethal alleles produce a recognizable phenotype in the heterozygote, as in the yellow mouse and Manx cat. Some lethal alleles are fully dominant and kill in one dose in the heterozygote. Some confer no detectable effect in the heterozygote at all, and the lethality is fully recessive. Furthermore, lethal alleles differ in the developmental stage at which they express their effects. Human lethals illustrate this very well. It has been estimated that we each carry a small number of recessive lethals in our genomes. The lethal effect is expressed in the homozygous progeny of a mating between two persons carrying the same recessive lethal in the heterozygous condition. Some lethals are expressed as deaths in utero, where they either go unnoticed, or are noticed as spontaneous abortions. Other lethals, such as those responsible for Duchenne muscular dystrophy, PKU, and cystic fibrosis, exert their effects in childhood. The time of death can even be in adulthood, as in Huntington's disease. The total of all the deleterious and lethal genes that are present in individuals is called *genetic load,* a kind of genetic burden that the population has to carry.

In some cases it is possible to trace the cascade of events that leads to death. How do lethal alleles kill? A common situation is that the allele causes a deficiency of some essential chemical reaction. The human disease PKU is a good example of this kind of deficiency. In other cases there is a structural defect. For example, a lethal allele of rats determines abnormal cartilage and the

Figure 4-8 Manx cat. All such cats are heterozygous for a dominant allele that causes no tail to form. The allele is lethal in homozygous condition. The dissimilar eyes are unrelated to taillessness. (Gérard Lacz/NHPA.)

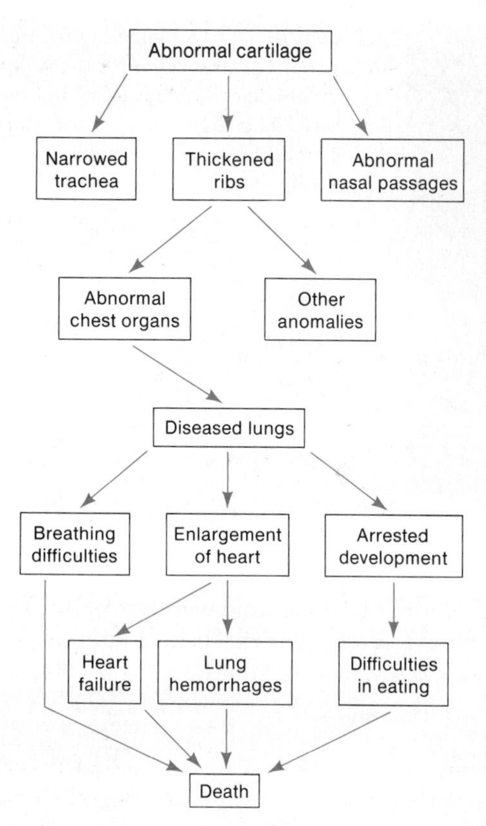

Figure 4-9 Diagram showing how one specific lethal allele causes death in rats. (From I. M. Lerner and W. J. Libby, *Heredity, Evolution, and Society,* 2d. ed. W. H. Freeman and Company, 1976; after H. Grüneberg.)

effect of this abnormality is expressed phenotypically in several different organs, resulting in lethal symptoms, as shown in Figure 4-9. Sickle-cell anemia is another example (Figure 1-3).

Whether an allele is lethal often depends on the environment in which the organism develops. Whereas certain alleles would be lethal in virtually any environment, others are viable in one environment but lethal in another. For example, remember that many of the phenotypes favored and selected by agricultural breeders would almost certainly be eliminated in nature as a result of competition with the members of the natural population. Modern grain varieties provide good examples; only careful nurturing by the farmer has maintained such phenotypes for our benefit.

In practical genetics, we commonly encounter situations in which expected Mendelian ratios are consistently skewed in one direction by reduced viability due to one allele. For example, in the cross $Aa \times aa$, we predict a progeny phenotypic ratio of 50 percent $A-$ and 50 percent aa, but we might consistently observe a ratio such as 55%:45% or 60%:40%. In such a case, the a allele is said to be **subvital,** or **semilethal,** because the lethality is expressed in only some individuals. Thus, lethality may range from 0 to 100 percent, depending on the gene itself, the rest of the genome, and the environment. We shall return to this topic later.

Several Genes Affecting the Same Character

We saw earlier that the identification of a major gene affecting a character does not mean that this is the *only* gene affecting that character. An organism is a highly complex machine in which all functions interact to a greater or lesser degree. At the level of genetic determination, the genes likewise can be regarded as cooperating. Therefore, a gene does not act in isolation; its effects depend not only on its own functions but also on the functions of other genes as well as on the environment. In many cases, genetic analysis can detect the complex interactions of major genes, and we now look at some examples. Typically, the key to an interaction is a **modified Mendelian ratio.**

Coat Color in Mammals

Studies of coat color in mammals reveal beautifully how different genes cooperate in the determination of one character. The mouse is a good mammal for genetic studies because it is small and thus easy to maintain in the laboratory and because its reproductive cycle is short; it is the best-studied mammal as regards the genetic determination of coat color. The genetic determination of coat color in other mammals closely parallels that of mice, and for this reason the mouse acts as a model system. We shall look at examples from other mammals as our discussion proceeds. At least five major genes interact to determine the coat color of mice: *A*, *B*, *C*, *D*, and *S*.

The *A* Gene. This gene determines the distribution of pigment in the hair. The wild-type allele *A* produces a phenotype called agouti. Agouti is an overall grayish color with a brindled or "salt-and-pepper" appearance. It is a common color of mammals in nature. The effect is caused by a band of yellow on the otherwise dark hair shaft. In the nonagouti phenotype (determined by the allele *a*), the yellow band is absent, so there is solid dark pigment throughout (Figure 4-10).

The lethal A^Y allele, discussed in the previous section, is another allele of this gene; it makes the entire shaft yellow. Still another allele is a^t, which results in a "black-and-tan" effect, a yellow belly with dark pigmentation elsewhere. For simplicity, we shall not include these two alleles in the following discussion.

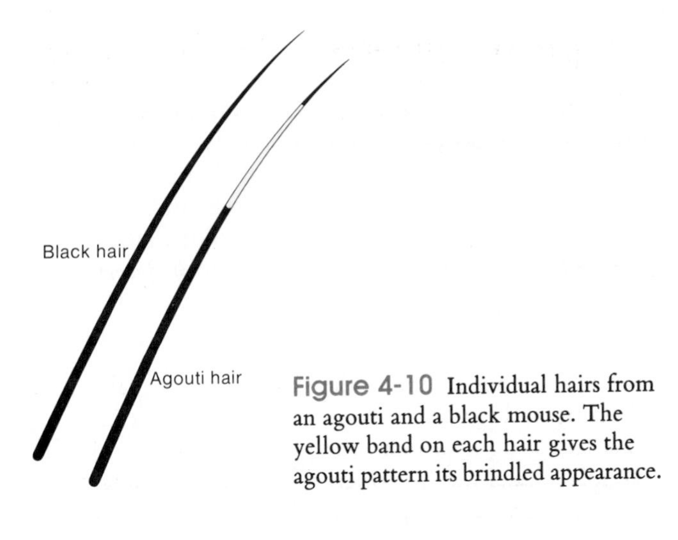

Figure 4-10 Individual hairs from an agouti and a black mouse. The yellow band on each hair gives the agouti pattern its brindled appearance.

(a)

(b)

Figure 4-11 Albinism in reptile and bird. In each case the phenotype is produced by a recessive allele that determines an inability to produce the dark pigment melanin in skin cells. (The normal allele determines ability to synthesize melanin.) (a) Rattlesnake. In this species the normal dark coloration is due entirely to melanin, so the albino allele results in a completely unpigmented appearance. (K. H. Switak/NHPA) (b) Penguin. In this species melanin normally makes most body feathers black, but the reddish-orange colors in the head feathers and beak are due to another pigment chemically unrelated to melanin. The recessive albino allele results in no melanin, but the reddish parts are unaffected and retain their normal coloration. (A.N.T./NHPA)

The *B* Gene. This gene determines the color of pigment. There are two major alleles of the *B* gene. The allele *B* gives the normal agouti color in combination with *A* but gives solid black with *aa*. The genotype *A* – *bb* gives a streaked brown color called cinnamon, and *aa bb* gives solid brown.

The following cross illustrates the inheritance pattern of the *A* and *B* genes:

AA bb (cinnamon) × *aa BB* (black)
or *AA BB* (agouti) × *aa bb* (brown)
↓
F_1 all *Aa Bb* (agouti)
Aa Bb (agouti) × *Aa Bb* (agouti)
↓
F_2 9 *A* – *B* – (agouti)
3 *A* – *bb* (cinnamon)
3 *aa B* – (black)
1 *aa bb* (brown)

The breeding of domestic horses seems to have eliminated the *A* allele that determines the agouti phenotype, although certain wild relatives of the horse do have this allele. The color we have called brown in mice is called chestnut in horses, and this phenotype also is recessive to black.

The *C* Gene. The wild-type allele *C* permits color expression, and the allele *c* prevents color expression. The *cc* constitution is said to be **epistatic** to the other color genes. The word epistatic literally means "standing upon"; in homozygous condition, the *c* allele "stands on" or blots out the expression of other genes concerned with coat color. The *cc* animals, lacking coat pigment, are called albinos. Albinos are common in many mammalian species and have also been reported among birds, snakes, and fish (Figure 4-11).

Epistatic genes produce interesting modified ratios, as seen in the following cross (in which both parents are *aa*):

$BB\,cc$ (albino) × $bb\,CC$ (brown)

or $BB\,CC$ (black) × $bb\,cc$ (albino)

↓

F_1 all $Bb\,Cc$ (black)

$Bb\,Cc$ (black) × $Bb\,Cc$ (black)

↓

F_2		
	9 $B{-}C{-}$ (black)	9
	3 $bb\,C{-}$ (brown)	3
	3 $B{-}cc$ (albino)	4
	1 $bb\,cc$ (albino)	

A phenotypic ratio of 9:3:4 is observed. This ratio is the signal for inferring gene interaction of the type called recessive epistasis. Recessive epistasis is also well illustrated by the three familiar colors of Labrador retriever dogs. The three colors, black, chocolate, and golden (Figure 4-12) are produced genetically in the following way. The alleles *B* and *b* are equivalent to those in mice and result in black and brown (chocolate) respectively. At another gene, the homozygous constitution *ee* is epistatic to both the *B*– and *bb* alternatives, resulting in the golden color. Therefore, to be black or brown, a dog must have the *E* allele. Whether a golden dog is *B*– or *bb* can be deduced sometimes by observing the color of the nose and lips because the epistasis of *ee* is effective mainly in the coat, an example of tissue-specific gene expression.

There is nothing mysterious about epistasis, despite the unusual F_2 ratios. Our example of the epistasis shown by mouse coat color makes perfect sense if we view the *C* allele as necessary for pigment synthesis, and the *B* and *b* alleles as the determinants of what color that pigment will be, with *B* determining black and *b* determining brown. If we then view *c* as an inactive form of *C*, then the mechanism of the epistasis becomes clear, because if no pigment is made, then it doesn't really matter whether the determinant for black or for brown is present, because these determinants simply will not be used. In fact, every time one gene is higher, or "upstream," in the genetic chain of command, we would expect there to be an epistatic effect on the genes lower in the hierarchy of command. Therefore, finding a case of epistasis as revealed by a modified Mendelian ratio can suggest hypotheses to the researcher about the sequence in which genes act.

We have already learned that the c^h (Himalayan) allele in rabbits determines that, at the extremities only, coat color is dark. This allele exists also in other mammals, including mice, which are also called Himalayan, and cats, which are called Siamese (Figure 4-13).

It should be pointed out here that the term epistasis is often used in a different way (mainly in population genetics) to describe *any* kind of gene interaction.

Message An epistatic gene allele of one gene eliminates expression of the alternative phenotypes of another gene, and inserts its own phenotype instead.

(a)

(b)

(c)

Figure 4-12 Coat color inheritance in Labrador retrievers. Two alleles of a pigment gene *B* and *b*, determine black **(a)** or brown **(b)** respectively. At a separate gene, *E* allows color deposition in the coat, and *ee* prevents deposition resulting in the golden phenotype **(c)**. This is a case of recessive epistasis. Thus the three homozygous genotypes are *BB EE* **(a)**, *bb EE* **(b)**, and *BB ee* or *bb ee* **(c)**. The dog in c is most likely *BB ee*—the animal still has the ability to make black pigment, as witnessed by the black nose and lips, but not to deposit this pigment in the hairs. The progeny of a dihybrid cross would produce a 9:3:4 ratio of black:brown:golden. (Anthony Griffiths)

(a)

(b)

Figure 4-13 Temperature-sensitive alleles of the *C* gene result in similar phenotypes in several different mammals. These alleles result in very much reduced or no synthesis of the dark pigment melanin in the skin covering warmer parts of the body. At lower temperatures, such as those found at the body extremities, melanin is synthesized, producing darker snout, ears, tail, and feet. (a) Himalayan mouse. (Anthony Griffiths) (b) Siamese cat (Walter Chandoha). Furthermore, Himalayan rabbits (not shown) are often sold as pets. All three are of genotype $c^h c^h$.

The *D* Gene. The *D* gene controls the intensity of pigment specified by the other coat-color genes. The genotypes *D D* and *D d* permit full expression of color in mice, but *d d* "dilutes" the color making it look "milky." The effect is due to an uneven distribution of pigment in the hair shaft. Dilute agouti, dilute cinnamon, dilute brown, and dilute black coats all are possible. A gene with such an effect is called a **modifier gene.** In the following cross, we assume that both parents are *a a C C*:

B B d d (dilute black) × *b b D D* (brown)

or *B B D D* (black) × *b b d d* (dilute brown)

↓

F_1 all *B b D d* (black)

B b D d (black) × *B b D d* (black)

↓

F_2 9 *B* – *D* – (black)

3 *B* – *d d* (dilute black)

3 *b b D* – (brown)

1 *b b d d* (dilute brown)

In horses, the *D* allele shows incomplete dominance. Figure 4-14 shows how dilution affects the appearance of chestnut and bay horses.

The *S* Gene. The *S* gene controls the presence or absence of spots. The genotype *S* – results in no spots, and *ss* produces a spotting pattern called piebald in both mice and horses. This pattern can be superimposed on any of the coat colors discussed earlier — with the exception of albino, of course.

By this time, the point of the discussion should be obvious. Normal coat appearance in wild mice is produced by a complex set of interacting genes determining pigment type, pigment distribution in the individual hairs, pigment distribution on the animal's body, and the presence or absence of pigment. Interacting genes determine most characters in any organism.

Figure 4-15 illustrates some of the pigment patterns in mice.

More Examples of Gene Interaction

Other organisms provide instructive examples of gene interaction. One of the simplest yet most striking is the inheritance of skin color in corn snakes *(Elaphe guttata)*. The natural color is a black-and-orange, repeating, camouflage pattern as shown in Figure 4-16a. The phenotype is produced by two separate pigments, both of which are under genetic control. One gene determines the orange pigment, and the alleles are *O* (presence of orange pigment) and *o* (absence of orange pigment). Another

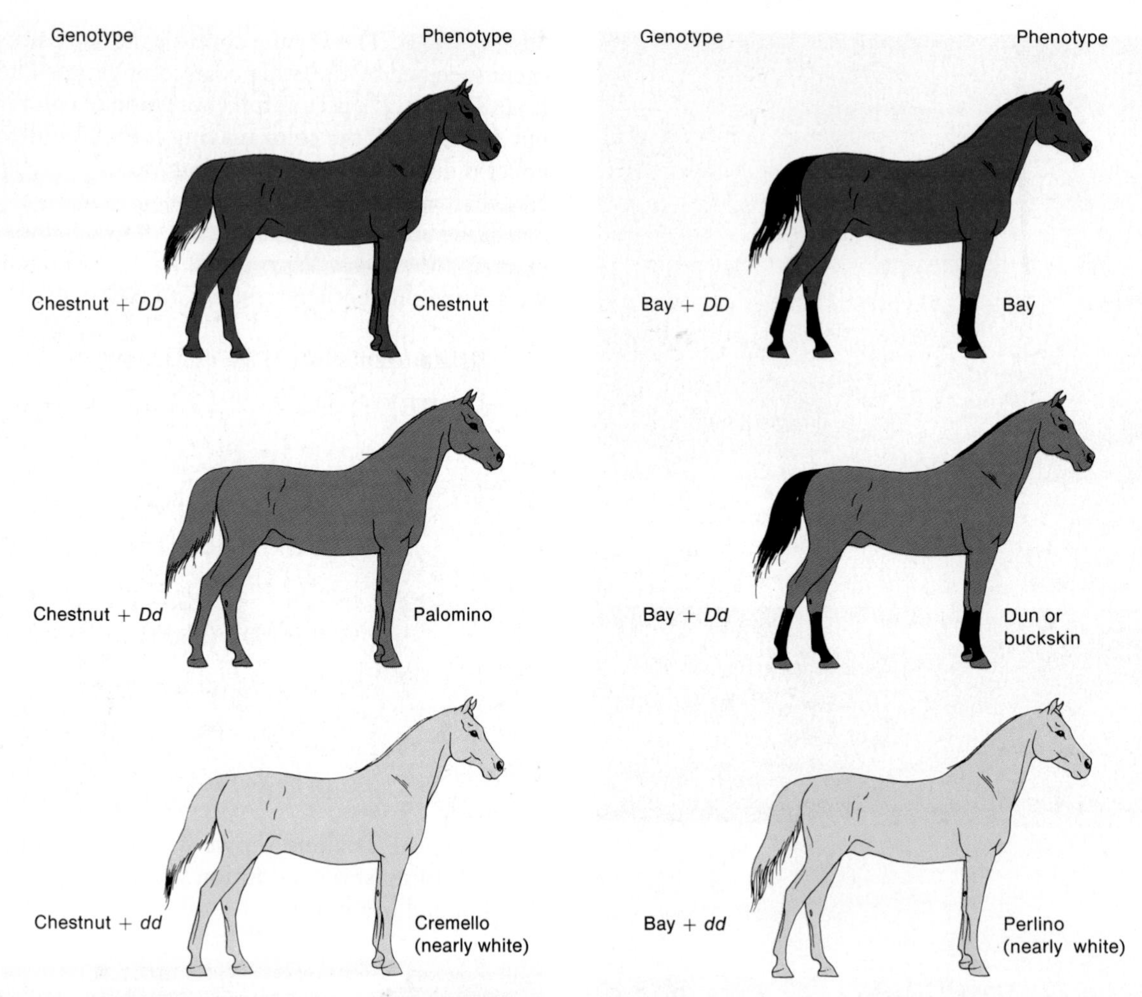

Figure 4-14 The modifying effect of the dilution allele on basic chestnut and bay genotypes in horses. Note the incomplete dominance shown by *D*. (From J. W. Evans et al., *The Horse*. W. H. Freeman and Company, 1977.)

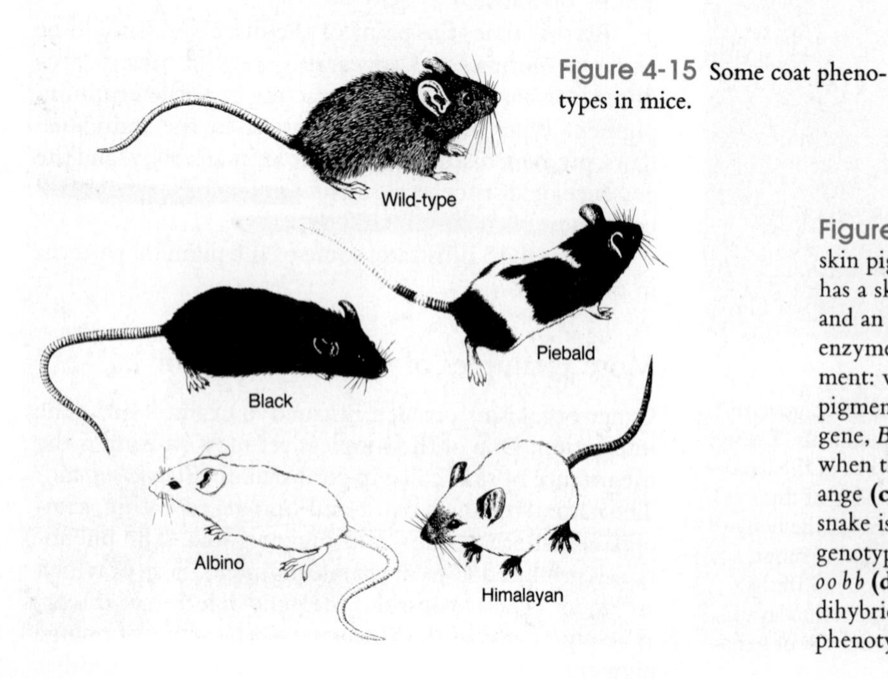

Figure 4-15 Some coat phenotypes in mice.

Figure 4-16 *(facing page)* Analysis of the genes for skin pigment in the corn snake. The wild type **(a)** has a skin pigmentation pattern made up of a black and an orange pigment. The gene *O* determines an enzyme in the synthetic pathway for orange pigment: when this enzyme is deficient (*oo*) no orange pigment is made and the snake is black **(b)**. Another gene, *B*, determines an enzyme for black pigment: when this enzyme is deficient (*bb*) the snake is orange **(c)**. When both enzymes are deficient, the snake is albino **(d)**. Hence the four homozygous genotypes are *OOBB* **(a)**, *ooBB* **(b)**, *OObb* **(c)** and *oobb* **(d)**. A cross of a × d or b × c would give a dihybrid wild-type F_1 and a 9:3:3:1 ratio of the four phenotypes in the F_2. (Anthony Griffiths)

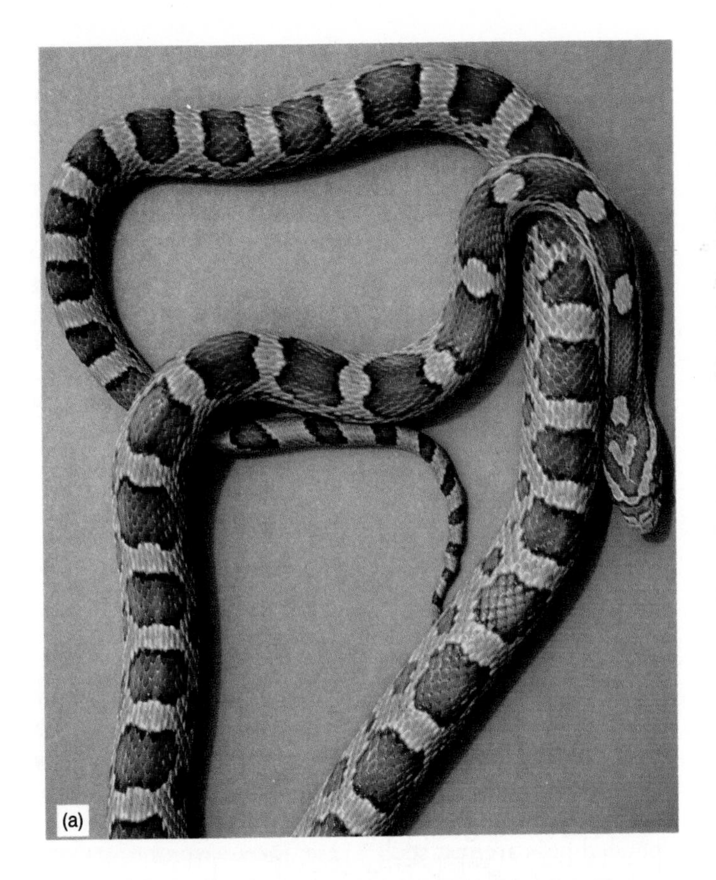
(a)

(b)

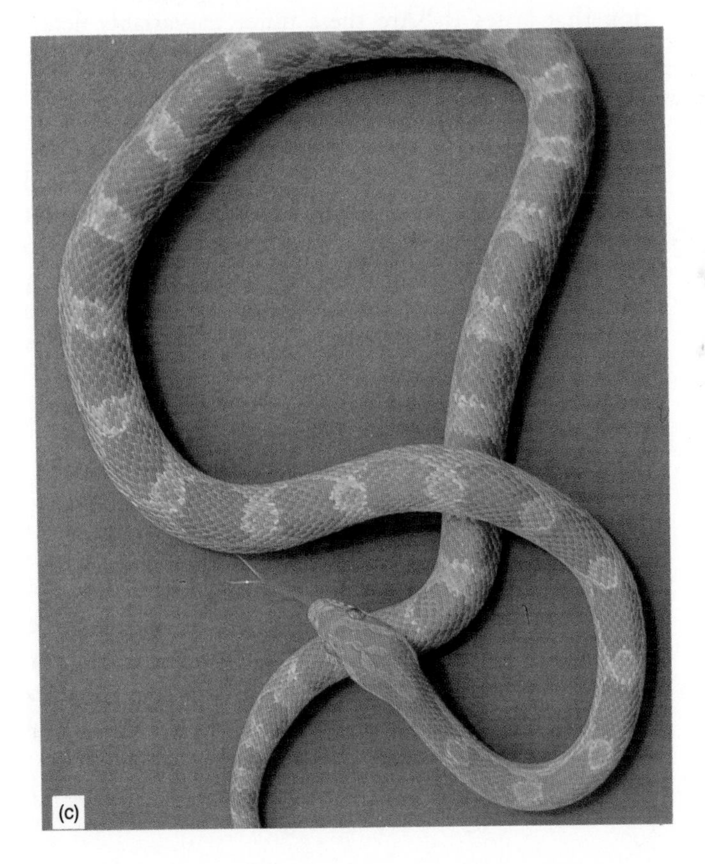
(c)

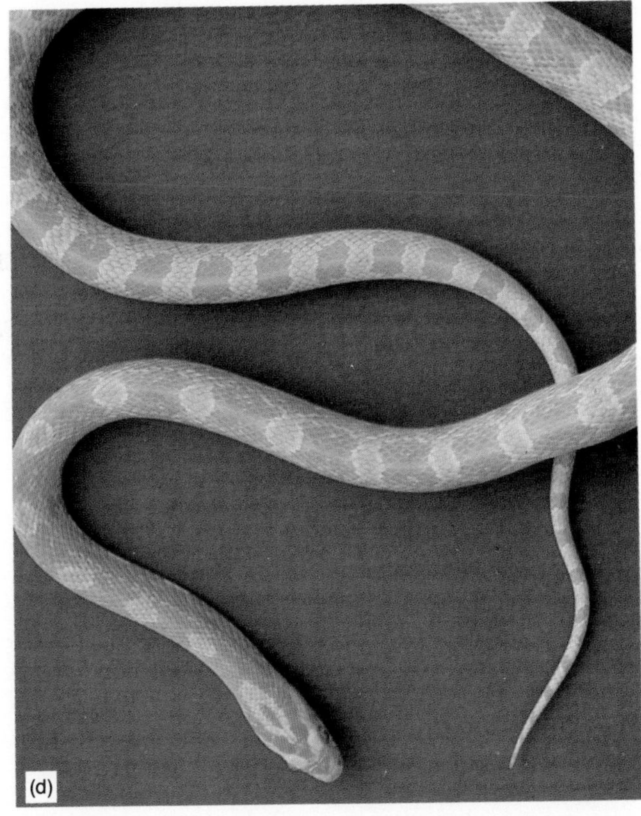
(d)

gene determines the black pigment, with alleles B (presence of black pigment) and b (absence of black pigment). The natural pattern is produced by the genotype $O{-}B{-}$, a snake that is $oo\,B{-}$ is black, and a snake that is $O{-}\,bb$ is orange (Figure 4-16c). The double homozygous recessive $oo\,bb$ is albino, as shown in the figure. Notice that the faint pink color of the albino is from yet another pigment, the hemoglobin of the blood that is visible through this snake's skin. The albino shows clearly that there is another element to the skin phenotype in addition to pigment, and this is the repeating motif in and around which pigment is deposited.

Since there are two genes in this system, we obtain a typical dihybrid inheritance pattern and the four unique phenotypes form a 9:3:3:1 ratio in the F_2. This is not really a modified Mendelian ratio, but the system shows clearly that the single character of skin coloration is produced by gene interaction. A typical pedigree might be as follows:

$OO\,bb$ (orange) × $oo\,BB$ (black)
↓
F_1 $OoBb$ (natural)
$Oo\,Bb$ (natural) × $OoBb$ (natural)
↓
F_2 $O{-}\,B{-}$ (natural) 9
$O{-}\,bb$ (orange) 3
$oo\,B{-}$ (black) 3
$oo\,bb$ (albino) 1

Peas provide a good example of another modified ratio that points to an important kind of gene interaction. When two specific white-petaled pure lines of peas are crossed, all of the F_1 have purple flowers. The F_2 produced by selfing shows both purple and white plants in a ratio of 9:7. How can these results be explained? By now, you should immediately suspect that the 9:7 ratio is a modification of the Mendelian 9:3:3:1 ratio. The explanation is that two different genes in the pea have similar effects on petal color. Let's represent the alleles of these genes by A, a, B, and b:

white line 1 × white line 2
$AA\,bb$ × $aa\,BB$
↓
F_1 all $Aa\,Bb$ (purple)
$Aa\,Bb$ (purple) × $Aa\,Bb$ (purple)
↓
F_2 9 $A{-}B{-}$ (purple) 9
3 $A{-}\,bb$ (white) }
3 $aa\,B{-}$ (white) } 7
1 $aa\,bb$ (white) }

Both genes affect petal color. Homozygosity for the recessive allele of *either* gene causes a plant to have white petals. To have the purple phenotype, a plant must have at least one dominant allele of *both* genes. This phenomenon is called **complementary gene action,** or **complementation**—a term that gives a good intuitive grasp of what is going on and the sense in which the two dominant alleles of different genes are uniting to produce a specific phenotype, in this example, purple petals. (Note that we might have mistakenly inferred purpleness to be a specific phenotype of one gene if, for example, we had crossed parents of genotypes $AA\,BB$ and $aa\,BB$.)

Message Complementation is the production of the dominant phenotype (generally wild type) when two separate genotypes determining similar recessive phenotypes come together in the same cell.

The two white lines of peas differ, even though they show exactly the same phenotype. Geneticists often have to deal with independently isolated lines that have the same phenotype. Generally such lines are abnormal recessive phenotypes derived from wild type. The white petaled peas are just such a case; most species of peas in nature have colored petals, but albino phenotypes arise spontaneously, and are nearly always recessive. When there are two or more such lines, the first obvious question that arises is "Are these different variants determined by recessive alleles of the same gene or not?". The complementation test provides an effective operational test of allelism. If the two recessive phenotypes are intercrossed and a wild-type phenotype is observed in the F_1, the parental genotypes have obviously complemented each other; the recessive alleles must be of different genes. On the other hand, what if we were to cross two independently obtained, recessive, albino lines and the F_1 and the F_2 were all albino? In this situation the evidence is consistent with the recessive determinants' being alleles of the same gene.

Message When independently derived genotypes producing similar recessive phenotypes fail to complement, we retain the hypothesis that the genetic determinants of the phenotypes are alleles of the same gene.

Now we have two tests for allelism; first, the monohybrid 3:1 ratio reveals a dominant and a recessive allele of the same gene, and second, lack of complementation in pairwise crosses of independently isolated lines reveals two recessive alleles of the same gene.

Genetic variants of foxgloves *(Digitalis purpurea)* provide excellent examples of gene interaction in the determination of the overall appearance of an organism. There are three important genes that interact to determine petal coloration. The first gene determines the synthesis of reddish anthocyanin pigment. The *M* allele of this gene stands for ability to synthesize, whereas *m* stands for inability to synthesize anthocyanin resulting in white petals. Figure 4-17 shows the phenotypes that we use in this discussion. The second gene is a modifier gene. One allele, *D*, stands for the synthesis of large amounts of anthocyanin (dark reddish), and *d* stands for low amounts (light reddish). The third gene affects pigment deposition. The allele *W* prevents pigment deposition in all parts of the petal except in the throat spots, whereas the recessive allele *w* allows deposition of pigment all over the petal. Thus these three genes control the ability to synthesize, the amount synthesized, and the ability for the pigment to be deposited in specific petal cells. We could consider a variety of dihybrid crosses and even a trihybrid cross, but we will consider only one sample cross, a dihybrid cross that illustrates a modified F_2 ratio not yet discussed.

Consider the cross between the two genotypes *MM DD ww* and *MM dd WW*. The phenotype of the first is dark reddish because it has the *D* modifier and the ability to deposit pigment. The second is white with reddish spots because although the plant has the ability to synthesize pigment (conferred by the allele *M*), the *W* allele prevents deposition except in the throat spots. Let us consider the usual type of pedigree but eliminate the *M* allele because it will be homozygous in all individuals.

(*MM*) *DD ww* × (*MM*) *dd WW*
(dark reddish) (white with reddish spots)
↓
F_1 *Dd Ww* (white with reddish spots)

Dd Ww × *Dd Ww*
(white with reddish spots) (white with reddish spots)
↓

F_2		
	9 *D– W* (white with reddish spots)	12
	3 *dd W–* (white with reddish spots)	
	3 *D– ww* (dark reddish)	3
	1 *dd ww* (light reddish)	1

Overall, a 12:3:1 phenotypic ratio is produced. This kind of interaction is called dominant epistasis because, as can

Figure 4-17 Pigment phenotypes in foxgloves, determined by three separate genes. *M* codes for an enzyme that synthesizes anthocyanin, the reddish pigment seen in these petals; *mm* produces no pigment and produces the phenotype albino with yellowish spots. *D* is an enhancer of anthocyanin, resulting in a darker pigment; *dd* does not enhance. At the third locus, *ww* allows pigment deposition in petals, but *W* prevents pigment deposition except in the spots, and so results in the white, spotted phenotype genotypes (and phenotypes) from left to right are *M – W – – –* (white with reddish spots), *mm – – – –* (white with yellowish spots), *M – ww dd* (light reddish), and *M – ww D –* (dark reddish). (Anthony Griffiths)

be seen from the F_2 results, the dominant allele *W* eliminates the two alternatives expressed by *D* and *d*, dark and light reddish, and replaces them with another phenotype, white with reddish spots.

Another important kind of interaction is **suppression** of one gene by another. This interaction is illustrated by the inheritance of the ability to produce a chemical called malvidin in the plant genus *Primula*. Malvidin production is determined by a single dominant allele *K*. However, the action of this dominant allele may be suppressed by a dominant suppressor *D*, an allele of a separate gene, as the following cross shows:

KK dd (malvidin) × *kk DD* (no malvidin)

↓

F_1 all *Kk Dd* (no malvidin)

Kk Dd (no malvidin) × *Kk Dd* (no malvidin)

↓

F_2
- 9 *K– D–* (no malvidin)
- 3 *kk D–* (no malvidin)
- 1 *kk dd* (no malvidin) } 13
- 3 *K– dd* (malvidin) 3

A suppressor gene may have an associated phenotype or it may—as in the malvidin example—have no detectable phenotypic effect other than the suppression.

Message A suppressor is a gene that eliminates the phenotypic expression of a specific allele of another gene.

Our final example of gene interaction introduces the concept that genes may be represented more than once in the genome. The example concerns the genes that control fruit shape in the plant called shepherd's purse, *Capsella bursa-pastoris*. Two different lines have fruits of different shapes: one is "round"; the other, "narrow." Are these two phenotypes determined by two alleles of a single gene? A cross between the two lines produces an F_1 with round fruit; this result is consistent with the hypothesis of determination by a pair of alleles. However, the F_2 shows a 15:1 ratio of round to narrow, and this ratio suggests a modification of the dihybrid 9:3:3:1 Mendelian ratio. The genetic control of fruit shape can be explained in terms of **duplicate genes** (Figure 4-18). Apparently, round fruits result from the presence of at least one dominant allele of either gene. The two genes appear to be identical in function. (Contrast this 15:1 ratio with the 9:7 ratio obtained from complementary genes, where *both* dominant genes are necessary to produce a specific phenotype.)

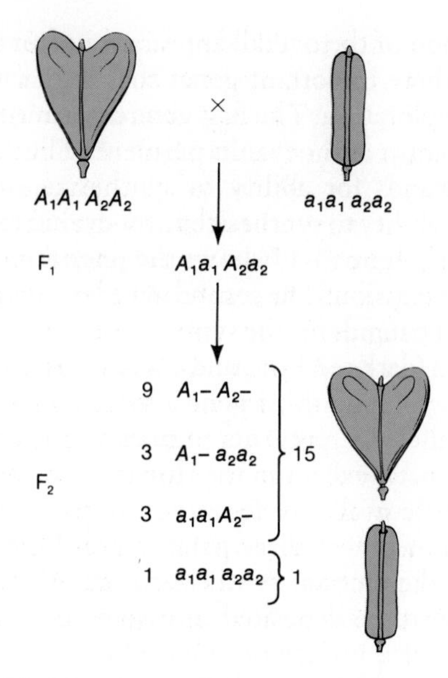

Figure 4-18 Inheritance pattern of duplicate genes controlling fruit shape in shepherd's purse. Either A_1 or A_2 can cause a round fruit.

Message Duplicate genes provide alternative genetic determination of a specific phenotype.

Table 4-3 summarizes the various types of gene interactions that modify Mendelian ratios. Note that so far

Table 4-3. Modified Phenotypic Ratios Produced by Gene Interaction

Type of gene interaction	9 *A – B –*	3 *A – bb*	3 *aa B –*	1 *aa bb*	Phenotypic ratio
None (four distinct phenotypes)	9	3	3	1	9:3:3:1
Complementary gene action	9	7			9:7
Dominant suppression by *A* of dominant allele *B*	12		3	1	13:3
Recessive epistasis of *aa* acting on *B* and *b* alleles	9	3	4		9:3:4
Dominant epistasis of *A* acting on *B* and *b* alleles	12		3	1	12:3:1
Duplicate genes	15			1	15:1

we have considered modified ratios of diploid organisms only. Gene interactions also can be detected in haploid organisms. We have already discussed a gene in the fungus *Neurospora* having an allele, *al*, that causes the asexual spores to be albino rather than the pink color produced by the wild-type allele of the same gene (see Figure 3-36). A cross *al* × al^+ gives $\frac{1}{2}$ *al* (albino) and $\frac{1}{2}$ al^+ (pink) progeny, conforming to Mendel's first law. Another interesting gene, *ylo*, gives yellow asexual spores (Figure 3-36), and the cross *ylo* × ylo^+ gives $\frac{1}{2}$ *ylo* (yellow) and $\frac{1}{2}$ ylo^+ (pink) progeny. When an *al* culture is crossed with a *ylo* culture, the resulting progeny are $\frac{1}{4}$ yellow, $\frac{1}{4}$ pink, and $\frac{1}{2}$ albino. How can we explain this result? The answer is a kind of epistasis: *al* and *ylo* are alleles of separate genes, each of which is necessary for the normal production of pink pigment. Writing out the genotypes in the cross shows how the epistasis works:

Haploid parental cultures	$al\ ylo^+$ (albino) × $al^+\ ylo$ (yellow)	
	↓	
Transient diploid	$al^+\ al\ ylo^+\ ylo$	
	↓ meiosis	
Progeny (cultures from sexual spores)	$\frac{1}{4}$ $al\ ylo$ (albino)	} $\frac{1}{2}$
	$\frac{1}{4}$ $al\ ylo^+$ (albino)	
	$\frac{1}{4}$ $al^+\ ylo$ (yellow)	$\frac{1}{4}$
	$\frac{1}{4}$ $al^+\ ylo^+$ (pink)	$\frac{1}{4}$

The data clearly suggest that al^+ and ylo^+ are both needed for production of pink pigment and that either *al* or *ylo* blocks this production. The epistatic mechanism is revealed by the albino phenotype shown by the *al ylo* genotype, because it suggests another example of an "upstream effect" like the one we saw in the example of mouse albinism. We can postulate from our genetic analysis that al^+ is needed for synthesis of pigment and that the alleles of the other gene determine whether that pigment is pink (ylo^+) or yellow (*ylo*). We shall see in later chapters that chemical studies generally confirm these ideas that come from genetic analysis.

> Message Modified Mendelian ratios reveal that a character is determined by the complex interaction of different genes.

Penetrance and Expressivity

The foregoing discussion shows that genes do not act in isolation. A gene does *not* determine a phenotype by acting alone; it does so only in conjunction with other genes and with the environment. Although geneticists do routinely ascribe a particular phenotype to a particular allele, we must remember that this is merely a convenient kind of shorthand designed to facilitate genetic analysis. This shorthand arises from the ability of geneticists to isolate individual components of a biological process and to study them as part of genetic dissection. Although this logical isolation is an essential aspect of genetics, the message of this chapter is that genes act in concert to produce the overall features of an organism.

In the preceding examples, the genetic basis of the dependence of one gene on another has been worked out from clear genetic ratios. In other situations, where the phenotype ascribed to a gene is known to be dependent on other factors but the precise inheritance of those factors has not been established, the terms *penetrance* and *expressivity* may be useful in describing the situation.

We have already encountered penetrance in the discussion of lethal alleles. **Penetrance** is defined as the percentage of individuals, with a given genotype who exhibit the phenotype associated with that genotype. For example, an organism may have a particular genotype but may not express the phenotype normally associated with that genotype because of modifiers, epistatic genes, or suppressors in the rest of the genome or because of a modifying effect of the environment. Penetrance can be used to measure such an effect when it is not known which of these types of modification underlies the effect.

On the other hand, **expressivity** describes the extent to which a given genotype is expressed phenotypically in an individual. Again, the lack of full expression may be due to the allelic constitution of the rest of the genome or to environmental factors. Figure 4-19 diagrams the distinction between penetrance and expressiv-

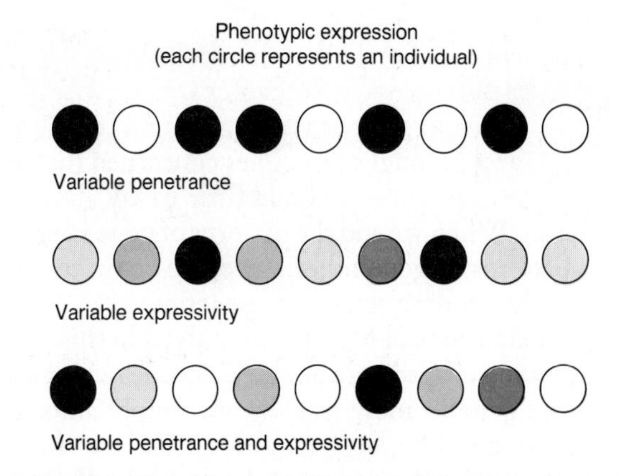

Figure 4-19 Diagram representing the effects of penetrance and expressivity through a hypothetical character "pigment intensity." In each row, all individuals have the same allele, say *P*, giving them the same "potential to produce pigment." However, effects deriving from the rest of the genome and from the environment may suppress or modify pigment production in an individual.

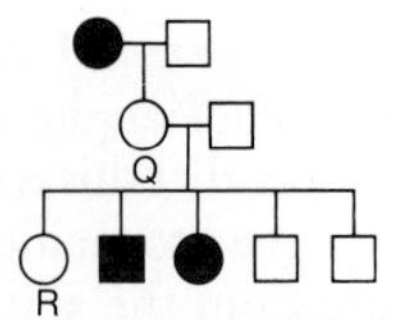

Figure 4-20 Lack of penetrance illustrated by a pedigree for a dominant allele. Individual Q must have the allele (because it was passed on to her progeny), but it was not expressed in her phenotype. An individual such as R cannot be sure that his or her genotype lacks the allele.

ity. Obviously, both penetrance and expressivity are integral to the concept of the norm of reaction, which was discussed in Chapter 1.

Human pedigree analysis and predictions in genetic counseling can often be thwarted by the phenomena of variable penetrance and expressivity. For example, if a disease-causing allele is not fully penetrant (as often is the case), it is difficult to give a clean genetic bill of health to any individual in a disease pedigree (for example, individual R in Figure 4-20). On the other hand, pedigree analysis can sometimes identify individuals who do not express but almost certainly do have a disease genotype (for example, individual Q in Figure 4-20).

Specific examples of variable expressivity are found in Figures 4-21 and 4-22 and also in Figure 2-21 and Problem 11 of Chapter 2.

Message The impact of a gene at the phenotype level depends not only on its dominance but also on the modifying effect of the rest of the genome and of the environment.

In conclusion, notice that the extensions of Mendelian analysis discussed in this chapter are mainly based on the complexities of gene expression, not on complexities of inheritance. Mendel's principles concerned the inheritance patterns of genes and had little to say about gene expression. When we add the patterns of gene expression to the patterns of gene inheritance, we begin to see the fabric of the modern science of genetics.

The extensions of Mendelian analysis in this chapter and following chapters are not the only ones encountered in routine genetic analysis; they are simply among the most common. The key recognition features for each extension — whether, for example, multiple allelism, incomplete dominance, epistasis, or variable expressivity — occur at the experimental level. The analyst must be constantly on the lookout for these and any other results that might indicate the uniqueness of any given situation. Such signals often lead to the discovery of new phenomena and to the opening up of new research areas.

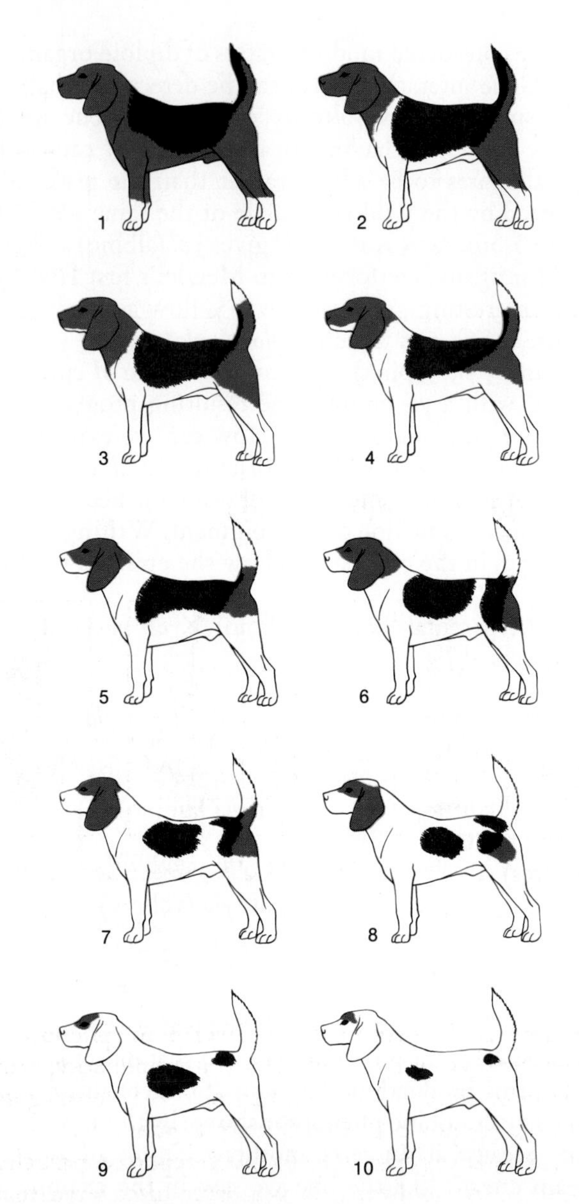

Figure 4-21 Variable expressivity shown by ten grades of piebald spotting in beagles. Each of these dogs has S^P, the allele responsible for piebald spots in dogs. (Adapted from Clarence C. Little, *The Inheritance of Coat Color in Dogs.* Cornell University Press, 1957, and from Giorgio Schreiber, *Journal of Heredity* 9, 1930, 403.)

Summary

Although Mendel's laws apply to all eukaryotic organisms, these laws are only a starting point for understanding heredity. Genes and chromosomes present many complexities in addition to those unraveled by Mendel.

Mendel observed full dominance in his experiments; we have added examples of incomplete dominance and

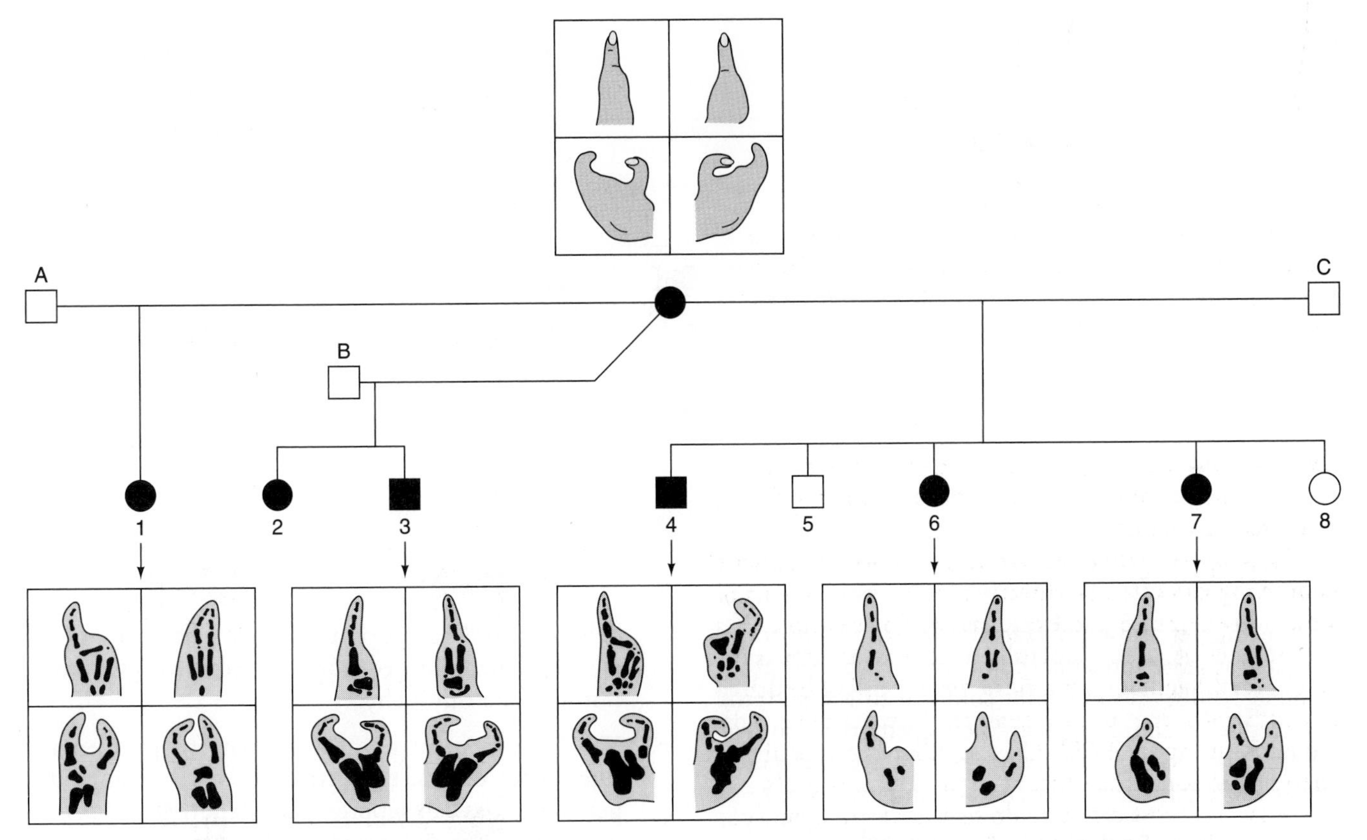

Figure 4-22 Variable expressivity in a pedigree of cleft hands and feet. Feet above, hands below.

codominance. In incomplete dominance, the phenotype of a heterozygote is anywhere between the phenotypes of the homozygotes. In codominance, the heterozygote shows the phenotypes of both homozygotes.

From his experiments, Mendel reported particles (genes) with two forms. It was later discovered that across a population a gene may have more than two forms. This situation is known as multiple allelism. A member of an allelic series may exhibit any type of dominance relative to the other members of the series.

One gene may affect more than one character—that is, it may be pleiotropic. The A^Y allele in mice is pleiotropic in affecting both coat color and viability.

A major gene affecting a character is not necessarily the only gene affecting that character; several genes may interact to affect a character. A good example of gene interaction is the coat color of mice, which is produced by a complex set of interacting genes that determine the presence or absence of pigment, pigment type, pigment distribution in the hair and pigment distribution on the animal.

Gene interaction often produces modified Mendelian ratios in the F_2. Some kinds of interaction have specific names, such as complementary gene action, epistasis, suppression, and duplicate gene action. Genes interact in both diploid and haploid organisms.

Two other important extensions of Mendelian analysis are the concepts of penetrance and expressivity. Penetrance is the percentage of individuals of a specific genotype who express the phenotype associated with that genotype. Expressivity refers to the degree to which a particular genotype is expressed as a phenotype. Variable penetrance and expressivity are caused by allelic and environmental variation.

Concept Map

Draw a concept map interrelating as many of the following terms as possible. Note that the terms are listed in no particular order.

dihybrid / environment / modified Mendelian ratio / gene / interaction / self-fertilization / 9:3:4 ratio / penetrance / recessive epistasis

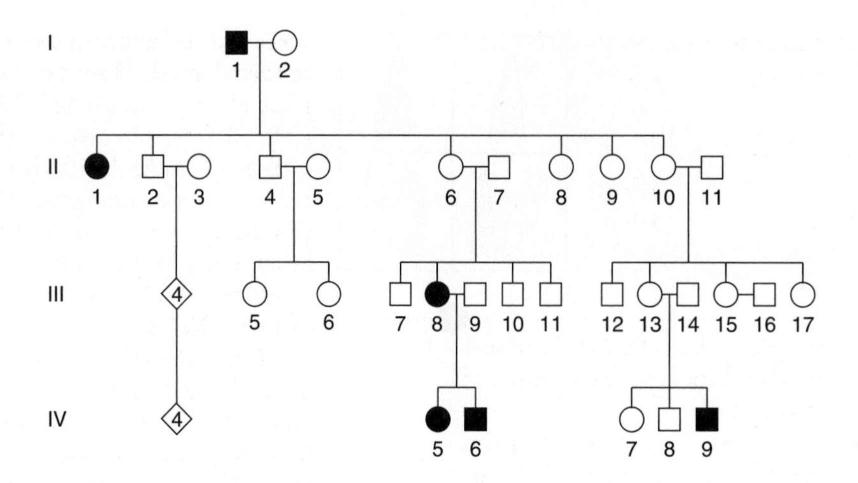

Chapter Integration Problem

Most pedigrees show polydactyly (Figure 2-18) to be inherited as a rare autosomal dominant, but the pedigrees of some families do not fully conform to the patterns expected for such inheritance. Such a pedigree is shown above. (The unshaded diamonds stand for the specified number of unaffected individuals of unknown sex.)

a. What irregularity does this pedigree show?

b. What genetic phenomenon does this pedigree illustrate?

c. Suggest a specific gene interaction mechanism that could produce such a pedigree, showing genotypes of pertinent family members.

Solution

a. The normal expectation for an autosomal dominant is for each affected individual to have an affected parent, but this is not seen in this pedigree, and this constitutes the irregularity. What are some possible explanations? Could some cases of polydactyly be caused by a different gene, one that is an X-linked dominant? This is not a useful suggestion, because we would still have to explain the absence of the condition in individuals II-6 and II-10. Furthermore, postulating recessive inheritance, whether autosomal or sex-linked, requires many persons in the pedigree to be heterozygotes, which is inappropriate because polydactyly is a rare condition.

b. This leaves us with the conclusion that polydactyly sometimes must be incompletely penetrant. We have learned in this chapter that some individuals who have the genotype for a particular phenotype do not express it. In this pedigree II-6 and II-10 seem to belong in this category; they must carry the polydactyly gene inherited from I-1 because they transmit it to their progeny.

c. We have seen in the chapter that environmental suppression of gene expression can cause incomplete penetrance, as can suppression by another gene. To give the requested genetic explanation, we must come up with a genetic hypothesis. What do we need to explain? The key is that I-1 passes the gene on to two types of progeny, represented by II-1, who expresses the gene, and by II-6 and II-10, who do not. (From the pedigree we cannot tell if the other children of I-1 have the gene or not.) Is genetic suppression at work? I-1 does not have a suppressor allele because he expresses polydactyly. So the only person a suppressor could come from is I-2. Furthermore, I-2 must be heterozygous for the suppressor gene because at least one of her children does express polydactyly. We have thus formulated the hypothesis that the mating in generation I must have been

$$\text{(I-1) } Pp\,ss \times \text{(I-2) } pp\,Ss$$

where S is the suppressor, and P is the allele responsible for polydactyly. From this hypothesis, we predict that the progeny will comprise the following four types if the genes assort independently:

Genotype	Phenotype	Example
$Pp\,Ss$	normal (suppressed)	II-6, II-10
$Pp\,ss$	polydactylous	II-1
$pp\,Ss$	normal	
$pp\,ss$	normal	

If S is rare, the matings and progenies of II-6 and II-10 are probably giving

Progeny genotype	Example
Pp Ss	III-13
Pp ss	III-8
pp Ss	
pp ss	

We cannot rule out the possibilities that II-2 and II-4 have the genotype Pp Ss and that by chance none of their descendants are affected.

Note that we have just used concepts from Chapter 1 (environmental effects), Chapter 2 (Mendelian inheritance), Chapter 3 (chromosomal locations of genes), and Chapter 4 (gene interactions).

Solved Problems

1. Beetles of a certain species may have green, blue, or turquoise wing covers. Virgin beetles were selected from a polymorphic laboratory population and mated to determine the inheritance of wing-cover color. The crosses and results were as follows:

Cross	Parents	Progeny
1	blue × green	all blue
2	blue × blue	$\frac{3}{4}$ blue:$\frac{1}{4}$ turquoise
3	green × green	$\frac{3}{4}$ green:$\frac{1}{4}$ turquoise
4	blue × turquoise	$\frac{1}{2}$ blue:$\frac{1}{2}$ turquoise
5	blue × blue	$\frac{3}{4}$ blue:$\frac{1}{4}$ green
6	blue × green	$\frac{1}{2}$ blue:$\frac{1}{2}$ green
7	blue × green	$\frac{1}{2}$ blue:$\frac{1}{4}$ green $\frac{1}{4}$ turquoise
8	turquoise × turquoise	all turquoise

a. Deduce the genetic basis of wing-cover color in this species.

b. Write the genotypes of all parents and progeny as completely as possible.

Solution

a. These data seem complex at first, but the inheritance pattern becomes clear if we consider the crosses one at a time. A general principle of solving such problems, as we have seen, is to begin by looking over all the crosses and by grouping the data to bring out the patterns.

One clue that emerges from an overview of the data is that all the ratios are one-gene ratios: there is no evidence of two separate genes being involved at all. How can such variation be explained with a single gene? The obvious answer is that there is variation for the single gene itself—that is, multiple allelism. Perhaps there are three alleles of one gene; let's call the gene w (for wing-cover color) and represent the alleles as w^g, w^b, and w^t. Now we have an additional problem, which is to determine the dominance of these alleles.

Cross (1) tells us something about dominance because the progeny of a blue × green cross are all blue; hence, blue appears to be dominant to green. This conclusion is supported by cross (5), because the green determinant must have been present in the parental stock to appear in the progeny. Cross (3) informs us about the turquoise determinants, which must have been present, although unexpressed, in the parental stock because there are turquoise wing covers in the progeny. So green must be dominant to turquoise. Hence, we have formed a model in which the dominance is $w^b > w^g > w^t$. Indeed, the inferred position of the w^t allele at the bottom of the dominance series is supported by the results of cross (7), where turquoise shows up in the progeny of a blue × green cross.

b. Now it is just a matter of deducing the specific genotypes. Notice that the question states that the parents were taken from a polymorphic population; this means that they could be either homozygous or heterozygous. A parent with blue wing covers, for example, might be homozygous ($w^b w^b$) or heterozygous ($w^b w^g$ or $w^b w^t$). Here, a little trial and error and common sense is called for, but by this stage, the question has essentially been answered and all that remains is to "cross the t's and dot the i's." The following genotypes explain the results. A dash indicates that the genotype may be either homozygous *or* heterozygous in having a second allele further down the allelic series.

Cross	Parents	Progeny
1	$w^b w^b \times w^g -$	$w^b w^g$ or $w^b -$
2	$w^b w^t \times w^b w^t$	$\frac{3}{4}\,w^b -:\frac{1}{4}\,w^t w^t$
3	$w^g w^t \times w^g w^t$	$\frac{3}{4}\,w^g -:\frac{1}{4}\,w^t w^t$
4	$w^b w^t \times w^t w^t$	$\frac{1}{2}\,w^b w^t:\frac{1}{2}\,w^t w^t$
5	$w^b w^g \times w^b w^g$	$\frac{3}{4}\,w^b -:\frac{1}{4}\,w^g w^g$
6	$w^b w^g \times w^g w^g$	$\frac{1}{2}\,w^b w^g:\frac{1}{2}\,w^g w^g$
7	$w^b w^t \times w^g w^t$	$\frac{1}{2}\,w^b -:\frac{1}{4}\,w^g w^t:\frac{1}{4}\,w^t w^t$
8	$w^t w^t \times w^t w^t$	all $w^t w^t$

2. The leaves of pineapples can be classified into three types: spiny (S), spiny tip (ST), and piping (nonspiny) (P). In crosses between pure strains followed by intercrosses of the F_1, the results shown below appeared.

	Phenotypes		
Cross	Parental	F_1	F_2
1	ST × S	ST	99 ST:34 S
2	P × ST	P	120 P:39 ST
3	P × S	P	95 P:25 ST:8 S

a. Assign gene symbols. Explain these results in terms of the genotypes produced and their ratios.

b. Using the model from part a, give the phenotypic ratios you would expect if you crossed: (1) the F_1 progeny from piping × spiny with the spiny parental stock, and (2) the F_1 progeny of piping × spiny with the F_1 progeny of spiny × spiny tip.

Solution

a. First, let's look at the F_2 ratios. We have clear 3:1 ratios in crosses 1 and 2, indicating single-gene segregations. Cross 3, however, shows a ratio that is almost certainly a 12:3:1 ratio. How do we know this? Well, there are simply not that many complex ratios in genetics, and trial and error brings us to the 12:3:1 quite quickly. In the 128 progeny total, the numbers of 96:24:8 are expected, but the actual numbers fit these expectations remarkably well.

One of the principles of this chapter is that modified Mendelian ratios reveal gene interactions. Cross 3 gives F_2 numbers appropriate for a modified dihybrid Mendelian ratio, so it looks as if we are dealing with a two-gene interaction. This seems the most promising place to start; we can go back to crosses 1 and 2 and try to fit them in later.

Any dihybrid ratio is based on the phenotypic proportions 9:3:3:1. Our observed modification groups them as follows:

$$\left.\begin{array}{l} 9\,A-B- \\ 3\,A-bb \end{array}\right\} \quad 12 \text{ piping}$$

$$3\,aa\,B- \qquad 3 \text{ spiny-tip}$$

$$1\,aa\,bb \qquad 1 \text{ spiny}$$

So without worrying about the name of the type of gene interaction (we are not asked to supply this anyway), we can already define our three pineapple leaf phenotypes in terms of the proposed allelic pairs A,a and B,b:

$$\text{piping} = A- \;\; (B,b \text{ irrelevant})$$

$$\text{spiny-tip} = aa\,B-$$

$$\text{spiny} = aa\,bb$$

What about the parents of cross 3? The spiny parent must be $aa\,bb$, and because the B gene is needed to produce F_2 spiny-tip individuals, the piping parent must be $AA\,BB$. (Note that we are *told* that all parents are pure, or homozygous.) The F_1 must therefore be $Aa\,Bb$.

Without further thought, we can write out cross 1 as follows:

$$aa\,BB \times aa\,bb \longrightarrow aa\,Bb \begin{cases} \tfrac{3}{4}\,aa\,Bb \\ \tfrac{1}{4}\,aa\,bb \end{cases}$$

Cross 2 can also be written out partially without further thought, using our arbitrary gene symbols:

$$AA\,-- \times aa\,BB \longrightarrow Aa\,B- \begin{cases} \tfrac{3}{4}\,A--- \\ \tfrac{1}{4}\,aa\,B- \end{cases}$$

We know that the F_2 of cross 2 shows single-gene segregation, and it seems certain now that the A,a allelic pair is involved. But the B allele is needed to produce the spiny tip phenotype, so all individuals must be homozygous BB:

$$AA\,BB \times aa\,BB \longrightarrow Aa\,BB \begin{cases} \tfrac{3}{4}\,A-BB \\ \tfrac{1}{4}\,aa\,BB \end{cases}$$

Notice that the two single-gene segregations in crosses 1 and 2 do not show that the genes are *not* interacting. What is shown is that the two-gene interaction is not *revealed* by these crosses—only by 3, in which the F_1 is heterozygous for both genes.

b. Now it is simply a matter of using Mendel's laws to predict cross outcomes:

(1) $Aa\,Bb \times aa\,bb \longrightarrow$ (independent assortment in a standard testcross)

$$\left.\begin{array}{l} \tfrac{1}{4}\,Aa\,Bb \\ \tfrac{1}{4}\,Aa\,bb \end{array}\right\} \tfrac{1}{2} \text{ piping}$$

$$\tfrac{1}{4}\,aa\,Bb \qquad \text{spiny tip}$$

$$\tfrac{1}{4}\,aa\,bb \qquad \text{spiny}$$

(2) $Aa\,Bb \times aa\,Bb \longrightarrow$

$$\tfrac{1}{2}\,Aa \begin{cases} \tfrac{3}{4}\,B- \longrightarrow \tfrac{3}{8} \\ \tfrac{1}{4}\,bb \longrightarrow \tfrac{1}{8} \end{cases} \Big\} \; \tfrac{1}{2} \text{ piping}$$

$$\tfrac{1}{2}\,aa \begin{cases} \tfrac{3}{4}\,B- \longrightarrow \tfrac{3}{8} \quad \text{spiny tip} \\ \tfrac{1}{4}\,bb \longrightarrow \tfrac{1}{8} \quad \text{spiny} \end{cases}$$

Problems

1. If a man of blood group AB marries a woman of blood group A whose father was of blood group O, what different blood groups can this man and woman expect their children to belong to?

2. Erminette fowls have mostly light-colored feathers with an occasional black one, giving a flecked appearance. A cross of two erminettes produced a total of 48 progeny, consisting of 22 erminettes, 14 blacks, and 12 pure whites. What genetic basis of the erminette pattern is suggested? How would you test your hypotheses?

3. Radishes may be long, round, or oval and they may be red, white, or purple. You cross a long, white variety with a round, red one and obtain an oval, purple F_1. The F_2 show nine phenotypic classes as follows: 9 long, red; 15 long, purple; 19 oval, red; 32 oval, purple; 8 long, white; 16 round, purple; 8 round, white; 16 oval, white; and 9 round, red.

   a. Provide a genetic explanation of these results. Be sure to define the genotypes, and show the constitution of parents, F_1, and F_2.

   b. Predict the genotypic and phenotypic proportions in the progeny of a cross between a long, purple radish and an oval, purple one.

4. In the multiple allele series that determines coat color in rabbits, $C^+ > C^{ch} > C^h$, dominance is from left to right as shown. In a cross of $C^+C^{ch} \times C^{ch}C^h$, what proportion of the progeny will be Himalayan? **(a)** 100 percent, **(b)** $\frac{3}{4}$, **(c)** $\frac{1}{2}$, **(d)** $\frac{1}{4}$, **(e)** 0 percent.

5. Black, sepia, cream, and albino are all coat colors of guinea pigs. Individual animals (not necessarily from pure lines) showing these colors were intercrossed; the results are tabulated below, where we are using the abbreviations A, albino; B, black; C, cream; and S, sepia, for the phenotypes:

Cross	Parental phenotypes	Phenotypes of Progeny			
		B	S	C	A
1	B × B	22	0	0	7
2	B × A	10	9	0	0
3	C × C	0	0	34	11
4	S × C	0	24	11	12
5	B × A	13	0	12	0
6	B × C	19	20	0	0
7	B × S	18	20	0	0
8	B × S	14	8	6	0
9	S × S	0	26	9	0
10	C × A	0	0	15	17

   a. Deduce the inheritance of these coat colors, using gene symbols of your own choosing. Show all parent and progeny genotypes.

   b. If the black animals in (7) and (8) are crossed, what progeny proportions can you predict using your model?

6. In a maternity ward, four babies become accidentally mixed up. The ABO types of the four babies are known to be O, A, B, and AB. The ABO types of the four sets of parents are determined. Indicate which baby belongs to each set of parents: **(a)** AB × O, **(b)** A × O, **(c)** A × AB, **(d)** O × O

7. Consider two blood polymorphisms that humans have in addition to the ABO system. Two alleles L^M and L^N determine the M, N, and MN blood groups. The dominant allele R of a different gene causes a person to have the Rh$^+$ (rhesus positive) phenotype whereas the homozygote for r is Rh$^-$ (rhesus negative). Two men took a paternity dispute to court, each claiming three children to be his own. The blood groups of the men, the children, and their mother were as follows:

Person	Blood group		
Husband	O	M	Rh$^+$
Wife's lover	AB	MN	Rh$^-$
Wife	A	N	Rh$^+$
Child 1	O	MN	Rh$^+$
Child 2	A	N	Rh$^+$
Child 3	A	MN	Rh$^-$

   From this evidence, can the paternity of the children be established?

8. On a fox ranch in Wisconsin, a mutation arose that gave a "platinum" coat color. The platinum color proved very popular with buyers of fox coats, but the breeders could not develop a pure-breeding platinum strain. Every time two platinums were crossed, some normal foxes appeared in the progeny. For example, the repeated matings of the same pair of platinums produced 82 platinum and 38 normal progeny. All other such matings gave similar progeny ratios. State a *concise* genetic hypothesis that accounts for these results.

9. Over a period of several years, Hans Nachtsheim investigated an inherited anomaly of the white blood cells of rabbits. This anomaly, termed the Pelger anomaly, involves an arrest of the segmentation of the nuclei of certain white cells. This anomaly does not appear to seriously inconvenience the rabbits.

   a. When rabbits showing the typical Pelger anomaly were mated with rabbits from a true-breeding normal stock, Nachtsheim counted 217 offspring showing the Pelger anomaly and 237 normal progeny. What appears to be the genetic basis of the Pelger anomaly?

b. When rabbits with the Pelger anomaly were mated to each other, Nachtsheim found 223 normal progeny, 439 showing the Pelger anomaly, and 39 extremely abnormal progeny. These very abnormal progeny not only had defective white blood cells but also showed severe deformities of the skeletal system; almost all of them died soon after birth. In genetic terms, what do you suppose these extremely defective rabbits represented? Why do you suppose there were only 39 of them?

c. What additional experimental evidence might you collect to support or disprove your answers to part b?

d. In Berlin, about one human in 1000 shows a Pelger anomaly of white blood cells very similar to that described in rabbits. The anomaly is inherited as a simple dominant, but the homozygous type has not been observed in humans. Can you suggest why, if you are permitted an analogy with the condition in rabbits?

e. Again by analogy with rabbits, what phenotypes and genotypes might be expected among the children of a man and woman who both show the Pelger anomaly?

(Problem 16 from A. M. Srb, R. D. Owen, and R. S. Edgar, *General Genetics,* 2d ed. W. H. Freeman and Company, 1965.)

10. Two normal-looking fruit flies were crossed and in the progeny there were 202 females and 98 males.

a. What is unusual about this result?

b. Provide a genetic explanation for this anomaly.

c. Provide a test of your hypothesis.

11. You have been given a virgin *Drosophila* female. You notice that the bristles on her thorax are much shorter than normal. You mate her with a normal male (with long bristles) and obtain the following F_1 progeny: $\frac{1}{3}$ short-bristled females, $\frac{1}{3}$ long-bristled females, and $\frac{1}{3}$ long-bristled males. A cross of the F_1 long-bristled females with their brothers gives only long-bristled F_2. A cross of short-bristled females with their brothers gives $\frac{1}{3}$ short-bristled females, $\frac{1}{3}$ long-bristled females, and $\frac{1}{3}$ long-bristled males. Provide a genetic hypothesis to account for all these results, showing genotypes in every cross.

12. A dominant allele *H* reduces the number of body bristles *Drosophila* flies have, giving rise to a "hairless" phenotype. In the homozygous condition, *H* is lethal. An independently assorting dominant allele *S* has no effect on bristle number except in the presence of *H*, in which case a single dose of *S* suppresses the hairless phenotype, thus restoring the hairy phenotype. However, *S* also is lethal in the homozygous (*SS*) condition.

a. What ratio of hairy to hairless individuals would you find in the live progeny of a cross between two hairy flies both carrying *H* in the suppressed condition?

b. When the hairless progeny are backcrossed with a parental hairy fly, what phenotypic ratio would you expect to find among their live progeny?

13. A pure-breeding strain of squash that produced disk-shaped fruits was crossed with a pure-breeding strain having long fruits. The F_1 had disk fruits, but the F_2 showed a new phenotype, sphere, and was composed of the following proportions:

disk	270
sphere	178
long	32

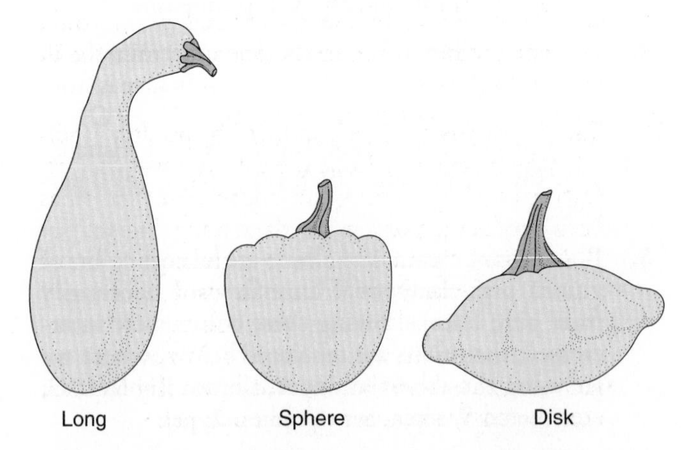

Propose an explanation for these results, and show the genotypes of P, F_1, and F_2 generations.

(Illustration from P. J. Russell, *Genetics,* 3d ed., Harper and Collins, 1992.)

14. Because snapdragons *(Antirrhinum)* possess the pigment anthocyanin, they have reddish purple petals. Two pure anthocyanin-less lines of *Antirrhinum* were developed, one in California and one in Holland. They looked identical in having no red pigment at all, manifested as white (albino) flowers. However, when petals from the two lines were ground up together in buffer in the same test tube,

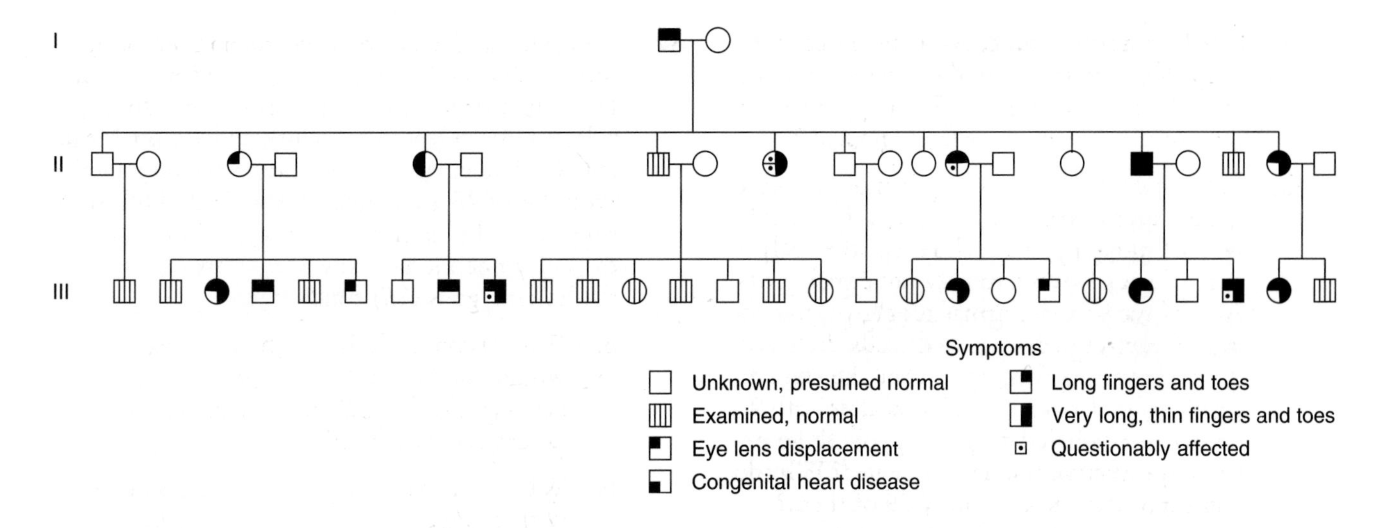

the solution, which appeared colorless at first, gradually turned red.

a. What control experiments should an investigator conduct before proceeding with further analysis.

b. What could account for the production of the red color in the test tube?

c. According to your explanation for part b, what would be the genotypes of the two lines?

d. If the two white lines are crossed, what would you predict would be the phenotypes of the F_1 and F_2?

15. The frizzle fowl is much admired by poultry fanciers. It gets its name from the unusual way that its feathers curl up, giving the impression that it has been (in the memorable words of animal geneticist F. B. Hutt) "pulled backwards through a knothole." Unfortunately, frizzle fowls do not breed true; when two frizzles are intercrossed they always produce 50 percent frizzles, 25 percent normal, and 25 percent with peculiar woolly feathers that soon fall out, leaving the birds naked.

a. Give a genetic explanation for these results, showing genotypes of all phenotypes, and provide a statement of how your explanation works.

b. If you wanted to mass produce frizzle fowls for sale, which types would be best to use as a breeding pair?

16. Marfan's syndrome is a disorder of the fibrous connective tissue, characterized by many symptoms including long, thin digits, eye defects, heart disease, and long limbs. (Flo Hyman, the American volleyball star, suffered from Marfan's syndrome. She died soon after a game from a ruptured aorta.)

a. Use the pedigree shown above to propose a mode of inheritance for Marfan's syndrome.

b. What genetic phenomenon is shown by this pedigree?

c. Speculate on a reason for such a phenomenon.

(Illustration from J. V. Neel and W. J. Schull, *Human Heredity*. University of Chicago Press, 1954.)

17. A snapdragon plant that bred true for white petals was crossed to a plant that bred true for purple petals, and all of the F_1 had white petals. The F_1 was selfed. Among the F_2, three phenotypes were observed in the following numbers:

white	240
solid purple	61
spotted purple	19
total	320

a. Propose an explanation of these results, showing genotypes of all generations (make up and explain your symbols).

b. A white F_2 plant was crossed to a solid purple F_2 plant, and the progeny were:

white	50%
solid purple	25%
spotted purple	25%

What were the genotypes of the F_2 plants crossed?

18. Most flour beetles are black, but several color variants are known. Crosses of pure-breeding parents produced the following results in the F_1 generation, and intercrossing the F_1 from each cross gave the ratios shown for the F_2 generation. The phenotypes are abbreviated Bl, black; Br, brown; Y, yellow; and W, white.

Cross	Parents	F_1	F_2
1	Br × Y	Br	3 Br:1 Y
2	Bl × Br	Bl	3 Bl:1 Br
3	Bl × Y	Bl	3 Bl:1 Y
4	W × Y	Bl	9 Bl:3 Y:4 W
5	W × Br	Bl	9 Bl:3 Br:4 W
6	Bl × W	Bl	9 Bl:3 Y:4 W

a. From these results deduce and explain the inheritance of these colors.

b. Write out the genotypes of each of the parents, the F_1, and the F_2 in all crosses.

19. Two albinos marry and have four normal children. How is this possible?

***20.** Plant breeders obtained three differently derived pure lines of white-flowered *Petunia* plants. They performed crosses and observed progeny phenotypes as follows:

Cross	Parents	Progeny
1	line 1 × line 2	F_1 all white
2	line 1 × line 3	F_1 all red
3	line 2 × line 3	F_1 all white
4	red F_1 × line 1	$\frac{1}{4}$ red:$\frac{3}{4}$ white
5	red F_1 × line 2	$\frac{1}{8}$ red:$\frac{7}{8}$ white
6	red F_1 × line 3	$\frac{1}{2}$ red:$\frac{1}{2}$ white

a. Explain these results, using gene symbols of your own choosing. (Show parental and progeny genotypes in each cross.)

b. If a red F_1 individual from cross 2 is crossed to a white F_1 individual from cross 3, what proportion of progeny will be red?

21. Consider production of flower color in the Japanese morning glory *(Pharbitis nil)*. Dominant alleles of either of two separate genes ($A - bb$ or $aaB -$) produce purple petals. $A - B -$ produces blue petals, and $aabb$ produces scarlet petals. Deduce the genotypes of parents and progeny in the following crosses:

Cross	Parents	Progeny
1	blue × scarlet	$\frac{1}{4}$ blue:$\frac{1}{2}$ purple:$\frac{1}{4}$ scarlet
2	purple × purple	$\frac{1}{4}$ blue:$\frac{1}{2}$ purple:$\frac{1}{4}$ scarlet
3	blue × blue	$\frac{3}{4}$ blue:$\frac{1}{4}$ purple
4	blue × purple	$\frac{3}{8}$ blue:$\frac{4}{8}$ purple:$\frac{1}{8}$ scarlet
5	purple × scarlet	$\frac{1}{2}$ purple:$\frac{1}{2}$ scarlet

22. Corn breeders obtained pure lines whose kernels turn sun-red, pink, scarlet, or orange when exposed to sunlight (normal kernels remain yellow in sunlight). Some crosses between these lines produced the following results. The phenotypes are abbreviated O, orange; P, pink; Sc, scarlet; and SR, sun-red.

	Phenotypes		
Cross	Parents	F_1	F_2
1	SR × P	all SR	66 SR:20 P
2	O × SR	all SR	998 SR:314 O
3	O × P	all O	1300 O:429 P
4	O × Sc	all Y	182 Y:80 O:58 Sc

Analyze the results of each cross, and provide a unifying hypothesis to account for *all* the results. (Explain all symbols you use.)

23. Many kinds of wild animals have the agouti coloring pattern, in which each hair has a yellow band around it (see Figure 4-10).

a. Black mice and other black animals do not have the yellow band; each of their hairs is all black. This absence of wild agouti pattern is called nonagouti. When mice of a true-breeding agouti line are crossed with nonagoutis, the F_1 is all agouti and the F_2 has a 3:1 ratio of agoutis to nonagoutis. Diagram this cross, letting A represent the allele responsible for the agouti phenotype and a, nonagouti. Show the phenotypes and genotypes of the parents, their gametes, the F_1, their gametes, and the F_2.

b. Another inherited color deviation in mice substitutes brown for the black color in the wild-type hair. Such brown-agouti mice are called cinnamons. When wild-type mice are crossed with cinnamons, the F_1 is all wild type and the F_2 has a 3:1 ratio of wild type to cinnamon. Diagram this cross as in part a, letting B stand for the wild-type black allele and b stand for the cinnamon brown allele.

c. When mice of a true-breeding cinnamon line are crossed with mice of a true-breeding nonagouti (black) line, the F_1 is all wild type. Use a genetic diagram to explain this result.

d. In the F_2 of the cross in part c, a fourth color called chocolate appears in addition to the parental cinnamon and nonagouti and the wild type of the F_1. Chocolate mice have a solid, rich-brown color. What is the genetic constitution of the chocolates?

e. Assuming that the Aa and Bb allelic pairs assort independently of each other, what would you expect to be the relative frequencies of the four color types in the F_2 described in part d? Diagram the cross of parts c and d, showing phenotypes and genotypes (including gametes).

f. What phenotypes would be observed in what proportions in the progeny of a backcross of F_1 mice from part c to the cinnamon parental stock? To the nonagouti (black) parental stock? Diagram these backcrosses.

g. Diagram a testcross for the F_1 of part c. What colors would result, and in what proportions?

h. Albino (pink-eyed white) mice are homozygous for the recessive member of an allelic pair Cc which assorts independently of the Aa and Bb pairs. Suppose that you have four different highly inbred (and therefore presumably homozygous) albino lines. You cross each of these lines with a true-breeding wild-type line, and you raise a large F_2 progeny from each cross. What genotypes for the albino lines would you deduce from the F_2 phenotypes shown below?

F_2 of line	Phenotypes of progeny: Wild type	Black	Cinnamon	Chocolate	Albino
1	87	0	32	0	39
2	62	0	0	0	18
3	96	30	0	0	41
4	287	86	92	29	164

(Problem 23 adapted from A. M. Srb, R. D. Owen, and R. S. Edgar, *General Genetics*, 2d ed. W. H. Freeman and Company, 1965.)

24. An allele A that is not lethal when homozygous causes rats to have yellow coats. The allele R of a separate gene that assorts independently produces a black coat. Together, A and R produce a grayish coat, whereas a and r produce a white coat. A gray male is crossed with a yellow female, and the F_1 is $\frac{3}{8}$ yellow, $\frac{3}{8}$ gray, $\frac{1}{8}$ black, and $\frac{1}{8}$ white. Determine the genotypes of the parents.

25. The genotype $rr\,pp$ gives fowl a single comb, $R-P-$ gives a walnut comb, $rr\,P-$ gives a pea comb, and $R-pp$ gives a rose comb.

a. What comb types will appear in the F_1 and in the F_2 in what proportions if single-combed birds are crossed with birds of a true-breeding walnut strain?

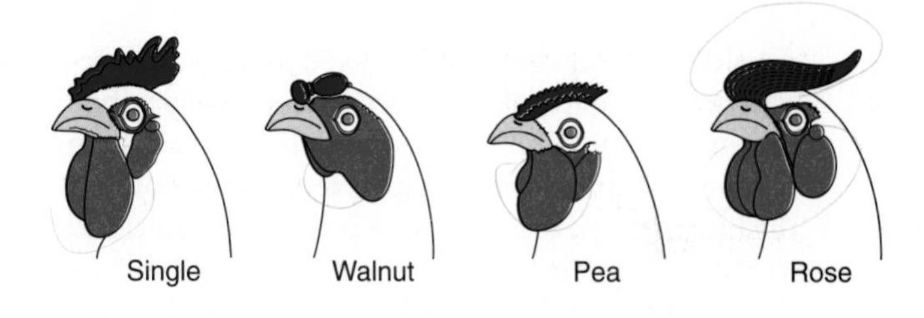

b. What are the genotypes of the parents in a walnut × rose mating from which the progeny are $\frac{3}{8}$ rose, $\frac{3}{8}$ walnut, $\frac{1}{8}$ pea, and $\frac{1}{8}$ single?

c. What are the genotypes of the parents in a walnut × rose mating from which all the progeny are walnut?

d. How many genotypes produce a walnut phenotype? Write them out.

26. The production of eye color pigment in *Drosophila* requires the dominant allele A. The dominant allele P of a second independent gene turns the pigment to purple but its recessive allele leaves it red. A fly producing no pigment has white eyes. Two pure lines were crossed with the following results:

P red-eyed female × white-eyed male

↓

F_1 Purple-eyed females
red-eyed males
$F_1 \times F_1$

↓

F_2 Both males and females: $\frac{3}{8}$ purple-eyed
$\frac{3}{8}$ red-eyed
$\frac{2}{8}$ white-eyed

Explain this mode of inheritance and show the genotypes of the parents, the F_1, and the F_2.

27. When true-breeding brown dogs are mated with certain true-breeding white dogs, all the F_1 pups are white. The F_2 progeny from some $F_1 \times F_1$ crosses were 118 white, 32 black, and 10 brown pups. What is the genetic basis for these results?

28. In corn, three dominant alleles, called A, C, and R, must be present to produce colored seeds. Genotypes $A-C-R-$ are colored: all others are colorless. A colored plant is crossed with three tester plants of known genotype. With tester $aa\,cc\,RR$, the colored plant produces 50 percent colored seeds; with $aa\,CC\,rr$, it produces 25 percent colored; and with $AA\,cc\,rr$, it produces 50 percent colored. What is the genotype of the colored plant?

29. The production of pigment in the outer layer of seeds of corn requires each of the three independently assorting genes A, C, and R to be represented by at least one dominant allele, as specified in problem 39. The dominant allele Pr of a fourth independently assorting gene is required to convert the biochemical precursor to a purple pigment, and its recessive allele pr makes the pigment red. Plants that do not produce pigment have yellow seeds. Consider a cross of a strain of genotype $AA\,CC\,RR\,pr\,pr$ with a strain of genotype $aa\,cc\,rr\,Pr\,Pr$.

- **a.** What are the phenotypes of the parents?
- **b.** What will be the phenotype of the F_1?
- **c.** What phenotypes, and in what proportions, will appear in the progeny of a selfed F_1?
- **d.** What progeny proportions do you predict from the test cross of an F_1?

30. Wild-type strains of the haploid fungus *Neurospora* can make their own tryptophan. An abnormal allele *td* renders the fungus incapable of making its own tryptophan. An individual of genotype *td* grows only when its medium supplies tryptophan. The allele *su* assorts independently of *td*; its only known effect is to suppress the *td* phenotype. Therefore, strains carrying both *td* and *su* do not require tryptophan for growth.

- **a.** If a *td su* strain is crossed with a genotypically wild-type strain, what genotypes are expected in the progeny, and in what proportions?
- **b.** What will be the ratio of tryptophan-dependent to tryptophan-independent progeny in the cross of part a?

31. The allele B gives mice a black coat, and b gives a brown one. The genotype ee for another, independently assorting gene prevents expression of B and b, making the coat color beige, whereas $E-$ permits expression of B and b. Both genes are autosomal. In the following pedigree, black symbols indicate a black coat, pink symbols indicate brown, and white symbols indicate beige.

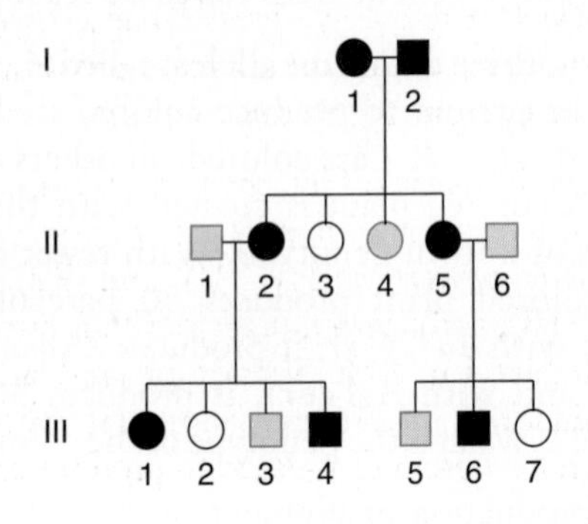

- **a.** What is the name given to the type of gene interaction in this example?
- **b.** What are the genotypes of the individuals in the pedigree? (If there are alternative possibilities, state them.)

32. Mice of the genotypes $AA\,BB\,CC\,DD\,SS$ and $aa\,bb\,cc\,dd\,ss$ are crossed. (These gene symbols are explained in the text of this chapter.) The progeny are intercrossed. What phenotypes will be produced in the F_2, and in what proportions?

33. Consider the genotypes of two lines of chickens: the pure-line Mottled Hondan is $ii\,DD\,MM\,WW$ and the pure-line Leghorn is $II\,dd\,mm\,ww$, where:

I = white feathers, i = colored feathers

D = duplex comb, d = simplex comb

M = bearded, m = beardless

W = white skin, w = yellow skin

These four genes assort independently. Starting with these two pure lines, what is the fastest and most convenient way of generating a pure line of birds that has colored feathers, simplex comb, yellow skin, and is beardless? Make sure you show:

- **a.** the breeding pedigree;
- **b.** the genotype of each animal represented;
- **c.** how many eggs to hatch in each cross, and why this number;
- **d.** why your scheme is the fastest and most convenient.

34. The following pedigree is for a dominant phenotype governed by an autosomal gene. What does this pedigree suggest about the phenotype, and what can you deduce about the genotype of individual A?

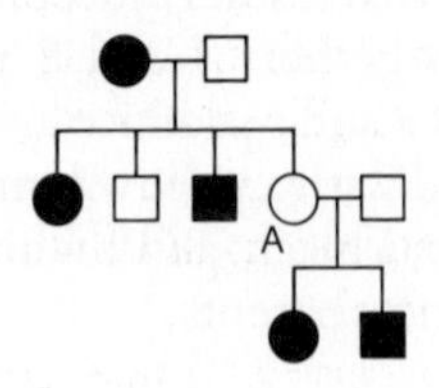

35. The genetic determination of petal coloration in foxgloves is given in Figure 4-19. Consider the following two crosses.

Cross	Parents	Progeny
1	dark reddish × white with yellowish spots	$\frac{1}{2}$ dark reddish: $\frac{1}{2}$ light reddish
2	white with yellowish spots × light reddish	$\frac{1}{2}$ white with reddish spots:$\frac{1}{4}$ dark reddish:$\frac{1}{4}$ light reddish

In each case, give the genotypes of parents and progeny with respect to the three genes.

36. A researcher crosses two white-flowered lines of *Antirrhinum plants* as follows and obtains the following results:

pure line 1 × pure line 2
↓
F_1 all white
$F_1 \times F_1$
↓
F_2 131 white
29 red

a. Deduce the inheritance of these phenotypes, using clearly defined gene symbols. Give the genotypes of the parents, F_1, and F_2.

b. Predict the outcome of crosses of the F_1 to each parental line.

37. In this problem, it is assumed that you know that enzymes are coded by genes—a topic not formally treated until Chapter 12. However, it is a good problem to try here, because it makes a strong connection between gene interactions and cell chemistry.

Assume that two pigments, red and blue, mix to give the normal purple color of petunia petals. Separate biochemical pathways synthesize the two pigments as shown in the top two rows of the diagram below. "White" refers to compounds that are not pigments. (Total lack of pigment results in a white petal.) Red pigment forms from a yellow intermediate that normally is at a concentration too low to color petals.

A third pathway whose compounds do not contribute pigment to petals normally does not affect the blue and red pathways, but if one of its intermediates ($white_3$) should build up in concentration, it can be converted to the yellow intermediate of the red pathway.

In the diagram, A to E represent enzymes; their corresponding genes, all of which are unlinked, may be symbolized by the same letters.

Pathway I $\cdots \longrightarrow white_1 \xrightarrow{E} blue$

Pathway II $\cdots \longrightarrow white_2 \xrightarrow{A} yellow \xrightarrow{B} red$

(C: $white_3$ → yellow)

Pathway III $\cdots \longrightarrow white_3 \xrightarrow{D} white_4$

Assume that wild-type alleles are dominant and code for enzyme function, and that recessive alleles represent lack of enzyme function. Deduce which combinations of true-breeding parental genotypes could be crossed to produce F_2 progenies in the following ratios:

a. 9 purple:3 green:4 blue

b. 9 purple:3 red:3 blue:1 white

c. 13 purple:3 blue

d. 9 purple:3 red:3 green:1 yellow

(NOTE: blue mixed with yellow makes green; assume that no mutations are lethal.)

38. The flowers of nasturtiums *(Tropaeolum majus)* may be single (S), double (D), or superdouble (Sd). Superdoubles are female sterile; they originated from a double-flowered variety. Crosses between varieties gave the progenies as listed in the following table, where pure means pure-breeding.

Cross	Parents	Progeny
1	Pure S × pure D	All S
2	Cross 1 F_1 × Cross 1 F_1	78 S:27 D
3	Pure D × Sd	112 Sd:108 D
4	Pure S × Sd	8 Sd:7 S
5	Pure D × Cross 4 Sd progeny	18 Sd:19 S
6	Pure D × Cross 4 S progeny	14 D:16 S

Using your own genetic symbols, propose an explanation for the above results, showing **(a)** all the genotypes in each of the six rows above and **(b)** the proposed origin of the superdouble.

***39.** In a certain species of fly, the normal eye color is red (R). Four abnormal phenotypes for eye color were found: two were yellow (Y1 and Y2), one was brown (B), and one was orange (O). A pure line was established for each phenotype and all possible combinations of the pure lines were crossed. Flies of each F_1 were intercrossed to produce an F_2. The

F_1s and F_2s are shown within the square below; the pure lines are given in the margins.

		Y1	Y2	B	O
Y1	F_1	all y	all r	all r	all r
	F_2	all y	9 r 7 y	9 r 4 y 3 b	9 r 4 o 3 y
Y2	F_1		all y	all r	all r
	F_2		all y	9 r 4 y 3 b	9 r 4 y 3 o
B	F_1			all b	all r
	F_2			all b	9 r 4 o 3 b
O	F_1				all o
	F_2				all o

a. Define your own symbols and show genotypes of all four pure lines.

b. Show how the F_1 phenotypes and the F_2 ratios are produced.

c. If you understand what a biochemical pathway is, show a biochemical pathway that explains the genetic results, indicating which gene controls which enzyme.

40. In common wheat, *Triticum aestivum,* kernel color is determined by multiply duplicated genes, each with a *R* and a *r* allele. Any number of *R* alleles will give red, and the complete lack of *R* alleles will give the white phenotype. In one cross between a red pure line and a white pure line, the F_2 was $\frac{63}{64}$ red and $\frac{1}{64}$ white.

a. How many *R* genes are segregating in this system?

b. Show genotypes of the parents, the F_1, and the F_2.

c. Different F_2 plants are backcrossed to the white parent. Give examples of genotypes that would give the following progeny ratios in such backcrosses, (1) 1 red:1 white, (2) 3 red:1 white, (3) 7 red:1 white.

***d.** What is the formula that generally relates the number of segregating genes to the proportion of red individuals in the F_2 in such systems?

41. The following pedigree shows the inheritance of deaf-mutism.

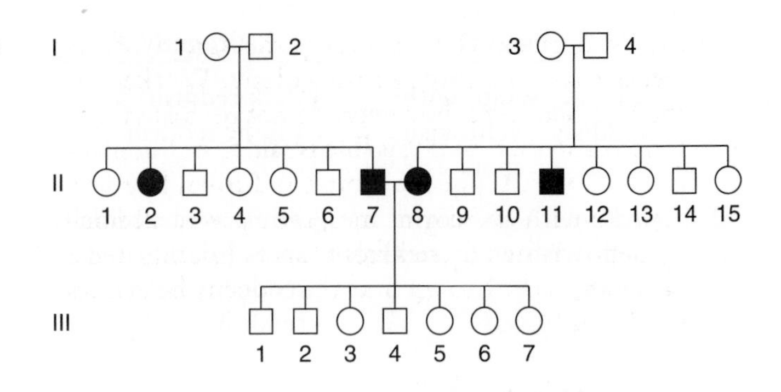

a. Provide an explanation for the inheritance of this rare condition in the two families in generations I and II, showing genotypes of as many individuals as possible using symbols of your own choosing.

b. Provide an explanation for the production of only normal individuals in generation III, making sure your explanation is compatible with the answer to part a.

42. The pedigree shown below is for blue sclera (bluish thin outer wall to the eye) and brittle bones:

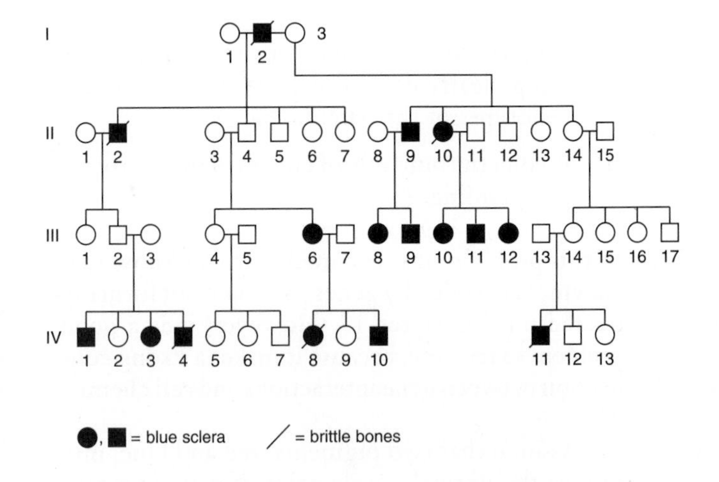

a. Are these two abnormalities caused by the same gene or separate genes? State your reasons clearly.

b. Is the gene or genes autosomal or sex-linked?

c. Does the pedigree show any evidence of incomplete penetrance or expressivity? If so, make the best calculations that you can of these measures.

43. Workers of the honeybee line known as Brown (nothing to do with color) show what is called "hygienic behavior," that is, they uncap hive compartments containing dead pupae, and then remove the dead pupae. This prevents the spread of infectious bacteria through the colony. Workers of the Van Scoy line, however, do not perform these actions and therefore this line is said to be "nonhygienic." When a queen from the Brown line was mated with Van Scoy drones, the F_1 were all nonhygienic. When drones from this F_1 inseminated a queen from the Brown line, the progeny behaviors were as follows:

$\frac{1}{4}$ hygienic

$\frac{1}{4}$ uncapping but no removing of pupae

$\frac{1}{2}$ nonhygienic

However, when the nonhygienic individuals were examined further, it was found that if the compartment of dead pupae were uncapped by the beekeeper, then about half of the individuals removed the dead pupae, but the other half did not.

a. Propose a genetic hypothesis to explain these behavioral patterns.

b. Discuss the data in terms of epistasis, dominance, and environmental interaction.

[NOTE: Workers are sterile, and all bees from one line carry the same alleles.]

5

Linkage I: Basic Eukaryotic Chromosome Mapping

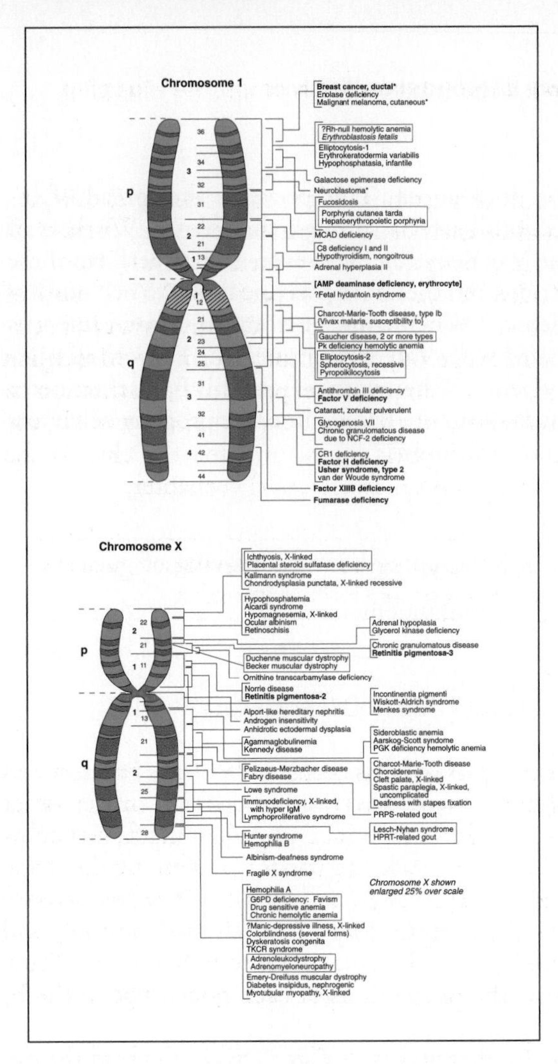

Map of the human chromosome 1 and the X chromosome. The genes have been positioned using several techniques, including those covered in this chapter and the next. (From the *Journal of NIH Research,* 1992.)

KEY CONCEPTS

Two genes close together on the same chromosome pair do not assort independently at meiosis.

Recombination produces genotypes with new combinations of parental alleles.

A pair of homologous chromosomes can exchange parts by crossing-over.

Recombination results from either independent assortment or crossing-over.

Gene loci on a chromosome can be mapped by measuring the frequencies of recombinants produced by crossing-over.

Interlocus map distances based on recombination measurements are roughly additive.

The occurrence of a crossover can influence the occurrence of a second crossover in an adjacent region.

We have already established the basic principles of segregation and assortment, and we have correlated them with chromosome behavior during meiosis. Thus, from the cross *AaBb* × *AaBb*, we expect a 9:3:3:1 ratio of phenotypes. As we learned from Bridges's study of nondisjunction (page 64), exceptions to simple Mendelian expectations can direct the experimenter's attention to new discoveries. Just such an exception observed in the progeny of a dihybrid cross provided the clue to the important concepts discussed in this chapter.

Message In genetic analysis, exceptions to predicted behavior often give important new insights.

The Discovery of Linkage

In the early years of this century, William Bateson and R. C. Punnett were studying inheritance in the sweet pea. They studied two genes: one affecting flower color (*P*, purple, and *p*, red), and the other affecting the shape of pollen grains (*L*, long, and *l*, round). They crossed pure lines *PPLL* (purple, long) × *ppll* (red, round), and selfed the F_1 *PpLl* heterozygotes to obtain an F_2. Table 5-1 shows the proportions of each phenotype in the F_2 plants.

The F_2 phenotypes deviated strikingly from the expected 9:3:3:1 ratio. What is going on? This does not appear to be explainable as a modified Mendelian ratio. Note that two phenotypic classes are larger than expected: the purple, long phenotype and the red, round phenotype. As a possible explanation for this, Bateson and Punnett proposed that the F_1 had actually produced more *PL* and *pl* gametes than would be produced by Mendelian independent assortment. Because these were the gametic types in the original pure lines, the researchers thought that physical **coupling** between the dominant alleles *P* and *L* and between the recessive alleles *p* and *l* might have prevented their independent assortment in the F_1. However, the researchers did not know what the nature of this coupling could be.

Table 5-1 Sweet Pea Phenotypes Observed in the F_2 by Bateson and Punnett

Phenotype (and genotype)	Number of Progeny	
	Observed	Expected from 9:3:3:1 Ratio
Purple, long $(P-L-)$	284	215
Purple, round $(P-ll)$	21	71
Red, long $(ppL-)$	21	71
Red, round $(ppll)$	55	24
	381	381

The confirmation of Bateson and Punnett's hypothesis had to await the development of *Drosophila* as a genetic tool. After the idea of coupling was first proposed, Thomas Hunt Morgan found a similar deviation from Mendel's second law while studying two autosomal genes in *Drosophila.* One of these genes affects eye color (*pr*, purple, and pr^+, red), and the other affects wing length (*vg*, vestigial, and vg^+, normal). The wild-type alleles of both genes are dominant. Morgan crossed *prpr vgvg* flies with $pr^+pr^+\ vg^+vg^+$ and then testcrossed the doubly heterozygous F_1 females: $pr^+pr\ vg^+vg$ ♀ × *prpr vgvg* ♂.

The use of the testcross is extremely important. Because one parent (the tester) contributes gametes carrying only recessive alleles, the phenotypes of the offspring reveal the gametic contribution of the other, doubly heterozygous parent. Hence, the analyst can concentrate on meiosis in one parent and forget about the other. This contrasts with the analysis of progeny from an F_1 self, where there are two sets of meioses to consider: one in the male parent and one in the female. Morgan's results follow; the alleles contributed by the F_1 female specify the F_2 classes:

$pr^+\ vg^+$	1339
$pr\ vg$	1195
$pr^+\ vg$	151
$pr\ vg^+$	154
	2839

Obviously, these numbers deviate drastically from the Mendelian prediction of a 1:1:1:1 ratio, and they indicate a coupling of genes. The two largest classes are the combinations $pr^+\ vg^+$ and *pr vg*, originally introduced by the homozygous parental flies. You can see that the testcross clarifies the situation. It directly reveals the allelic combinations in the gametes from one sex in the F_1, thus clearly showing the coupling that could only be inferred from Bateson and Punnett's F_1 self. The testcross also reveals something new: there is approximately a 1:1 ratio between the two parental types and also between the two nonparental types.

Now let us consider what may be learned by repeating the crossing experiments but changing the combinations of alleles contributed as gametes by the homozygous parents in the first cross. In this cross, each parent was homozygous for one dominant allele and for one recessive allele. Again F_1 females were testcrossed:

P $pr^+pr^+\ vgvg$ × $prpr\ vg^+vg^+$

↓

F₁ $pr^+pr\ vg^+vg$

$pr^+pr\ vg^+vg$ ♀ × *prpr vgvg* ♂

The following progeny were obtained from the testcross:

$pr^+ vg^+$	157
$pr\ vg$	146
$pr^+ vg$	965
$pr\ vg^+$	1067
	2335

Again, these results are not even close to a 1:1:1:1 Mendelian ratio. Now, however, the largest classes are those that have one dominant allele or the other rather than, as before, two dominant alleles or two recessives. But notice that once again the allelic combinations that were originally contributed to the F_1 by the parental flies provide the most frequent classes in the testcross progeny. In the early work on coupling, Bateson and Punnett coined the term **repulsion** to describe this situation, because it seemed to them that in this case the nonallelic dominant alleles "repelled" each other — the opposite of the situation in coupling, where the dominant alleles seemed to "stick together." How can we explain these two phenomena: coupling and repulsion?

Morgan suggested that the genes governing both phenotypes are located *on the same pair of homologous chromosomes.* Thus, when *pr* and *vg* are introduced from one parent, they are physically located on the same chromosome, whereas pr^+ and vg^+ are on the homologous chromosome from the other parent (Figure 5-1). This hypothesis also explains repulsion. In that case, one parental chromosome carries *pr* and vg^+ and the other carries pr^+ and *vg*. Repulsion, then, is just another case of coupling: in this case, the dominant allele of one gene is coupled with the recessive allele of the other gene. This hypothesis explains why allelic combinations from P remain together, but how do we explain the appearance of nonparental combinations?

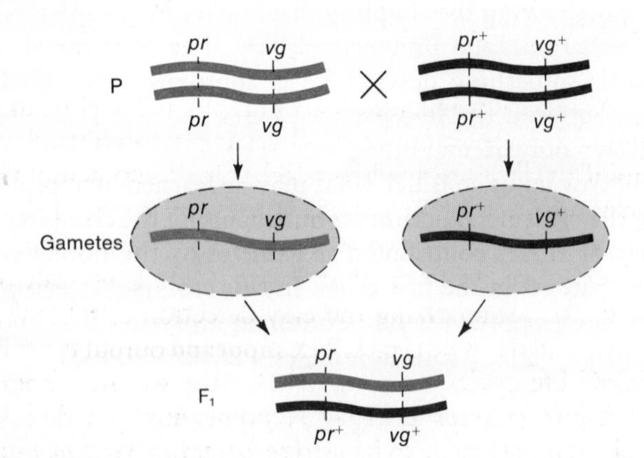

Figure 5-1 Simple inheritance of two pairs of alleles located on the same chromosome pair.

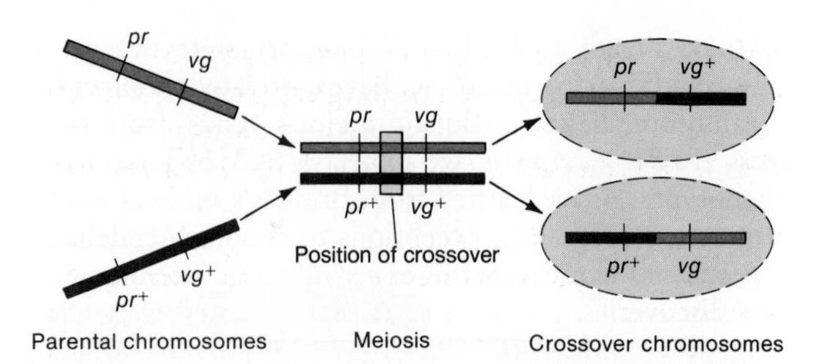

Figure 5-2 Crossing-over during meiosis. An individual receives one homolog from each parent. The exchange of parts by crossing-over may produce gametic chromosomes whose allelic combinations differ from the parental combinations.

Morgan suggested that when homologous chromosomes pair during meiosis, the chromosomes occasionally exchange parts during a process called crossing-over. Figure 5-2 illustrates this physical exchange of chromosome segments. The original arrangements of alleles on the two chromosomes are called the **parental** combinations. The two new combinations are called **crossover products** or **recombinants.**

Morgan's hypothesis that homologs may exchange parts may seem a bit farfetched. Is there any cytologically observable process that could account for crossing-over? We saw in Chapter 3 that during meiosis, when duplicated homologous chromosomes pair with each other, two nonsister chromatids often appear to cross each other. This is diagrammed in Figure 5-3. Recall that the resulting cross-shaped structure is called a chiasma. To Morgan, the appearance of the chiasmata visually corroborated the concept of crossing-over. (Note that the chiasmata seem to indicate that it is chromatids, not unduplicated chromosomes, that cross over. We shall return to this point later.) For the present, let's accept Morgan's interpretation that chiasmata are the cytological counterparts of crossovers and leave the proof until Chapter 19. Note that Morgan did not arrive at this interpretation out of nowhere; he was looking for a *phys-*

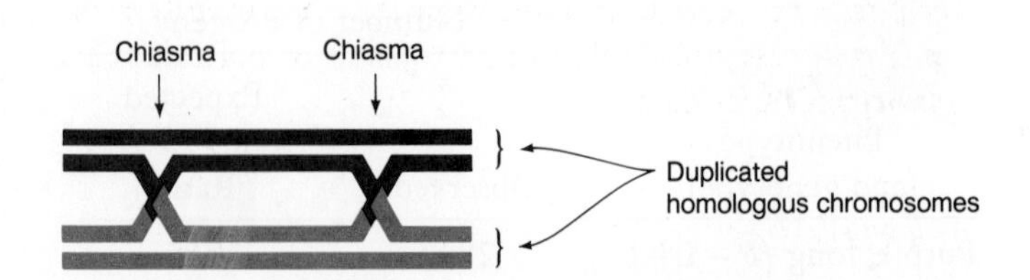

Figure 5-3 Diagrammatic representation of chiasmata at meiosis. Each line represents a chromatid of a pair of synapsed chromosomes.

ical explanation for his *genetic* results. His achievement in correlating the results of breeding experiments with cytological phenomena thus emphasizes the importance of the chromosome theory as a powerful basis for research.

Message Chiasmata are the visible manifestations of crossovers.

Data like those just presented, showing coupling and repulsion in testcrosses and in F_1 selfs, are commonly encountered in genetics. Clearly, results of this kind represent a departure from independent assortment. Such exceptions, in fact, constitute a major addition to Mendel's view of the genetic world.

Message When two genes are close together on the same chromosome pair, they do not assort independently.

The residing of genes on the same chromosome pair is termed **linkage.** Two genes on the same chromosome pair are said to be linked. It is also proper to refer to the linkage of specific alleles: for example, in one $AaBb$ individual, A might be linked to b; a would then of necessity be linked to B. These terms graphically allude to the existence of a physical entity linking the genes—that is, the chromosome itself! You may wonder why we refer to such genes as "linked" rather than "coupled"; the answer is that the words coupling and repulsion are now used to indicate two different types of linkage conformation in a double heterozygote, as follows:

Coupling conformation $\dfrac{pr \quad vg}{pr^+ \quad vg^+}$

Repulsion conformation $\dfrac{pr \quad vg^+}{pr^+ \quad vg}$

In other words, coupling refers to the linkage of two dominant or two recessive alleles, whereas repulsion indicates that dominant alleles are linked with recessive alleles. To ascertain whether a double heterozygote is in coupling or repulsion conformation, an investigator must testcross the double heterozygote or consider the genotypes of its parents.

Recombination

We have already introduced the term recombination. This term is widely used in many areas of practical and theoretical genetics, so it is absolutely necessary to be clear about its meaning at this stage. Recombination has been observed in a variety of situations in addition to meiosis, but for the present, let's define it in relation to meiosis. **Meiotic recombination** is any meiotic process that generates a haploid product with a genotype that differs from both haploid genotypes that constituted the meiotic diploid cell. The product of meiosis so generated is called a **recombinant.** This definition makes the important point that we detect recombination by comparing the *output* genotypes of meiosis and the *input* genotypes (Figure 5-4). The input genotypes are the two haploid genotypes that combined to make the genetic constitution of the meiocyte, the diploid cell that undergoes meiosis.

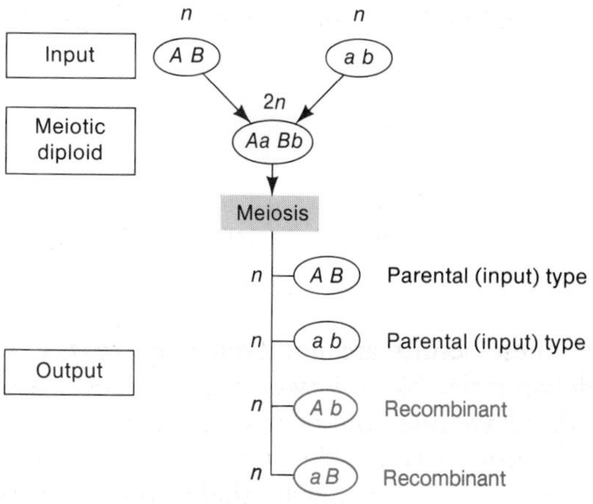

Figure 5-4 Recombinants are those products of meiosis with allelic combinations different from those of the haploid cells that formed the meiotic diploid.

Message During meiosis recombination generates haploid genotypes differing from the haploid parental genotypes.

Meiotic recombination is a part of both haploid and diploid life cycles; however, detecting recombinants in haploid cycles is easy whereas detecting them in diploid cycles takes work. The input and output types in haploid cycles are the genotypes of individuals and may thus be inferred directly from phenotypes. Figure 5-4 can be viewed as summarizing the easy detection of recombinants in haploid life cycles. The input and output types in diploid life cycles are gametes. Because we must know the input gametes to detect recombinants in a diploid cycle, it is preferable to have pure-breeding parents. Furthermore, we cannot detect recombinant output gametes directly: we must testcross the diploid individual and

Interchromosomal

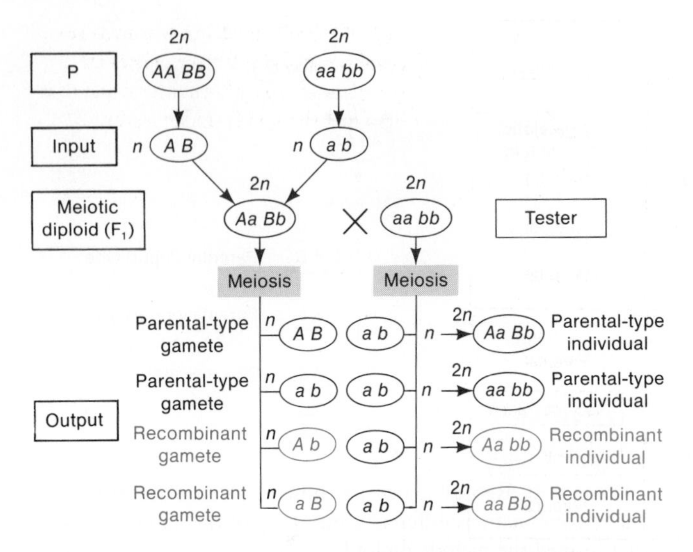

Figure 5-5 The detection of recombination in diploid organisms. Note that Figure 5-4 is actually a part of this figure. Recombinant products of a diploid meiosis are most readily detected in a cross of a heterozygote to a recessive tester.

observe its progeny (Figure 5-5). If a testcross offspring is shown to have been constituted from a recombinant product of meiosis, it too is called a recombinant. Notice again that the testcross allows us to concentrate on *one* meiosis and avoid ambiguity. From a self of the F_1 in Figure 5-5, for example, a recombinant *A A B b* offspring cannot be distinguished from *AA BB* without further crosses.

There are two kinds of recombination, and they produce recombinants by completely different methods. But a recombinant is a recombinant, so how can we decide which type of recombination accounts for any particular progeny? The answer lies in the *frequency* of recombinants, as we shall soon see. First, though, let's consider the two types of recombination: interchromosomal and intrachromosomal.

Interchromosomal Recombination

Mendelian independent assortment brings about **interchromosomal recombination** (Figure 5-6). In a testcross the two recombinant classes always make up 50 percent of the progeny; that is, there are 25 percent of each recombinant type among the progeny. If we observe this frequency, we can infer that the two genes under study assort independently.

The simplest interpretation of these frequencies is that the two genes are on separate chromosome pairs. However, as has already been hinted, genes that are far apart on the *same* chromosome pair can act virtually independently and produce the same result. In Chapter 6, we shall see how a large amount of crossing-over effectively unlinks distantly linked genes.

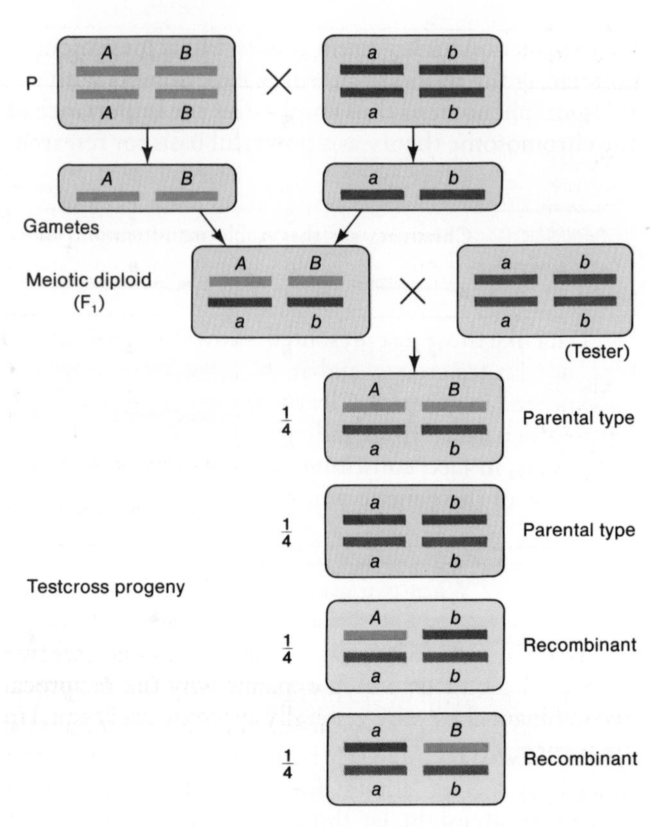

Figure 5-6 Interchromosomal recombination, which always produces a recombinant frequency of 50 percent. This diagram shows two chromosome pairs of a diploid organism with *A* and *a* on one pair and *B* and *b* on the other. Note that we could represent the haploid situation by removing the part marked P and the testcross.

Intrachromosomal Recombination

Crossing-over produces **intrachromosomal recombination.** Any two nonsister chromatids can cross over. (We shall show proof of this in Chapter 6.) Of course, there is not a crossover between two specific genes in all meioses, but when there is, one-half of the products of *that* meiosis are recombinant, as shown in Figure 5-7. Meiosis with no crossover between the genes under study produce only parental genotypes for these genes.

The sign of intrachromosomal recombination is a recombinant frequency of less than 50 percent. The physical linkage of parental gene combinations prevents the independent assortment of two genes that generates the recombinant frequency of 50 percent (Figure 5-8). We saw an example of this situation in Morgan's data (page 119), where the recombinant frequency was $(151 + 154) \div 2839 = 10.7$ percent. This is obviously much less than the 50 percent we would expect with independent assortment. What about recombinant frequencies greater than 50 percent? The answer is that such frequencies are *never* observed, as we shall see in Chapter 6.

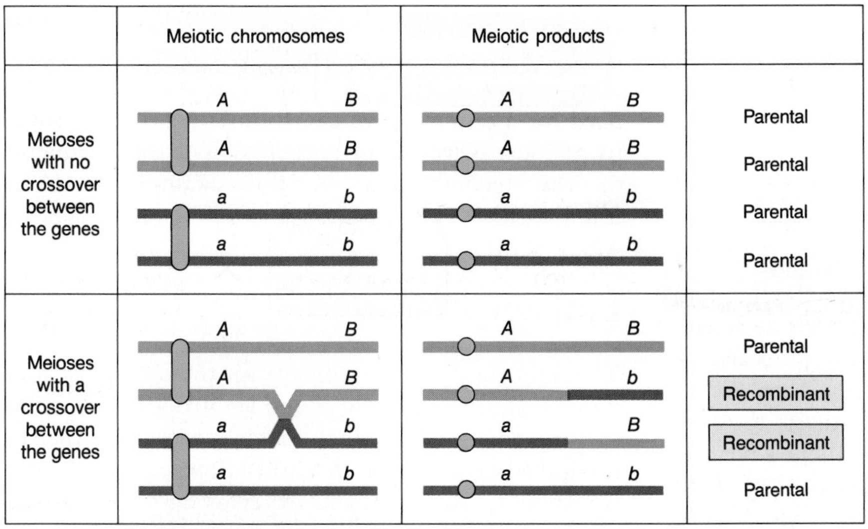

	Meiotic chromosomes	Meiotic products	
Meioses with no crossover between the genes	A B / A B / a b / a b	A B	Parental
		A B	Parental
		a b	Parental
		a b	Parental
Meioses with a crossover between the genes	A B / A B / a b / a b	A B	Parental
		A b	Recombinant
		a B	Recombinant
		a b	Parental

Figure 5-7 Intrachromosomal recombinants arise from meioses in which nonsister chromatids cross over between the genes under study.

Note in Figure 5-7 that crossing-over generates two reciprocal products, which explains why the reciprocal recombinant classes are generally approximately equal in frequency.

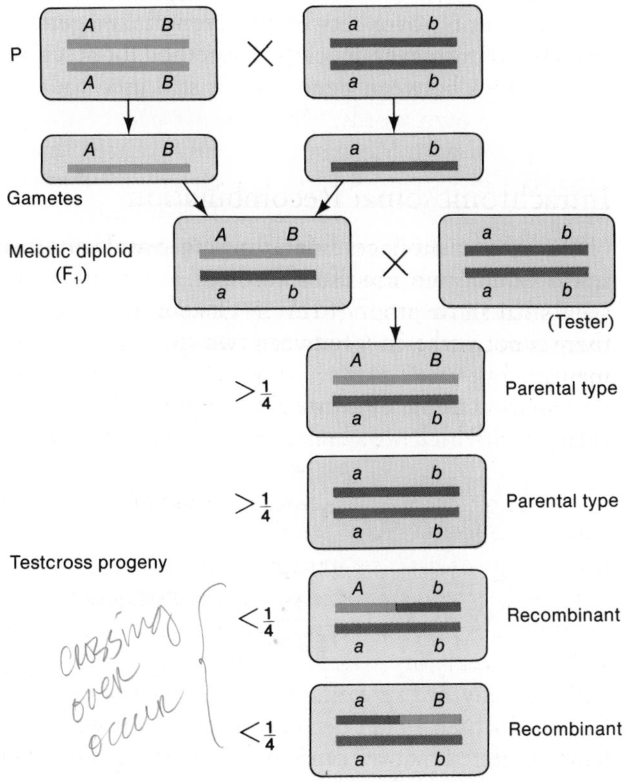

Figure 5-8 Intrachromosomal recombination. Notice that the frequencies of the recombinants add up to less than 50 percent.

Message A recombinant frequency significantly less than 50 percent shows that the genes are linked. A recombinant frequency of 50 percent generally means that the genes are unlinked on separate chromosomes.

The remainder of this chapter deals with intrachromosomal recombination and its causative crossovers.

Linkage Symbolism

Our symbolism for describing crosses becomes cumbersome with the introduction of linkage. We can depict the genetic constitution of each chromosome in the *Drosophila* cross as in the following example:

$$\frac{pr \quad vg}{pr^+ \quad vg^+}$$

where each line represents a chromosome; the alleles above are on one chromosome, and those below are on the other chromosome. A crossover is represented by placing an X between the two chromosomes, so that

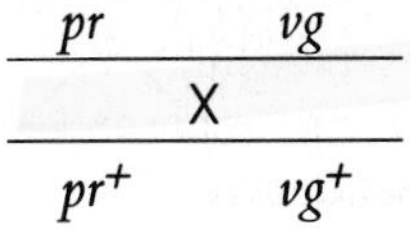

$$\frac{pr \quad vg}{\quad X \quad}\,\frac{}{pr^+ \quad vg^+}$$

is the same as

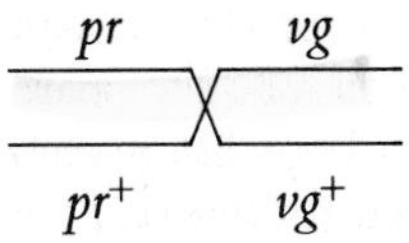

$$\frac{pr \quad vg}{pr^+ \quad vg^+}$$

We can simplify the genotypic designation of linked genes by drawing a single line, with the genes on each side being on the same chromosome; now our symbol is

$$\frac{pr \quad vg}{pr^+ \quad vg^+}$$

But this is still inconvenient for typing and writing, so let's tip the line to give us $pr\ vg\ /\ pr^+\ vg^+$, still keeping the genes of one chromosome on one side of the line and those of its homolog on the other. We always designate linked genes on each side in the same order; it is always $a\ b\ /\ a\ b$, never $a\ b\ /\ b\ a$. The rule that genes are always written in the same order permits geneticists to use a shorter notation in which the wild-type allele is written with a plus sign alone. In this notation the genotype $pr\ vg\ /\ pr^+\ vg^+$ becomes $pr\ vg\ /\ +\ +$. You may see this notation in other books or in research papers.

Now if we reconsider the results obtained by Bateson and Punnett, we can easily explain the coupling phenomenon by using the concept of linkage. Their results are complex because they did not do a testcross. See if you can derive estimated numbers for recombinant and parental types in the gametes.

Linkage of Genes on the X Chromosome

Until now, we have been considering recombination of autosomal genes. What are the consequences of nonsister chromatids of the X chromosome crossing over between two genes of interest? Recall that a human, or *Drosophila,* female produces male progeny hemizygous for the genes of the X chromosome, so that the genotype of the gamete that a mother contributes to her son is the sole determinant of the son's phenotype. Let's consider an example in which we first observe the F_1 progeny from the mating of two *Drosophila* flies and then the F_2 progeny from intercrossing the F_1. We use here the following symbols: y and y^+ for the alleles governing yellow body and brown body, respectively; w and w^+ for alleles for white eye and red eye; and Y for the Y chromosome.

P $y\,w^+ / y\,w^+$ ♀ × $y^+ w$ / Y ♂

F_1 $y\,w^+ / y^+ w$ ♀ × $y\,w^+$ / Y ♂

The numbers of F_2 males in the phenotypic classes are

$y\ w$	43	recombinant
$y^+ w$	2146	parental
$y\ w^+$	2302	parental
$y^+ w^+$	22	recombinant
	4513	

Because the F_2 males obtain only a Y chromosome from the F_1 males, these classes reflect perfectly the products of meiosis in the F_1 females. Notice that this eliminates the need for a testcross; we can follow meiosis in a single parent, just as we can in a testcross. The total frequency of the recombinants in this example is $(43 + 22) \div 4513 = 1.4$ percent.

Linkage Maps

The frequency of recombinants for the *Drosophila* autosomal genes we studied (*pr* and *vg*) was 10.7 percent of the progeny—a frequency much greater than that for the linked genes on the X chromosome, studied above. Apparently, the amount of crossing-over between various linked genes differs. Indeed, there is no reason to expect that chromatids would cross over between different linked genes with the same frequency. As Morgan studied more linked genes, he saw that the proportion of recombinant progeny varied considerably, depending on which linked genes were being studied, and he thought that these variations in crossover frequency might somehow reflect the actual distances separating genes on the chromosomes. Morgan assigned the study of this problem to a student, Alfred Sturtevant, who (like Bridges) became a great geneticist. Morgan asked Sturtevant, still an undergraduate at the time, to make some sense of the data on crossing-over between different linked genes. In one night, Sturtevant developed a method for describing relationships between genes that is still used today. In Sturtevant's own words, "In the latter part of 1911, in conversation with Morgan, I suddenly realized that the variations in strength of linkage, already attributed by Morgan to differences in the spatial separation of genes, offered the possibility of determining sequences in the linear dimension of a chromosome. I went home and spent most of the night (to the neglect of my undergraduate homework) in producing the first chromosome map."

As an example of Sturtevant's logic, consider a testcross from which we obtain the following results:

$pr\ vg\ /\ pr\ vg$	165	Parental
$pr^+\ vg^+\ /\ pr\ vg$	191	
$pr\ vg^+\ /\ pr\ vg$	23	Recombinant
$pr^+\ vg\ /\ pr\ vg$	21	
	400	

The progeny in this example represent 400 female gametes, of which 44 (11 percent) are recombinant. Sturtevant suggested that we can use the percentage of recombinants as a quantitative index of the linear distance between two genes on a genetic map, or **linkage map,** as it is sometimes called.

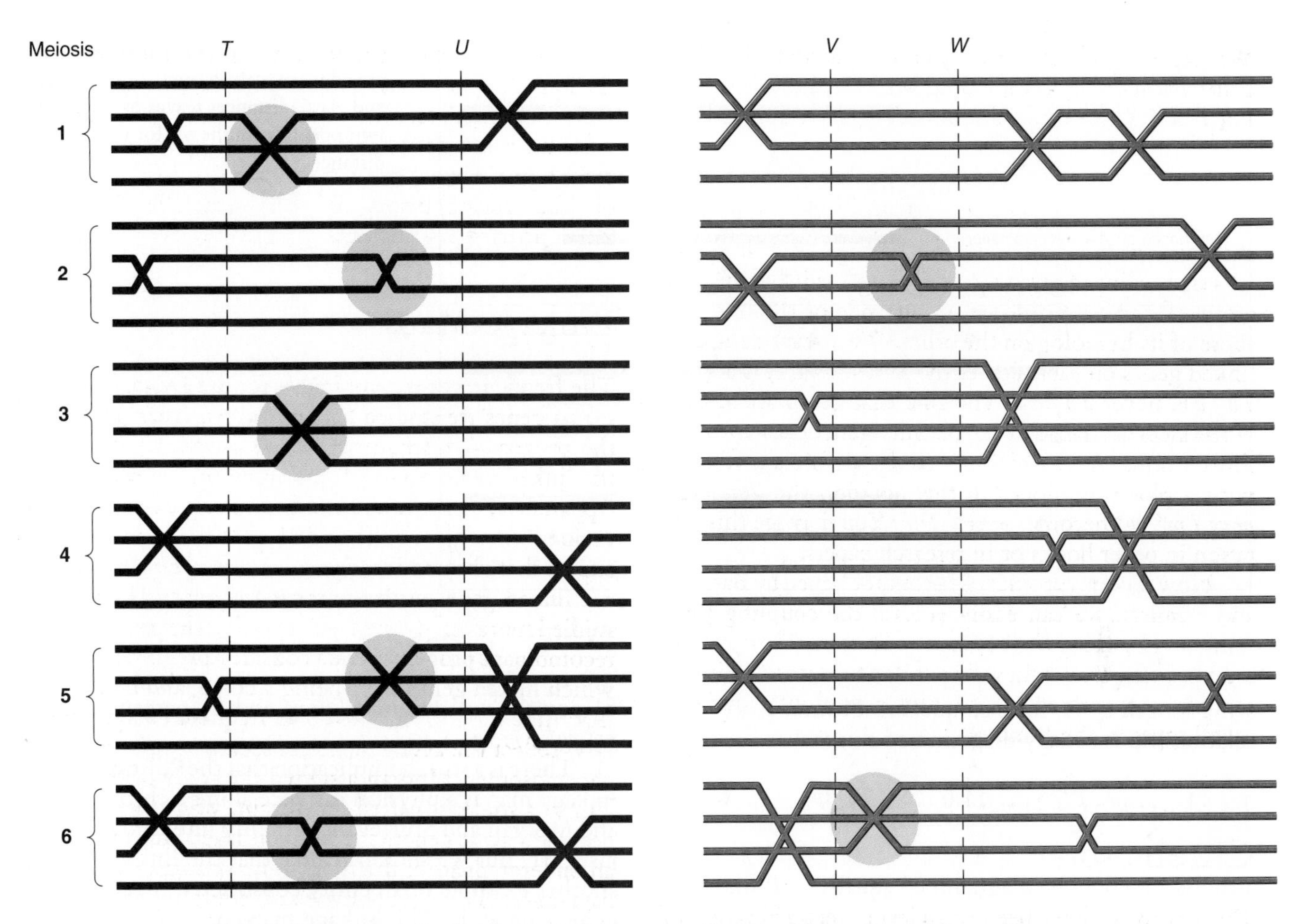

Figure 5-9 Proportionality between chromosome distance and recombinant frequency. During every meiosis, chromatids cross over at random along the chromosome. The two genes *T* and *U* are farther apart on a chromosome than *V* and *W*. Chromatids cross over between *T* and *U* in a larger proportion of meioses than between *V* and *W*, so the recombinant frequency for *T* and *U* is higher than that for *V* and *W*. As we will learn later in the chapter, a crossover can occur between any two nonsister chromatids.

The basic idea here is quite simple. Imagine two specific genes positioned a certain fixed distance apart. Now imagine random crossing-over along the paired homologs. In some meiotic divisions, nonsister chromatids cross over by chance in the chromosomal region between these genes; from these meioses, recombinants are produced. In other meiotic divisions, there are no crossovers between these genes; no recombinants result from these meioses. Sturtevant postulated a rough proportionality: the greater the distance between the linked genes, the greater the chance that nonsister chromatids would cross over in the region between the genes and, hence, the greater the proportion of recombinants that would be produced. Thus, by determining the frequency of recombinants, we can obtain a measure of the map distance between the genes (Figure 5-9). In fact, we can define one **genetic map unit (m.u.)** as that distance between genes for which one product of meiosis out of 100 is recombinant. Put another way, a **recombinant frequency (RF)** of 0.01 (or 1 percent) is defined as 1 m.u. (A map unit is sometimes referred to as a centimorgan (cM) in honor of Thomas Hunt Morgan.)

A direct consequence of the way map distance is measured is that if 5 map units (5 m.u.) separate genes *A* and *B* whereas 3 m.u. separate genes *A* and *C*, then *B* and *C* should be either 8 or 2 m.u. apart (Figure 5-10). Sturtevant found this to be the case. In other words, his analysis strongly suggested that genes are arranged in some linear order.

The place on the map where a gene is located—and on the chromosome—is called the **gene locus** (plural, **loci**). The locus of the eye-color gene and the locus of the

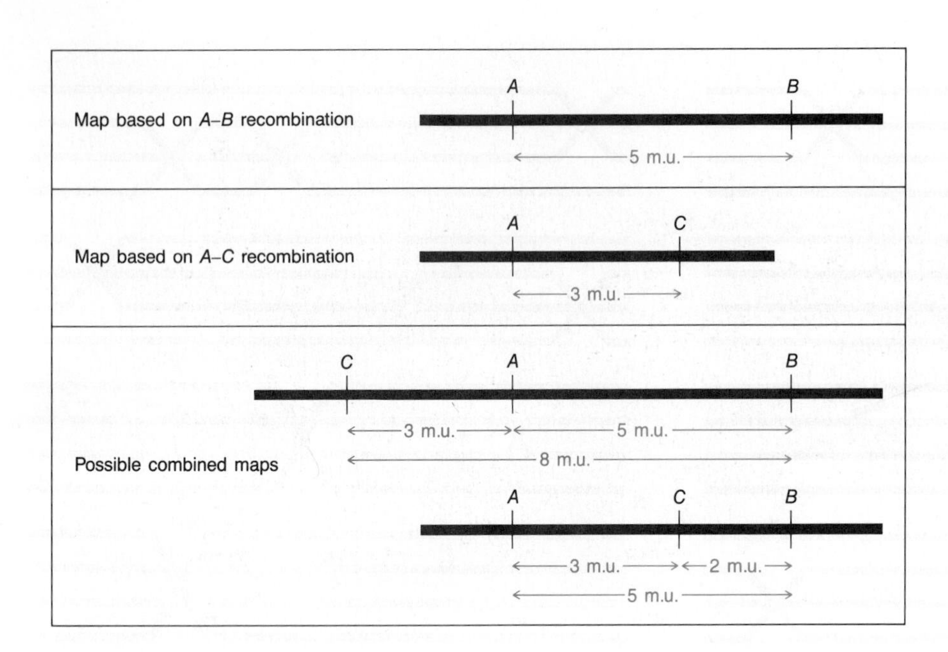

Figure 5-10 Because map distances are additive, calculation of the $A-B$ and $A-C$ distances leaves us with the two possibilities shown for the $B-C$ distance.

wing-length gene, for example, are 11 m.u. apart. The relationship is usually diagrammed this way:

pr ——— 11.0 ——— *vg*

although it could be diagrammed equally well like this:

pr^+ ——— 11.0 ——— vg^+

or like this:

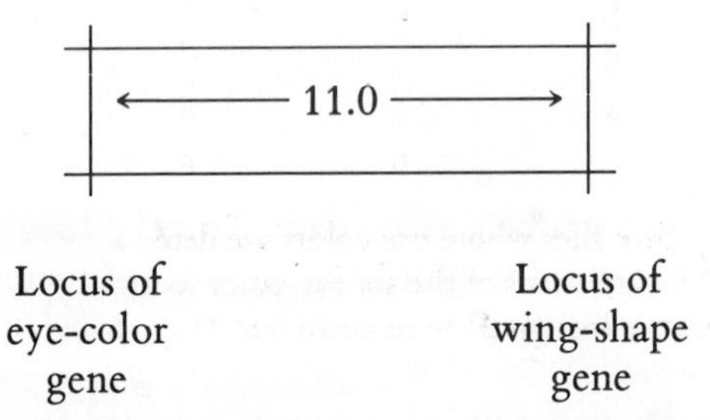

Usually we refer to the locus of this eye-color gene in shorthand as the "*pr* locus," after the first discovered non-wild-type allele, but we mean the place on the chromosome where any allele of this gene will be found.

Given a genetic distance in map units, we can predict frequencies of progeny in different classes. For example, in the progeny from a testcross of a female $pr\ vg / pr^+\ vg^+$ heterozygote, we know there will be 11 percent recombinants, of which $5\frac{1}{2}$ percent will be $pr\ vg^+ / pr\ vg$ and $5\frac{1}{2}$ percent will be $pr^+\ vg / pr\ vg$; of the progeny from a testcross of a female $pr\ vg^+ / pr^+\ vg$ heterozygote, $5\frac{1}{2}$ percent will be $pr\ vg / pr\ vg$ and $5\frac{1}{2}$ percent will be $pr^+\ vg^+ / pr\ vg$.

There is a strong implication that the "distance" on a linkage map is a physical distance along a chromosome, and Morgan and Sturtevant certainly intended to imply just that. But we should realize that the linkage map is another example of an entity constructed from a purely genetic analysis. The linkage map could have been derived without even knowing that chromosomes existed. Furthermore, at this point in our discussion, we cannot say whether the "genetic distances" calculated by means of recombinant frequencies in any way reflect actual physical distances on chromosomes, although cytogenetic and molecular analysis has shown that genetic distances are, in fact, roughly proportional to chromosome distances. Nevertheless, it must be emphasized that the hypothetical structure (the linkage map) was developed with a very real structure (the chromosome) in mind. In other words, the chromosome theory provided the framework for the development of linkage mapping.

Message Recombination between linked genes can be used to map their distance apart on the chromosome. The unit of mapping (1 m.u.) is defined as a recombinant frequency of 1 percent.

The stage of analysis that we have reached in our discussion is well illustrated by linkage maps of the screwworm *(Cochliomyia hominivorax)*. The larval stage of this insect — the worm — is parasitic on mammalian wounds and is a costly pest of livestock in some parts of

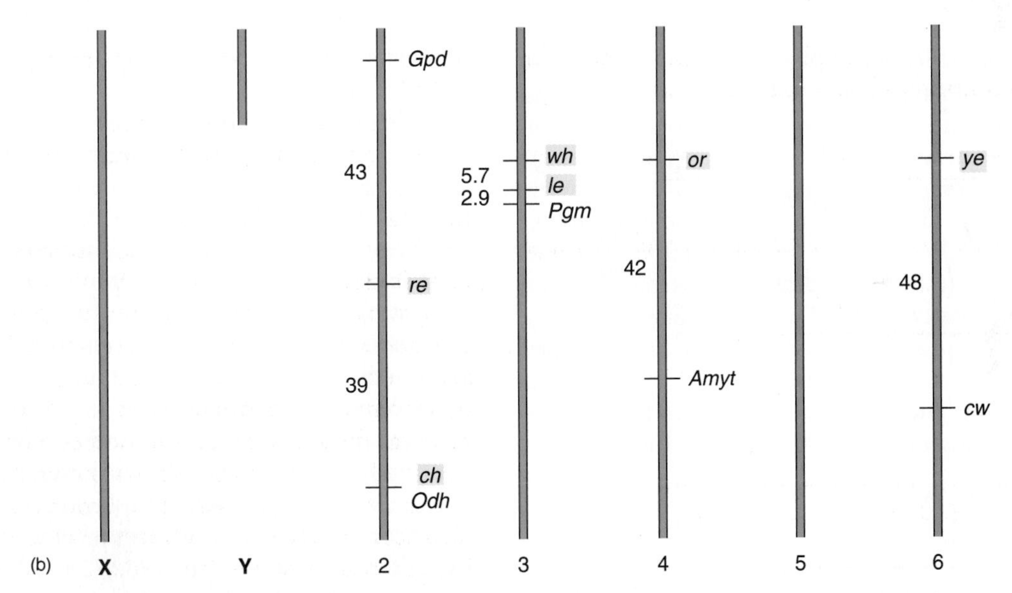

Figure 5-11 (a) Wild-type adult of screwworm and six flies whose eye colors are determined by alleles at six different autosomal loci. (b) Linkage maps of the six eye-color loci and five other loci of screwworms. The numbers between the loci give the recombinant frequencies. (D. B. Taylor, USDA)

the world. A genetic system of population control has been proposed, of a type that has been successful in other insects. In order to accomplish this goal, an understanding of the basic genetics of the insect is needed, and one important part of this is to prepare a map of the chromosomes. This animal has six chromosome pairs, and mapping has begun recently.

The job of general mapping starts by finding and analyzing as many variant phenotypes as possible. The adult stage of this insect is a fly, and geneticists have found phenotypic variants among screwworm flies. They found flies of six different eye colors, all different from the brown-eyed, wild-type flies, as Figure 5-11a shows. They also found five variant phenotypes for some other characters. Eleven mutant alleles were shown to determine the 11 variant phenotypes, each at a different autosomal locus. Pure lines of each phenotype were intercrossed to generate dihybrid F_1s, and then these were testcrossed. This revealed the set of four linkage groups shown in Figure 5-11b. Notice that the *ye* and *cw* loci are shown tentatively linked although the recombinant frequency is not significantly different from 50 percent.

A linkage analysis such as the above cannot assign linkage groups to specific chromosomes; this must be done using cytogenetic techniques to be discussed in Chapter 8. In the present example, such cytogenetic techniques have allowed the linkage groups to be correlated with the chromosomes previously numbered as shown in the figure.

Three-Point Testcross

So far, we have looked at linkage in crosses of double heterozygotes to doubly recessive testers. The next level of complexity is a cross of a triple heterozygote to a triply recessive tester. This kind of cross, called a **three-point testcross,** illustrates the standard kind of approach used in linkage analysis. We will consider two examples of such crosses here.

First, we shall focus on three *Drosophila* genes that have the non-wild-type alleles *sc* (short for scute, or loss of certain thoracic bristles), *ec* (short for echinus, or roughened eye surface), and *vg* (short for vestigial wing). We can cross *sc sc ec ec vg vg* triply recessive flies with wild-type flies to generate triple heterozygotes, $sc\,sc^+\,ec\,ec^+\,vg\,vg^+$. We analyze recombination in these heterozygotes by testcrossing heterozygous females with triply recessive tester males. The results of such a testcross follow. The progeny are listed as gametic genotypes derived from the heterozygous females. Eight gametic types are possible, and these were counted in the following numbers in a sample of 1008 progeny flies:

sc	ec	vg	235
sc^+	ec^+	vg^+	241
sc	ec	vg^+	243
sc^+	ec^+	vg	233
sc	ec^+	vg	12
sc^+	ec	vg^+	14
sc	ec^+	vg^+	14
sc^+	ec	vg	16
			1008

The systematic way to analyze such crosses is to calculate all possible recombinant frequencies, but it is always worthwhile to inspect the data for obvious patterns before doing this. At first glance, we can note in the above data that there is a considerable deviation from the 1 : 1 : 1 : 1 : 1 : 1 : 1 : 1 ratio that is expected if the genes are all unlinked. So let's begin to calculate recombinant frequency values, taking the loci a pair at a time. Starting with the *sc* and *ec* loci (ignoring the *vg* locus for the time being), we determine which of the gametic genotypes are recombinant for *sc* and *ec*. Because we know that the heterozygotes were established from *sc ec* and $sc^+\,ec^+$ gametes, we know that the recombinant products of meiosis must be $sc\,ec^+$ and $sc^+\,ec$. We note from the list that there are 12 + 14 + 14 + 16 = 56 of these types; therefore, RF = (56/1008) × 100 = 5.5 m.u. This tells us that these loci must be linked on the same chromsome, as follows:

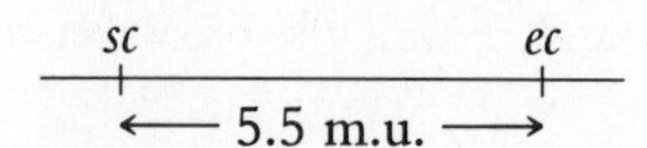

Now let's look at recombination between the *sc* and the *vg* loci. The "input" parental genotypes were *sc vg* and $sc^+\,vg^+$, so we must calculate the frequency of $sc\,vg^+$ and $sc^+\,vg$ progeny types (this time, we ignore *ec*). We see that there are 243 + 233 + 14 + 16 = 506 recombinants; because 506/1008 is very close to an RF of 50 percent, we conclude that the *sc* and *vg* loci are not linked and probably not on the same chromosome. We can summarize the linkage relationship as follows:

sc ——— 5.5 m.u. ——— *ec*   *vg*

Now you should see that the *ec* and *vg* loci must also be unlinked. Adding up the recombinants in the list and calculating the RF will confirm this. (Try it.)

A second example, using some other loci of *Drosophila,* will introduce some more important genetic concepts. Here the non-wild-type alleles are *v* (vermilion eyes), *cv* (crossveinless, or absence of a crossvein on the wing), and *ct* (cut, or snipped wing edges). This time the parental stocks are homozygous doubly recessive flies of genotype $v^+v^+\,cv\,cv\,ct\,ct$ and homozygous singly recessive flies of genotype $v\,v\,cv^+cv^+\,ct^+ct^+$. From this cross, triply heterozygous progeny of genotype $v\,v^+\,cv\,cv^+\,ct\,ct^+$ are obtained, and females of this genotype are testcrossed to triple recessives of genotype *v v cv cv ct ct*. The female gametic genotypes determining the eight progeny types from this testcross are shown here, with their numbers out of a total sample of 1448 flies:

v	cv^+	ct^+	580
v^+	cv	ct	592
v	cv	ct^+	45
v^+	cv^+	ct	40
v	cv	ct	89
v^+	cv^+	ct^+	94
v	cv^+	ct	3
v^+	cv	ct^+	5
			1448

Once again, the standard recombination approach is called for, but we must be careful in our classification of parental and recombinant types. Note that the parental input genotypes for the triple heterozygotes are v^+ *cv ct* and *v* cv^+ ct^+; we must take this into consideration when we decide what constitutes a recombinant.

Starting with the *v* and *cv* loci, we see that the recombinants are of genotype *v cv* and v^+ cv^+ and that there are $45 + 40 + 89 + 94 = 268$ of these. Out of a total of 1448 flies, this gives an RF of 18.5 percent.

For the *v* and *ct* loci, the recombinants are *v ct* are v^+ ct^+. There are $89 + 94 + 3 + 5 = 191$ of these among 1448 flies, so that RF = 13.2 percent.

For *ct* and *cv*, the recombinants are *cv* ct^+ and cv^+ *ct*. There are $45 + 40 + 3 + 5 = 93$ of these among the 1448, so that RF = 6.4 percent.

Obviously, all of the loci are linked on the same chromosome, because the RF values are all considerably less than 50 percent. Because the *v* and *cv* loci show the largest RF value, they must be the farthest apart; therefore, the *ct* locus must be between them. A map can be drawn as follows:

v ——— *ct* ——— *cv*

← 13.2 m.u. →← 6.4 m.u. →

Note several important points here. The first is that we have deduced a different gene order from that of our listing of the progeny genotypes. As the point of the exercise was to determine the linkage relationships of these genes, the original listing was of necessity arbitrary; the order simply was not known before the data were analyzed.

Second, we have definitely established that *ct* is between *v* and *cv* and what the distances are between *ct* and these loci in map units. But we have arbitrarily placed *v* to the left and *cv* to the right; the map could equally well be inverted!

A third point to note is that the two smaller map distances, 13.2 m.u. and 6.4 m.u., add up to 19.6 m.u., which is greater than 18.5 m.u., the distance calculated for *v* and *cv*. Why is this so? The answer to this question lies in the way in which we have analyzed the two rarest classes in our classification of recombination for the *v* and *cv* loci. Now that we have the map, we can see that these two rare classes are in fact double recombinants, arising from two crossovers (Figure 5-12). However, we did not count the *v ct* cv^+ and v^+ ct^+ *cv* genotypes when we calculated the RF value for *v* and *cv*; after all, with regard to *v* and *cv*, they are parental combinations (*v* cv^+ and v^+ *cv*). In the light of our map, however, we see that this led to an underestimate of the distance between the *v* and *cv* loci. Not only should we have counted the two rarest classes, we should have counted each of them twice because each represents a double recombinant class! Hence, we can correct the value by adding the numbers $45 + 40 + 89 + 94 + 3 + 3 + 5 + 5 = 284$. Out of the total of 1448, this is exactly 19.6 percent, which is identical to the sum of the two component values.

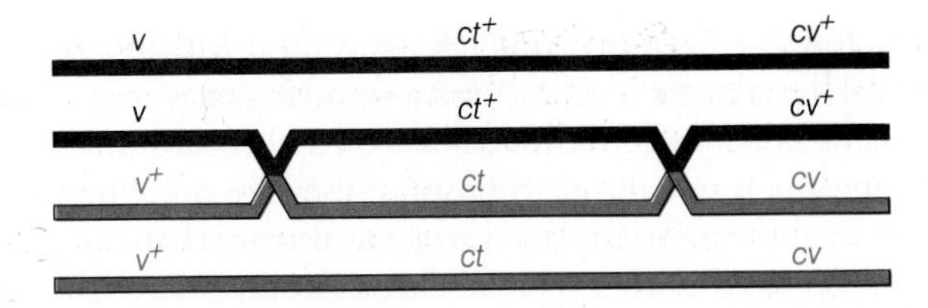

Figure 5-12 An example of a double crossover. Notice that a double crossover produces double recombinant chromatids that have the parental allelic combinations at the outer loci.

Now that we have had some experience with the data of this cross, we can look back at the progeny listing and see that it is usually possible to deduce gene order by inspection, without a recombinant frequency analysis. Only three gene orders are possible, each with a different gene in the middle position. It is generally true that the double recombinant classes are the smallest ones. Only one order should be compatible with the smallest classes having been formed by double crossovers, as shown in Figure 5-13. Only one order gives double recombinants of genotype *v ct* cv^+ and v^+ ct^+ *cv*. Notice in passing that the ability to detect double crossovers depends on having a heterozygous gene between the two crossovers; if the mothers of these progeny had not been heterozygous *ct* ct^+, we could never have identified the double recombinant classes.

Possible gene orders	Double recombinant chromatids
v ct+ cv+ / v+ ct cv	v ct cv+ / v+ ct+ cv
ct+ v cv+ / ct v+ cv	ct+ v+ cv+ / ct v cv
ct+ cv+ v / ct cv v+	ct+ cv v / ct cv+ v+

Figure 5-13 With three genes, only three gene orders are possible. Double crossovers create unique double recombinant genotypes for each gene order. Only the first possibility is compatible with the data in the text.

Finally, note that linkage maps merely map the loci in relation to each other, using standard map units. We do not know where the loci are on a chromosome—or even which specific chromosome they are on. The linkage map is essentially an abstract construct that can only be correlated with a specific chromosome and with specific chromosome regions by applying special kinds of cytogenetic analyses, as we shall see in Chapter 8.

Message Three (and higher) point testcrosses enable linkage between three (or more) genes to be evaluated in one cross.

Interference

The detection of the double recombinant classes shows that double crossovers must occur. Knowing this, one of the first questions that we may think of is whether the crossovers in adjacent chromosome regions are independent or whether they interact with each other in some way. For example, we might ask if a crossover in one region affects the likelihood of there being a crossover in an adjacent region. It turns out that often the answer is yes, and the interaction is called **interference.**

The analysis can be approached in the following way. If the crossovers in the two regions are independent, then according to the product rule (see page 27), the frequency of double recombinants would equal the product of the recombinant frequencies in the adjacent regions. In the *v–ct–cv* recombination data, the *v–ct* RF value is 0.132 and the *ct–cv* value is 0.064, so double recombinants might be expected at the frequency $0.132 \times 0.064 = 0.0084$ (0.84 percent) if there is independence. In the sample of 1448 flies, $0.0084 \times 1448 = 12$ double recombinants are expected. But the data show that only 8 were actually observed. If this deficiency of double recombinants were consistently observed, it would show us that the two regions are not independent and suggest that the distribution of crossovers favors singles at the expense of doubles. In other words, there is some kind of interference: a crossover reduces the probability of a crossover in an adjacent region.

Interference is quantified by first calculating a term called the **coefficient of coincidence (c.o.c.),** which is the ratio of observed to expected double recombinants and then subtracting this value from 1. Hence

Interference (I) = 1 − c.o.c. =

$$1 - \left[\frac{\text{Observed frequency or number of double recombinants}}{\text{Expected frequency or number of double recombinants}}\right]$$

In our example

$$I = 1 - \tfrac{8}{12} = \tfrac{4}{12} = \tfrac{1}{3}, \text{ or } 33\%$$

In some regions, there are never any observed double recombinants. In these cases, c.o.c. = 0, so I = 1 and interference is complete. Most of the time, the interference values that are encountered in mapping chromosome loci are between 0 and 1, but in certain special situations, observed doubles exceed expected, giving negative interference values.

Recombination analysis relies so heavily on three-point testcrosses and extended versions of them that it is worth making a step-by-step summary of the analysis, ending with an interference calculation. We shall use numerical values from the data on the *v*, *ct*, and *cv* loci.

1. Calculate recombinant frequencies for each pair of genes:

$$v - cv = 18.5\%$$
$$cv - ct = 6.4\%$$
$$ct - v = 13.2\%$$

2. Represent linkage relationships in a linkage map:

v —— 13.2 m.u. —— *ct* —— 6.4 m.u. —— *cv*

3. Determine the double recombinant classes.
4. Calculate the frequency and number of double recombinants expected if there is no interference:

$$\text{Expected frequency} = 0.132 \times 0.064 = 0.0084$$
$$\text{Expected number} = 0.0084 \times 1448 = 12$$

5. Calculate interference:

$$\text{Observed number of double recombinants} = 8$$
$$\text{Expected number of double recombinants} = 12$$
$$\therefore I = 1 - \tfrac{8}{12} = \tfrac{4}{12} = 0.33, \text{ or } 33\%$$

You may have wondered why we always used heterozygous females for testcrosses in our examples of linkage in *Drosophila*. When $pr\ vg / pr^+\ vg^+$ males are crossed with *pr vg* / *pr vg* females, only $pr\ vg / pr^+\ vg^+$ and *pr vg* / *pr vg* progeny are recovered. This result shows that there is no crossing-over in *Drosophila* males. However, this absence of crossing-over in one sex is limited to certain species; it is not the case for males of all species (or for the heterogametic sex). In other organisms, there is crossing-over in XY males and in WZ females. The

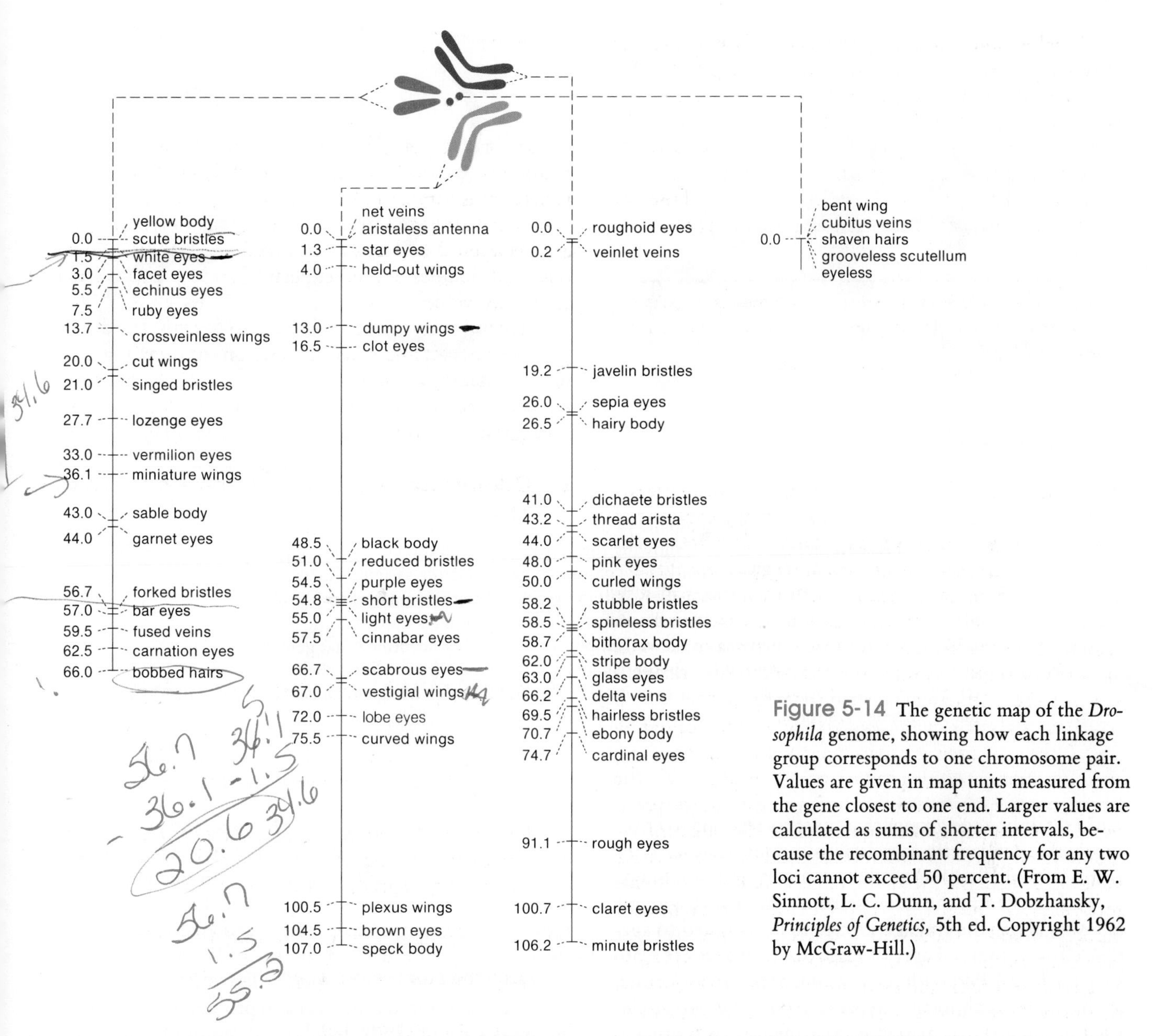

Figure 5-14 The genetic map of the *Drosophila* genome, showing how each linkage group corresponds to one chromosome pair. Values are given in map units measured from the gene closest to one end. Larger values are calculated as sums of shorter intervals, because the recombinant frequency for any two loci cannot exceed 50 percent. (From E. W. Sinnott, L. C. Dunn, and T. Dobzhansky, *Principles of Genetics,* 5th ed. Copyright 1962 by McGraw-Hill.)

reason for the absence of crossing-over in *Drosophila* males is that they have an unusual prophase I, with no synaptonemal complexes.

Incidentally, there is a recombination difference between human sexes as well. Women show higher recombinant frequencies for the same loci than do men. This is shown in a map in a later chapter (Figure 15-25, page 558).

Linkage maps are an essential aspect of the experimental genetic study of any organism. They are the prelude to any serious piece of genetic manipulation. Many organisms have had their chromosomes intensively mapped in this way. The resultant maps represent a vast amount of genetic analysis generally achieved by collaborative efforts of research groups throughout the world. Figures 5-14 and 5-15 show two examples of linkage maps: one from *Drosophila* and one from the tomato. The *Drosophila* genome is one of the most intensively mapped of all model genetic organisms. The map in Figure 5-14 shows only a fraction of the known loci. Tomatoes, also, have been interesting from the perspectives of both basic and applied genetic research, and the tomato genome is one of the best mapped of plants.

The different panels of Figure 5-15 illustrate some of the stages of understanding through which research arrives at a comprehensive map. First, although chromosomes are visible under the microscope, there is initially no way to locate genes on them. However, the chromosomes can be individually identified and numbered, based on their inherent landmarks such as staining patterns and centromere positions, as has been done in parts a and b. Next, analysis of recombinant frequencies gener-

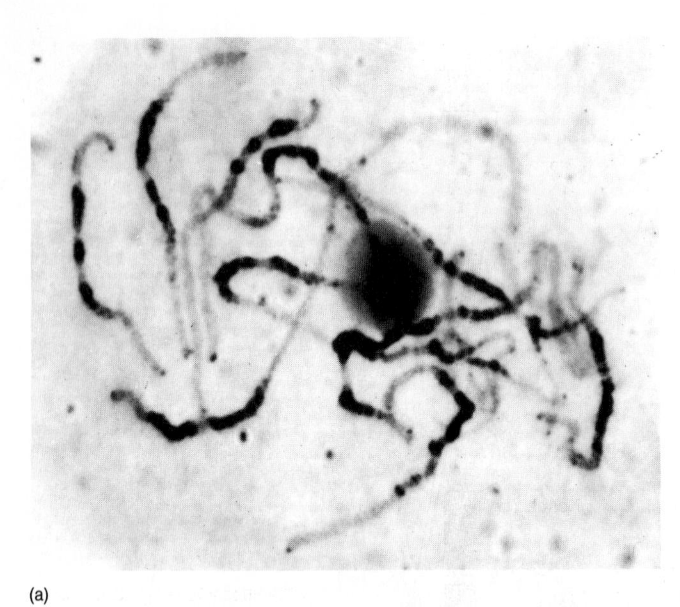

(a)

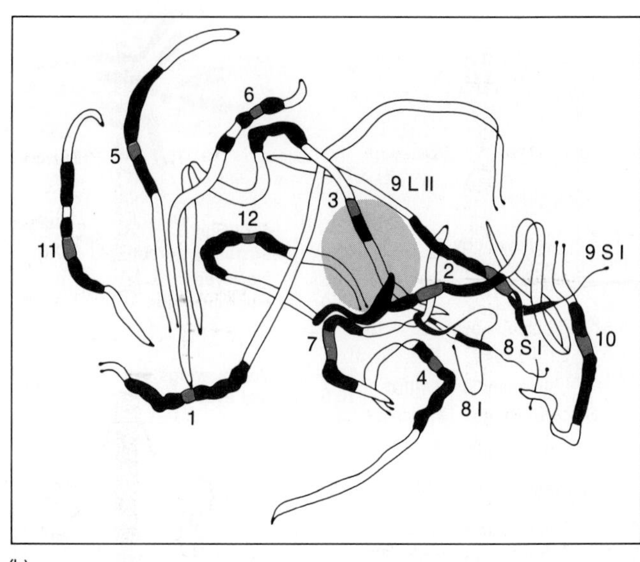

(b)

Figure 5-15 Mapping the chromosomes of tomatoes. (a) Photomicrograph of a meiotic prophase I (pachytene) from anthers, showing the 12 pairs of chromosomes as they appear under the microscope. (b) The currently used chromosome numbering system. The centromeres are colored and the flanking, densely staining regions (heterochromatin) are shown in black. (c) A linkage map made in 1952 showing the linkage groupings known at the time. Each locus is flanked by drawings of the variant phenotype that first identified that genetic locus (to the right or above) and the appropriate normal phenotype (to the left or below). Interlocus map distances are shown in map units. (Parts a and b from C. M. Rick, "The Tomato," *Scientific American*, 1978; part c from L. A. Butler.)

ates a set of linkage groups that must correspond to chromosomes, but specifc correlations cannot necessarily be made with the numbered chromosomes. At some stage, as we discussed in the screwworm example, cytogenetic analyses allow the linkage groups to be assigned to specific chromosomes. Part c of Figure 5-15 shows a tomato map made in 1952, with the linkages of the genes known at that time. Each locus is represented by the two alleles used in the original mapping experiments. As more and more loci became known, they were mapped in relation to the loci shown in the figure, so today the map contains hundreds of loci. Some of the chromosome numbers shown in part c are tentative and do not correspond to the modern chromosome numbering system.

The χ^2 Test

We come now to a subject that is particularly relevant to the detection of linkage. We have seen that the functional test for the presence or absence of linkage is based on the relative frequencies of the meiotic product types. If there is no linkage between the *A* and *B* loci, then the four product-cell types *A B*, *a b*, *A b*, and *a B* are produced in a 1:1:1:1 ratio, which fixes the recombinant frequency at 50 percent. If there is linkage, the proportions of the four types deviate from the 1:1:1:1 ratio and two types, the recombinants comprise < 50 percent of the whole. In practical terms, the question "Is this a 1:1:1:1 ratio?" must be answered in any particular case. An answer of "yes" indicates the absence of linkage, and "no" indicates linkage. "Obvious" departures from the 1:1:1:1 ratio present no decision problems, but smaller departures are tricky to handle and require further analytic approaches.

What is the precise problem here? An example will be helpful. A double heterozygote in coupling conformation produces 500 meiotic products, distributed as follows:

A B	140
a b	135
A b	110
a B	115

By applying the recombinant-frequency test, we find 225 recombinants, or 45 percent. This is admittedly less than 50 percent, but not convincingly so. The skeptic would say, "This is merely a chance deviation from a

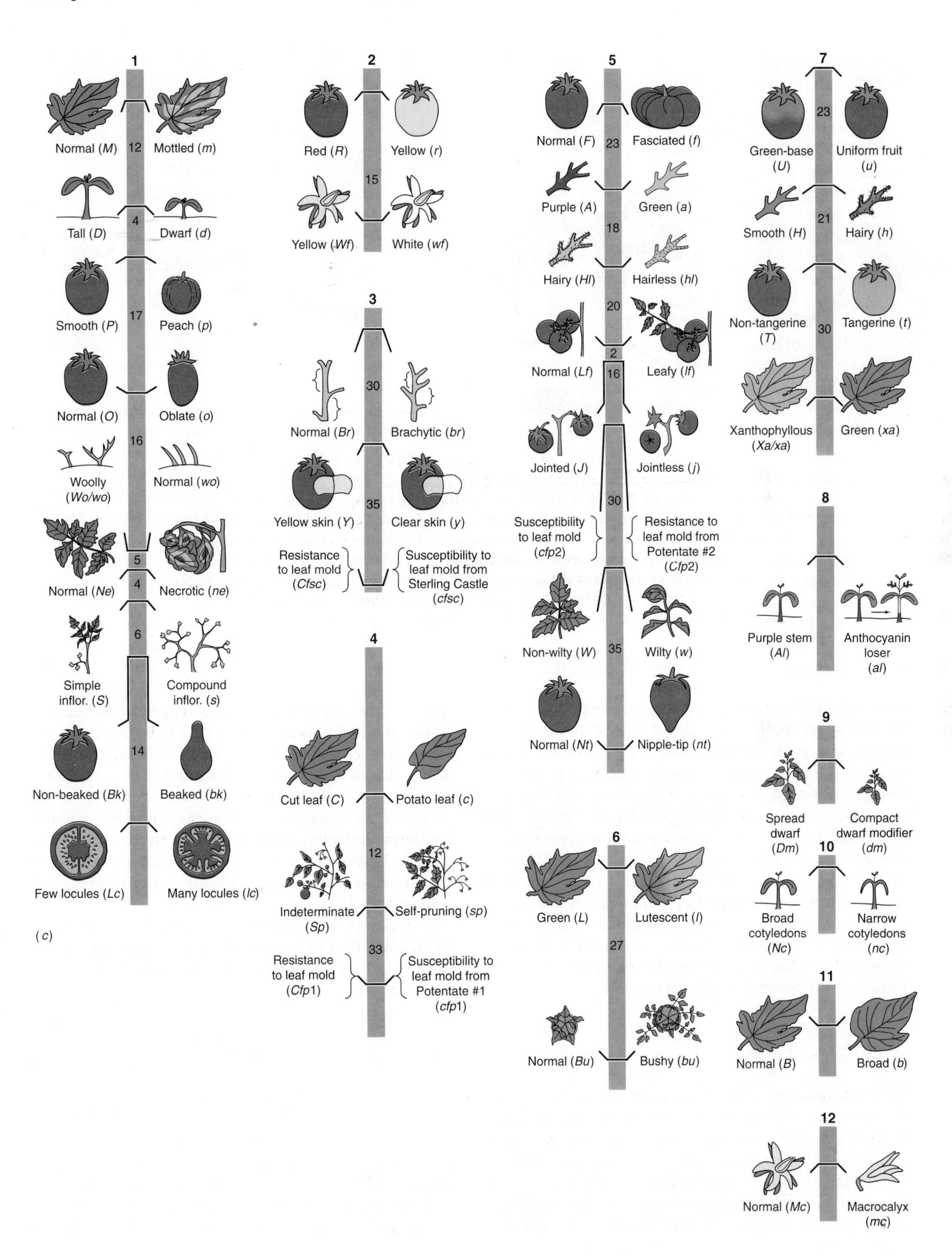
1
Normal (M)
12
Mottled (m)
Tall (D)
4
Dwarf (d)
Smooth (P)
17
Peach (p)
Normal (O)
Oblate (o)
16
Woolly (Wo/wo)
Normal (wo)
5
Normal (Ne)
4
Necrotic (ne)
6
Simple inflor. (S)
Compound inflor. (s)
14
Non-beaked (Bk)
Beaked (bk)
Few locules (Lc)
Many locules (lc)
2
Red (R)
Yellow (r)
15
Yellow (Wf)
White (wf)
3
30
Normal (Br)
Brachytic (br)
35
Yellow skin (Y)
Clear skin (y)
Resistance to leaf mold (Cfsc)
Susceptibility to leaf mold from Sterling Castle (cfsc)
4
Cut leaf (C)
Potato leaf (c)
12
Indeterminate (Sp)
Self-pruning (sp)
33
Resistance to leaf mold (Cfp1)
Susceptibility to leaf mold from Potentate #1 (cfp1)
5
Normal (F)
23
Fasciated (f)
Purple (A)
Green (a)
18
Hairy (Hl)
Hairless (hl)
20
2
Normal (Lf)
16
Leafy (lf)
Jointed (J)
Jointless (j)
30
Susceptibility to leaf mold (cfp2)
Resistance to leaf mold from Potentate #2 (Cfp2)
Non-wilty (W)
35
Wilty (w)
Normal (Nt)
Nipple-tip (nt)
6
Green (L)
Lutescent (l)
27
Normal (Bu)
Bushy (bu)
7
23
Green-base (U)
Uniform fruit (u)
21
Smooth (H)
Hairy (h)
30
Non-tangerine (T)
Tangerine (t)
Xanthophyllous (Xa/xa)
Green (xa)
8
Purple stem (Al)
Anthocyanin loser (al)
9
Spread dwarf (Dm)
Compact dwarf modifier (dm)
10
Broad cotyledons (Nc)
Narrow cotyledons (nc)
11
Normal (B)
Broad (b)
12
Normal (Mc)
Macrocalyx (mc)

(c)

1:1:1:1 ratio! If you repeatedly sampled 500 marbles blindly from a sack containing equal numbers of red, blue, yellow, and green, you would get this big a deviation from the 1:1:1:1 ratio fairly often." What should we decide? The χ^2 test can help in this kind of predicament.

In general, the χ^2 test tells us how often observations deviate from expectations purely on the basis of chance. The procedure is as follows:

1. *State a simple hypothesis that gives a precise expectation.* In our example, the best hypothesis is "lack of linkage," which yields an expected 1:1:1:1 ratio. This **null hypothesis** is obviously better than a hypothesis of linkage because, if we postulate linkage, we do not know what to choose as an RF value. If we disprove "lack of linkage," then the genes must be linked.
2. *Calculate χ^2.* The statistic χ^2 is always calculated from actual numbers—never from percentages, fractions, or decimal fractions. In fact, part of the usefulness of the χ^2 test is that it takes sample size into consideration. The sample comprises several classes, with O representing the observed number in a class and E representing the corresponding expected number based on the hypothesis. The formula for calculating χ^2 is as follows:

$$\chi^2 = \text{total of } \frac{(O-E)^2}{E} \text{ for all classes}$$

We would set up the calculation of our example as shown in Table 5-2.

3. *Estimate p.* Using χ^2, we estimate the probability p of obtaining the observed results if the null hypothesis is correct. Before this can be done, however, we must compute another item: the number of **degrees of freedom (df).** In the present context, the number of degrees of freedom can be simply defined as

$$\text{df} = (\text{number of classes} - 1)$$

In our example

$$\text{df} = (4 - 1) = 3$$

We now turn to the table of χ^2 values (Table 5-3), which will give us our p values if we plug in our computed χ^2 and df values. Looking along the df = 3 line, we find that our χ^2 value of 5.2 lies between the p values of 0.5 (50 percent) and 0.1 (10 percent). Hence, we can conclude that our p value is just greater than 10 percent (0.1). This means that if our null hypothesis is correct, a deviation *at least* as great as that observed is expected about 10 percent of the time.

Table 5-2 Calculation of χ^2

Class	O	E	$(O-E)^2$	$\frac{(O-E)^2}{E}$
$A\,B$	140	125	225	1.8
$a\,b$	135	125	100	0.8
$A\,b$	110	125	225	1.8
$a\,B$	115	125	100	0.8
	500	500		$\chi^2 = 5.2$

4. *Reject or accept the hypothesis.* How do we know which p value is too low to be acceptable? Scientists in general arbitrarily use the 5 percent level. Thus, any p value less than 5 percent results in rejection of the hypothesis, and any p value greater than 5 percent results in acceptance of the hypothesis. In the case of acceptance, of course, the hypothesis is not proved, merely possible. In the case of rejection, the hypothesis is not disproved, merely improbable. Now that we have accepted our null hypothesis of "lack of linkage," we must live with it and acknowledge that the skeptic had a point; in the sample size used an RF of 45 percent did not constitute a case for linkage.

In this way, the χ^2 test helps us decide between linkage and absence of linkage. In fact, a major use of the χ^2 test in genetics is in the determination of linkage. But there are several other situations in which the χ^2 test is useful, and these also test actual results against simple expectations of a hypothesis. In our linkage example, we used χ^2 to test observed data against an expected 1:1:1:1 ratio, but χ^2 is ideal for testing deviations for any genetic ratio: 3:1, 9:3:3:1, 9:7, 1:1, and so on.

However, there are pitfalls. Suppose, for example, you believe you have identified a major gene that affects a specific biological function. You testcross a presumed heterozygote $A\,a$ to homozygous recessive individual $a\,a$, expecting a 1:1 phenotypic ratio in the progeny. Out of 200 progeny, 116 are $A\,a$ and 84 are $a\,a$. Here $\chi^2 = (16^2 \div 100) + (16^2 \div 100) = 5.12$ with 1 df. The p value is 2.5 percent, so you reject the null hypothesis that a single pair of alleles determines the phenotypes; you might be persuaded to consider an alternative hypothesis involving more than one gene. Actually, however, you must recall that the 1:1 ratio is expected only if there is a single pair of alleles *of equal viability* segregating. With the rejection of the null hypothesis, you must reject the notion *either* of a single allele pair *or* of the equal viability of alleles (or both). The χ^2 test cannot tell you which portion of a compound null hypothesis to reject, so care must be taken to determine all of the assumptions in a null hypothesis if error is to be avoided in the application of statistical testing. Another pitfall is to believe that the

Table 5-3. Critical Values of the χ^2 Distribution

df \ p	0.995	0.975	0.9	0.5	0.1	0.05	0.025	0.01	0.005	df
1	.000	.000	0.016	0.455	2.706	3.841	5.024	6.635	7.879	1
2	0.010	0.051	0.211	1.386	4.605	5.991	7.378	9.210	10.597	2
3	0.072	0.216	0.584	2.366	6.251	7.815	9.348	11.345	12.838	3
4	0.207	0.484	1.064	3.357	7.779	9.488	11.143	13.277	14.860	4
5	0.412	0.831	1.610	4.351	9.236	11.070	12.832	15.086	16.750	5
6	0.676	1.237	2.204	5.348	10.645	12.592	14.449	16.812	18.548	6
7	0.989	1.690	2.833	6.346	12.017	14.067	16.013	18.475	20.278	7
8	1.344	2.180	3.490	7.344	13.362	15.507	17.535	20.090	21.955	8
9	1.735	2.700	4.168	8.343	14.684	16.919	19.023	21.666	23.589	9
10	2.156	3.247	4.865	9.342	15.987	18.307	20.483	23.209	25.188	10
11	2.603	3.816	5.578	10.341	17.275	19.675	21.920	24.725	26.757	11
12	3.074	4.404	6.304	11.340	18.549	21.026	23.337	26.217	28.300	12
13	3.565	5.009	7.042	12.340	19.812	22.362	24.736	27.688	29.819	13
14	4.075	5.629	7.790	13.339	21.064	23.685	26.119	29.141	31.319	14
15	4.601	6.262	8.547	14.339	22.307	24.996	27.488	30.578	32.801	15

χ^2 test shows the truth. It does not; obviously wide deviations do sometimes occur, and you might have one of them in the experiment you are conducting.

Message The χ^2 test is used to test experimental results against the expectations derived from a precise hypothesis. The test generates a p value that is the probability of obtaining by chance a specific deviation at least as great as the one observed, assuming that the hypothesis is correct.

Early Thoughts on the Nature of Crossing-Over

The idea that intrachromosomal recombinants were produced by some kind of exchange of material between homologous chromosomes was a compelling one. But experimentation was necessary to test this idea. One of the first steps was to correlate the appearance of a genetic recombinant with an exchange of parts of chromosomes. Several investigators approached this problem in the same way. In 1931, Harriet Creighton and Barbara McClintock were studying two loci of chromosome 9 of corn: one affecting seed color (*C*, colored; *c*, colorless), and the other affecting endosperm composition (*Wx*, waxy; *wx*, starchy). Furthermore, the chromosome carrying *C* and *Wx* was unusual in that it carried a large, densely staining element (called a knob) on the *C* end and a longer piece of chromosome on the *Wx* end; thus, the heterozygote was

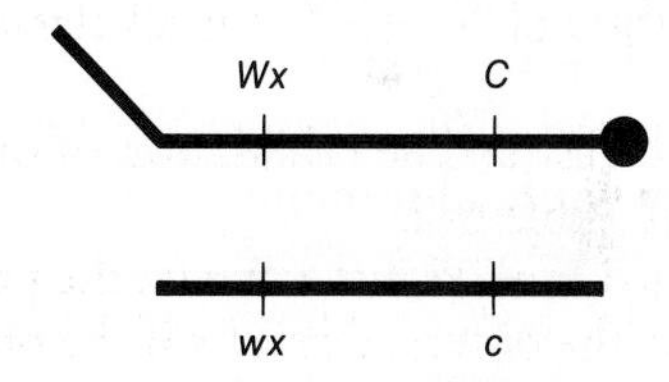

When they compared the chromosomes of genetic recombinants with those of parental type progeny, Creighton and McClintock found that all the parental types retained the parental chromosomal arrangements, whereas all the recombinants were

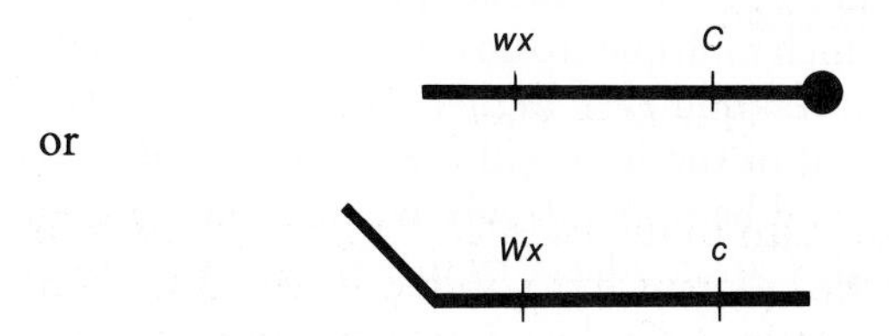

Thus, they correlated the genetic and cytological events of intrachromosomal recombination. The chiasmata appeared to be the sites of the exchange, but the final proof of this did not come until 1978.

But what is the mechanism of chromosome exchange in a crossover event? Is it a breakage and reunion

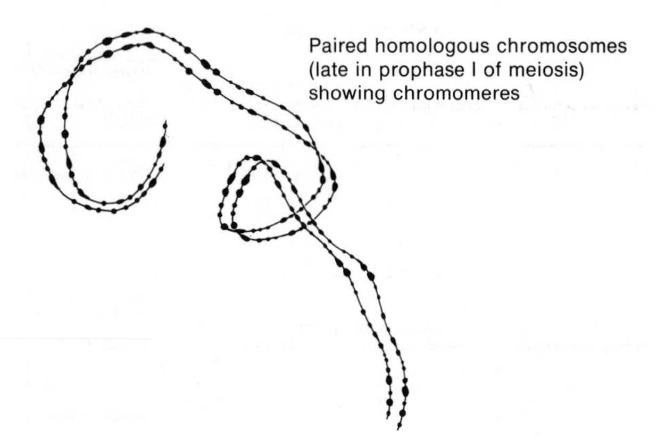

Paired homologous chromosomes
(late in prophase I of meiosis)
showing chromomeres

Figure 5-16 Diagrammatic representation of the chromomeres that are important in Belling's copy-choice model.

process, like splicing sound tapes? Or is it a process that only *appears* to consist of a breakage and a reunion? An idea that favors the latter explanation was proposed in 1928 by John Belling, who studied meiosis in plant chromosomes and observed bumps along the chromosome (the chromomeres), which he thought might correspond to genes (Figure 5-16). Belling visualized the genes as beads strung together with some nongenic linking substance. He reported that during prophase I of meiosis, chromomeres duplicate, so that newly made chromomeres are attached to the originals (Figure 5-17). After duplication, the newly formed chromomeres are fastened together, but because all the chromomeres are tightly juxtaposed, the linking elements could switch from a newly made chromomere on one homologous chromosome to an adjacent one on the other homolog. Belling's model became known as the **copy-choice model** of crossing-over. You can see how it can generate a crossover chromatid that might appear to have arisen from a physical breakage and reunion of chromosomes (Figure 5-18). You can also see that it suggests that only the newly made chromomeres (and hence newly made chromatids) would recombine, so that any multiple crossover could involve only two chromatids.

Thus, the copy-choice model predicts that in every meiosis in which multiple crossovers occur (for example, double crossovers and even higher multiples), only two chromatids out of the four will ever be involved. This prediction could be tested if only we had some way of recovering all four products from individual meioses.

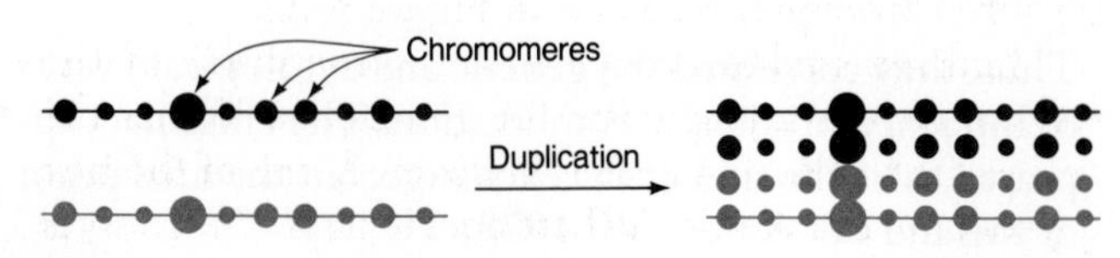

Figure 5-17 Duplication of the chromomeres according to Belling's model.

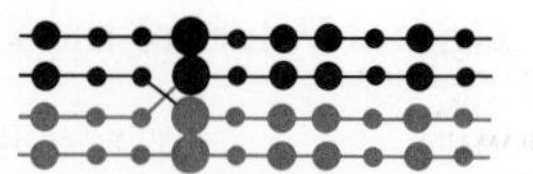

Figure 5-18 The hooking together of the newly synthesized chromomeres according to Belling's model. Joining usually forms parental combinations, but sometimes a switch can occur.

Each group of four could then be examined for the presence of multiple recombinants. In any group of four that contains at least one multiple recombinant, there must be two accompanying parental types for the copy-choice model to be possible. Luckily, several haploid organisms are suited to the recovery of all four products of a single meiosis. These organisms are some fungi and some unicellular algae. We consider the full analysis of these groups of four (called *tetrad analysis*) at length in Chapter 6. But we need them right now to answer one specific question: What companion genotypes are found in tetrads containing multiple recombinants? Several types of such tetrads are possible, but one interesting situation (often encountered in tetrad analysis) is illustrated in the following diagram, in which the genes are linked in the order shown:

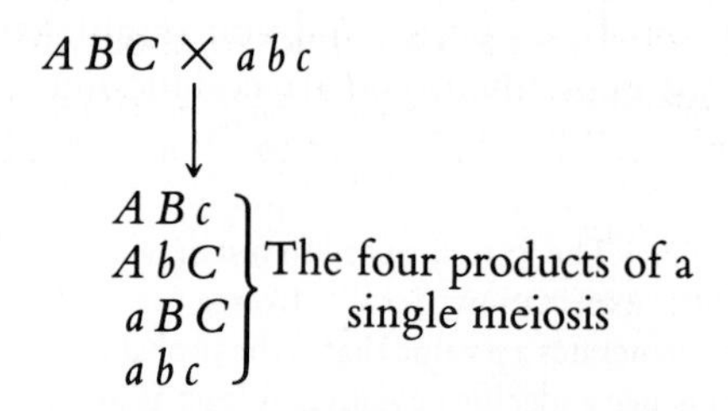

Note that the homologous chromosomes recombined between the first two loci (*A* and *B*) and also recombined in the same meiosis between loci *B* and *C*; there has been a double-exchange and one of the two products, *A b C*, is a double recombinant. *But* more than two chromatids must have been involved; in fact, *three must* have been involved (Figure 5-19). Thus, Belling's suggestion (in which only two chromatids could ever be involved) is most unlikely. (We shall see some *positive* evidence for the model of breakage and reunion in Chapter 19.)

Message The breakage and reunion model of chromosome crossing-over wins by default.

The ability to isolate the four products of meiosis in fungi and algae also clears up another mystery, about whether crossing-over occurs at the two-strand (two-chromosome) stage or at the four-strand (four-chroma-

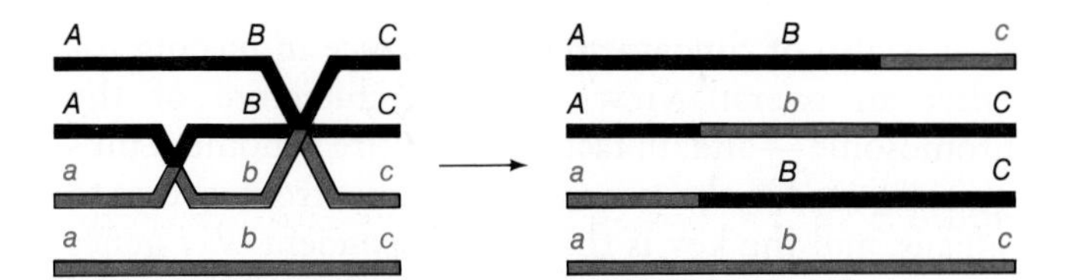

Figure 5-19 One of the several possible types of double-crossover tetrads that are regularly observed. Note that more than two chromatids exchanged parts. This evidence makes Belling's copy-choice model very unlikely.

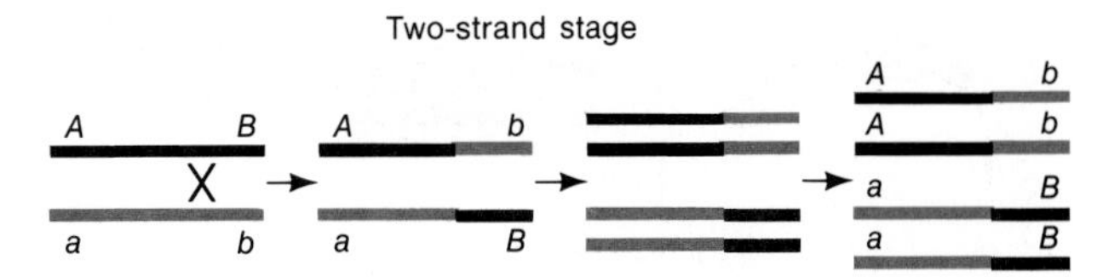

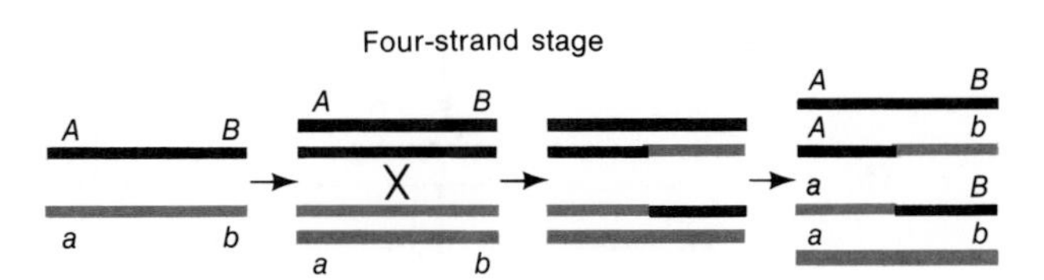

Figure 5-20 Tetrad analysis provides evidence that enabled geneticists to decide whether crossing-over occurs at the two-strand (two-chromosome) or at the four-strand (four-chromatid) stage of meiosis. Because more than two different products of a single meiosis can be seen in some tetrads, crossing-over cannot occur at the two-strand stage.

tid) stage. If it occurs at the two-strand stage (before replication), there can never be more than two different products of a given meiosis. If it occurs at the four-strand stage (after replication), up to four different products of meiosis are possible (Figure 5-20). In fact, four different products of a single meiosis are regularly observed, showing that crossing-over occurs at the four-strand stage of meiosis.

Crossing-over is a remarkably precise process. The synapsis and exchange of chromosomes is such that no segments are lost or gained, and four complete chromosomes emerge in a tetrad. At present, however, the actual mechanisms of crossing-over and interference are still somewhat of an enigma, although several attractive models do exist and a lot is known about the sorts of chemical reactions that might ensure precision at the molecular level. We shall return to these points in Chapter 19.

Linkage Mapping by Recombination in Humans

Although humans have thousands of autosomally inherited conditions, producing human linkage maps by the kind of analysis we have pursued in this chapter has not been easy. There are several reasons for this. First, the human genome is immense, with a large number of chromosomes and large distances between genes. Also, the types of crosses that are informative in linkage analysis, especially testcrosses of dihybrids, cannot be made for experimental purposes and are difficult to identify from data on marriages. Then, of course, human families are generally not large enough to provide the sample sizes needed to distinguish linkage from independent assortment. This problem can be partly overcome by combining data from many similar families. Despite the problems, considerable progress has been made in mapping the human genome. Part of the success is the recently developed technique of using as markers naturally occurring variation in DNA ("DNA polymorphisms"). Although these DNA markers generally have no phenotype at the organismal level, they are extremely numerous, and act as convenient reference points or "mileposts" for the overall mapping process. Another advance that has facilitated mapping of the human genome is a special cell culture method for assigning genes to parts of chromosomes. We shall discuss this in the next chapter. Other special techniques, especially molecular ones, have also played a major part. We shall encounter these in later chapters too.

The X chromosome, however, has been far more amenable to mapping by recombination analysis than have the autosomes. The reason is that males are hemizygous for X-linked genes, and, just as we did for *Drosophila,* if we look just at male progeny of a dihybrid female, we are effectively sampling her gametic output. Consider the following situation involving the rare recessive alleles for defective sugar processing (*g*) and, at another locus, for color blindness (*c*). A doubly affected male (*c g* / Y) marries a normal woman (who is almost certainly *C G* / *C G*). The daughters of this mating are coupling-conformation heterozygotes. When they marry, they will almost certainly marry normal men (*C G* / Y), and their male children will provide an opportunity for researchers to study the frequency of recombinants (Figure 5-21). The total frequency of recombinants may be estimated from the phenotypes in pedigrees of such situations. A human linkage map based on such techniques is shown in Figure 5-22.

Linkage studies occupy a large proportion of the routine day-to-day activities of geneticists. When the first variant allele of a gene is discovered, one of the first questions to be asked is "Where does it map?" Not only is this knowledge a necessary component in the engineering and maintenance of genetic stocks for research use,

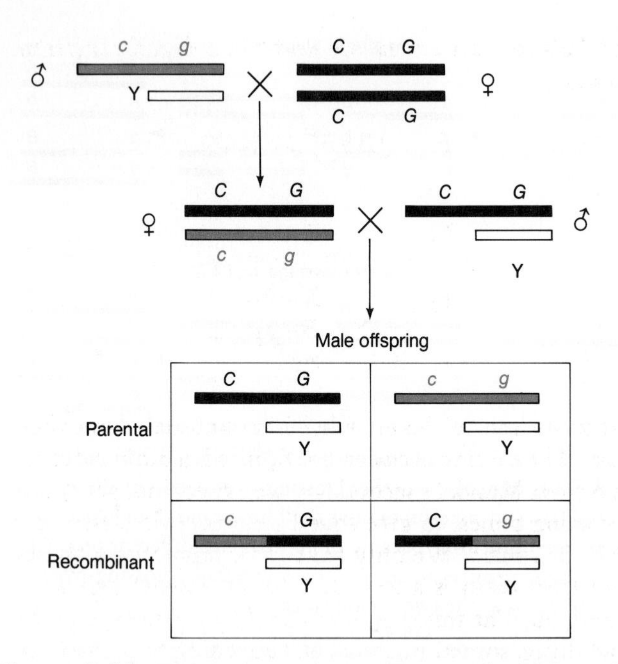

Figure 5-21 The phenotypic proportions in male children of women heterozygous for two X-linked genes can be used to calculate recombinant frequency. Thus, the X chromosome can be mapped by combining such pedigrees.

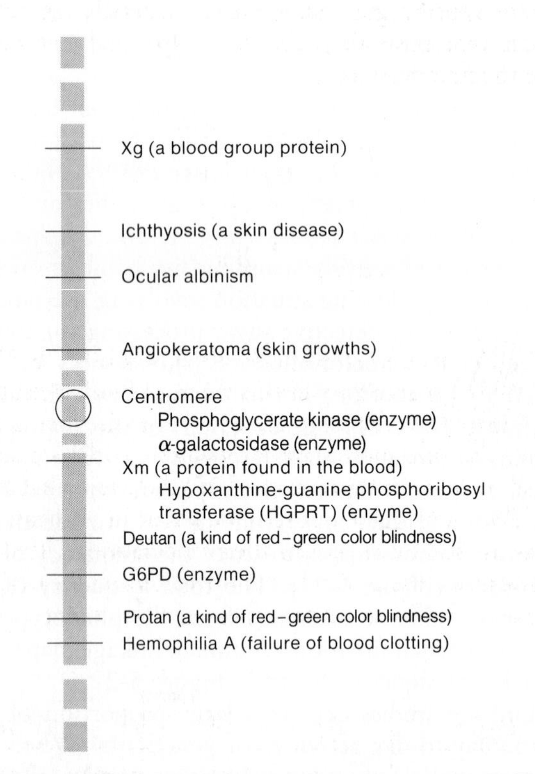

Figure 5-22 A map of the human X chromosome, derived from the analysis of recombinant frequencies. (From W. F. Bodmer and L. L. Cavalli-Sforza, *Genetics, Evolution, and Man*. Copyright 1976 by W. H. Freeman and Company.)

but it is also of fundamental importance in piecing together an overall view of the architecture of the chromosome—and, in fact, of the entire genome. Linkage is detected in the same way in studies from viruses to humans, and the key is the nonindependent way genes are transmitted from generation to generation. Look for this key when we discuss linkage in following chapters.

Summary

After making dihybrid crosses of sweet pea plants, William Bateson and R. C. Punnett discovered deviations from the 9 : 3 : 3 : 1 ratio of phenotypes expected in the F_2 generation. The parental gametic types outnumbered the other two classes. Later, in his studies of two different autosomal genes in *Drosophila,* Thomas Hunt Morgan found a similar deviation from Mendel's law of independent assortment. Morgan postulated that the two genes were located on the same pair of homologous chromosomes. This relationship is called linkage.

Linkage explains why the parental gene combinations stay together but not how the nonparental combinations arise. Morgan postulated that during meiosis there may be a physical exchange of chromosome parts by a process now called crossing-over, or intrachromosomal recombination. Thus, there are two types of meiotic recombination. Interchromosomal recombination is achieved by Mendelian independent assortment and results in a recombinant frequency of 50 percent. Intrachromosomal recombination occurs when the physical linkage of the parental gene combinations prevents their independent assortment and results in a recombinant frequency of less than 50 percent.

As Morgan studied more linked genes, he discovered many different values for recombinant frequency and wondered if these reflected the actual distances between genes on a chromosome. Alfred Sturtevant, a student of Morgan's, developed a method of determining the distance between genes on a linkage map, based on the percentage of recombinants. A linkage map is another example of a hypothetical entity based on genetic analysis.

Although the basic test for linkage is deviation from the 1 : 1 : 1 : 1 ratio of progeny types in a testcross, such a deviation may not be all that obvious. The χ^2 test, which tells how often observations deviate from expectations purely by chance, can help us determine whether loci are linked. The χ^2 test has other applications in genetics in the testing of observed against expected events.

Several theories about how recombinant chromosomes are generated have been set forth. We now know that crossing-over is the result of a physical breakage and a reunion of chromosome parts and that it occurs at the four-strand stage of meiosis.

Concept Map

Draw a concept map interrelating as many of the following terms as possible. Note that the terms are listed in no particular order.

crossing-over / chromatids / mapping / chromosomes / testcross / linkage / map units / recombination / additivity / independent assortment

Chapter Integration Problem

In *Drosophila,* the alleles *P* and *p* determine red and purple eyes, respectively. The alleles *B* and *b* determine brown and black body, respectively. A geneticist crossed a purple-eyed, brown-bodied female from a pure line with a red-eyed, black-bodied male, also from a pure line. The F_1 flies all had red eyes and brown bodies. The geneticist produced an F_2 by interbreeding the F_1 flies. The F_2 had the following composition:

brown, red	684
brown, purple	343
black, red	341
	1368

What do these data tell us about the chromosomal location of the genes?

Solution

When we think about chromosomal location, which of the phenomena we have studied so far come to mind? First, there is the distinction between autosomal and sex-linked inheritance. Then there are the alternative possibilities of linkage or independent assortment. There isn't much else that we should worry about at this stage. How can we sort out these possibilities? Let's start with sex linkage. The characteristic feature of sex linkage is that there is an inheritance pattern that is correlated with sex in some way. There is no evidence of this in our data; as we were given no information to the contrary, we must assume that equal numbers of males and females were obtained in each class. In the genetics of *Drosophila,* there is, however, a fact touching on the sex of the flies that must always be kept in mind; as we learned in this chapter, during meiosis in *Drosophila* males, there is no crossing over. We will put that notion on the back burner for the moment; it might become useful later.

So we have concluded that both genes must be autosomal. Now we can reconstruct the crosses using gene symbols to restate what we know in a slightly different format.

$$\begin{array}{ll} P & BBpp \times bbPP \\ & \quad\downarrow \\ F_1 & BbPp \\ & BbPp \times BbPp \\ F_2 & B{-}P{-} \\ & B{-}pp \\ & bbP{-} \end{array}$$

Now we can see better what is so unusual about these data. The mating between the F_1 flies is a dihybrid cross, and from Mendel's second law we expect independently assorting genes, to give four F_2 phenotypic classes in a 9:3:3:1 ratio. We only have three phenotypic classes, and their ratio is a 2:1:1. Clearly, something else is going on. The missing class is the *bb pp* genotype, so for one thing, something must be happening to prevent the production of this class.

Remembering the discussion of lethality in Chapter 4, we might speculate that the *bb pp* genotype is lethal. This is unlikely, however, because that would leave us with the 9:3:3 part of the ratio, which would reduce to 3:1:1, not 2:1:1. But we have just considered another phenomenon that can keep genes from assorting freely to produce specific genotypes, that is, linkage, so let us explore that possibility. For this hypothesis we could write the first cross as

$$\begin{array}{ll} \text{P} & \dfrac{B\,p}{B\,p} \times \dfrac{b\,P}{b\,P} \\ & \qquad\downarrow \\ F_1 & \qquad\dfrac{B\,p}{b\,P} \end{array}$$

If this is true, what ratios do we expect in the F_2? We might think that to answer this we must know the number of map units that separate the two loci. That is reasonable, but we must start somewhere, so let's assume that 20 map units separate the loci, then see what happens. (Don't forget about trial and error in analysis; scientists use it a lot as part of their calculations.) The expectations for the F_2 can be set up as follows, bearing in mind that there is no crossing-over in the males:

Eggs		*Sperm* *B p*	50%	*b P*	50%
B P	10%	*BB Pp*	5%	*Bb PP*	5%
b p	10%	*Bb pp*	5%	*bb Pp*	5%
B p	40%	*BB pp*	20%	*Bb Pp*	20%
b P	40%	*Bb Pp*	20%	*bb PP*	20%

Overall, the phenotypic proportions are

$B-P-$	$5+5+20+20$	$=50\%$
$B-pp$	$5+20$	$=25\%$
$bbP-$	$5+20$	$=25\%$
$bbpp$		$=0\%$

This gives us precisely the ratio that we need, 2:1:1! How did this happen? Were we just lucky in choosing the 20 percent recombinant frequency? The answer is no, it doesn't matter what frequency we might have chosen, the same result would have prevailed (try it).

In conclusion then, we can say that the ratio was produced by the inheritance of linked genes in repulsion phase in the parental strains, but we cannot determine how far apart they are. Precise mapping would require a testcross.

Although this was a tricky problem, the path we took through it has shown how a variety of different concepts can be used to rule out various possibilities and arrive at a solution that works.

Solved Problems

1. The allele b gives *Drosophila* flies a black body and b^+ gives brown, the wild-type phenotype. The allele wx of a separate gene gives waxy wings and wx^+ gives nonwaxy, the wild-type phenotype. The allele cn of a third gene gives cinnabar eyes and cn^+ gives red, the wild-type phenotype. A female heterozygous for these three genes is testcrossed, and 1000 progeny are classified as follows: 5 wild type; 6 black, waxy, cinnabar; 69 waxy, cinnabar; 67 black; 382 cinnabar; 379 black, waxy; 48 waxy; and 44 black, cinnabar.

a. Explain these numbers.

b. Draw the alleles in their proper positions on the chromosomes of the triple heterozygote.

c. If it is appropriate according to your explanation, calculate interference.

d. If two triple heterozygotes of the above type were crossed, what proportion of progeny would be black, waxy? (Remember that there is no crossing-over in *Drosophila* males.)

Solution

a. One of the general pieces of advice given earlier is to be methodical. Here it is a good idea to write out the genotypes that may be inferred from the phenotypes. The cross is a testcross of type

$$b^+b\ wx^+wx\ cn^+cn \times bb\ wxwx\ cncn$$

Notice that there are distinct pairs of progeny classes in terms of frequency. Already, we can guess that the two largest classes represent parental chromosomes, that the two classes of about 68 represent single crossovers in one region, that the two classes of about 45 represent single crossovers in the other region, and that the two classes of about 5 represent double crossovers. Note that a progeny group may be specified by listing only the non-wild-type phenotypes. We can write out the progeny as classes derived from the female's gametes, grouped as follows:

b^+	wx^+	cn	382
b	wx	cn^+	379
b^+	wx	cn	69
b	wx^+	cn^+	67
b^+	wx	cn^+	48
b	wx^+	cn	44
b	wx	cn	6
b^+	wx^+	cn^+	5
			1000

Writing the classes out this way confirms that the pairs of classes are in fact reciprocal genotypes arising from zero, one, or two crossovers.

At first, because we do not know the parents of the triple heterozygous female, it looks as if we cannot apply the definition of recombination in which gametic genotypes are compared with the two input genotypes that form an individual. But on reflection, the only parental types that make sense in terms of the data presented are $b^+b^+\ wx^+wx^+\ cncn$ and $bb\ wxwx\ cn^+cn^+$ because these are still the most common classes.

Now we can calculate the recombinant frequencies. For

$b-wx$, the RF

$$= \frac{69+67+48+44}{1000} = 22.8\%$$

$$b-cn\text{, the RF} = \frac{48+44+6+5}{1000} = 10.3\%$$

$$wx-cn\text{, the RF} = \frac{69+67+6+5}{1000} = 14.7\%$$

The map is therefore

b ——— cn ——— wx

← 10.3 m.u. →← 14.7 m.u. →

b. The parental chromosomes in the triple heterozygote were

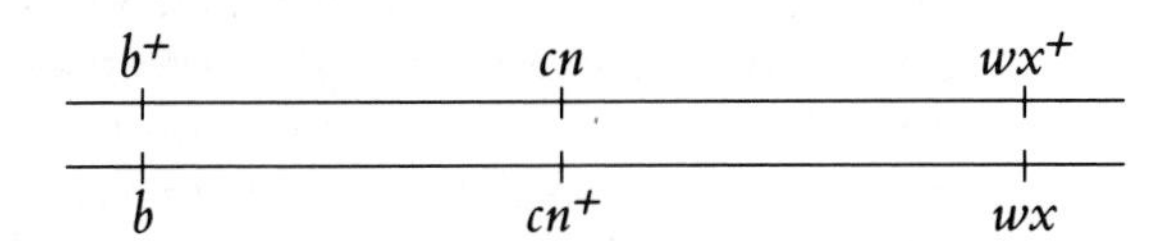

c. The expected number of double recombinants is $0.103 \times 0.147 \times 1000 = 15.141$. The observed number is $6 + 5 = 11$, so interference can be calculated as $I = 1 - 11/15.141 = 1 - 0.726 = 0.274 = 27.4$ percent.

d. Here we are asked to use the newly gained knowledge of the linkage of these genes to predict the outcome of a cross. Note that we are not dealing with a testcross here. We are asked for the expected proportion of black, waxy flies among the progeny of two triple heterozygotes. Because there is no crossing-over in the male, one-half of his gametes are $b\ cn^+\ wx$ and the other half are $b^+\ cn\ wx^+$ but the latter are incapable of contributing to the black, waxy phenotype. In the female, two gametic types contribute to black, waxy offspring: $b\ cn^+\ wx$ and $b\ cn\ wx$. The contriutions of the parents to the black, waxy phenotype are shown here:

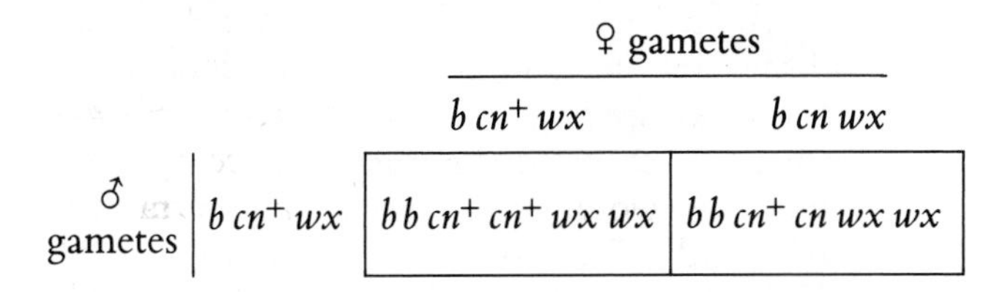

		♀ gametes	
		$b\ cn^+\ wx$	$b\ cn\ wx$
♂ gametes	$b\ cn^+\ wx$	$b\,b\ cn^+\ cn^+\ wx\ wx$	$b\,b\ cn^+\ cn\ wx\ wx$

The chromosome carrying $b\ cn^+\ wx$ is a nonrecombinant parental type. There is a total of $382 + 379/1000 = 76.1$ percent of nonrecombinant parental chromosomes, half of which, or 38.05 percent, carry $b\ cn^+\ wx$. The frequency of $b\ cn\ wx$ is $11 \div 2/1000 = 0.55$ percent. In total, then, $38.05 + 0.55 = 38.6$ percent of the female's gametes are capable of forming a black, waxy offspring; one-half of these fuse with a $b\ cn^+\ wx$ gamete from the male, so the frequency of black, waxy offspring is $38.6/2 = 19.3$ percent.

In summary, the appropriate gametic fusion can be shown this way:

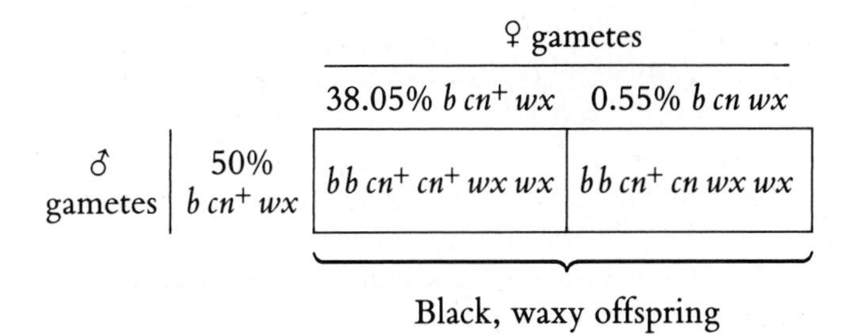

		♀ gametes	
		38.05% $b\ cn^+\ wx$	0.55% $b\ cn\ wx$
♂ gametes	50% $b\ cn^+\ wx$	$b\,b\ cn^+\ cn^+\ wx\ wx$	$b\,b\ cn^+\ cn\ wx\ wx$

Black, waxy offspring

2. A human pedigree shows persons affected with the rare nail-patella syndrome (misshapen nails and kneecaps) and also gives the ABO blood-group genotype of each individual. Both loci concerned are autosomal. Study the pedigree below.

a. Is the nail-patella syndrome a dominant or recessive phenotype? Give reasons to support your answer.

b. Is there evidence of linkage between the nail-patella gene and the gene for ABO blood type, as judged from this pedigree? Why or why not?

c. If there is evidence of linkage, then draw the alleles on the relevant homologs of the grandparents. If there is no evidence of linkage, draw the alleles on two homologous pairs.

d. According to your model, which descendants represent recombinants?

e. What is the best estimate of RF?

f. If man III-1 marries a normal woman of blood type O, what is the probability that their first child will be blood type B with nail-patella syndrome?

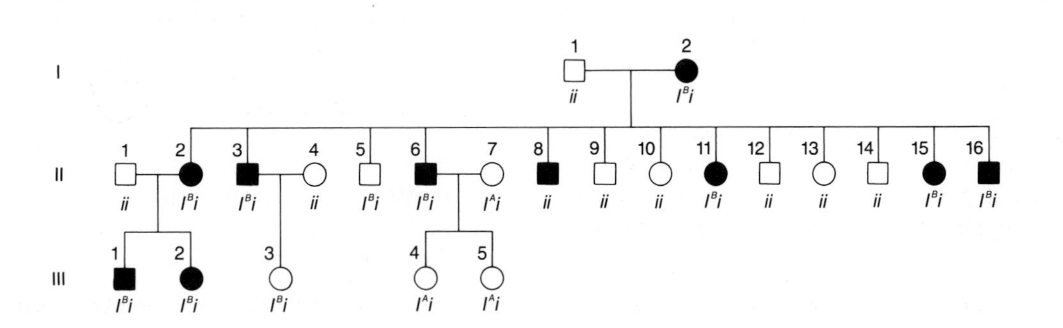

Solution

a. Nail-patella syndrome is most likely dominant. We are told that it is a rare abnormality, so it is unlikely that the unaffected people marrying into the family carry a presumptive recessive allele for nail-patella syndrome. Let N be the causative allele. Then all people with the syndrome are heterozygotes Nn because all (probably including the grandmother, too) result from a mating to an nn normal individual. Notice that the syndrome appears in all three successive generations—another indication of dominant inheritance.

b. There is evidence of linkage. Notice that most of the affected persons—those that carry the N allele—also carry the I^B allele; most likely, these alleles are linked on the same chromosome.

$$\frac{n \quad i}{n \quad i} \times \frac{N \quad I^B}{n \quad i}$$

(The grandmother must carry both recessive alleles in order to produce offspring of genotype $i\,i$ and $n\,n$.)

d. Notice that the grandparental mating is equivalent to a testcross, so the recombinants in generation II are

#5: $n\,I^B / n\,i$ and #8: $N\,i / n\,i$

whereas all others are nonrecombinants, being either $N\,I^B / n\,i$ or $n\,i / n\,i$.

e. Notice that the grandparental cross and the first two crosses in generation II are identical and, of course, are all testcrosses. Three of the total 16 progeny are recombinant (II-5, II-8, and III-3). This gives a recombinant frequency of RF = $\frac{3}{16}$ = 18.8 percent. (We cannot include the cross of II-6 × II-7, because the progeny cannot be designated as recombinant or not.)

f.

(III-1 ♂) $\frac{N \quad I^B}{n \quad i} \times \frac{n \quad i}{n \quad i}$ (normal type O ♀)

↓

Gametes

	Gamete	Frequency	
81.2%	$N\,I^B$	40.6%	←Nail-patella, blood type B
	$n\,i$	40.6%	
18.8%	$N\,i$	9.4%	
	$n\,I^B$	9.4%	

The two parental classes are always equal, and so are the two recombinant classes. Hence, the probability that the first child will have nail-patella syndrome and blood type B is 40.6 percent.

Problems

1. A plant of genotype

$$\frac{A \quad B}{a \quad b}$$

is testcrossed to

$$\frac{a \quad b}{a \quad b}$$

If the two loci are 10 m.u. apart, what proportion of progeny will be $A\,a\,B\,b$?

2. The A locus and the D locus are so tightly linked that no recombination is ever observed between them. If $A\,A\,d\,d$ is crossed to $a\,a\,D\,D$, and the F_1 is intercrossed, what phenotypes will be seen in the F_2 and in what proportions?

3. The R and S loci are 35 m.u. apart. If a plant of genotype

$$\frac{R \quad S}{r \quad s}$$

is selfed, what progeny phenotypes will be seen and in what proportions?

4. The cross $E\,E\,F\,F \times e\,e\,f\,f$ is made, and the F_1 is then backcrossed to the recessive parent. The progeny genotypes are inferred from the phenotypes. The progeny genotypes, written as the gametic contributions of the heterozygous parent, are in the following proportions:

$E\,F$	$\frac{2}{6}$
$E\,f$	$\frac{1}{6}$
$e\,F$	$\frac{1}{6}$
$e\,f$	$\frac{2}{6}$

Explain these results.

5. A strain of *Neurospora* with the genotype $H\,I$ is crossed with a strain with the genotype $h\,i$. One-half of the progeny are $H\,I$, and one-half are $h\,i$. Explain how this is possible.

6. A female animal with genotype $A\,a\,B\,b$ is crossed with a double-recessive male ($a\,a\,b\,b$). Their progeny includes 442 *Aa Bb*, 458 *aa bb*, 46 *Aa bb*, and 54 *aa Bb*. Explain these results.

7. If *AA BB* is crossed to *aa bb*, and the F_1 is testcrossed, what percent of the testcross progeny will be *aa bb* if the two genes are **(a)** unlinked; **(b)** completely linked (no crossing-over at all); **(c)** 10 map units apart; **(d)** 24 map units apart?

8. In a haploid organism, the *C* and *D* loci are 8 m.u. apart. From a cross *C d* × *c D*, give the proportion of each of the following progeny classes: **(a)** *C D*, **(b)** *c d*, **(C)** *C d*, **(d)** all recombinants.

9. A fruit fly of genotype *B R* / *b r* is testcrossed to *b r* / *b r*. In 84 percent of the meioses, there are no chiasmata between the linked genes; in 16 percent of the meioses, there is one chiasma between the genes. What proportion of the progeny will be *Bb rr*? **(a)** 50 percent, **(b)** 4 percent, **(c)** 84 percent, **(d)** 25 percent, **(e)** 16 percent?

10. An individual heterozygous for four genes, *Aa Bb Cc Dd*, is testcrossed to *aa bb cc dd*, and 1000 progeny are classified by the gametic contribution of the heterozygous parent as follows:

a B C D	42
A b c d	43
A B C d	140
a b c D	145
a B c D	6
A b C d	9
A B c d	305
a b C D	310

a. Which genes are linked?

b. If two pure-breeding lines were crossed to produce the heterozygous individual, what were their genotypes?

c. Draw a linkage map of the linked genes, showing the order and the distances in map units.

d. Calculate an interference value, if appropriate.

11. The squirting cucumber, *Echballium elaterium*, has two separate sexes (it is dioecious). The sexes are determined, not by heteromorphic sex chromosomes, but by the alleles of two genes. The alleles at the two loci govern sexual phenotypes as follows: *M* determines male fertility; *m* determines male sterility; *F* determines female sterility; *f* determines female fertility. In populations of this plant, individuals can be male (approximately 50 percent) or female (approximately 50 percent). In addition, a hermaphroditic type is found, but only at a very low frequency. The hermaphrodite has male and female sex organs on the same plant.

a. What must be the full genotype of a male plant? (Indicate linkage relations of the genes.)

b. What must be the full genotype of a female plant? (Indicate linkage relations of the genes.)

c. How does the population maintain an approximately equal proportion of males and females?

d. What is the origin of the rare hermaphrodite?

e. Why are hermaphrodites rare?

***12.** There is an autosomal gene *N* in humans that causes abnormalities in nails and patellae (kneecaps), called the nail-patella syndrome. Consider marriages in which one partner has the nail-patella syndrome and blood type A and the other partner has normal nail-patella and blood type O. These marriages produce some children who have both the nail-patella syndrome and blood type A. Assume that unrelated children from this phenotypic group mature, intermarry, and have children. Four phenotypes are observed in the following percentages in this second generation:

nail-patella syndrome, blood type A	66%
normal nail-patella, blood type O	16%
normal nail-patella, blood type A	9%
nail-patella syndrome, blood type O	9%

Fully analyze these data, explaining the relative frequencies of the four phenotypes.

***13.** Using the data obtained by Bateson and Punnett (Table 5-1), calculate the map distance (in m.u.) separating the color and shape genes. (This will require some trial and error.)

14. You have a *Drosophila* line that is homozygous for autosomal recessive alleles *a*, *b*, and *c*, linked in that order. You cross females of this line with males homozygous for the corresponding wild-type alleles. You then cross the F_1 heterozygous males with their heterozygous sisters. You obtain the following F_2 phenotypes (where letters denote recessive phenotypes and pluses denote wild-type phenotypes): 1364 + + +, 365 a b c, 87 a b +, 84 + + c, 47 a + +, 44 + b c, 5 a + c, and 4 + b +.

***a.** What is the recombinant frequency between *a* and *b*? Between *b* and *c*?

b. What is the coefficient of coincidence?

15. R. A. Emerson crossed two different pure-breeding lines of corn and obtained a phenotypically wild-type F_1 that was heterozygous for three alleles that determine recessive phenotypes: *an* determines anther; *br*, brachytic; and *f*, fine. He testcrossed the F_1 to a tester that was homozygous recessive for the three genes and obtained these progeny phenotypes: 355 anther; 339 brachytic, fine; 88 completely wild-type; 55 anther, brachytic, fine; 21 fine; 17 anther, brachytic; 2 brachytic; 2 anther, fine.

a. What were the genotypes of the parental lines?

b. Draw a linkage map for the three genes (include map distances).

c. Calculate the interference value.

16. Chromosome 3 of corn carries three loci having alleles b and b^+, v and v^+, and lg and lg^+. The corresponding recessive phenotypes are abbreviated b (for plant-color booster), v (for virescent), and lg (for liguleless); pluses denote wild-type phenotypes. A testcross of triple recessives with F_1 plants heterozygous for the three genes yields the following progeny phenotypes: 305 + v lg, 275 b + +, 128 b + lg, 112 + v +, 74 + + lg, 66 b v +, 22 + + +, and 18 b v lg. Give the gene sequence on the chromosome, the map distances between genes, and the coefficient of coincidence.

17. Groodies are useful (but fictional) haploid organisms that are pure genetic tools. A wild-type groody has a fat body, a long tail, and flagella. Non-wild-type lines are known that have thin bodies, or are tailless, or do not have flagella. Groodies can mate with each other (although they are so shy that we do not know how) and produce recombinants. A wild-type groody mates with a thin-bodied groody lacking both tail and flagella. The 1000 baby groodies produced are classified as shown in the following figure.

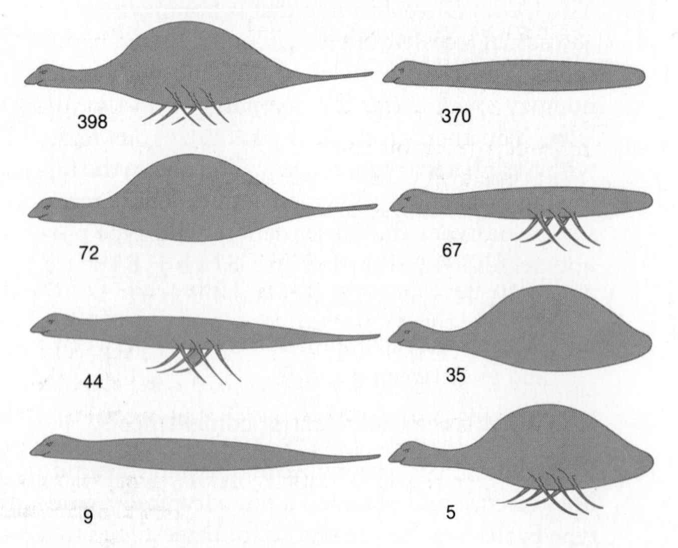

Assign genotypes, and map the three genes.

(Problem 17 from Burton S. Guttman.)

18. Assume that three pairs of alleles are found in *Drosophila:* x^+ and x, y^+ and y, and z^+ and z. As shown by the symbols, each non-wild-type allele is recessive to its wild-type allele. A cross between females heterozygous at these three loci and wild-type males yields the following progeny, classified by the gametic genotypes contributed by the females: 1010 $x^+ y^+ z^+$ females; 430 $x\, y^+ z$ males; 441 $x^+ y\, z^+$ males; 39 $x\, y\, z$ males; 32 $x^+ y^+ z$ males; 30 $x^+ y^+ z^+$ males; 27 $x\, y\, z^+$ males; 1 $x^+ y\, z$ male; and 0 $x\, y^+ z^+$ males.

a. In what chromosome of *Drosophila* are these genes carried?

b. Draw the relevant chromosomes in the heterozygous female parent, showing the arrangement of the alleles.

c. Calculate the map distances between the genes and the coefficient of coincidence.

19. From the five sets of data given in the following table determine the order of genes by inspection—that is, without calculating recombination values. Recessive phenotypes are symbolized by lowercase letters and dominant phenotypes by pluses.

Phenotypes observed in 3-point testcross	Data set 1	2	3	4	5
+ + +	317	1	30	40	305
+ + c	58	4	6	232	0
+ b +	10	31	339	84	28
+ b c	2	77	137	201	107
a + +	0	77	142	194	124
a + c	21	31	291	77	30
a b +	72	4	3	235	1
a b c	203	1	34	46	265

20. From the phenotypic data given below for two three-point testcroses involving (1) a, b, and c, and (2) b, c, and d, determine the sequence of the four genes a, b, c, and d, and the three map distances between them. Recessive phenotypes are symbolized by lower-case letters and dominant phenotypes by pluses.

(1)		(2)	
+ + +	669	b c d	8
a b +	139	b + +	441
a + +	3	b + d	90
+ + c	121	+ c d	376
+ b c	2	+ + +	14
a + c	2,280	+ + d	153
a b c	653	+ c +	64
+ b +	2,215	b c +	141

21. In *Drosophila,* the allele dp^+ determines long wings and *dp* determines short ("dumpy") wings. At a separate locus e^+ determines gray body and *e* determines ebony body. Both loci are autosomal. The following crosses were made, starting with pure-breeding parents:

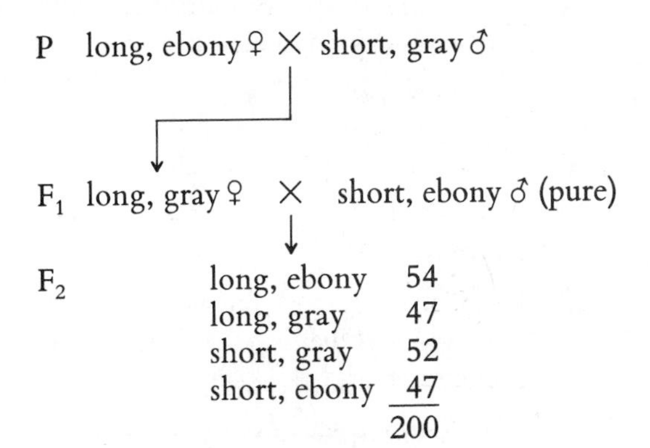

Use the χ^2 test to determine if these loci are linked. In doing so, state **(a)** hypothesis, **(b)** calculation of χ^2, **(c)** *p* value, **(d)** what the *p* value means, **(e)** conclusion, **(f)** inferred chromsomal constitutions of parents, F_1, tester, and progeny.

22. Two strains of haploid yeast each carrying two non-wild-type alleles were crossed to determine gene linkage. Strain A had *asp* and *gal*, which caused it to require aspartate and be unable to utilize galactose; strain B had *rad* and *aro*, which caused it to be sensitive to radiation and to require aromatic amino acids. Haploid progeny were obtained and tested for genotype with the following results (where pluses denote wild-type alleles):

Genotype	Frequency
asp gal + +	0.136
asp + + *rad*	0.136
asp gal + *rad*	0.064
asp + + +	0.064
+ *gal aro* +	0.136
+ + *aro rad*	0.136
+ *gal aro rad*	0.064
+ + *aro* +	0.064
asp gal aro +	0.034
asp + *aro rad*	0.034
asp gal aro rad	0.016
asp + *aro* +	0.016
+ *gal* + +	0.034
+ + + *rad*	0.034
+ *gal* + *rad*	0.016
+ + + +	0.016
	1.000

a. Calculate the six recombinant frequencies.

b. Draw a linkage map to illustrate positions of the four genetic loci and label in map units.

23. A geneticist puts a female mouse from a line that breeds true for wild-type eye and body color with a male mouse from a pure line having apricot eyes (determined by allele *a*) and gray coats (determined by allele *g*). The mice mate and produce an F_1 that is all wild type. The F_1 mice intermate and produce an F_2 of the following composition:

females	all wild type	
males	wild type	45%
	apricot, gray	45%
	gray	5%
	apricot	5%

a. Explain these frequencies.

b. Give the genotypes of the parents, and of both sexes of the F_1 and F_2.

24. The mother of a family with ten children has blood type Rh^+. She also has a very rare condition (elliptocytosis, phenotype E) that causes red blood cells to be oval rather than round in shape but that produces no adverse clinical effects. The father is Rh^- (lacks the Rh^+ antigen) and has normal red cells (phenotype e). The children are 1 Rh^+ e, 4 Rh^+ E, and 5 Rh^- e. Information is available on the mother's parents, who are Rh^+ E and Rh^- e. One of the ten children (who is Rh^+ E) marries someone who is Rh^+ e, and they have an Rh^+ E child.

a. Draw the pedigree of this whole family.

b. Is the pedigree in agreement with the hypothesis that the Rh^+ allele is dominant and Rh^- is recessive?

c. What is the mechanism of transmission of elliptocytosis?

***d.** Could the genes governing the E and Rh phenotypes be on the same chromosome? If so, estimate the map distance between them, and comment on your result.

25. The father of Mr. Spock, first officer of the starship *Enterprise,* came from planet Vulcan; Spock's mother came from Earth. A Vulcan has pointed ears (determined by allele *P*), adrenals absent (determined by *A*), and a right-sided heart (determined by *R*). All of these alleles are dominant to normal Earth alleles. The three loci are autosomal, and they are linked as shown in this linkage map:

P ——— *A* ——— *R*

← 15 m.u. →← 20 m.u. →

If Mr. Spock marries an Earth woman and there is no (genetic) interference, what proportion of their children

a. will have Vulcanian phenotypes for all three characters?

b. will have Earth phenotypes for all three characters?

c. will have Vulcanian ears and heart but Earth adrenals?

d. will have Vulcanian ears but Earth heart and adrenals?

(Problem 25 from D. Harrison, *Problems in Genetics*, Addison-Wesley, 1970.)

26. In a certain diploid plant, the three loci A, B, and C are linked as follows:

A		B		C
←	20 m.u.	→←	30 m.u.	→

One plant is available to you (call it the parental plant). It has the constitution $A\,b\,c\,/\,a\,B\,C$.

a. Assuming no interference, if the plant is selfed, what proportion of the progeny will be of the genotype $a\,b\,c\,/\,a\,b\,c$?

b. Again assuming no interference, if the parental plant is crossed with the $a\,b\,c\,/\,a\,b\,c$ plant, what genotypic classes will be found in the progeny? What will be their frequencies if there are 1000 progeny?

***c.** Repeat part b, this time assuming 20 percent interference between the regions.

27. From several crosses of the general type $A A\,B B \times a a\,b b$ the F_1 individuals of type $A a\,B b$ were testcrossed to $a a\,b b$. The results are shown below.

Testcross of F_1 from cross	Testcross progeny $A a\,B b$	$a a\,b b$	$A a\,b b$	$a a\,B b$
1	310	315	287	288
2	36	38	23	23
3	360	380	230	230
4	74	72	50	44

For each set of progeny, use the χ^2 test to decide if there is evidence of linkage.

28. Certain varieties of flax show different resistances to specific races of the fungus called flax rust. For example, the flax variety 77OB is resistant to rust race 24 but susceptible to rust race 22, whereas flax variety Bombay is resistant to rust race 22 and susceptible to rust race 24. When 77OB and Bombay were crossed, the F_1 hybrid was resistant to both rust races. When selfed, the F_1 produced an F_2 containing the phenotypic proportions shown below, where R stands for resistant and S stands for susceptible.

		Rust race 22 R	S
Rust race 24	R	184	63
	S	58	15

a. Propose a hypothesis to account for the genetic basis of resistance in flax to these particular rust races. Make a concise statement of the hypothesis, and define any gene symbols you use. Show your proposed genotypes of the 77OB, Bombay, F_1, and F_2 flax plants.

b. Test your hypothesis, using the χ^2 test. Give the expected values, the value of χ^2 (to two decimal places), and the appropriate probability value. Explain exactly what this value means. Do you accept or reject your hypothesis on the basis of the χ^2 test?

(Problem 28 adapted from M. Strickberger, *Genetics*, Macmillan, 1968.)

29. Here are two pedigrees in which a vertical bar in a symbol stands for steroid sulfatase deficiency, and a horizontal bar stands for ornithine transcarbamylase deficiency.

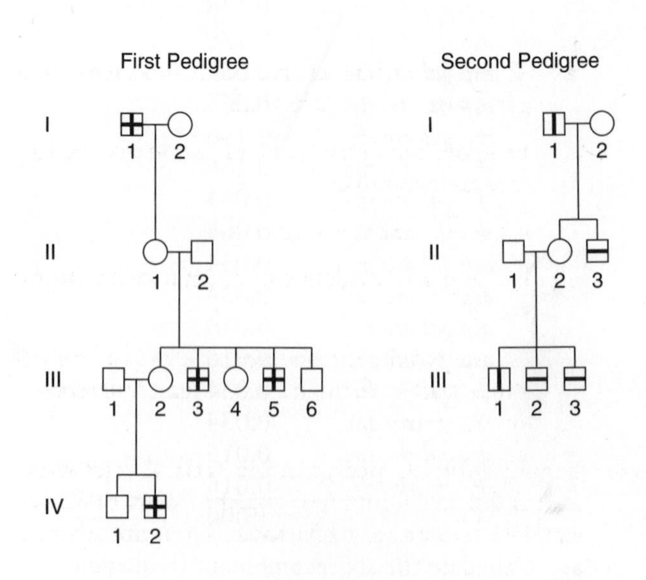

a. Is there any evidence in these pedigrees that the genes determining the deficiencies are linked?

b. If the genes are linked, is there any evidence in the pedigree of crossing-over between them?

c. Draw genotypes of these individuals as far as possible.

30. The following pedigree shows a family with two rare abnormal phenotypes, blue sclerotics (a brittle bone defect), represented by the black borders to the symbols, and hemophilia, represented by the black centers to the symbols. Individuals represented by completely black symbols have both disorders.

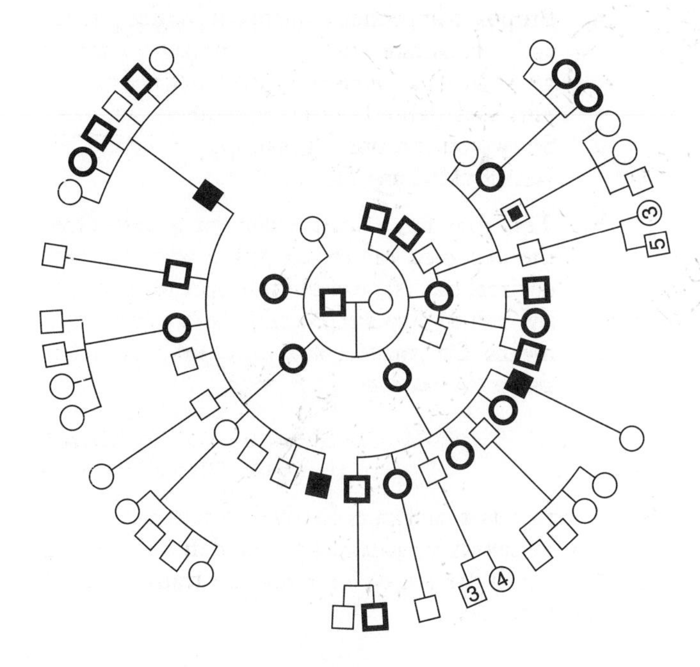

a. What pattern of inheritance is shown by each condition in this pedigree?

b. Provide the genotypes of as many family members as possible.

c. Is there evidence of linkage?

d. Is there any evidence of independent assortment?

e. Can any of the members be judged as recombinants (that is, formed from at least one recombinant gamete)?

31. In the following pedigree, the vertical lines stand for protan colorblindness, and the horizontal lines stand for deutan colorblindness. These are separate conditions causing different misperceptions of colors; each is determined by a separate gene.

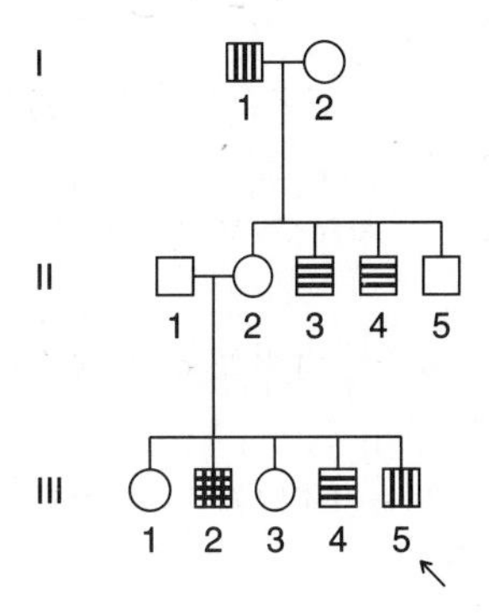

a. Does the pedigree show any evidence that the genes are linked?

b. If there is linkage, does the pedigree show any evidence of crossing-over? Explain both of your answers with the aid of the diagram.

c. Can you calculate a value for the recombination between these genes? Is this an intrachromosomal or an interchromosomal type of recombination?

***32.** The human genes for colorblindness and for hemophilia are both on the X chromosome, and they show a recombinant frequency of about 10 percent. Linkage of a pathological gene to a relatively harmless one can be used for genetic prognosis. Here is part of a more extensive pedigree. Blackened symbols indicate that the persons had hemophilia and crosses indicate colorblindness.

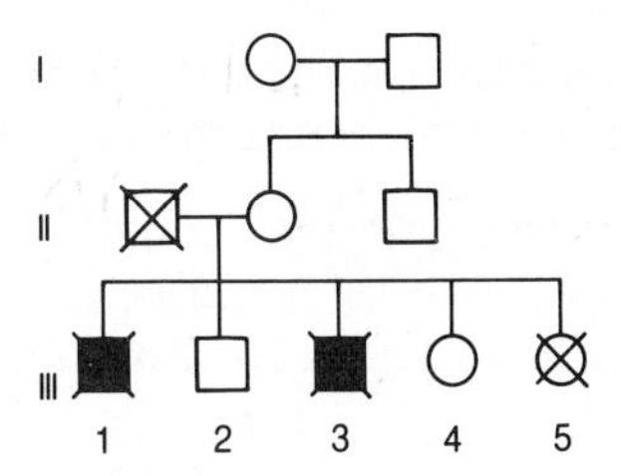

What information could be given to the women III 4 and III 5 as to the likelihood of their having sons with hemophilia?

(Problem 32 adapted from J. F. Crow, *Genetics Notes: An Introduction to Genetics,* Burgess Publishing, 1983.)

33. A geneticist mapping the genes A, B, C, D, and E makes two three-point testcrosses. The first cross of pure lines is

$$AA\,BB\,CC\,DD\,EE \times aa\,bb\,CC\,dd\,EE$$

The geneticist crosses the F_1 with a recessive tester and classifies the progeny by the gametic contribution of the F_1:

A B C D E	316
a b C d E	314
A B C d E	31
a b C D E	39
A b C d E	130
a B C D E	140
A b C D E	17
a B C d E	13
	1000

The second cross of pure lines is

$$AA\,BB\,CC\,DD\,EE \times aa\,BB\,cc\,DD\,ee$$

The geneticist crosses the F_1 from this cross with a recessive tester and obtains

A B C D E	243
a B c D e	237
A B c D e	62
a B C D E	58
A B C D e	155
a B c D E	165
a B C D e	46
A B c D E	34
	1000

The geneticist also knows that genes *D* and *E* assort independently.

a. Draw a map of these genes, showing distances in map units wherever possible.

b. Is there any evidence of interference?

6

Linkage II: Special Eukaryotic Chromosome Mapping Techniques

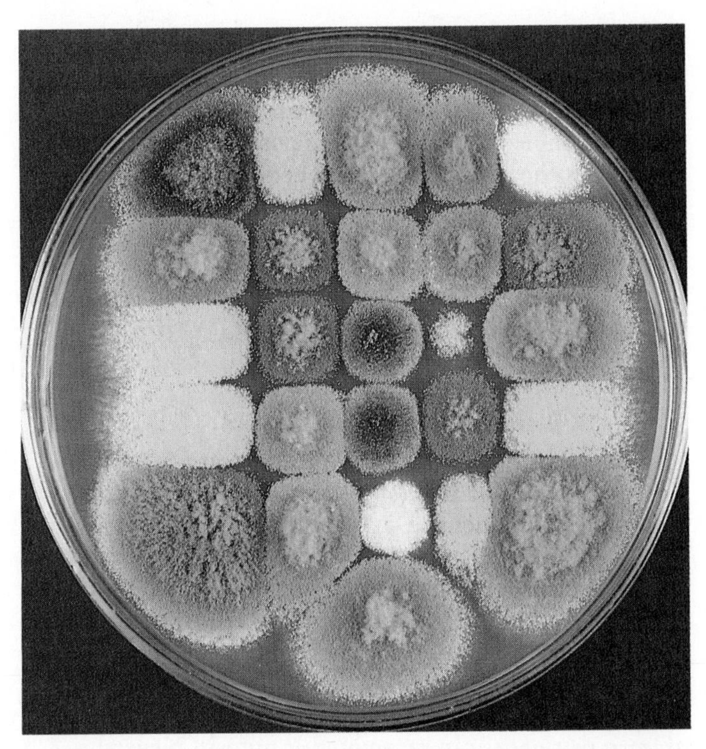

Some of the color variants of the Ascomycete fungus *Aspergillus,* normally dark green. The alleles producing such colors have been useful as markers in the study of mitotic crossing-over, one of the subjects of this chapter. Such colors are usually pigments that are intermediates in the biochemical reaction sequences terminating with the green pigment. (J. Peberdy, Department of Life Sciences, University of Nottingham, England)

KEY CONCEPTS

Double and higher multiple crossovers cause underestimates of map distances calculated from recombinant frequencies.

Special mapping formulas provide accurate map distances corrected for the effect of multiple crossovers.

Analysis of individual meioses can position centromeres on genetic maps.

Analysis of individual meioses provides insight into the mechanisms of gene segregation and recombination during meiosis.

Occasionally chromosomes cross over and assort in diploid cells undergoing *mitosis.*

Hybrid cells produced by fusing human and rodent cells have provided a powerful new technique for determining the chromosomal locations of human genes.

In Chapter 5 we learned the basic method for mapping eukaryotic chromosomes. According to that method, the investigator measures the recombinant frequency in a random sample of products of meiosis, and from this determines the degree of linkage between the genes in question. In this chapter we will extend the analysis beyond the three key concepts of the previous sentence:

recombinant frequency,

random products, and

meiosis

First, recombinant frequency. We will learn that measuring recombinant frequency does not always give an accurate measurement of map distance, and sometimes corrections have to be made. Second, random products. Some organisms, because of the way they complete their life cycles, allow the investigator to study the products of individual meioses, and this permits insightful studies of recombination that are not possible using random meiotic products. Third, meiosis. It might come as a surprise to learn that genes can recombine at mitosis. Nevertheless, in some cases they do, and this opens up some important mapping methodologies in those situations.

Mapping chromosomes is one of the central activities of geneticists. The goal of genetics, after all, is to understand the structure, function, and evolution of the genome, and, clearly, knowing the locations of genes is central to this task. The concepts in this chapter are all part of the everyday vocabulary of genetics. In the same way that earlier we extended Mendel's basic rules of segregation, now we have to extend the basic rules of mapping.

Accurate Calculation of Large Map Distances

In Chapter 5 we saw that the basic genetic method of measuring map distance is based on recombinant frequency (RF). A genetic map unit (m.u.) was defined as a recombinant frequency of 1 percent. This is a useful fundamental unit that has stood the test of time and is still used in genetics. However, the larger the recombinant frequency, the less accurate it is as a measure of map distance. In fact, map units calculated from larger recombinant frequencies are smaller than map units calculated from smaller recombinant frequencies! How can this be demonstrated? We have already encountered the effect in examples from the previous chapter. Typically when measuring recombination between three linked loci, the sum of the two internal recombinant frequencies is greater than the recombinant frequency between the outside loci, as shown in the following diagram.

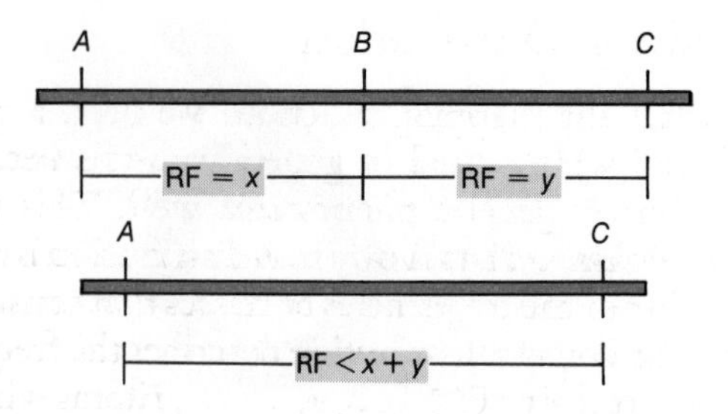

Using such data, what is the most accurate estimate that can be made of map distance between the two outside loci A and C? The answer is that $x + y$ is the best estimate, and more accurate than the smaller overall $A-C$ value. This gives us a useful mapping principle, which is that the best estimates of map distance are obtained from the sum of the distances calculated for shorter intervals.

However, what if we have no intervening marker loci available to measure recombination in shorter intervals? This would be a situation commonly encountered when beginning to map a new experimental organism, or in cases where the genome is huge as it is in human beings. For example, in the above diagram, what if there were no known B locus? Would we have to make do with the map distance value obtained directly from the $A-C$ recombination frequency? Furthermore, what about the shorter intervals themselves? If there were other loci between A and B, and between C and D, then we might obtain even better estimates of the A to D distance. Luckily, there is a way of taking any recombinant frequency and performing a calculation to make it a more accurate measure of map distance, without studying shorter and shorter intervals.

Before we consider the calculation, let's think about the reason why larger RF values are less accurate measures of map distance. We have already encountered the culprit: multiple crossovers. In the previous chapter we saw that double crossovers often lead to a parental arrangement of alleles and therefore the resulting meiotic products are not counted when measuring recombinant frequency. The same is true for other types of multiple crossovers: triples, quadruples, and so on. So it is easy to see that these automatically lead to an underestimate of map distance, and since the multiples are expected to be relatively more common over longer regions, we can see why the problem is worse for larger recombinant frequencies.

How can we take these multiple crossovers into account when calculating map distances? What we need is a mathematical function that accurately relates recombination to map distance. In other words, what we need is a **mapping function.**

Message The mapping function estimates a map distance better than do recombinant frequencies alone.

The Poisson Distribution

To calculate the mapping function, we need a mathematical tool widely used in genetic analysis because it describes many genetic phenomena well. This tool is called the **Poisson distribution.** A distribution is merely a description of the frequencies of classes that constitute a sample. The Poisson distribution describes the frequency of classes containing 0, 1, 2, 3, 4, . . . , *i* items when the average number of items per sample is small in relation to the total number of items possible. For example, the *possible* number of tadpoles obtainable in a single dip of a net in a pond is quite large, but most dips yield only one or two or none. The number of dead birds on the side of the highway is potentially very large, but in a sample kilometer the number is usually small. Such samplings are described well by the Poisson distribution.

Let's consider a numerical example. Suppose we randomly distribute 100 one-dollar bills to 100 students in a lecture room, perhaps by scattering them over the class from some vantage point near the ceiling. The average (or mean) number of bills per student is 1.0, but common sense tells us that it is very unlikely that each of the 100 students will capture one bill. We would expect a few lucky students to grab three or four bills each and quite a few students to come up with two bills each. However, we would expect most students to get either one bill or none. The Poisson distribution provides a quantitative prediction of the results.

In this example, the item being considered is the capture of a bill by a student. We want to divide the students into classes according to the number of bills each captures; then find the frequency of each class. Let m represent the mean number of items (here, $m = 1.0$ bill per student). Let i represent the number for a particular class (say, $i = 3$ for those students who get three bills each). Let $f(i)$ represent the frequency of the i class — that is, the proportion of the 100 students who each capture i bills. The general expression for the Poisson distribution states that

$$f(i) = \frac{e^{-m} m^i}{i!}$$

where e is the base of natural logarithms ($e \cong 2.7$) and ! is the factorial symbol (as examples, $3! = 3 \times 2 \times 1 = 6$ and $4! = 4 \times 3 \times 2 \times 1 = 24$; by definition, $0! = 1$). When computing $f(0)$, recall that any number raised to the power of 0 is defined as 1. Table 6-1 gives values of e^{-m} for m values from 0.000 to 1.000.

Table 6-1. Values of e^{-m} for m Values of 0 to 1

m	e^{-m}	m	e^{-m}	m	e^{-m}	m	e^{-m}
0.000	1.00000	0.250	0.77880	0.500	0.60653	0.750	0.47237
0.010	0.99005	0.260	0.77105	0.510	0.60050	0.760	0.46767
0.020	0.98020	0.270	0.76338	0.520	0.59452	0.770	0.46301
0.030	0.97045	0.280	0.75578	0.530	0.58860	0.780	0.45841
0.040	0.96079	0.290	0.74826	0.540	0.58275	0.790	0.45384
0.050	0.95123	0.300	0.74082	0.550	0.57695	0.800	0.44933
0.060	0.94176	0.310	0.73345	0.560	0.57121	0.810	0.44486
0.070	0.93239	0.320	0.72615	0.570	0.56553	0.820	0.44043
0.080	0.92312	0.330	0.71892	0.580	0.55990	0.830	0.43605
0.090	0.91393	0.340	0.71177	0.590	0.55433	0.840	0.43171
0.100	0.90484	0.350	0.70469	0.600	0.54881	0.850	0.42741
0.110	0.89583	0.360	0.69768	0.610	0.54335	0.860	0.42316
0.120	0.88692	0.370	0.69073	0.620	0.53794	0.870	0.41895
0.130	0.87810	0.380	0.68386	0.630	0.53259	0.880	0.41478
0.140	0.86936	0.390	0.67706	0.640	0.52729	0.890	0.41066
0.150	0.86071	0.400	0.67032	0.650	0.52205	0.900	0.40657
0.160	0.85214	0.410	0.66365	0.660	0.51685	0.910	0.40252
0.170	0.84366	0.420	0.65705	0.670	0.51171	0.920	0.39852
0.180	0.83527	0.430	0.65051	0.680	0.50662	0.930	0.39455
0.190	0.82696	0.440	0.64404	0.690	0.50158	0.940	0.39063
0.200	0.81873	0.450	0.63763	0.700	0.49659	0.950	0.38674
0.210	0.81058	0.460	0.63128	0.710	0.49164	0.960	0.38289
0.220	0.80252	0.470	0.62500	0.720	0.48675	0.970	0.37908
0.230	0.79453	0.480	0.61878	0.730	0.48191	0.980	0.37531
0.240	0.78663	0.490	0.61263	0.740	0.47711	0.990	0.37158
						1.000	0.36788

SOURCE: F. James Rohlf and Robert R. Sokal, *Statistical Tables*, 2d ed. W. H. Freeman and Company, 1981.

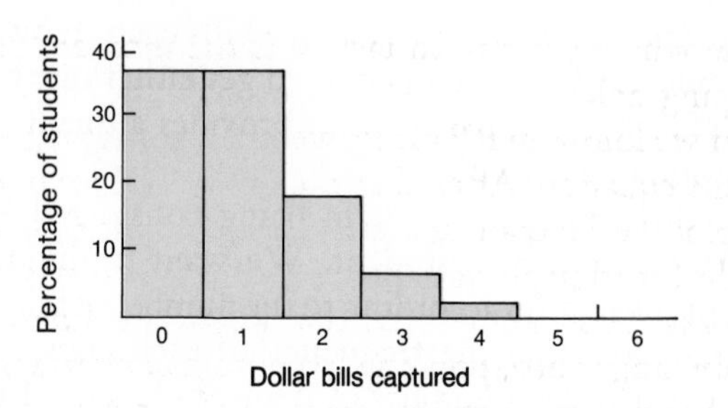

Figure 6-1 Poisson distribution for a mean of 1.0, illustrated by a random distribution of dollar bills to students.

In our example, $m = 1.0$. Using Table 6-1, we compute the frequencies of the classes of students capturing 0, 1, 2, 3, and 4 bills as follows:

$$f(0) = \frac{e^{-1}\,1^0}{0!} = \frac{e^{-1}}{1} = 0.368$$

$$f(1) = \frac{e^{-1}\,1^1}{1!} = \frac{e^{-1}}{1} = 0.368$$

$$f(2) = \frac{e^{-1}\,1^2}{2!} = \frac{e^{-1}}{2 \times 1} = \frac{e^{-1}}{2} = 0.184$$

$$f(3) = \frac{e^{-1}\,1^3}{3!} = \frac{e^{-1}}{3 \times 2 \times 1} = \frac{e^{-1}}{6} = 0.061$$

$$f(4) = \frac{e^{-1}\,1^4}{4!} = \frac{e^{-1}}{4 \times 3 \times 2 \times 1} = \frac{e^{-1}}{24} = 0.015$$

Figure 6-1 shows a histogram of this distribution. We predict that about 37 students will capture no bills, about 37 will capture one bill, about 18 will capture two bills, about 6 will capture three bills, and about 2 will capture four bills. This accounts for all 100 students; in fact, you can verify that the Poisson distribution yields $f(5) = 0.003$, which makes it likely that no student in this sample of 100 will capture five bills.

Similar distributions may be developed for other m values. Some are shown in Figure 6-2 as curves instead of bar histograms.

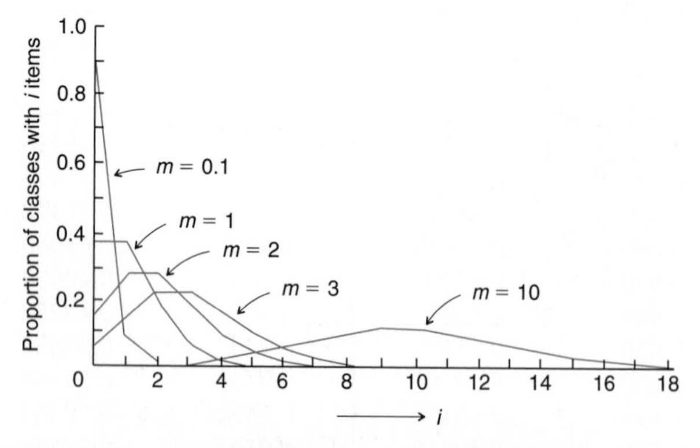

Figure 6-2 Poisson distributions for five different mean values: m is the mean number of items per sample, and i is the actual number of items per sample. (Modified from R. R. Sokal and F. J. Rohlf, *Introduction to Biostatistics*. W. H. Freeman and Company, 1973.)

Derivation of the Mapping Function

The Poisson distribution describes the distribution of crossovers along a chromosome during meiosis. In any chromosomal region, the actual number of crossovers is probably small in relation to the total number of possible crossovers in that region. If we knew the *mean* number of crossovers in the region per meiosis, we could calculate the distribution of meioses with zero, one, two, three, four, and more multiple crossovers. This is unnecessary in the present context because, as we shall see, the only class we are really interested in is the zero class. We want to correlate map distances with observable RF values. Meioses in which there are one, two, three, four, or *any* finite number of crossovers per meiosis all behave similarly in that they produce an RF of 50 percent *among the products of those meioses,* whereas the meiosis with no crossovers produce an RF of 0 percent. Consequently, the determining force in actual RF values is the ratio of class zero to the rest. To see how this can be so, consider a series of meioses in which nonsister chromatids cross over not at all, once, and twice, as shown in Figure 6-3. We get recombinant products *only* from meioses with at least one crossover in the region, and *always* precisely one-half of the products of such meioses are recombinant. We see then that the real determinant of the RF value is the size of the zero crossover class in relation to the rest.

As noted in Figure 6-3, we consider only crossovers between nonsister chromatids; **sister-chromatid exchange** is thought to be rare at meiosis. If it occurs, it can be shown to have no net effect in most meiotic analyses.

At last, we can derive the mapping function. Recombinants make up one-half of the products of those meioses having at least one crossover in the region. The proportion of meioses with at least one crossover is 1 minus the fraction with zero crossovers. The zero-class frequency will be

$$\frac{e^{-m}\,m^0}{0!}$$

which equals

$$e^{-m}$$

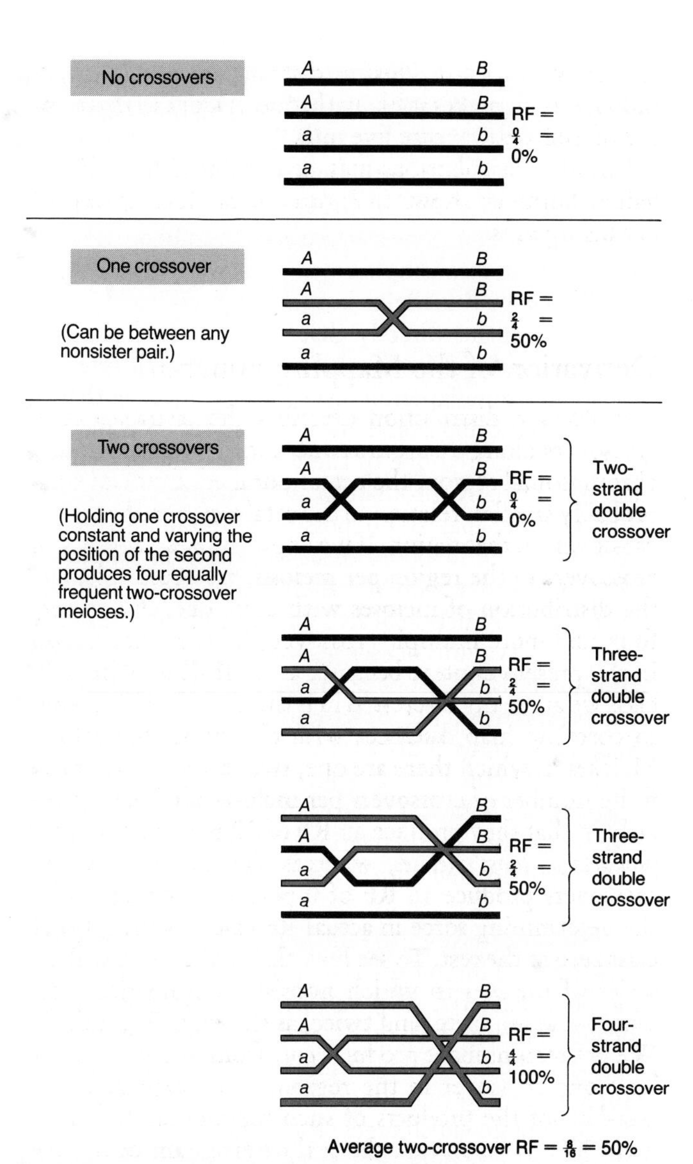

Figure 6-3 Demonstration that the average RF is 50 percent for meioses in which the number of crossovers is not zero. Recombinant chromatids are red. Two-strand double crossovers produce all parental types so all the chromatids are black. Note that all crossovers are between nonsister chromatids. Try the triple crossover class yourself.

so the mapping function can be stated as

$$RF = \tfrac{1}{2}(1 - e^{-m})$$

This formula relates recombinant frequency to m, the mean number of crossovers. Since the whole concept of mapping by genetics is based on the occurrence of crossovers, and proportionality between crossover frequency and the physical size of a chromosomal region, then you can see that m is probably the most fundamental variable in the whole process. In fact, m is the ultimate genetic mapping unit.

If we know an RF value, we can calculate m by solving the equation. After obtaining many values of m, we can plot the function as a graph, as in Figure 6-4. Viewing the function plotted as a graph should help us see how it works. First, notice that the function is linear for a certain range corresponding to very small m values. (Remember that m is our best measure of genetic distance.) Therefore, RF is a good measure of distance where the dashed line coincides with the function in Figure 6-4. In this region, the map unit defined as 1 percent RF has real meaning. Therefore, let's use this region of the curve to define corrected map units by considering some small values of m:

For $m = 0.05$, $e^{-m} = 0.95$, and
$$RF = \tfrac{1}{2}(1 - 0.95) = \tfrac{1}{2}(0.05) = \tfrac{1}{2} \times m$$

For $m = 0.10$, $e^{-m} = 0.90$, and
$$RF = \tfrac{1}{2}(1 - 0.90) = \tfrac{1}{2}(0.10) = \tfrac{1}{2} \times m$$

We see that $RF = m/2$, and this relation defines the dashed line in Figure 6-4. It allows us to translate m values into corrected map units. Expressing m as a percentage, we see that an m of 100 percent (=1) is the equivalent of 50 corrected map units. So an m value of 1 is the equivalent of 50 corrected map units, and we can express the horizontal axis of Figure 6-4 in our new map units. Now we can notice from the graph that two loci separated by 150 corrected map units show an RF of only 50 percent. We can use the graph of the function to convert any RF into map distance simply by drawing a horizontal line from the RF value to the curve and dropping a perpendicular to the map-unit axis—a process equivalent to using the equation $RF = \tfrac{1}{2}(1 - e^{-m})$ to solve for m.

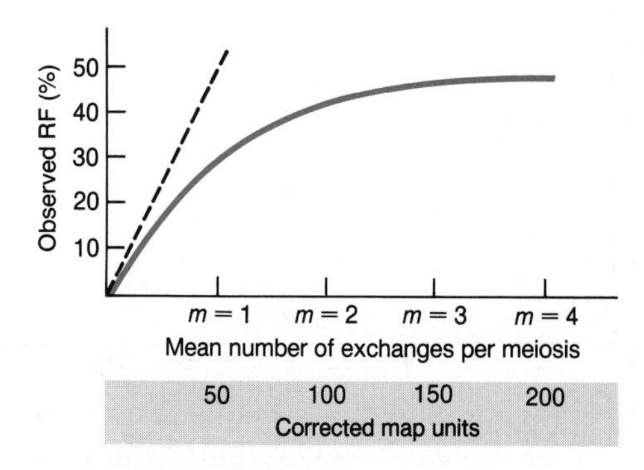

Figure 6-4 The blue line gives the mapping function in graphic form. Where the blue curve and the dashed line coincide, the function is linear and RFs estimate correct map distances well.

Let's consider a specific numerical example of the use of the mapping function. Suppose we get an RF of 27.5 percent. How many corrected map units does this represent?

$$0.275 = \tfrac{1}{2}(1 - e^{-m})$$

$$0.55 = 1 - e^{-m}$$

Therefore

$$e^{-m} = 1 - 0.55 = 0.45$$

From e^{-m} tables (or by using a calculator), we find that $m = 0.8$, which is the equivalent of 40 corrected map units. If we had been happy to accept 27.5 percent RF as representing 27.5 map units, we would have been considerably underestimating the distance between the loci.

Message To estimate map distances most accurately, put RF values through the mapping function. Alternatively, add distances that are each short enough to be in the region where the mapping function is linear.

A corollary of the second statement of this Message is that for organisms for which the chromosomes are already well mapped, such as *Drosophila,* a geneticist seldom needs to calculate from the map function to place newly discovered genes on the map. This is because the map is already divided into small, marked regions by known loci. However, when the process of mapping has just begun in a new organism or when the available genetic markers are sparsely distributed (as in human maps), the corrections provided by the function are needed.

Notice that no matter how far apart two loci are on a chromosome, we never observe an RF value of greater than 50 percent. Consequently, an RF value of 50 percent would leave us in doubt about whether two loci are linked or are on separate chromosomes. Stated another way, as m gets larger, e^{-m} gets smaller and RF approaches $\frac{1}{2}(1 - 0) = \frac{1}{2} \times 1 = 50$ percent. This is an important point: RF values of 100 percent can never be observed, no matter how far apart the loci are!

Analysis of Single Meioses

In Chapter 5 we briefly encountered a type of organism in which the products of single meioses remain together as groups of four cells. Such a group of four cells is called a **tetrad,** a term based on the Greek word for four. Tetrads can be isolated only in certain fungi and single-celled algae.

You might be wondering why the study of individual meioses is important; after all we have just developed some of the basic analytical rules of genetics in the previous chapters without using them. The first answer to this question is that by studying individual meioses the geneticist can make *direct* observations on the behavior of genes in meiotic processes, with less need for inferences. Consider, for example, the inference that we implicitly make when studying allele segregation in a heterozygote, say $A\,a$. When we observe equal numbers of A and a alleles in a random sample of gametes, we attribute this to segregation of the alleles in individual meioses. Tetrad analysis confirms directly that this is indeed the case. This kind of directness is most useful in genetic research. In addition, tetrads permit several kinds of studies that are not possible using conventional analysis. We have already encountered one of these studies, the determination of how many chromatids can be involved in crossing-over. Another analysis not possible with random meiotic products is the mapping of centromeres in relation to other loci; this fundamental mapping method eventually led to the construction of artificial chromosomes in yeast. Lastly, observations of exceptional allele segregations in tetrad analysis led to one of the central molecular models of crossing-over, the heteroduplex (or hybrid) DNA model, and we will discuss this in detail in Chapter 19.

But do fungi and algae have typical eukaryotic chromosomes with typical eukaryotic behavior? The answer is generally yes, so discoveries made in these model systems can be applied with reasonable confidence to plants and animals.

Notice that fungi and unicellular algae are microbial eukaryotes. This means that just like the prokaryotic microbes, bacteria, they can be propagated easily and cheaply, take up little space, and can be studied in large numbers. These are also useful properties in research. Moreover, recall that fungi and algae are haploid. This has two basic advantages for genetic studies. First, because there is only one chromosome set, dominance and recessiveness normally do not complicate gene expression, and the phenotype is a direct reflection of the genotype. Second, in addition to the usefulness of tetrad analysis, even random meiotic product analysis is easier than in diploids. Recall that in the haploid life cycle the two haploid parental cells unite to form the diploid meiocyte, so for the geneticist this is the only meiosis to worry about. In contrast, in diploids the analyst must consider meioses in both parents. As an example, let's look at how easy it is to study linkage in haploids. We can make a cross such as $a^+ b^+ \times a\,b$, and test a random sample of meiotic products. This is straightforward because each product develops directly into a haploid progeny individual. We might find the following genotypic frequencies in the progeny:

$a^+ b^+$	45%
$a\,b$	45%
$a\,b^+$	5%
$a^+ b$	5%

From these data, it is easy to compute the RF value of 10 percent, indicating linkage of the two genes 10 map units apart.

In summary, tetrads provide a direct method for the genetic analysis of individual meioses; the method can be used to examine processes that are not accessible by studying random meiotic products. Meiosis is one of the central processes of eukaryotic biology. Although a great deal is known about meiosis, there is still much to be learned about the mechanisms of such processes as pairing, crossing-over, and segregation.

Message Tetrads are ideal structures for studying any aspects of the genetics of meiosis.

In some eukaryotic organisms in which tetrads can be isolated and analyzed, the cells that represent the four products of meiosis develop directly into four sexual spores (see Figure 6-5). In other such organisms, each of the four products of meiosis undergoes an additional mitotic division, yielding a group of eight cells called an **octad.** However, an octad is simply a double tetrad com-

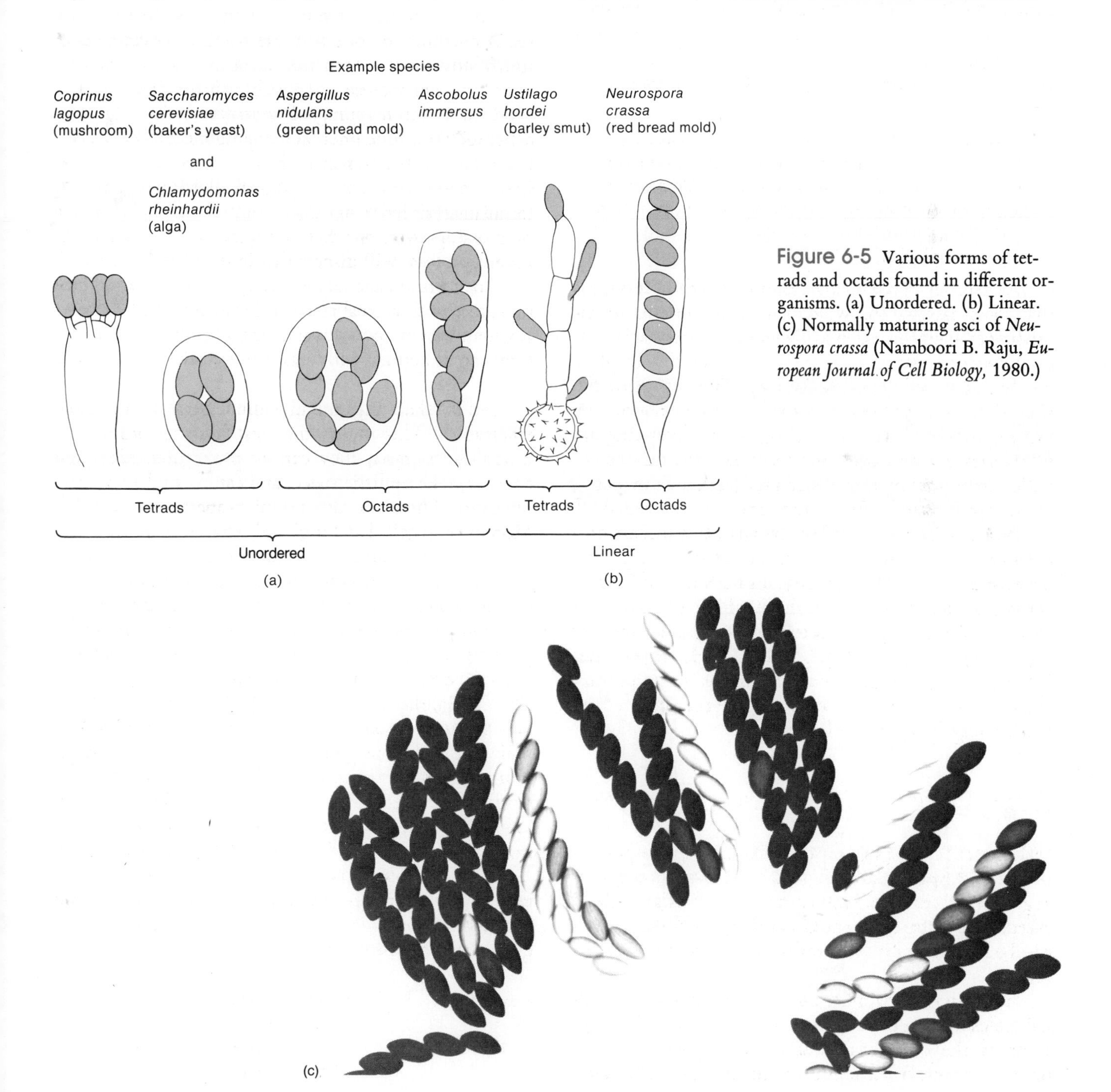

Figure 6-5 Various forms of tetrads and octads found in different organisms. (a) Unordered. (b) Linear. (c) Normally maturing asci of *Neurospora crassa* (Namboori B. Raju, *European Journal of Cell Biology,* 1980.)

posed of four spore pairs. The members of a spore pair are identical, being mitotic daughter cells of one of the four products of meiosis.

The sexual spores, whether four or eight in number, can be found in a variety of arrangements. In some species, the spores are found in a jumbled arrangement called an **unordered tetrad,** shown in Figure 6-5(a). In other species, the spores are arranged in a striking linear arrangement called a **linear tetrad,** shown in Figure 6-5(b). Unordered and linear tetrads each have their specific uses in genetic analysis. We shall deal with the linear types first.

Linear Tetrad Analysis

How are linear tetrads produced? The key fact is that the spindles of the first and second meiotic divisions and of the post-meiotic mitosis are positioned end-to-end in the long ascus sac and do not overlap. The reason is probably related to the fact that these divisions occur in a tubelike structure, which physically prevents the spindles from overlapping. In any case, the absence of spindle overlap means that the nuclei are laid out in a straight array, and the lineage of each of the eight nuclei of the final octad can therefore be traced back through meiosis, as shown in Figure 6-6.

As you might expect, linear tetrads beautifully illustrate the segregation and independent assortment of genes at meiosis. However, linear tetrads are ideally suited to a special kind of analysis that is not possible in most organisms: **centromere mapping,** the locating of centromeres in relation to gene loci on the chromosomes. The centromere is a fascinating region of the chromosome that interacts with the spindle fibers and ensures proper chromosome movement during nuclear division. When this process fails, the daughter cells have abnormal chromosome numbers, which can lead to death or phenotypic abnormality. For example, abnormal chromosome numbers in humans have produced a major class of genetic diseases (which we shall discuss in Chapter 9). Therefore, the centromere has been the subject of considerable attention in genetics.

How does centromere mapping work? We shall use the fungus *Neurospora crassa* as an example. This fungus belongs to a group called Ascomycetes. In these fungi, a membranous sac encloses the sexual spores; together, the spores and sac are called an ascus (see Figure 6-5 for examples). *Neurospora* produces linear octads. In its simplest form, centromere mapping considers a gene locus and asks how far this locus is from its centromere. We could pick any *Neurospora* locus to illustrate this technique, but we will use the **mating-type locus** because it was one of the first to be used in such an analysis. This locus has two alleles, which are represented as *A* and *a* even though neither is dominant or recessive. These alleles determine the *A* and *a* mating types. Although the mating-type phenomenon is interesting in itself, here we are merely using the locus as a genetic marker to illustrate the analysis and we will not be concerned with its function.

Centromere mapping is based on the fact that a meiosis in which nonsister chromatids cross over between the centromere and a heterozygous locus produces a different allele pattern in the octad than a meiosis in which nonsister chromatids do not cross over in that region. Figures 6-7 and 6-8 show examples of these two

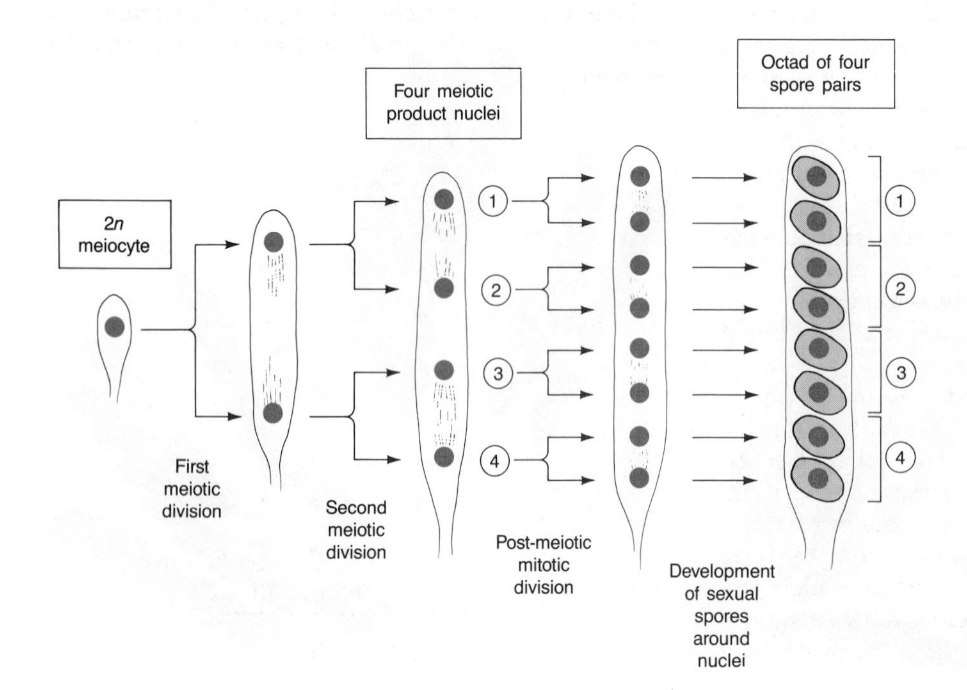

Figure 6-6 Meiosis and post-meiotic mitosis in a linear tetrad. The nuclear spindles do not overlap at any division, so that nuclei never pass each other in the sac. The resulting eight nuclei are laid down in a linear array that can be traced back through the divisions.

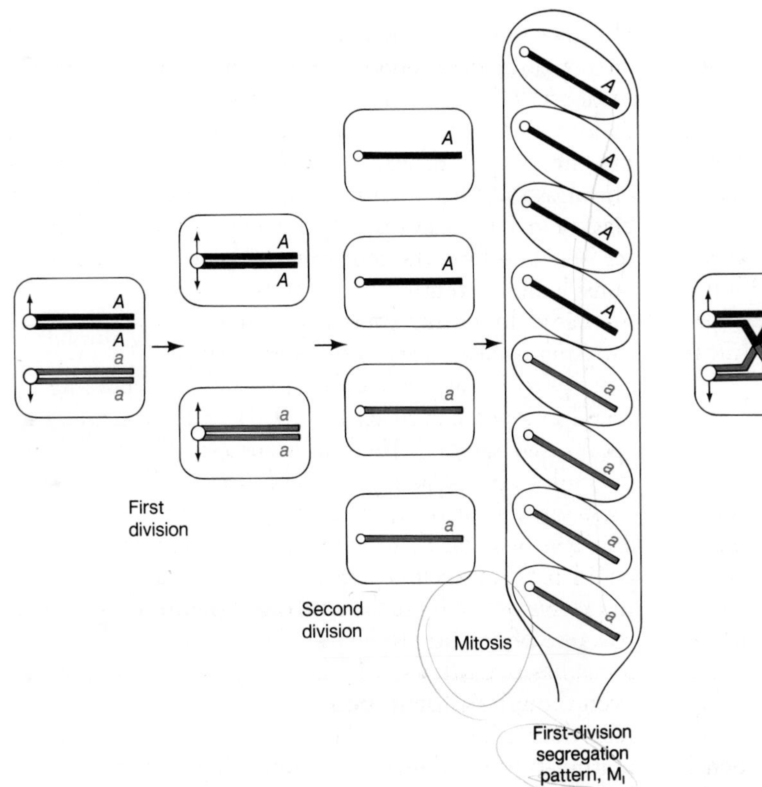

Figure 6-7 *A* and *a* segregate into separate nuclei at the first meiotic division when there is no crossover between the centromere and the locus. The resultant allele pattern in the octad is called a first-division segregation pattern.

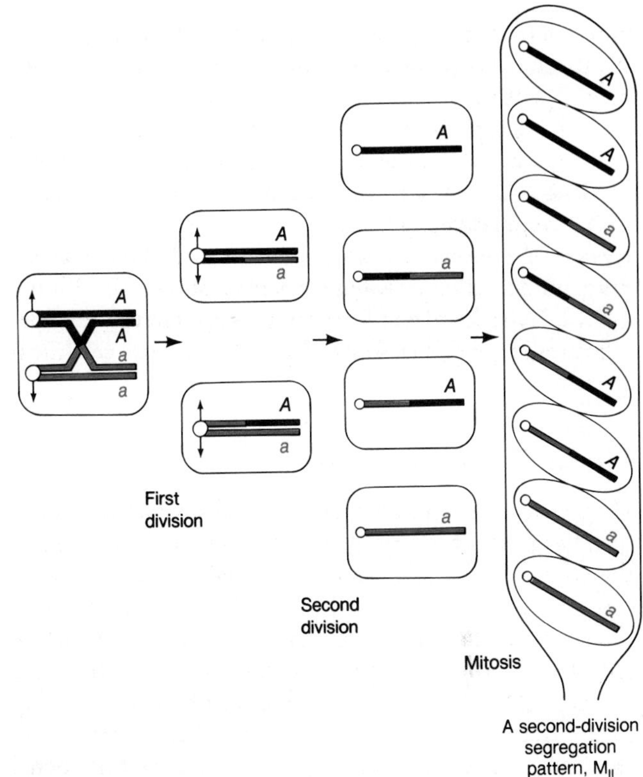

Figure 6-8 *A* and *a* segregate into separate nuclei at the second meiotic division when there is a crossover between the centromere and the locus.

possibilities. The simpler pattern, shown in Figure 6-7, arises when there is no crossover. This pattern is typified by all of the products at one end of the tetrad or octad carrying one allele and the products at the other end carrying the other allele. You can see from the diagram how this pattern is produced: because there is no spindle overlap, the *A*-bearing nuclei and the *a*-bearing nuclei never pass each other in the ascus. Also notice that although the *A* and *a* alleles are together in the diploid meiocyte nucleus, the first meiotic division cleanly segregates the *A* and *a* alleles and these alleles remain separate throughout the second division of meiosis. This gives rise to the term **first-division segregation,** and the allele pattern in the spores is called a *first-division segregation pattern* or **M$_I$ pattern.**

When nonsister chromatids do cross over between the centromere and the locus (see Figure 6-8), *A* and *a* are still together in the nuclei at the end of the first division of meiosis. There has been no first-division segregation. However, the second meiotic division does move the *A* and *a* alleles into separate nuclei, giving rise to the term **second-division segregation.** The resulting allele pattern in the tetrad or octad is called a *second-division segregation pattern* or **M$_{II}$ pattern.** This pattern is typified by any arrangement in which a half tetrad or a half octad contains both alleles.

Now let's look at the following experimental results and interpret them in light of these ideas. Remember, the cross was *A* × *a*. The octad patterns obtained were the following:

Octads					
A	*a*	*A*	*a*	*A*	*a*
A	*a*	*A*	*a*	*A*	*a*
A	*a*	*a*	*A*	*a*	*A*
A	*a*	*a*	*A*	*a*	*A*
a	*A*	*A*	*a*	*a*	*A*
a	*A*	*A*	*a*	*a*	*A*
a	*A*	*a*	*A*	*A*	*a*
a	*A*	*a*	*A*	*A*	*a*
126	132	9	11	10	12

Total = 300

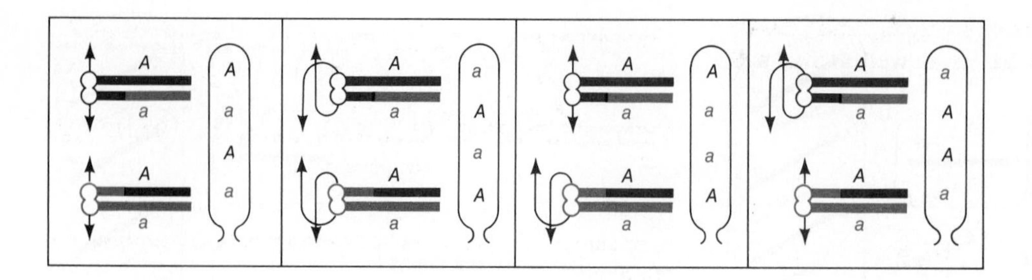

Figure 6-9 Four second-division segregation ascus patterns are equally frequent in linear asci because centromeres attach to spindles at the second meiotic division at random as regards whether they will be pulled "up" or "down."

The first two octads have the M_I segregation pattern, with the first octad being simply an upside-down version of the second. Notice that these two are more-or-less equal in frequency (126 versus 132). This equality simply reflects that centromeres attach to spindles at the first meiotic division at random as regards whether the arrangement will pull *A* "up" and *a* "down" or *a* "up" and *A* "down." We can deduce that $126 + 132 = 258$ meioses out of 300, or 86 percent of meioses, had no crossover in the region between the mating-type gene and the centromere.

The remaining four octads all have both *A* and *a* present in half octads; hence by the above definition *A* and *a* segregated not at the first but at the second division of meiosis. What accounts for these four different variations on the same basic M_{II} theme? Once again, it is the randomness of spindle attachment, not only at the first but also at the second division of meiosis. This is illustrated in Figure 6-9. The M_{II} patterns in our example total $9 + 11 + 10 + 12 = 42$, or 14 percent, and show that nonsister chromatids crossed over between the mating-type locus and the centromere in 14 percent of the meioses.

Well, we have measured the M_{II} pattern frequency at 14 percent in this example. Does this mean that the mating-type locus is 14 map units from the centromere? The answer is no, but this value can be used to calculate the number of map units. The 14 percent value is a percentage of meioses, and this is not the way map units are defined. Map units are defined in terms of the percentage of recombinant chromosomes issuing from meiosis. Figure 6-10 shows that when a crossover is in the region spanned by the centromere and the locus, then only one-half of the chromosomes issuing from that meiosis will be recombinant. So to specify the length of the region in map units, it is necessary to divide the M_{II} pattern frequency by 2. In our example, the distance of the mating-type locus from the centromere is therefore $14 \div 2 = 7$ map units.

Message To calculate the distance of a locus from its centromere in map units, measure the percentage of tetrads showing second-division segregation patterns for that locus and divide by 2.

The above analysis can be extended to any number of heterozygous pairs segregating in a cross. Let's consider meiosis in a *Neurospora* diploid meiocyte of genotype $a\,a^+\,b\,b^+$ as an example. We do not know if the two loci are linked; nor do we know how they are located relative to centromeres. There are only limited possibilities, however:

1. The loci are on separate chromosomes.
2. The loci are on opposite sides of the centromere on the same chromosome.
3. The loci are on the same side of the centromere on the same chromosome.

The first two possibilities would both show independence of the M_{II} patterns for both loci. The third possibility is more interesting in that a crossover in the region between the centromere and the locus closest to it would produce coinciding M_{II} patterns for the two loci in an ascus (barring rare double crossovers). Hence, if *a* is closer to the centromere than *b*, for example, we would expect most M_{II} patterns for *a* to coincide with M_{II} patterns for *b* (Figure 6-11). Such coincidence of M_{II} patterns provides clues about linkage between loci.

Message Coincidence of second-division segregation patterns indicates that the loci are on the same chromosome arm.

The farther a locus is from a centromere, the greater will be the M_{II} frequency. But the M_{II} frequency never reaches 100 percent; in fact, the theoretical maximum is 67 percent, or $\frac{2}{3}$. The reason for this is that multiple crossovers (especially doubles) become more and more

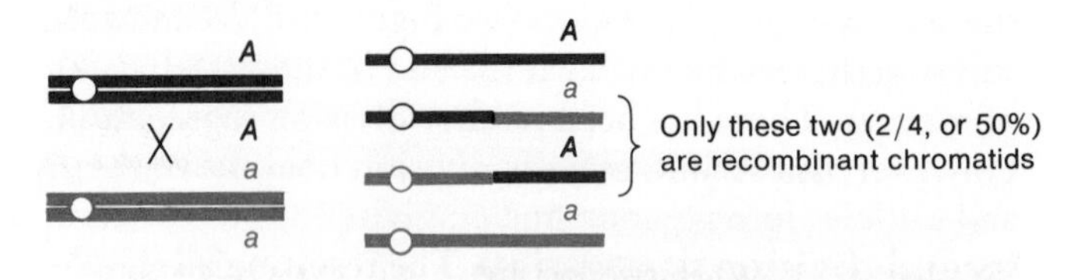

Figure 6-10 Only one-half of the chromatids from a meiosis with a single crossover are recombinant.

prevalent as the interval becomes larger, and double crossovers can generate M_I patterns as well as M_{II} patterns. For example:

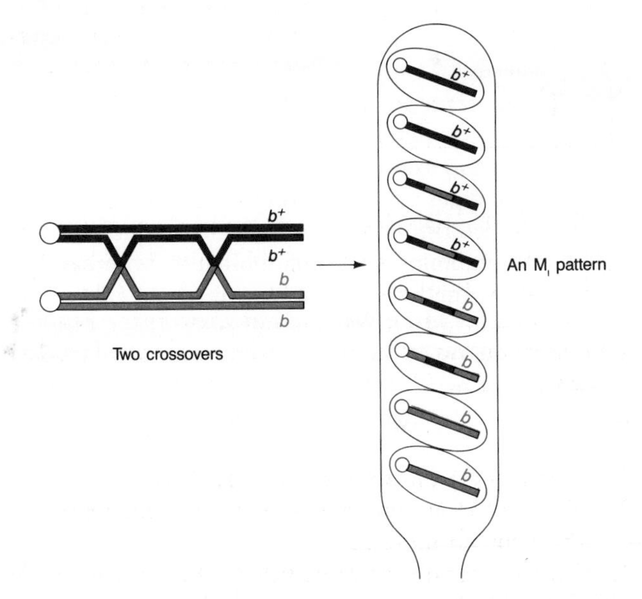

One easy way to calculate the maximum M_{II} frequency is by the following "thought experiment." Consider a heterozygous locus, b^+b, that is so far from the

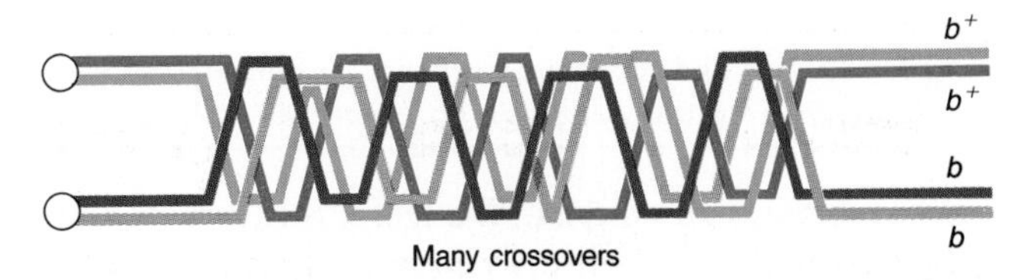

Figure 6-12 When a locus is far from the centromere, a large number of crossovers effectively unlink the locus from the centromere.

centromere that there are many crossovers in the intervening region (Figure 6-12). These many crossovers effectively uncouple the locus from the centromere, and the two b alleles and the two b^+ alleles can end up in the tetrad in any arrangement. We can simulate this by considering how many different ways there are of dropping four marbles (two b and two b^+) into a test tube (follow this with Figure 6-13). The first marble can be b^+ or b; it makes no difference. Let's assume that it is b^+. We then have two b's and one b^+ left. The next marble determines if the pattern will be M_I or M_{II}. If the marbles are dropped at random into the tube over and over again, one-third of the time, the second marble will be b^+; thus, the third and fourth marbles must be b, and an M_I pattern is generated. The other two-thirds of the time, the second marble will be b, generating an M_{II} pattern. Therfore, we see that even with this very large number of crossovers, the $\frac{2}{3}$ frequency of M_{II} patterns can never be exceeded.

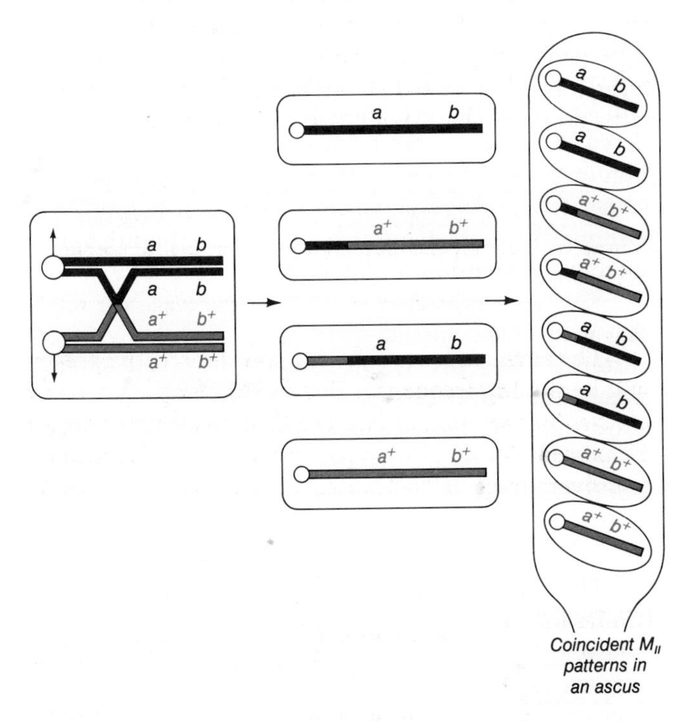

Figure 6-11 When two loci are on the same chromosome arm, a crossover between the centromere and the locus closest to it produces a coincident M_{II} pattern for both loci.

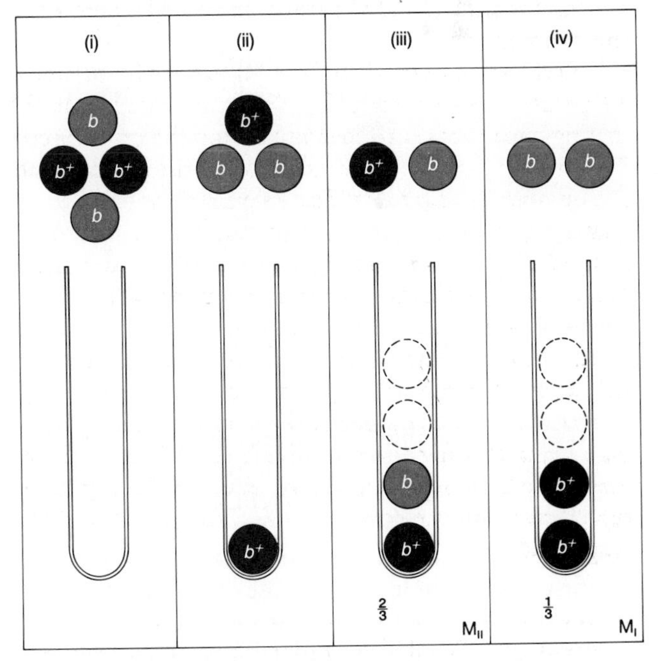

Figure 6-13 Demonstration of the limiting M_I and M_{II} segregation pattern frequencies of $\frac{1}{3}$ and $\frac{2}{3}$ in linear tetrad analysis. (See text for details.)

Message The maximum frequency possible for second-division segregation patterns is 66.7 percent.

This raises an enigma. We feel intuitively that this M_{II} maximum should directly equate with the 50 percent RF maximum observed for very large map distances, but $66.7 \div 2 = 33.3$ percent, which would be 33.3 m.u. The villain, once again, is the existence of multiple crossovers. We could derive a map function for M_{II} patterns, but in practice, it is simpler to stick to the analysis of smaller intervals and to recognize that the larger M_{II} frequencies provide increasingly inaccurate estimates of distances between centromeres and loci.

Unordered Tetrad Analysis

In unordered tetrads or octads, the spores are in no particular sequence, so that centromere mapping studies of the type just discussed are not possible. Nevertheless, unordered tetrads are generally useful in analyzing many aspects of meiotic segregation and recombination. Some of these uses will be explored in the problem sets. We will illustrate their use here with one example, which shows how they can be used as an alternative to a mapping function to correct for double crossovers. In this example, we will see again the directness of the tetrad approach, avoiding some of the assumptions of the mathematical method.

Let us assume that we are studying two yeast loci *a* and *b* and we want to determine if they are linked and, if so, calculate a corrected map distance. Yeast produces unordered tetrads. How can we use these tetrads to explore linkage? First let's see what different types of unordered tetrad can be expected in such a cross. It turns out that in *any* dihybrid cross, with or without linkage, there are only three possible unordered tetrad types, and they are defined as follows, using a cross of $a\ b \times a^+\ b^+$:

	Parental ditype (PD)	Nonparental ditype (NPD)	Tetratype (T)
Spores	$a\ b$	$a\ b^+$	$a\ b$
	$a\ b$	$a\ b^+$	$a\ b^+$
	$a^+\ b^+$	$a^+\ b$	$a^+\ b$
	$a^+\ b^+$	$a^+\ b$	$a^+\ b^+$

Remember that these asci are unordered, so even though the first type might look like a case of M_I segregation for both loci, that is not the case: the spores could have been written equally well in any order. These asci merely have been classified according to whether they contain two genotypes (*di*types) or four genotypes (*tetra*types, represented by T). Within the ditype class, both genotypes can be either parental (PD) or nonparental (NPD). You might try writing out a few asci to convince yourself that no types other than PD, NPD, and T are possible.

Now, what do these types tell us about linkage? First it should be stated that only the three types PD, NPD, and T are obtained no matter whether the loci are linked or not. If presented with the frequencies of these three types, could one determine if two loci are linked? You will notice from examining the three types that only the NPD and T types contain recombinants so these are key classes in determining recombinant frequency. The NPD class contains only recombinants, whereas only half the spores in the T class are recombinant. Therefore, we could write a formula for determining RF in tetrads, where T and NPD represent the percentages of those classes:

$$\mathrm{RF} = \tfrac{1}{2}\mathrm{T} + \mathrm{NPD}$$

If this formula gives a frequency of 50 percent, then we know the loci must be unlinked, and correspondingly if the RF is less than 50 percent, then the genes must be linked and we could use that value to represent the number of map units between them. However, just as with other linkage analyses we have studied earlier in the chapter, this is an underestimate because it does not consider double recombinants and other higher level crossovers. Calculations based on other aspects of unordered tetrad frequencies can, however, provide a value corrected for doubles. First we need to determine what the PD, NPD, and T classes represent for linked loci. Let us assume that *a* and *b* are linked. If we assume that individual meioses can have no crossovers (NCO), a single crossover (SCO), or a double crossover (DCO) in the *a* to *b* region, then we can represent the classes of unordered asci that emerge from such meioses as shown in Figure 6-14. Of course, triples and higher numbers of crossovers might occur, but we may assume that these are rare and therefore negligible. Notice that Figure 6-14 is merely an extension of Figure 6-3.

The key to the analysis is the NPD class, the tetrad that arises only from a double crossover involving all four chromatids. Because we are assuming that double crossovers occur randomly between the chromatids, we can also assume that the frequencies of the four DCO classes are equal. This means that the NPD class should contain $\frac{1}{4}$ of the DCOs, and we can estimate that

$$\mathrm{DCO} = 4\mathrm{NPD}$$

The single crossover class can also be calculated by a similar kind of reasoning. Notice that tetratype (T) asci can result from either single crossover or double crossover meioses. But we can estimate the component of the T class that comes from DCO meioses to be 2NPD. Hence, the size of the SCO class can be stated as

$$\mathrm{SCO} = \mathrm{T} - 2\mathrm{NPD}$$

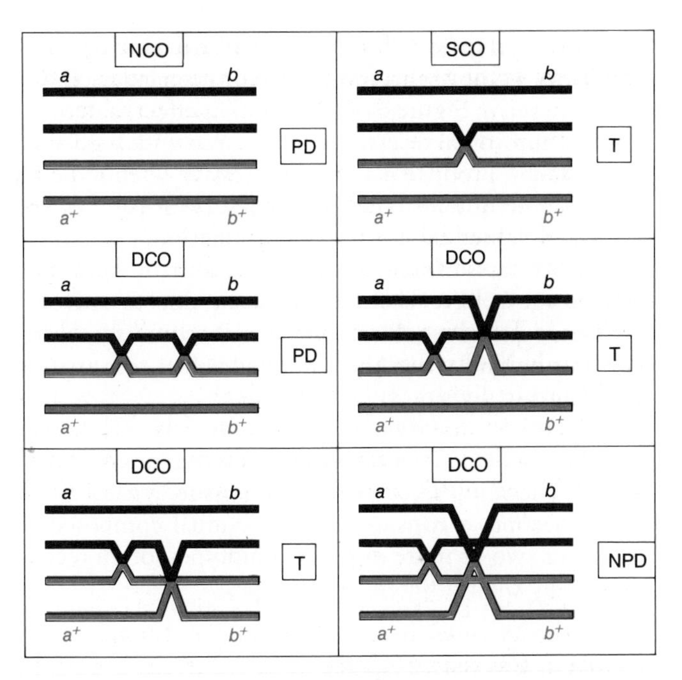

Figure 6-14 The ascus classes produced by crossovers between linked loci. NCO = noncrossover meioses; SCO = single crossover meioses; DCO = double crossover meioses.

Now that we have estimated the sizes of the SCO and DCO classes, the noncrossover class can be estimated as

$$\text{NCO} = 1 - (\text{SCO} + \text{DCO})$$

Thus, we have estimated the distribution of NCO, SCO, and DCO classes in this marked region. We can use these values to derive a value for m, the mean number of crossovers in this region. We calculate the value of m simply by taking the sum of the SCO class plus *twice* the DCO class (because this class contains two crossovers). Hence

$$\begin{aligned} m &= (\text{T} - 2\text{NPD}) + 2(4\text{NPD}) \\ &= \text{T} + 6\text{NPD} \end{aligned}$$

In the mapping-function section, we saw that to convert an m value to map units, it must be multiplied by 50. So

$$\text{map distance} = 50(\text{T} + 6\text{NPD}) \text{ m.u.}$$

Let's assume that in our hypothetical cross of $a\,b \times a^+\,b^+$, the observed frequencies of the ascus classes are 56 percent PD, 41 percent T, and 3 percent NPD. Using the formula, the distance between the a and b loci is

$$\begin{aligned} 50[0.41 + (6 \times 0.03)] &= 50(0.59) \\ &= 29.5 \text{ m.u.} \end{aligned}$$

Let us compare this value with that obtained directly from the RF. Recall that the formula is

$$\text{RF} = \tfrac{1}{2}\text{T} + \text{NPD}$$

In our example

$$\begin{aligned} \text{RF} &= 0.205 + 0.03 \\ &= 0.235, \text{ or } 23.5 \text{ m.u.} \end{aligned}$$

This is 6 m.u. less than the estimate we obtained using the map-distance formula because we could not correct for double crossovers in the RF analysis.

> Message Unordered tetrads can be used to study several aspects of meiotic genetics, including the distribution of single and double crossovers, which can be used to calculate accurate map distances.

Chromatid Interference

Does a crossover between two nonsister chromatids affect the probability of a second crossover between the *same* chromatids? This phenomenon is called **chromatid interference,** an effect of crossover interaction *across* the four chromatids, in contrast to the interference we encountered *along* the chromosome, which is sometimes called **chromosome interference.** As usual, the best hypothesis to test is a null hypothesis, which is that the second crossover occurs randomly between any pair of nonsister chromatids. This hypothesis means that the four possibilities, which are indicated in Figure 6-15, are equally frequent. As two of the equally frequent possibilities produce three-strand doubles, we can predict from the null hypothesis that second crossovers generate a 1 : 2 : 1 ratio of 2-strand : 3-strand : 4-strand doubles. If we observe a statistically significant deviation from this ratio, we reject the null hypothesis and conclude that there is chromatid interference. In fact, geneticists find very little or no consistent evidence for interference of this sort when second crossovers in fungal asci are analyzed.

Note that the RF limit of 50 percent applies only when there is no chromatid interference. Chromatid in-

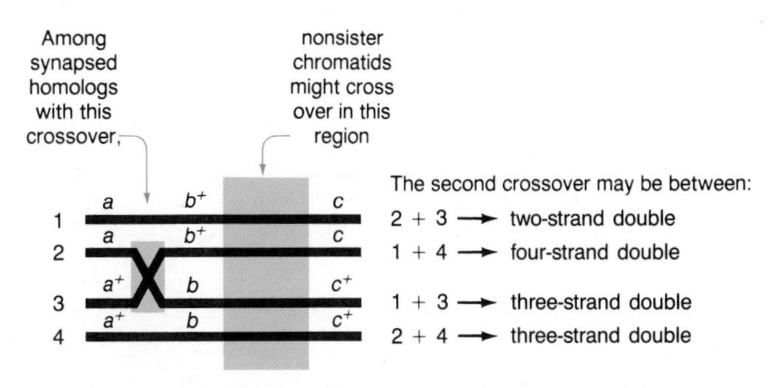

Figure 6-15 Double crossovers in the region of three heterozygous loci can produce several distinctive patterns of alleles in the ascus, depending on which strands cross over.

terference would favor the production of NPD tetrads, in which all the products of meiosis are recombinant, at the expense of PD and T classes. Hence, the RF could rise above 50 percent to a value dependent on the degree of chromatid interference.

Mitotic Segregation and Recombination

We normally think of segregation and recombination in connection with meioses, but they do occur (although less frequently) during mitosis. Mitotic segregation and recombination can easily be demonstrated if the genetic system is appropriately chosen.

Mitotic Segregation

In genetics, **segregation** refers to the separation of two alleles constituting a heterozygous genotype into two phenotypically distinguishable individuals or cells. We have seen segregation repeatedly, of course, in our meiotic analyses based on Mendel's first law. However, the alleles of a heterozygote occasionally can be seen to segregate when the heterozygous cell undergoes *mitotic* division. The following example clarifies the way in which geneticists detect mitotic segregation.

In the 1930s, Calvin Bridges was observing *Drosophila* females that were genotypically MM^+ (M is a dominant allele that produces a phenotype of slender bristles). Surprisingly, some females had a patch, or **sector,** of wild-type bristles on a body of predominantly M phenotype. Thus, the alleles of the heterozygote were showing segregation at the phenotypic level. Bridges concluded that this segregation was the result of **mitotic nondisjunction,** a type of chromosome separation failure that is diagrammed in Figure 6-16. In a similar effect, heterozygotes of autosomal recessive alleles paired with wild-type alleles ($a\,a^+$), produce patches of recessive phenotype on backgrounds of wild-type phenotype. These patches also can be explained by mitotic nondisjunction.

Other cases of segregation in the somatic tissue of a heterozygote have been found to be due to **mitotic chromosome loss.** Here one chromosome somehow gets left behind when the daughter nuclei reconstitute after mitotic division (Figure 6-17).

Geneticists find two other terms useful in relation to such phenomena. First, **variegation** is the coexistence of different-looking sectors of somatic tissue, whatever the cause. Second, a **mosaic** is an individual composed of tissues of two or more different genotypes, often recognizable by their different phenotypes.

Mitotic Crossing-Over

In 1936, Curt Stern showed that sectors in a mosaic are not always the result of mitotic nondisjunction or chromosome loss. Working with the *Drosophila* sex-linked genes y (which codes for yellow body) and sn (singed, because it codes for short, curly bristles), Stern made a cross $y^+ sn\,/\,y^+ sn \times y\,sn^+/\,\mathrm{Y}$. The female progeny were predominantly wild type in appearance, as expected from their $y^+ sn\,/\,y\,sn^+$ genotypes. Some females had sectors of yellow tissue or of singed tissue; these could be explained by nondisjunction or chromosome loss. Other females showed **twin spots.** A twin spot, in this example, is two adjacent sectors — one of yellow tissue and one of

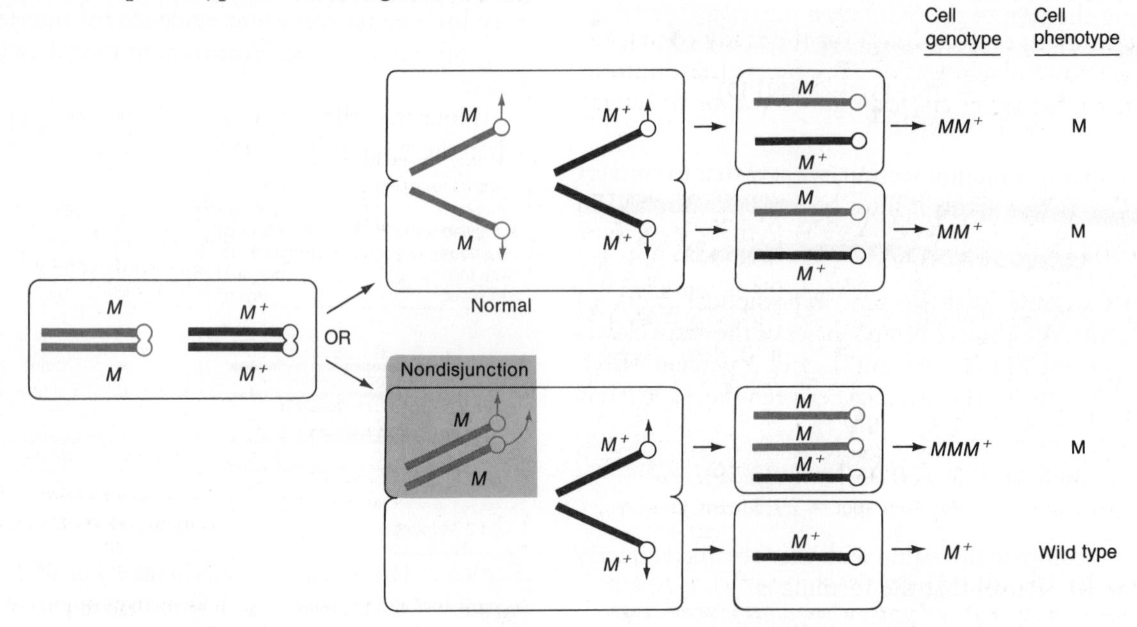

Figure 6-16 Mitotic nondisjunction can lead to phenotypic segregation.

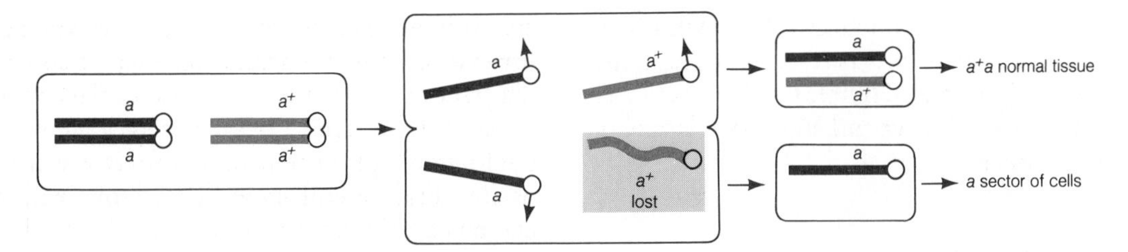

Figure 6-17 Chromosome loss at mitotic division can lead to phenotypic segregation.

singed tissue — in a background of wild-type tissue (Figure 6-18). Stern noticed that the twin spots were too common to be chance juxtapositions of single spots. Because of this, and because the twin sectors of a twin spot were always adjacent, he reasoned that the twin sectors must be reciprocal products of the same event. That event, he realized, must have begun when, by chance, the homologous parental chromosomes had come to lie in a pairing conformation. From such a conformation, Stern postulated, chromatids of the different homologs must have crossed over between the *sn* locus and the centromere. Figure 6-19 diagrams this crossover.

Twin spots have been observed in other diploid cells, including those of plants. All of these observations can be explained by mitotic crossing-over, which seems to take place quite regularly (although rarely) in all eukaryotes.

The definition of mitotic recombination is similar to that of meiotic recombination: **mitotic recombination** is any mitotic process that generates a diploid daughter cell with a combination of alleles different from that in the diploid parental cell. Compare this definition with that of meiotic recombination on page 121.

Fungal Detection Systems

Geneticists often employ fungi for the study of mitotic recombination and segregation. To observe these mitotic phenomena, the geneticist must generate diploid fungal cells because a haploid cell does not provide the opportunity for two genomes to recombine. Diploids form spontaneously in many fungi. The fungus we shall examine is *Aspergillus,* a greenish mold. *Aspergillus* is highly suitable for mitotic analysis. The aerial hyphae of *Aspergillus* produce long chains of **conidia** (asexual spores). Each conidium has a single nucleus, and the phenotype of any individual spore is dependent only on the genotype of its own nucleus. This makes certain kinds of selective techniques possible. If two haploid strains are mixed, the hyphae fuse; both types of nuclei are then present in a common cytoplasm. Such a strain is called a **heterokaryon.** Note that heterokaryons, like other strains, produce uninucleate conidia.

Consider the following parts of two chromosome pairs of a heterokaryon:

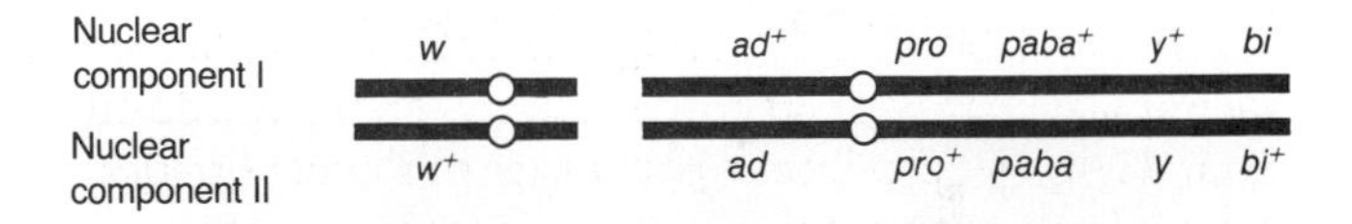

The alleles *ad*, *pro*, *paba*, and *bi* are all recessive to their wild-type counterparts, and each confers a requirement for a specific chemical supplement to permit growth.

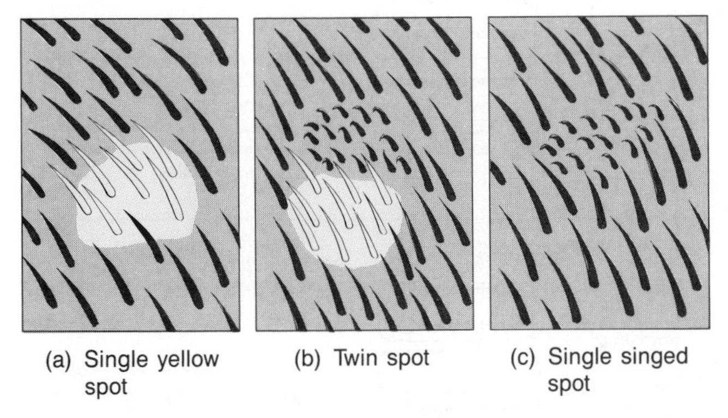

Figure 6-18 Unexpected sectors of body-surface phenotypes of *Drosophila* genotype $y^+\ sn\ /\ y\ sn^+$, where *sn* represents singed bristles and *y* represents yellow body.

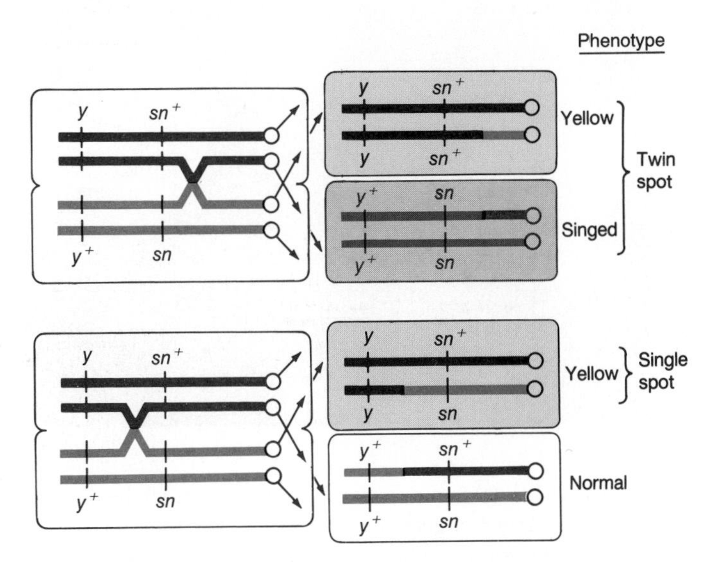

Figure 6-19 A mitotic crossover can lead to phenotypic segregation of the type shown in Figure 6-18.

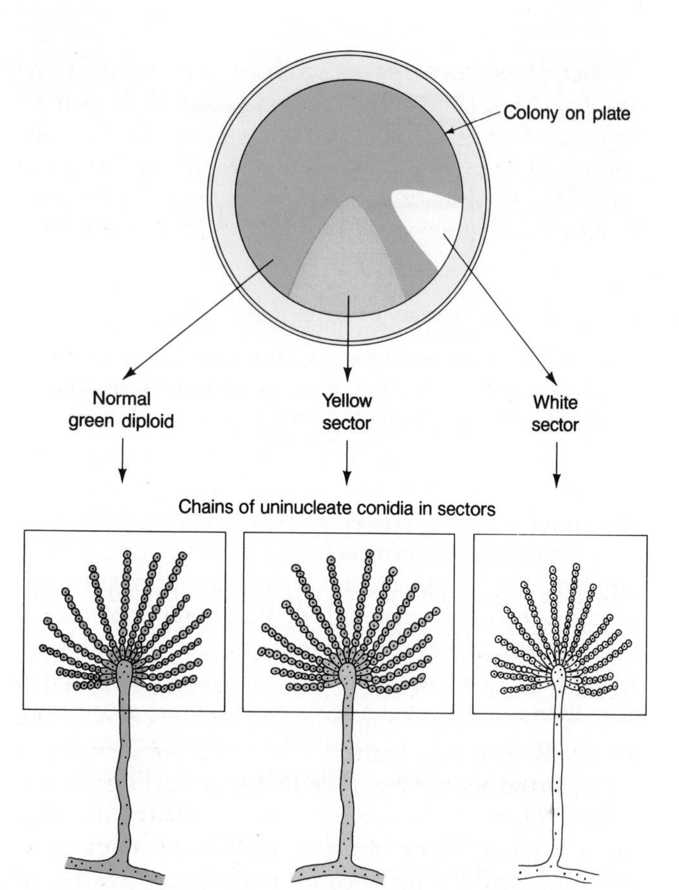

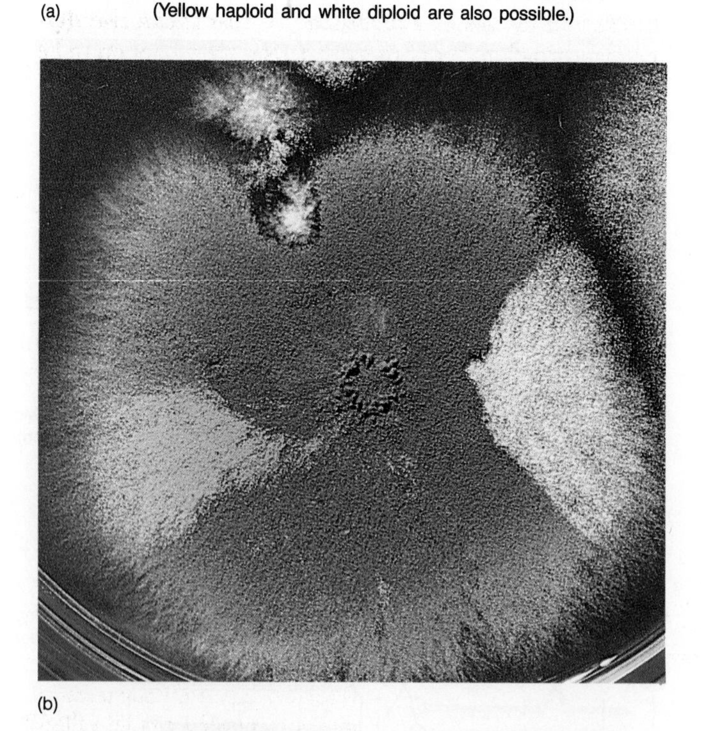

Figure 6-20 (a) Some sectors showing segregation in an *Aspergillus* diploid of genotype $w^+ w\, y^+ y$, where *w* and *y* cause conidia to be white and yellow, respectively. Haploids are smaller. (b) Photograph of a white and a yellow sector in a diploid colony. (Part b from Etta Käfer.)

Because the heterokaryon has a wild-type allele of each locus, it does not require any supplement for growth. The recessive alleles *w* and *y* cause conidia to be white and yellow, respectively. Because each conidium produced by a heterokaryon is uninucleate, it is either yellow or white. The heterokaryon itself looks yellowish-white and has a kind of "pepper-and-salt" appearance.

Green sectors appear in some heterokaryons. Green is the normal wild-type color of the fungus, and green coloration in the present example reveals that a diploid nucleus has formed spontaneously and has multiplied to form the sector. The presence of the dominant y^+ and w^+ alleles in the same nucleus produces the wild-type coloration. Diploid conidia can be removed from a green sector, and a diploid culture can be grown from it. Like the heterokaryon, the diploid cells require no growth supplements because they have the wild-type alleles. When the cultured diploid is fully grown, rare sectors producing either white or yellow conidia can be observed. Some of these sectors are diploid (recognizable because of their large diameter conidia) and some are haploid (with small diameter conidia). Two types are particularly suitable for illustrating the phenomena at work: white haploid sectors, and yellow diploid sectors (Figure 6-20).

Haploid White Sectors. If haploid white sectors are isolated and tested, almost exactly one-half of them prove to have the genotype $w\ ad^+\ pro\ paba^+\ y^+\ bi$ and one-half have the genotype $w\ ad\ pro^+\ paba\ y\ bi^+$. Thus, the original diploid nucleus has somehow become haploid (a process known as **haploidization**), presumably through the progressive loss of one member of each chromosome pair. By looking at only white haploid conidia, we automatically select for the *w*-bearing chromosome. In one-half of the *w* sectors, the $ad^+\ pro\ paba^+\ y^+\ bi$ chromosome is retained; in the other half, the $ad\ pro^+\ paba\ y\ bi^+$ chromosome is retained (Figure 6-21). In this way, the recessive color alleles can be used to derive linkage information because haploidization is similar to independent

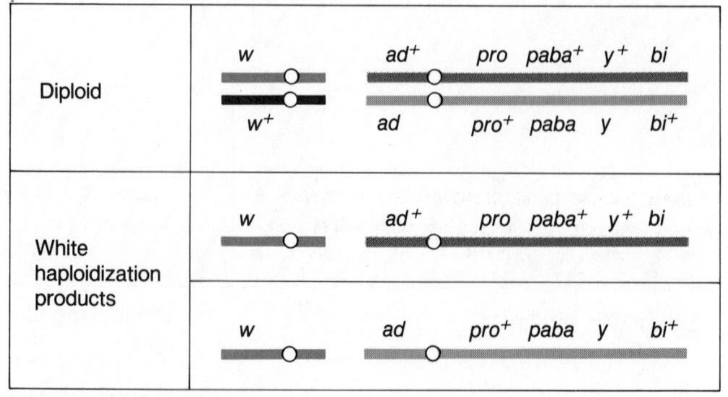

Figure 6-21 Genotypes of an *Aspergillus* diploid and two white haploidization products derived from it.

assortment. You can see that we first select the chromosome bearing the color marker and then observe which genes are retained with it and which are independent of it, and in what groupings.

Diploid Yellow Sectors. When a number of diploid yellow sectors are examined, they usually prove to contain recombinant chromosomes. For example, one sector type was yellow and also required "paba" (*para*-aminobenzoic acid) for growth. Remember that mitotic crossing-over can make heterozygous loci homozygous, so mitotic crossing-over explains this type (Figure 6-22). Notice that we must follow two spindle fibers to each pole in mitotic analysis. Although one chromatid pair would normally *not* lie adjacent to its homologous chromatid pair, by chance this *has* happened. This yellow, paba-requiring diploid arose from a mitotic exchange that obviously took place in the centromere-to-*paba* region; other yellow diploids would arise from mitotic exchanges in the *paba*-to-*y* region, but these would not require paba for growth. The relative frequencies of these two types would provide a kind of mitotic linkage map for that region. Such mapping can be carried out in several fungi. As expected, gene order in mitotic maps corresponds to gene order in meiotic maps, but, unexpectedly, the relative sizes of many of the intervals are found to differ when meiotic and mitotic maps are compared. The reason for this is not known. Note that every locus from the point of the crossover to the end of the chromosome arm is made homozygous by a mitotic crossover. This in itself can provide useful mapping information. Because mitotic recombination experiments are done in fast-growing vegetative fungal cultures, the results can be obtained much more quickly than a meiotic analysis, which requires the slower sexual phase.

> **Message** A mitotic crossover can result in the coincident homozygosity of all heterozygous gene pairs distal to the crossover, and such events may be used to deduce gene order and map distances.

In this section we have encountered several different processes that can cause a heterozygous allele pair to segregate during mitosis and thus code for different phenotypes. In later chapters we shall encounter other processes that can produce such variegation, but for the time being we can summarize as in the following message.

> **Message** Because of mitotic nondisjunction, chromosome loss, or crossing-over, a heterozygous pair of alleles may segregate, resulting in tissue variegation (phenotypically different sectors in one tissue).

Mapping Human Chromosomes

We have seen that human beings are not ideal subjects for traditional genetic analysis. Until the 1960s, geneticists had to rely on the study of family pedigrees to deduce linkage. Then, however, a technique was developed that has revolutionized the mapping of both sex-linked and autosomal genes in humans. This technique, which is called **somatic-cell hybridization** uses cells grown in culture (Figure 6-23). A virus called the Sendai virus has a useful property that makes this mapping technique possible. Normally, a virus has a specific point at which it attaches to and penetrates a host cell. Each Sendai virus has several points of attachment, so that it can simultaneously attach to two different cells if they happen to be close together. A virus, though, is very small in comparison with a cell (similar to the comparison between the earth and the sun), so that the two cells to which the virus is attached are held very close together indeed. In fact, in many cases, the membranes of the two cells fuse together and the two cells become one—a binucleate heterokaryon.

If suspensions of human and mouse cells are mixed together in the presence of Sendai virus which has been inactivated by ultraviolet light, the virus can mediate fusion of the two kinds of cells (Figure 6-24). Once the cells fuse, the nuclei can fuse to form a uninucleate cell

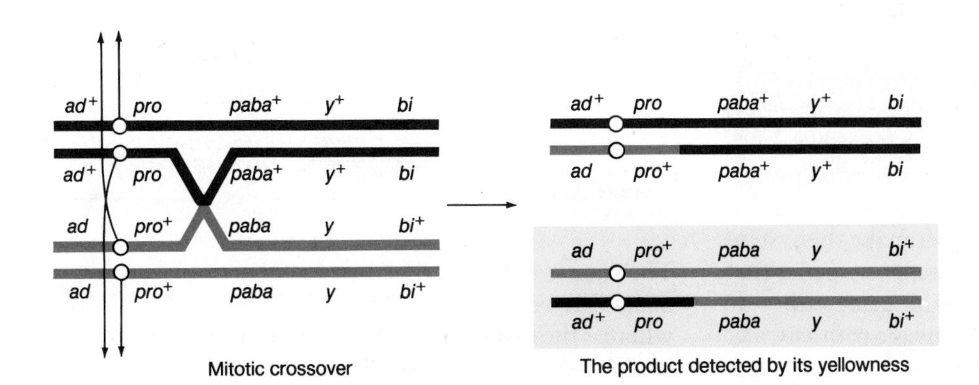

Figure 6-22 A mitotic crossover can produce a diploid yellow sector in the *Aspergillus* diploid shown. Note that the crossover produces homozygosity at all heterozygous loci beyond the crossover.

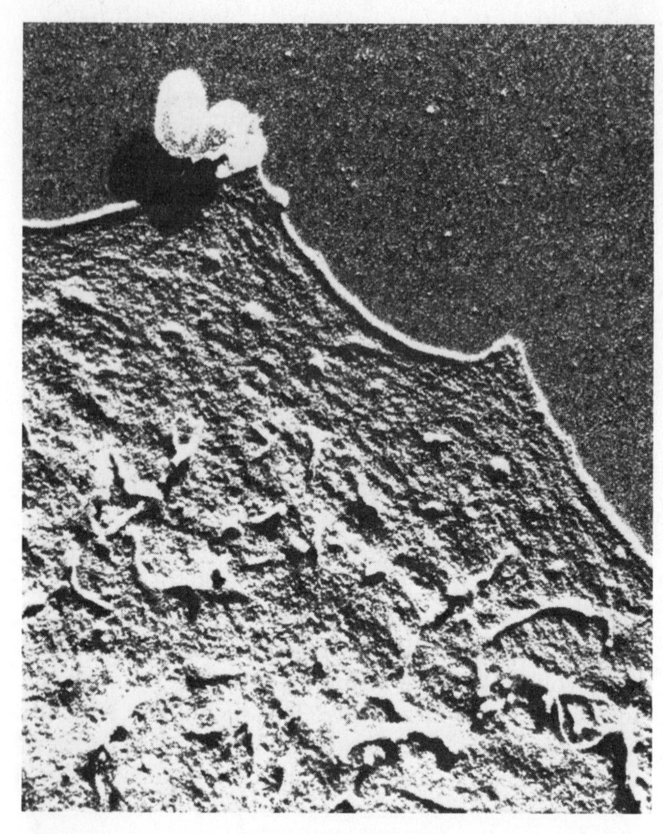

Figure 6-23 A human epithelial cell contrasted in size with two bacterial cells. The wavy line is the edge of the human cell, of which only about one-eighth is shown. The spherical images are *Escherichia coli* bacteria from the human intestine. (From Jack D. Griffith.)

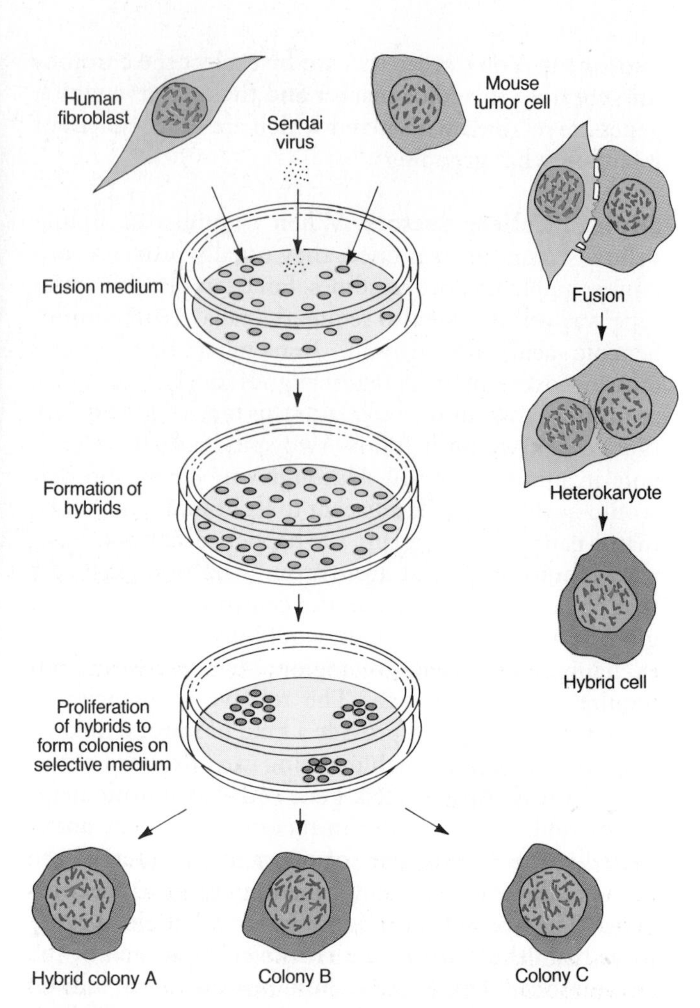

Figure 6-24 Cell-fusion techniques applied to human and mouse cells produce colonies, each of which contains a full mouse genome plus a few human chromosomes (blue). A fibroblast is a cell of fibrous connective tissue. (From F. H. Ruddle and R. S. Kucherlapati, "Hybrid Cells and Human Genes." *Scientific American,* 1974.)

line. Because the mouse and human chromosomes are very different in number and shape, the hybrid cells can be readily recognized. However, for unknown reasons, the human chromosomes are gradually eliminated at random from the hybrid as the cells divide (perhaps this is analogous to haploidization in *Aspergillus*). This process can be arrested to encourage the formation of a stable partial hybrid in the following way. The mouse cells that are used can be made genetically deficient in some function (usually a nutritional one) so that the function must be supplied by the human genome if the cells are to grow. This selective technique usually results in the maintenance of hybrid cells that have a complete set of mouse chromosomes and a small number of human chromosomes, which vary in number and type from hybrid to hybrid but which always include the nutritionally sufficient human chromosome.

Let's look at the specific genes that make the system work. The most popular selective system uses genes that synthesize DNA. In cells DNA can be made either de novo ("from scratch") or through a salvage pathway that uses molecular skeletons already available. The selective technique adds a chemical, aminopterin, that blocks the de novo pathway, restricting DNA synthesis to the salvage pathway. Two essential salvage enzymes are relevant to the system, as shown in the following two reactions:

$$\text{thymine} \xrightarrow{\text{thymidine kinase (TK)}} \text{thymidilic acid (a DNA precursor)}$$

$$\text{hypoxanthine} \xrightarrow[\text{(HGPRT)}]{\text{hypoxanthine-guanine phosphoribosyl transferase}} \text{inosinic acid (a DNA precursor)}$$

The mouse cell line to be fused is genetically unable to make TK because it is homozygous for the allele tk^-, whereas the human cell line is genetically unable to make HGPRT because it is homozygous at another locus for

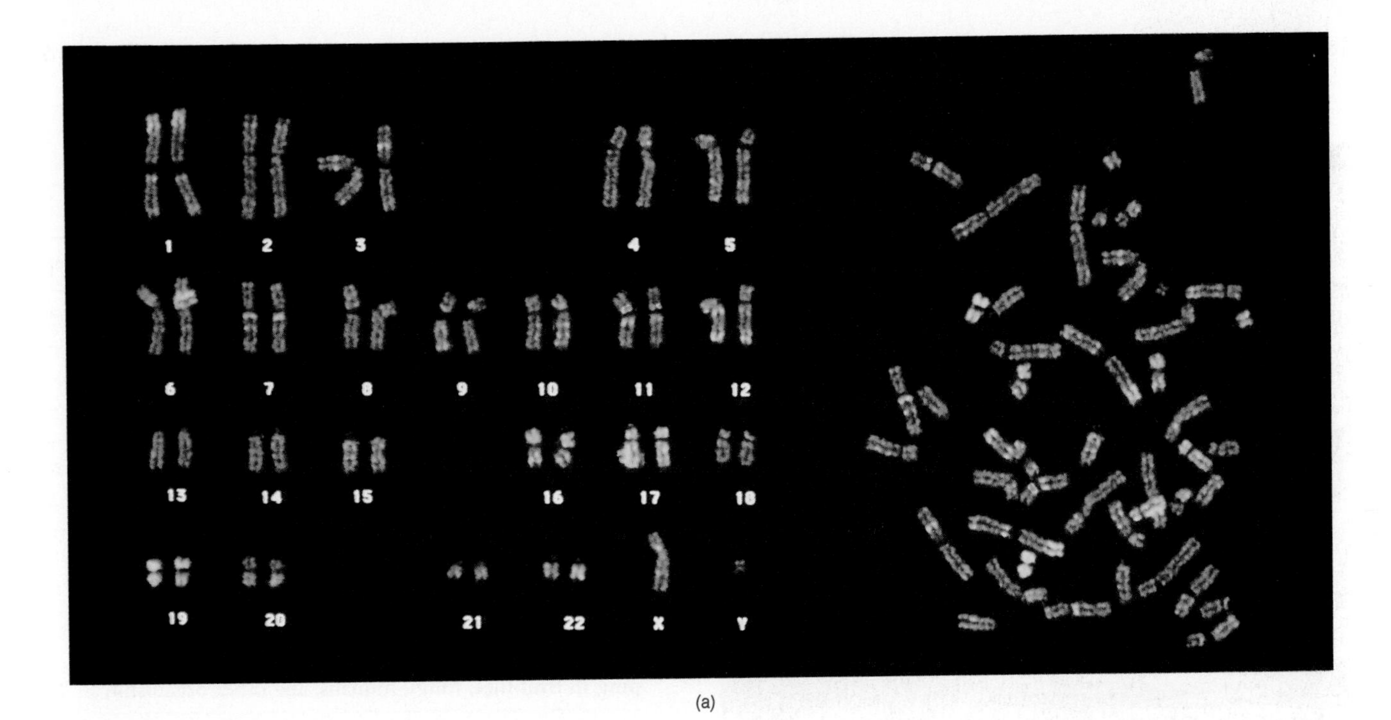

(a)

the allele *hgprt*$^-$. So the phenotypes of the two fusing cell lines are

mouse:	TK−	HGPRT+
human:	TK+	HGPRT−

Because each is deficient for one enzyme, neither the mouse nor the human cells are able to make DNA individually. In the hybrid cells, however, *tk*$^+$ allele complements the *hgprt*$^+$ allele, so that the cells can make both enzymes and DNA and hence can proliferate. Most human chromosomes are eliminated from the hybrid cell cultures because their loss has no effect on the cultures' ability to grow. But to continue to grow, a hybrid culture must retain at least one of the human chromosomes that carries the *tk*$^+$ allele.

Luckily, the progressive elimination of the human chromosomes from the fused cell lines can be followed under the microscope because mouse chromosomes can easily be distinguished from human chromosomes. This microscopic observation is facilitated by stains such as quinacrine and Giemsa that reveal a pattern of "banding" within the chromosomes. The size and the position of these bands vary from chromosome to chromosome but the banding patterns are highly specific and constant for each chromosome. Thus, it is relatively easy to identify the human chromosomes that are present in any hybrid cell (Figure 6-25). Different hybrid cells are grown separately into lines; eventually a bank of lines is produced that contains, in total, all the human chromosomes.

The mapping technique works as follows. If the human chromosome set is homozygous for a genetic marker (such as an allele that controls a specific cell-surface antigen, drug resistance, a nutritional requirement, or a protein variant), then the presence or absence of this genetic marker in each line of hybrid cells can be

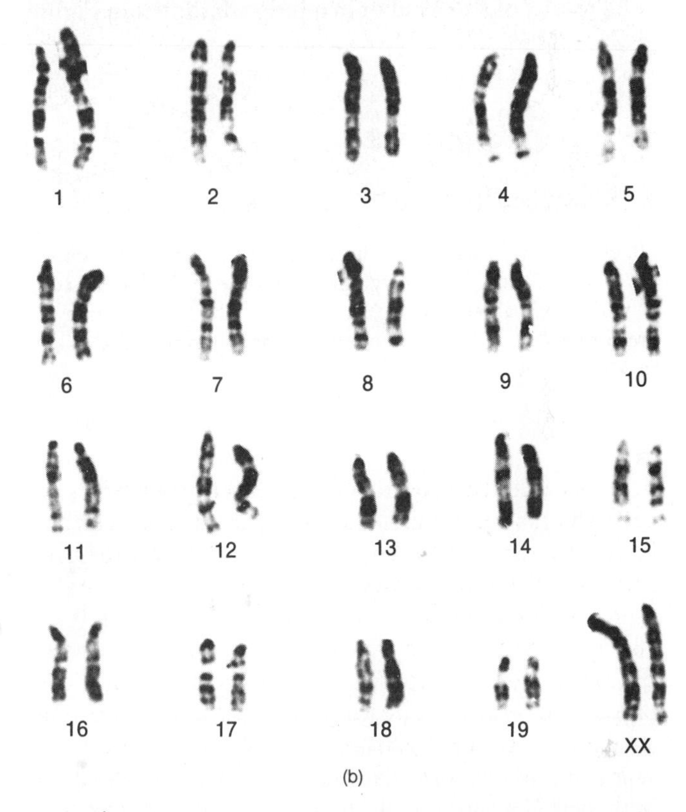

(b)

Figure 6-25 (a) Stained human chromosomes. Under the microscope the chromosomes appear as a jumbled cluster, as shown to the right. This array is photographed, the individual chromosomes are cut out of the photograph and then grouped by size and banding pattern, as shown at left. The chromosome set of a male is shown. (b) The chromosomes of a female mouse, shown for comparison. To the experienced eye, the mouse chromosomes can be easily distinguished from human chromosomes, as required in the human–rodent cell-hybrid technique of gene localization. (Part a from David Ward, Yale University School of Medicine. Part b from Jackson Laboratory, Bar Harbor, Maine.)

correlated with the presence or absence of certain human chromosomes in each line. Data of this sort are presented in Table 6-2, where + means presence and − means absence. We can see that in the different hybrid cell lines, genes 1 and 3 are always present or absent together. We can conclude, then, that they are linked. Furthermore, the presence or absence of genes 1 and 3 is directly correlated with the presence or absence of chromosome 2, so we can assume that these genes are located on chromosome 2. By the same reasoning, gene 2 must be on chromosome 1, but the location of gene 4 cannot be assigned.

Large numbers of human genes have now been localized to specific chromosomes in this way. To make a linkage map showing the order and distances between these genes, we must perform other manipulations; for example, we might correlate the loss or retention of variously sized bits of a specific chromosome with the presence or absence of the genetic markers being tested. The results of this kind of mapping are shown in Figure 6-26.

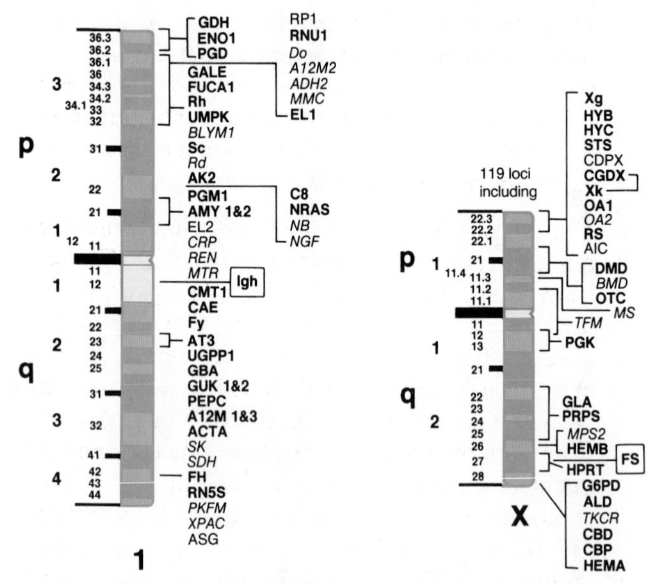

Figure 6-26 Maps of the human chromosome 1 and the X chromosome. The loci are labeled with code names. Many code names are based on the protein specified by the gene; others represent marker loci identified only as DNA variants. Such loci with some molecular tag can be mapped well by somatic cell hybridization methods. Some of these loci are interesting in themselves, and some are useful as chromosomal landmarks to which other loci may be mapped. For autosomes such as chromosome 1, approximately $\frac{2}{3}$ of the loci have been assigned by somatic hybridization, and the remainder from pedigree analysis. Compare these maps with those in the Chapter 5 opener, which list genes that determine various disorders and other phenotypes that map to these chromosomes. Protein variants are the basis of most of these conditions. p and q designate the short and long arms, respectively. (From V. A. McKusick, *Genetic Maps*, Vol. 2. Edited by S. J. O'Brien. Cold Spring Harbor, 1982.)

Table 6-2. Comparison of Five Hybrid Lines

		Hybrid Cell Lines				
		A	B	C	D	E
Human genes	1	+	−	−	+	−
	2	−	+	−	+	−
	3	+	−	−	+	−
	4	+	+	+	−	−
Human chromosomes	1	−	+	−	+	−
	2	+	−	−	+	−
	3	−	−	−	+	+

Message Mitotic recombination and segregation can provide mapping information to supplement meiotic mapping in fruit flies, fungi, humans, and other organisms.

Summary

Because there are likely to be multiple crossovers between loci, map distance between such loci is not linearly related to recombinant frequency. A mathematical relationship between distance along a chromosome and recombinant frequency is called the mapping function. We use the Poisson distribution to calculate the mapping function.

Another useful genetic tool is tetrad analysis in those fungi and single-celled algae in which the four products of each meiosis are held together in a kind of bag. Tetrad analysis provides the opportunity to test directly some of the assumptions of the chromosome theory of heredity, to map centromeres as genetic loci, to investigate the possibility of chromatid interference, and to examine the mechanisms of crossing-over.

Tetrads may be linear or unordered. Analysis of linear tetrads enables us to map loci in relation to their centromeres and to each other. From crosses in which the parents differ at two linked loci, the asci in a linear or unordered tetrad may be classified as parental ditype (PD), nonparental ditype (NPD), and tetratype (T). Unordered tetrad analysis can be used to correct for multiple crossovers in mapping studies.

Although segregation and recombination are normally thought of as meiotic phenomena, alleles do occasionally segregate and recombine during mitosis. Mitotic segregation was first identified in the 1930s, when Calvin Bridges observed patches of M^+ bristles on the body of a female *Drosophila* of predominantly M phenotype. He concluded that the patches were the result of abnormal chromosome segregation during mitosis. Around the same time, Curt Stern observed twin spots in *Drosophila* and assumed they must be the reciprocal products of mitotic crossing-over. Fungi are extensively used to study mitotic segregation and recombination.

Humans are generally unsuitable subjects for traditional mapping analysis, but somatic-cell hybridization has provided another approach. When the Sendai virus is used to fuse human and mice cells the nuclei also fuse, and then human chromosomes are progressively reduced. Chromosomal assignments of human genetic markers present are obtained by correlating the retention of these markers with human chromosomes in the reduced hybrids.

Concept Map

Draw a concept map interrelating as many of the following terms as possible. Note that the terms are listed in no particular order.

map distance / map function / linkage map / recombinant frequency / tetrad analysis / centromere mapping / second division segregation / correction for double crossovers / meiosis

Chapter Integration Problem

Strains of the haploid fungus *Neurospora* bearing the *am* allele do not grow unless alanine is added to the medium. Normal wild-type strains grow without alanine supplementation. Linked to the *am* locus is a suppressor gene *ssu*. The *ssu* allele suppresses the alanine requirement, and *am ssu* strains do not need alanine supplementation. The *ssu* allele has no effect on the wild-type am^+ allele, and furthermore the wild-type allele ssu^+ has no known effect on the *am* locus alleles.

It has been calculated that triple and higher crossovers are negligible in the region between *am* and *ssu*. Meioses thus have no crossover (41 percent), one crossover (39 percent), or two crossovers (20 percent). From this information answer the following questions:

a. In what way is the genetic system described dependent on the environment?

b. What is the recombinant frequency (RF value)?

c. If ascospores from the cross *am ssu* × $am^+ ssu^+$ are sampled at random, what proportion will require alanine to grow?

d. What is the distance between the loci in corrected map units?

e. If asci from the cross *am ssu* × $am^+ ssu^+$ are analyzed as unordered tetrads, what proportion will show the following five phenotypic patterns:

Pattern	Number of ascospores not requiring alanine	Number of ascospores requiring alanine
(1)	8	0
(2)	6	2
(3)	4	4
(4)	2	6
(5)	0	8

Solution

a. The only aspect of the environment that is mentioned is alanine as part of the growth medium. If there is alanine in the medium, we cannot identify the two basic phenotypes, the growth phenotype and the no-growth phenotype. Therefore, the ability to study and analyze this interesting case of gene interaction is environmentally dependent, because the starting point for any piece of genetic analysis is variation, and without it there would be no study. Another point that is relevant is that the *am* allele would probably be lethal in a natural environment; it is only because scientists propagate such phenotypes (in this case on special medium—that is, in a special environment) that we have them for study.

b. The chapter has provided us with the formula for calculating the RF from tetrad analysis. We have learned that RF = T/2 + NPD. However, we do not know what the values of T and NPD are, so they have to be calculated. We have been given the frequencies of meioses with 0, 1, and 2 crossovers, and we have learned how PD, T, and NPD are produced from such meioses. We thus have what we need to calculate the required frequencies. All asci with no crossovers are PD, so this contributes 41 percent to the PD class. Single crossovers are also easy; they produce only T asci, so this contributes 39 percent to the T class. Double crossovers are a bit more complicated, but we have seen that they produce $\frac{1}{4}$ PD, $\frac{1}{2}$ T and $\frac{1}{4}$ NPD, which gives us 5 percent PD, 10 percent T, and 5 percent NPD. Adding these, we get 46 percent PD, 49 percent T, and 5 percent NPD. Now we can calculate the RF value, which is $\frac{49}{2} + 5 =$ 29.5 percent.

c. We now expect 29.5 percent recombinants in the ascospores from the cross, comprising the two recombinant classes, *am* ssu^+ and am^+ *ssu*, each being 14.75 percent. The two parental classes are *am ssu* and $am^+ ssu^+$ and we expect them to make up the difference 100 − 29.5 = 70.5 percent, each being 35.25 percent. Which of these four genotypes will require alanine? Obviously, $am^+ ssu^+$ will not, neither will *am ssu*. The genotype $am^+\ ssu$ will also not require alanine because it has the wild-type allele at the *am* locus, and we have been told that *ssu* does not affect that allele. However, $am\ ssu^+$ does not have the suppressor allele, and will require alanine, so the answer is 14.75 percent, the frequency of this class.

d. We have learned in this chapter that if we know the frequency of PD, T, and NPD in an unordered tetrad analysis, we can use these values to correct for the effect of double crossovers, which always tend to cause an underestimation of map distance. The formula is $50(T + 6NPD)$. Applying our calculated values we get $50(0.49 + 0.30)$, which equals 39.5 corrected map units.

e. We have done most of the reasoning to answer this part because we have deduced that the only genotype that will confer a requirement for alanine is the genotype $am\ ssu^+$. Since we know there are only three ascus types in an unordered tetrad analysis of two heterozygous loci (PD, T, and NPD), we simply need to deduce how many $am\ ssu^+$ ascospores will be seen in each. In PD asci there will be no $am\ ssu^+$ genotypes, so these asci will give us an 8:0 ratio. Tetratype (T) asci will contain one spore pair of genotype $am\ ssu^+$, so these will give us a 6:2 ratio. The NPD class will be composed of four $am\ ssu^+$ ascospores and four $am^+\ ssu$ ascospores, giving a 4:4 ratio. And that covers all the possibilities, so we can answer that the proportions for the table are (1) 46 percent, (2) 49 percent, (3) 5 percent, (4) 0 percent, (5) 0 percent.

In summary, notice the concepts we used from previous chapters: Mendelian segregation, chromosomal inheritance, gene interaction, lethal effects, environmental effects, crossing-over, and mapping.

Solved Problems

1. A cross is made between a haploid strain of *Neurospora,* of genotype $nic^+\ ad$ and another haploid strain of genotype $nic\ ad^+$. From this cross, a total of 1000 linear asci are isolated and categorized as follows:

1	2	3	4	5	6	7
$nic^+\ ad$	$nic^+\ ad^+$	$nic^+\ ad^+$	$nic^+\ ad$	$nic^+\ ad$	$nic^+\ ad^+$	$nic^+\ ad^+$
$nic^+\ ad$	$nic^+\ ad^+$	$nic^+\ ad^+$	$nic^+\ ad$	$nic^+\ ad$	$nic^+\ ad^+$	$nic^+\ ad^+$
$nic^+\ ad$	$nic^+\ ad^+$	$nic^+\ ad$	$nic\ ad$	$nic\ ad^+$	$nic\ ad$	$nic\ ad$
$nic^+\ ad$	$nic^+\ ad^+$	$nic^+\ ad$	$nic\ ad$	$nic\ ad^+$	$nic\ ad$	$nic\ ad$
$nic\ ad^+$	$nic\ ad$	$nic\ ad^+$	$nic^+\ ad^+$	$nic^+\ ad$	$nic^+\ ad^+$	$nic^+\ ad$
$nic\ ad^+$	$nic\ ad$	$nic\ ad^+$	$nic^+\ ad^+$	$nic^+\ ad$	$nic^+\ ad^+$	$nic^+\ ad$
$nic\ ad^+$	$nic\ ad$	$nic\ ad$	$nic\ ad^+$	$nic\ ad^+$	$nic\ ad$	$nic\ ad^+$
$nic\ ad^+$	$nic\ ad$	$nic\ ad$	$nic\ ad^+$	$nic\ ad^+$	$nic\ ad$	$nic\ ad^+$
808	1	90	5	90	1	5

Map the *ad* and *nic* loci in relation to centromeres and to each other.

Solution

What principles can we draw on to solve this problem? It is a good idea to begin by doing something straightforward, which is to calculate the two locus-to-centromere distances. We do not know if the *ad* and the *nic* loci are linked but we do not need to know. The frequencies of the M_{II} patterns for each locus give the distance from locus to centromere. (We can worry about whether it is the same centromere later.)

Remember that an M_{II} pattern is any pattern that is not two blocks of four. Let's start with the distance between the *nic* locus and the centromere. All we have to do is add the ascus types 4, 5, 6, and 7, because these are all M_{II} patterns for the *nic* locus. The total is $5 + 90 + 1 + 5 = 101$, or 10.1 percent. In this chapter, we have seen that to convert this to map units, we must divide by 2, which gives 5.05 m.u.

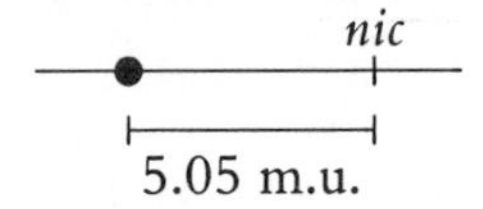

Now we do the same thing for the *ad* locus. Here the total of the M_{II} patterns is given by types 3, 5, 6, and 7 and is $90 + 90 + 1 + 5 = 186$, or 18.6 percent, which is 9.3 m.u.

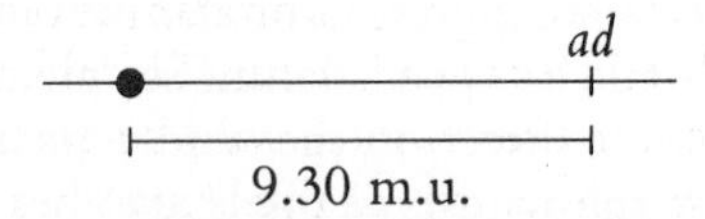

Now we have to put these two together and decide between the following alternatives, all of which are compatible with the above locus-to-centromere distances:

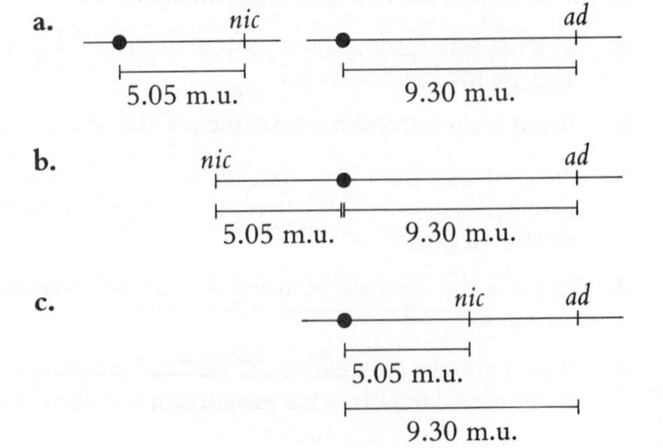

Here a combination of common sense and simple analysis tells us which alternative is correct. First, an inspection of the asci reveals that the most common single type is the one labeled 1, which contains more than 80 percent of all the asci. This type contains only *nic*$^+$ *ad* and *nic ad*$^+$ genotypes, and they are *parental* genotypes. So we know that recombination is quite low and the loci are certainly linked. This rules out alternative a.

Now consider alternative c; if this were correct, a crossover between the centromere and the *nic* locus would generate not only an M_{II} pattern for that locus but also an M_{II} pattern for the *ad* locus, because it is farther from the centromere than *nic*. The ascus pattern produced by alternative c should be

nic$^+$ *ad*
nic$^+$ *ad*
nic ad$^+$
nic ad$^+$
→
nic$^+$ *ad*
nic$^+$ *ad*
nic ad$^+$
nic ad$^+$
nic$^+$ *ad*
nic$^+$ *ad*
nic ad$^+$
nic ad$^+$

Remember that the *nic* locus shows M_{II} patterns in asci 4, 5, 6, and 7 (a total of 101 asci); of these, type 5 is the very one we are talking about and contains 90 asci. Therefore, alternative c appears to be correct because ascus type 5 comprises about 90 percent of the M_{II} asci for the *nic* locus. This relationship would not hold if alternative b were correct, because crossovers on either side of the centromere would generate the M_{II} patterns for the *nic* and the *ad* loci independently.

Is the map distance from *nic* to *ad* simply 9.30 − 5.05 = 4.25 m.u.? Close, but not quite. The best way of calculating map distances between loci is always by measuring the recombinant frequency (RF). We could go through the asci and count all the recombinant ascospores, but it is simpler to use the formula RF = $\frac{1}{2}$T + NPD. Even though the asci are linear, they can still be scored PD, NPD, and T. The T asci are classes 3, 4 and 7, and the NPD asci are classes 2 and 6. Hence, RF + [$\frac{1}{2}$(100) + 2]/1000 = 5.2 percent or 5.2 m.u., and a better map is

nic *ad*
5.05 m.u. 5.2 m.u.
10.25 m.u.

The reason for the underestimate of the *ad*-to-centromere distance calculated from the M_{II} frequency is the occurrence of double crossovers, which can produce an M_I pattern for *ad*, as in ascus type 4:

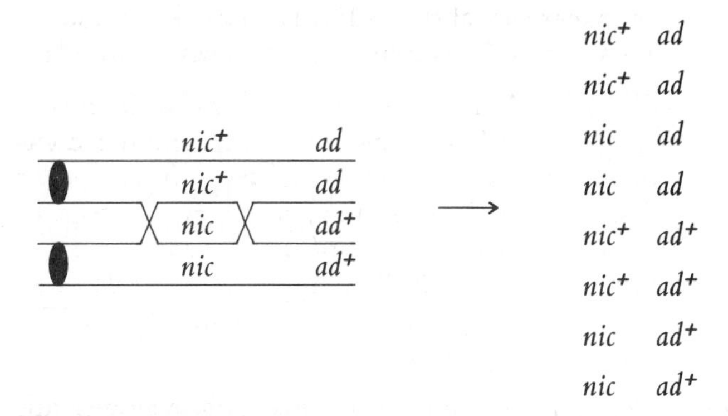

2. In *Aspergillus*, the recessive chromosome VI alleles *leu1*, *met5*, *thi3*, *pro2*, and *ad2* confer requirements for leucine, methionine, thiamine, proline, and adenine, respectively, whereas their wild-type alleles confer no such requirements. At the tip of chromosome VI there is a locus with a recessive allele *su* that suppresses *ad2*, so that strains expressing *ad2* and *su* require no adenine. A diploid strain is made by combining the following haploid genotypes, where the loci, with the exception of *su*, are written in no particular order:

su leu1 met5 thi3 pro2 ad2
and
su$^+$ *leu1*$^+$ *met5*$^+$ *thi3*$^+$ *pro2*$^+$ *ad2*

Diploid asexual spores were spread on a medium containing all supplements except adenine. Most spores did not grow due to the recessiveness of *su*, but a few did grow into colonies; 100 of these were removed and tested for the phenotypes of the other markers. The following four classes were found (note that these are diploid phenotypes, not haploid genotypes):

(1)	su	LEU+	MET−	THI+	PRO+	AD−	60
(2)	su	LEU−	MET−	THI+	PRO+	AD−	25
(3)	su	LEU+	MET+	THI+	PRO+	AD−	10
(4)	su	LEU−	MET−	THI−	PRO+	AD−	5

a. Explain the production of these four classes and their relative amounts.

b. Why do you think that no colonies expressing *pro2* were recovered?

Solution

a. The experiment concerns the behavior of diploid cells at mitosis, so the four classes must be explained by a mitotic mechanism. It seems likely that the *su* allele in all classes has been made homozygous *su su*, because only in that condition can it suppress an *ad2 ad2* homozygote to permit growth without adenine. This then was the basis of the original selection of the 100 colonies. But evidently other alleles have become homozygous too—and in different combinations in different classes.

Chapter 6 demonstrates that mitotic crossing-over can produce homozygosity in any recessive allele that is distal to the crossover. Can crossovers in different regions of chromosome VI explain the different classes? Inspection of the classes shows that homozygosity can be produced for either *su* alone (class 3), or *su* and *met* (class 1), or *su, met* and *leu* (class 2), or *su, met, leu,* and *thi* (class 4). This virtually dictates to us that the order must be *su–met–leu–thi*–centromere. The crossovers responsible for homozygosity must then be in the following regions.

su——*met*——*leu*——*thi*——centromere

class 3 class 1 class 2 class 4

For example, the following crossover is necessary to produce class 2:

su met leu thi

su met leu thi
su$^+$ *met*$^+$ *leu*$^+$ *thi*$^+$

su$^+$ *met*$^+$ *leu*$^+$ *thi*$^+$

su met leu thi
su met leu thi$^+$

The relative sizes of the classes must reflect the relative distances separating the loci as follows:

su met *leu* *thi*

10 60 25 5

Note that these units are relative proportions and are not the same as meiotic map units.

b. We are told that *pro2* is on chromosome VI, yet it is never made homozygous. Note the importance of the fact that to select for mitotic crossovers, we are making use of an allele *su* which is at the tip of one arm. Recessive alleles in the other arm would not be simultaneously made homozygous by these crossovers, so *pro2* is probably in the other arm. We have no way of knowing where *ad2* is, because it is homozygous from the outset.

Problems

1. The *Neurospora* cross *al-2*$^+$ × *al-2* is made. A linear tetrad analysis reveals that the second-division segregation frequency is 8 percent.

a. Draw two examples of second-division segregation patterns in this cross.

b. What can the 8 percent value be used to calculate?

2. From the fungal cross *arg-6 al-2* × *arg-6*$^+$ *al-2*$^+$, what will the spore genotypes be in unordered tetrads that are **(a)** parental ditypes? **(b)** tetratypes? **(c)** nonparental ditypes?

3. For a certain chromosomal region, the mean number of crossovers at meiosis is calculated to be two per meiosis. In that region, what proportion of meioses are predicted to have **(a)** no crossovers? **(b)** one crossover? **(c)** two crossovers?

4. In a *Drosophila* of genotype

y *sn*
2 1
y$^+$ *sn*$^+$

y determines yellow body and *y*$^+$ determines brown body; *sn* determines singed hairs and *sn*$^+$ unsinged. What is the detectable outcome if there is a mitotic crossover during development in **(a)** Region 1? **(b)** Region 2?

5. In a haploid yeast, a cross between *arg*$^-$ *ad*$^-$ *nic*$^+$ *leu*$^+$ and *arg*$^+$ *ad*$^+$ *nic*$^-$ *leu*$^-$ produces haploid sexual spores, and 20 of these are isolated at random. When the resulting cultures are tested on various media, they give the results shown below, where Arg means arginine; Ad, adenine; Nic, nicotinamide; Leu, leucine; +, growth; and − no growth.

	Minimal Medium Plus			
Culture	Arg, Ad, Nic	Arg, Ad, Leu	Arg, Nic, Leu	Ad, Nic, Leu
1	+	+	−	−
2	−	−	+	+
3	−	+	−	+
4	+	−	+	−
5	−	−	+	+
6	+	+	−	−
7	+	+	−	−
8	−	−	+	+
9	+	−	+	−
10	−	+	−	+
11	−	+	−	+
12	+	−	+	−
13	+	+	−	−
14	+	−	+	−
15	−	+	−	+
16	+	−	−	−
17	+	+	−	−
18	−	−	+	+
19	+	+	−	−
20	−	+	−	+

a. What can you say about linkage among these genes?

b. What is the origin of culture 16?

6. Every Friday night, genetics student Jean Allele, exhausted by her studies, goes to the students' union bowling lane to relax. But even there, she is haunted by her genetics studies. The rather modest bowling lane has only four bowling balls: two red and two blue. These are bowled at the pins and are then collected and returned down the chute in random order, coming to rest at the end stop. Over the evening, Jean notices familiar patterns of the four balls as they come to rest at the stop. Compulsively, she counts the different patterns. What patterns did she see, what were their frequencies, and what is the relevance of this matter to genetics? (This is not a trivial question.)

7. **a.** Use the mapping function to calculate the corrected map distance between loci having a recombinant frequency of 20 percent. Remember that an m value of 1 is equal to 50 corrected map units.

b. If you obtain an RF value of 45 percent in one experiment, what can you say about linkage? (The actual numbers you observed were 58 and 52 parental types and 47 and 43 recombinant types out of 200 progeny.)

***8.** In a tetrad analysis, the linkage arrangement of the p and q loci is as follows.

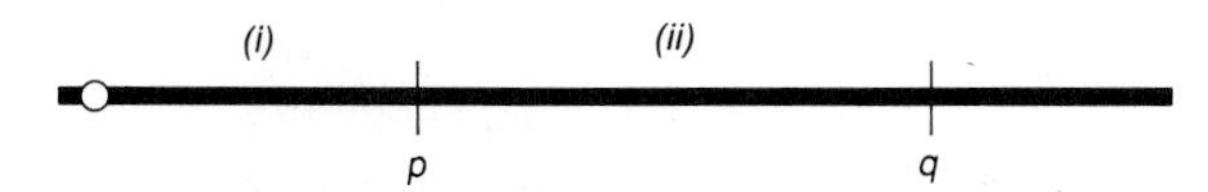

Assume that

- In region (i), there is no crossover in 88 percent of meioses and a single crossover in 12 percent of meioses.
- In region (ii), there is no crossover in 80 percent of meioses and a single crossover in 20 percent of meioses.
- There is no interference (in other words, the situation in one region doesn't affect what is going on in the other region).

What proportions of tetrads will be of the following types? **(a)** M_IM_I, PD; **(b)** M_IM_I, NPD; **(c)** M_IM_{II}, T; **(d)** $M_{II}M_I$, T; **(e)** $M_{II}M_{II}$, PD; **(f)** $M_{II}M_{II}$, NDP; **(g)** $M_{II}M_{II}$, T. (NOTE: Here the M pattern written first is the one that pertains to the p locus.) HINT: The easiest way to do this problem is to start by calculating the frequencies of asci with crossovers in both regions, region 1, region 2, and in neither region. Then determine what M_I and M_{II} patterns result.

9. The following cross is made in *Neurospora*: $a^+ b^+ c^+ d^+ \times a\,b\,c\,d$ (loci a, b, c, and d are linked in the order written). Construct crossover diagrams to illustrate how the following unordered ascus patterns could arise:

a.	**b.**	**c.**
$a^+ b^+ c\ d^+$	$a^+ b\ c\ d$	$a^+ b^+ c\ d^+$
$a\ b\ c\ d^+$	$a^+ b^+ c^+ d$	$a^+ b\ c^+ d$
$a^+ b\ c^+ d$	$a\ b\ c\ d^+$	$a\ b^+ c\ d^+$
$a\ b^+ c^+ d$	$a\ b^+ c^+ d^+$	$a\ b\ c^+ d$

d.	**e.**	**f.**
$a^+ b\ c^+ d$	$a^+ b\ c\ d$	$a^+ b\ c^+ d$
$a^+ b\ c^+ d$	$a\ b\ c\ d^+$	$a^+ b^+ c^+ d$
$a\ b^+ c\ d^+$	$a^+ b^+ c^+ d$	$a\ b\ c\ d^+$
$a\ b^+ c\ d^+$	$a\ b^+ c^+ d^+$	$a\ b^+ c\ d^+$

g.	**h.**
$a^+ b\ c^+ d^+$	$a^+ b\ c^+ d$
$a^+ b\ c^+ d^+$	$a\ b^+ c\ d^+$
$a\ b^+ c\ d$	$a^+ b\ c^+ d$
$a\ b^+ c\ d$	$a\ b^+ c\ d^+$

10. A geneticist studies 11 different pairs of *Neurospora* loci by making crosses of the type $a\,b \times a^+ b^+$ and then analyzing 100 linear asci from each cross. For the convenience of making a table, the geneticist organizes the data as if all 11 pairs of loci had the same designation—a and b—as shown below:

Cross	Number of asci of type						
	$a\,b$ $a\,b$ $a^+ b^+$ $a^+ b^+$	$a\,b^+$ $a\,b^+$ $a^+ b$ $a^+ b$	$a\,b$ $a\,b^+$ $a^+ b^+$ $a^+ b$	$a\,b$ $a^+ b$ $a^+ b^+$ $a\,b^+$	$a\,b$ $a^+ b^+$ $a^+ b^+$ $a\,b$	$a\,b^+$ $a^+ b$ $a^+ b$ $a\,b^+$	$a\,b^+$ $a^+ b$ $a^+ b^+$ $a\,b$
1	34	34	32	0	0	0	0
2	84	1	15	0	0	0	0
3	55	3	40	0	2	0	0
4	71	1	18	1	8	0	1
5	9	6	24	22	8	10	20
6	31	0	1	3	61	0	4
7	95	0	3	2	0	0	0
8	6	7	20	22	12	11	22
9	69	0	10	18	0	1	2
10	16	14	2	60	1	2	5
11	51	49	0	0	0	0	0

For each cross, map the loci in relation to each other and to centromeres.

11. In *Neurospora*, the a locus is 5 m.u. from the centromere on chromosome 1. The b locus is 10 m.u. from the centromere on chromosome 7. From the cross of $a\,b^+ \times a^+ b$, determine the frequencies of the following: **(a)** parental ditype asci, **(b)** nonparental ditype asci, **(c)** tetratype asci, **(d)** recombinant ascospores, **(e)** wild-type ascospores. (NOTE: Don't bother with mapping-function complications here.)

12. Three different crosses in *Neurospora* are analyzed on the basis of unordered tetrads. Each cross combines a different pair of linked genes. The results follow:

Cross	Parents	Parental ditypes (%)	Tetra-types (%)	Non-parental ditypes (%)
1	$a\,b^+ \times a^+ b$	51	45	4
2	$c\,d^+ \times c^+ d$	64	34	2
3	$e\,f^+ \times e^+ f$	45	50	5

For each cross, calculate:

a. The frequency of recombinants (RF).

b. The uncorrected map distance, based on RF.

c. The corrected map distance, based on tetrad frequencies.

13. A geneticist crosses two yeast strains differing at the linked loci *ura3* (which governs uracil requirement) and *lys4* (which governs lysine requirement):

$$ura3\ lys4^+ \times ura3^+\ lys4$$

The geneticist isolates and classifies 300 *unordered* tetrads as follows:

$ura3\ lys4^+$	$ura3\ lys4$	$ura3\ lys4^+$
$ura3^+\ lys4$	$ura3\ lys4$	$ura3\ lys4^+$
$ura3\ lys4$	$ura3^+\ lys4^+$	$ura3^+\ lys4$
$ura3^+\ lys4^+$	$ura3^+\ lys4^+$	$ura3^+\ lys4$
138	12	150

a. What is the recombinant frequency?

b. If it is assumed that there may be zero, one, or two (never more) crossovers between these loci at meiosis, what are the percentages of zero, one, and two crossover meioses?

c. What is the distance between the loci in map units *corrected* for double crossovers?

***14.** For an experiment with haploid yeast, you have two different cultures. Each will grow on minimal medium to which arginine has been added, but neither will grow on minimal medium alone. (Minimal medium is inorganic salts plus sugar.) Using appropriate methods, you induce the two cultures to mate. The diploid cells then divide meiotically and form unordered tetrads. Some of the ascospores will grow on minimal medium. You classify a large number of these tetrads for the phenotypes ARG− (arginine requiring) and ARG+ (arginine independent) and record the following data:

Segregation of ARG−: ARG+	Frequency (%)
4:0	40
3:1	20
2:2	40

a. Using symbols of your own choosing, assign genotypes to the two parental cultures. For each of the three kinds of segregation, assign genotypes to the segregants.

b. If there is more than one locus governing arginine requirement, are these loci linked?

***15.** Four histidine loci are known in *Neurospora*. As shown here, each of the four loci is located on a different chromosome.

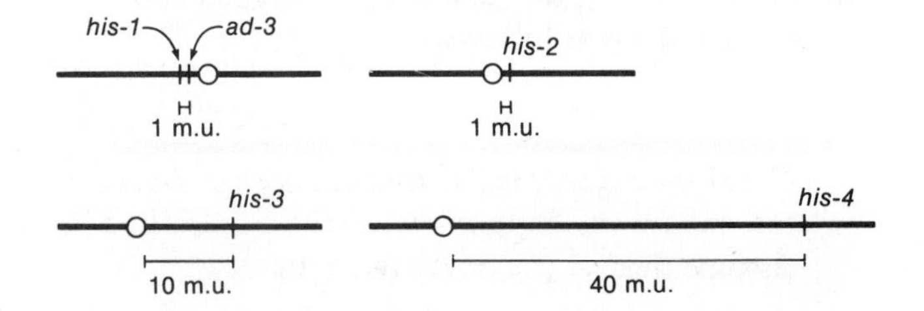

In your experiment, you begin with an *ad-3* line from which you recover a cell that also requires histidine. Now you wish to determine which of the four histidine loci is involved. You cross the *ad-3 his-?* strain with a wild type $(ad\text{-}3^+\ his\text{-}1^+\ his\text{-}2^+\ his\text{-}3^+\ his\text{-}4^+)$ and analyze ten unordered tetrads: two are PD, six are T, and two are NPD. From this result, which of the four *his* loci is most probably the one that changed from his^+ to *his*?

(Problem 15 from Luke deLange.)

16. The haploid ascomycete fungus *Sordaria* has ascospores that are normally black. Two ascospore-color variants (mutants) are isolated. When mutant 1 is crossed to a wild type, the asci produced contain four black spores and four white spores; when mutant 2 is crossed to a wild type, the asci produced contain four black spores and four tan spores. When mutants 1 and 2 are intercrossed, some asci contain four black and four white spores, some asci contain four tan and four white spores, and some asci contain four white and two black and two tan spores. Explain these results giving

a. genotypes underlying the three phenotypes;

b. an explanation of the types of asci produced in the crosses to wild type and the intercross.

***17.** *Drosophila melanogaster* can have two entire X chromosomes attached to the same centromere:

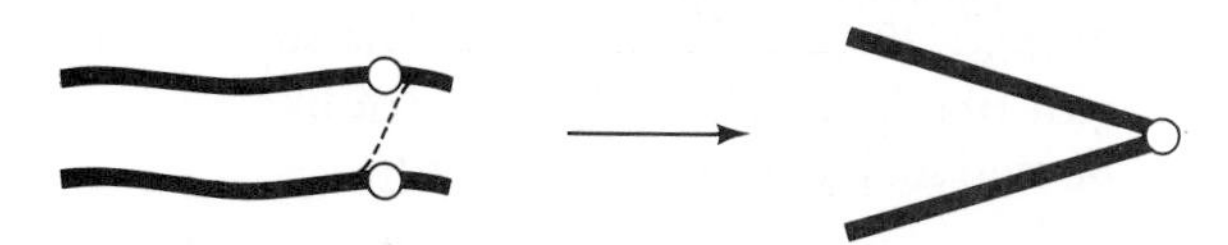

The joined X chromosomes behave as a single chromosome, called an attached X. Work out the inheritance of sex chromosomes in crosses of attached-X-bearing females with normal males. During meiosis, an attached-X chromosome duplicates and segregates as follows:

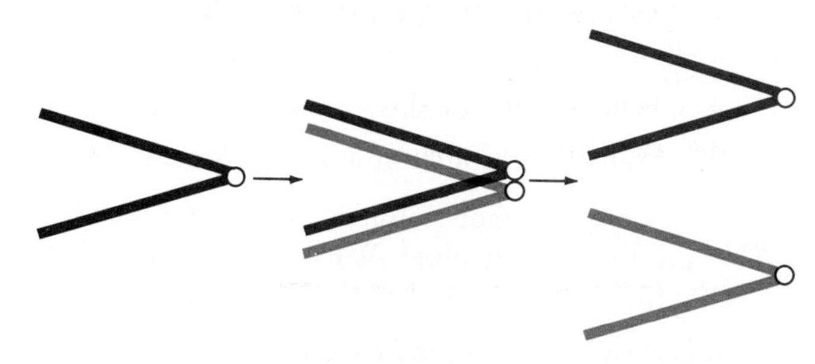

Nonsister chromatids of attached-x chromosomes can cross over. Suppose you have an attached-X chromosome of the following genotype:

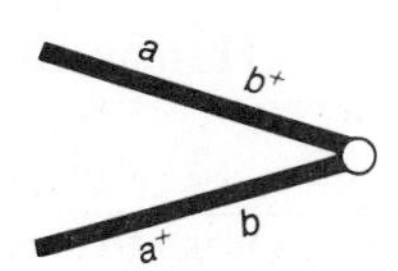

Diagram all the possible genotypic products and their phenotypes **(a)** from single crossovers between *a* and *b* and between *b* and the centromere, and **(b)** from double exchanges between *a* and *b* and between *b* and the centromere. Make sure you consider all possibilities. Describe how such a procedure represents a kind of tetrad analysis.

18. In *Neurospora,* the mating-type locus, which has alleles *a* and *A*, is 5 m.u. from the centromere on chromosome 1, and the cycloheximide resistance locus, which has alleles *c* and *C*, is 8 m.u. to the other side of the centromere on the same chromosome. What proportion of *unordered* tetrads from the cross *A C* × *a c* will be of the type

a.	**b.**	**c.**
A c	*A C*	*a C*
A C	*A C*	*a C*
a c	*a c*	*A c*
a C	*a c*	*A c*

19. A cross was made in the fungus *Sordaria,*

$$un\ cyh \times un^{+}\ cyh^{+}$$

and 100 linear tetrads were isolated. There proved to be six classes in the following proportions:

1	2	3
$un\ cyh$	$un\ cyh$	$un\ cyh$
$un^+\ cyh$	$un\ cyh^+$	$un\ cyh$
$un\ cyh^+$	$un^+\ cyh$	$un^+\ cyh^+$
$un^+\ cyh^+$	$un^+\ cyh^+$	$un^+\ cyh^+$
15	29	47

4	5	6
$un\ cyh$	$un\ cyh^+$	$un\ cyh$
$un^+\ cyh^+$	$un^+\ cyh$	$un^+\ cyh^+$
$un\ cyh$	$un\ cyh^+$	$un\ cyh^+$
$un^+\ cyh^+$	$un^+\ cyh$	$un^+\ cyh$
2	2	5

a. Map the genes in relation to each other and to centromeres. Show distances in uncorrected map units.

b. Draw six crossover diagrams, one to explain the origin of each class.

c. Which type of tetrad is missing and why do you think it is?

d. Why is class 6 more common than class 4 or 5?

20. *Neurospora*'s chromosome 4 carries the *leu3* locus, located near the centromere; its alleles always segregate at the first division. On the other arm of chromosome 4, the *cys2* locus is 8 m.u. away from the centromere. If we cross a *leu3*$^+$ *cys2* strain and a *leu3 cys2*$^+$ strain, and if we ignore double and other multiple crossovers, what will we expect as the frequencies of the following seven classes of linear tetrad (where l = *leu3* and c = *cys2*)?

a.	**b.**	**c.**	**d.**
$l\ c$	$l\ c^+$	$l\ c$	$l\ c$
$l\ c$	$l\ c^+$	$l\ c^+$	$l^+\ c$
$l^+\ c^+$	$l^+\ c$	$l^+\ c^+$	$l^+\ c^+$
$l^+\ c^+$	$l^+\ c$	$l^+\ c$	$l\ c^+$

e.	**f.**	**g.**
$l\ c$	$l\ c^+$	$l\ c^+$
$l^+\ c^+$	$l^+\ c$	$l^+\ c$
$l^+\ c^+$	$l^+\ c$	$l^+\ c^+$
$l\ c$	$l\ c^+$	$l\ c$

21. Consider the following *Drosophila* alleles:

g codes for gray body (wild type is black)

c codes for curly bristles (wild type is straight)

s codes for stippled body (wild type is smooth)

These autosomal loci are arranged in the order *g*–*c*–*s*–centromere. A homozygous gray, stippled strain was crossed to a homozygous curly strain and the F_1 were generally wild type in their phenotypic appearance, but microscopic examination of the flies revealed that a few rare individuals had one of three basic patterns. The patterns are

a. a gray stippled sector next to a curly sector

b. a gray sector next to a curly sector

c. solitary gray sectors

Explain the origin of these three patterns with a single unifying hypothesis.

22. Previous experiments on the fungus, *Aspergillus*, had shown that the loci *ad, col, phe, pu, sm*, and *w* are all on the same chromosome, but their order was not known. A diploid was constructed with the following genotype: $ad\ col\ phe\ pu\ sm\ w\ /\ ad^+\ col^+\ phe^+\ pu^+\ sm^+ w^+$ (the gene order is written alphabetically). When this diploid was cultured, white diploid sectors (*ww*) were observed and isolated. Upon further testing, these were found to be of the following *phenotypes* (none of the white diploid sectors expressed *col*):

$ad\ phe\ pu\ sm$	41%
$ad\ phe^+\ pu\ sm^+$	30%
$ad\ phe^+\ pu\ sm$	24%
$ad^+\ phe^+\ pu\ sm^+$	5%

a. What is the likely origin of these sectors?

b. What is the relative order of the six genes and the centromere?

c. Give relative map distances where possible.

d. Why was *col* not expressed by the white diploid cells?

23. An incompletely dominant allele of soybeans, *Y*, causes yellowish leaves in Y^+Y heterozygotes. However, heterozygotes regularly show rare patches of green adjacent to a patch of very pale yellow, all in the yellowish background. Propose an explanation for these rare patches.

24. You isolate white haploidized sectors from an *Aspergillus* diploid that is $w^+w\ a^+a\ b^+b\ c^+c$ and score them for *a, b*, and *c*. You find that 25 percent are $w\ a\ b\ c$, 25 percent are $w\ a^+\ b^+\ c^+$, 25 percent are $w\ a\ b^+\ c$, and 25 percent are $w\ a^+\ b\ c^+$. What linkage relations can you deduce from these frequencies? Sketch your conclusions.

25. You know that two loci, *y* and *ribo*, are linked in *Aspergillus*, but you do not know their locations relative to the centromere. You have a diploid culture of genotype $y^+\ ribo^+\ /\ y\ ribo$ that is green and that grows without riboflavin supplementation. You notice some yellow sectors in the culture and study them. They are diploid and you discover that 80 percent of them can grow without riboflavin whereas 20 percent require media supplemented with riboflavin. What is the most likely order of the two loci and the centromere?

26. An *Aspergillus* diploid is $pro^+ pro\ fpa^+ fpa\ paba^+\ paba$; *pro* is a recessive allele for proline requirement, *fpa* is a recessive allele for fluorophenylalanine resistance, and *paba* is a recessive allele for *para*-aminobenzoic acid (paba) requirement. When conidia are plated on fluorophenylalanine, only resistant colonies develop. Of 154 *diploid* resistant colonies, 35 do not require proline or paba, 110 require paba, and 9 require both.

a. What do these figures indicate?

b. Sketch your conclusions in the form of a map.

c. Some resistant colonies (not the ones described) are haploid. What would you predict their genotype to be?

27. A geneticist succeeds in maintaining three colonies of human–mouse hybrid cells. The only human chromosomes retained by the hybrid cells are those indicated by pluses below:

Hybrid colony	Human chromosome 1	2	3	4	5	6	7	8
A	+	+	+	+	−	−	−	−
B	+	+	−	−	+	+	−	−
C	+	−	+	−	+	−	+	−

The geneticist tests each of the colonies for the presence of five enzymes (α, β, γ, δ, and ϵ) with the following results: α is active only in colony C, β is active in all three colonies, γ is active only in colonies B and C, δ is active only in colony B, and ϵ shows no activity in any colony. What can the geneticist conclude about the locations of the genes responsible for these enzyme activities?

28. Consider the following set of eight hybridized

human-mouse cell lines:

Cell line	Chromosome 1	2	6	9	12	13	17	21	X
A	+	+	−	q	−	p	+	+	+
B	+	−	p	+	−	+	+	−	−
C	−	+	+	+	p	−	+	−	+
D	+	+	−	+	+	−	q	−	+
E	p	−	+	−	q	−	+	+	q
F	−	p	−	−	q	−	+	+	p
G	q	+	−	+	+	+	+	−	−
H	+	q	+	−	−	q	+	−	+

Each cell line may carry an intact (numbered) chromosome (+), only its long arm (q), only its short arm (p), or it may lack the chromosome (−).

The following human enzymes were tested for their presence (+) or absence (−) in cell lines A – H:

Enzyme	Cell line A	B	C	D	E	F	G	H
steroid sulfatase	+	−	+	+	−	+	−	+
phosphoglucomutase-3	−	−	+	−	+	−	−	+
esterase D	−	+	−	−	−	−	+	+
phosphofructokinase	+	−	−	−	+	+	−	−
amylase	+	+	−	+	+	−	−	+
galactokinase	+	+	+	+	+	+	+	+

Identify the chromosome carrying each enzyme locus. Where possible, identify the chromosome arm.

(Problem 28 from L. A. Snyder, D. Freifelder, and D. L. Hartl, *General Genetics.* Jones and Bartlett, 1985.)

7

Gene Mutation

The phenotype produced by an unstable mutation in *Zinnia.* The mutation eliminates red pigment, resulting in white tissue. However during development the mutation frequently reverts back to the normal allele, which permits synthesis of red pigment. Stripes are produced because cell division in the petals takes place mainly on the long axis, so the daughters of the revertant cells tend to be arranged longitudinally. (From M. A. L. Smith, Department of Horticulture, University of Illinois; see *Journal of Heredity* 80, 1989.)

KEY CONCEPTS

Mutation is the process whereby genes change from one allelic form to another.

▪

Mutations in germ-line cells can be transmitted to progeny, but somatic mutations cannot.

▪

Selective systems make it easier to obtain mutations.

▪

Mutagens are agents that increase normally low rates of mutation.

▪

Genes mutate randomly, at any time and in any cell of an organism.

▪

A biological process can be dissected genetically if mutations that affect that process can be obtained. Each gene identified by a mutation identifies a separate component of the process.

Genetic analysis would not be possible without **variants** — organisms that differ in a particular character. We have considered many examples in which we could analyze differing phenotypes for particular characters. Now we consider the origin of the variants. How, in fact, do genetic variants arise?

The simple answer to this question is that organisms have an inherent tendency to undergo change from one hereditary state to another. Such hereditary change is called **mutation.** Geneticists recognize two different levels at which mutation takes place. In a **gene mutation,** an allele of a gene changes, becoming a different allele. Because such a change occurs within a single gene and maps to one chromosomal locus ("point"), a gene mutation is sometimes called a **point mutation.** In the other level of hereditary change — **chromosome mutation** — segments of chromosomes, whole chromosomes, or even entire sets of chromosomes change. Gene mutation is not necessarily involved in such a process; the effects of chromosome mutation are due more to the new arrangements of chromosomes and of the genes they contain. Nevertheless, some chromosome mutations, in particular those proceeding from chromosome breaks, are accompanied by gene mutations caused by the disruption at the breakpoint. In this chapter, we explore gene mutation; in Chapters 8 and 9, we shall consider chromosome mutation.

To consider change, we must have a fixed reference point, or standard. In genetics, the wild type provides the standard. Remember that the wild-type allele may be either a form found in nature or a form commonly used as characterizing a standard laboratory stock. Any change away from the wild-type allele is called **forward mutation;** any change to the wild-type allele is called **reverse mutation** (or reversion or back mutation). For example,

$$\left.\begin{array}{r} a^+ \longrightarrow a \\ D^+ \longrightarrow D \end{array}\right\} \text{forward mutation}$$

$$\left.\begin{array}{r} a \longrightarrow a^+ \\ D \longrightarrow D^+ \end{array}\right\} \text{reverse mutation}$$

The non-wild-type allele of a gene is sometimes referred to as a mutation. To use the same word for the process and the product may seem confusing at first, but in practice little confusion arises. Thus, we can speak of a dominant mutation (such as *D* above) or a recessive mutation (such as *a*). Bear in mind how arbitrary these gene states are; the wild-type of today may have been a mutation in the evolutionary past, and vice versa.

Another useful term is **mutant.** This is, strictly speaking, an adjective and should properly precede a noun. A mutant organism or cell is one whose changed phenotype is attributable to the possession of a mutation. Sometimes the noun is left unstated; in this case, a mutant always means an individual or cell with a phenotype that shows that it bears a mutation.

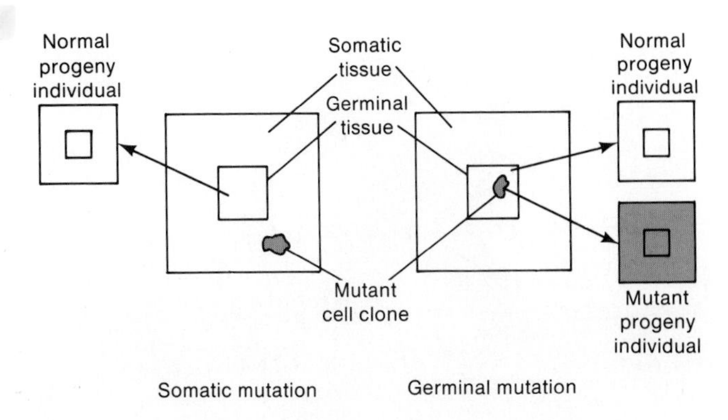

Figure 7-1 Somatic mutations are not transmitted to progeny, but germinal mutations may be transmitted to some or all progeny.

Two other useful terms are **mutation event,** which is the actual occurrence of a mutation, and **mutation frequency,** the proportion of mutations in a population of cells or individuals.

Somatic Versus Germinal Mutation

Genes or chromosomes can mutate in either somatic or germinal tissue. We thus speak of somatic and germinal mutations, which are diagrammed in Figure 7-1.

Somatic Mutation

A **somatic mutation** — a mutation in developing somatic tissue — can lead to a population of identical mutant cells, all of which have descended from the cell that mutated. A population of identical cells derived asexually from one progenitor cell is called a **clone.** Because the members of a clone tend to stay in the same vicinity, a mutation is often phenotypically expressed as a mutant **sector.** The earlier in development the mutation event, the larger the mutant clone will be (Figure 7-2). Mutant

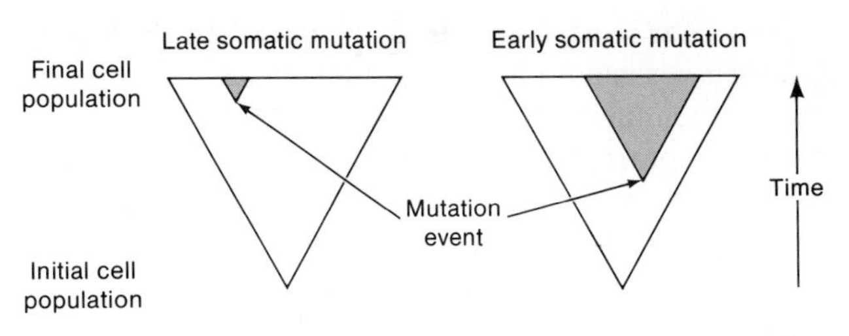

Figure 7-2 Early mutation produces a larger proportion of mutant cells in the growing population than does later mutation.

Figure 7-3 Somatic mutation in the red delicious apple. The mutant allele determining the golden color arose in a flower's ovary wall, which eventually developed into the fleshy part of the apple. The seeds would not be mutant, and would give rise to red-appled trees. (Note that, in fact, the golden delicious apple originally arose as a mutant branch on a red delicious tree.) (Anthony Griffiths)

Figure 7-4 A mutation to an allele for white petals that arose originally in somatic tissue, but eventually became part of germinal tissue and could be transmitted through seeds. The mutation arose in the primordium of a side branch of the rose. The branch grew long and eventually produced flowers. (From Harper Horticultural Slide Library.)

sectors sometimes can be identified by eye if their phenotype contrasts visually with the phenotype of the surrounding wild-type cells (Figure 7-3).

What about transmitting a somatic mutation to progeny? By definition, this is not possible. However, note that if we take a plant cutting from tissue that includes a mutant clone, the plant that grows from the cutting may develop mutant germinal tissue and transmit the mutant gene to progeny. Hence, what arose as a somatic mutation can be transmitted sexually. A similar situation is shown in Figure 7-4.

Any method for the detection of somatic mutation must be able to rule out the possibility that the sector is due to mitotic segregation or recombination (Chapter 6). If the individual is a homozygous diploid, such sectoring is almost certainly due to mutation.

Germinal Mutation

A **germinal mutation** occurs in tissue that ultimately will form sex cells. Then, if the mutant sex cells fertilize zygotes the mutation will be passed on to the next generation. Of course, an individual of perfectly normal phenotype and of normal ancestry can harbor undetected mutant sex cells. These mutations can be detected only if they are included in a zygote (Figures 7-5 and 7-6). You will remember from Chapter 3 that the X-linked hemophilia mutation in the European royal families is thought

Figure 7-5 Germinal mutation determining white petals in viper's bugloss *(Echium vulgare)*. A recessive germinal mutation *a* arose in an *AA* blue plant of the previous generation, making its germinal tissue *Aa*. Upon selfing, the mutation was transmitted to progeny, some of which were *aa* and expressed the mutant phenotype. (Anthony Griffiths)

Figure 7-6 A mutation to an allele determining curled ears arose in the germ line of a normal straight-eared cat, and was expressed in progeny such as the individual shown here. This mutation arose in a population in Lakewood, California, in 1981, and it is an autosomal dominant. (From R. Robinson, *Journal of Heredity* 80, 1989, 474.)

to have arisen in the germ cells of Queen Victoria or one of her parents. The mutation was expressed only in her male descendants.

The experimental detection of germinal mutation depends on the ability to rule out meiotic segregation and recombination as possible causes of phenotypic differences between parents and offspring.

Message Before a new heritable phenotype can be attributed to mutation, both segregation *and* recombination must be ruled out. This is true for both somatic and germinal mutations.

Mutant Types

The phenotypic consequences of mutation may be so subtle as to require refined biochemical techniques to detect a difference from the phenotype conferred by the wild-type allele. Alternatively, the mutation may be so severe as to produce gross morphological defects or death. A rough classification follows, based only on the ways in which the mutations are recognized. This is by no means a complete classification.

Morphological Mutations. *Morph* means "form." Morphological mutations affect the outwardly visible properties of an organism, such as shape, color, or size. Albino ascospores in *Neurospora,* curly wings in *Drosophila,* and dwarf peas are all morphological mutations. Some examples of morphological mutants are shown in Figure 7-7.

Lethal Mutations. A new lethal mutant allele is recognized by its effects on the survival of the organism. Sometimes a primary cause of death from a lethal mutation is easy to identify (for example, in certain blood abnormalities). But often the mutant allele is recognizable *only* by its effects on viability. An example of a lethal mutation is shown in Figure 7-8.

Conditional Mutations. In the class of conditional mutations, a mutant allele expresses the mutant phenotype under a certain environmental condition called the **restrictive condition** but expresses a wild-type phenotype under another condition called the **permissive condition.** Geneticists have studied many temperature-conditional mutations. For example, certain *Drosophila* mutations are known as "dominant heat-sensitive lethals." Heterozygotes (say, H^+H) are wild-type at 20°C (the permissive condition) but die if the temperature is raised to 30°C (the restrictive condition).

Many mutant organisms are less vigorous than their normal counterparts and thus more troublesome as experimental subjects. For this reason, conditional mutants are useful because they can be grown under permissive conditions and then shifted to restrictive conditions for study. Another advantage is that by shifting from permissive to restrictive conditions, a **temperature-sensitive period** can be identified, which can provide clues about when and how the particular gene acts.

Biochemical Mutations. Generally, cell cultures are the basic material for the study of biochemical mutations, which are identified by the loss or change of some biochemical function of the cells. This change typically results in an inability to grow and proliferate. In many cases, however, growth of a mutant cell can be restored by supplementing the growth medium with a specific nutrient. Biochemical mutations have been extensively analyzed in microorganisms. Microorganisms, by and large, are **prototrophic:** they are nutritionally self-sufficient and can exist on a substrate of simple inorganic salts and an energy source; such a growth medium is called a **minimal medium.** Biochemical mutants, however, often are **auxotrophic:** they must be supplied certain nutrients if they are to grow. For example, a certain class of biochemically mutant fungi will not grow unless supplied with adenine. Mutant alleles at an *ad* locus determine the auxotrophic, adenine-requiring

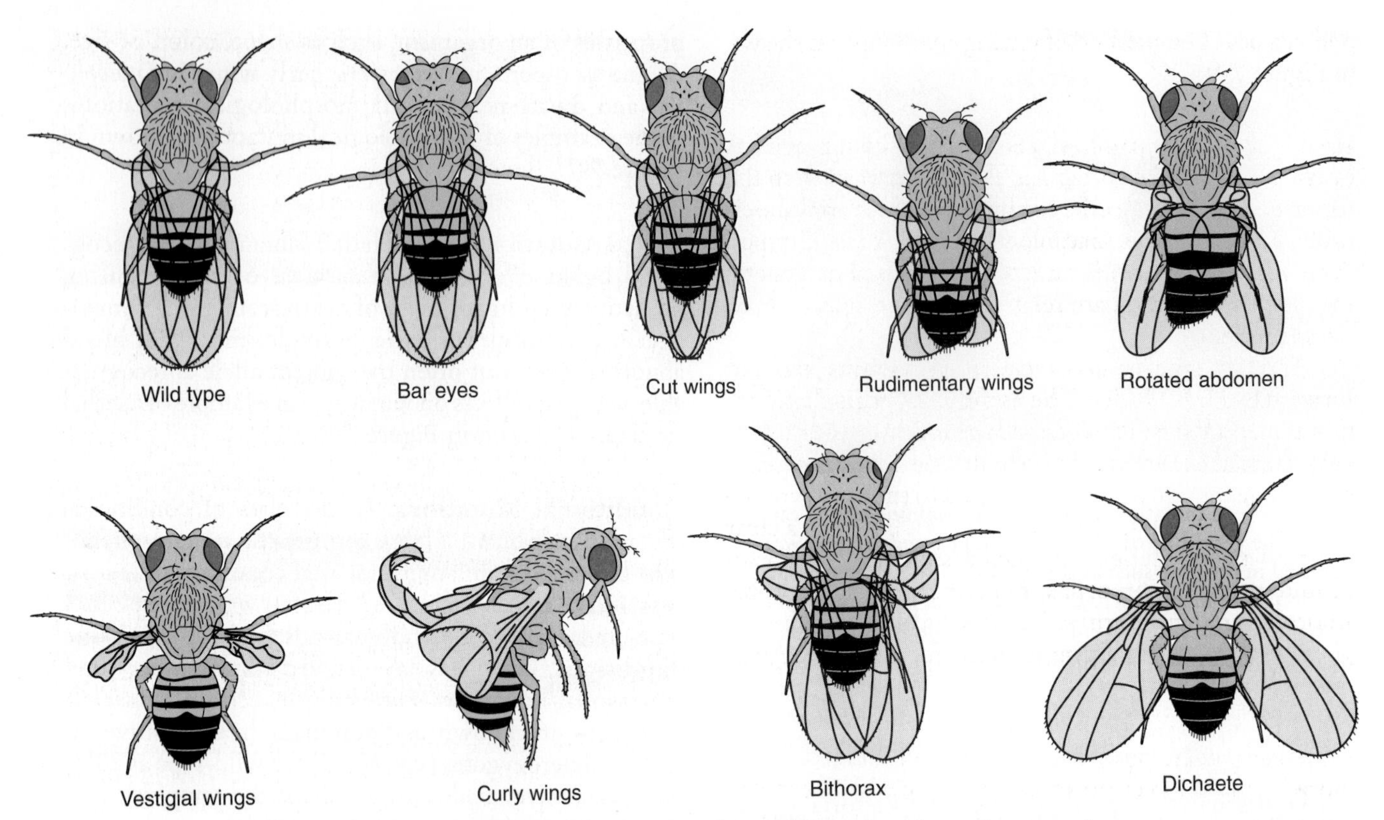

Figure 7-7 Eight morphological mutations of *Drosophila,* and wild type for comparison. Most of the mutant phenotypes are self-explanatory; bithorax is an abnormality of the thorax featuring small wings instead of balancers; the most prominent feature of dichaete is that the wings are held at 45 degrees to the body.

Figure 7-8 Phenotypes of (a) wild type and (b) a mutation affecting plumage of Japanese quail. This mutation arose in a laboratory colony of quail, and could be maintained as an interesting subject for genetic analysis. However, if such a mutation had arisen in nature it would almost certainly be lethal. (Janet Fulton)

phenotypes. The method of testing auxotrophs is shown in Figure 7-9.

Resistance Mutations. A resistance mutation confers on the mutant cell or organism the ability to grow in the presence of some specific inhibitor, such as cycloheximide, or a pathogen, that blocks growth of wild types. Such mutations have been extensively used as genetic markers because they are relatively easy to select for, as we shall see.

Another type of classification of mutants was put forward by H. J. Muller. The usefulness of this classification is that it focuses on the mode of action of the mutant gene, and therefore can provide insight into how genes work. The mutant actions are related to the wild type, or standard, and the classification only applies where a clear standard exists. Muller suggested designating the classes of mutants as hypomorphs, hypermorphs, neomorphs, amorphs, and antimorphs. These words can also be applied to the underlying mutations, for example hypomorphic mutation.

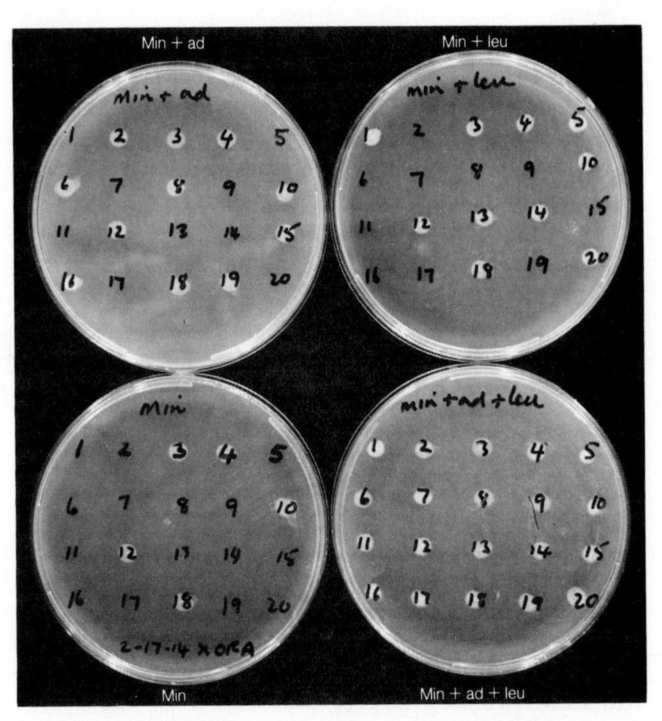

Figure 7-9 Testing strains of *Neurospora crassa* for auxotrophy and prototrophy. In this experiment the test utilizes 20 progeny from a cross of an adenine-requiring auxotroph and a leucine-requiring auxotroph. Genotypically, the cross was *ad* leu^+ × ad^+ *leu*, and the progeny can carry any of the four possible combinations of these alleles. To test the progeny, the geneticist attempts to grow cells on various kinds of gelled media in petri dishes. The media are minimal medium (Min) with either adenine (ad, *top left*), leucine (leu, *top right*), neither (*bottom left*), or both (*bottom right*). Growth appears as a small circular colony (white in the photograph). Any culture growing on minimal must be ad^+ leu^+, one growing on adenine and no leucine must be leu^+, and one on leucine and no adenine must be ad^+. All should grow on adenine plus leucine; it is a kind of control to check viability. As examples, culture 8 must be *ad* leu^+, 9 must be *ad leu*, 10 must be ad^+ leu^+ and 13 must be ad^+ *leu*. (Anthony Griffiths)

Hypomorphs. Hypo means "less than," and a hypomorphic mutation creates an allele that determines some phenotypic measure less than wild type. A hypomorph does the same thing as its wild-type counterpart but does it less efficiently. For example, the *Drosophila* mutation *e* gives an eosin eye, which is light-red rather than the wild-type bright red. Geneticists noticed that these eyes appear to produce less pigment than the wild type. The geneticists confirmed this idea by adding extra fragments of chromosome containing the *e* mutant allele, using special chromosomal engineering techniques to be covered in later chapters. For example, the eye color of an *ee* fly that has received an extra *e*-bearing fragment to make it *eee* is almost wild type. In microbial genetics, hypomorphic mutants are sometimes said to be "leaky," which draws attention to the fact that they have a partly wild-type function.

Hypermorphs. Hyper means "more than," so hypermorphic mutant alleles do a job in excess of the wild type. Good examples are to be found in industrial microbiology; geneticists regularly isolate mutant strains that overproduce some industrially important chemical. For example, they isolate hypermorphs of the fungus *Penicillium* that produce large amounts of the medically important antibiotic penicillin.

Amorphs. Some mutations create alleles whose phenotypes are the complete absence of wild-type function. They could be thought of as extreme hypomorphs but are classified separately as amorphs. In microbial genetics, amorphs are often called "null" mutations. They are particularly easy to identify at the chemical level from the absence of some chemical (for example a protein) that is normally found in the wild type.

Antimorphs. Some mutations seem to antagonize the wild type. These antimorphic mutations appear to have the opposite effect of the wild type as can be seen by considering the ebony mutation of *Drosophila.* The ebony (*b*) allele has the effect of producing a darker fly. From a series of genotypes and corresponding phenotypes, we can see the antimorphic action of *b*: b^+b^+ determines the lightest phenotype, whereas that from b^+b is darker, that from b^+b plus an extra chromosomal fragment bearing *b* is darker still, and the phenotype of *bb* is the darkest of all. In other words, the antimorphic and wild-type alleles pull in opposite directions, or titrate each other out.

Neomorphs. Some mutations, called neomorphic by Muller, create alleles that do something completely different from the corresponding wild-type allele. The *Drosophila* mutation *Hw*, which produces hairy wings, is a good example. *HwHw* gives a hairier phenotype than $HwHw^+$, but adding a fragment containing Hw^+ to the *HwHw* genotype has no effect. In other words, it is the number of doses of *Hw* that is important; Hw^+ has no effect and behaves like an amorphic allele when combined with *Hw*.

Like any classification system, this one has its gray areas and overlaps. Its usefulness is directing our attention to the way in which a mutation changes a function.

The Occurrence of Mutations

Mutation is a biological process that has characterized life from its beginning. As such, it is certainly fascinating and worthy of study. Mutant alleles such as the ones mentioned in the previous sections obviously are invaluable in the study of the process of mutation itself. For instance, they allow us to measure the frequency of mutation. In this connection, they are used as genetic markers, or representative genes: their precise function is not particularly important, except as a way to follow the process.

In modern genetics, however, mutant genes have another role in which their precise function *is* important. We have already referred (in Chapter 2) to genetic dissection as an established approach to biological analysis. Mutant genes can be used as probes to disassemble the constituent parts of a biological function and to examine their workings and interrelationships. Thus, it is of considerable interest to a biologist studying a particular function to have as many mutant forms that affect the function as possible. This has led to "mutant hunts" as an important prelude to any genetic dissection in biology. To identify a genetic variant is to identify a component of the biological process. We explore this idea further later in the chapter.

Message Mutations serve two research purposes: geneticists use mutations to study the process of mutation itself and to genetically dissect biological functions.

Mutation Detection Systems

The tremendous stability and constancy of form of species from generation to generation suggest that mutation must be a rare process. This supposition has been confirmed, creating a problem for the geneticist who is trying to demonstrate mutation.

The prime need is for a detection system—a set of circumstances in which a mutant allele will make its presence known at the phenotypic level. Such a system ensures that the rare mutations that might occur will not be missed.

One of the main considerations here is that of dominance. The system must be set up so that mutations that create new recessive alleles, which are the most common type, will not be masked by a dominant wild-type allele. (Dominant mutations are less of a problem.) As an example, we can use one of the first detection systems ever set up—the one used by Lewis Stadler in the 1920s to study mutation in corn from *C*, expressed phenotypically as a colored kernel, to *c*, expressed as a white kernel. Here we are dealing with the phenotype of the endosperm of the seed. If you refer to Figure 3-40 (page 78), you will see that this tissue forms when two identical haploid female nuclei fuse with one haploid nucleus from the male pollen cell. Hence, the tissue has three chromosome sets (it is $3n$). One dominant *C* allele in combination with two *c* alleles causes the kernel to be colored.

Stadler crossed *cc* ♀ × *CC* ♂ and examined thousands of individual kernels on the corn ears that resulted from this cross. Each kernel is a progeny individual. In the absence of mutation, every kernel is *Ccc* and shows the colored phenotype. Therefore, a white kernel reveals that *C* mutated to *c* in a male reproductive cell of the *CC* parent. The system has thus detected a germinal mutation. Although laborious, this is a very straightforward and reliable method of mutant detection (Figure 7-10).

This basic system can be extended to as many loci as can be conveniently made heterozygous in the same cross. For example

$$a^+a^+\ b^+b^+\ c^+c^+\ d^+d^+ \times aa\ bb\ cc\ dd$$

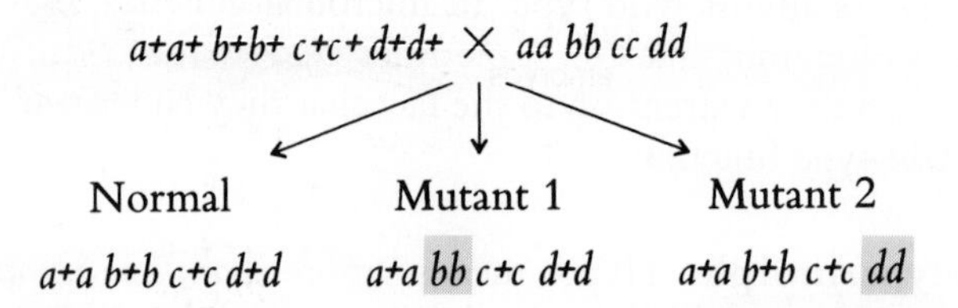

By increasing the number of loci under study, the investigator increases the likelihood of detecting a mutation in the experiment.

Stadler's method for detecting germinal mutations is now called the **specific locus test.** The same principle can be applied to somatic mutations too. Once again, to increase the likelihood of finding and quantifying mutations, the number of loci under study can be increased to any level that is practically feasible. Let's look at an example in which detectable mutations affect the coat colors of mice. The phenotypes and genes responsible for them are leaden (*ln*), pallid (*pa*), brown (*b*), chinchilla

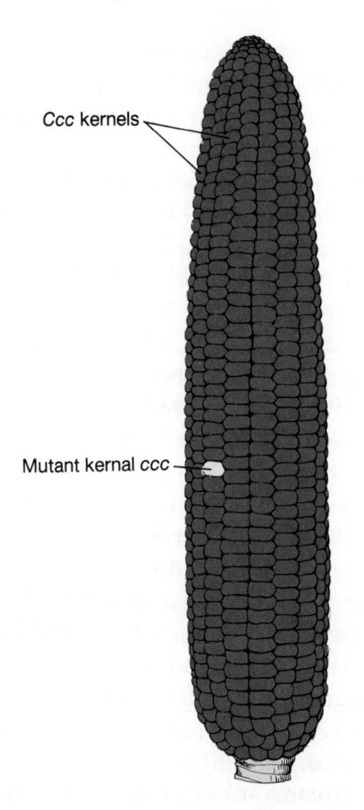

Figure 7-10 The detection system for mutations at a specific locus of corn. The *C* allele determines the presence of a purple pigment in kernels, whereas *c* results in none. The geneticist makes the cross cc♀ × CC♂, and $C \rightarrow c$ mutations in the male germ line show up as unpigmented kernels on the cobs.

(*ch*), pink-eyed dilution (*p*), dilution (*d*), and pearl (*pe*). Mice from pure lines of the following genotypes are crossed:

$$ln\,ln\;pa\,pa\;b^+b^+\;ch^+ch^+\;p^+p^+\;d^+d^+\;pe\,pe \times ln^+ln^+\;pa^+pa^+\;b\,b\;ch\,ch\;p\,p\;d\,d\;pe^+pe^+$$

and the females bear embryos of the multiply heterozygous genotype

$$ln\,ln^+\;pa\,pa^+\;b\,b^+\;ch\,ch^+\;p\,p^+\;d\,d^+\;pe\,pe^+$$

Because all the mutant alleles are recessive, the coats of the F_1 are expected to be wild type. However, somatic mutations from wild type to mutant at any of the heterozygous loci cause mutant sectors on the coat (Figure 7-11). The frequency of these mutant sectors increases dramatically if the investigator injects a mutation-inducing chemical into the uterus of the pregnant mother at the eighth day of pregnancy. This exposes the developing embryos to the injected chemical. The progeny show many more mutant sectors than do mice that were not exposed to the mutagen as embryos. The potency of the chemical can be quantified simply by counting the number of mutant sectors and comparing this number with that observed in untreated animals.

In the plant *Tradescantia,* an effective somatic specific locus test has been developed using a single heterozygous

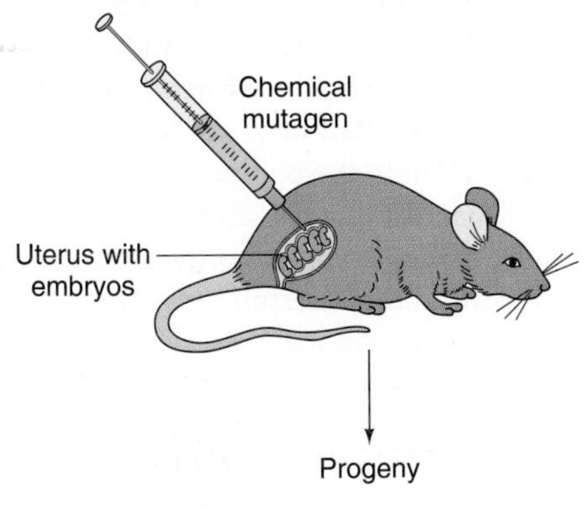

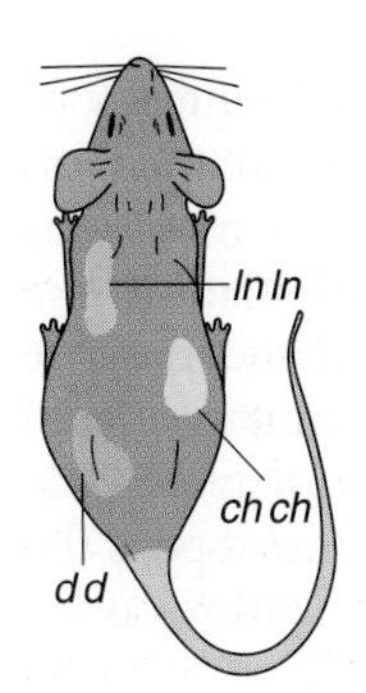

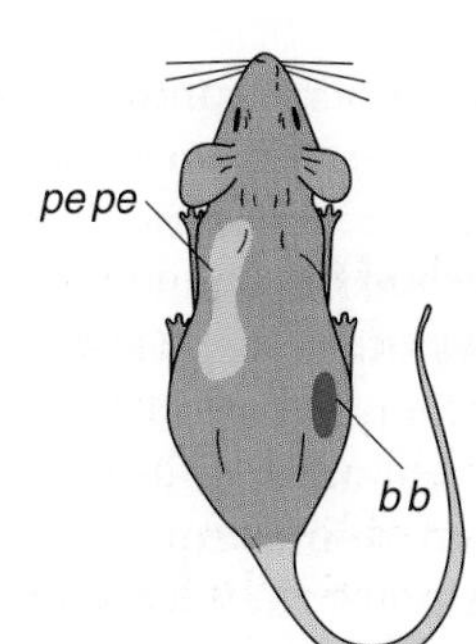

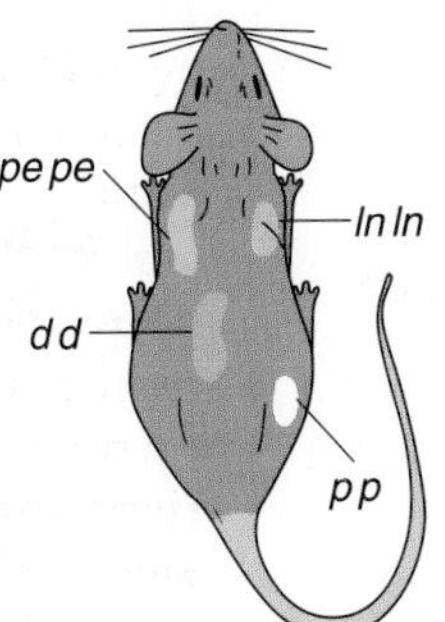

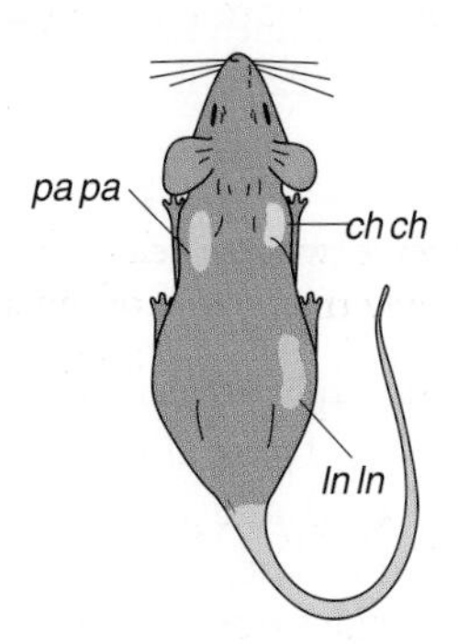

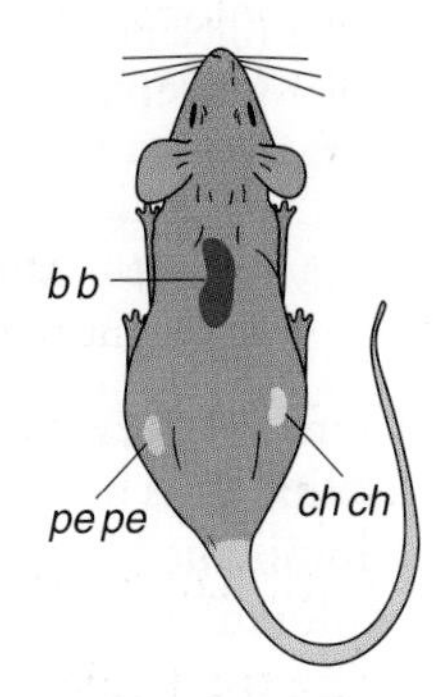

Genotype of progeny is $ln\,ln^+\;pa\,pa^+\;b\,b^+\;ch\,ch^+\;p\,p^+\;d\,d^+\;pe\,pe^+$
Phenotype of progeny is wild type with mutant sectors

Figure 7-11 A detection system for recessive somatic mutations at seven coat-color loci in mice. The cross $ln\,ln\;pa\,pa\;b^+b^+\;ch^+ch^+\;p^+p^+\;d^+d^+\;pe\,pe \times ln^+ln^+\;pa^+pa^+\;b\,b\;ch\,ch\;p\,p\;d\,d\;pe^+pe^+$ results in progeny heterozygous for all seven genes and predominantly wild type in appearance. Any somatic mutation from wild type to mutant at any of the loci causes a mutant sector in the coat of the offspring. The mutant colors are leaden (*ln*), pallid (*pa*), brown (*b*), pink-eyed dilution (*p*), dilution (*d*), and pearl (*pe*). The frequency of mutations can be increased by administering to developing embryos a chemical that is known to produce mutations (a chemical mutagen).

(a)

Anther

(b)

Figure 7-12 (a) Stamen of a *Tradescantia* plant heterozygous for P and p alleles. (b) In the chains that constitute the lateral hairs, some cells (marked by arrows) are mutant. The darker color (blue pigmentation) is determined by P and the paler color (pink pigmentation) by pp. (Part a from Runk/Schoenberger/Grant Heilman.)

gene. A vegetatively propagated line of *Tradescantia* called 02 is heterozygous Pp. These alleles determine the dominant blue (dark colored) and recessive pink (pale colored) pigmentation phenotypes. The pigmentation is expressed in the petals and the stamens. The stamens of this plant have hairs that are chains of single cells. Millions of single cells can easily be scanned for pink cells whose color reveals that a P allele has mutated, becoming a p allele, in somatic tissue (Figure 7-12).

Investigators studying human genetics detect germinal mutation by the sudden appearance of an abnormal phenotype in a pedigree in which there is no previous record of the abnormality. Dominant mutations are more accessible to such detection because heterozygous genotypes express the abnormal dominant phenotypes. An example is given in Figure 7-13, which contains a pedigree for neurofibromatosis, an autosomal dominant disorder characterized by abnormal skin pigmentation ("cafe au lait" spots) and by numerous tumors, called neurofibromas, that associate with the peripheral or central nervous system. The tumors are visible on the surface of the body as shown in Figure 7-13. The pedigree shows an ancestry free of neurofibromatosis, with the disorder arising suddenly in the children of one couple in generation IV. The mutation must have arisen in the germinal tissue of either the mother or the father. It so happens that both neurofibromatosis and achondroplasia have very high mutation frequencies in humans, so a large proportion of cases are from mutation. At the other end of the spectrum, Huntington's disease has the lowest mutation frequency, so most cases of this disease are inherited from previous generations. In fact, most cases of Huntington's disease in North America can be traced to two immigrant families.

A human recessive mutation is more difficult to demonstrate; because of its recessiveness, the mutation is not expressed in the heterozygous state and can go unnoticed for generations. X-linked recessive mutations can be spotted more easily than autosomal recessive mutations. We have already discussed such a mutation to hemophilia in the European royal families. The disease was not present in the pedigree until shown by one of the sons of Queen Victoria, and therefore the mutation event must have occurred either in the germinal tissue of Queen Victoria herself, or in the germinal tissue of one of her parents. Another similar case is shown in the hemophilia pedigree in Figure 7-14. It is possible that the allele was present in the female ancestors, but the absence of hemophilia in a total of 11 male predecessors makes this possibility unlikely.

How can one obtain some idea of the frequency of autosomal recessive mutation events? The answer begins with the idea that for any particular condition, mutations are constantly occurring, entering the population like drops from a dripping faucet. On the other hand, they are removed from the population by reverse mutation, and by selection, which are like leaks in the container. The two tendencies come into equilibrium, which opens the way for calculation of the frequency of mutation, as we shall see in the population genetics chapter.

Haploids have a great advantage over diploids in mutation studies. The system of detecting haploid muta-

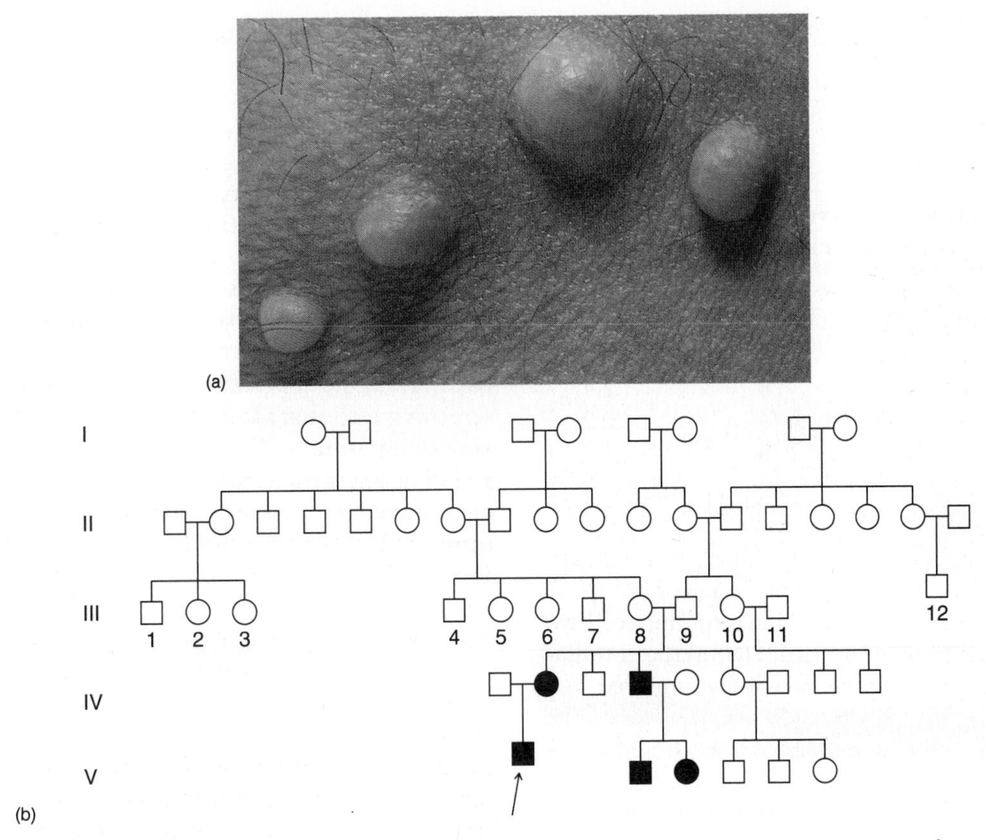

Figure 7-13 Mutation to neurofibromatosis. (a) Neurofibromas. (b) Pedigree showing that the neurofibromatosis mutation must have been in the germ line of III8 or III9. (Part a from Michael English/Custom Medical Stock.)

tions is quite straightforward: any newly arisen recessive allele announces its presence unhampered by a dominant partner allele. In fact, the question of dominance or recessiveness need never arise: there is what amounts to a built-in detection facility. In some cases, a direct identification of mutants is possible. In *Neurospora,* for example, auxotrophic adenine-requiring mutations have been found to map at several loci. One of these loci (*ad-3*) is unique in that auxotrophic mutant cells accumulate a purple pigment when grown on a low concentration of adenine. Thus, auxotrophic mutations of this gene may be detected simply by allowing single asexual spores to grow into colonies on a medium with limited adenine. The purple mutant colonies can be identified easily among the normal white colonies.

What about other auxotrophs? Usually, there are no visual pleiotropic effects, as are exhibited with *ad-3*. We will examine the most commonly used detection technique, called replica plating, later in this chapter.

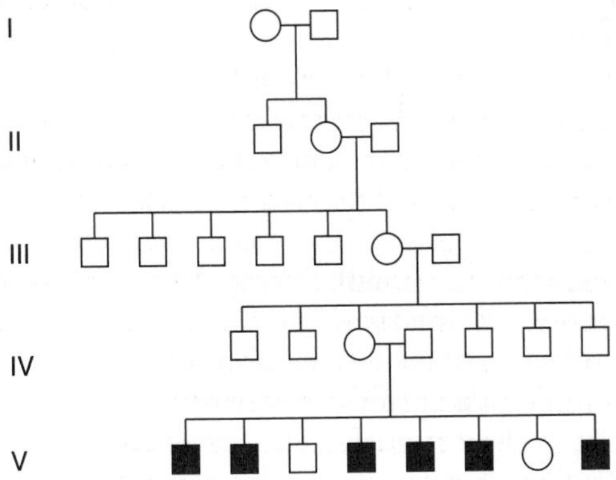

Figure 7-14 Mutation to X-linked recessive hemophila.

How Common Are Mutations?

A detection system must be designed before an investigator can find mutations. If a detection system is available, the investigator can set out on a mutant hunt. One thing will be readily apparent: mutations are in general very rare. This is shown in Table 7-1, which presents some data Stadler collected while working with several corn loci. Mutation studies of this sort are a lot of work! Counting a million of *anything* is no small task. Another feature shown by these data is that different genes seem to generate different frequencies of mutations; a 500-fold range is seen in the corn results. Obviously, one of the prime requisites of mutation analysis is to be able to

Table 7-1. Forward-Mutation Frequencies at Some Specific Corn Loci

Gene	Number of gametes tested	Number of mutations	Average number of mutations per million gametes
$R \longrightarrow r$	554,786	273	492.0
$I \longrightarrow i$	265,391	28	106.0
$Pr \longrightarrow pr$	647,102	7	11.0
$Su \longrightarrow su$	1,678,736	4	2.4
$Y \longrightarrow y$	1,745,280	4	2.2
$Sh \longrightarrow sh$	2,469,285	3	1.2
$Wx \longrightarrow wx$	1,503,744	0	0.0

measure the tendency of different genes to mutate. Two terms are commonly used to quantify mutation: mutation frequency, which we have already encountered, and mutation rate. The **mutation rate** is a measure of the basic tendency of a gene to mutate. It is expressed as the number of mutations occurring in a unit that measures the opportunity to mutate. The units of opportunity are commonly a cell generation span, an organismal generation span, or a cell division. You can see that these are all biological units of time.

Consider the lineage of cells in Figure 7-15. Obviously, there has been only one mutation event (M), so the numerator of a mutation rate is established. But what can we use as the denominator? The total opportunity for mutation—the "time" element—may be represented either by the total number of lines in Figure 7-15 (14 total generation spans) or, alternatively, by the total number of cell divisions (7). We might state the mutation rate, for example, as one mutation per seven cell divisions.

The **mutation frequency** is the frequency at which a specific kind of mutation (or mutant) is found in a population of cells or individuals. The cell population can be gametes, asexual spores, or almost any other cell type. In our example in Figure 7-15, the mutation frequency in the final population of eight cells would be $\frac{2}{8} = 0.25$.

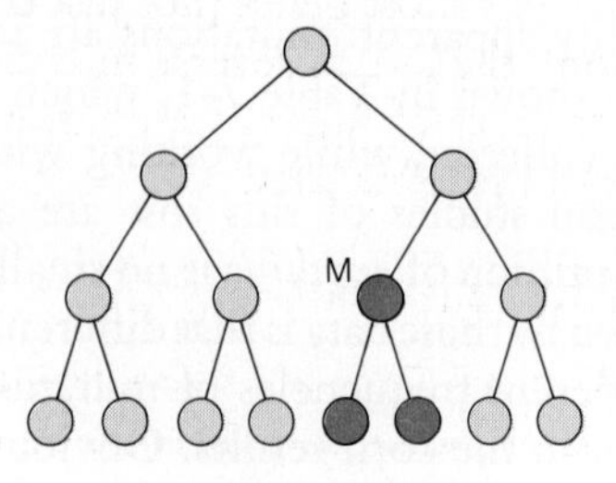

Figure 7-15 A simple cell lineage, showing a mutation at M.

Some mutation rates and frequencies are shown in Table 7-2.

Selective Systems

The rarity of mutations is a problem if an investigator is trying to amass a collection of a specific type of mutation for genetic study. Geneticists respond to this problem in two ways. One approach is to use a **selective system,** a screening technique designed to separate the desired mutant types from wild-type individuals. The other approach is to increase the mutation rate using **mutagens,** agents that have the biological effect of inducing mutations above the background, or spontaneous, rate.

Message Selective systems and mutagens offer two ways to improve the recovery of rare mutations.

Most of the examples of selective systems presented here are in microorganisms. This doesn't mean that selective systems are impossible for studies of more complex organisms, but merely that selection can be used to much better advantage in microbes. A million spores or bacterial cells are easy to produce; a million mice, or even a million fruit flies, require a large-scale commitment of money, time, and laboratory space. Microbes appear frequently in the discussion that follows, so a few words on microbe culture and routine microbial manipulation are appropriate here.

Microbial Techniques

The microbes that we consider in this book are bacteria, fungi, and unicellular algae, all of which are haploid. But whereas fungi and algae are eukaryotic (have genomes organized into one or more chromosomes in a nucleus surrounded by a nuclear membrane), bacteria are **prokaryotic** (their chromosomes are not enclosed in a separate compartment of any kind).

In liquid culture, prokaryotic organisms proliferate as suspensions of individual cells. Each starting cell goes through repeated cell divisions, so that, from each, a series of 2, 4, 8, 16, . . . descendant cells is produced. This exponential growth is limited by the availability of nutrients in the medium, but a dense suspension of millions of separate cells is soon produced. Such fungi as yeasts follow this growth pattern exactly, but the mycelial fungi vary it slightly. The descendant cells of mycelial fungi remain attached as long chains called hyphae; thus, a liquid culture started from asexual spores eventu-

Table 7-2. Mutation Rates or Frequencies in Various Organisms

Organism	Mutation	Value	Units
Bacteriophage T2 (bacterial virus)	Lysis inhibition $r \rightarrow r^+$	1×10^{-8}	*Rate:* mutant genes per gene replication
	Host range $h^+ \rightarrow h$	3×10^{-9}	
Escherichia coli (bacterium)	Lactose fermentation $lac \rightarrow lac^+$	2×10^{-7}	*Rate:* mutant cells per cell division
	Histidine requirement $his^- \rightarrow his^+$	4×10^{-8}	
	$his^+ \rightarrow his^-$	2×10^{-6}	
Chlamydomonas reinhardtii (alga)	Streptomycin sensitivity $str^s \rightarrow str^r$	1×10^{-6}	
Neurospora crassa (fungus)	Inositol requirement $inos^- \rightarrow inos^+$	8×10^{-8}	*Frequency* per asexual spore
	adenine requirement $ad^- \rightarrow ad^+$	4×10^{-8}	
Corn	See Table 7-2		
Drosophila melanogaster (fruit fly)	Eye color $W \rightarrow w$	4×10^{-5}	*Frequency* per gamete
Mouse	Dilution $D \rightarrow d$	3×10^{-5}	
Human			
to autosomal dominants	Huntington's chorea	0.1×10^{-5}	
	Nail-patella syndrome	0.2×10^{-5}	
	Epiloia (predisposition to type of brain tumor)	$0.4-0.8 \times 10^{-5}$	
	Multiple polyposis of large intestine	$1-3 \times 10^{-5}$	
	Achondroplasia (dwarfism)	$4-12 \times 10^{-5}$	
	Neurofibromatosis (predisposition to tumors of nervous system)	$3-25 \times 10^{-5}$	
to X-linked recessives	Hemophilia A	$2-4 \times 10^{-5}$	
	Duchenne's muscular dystrophy	$4-10 \times 10^{-5}$	
bone-marrow tissue-culture cells	Normal $\rightarrow$ azaguanine resistance	7×10^{-4}	*Rate:* mutant cells per cell division

SOURCE: R. Sager and F. J. Ryan, *Heredity*, John Wiley, 1961.

ally looks like tapioca pudding, with small fuzzy balls of hyphae in suspension, each originating from one spore.

In solid culture, usually on an agar-gel surface, descendant cells again tend to stay together, so that colonies are produced. When a suspension of cells is spread on the surface of a plate of culture medium, one colony develops from each original cell in the suspension.

What phenotypes of microorganisms may be examined? Morphological mutations affecting color, shape, and size of colony create useful but limited ranges of phenotypes. Other characters have been far more useful — as examples, auxotrophic phenotypes (can these cells grow without this specific supplement?), resistance phenotypes (can these cells grow in the presence of this growth inhibitor?), and substrate-utilization phenotypes (can these cells utilize this sugar as an energy source, as wild types can?).

We now return to selective systems and some examples thereof.

Reverse Mutation of Auxotrophs

For the detection of the reverse mutation of auxotrophy to prototrophy, there is a direct selective system. Take an adenine-requiring auxotroph, for example. Grow a culture of this strain on an adenine-containing medium. Then plate some of the cells on a solid medium containing no adenine. The only cells that can grow and divide on this medium are adenine prototrophs, which must have arisen by reverse mutation in the original culture (Figure 7-16). For most genes (not just those concerned with nutrition), the rate of reverse mutation is generally lower than the rate of forward mutation. We shall explore the reason for this later.

Filtration Enrichment

Filamentous fungi such as *Neurospora* and *Aspergillus* grow as a mass of branching threads. Fungal geneticists can make use of this property when selecting auxo-

Figure 7-26 *Potentilla fruticosa* wild type (center) and horticultural varieties arranged in a circle. All the attractive novel shapes, colors, patterns and sizes arose through mutations, which were then selected and bred by the horticulturalists. Most flowering plants in our gardens and parks and most food plants in use today have been produced by such a procedure, beginning with either spontaneous or induced mutations. (Anthony Griffiths)

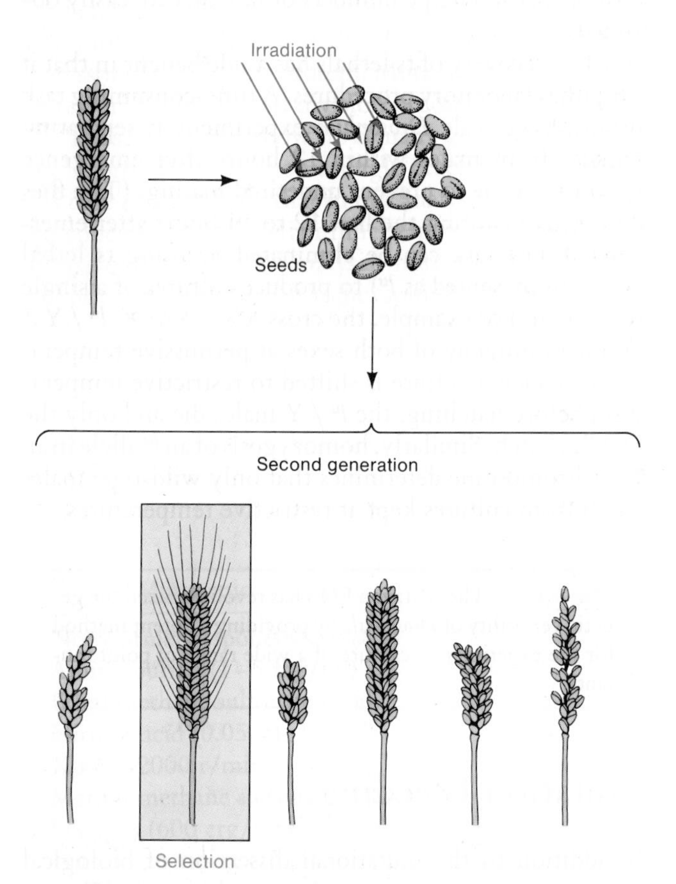

Figure 7-27 Mutation breeding in crops. (From Björn Sigurbjornsson, "Induced Mutations in Plants," *Scientific American,* 1970.)

is to make a hybrid and then to select the desired recombinants from the progeny generations. That approach makes use of the variation naturally found between available stocks or isolates from nature. Figure 7-26 shows an example. Another way to generate variability for selection is to treat with a mutagen. In this way, the variability is increased through human intervention.

A variety of procedures may be used. Pollen may be mutagenized and then used in pollination. Dominant mutations then appear in the next generation, and further generations of selfing reveal recessives. Alternatively, seeds may be mutagenized. A cell in the enclosed embryo of a seed may become mutant, and then it may become part of germinal tissue or somatic tissue. If the mutation is in somatic tissue, any dominant mutations will show up in the plant derived from that seed, but this will be the end of the road for such mutations. Germinal mutations will show up in later generations, where they can be selected as appropriate. Figure 7-27 summarizes mutation breeding.

In this chapter, we have seen how gene mutation not only is of biological interest in itself but can also be put to a variety of experimental and practical uses. In the next chapter, we shall see that much the same kind of statement can be made about chromosome mutation.

Summary

Genes may mutate in somatic cells or germinal cells. Within these two categories, there are several kinds of mutations, including morphological, lethal, conditional, biochemical, and resistance mutations.

Mutations can be used to study the process of mutation itself or to permit genetic dissection of biological functions. In order to carry out such studies, however, it is necessary to have a system for detecting mutant alleles at the phenotypic level. In diploids, dominant mutations are easily detected; recessive mutations, on the other hand, may never be manifested in the phenotype. For this reason, detection systems are much more straightforward in haploids, where the question of dominance does not arise.

Somatic mutation in oncogenes is the major cause of cancer.

Because spontaneous mutations are rare, geneticists use selective systems or mutagens (or both) to obtain mutations. Selective systems automatically distinguish between mutant and nonmutant states. Mutagens are valuable not only in studying the mechanisms of mutation but also in inducing mutations to be used in other genetic studies. In addition, mutagens are frequently used in crop breeding.

Concept Map

Draw a concept map interrelating as many of the following terms as possible. Note that the terms are listed in no particular order.

mutation / allele / reversion / wild type / recessive / somatic / mutagen / germinal / genetic dissection / progeny

Chapter Integration Problem

Some recessive mutations in maize affect the color, texture or shape of the kernel. To discuss these mutations in general, we'll use m to represent the recessive allele. These mutations can be detected in a straightforward manner by crossing an m^+m^+ male with an $m\,m$ female. Any new mutations are found by visually scanning many seeds, in other words millions of kernels on thousands of corncobs. Finding a seed with the recessive phenotype shows that the gene mutated in the male. When maize geneticists sow seeds with the mutant phenotype and self the resulting plants, they find one or the other of two kinds of results:

If the mutation was spontaneous, the progeny from the selfing are generally all mutant.

If the mutation followed treatment of the pollen with a mutagen, the progeny from the selfing are generally $\frac{3}{4}$ wild type and $\frac{1}{4}$ mutant.

Provide an explanation for these two different outcomes.

Solution

As usual, if we restate the results in a slightly different way, it gives us a clue as to what is going on. The difference between outcomes seems to reflect some difference between spontaneous and induced mutation. From a spontaneous mutation, we obtain an individual that seems to be of genotype $m\,m$ because it breeds true for the mutant phenotype. From an induced mutation, we obtain an individual that seems, from the 3:1 phenotypic ratio upon selfing, to have an m^+m genotype. Clearly, the latter deduction does not mesh with the observation of a mutant phenotype. This is the paradoxical case. How is it possible for a seed to have a recessive mutant phenotype but to mature into a plant that produces the selfed progeny expected from a heterozygote?

The moral of this example is that the experimenter should always keep in mind the peculiarities of the organism that is being used. What do we know about corn that might be relevant to this situation? The explanation hinges on mutational events in and around the pollen grain in the anthers of the plant. Recall from Chapter 3, that the mature pollen grain of maize is not just a cell with a single haploid nucleus, but that there are three important components: a haploid tube nucleus and two haploid nuclei that act as sperm. One of these sperm fuses with one female nucleus to create the zygote, and the other sperm fuses with two other female nuclei to create the triploid endosperm. When mature pollen cells are treated with a mutagen, it is likely that only one of the sperm nuclei mutates in any one cell; the only mutations we see are those in sperm that produce the endosperm, because it is the endosperm that forms the bulk of the mature kernel. This is the solution to the problem, but let's write out the genotypes: The endosperm must be $m\,m\,m$, and the embryo must be m^+m. This is why selfing produces a 3:1 ratio.

What about the spontaneous mutations? A gene can mutate at any time. There is only a relatively short time between the mitosis that produces the three nuclei of a mature pollen cell and fertilization. A gene is thus most likely to mutate during the relatively long time that precedes the pollen cell's maturation, possibly even a long time before, in tissue that is still somatic. If a gene does mutate during this time to become allele m, the pollen cell's mitoses will produce three nuclei of genotype m. When such a cell unites with a gamete from the $m\,m$ female, the endosperm is $m\,m\,m$ and the embryo is $m\,m$.

In summary, we have used concepts from the discussion of germinal and somatic mutation, and of detection systems (all from this chapter); and from previous chapters we have used our knowledge of dominance relations, Mendelian genetics, mitosis and meiosis, and organismal life cycles.

Solved Problems

1. A certain plant species normally has the purple pigment anthocyanin dispersed throughout the plant. This makes the green parts of the plant look brown (green plus purple) and the parts of the flower that lack chlorophyll (petals, ovary, anthers) look bright purple. The allele A is essential for anthocyanin production, and, in a homozygote, the recessive alleles a result in a plant that lacks anthocyanin. An interesting new allele called a^u arises, which is unstable. The a^u allele reverts to A at a frequency thousands of times greater than regular (stable) a alleles do.

   a. What phenotype would you expect in plants of genotype (1) $a^u a^u$; (2) $a^u a$; (3) $A\,a^u$?

   b. How can you confirm that the reversions are true mutations?

Solution

a. (1) We have seen that mutations tend to be very rare; in the normal course of events, they are seldom observed unless they are seriously sought out. However, an allele that is reverting as frequently as a^u is likely to announce its reversion behavior prominently if a proper detection system is available. In a plant containing billions of somatic cells, many cells will undoubtedly revert at some stage of development. As development proceeds, each initial revertant cell will give rise to a clone of revertant cells that should be visible as a purple or brown sector. Therefore, a plant of $a^u a^u$ genotype should have white flowers with purple sectors of varying size and photosynthetic tissues that are basically green with brown sectors.

(2) Each sector is a clone derived from a single cell with a^u reverting to A. The allele a^u is expected to be dominant over a, because a is essentially inactive and will not prevent a^u from reverting. Hence, $a^u a$ plants will look the same as $a^u a^u$, but with fewer purple or brown sectors because there are half the chances for reversion.

(3) Because A produces pigment in all cells, $A a^u$ will be indistinguishable from AA. Even though a^u is reverting, the reversions will not show up—a detection system is lacking. Such plants will be purple-brown throughout.

b. How can we prove that a pigmented sector is due to a mutation? It is worth mentioning that some spotted patterns (as in, for example, Dalmatian dogs) are not caused by mutant clones but rather by developmental effects. So spots do not necessarily indicate a mutation—a skeptic could invoke some other cause.

The key is that genes are hereditary units and mutant alleles should be transmitted to subsequent generations of cells or individuals. How can we detect the transmission of revertant A alleles to progeny cells or individuals? If we were able to take or "scrape" a few cells out of a sector and grow them into a plant, and that plant turned out to be purple, then the reversion hypothesis would be proved. Well, the technology for doing such an experiment is available. Is there any other way? If revertant sectors appear in the flower, then presumably the germinal tissue (anthers and ovaries) should sometimes be part of a sector. Therefore, we could collect pollen from such a flower and pollinate a plant of genotype aa. If there is some pollen carrying a revertant A, then the progeny from this pollen should be Aa and purple-brown throughout. Finding some progeny of this phenotype would prove the reversion is a true mutation because there is nowhere else the A gene could have come from.

2. A mutation experiment is performed on the *tryp4* locus in yeast. A *tryp4* mutation confers a requirement for the amino acid tryptophan. A *tryp4* mutant allele named *tryp4-1* is known to be revertible; the experimenters want to measure reversion frequency in a population of haploid cells. A culture of mating type α and of genotype *tryp4-1* is grown and 10 million cells are plated on a medium lacking tryptophan; 120 colonies are obtained. The genotypes of these prototrophic colonies are checked by crossing each one to a wild-type culture of mating type *a*. Based on the results of these crosses, it is found that the prototrophic colonies were of two types: two-thirds of type 1 and one-third of type 2:

Type 1α × wild-type *a* ⟶
progeny all tryptophan independent

Type 2α × wild-type *a* ⟶
$\frac{3}{4}$ progeny tryptophan independent
$\frac{1}{4}$ progeny tryptophan requiring

a. Propose a genetic explanation for the two types.

b. Calculate the frequency of revertants.

Solution

a. The technique of plating on a medium lacking tryptophan should be a selection system for *tryp4*$^+$ revertants, because they do not require tryptophan for growth. When backcrossed to a wild type (*tryp4*$^+$), all progeny of revertants should be tryptophan independent. This behavior is exhibited by type 1 colonies, so these prototrophs are revertants.

Now what about the type 2 colonies? The fact that some progeny are tryptophan requiring shows that the *tryp4-1* mutation could not have genuinely reverted. Rather, the requirement for tryptophan appears to have been merely masked or suppressed. We have already studied several examples of the suppression of a mutant allele by a new mutation at a separate locus. But even if we had not remembered about these examples, the 3 : 1 ratio in a haploid organism should provide the clue that two independent loci are involved. Let's designate the suppressor mutation as *su* and its inactive wild-type allele as *su*$^+$. Type 2 colonies must be *tryp4-1 su*, and the wild types are *tryp4*$^+$ *su*$^+$. A cross of these strains produces

$\frac{1}{2}$ *tryp4-1* → $\frac{1}{2}$ *su+* ⟶ $\frac{1}{4}$ tryptophan-requiring cells
$\frac{1}{2}$ *tryp4-1* → $\frac{1}{2}$ *su*
$\frac{1}{2}$ *tryp4+* → $\frac{1}{2}$ *su+*
$\frac{1}{2}$ *tryp4+* → $\frac{1}{2}$ *su*
(last three combined) $\frac{3}{4}$ tryptophan-independent cells

(Notice that *su* has no effect on the *tryp4*$^+$ allele.)

b. We are told that two-thirds of the colonies are type 1 or true revertants. Thus, there are $120 \times \frac{2}{3} = 80$ revertants out of a total of 10^7, or a revertant frequency of 8×10^{-6} cells.

Problems

1. A certain species of plant produces flowers with petals that are normally blue. Plants with the mutation *w* produce white petals. In a plant of genotype *ww*, one *w* allele reverts during the development of a petal. What detectable outcome would this reversion produce in the resulting petal?

2. *Penicillium* (a commercially important filamentous fungus) normally can synthesize its own leucine (an amino acid). How would you go about selecting mutants that are leucine-requiring (that cannot synthesize their own leucine)? (NOTE: Like many filamentous fungi, *Penicillium* produces profuse numbers of asexual spores.)

3. How would you select revertants of the yeast allele *pro-1*? This allele confers an inability to synthesize the amino acid proline, which can be synthesized by wild-type yeast and which is necessary for growth.

4. How would you use the replica-plating technique to select arginine-requiring mutants of haploid yeast?

5. Using the filtration-enrichment technique, you do all your filtering with a minimal medium and do your final plating on a complete medium that contains every known nutritional compound. How would you find out what *specific* nutrient is required? After replica plating onto every kind of medium supplement known to science, you still can't identify the nutritional requirement of your new yeast mutant. What could be the reason(s)?

6. An experiment is initiated to measure the reversion rate of an *ad-3* mutant allele in haploid yeast cells. Each of 100 tubes of a liquid, adenine-containing medium are inoculated with a very small number of mutant cells and incubated until there are 10^6 cells per tube. Then the contents of each tube are spread over a separate plate of solid medium containing no adenine. The plates are observed after one day, and colonies are seen on 63 plates. Calculate the reversion rate of this allele per cell division.

7. Suppose that you want to determine whether caffeine induces mutations in higher organisms. Describe how you might do this (include control tests).

8. In corn, alleles at a single locus determine presence (*Wx*) or absence (*wx*) of amylose in the cell's starch. Cells that have *Wx* stain blue with iodine; those that have only *wx* stain red. Design a system for studying the frequency of rare mutations from *Wx* → *wx* without using acres of plants. (HINT: You might start by thinking of an easily studied cell type.)

9. Suppose that you cross a single male mouse from a homozygous wild-type stock with several virgin females homozygous for the allele that determines black coat color. The F_1 consists of 38 wild-type mice and five black mice. Both sexes are represented equally in both phenotypic classes. How can you explain this result?

10. A man employed for several years in a nuclear power plant becomes the father of a hemophiliac boy. There is no hemophilia in the extensive pedigrees of the man's ancestors and his wife's ancestors. Another man, also employed for several years in the same plant, has an achondroplastic dwarf child—a condition nowhere recorded in his ancestry or in that of his wife. Both men sue their employer for damages. As a geneticist, you are asked to testify in court. What do you say about each situation? (NOTE: Hemophilia is an X-linked recessive; achondroplasia is an autosomal dominant.)

11. In a large maternity hospital in Copenhagen, there were 94,075 births. Ten of the infants were achondroplastic dwarfs. (Achondroplasia is an autosomal dominant, showing full penetrance.) Only two of the dwarfs had a dwarf parent. What is the mutation frequency to achondroplasia in gametes? Do you have to worry about reversion rates in this problem? Explain.

12. One of the jobs of the Hiroshima-Nagasaki Atomic Bomb Casualty Commission was to assess the genetic consequences of the blast. One of the first things they studied was the sex ratio in the offspring of the survivors. Why do you suppose they did this?

13. Many organisms present examples of unstable recessive alleles that revert to wild types at very high frequencies. Different unstable alleles often revert in different ways. One gene in corn (C), which produces a reddish pigment in the kernels, has several unstable null alleles. The unstable allele c^{m1} reverts late in the development of the kernel and at a very high rate. Another unstable allele, c^{m2}, reverts earlier in the development of the kernel and at a lower rate. Assuming that lack of pigment leaves the cell looking yellowish, what phenotype do you predict in plants of genotype:

a. $C\,c^{m1}$
b. $C\,c^{m2}$
c. $c^{m1}\,c^{m1}$
d. $c^{m2}\,c^{m2}$
e. $c^{m1}\,c$ (c is a stable mutant allele)
f. $c^{m2}\,c$
g. $c^{m1}\,c^{m2}$

14. A new dominant mutation arises in *Drosophila* and is tested in combination with other alleles at the same locus. The genotypes are ranked according to the degree to which mutation A is expressed.

$$AA > A/\text{del} > AA^{+} > A^{+}A^{+}$$

(A/del indicates that the A allele is on one homolog, and the A locus is completely deleted on the other homolog.) Which of Muller's mutational classes do you think A is in? State your reasons.

15. The mutant allele w^{ap}, which governs a *Drosophila* eye color called apricot, resides at the same locus as w, which determines white eyes. The apricot mutation is tested and the following results are obtained: $w^{ap}w^{ap}$ gives a more normal eye pigmentation than $w^{ap}w$, which gives a more normal phenotype than ww. Which of Muller's mutational classes does apricot fall into?

16. Outline the precise steps that you would take to find the chromosomal locus of a new recessive mutation that you have just found in tomatoes. Show all steps including what crosses you would make, which lines you would use, how you would obtain the lines, how much field or greenhouse space is needed, how long it would take, etc. (You may refer to the tomato map, which is shown in Figure 5-15.)

17. A *Neurospora* strain unable to synthesize arginine produces a revertant, arginine-independent colony. A cross is made between the revertant and a wild-type strain. What proportion of the progeny from this cross would be arginine independent if

a. the reversion precisely reversed the original change that produced the arg^- mutant allele?

b. the revertant phenotype was produced by a mutation in a second gene located on a different chromosome (the new mutation suppresses arg^-)?

c. a suppressor mutation occurred in a second gene located 10 map units from the arg^- locus on the same chromosome?

18. A certain haploid fungus is normally red due to a carotenoid pigment. Mutants were obtained that were different colors due to the presence of different pigments: orange (o^-), pink (p^-), white (w^-), yellow (y^-), and beige (b^-). Each phenotype was inherited as if a single gene mutation governed it. To determine what these mutations signified, double mutants were constructed with all possible combinations, and the results were as follows:

	p^-	w^-	y^-	b^-
o^-	pink	white	yellow	beige
p^-	—	white	pink	pink
w^-		—	white	white
y^-			—	yellow

Interpret this table as follows, using the first entry as an example: the double mutant $o^-\,p^-$ has a pink phenotype.

a. What do the results of the table mean? Explain in detail.

b. What phenotypes would result if the mutants were paired as heterokaryons in all the combinations possible?

c. What phenotypic proportions would be found in the progeny of a cross between an $o^-\,p^-$ double mutant and a wild type, if the loci are linked 16 map units apart?

8

Chromosome Mutation I: Changes in Chromosome Structure

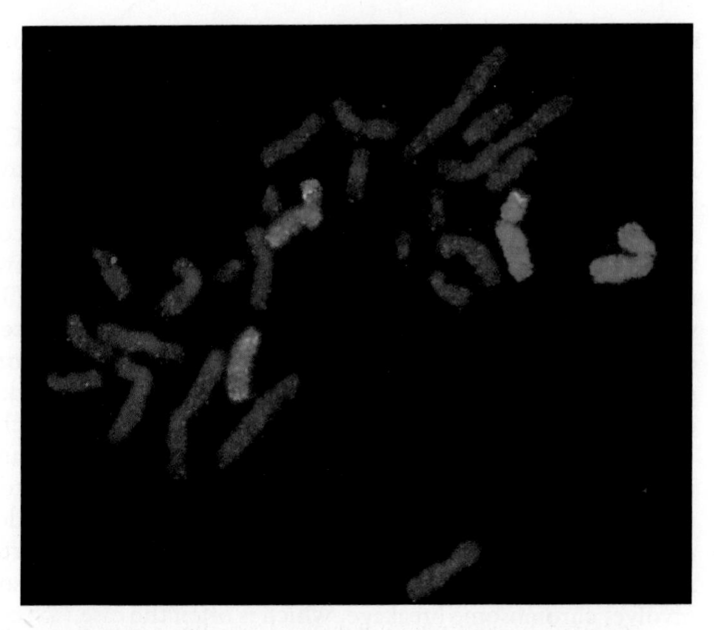

A reciprocal translocation demonstrated by a technique called chromosome painting. A suspension of chromosomes from many cells is passed under an electronic device that sorts them by size. DNA is extracted from individual chromosomes, joined to one of several fluorescent dyes, then added to partially denatured chromosomes on a slide. The fluorescent DNA "finds" its own chromosome and binds along its length to paint it. In this preparation a light blue and a pink dye have been used to paint different chromosomes. The preparation shows one normal pink chromosome, one normal blue, and two that have exchanged their tips. (Lawrence Berkeley Laboratory)

KEY CONCEPTS

Chromosomes have many landmarks that can be used for the detection of rearrangements.

▪

Owing to the strong meiotic pairing affinity of homologous regions, diploids with one standard and one rearranged chromosome set produce pairing structures that have shapes and properties unique to that rearrangement.

▪

A deletion in one chromosome set is generally deleterious as a result of gene imbalance and the unmasking of deleterious alleles in the other chromosome set.

▪

Duplications can lead to gene imbalance but also provide extra material for evolutionary divergence.

▪

Heterozygous inversions have reduced fertility and reduced recombination in the region spanned by the inversion.

▪

A heterozygous translocation has 50 percent sterility and linkage of genes on the chromosomes involved in the translocation.

Chromosome mutation is the name given to the processes of change that results in rearranged chromosome parts, or abnormal numbers of individual chromosomes, or abnormal numbers of chromosome sets. As with gene mutation, the term chromosome mutation is applied both to the process and the product, so the novel genomic arrangements may be called chromosome mutations. Sometimes these genomic modifications can be detected microscopically, sometimes by genetic analysis, and sometimes by both. In contrast, gene mutations are never detectable microscopically; a chromosome bearing a gene mutation looks the same under the microscope as one carrying the wild-type allele.

Many chromosome mutations lead to abnormalities in cell and organismal function. There are two basic reasons for this. First, the chromosome mutations can result in abnormal effects by virtue of abnormal gene number or position. Second, if chromosome mutation involves chromosome breakage, which is often the case, then the break may occur in the middle of a gene, thereby disrupting its function.

Chromosome mutations are important at several different levels of biology. First, in research, they provide ways of synthesizing special arrangements of genes, uniquely suited to answer certain biological questions. Second, chromosome mutations are important at the applied level, especially in medicine, and in plant and animal breeding. Finally, chromosome mutations have been instrumental in shaping genomes as part of the evolutionary process.

The study of normal and abnormal chromosome sets and their genetic properties is called **cytogenetics,** a discipline that combines cytology with genetics. Cytogenetics uses the normal genome as its standard or wild type. A cytogeneticist must be familiar with the normal features and landmarks of the genome before being able to detect any changes that have occurred. Therefore we begin our discussion with the topography of chromosomes.

The Topography of Chromosomes

The following features of chromosomes are commonly used as landmarks in cytogenetic analysis.

Chromosome Size. The chromosomes of a single genome may differ considerably in size. In the human genome, for example, there is about a three- to fourfold range in size from chromosome 1 (the biggest) to chromosome 21 (the smallest), as we saw in Figure 6-26. In studying the chromosomes of some species, a cytogeneticist may have difficulty identifying individual chromosomes by size alone, but may be able to group chromosomes of similar size. A change may then be pinpointed in, for example, "one of the chromosomes in size group A."

Centromere Position. The centromere is the structure to which spindle fibers attach. The centromere region usually appears to be constricted, and the position of this constriction defines the ratio between the lengths of the two chromosome arms; this ratio is a useful characteristic (Table 8-1). Centromere positions can be categorized as **telocentric** (at one end), **acrocentric** (off center), or **metacentric** (in the middle). The centromere position determines not only arm ratio but also the shapes of chromosomes as they migrate to opposite poles during anaphase. These anaphase shapes range from a rod to a J to a V (Figure 8-1). In some organisms, such as the Lepidoptera, centromeres are "diffuse," so that spindle fibers attach all along the chromosome. When such a chromosome breaks, both parts can still migrate to the poles. In contrast, a break in a chromosome with a single centromere creates a fragment that cannot move to the pole because it has no centromere. Chromosome fragments lacking a centromere are known as **acentric** (Figure 8-2).

The molecular structure of the centromeres of certain organisms is now known, as is the structure of the tips of the chromosomes, the **telomeres.** Although not morphologically distinct, the telomeres have a unique molecular structure that is crucial to normal chromosome behavior. Such knowledge about molecular structure has permitted a new line of research in yeast in which completely novel chromosomes can be made by splicing together the DNA of a functional centromere, two telomeres, and any other genetic material of interest to the experimenter.

Position of Nucleolar Organizers. Different organisms are differently endowed with nucleoli, which

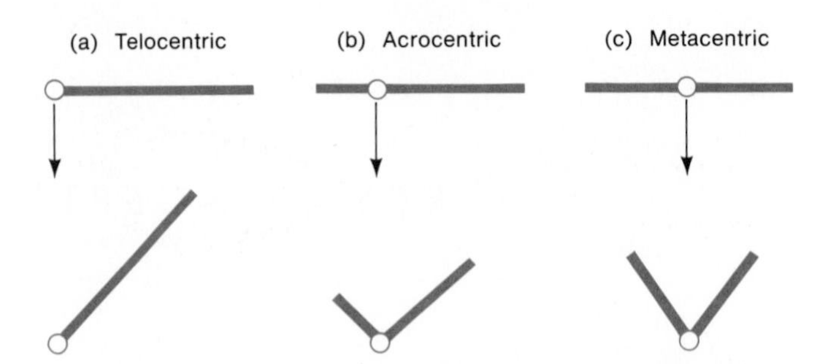

Figure 8-1 The classification of chromosomes by the position of the centromere. A telocentric chromosome has its centromere at one end; when the chromosome moves toward one pole of the cell during the anaphase of cellular division, it appears as a simple rod. An acrocentric chromosome has its centromere somewhere between the end and the middle of the chromosome; during anaphase movement, the chromosome appears as a J. A metacentric chromosome has its centromere in the middle and appears as a V during anaphase.

Table 8-1. Human Chromosomes

Group	Number	Diagrammatic representation	Relative length[a]	Centromeric index[b]
Large chromosomes				
A	1		8.4	48 (M)
	2		8.0	39
	3		6.8	47 (M)
B	4		6.3	29
	5		6.1	29
Medium chromosomes				
C	6		5.9	39
	7		5.4	39
	8		4.9	34
	9		4.8	35
	10		4.6	34
	11		4.6	40
	12		4.7	30
D	13		3.7	17 (A)
	14		3.6	19 (A)
	15		3.5	20 (A)
Small chromosomes				
E	16		3.4	41
	17		3.3	34
	18		2.9	31
F	19		2.7	47 (M)
	20		2.6	45 (M)
G	21		1.9	31
	22		2.0	30
Sex chromosomes				
	X		5.1 (group C)	40
	Y		2.2 (group G)	27 (A)

a. Percentage of the total combined length of a haploid set of 22 autosomes.

b. Percentage of a chromosome's length spanned by its short arm. The four most metacentric chromosomes are indicated by an (M); the four most acrocentric by an (A).

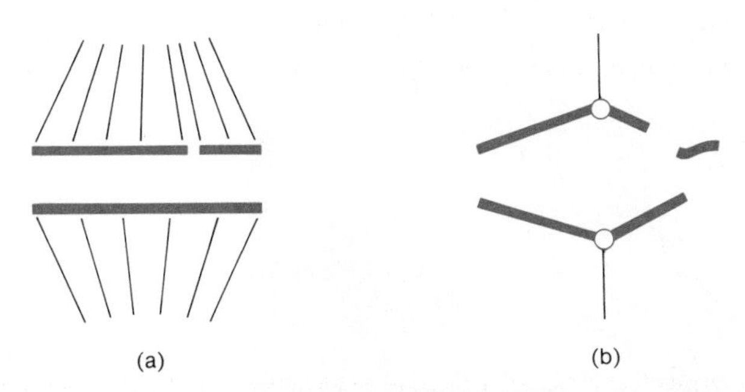

Figure 8-2 Spindle-fiber attachments to chromosomes. (a) A diffuse centromere with many spindle fibers attaching along the length of a single chromosome. This arrangement is rare. A break in the chromosome does not result in the loss of any chromosome material. (b) Spindle-fiber attachment to a single centromere (the common arrangement). In this case, a break in the chromosome results in an acentric fragment, which is lost during cellular division.

range in number from one to many. The nucleoli contain ribosomal RNA, an important component of ribosomes. The diploid cells of many species have two nucleoli. The nucleoli reside next to secondary constrictions of the chromosomes, called **nucleolar organizers,** which have highly specific positions in the chromosome set. Nucleolar organizers contain the genes for ribosomal RNA. Their positions, like those of centromeres, are landmarks for cytogenetic analysis.

Chromomere Patterns. We have already encountered chromomeres in Chapter 5 (page 136). They are visible during certain stages of cell division as beadlike, localized thickenings of the chromosome.

Heterochromatin Patterns. When a chromosome is stained with standard reagents that react with DNA, such as Feulgen stain, distinct regions with different staining characteristics are visually revealed. Densely staining regions are designated **heterochromatin;** poorly staining regions are said to be **euchromatin.** The distinction is thought to reflect the degree of compactness of the DNA in the chromosome. Heterochromatin can be either **constitutive** or **facultative.** The constitutive type is a permanent feature of a specific chromosome location and is, in this sense, a hereditary feature; Figure 5-15 shows good examples of constitutive heterochromatin in the tomato. The facultative type of heterochromatin can be either present or absent at any particular chromosomal location. The patterns of heterochromatin and euchromatin along a chromosome constitute good cytogenetic markers.

Banding Patterns. Special staining procedures have revealed several sets of intricate chromosome bands in a wide range of organisms. These are Q bands (produced by quinacrine hydrochloride), G bands (produced by Giemsa stain), and R bands (reversed Giemsa). An example of G bands in human chromosomes is shown in Figure 6-25a.

A rather specialized kind of banding, which has been used extensively by cytogeneticists for many years, characterizes the so-called giant chromosomes in certain organs of the Diptera. In 1881, E. G. Balbiani recorded peculiar structures in the nuclei of certain secretory cells of two-winged flies. These structures were long and sausage-shaped and were marked by swellings and cross

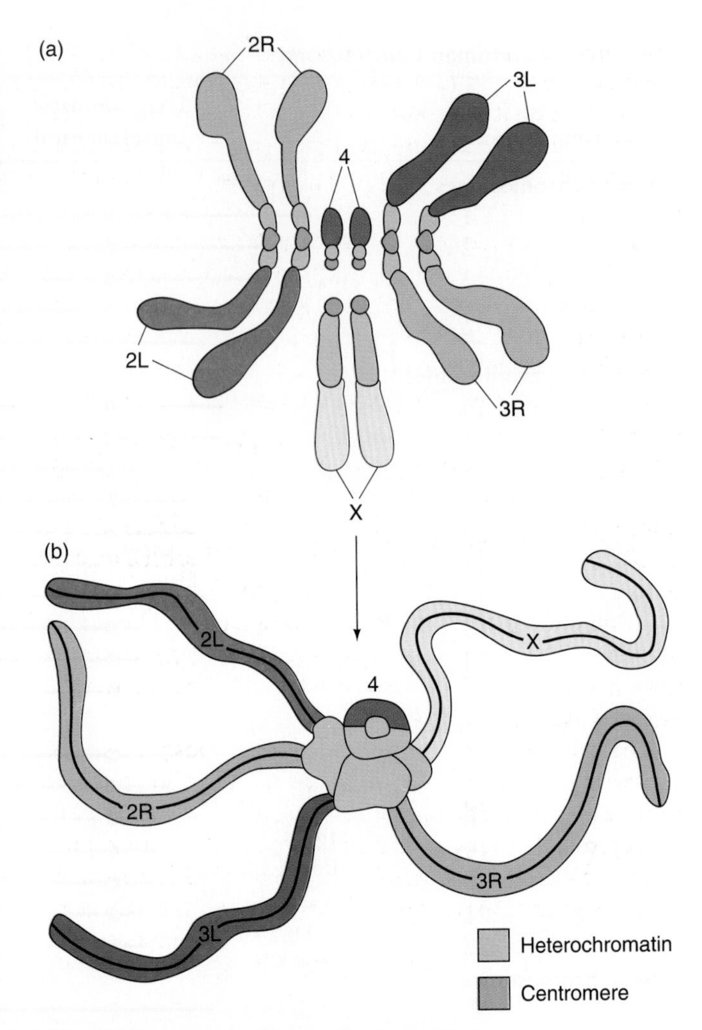

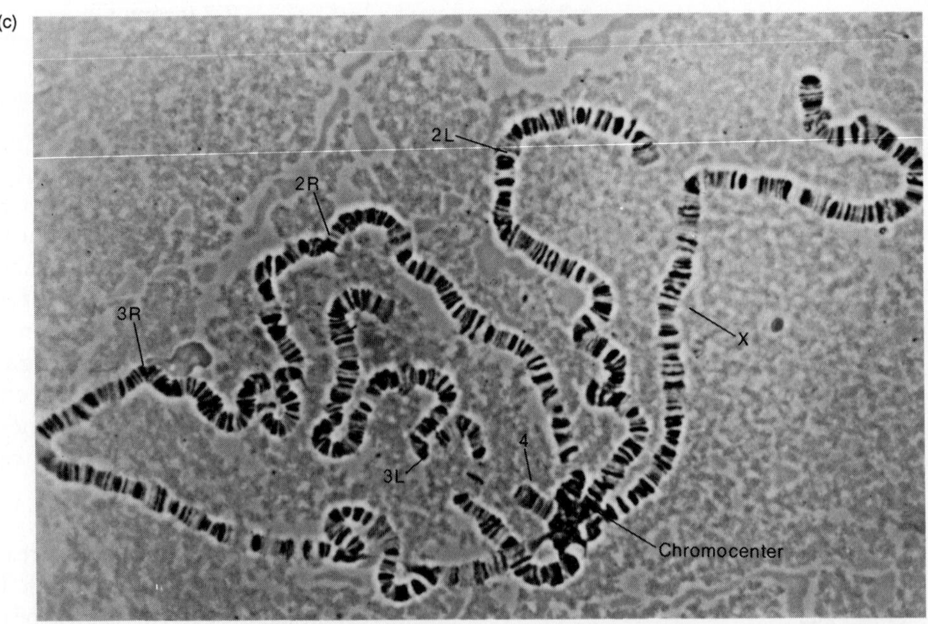

Figure 8-3 Polytene chromosomes form a chromocenter in a *Drosophila* salivary-gland. (a) Mitotic metaphase chromosomes, with arms represented by different shades. (b) Heterochromatin coalesces to form the chromocenter. (c) Photograph of polytene chromosomes. (Tom Kaufman.)

striations. Unfortunately, Balbiani did not recognize them as chromosomes, and his report remained buried in the literature. It was not until 1933 that Theophilus Painter, Ernst Heitz, and H. Bauer rediscovered them and realized that these structures are in fact chromosomes.

In secretory tissues, such as the Malpighian tubules, rectum, gut, footpads, and salivary glands of the Diptera, the chromosomes replicate their genetic material many times without actually separating into distinct chromatids. Thus, as a chromosome increases in replicas, it elongates and thickens. The bundle of replicas is called a **polytene chromosome.** *Drosophila melanogaster,* for example, has an n number of 4, but only four polytene chromosomes are seen in the cells of such tissues because homologs are tightly paired. Furthermore, all four chromosomes are joined at the **chromocenter,** which represents a coalescence of the heterochromatic areas around the centromeres of all four chromosome pairs. The chromocenter of *Drosophila* salivary-gland chromosomes is shown in Figure 8-3, where L and R stand for arbitrarily assigned left and right arms.

Along the length of a polytene chromosome, characteristic stripes called **bands** can be identified; these bands differ in width and morphology. In addition, there are regions that may at times appear swollen **(puffs)** or greatly distended **(Balbiani rings);** these are presumed to correspond to regions of RNA synthesis. The polytene chromosomes that correspond to specific linkage groups in *Drosophila* have been determined through the use of special chromosome mutations. Such mutations, as we shall see, have also been useful in the specific localization of genes along the chromosomes. As we shall see in Chapter 16, recent molecular studies have shown that a chromosomal region of *Drosophila* has more genes than there are polytene bands, so there is not a one-to-one correspondence of bands and genes as was once believed. Similarly, the significance of the banding patterns of the chromosomes of vertebrates is not clear, but the bands probably reflect nucleotide content. It is known that most of the active genes reside in the light bands.

Using various landmarks of chromosomes collectively, cytogeneticists can distinguish each of the chromosomes of many species. Figure 8-4 shows, for example, a diagram of the chromosomes of corn. Notice how the landmarks make each of the ten chromosomes of corn unique.

> Message Such features as size, arm ratio, heterochromatin, number and position of thickenings, and the number and location of nucleolar organizers identify the individual chromosomes within the set that characterizes a species.

Now that we can distinguish chromosomes topographically, we have what we need for learning about the chromosome mutations that are the subject of this chapter. Mutations changing chromosome size and the relative locations of the landmarks — in short, mutations in

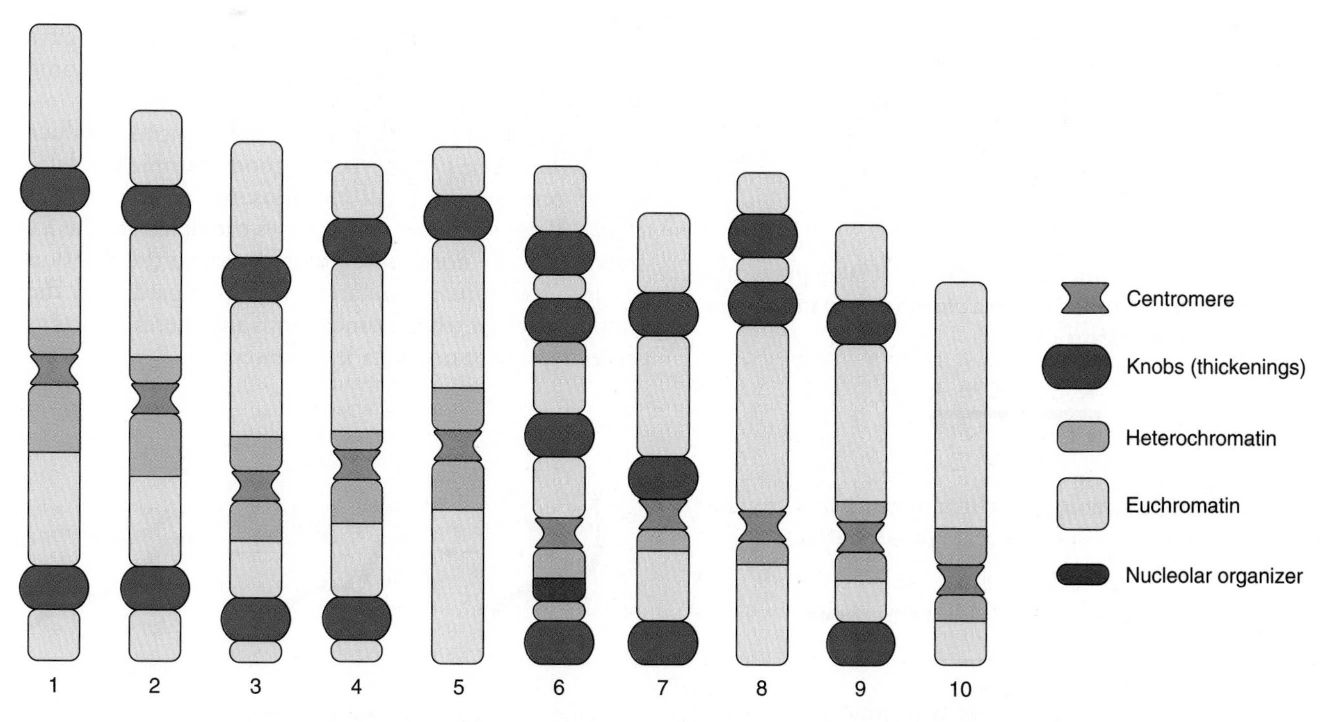

Figure 8-4 The landmarks that distinguish the chromosomes of corn.

structure — are the subject of the rest of this chapter. We shall learn about mutations affecting chromosome number in Chapter 9.

Let's consider two important features of chromosome behavior that are useful in understanding structural chromosome mutations. One is that during prophase I of meiosis, homologous regions of chromosomes have a strong pairing affinity and contort if necessary, in order to pair. Because of this, many curious structures may be seen in a cell that has one standard set of chromosomes and one aberrant set. Polytene chromosome homologs may also contort to pair, and equivalent shapes result. The other important feature is that changes in structure usually involve chromosome breakage, and the broken chromosome ends are highly "reactive," tending strongly to join with other broken ends. The telomeres (the regular chromosome ends), however, do not tend to join.

Types of Changes in Chromosome Structure

A chromosome may mutate by losing a segment. To discuss this and other mutational events, let's use letters to represent arbitrary chromosome regions, each of which contains many genes. For example, a chromosome may lose a region, shown here as region B:

A B C D E F → A C D E F

This type of change is called a **deletion,** or a **deficiency.** The reciprocal change is a **duplication:**

A B B C D E F

A segment of a chromosome can rotate 180 degrees and rejoin the chromosome. When a chromosome mutates in this way, an **inversion** results:

A E D C B F

Finally, two nonhomologous chromosomes may mutate by exchanging parts to produce a **translocation:**

A B C D J K   G H I E F

Chromosomes commonly mutate to produce these types of chromosomal aberrations and we shall examine their genetic and cytological properties. Collectively, this class of aberrations is known as **chromosomal rearrangements.**

Deletions

Obviously, a chromosomal rearrangement begins with a break in the linear continuity of a chromosome. Deletions and duplications can be produced by the same event if two homologs break simultaneously at different points. We can see how this might happen if we picture two homologs first overlapping and then breaking and rejoining, as shown in Figure 8-5. However, spontaneous duplications and deletions are rarely reciprocal products of the same event.

If a deletion is made homozygous (that is, if both homologs have the same deletion), then the combination is generally lethal. This suggests that most regions of the chromosomes are essential for normal viability and that complete elimination of any segment from the genome is deleterious. Even individuals heterozygous for a deletion — those with one normal homolog and one that carries the deletion — may not survive because the genome has been "fine-tuned" during evolution to require a specific balance of most genes and the deletion upsets this balance. Nevertheless, some small deletions are viable in combination with a normal homolog. If meiotic chromosomes in heterozygous deletions can be examined, the region of the deletion can be determined by the failure of the corresponding segment on the normal homolog to pair, resulting in a **deletion loop** (Figure 8-6a). In insects deletion loops are also detected in the polytene chromosomes, in which the homologs are fused (Figure 8-6b). Deletions can be assigned to specific chromosome locations by this technique.

Deletions of some chromosome regions produce their own unique phenotypes. A good example is a deletion of one specific small chromosome region of *Drosophila.* When one homolog carries the deletion, the fly shows a unique notch-wing phenotype, so the deletion acts like a dominant mutation in this regard. But the deletion is lethal when homozygous and therefore acts as a recessive in regard to its lethal effect.

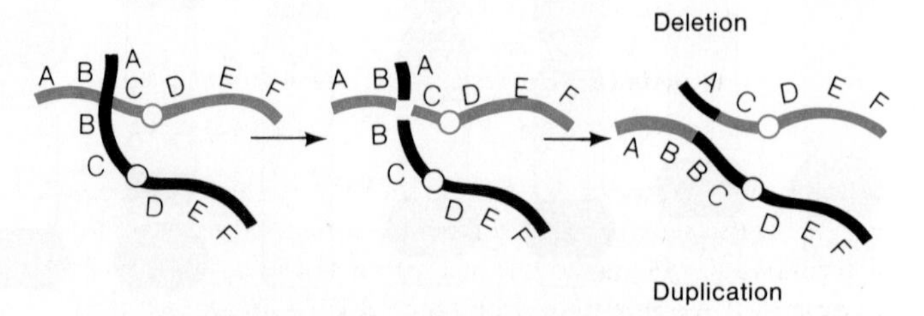

Figure 8-5 One possible way of producing deletions and duplications: by the reunion of broken homologs.

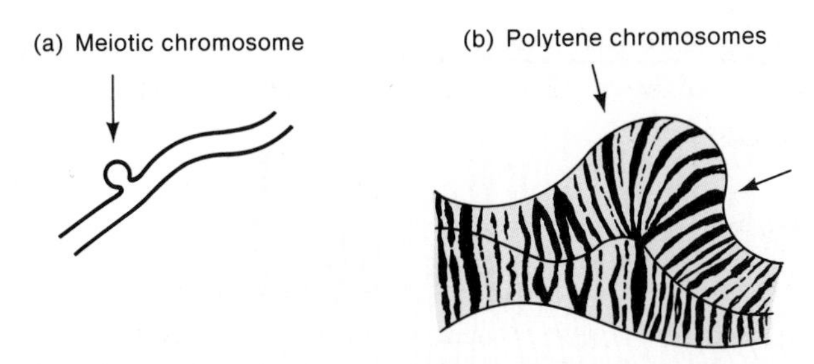

Figure 8-6 Looped configurations in a *Drosophila* deletion heterozygote. During the meiotic pairing, the normal homolog forms a loop. The genes in this loop have no alleles to synapse with. (b) Because polytene chromosomes in *Drosophila* have specific banding patterns, we can infer which bands are missing from the homolog with the deletion by observing which bands appear in the loop of the normal homolog.

An effective mutagen for inducing chromosomal rearrangements of all kinds is ionizing radiation. This kind of radiation, of which X rays and γ rays are examples, is highly energetic and causes chromosome breaks. The way in which the breaks rejoin determines the kind of rearrangement produced. Two types of deletion are possible: a single break can cause a **terminal deletion** and two breaks can produce an **interstitial deletion** (Figure 8-7).

But what about the genetic properties of deletions in general? When we lack useful cytological approaches—such as detecting deletions by observing the salivary-gland chromosomes of *Drosophila*—can we proceed anyway? The answer is "yes." There are genetic criteria for inferring the presence of a deletion. One is the failure of the chromosome to survive as a homozygote, but this criterion, of course, could also be produced by any lethal mutation. Another is the suppression of crossing-over in the region spanning the deletion, but again other aberrations could cause this condition and small deletions may have only minor effects on crossing-over. The most useful criterion is that chromosomes with deletions can never revert to a normal condition.

Another reliable criterion for inferring the presence of a deletion is the phenotypic expression of recessive alleles on a normal chromosome when the region in which their wild-type counterparts are located has been deleted from the homolog. Consider, for example, the deletion of two out of six loci:

$a\ b\ c\ d\ e\ f$

$a^+ \quad d^+\ e^+\ f^+$

Phenotype is $a^+\ b\ c\ d^+e^+f^+$

This unmasking effect is relevant to the general inviability of deletions. It is known that many diploid organisms harbor recessive deleterious (or even lethal) mutations, sheltered by their normal dominant alleles. Deletions might lead to the expression of such recessives by removing sections containing the normal alleles. Such **pseudodominance,** the expression of a recessive allele when present in a single dose, also allows the physical localization of the allele's locus by plotting the locations of the deletions that can cause the allele to be pseudodominant. This kind of plotting enables a correlation to be made between the genetic map (based on linkage analysis) and the cytological map (based on pseudodominance in deletion heterozygotes). By and large, where this has been done, the maps correspond well—a satisfying cytological endorsement of a purely genetic creation. An example from the fruit fly *Drosophila* is shown in Figure 8-8. In this diagram, the recombination map is shown above, marked with distances in map units from the left end. The horizontal bars below the chromosome show the extent of the deletions identified to the left. The mutation prune *(pn)* shows pseudodominance only with deletion 264-38, which determines its location in the 2D-4 to 3A-2 region. However, *fa* shows pseudodominance with all but two deletions, so its position can be pinpointed to band 3C-7.

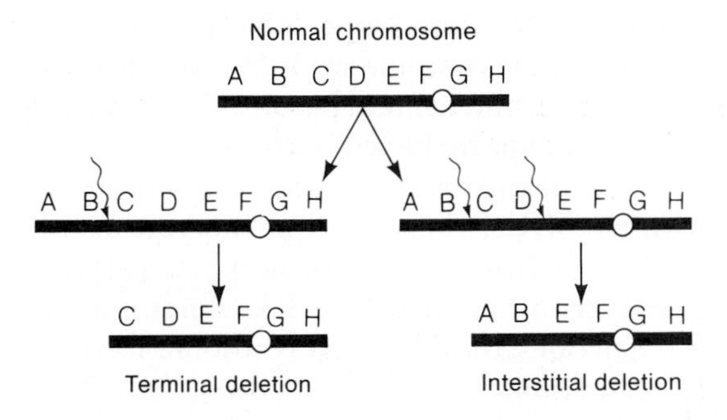

Figure 8-7 Terminal and interstitial deletions. Chromosomes can be broken when struck by ionizing radiation (wavy arrows). A terminal deletion is the loss of the end of a chromosome. An interstitial deletion results after two breaks are induced if the terminal portion (AB) rejoins the main body of the chromosome while the acentric fragment (CD) is lost.

Message Chromosome maps made by analyzing deletions agree well with linkage maps made by analyzing recombinants.

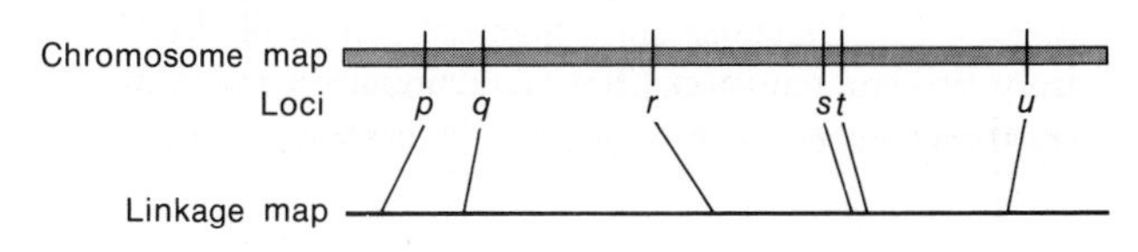

Moreover, pseudodominance can be used to map a small deletion that cannot be visualized microscopically. Let's consider an X chromosome in *Drosophila* that carries a

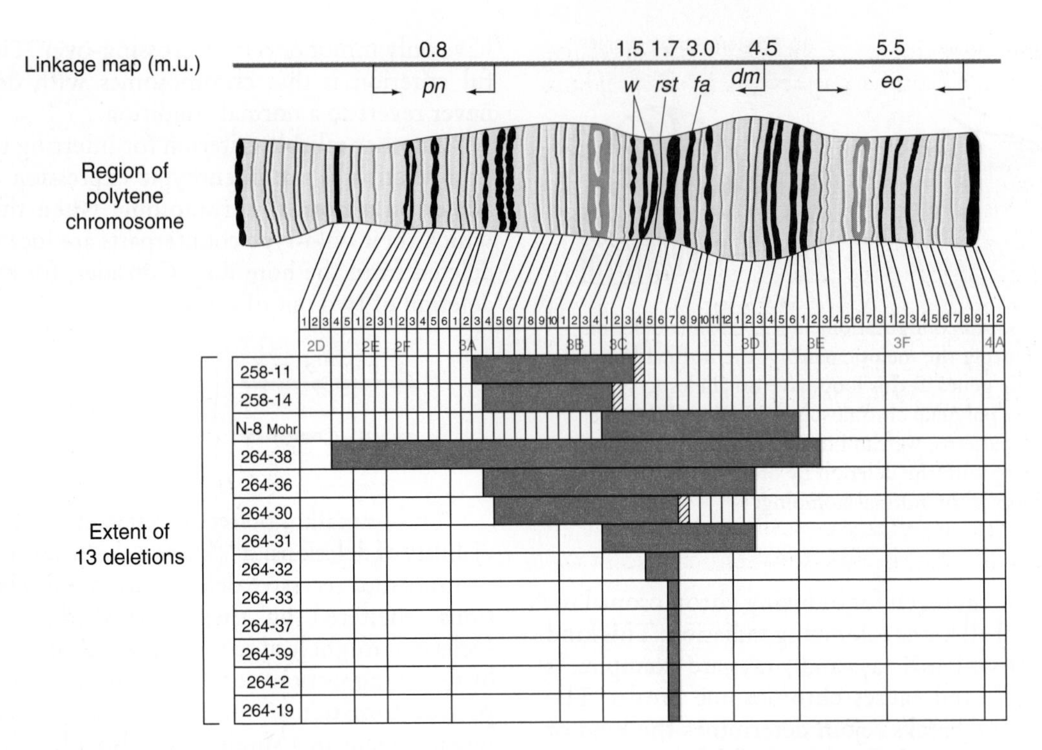

Figure 8-8 Locating genes to chromosomal regions by observing pseudodominance in *Drosophila* heterozygous for deletion and normal chromosomes. The red bars show the extent of the deleted segments in 13 deletions. All recessive alleles spanned by a deletion will be expressed.

recessive lethal suspected of being a deletion; we call this chromosome "X?." We can cross X?-bearing females with males carrying recessive alleles of loci on that chromosome. For example, a map of loci in the tip region is

y	*dor*	*br*	*gt*	*swa*	*w*	*rst*	*vt*
0.3	0.3	0.3	0.4	0.2	0.2	0.6	

Suppose we obtain all wild-type flies in crosses between X?/X females and males carrying *y*, *dor*, *br*, *gt*, *rst*, and *vt* but obtain pseudodominance of *swa* and *w* with X? (that is, X?/*swa* shows the recessive swa phenotype and X?/*w* shows the recessive w phenotype). Then we have good genetic evidence for a deletion of the chromosome that includes at least the *swa* and *w* loci but not *gt* or *rst*.

Message Deletions are recognized genetically by (1) lack of reverse mutation, (2) pseudodominance, and (3) recessive lethality, and, cytologically, by (4) deletion loops.

Clinicians regularly find deletions in human chromosomes. In most cases the deletions are relatively small, but they nevertheless have an adverse phenotypic effect when carried in the heterozygous condition. This is somewhat surprising; how can a deletion of only one or two bands of a chromosome produce severe abnormalities when the other homolog appears to carry all the normal genetic material necessary for normal development? Presumably the balance between regions is absolutely crucial, and even minor perturbations of this can cause abnormality.

Deletions of specific human chromosome regions cause unique syndromes of phenotypic abnormalities. An example is the "cri du chat" syndrome, caused by a deletion at the tip of the short arm of chromosome 5 (Figure 8-9). It is the convention to call the short arm of a chromosome p, and the long arm q. The specific bands deleted in cri du chat syndrome are 5p15-2 and 5p15-3, the two most distal bands identifiable on 5p. The most characteristic phenotype in the syndrome is the one that gives its name, the distinctive catlike mewing cries made by infants with this deletion. Other phenotypic manifestations of the syndrome are microencephaly (abnormally small head) and a moonlike face. Like syndromes caused by other deletions, the cri du chat syndrome also includes mental retardation.

Most human deletions, such as those we have just considered, arise from a new germinal mutation in one of the parents of an affected person, and no sign of the deletions are found in the somatic chromosomes of the parents. However, as we shall see in a later section, some human deletions are produced by the behavior of other rearrangements at meiosis. Cri du chat syndrome, for example, can result from a translocation in the previous generation.

Geneticists have mapped human genes from deletions by using a molecular technique called in situ hybridization. This will be explained in detail in a later chapter, but for now we can cover the basics to show the usefulness of deletions. If an interesting gene or other DNA fragment has been isolated using modern molecular technology, it can be tagged with radioactivity or a chemical label, and then added to a chromosomal preparation under the microscope. In such a situation, the DNA recognizes and physically binds to its normal chromosomal counterpart by nucleotide pairing; the DNA announces its presence as a spot of radioactivity or dye. The precise location of such spots is difficult to correlate with specific bands, but the deletion technique comes to the rescue. If a deletion happens to span the locus in question, no spot will appear when the test is run with the chromosome carrying the deletion because the region for binding simply is not present (Figure 8-10). By saving cell lines from patients with deletions, geneticists develop test panels of overlapping deletions spanning specific chromosomal regions, and these can be used to pinpoint a gene's position. An example from chromosome 11 is shown in Figure 8-11. The extent of the deletions in the test panel are shown as vertical bars, and the coded DNA fragments under test are shown at the right. By showing that fragment 270, for example, failed to bind to deletions 35, 8, 10, 7, 9, 23, 24, A2, 27A, and 4D, but did bind to the other deletions, it can be inferred that this piece of DNA originally came from the region spanned by 11q13.5 and 11q21. Notice that many of the deletions we have discussed are drawn as terminal. It is unlikely that they are truly terminal, because a telomere at the tip is necessary for normal chromosome replication.

Chromosome mutations often arise in cancer cells, and we shall see several cases in this chapter and the next.

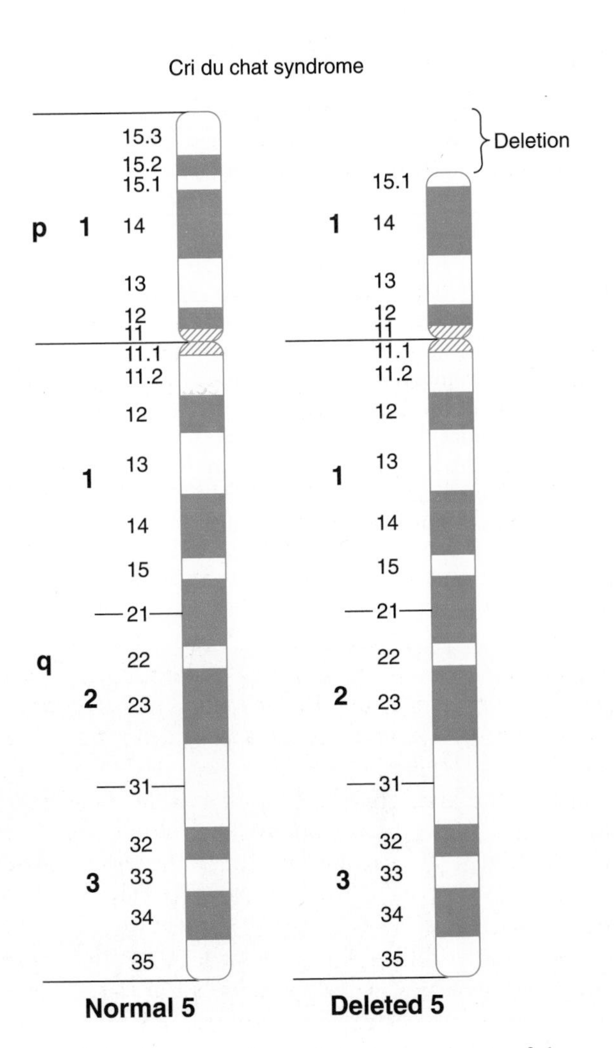

Figure 8-9 The cause of the cri du chat syndrome of abnormalities in humans is loss of the tip of the short arm of one of the homologs of chromosome 5.

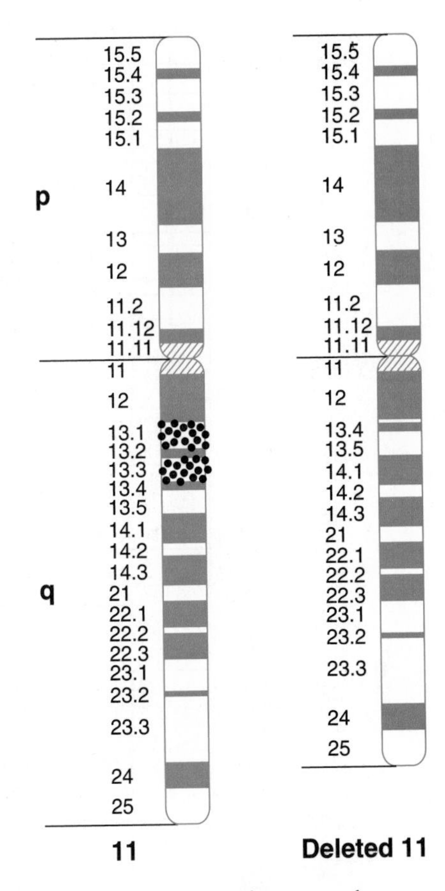

Figure 8-10 Radioactive spots show up only on one chromosome 11, because the other one has a deletion in the region where the radioactive DNA binds.

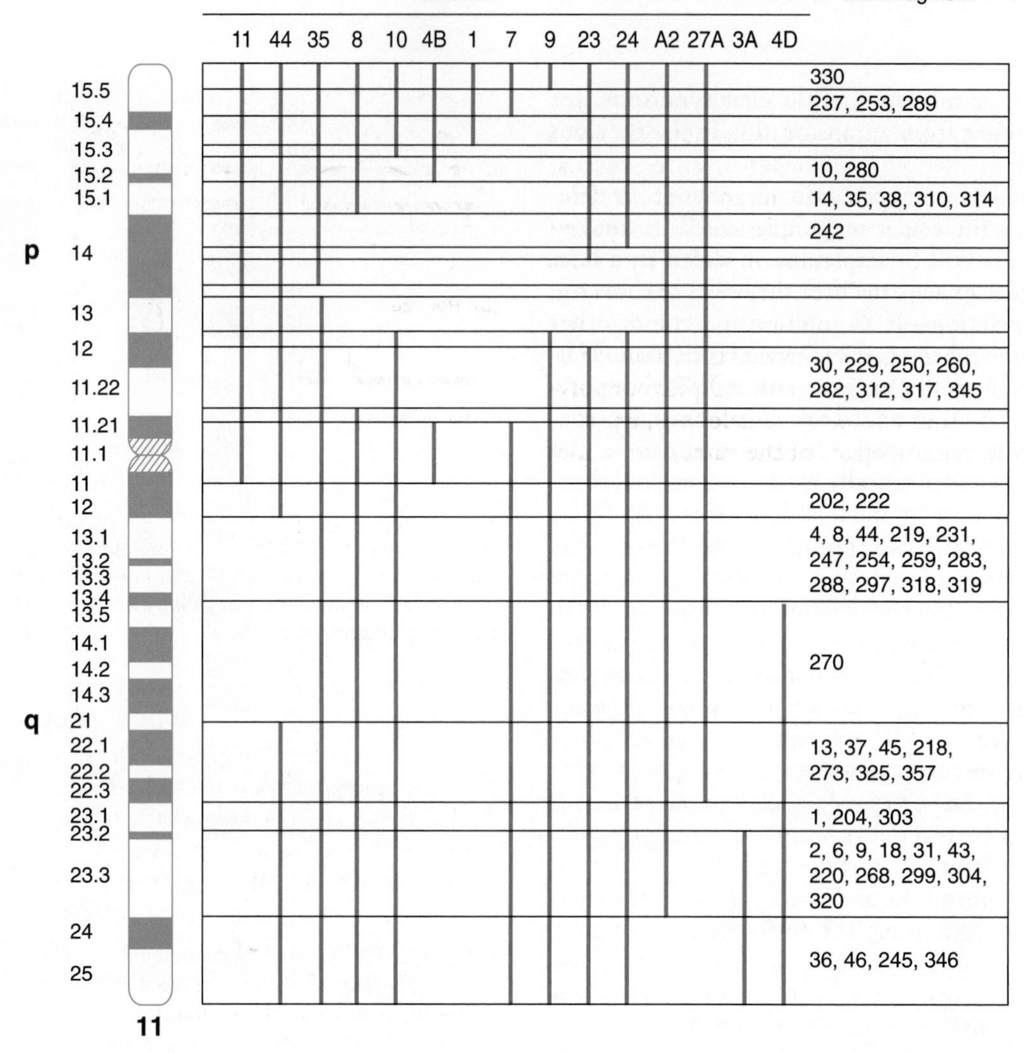

Figure 8-11 Human DNA fragments mapped to regions of chromosome 11 by their failure to bind to particular deletions. The red bars show the extent of the deletions, and the DNA fragments that were mapped are identified at the right. Notice that fragment 270, for example, failed to bind to deletions 35, 8, 10, 7, 9, 23, 24, A2, 27A, and 4D, but did bind to the other deletions. The region spanned by 11q13.5 and 11q21 is missing from all the deletions that did not bind 270; 270 was thus inferred to lie within this region. (From Y. Nakamura.)

As an example, Figure 8-12 shows some deletions consistently found in solid tumors. Not all the cells in a tumor show the deletion indicated, and often a mixture of different chromosome mutations can be found in one tumor. The contribution of such changes to the cancer phenotype is not understood.

An interesting difference between animals and plants is revealed by deletions. A male animal that is heterozygous for a deletion chromosome and a normal one produces functional sperm carrying each of the two chromosomes in approximately equal numbers. In other words, sperm seem to function to some extent regardless of their genetic content. In diploid plants, on the other hand, the pollen produced by a deletion heterozygote is of two types: functional pollen carrying the normal

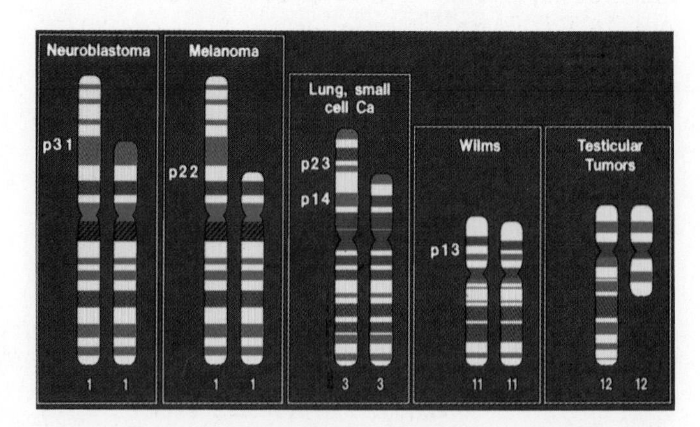

Figure 8-12 Deletions found consistently in several different types of solid tumors in humans. Band numbers represent recurrent breakpoints. (Jorge Yunis)

chromosome, and nonfunctional (or aborted) pollen carrying the deficient homolog. Thus, pollen cells seem to be sensitive to changes in *amount* of chromosomal material, and this sensitivity might act to weed out deletions. The situation is somewhat different for polyploid plants, which are far more tolerant of pollen deletions. This tolerance is due to the fact that there are several chromosome sets even in the pollen, and the loss of a segment in one of these sets is less crucial than it would be in a haploid pollen cell. Ovules in either diploid or polyploid plants also are quite tolerant of deletions, presumably due to the nurturing effect of the surrounding maternal tissues.

Duplications

The processes of chromosome mutation sometimes produce an extra copy of some chromosome region. In considering a haploid organism, which has one chromosome set, it is easy to see why such a product is called a duplication because the region is now present in duplicate. The duplicate regions can be located adjacent to each other, or one can be in its normal location and the other in a novel location on a different part of the same chromosome or even on another chromosome. In a diploid organism, the chromosome set containing the duplication is generally present together with a standard chromosome set. The cells of such an organism will, of course, have three copies of the chromosome region in question, but nevertheless such duplication heterozygotes are generally referred to as duplications because they carry the product of one duplication event.

Duplication heterozygotes also show interesting pairing structures at meiosis or in salivary gland chromosomes. The precise structure that forms depends on the type of duplication. For the present we shall discuss only adjacent duplications. These can be **tandem,** as in the example A B C B C D, or **reverse** as in the case A B C C B D. The pairing structures in heterozygotes for adjacent duplications are shown in Figure 8-13.

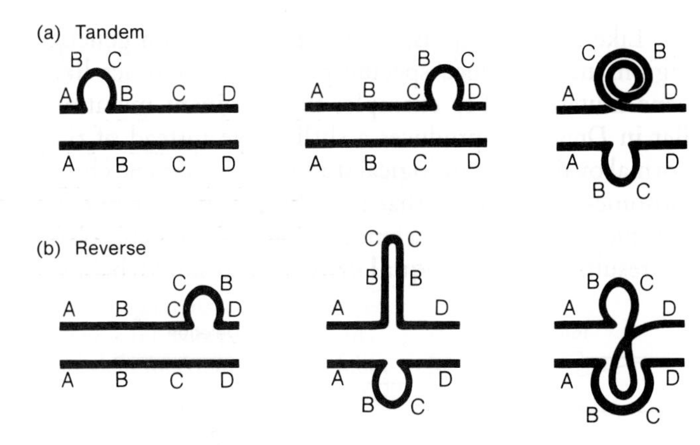

Figure 8-13 Possible pairing configurations in heterozygotes of a standard chromosome and a side-by-side duplication. Duplicated segments may be (a) in tandem or (b) in reverse order. A particular duplicated segment may assume different configurations in different meioses.

Once a tandem duplication arises in an individual, then by inbreeding it is possible that some descendants will be duplication homozygotes, carrying a total of four copies of the duplicated chromosome region. Such individuals present another interesting feature of duplications, which is the possibility of asymmetrical pairing, as shown in Figure 8-14. The crossing-over at meiosis of such asymmetrically paired regions may create a tandem triplication of the chromosome region.

The extra region of a duplication is free to undergo gene mutation because the necessary basic functions of the region will be provided by the other copy. This provides an opportunity for divergence in the function of the duplicated genes, which could be advantageous in genome evolution. Indeed, from situations in which different gene products with related functions can be compared, such as the globins (to be discussed later), there is good evidence that these products arose as duplicates of each other.

Message Duplications supply additional genetic material capable of evolving new functions.

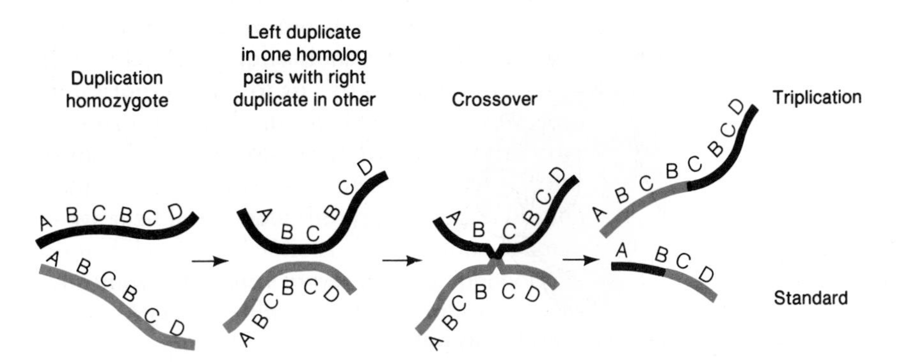

Figure 8-14 Generation of higher orders of duplications by asymmetric pairing followed by crossing-over in a duplication homozygote.

ments of DNA from their own species, and this DNA can insert into the linear structure of a chromosome. It can insert either at its regular locus, replacing the resident sequence, or at a completely different site, called an **ectopic site.** Insertion of the DNA at an ectopic site creates a duplication of the fragment that was taken up. This is a direct and convenient way of making small duplications.

Inversions

If two breaks occur in one chromosome, sometimes the region between the breaks rotates 180 degrees before rejoining with the two end fragments. This creates a chromosomal mutation called an **inversion.** Unlike deletions and duplications, inversions involve no change in the overall amount of the genetic material, so inversions are generally viable and show no particular abnormalities at the phenotypic level. In some cases one of the chromosome breaks is within a gene of essential function, and then that breakpoint acts as a lethal gene mutation linked to the inversion. In such a case the inversion could not be made homozygous. However many inversions can be made homozygous, and furthermore inversions can be detected in haploid organisms, so in these cases the breakpoint is clearly not in an essential region.

Most analyses of inversions use heterozygous inversions, diploids in which one chromosome has standard sequence and one carries the inversion. Microscopic observation of meioses in inversion heterozygotes reveals the location of the inverted segment (Figure 8-18) because one chromosome twists once at the ends of the inversion to pair with the other, untwisted chromosome; in this way the paired homologs form an **inversion loop** (Figure 8-18).

The location of the centromere relative to the inverted segment determines the genetic behavior of the chromosome. If the centromere is outside the inversion, then the inversion is said to be **paracentric,** whereas inversions spanning the centromere are **pericentric:**

Paracentric A B C D E F → A B E D C F

Pericentric A B C D E F → A D C B E F

How do inversions behave genetically? Crossing-over within the inversion loop of a paracentric inversion connects homologous centromeres in a **dicentric bridge** while also producing an acentric fragment — one without a centromere. Then, as the chromosomes separate during anaphase I, the centromeres remain linked by the bridge. This orients the centromeres so that the noncrossover chromatids lie farthest apart. The acentric fragment cannot align itself or move and, consequently, is lost. Tension eventually breaks the bridge, forming two chromosomes with terminal deletions (Figure 8-19). The gametes containing such deleted chromosomes may be inviable, but even if viable the zygotes they eventually form will be inviable. Hence, a crossover event, which normally generates the recombinant class of meiotic products, instead produces lethal products. The overall result is a smaller recombinant frequency. In fact for genes within the inversion the RF is zero. For genes flanking the inversion, the RF is reduced in proportion to the relative size of the inversion.

Inversions affect recombination in another way too. Inversion heterozygotes often have mechanical pairing problems in the region of the inversion; this reduces the frequency of crossing-over and hence the recombinant frequency in the region.

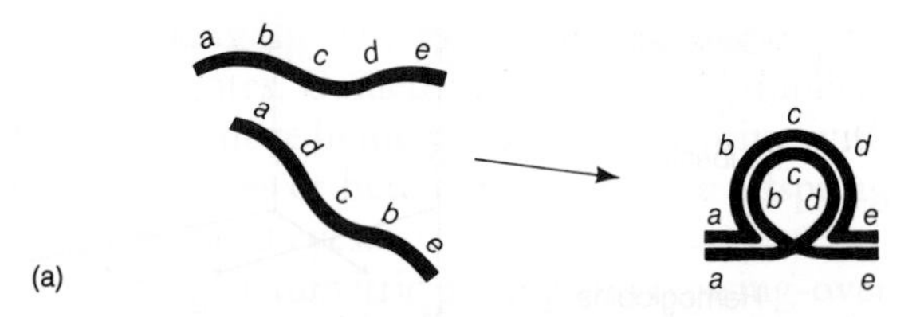

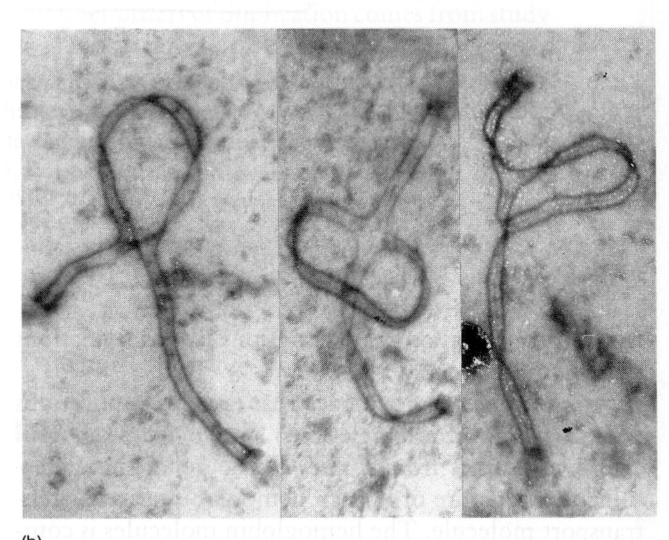

Figure 8-18 The chromosomes of inversion heterozygotes pair in a loop at meiosis. (a) Diagrammatic representation; each chromosome is actually a pair of sister chromatids. (b) Electron micrograph of synaptonemal complexes at prophase I of meiosis in a mouse heterozygous for a paracentric inversion. Three different meiocytes are shown. (Part b from M. J. Moses, Department of Anatomy, Duke Medical Center.)

Message Two mechanisms reduce the number of recombinant products among the progeny of inversion heterozygotes: elimination of the products of crossovers in the inversion loop and inhibition of pairing in the region of the inversion.

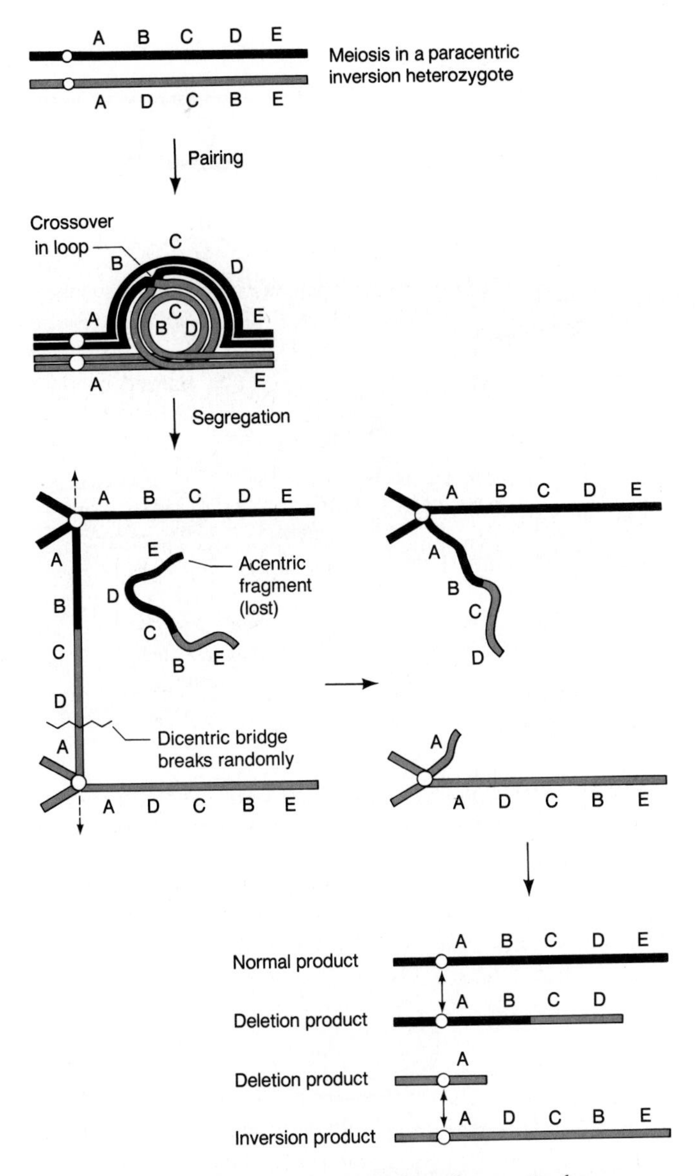

Figure 8-19 Meiotic products resulting from a single crossover within a paracentric inversion loop. Two nonsister chromatids cross over within the loop.

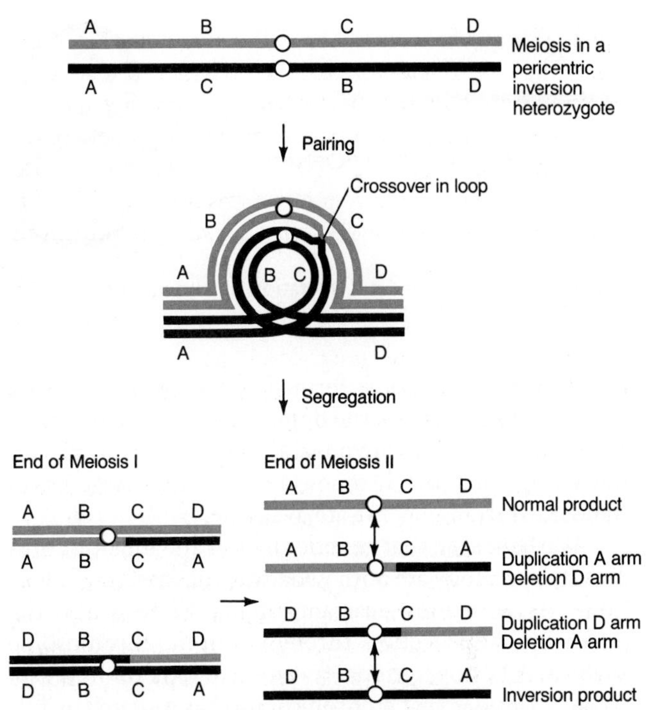

Figure 8-20 Meiotic products resulting from a meiosis with a single crossover within a pericentric inversion loop.

The net genetic effect of a pericentric inversion is the same as that of a paracentric one — crossover products are not recovered — but for different reasons. In a pericentric inversion, because the centromeres are contained within the inverted region, the chromosomes that have crossed over disjoin in the normal fashion, without the creation of a bridge. However, the crossover produces chromatids that contain a duplication and a deficiency for different parts of the chromosome (Figure 8-20). In this case, if a nucleus carrying a crossover chromosome is fertilized, the zygote dies because of its genetic imbalance. Again, the result is the selective recovery of noncrossover chromosomes in viable progeny.

It is worth adding a note about homozygous inversions. In such cases the homologous inverted chromosomes pair and cross over normally, there are no bridges, and the meiotic products are viable. However, there is an interesting effect, which is that the linkage map will show the inverted gene order.

Geneticists use inversions to create duplications of specific chromosome regions for various experimental purposes. For example, consider a heterozygous pericentric inversion with one breakpoint at the tip (T) of the chromosome, as shown in Figure 8-21. A crossover in the loop produces a chromatid type in which the entire left arm is duplicated; if the tip is nonessential, a duplication stock is generated for investigation. Another way to make a duplication (and a deficiency) is to use two paracentric inversions with overlapping breakpoints (Figure

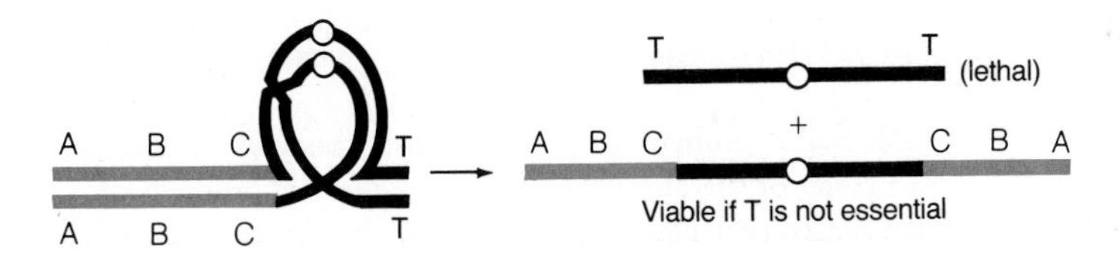

Figure 8-21 Generation of a viable nontandem duplication from a pericentric inversion close to a dispensable chromosome tip.

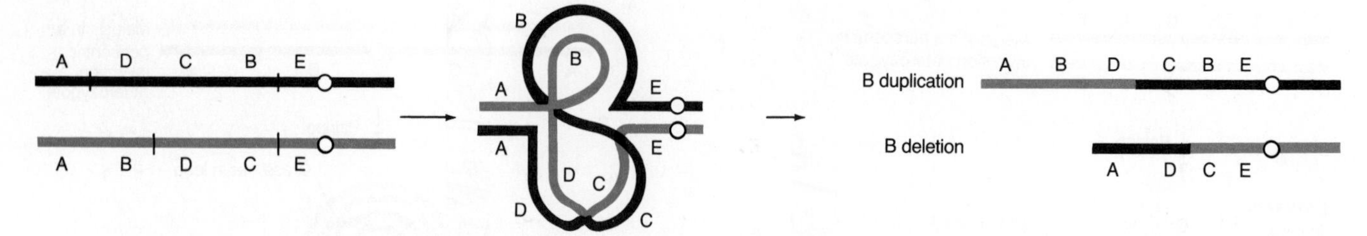

Figure 8-22 Generation of a nontandem duplication by crossing-over between two overlapping inversions.

8-22). A complex loop is formed, and a crossover within the inversion produces the duplication and the deletion. These manipulations are possible only in organisms with thoroughly mapped chromosomes for which large sets of standard rearrangements are available.

We have seen that genetic analysis and meiotic chromosome cytology are both good ways of detecting inversions. As with most rearrangements, there is also the possibility of detection through mitotic chromosome analysis. A key operational feature is to look for new arm ratios. Consider that a chromosome has mutated as follows:

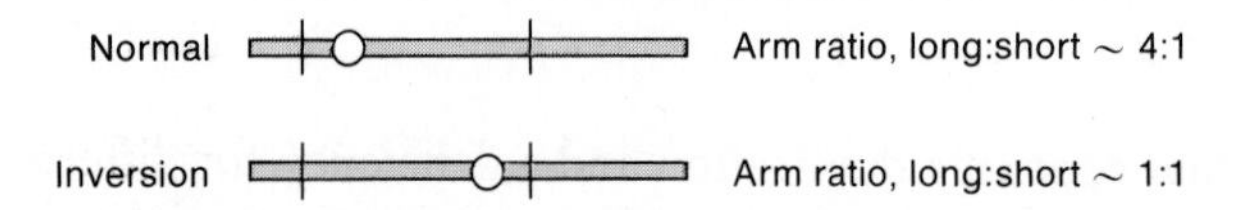

Note that the ratio of the long to the short arm has been changed from about 4 to about 1 by the pericentric inversion. Paracentric inversions do not alter the arm ratio but they may be detected microscopically if banding or other chromosome landmarks are available.

Message The main diagnostic features of inversions are reduction of recombinant frequency, inversion loops, and partial sterility from unbalanced or deleted meiotic products, all in individuals heterozygous for inversions. Some inversions may be directly observed as an inverted arrangement of chromosomal landmarks.

Inversions are found in about 2 percent of humans. The heterozygous inversion carriers generally show no adverse phenotype, but produce the expected array of abnormal meiotic products from crossing-over in the inversion loop. Let us consider pericentric inversions as an example. Individuals heterozygous for pericentric inversions produce offspring with the duplication–deletion chromosomes predicted; these offspring show varying degrees of abnormalities depending on the lengths of the chromosome regions involved. Some phenotypes caused by duplication–deletion chromosomes are so abnormal

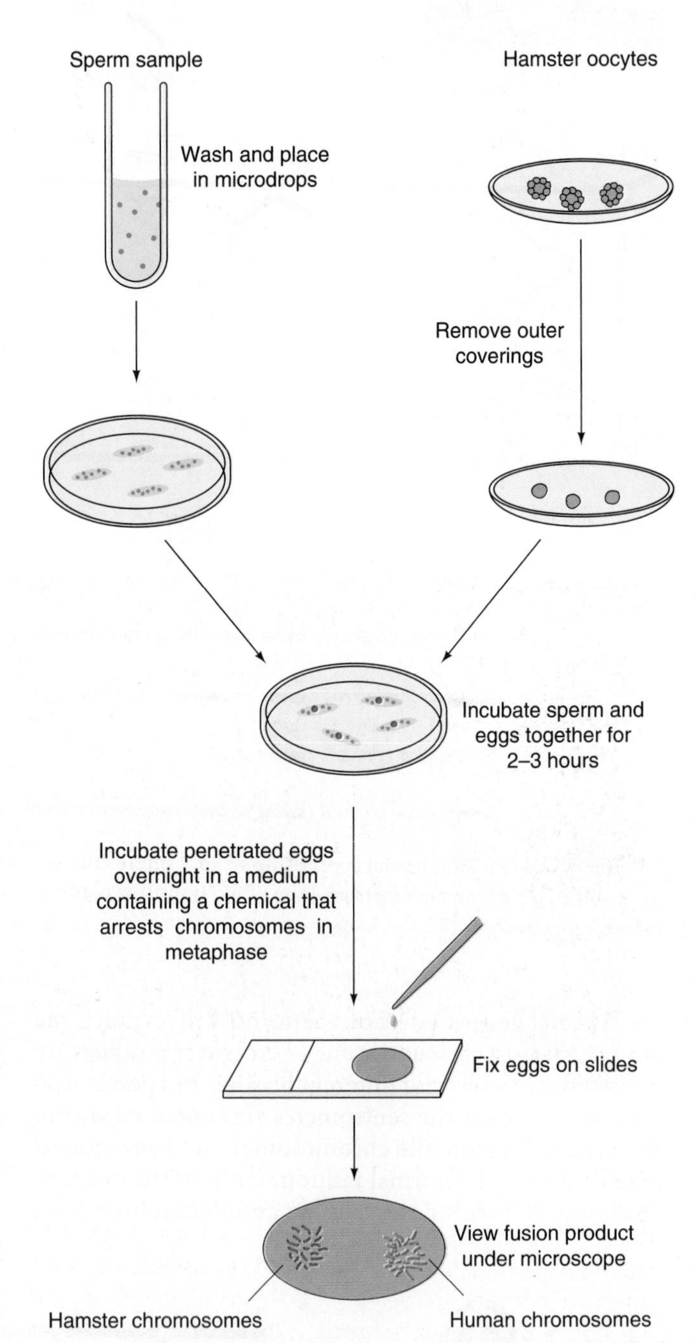

Figure 8-23 Human sperm and hamster oocytes are fused to permit study of the chromosomes in the meiotic products of human males. (Adapted from original art by Renée Martin.)

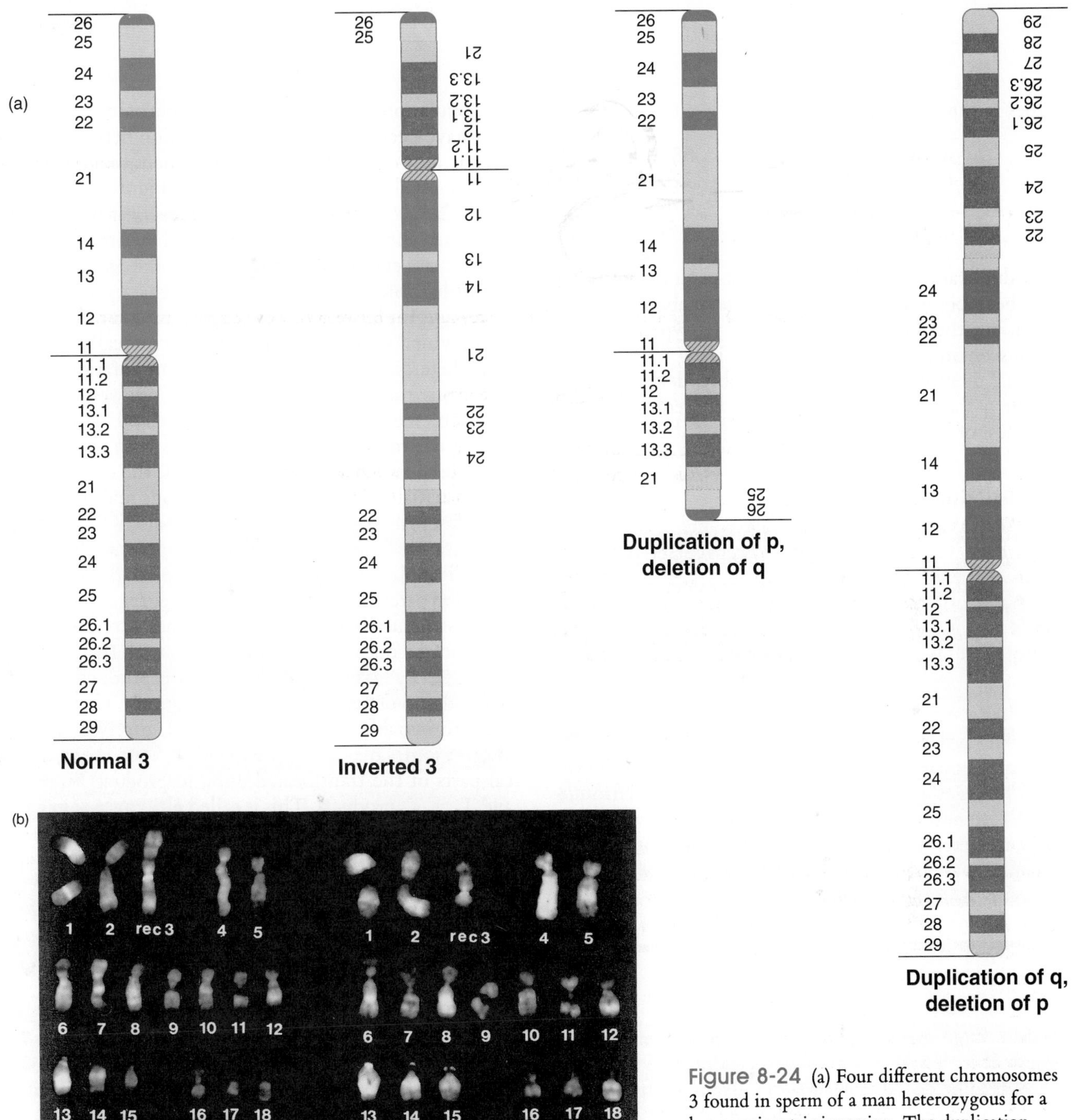

Figure 8-24 (a) Four different chromosomes 3 found in sperm of a man heterozygous for a large pericentric inversion. The duplication–deletion types result from a crossover in the inversion loop. (b) Two complete sperm chromosome sets containing the two duplication–deletion types (labeled rec 3). (Renée Martin)

as to be incapable of survival to birth, and are lost as spontaneous abortions. However, there is a way to study the abnormal meiotic products that does not depend on survival to term. Human sperm placed in contact with unfertilized eggs of the golden hamster penetrate the eggs but fail to fertilize them. The sperm nucleus does not fuse with the egg nucleus, and if the cell is prepared for cytogenetic examination, the human chromosomes are seen as a distinct group (Figure 8-23). This makes it possible to study directly the chromosomal products of a male meiosis, and is particularly useful in the study of meiotic products of men with chromosome mutations.

In one case, a man heterozygous for an inversion of chromosome 3 was subjected to sperm analysis. This was a large inversion with a high potential for crossing-over in the loop. Four chromosome 3 types were represented in the man's sperm — normal, inversion, and two recombinant types (Figure 8-24). The sperm contained the four types in the following frequencies:

normal	38%
inversion	32%
duplication q, deletion p	17%
duplication p, deletion q	13%

The duplication q – deletion p recombinant chromosome had been observed previously in several abnormal children, but the duplication p – deletion q type had never been seen and probably zygotes receiving it are too abnormal to survive to term. Presumably, deletion of the larger q fragment has more severe consequences than deletion of the smaller p fragment.

Translocations

When two nonhomologous chromosomes mutate by exchanging parts, the resulting chromosomal rearrangements are translocations. Here we consider **reciprocal translocations,** the most common kind of translocation. A segment from one chromosome is exchanged with a segment from another nonhomologous one, so that two translocation chromosomes are generated simultaneously.

The exchange of chromosome parts between nonhomologs establishes new linkage relationships. These new linkages are revealed if the translocated chromosomes are homozygous and, as we shall see, even when they are heterozygous. Furthermore, translocations may drastically alter the size of a chromosome as well as the position of its centromere. For example,

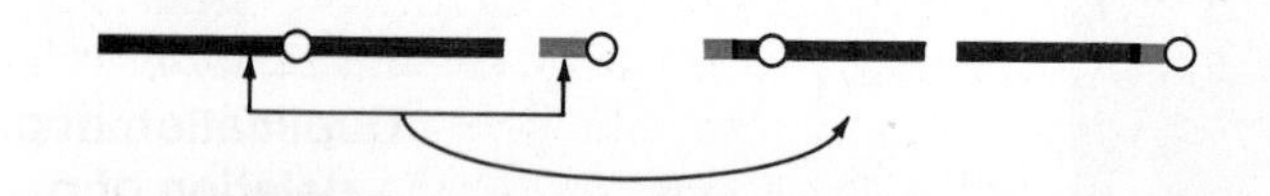

Here a large metacentric chromosome is shortened by one-half its length to an acrocentric one, and the small chromosome becomes a large one. Examples from natural populations are known in which chromosome numbers have been changed by translocation between acrocentric chromosomes and the subsequent loss of the resulting small chromosome elements (Figure 8-25).

In heterozygotes having two translocated chromosomes and their normal counterparts, there are important genetic and cytological effects. Again, the pairing affinities of homologous regions dictate a characteristic configuration for chromosomes synapsed in meiosis. Figure 8-26, which illustrates meiosis in a reciprocally translocated heterozygote, shows that the configuration is that of a cross.

Remember, the configuration presented in the figure lies on the metaphase equatorial plate with the spindle fibers perpendicular to the page. Thus, the centromeres would actually migrate up out of the page or down under it. Homologous paired centromeres disjoin, whether or not a translocation is present. Because Mendel's second law still applies to *different paired centromeres,* there are two common patterns of disjunction. The segregation of each of the structurally normal chromosomes with one of the translocated ones (T_1 with N_2 and T_2 with N_1) is called **adjacent-1 segregation.** Both meiotic products are duplicated and deficient for different regions. These products are inviable. On the other hand, the two normal chromosomes may segregate together, as do the reciprocal parts of the translocated ones, to produce $N_1 + N_2$ and $T_1 + T_2$ products. This is called **alternate segregation.** These products are viable. Since the adjacent-1 and alternate segregation patterns are equally frequent, approximately 50 percent of the products are viable and 50 percent inviable, a condition known as **semisterility.** (There is another event, called **adjacent-2 segregation,** in which homologous centromeres migrate to the same pole, but in general this is rare.)

Semisterility, or "half-sterility," diagnoses translocation heterozygotes. However, semisterility is defined differently for plants and animals. In plants, the 50 percent unbalanced meiotic products from the adjacent-1

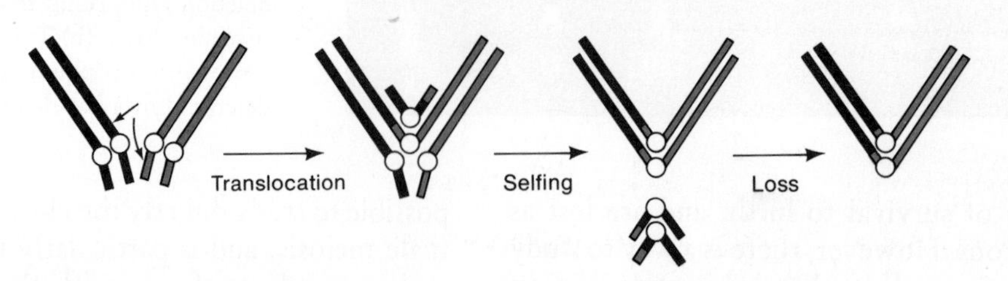

Figure 8-25 Genome restructuring by translocations. Small arrows indicate breakpoints in one homolog of each of two pairs of acrocentric chromosomes. The resulting fusion of the breaks yields one short and one long metacentric chromosome. If, as in plants, self-fertilization (selfing) takes place, an offspring could be formed with only one pair of long and only one pair of short metacentric chromosomes. Under appropriate conditions, the short metacentric chromosome may be lost. Thus, we see a conversion from two acrocentric pairs of chromosomes to one pair of metacentrics.

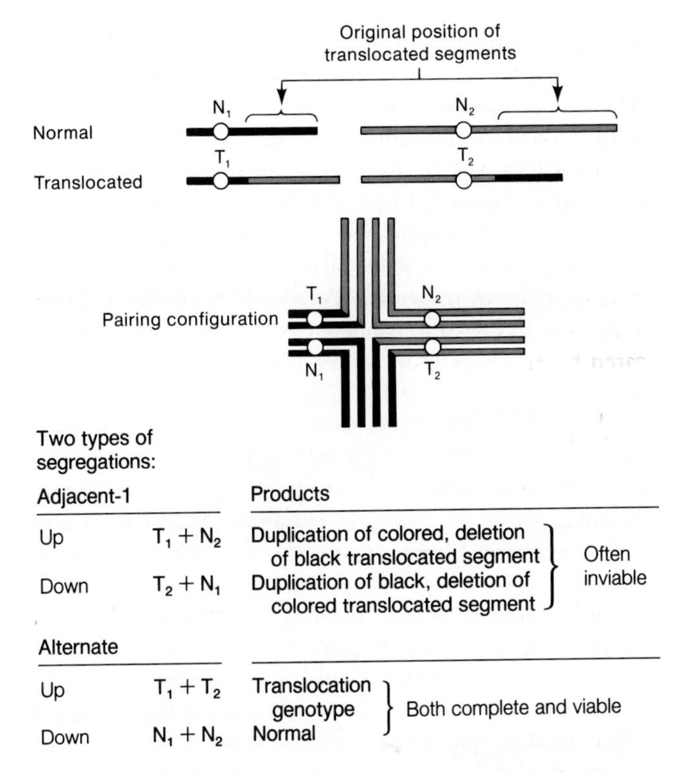

Figure 8-26 The meiotic products resulting from the two most commonly encountered chromosome segregation patterns in a reciprocal translocation heterozygote.

Figure 8-27 Photo of normal and aborted pollen of a semisterile corn plant. The clear pollen grains contain chromosomally unbalanced meiotic products of a reciprocal translocation heterozygote. The opaque pollen grains, which contain either the complete translocation genotype or normal chromosomes, are functional in fertilization and development. (William Sheridan)

segregation generally abort at the gametic stage (Figure 8-27). In animals, however, the duplication–deletion products are viable as gametes, but lethal to the zygote.

Remember also that heterozygotes for the other rearrangements such as deletions and inversions may be partly sterile; but the precise 50 percent reduction in viable gametes or zygotes is usually a good diagnostic clue for a translocation.

Message Translocations, inversions, and deletions produce partial sterility by generating unbalanced meiotic products that may themselves die or that may cause zygotes to die.

Genetically, markers on nonhomologous chromosomes appear to be linked if these chromosomes are involved in a translocation and the loci are close to the translocation breakpoint. Figure 8-28 shows a situation in which a translocation heterozygote has been established by crossing an *aa bb* individual with a translocation homozygote bearing the wild-type alleles. We shall assume that *a* and *b* are close to the translocation breakpoint. On testcrossing the heterozygote, the only viable progeny are those bearing the parental genotypes, so linkage is seen between loci that were originally on different chromosomes. In fact, if all four arms of the meiotic pairing structure are genetically marked, recombination studies should result in a cross-shaped linkage

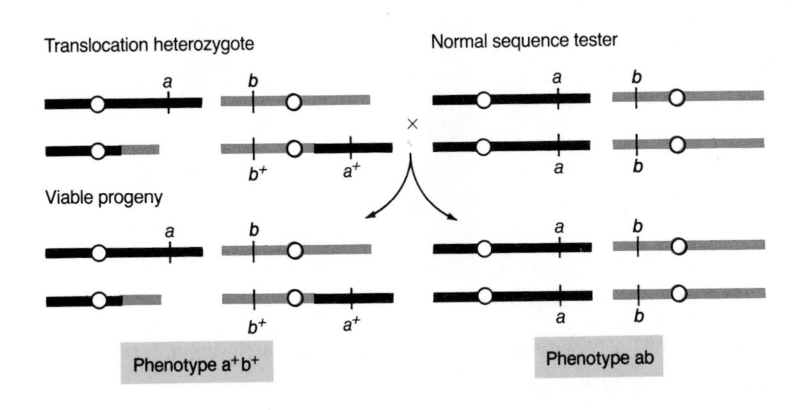

Figure 8-28 When a translocated fragment carries a marker gene, this marker can show linkage to genes on the other chromosome because the recombinant genotypes (in this case $a^+ b$ and $a\ b^+$) tend to be in duplication–deletion gametes and do not survive.

map. Apparent linkage of genes known to be on separate chromosomes is a genetic giveaway for the presence of a translocation.

Message Reciprocal translocations are diagnosed genetically by semisterility and by the apparent linkage of genes known to be on separate chromosomes.

The Importance of Translocations in Human Affairs. Translocations are economically important. In agriculture, translocations in certain crop strains can reduce yields considerably due to the number of unbalanced zygotes that form. On the other hand, translocations are potentially useful: it has been proposed that insect pests could be controlled by introducing translocations into their wild populations. According to the proposal, 50 percent of the offspring of crosses between insects carrying the translocation and wild types would die, and $\frac{10}{16}$ of the progeny of crosses between translocation-bearing insects would die.

Translocations in humans are always carried in the heterozygous state. An example involving chromosomes 5 and 11 is shown in Figure 8-29. The offspring of the person with this particular translocation had a duplication of 11q and a deletion of 5p. These children showed symptoms of both cri du chat syndrome, which is caused by the deletion of 5p, and the syndrome associated with the duplication of 11q. The reciprocal duplication–deficiency chromosome was not observed.

Down syndrome is usually caused by the presence of an extra chromosome 21 that failed to segregate from its homolog at meiosis (see Chapter 9). However, Down syndrome can also arise in the progeny of an individual heterozygous for a translocation involving chromosome 21. The heterozygous person is a phenotypically normal carrier. During meiosis, an adjacent-1 segregation produces gametes carrying duplicated parts of chromosome 21 and possibly a deficiency for some part of the other chromosome involved in the translocation. The extra segment of chromosome 21 is the cause of Down syndrome. One-half of the normal children of a carrier, on the average, are carriers (Figure 8-30). Under the mechanism just outlined, there should be a high recurrence rate in the pedigree of the affected family: a translocation heterozygote can repeatedly produce children with Down syndrome, and of course the carriers can do the same in subsequent generations. The other mechanism is sporadic and does *not* show this recurrence within a family. The factors responsible for numerous other heredi-

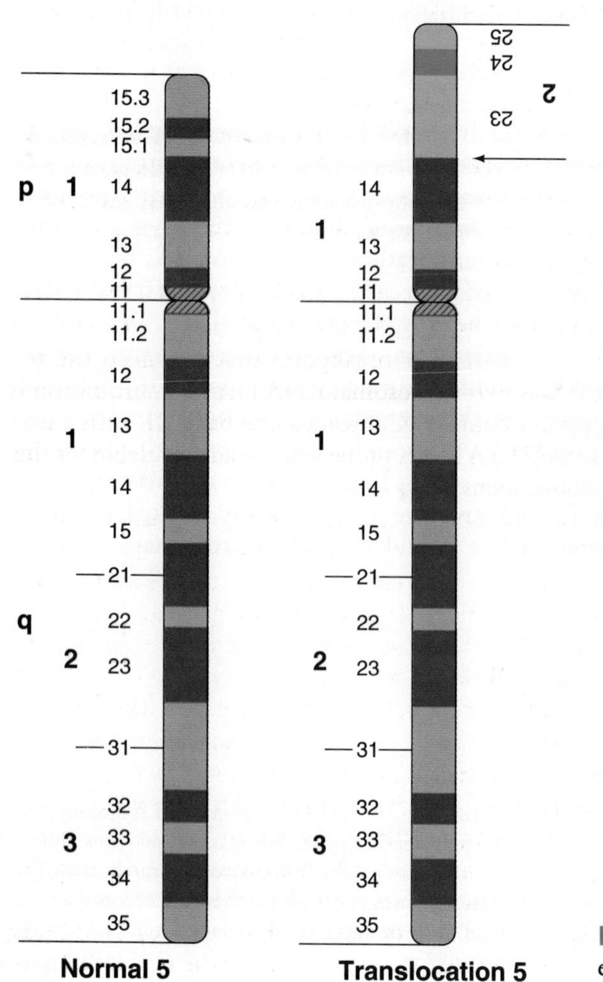

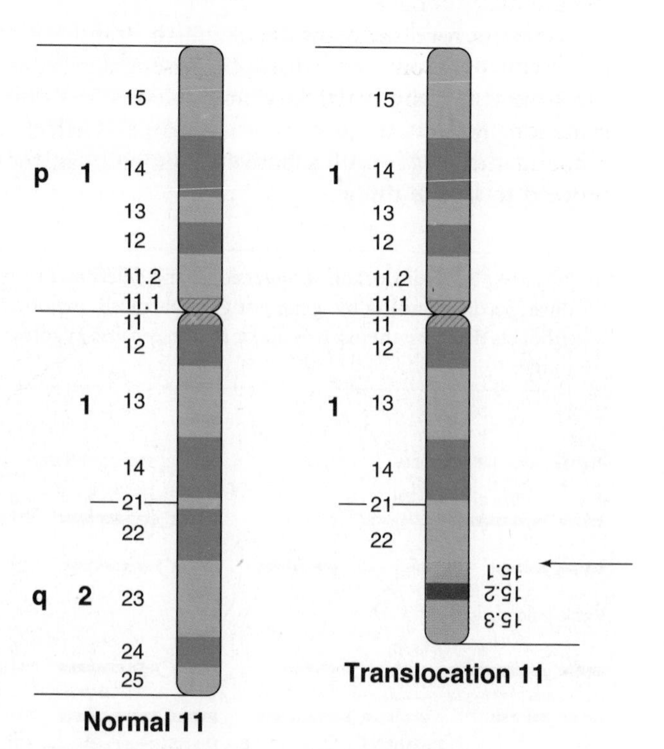

Figure 8-29 A human translocation heterozygote involving reciprocal exchange of 5p and 11q (5p15;11q23).

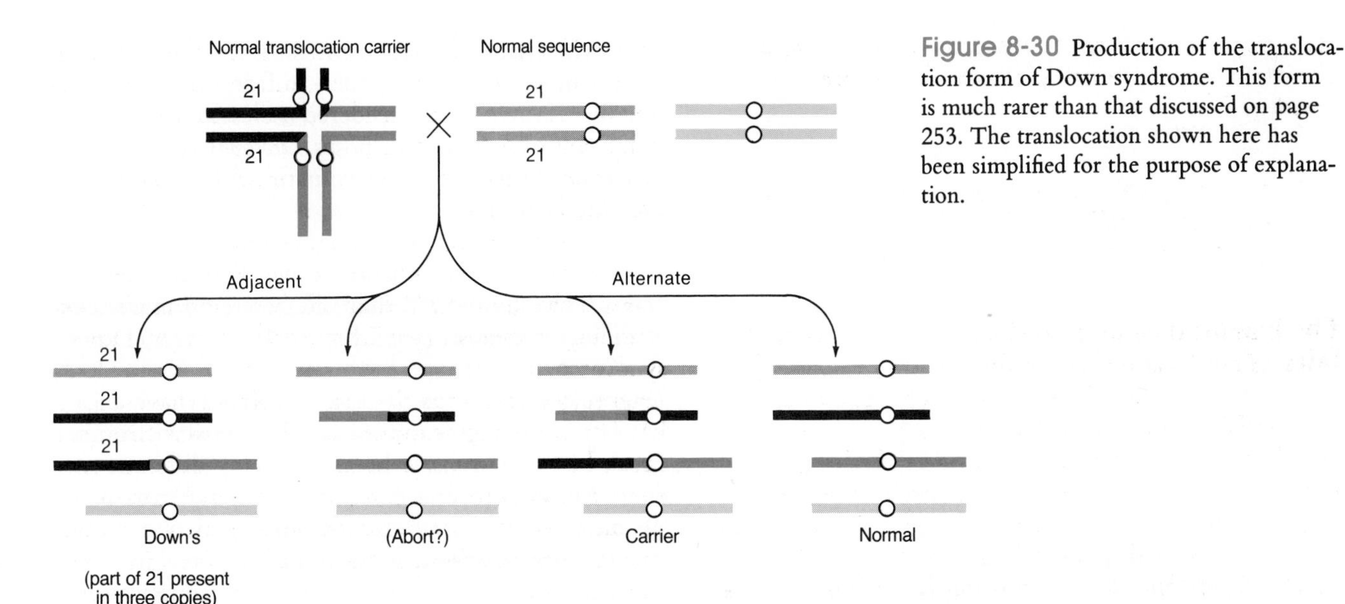

Figure 8-30 Production of the translocation form of Down syndrome. This form is much rarer than that discussed on page 253. The translocation shown here has been simplified for the purpose of explanation.

tary disorders have been traced to translocation heterozygosity in the parents.

Translocations also appear in cancer cells, and some examples are shown for solid tumors in Figure 8-31. In solid tumors translocations are not as common as deletions. As with other rearrangements in cancer cells, the involvement with the cancer phenotype generally is not clear. However, in a later section we shall study an example in which the relocation of a specific oncogene seems to be causally connected to cancer.

As is true for other rearrangements, the translocation breakpoints can sometimes disrupt an essential gene, and the gene is thereby inactivated and behaves as a point mutation. Molecular geneticists can use this effect to pinpoint the location of a human gene and can then proceed to isolate the gene. For example, information

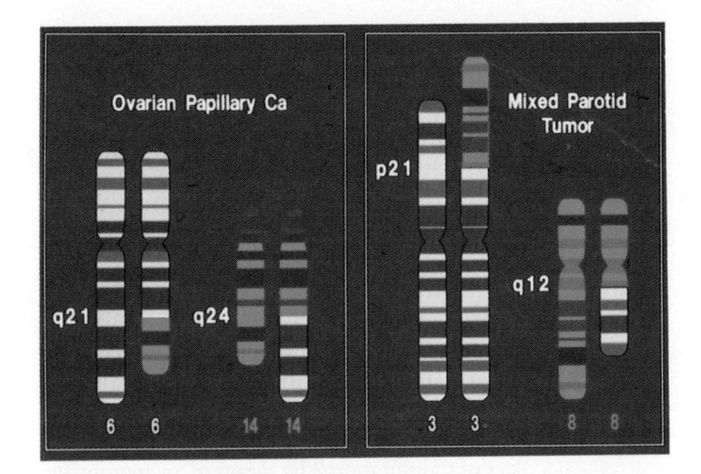

Figure 8-31 Translocations found consistently in several different types of solid tumors in humans. Band numbers indicate breakpoints. (Jorge Yunis)

from translocations helped in the isolation of the gene for Duchenne muscular dystrophy. Some rare cases of Duchenne muscular dystrophy were females who were also heterozygous for translocations between the X chromosome and a variety of different autosomes. The X chromosome breakpoint was always in the band Xp21, so the gene for muscular dystrophy, already known to be X-linked, was evidently in this band and had been disrupted by the break. The hunt for the gene could begin by focusing on that area. In passing, note that the expression of the mutant phenotype in a female must have been because the normal X chromosome was inactivated (Figure 8-32 and page 71).

One specific autosomal breakpoint proved to be useful in providing a molecular "tag" for the Duchenne gene. The particular breakpoint that advanced the research was in the ribosomal RNA locus on chromosome 21. (Don't confuse X chromosome band 21 with chromosome 21.) A DNA probe was already available for the ribosomal locus. It was reasoned that, in the X/21 translocation, the Duchenne gene must be disrupted and attached next to the ribosomal RNA gene. The geneticists therefore used the ribosomal probe to isolate a DNA segment with part of the Duchenne gene on it.

Translocations were also used to isolate the human gene for neurofibromatosis. Once again, the critical chromosomal material came from persons who had not only the disease but also translocations. The translocations all had one of their breakpoints in chromosome 17, in a band close to the centromere. Hence it appeared that this must be the locus of the neurofibromatosis gene, which had been disrupted by the translocation breakage. Subsequent analysis showed that the chromosome 17 breakpoints were not identical, and their positions helped to map the region occupied by the neurofibroma-

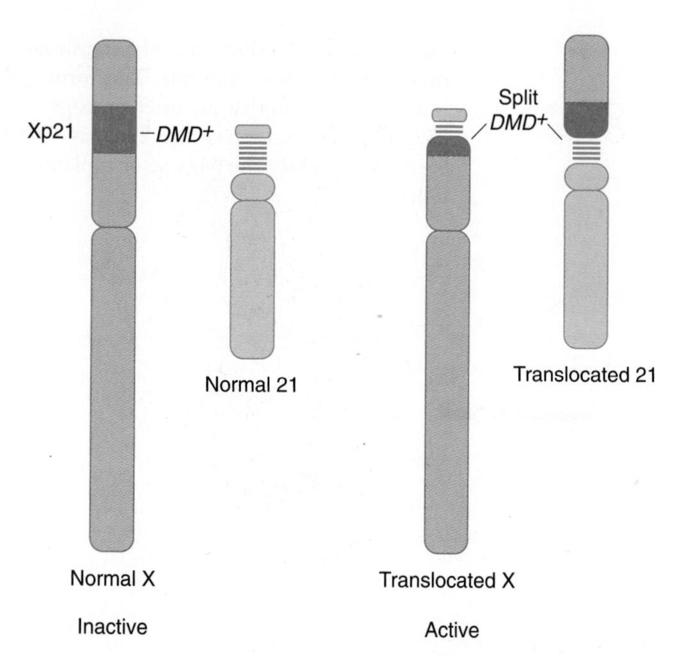

Figure 8-32 Diagram of the chromosomes of a woman with Duchenne muscular dystrophy and heterozygous for a reciprocal translocation between the X chromosome and chromosome 21. The translocation breakpoint disrupted one DMD^{+} allele rendering it nonfunctional, and X chromosome inactivation has made its undisrupted partner DMD^{+} allele also nonfunctional.

tosis gene. Isolation of DNA fragments from this region eventually led to the recovery of the gene itself.

The Use of Translocations in Producing Duplications and Deletions. Geneticists regularly need to make specific duplications or deletions to answer specific experimental questions. We have seen that they use inversions to do this, and now we shall see how translocations also can be used for the same purpose. Let's take *Drosophila* as an example. For reasons that are as yet unclear, heterochromatin near the centromere is physically extensive but contains few genes. In fact, for a long time, heterochromatin was considered useless and inert material. In any case, for our purposes, *Drosophila* can tolerate a loss or an excess of heterochromatin with little effect on viability or fertility.

Now let's select two different reciprocal translocations of the same two chromosomes. Each translocation has a breakpoint somewhere in heterochromatin, and each has another breakpoint in euchromatin on one side or the other of the region we want to duplicate or delete (Figure 8-33). It can be seen that if we have a large collection of translocations having one heterochromatic break and euchromatic breaks at many different sites, then duplications and deletions for many parts of the genome can be produced at will for a variety of experimental purposes. More generally, if one breakpoint of a translocation is near a dispensable tip, then duplication or deletion of this tip can be ignored, and the translocation can be used as a way of generating duplications or deficiencies for the other translocated segment.

Position-Effect Variegation. In previous chapters, we considered several mechanisms of generating variegation in the somatic cells of a multicellular organism. These mechanisms were somatic segregation, somatic crossover, and somatic mutation. Another cause of variegation is associated with translocations and is called **position-effect variegation.**

The locus for white eye color in *Drosophila* is near the tip of the X chromosome. Consider a translocation in which the tip of an X chromosome carrying w^{+} is relocated next to the heterochromatic region, say that of chromosome 4. Position-effect variegation is observed in flies that are heterozygotes for such a translocation, with the normal X chromosome carrying the recessive allele w. The expected eye phenotype is red because the wild-

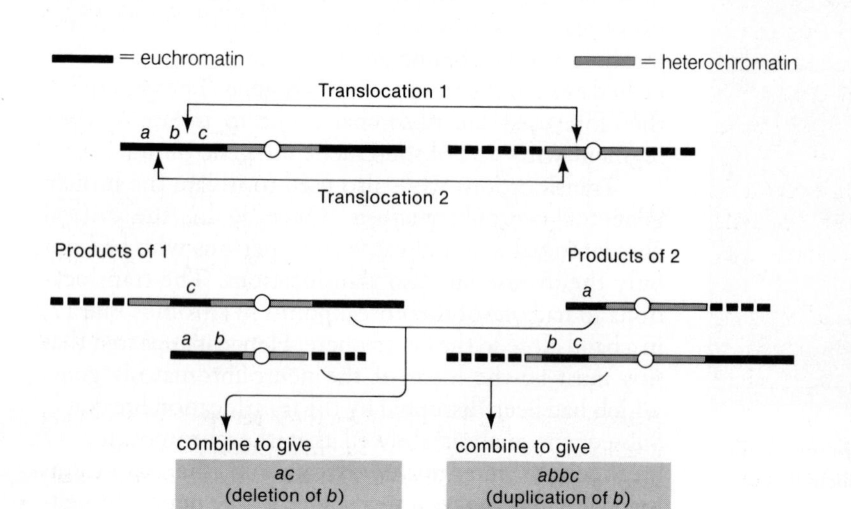

Figure 8-33 Using translocations with one breakpoint in heterochromatin to produce a duplication and a deletion. If the upper product of translocation 1 is combined with the upper product of translocation 2 by means of an appropriate mating, a deletion of *b* results. If the lower products of the two translocations are combined, the genotype *a b b c* is produced.

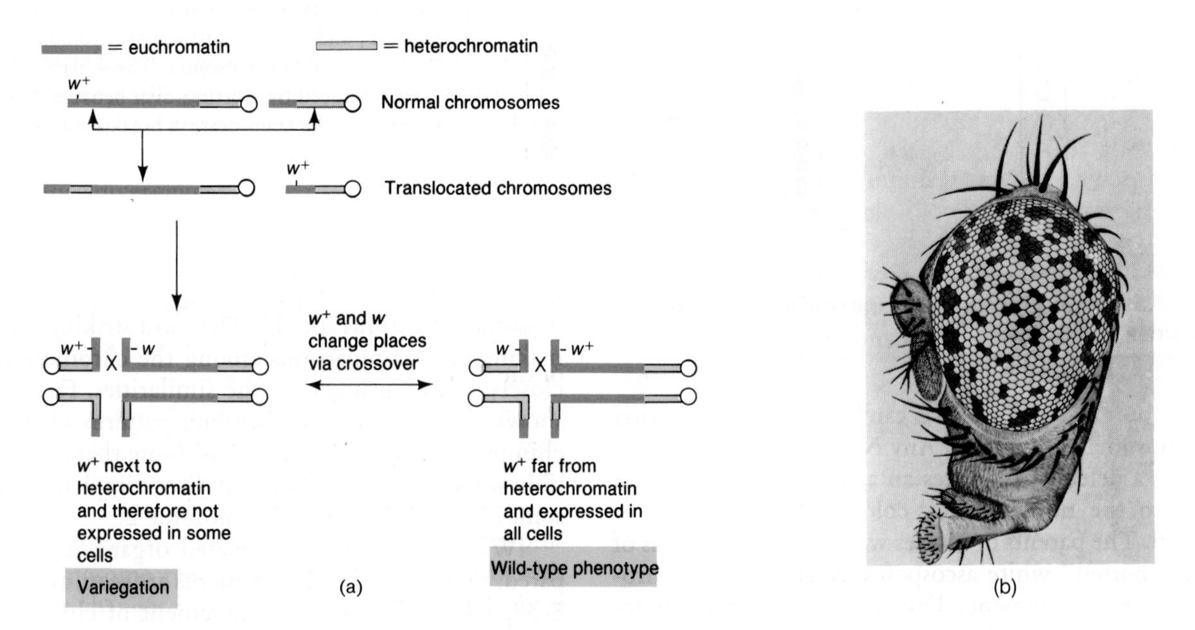

Figure 8-34 Position-effect variegation. (a) The translocation of w^+ to a position next to heterochromatin causes the w^+ function to fail in some cells, producing position-effect variegation. (b) A *Drosophila* eye showing the position-effect variegation. (Part b from Randy Mottus.)

type allele is dominant to *w*. However, in such cases the observed phenotype is a variegated mixture of red and white eye facets. How can we explain the white areas? We could suppose that when the chromosomes broke and rejoined in the translocation, the w^+ allele was somehow changed to a state that made it more mutable in somatic cells; so the white eye tissue reflects cells in which w^+ has mutated to *w*.

In 1972, Burke Judd tested this hypothesis by recombining the w^+ allele out of the translocation and onto a normal X chromosome and by recombining a *w* allele from the normal X chromosome onto the translocation (Figure 8-34). Judd found that when the w^+ allele on the translocation was crossed onto a normal X chromosome and *w* was then inserted into the translocation, the eye color was red; so obviously the w^+ allele was not defective. When he crossed a w^+ allele back onto the translocation, the phenotype was again variegated. Thus, we can conclude that, for some reason, the w^+ allele in the translocation is not expressed in some cells, thereby allowing the expression of *w*. This kind of variegation is called position-effect variegation because the unstable expression of a gene reflects its position in a rearrangement.

Such position effects can affect the genes that cause cancer. For example, most cases of Burkitt's lymphoma, a cancer of certain human antibody-producing cells called B cells, are caused by the relocation of an oncogene to a position next to a region that normally enhances the production of antibodies (Figure 8-35). The oncogene is then activated, resulting in cancer.

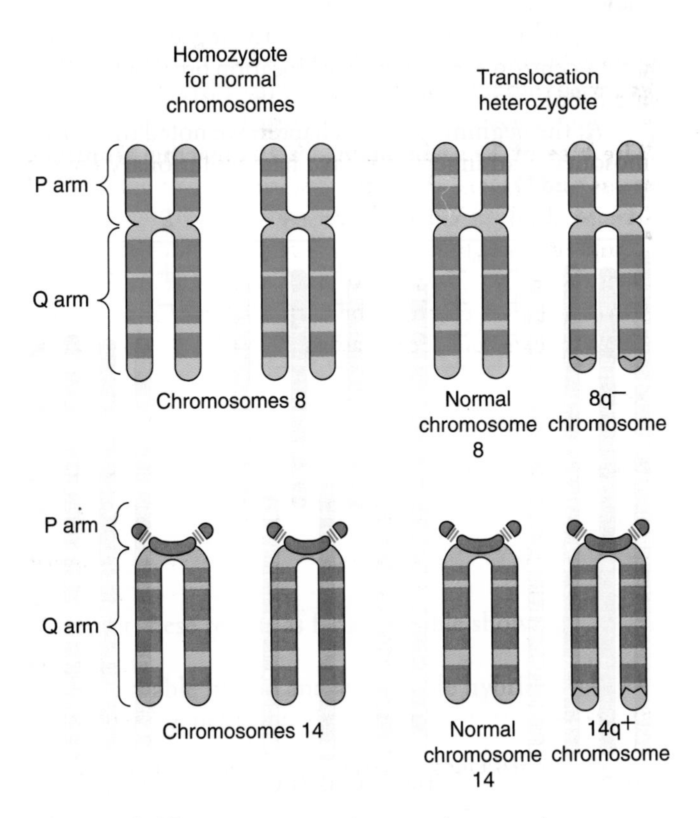

Figure 8-35 Reciprocal translocations between chromosomes 8 and 14 causes most cases of Burkitt's lymphoma. An oncogene on the tip of chromosome 8 becomes relocated next to an antibody gene enhancer region on chromosome 14.

b. You find that this fly has the recessive alleles *bw* (brown eye) and *e* (ebony body) on the nontranslocated chromosomes 2 and 3, respectively, and wild-type alleles on the translocated chromosomes. The fly is mated with a female that has normal chromosomes and is homozygous for *bw* and *e*. What type of offspring would you expect and in what ratio? (HINT: Remember that zygotes that have an extra chromosome arm or that are deficient for one chromosome do not survive. There is no crossing-over in *Drosophila* males.)

25. An *insertional* translocation consists of a piece from the center of one chromosome inserted into the middle of another (nonhomologous) chromosome. Thus

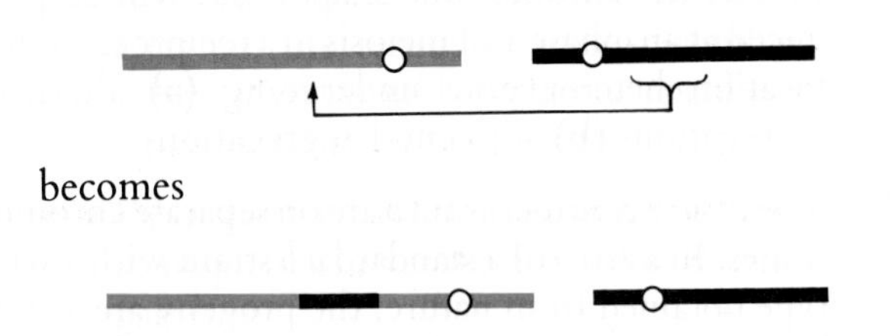

How will genomes that are heterozygous for such translocations pair at meiosis? What spore abortion patterns will be produced and in what relative proportions in a *Neurospora* cross between a wild type and an insertional translocation? (HINT: Remember that spores with chromosomal duplications survive and become dark, while those with deficiencies are light spored and inviable.)

26. Two auxotrophic mutations in *Neurospora*, *ad-3* and *pan-2*, are located on chromosomes 1 and 6, respectively. An unusual *ad-3* line arises in the laboratory, giving the following results:

	Ascospore appearance	RF between *ad-3* and *pan-2*
1. normal *ad-3* × normal *pan-2*	all black	50%
2. abnormal *ad-3* × normal *pan-2*	about $\frac{1}{2}$ black and $\frac{1}{2}$ white (inviable)	1%

3. Of the black spores from cross 2, about one-half were completely normal and one-half repeated the same behavior as the original abnormal *ad-3* strain.

Explain all three results with the aid of clearly labeled diagrams. (NOTE: In *Neurospora*, ascospores with extra chromosomal material survive and are the normal black color, whereas ascospores lacking any chromosome region are white and inviable.)

27. Corn plants that carry the allele *P* produce purple leaves, and plants homozygous for *p* produce green leaves. The *P* locus is 20 m.u. from the centromere on the long arm of chromosome 9. A homozygous *pp* plant is crossed to a *PP* plant homozygous for a reciprocal translocation that has one breakpoint 30 m.u. from the centromere on the long arm of chromosome 9. The F_1 is obtained and backcrossed to the *pp* parent. Predict what proportions of progeny from the backcross will be

a. green, semisterile

b. green, fully fertile

c. purple, semisterile

d. purple, fully fertile?

e. If the F_1 is selfed, what proportion of progeny will be green, fully fertile?

28. Chromosomally normal corn plants have a *p* locus on chromosome 1 and an *s* locus on chromosome 5.

P gives dark green leaves; *p*, pale green
S gives large ears; *s*, shrunken ears

An original plant of genotype *Pp Ss* has the expected phenotype (large ears, dark green) but gives unexpected results in crosses as follows:

- On selfing, fertility is normal, but the frequency of *pp ss* types is $\frac{1}{4}$ (not $\frac{1}{16}$, as expected).
- When crossed to a normal tester of genotype *pp ss*, the F_1 progeny are $\frac{1}{2}$ *Pp Ss* and $\frac{1}{2}$ *pp ss*; fertility is normal.
- When an F_1 *Pp Ss* plant is crossed to a normal *pp ss* tester, it proves to be semisterile, but again the progeny are $\frac{1}{2}$ *Pp Ss* and $\frac{1}{2}$ *pp ss*.

Explain these results, showing the full genotypes of the original plant, the tester, and the F_1 individuals. How would you test your hypothesis?

29. A corn plant of genotype *prpr* that has standard chromosomes is crossed with a *PrPr* plant that is homozygous for a reciprocal translocation between chromosomes 2 and 5. The F_1 is semisterile and phenotypically Pr (a seed color). A backcross to the parent with standard chromosomes gives 764 semisterile Pr; 145 semisterile pr; 186 normal Pr; and

727 normal pr. What is the map distance between the *Pr* locus and the translocation point?

30. A reciprocal translocation of the following types is obtained in *Neurospora:*

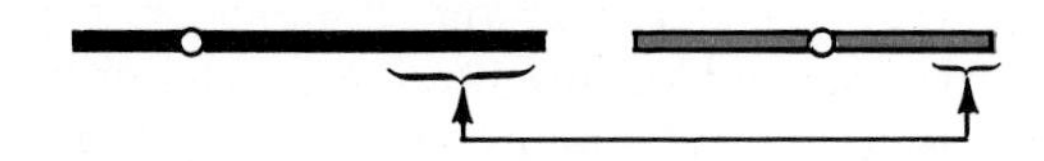

The following cross is then made:

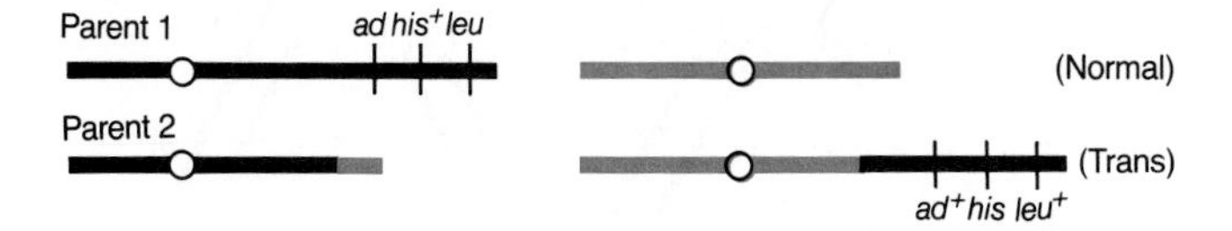

Assume that the small, colored piece of the chromosome involved in the translocation does not carry any essential genes. How would you select products of meiosis that are duplicated for the translocated part of the solid chromosome?

31. A male rat that is phenotypically normal shows reproductive anomalies when compared with normal male rats, as shown in Table 8-2. Propose a genetic explanation of these unusual results, and say how your idea could be tested.

32. Let's say you undertake a study of hybridized human and mouse cells because you want to map part of human chromosome 17. Three loci on this chromosome, *a*, *b*, and *c*, are concerned with making the compounds a, b, and c—all of which are essential for growth and present in mice as well as humans. You fuse mouse cells that are phenotypically $a^- b^- c^-$ with $a^+ b^+ c^+$ human cells. Assume that you find a hybrid in which the only human component is the right arm of chromosome 17 (17R), translocated by some unknown mechanism to a mouse chromosome. The hybrid can make the compounds a, b, and c. Then you treat cells with adenovirus, which causes chromosome breaks. Assume that you can isolate 200 lines in which bits of the translocated 17R have been clipped off. You test these lines for the ability to make a, b, and c, and obtain the following results:

Number	Can Make
0	a only
0	b only
12	c only
0	a and b only
80	b and c only
0	a and c only
60	a, b, and c
48	nothing

a. How would these different types arise?

b. Are *a*, *b*, and *c* all located on the right arm of chromosome 17? If so, draw a map indicating their relative positions.

c. How would banding patterns help you in your analysis?

33. During ascus formation in *Neurospora* any ascospore with a chromosomal deletion aborts and appears white, and any with a duplication survives as a dark spore. What patterns of ascospore abortion would you observe, and in what approximate proportions, in octads from *Neurospora* crosses between a strain with normal chromosomes and one with

a. a reciprocal translocation

b. a pericentric inversion

c. a paracentric inversion

Be sure to consider the effect of crossing-over in each one.

Table 8-2.

Mating	Embryos (mean no.): Implanted in uterine wall	Embryos (mean no.): Degenerating after implantation	Embryos (mean no.): Normal	Degeneration (%)
exceptional ♂ × normal ♀	8.7	5.0	3.7	57.5
normal ♂ × normal ♀	9.5	0.6	8.9	6.5

34. Suppose that you are studying the cytogenetics of five closely related species of *Drosophila*. The figure shown below gives the arrangement of loci (the letters indicate loci that are identical in all five species) and the chromosomal patterns that you find in each species. Show how these species probably evolved from each other, describing the changes occurring at each step. (NOTE: Be sure to compare gene order carefully.)

***35.** Show how the $\frac{10}{16}$ ratio on page 224 was derived.

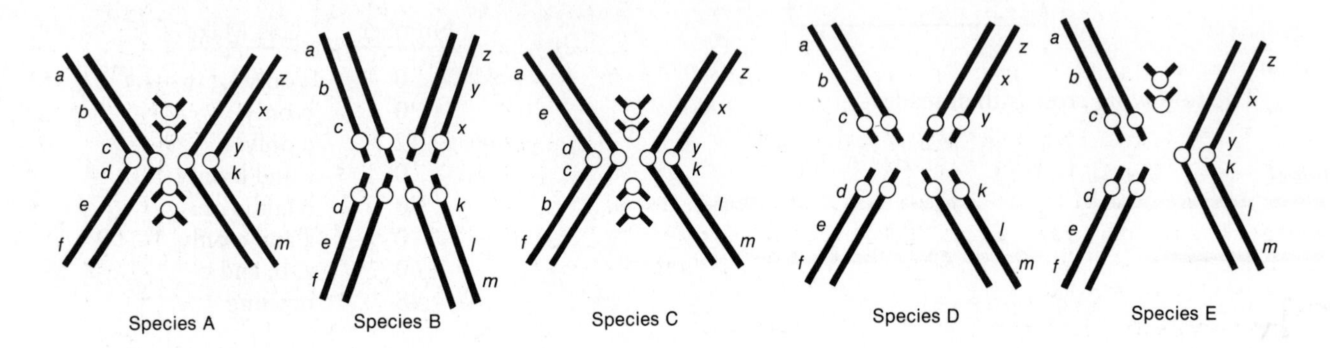

9

Chromosome Mutation II: Changes in Number

A geneticist examining an orchid with multiple chromosome sets (a polyploid). Compare the dimensions of the polyploid with those of its diploid parents shown in the foreground. (Runk/Schoenberger/Grant Heilman)

KEY CONCEPTS

Organisms with multiple chromosome sets (polyploids) are generally larger than diploid organisms, but meiotic pairing anomalies make some polyploid organisms sterile.

An even number of polyploid sets is generally more likely to result in fertility. Then the single-locus segregation ratios are different from those of diploids.

Crosses between two different species followed by the doubling of the chromosome number in the hybrid produces a special kind of fertile interspecific polyploid.

Variants in which a single chromosome has been gained or lost generally arise by nondisjunction (abnormal chromosome segregation at meiosis or mitosis).

Such variants tend to be sterile and show the abnormalities attributable to gene imbalance.

When fertile, such variants show abnormal gene segregation ratios for the misrepresented chromosome only.

The second major type of chromosome mutation is change in chromosome number. The topic of change in chromosome number might initially seem esoteric, and relevant only to geneticists. However, there are few aspects of genetics that impinge on human affairs quite so directly as this one. Chromosome numbers change spontaneously as accidents within cells, and this process has been going on as long as there has been life on the planet. In fact, changes in chromosome number have been instrumental in molding genomes during evolution. For examples of such changes, we have to look no farther than the food on our dining tables because many of the plants (and some of the animals) that we eat arose through spontaneous changes in chromosome number during the evolution of those species. Today breeders emulate this process by manipulating chromosome number to improve productivity or some other useful feature of the organism. But perhaps the main relevance of chromosome numbers focuses on members of our own species in that a large proportion of genetically determined ill health in humans is caused by abnormal chromosome numbers.

In this chapter we shall investigate the processes that produce new chromosome numbers, the diagnostic tests for detecting such changes, and the properties of cells and individuals carrying the different kinds of variant chromosomal complements. As with any area of cytogenetics, the techniques are a combination of genetics and microscopy. Changes in chromosome number are usually classified into two types, changes in whole chromosome sets and changes in parts of chromosome sets, and these two types are dealt with in the two following sections.

Euploidy

The number of chromosomes in a basic set is called the **monoploid number** (x). Organisms with multiples of the monoploid number of chromosomes are called **euploid.** Euploid types that have more than two sets of chromosomes are called **polyploid.** Thus, $1x$ is **monoploid,** $2x$ is **diploid,** and the polyploid types are $3x$ **(triploid),** $4x$ **(tetraploid),** $5x$ **(pentaploid),** $6x$ **(hexaploid),** and so on. The haploid number (n), which we have already used extensively, refers strictly to the number of chromosomes in gametes. In most animals and many plants that we are familiar with, the haploid number and monoploid number are the same. Hence, n or x (or $2n$ or $2x$) can be used interchangeably. However, in certain plants such as modern wheat, n and x are different. Wheat has 42 chromosomes, but careful study reveals that it is hexaploid, with six rather similar but not identical sets of seven chromosomes. Hence, $6x = 42$ and $x = 7$. However, the gametes of wheat contain 21 chromosomes, so $n = 21$ and $2n = 42$.

Monoploids

Male bees, wasps, and ants are monoploid. In the normal life cycles of these insects, males develop parthenogenetically—that is, they develop from unfertilized eggs. In most species, however, monoploid individuals are abnormal, arising in natural populations as rare aberrations. Throughout the chapter, we shall often focus on individuals having chromosome numbers abnormal for their species because this approach serves our interest in genetic analysis.

The germ cells of a monoploid cannot proceed through meiosis normally because the chromosomes have no pairing partners. Thus, monoploids are characteristically sterile. (Male bees, wasps, and ants bypass meiosis in forming gametes; here, mitosis produces the gametes.) If a monoploid cell does undergo meiosis, the single chromosomes segregate randomly, and the probability of all chromosomes going to one pole is $\frac{1}{2}^{x-1}$, where x is the number of chromosomes. This will determine the frequency of viable (whole-set) gametes, obviously a vanishingly small number if x is large.

Monoploids play a major role in modern approaches to plant breeding. Diploidy is an inherent nuisance when

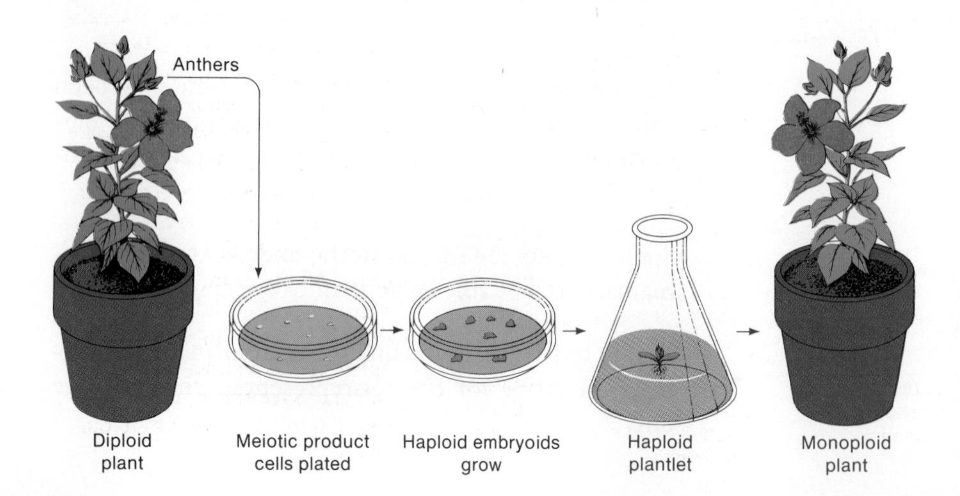

Figure 9-1 Generating monoploid plants by tissue culture. Pollen grains (haploid) are treated so that they will grow and placed on agar plates containing certain plant hormones. Under these conditions, haploid embryoids will grow into monoploid plantlets. After being moved to a medium containing different plant hormones, these plantlets will grow into mature monoploid plants with roots, stems, leaves, and flowers.

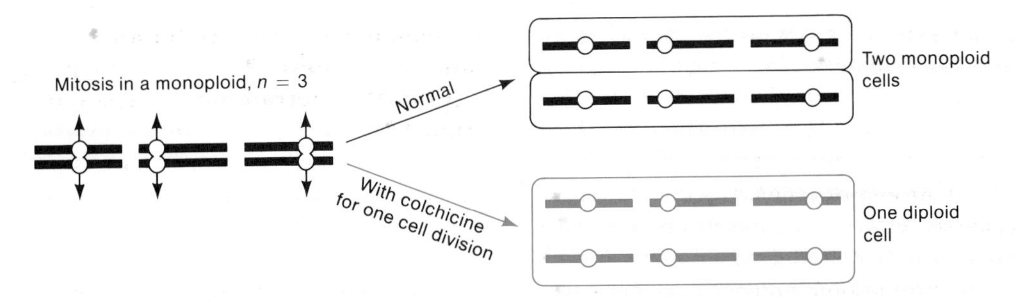

Figure 9-2 Using colchicine to generate a diploid from a monoploid. Colchicine added to mitotic cells during metaphase and anaphase disrupts spindle-fiber formation, preventing the separation of chromatids after the centeromere is split. A single cell is created that contains pairs of identical chromosomes that are homozygous at all loci.

breeders want to induce and select new gene mutations that are favorable and to find new combinations of alleles at different loci. New recessive mutations must be made homozygous before they can be expressed; favorable allelic combinations in heterozygotes are broken up by meiosis. Monoploids provide a way around some of these problems. In some plants, monoploids may be artificially derived from the products of meiosis in the plant's anthers. A cell destined to become a pollen grain may instead be induced by cold treatment to grow into an embryoid, a small dividing mass of cells. The embryoid may be grown on agar to form a monoploid plantlet, which can then be potted in soil and allowed to mature (Figure 9-1).

Plant monoploids may be exploited in several ways. In one, they are first examined for favorable traits or allelic combinations, which may arise from heterozygosity already present in the parent or induced in the parent by mutagens. The monoploid can then be subjected to chromosome doubling to achieve a completely homozygous diploid with a normal meiosis, capable of providing seed. How is this achieved? Quite simply, by the application of a compound called **colchicine** to meristematic tissue. Colchicine—an alkaloid drug extracted from the autumn crocus—inhibits the formation of the mitotic spindle, so that cells with two chromosome sets are produced (Figure 9-2). These cells may proliferate to form a sector of diploid tissue that can be identified cytologically.

Another way in which the monoploid may be used is to treat its cells basically like a population of haploid organisms in a mutagenesis-and-selection procedure. The cells are isolated, their walls are removed by enzymatic treatment, and they are treated with mutagen. They are then plated on a selective medium (perhaps a toxic compound normally produced by one of the plant's parasites or a herbicide) to select resistant cells. Resistant plantlets eventually grow into haploid plants, which can then be doubled (using colchicine) into a pure-breeding, diploid, resistant type (Figure 9-3).

These are powerful techniques that can circumvent the normally slow process of what is basically meiotic plant breeding. The techniques have been successfully applied to several important crop plants, such as soybeans and tobacco. This is, of course, another aspect of somatic-cell genetics in higher organisms.

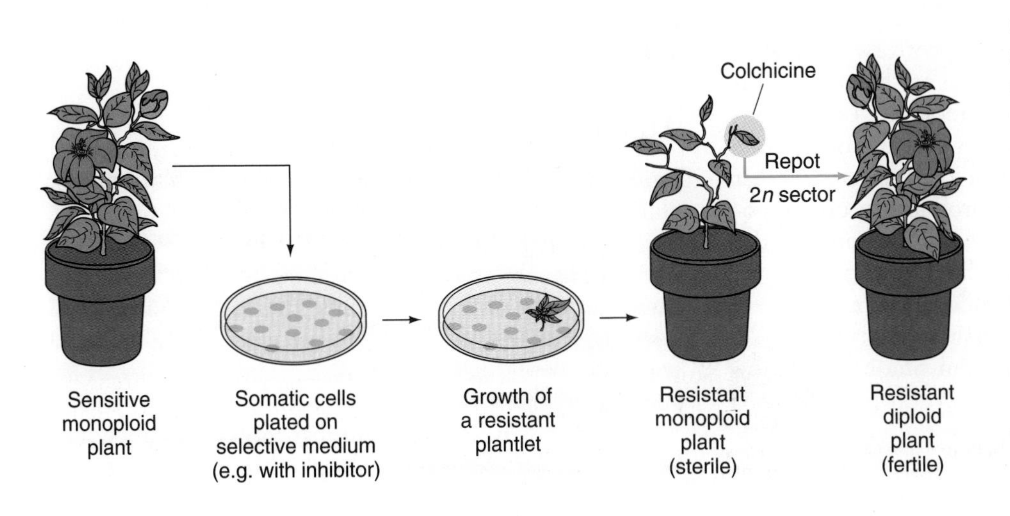

Figure 9-3 Using microbial techniques in plant engineering. The cell walls of haploid cells are removed enzymatically. The cells are then exposed to a mutagen and plated on an agar medium containing a selective agent, such as a toxic compound produced by a plant parasite. Only those cells containing a resistance mutation that allows them to live within the presence of this toxin will grow. After treatment with the appropriate plant hormones, these cells will grow into mature monoploid plants and, with proper colchicine treatment, can be converted into homozygous diploid plants.

The anther technique for producing monoploids does not work in all organisms or in all genotypes of an organism. Another useful technique has been developed in barley, an important crop plant. Diploid barley, *Hordeum vulgare,* may be fertilized by pollen from a diploid wild relative called *Hordeum bulbosum.* During the ensuing somatic cell divisions, however, the chromosomes of *H. bulbosum* are eliminated from the zygote while all the chromosomes of *H. vulgare* are retained, resulting in a haploid embryo. (The haploidization process appears to be caused by a genetic incompatibility between the chromosomes of the different species.) The chromosomes of the resulting haploids can be doubled with colchicine. This approach has led to the rapid production and widespread planting of several new barley varieties, and it is being used successfully in other species too.

Message To create new plant genotypes, geneticists produce monoploids and then double the chromosomes to form fertile, homozygous, diploid lines.

Polyploids

Once we are into the realm of polyploids, we must distinguish between **autopolyploids,** which are composed of multiple sets from within one species, and **allopolyploids,** which are composed of sets from different species. Allopolyploids form only between closely related species; however, the different chromosome sets are **homeologous** (only partially homologous)—not fully homologous, as they are in autopolyploids.

Triploids

Triploids are usually autopolyploids. They arise in nature or are constructed by geneticists from the cross of a $4x$ (tetraploid) and a $2x$ (diploid). The $2x$ and the x gametes unite to form a $3x$ triploid.

Triploids are characteristically sterile. The problem, like that of monoploids, involves pairing at meiosis. We see in Figure 9-4 that the net result of the pairing possibilities is an unbalanced segregation with two chromosomes going in one direction and one in the other. This happens for every chromosome threesome. Let us consider a hypothetical triploid species that has only three chromosomes in a set (most real organisms have far more). One-half of the meiotic products in our hypothetical species will have only one chromosome of the first set. This is also true for the other two sets when we consider each individually. The probability that a gamete will receive exactly one copy of all three sets is $\frac{1}{2} \times \frac{1}{2} \times \frac{1}{2}$, or $\frac{1}{8}$. The probability that a random union will bring together two such gametes to form a zygote is $\frac{1}{8} \times \frac{1}{8}$, or $\frac{1}{64}$. Notice that this zygote would have the chromosomal complement of the diploid ancestral species rather than

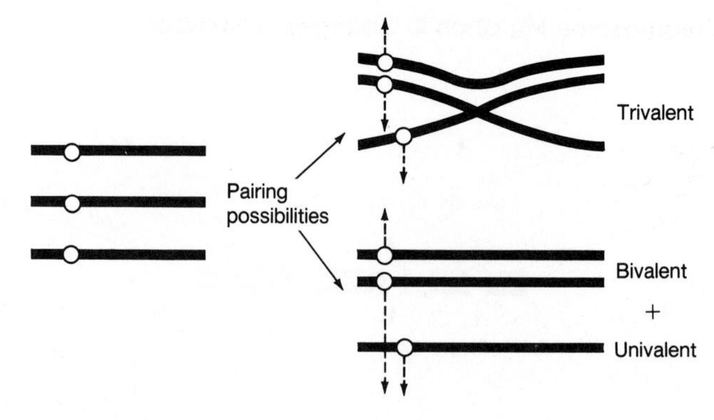

Figure 9-4 Two possibilities for the pairing of three homologous chromosomes before the first meiotic division in a triploid. Notice that the outcome will be the same in both cases: one resulting cell will receive two chromosomes and the other will receive just one. The probability that the latter cell can become a functional haploid gamete is very small, however, because to do so it would have also to receive only one of the three homologous chromosomes of every other set in the organism. Note that each chromosome is really a pair of chromatids.

being a new individual with the triploid complement. There is also one chance in 64 that a zygote will have the chromosomal complement of the tetraploid parent. All other zygotes would be chromosomally unbalanced and for this reason inviable or, as we will see later in the chapter, abnormal. Perhaps in this hypothetical case we would say the triploid species was of greatly reduced fertility, rather than calling it sterile.

Let us now extend our probabilistic reasoning to a chromosomally more realistic organism. The bananas that are widely available commercially are triploids with 11 chromosomes in each set. The probability that a gamete from such a triploid has exactly one chromosome from each set is $\frac{1}{2048}$ and there is only one chance in five million that two such gametes will unite to form a zygote. So saying that triploid bananas are sterile is not far off the mark!

The most obvious expression of the sterility of bananas is that there are no seeds in the fruit that we eat. Another example of the commercial exploitation of triploidy in plants is the production of triploid watermelons. For the same reasons presented in our discussion of bananas, triploid watermelons are seedless, a phenotype favored by some for its convenience.

Autotetraploids

Autotetraploids arise naturally, by the spontaneous accidental doubling of a $2x$ genome to a $4x$ genome, and also artificially, through the use of colchicine. Autotetraploid plants are advantageous as commercial crops because, as with other polyploids, the larger number of chromosome sets often leads to increased size. Cell size, fruit size, flower size, stomata size, and so on can all be larger in the polyploid (Figure 9-5).

Because 4 is an even number, autotetraploids can have a regular meiosis, although this is by no means

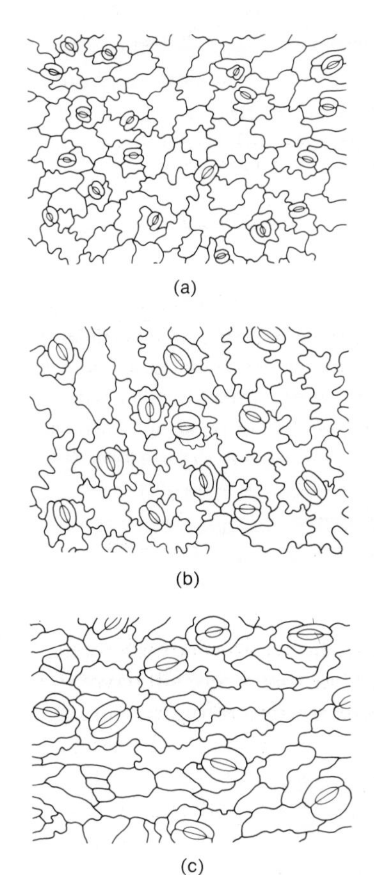

Figure 9-5 Epidermal leaf cells of tobacco plants, showing an increase in cell size, particularly evident in stomata size, with an increase in autopolyploidy. (a) Diploid, (b) tetraploid, (c) octoploid. (Parts a, b, c from W. Williams, *Genetic Principles and Plant Breeding,* Blackwell Scientific Publications, Ltd.)

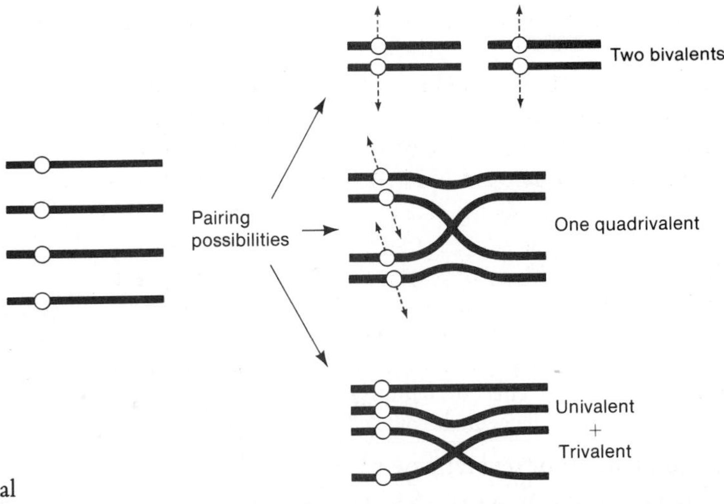

Figure 9-6 Meiotic pairing possibilities in tetraploids. (Each chromosome is really two chromatids.) The four homologous chromosomes may pair as two bivalents or as a quadrivalent. Both possibilities can yield functional gametes. However, the four chromosomes may also pair in a univalent–trivalent combination, yielding nonfunctional gametes. A specific tetraploid can show one or more of these pairings.

always the case. The crucial factor is how the four chromosomes of each set pair and segregate. There are several possibilities, as shown in Figure 9-6. Paired chromosomes of the type found in diploids are called **bivalents.** Pairings of three chromosomes are called **trivalents,** four are **quadrivalents,** and one unpaired chromosome is a **univalent.** In tetraploids, the two-bivalent and the quadrivalent pairing modes tend to be most regular in segregation, but even here there is no guarantee of a $2 \leftrightarrow 2$ segregation. If all chromosome sets segregate $2 \leftrightarrow 2$ as they do in some species, then the gametes will be functional and a formal genetic analysis can be developed for such autotetraploids.

Let's consider the genetics of a fertile tetraploid. We can consider an experiment in which colchicine is used to double the chromosomes of an *Aa* plant to form an *AAaa* autotetraploid, which we will assume shows $2 \leftrightarrow 2$ segregation. We now have a further concern because polyploids such as tetraploids give different segregation ratios in their progeny, depending on whether or not the locus in question is tightly linked to the centromere. First, we consider a centromeric gene. The three possible pairing and segregation patterns are presented in Figure 9-7; these occur by chance and with equal frequency. As the figure shows, the 2*x* gametes are *Aa*, *AA*, or *aa*, produced in a ratio of 8:2:2, or 4:1:1. If such a plant is selfed, the probability of an *aaaa* phenotype in

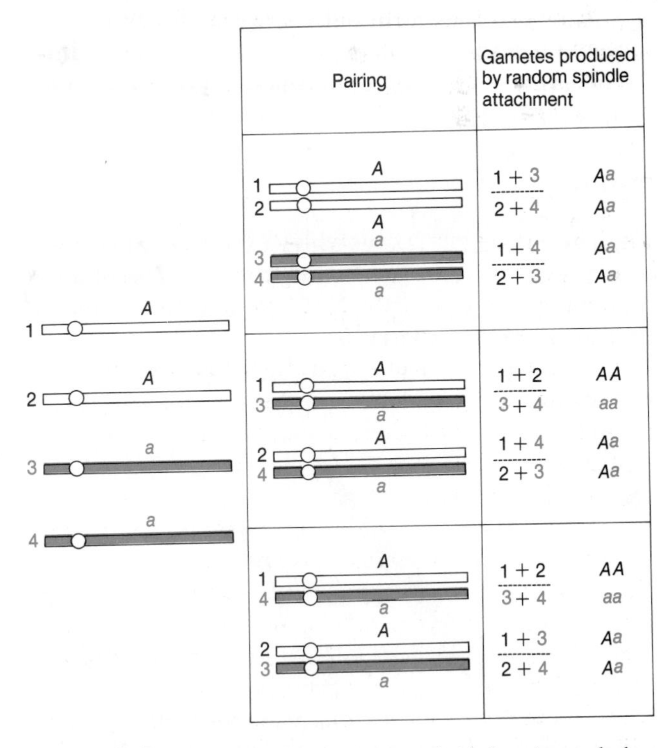

Figure 9-7 Gene segregation in a tetraploid showing orderly pairing by bivalents. (Each chromosome is really two chromatids.) The locus is assumed to be close to the centromere. Self-fertilization could yield a variety of genotypes, including *aaaa.*

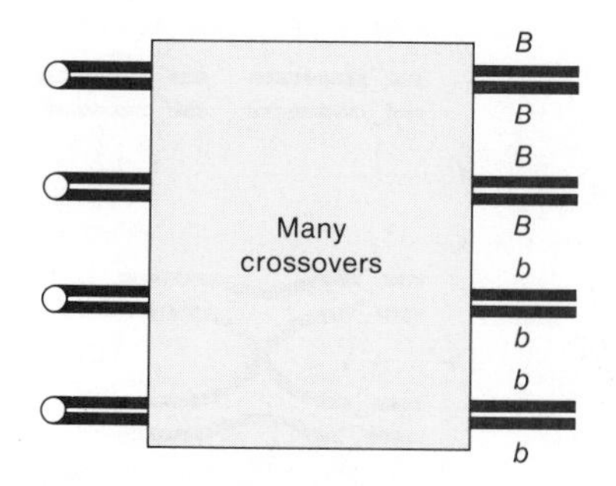

Figure 9-8 Highly diagrammatic representation of a tetraploid meiosis involving a heterozygous locus distant from the centromere. The net effect of multiple crossovers in such a long region will be that the genes become effectively unlinked from their original centromeres. Genes are packaged two at a time into gametes, much as two balls may be grabbed at random from a bag containing eight balls: four of one kind, and four of another.

the offspring is $\frac{1}{6} \times \frac{1}{6} = \frac{1}{36}$. In other words, a 35:1 phenotypic ratio will be observed if *A* is fully dominant over three *a* alleles.

If, in the same kind of plant, a genetic locus having the alleles *B* and *b* is very far removed from the centromere, crossing-over must be considered. This forces us to think in terms of chromatids instead of chromosomes; there are four *B* chromatids and four *b* chromatids (Figure 9-8). Because the number of crossovers in such a long region will be large, the genes will become effectively unlinked from their original centromeres. The packaging of genes two at a time into gametes is very much like grabbing two balls at random from a bag of eight balls: four of one kind, and four of another. The probability of picking two *b* genes is then

$$\frac{4}{8}\text{ (the first one)} \times \frac{3}{7}\text{ (the second one)} = \frac{12}{56} = \frac{3}{14}$$

So, in a selfing, the probability of a *bbbb* phenotype is $\frac{3}{14} \times \frac{3}{14} = \frac{9}{196} \cong \frac{1}{22}$. Hence, there will be a 21:1 phenotypic ratio of *B* – – –: *bbbb*. For genetic loci of intermediate position, intermediate ratios will, of course, result.

Allopolyploids

The "classic" allopolyploid was synthesized by G. Karpechenko in 1928. He wanted to make a fertile hybrid that would have the leaves of the cabbage *(Brassica)* and the roots of the radish *(Raphanus)*. Each of these species has 18 chromosomes, and they are related closely enough to allow intercrossing. A viable hybrid progeny individual was produced from seed. However, this hybrid was functionally sterile because the nine chromosomes from the cabbage parent were different enough from the radish chromosomes that pairs did not synapse and disjoin normally:

Radish $2n_1 = 18,\ n_1 = 9$ → $n_1 + n_2 = 18$ Sterile hybrid ← Cabbage $n_2 = 9,\ 2n_2 = 18$

However, one day a few seeds were in fact produced by this (almost!) sterile hybrid. On planting, these seeds produced fertile individuals with 36 chromosomes. All of these individuals were allopolyploids. They had apparently been derived from spontaneous, accidental chromosome doubling to $2n_1 + 2n_2$ in the sterile hybrid, presumably in tissue that eventually became germinal and underwent meiosis. Thus, in $2n_1 + 2n_2$ tissue, there is a pairing partner for each chromosome and balanced gametes of the type $n_1 + n_2$ are produced. These fuse to give $2n_1 + 2n_2$ allopolyploid progeny, which are also fertile. This kind of allopolyploid is sometimes called an **amphidiploid** (Figure 9-9). (Unfortunately for Karpe-

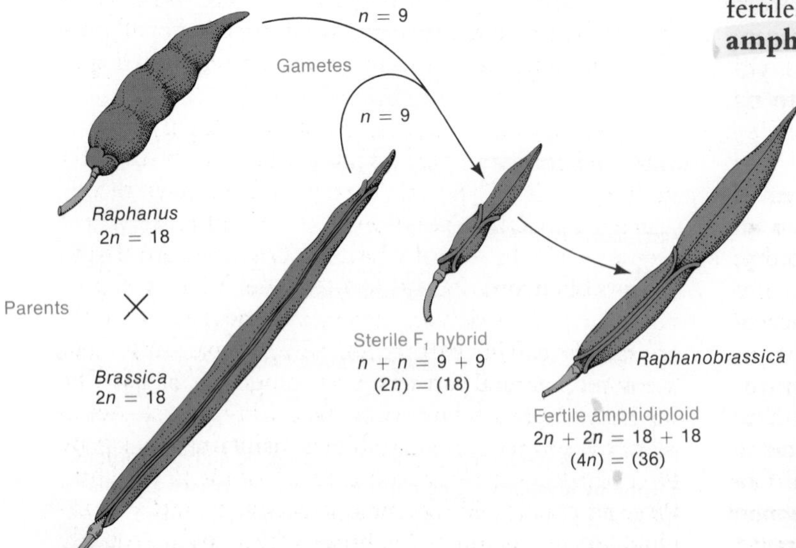

Figure 9-9 The origin of the amphidiploid *(Raphanobrassica)* formed from cabbage *(Brassica)* and radish *(Raphanus)*. The fertile amphidiploid arose in this case from spontaneous doubling in the $2n = 18$ sterile hybrid. Colchicine can be used to promote doubling. (From A. M. Srb, R. D. Owen, and R. S. Edgar, *General Genetics*, 2d ed. Copyright 1965 by W. H. Freeman and Company. After G. Karpechenko, *Z. Indukt. Abst. Vererb.* 48, 1928, 27.)

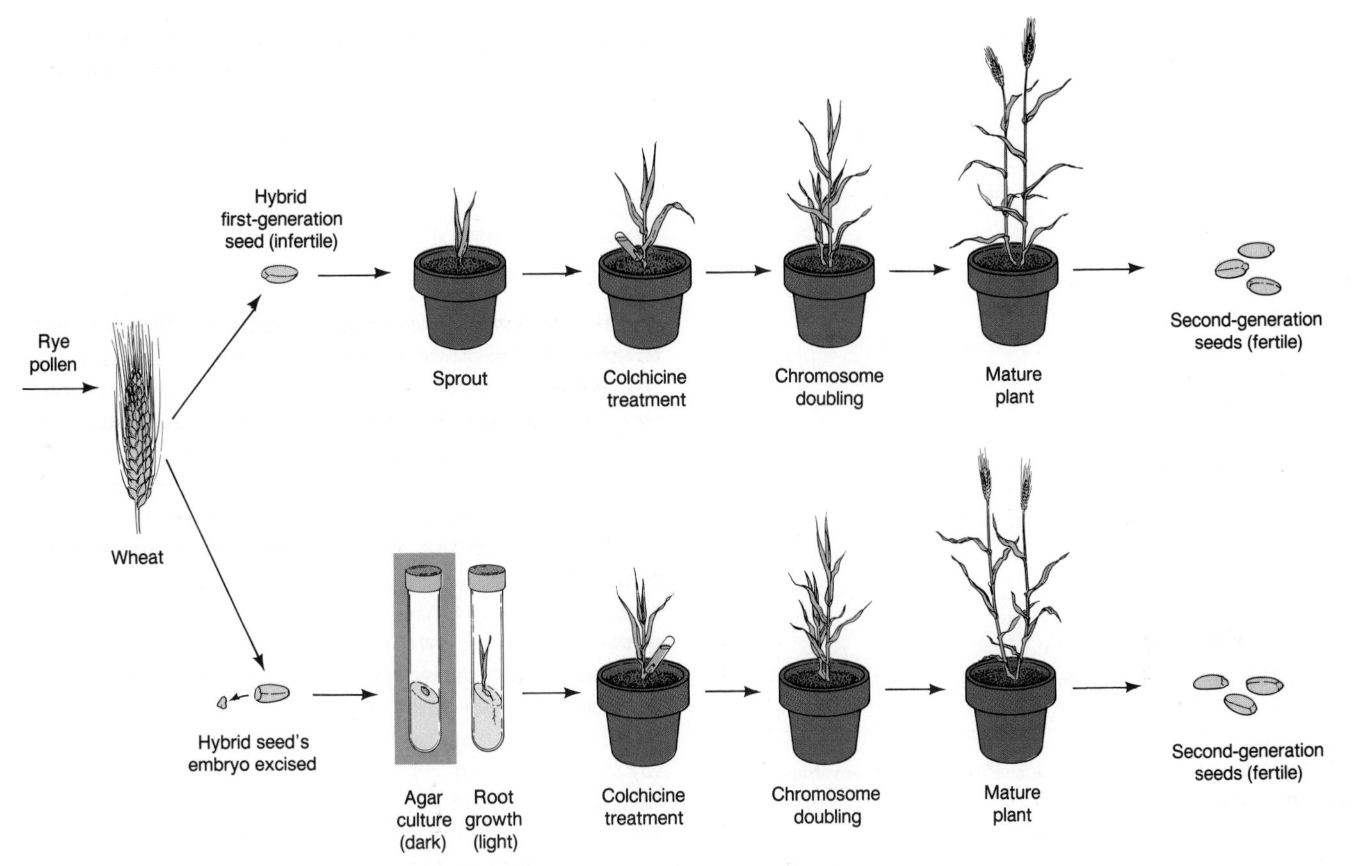

Figure 9-10 Techniques for the production of the amphidiploid *Triticale.* If the hybrid seed does not germinate, then tissue culture *(below)* may be used to obtain a hybrid plant. (From Joseph H. Hulse and David Spurgeon, "Triticale." Copyright 1974 by *Scientific American.*)

chenko, his amphidiploid had the roots of a cabbage and the leaves of a radish.)

When the allopolyploid was crossed to either parental species, sterile offspring resulted. The offspring of the cross to radish were $2n_1 + n_2$, constituted from an $n_1 + n_2$ gamete from the allopolyploid and an n_1 gamete from the radish. Obviously, the n_2 chromosomes had no pairing partners, so sterility resulted. Consequently, Karpechenko had effectively created a new species, with no possibility of gene exchange with its parents. He called his new species *Raphanobrassica.*

Today, allopolyploids are routinely synthesized in plant breeding. Instead of waiting for spontaneous doubling to occur in the sterile hybrid, colchicine is added to induce doubling. The goal of the breeder obviously is to combine some of the useful features of both parental species into one type. This kind of endeavor is very unpredictable, as Karpechenko learned. In fact, only one synthetic amphidiploid has ever been widely used. This is *Triticale,* an amphidiploid between wheat (*Triticum,* $2n = 6x = 42$) and rye (*Secale,* $2n = 2x = 14$). *Triticale* combines the high yields of wheat with the ruggedness of rye. A massive international *Triticale* testing program is now under way, and many breeders have great hopes for the future of this artificial amphidiploid. Figure 9-10 shows the procedure for synthesizing *Triticale.*

In nature, allopolyploidy seems to have been a major force in speciation of plants. There are many different examples. One particularly satisfying one is shown by the genus *Brassica,* as illustrated in Figure 9-11. Here three different parent species have hybridized in all possible pair combinations to form new amphidiploid species.

A particularly interesting natural allopolyploid is bread wheat, *Triticum aestivum* ($2n = 6x = 42$). By studying various wild relatives, geneticists have reconstructed a probable evolutionary history of bread wheat (Figure 9-12). In a bread wheat meiosis, there are always 21 pairs of chromosomes. Furthermore, it has been possible to establish that any given chromosome has only one specific pairing partner (homologous pairing)—not five other potential partners (homeologous pairing). The suppression of such homeologous pairing (which would make the species more unstable) is maintained by a gene *Ph* on the long arm of chromosome 5 of the B set. Thus, *Ph* ensures a diploid-like meiotic behavior for this hexaploid species. Without *Ph,* bread wheat could probably never have arisen. It is interesting to speculate as to

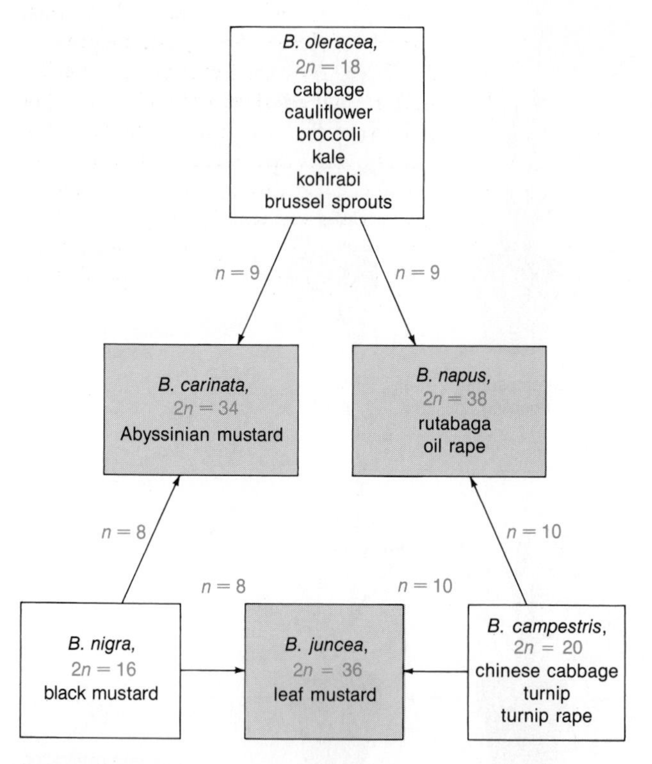

Figure 9-11 A species triangle, showing how amphidiploidy has been important in the production of new species of *Brassica.*

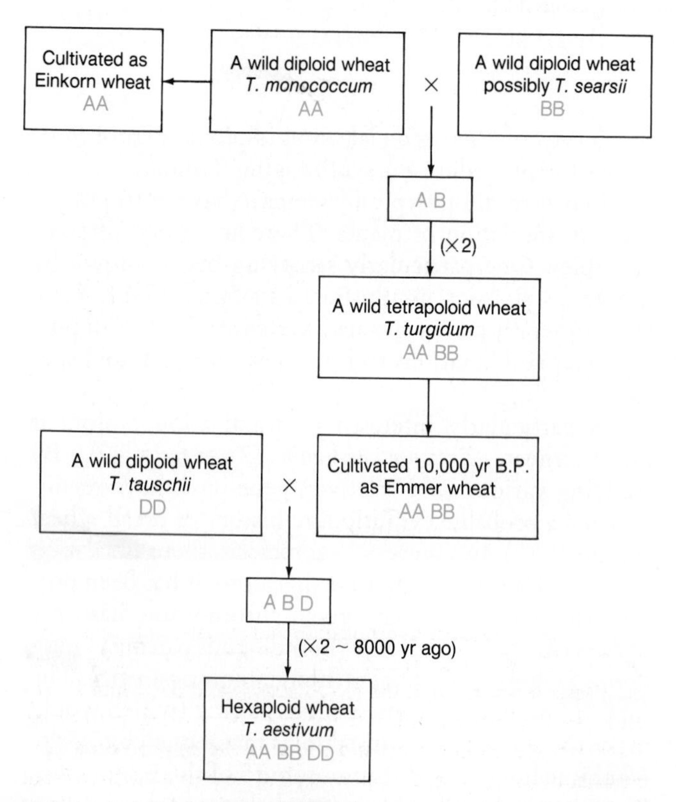

Figure 9-12 Diagram of the proposed evolution of modern hexaploid wheat involving amphidiploid production at two points. A, B, and D are different chromosome sets.

whether Western civilization could have begun or progressed without this species—in other words, without *Ph.*

Somatic Allopolyploids from Cell Hybridization

Another innovative approach to plant breeding is to try to make allopolyploid-like hybrids by asexual cell-fusion methods. Theoretically, such a technique would permit the combination of widely differing parental species. The technique does indeed work, but the only allopolyploids that have been produced so far can also be made by the sexual methods we have considered already. In the cell-fusion procedure, cell suspensions of the two parental species are prepared and stripped of their cell walls by special enzyme treatments. The stripped cells are called **protoplasts.** The two protoplast suspensions are then combined with polyethylene glycol, which enhances protoplast fusion. The parental cells and the fused cells proliferate on an agar medium to form colonies (in much the same way as microbes). If these colonies, or calluses as they are called, are examined, a fair percentage of them are found to be allopolyploid-like hybrids with chromosome numbers equal to the sum of the parental numbers. Thus, not only do the protoplast cell membranes fuse to form a kind of heterokaryon, but the nuclei fuse too.

Another good example of an allopolyploid-like hybrid is commercial tobacco, *Nicotiana tabacum,* which has 48 chromosomes. This species of tobacco was originally found in nature as a spontaneous amphidiploid. The two probable parents are *N. sylvestris* and *N. tomentosiformis,* each of which has 24 chromosomes. A sexual cross between *N. tabacum* and either of these two probable parents gives a 36-chromosome hybrid containing 12 chromosome pairs plus 12 unpaired chromosomes. A cross between *N. sylvestris* and *N. tomentosiformis* yields a 24-chromosome hybrid in which there is no pairing at all. Hence, it appears that part of the *N. tabacum* genome is from *N. sylvestris* and part is from *N. tomentosiformis.* This amphidiploid can be re-created either sexually, by using colchicine as described previously, or somatically by cell fusion. When cells of the prospective parental species are fused, a 48-chromosome hybrid cell line is produced from which plants identical in behavior to *N. tabacum* may be grown. (Note that in the latter method, colchicine is not required because the fusion product is already amphidiploid.)

The recovery of somatic hybrids may be enhanced if a selective system is available. In one example in *N. tabacum,* two monoploid lines were fused to form a diploid hybrid culture using complementation as the selection system. The first line had whitish, light-sensitive leaves due to a recessive mutation *w.* The other had yellowish,

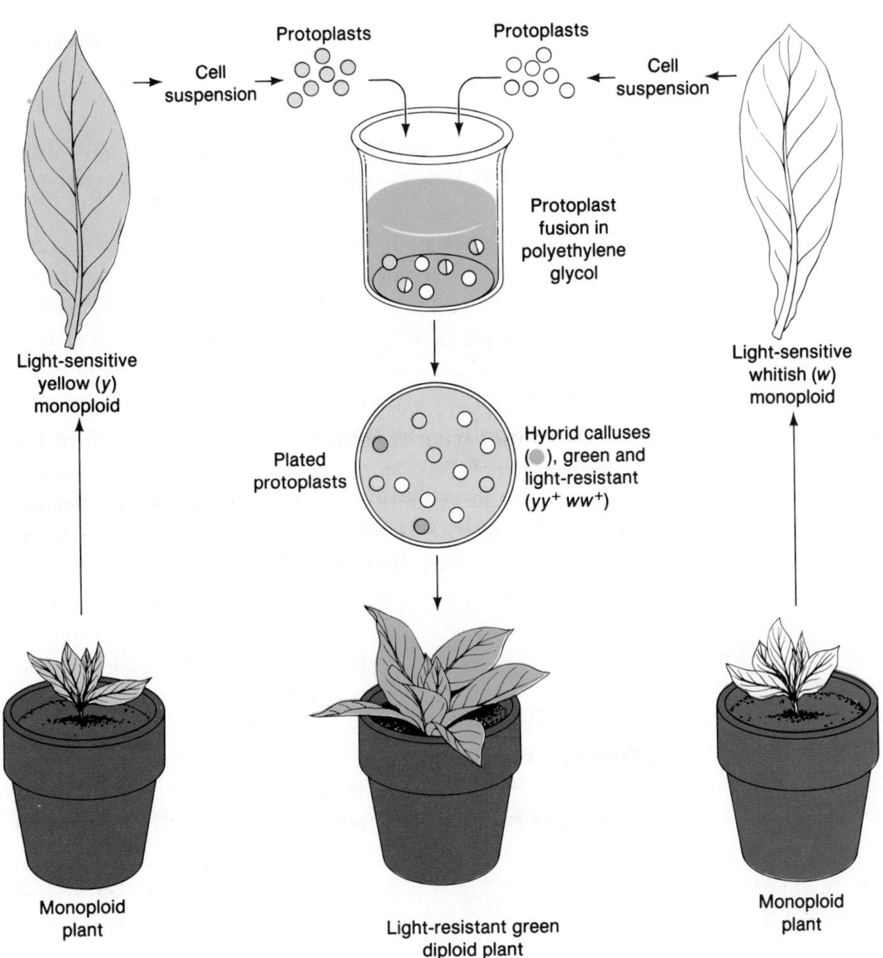

Figure 9-13 Creating a hybrid of two monoploid lines of *Nicotiana tabacum* by cell fusion. One line has light-sensitive yellowish leaves, and the other has light-sensitive whitish leaves. Protoplasts are produced by enzymatically stripping the cell walls from the leaf cells of each strain. The protoplasts can fuse, as indicated; those that fuse as hybrids can be grown into calluses that are light resistant as a result of recessiveness of the parental genotypes. The calluses, under the appropriate hormone regime, can be grown into green diploid plants.

light-sensitive leaves due to a mutation y at a separate locus. When the cells were combined in a petri dish the diploid $w^+w\ y^+y$ calluses could be selected by their resistance to light and their normal green color. The calluses can be grown into plantlets, which then are either grafted onto a mature plant to develop or potted themselves. The protocol for this experiment is illustrated in Figure 9-13.

Two allotetraploids of *Petunia,* one produced by sexual hybridization and the other by somatic hybridization, are compared in Figure 9-14. The figure shows that the two are identical in appearance and produce the same range of progeny types.

Message Allopolyploids can be synthesized by crossing related species and doubling the chromosomes of the hybrid or by asexually fusing the cells of different species.

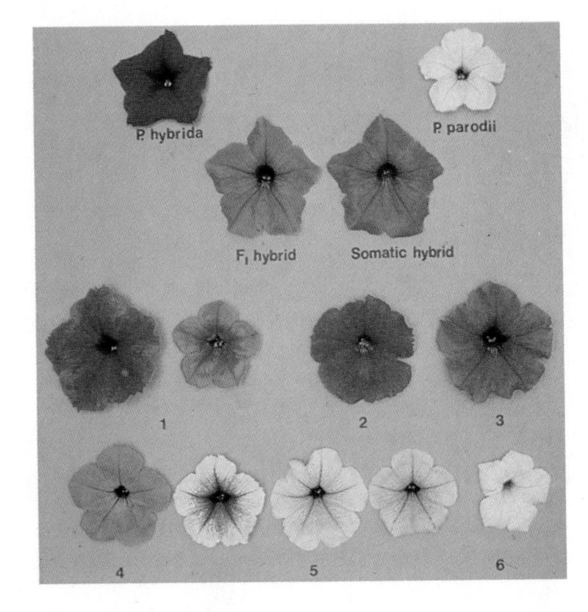

Figure 9-14 Sexual hybridization and somatic hybridization produce identical *Petunia* allotetraploids. The two parental lines are illustrated at the top, the red *P. hybrida* and the white *P. parodii.* The F_1 allotetraploid in the center row resulted from crossing the two diploid parents and doubling the chromosome number of the hybrid; the allotetraploid next to it resulted from fusion of somatic cells of the two parental lines. The two allotetraploids produce the same range of progeny, illustrated. The numbers identify plants used in further experiments. (J. B. Power)

Polyploidy in Animals

You may have noticed that most of the discussion of polyploidy so far has concerned plants. Indeed, polyploidy is more common in plants than in animals, but nevertheless there are many cases of polyploid animals. Examples are found in flatworms, leeches, and brine shrimps. In these cases reproduction is by **parthenogenesis,** the development of a special type of unfertilized egg into an embryo, without the need for fertilization. However, examples are not confined to these so-called lower forms. Polyploid amphibians and reptiles are surprisingly common. These show several modes of reproduction. Polyploid frogs and toads have males and females participating in their sexual cycles, whereas polyploid salamanders and lizards are parthenogenetic.

Some fish are also polyploid, and in two cases it appears that a single polyploid event has given rise to an entire taxonomic family in evolution. This situation contrasts with that in amphibians and reptiles because in those cases the polyploids all have closely related diploid species and hence the polyploid events do not seem to have been important in the evolution of the group as a whole. The Salmonidae family of fishes (containing salmon and trout) is a familiar example of a group that appears to have originated through polyploidy. Salmonids have twice as much DNA as related fish. Different salmonid species have different chromosome numbers, but it has been discovered that the group has an almost invariant number of chromosome arms (some fused in some species) that is twice the number of arms in related groups. Hence, the evidence points to the salmonids having evolved from a single event that gave rise to a tetraploid.

You might also be interested to know that the sterility of triploids has been commercially exploited in animals as well as plants. Triploid oysters have been developed and these have a commercial advantage over their diploid relatives. The diploids go through a spawning season when they are unpalatable, but triploids, because of their sterility, do not spawn and are palatable the whole year round.

Human polyploid zygotes do arise through various kinds of mistakes in cell division. Most die in utero. Occasionally, triploid babies are born, but none survive.

Aneuploidy

As noted at the beginning of the chapter, changes in parts of chromosome sets are the second category of chromosome mutations involving chromosome number. The number of one or more chromosomes may change dur-

Figure 9-15 The nullisomics of wheat. Although nullisomics are usually lethal in regular diploids, organisms like wheat, which "pretends" to be diploid but is hexaploid, can tolerate nullisomy. Nullisomics, however, are less vigorous growers. (E. R. Sears)

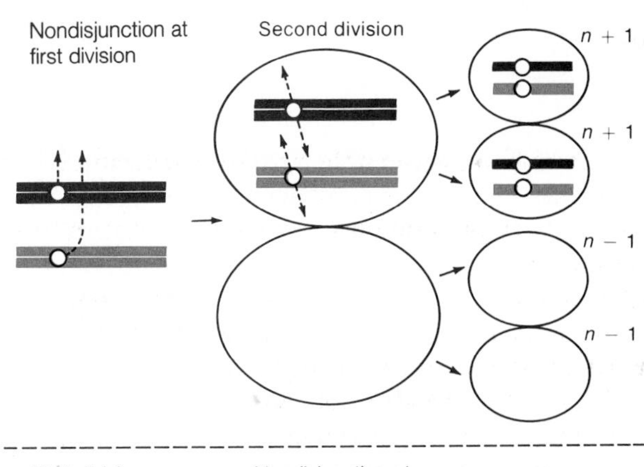

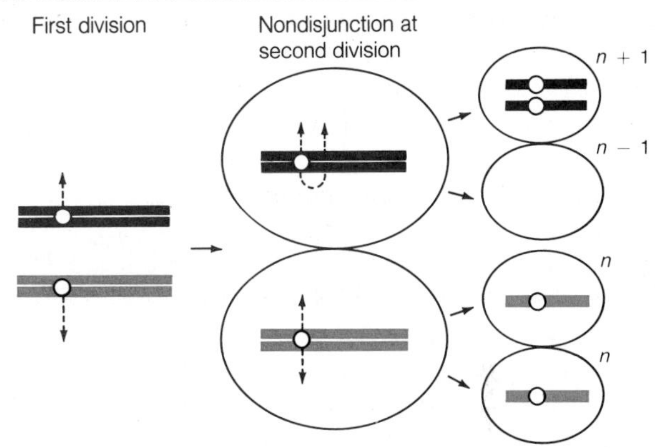

Figure 9-16 The origin of aneuploid gametes by nondisjunction at the first or second meiotic division.

ing the formation of a new individual. Such a change creates an individual called an **aneuploid,** which means "not euploid." Chromosomes may be added, so that the individual has one or more extra chromosomes, or they may be subtracted, so that the individual lacks chromosomes that are present in its ancestral genome. We indicate that one chromosome has been added or subtracted in a particular diploid individual by writing $2n + 1$ or $2n - 1$, where $2n$ is the number of chromosomes in a normal diploid of the species. A predominantly diploid organism that is $2n - 1$ is a **monosomic,** $2n + 1$ is a **trisomic,** $2n - 2$ is a **nullisomic** (two homologs have been lost), and $2n + 1 + 1$ is a **double trisomic.** An $n + 1$ aneuploid in a haploid species is called a **disomic.** It can be seen that these terms refer to the number of copies of particular chromosomes that are present in the aneuploid.

Nullisomics ($2n - 2$)

Although nullisomy is a lethal condition in diploids, an organism like bread wheat (which behaves meiotically like a diploid although it is a hexaploid) can tolerate nullisomy. The four homeologous chromosomes apparently compensate for a missing pair of homologs. In fact, all of the possible 21 bread wheat nullisomics have been produced; these are illustrated in Figure 9-15. Their appearances differ from the normal hexaploids; furthermore, most of the nullisomics grow less vigorously.

Monosomics ($2n - 1$)

Monosomic chromosome complements are generally deleterious for two main reasons. First, the missing chromosome grossly disturbs the overall balance of chromosomes, which has been carefully put together during evolution and which is necessary to fine tune cellular homeostasis. Second, having a chromosome missing causes any deleterious recessive allele on the single chromosome to be hemizygous and thus to be directly expressed phenotypically. Notice that these are the same effects produced by deletions.

Nondisjunction during mitosis or meiosis produces gametes that are the source of monosomics, trisomics ($2n + 1$), and other aneuploids. Disjunction is the normal separation of chromosomes or chromatids to opposite poles at nuclear division. **Nondisjunction** is a failure of this process and ends up with two going to one pole and none to the other. The chromosomes may fail to disjoin at either the first or second division (Figure 9-16). Either way, $n + 1$ and $n - 1$ gametes are produced. If an $n - 1$ gamete is fertilized by an n gamete, a monosomic ($2n - 1$) zygote is produced. An $n + 1$ and an n gamete yield a trisomic $2n + 1$; an $n + 1$ and an $n + 1$ produce a tetrasomic if the same chromosome is involved or a double trisomic if different chromosomes are involved, and so on.

In *Neurospora* (a haploid), an $n - 1$ meiotic product aborts and does not darken like a normal ascospore; so we may detect M_I and M_{II} nondisjunctions by observing asci with 4:4 and 6:2 ratios of normal to aborted spores, respectively, as shown below.

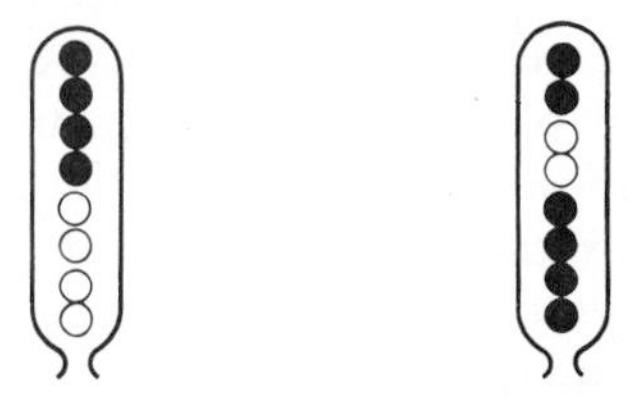

Diagram the chromosome content of the various spores to convince yourself of the relation of the spore pattern to nondisjunction. What ascus genotypes are produced for loci on the aneuploid chromosomes?

In humans, a sex-chromosome monosomic complement of 44 autosomes + 1X produces a phenotype known as Turner syndrome. Affected people have a characteristic, easily recognizable phenotype: they are sterile females, are short in stature, and often have a web of skin extending between the neck and shoulders (Figure 9-17). Although their intelligence is near normal, some of their specific cognitive functions are defective.

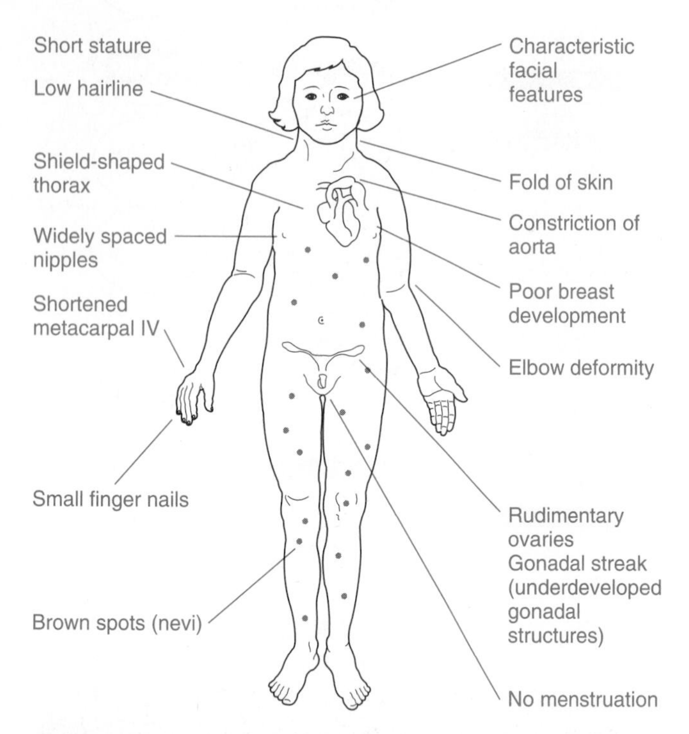

Figure 9-17 Characteristics of Turner syndrome, which results from having a single X chromosome (XO). (Adapted from F. Vogel and A. G. Motulsky, *Human Genetics*, Springer-Verlag, 1982.)

About 1 in 5000 female births have this monosomic chromosomal complement. Monosomics for all human autosomes die in utero.

Geneticists have used viable plant nullisomics and monosomics to identify the chromosomes that carry the loci of newly found recessive alleles. For example, a geneticist may obtain different monosomic lines, each of which lacks a different chromosome. Homozygotes for the new allele are crossed with each monosomic line, and the progeny of each cross are inspected for expression of the recessive phenotype. The phenotype appears in some of the progeny of the parent monosomic for the locus-bearing chromosome and thus identifies it. Figure 9-18 shows that this test works because meiosis in the monosomic parent produces some gametes that lack the chromosome bearing the locus. When one of these gametes forms a zygote, the single chromosome contributed by the other homozygous recessive parent determines the phenotype.

Genetic analysis of humans occasionally reveals a similar unmasking of a recessive phenotype by an $n - 1$ gamete. For example, two people whose vision is normal may produce a daughter who has Turner syndrome and who is also red-green colorblind. This shows that the allele for red-green colorblindness is recessive, that it is located on the X chromosome of the mother, and that it was in a meiosis in the father that chromosomes failed to disjoin. (Can you see why?)

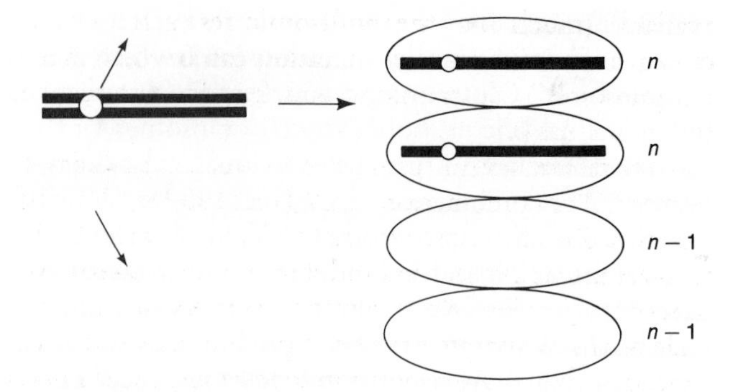

Figure 9-18 Meiosis in which the chromosome of interest is monosomic. Two of the resulting gametes are haploid (n); the other two gametes contain a set lacking a chromosome ($n - 1$).

Trisomics ($2n + 1$)

The trisomic condition is also one of chromosomal imbalance and can result in abnormality or death. However, there are many examples of viable trisomics. You might remember that we studied the trisomics of the Jimson weed *Datura* in Figure 3-8. Furthermore, trisomics can be fertile. When cells from some trisomic organisms are observed under the microscope at the time of meiotic chromosome pairing, the trisomic chromosomes are seen to form a trivalent, a paired group of three, while the other chromosomes form regular bivalents. What genetic ratios might we expect for genes on the trisomic chromosome? Let us consider a gene A that is close to the centromere on that chromosome, and let us assume that the genotype is $A a a$. Furthermore, if we postulate that two of the centromeres disjoin to opposite poles as in a normal bivalent, and that the other chromosome passes randomly to either pole, then we can predict the three equally frequent segregations shown in Figure 9-19. These segregations result in an overall gametic ratio of $1A:2Aa:2a:1aa$. This ratio and the one corresponding to AAa are observed in practice. If a trisomic tester set is

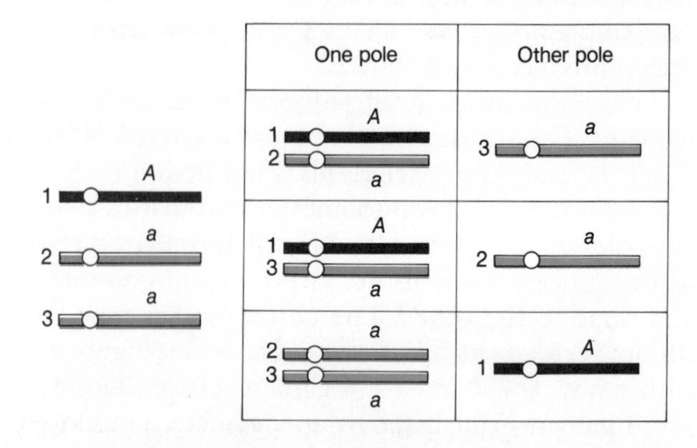

Figure 9-19 Genotypes of the meiotic products of an Aaa trisomic. Three segregations are equally likely.

available (much like the nullisomic tester set we discussed earlier) then a new mutation can be located to a chromosome by determining which of the testers gives the special ratio.

There are several examples of viable trisomics in humans. The combination XXY (1 in 1000 male births) results in Klinefelter syndrome, males with lanky builds who are mentally retarded and sterile (Figure 9-20). Another combination, XYY, also occurs in about 1 in 1000 male births. Attempts have been made to link the XYY condition with a predisposition toward violence. This is still hotly debated, although it is now clear that an XYY condition in no way guarantees such behavior. Nevertheless, several enterprising lawyers have attempted to use the XYY genotype as grounds for acquittal or compassion in crimes of violence. The XYY males are usually fertile. Their meioses are of the XY type; the extra Y is not transmitted, and their gametes contain either X or Y, never YY or XY.

The most common type of viable human aneuploid is Down syndrome (Figure 9-21), occurring at a frequency of about 0.15 percent of all live births. We have already encountered the translocation form of Down syndrome in Chapter 8. However, by far the most com-

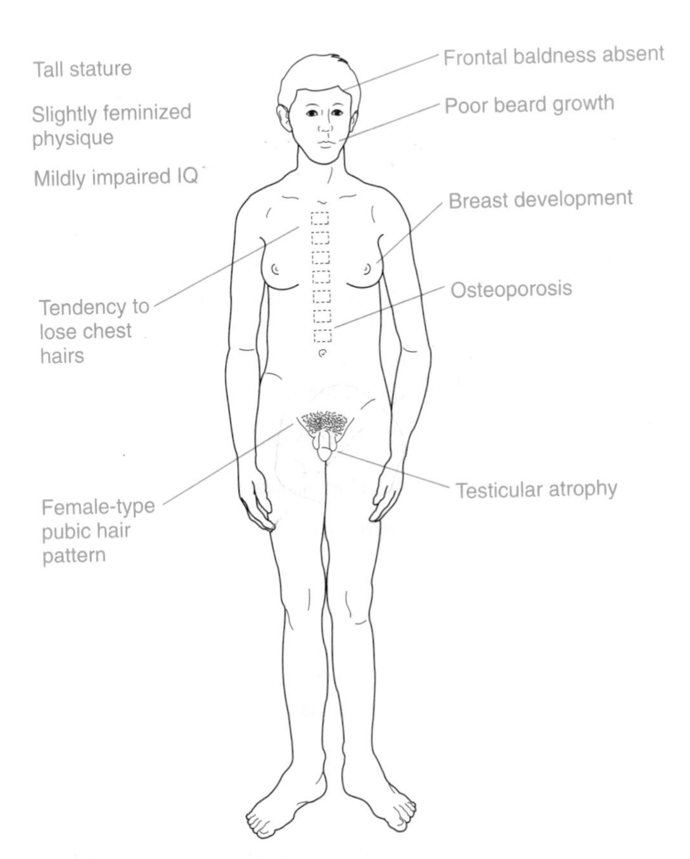

Figure 9-20 Characteristics of Klinefelter syndrome (XXY). (Adapted from F. Vogel and A. G. Motulsky, *Human Genetics*, Springer-Verlag, 1982.)

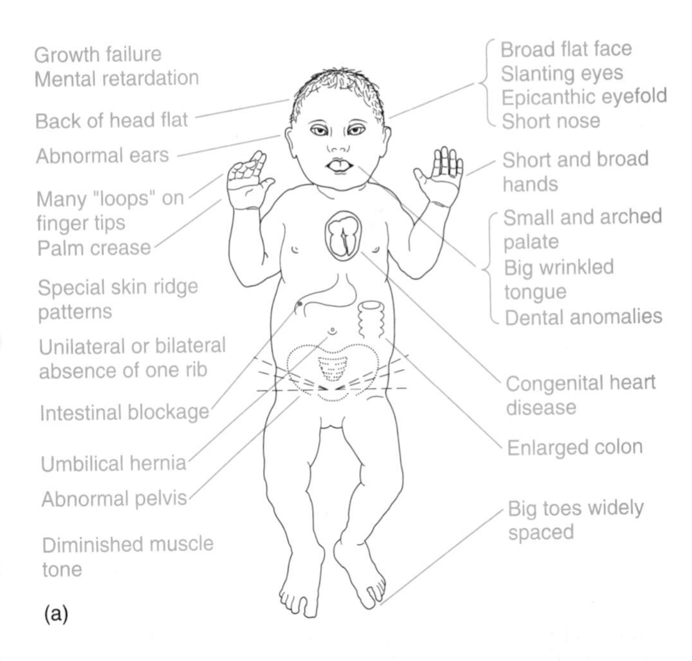

Figure 9-21 Characteristics of Down syndrome (trisomy 21). (a) Diagrammatic representation of the syndrome in an infant. (b) Athletes with Down syndrome. (Part a adapted from F. Vogel and A. G. Motulsky, *Human Genetics*, Springer-Verlag, 1982; part b from Bob Daemmrich/The Image Works.)

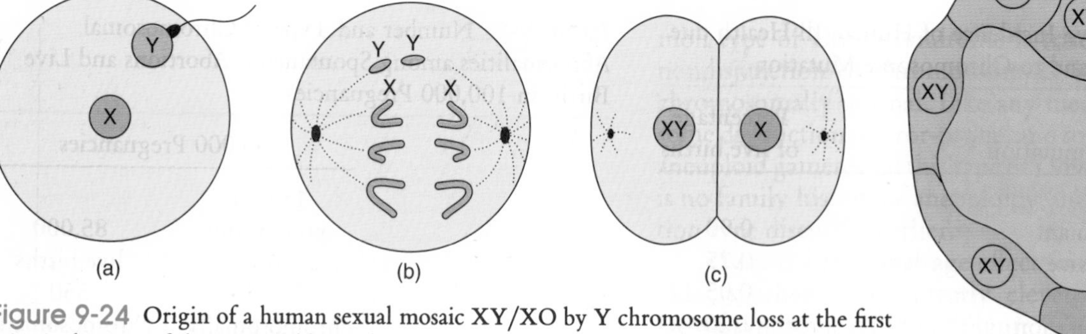

Figure 9-24 Origin of a human sexual mosaic XY/XO by Y chromosome loss at the first mitotic division of the zygote. (a) Fertilization. (b) Chromosome loss. (c) Resulting male and female cells. (d) Mosaic blastocyst. (Adapted from C. Stern, *Principles of Human Genetics,* 3d ed. W. H. Freeman and Company, 1973.)

dentally, there is no evidence that would suggest these aberrations are produced by environmental insult to our reproductive systems, or that the frequency of the aberrations is increasing.

Somatic Aneuploids

Aneuploid cells can arise spontaneously in somatic tissue or in cell culture. In such cases, the initial result is a genetic mosaic of cell types. Good examples are provided by certain conditions in humans.

Sexual mosaics—people whose bodies are a mixture of male and female tissue—provide the first example. One type of sexual mosaic, XO/XYY, can be explained by postulating an XY zygote in which the Y chromosomes fail to disjoin at an early mitotic division, so that both go to one pole:

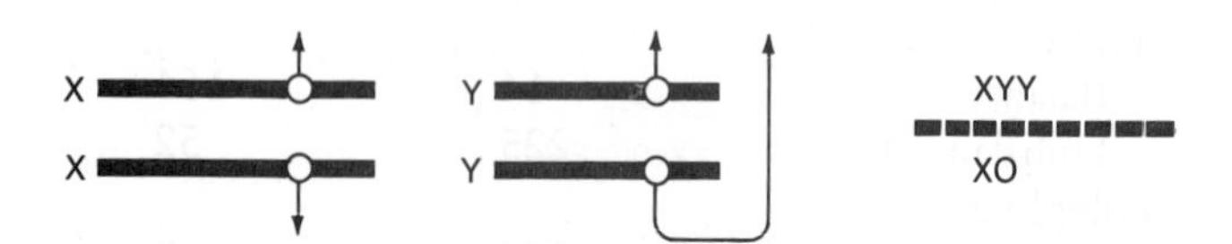

The phenotypic sex of such individuals depends on where the male and female sectors end up in the body. In this case, nondisjunction at a later mitotic division produces a three-way mosaic XY/XO/XYY, which contains a clone of normal male cells. Other sexual mosaics have different explanations; as examples, XO/XY is probably due to chromosome loss in a male zygote (Figure 9-24) and XX/XY is probably the result of a double fertilization (fused twins). In general, sexual mosaics are called gynandromorphs. Geneticists working with many species of experimental animals occasionally find gynandromorphs among their stocks. A classic example is the *Drosophila* shown in Figure 9-25. In this case the zygote started out as a female with the genotype $w^+ \ m^+ \ / \ w \ m$ for two X-linked genes. Loss of the wild-type X chromosome at the first mitotic division resulted in the two cell lines and ultimately in a fly differing from one side to the other in sex, eye color, and size of wing. A similar gynandromorph in the Io moth is shown in Figure 9-26.

Somatic aneuploidy and its resulting mosaics are often observed in association with cancer. People suffering from chronic myeloid leukemia (CML), a cancer of the white blood cells, frequently harbor cells containing the so-called Philadelphia chromosome. This chromosome was once thought to represent an aneuploid condition, but it is now known to be a translocation product in which part of the long arm of chromosome 22 attaches to the long arm of chromosome 9. However, CML patients often show aneuploidy in addition to the Philadelphia chromosome. In one study of 67 people with CML, 33 proved to have an extra Philadelphia chromosome and the remainder had various aneuploidies; the most com-

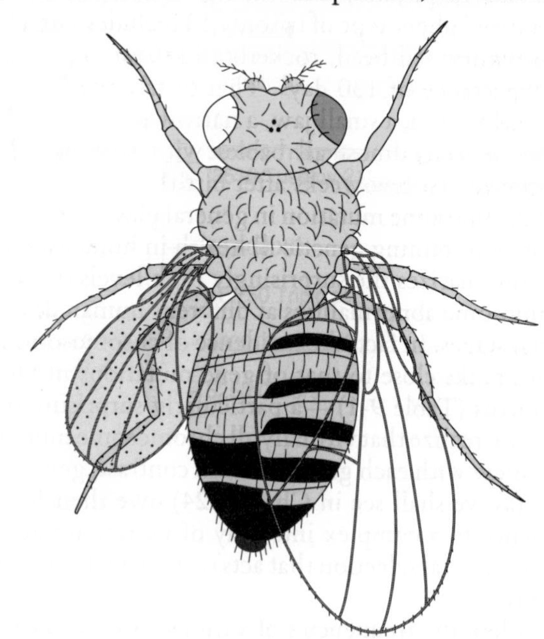

Figure 9-25 A bilateral gynandromorph of *Drosophila.* The zygote was $w^+ \ m^+ \ / \ w \ m$ but loss of the $w^+ \ m^+$ chromosome in the first mitotic division produced a fly that was $\frac{1}{2}$ O / $w \ m$ and male (left) and $\frac{1}{2}$ $w^+ \ m^+ \ / \ w \ m$ and female (right). Mutant allele w gives white eye, and m gives miniature wing.

Figure 9-26 A bilateral gynandromorph in the Io moth, *Automeris io io.* One half of the body is female and happens to carry the sex chromosome mutation "broken eye"; the other half of the body is male and carries the normal allele of the broken-eye gene. This moth resulted from an event similar to that which produced the *Drosophila* gynandromorph in Figure 9-25. The color and size differences are both sex differences. (T. R. Manley)

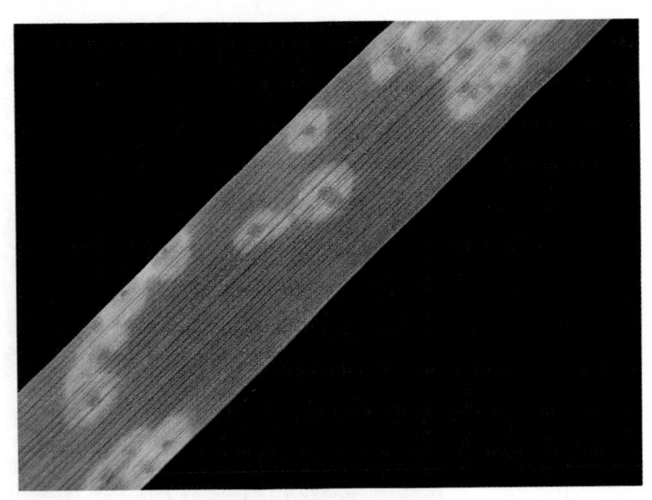

Figure 9-27 A wheat leaf infected by rust fungus, which derives its name from the rust-colored patches of fungal spores produced at the infection centers. Various kinds of rust fungi are pathogens on many important crops species including cereals, pines, and coffee. Rust infection results in billions of dollars of damage yearly, and a large part of plant breeding is directed at producing genetically resistant plant lines. (V. A. Wilmot/Visuals Unlimited)

mon aneuploidy was trisomy for the long arm of chromosome 17, which was detected in 28 people. Of 58 people with acute myeloid leukemia, 21 were shown to have aneuploidy for chromosome 8; 16 for chromosome 9; and 10 for chromosome 21. In another study of 15 patients with intestinal tumors, 12 had cells with abnormal chromosomes, at least some with trisomy for chromosome 8, 13, 15, 17, or 21. Of course, such studies merely established correlations, and it is not clear whether the abnormalities are best thought of as a cause or as an effect of cancer.

Message Aneuploids are produced by nondisjunction or some other type of chromosome misdivision at either meiosis or mitosis.

Chromosome Mechanics in Plant Breeding

Some of the material covered in the present chapter may seem somewhat esoteric. The purpose of this section, then, is to provide convincing evidence that the details covered here are of immense significance in the genetic engineering that is so necessary to produce and maintain new crop types in our hungry world. The sole experiment to be described, performed by E. R. Sears in the 1950s, concerns the transfer of a gene for leaf-rust resistance from a wild grass, *Aegilops umbellulata,* to bread wheat, which is highly susceptible to this disease, to offset a major problem in the wheat industry (Figure 9-27). This is a classic experiment of its kind.

The first problem that Sears encountered was that these two species (Figure 9-28) are not interfertile, so the feat of gene transfer seemed impossible. Sears sidestepped this problem with a **bridging cross,** in which he crossed *A. umbellulata* to a wild relative of bread wheat called emmer, *Triticum dicoccoides.* (Follow the process in Figure 9-29). *A. umbellulata* is a diploid, $2n = 2x = 14$. We shall call its chromosome sets CC. *T. dicoccoides* is a tetraploid, $2n = 4x = 28$, with sets AA BB. From this cross, the resulting sterile hybrid ABC was doubled into a fertile amphidiploid AA BB CC with 42 chromosomes. This amphidiploid was fertile in crosses with bread wheat *(T. aestivum),* which is represented as $2n = 6x = 42$, AA BB DD. The offspring, AA BB CD, were almost completely sterile due to pairing irregularities between the C and D sets. However, backcrosses to wheat did produce a few rare seeds, some of which grew into plants that were resistant to leaf rust, like the wild parent in the first cross. Some of these were almost the desired type, having 43 chromosomes (42 of which were wheat, plus one *Aegilops* chromosome bearing the resistance gene). Thus, in the AA BB CD hybrid, some aberrant form of chromosome assortment had produced a gamete with 22 chromosomes: ABD plus one from the C group.

Thus far, we have dealt almost exclusively with genes that are packed into chromosomes and enclosed within the nuclei of eukaryotic organisms. However, a very large part of the history of genetics and current genetic analysis (particularly molecular genetics) is concerned with prokaryotic organisms, which have no distinct nuclei, and with viruses. Although **viruses** share some of the definitive properties of organisms, many biologists regard viruses as distinct entities that in some sense are not fully alive. They are not cells; they cannot grow or multiply alone. To reproduce, they must parasitize living cells and use their metabolic machinery. Nevertheless, viruses do have hereditary properties that can be subjected to genetic analysis. Genetic analysis of bacteria and their viruses has yielded key insights into the nature and structure of the genetic material, the genetic code, and mutation.

The **prokaryotes** are the blue-green algae, now classified as "cyanobacteria," and the bacteria. The best studied **viruses**—those that parasitize bacteria—are called **bacteriophages** or, simply, **phages.** Pioneering work with bacteriophages has led to a great deal of recent research on tumor-causing viruses and other kinds of animal and plant viruses.

Compared to eukaryotes, prokaryotic organisms and viruses have simple chromosomes that are not contained within a nuclear membrane. Because they are monoploid, these chromosomes do not undergo meiosis, but they do go through stages analogous to meiosis. The approach to the genetic analysis of recombination in these organisms is surprisingly similar to that for eukaryotes.

The opportunity for genetic recombination in bacteria can arise in several different ways, as this chapter will detail. In the first process we'll examine, **conjugation,** one bacterial cell transfers DNA segments to another cell via direct cell-to-cell contact. A bacterial cell can also pick up a piece of DNA from the environment and incorporate this DNA into its own chromosome; this procedure is called **transformation.** In addition, certain bacterial viruses can pick up a piece of DNA from one bacterial cell and inject it into another, where it can be incorporated into the chromosome, in a process known as **transduction.**

Working with Microorganisms

Bacteria can be grown in a liquid medium or on a solid surface, such as an agar gel, as long as basic nutritive ingredients are supplied. In a liquid medium, the bacteria divide by binary fission: they multiply geometrically until the nutrients are exhausted or until toxic factors (waste products) accumulate to levels that halt the population growth. A small amount of such a liquid culture can be pipetted onto a petri plate containing an agar medium and spread evenly on the surface with a sterile spreader, in a process called **plating** (Figure 10-1). Each cell then reproduces by fission. Because the cells are immobilized in the gel, all of the daughter cells remain together in a clump. When this mass reaches more than 10^7 cells, it becomes visible to the naked eye as a **colony.** If the initially plated sample contains very few cells, then each distinct colony on the plate will be derived from a single original cell. Members of a colony that share a single genetic ancestor are known as **clones.**

As we discussed in Chapter 7, wild-type bacteria are **prototrophic:** they can grow colonies on **minimal medium**—a substrate containing only inorganic salts, a carbon source for energy, and water. Mutant clones can be identified because they are **auxotrophic:** they will not grow unless the medium contains one or more specific nutrients—say, adenine, or threonine and biotin. Furthermore, wild types are susceptible to certain inhibitors, such as streptomycin, whereas **resistant mutants** can form colonies despite the presence of the inhibitor. These properties allow the geneticist to distinguish different phenotypes among plated colonies.

For many characters, the phenotype of a clone can be determined readily through visual inspection or simple

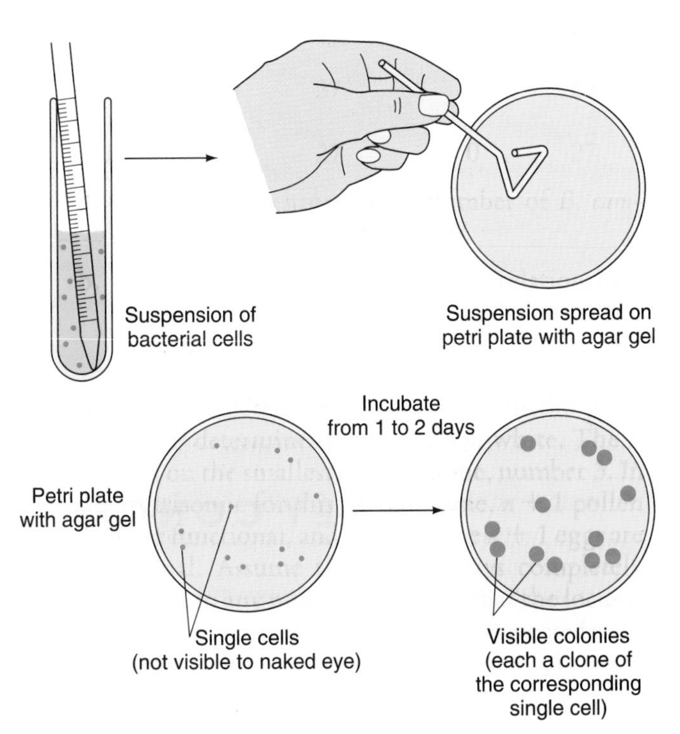

Figure 10-1 Methods of growing bacteria in the laboratory. A few bacterial cells that have been grown in a liquid medium containing nutrients can be spread on an agar medium also containing the appropriate nutrients. Each of these original cells will divide many times by binary fission and eventually give rise to a colony. All cells in a colony, being derived from a single cell, will have the same genotype and phenotype.

Table 10-1. Some Genotypic Symbols Used in Bacterial Genetics

Symbol	Character or phenotype associated with symbol
bio$^-$	Requires biotin added as a supplement to minimal medium
arg$^-$	Requires arginine added as a supplement to minimal medium
met$^-$	Requires methionine added as a supplement to minimal medium
lac$^-$	Cannot utilize lactose as a carbon source
gal$^-$	Cannot utilize galactose as a carbon source
str^r	Resistant to the antibiotic streptomycin
str^s	Sensitive to the antibiotic streptomycin

NOTE: Minimal medium is the basic synthetic medium for bacterial growth, without nutrient supplements.

chemical tests. This phenotype can then be assigned to the original cell of the clone, and the frequencies of various phenotypes in the pipetted sample can be determined. Table 10-1 lists some bacterial phenotypes and their genetic symbols. Also refer to Chapter 7 for a further review of microbial genetic techniques.

Bacterial Conjugation

This and the following sections describe the discovery of gene transfer in bacteria and explain different types of gene transfer and their use in bacterial genetics. First, we'll consider **conjugation,** a process by which certain bacterial cells can transfer DNA to a second cell with which they make contact. Before we examine the discovery of conjugation, let's summarize what we now know.

The Remarkable Properties of F

The ability to transfer DNA by conjugation is dependent on the presence of a cytoplasmic entity termed the **fertility factor,** or **F**. Cells carrying F are termed **F$^+$**; cells without F are **F$^-$**. F is a small, circular DNA element that acts like a minichromosome. F contains approximately 100 genes; these give F several important properties:

1. F can replicate its DNA, which allows F to be maintained in a cellular population that is dividing (Figure 10-2a).
2. Cells carrying F produce **pili** (singular, pilus)—minute proteinaceous tubules that allow the F$^+$ cells to attach to other cells and maintain contact with them (Figure 10-2b).
3. F$^+$ cells can transfer the newly synthesized copy of the circular F genome to a recipient (F$^-$) cell that lacks such a genome (Figure 10-2c); note that a copy of F always remains behind in the donating cell. When a donor cell transfers a copy of its cytoplasmic F to an F$^-$ cell, the recipient cell also becomes an F$^+$ cell, because it now contains a circular F genome.
4. F$^+$ cells are usually inhibited from making contact with other F$^+$ cells and do not usually transfer the F genome to F$^+$ cells.
5. Occasionally, F leaves the cytoplasm and integrates itself into the host bacterial chromosome, as dia-

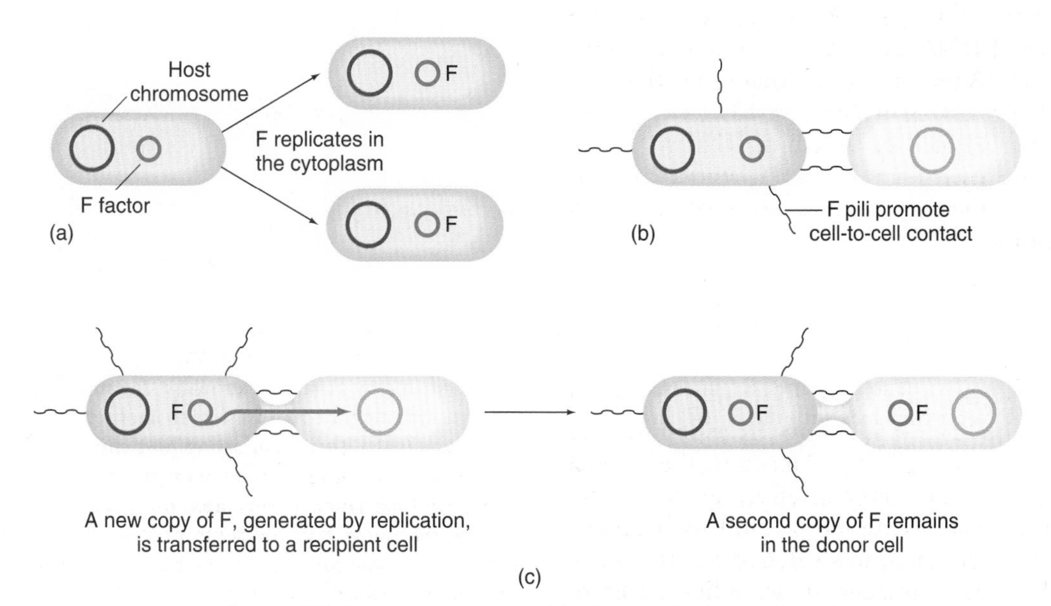
Host chromosome
F replicates in the cytoplasm
F factor
F
F
(a)
F pili promote cell-to-cell contact
(b)
F
F
F
F
A new copy of F, generated by replication, is transferred to a recipient cell
A second copy of F remains in the donor cell
(c)

Figure 10-2 Some properties of the fertility (F) factor of *E. coli.*

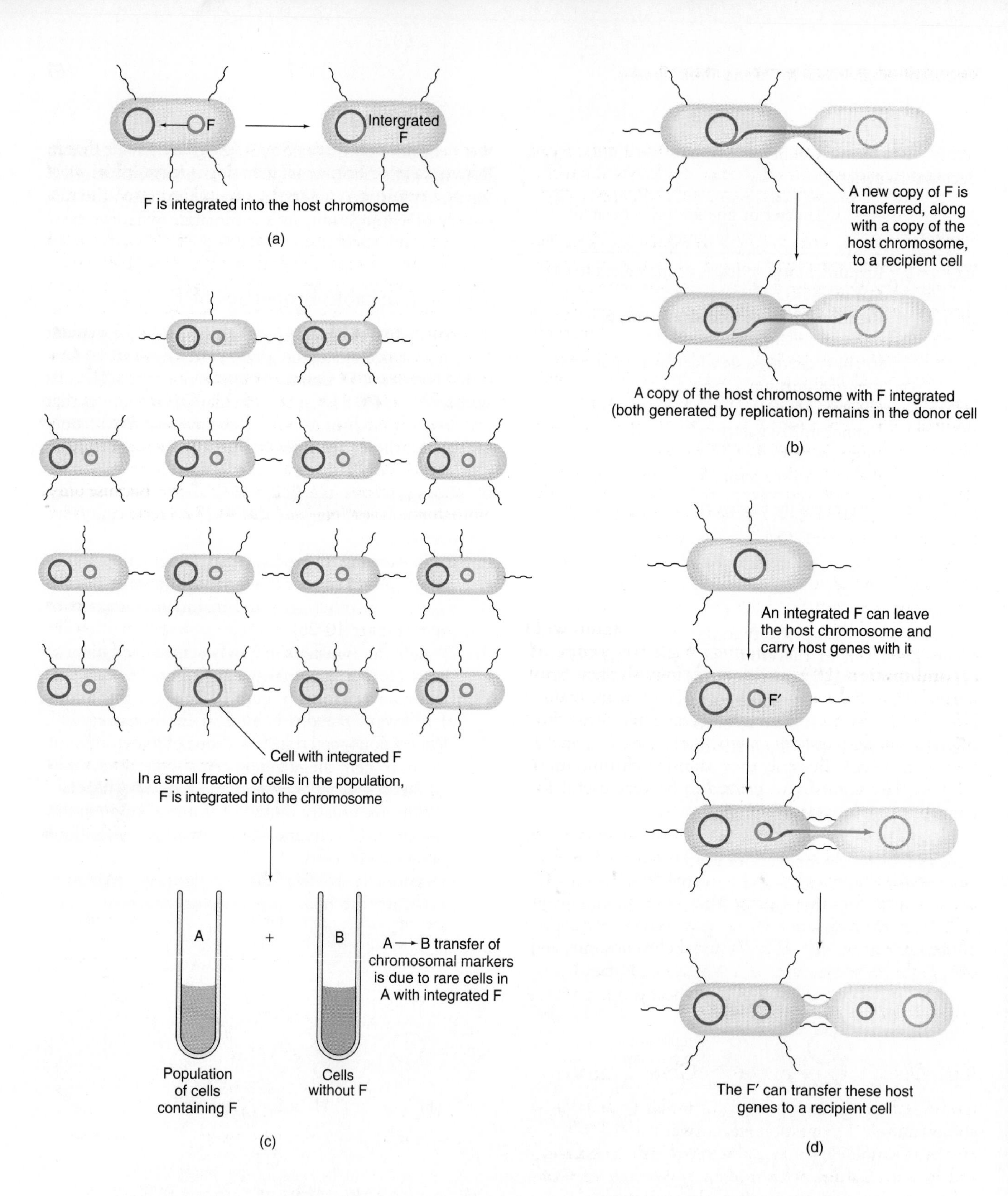

Figure 10-3 The transfer of *E. coli* chromosomal markers mediated by F. (a) Occasionally, the independent F factor combines with the *E. coli* chromosome. (b) When the integrated F transfers to another *E. coli* cell during conjugation, it carries along any *E. coli* DNA that is attached, thus transferring host chromosomal markers to a new cell. (c) In a population of F^+ cells, a few cells will have F integrated into the chromosome; these few cells can transfer chromosomal markers. Therefore, when a population of F^+ cells (tube A) is mixed with a population of F^- cells (tube B), a few B cells will show the acquisition of markers from A. (d) Occasionally, the integrated F can leave the chromosome and return to the cytoplasm. In rare cases F can carry host genes with it, incorporating them into the circular F, which is now termed an F′. The F′ can transfer these genes at high efficiency to other cells, since they are part of the F′ genome.

grammed in Figure 10-3a. When this occurs, F can also transfer host chromosomal markers to the recipient cell along with its own DNA (Figure 10-3b).

The linked transfer of chromosomal markers has some interesting consequences. First, in any population of cells containing the F factor, F will integrate into the chromosomes of a small fraction of cells (Figure 10-3c). These few cells can now transfer chromosomal markers to a second strain. Although the level of transfer is small, it is detectable because the transferred chromosomal markers produce genetic recombinants in the second strain of cells. This phenomenon is what led to the initial discovery of gene transfer by conjugation.

It is possible to isolate the specific cells in the bacterial population that have the F factor integrated into the host chromosome and to cultivate pure strains derived from these cells. In such strains, every cell donates chromosomal markers during F transfer, so that the frequency of recombinants for these strains is much higher than it is for cells in the original population, where the F factor is nearly always in the cytoplasm. Therefore, strains with an integrated F factor are termed **high frequency of recombination (Hfr)** strains to distinguish them from normal F^+ strains, which contain only a few individual Hfr cells that can transfer chromosomal markers and thus display a low frequency of recombination for the population as a whole. Because they transfer chromosomal markers, Hfr strains have proved to be very useful for genetic mapping, as we shall see later on.

F occasionally leaves the chromosome of an Hfr cell and moves back to the cytoplasm, carrying a few host chromosomal genes with it. This modified F, called **F′**, can now transfer these specific host genes to a recipient cell. Thus, the recipient cell can now contain two copies of the same gene, one on its bacterial chromosome, and one on the newly transferred cytoplasmic F′ factor.

Let's now look at how some of these processes were discovered by pioneers in the field.

The Discovery of Bacterial Gene Transfer

Do bacteria possess any processes similar to sex and recombination? The question was answered in 1946 by the elegantly simple experimental work of Joshua Lederberg and Edward Tatum, who studied two strains of *Escherichia coli* with different nutritional requirements. Strain A would grow on a minimal medium only if the medium were supplemented with methionine and biotin; strain B would grow on a minimal medium only if it were supplemented with threonine, leucine, and thiamine. Thus, we can designate strain A as *met⁻ bio⁻ thr⁺ leu⁺ thi⁺* and strain B as *met⁺ bio⁺ thr⁻ leu⁻ thi⁻*. (It is perhaps presumptuous even to use gene symbols to designate these phenotypes because it had not been proved at the time that they were determined by single genes.) Note that in this instance no linkage relationship is intended in writing the symbols in a specific order. The symbolism is merely a convenience.

In their experiment, Lederberg and Tatum plated bacteria into dishes containing only unsupplemented minimal medium. Some of the dishes were plated with only strain A bacteria, some with only strain B bacteria, and some with a mixture of strain A and strain B bacteria that had been incubated together for several hours in a liquid medium containing all of the supplements (Figure 10-4). The plates that received the mixture of strain A and strain B produced growing colonies with a frequency of one in every 10,000,000 cells plated (in scientific notation, 1×10^{-7}), whereas no colonies arose on plates containing either strain A or strain B alone. Because only prototrophic (*met⁺ bio⁺ thr⁺ leu⁺ thi⁺*) bacteria can grow

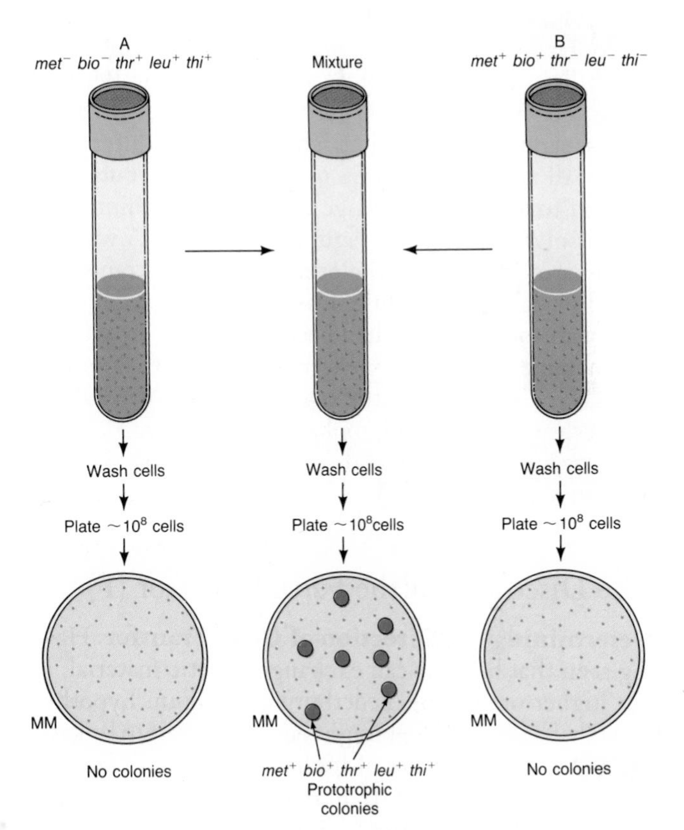

Figure 10-4 Demonstration by Lederberg and Tatum of genetic recombination occurring between bacterial cells. Cells of type A or type B cannot grow on an unsupplemented (minimal) medium (MM), because A and B each carry mutations that cause the inability to synthesize constituents needed for cell growth. When A and B are mixed for a few hours and then plated, however, a few colonies appear on the agar plate. These colonies derive from single cells in which an exchange of genetic material has occurred; they are therefore capable of synthesizing all of the required constituents of metabolism.

on a minimal medium, this observation suggested that some form of recombination of genes had occurred between the genomes of the two strains.

Now, you might object: "What about the possibility that these wild-type colonies were produced by mutation?" The answer is that if the results were due to mutation, then wild-type colonies should have appeared when the strain A and strain B bacteria were plated *by themselves* onto the minimal medium. But they didn't. Note too that more than one mutation would be required to convert each strain to prototrophy (the ability to grow on minimal medium without added nutrients). Therefore, we can conclude that the wild-type colonies were most likely produced by an exchange of genetic material between the two strains.

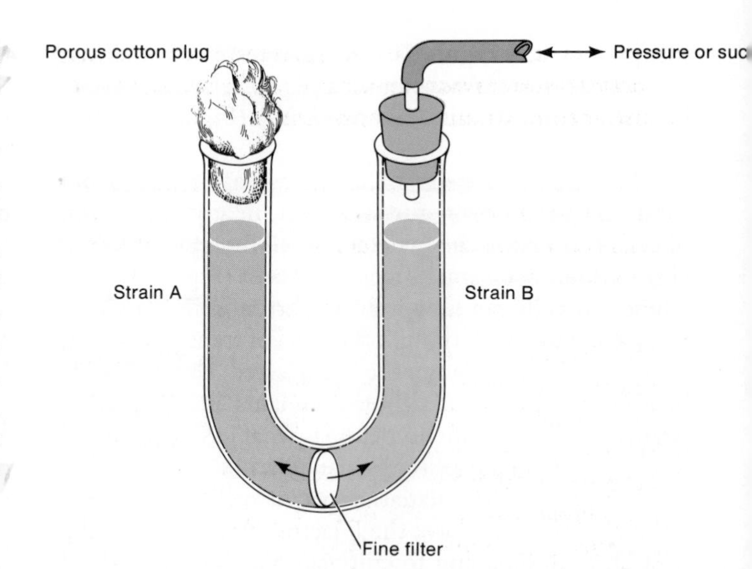

Figure 10-5 Experiment demonstrating that physical contact between bacterial cells is needed for genetic recombination to occur. A suspension of a bacterial strain unable to synthesize certain nutrients is placed in one arm of a U tube. A strain genetically unable to synthesize different required metabolites is placed in the other side. Liquid may be transferred between the arms by the application of pressure or suction, but bacterial cells cannot pass through the center filter. After several hours of incubation, the cells are plated, but no colonies grow on the minimal medium.

Requirement for Physical Contact

It could be suggested that the cells of the two strains do not really exchange genes but instead leak substances that the other cells can absorb and use for growing. This possibility of "cross-feeding" was ruled out by Bernard Davis. He constructed a U tube in which the two arms were separated by a fine filter. The pores of the filter were too small to allow bacteria to pass through but large enough to allow easy passage of the fluid medium and any dissolved substances (Figure 10-5). Strain A was put in one arm; strain B in the other. After the strains had been incubated for a while, Davis tested the content of each arm to see if cells had become able to grow on a minimal medium, and none were found. In other words, *physical contact* between the two strains was needed for wild-type cells to form. It looked as though some kind of gene transfer was involved, and genetic recombinants were indeed produced.

The Discovery of the Fertility Factor (F)

Determining the Direction of Gene Transfer. Having seen that bacteria can exchange genetic material, as can higher organisms, Lederburg and Tatum hypothesized that bacterial genes might be organized into linkage groups, as are the genes of higher organisms. They extended their work with strain A and strain B in attempts to support their hypothesis. Their data clearly showed linkage, because the allele frequencies they obtained from A × B crosses showed great departures from the 1 : 1 ratio typical of independent gene assortment. However, their results were unclear, in part because they did not know the direction of genetic transfer.

In 1953, William Hayes exploited the properties of the antibiotic streptomycin to determine that genetic transfer occurred in one direction in the types of crosses carried out. Hayes verified the results of Lederberg and Tatum, using a similar cross:

Strain A		Strain B
$met^- \ thr^+ \ leu^+ \ thi^+$	×	$met^+ \ thr^- \ leu^- \ thi^-$

However, he treated one of the strains with streptomycin, which prevents cell division and subsequently kills cells but does permit mating to continue for a short period of time. The streptomycin can be washed out after it takes effect on the initial strain, so that a second strain will be unexposed to the drug. When Hayes treated strain A with the streptomycin, washed out the streptomycin, mixed in strain B, and then plated the culture on a minimal medium, he obtained the same frequency of colonies he had obtained in the control experiment with both strains untreated. However, when he treated strain B with streptomycin, washed out the streptomycin, mixed in strain A, and then plated the culture on a minimal medium, he obtained no colonies.

This experiment showed that all the recombinants detected in these genetic transfer experiments took place in strain B. Therefore, the production of colonies of strain B—but not of strain A—on the selective minimal medium was required. The obvious interpretation is that the genetic transfer was not reciprocal but occurred only by transfer from strain A to strain B.

Message The transfer of genetic material in *E. coli* is not reciprocal. One cell acts as the **donor,** and the other cell acts as the **recipient.**

In Hayes's experimental strains, the donor could still transmit genes after exposure to streptomycin, but the recipient could not divide to produce colonies if it had been exposed to streptomycin. This kind of unidirectional transfer of genes was originally analogized to a sexual difference, with the donor being termed "male" and the recipient "female." Although the terms "male" and "female" still persist, it should be stressed that this type of gene transfer is not sexual reproduction. In *bacterial gene transfer,* one organism receives genetic information from a donor; the recipient is changed by that information. In *sexual reproduction,* two organisms donate equally (or nearly so) to the formation of a new organism, but only in exceptional cases is either of the donors changed.

Loss and Regain of Ability to Transfer. By accident, Hayes discovered a variant of his original A strain (male) that would not produce recombinants on crossing with the B strain (female). Apparently, the A males had lost the ability to transfer genetic material and had changed into females. In his analysis of this "sterile" variant, Hayes realized that the fertility of *E. coli* could be lost and regained rather easily.

In further studies, Hayes isolated and cultured a streptomycin-resistant mutant of the sterile A variant. He then mixed these sterile A str^r cells with fertile A male str^s cells. He found that as many as one-third of the A str^r cells had again become fertile, which he judged by their ability to transfer genetic markers to B females. Hayes suggested that maleness, or donor ability, is itself a hereditary state imposed by a **fertility factor (F).** Females lack F and therefore are recipients. Thus, females he designated F^-, and males F^+.

Transfer of F during Conjugation. Recombinant genotypes for marker genes occurred relatively rarely in bacterial crosses, Hayes noted, but the F factor apparently was transmitted effectively during physical contact, or **conjugation,** since Hayes recovered male fertile str^r cells very soon after mixing F^- str^r cells with F^+ str^s cells. There seemed to be a kind of **"infectious transfer"** of the F factor that was taking place far more quickly than the exchange of genetic markers by recombination. The physical nature of the F factor was elucidated much later, but these early experiments clearly showed that it involved some kind of errant particle not closely tied to the genetic markers.

Hfr Strains

An important breakthrough came when Luca Cavalli-Sforza obtained a new kind of male from an F^+ strain (which we now know resulted from the integration of F into the chromosome). On crossing with F^- females, this new strain produced 1000 times more recombinants for genetic markers than did a normal F^+ strain. Cavalli-Sforza designated this derivative an **Hfr** strain to indicate a high frequency of recombination. (The derivative later became known as Hfr C to distinguish it from a similar strain, Hfr H, found by Hayes.) In cells resulting from an $F^+ \times F^-$ cross, a large proportion of the F^- parents were converted to F^+ by infectious transfer of the F factor. However, in Hfr $\times$ F^- crosses, virtually none of the F^- parents were converted to F^+ or to Hfr. Thus, infectious transfer of F did not occur in these crosses since, as we now know, F was no longer in the cytoplasm.

Determining Linkage from Interrupted-Mating Experiments

The exact nature of Hfr strains became clearer in 1957 when Elie Wollman and François Jacob investigated the pattern of transmission of Hfr genes to F^- cells during a cross. They crossed Hfr $str^s\ a^+\ b^+\ c^+\ d^+$ with F^- $str^r\ a^-\ b^-\ c^-\ d^-$. At specific time intervals after mixing, they removed samples. Each sample was put into a kitchen blender for a few seconds to disrupt the mating cell pairs and then was plated onto a medium containing streptomycin to kill the Hfr donor cells. This is called an **interrupted-mating** procedure. The str^r cells then were tested for the presence of marker alleles from the donor. Those str^r cells bearing donor marker alleles must have been involved in conjugation; such cells are called **exconjugants.** Figure 10-6a shows a plot of the results; azi^r, ton^r, lac^+, and gal^+ correspond to the a^+, b^+, c^+, and d^+ mentioned in our generalized description of the experiment. Figure 10-6b portrays the transfer of markers.

The most striking thing about these results is that each donor allele first appeared in the F^- recipients at a specific time after mating began. Furthermore, the donor alleles appeared in a specific sequence. Finally, the maximal yield of cells containing a specific donor allele was smaller for the donor markers that entered later. Putting all these observations together, Wollman and Jacob concluded that gene transfer occurs from a fixed point on the donor chromosome, termed the **origin,** or **O**, and continues in a linear fashion.

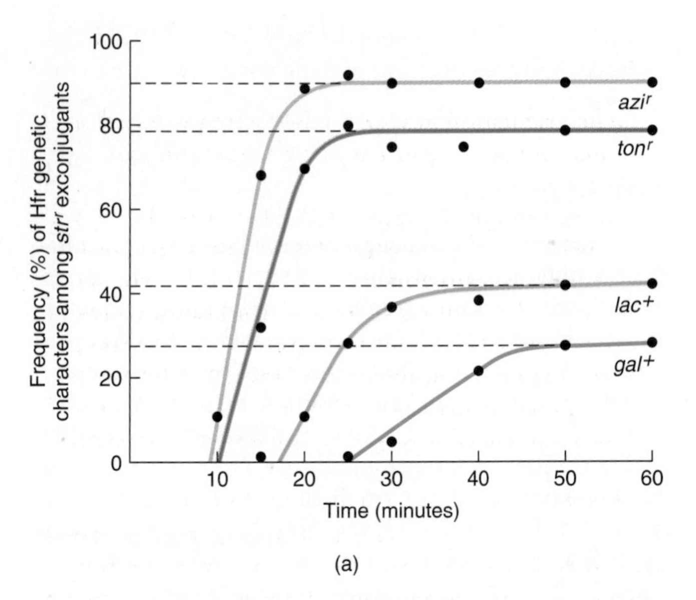

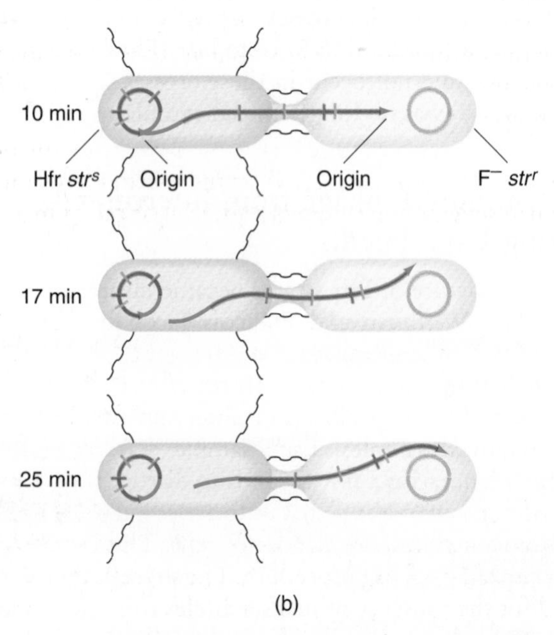

Figure 10-6 Interrupted-mating conjugation experiments with *E. coli.* F^- cells that are str^r are crossed with Hfr cells that are str^s. The F^- cells have a number of mutations (indicated by the genetic markers *azi*, *ton*, *lac*, and *gal*) that prevent them from carrying out specific metabolic steps. However, the Hfr cells are capable of carrying out all of these steps. At different times after the cells are mixed, samples are withdrawn, disrupted in a blender to break conjugation between cells, and plated on media containing streptomycin. The antibiotic kills the Hfr cells but allows the F^- cells to grow and to be tested for their ability to carry out the four metabolic steps. (a) A plot of the frequency of recombinants for each metabolic marker as a function of time after mating. Transfer of the donor allele for each metabolic step obviously depends on how long conjugation is allowed to continue. (b) A schematic view of the transfer of markers over time. (Part a modified from E. L. Wollman, F. Jacob, and W. Hayes, *Cold Spring Harbor Symposia on Quantitative Biology* 21, 1956, 141.)

Message The Hfr chromosome, originally circular, unwinds and is transferred to the F^- cell in a linear fashion. The unwinding and transfer begin from a specific point at one end of the integrated F, called the **origin,** or **O**. The farther a gene is from O, the later it is transferred to the F^-; the transfer process most likely will stop before the farthermost genes are transferred.

Wollman and Jacob realized that it would be easy to construct linkage maps from the interrupted-mating results, using as a measure of "distance" the times at which the donor alleles first appear after mating. The units of distance in this case are minutes. Thus, if b^+ begins to enter the F^- cell 10 minutes after a^+ begins to enter, then a^+ and b^+ are 10 units apart (Figure 10-7). Like the maps based on crossover frequencies, these linkage maps are purely genetic constructions; at the time, they had no known physical basis.

Chromosome Circularity and Integration of F

When Wollman and Jacob allowed Hfr × F^- crosses to continue for as long as two hours before blending, they found that a few of the exconjugants were converted into Hfr. In other words, an important part of F (the terminal portion now known to confer maleness, or donor ability), was eventually being transmitted, but at a very low efficiency, and it apparently was transmitted as the last element of the linear chromosome. We now have the following map, in which the arrow indicates the process of transfer, beginning with O:

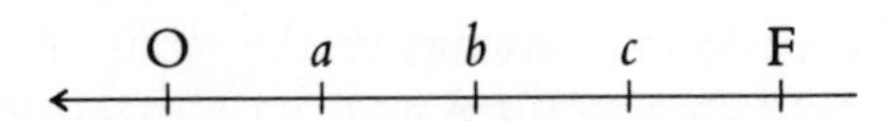

However, when several different Hfr linkage maps were derived by interrupted-mating and "time-of-entry" studies using different, separately derived Hfr strains, the maps differed from strain to strain:

Hfr H	O *thr pro lac pur gal his gly thi* F
1	O *thr thi gly his gal pur lac pro* F
2	O *pro thr thi gly his gal pur lac* F
3	O *pur lac pro thr thi gly his gal* F
AB 312	O *thi thr pro lac pur gal his gly* F

At first glance, there seems to be a random reshuffling of genes. However, a pattern does exist; the genes are not thrown together at random in each strain. For example, note that in every case the *his* gene has *gal* on one side and *gly* on the other. Similar statements can be made about

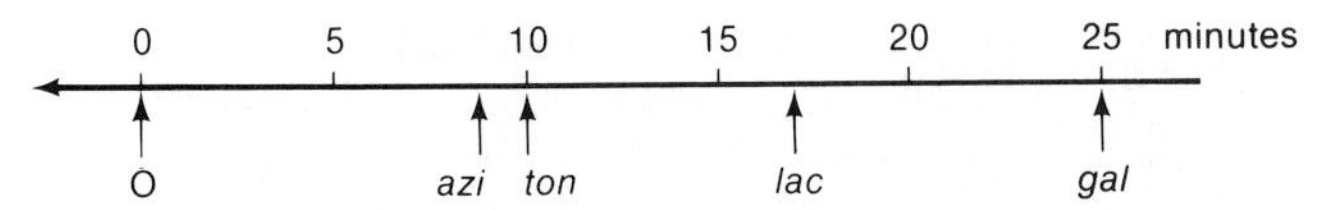

Figure 10-7 Chromosome map from Figure 10-6. A linkage map can be constructed for the *E. coli* chromosome from interrupted-mating studies, using the time at which the donor alleles first appear after mating. The units of distance are given in minutes; arrowhead at left indicates the direction of transfer of the donor alleles.

each gene, except when it appears at one end or the other of the linkage map. The order in which the genes are transferred is not constant. In two Hfr strains, for example, the *his* gene is transferred before the *gly* gene (*his* is closer to O), but in three strains the *gly* gene is transferred before the *his* gene.

How can we account for these unusual results? Alan Campbell proposed a startling hypothesis: Suppose that in an F^+ male, F is a small cytoplasmic element (and therefore easily transferred to an F^- cell on conjugation). If the *chromosome* of the F^+ male were a *ring*, any of the linear Hfr chromosomes could be generated simply by inserting F into the ring at the appropriate place and in the appropriate orientation (Figure 10-8).

Several conclusions—later confirmed—follow from this hypothesis.

1. The orientation in which F is inserted would determine the polarity of the Hfr chromosome, as indicated in Figure 10-8a.
2. At one end of the integrated F factor would be the **origin,** where transfer of the Hfr chromosome begins; the **terminus** at the other end of F would not be transferred unless all of the chromosome had been transferred. Since the chromosome often breaks before all of it is transferred, and since the F terminus is what confers maleness, then only a small fraction of the recipient cells would be converted to male cells.

How, then, might F integration be explained? Wollman and Jacob suggested that some kind of crossover event between F and the F^+ chromosome might generate the Hfr chromosome. Campbell then came up with a brilliant extension of that idea. He proposed that if F, like the chromosome, were circular, then a crossover between the two rings would produce a single larger ring with F inserted (Figure 10-9a).

Now suppose that F consists of three different regions, as shown in Figure 10-9b. If the bacterial chromosome had several homologous regions that could match

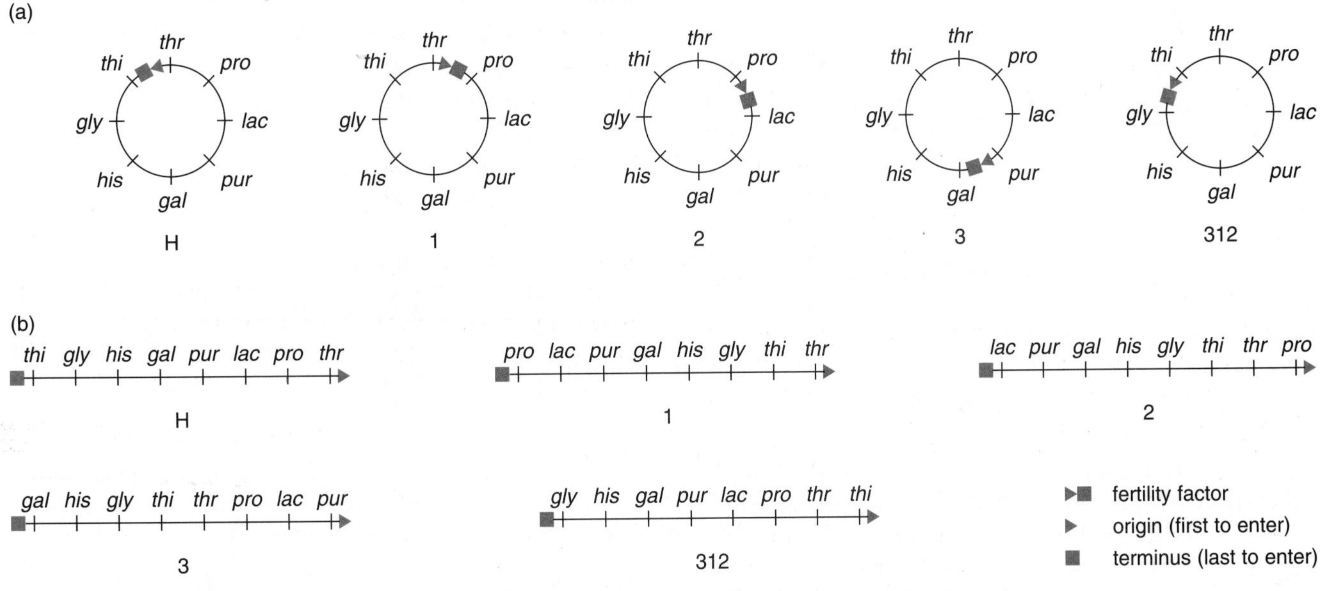

Figure 10-8 Circularity of the *E. coli* chromosome. (a) Through the use of different Hfr strains (H, 1, 2, 3, 312) that have the fertility factor inserted into the chromosome at different points and in different directions, interrupted-mating experiments indicate that the chromosome is circular. The mobilization point (origin) is shown for each strain. (b) The linear order of transfer of markers for each Hfr strain; arrowheads indicate the origin and direction of transfer.

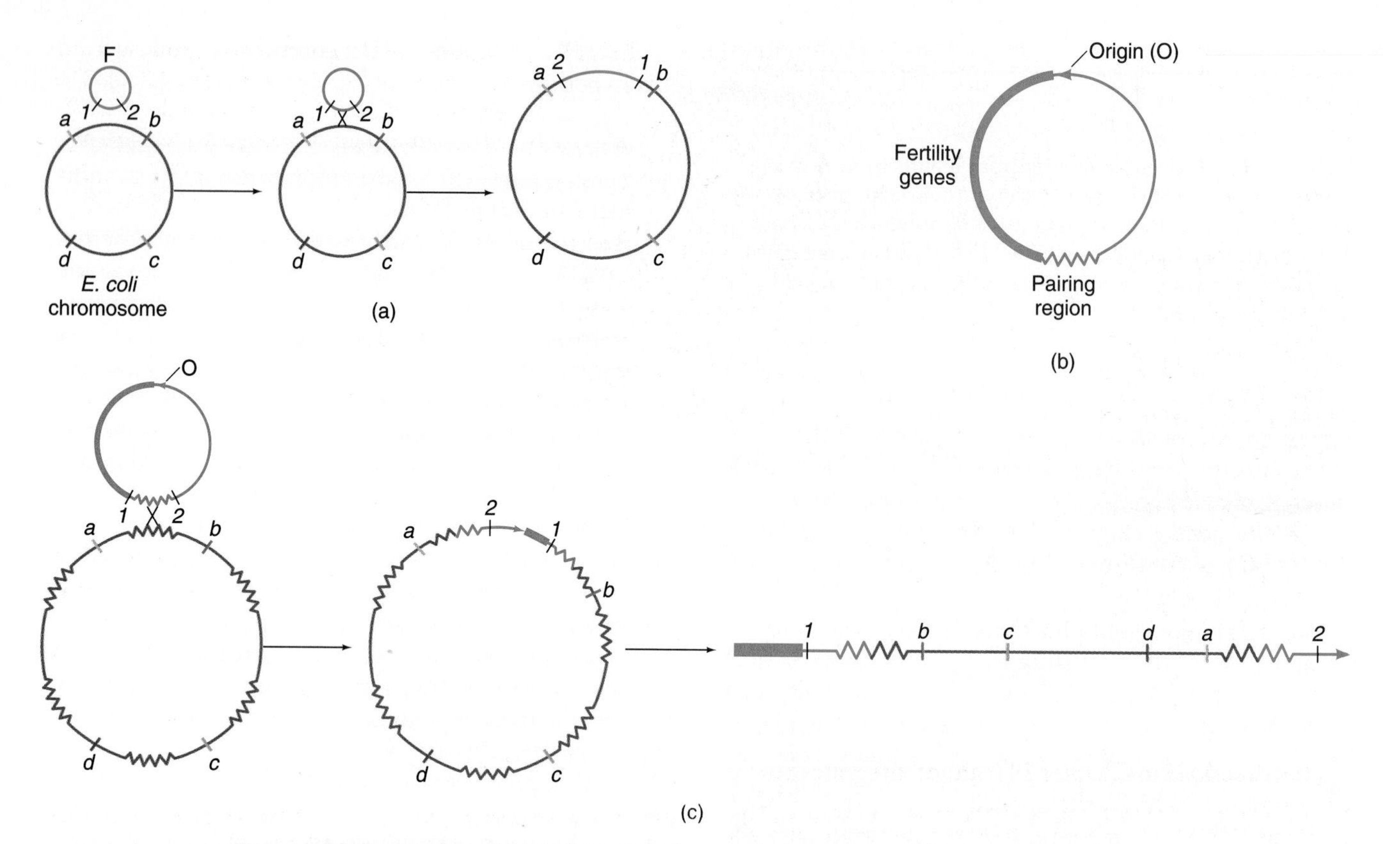

Figure 10-9 Insertion of the F factor into the *E. coli* chromosome. (a) Crossing-over mechanism for the attachment of F. (Hypothetical markers 1 and 2 are shown on F to depict the direction of insertion.) (b) Hypothetical regions in F. The origin (O) is the mobilization point where insertion begins; the pairing region is homologous with a region on the *E. coli* chromosome; fertility genes are responsible for the F^+ phenotype. (c) Model for the insertion of F. An area of F may have regions of pairing homology for different regions of the circular *E. coli* chromosome. Crossover between F and the chromosome could thus insert F into various points on the chromosome. (Jagged lines represent possible regions of homology; the arrowhead indicates the direction of transfer from the origin, O.) In this example, the Hfr cell created by the insertion of F would transfer its genes in the order *a, d, c, b*. What would be the order of transfer if F inserted at the other homologous sites?

up with the pairing region of F, then different Hfr chromosomes could easily be generated by crossovers at these different sites (Figure 10-9c).

Chromosomal and F circularity were wildly implausible concepts initially, inferred solely from the genetic data; confirmation of their physical reality came only a number of years later. The direct-crossover model of integration was also subsequently confirmed.

Episomes

The fertility factor thus exists in two states: (1) as a free cytoplasmic element F that is easily transferred to F^- recipients and (2) as an integrated part of a circular chromosome that is transmitted only very late in conjugation. The word **episome** (literally, "additional body") was coined for a genetic particle having such a pair of states. A cell containing F in the first state is called an F^+ cell, a cell containing F in the second state is an Hfr cell, and a cell lacking F is an F^- cell.

Message An **episome** is a genetic factor in bacteria that can exist either as an element in the cytoplasm or as an integrated part of a chromosome. The F factor is an episome.

Infectious elements other than F have been found in *E. coli* and other bacteria. When F or other factors exist in the cytoplasmic state they are termed **plasmids.** Some plasmids (including many used in the recombinant DNA

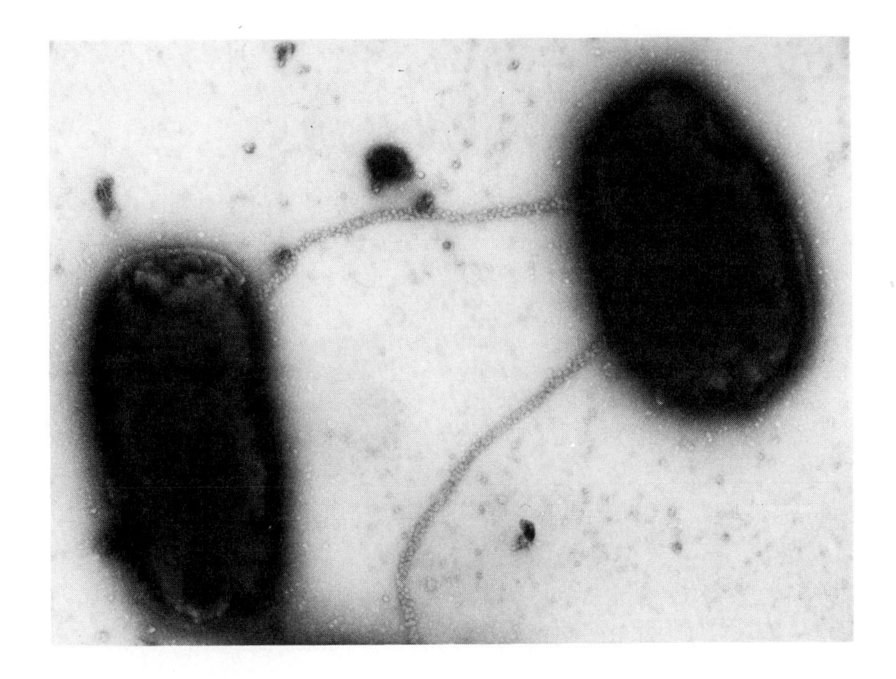

Figure 10-10 Electron micrograph of a cross between two *E. coli* cells (×34,300). The pili of the Hfr cell (not present in the F^- cell) have been visualized through the addition of viruses that attach specifically to them. (David P. Allison, Biology Division, Oak Ridge National Laboratory.)

studies described in Chapter 14) cannot integrate into the chromosome, while others, such as temperate phages (see page 286), are episomes that can integrate into the bacterial chromosome as the F factor does. Plasmids are of central importance in genetic engineering; we consider them in more detail later in this book.

Mechanics of Transfer

Does an Hfr cell die after donating its chromosome to an F^- cell? The answer is no (unless the culture is treated with streptomycin). The Hfr chromosome replicates while it is transferring a single strand to the F^- cell; this ensures a complete chromosome for the donor cell after mating. The transferred strand is replicated in the recipient cell, and donor genes may become incorporated in the recipient's chromosome through crossovers, creating a recombinant cell. Otherwise, transferred fragments of DNA in the recipient are lost during cell division.

We assume that the F^- chromosome is also circular, because the recipient F^- cell, if it receives the F factor, from an F^+ cell, is readily converted into an F^+ cell from which an Hfr cell can be derived.

The picture emerges of a circular Hfr chromosome unwinding a copy of itself, which is then transferred in a linear fashion into the F^- cell. How is the transfer achieved? Electron-microscope studies show that Hfr and F^+ cells have fibrous structures, **F pili,** protruding from their cell walls, as shown in Figure 10-10. The F pili facilitate cell-to-cell contact, during which DNA transfer occurs, apparently through pores in the F^-, although the exact mechanism of transfer is still unclear.

The *E. coli* Conjugation Cycle

We can now summarize the various aspects of the conjugation cycle in *E. coli* (Figure 10-11). We'll review the conjugation cycle in terms of the differences between F^-, F^+, and Hfr cells, since these differences epitomize the cycle.

F^- strains do not contain the F factor and cannot transfer DNA by conjugation. They are, however, recipients of DNA transferred from F^+ or Hfr cells by conjugation.

F^+ cells contain the F factor in the cytoplasm and therefore can transfer F in a highly efficient manner to F^- cells during conjugation.

Hfr cells have F integrated into the bacterial chromosome, not in the cytoplasm.

Chromosomal markers are transferred in a strain of F^+ cells because in any population of F^+ cells, a small fraction of cells (about 1 in 1000) have been converted to Hfr cells by the integration of F into the bacterial chromosome. Since conjugation experiments are usually carried out by mixing 10^7 to 10^8 cells comprising prospective donors and recipients, the population will contain various different Hfr cells derived from independent integrations of F into the chromosome at various different sites. Therefore, when chromosomal markers are transferred by different cells in the population, transfer will start at different points on the chromosome. This results in an approximately equal transfer of markers all around the chromosome, although at a low frequency.

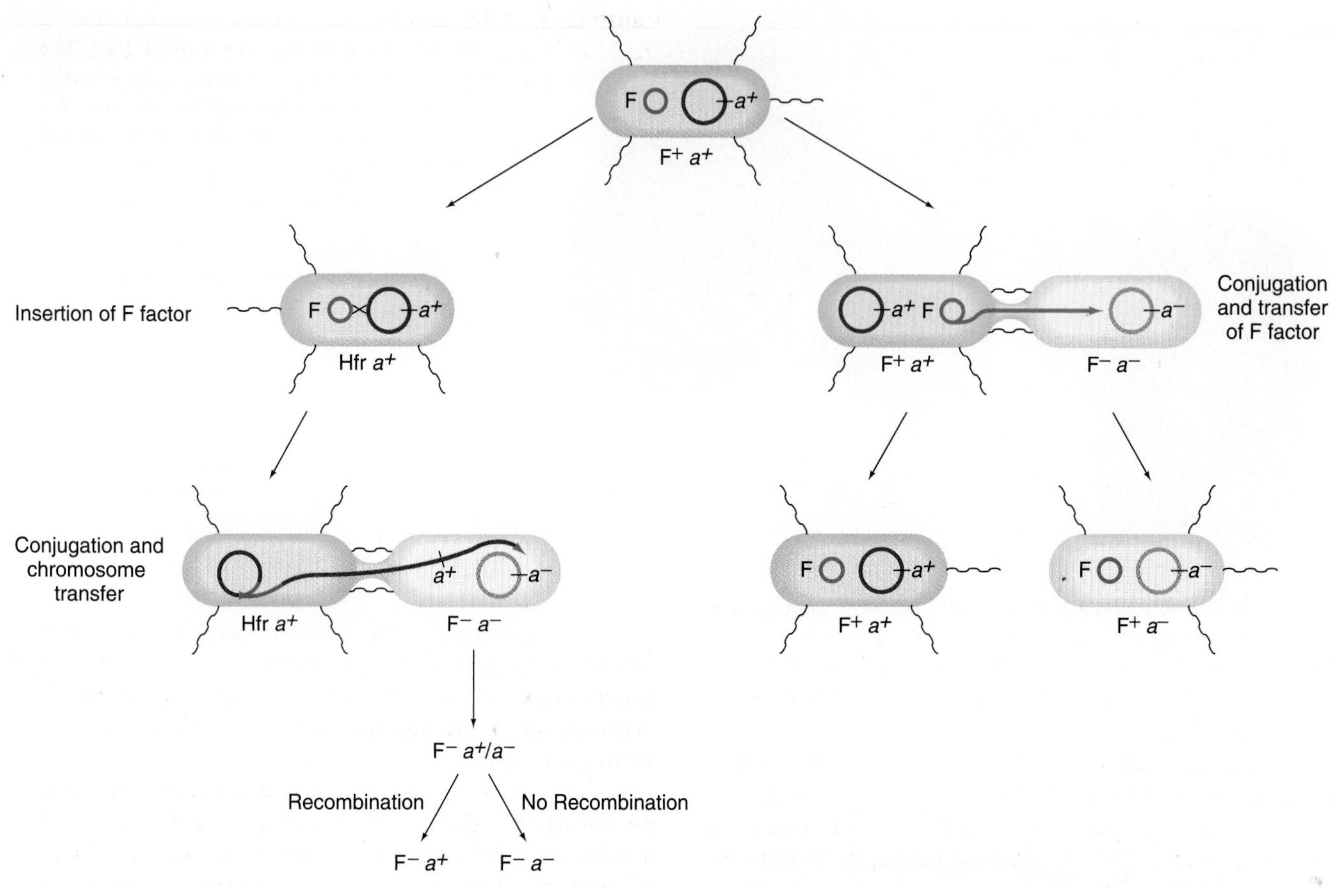

Figure 10-11 Summary of the various events that occur in the conjugational cycle of *E. coli.*

This type of F^+-mediated transfer is what Lederberg and Tatum observed when they discovered gene transfer in bacteria.

Each of the Hfr cells in an F^+ population with an integrated F factor can be the source of a new Hfr strain if it is isolated and used to start a clone.

Hfr strains are derived from a clone of Hfr cells in which a specific integration of F into the bacterial chromosome has occurred. Therefore, all the cells in any given Hfr strain have F integrated into the chromosome at exactly the same point.

Hfr populations transfer chromosomal markers to F^- cells at a high rate compared to F^+ populations, since only a fraction of cells in an F^+ population have F integrated into the chromosome. Further, in any given Hfr strain the markers are transferred from a fixed point in a specific order. This also contrasts with F^+ populations, where the Hfr cells transfer chromosomal markers in no particular fixed order, since the F factor integrates into the chromosome at different points in different F^+ cells.

In an Hfr × F^- cross, the F^- is not converted to Hfr or to F^+, except in very rare cases, because the Hfr chromosome nearly always breaks before the F terminus is transferred to the F^- cell.

Bacterial Recombination and Mapping the *E. coli* Chromosome

Recombination between Marker Genes after Transfer

Thus far, we have studied only the process of the transfer of genetic information between individuals in a cross. This transfer is inferred from the existence of recombinants produced from the cross. However, before a stable recombinant can be produced, the transferred genes must be integrated or incorporated into the recipient's genome by an exchange mechanism. We now consider some of the special properties of this exchange event.

Genetic exchange in prokaryotes does not take place between two whole genomes (as it does in eukaryotes); rather, it takes place between one *complete* genome, derived from F^-, called the **endogenote,** and an *incomplete* one, derived from the donor, called the **exogenote.** What we have in fact is a partial diploid, or **merozygote.** Bacterial genetics is merozygous genetics. Figure 10-12a is a diagram of a merozygote.

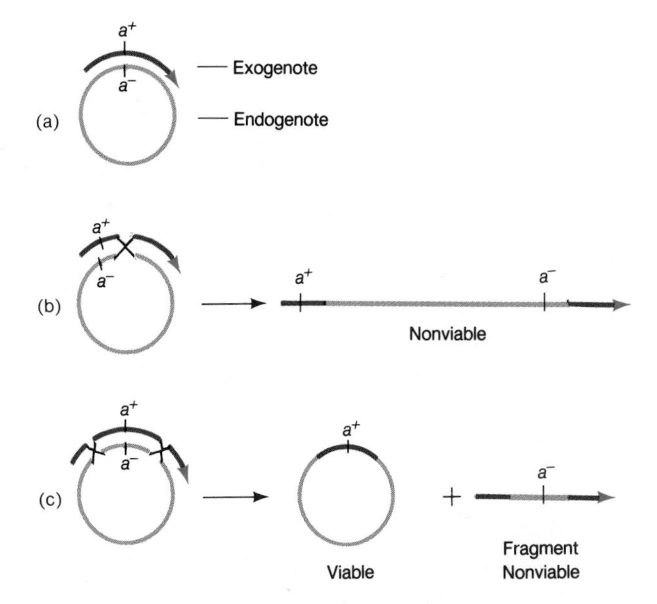

Figure 10-12 Crossover between exogenote and endogenote in a merozygote. (a) The merozygote. (b) A single crossover leads to a partially diploid linear chromosome. (c) An even number of crossovers leads to a ring plus a linear fragment.

A single crossover would not be very useful in generating viable recombinants, because the ring is broken to produce a strange, partially diploid linear chromosome (Figure 10-12b). To keep the ring intact, there must be an even number of crossovers (Figure 10-12c). The fragment produced in such a crossover is only a partial genome, which is generally lost during subsequent cell growth. Hence, both reciprocal products of recombination do not survive—only one does. A further unique property of bacterial exchange, then, is that we must forget about reciprocal exchange products in most cases.

Message In the merozygous genetics of bacteria, we generally are concerned with double crossovers and we do not expect reciprocal recombinants.

The Gradient of Transfer

Only partial diploids exist in the merozygote. Some genes don't even get into the act! To better appreciate this, let's look again at the consequences of gene transfer. Note how only a fragment of the donor chromosome appears in the recipient [Figure 10-12(c)]. This is because there is spontaneous breakage of the mating pairs, so that the entire chromosome is rarely transferred. The spontaneous breakage can occur at any time after transfer begins. This creates a natural **gradient of transfer,** which makes it less and less likely that a recipient cell will receive later and later genetic markers. ("Later" here refers to markers that are increasingly farther from the origin and hence are donated later in the order of markers transferred). For example, in a cross of Hfr-donating markers in the order *met, arg, leu,* we would expect a distribution of fragments such as the one represented here:

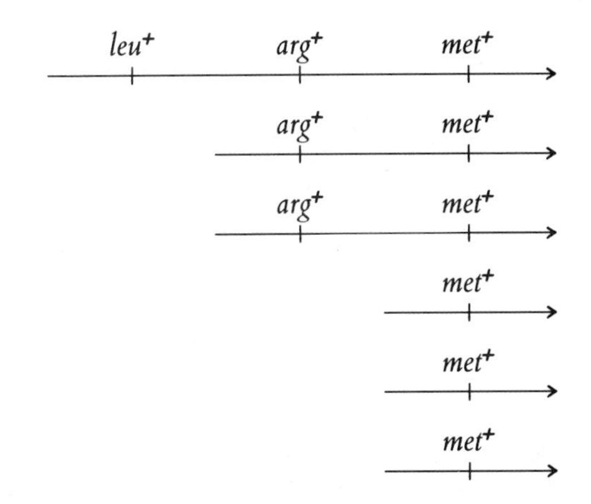

Note that many more fragments contain the *met* locus than the *arg* locus and that the *leu* locus is present on only one fragment. It is easy to see that the closer the marker is to the origin, the greater the chance it will be transferred during conjugation.

The concept of the gradient of transfer is the same as the one described earlier for interrupted matings, except that here we are allowing the natural disruption of mating pairs to occur instead of interrupting the pairs mechanically.

Determining Gene Order from the Gradient of Transfer

We can use the natural gradient of transfer to establish the order of genetic markers, provided we select for an early marker which enters before the markers we are ordering. Let's see how this works. Suppose that we use an Hfr strain which donates markers in the order *met, arg, aro, his.* In a cross of an Hfr that is $met^+\ arg^+\ aro^+\ his^+\ str^s$ with an F^- that is $met^-\ arg^-\ aro^-\ his^-\ str^r$ recombinants are selected that can grow on a minimal medium without methionine but with arginine, aromatic amino acids, histidine, and in the presence of streptomycin. Here we are selecting for recombinants in the F^- strain that are met^+ in a cross in which the *met* locus is transferred as the earliest marker. We can then score for inheritance of the other markers present in the Hfr. A typical result would be

$$met^+ = 100\%$$
$$arg^+ = 60\%$$
$$aro^+ = 20\%$$
$$his^+ = 4\%$$

Note how the frequency of inheritance corresponds to the order of transfer. This is because the frequency of inheritance reflects the frequency of transfer. For this method to work, it is crucial that it be applied only to genetic markers that enter after the selected marker—in this case, after *met.*

Higher-Resolution Mapping by Recombinant Frequency in Bacterial Crosses

While interrupted-mating experiments and the natural gradient of transfer can give us a rough set of gene locations over the entire map, other methods are needed to obtain a higher resolution between marker loci that are close together. Here we consider one approach to the problem of using the frequency of recombinants to measure linkage.

Previous attempts to measure linkage in conjugational crosses were hindered by the failure to understand that only fragments of the chromosome are transferred and that the gradient of transfer produces a bias toward the inheritance of early markers. In order to measure linkage and to attach any meaning to a calculated map distance, it is necessary to produce a situation in which every marker has an equal chance at being transferred, so that the recombinant frequencies are dependent only on the distance between the relevant genes.

Suppose that we consider three markers: *met, arg,* and *leu.* If the order is *met, arg, leu,* and if *met* is transferred first and *leu* last, then we really want to set up the situation diagrammed below to calculate map distances separating these markers:

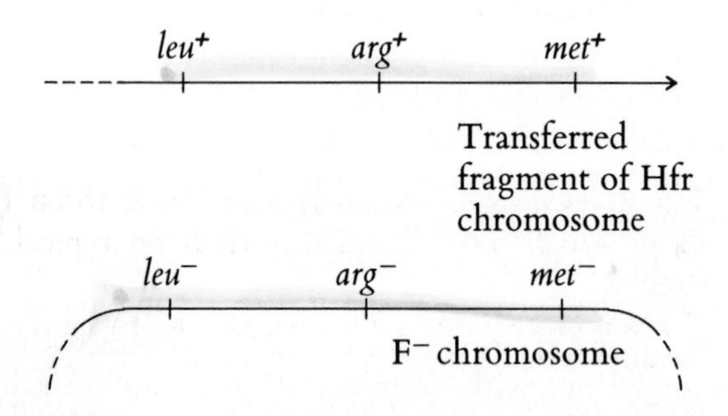

In this case, we have to arrange to select the *last* marker to enter, which in this case is *leu.* Why? Because if we select for the last marker, then we know that every cell that received fragments containing the last marker also received the earlier markers, namely *arg* and *met,* on the same fragments. We can then proceed to calculate map distance in the classic manner, where 1 map unit (m.u.) is equal to one percent crossovers in the respective interval on the map. In practice, this is done by calculating, among the total recovered recombinants, the percentage of recombinants produced by crossovers between two markers. Let's look at an example.

Sample Cross

In the cross of the Hfr strain just described (met^+ arg^+ leu^+ str^s) with an F^- that is met^- arg^- leu^- str^r, we would select leu^+ recombinants and then examine them for the *arg* and *met* markers. In this case, the *arg* and *met* markers are called the **unselected markers.** Figure 10-13 depicts the types of crossover events expected. Note how two crossover events are required to incorporate part of the incoming fragment into the F^- chromosome. One crossover must be on each side of the selected *(leu)* marker. Thus, in Figure 10-13, one crossover must be on the left side of the *leu* marker and the second must be on the right side. Suppose that the map distance between each marker is 5 m.u. (5 percent recombination). In 5 percent of the total leu^+ recombinants, the second crossover occurs between *leu* and *arg* (Figure 10-13a); in another 5 percent of the cases, the second crossover occurs between *leu* and *met* (Figure 10-13b). We would then expect 90 percent of the selected leu^+ recombinants to be arg^+ met^+, because the second crossover occurs outside the *leu–arg–met* interval (Figure 10-13c) in 90 percent of the cases. We would also expect 5 percent of the leu^+ recombinants to be arg^- met^-, resulting from a crossover between *leu* and *arg,* and 5 percent of the leu^+ recombinants to be arg^+ met^-, resulting from a crossover between *arg* and *met.* In reality, then, we are simply determining the percentage of the time that the second crossover occurs in each of the three possible intervals.

In a cross such as the one just described, one class of potential recombinants requires an additional two crossover events (Figure 10-13d). In this case, the leu^+ arg^- met^+ recombinants would require four crossovers instead of two. These recombinants are rarely recovered, because their frequency is sharply reduced compared to the other classes of recombinants.

Deriving Gene Order by Reciprocal Crosses

In many cases, two loci are so close together that it becomes difficult to order them in relation to a third locus. For example, consider three loci linked in this way:

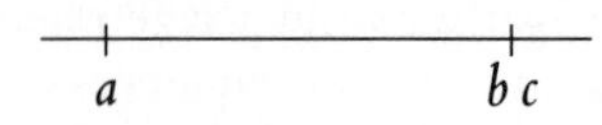

Here *b* and *c* are too closely linked to separate easily. The recombinant frequency (RF) between *a* and *b* under most experimental conditions will be more or less the same as the RF between *a* and *c*. Is the order *a, b, c* or *a, c, b*? One often-used technique for finding out is to make a pair of reciprocal crosses, using the same marker genotypes for the donor in one cross and for the recipient in the other cross. Figure 10-14 shows the crossover events needed to generate prototrophs from the cross

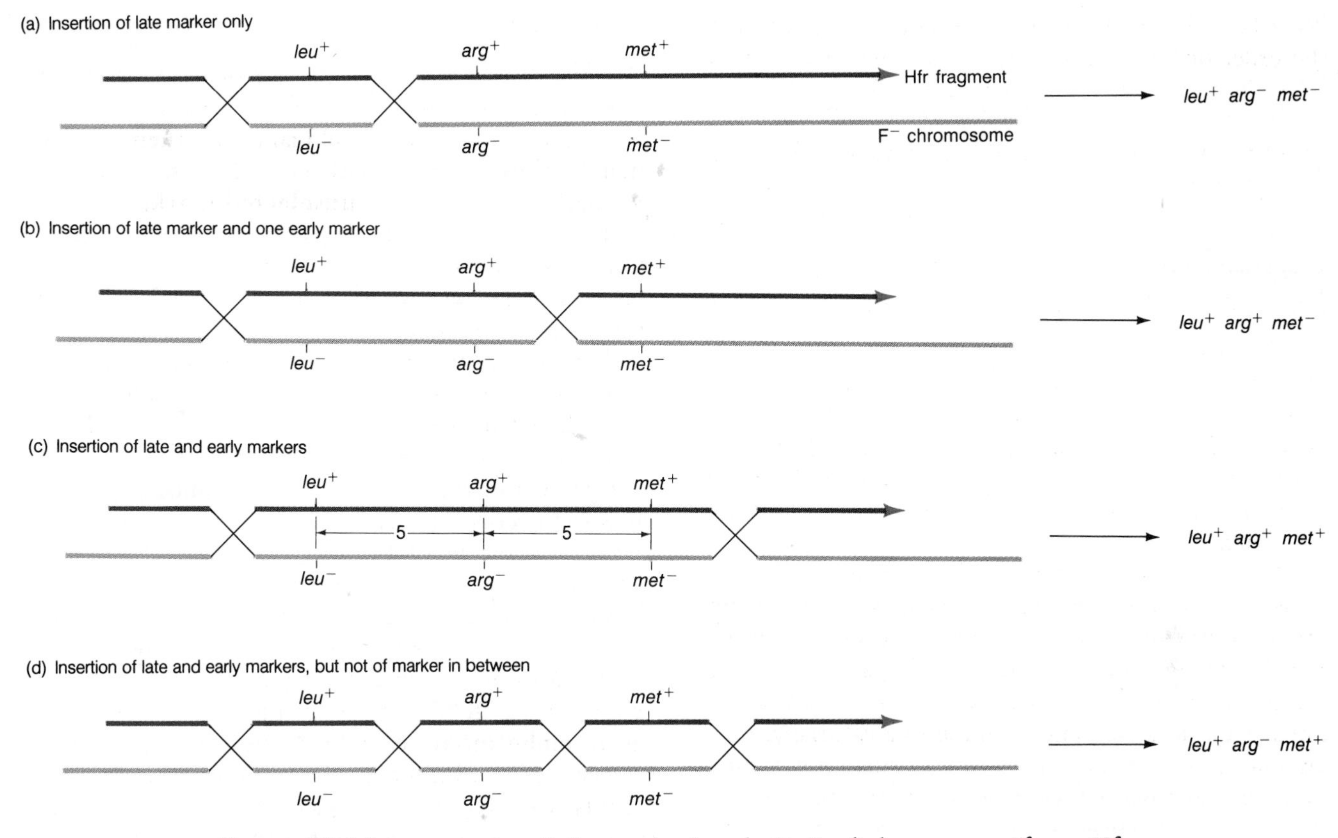

Figure 10-13 Incorporation of a late marker into the F$^-$ *E. coli* chromosome. After an Hfr cross, selection is made for the *leu*$^+$ marker, which is donated late. The early markers (*arg*$^+$ and *met*$^+$) may or may not be inserted, depending on the site where recombination between the Hfr fragment and the F$^-$ chromosome occurs. If, as indicated in (c), the distance between *leu*$^+$ and *arg*$^+$ is 5 m.u. and the distance between *arg*$^+$ and *met*$^+$ is 5 m.u., then crossovers will occur in each of these intervals 5 percent of the time (a and b), resulting in the *leu*$^+$ *arg*$^-$ *met*$^-$ (a) and *leu*$^+$ *arg*$^+$ *met*$^-$ (b) recombinants. Crossovers occurring outside of these intervals can result in *leu*$^+$ *arg*$^+$ *met*$^+$ recombinant (c). The *leu*$^+$ *arg*$^-$ *met*$^+$ recombinant requires an additional two crossovers (d).

$a\ b\ c^+ \times a^+\ b^+\ c$, depending on the order of the loci. If the reciprocal crosses give dramatically different frequencies of wild-type survivors on a minimal medium, then we know that the order is *a, c, b*. If we observe no difference in frequencies between the reciprocal crosses, then the order must be *a, b, c*. Once again, this same principle can be used in other bacterial and phage mapping systems.

Infectious Marker-Gene Transfer by Episomes

Edward Adelberg's work led to the discovery of gene transfer at high frequency by episomes. When he began his recombination experiments in 1959, the particular Hfr strain he used kept producing F$^+$ cells, so the recombination frequencies were not very large. Adelberg called this particular fertility factor F′ to distinguish it from the normal F, for the following reasons:

1. The F′-bearing F$^+$ strain reverted back to an Hfr strain much more frequently than do typical F$^+$ strains.
2. F′ always integrated at the *same place* to give back the original Hfr chromosome. (Remember that randomly selected Hfr derivatives from F$^+$ males have origins at many different positions.)

How could these properties of F′ be explained? The answer came from the recovery of an F′ from an Hfr strain in which the *lac*$^+$ locus was near the end of the Hfr chromosome (since it was transferred very late). Using this Hfr *lac*$^+$ strain, François Jacob and Adelberg found an F$^+$ derivative that transferred *lac*$^+$ to F$^-$ *lac*$^-$ recipients at a very high frequency. Furthermore, the recipients that became F$^+$ *lac*$^+$ phenotypically occasionally produced F$^-$ *lac*$^-$ daughter cells, at a frequency of 1×10^{-3}. Thus, the genotype of these recipients appeared to be F *lac*$^+$/*lac*$^-$.

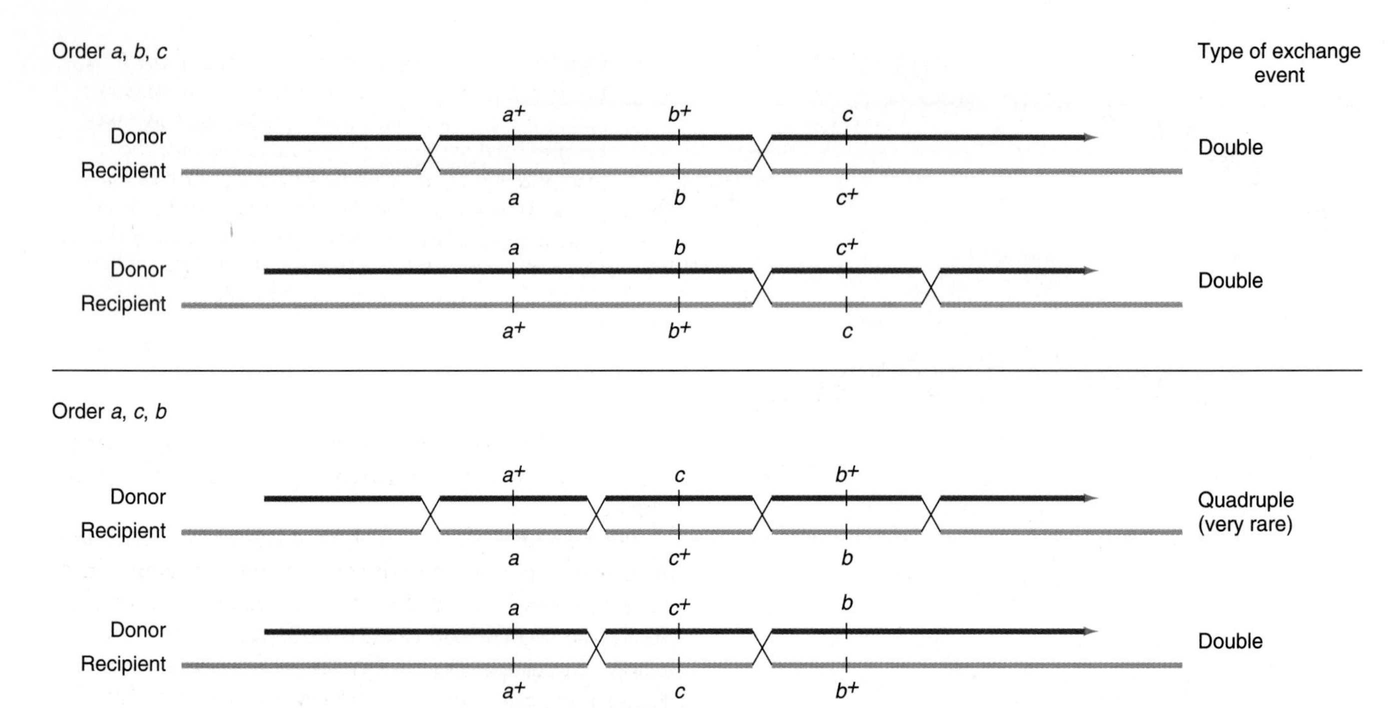

Figure 10-14 Inferring gene order from the relative frequencies of recombinants in reciprocal crosses involving parental genotypes $a\ b\ c^+ \times a^+\ b^+\ c$. If the order is *a*, *b*, *c*, then the frequency of $a^+\ b^+\ c^+$ progeny from the reciprocal crosses will be approximately equal. However, if the order is *a*, *c*, *b*, then the frequencies will be quite different.

Now we have the clue: F′ is a cytoplasmic element that carries a part of the bacterial chromosome. In fact, it is nothing more than F with a piece of the host chromosome incorporated. Its origin and reintegration can be visualized as shown in Figure 10-15. This F′ is known as F′-*lac*, because the piece of host chromosome it picked up has the *lac* gene on it. F′ factors have been found carrying many different chromosomal genes and have been named accordingly. For example, F′ factors carrying *gal* or *trp* are called F′-*gal* and F′-*trp*, respectively. Because F lac^+/lac^- cells are lac^+ in phenotype, we know that lac^+ is dominant over lac^-. As we shall see later, in Chapter 17, the dominant–recessive relationship between alleles can be a very useful bit of information in interpreting gene function.

Partial diploidy, called **merodiploidy,** for specific segments of the genome can be made with an array of F′ derivatives from Hfr strains. The F′ cells can be selected by looking for the infectious transfer of normally late genes in a specific Hfr strain. The creation of partial diploids by F′ elements is called **sexduction,** or **F′-duction.** Some F′ strains can carry very large parts (up to one-quarter) of the bacterial chromosome; if appropriate markers are used, the merozygotes generated can be used for recombination studies.

Message During conjugation between an Hfr donor and an F^- recipient, the genes of the donor are transmitted *linearly* to the F^- cell, via the bacterial chromosome, with the inserted fertility factor transferring last.

During conjugation between an F^+ donor carrying an F′ plasmid and an F^- recipient, a specific part of the donor genome may be transmitted *infectiously* to the F^- cell, via the plasmid. The transmitted part was originally adjacent to the F locus in an Hfr strain from which the F^+ was derived.

Bacterial Transformation

Some bacteria have another method of transferring DNA and producing recombinants that does not require conjugation. The conversion of one genotype into another by the introduction of exogenous DNA (that is, bits of DNA from an external source) is termed **transformation.** Transformation was discovered in *Streptococcus pneumoniae* in 1928 by Frederick Griffith, and in 1944 Oswald T. Avery, Colin M. McLeod, and Maclyn McCarty demonstrated that the "transforming principle" was DNA. Both results are milestones in the elucidation of the molecular nature of genes. We consider this work in more detail in Chapter 11.

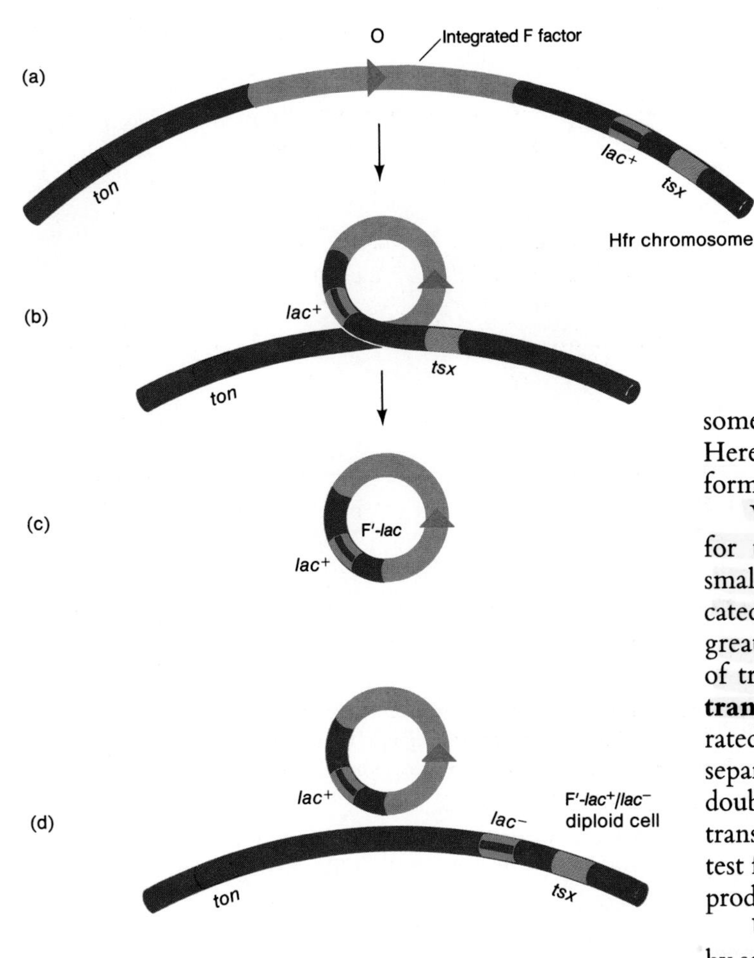

Figure 10-15 Origin and reintegration of the F′ factor, in this case, F′-*lac*. (a) F is inserted in an Hfr strain between the *ton* and lac^+ alleles. (b,c) Abnormal "outlooping" and separation of F occurs to include the *lac* locus, producing the F′-lac^+ particle. (d) An F lac^+/lac^- partial diploid is produced by the transfer of the F′-*lac* particle to an F^- lac^- recipient. (From G. S. Stent and R. Calendar, *Molecular Genetics,* 2d ed. Copyright 1978 by W. H. Freeman and Company, New York.)

After it was shown that DNA is the agent that determines the polysaccharide character of *S. pneumoniae,* transformation was demonstrated for other genes, such as those for drug resistance (Figure 10-16). The transforming principle, exogenous DNA, is incorporated into the bacterial chromosome by a breakage-and-insertion process analogous to that depicted for Hfr × F^- crosses in Figure 10-12. (Note, however, that in *conjugation,* DNA is transferred from one living cell to another through the F pilus, whereas in *transformation,* isolated pieces of external DNA are taken up by a cell.) Thus, if radioactively labeled DNA from an arg^+ bacterial culture is added to unlabeled arg^- cells, the arg^+ transformants (selected by plating on a minimal medium without arginine) can be shown to contain some of the radioactivity. We examine a molecular model of this process in Chapter 19. For now, let's consider transformation simply as a genetic tool.

Linkage Information from Transformation

Transformation has been a very handy tool in several areas of bacterial research. We learn later how it is used in some of the modern techniques of genetic engineering. Here we examine its usefulness in providing linkage information.

When DNA (the bacterial chromosome) is extracted for transformation experiments, some breakage into smaller pieces is inevitable. If two donor genes are located close together on the chromosome, then there is a greater chance that they will be carried on the same piece of transforming DNA and hence will cause a **double transformation.** Conversely, if genes are widely separated on the chromosome, then they will be carried on separate transforming segments and the frequency of double transformants will equal the product of the single-transformation frequencies. Thus, it should be possible to test for close linkage by testing for a departure from the product rule.

Unfortunately, the situation is made more complex by several factors—the most important of which is that not all cells in a population of bacteria are **competent,** or able to be transformed. Because single transformations are expressed as proportions, the success of the product

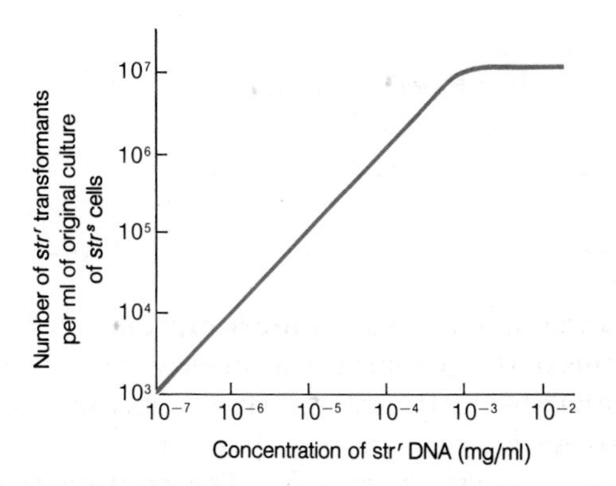

Figure 10-16 The genetic transfer of streptomycin resistance (str^r) to streptomycin-sensitive (str^s) cells of *E. coli.* The recovery of str^r transformants among str^s cells depends on the concentration of str^r DNA. (From G. S. Stent and R. Calendar, *Molecular Genetics,* 2d ed. Copyright 1978 by W. H. Freeman and Company, New York.)

rule obviously depends on the absolute size of these proportions. There are ways of calculating the proportion of competent cells, but we need not detour into that subject now. You can sharpen your skills in transformation analysis in one of the problems at the end of the chapter, which assumes 100 percent competence.

Bacteriophage Genetics

Infection of Bacteria by Phages

Many bacteriophages play an important role in gene transfer, as we will see below. Most bacteria are susceptible to attack by bacteriophages, which literally means "eaters of bacteria." A phage consists of a nucleic acid "chromosome" (DNA or RNA) surrounded by a coat of protein molecules. One well-studied set of phage strains are identified as T1, T2, and so on. Figures 10-17 and 10-18 show the complicated structure of a phage belonging to the class called **T-even phages** (T2, T4, and so on).

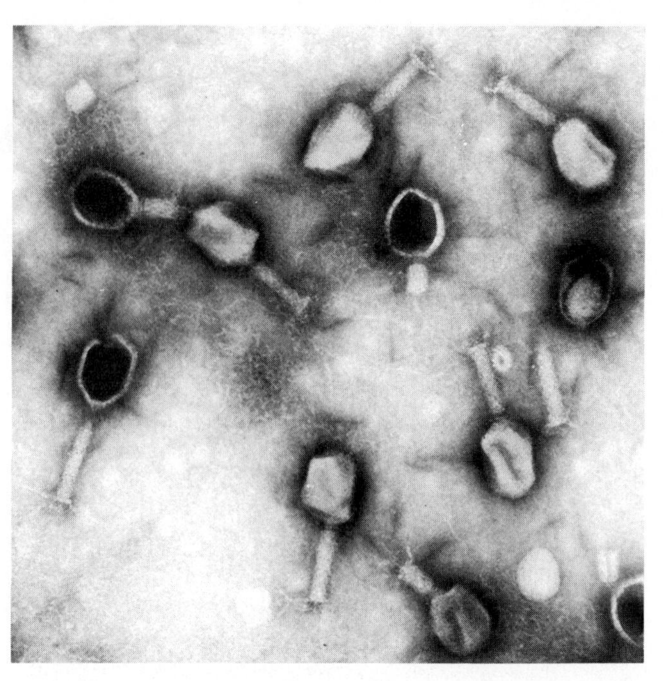

Figure 10-18 Mature particles of the *E. coli* phage T4 (×97,500). (Grant Heilman.)

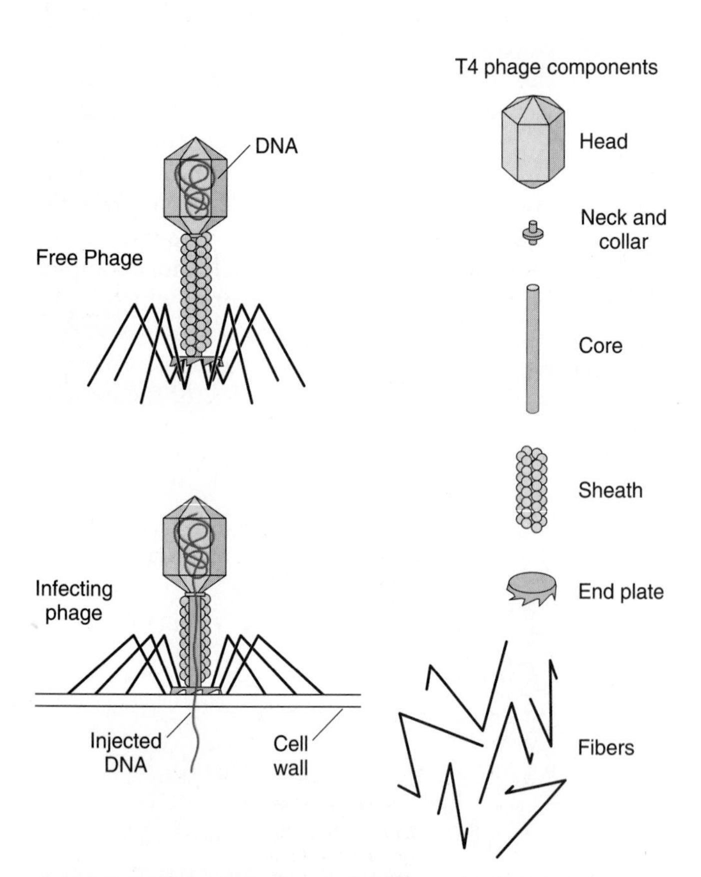

Figure 10-17 Phage T4, shown in its free state and in the process of infecting an *E. coli* cell. The infecting phage injects DNA through its core structure into the cell. On the right, a phage has been diagrammatically exploded to show its highly ordered three-dimensional structure. (Modified from R. S. Edgar and R. H. Epstein, "The Genetics of a Bacterial Virus." Copyright 1965 by Scientific American, Inc. All rights reserved.)

During infection, phage attaches to a bacterium and injects its genetic material into the bacterial cytoplasm (Figure 10-19a). The phage genetic information then takes over the machinery of the bacterial cell by turning off the synthesis of bacterial components and redirecting the bacterial synthetic material to make more phage components (Figure 10-19b). (The use of the word "information" is interesting in this connection; it literally means "to give form." And of course, that is precisely the role of the genetic material: to provide blueprints for the construction of form. In the present discussion, the form is the elegantly symmetrical structure of the new phages.) Ultimately, many phage descendants are released when the bacterial cell wall breaks open. This breaking-open process is called **lysis.**

But how can we study inheritance in phages when they are so small that they are visible only under the electron microscope? In this case, we cannot produce a visible colony by plating, but we can produce a visible manifestation of an infected bacterium by taking advantage of several phage characters. Let's look at the consequences of a phage's infecting a single bacterial cell. Figure 10-20 shows the sequence of events in the infectious cycle that leads to the release of progeny phages from the lysed cell. After lysis, the progeny phages infect neighboring bacteria. This is an exponentially explosive phe-

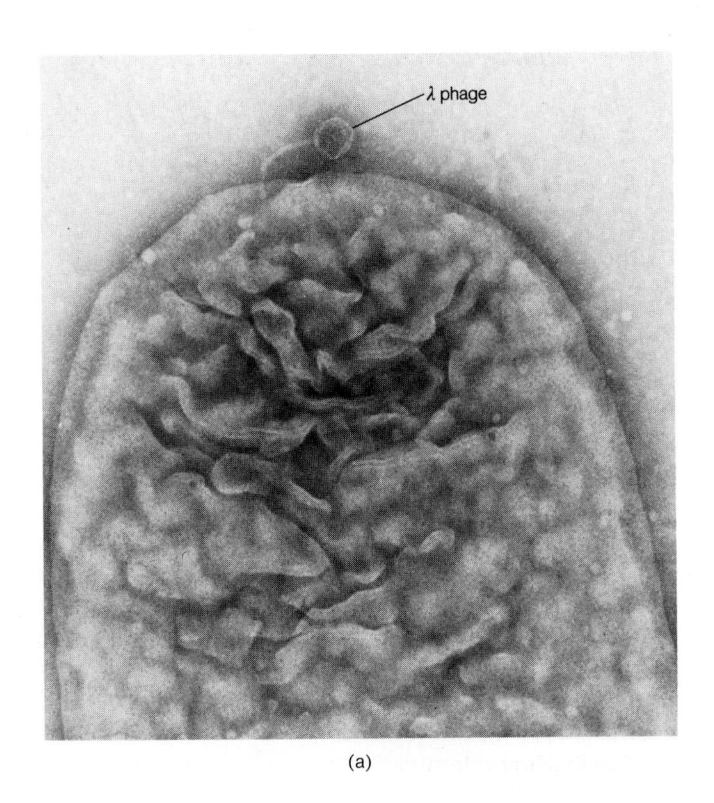

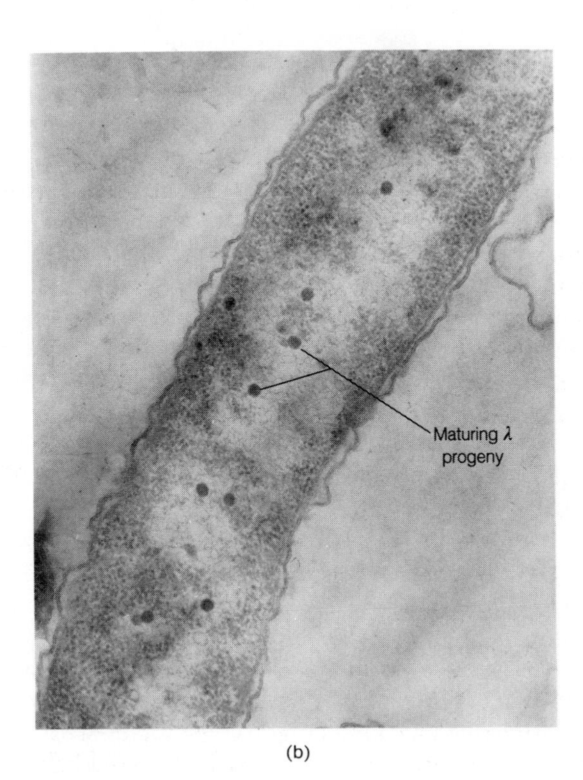

Figure 10-19 (a) Bacteriophage λ attached to an *E. coli* cell and injecting its genetic material. (b) Progeny particles of phage λ maturing inside an *E. coli* cell. (Jack D. Griffith.)

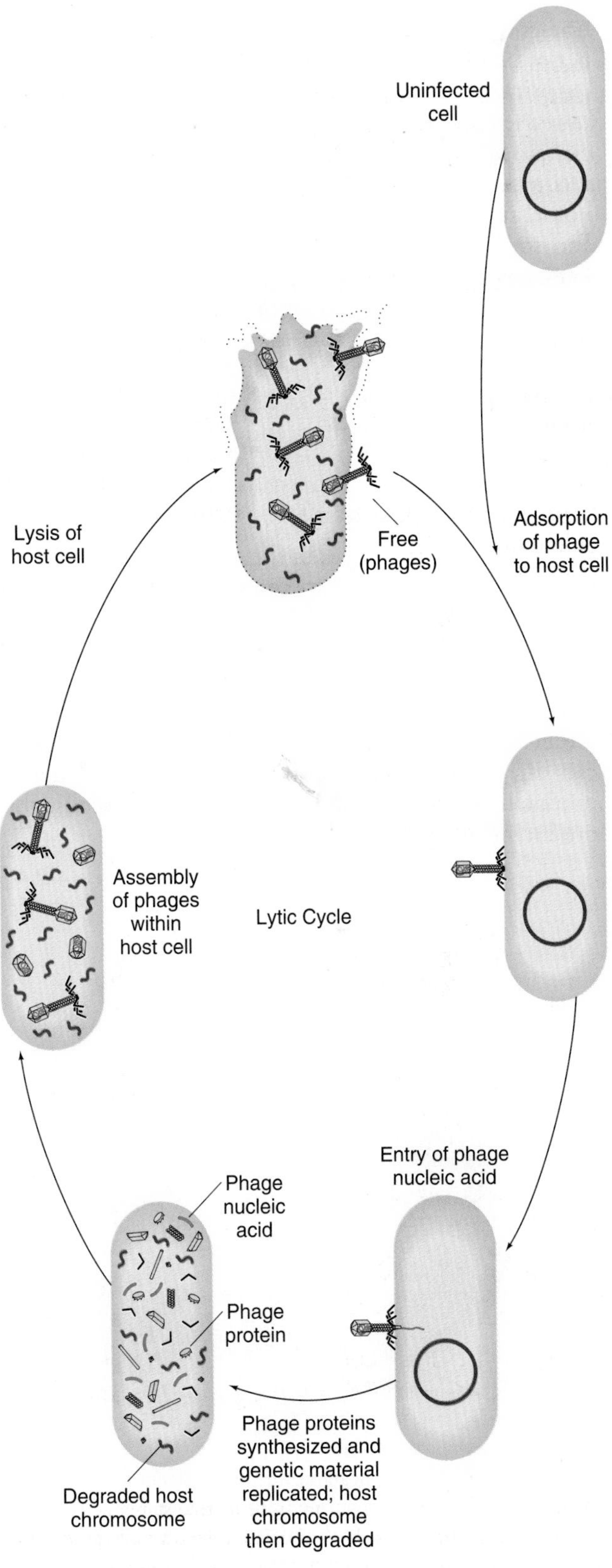

Figure 10-20 A generalized bacteriophage lytic cycle. (Adapted from J. Darnell, H. Lodish and D. Baltimore, *Molecular Cell Biology.* Copyright 1986 by W. H. Freeman and Company, New York.)

nomenon (it causes an exponential increase in the number of lysed cells). Within 15 hours after the start of an experiment of this type, the effects are visible to the naked eye: a clear area, or **plaque,** is present on the opaque lawn of bacteria on the surface of a plate of solid medium (Figure 10-21). Depending on the phage genome, such plaques can be large or small, fuzzy or sharp, and so forth. Thus, **plaque morphology** is a phage character that can be analyzed.

Another phage phenotype that can be analyzed genetically is **host range.** Certain strains of bacteria are immune to adsorption (attachment) or injection by phages. Phages, in turn, may differ in the spectra of bacterial strains they can infect and lyse.

The Phage Cross

Can we cross two phages in the same way we cross two bacterial strains? A phage cross can be illustrated by a cross of T2 phages originally studied by Alfred Hershey. The genotypes of the two parental strains of T2 phage in Hershey's cross were $h^- r^+ \times h^+ r^-$. The alleles are identified by the following characters: h^- can infect two different *E. coli* strains (which we can call strains 1 and 2); h^+ can infect only strain 1; r^- rapidly lyses cells, thereby producing large plaques; and r^+ slowly lyses cells, thus producing small plaques.

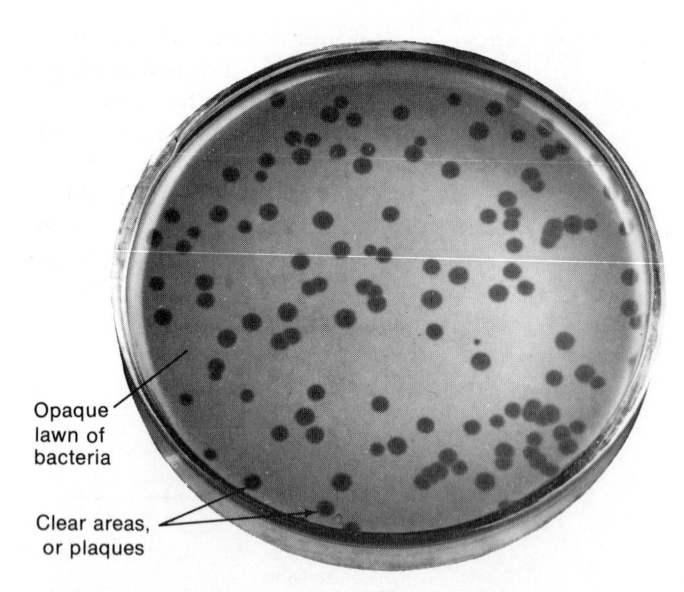

Figure 10-21 The appearance of phage plaques. Individual phages are spread on an agar medium that contains a fully grown "lawn" of *E. coli.* Each phage infects one bacterial cell, producing 100 or more progeny phages that burst the *E. coli* cell and infect neighboring cells. They in turn are exploded with progeny, and the continuing process produces a clear area, or plaque, on the opaque lawn of bacterial cells. (From G. S. Stent, *Molecular Biology of Bacterial Viruses.* Copyright 1963 by W. H. Freeman and Company, New York.)

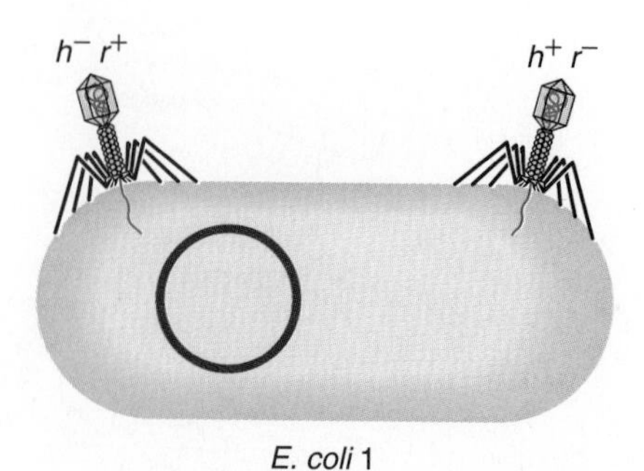

Figure 10-22 A double infection of *E. coli* by two phages.

In the cross, *E. coli* strain 1 is infected with both parental T2 phage genotypes at a phage : bacteria concentration (called **multiplicity of infection**) that is high enough to ensure that a large percentage of cells are simultaneously infected by both phage types. This kind of infection (Figure 10-22) is called a **mixed infection,** or a **double infection.** The phage lysate (the progeny phage) is then analyzed by spreading it onto a bacterial lawn composed of a mixture of *E. coli* strains 1 and 2. Four plaque types are then distinguishable (Figure 10-23 and Table 10-2). These four genotypes can be scored easily as parental ($h^- r^+$ and $h^+ r^-$) and recombinant, and a recombinant frequency can be calculated as follows:

$$\text{RF} = \frac{(h^+ r^+) + (h^- r^-)}{\text{total plaques}}$$

If we assume that entire phage genomes recombine, then we are not faced with a merozygous situation, as in a bacterial cross. Presumably, then, single exchanges can occur and produce viable reciprocal products. Nevertheless, phage crosses are subject to complications. First, several rounds of exchange can potentially occur within the host: a recombinant produced shortly after infection may undergo further exchange at later times. Second, recombination can occur between genetically similar phages as well as between different types. Thus, $P_1 \times P_1$ and $P_2 \times P_2$ occur in addition to $P_1 \times P_2$ (P_1 and P_2 refer to phage 1 and phage 2, respectively). For both of these

Table 10-2. Progeny-phage Plaque Types from Cross $h^- r^+ \times h^+ r^-$

Phenotype	Inferred genotype
Clear and small	$h^- r^+$
Cloudy and large	$h^+ r^-$
Cloudy and small	$h^+ r^+$
Clear and large	$h^- r^-$

NOTE: Clearness is produced by the h^- allele, which allows infection of *both* bacterial strains in the lawn; cloudiness is produced by the h^+ allele, which limits infection to the cells of strain 1.

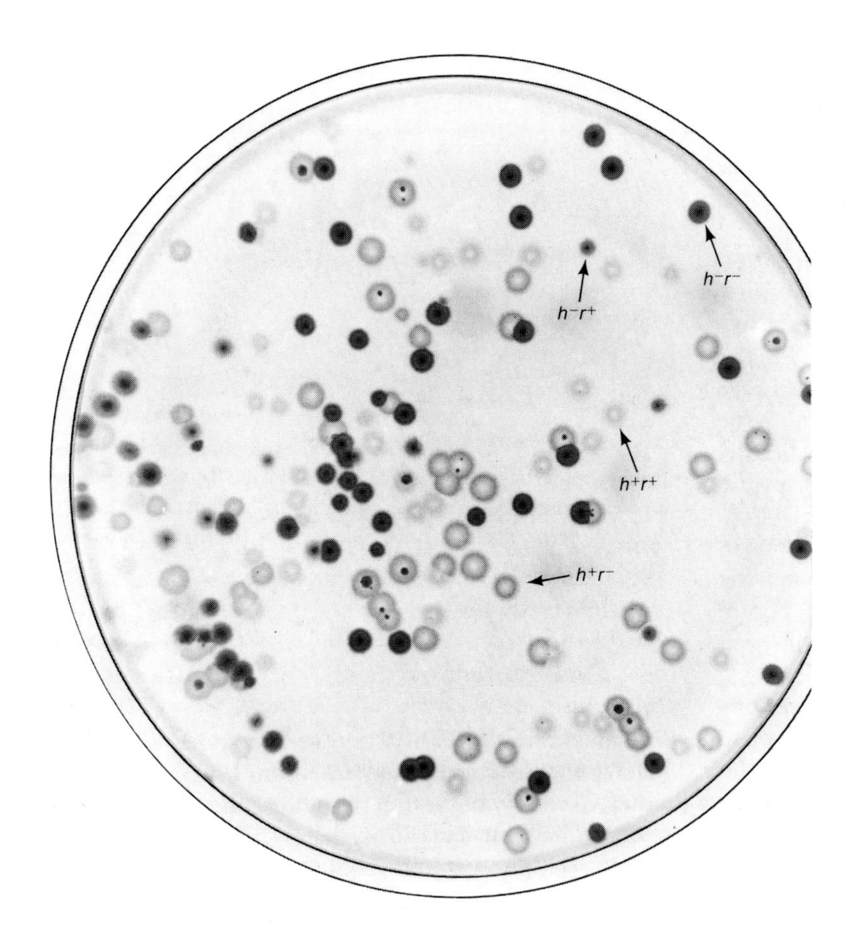

Figure 10-23 Plaque phenotypes produced by progeny of the cross $h^- r^+ \times h^+ r^-$. Enough phages of each genotype are added to ensure that most bacterial cells are infected with at least one phage of each genotype. After lysis, the progeny phages are collected and added to an appropriate *E. coli* lawn. Four plaque phenotypes can be differentiated, representing two parental types and two recombinants. (From G. S. Stent, *Molecular Biology of Bacterial Viruses.* Copyright 1963 by W. H. Freeman and Company, New York.)

reasons, recombinants from phage crosses are a consequence of a *population* of events rather than defined, single-step exchange events. Nevertheless, *all other things being equal,* the RF calculation does represent a valid index of map distance in phages.

Circularity of the T2 Genetic Map

Hershey obtained several different T2 strains with the rapid-lysis phenotype; he called their genotypes *r1*, *r2*, and so forth (in the order of discovery). Let's indicate three different *r* strains as r_a, r_b, and r_c, each with a mutation in a different gene, and make the cross $r_x^- h^+ \times r_x^+ h^-$, where r_x represents one of the three *r* genes. Table 10-3 shows the results. We can construct a linkage map in just the same way that Sturtevant constructed the original eukaryotic maps in *Drosophila.* The parental types ($r^- h^+$ and $r^+ h^-$) occur with the highest frequencies, although not with equal frequency. The two recombinant classes, however, are equally frequent. We can construct linkage maps for each cross (Figure 10-24a). The different recombination values indicate that the loci for the three *r* genes are in different places on the chromosome, so there are four possible linkage maps (Figure 10-24b).

Table 10-3. Frequency of Progeny-phage Types in Crosses Involving Several *r* Mutants and an *h* Mutant

	Percentage of each genotype			
Cross	$r^- h^+$	$r^+ h^-$	$r^+ h^+$	$r^- h^-$
$r_a^- h^+ \times r_a^+ h^-$	34.0	42.0	12.0	12.0
$r_b^- h^+ \times r_b^+ h^-$	32.0	56.0	5.9	6.4
$r_c^- h^+ \times r_c^+ h^-$	39.0	59.0	0.7	0.9

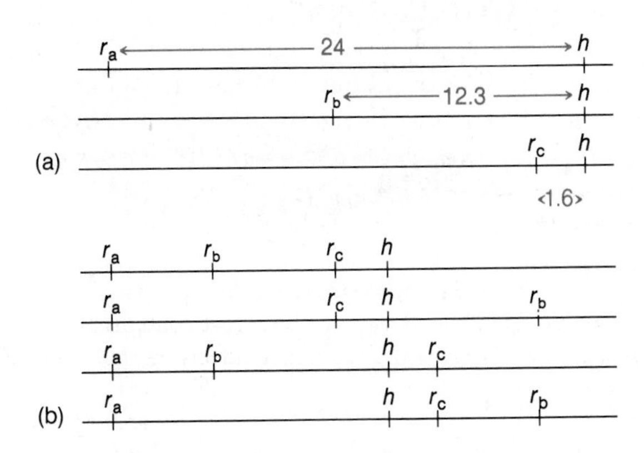

Figure 10-24 (a) Distances between gene pairs for each cross given in Table 10-3. (b) The various possible linkage relationships inferred from the distances in (a).

Can we distinguish among these alternatives? First, let's take only r_b, r_c, and h and ask whether the order is r_c, h, r_b or h, r_c, r_b. We can make the cross $r_c^- r_b^+ \times r_c^+ r_b^-$ and compare the RF with the value of 12.3 obtained for the $r_b - h$ interval. From this comparison, we find that h is located between r_c and r_b (r_c, h, r_b).

Now we ask whether r_a lies on the side of h next to r_b or on the side next to r_c. The data from crosses of r_a with r_b and r_c do not provide a clear-cut answer. After intensive genetic mapping with many different strains of T2, the answer turns out to be that *both* alternative maps are correct. How can both r_a, r_c, h, r_b and r_c, h, r_b, r_a be correct? We have encountered a similar situation before, and the answer is similar. Like the bacterial map, the linkage map of T2 is circular, as shown in Figure 10-25a. After infection, the linear phage genome forms a circle in order to replicate the chromosome. The total genetic length of the T2 linkage map is about 1500 m.u. Figure 10-25b shows a more complex map for another T-even phage.

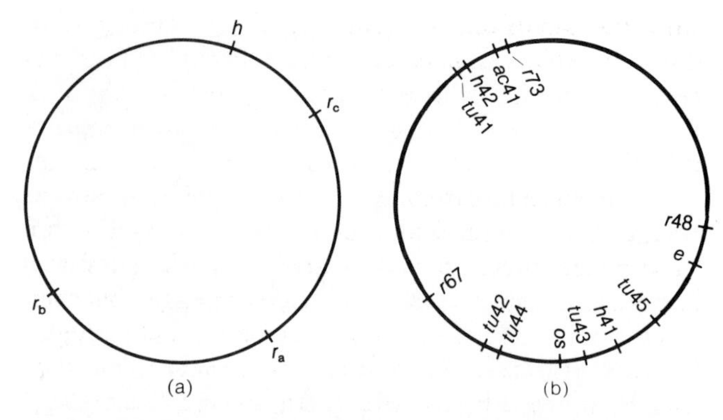

Figure 10-25 Circular maps for T-even phages. (a) Simple linkage map for phage T2. (b) Map of phage T4, inferred from crosses of nonlethal mutants *r* (rapid lysis), *h* (host range), *ac* (acridine resistance), *tu* (turbid plaques), *os* (resistance to osmotic shock), and *e* (lysis defective).

In the bacterial and phage experiments we have considered, the data are initially confusing. However, once the novel conditions for "crossing" are understood, the analysis of recombination and the construction of linkage maps are fairly straightforward procedures.

Message Recombination between phage chromosomes can be studied by bringing the parental chromosomes together in one host cell through mixed infection. Progeny phages can be examined for parental versus recombinant genotypes.

Lysogeny

In the 1920s, long before *E. coli* became the favorite organism of microbial geneticists, some interesting results were obtained in the study of phage infections of *E. coli.* Some bacterial strains were found to be resistant to infection by certain phages, but these resistant bacteria would cause lysis of nonresistant bacteria when the two bacterial strains were mixed together. The resistant bacteria that induced lysis in other cells were said to be **lysogenic bacteria,** or **lysogens.** When nonlysogenic bacteria were infected with phages derived from a lysogenic strain, a small fraction of the infected cells do not lyse but instead became lysogenic themselves.

Apparently, the lysogenic bacteria could somehow "carry" the phages while remaining immune to their lysing action. Initially, little attention was paid to this phenomenon after some studies seemed to show that the lysogenic bacteria were simply contaminated with external phages that could be removed by careful purification. However, in the mid-1940s, André Lwoff examined lysogenic strains of *Bacillus megaterium* and followed the behavior of a lysogenic strain through many cell divisions. Carefully observing his culture, he separated each pair of daughter cells immediately after division. One cell was put into a culture; the other was observed until it divided. In this way, Lwoff obtained 19 cultures representing 19 generations (19 consecutive cell divisions). All 19 cultures were lysogenic, but tests of the medium showed no free phage at any time during these divisions, thereby confirming that lysogenic behavior is a character that persists through reproduction in the absence of free phage.

On rare occasions, Lwoff observed spontaneous lysis in his cultures. When the medium was spread on a lawn of nonlysogenic cells after one of these spontaneous lyses, plaques appeared, showing that free phages had been released in the lysis. Lwoff was able to propose a hypothesis to explain all his observations: each bacterium of the lysogenic strain contains a noninfective factor that is passed from bacterial generation to generation, but this factor occasionally gives rise to the production of infective phage (without the presence of free phage in the medium). Lwoff called this factor the **prophage** because it somehow seemed to be able to *induce* the formation of a "litter" of infective phage. Later studies showed that a variety of agents, such as ultraviolet light or certain chemicals, could activate the prophage, inducing lysis and infective phage release in a large fraction of a population of lysogenic bacteria.

We now know exactly how Lwoff's observations occur. A lysogenic bacterium contains a prophage, which somehow protects the cell against additional infection, or **superinfection,** from free phages and which is duplicated and passed on to daughter cells during division. In a small fraction of the lysogenic cells, the prophage is induced, or activated, producing infective phage. This process robs the cell of its protection against the phage; it lyses and releases infective phage into the medium, thus infecting any nonlysogenic cells present in the culture.

Phages can be categorized into two types. **Virulent phages** have an infectious cycle that is always **lytic** — for these phages there are no lysogenic bacteria. (Resistant bacterial mutants may exist for virulent phages, but their resistance is not due to lysogeny.) **Temperate phages** follow a lytic cycle under some circumstances, but they usually initiate a **lysogenic cycle,** in which the phage exists as a prophage within the bacterial cell. In this case, the lysogenic bacterium becomes resistant to superinfection, an "immunity" conferred by the presence of the prophage, which is transmitted genetically through many bacterial generations. Temperate phages also cause lysis when the prophage is **induced,** or activated. Figure 10-26 diagrams the lytic and lysogenic infectious cycles of a typical temperate phage.

Message Virulent phages cannot become prophages; they are always lytic. Temperate phages can exist within the bacterial cell as prophages, allowing their hosts to survive as lysogenic bacteria; they are also capable of direct bacterial lysis.

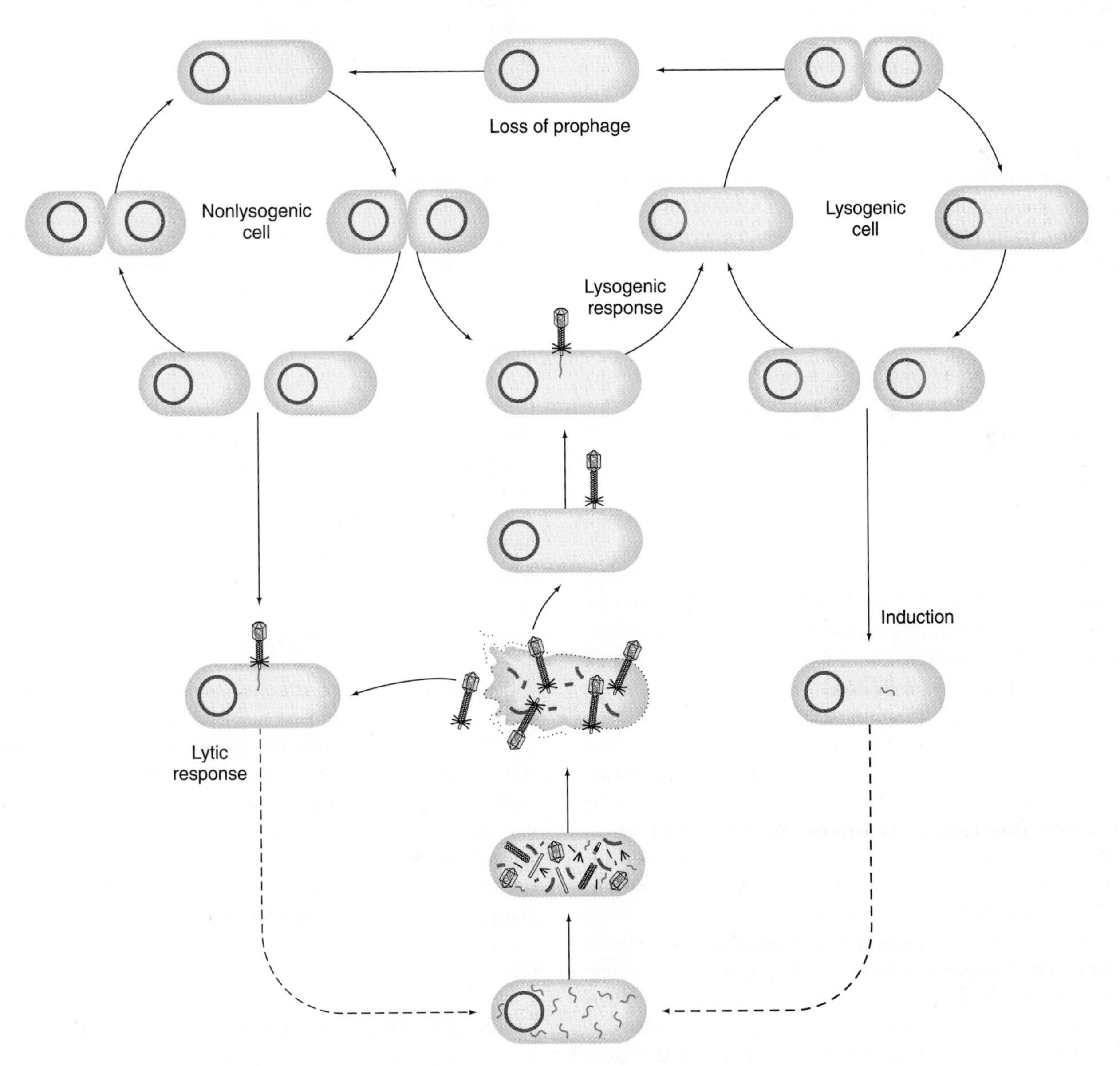

Figure 10-26 Alternative cell cycles of a temperate phage and its host. (Adapted from A. Lwoff, *Bacteriological Reviews* 17, 1953, 269.)

The Genetic Basis of Lysogeny

What is the nature of the prophage? On induction, the prophage is capable of directing the production of complete mature phage, so all of the phage genome must be present in the prophage. But is the prophage a small particle free in the bacterial cytoplasm—a plasmid—or is it somehow associated with the bacterial genome? Fortuitously, the original strain of *E. coli* used by Lederberg and Tatum (page 269) proved to be lysogenic for a temperate phage called **lambda (λ)**. Phage λ (Figure 10-27) has become the most intensively studied and best characterized phage. Crosses between F^+ and F^- cells have yielded interesting results. It turns out that $F^+ \times F^-(\lambda)$ crosses yield recombinant lysogenic recipients, whereas the reciprocal cross $F^+(\lambda) \times F^-$ almost never gives lysogenic recombinants.

These results became more understandable when Hfr strains were discovered. In the cross Hfr $\times$ $F^-(\lambda)$, lysogenic F^- exconjugants with Hfr genes are readily recovered. However, in the reciprocal cross Hfr(λ) $\times$ F^-, the early genes from the Hfr chromosome are recovered among the exconjugants, but recombinants for late markers (those expected to transfer after a certain time in mating) are not recovered. Furthermore, lysogenic exconjugants are almost never recovered from this reciprocal cross. What is the explanation? The observations make sense if the λ prophage is behaving like a bacterial gene locus (that is, like part of the bacterial chromosome). In interrupted-mating experiments, the λ prophage always enters the F^- cell at a specific time, closely linked to the *gal* locus. Thus, we can assign the λ prophage to a specific locus next to the *gal* region.

In the cross of a lysogenic Hfr with a nonlysogenic (nonimmune) F^- recipient, the entry of the λ prophage into the nonimmune cell immediately triggers the prophage into a lytic cycle; this is called **zygotic induction.** But in the cross Hfr(λ) $\times$ $F^-(\lambda)$, any recombinants are readily recovered (that is, no induction of the prophage, and consequently lysis, occurs). It would seem that the cytoplasm of the F^- cell must exist in two different states (depending on whether or not the cell contains a λ prophage), so that contact between an entering prophage and the cytoplasm of a nonimmune cell immediately induces the lytic cycle. We now know that a cytoplasmic factor specified by the prophage represses the multiplication of the virus. Entry of the prophage into a nonlysogenic environment immediately dilutes this repressing factor, and therefore the virus reproduces. But if the virus specifies the repressing factor, then why doesn't the virus shut itself off again? Clearly it does, because a fraction of infected cells do become lysogenic. There is a race between the λ gene signals for reproduction and those specifying a shutdown. The model of a phage-directed cytoplasmic repressor nicely explains the immunity of the lysogenic bacteria, because any superinfecting phage would immediately encounter a repressor and be inactivated. We discuss this model in more detail in Chapter 17.

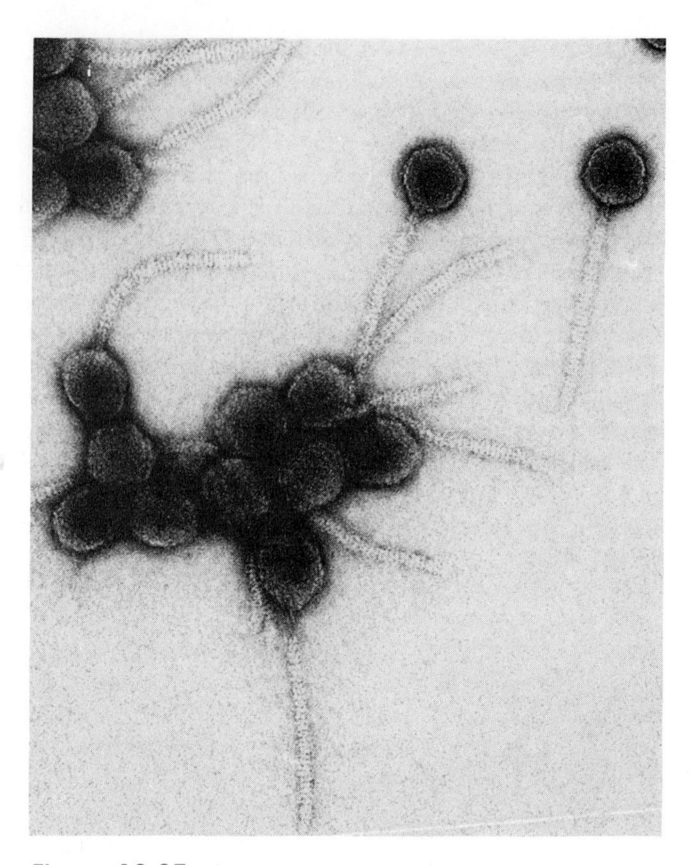

Figure 10-27 Electron micrograph of λ phages. (Jack D. Griffith.)

Prophage Attachment

How is the prophage attached to the bacterial genome? Allan Campbell proposed in 1962 that λ attaches to the bacterial chromosome by a reciprocal crossover between the circular λ chromosome and the circular *E. coli* chromosome, as shown in Figure 10-28. The crossover point would occur between a specific site in λ, the **λ attachment site,** and a site in the bacterial chromosome located between the genes *gal* and *bio*, since λ integrates at that position in the *E. coli* chromosome.

One attraction of Campbell's proposal is that it allows predictions that geneticists can test, using phage λ:

1. Integration of the prophage into the *E. coli* chromosome should increase the genetic distance between flanking bacterial markers, as can be seen in Figure 10-28 for *gal* and *bio*. In fact, studies show that time-

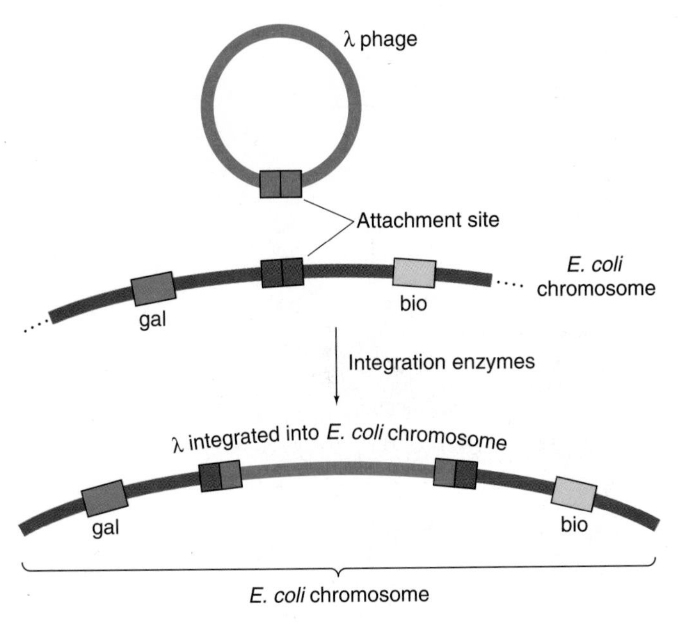

Figure 10-28 Campbell's model for the integration of phage λ into the *E. coli* chromosome. Reciprocal recombination occurs between a specific attachment site on the circular λ DNA and a specific region on the bacterial chromosome between the *gal* and *bio* genes.

of-entry or recombination distances between the bacterial genes *are* increased by lysogeny.

2. Deleting bacterial segments adjacent to the prophage site should delete phage genes at least some of the time. Experimental studies also confirm this prediction.

The phenomenon of lysogeny is a very successful way for a temperate phage to avoid eating itself out of house and home. Lysogenic cells can perpetuate and carry the phages around. We consider lysogeny and the integration of the λ phage into the host chromosome in more detail in Chapter 17.

Transduction

Some phages are able to "mobilize" bacterial genes and carry them from one bacterial cell to another through the process of **transduction.** Thus, transduction joins the battery of modes of genetic transfer in bacteria—along with conjugation, infectious transfer of episomes, and transformation.

There are two kinds of transduction: generalized and specialized. Generalized transducing phage can carry any part of the chromosome, whereas, **specialized** transducing phages carry only restricted parts of the bacterial chromosome.

The Discovery of Transduction

In 1951, Joshua Lederberg and Norton Zinder were testing for recombination in the bacterium *Salmonella typhimurium,* using the techniques that had been successful with *E. coli.* The researchers used two different strains: one was *phe⁻ trp⁻ tyr⁻*, and the other was *met⁻ his⁻*. (We won't worry about the nature of these markers except to note that the mutant alleles confer nutritional requirements.) When either strain was plated on a minimal medium, no wild-type cells were observed. However, after mixing the two strains, wild-type cells occurred at a frequency of about 1 in 10^5. Thus far, the situation seems similar to that for recombination in *E. coli.*

However, in this case, the researchers also recovered recombinants from a U-tube experiment, in which cell contact (conjugation) was prevented by a filter separating the two arms. By varying the size of the pores in the filter, they found that the agent responsible for recombination was about the size of the virus P22, a known temperate phage of *Salmonella.* Further studies have supported the suggestion that the vector of recombination is indeed P22. The filterable agent and P22 are identical in properties of size, sensitivity to antiserum, and immunity to hydrolytic enzymes. Thus, Lederberg and Zinder, instead of confirming conjugation in *Salmonella,* had discovered a new type of gene transfer mediated by a virus. They called this process **transduction.** During the lytic cycle, some virus particles somehow pick up bacterial genes that are then transferred to another host, where the virus inserts its contents. Transduction has subsequently been shown to be quite common among both temperate and virulent phages.

Transducing Phages and Generalized Transduction

How are transducing phages produced? In 1965, K. Ikeda and J. Tomizawa threw light on this question in some experiments on the temperate *E. coli* phage P1. They found that when a nonlysogenic donor cell is lysed by P1, the bacterial chromosome is broken up into small pieces. Occasionally, the forming phage particles mistakenly incorporate a piece of the bacterial DNA into a phage head in place of phage DNA. This is the origin of the transducing phage. A similar process can occur when the prophage of P1 is induced.

Because the phage coat proteins determine a phage's ability to attack a cell, transducing phages can bind to a bacterial cell and inject their contents, which now happen to be donor bacterial genes. When a transducing phage injects its contents into a recipient cell, a merodiploid situation is created in which the transduced bacterial genes can be incorporated by recombination (Figure 10-29). Because any of the host markers can be trans-

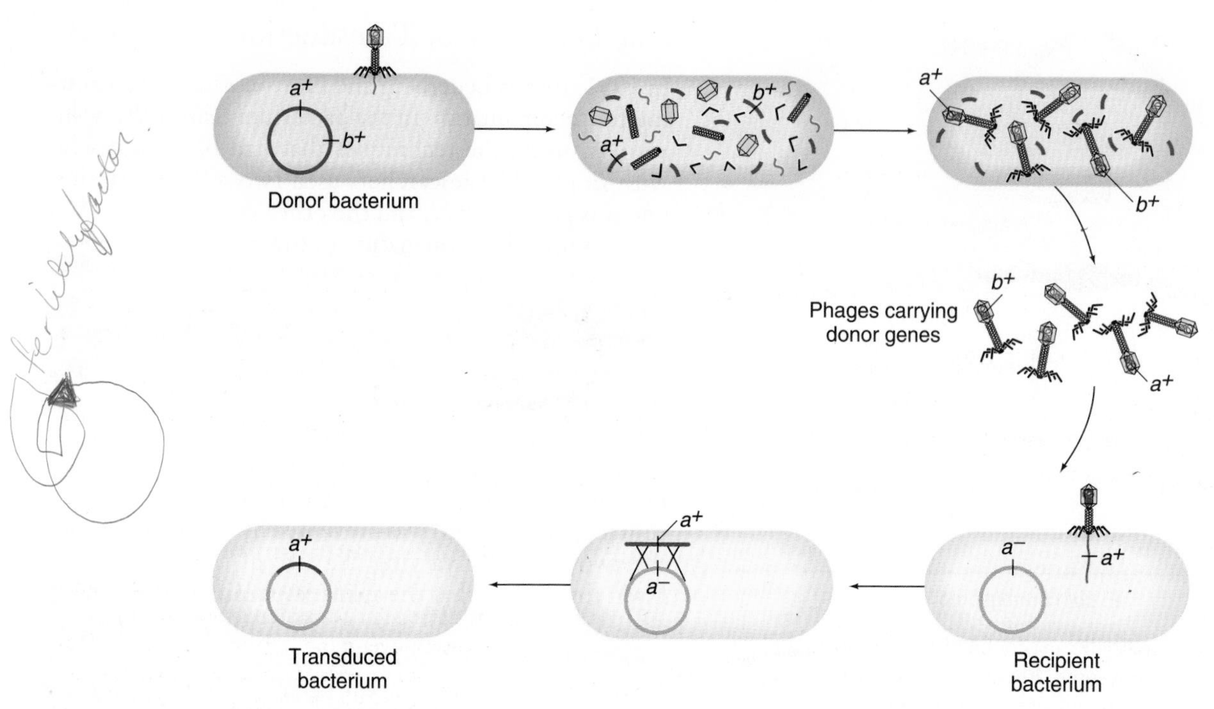

Figure 10-29 The mechanism of generalized transduction. In reality, only a very small minority of phage progeny (1 in 10,000) carries donor genes.

duced, this type of transduction is termed **generalized transduction.**

Phages P1 and P22 both belong to a phage group that shows generalized transduction (that is, they transfer virtually any gene of the host chromosome). As prophages, P22 probably inserts into the host chromosome and P1 remains free like a large plasmid. But both transduce by faulty headstuffing during lysis.

Specialized Transduction

We look now at the process of **specialized transduction,** in which only certain host markers can be transduced.

Lambda (λ) is a good example of a specialized transducer. As a prophage, λ always inserts between the *gal* region and the *bio* region of the *E. coli* host chromosome. In transduction experiments, λ can transduce only the *gal* and *bio* genes. Let's visualize the mechanism of λ transduction:

In Figure 10-30a, we see a schematic representation of the production of a lysogen, or lysogenic bacterium. (The actual recombination between regions of λ and the bacterial chromosome is catalyzed by a specific enzyme system, described more fully in Chapter 19.) The phage and bacterial integration regions are not completely identical, so the integration of λ results in two hybrid integration sites, as shown.

We can induce the lytic cycle with ultraviolet light and produce a lysate (progeny phage population). The normal outlooping of the prophage restores the original phage integration site (Figure 10-30b, i). These phages can integrate normally (as in Figure 10-30a) on subsequent infection of a strain that is not lysogenic for λ.

Very rarely, abnormal outlooping can result in phage particles that now carry the *gal* gene (Figure 10-30b, ii). These particles are **defective** in that some phage genes have been left behind in the host; consequently, they are called **λdgal** (λ-defective *gal*). The λ*dgal* particle has a λ protein coat and tail fibers and can infect bacteria, but it is also defective in its integration site. The hybrid integration site left in λ*dgal* does not provide a correct substrate for the phage-specific enzyme that promotes recombination between the phage and bacterial integration sites. Therefore, efficient integration cannot occur in a single infection of *E. coli* by λ*dgal*. However, coinfection with a wild-type λ phage results in efficient integration of the λ*dgal* phage (Figure 10-13c, i); the second, wild-type phage is often termed a **helper phage**. In general, a helper phage provides something that is required for integration into the host chromosome.

In practice, the lysate of λ produced originally (Figure 10-30b) can be used to infect a gal^- recipient culture that is nonlysogenic for λ. These infected cells are plated on a minimal medium, and rare gal^+ transductants are the only ones to grow and produce colonies. A small per-

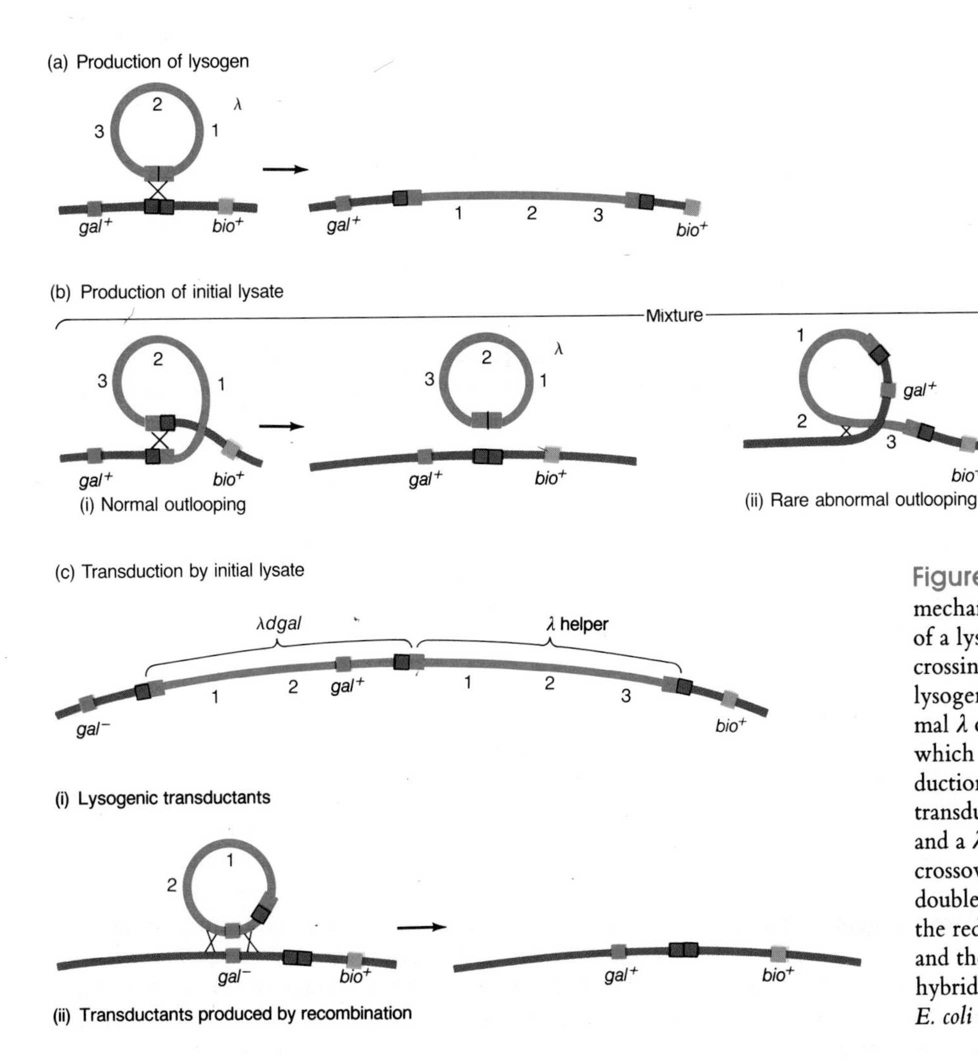

Figure 10-30 Specialized transduction mechanisms in phage λ. (a) The production of a lysogenic bacterium takes place by crossing-over in a specialized region. (b) The lysogenic bacterial culture can produce normal λ or, rarely, an abnormal particle, *λdgal*, which is the transducing particle. (c) Transduction by the mixed lysate can produce gal^+ transductants by the coincorporation of *λdgal* and a λ helper phage or, more rarely, by crossovers flanking the *gal* gene. The purple double squares are bacterial integration sites, the red double squares are λ integration sites, and the pairs of purple and red squares are hybrid integration sites, derived partly from *E. coli* and partly from λ.

centage of these transductants result from recombination between the *gal* regions of the *λdgal* and the recipient chromosome (Figure 10-30c, ii). The vast majority of the transductants (Figure 10-30c, i) are double lysogens: If such a culture is lysed and used as a gal^+ donor in transduction, a very high frequency of transduction is obtained. This lysate is called a **high-frequency transduction (HFT)** lysate. HFT lysates contain a significant fraction of *λdgal* specialized-transducing phage, since each lysogen already contained a *λdgal* phage, and induction results in the excision and propagation of both *λdgal* and helper phage. This is in contrast to the original single lysogen (Figure 10-30a), which on induction yielded *λdgal* phage (Figure 10-30b, i) at a very low frequency.

Specialized transduction can only mobilize small regions of the bacterial genome that flank the prophage. Transduction by λ can mobilize either the *gal* or the *bio* genes, since these are the genes which can be incorporated into λ by rare, aberrant excision events. Specialized transduction is useful for moving genes from one bacterial strain to another.

Linkage Data from Transduction

Generalized transduction allows us to derive linkage information about bacterial genes when markers are close enough that the phage can pick them up and transduce them in a single piece of DNA. For example, suppose we wanted to find the linkage between *met* and *arg* in *E. coli*. We might set up a cross of a $met^+ arg^+$ strain with a $met^- arg^-$ strain. We could grow phage P1 on the donor $met^+ arg^+$ strain, allow P1 to infect the $met^- arg^-$ strain, and select for met^+ colonies. Then, we could note the percentage of met^+ colonies that became arg^+. Strains transduced to both met^+ and arg^+ are called **cotransductants.**

Linkages usually are expressed as cotransduction frequencies (Figure 10-31). The greater the contransduction frequency, the closer two genetic markers are.

We can estimate the size of the piece of host chromosome that a phage can pick up by using P1 phage in the following type of experiment:

$$\text{donor } leu^+\ thr^+\ azi^r \longrightarrow \text{recipient } leu^-\ thr^-\ azi^s$$

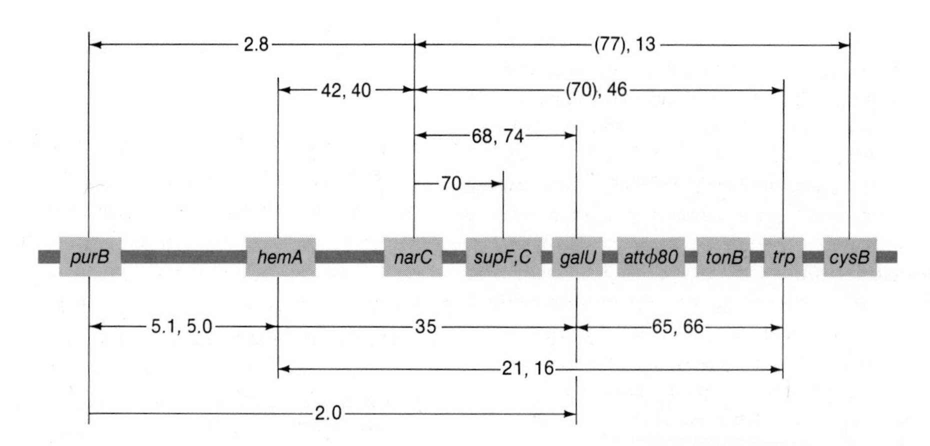

Figure 10-31 Genetic map of the *purB* to *cysB* region of *E. coli* determined by P1 cotransduction. The numbers given are the averages in percent for cotransduction frequencies obtained in several experiments. Where transduction crosses were performed in both directions, the head of each arrow points to the selective marker with the corresponding linkage nearest to each arrow. The values in parentheses are considered unreliable owing to interference from the nonselective marker. (Redrawn from J. R. Guest, *Molecular and General Genetics* 105, 1969, 285.)

We can select for one or more donor markers in the recipient and then (in true merozygous genetics style) look for the presence of the other unselected markers, as outlined in Table 10-4. Experiment 1 in the table tells us that *leu* is relatively close to *azi* and distant from *thr*, leaving us with two possibilities:

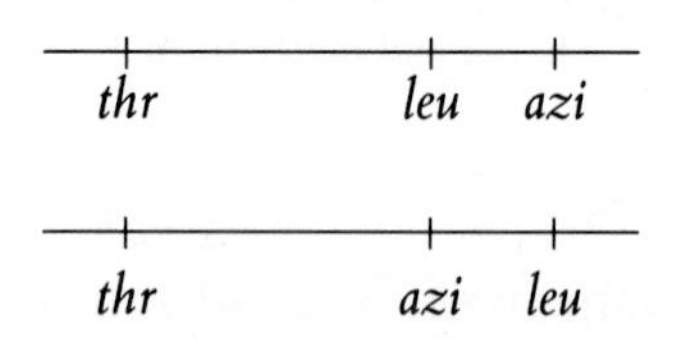

Experiment 2 tells us that *leu* is closer to *thr* than *azi* is, so that the map must be

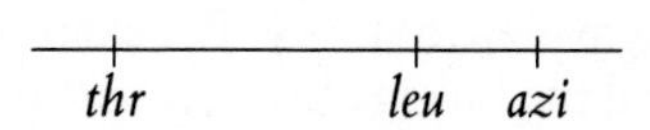

By selecting for thr^+ and leu^+ in the transducing phage in experiment 3, we see that the transduced piece of genetic material never includes the *azi* locus.

Table 10-4. Accompanying Markers in Specific P1 Transductions

Experiment	Selected marker	Unselected markers
1	leu^+	50% are azi^r; 2% are thr^+
2	thr^+	3% are leu^+; 0% are azi^r
3	leu^+ and thr^+	0% are azi^r

If enough markers were studied to produce a more complete linkage map, we could estimate the size of a transduced segment. Such experiments indicate that P1 cotransduction occurs within approximately 1.5 minutes of the *E. coli* chromosome map (1 minute equals the length of chromosome transferred by an Hfr in one minute's time at 37°C).

Linkage in the Discovery of Transduction

At this point, we can figure out how Lederberg and Zinder were able to detect transduction in their experiment. Recall that they utilized two multiply marked strains. One strain required phenylalanine, tryptophan, and tyrosine for growth; the second required methionine and histidine. Based on what we now know about P22- and P1-mediated generalized transduction, we would not expect to restore either of these strains to the wild type unless the markers were so closely linked that they could be carried on the same transducing particle. And yet, that is precisely what happened in the experiment. The met^- his^- strain was the source of the P22 transducing phage. This strain donated the wild-type markers that replaced the phe^- trp^- tyr^- markers. These markers are all part of the same pathway and are very closely linked. (In fact, we now know that the phenylalanine and tyrosine requirements are caused by the same mutation, although the tryptophan requirement is due to a mutation in a different gene.) If Lederberg and Zinder had used a different set of markers, they might never have discovered transduction!

> Message Transduction occurs when newly forming phages acquire host genes and transfer them to other bacterial cells. **Generalized transduction** can transfer any host gene. It occurs when phage packaging accidentally incorporates bacterial DNA instead of phage DNA. This can happen in the lytic cycle of some virulent and temperate phages or when a temperate prophage is induced to lysis. **Specialized transduction** occurs only during prophage induction. It is due to faulty separation of the prophage from the bacterial chromosome, so that the new phage includes both phage and bacterial genes. The transducing phage can transfer only specific host genes because the prophage of any given temperate strain is usually inserted at one specific bacterial locus.

Mapping of Bacterial Chromosomes

Some very detailed chromosomal maps for bacteria have been obtained by combining the mapping techniques of interrupted mating, recombination mapping, transformation, and transduction. Today, new genetic markers are typically mapped first into a segment of about 10 to 15 map minutes by using a series of Hfr strains that transfer from different points around the chromosome. This allows the selection of markers within the interval to be used for P1 cotransduction.

By 1963, the *E. coli* map (Figure 10-32) already detailed the positions of approximately 100 genes. After 27

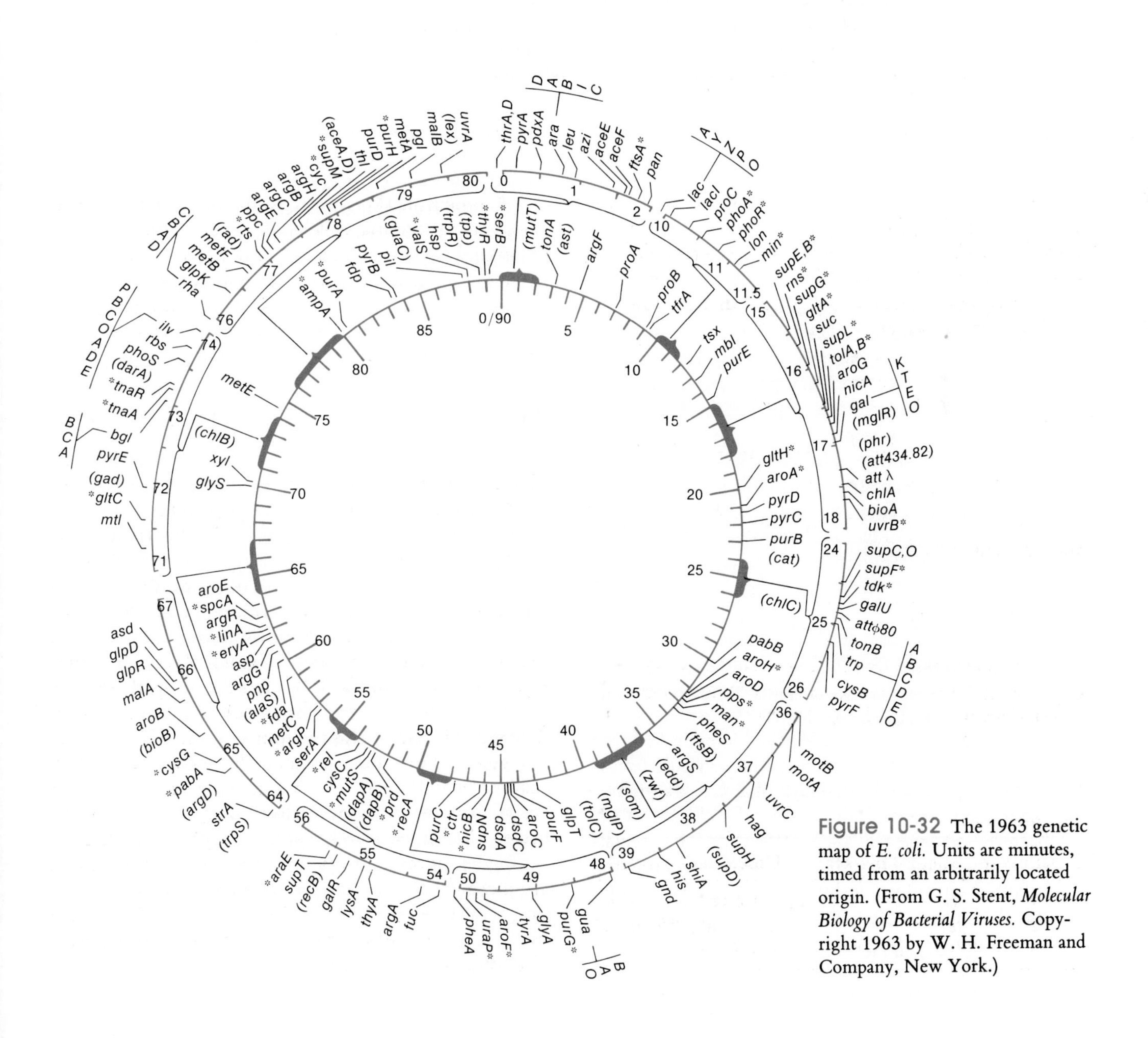

Figure 10-32 The 1963 genetic map of *E. coli*. Units are minutes, timed from an arbitrarily located origin. (From G. S. Stent, *Molecular Biology of Bacterial Viruses*. Copyright 1963 by W. H. Freeman and Company, New York.)

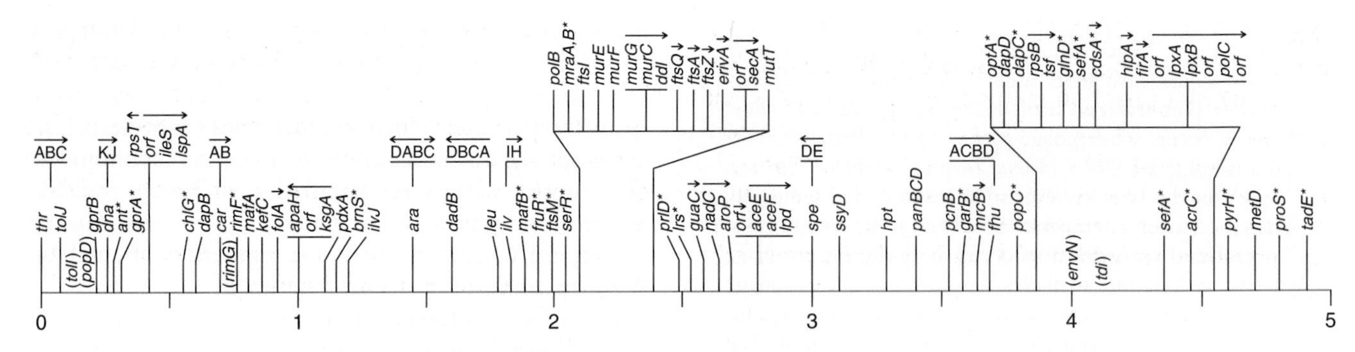

Figure 10-33 Linear scale drawing of a 5-minute portion of the 100-minute 1990 *E. coli* linkage map. Markers in parentheses are not precisely mapped; those marked by asterisks are more precisely mapped than those with parentheses, but are still not exactly known. Arrows above genes and groups of genes indicate the direction of transcription of these loci. From B.7 J. Bachmann, "Linkage Map of *Escherichia coli* K-12, Edition 8." *Microbiological Reviews* 54, 1990, 130–197.

years of further refinement, the 1990 map depicts the positions of more than 1400 genes! Figure 10-33 shows a 5-minute portion of the 1990 map (which is adjusted to a scale of 100 minutes). The complexity of these maps illustrates the power and sophistication of genetic analysis at its best.

Bacterial Gene Transfer in Review

1. Gene transfer in bacteria can be achieved through conjugation, transformation, and viral transduction.
2. The inheritance of genetic markers via the conjugative transfer of DNA by Hfr strains, the transformation of portions of the donor chromosome, and generalized transduction all share one important property. Each process introduces a DNA fragment into the recipient cell; then a double crossover event must occur if the fragment is to be incorporated into the recipient genome and subsequently inherited. Unincorporated fragments cannot replicate and are diluted out and lost from the population of daughter cells.
3. The conjugative transfer of F′ factors that carry bacterial genes and the specialized transduction of certain genetic markers are similar processes in that a specific and limited set of bacterial genes in each case is efficiently introduced into the recipient cell. Inheritance does not require normal recombination, as in the case of the inheritance of DNA fragments. After the F′ transfer, the F′ factor replicates in the bacterial cytoplasm as a separate entity. The specialized transducing phage DNA is recombined into the bacterial chromosome by a recombination system specific for that phage. In both cases, a partial diploid (merodiploid) results, because each process allows the inheritance of the transferred gene and also of the recipient's counterpart.

Summary

Advances in microbial genetics within the past four decades have provided the foundation for recent advances in molecular biology (discussed in the next several chapters). Early in this period, it was discovered that gene transfer and recombination occur between certain different strains of bacteria. In bacteria, however, genetic material is passed in only one direction—from a donor cell (F^+ or Hfr) to a recipient cell (F^-). Donor ability is determined by the presence in the cell of a fertility (F) factor acting as an episome.

On occasion, the F factor present in a free state in F^+ cells can integrate into the *E. coli* chromosome and form an Hfr cell. When this occurs, gene transfer and subsequent recombination take place. Furthermore, since the F factor can insert at different places on the host chromosome, investigators were able to show that the *E. coli* chromosome is a single circle, or ring. Interruptions of the transfer at different times has provided geneticists with a new method for constructing a linkage map of the single chromosome of *E. coli* and other similar bacteria.

Genetic traits can also be transferred from one bacterial cell to another in the form of purified DNA. This process of transformation in bacterial cells was the first demonstration that DNA is the genetic material. For transformation to occur, DNA must be taken into a recipient cell and recombination between a recipient chromosome and the incorporated DNA then must take place.

Bacteria can also be infected by bacteriophages. In one method of infection, the phage chromosome may enter the bacterial cell and, using the bacterial metabolic machinery, produce progeny phage that burst the host bacteria. The new phages can then infect other cells. If two phages of different genotypes infect the same host, recombination between their chromosomes can take place during this lytic process. Mapping the genetic loci through these recombinational events has led to the discovery that some phage chromosomes also are circular.

In another infection method, lysogeny, the injected phage lies dormant in the bacterial cell. In many cases, this dormant phage (the prophage) incorporates into the host chromosome and replicates with it. Either spontaneously or under appropriate stimulation, the prophage can arise from its latency and can lyse the bacterial host cell.

Phages can carry bacterial genes from a donor to a recipient. In generalized transduction, random host DNA is incorporated alone into the phage head during lysis. In specialized transduction, faulty outlooping of the prophage from a unique chromosomal locus results in the inclusion of specific host genes as well as phage DNA in the phage head.

Figure 10-34 summarizes the processes of conjugation, transformation, and transduction.

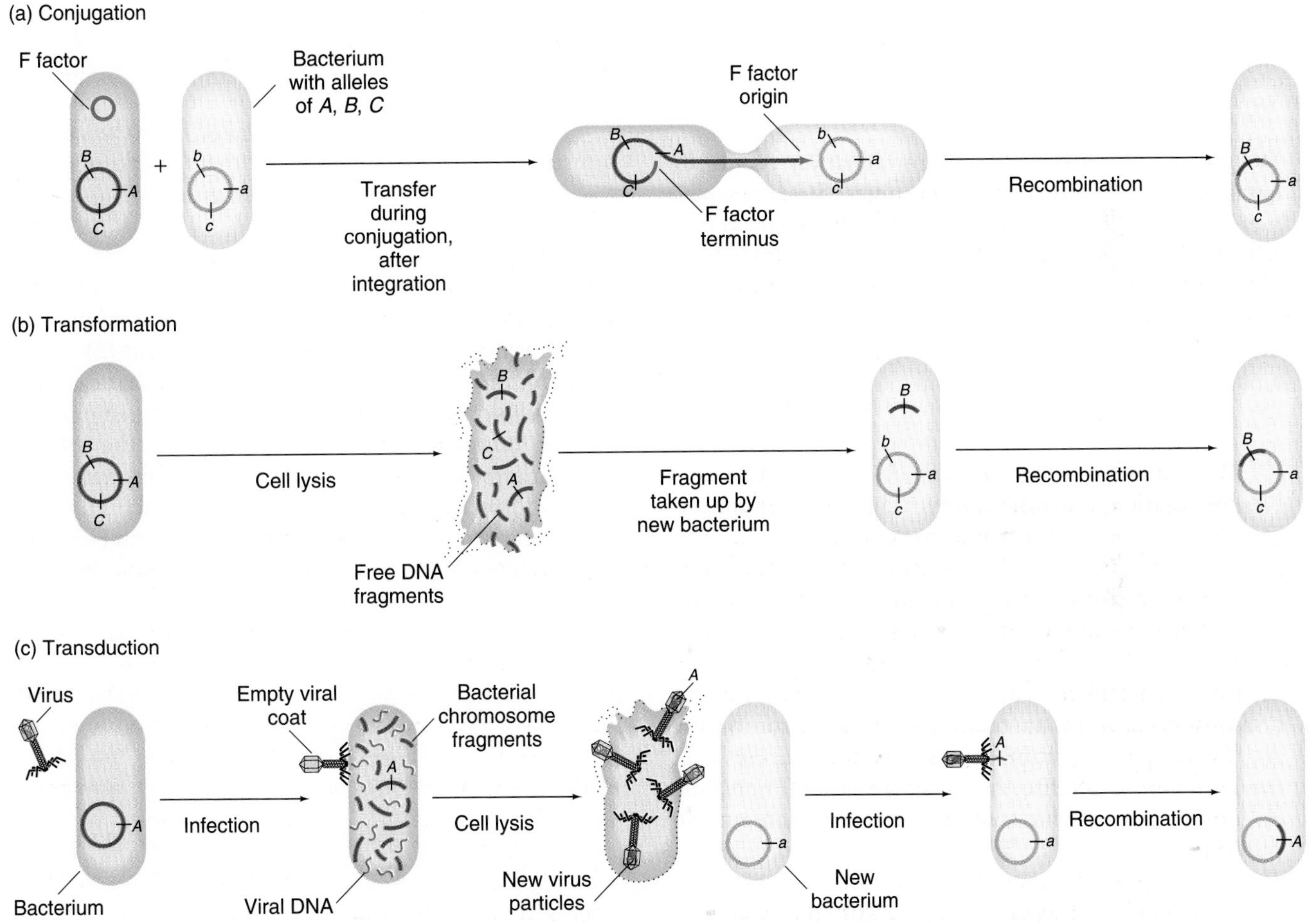

Figure 10-34 Recombination processes in bacteria. Bacterial recombination requires that a bacterial cell receive an allele obtained from another cell. (a) In conjugation, a cytoplasmic element such as the fertility factor (F) integrates into the chromosome of a bacterial cell. During cell-to-cell contact, the integrated factor can transfer part or all of that chromosome to another cell whose chromosome carries alleles of genes on the transferred chromosome. The transferred segment recombines with a homologous segment in the recipient cell's chromosome; in the example shown here, allele *B* thereby replaces allele *b*. (b) In transformation, a DNA segment bearing a particular allele is taken up from the environment by a cell whose chromosome carries a matching allele; the alleles (in our example, *B* and *b*) are then exchanged by homologous recombination. (c) In transduction, after a phage has infected a bacterial cell, one of the newly forming phage particles picks up a bacterial DNA segment instead of viral DNA. When this phage particle infects another cell it injects its bacterial DNA, which recombines with a homologous segment in the second cell, thereby exchanging any corresponding alleles (in our example, *A* and *a*).

Concept Map

Draw a concept map interrelating as many of the following terms as possible. Note that the terms are listed in no particular order.

bacteria / conjugation / recombination / F plasmid / Hfr / F⁻ / donor / recipient / interrupted mating / chromosome map / pilus / merozygote / gene

Chapter Integration Problem

We saw in Chapter 5 how recombination occurs by breakage and reunion of chromosomes. Suppose a cell were unable to carry out breakage and reunion by generalized recombination (*rec*⁻). How would this cell behave as a recipient in generalized and in specialized transduction? First compare each type of transduction and then determine the effect of the *rec*⁻ mutation on the inheritance of genes by each process.

Solution

Generalized transduction involves the incorporation of chromosomal fragments into phage heads, which then infect recipient strains. Fragments of the chromosome are incorporated randomly into phage heads, so that any marker on the bacterial host chromosome can be transduced to another strain by generalized transduction. By contrast, specialized transduction involves the integration of the phage at a specific point on the chromosome and the rare incorporation of chromosomal markers near the integration site into the phage genome. Therefore, only those markers that are near the specific integration site of the phage on the host chromosome can be transduced.

Inheritance of markers occurs by different routes in generalized and specialized transduction. A generalized transducing phage injects a fragment of the donor chromosome into the recipient. This fragment must be incorporated into the recipient's chromosome by recombination, using the recipient recombination system. Therefore, a *rec*⁻ recipient will not be able to incorporate fragments of DNA and cannot inherit markers by generalized transduction. On the other hand, the major route for the inheritance of markers by specialized transduction involves integration of the specialized transducing particle into the host chromosome at the specific phage integration site. This integration, which sometimes requires an additional wild-type (helper) phage, is mediated by a phage-specific enzyme system that is independent of the normal recombination enzymes. Therefore, a *rec*⁻ recipient can still inherit genetic markers by specialized transduction.

Solved Problems

1. In *E. coli,* four Hfr strains donate the following genetic markers shown in the order donated:

strain 1:	Q	W	D	M	T
strain 2:	A	X	P	T	M
strain 3:	B	N	C	A	X
strain 4:	B	Q	W	D	M

All of these Hfr strains are derived from the same F⁺ strain. What is the order of these markers on the circular chromosome of the original F⁺?

Solution

Recall the two-step approach that works well: (1) determine the underlying principle, and (2) draw a diagram. Here the principle is clearly that each Hfr strain donates genetic markers from a fixed point on the circular chromosome and that the earliest markers are donated with the highest frequency. Since not all markers are donated by each Hfr, only the early markers must be donated for each Hfr. Each strain allows us to draw the following circles:

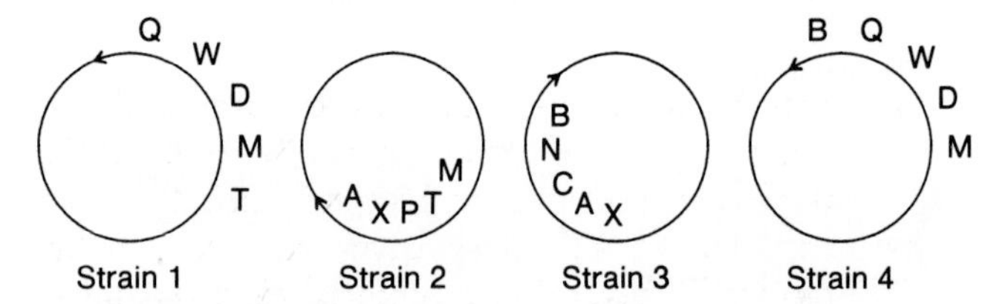

From this information, we can consolidate each circle into one circular linkage map of the order Q, W, D, M, T, P, X, A, C, N, B, Q.

2. In an Hfr × F⁻ cross, *leu*⁺ enters as the first marker, but the order of the other markers is unknown. If the Hfr is wild-type and the F⁻ is auxotrophic for each marker in question, what is the order of the markers in a cross where *leu*⁺ recombinants are selected if 27 percent are *ile*⁺, 13 percent are *mal*⁺, 82 percent are *thr*⁺, and 1 percent are *trp*⁺?

Solution

Recall that spontaneous breakage creates a natural gradient of transfer, which makes it less and less likely for a recipient to receive later and later markers. Because we have selected for the earliest marker in this cross, the frequency of recombinants is a function of the order of entry for each marker. Therefore, we can immediately determine the order of the genetic markers simply by looking at the percentage of recombinants for any marker among the leu^+ recombinants. Because the inheritance of thr^+ is the highest, this must be the first marker to enter after *leu*. The complete order is *leu, thr, ile, mal, trp*.

3. A cross is made between an Hfr that is $met^+\ thr^+\ pur^+$ and an F^- that is $met^-\ thi^-\ pur^-$. Interrupted-mating studies show that met^+ enters the recipient last, so that met^+ recombinants are selected on a medium containing supplements that satisfy only the *pur* and *thi* requirements. These recombinants are tested for the presence of the thi^+ and pur^+ alleles. The following numbers of individuals are found for each genotype:

Genotype	Number
$met^+\ thi^+\ pur^+$	280
$met^+\ thi^+\ pur^-$	0
$met^+\ thi^-\ pur^+$	6
$met^+\ thi^-\ pur^-$	52

a. Why was methionine (met) left out of the selection medium?

b. What is the gene order?

c. What are the map distances in recombination units?

Solution

a. Methionine was left out of the medium to allow selection for met^+ recombinants, because met^+ is the last marker to enter the recipient. This ensures that all of the loci we are considering in the cross will have already entered each recombinant that we analyze.

b. Here it is helpful to diagram the possible gene orders. Since we know that *met* enters the recipient last, there are only two possible gene orders if the first marker enters on the right: *met, thi, pur* or *met, pur, thi*. How can we distinguish between these two orders? Fortunately, one of the four possible classes of recombinants requires two additional crossovers. Each possible order predicts a different class that arises by four crossovers rather than two. For instance, if the order were *met, thi, pur*, then $met^+\ thi^-\ pur^+$ recombinants would be very rare. On the other hand, if the order were *met, pur, thi*, then the four-crossover class would be $met^+\ pur^-\ thi^+$. From the information given in the table, it is clear that the $met^+\ pur^-\ thi^+$ class is the four-crossover class and therefore that the gene order *met, pur, thi* is correct.

c. Refer to the following diagram:

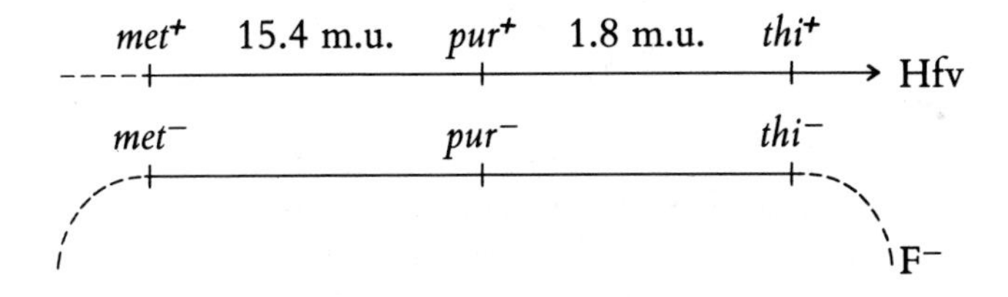

To compute the distance between *met* and *pur*, we compute the percentage of $met^+\ pur^-\ thi^-$, which is $52/338 = 15.4$ m.u. The distance between *pur* and *thi* is, similarly, $6/338 = 1.8$ m.u.

4. Compare the mechanism of transfer and inheritance of the lac^+ genes in crosses with Hfr, F^+, and $F'\ lac^+$ strains. How would an F^- cell that cannot undergo normal homologous recombination *(rec^-)* behave in crosses with each of these three strains? Would the cell be able to inherit the lac^+ genes?

Solution

Each of these three strains donates genes by conjugation. In the cases of the Hfr and F^+ strains, the lac^+ genes on the host chromosome are donated. In the Hfr strain, the F factor is integrated into the chromosome in every cell, so that efficient donation of chromosomal markers can occur, particularly if the marker is near the integration site of F and is donated early. The F^+ cell population contains a small percentage of Hfr cells, in which F is integrated into the chromosome. These cells are responsible for the gene transfer displayed by cultures of F^+ cells. In the cases of Hfr- and F^+-mediated gene transfer, inheritance requires the incorporation of a transferred fragment by recombination (recall that two crossovers are needed) into the F^- chromosome. Therefore, an F^- strain that cannot undergo recombination cannot inherit donor chromosomal markers even though they are transferred by Hfr strains or Hfr cells in F^+ strains. The fragment cannot be incorporated into the chromosome by recombination. Since these fragments do not possess the ability to replicate within the F^- cell, they are rapidly diluted out during cell division.

Unlike Hfr cells, F' cells transfer genes carried on the F' factor, a process that does not require chromosome transfer. In this case, the lac^+ genes are linked to the F' and are transferred with the F' at a high efficiency. In the F^- cell, no recombination is required, because the F′-*lac*$^+$ strain can replicate and be maintained in the dividing F$^-$ cell population. Therefore, the *lac*$^+$ genes are inherited even in a *rec*$^-$ strain.

Problems

1. A microbial geneticist isolates a new mutation in *E. coli* and wishes to map its chromosomal location. She uses interrupted-mating experiments with Hfr strains and generalized transduction experiments with phage P1. Explain why each technique, by itself, is insufficient for accurate mapping.

2. In *E. coli,* four Hfr strains donate the following markers, shown in the order donated:

strain 1:	M	Z	X	W	C
strain 2:	L	A	N	C	W
strain 3:	A	L	B	R	U
strain 4:	Z	M	U	R	B

All of these Hfr strains are derived from the same F$^+$ strain. What is the order of these markers on the circular chromosome of the original F$^+$?

3. Four *E. coli* strains of genotype $a^+ b^-$ are labeled 1, 2, 3, and 4. Four strains of genotype $a^- b^+$ are labeled 5, 6, 7, and 8. The two genotypes are mixed in all possible combinations and (after incubation) are plated to determine the frequency of $a^+ b^+$ recombinants. The following results are obtained, where M = many recombinants, L = low numbers of recombinants, and 0 = no recombinants.

	1	2	3	4
5	0	M	M	0
6	0	M	M	0
7	L	0	0	M
8	0	L	L	0

On the basis of these results, assign a sex type (either Hfr, F$^+$, or F$^-$) to each strain.

4. An Hfr strain of genotype $a^+ b^+ c^+ d^- str^s$ is mated with a female strain of genotype $a^- b^- c^- d^+ str^r$. At various times, the culture is shaken vigorously to separate mating pairs. The cells are then plated on agar of the following three types, where nutrient A allows the growth of a^- cells; nutrient B, of b^- cells; nutrient C, of c^- cells; and nutrient D, of d^- cells (a plus indicates the presence of streptomycin and a nutrient, and a minus indicates its absence):

Agar type	str	A	B	C	D
1	+	+	+	−	+
2	+	−	+	+	+
3	+	+	−	+	+

a. What donor genes are being selected on each type of agar?

b. Table 10-5 shows the number of colonies on each type of agar for samples taken at various times after the strains are mixed. Use this information to determine the order of the genes *a*, *b*, and *c*.

c. From each of the 25-minute plates, 100 colonies are picked and transferred to a dish containing agar with all of the nutrients except D. The numbers of colonies that grow on this medium are 89 for the sample from agar type 1, 51 for the sample from agar type 2, and 8 for the sample from agar type 3. Using these data, fit gene *d* into the sequence of *a*, *b*, and *c*.

d. At what sampling time would you expect colonies to first appear on agar containing C and streptomycin but no A or B?

Table 10-5.

Time of sampling (minutes)	Number of colonies on agar of type		
	1	2	3
0	0	0	0
5	0	0	0
7.5	100	0	0
10	200	0	0
12.5	300	0	75
15	400	0	150
17.5	400	50	225
20	400	100	250
25	400	100	250

(Problem 4 is from D. Freifelder, *Molecular Biology and Biochemistry.* Copyright 1978 by W. H. Freeman and Company, New York.)

5. You are given two strains of *E. coli.* The Hfr strain is $arg^+ ala^+ glu^+ pro^+ leu^+ T^s$; the F$^-$ strain is $arg^- ala^- glu^- pro^- leu^- T^r$. The markers are all nutritional except *T*, which determines sensitivity or resistance to phage T1. The order of entry is as given, with arg^+ entering the recipient first and T^s last. You find that the F$^-$ strain dies when exposed to penicillin (pen^s) but the Hfr strain does not (pen^r).

How would you locate the locus for pen on the bacterial chromosome with respect to *arg*, *ala*, *glu*, *pro*, and *leu*? Formulate your answer in logical, well-explained steps and draw explicit diagrams where possible.

6. A cross is made between two *E. coli* strains: Hfr arg^+ bio^+ leu^+ × F^- art^- bio^- leu^-. Interrupted-mating studies show that arg^+ enters the recipient last, so that arg^+ recombinants are selected on a medium containing *bio* and *leu* only. These recombinants are tested for the presence of bio^+ and leu^+. The following numbers of individuals are found for each genotype:

arg^+ bio^+ leu^+	320
arg^+ bio^+ leu^-	8
arg^+ bio^- leu^+	0
arg^+ bio^- leu^-	48

a. What is the gene order?

b. What are the map distances in recombination units?

7. You make the following *E. coli* cross: Hfr Z_1^- ade^+ str^s × F^- Z_2^- ade^- str^r, in which *str* determines resistance or sensitivity to streptomycin, *ade* determines adenine requirement for growth, and Z_1 and Z_2 are two very close sites having Z^- alleles that cause an inability to use lactose as an energy source. After about an hour, the mixture is plated on a medium containing streptomycin, with glucose as the energy source. Many of the ade^+ colonies that grow are found to be capable of using lactose. However, hardly any of the ade^+ colonies from the reciprocal cross Hfr Z_2^- ade^+ str^s × F^- Z_1^- ade^- str^r are found to be capable of using lactose. What is the order of the Z_1 and Z_2 sites in relation to the *ade* locus? (Note that the *str* locus is terminal.)

8. Jacob selected eight closely linked lac^- mutations (called *lac*-1 through *lac*-8) and then attempted to order the mutations with respect to the outside markers *pro* (proline) and *ade* (adenine) by performing a pair of reciprocal crosses for each pair of *lac* mutants:

Cross A Hfr pro^- *lac-x* ade^+ × F^- pro^+ *lac-y* ade^-

Cross B Hfr pro^- *lac-y* ade^+ × F^- pro^+ *lac-x* ade^-

In all cases, prototrophs were selected by plating on a minimal medium with lactose as the only carbon source. Table 10-6 shows the number of colonies in the two crosses for each pair of mutants. Determine the relative order of the mutations.

Table 10-6.

x	*y*	Cross A	Cross B	*x*	*y*	Cross A	Cross B
1	2	173	27	1	8	226	40
1	3	156	34	2	3	24	187
1	4	46	218	2	8	153	17
1	5	30	197	3	6	20	175
1	6	168	32	4	5	205	17
1	7	37	215	5	7	199	34

9. Linkage maps in an Hfr bacterial strain are calculated in units of minutes (the number of minutes between genes indicates the length of time it takes for the second gene to follow the first during conjugation). In making such maps, microbial geneticists assume that the bacterial chromosome is transferred from Hfr to F^- at a constant rate. Thus, two genes separated by 10 minutes near the origin end are assumed to be the same physical distance apart as two genes separated by 10 minutes near the F-attachment end. Suggest a critical experiment to test the validity of this assumption.

10. In the cross Hfr aro^+ arg^+ ery^r str^s × F^- aro^- arg^- ery^s str^r, the markers are transferred in the order given (with aro^+ entering first), but the first three genes are very close together. Exconjugants are plated on a medium containing str (streptomycin, to contraselect Hfr cells), ery (erythromycin), arg (arginine), and aro (aromatic amino acids). The following results are obtained for 300 colonies from these plates isolated and tested for growth on various media: on ery only, 263 strains grow; on ery + arg, 264 strains grow; on ery + aro, 290 strains grow; on ery + arg + aro, 300 strains grow.

a. Draw up a list of genotypes, and indicate the number of individuals in each.

b. Calculate the recombination frequencies.

c. Calculate the ratio of the size of the *arg*-to-*aro* region to the size of the *ery*-to-*arg* region.

11. A particular Hfr strain normally transmits the pro^+ marker as the last one during conjugation. In a cross of this strain with an F^- strain, some pro^+ recombinants are recovered early in the mating process. When these pro^+ cells are mixed with F^- cells, the majority of the F^- cells are converted to pro^+ cells that also carry the F factor. Explain these results.

12. F′ strains in *E. coli* are derived from Hfr strains. In some cases, these F′ strains show a high rate of integration back into the bacterial chromosome. Furthermore, the site of integration often is the same site that the sex factor occupied in the original Hfr strain (before production of the F′ strains). Explain these results.

13. You have two *E. coli* strains, F^- str^r ala^- and Hfr str^s ala^+, in which the F factor is inserted close to ala^+. Devise a screening test to detect F′ ala^+ sexductants.

14. Five Hfr strains A through E are derived from a single F^+ strain of *E. coli.* The following chart shows the entry times of the first five markers into an F^- strain when each is used in an interrupted-conjugation experiment:

A		B		C		D		E	
mal^+	(1)	ade^+	(13)	pro^+	(3)	pro^+	(10)	his^+	(7)
str^s	(11)	his^+	(28)	met^+	(29)	gal^+	(16)	gal^+	(17)
ser^+	(16)	gal^+	(38)	xyl^+	(32)	his^+	(26)	pro^+	(23)
ade^+	(36)	pro^+	(44)	mal^+	(37)	ade^+	(41)	met^+	(49)
his^+	(51)	met^+	(70)	str^s	(47)	ser^+	(61)	xyl^+	(52)

a. Draw a map of the F^+ strain, indicating the positions of all genes and their distances apart in minutes.

b. Show the insertion point and orientation of the F plasmid in each Hfr strain.

c. In using each of these Hfr strains, state which gene you would select to obtain the highest proportion of Hfr exconjugants.

15. *Streptococcus pneumoniae* cells of genotype str^s mtl^- are transformed by donor DNA of genotype str^r mtl^+ and (in a separate experiment) by a mixture of two DNAs with genotypes str^r mtl^- and str^s mtl^+. Table 10-7 shows the results.

a. What does the first line of the table tell you? Why?

b. What does the second line of the table tell you? Why?

Table 10-7.

Transforming DNA	Percentage of cells transformed to		
	str^r mtl^-	str^s mtl^+	str^r mtl^+
str^r mtl^+	4.3	0.40	0.17
str^r mtl^- + str^s mtl^+	2.8	0.85	0.0066

Table 10-8.

Drug(s) added	Number of colonies	Drug(s) added	Number of colonies
None	10,000	BC	51
A	1156	BD	49
B	1148	CD	786
C	1161	ABC	30
D	1139	ABD	42
AB	46	ACD	630
AC	640	BCD	36
AD	942	ABCD	30

16. A transformation experiment is performed with a donor strain that is resistant to four drugs: A, B, C, and D. The recipient is sensitive to all four drugs. The treated recipient-cell population is divided up and plated on media containing various combinations of the drugs. Table 10-8 shows the results.

a. One of the genes obviously is quite distant from the other three, which appear to be tightly (closely) linked. Which is the distant gene?

b. What is the probable order of the three tightly linked genes?

(Problem 16 is from Franklin Stahl, *The Mechanics of Inheritance*, 2d ed. Copyright 1969, Prentice-Hall, Englewood Cliffs, New Jersey. Reprinted by permission.)

17. Recall that in Chapter 5 we discussed the possibility that a crossover event may affect the likelihood of another crossover. In the bacteriophage T4, gene *a* is 1.0 m.u. from gene *b*, which is 0.2 m.u. from gene *c*. The gene order is *a, b, c*. In a recombination experiment, you recover five double crossovers between *a* and *c* from 100,000 progeny viruses. Is it correct to conclude that interference is negative? Explain your answer.

18. You have infected *E. coli* cells with two strains of T4 virus. One strain is minute *(m)*, rapid lysis, *(r)*, and turbid *(tu)*; the other is wild-type for all three markers. The lytic products of this infection are plated and classified. Of 10,342 plaques, the following numbers are classified as each genotype:

m r tu	3467	*m* + +	520
+ + +	3729	+ *r tu*	474
m r +	853	+ *r* +	172
m + *tu*	162	+ + *tu*	965

a. Determine the linkage distances between *m* and *r*, between *r* and *tu*, and between *m* and *tu*.

b. What linkage order would you suggest for the three genes?

c. What is the coefficient of coincidence in this cross, and what does it signify?

(Problem 18 is reprinted with the permission of Macmillan Publishing Co., Inc., from Monroe W. Strickberger, *Genetics.* Copyright 1968 by Monroe W. Strickberger.)

19. Using P22 as a generalized transducing phage grown on a $pur^+ \, pro^+ \, his^+$ bacterial donor, a recipient strain of genotype $pur^- \, pro^- \, his^-$ is infected and incubated. Afterwards, transductants for pur^+, pro^+, and his^+ are selected individually in experiments I, II, and III, respectively.

a. What media are used for these selection experiments?

b. The transductants are examined for the presence of unselected donor markers, with the following results:

I		II		III	
$pro^- \, his^-$	87%	$pur^- \, his^-$	43%	$pur^- \, pro^-$	21%
$pro^+ \, his^-$	0%	$pur^+ \, his^-$	0%	$pur^+ \, pro^-$	15%
$pro^- \, his^+$	10%	$pur^- \, his^+$	55%	$pur^- \, pro^+$	60%
$pro^+ \, his^+$	3%	$pur^+ \, his^+$	2%	$pur^+ \, pro^+$	4%

What is the order of the bacterial genes?

c. Which two genes are closest together?

d. On the basis of the order you proposed in part c, explain the relative proportions of genotypes observed in experiment II.

(Problem 19 is from D. Freifelder, *Molecular Biology and Biochemistry.* Copyright 1978 by W. H. Freeman and Company, New York.)

20. Although most λ-mediated gal^+ transductants are inducible lysogens, a small percentage of these transductants in fact are not lysogens (that is, they contain no integrated λ). Control experiments show that these transductants are not produced by mutation. What is the likely origin of these types?

21. An $ade^+ \, arg^+ \, cys^+ \, his^+ \, leu^+ \, pro^+$ bacterial strain is known to be lysogenic for a newly discovered phage, but the site of the prophage is not known. The bacterial map is

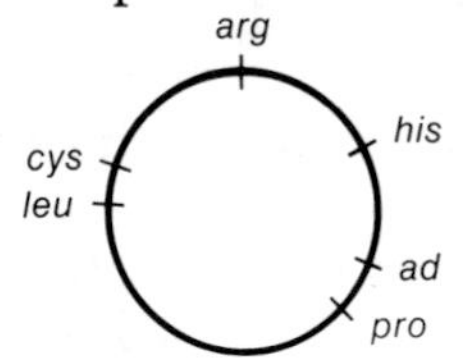

Table 10-9.

Medium	ade	arg	cys	his	leu	pro	Presence of colonies
	Nutrient supplementation in medium						
1	−	+	+	+	+	+	N
2	+	−	+	+	+	+	N
3	+	+	−	+	+	+	C
4	+	+	+	−	+	+	N
5	+	+	+	+	−	+	C
6	+	+	+	+	+	−	N

NOTE: + indicates the presence of a nutrient supplement; − indicates supplement not present. N indicates no colonies; C indicates colonies present.

The lysogenic strain is used as a source of the phage, and the phages are added to a bacterial strain of genotype $ade^- \, arg^- \, cys^- \, his^- \, leu^- \, pro^-$. After a short incubation, samples of these bacteria are plated on six different media, with the supplementations indicated in Table 10-9. The table also shows whether or not colonies were observed on the various media.

a. What genetic process is at work here?

b. What is the approximate locus of the prophage?

22. You have two strains of λ that can lysogenize *E. coli;* the following figure shows their linkage maps:

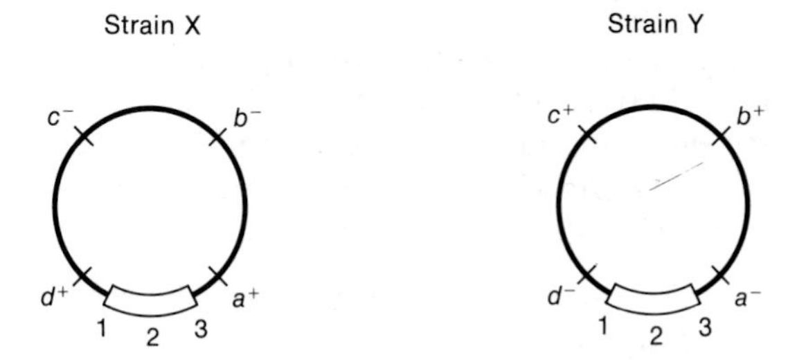

The segment shown at the bottom of the chromosome, designated 1–2–3, is the region responsible for pairing and crossing-over with the *E. coli* chromosome. (Keep the markers on all your drawings.)

a. Diagram the way in which λ strain X is inserted into the *E. coli* chromosome (so that the *E. coli* is lysogenized).

b. It is possible to superinfect the bacteria that are lysogenic for strain X by using strain Y. A certain percentage of these superinfected bacteria become "doubly" lysogenic (that is, lysogenic for both strains). Diagram how this will occur. (Don't worry about how double lysogens are detected.)

c. Diagram how the two λ prophages can pair.

d. It is possible to recover crossover products between the two prophages. Diagram a crossover event and the consequences.

23. You have three strains of *E. coli*. Strain A is F′ *cys*$^+$ *trp1/cys*$^+$ *trp1* (that is, both the F′ and the chromosome carry *cys*$^+$ and *trp1*, an allele for tryptophan requirement). Strain B is F$^-$ *cys*$^-$ *trp2* Z (this strain requires cysteine for growth and carries *trp2*, another allele causing a tryptophan requirement; strain B also is lysogenic for the generalized transducing phage Z). Strain C is F$^-$ *cys*$^+$ *trp1* (it is an F$^-$ derivative of strain A that has lost the F′).

a. How would you determine whether *trp1* and *trp2* are alleles of the same locus? (Describe the crosses and the results expected.)

b. Suppose that *trp1* and *trp2* are nonallelic and that the *cys* locus is cotransduced with the *trp* locus. Using phage Z to transduce genes from strain C to strain B, how would you determine the genetic order of *cys*, *trp1*, and *trp2*?

24. A generalized transducing phage is used to transduce an $a^- b^- c^- d^- e^-$ recipient strain of *E. coli* with an $a^+ b^+ c^+ d^+ e^+$ donor. The recipient culture is plated on various media with the results shown in Table 10-10. (Note that a^- determines a requirement for A as a nutrient, and so forth.) What can you conclude about the linkage and order of the genes?

Table 10-10.

Compounds added to minimal medium	Presence (+) or absence (−) of colonies
C D E	−
B D E	−
B C E	+
B C D	+
A D E	−
A C E	−
A C D	−
A B E	−
A B D	+
A B C	−

25. In a generalized transduction system using P1 phage, the donor is *pur*$^+$ *nad*$^+$ *pdx*$^-$ and the recipient is *pur*$^-$ *nad*$^-$ *pdx*$^+$. The donor allele *pur*$^+$ is initially selected after transduction, and 50 *pur*$^+$ transductants are then scored for the other alleles present. The results follow:

Genotype	Number of colonies
nad$^+$ *pdx*$^+$	3
nad$^+$ *pdx*$^-$	10
nad$^-$ *pdx*$^+$	24
nad$^-$ *pdx*$^-$	13
	50

a. What is the cotransduction frequency for *pur* and *nad*?

b. What is the cotransduction frequency for *pur* and *pdx*?

c. Which of the unselected loci is closest to *pur*?

d. Are *nad* and *pdx* on the same side or on opposite sides of *pur*? Explain. (Draw the exchanges needed to produce the various transformant classes under either order to see which requires the minimum number to produce the results obtained.)

26. In a generalized transduction experiment, phages are collected from an *E. coli* donor strain of genotype *cys*$^+$ *leu*$^+$ *thr*$^+$ and used to transduce a recipient of genotype *cys*$^-$ *leu*$^-$ *thr*$^-$. Initially, the treated recipient population is plated on a minimal medium supplemented with leucine and threonine. Many colonies are obtained.

a. What are the possible genotypes of these colonies?

b. These colonies are then replica-plated onto three different media: (1) minimal plus threonine only, (2) minimal plus leucine only, and (3) minimal. What genotypes could, in theory, grow on these three media?

c. It is observed that 56 percent of the original colonies grow on (1), 5 percent grow on (2), and no colonies grow on (3). What are the actual genotypes of the colonies on (1), (2), and (3)?

d. Draw a map showing the order of the three genes and which of the two outer genes is closer to the middle gene.

***27.** In 1965, Jon Beckwith and Ethan Signer devised a method of obtaining specialized transducing phages carrying the *lac* region. In a two-step approach, the researchers first "transposed" the *lac* genes to a new region of the chromosome and then isolated the specialized transducing particles. They noted that the integration site for the temperate phage $\phi 80$ (a relative of phage λ), designated *att80,* was located near one of the genes involved in conferring resistance to the virulent phage T1, termed *tonB:*

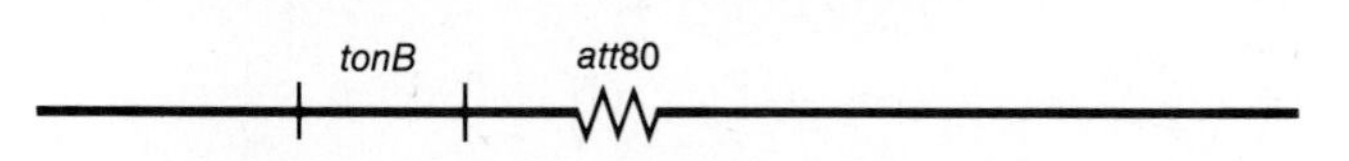

Beckwith and Signer used an F′-*lac* episome that could not replicate at high temperatures in a strain carrying a deletion of the *lac* genes. By forcing the cell to remain *lac*$^+$ at high temperatures, the researchers could select strains in which the episome had integrated into the chromosome, thereby allowing the F′-*lac* to be maintained at high temperatures. By combining this selection with a simultaneous selection for resistance to T1 phage infection, they found that the only survivors were cells in which the F′-*lac* had integrated into the *tonB* locus, as shown in the accompanying figure. Can you see why?

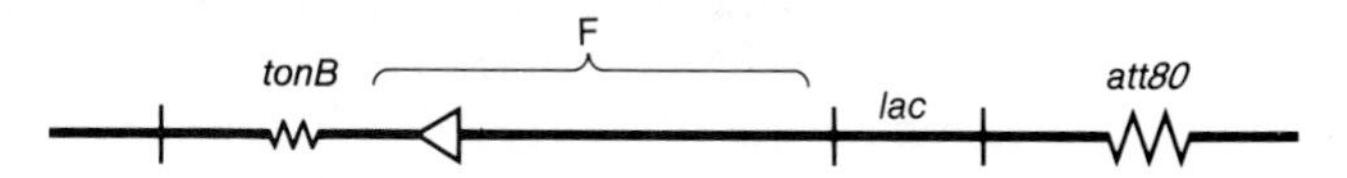

This placed the *lac* region near the integration site for phage $\phi 80$. Describe the subsequent steps that the researchers must have followed to isolate the specialized transducing particles of phage $\phi 80$ that carried the *lac* region.

11 The Structure of DNA

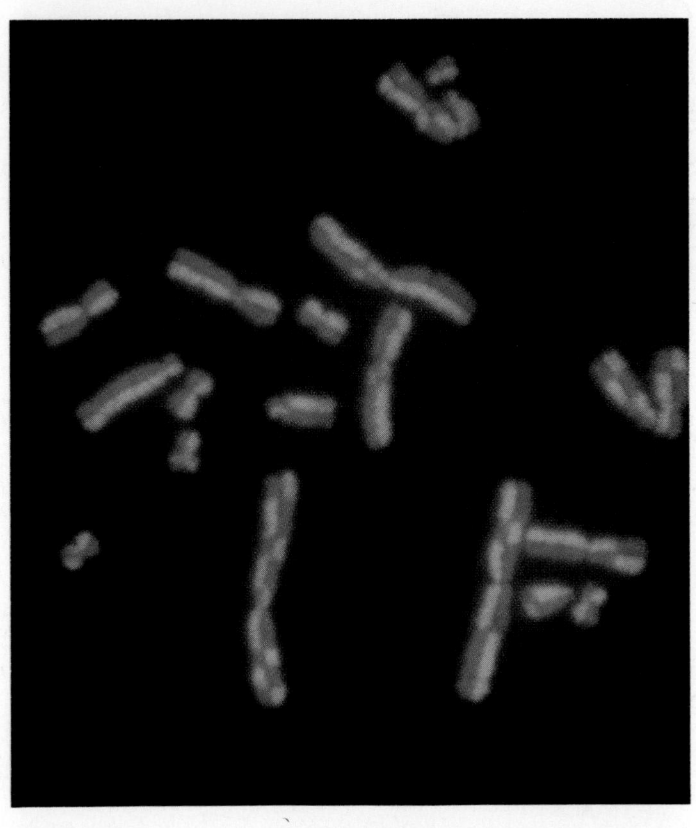

Harlequin chromosomes. (Sheldon Wolff/University of California, San Francisco)

KEY CONCEPTS

Bacterial cells that express one phenotype can be transformed into cells that express a different phenotype; the transforming agent is DNA.

Experiments with labeled T2 phage have established that DNA is the hereditary material.

James Watson and Francis Crick showed that the structure of DNA is a double helix, in which each helix is a chain of nucleotides held together by phosphodiester bonds and in which specific hydrogen bonds are formed by pairs of bases.

The DNA structure suggests that the fidelity of replication can be ensured if the complementary base of each base is specified by hydrogen bonding.

The replication of DNA is semiconservative in that each daughter duplex contains one parental and one newly synthesized strand.

Many of the enzymes involved in DNA synthesis in bacteria have been characterized.

Until now, we have looked at genes as abstract entities that somehow control hereditary traits. Through purely genetic analysis, we have studied the inheritance of different genes. But what about the physical nature of the gene? This question puzzled scientists for many years until it was realized that genes are composed of deoxyribonucleic acid (abbreviated *DNA*) and that DNA has a fascinating structure.

The elucidation of the structure of DNA in 1953 by James Watson and Francis Crick was one of the most exciting discoveries in the history of genetics. It paved the way for the understanding of gene action and heredity in molecular terms. Before we see how the solution of DNA structure was achieved, let's review what was known about genes and DNA at the time that Watson and Crick began their historic collaboration:

1. Genes—the hereditary "factors" described by Mendel—were known to be associated with specific character traits, but their physical nature was not understood.
2. The one-gene–one-enzyme theory (described more fully in Chapter 12) postulated that genes control the structure of proteins.
3. Genes were known to be carried on chromosomes.
4. The chromosomes were found to consist of DNA and protein.
5. Research by Frederick Griffith and, subsequently, by Oswald Avery and his coworkers pointed to DNA as the genetic material. These experiments, described here, showed that bacterial cells that express one phenotype can be transformed into cells that express a different phenotype and that the transforming agent is DNA.

DNA: The Genetic Material

The Discovery of Transformation

A puzzling observation was made by Frederick Griffith in the course of experiments on the bacterium *Streptococcus pneumoniae* in 1928. This bacterium, which causes pneumonia in humans, is normally lethal in mice. However, different strains of this bacterial species have evolved that differ in virulence (in the ability to cause disease or death). In his experiments, Griffith used two strains that are distinguishable by the appearance of their colonies when grown in laboratory cultures. In one strain, a normal virulent type, the cells are enclosed in a polysaccharide capsule, giving colonies a smooth appearance; hence, this strain is labeled *S*. In Griffith's other strain, a mutant nonvirulent type that grows in mice but is not lethal, the polysaccharide coat is absent, giving colonies a rough appearance; this strain is called *R*.

Griffith killed some virulent cells by boiling them and injected the heat-killed cells into mice. The mice survived, showing that the carcasses of the cells do not cause death. However, mice injected with a mixture of heat-killed virulent cells and live nonvirulent cells did die. Furthermore, live cells could be recovered from the dead mice; these cells gave smooth colonies and were virulent on subsequent injection. Somehow, the cell debris of the boiled S cells had converted the live R cells into live S cells. The process is called **transformation.** Griffith's experiment is summarized in Figure 11-1.

This same basic technique was then used to determine the nature of the *"transforming principle"*—the agent in the cell debris that is specifically responsible for transformation. In 1944, Oswald Avery, C. M. MacLeod, and M. McCarty separated the classes of molecules found in the debris of the dead S cells and tested them for transforming ability, one at a time. These tests showed first that the polysaccharides themselves do not transform the rough cells. Therefore, the polysaccharide coat, although undoubtedly concerned with the pathogenic reaction, is only the phenotypic expression of virulence. In screening the different groups, Avery and his colleagues found that only one class of molecules, DNA, induced transformation of R cells (Figure 11-2). They deduced that DNA is the agent that determines the polysaccharide character and hence the pathogenic character. Furthermore, it seemed that providing R cells with S DNA was tantamount to providing these cells with S genes!

> Message The demonstration that DNA is the transforming principle was the first demonstration that genes are composed of DNA.

The Hershey-Chase Experiment

The experiments conducted by Avery and his colleagues were definitive, but many scientists were very reluctant to accept DNA (rather than proteins) as the genetic material. The clincher was provided in 1952 by Alfred Hershey and Martha Chase using the phage (virus) T2. They reasoned that phage infection must involve the introduction (injection) into the bacterium of the specific information that dictates viral reproduction. The phage is relatively simple in molecular constitution. Most of its structure is protein, with DNA contained inside the protein sheath of its "head."

Phosphorus is not found in proteins but is an integral part of DNA; conversely, sulfur is present in proteins but never in DNA. Hershey and Chase incorporated the radioisotope of phosphorus (^{32}P) into phage DNA and that of sulfur (^{35}S) into the proteins of a separate phage culture. They then used each phage culture independently

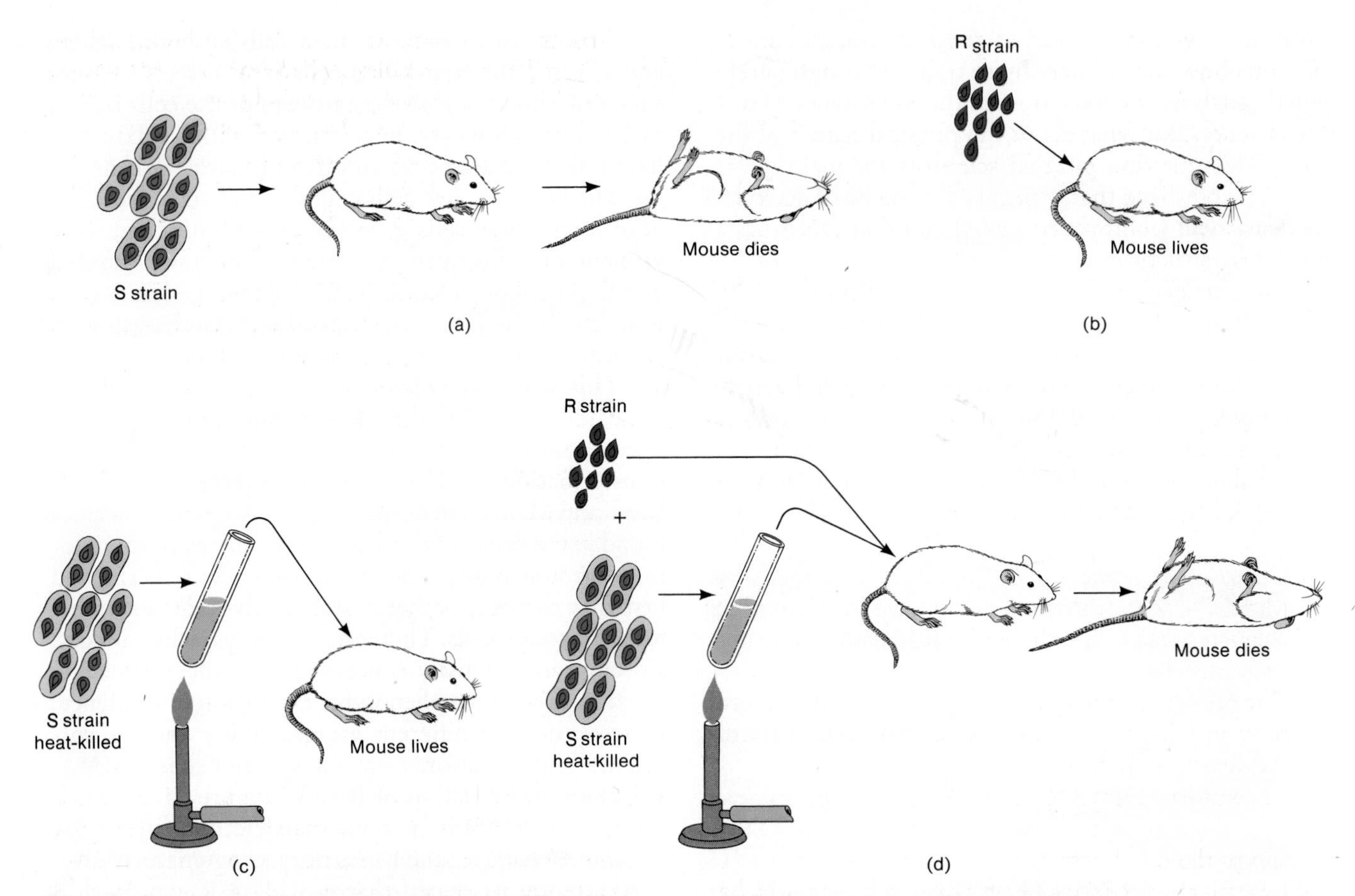

Figure 11-1 The first demonstration of bacterial transformation. (a) Mouse dies after injection with the virulent S strain. (b) Mouse survives after injection with the R strain. (c) Mouse survives after injection with heat-killed S strain. (d) Mouse dies after injection with a mixture of heat-killed S strain and live R strain. The heat-killed S strain somehow transforms the R strain to virulence. Parts (a), (b), and (c) act as control experiments for this demonstration. (From G. S. Stent and R. Calendar, *Molecular Genetics,* 2d ed. Copyright 1978 by W. H. Freeman and Company. After R. Sager and F. J. Ryan, *Cell Heredity.* John Wiley, 1961.)

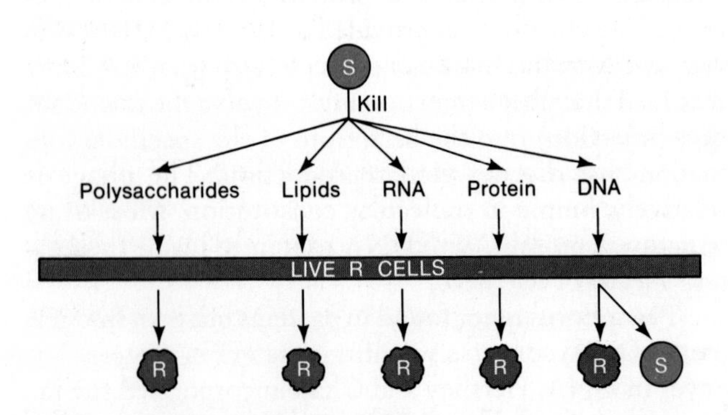

Figure 11-2 Demonstration that DNA is the transforming agent. DNA is the only agent that produces smooth (S) colonies when added to live rough (R) cells.

to infect *E. coli* with many virus particles per cell. After sufficient time for injection to occur, they sheared the empty phage carcasses (called *ghosts*) off the bacterial cells by agitation in a kitchen blender. They used centrifugation to separate the bacterial cells from the phage ghosts and then measured the radioactivity in the two fractions. When the ^{32}P-labeled phages were used, most of the radioactivity ended up inside the bacterial cells, indicating that the phage DNA entered the cells. ^{32}P can also be recovered from phage progeny. When the ^{35}S-labeled phages were used, most of the radioactive material ended up in the phage ghosts, indicating that the phage protein never entered the bacterial cell (Figure 11-3). The conclusion is inescapable: DNA is the hereditary material; the phage proteins are mere structural packaging that is discarded after delivering the vital DNA to the bacterial cell.

Why such reluctance to accept this conclusion? DNA was thought to be a rather simple chemical. How could all the information about an organism's features be stored in such a simple molecule? How could such information be passed on from one generation to the next? Clearly, the genetic material must have both the ability to encode specific information and the capacity to duplicate that information precisely. What kind of structure could allow such complex functions in so simple a molecule?

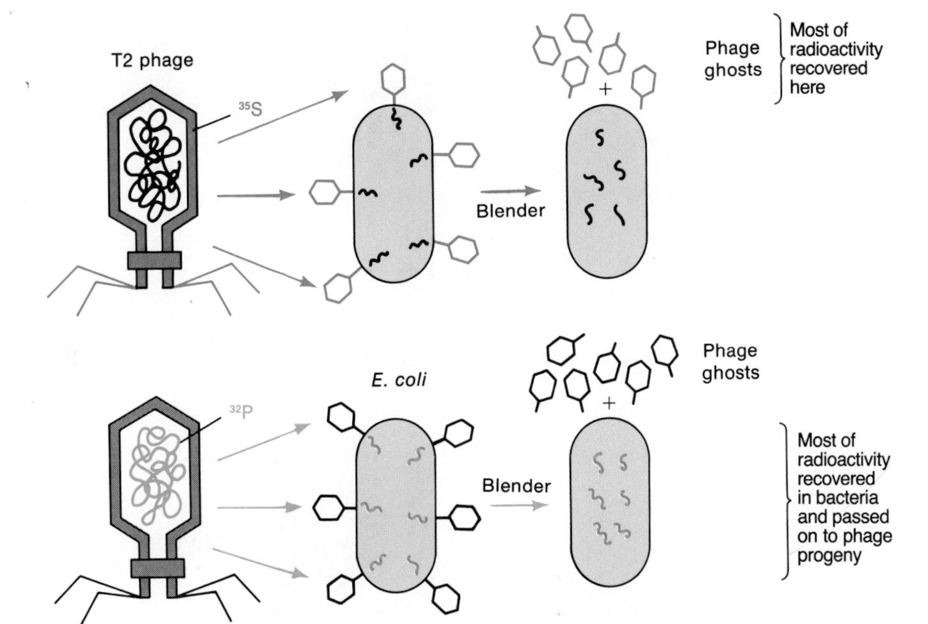

Figure 11-3 The Hershey-Chase experiment, which demonstrated that the genetic material of phage is DNA, not protein. The experiment uses two sets of T2 bacteriophage. In one set, the protein coat is labeled with radioactive sulfur (^{35}S) not found in DNA. In the other set, the DNA is labeled with radioactive phosphorus (^{32}P) not found in protein. Only the ^{32}P is injected into the *E. coli,* indicating that DNA is the agent necessary for the production of new phages.

The Structure of DNA

Although the DNA structure was not known, the basic building blocks of DNA had been known for many years. The basic elements of DNA had been isolated and determined by partly breaking up purified DNA. These studies showed that DNA is composed of only four basic molecules called **nucleotides,** which are identical except that each contains a different nitrogen base. Each nucleotide contains phosphate, sugar (of the deoxyribose type), and one of the four bases (Figure 11-4). When the phosphate group is not present, the base and the deoxyribose form a **nucleoside** rather than a nucleotide. The four bases are **adenine, guanine, cytosine,** and **thymine.** The full chemical names of the nucleotides are deoxyadenosine 5′-monophosphate (or deoxyadenylate, or dAMP), deoxyguanosine 5′-monophosphate (or deoxyguanylate, or dGMP), deoxycytidine 5′-monophosphate (or deoxycytidylate, or dCMP), and deoxythymidine 5′-monophosphate (or deoxythymidylate, or dTMP). However, it is more convenient just to refer to each nucleotide by the abbreviation of its base (A, G, C, and T, respectively). Two of the bases, adenine and guanine, are similar in structure and are called **purines.** The other two bases, cytosine and thymine, also are similar and are called **pyrimidines.**

After the central role of DNA in heredity became clear, many scientists set out to determine the exact structure of DNA. How can a molecule with such a limited range of different components possibly store the vast range of information about all the protein primary structures of the living organism? The first to succeed in putting the building blocks together and finding a reasonable DNA structure — Watson and Crick in 1953 — worked from two kinds of clues. First, Rosalind Franklin and Maurice Wilkins had amassed X-ray diffraction data on DNA structure. In such experiments, X rays are fired at DNA fibers, and the scatter of the rays from the fiber is observed by catching them on photographic film, where the X rays produce spots. The angle of scatter represented by each spot on the film gives information about the position of an atom or certain groups of atoms in the DNA molecule. This procedure is not simple to carry out (or to explain), and the interpretation of the spot patterns is very difficult. The available data suggested that DNA is long and skinny and that it has two similar parts that are parallel to one another and run along the length of the molecule. The X-ray data showed the molecule to be helical (spiral-like). Other regularities were present in the spot patterns, but no one had yet thought of a three-dimensional structure that could account for just those spot patterns.

The second set of clues available to Watson and Crick came from work done several years earlier by Erwin Chargaff. Studying a large selection of DNAs from different organisms (see Table 11-1), Chargaff established certain empirical rules about the amounts of each component of DNA:

1. The total amount of pyrimidine nucleotides (T + C) always equals the total amount of purine nucleotides (A + G).

Purine nucleotides

Phosphate

← Nitrogen base (adenine, A)

← Deoxyribose sugar

Deoxyadenosine 5′-phosphate (dAMP)

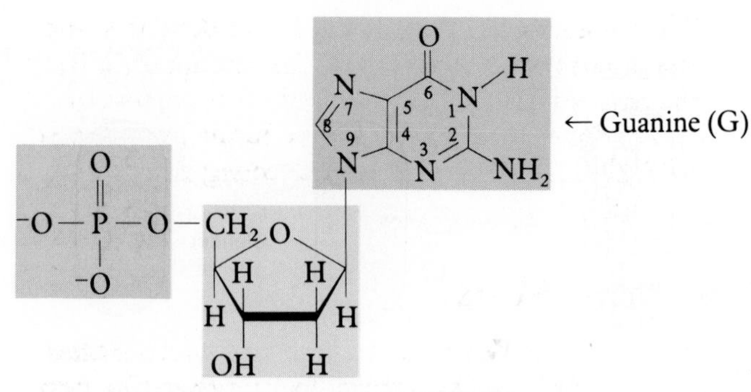

Deoxyguanosine 5′-phosphate (dGMP)

Pyrimidine nucleotides

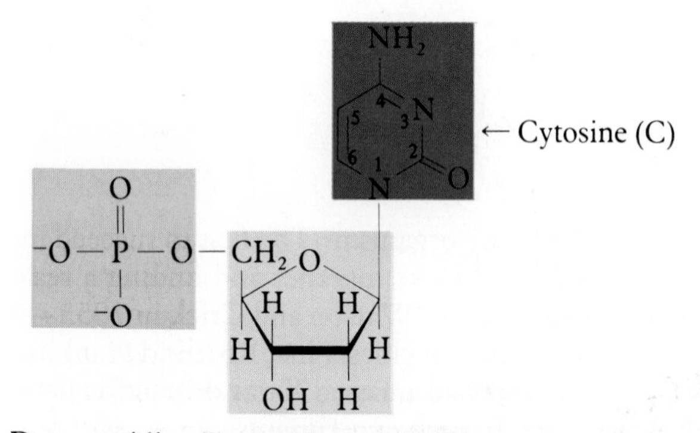

Deoxycytidine 5′-phosphate (dCMP)

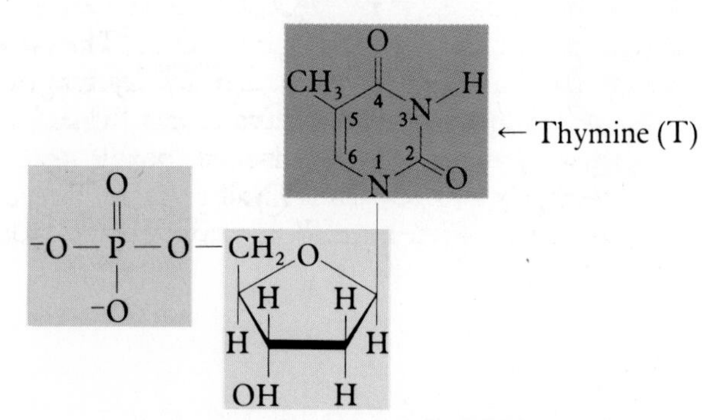

Deoxythymidine 5′-phosphate (dTMP)

Figure 11-4 Chemical structure of the four nucleotides (two with purine bases and two with pyrimidine bases) that are the fundamental building blocks of DNA. The sugar is called *deoxyribose* because it is a variation of a common sugar, ribose, that has one more oxygen atom.

Table 11-1. Molar Properties of Bases* in DNAs from Various Sources

Organism	Tissue	Adenine	Thymine	Guanine	Cytosine	$\frac{A+T}{G+C}$
Escherichia coli (K12)	—	26.0	23.9	24.9	25.2	1.00
Diplococcus pneumoniae	—	29.8	31.6	20.5	18.0	1.59
Mycobacterium tuberculosis	—	15.1	14.6	34.9	35.4	0.42
Yeast	—	31.3	32.9	18.7	17.1	1.79
Paracentrotus lividus (sea urchin)	Sperm	32.8	32.1	17.7	18.4	1.85
Herring	Sperm	27.8	27.5	22.2	22.6	1.23
Rat	Bone marrow	28.6	28.4	21.4	21.5	1.33
Human	Thymus	30.9	29.4	19.9	19.8	1.52
Human	Liver	30.3	30.3	19.5	19.9	1.53
Human	Sperm	30.7	31.2	19.3	18.8	1.62

* Defined as moles of nitrogenous constituents per 100 g-atoms phosphate in hydrolysate.

SOURCE: E. Chargaff and J. Davidson, eds., *The Nucleic Acids*. Academic Press, 1955.

2. The amount of T always equals the amount of A, and the amount of C always equals the amount of G. But the amount of A + T is not necessarily equal to the amount of G + C, as can be seen in the last column of Table 11-1. This ratio varies among different organisms.

The Double Helix

The structure that Watson and Crick derived from these clues is a **double helix,** which looks rather like two interlocked bedsprings. Each bedspring (helix) is a chain of nucleotides held together by **phosphodiester bonds,** in which a phosphate group forms a bridge between —OH groups on two adjacent sugar residues. The two bedsprings (helices) are held together by **hydrogen bonds,** in which two electronegative atoms "share" a proton, between the bases. Hydrogen bonds occur between hydrogen atoms with a small positive charge and acceptor atoms with a small negative charge. For example

Each hydrogen atom in the NH_2 group is slightly positive (δ^+) because the nitrogen atom tends to attract the electrons involved in the N—H bond, thereby leaving the hydrogen atom slightly short of electrons. The oxygen atom has six unbonded electrons in its outer shell, making it slightly negative (δ^-). A hydrogen bond forms between one H and the O. Hydrogen bonds are quite weak (only about 3 percent of the strength of a covalent chemical bond), but this weakness (as we shall see) plays an important role in the function of the DNA molecule in heredity. One further important chemical fact: the hydrogen bond is much stronger if the participating atoms are "pointing at each other" in the ideal orientations.

The hydrogen bonds are formed by pairs of bases and are indicated by dotted lines in Figure 11-5, which shows a part of this paired structure with the helices uncoiled. Each base pair consists of one purine base and one pyrimidine base, paired according to the following rule: G pairs with C, and A pairs with T. In Figure 11-6, a simplified picture of the coiling, each of the base pairs is represented by a "stick" between the "ribbons," or so-called sugar-phosphate backbones of the chains. In Figure 11-5, note that the two backbones run in opposite directions; they

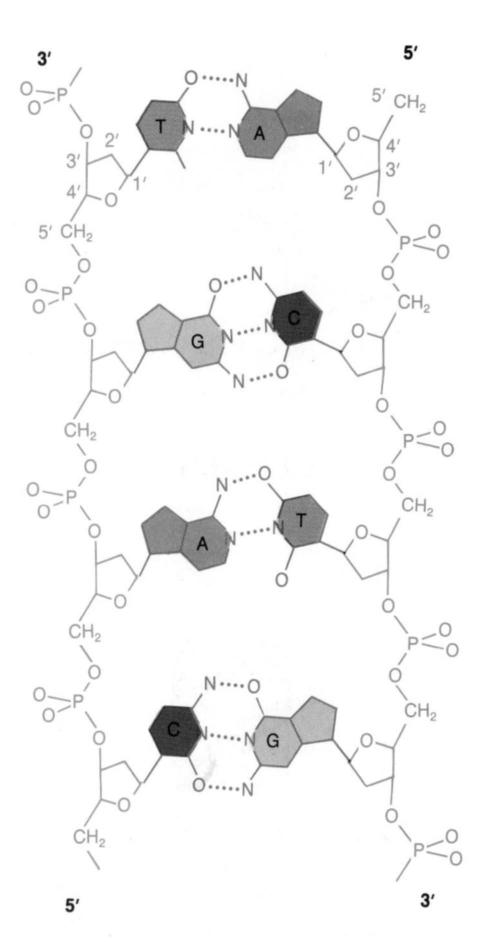

Figure 11-5 The DNA double helix, unrolled to show the sugar-phosphate backbones (blue) and base-pair rungs (red). The backbones run in opposite directions; the 5′ and 3′ ends are named for the orientation of the 5′ and 3′ carbon atoms of the sugar rings. Each base pair has one purine base, adenine (A) or guanine (G), and one pyrimidine base, thymine (T) or cytosine (C), connected by hydrogen bonds *(dotted lines)*. (From R. E. Dickerson, "The DNA Helix and How It is Read." Copyright 1983 by Scientific American, Inc. All rights reserved.)

are thus said to be **antiparallel,** and (for reasons apparent in the figure) one is called the 5′ → 3′ strand and the other the 3′ → 5′ strand.

The double helix accounted nicely for the X-ray data and also tied in very nicely with Chargaff's data. Studying models they made of the structure, Watson and Crick realized that the observed radius of the double helix (known from the X-ray data) would be explained if a purine base always pairs (by hydrogen bonding) with a pyrimidine base (Figure 11-7). Such pairing would account for the (A + G) = (T + C) regularity observed by Chargaff, but it would predict four possible pairings: T · · · A, T · · · G, C · · · A, and C · · · G. Chargaff's data, however, indicate that T pairs only with A and C

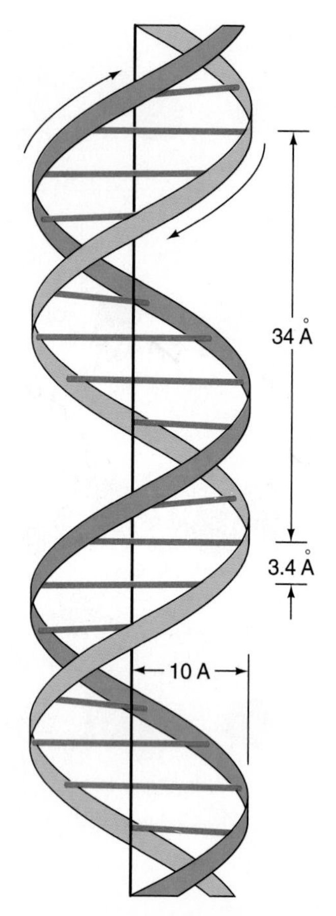

Figure 11-6 A simplified model showing the helical structure of DNA. The sticks represent base pairs, and the ribbons represent the sugar-phosphate backbones of the two antiparallel chains. The various measurements are given in angstroms (1 Å = 0.1 nm).

Pyrimidine + pyrimidine: DNA too thin

Purine + purine: DNA too thick

Purine + pyrimidine: thickness compatible with X-ray data

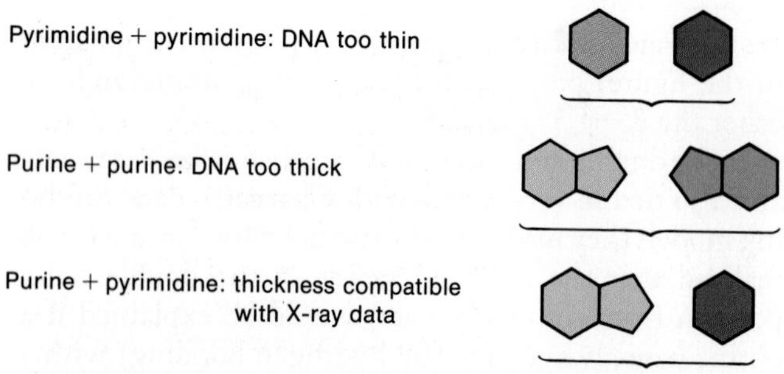

Figure 11-7 The pairing of purines with pyrimidines accounts exactly for the diameter of the DNA double helix determined from X-ray data. (From R. E. Dickerson, "The DNA Helix and How It Is Read." Copyright 1983 by Scientific American, Inc. All rights reserved.)

pairs only with G. Watson and Crick showed that only these two pairings have the necessary complementary "lock-and-key" shapes to permit efficient hydrogen bonding (Figure 11-8).

Note that the G–C pair has three hydrogen bonds, whereas the A–T pair has only two. We would predict that DNA containing many G–C pairs would be more stable than DNA containing many A–T pairs. In fact, this prediction is confirmed. We now have a neat explanation for Chargaff's data in terms of DNA structure (Figure 11-9). We also have a structure that is consistent with the X-ray data.

Three-Dimensional View of the Double Helix

In three dimensions, the bases actually form rather flat structures, and these flat bases partially stack on top of one another in the twisted structure of the double helix. This stacking of bases adds tremendously to the stability of the molecule by excluding water molecules from the spaces between the base pairs. (This phenomenon is very much like the stabilizing force that you can feel when you squeeze two plates of glass together underwater and then try to separate them.) Subsequently, it was realized that there were two forms of DNA in the fiber analyzed by diffraction. The **A form** is less hydrated than the **B form** and is more compact. Figure 11-10 shows both the A and the B forms in a schematic three-dimensional drawing. Note the stacking of bases as represented in this diagram. It is believed that the B form of DNA is the form found most frequently in living cells.

The stacking of the base pairs in the double helix results in two grooves in the sugar-phosphate backbones. These are termed the **major** and **minor** grooves and can be readily seen in the space-filling (three-dimensional) model in Figure 11-9.

Implications of DNA Structure

Elucidation of the structure of DNA caused a lot of excitement in genetics (and in all areas of biology) for two basic reasons. First, the structure suggests an obvious way in which the molecule can be **duplicated,** or **replicated,** since each base can specify its complementary base by hydrogen bonding. This essential property of a genetic molecule had been a mystery until this time. Second, the structure suggests that perhaps the *sequence* of nucleotide pairs in DNA is dictating the sequence of amino acids in the protein organized by that gene. In other words, some sort of **genetic code** may write information in DNA as a sequence of nucleotide pairs and then translate it into a different language of amino acid sequences in protein.

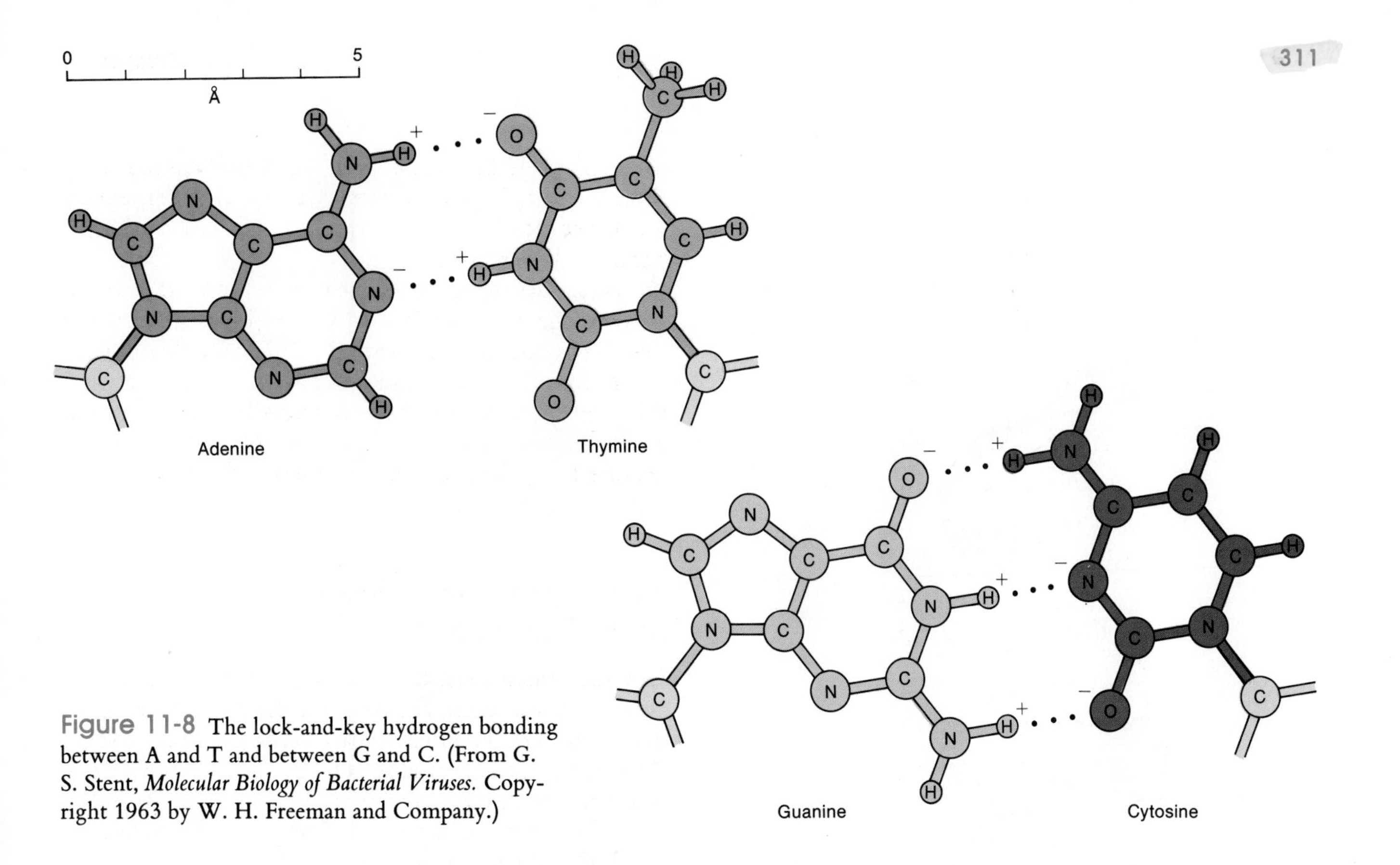

Figure 11-8 The lock-and-key hydrogen bonding between A and T and between G and C. (From G. S. Stent, *Molecular Biology of Bacterial Viruses.* Copyright 1963 by W. H. Freeman and Company.)

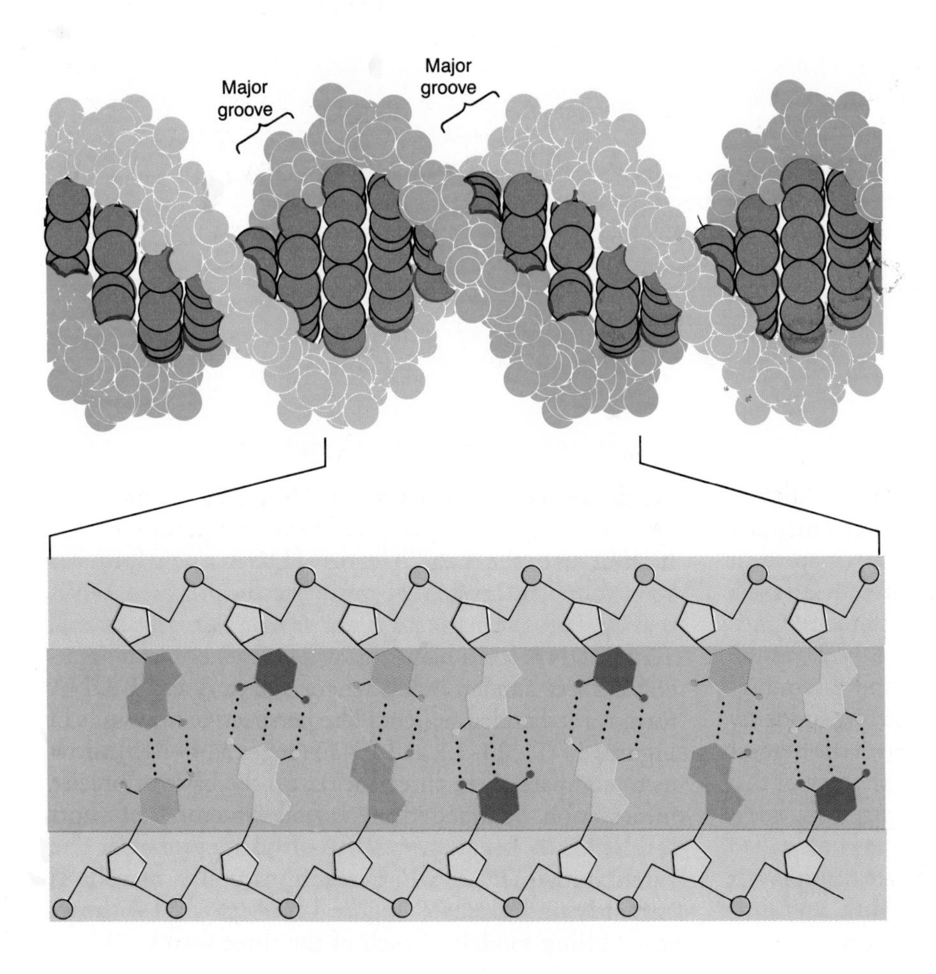

Figure 11-9 *Top:* a space-filling model of the DNA double helix. *Bottom:* an unwound representation of a short stretch of nucleotide pairs, showing how A–T and G–C pairing produces the Chargaff ratios. This model is of one of several forms of DNA, termed the B form (see Figure 11-10). (Space-filling model from C. Yanofsky, "Gene Structure and Protein Structure." Copyright 1967 by Scientific American, Inc. All rights reserved. Unwound structure based on A. Kornberg, "The Synthesis of DNA." Copyright 1968 by Scientific American, Inc. All rights reserved.)

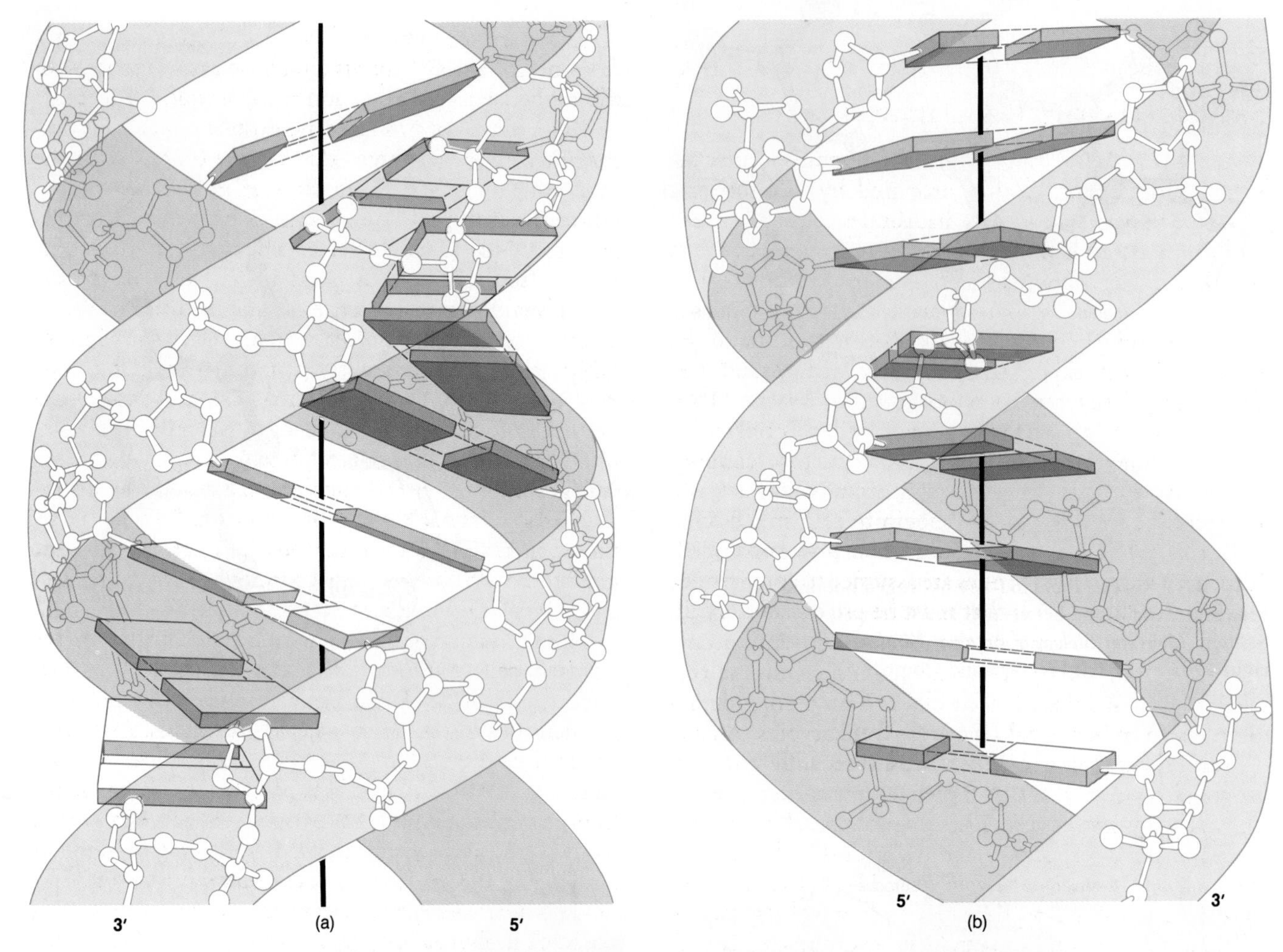

Figure 11-10 Schematic drawings of the structures of two forms of DNA. (a) A DNA. (b) B DNA. Analyses of DNA fibers by Struther Arnott and others provided the conceptual basis for these drawings. The sugar-phosphate backbones of the double helix are represented as ribbons and the runglike base pairs connecting them as planks. In A DNA, the base pairs are tilted and are pulled away from the axis of the double helix. In B DNA, on the other hand, the base pairs sit astride the helix axis and are perpendicular to it.

This basic information about DNA is now familiar to almost anyone who has read a biology text in elementary or high school, or even magazines and newspapers. It may seem trite and obvious, but try to put yourself back into the scene in 1953 and imagine the excitement! Until then, the evidence that the uninteresting DNA was the genetic molecule seemed disappointing and discouraging. But the Watson-Crick structure of DNA suddenly opened up the possibility of explaining two of the biggest "secrets" of life. James Watson has told the story of this discovery (from his own point of view, strongly questioned by others involved) in a fascinating book called *The Double Helix,* which reveals the intricate interplay of personality clashes, clever insights, hard work, and simple luck in such important scientific advances.

Alternative Structures

In addition to the A and B forms of DNA, a new form has been found in crystals of synthetically prepared DNA that contain alternating G's and C's on the same strand. This **Z DNA** form has a zigzag-like backbone and generates a left-handed helix, whereas both A and B DNA form right-handed helices. The perspective drawings in Figures 11-11, 11-12, and 11-13 (pages 314–315) allow us to compare these three forms, on the basis of precise information obtained from crystal structures of short synthetic DNA sequences called **oligonucleotides.** The extent to which Z DNA occurs naturally in cells is currently under debate. Figure 11-14 (page 314) shows space-filling models of each of the three forms.

Replication of DNA

Semiconservative Replication

Figure 11-15 (page 316) diagrams the possible mechanism for DNA replication proposed by Watson and Crick. The sugar-phosphate backbones are represented by lines, and the sequence of base pairs is random. Let's imagine that the double helix is like a zipper that unzips, starting at one end (the bottom in this figure). We can see that if this zipper analogy is valid, the unwinding of the two strands will expose single bases on each strand. Because the pairing requirements imposed by the DNA structure are strict, each exposed base will pair only with its complementary base. Because of this base complementarity, each of the two single strands will act as a **template,** or mold, and will begin to reform a double helix identical with the one from which it was unzipped. The newly added nucleotides are assumed to come from a pool of free nucleotides that must be present in the cell.

If this model is correct, then each daughter molecule should contain one parental nucleotide chain and one newly synthesized nucleotide chain. This prediction has been tested in both prokaryotes and eukaryotes. A little thought shows that there are at least three different ways in which a parental DNA molecule might be related to the daughter molecules. These hypothetical modes are called semiconservative (the Watson-Crick model), conservative, and dispersive (Figure 11-16; page 316). In **semiconservative** replication, each daughter duplex contains one parental and one newly synthesized strand. However, in **conservative** replication, one daughter duplex consists of two newly synthesized strands, and the parent duplex is conserved. **Dispersive** replication results in daughter duplexes that consist of strands containing only *segments* of parental DNA and newly synthesized DNA.

The Meselson-Stahl Experiment

In 1958, Matthew Meselson and Franklin Stahl set out to distinguish among these possibilities. They grew *E. coli* cells in a medium containing the heavy isotope of nitrogen (^{15}N) rather than the normal light (^{14}N) form. This isotope was inserted into the nitrogen bases, which then were incorporated into newly synthesized DNA strands. After many cell divisions in ^{15}N, the DNA of the cells were well labeled with the heavy isotope. The cells were then removed from the ^{15}N medium and put into a ^{14}N medium; after one and two cell divisions, samples were taken. DNA was extracted from the cells in each of these samples and put into a solution of cesium chloride (CsCl) in an ultracentrifuge.

If cesium chloride is spun in a centrifuge at tremendously high speeds (50,000 rpm) for many hours, the cesium and chloride ions tend to be pushed by centrifugal force toward the bottom of the tube. Ultimately, a gradient of Cs^+ and Cl^- ions is established in the tube, with the highest ion concentration at the bottom. Molecules of DNA in the solution also are pushed toward the bottom by centrifugal force. But as they travel down the tube, they encounter the increasing salt concentration, which tends to push them back up owing to the buoyancy of DNA (or its tendency to float). Thus, the DNA finally "settles" at some point in the tube where the centrifugal forces just balance the buoyancy of the molecules in the cesium chloride gradient. The buoyancy of DNA depends on its density (which in turn reflects the ratio of G–C to A–T base pairs). The presence of the heavier isotope of nitrogen changes the buoyant density of DNA. The DNA extracted from cells grown for several generations on ^{15}N medium can readily be distinguished from the DNA of cells grown on ^{14}N medium by the equilibrium position reached in a cesium chloride gradient. Such samples are commonly called "heavy" and "light" DNA, respectively.

Meselson and Stahl found, one generation after the "heavy" cells were moved to ^{14}N medium, the DNA formed a single band of an intermediate density between the densities of the heavy and light controls. After two generations in ^{14}N medium, the DNA formed two bands: one at the intermediate position, the other at the light position (Figure 11-17; page 316). This result would be expected from the semiconservative mode of replication; in fact, the result is compatible *only* with this mode *if* the experiment begins with chromosomes composed of individual double helices (Figure 11-18; page 317).

Autoradiography

The Meselson-Stahl experiment on *E. coli* was essentially duplicated in 1958 by Herbert Taylor on the chromosomes of bean root-tip cells, using a cytological technique. Taylor put root cells into a solution containing tritiated thymidine ([^3H]-thymidine) — the thymine nucleotide labeled with a radioactive hydrogen isotope called *tritium.* He allowed the cells to undergo mitosis in this solution, so that the [^3H]-thymidine could be incorporated into DNA. He then washed the tips and transferred them to a solution containing nonradioactive thymidine. Addition of colchicine to such a preparation inhibits the spindle apparatus so that chromosomes in metaphase fail to separate and sister chromatids remain "tied together" by the centromere.

The cellular location of ^3H can be determined by **autoradiography.** As ^3H decays, it emits a beta particle (an energetic electron). If a layer of photographic emulsion is spread over a cell that contains ^3H, a chemical reaction takes place wherever a beta particle strikes the emulsion. The emulsion can then be developed like a photographic print, so that the emission track of the beta

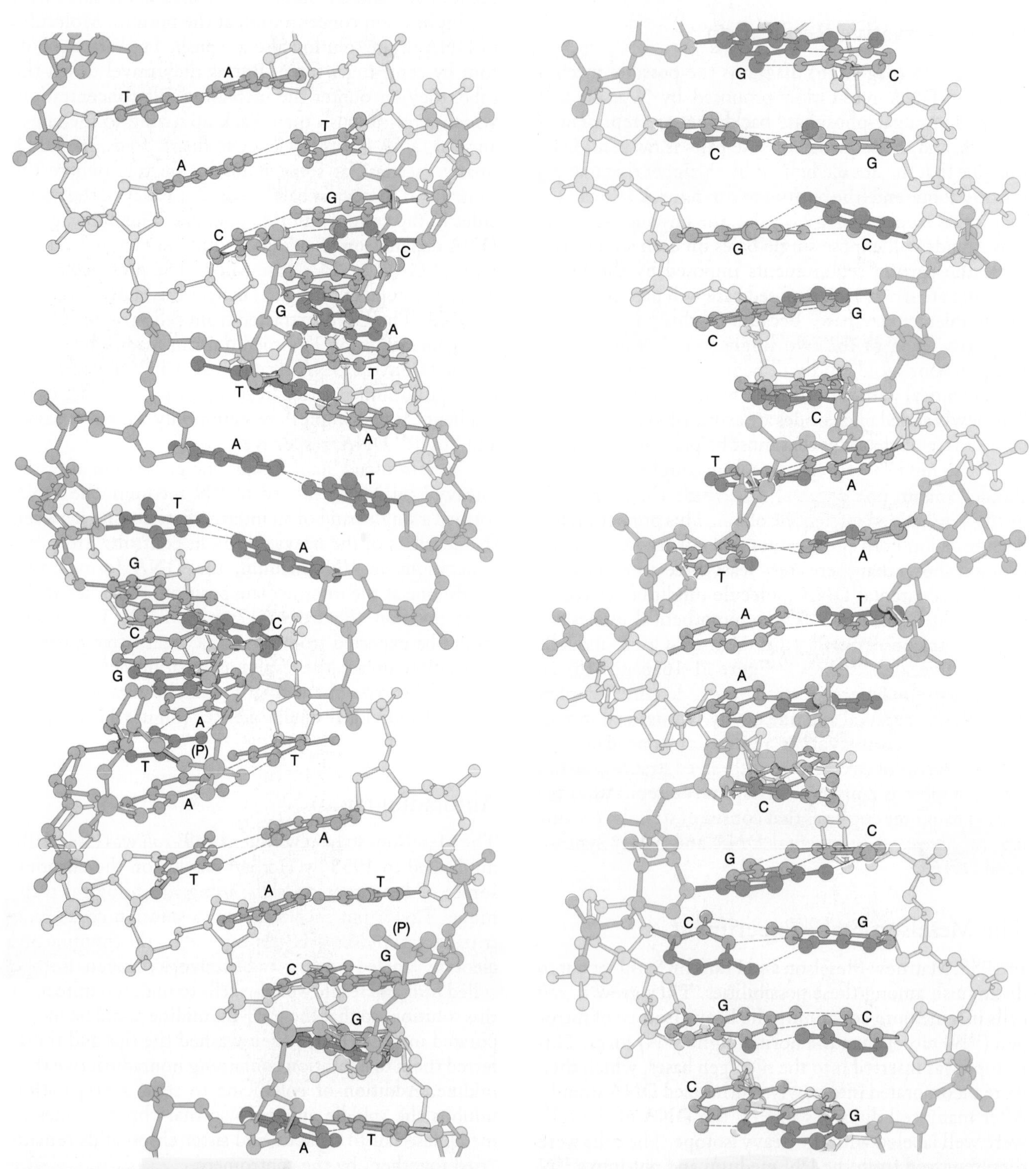

Figure 11-11 The A DNA helix. This perspective drawing was generated from repetition of the central six bases of the octamer GGTATACC. Note how phosphate groups (P) on opposite chains face each other across the major groove. (From R. E. Dickerson, "The DNA Helix and How It Is Read." Copyright 1983 by Scientific American, Inc. All rights reserved.)

Figure 11-12 The B DNA helix. This perspective drawing was generated from repetition of the central 10 base pairs of the dodecamer CGCGAATTCGCG. Note the large twist relative to A DNA and how the twist improves the stacking of the bases along each backbone chain. (From R. E. Dickerson, "The DNA Helix and How It Is Read." Copyright 1983 by Scientific American, Inc. All rights reserved.)

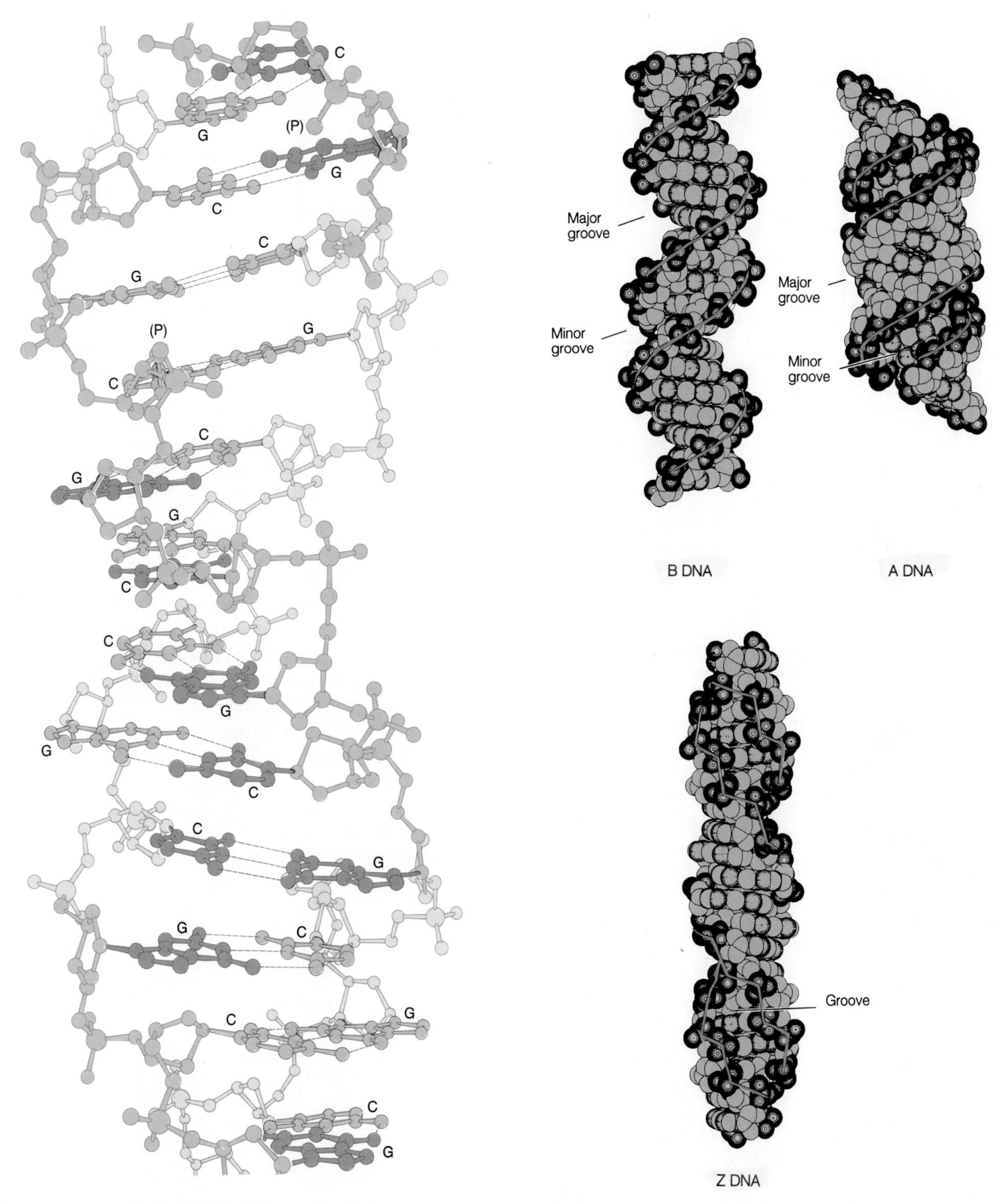

Figure 11-13 The Z DNA helix. This drawing depicts the structure as a left-handed helix of alternating guanines and cytosines, generated from the central four base pairs of CGCGCG. Phosphate groups on different chains now face each other across the deep minor groove. (From R. E. Dickerson, "The DNA Helix and How It Is Read." Copyright 1983 by Scientific American, Inc. All rights reserved.)

Figure 11-14 Structure of B DNA *(left)*, A DNA *(center)* and Z DNA *(right)*. (From A. Kornberg and T. A. Baker, *DNA Replication,* 2d ed. Copyright 1992 by W. H. Freeman and Company.)

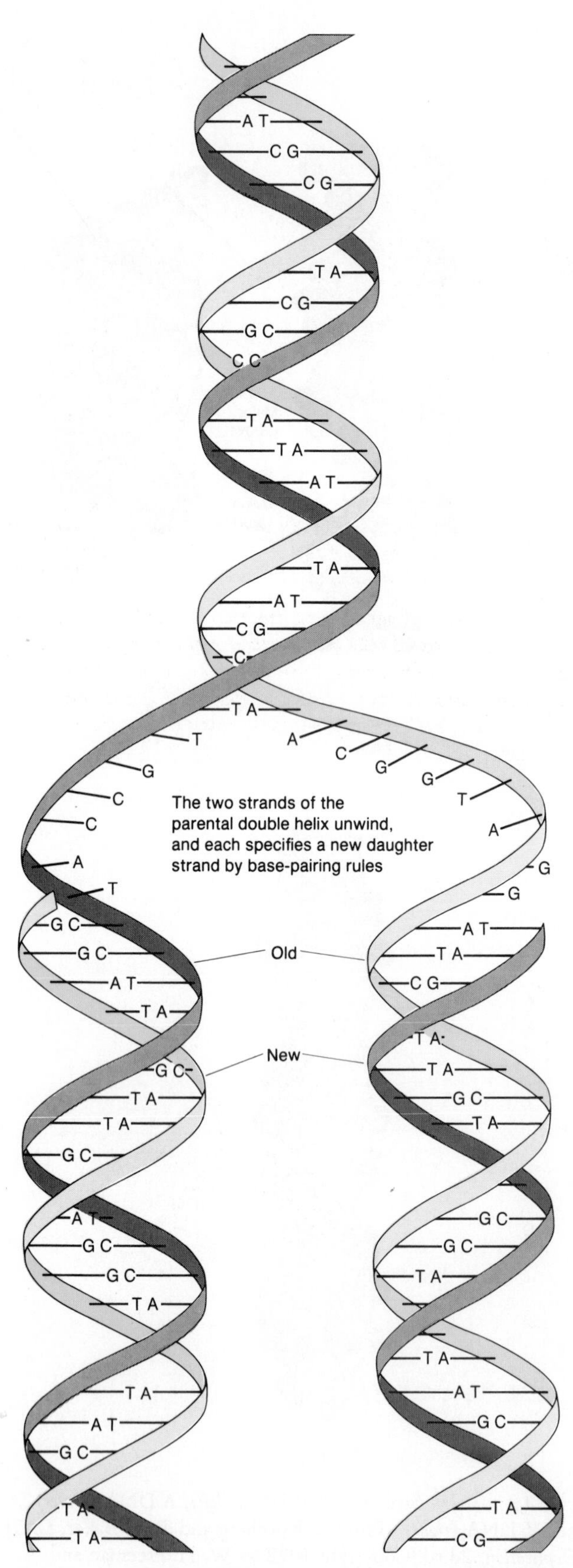

Figure 11-15 The model of DNA replication proposed by Watson and Crick is based on the hydrogen-bonding specificity of the base pairs. Complementary strands are shown in different colors.

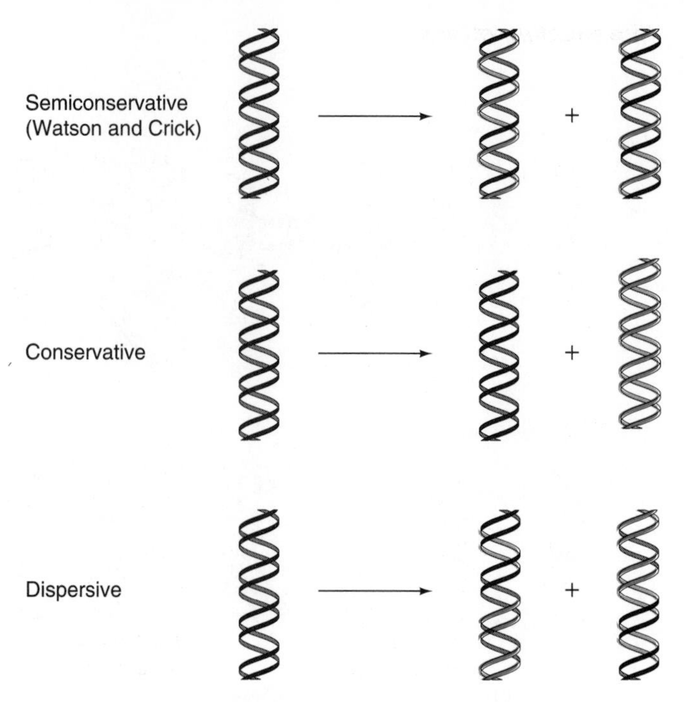

Figure 11-16 Three alternative patterns for DNA replication. The Watson-Crick model would produce the first (semiconservative) pattern. Red lines represent the newly synthesized strands.

particle appears as a black spot or grain. The cell can also be stained, so that the structure of the cell is visible, to identify the location of the radioactivity. In effect, autoradiography is a process in which radioactive cell structures "take their own pictures."

Figure 11-19 shows the results observed when colchicine is added during the division in [^{3}H]-thymidine or during the subsequent mitotic division. It is possible to interpret these results by representing each chromatid as a single DNA molecule that replicates semiconservatively (Figure 11-20).

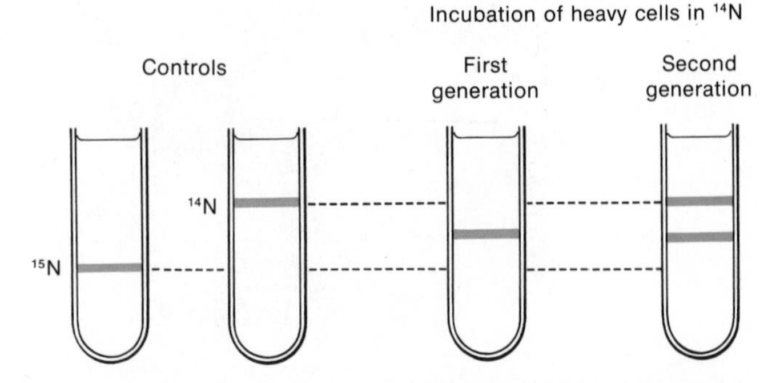

Figure 11-17 Centrifugation of DNA in a cesium chloride (CsCl) gradient. Cultures grown for many generations in ^{15}N and ^{14}N media provide control positions for "heavy" and "light" DNA bands, respectively. When the cells grown in ^{15}N are transferred to a ^{14}N medium, the first generation produces an intermediate DNA band and the second generation produces two bands: one intermediate and one light.

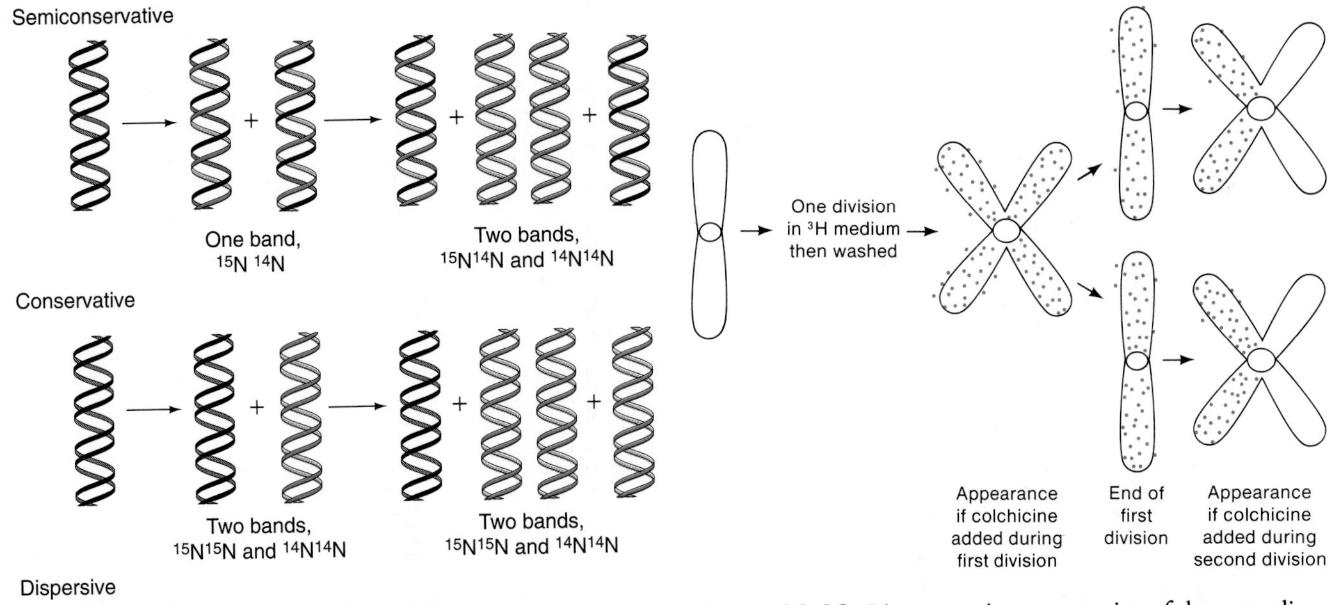

Figure 11-18 Only the semiconservative model of DNA replication predicts results like those shown in Figure 11-17: a single intermediate band in the first generation and one intermediate and one light band in the second generation. (See Figure 11-16 for explanations of symbols.)

Figure 11-19 Diagrammatic representation of the autoradiography of chromosomes from cells grown for one cell division in the presence of the radioactive hydrogen isotope ^{3}H (tritium) and then grown in a nonradioactive medium for a second mitotic division. Each dot represents the track of a particle of radioactivity.

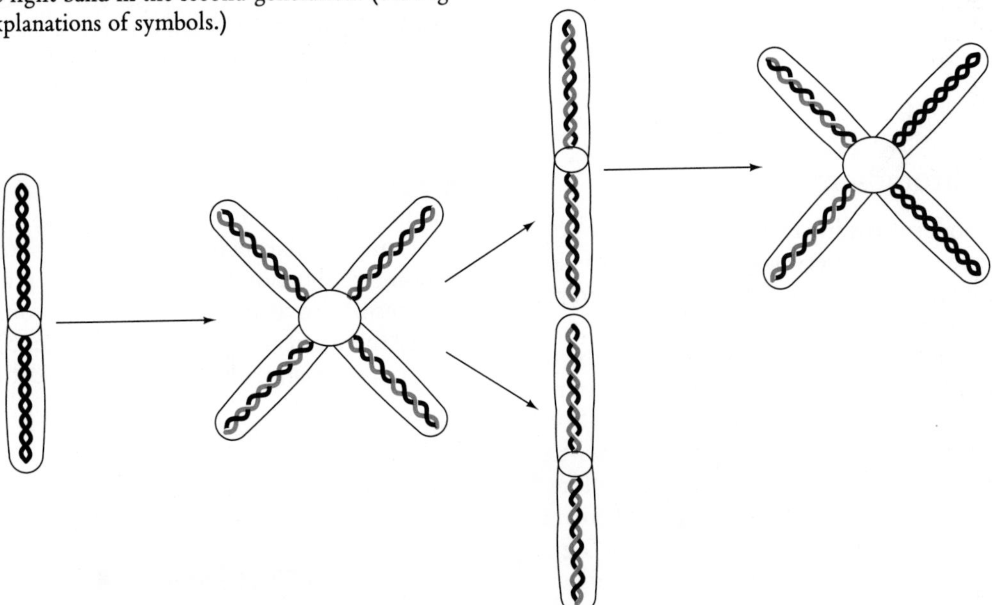

Figure 11-20 An explanation of Figure 11-19 at the DNA level. Red lines represent radioactive strands. In the second replication (which occurs in nontritiated solution), both the ^{3}H strand and the nontritiated strand incorporate nonradioactive nucleotides, yielding one hybrid and one nontritiated chromatid.

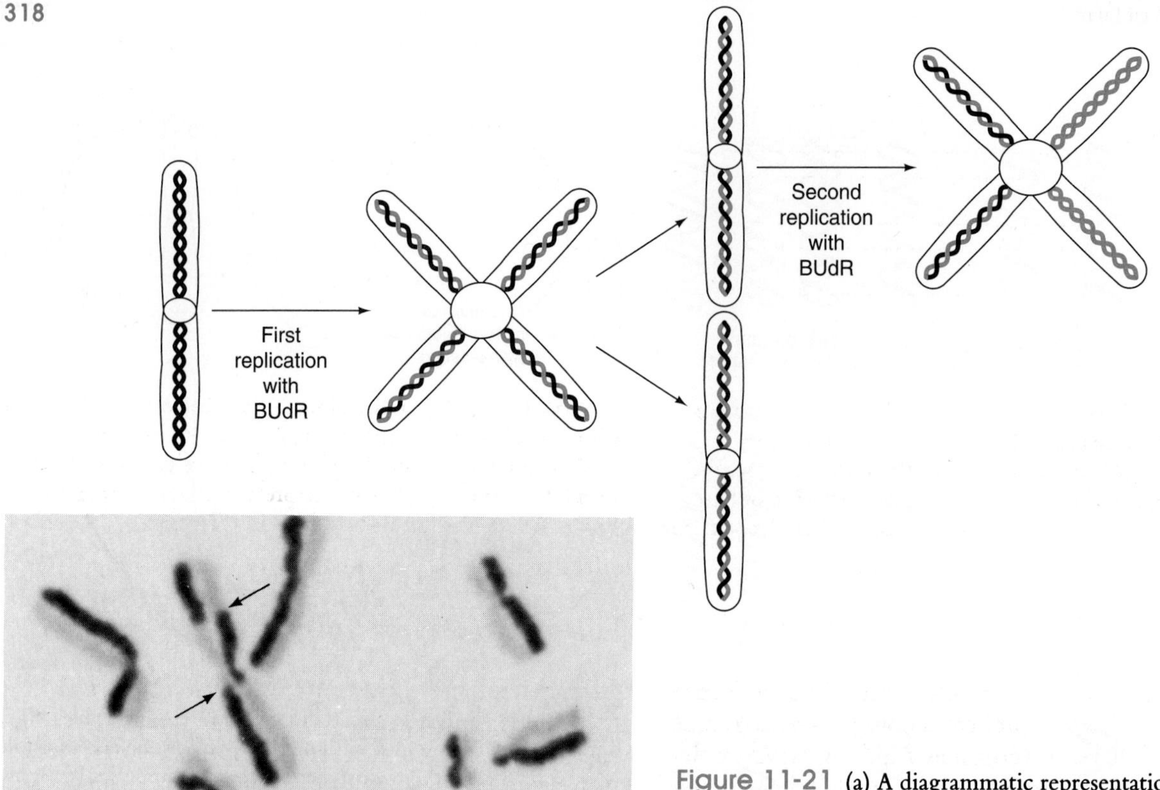

(b)

Figure 11-21 (a) A diagrammatic representation of the production of harlequin chromosomes. The procedure involves letting the chromosomes go through two rounds of replication in the presence of bromodeoxyuridine (BUdR), which replaces thymidine in the newly synthesized DNA. The chromosomes then are stained with a fluorescent dye and Giemsa stain, producing the appearance shown. (The red lines represent the BUdR-substituted strands.) (b) Photograph of harlequin chromosomes in a Chinese hamster ovary (CHO) cell. The chromatids with two strands containing BUdR are light in this photo, while those with one BUdR strand and one original strand are dark. A chromosome at the top (see arrows) has two sister chromatid exchanges. (Photograph courtesy of Sheldon Wolff and Judy Bodycote.)

Harlequin Chromosomes

Using a more modern staining technique, it is now possible to visualize the semiconservative replication of chromosomes at mitosis without the aid of autoradiography. In this procedure, the chromosomes are allowed to go through two rounds of replication in bromodeoxyuridine (BUdR). The BUdR labeling pattern, shown in Figure 11-21a is the reciprocal of that of Figure 11-20, since the BUdR is used in both replications, rather than being replaced by normal thymidine for the second replication as in the autoradiographic process. The chromosomes are then stained with fluorescent dye and Giemsa stain; this process distinguishes hybrid chromatids with one BUdR-containing strand and one original strand (dark stain) from those in which both strands contain BUdR (no stain), and generates so-called **harlequin chromosomes** (Figure 11-21b). (Note, in passing, that harlequin chromosomes are particularly favorable for the detection of sister chromatid exchange at mitosis; two examples are seen in Figure 11-21b).

Using similar techniques, Taylor showed that chromosome replication at meiosis also is semiconservative. This result drove another nail in the coffin of the copy-choice theory of crossing-over (Chapter 5), which would require *conservative* chromosome replication at meiosis.

Chromosome Structure

Figures 11-19 and 11-20 bring up one of the remaining great unsolved questions of genetics: is a eukaryotic chromosome basically a single DNA molecule surrounded by a protein matrix? Two things strongly suggest that this is, in fact, the case. First, if there were many DNA molecules in the chromosome (whether they were side by side, end to end, or randomly oriented), it would be almost impossible for the chromosome to replicate semiconservatively (with all the label going

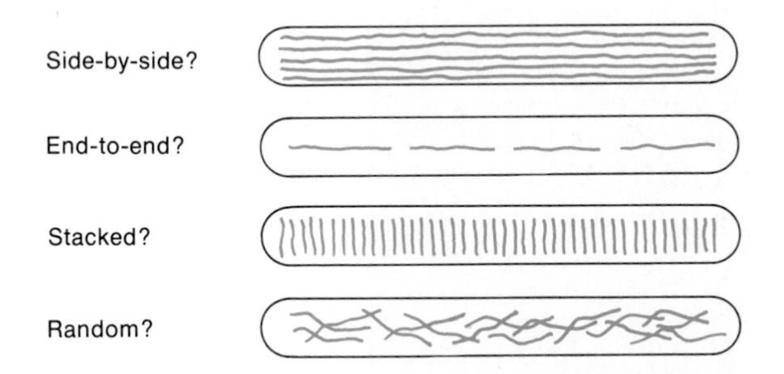

Figure 11-22 Some theoretical alternative packing arrangements of DNA in a eukaryotic chromosome. All of these models are hard to reconcile with the data supporting a semiconservative model of DNA replication. The question of the packaging of a long DNA molecule is considered in Chapter 16.

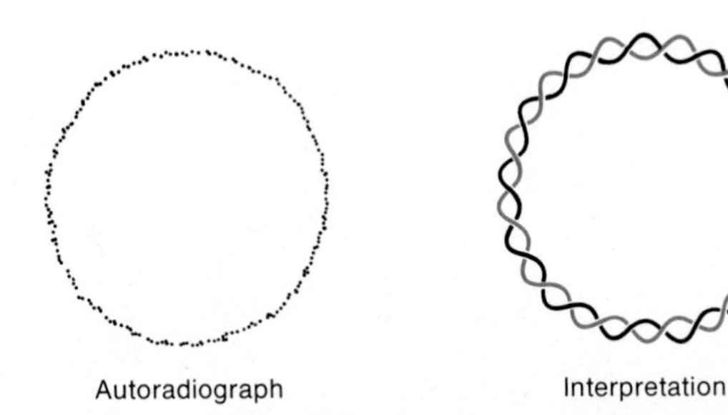

Figure 11-23 *Left:* autoradiograph of a bacterial chromosome after one replication in tritiated thymidine. According to the semiconservative model of replication, one of the two strands should be radioactive. *Right:* interpretation of the autoradiograph. The red line represents the tritiated strand.

into one chromatid, as in Taylor's results). Look at Figure 11-22, and try to figure out how it could be done. Recent studies on isolated chromosomes and long DNA molecules are consistent with the suggestion that *each chromatid is a single molecule of DNA.* That makes a very long molecule. There is enough DNA in a single human chromosome, for example, to stretch out an inch or two and enough DNA in a single nucleus to stretch one meter. (This raises another interesting problem: how is this long molecule packed into the chromosome to permit easy replication?) The second fact supporting a single-molecule hypothesis is that DNA and genes behave as though they are attached end to end in a single string or thread that we call a *linkage group.* All genetic linkage data (Chapter 5) tell us that we need nothing more than a single linear array of genes per chromosome to explain the genetic facts.

As just mentioned, there is far too much DNA in a chromosome for it to extend linearly along the chromosome. It must be packed very efficiently into the chromosome. Current thinking (supported by good microscopic evidence) tends toward a process of coiling and supercoiling of the DNA. Twist a rubber band with your fingers and notice the way it coils; chromosomes may be like this. We return to these questions in Chapter 16.

The Replication Fork

A prediction of the Watson-Crick model of DNA replication is that a replication zipper, or **fork,** will be found in the DNA molecule during replication. In 1963, John Cairns tested this prediction by allowing replicating DNA in bacterial cells to incorporate tritiated thymidine. Theoretically, each newly synthesized daughter molecule should then contain one radioactive ("hot") strand and another nonradioactive ("cold") strand. After varying intervals and varying numbers of replication cycles in a "hot" medium, Cairns extracted the DNA from the cells, put it on a slide, and autoradiographed it for examination under the electron microscope. After one replication cycle in [^{3}H]-thymidine, rings of dots appeared in the autoradiograph. Cairns interpreted these rings as shown in Figure 11-23. It is also apparent from this figure that the bacterial chromosome is circular—a fact that also emerged from genetic analysis described earlier (Chapter 10).

During the second replication cycle, the forks predicted by the model were indeed seen. Furthermore, the density of grains in the three segments was such that the interpretation shown in Figure 11-24 could be made. Cairns saw all sizes of these moon-shaped, autoradiographic patterns, corresponding to the progressive movement of the replication zipper, or fork, around the ring. Structures of the sort shown in Figure 11-24 are called **theta (θ) structures.**

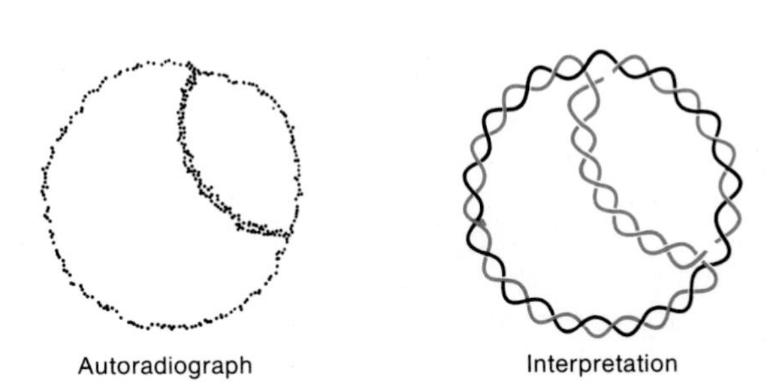

Figure 11-24 *Left:* autoradiograph of a bacterial chromosome during the second round of replication in tritiated thymidine. In this theta θ structure, the newly replicated double helix that crosses the circle could consist of two radioactive strands (if the parental strand was the radioactive one). *Right:* the double thickness of the radioactive tracing on the autoradiogram appears to confirm the interpretation shown here. The red helices represent the "hot" strands.

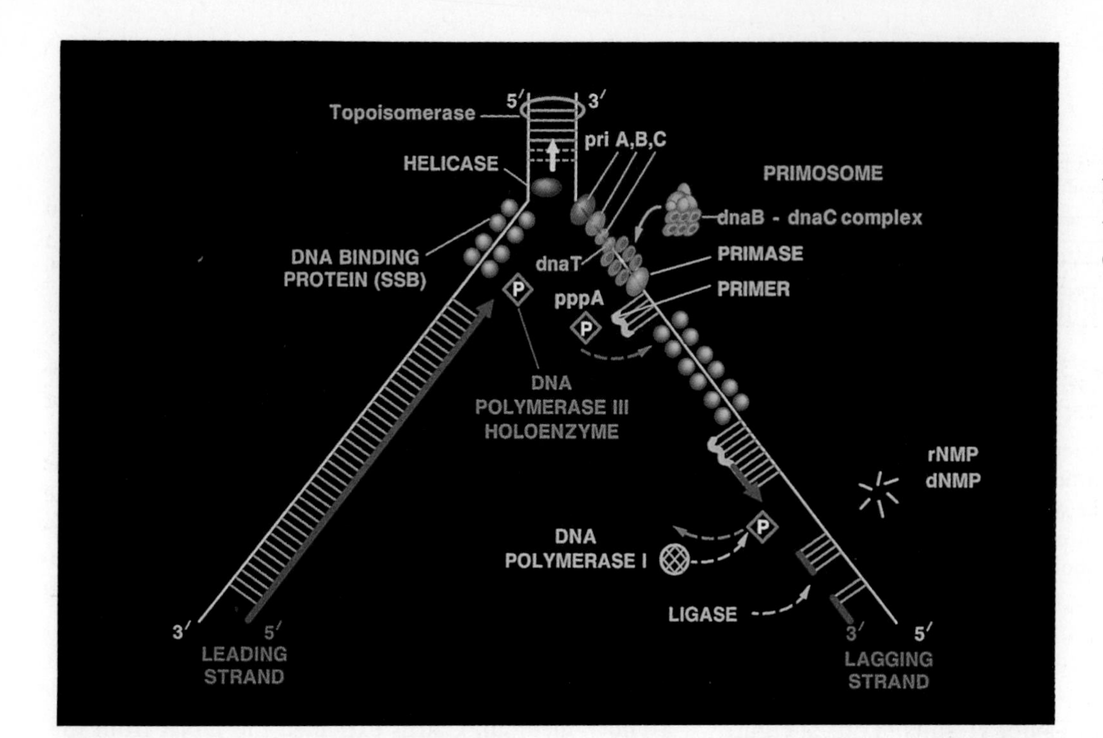

Figure 11-25 DNA replication fork. (From A. Kornberg and T. A. Baker, *DNA Replication,* 2d ed. Copyright 1992 by W. H. Freeman and Company.)

Mechanism of DNA Replication

Watson and Crick first reasoned that complementary base pairing provides the basis of fidelity in DNA replication, that is, that each base in the template strand dictates the complementary base in the new strand. However, we now know that the process of DNA replication is very complex and requires the participation of many different components. Let's examine each of these components and see how they fit together to produce our current picture of DNA synthesis in *Escherichia coli,* the best-studied cellular replication system. In the previous section we introduced the concept of the replication fork. Figure 11-25 gives a detailed schematic view of fork movement during DNA replication; we can refer to this illustration as we consider each component of the process.

DNA Polymerases

In the late 1950s, Arthur Kornberg successfully identified and purified the first DNA polymerase, an enzyme that catalyzes the replication reaction

$$\begin{matrix}\text{Primer}\\ \text{(parental)} +\\ \text{DNA}\end{matrix} \left\{\begin{matrix}\text{dATP}\\ +\\ \text{dGTP}\\ +\\ \text{dCTP}\\ +\\ \text{dTTP}\end{matrix}\right\} \xrightarrow[\text{polymerase}]{\text{DNA}} \text{progeny DNA}$$

This reaction works only with the triphosphate forms of the nucleotides (such as deoxyadenosine triphosphate, or dATP). The total amount of DNA at the end of the reaction can be as much as 20 times the amount of original input DNA, so most of the DNA present at the end must be progeny DNA. Figure 11-26 depicts the chain elongation reaction, or **polymerization** reaction catalyzed by DNA polymerases. We now know that there are three DNA polymerases in *Escherichia coli.* The first enzyme Kornberg purified is called **DNA polymerase I,**

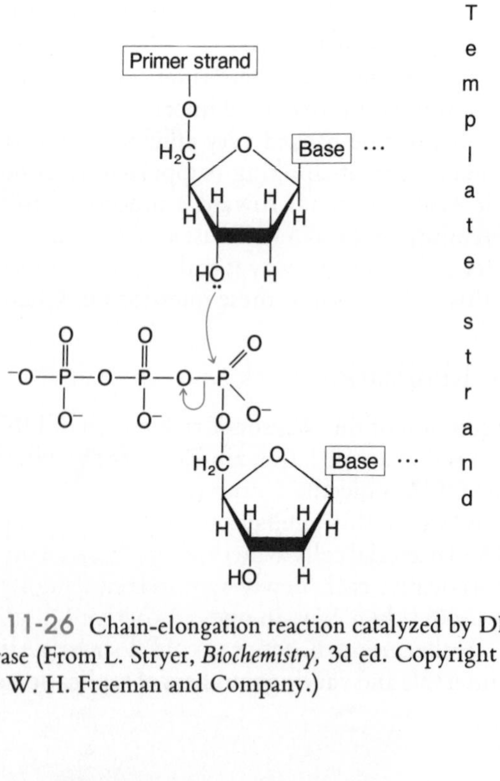

Primer strand

Base ···

Base ···

Template strand

Figure 11-26 Chain-elongation reaction catalyzed by DNA polymerase (From L. Stryer, *Biochemistry,* 3d ed. Copyright 1988 by W. H. Freeman and Company.)

or **pol I.** This enzyme has three activities, which appear to be located in different parts of the molecule:

1. A polymerase activity, which catalyzes chain growth in the $5' \rightarrow 3'$ direction
2. A $3' \rightarrow 5'$ exonuclease activity, which removes mismatched bases
3. A $5' \rightarrow 3'$ exonuclease activity, which degrades double-stranded DNA

Subsequently, two additional polymerases, pol II and pol III, were identified in *E. coli.* Pol II may operate to repair damaged DNA, although no particular role has been assigned to this enzyme. Pol III, together with pol I, is involved in the replication of *E. coli* DNA (Figure 11-25). The complete complex, or **holoenzyme,** of pol III contains at least 20 different polypeptide subunits, although the catalytic "core" consists of only three subunits, alpha (α), epsilon (ϵ), and theta (θ). The pol III complex will complete the replication of single-stranded DNA if there is at least a short segment of duplex already present. The short oligonucleotide that creates the duplex is termed a **primer.**

Origins of Replication

E. coli replication begins from a fixed **origin,** but then proceeds **bidirectionally** (with moving forks at both ends of the replicating piece), as shown in Figure 11-27, ending at a site called the **terminus.** The unique origin is termed ***oriC*** and is located at 83 min on the genetic map. It is 245 bp long and has several components, as illustrated in Figure 11-28. First, there is a side-by-side, or tandem, set of 13-bp sequences, which are nearly identical. There is also a set of binding sites for a protein, the dnaA protein. An initial step in DNA synthesis is the unwinding of the DNA at the origin in response to binding of the dnaA protein. The consequences of bidirectional replication can be seen in Figure 11-29, which gives a larger view of DNA replication.

Priming DNA Synthesis

DNA polymerases can extend a chain but cannot start a chain. Therefore, as mentioned above, DNA synthesis must first be initiated with a primer, or short oligonucleotide that generates a segment of duplex DNA. The

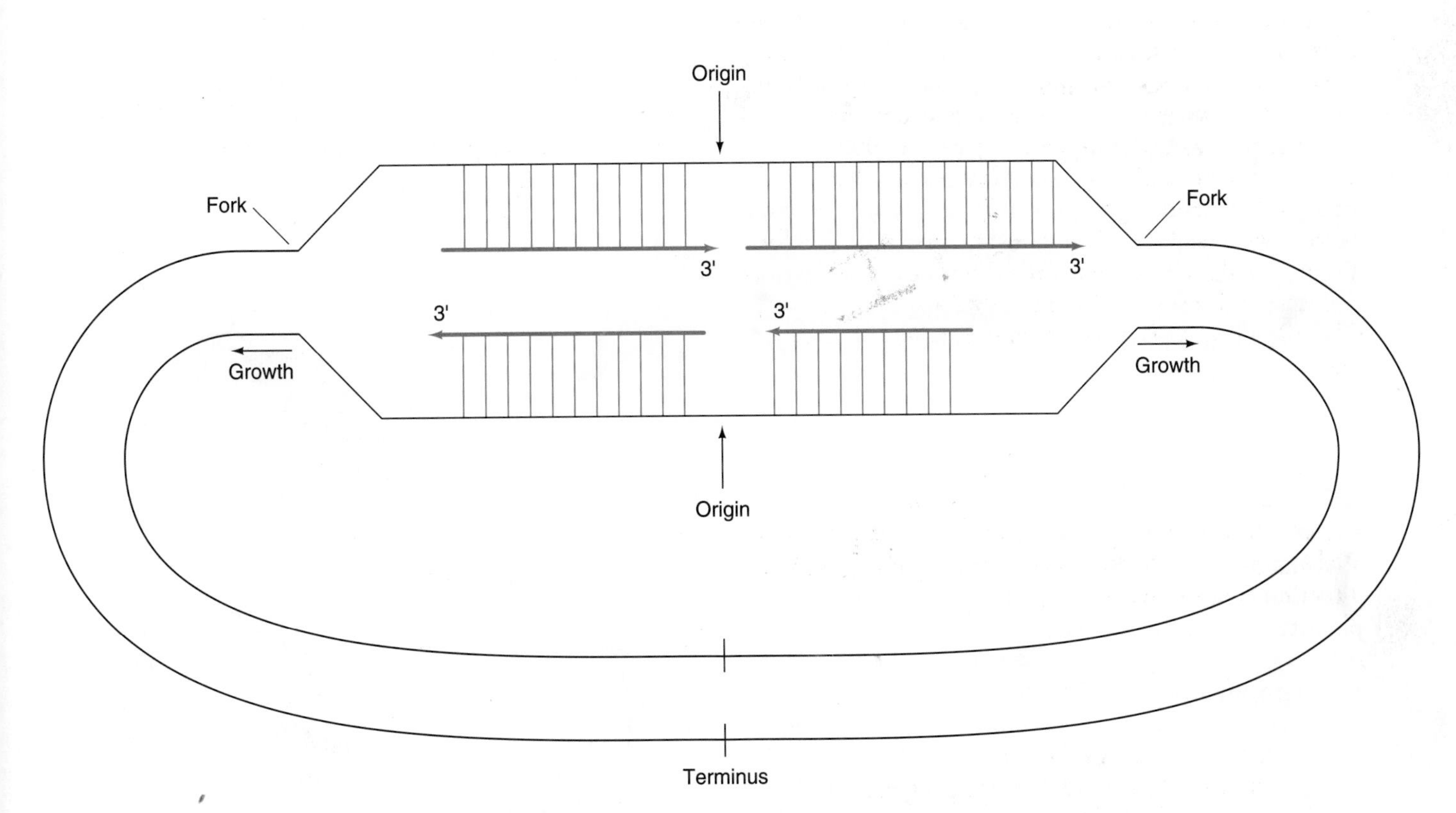

Figure 11-27 Diagrammatic representation of DNA replication proceeding in both directions from the origin (Modified from A. Kornberg, *DNA Synthesis.* Copyright 1974 by W. H. Freeman and Company).

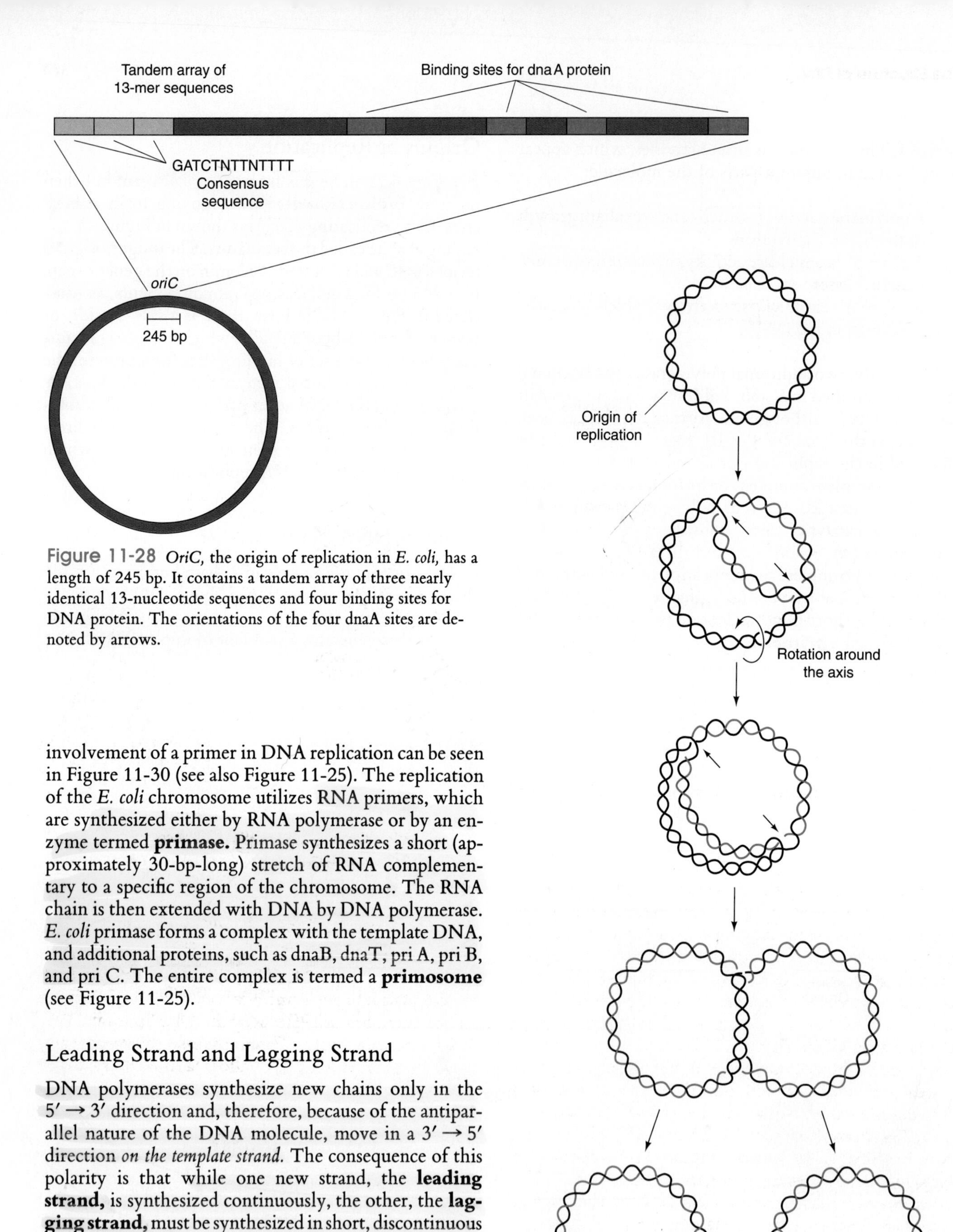

Figure 11-28 *OriC,* the origin of replication in *E. coli,* has a length of 245 bp. It contains a tandem array of three nearly identical 13-nucleotide sequences and four binding sites for DNA protein. The orientations of the four dnaA sites are denoted by arrows.

Figure 11-29 Bidirectional replication of a circular DNA molecule.

involvement of a primer in DNA replication can be seen in Figure 11-30 (see also Figure 11-25). The replication of the *E. coli* chromosome utilizes RNA primers, which are synthesized either by RNA polymerase or by an enzyme termed **primase.** Primase synthesizes a short (approximately 30-bp-long) stretch of RNA complementary to a specific region of the chromosome. The RNA chain is then extended with DNA by DNA polymerase. *E. coli* primase forms a complex with the template DNA, and additional proteins, such as dnaB, dnaT, pri A, pri B, and pri C. The entire complex is termed a **primosome** (see Figure 11-25).

Leading Strand and Lagging Strand

DNA polymerases synthesize new chains only in the $5' \rightarrow 3'$ direction and, therefore, because of the antiparallel nature of the DNA molecule, move in a $3' \rightarrow 5'$ direction *on the template strand.* The consequence of this polarity is that while one new strand, the **leading strand,** is synthesized continuously, the other, the **lagging strand,** must be synthesized in short, discontinuous segments, as can be seen in Figure 11-31 (see also Figure 11-25). This is because addition of nucleotides along the template for the lagging strand must proceed toward the template's 5′ end (since replication *always* moves along the template in a $3' \rightarrow 5'$ direction, so that the new strand can grow $5' \rightarrow 3'$). Thus, the new strand must grow in a

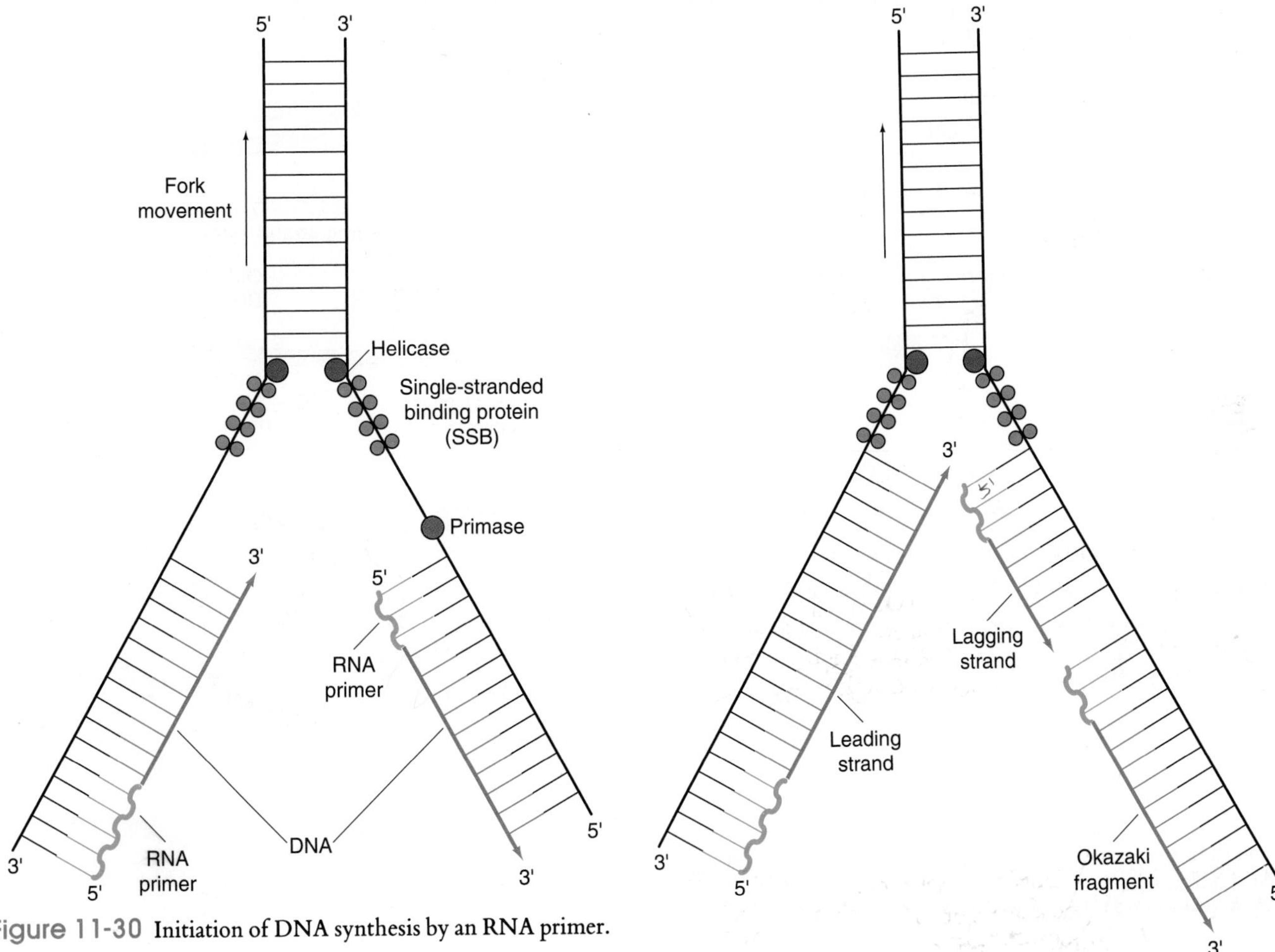

Figure 11-30 Initiation of DNA synthesis by an RNA primer.

Figure 11-31 DNA synthesis proceeds by continuous synthesis on the leading strand and discontinuous synthesis on the lagging strand.

direction opposite to the movement of the replication fork. As fork movement exposes a new section of lagging-strand template, a new lagging-strand fragment is begun and proceeds away from the fork until it is stopped by the preceding fragment. In *E. coli,* pol III carries out most of the DNA synthesis on both strands, and pol I fills in the gaps left in the lagging strand, which are then sealed by the enzyme **DNA ligase.** DNA ligases join broken pieces of DNA by catalyzing the formation of a phosphodiester bond between the 5′ phosphate end of a hydrogen-bonded nucleotide and an adjacent 3′ OH group, as shown in Figure 11-32. It is the only enzyme that can seal DNA chains. Figure 11-33 shows the lagging-strand synthesis and gap repair in detail. The primers for the discontinuous synthesis on the lagging strand are synthesized by primase [step (a)]. The primers are extended by DNA polymerase [step (b)] to yield DNA fragments that were first detected by Reiji Okazaki, and which are termed **Okazaki fragments.** The 5′ → 3′ exonuclease activity of pol I removes the primers [step (c)] and fills in the gaps with DNA, which are sealed by DNA ligase [step (d)]. One proposed mechanism that allows the same dimeric holoenzyme molecule to participate in both leading- and lagging-strand synthesis is shown in Figure 11-34. Here, the looping of the template for the lagging strand allows a single pol III dimer to generate both daughter strands. After approximately 1000 base pairs, pol III would release the segment of lagging-strand duplex and allow a new loop to be formed.

Helicases and Topoisomerases

Helicases are enzymes that disrupt the hydrogen bonds which hold the two DNA strands together in a double helix. Hydrolysis of ATP drives the reaction. Among *E. coli* helicases are the dnaB protein and the rep protein. The rep protein may help to unwind the double helix ahead of the polymerase (refer to Figure 11-25). The unwound DNA is stabilized by the SSB (single-stranded binding) protein, which binds to the single-stranded DNA and retards reformation of the duplex.

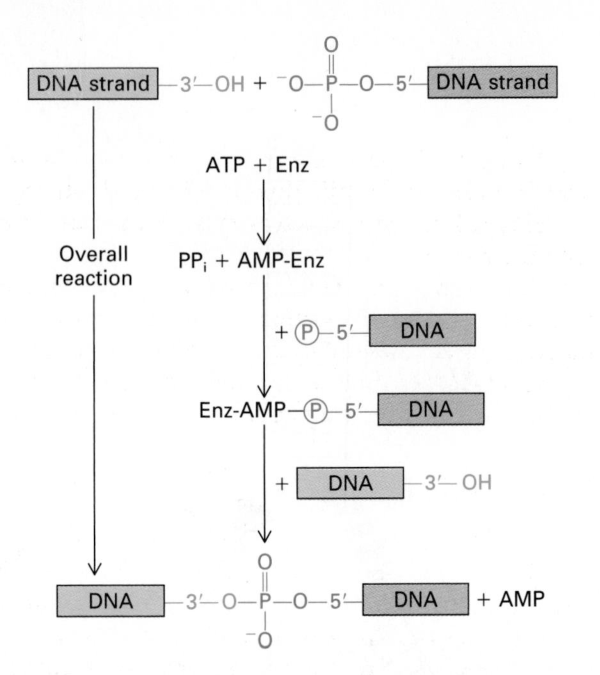

Figure 11-32 The reaction catalyzed by DNA ligase (Enz) joins the 3′ OH end of one fragment to the 5′ phosphate of the adjacent fragment (From J. Darnell, H. Lodish, and D. Baltimore, *Molecular Cell Biology,* 2d ed. Copyright 1990 by Scientific American Books, Inc.)

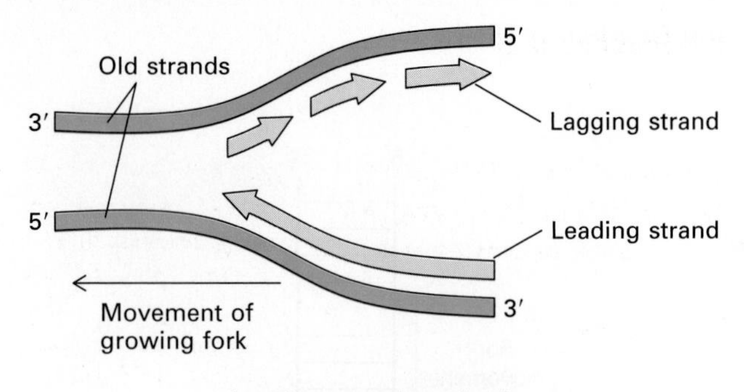

Lagging-strand synthesis

(a) RNA oligonucleotides (primer) copied from DNA

(b) DNA polymerase elongates RNA primers with new DNA

(c) DNA polymerase removes 5′ RNA at end of neighboring fragment and fills gap

(d) DNA ligase joins adjacent fragments

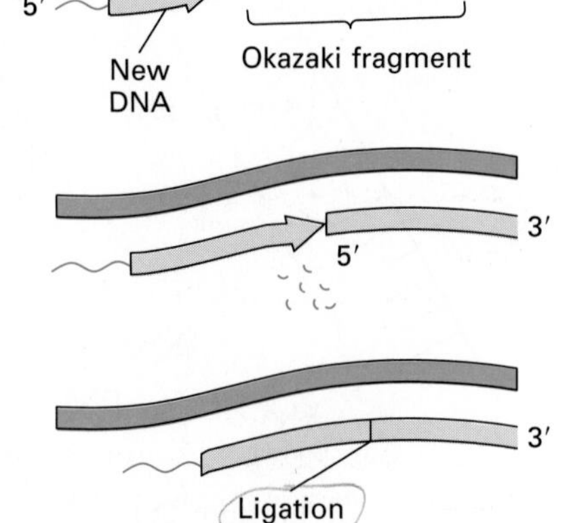

Figure 11-33 The overall structure of a growing fork (top) and steps in synthesis of the lagging strand. (From J. Darnell, H. Lodish, and D. Baltimore, Molecular Cell Biology, 2d ed. Copyright 1990 by Scientific American Books, Inc.)

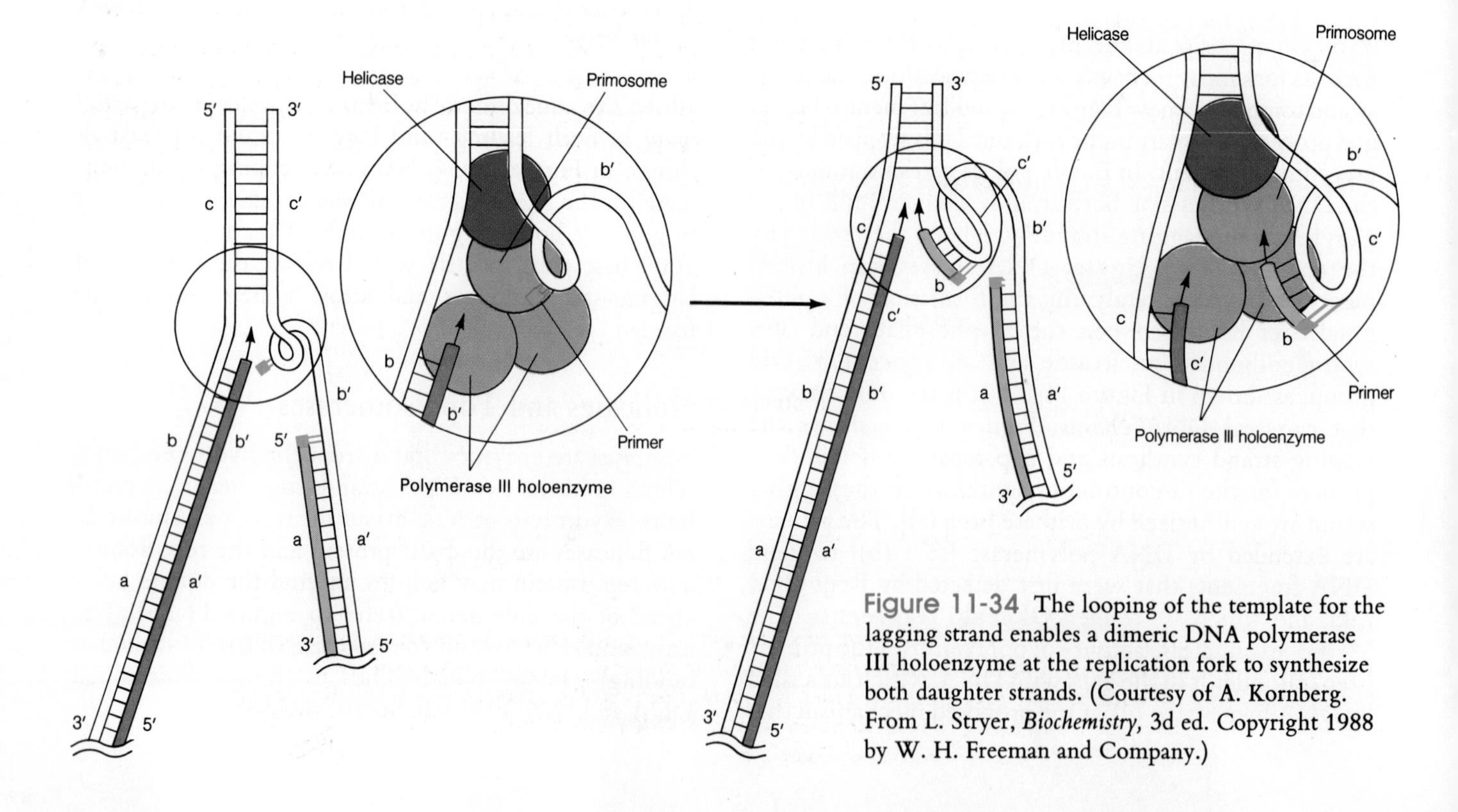

Figure 11-34 The looping of the template for the lagging strand enables a dimeric DNA polymerase III holoenzyme at the replication fork to synthesize both daughter strands. (Courtesy of A. Kornberg. From L. Stryer, *Biochemistry,* 3d ed. Copyright 1988 by W. H. Freeman and Company.)

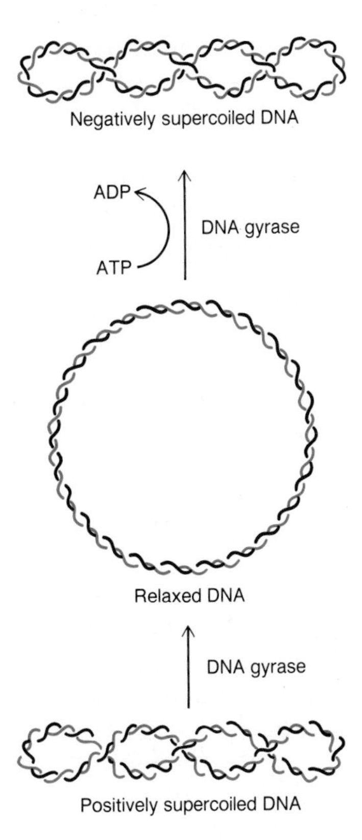

Figure 11-35 DNA gyrase–catalyzed supercoiling. Replicating DNA generates "positive" supercoils, depicted at the bottom of the diagram, as a result of rapid rotation of the DNA at the replication fork. DNA gyrase can nick and close phosphodiester bonds, relieving the supercoiling, as shown here (relaxed DNA). Gyrase can also generate supercoils twisted in the opposite direction, termed *negative* supercoils; this arrangement facilitates the unwinding of the helix. (Modified from L. Stryer, *Biochemistry,* 3d ed. Copyright 1988 by W. H. Freeman and Company.)

The action of helicases during DNA replication generates twists in the circular DNA that need to be removed to allow replication to continue. We should appreciate that circular DNA can be twisted and coiled, much like the extra coils that can be introduced into a rubber band. This **supercoiling** can be created or relaxed by enzymes termed **topoisomerases,** an example of which is DNA gyrase (Figure 11-35). Topoisomerases can also induce *(catenate)* or remove *(decatenate)* knots, or links in a chain. There are two basic types of isomerases. Type I enzymes induce a single-stranded break into the DNA duplex. Type II enzymes cause a break in both strands. In *Escherichia coli,* topo I and topo III are examples of type I enzymes, whereas gyrase is an example of a type II enzyme.

Untwisting of the DNA strands to open the replication fork causes extra twisting at other regions, and the supercoiling releases the strain of the extra twisting (see Figure 11-36). During replication, gyrase is needed to remove positive supercoils ahead of the replication fork.

Exonuclease Editing

Both DNA polymerase I and DNA polymerase III also possess $3' \rightarrow 5'$ exonuclease activity, which serves a "proofreading" and "editing" function by searching for mismatched bases that were inserted erroneously during polymerization and excising them. The proofreading activity of pol III is in the ϵ subunit, which must be bound to α for full proofreading activity (see Figure 11-37). Strains lacking a functional ϵ have a higher mutation rate (see Chapter 17). Figure 11-38 shows the excision of a cytosine residue that has erroneously been paired with an adenine. As can be seen, hydrolysis occurs at the 5′ end of the mismatched base; removal of the incorrect base leaves a 3′ OH group on the preceding base, which is then free to continue the growing strand by accepting the correct nucleotide triphosphate (thymidine, in this case).

Note that this exonuclease activity occurs at the 3′ end of the growing strand (and is therefore $3' \rightarrow 5'$). The coordination of exonuclease activity with strand growth helps to explain why replication occurs in the $5' \rightarrow 3'$ direction. As we saw earlier, new bases are added when the 3′ OH on the terminal deoxyribose of the growing strand attacks the high-energy phosphate of the nucleotide triphosphate that is being added (see Figure 11-26). Chain growth is thus $5' \rightarrow 3'$. It is conceivable that replication could occur $3' \rightarrow 5'$ (in Figure 11-26, the 5′ triphosphate at the bottom would be the last base on the

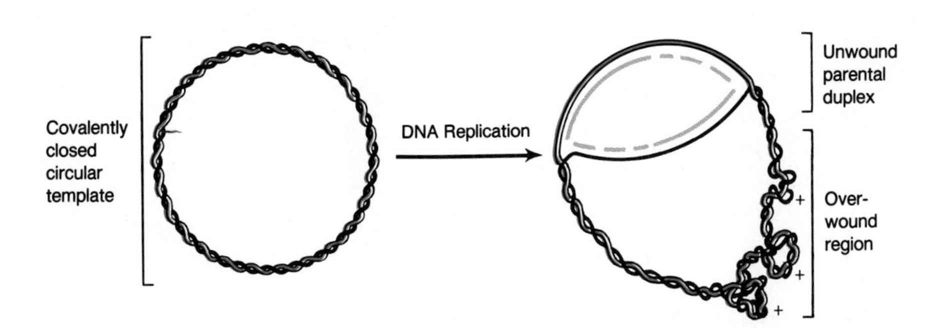

Figure 11-36 Swivel function of topoisomerase during replication. Extra-twisted (positively supercoiled) regions accumulate ahead of the fork as the parental strands separate for replication. A topoisomerase is required to remove these regions, acting as a swivel to allow extensive replication. (From A. Kornberg and T. A. Baker, *DNA Replication,* 2d ed. Copyright 1992 by W. H. Freeman and Company.)

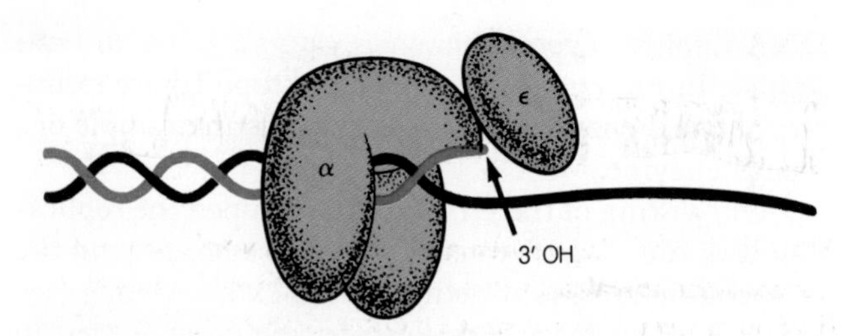

Figure 11-37 Proofreading by the pol III α-ε complex (From A. Kornberg and T. A. Baker, *DNA Replication,* 2d ed. Copyright 1992 by W. H. Freeman and Company.)

chain, and the 3′ OH that attacks it would be on the free nucleotide triphosphate about to be added to the strand). However, if replication occurred in this direction, exonuclease excisions would occur at the 5′ end of the strand. When a mismatched base was removed, a 5′ OH would be left at the end of the growing strand. The 3′ OH of an incoming nucleotide triphosphate would thus be facing this 5′ OH instead of the high-energy 5′ triphosphate necessary for bond formation. No bond would form and strand growth would stop. Therefore, replication does not occur in the 3′ → 5′ direction.

DNA Replication in Eukaryotes

Although many of the details of DNA replication are similar in both prokaryotes and eukaryotes, there are a number of differences. For instance, in bacteria such as *E.*

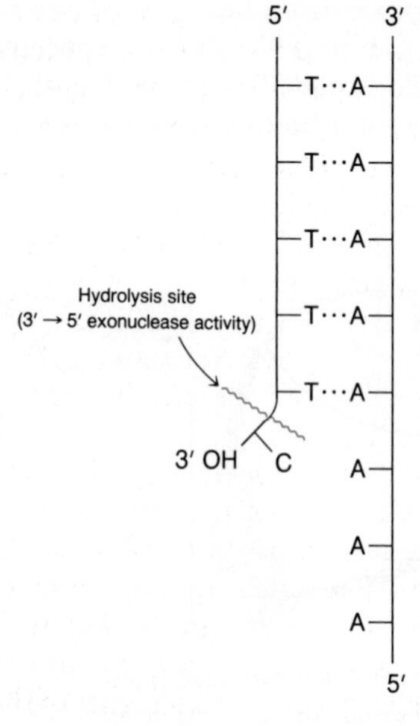

Figure 11-38 3′ → 5′ exonuclease action of DNA polymerase III.

coli the replication-division cycle usually requires 40 minutes, but in eukaryotes it can vary from 1.4 hours in yeast to 24 hours in cultured animal cells, and may last 100 to 200 hours in some cells. Eukaryotes have to solve the problem of coordinating the replication of more than one chromosome, as well as replicating the complex structure of the chromosome itself (see Chapter 16 for a description of chromosome structure).

The Cell Cycle

The eukaryotic **cell cycle** can be divided into four distinct phases, pictured schematically in Figure 11-39. These are

1. **Gap 1 (G1) phase.** The decision to replicate is made during this phase. Cells synthesize many of the factors and enzymes needed for replication. A commitment to proliferate is made when the cell proceeds past the "restriction point" (R), after which the programmed set of events cannot be reversed. Cells must reach a certain size to be able to enter the division cycle. Cells not entering the division cycle become quiescent (the **G0 phase**).
2. **Synthesis (S) phase.** Initiation of DNA synthesis occurs at multiple origins (see below), and the genome is replicated.
3. **Gap 2 (G2) phase.** The cell prepares for entry into mitosis. Chromosomes condense, and cytoplasmic factors important for mitosis are synthesized.
4. **Mitosis (M) phase.** Chromosomes are segregated into daughter cells.

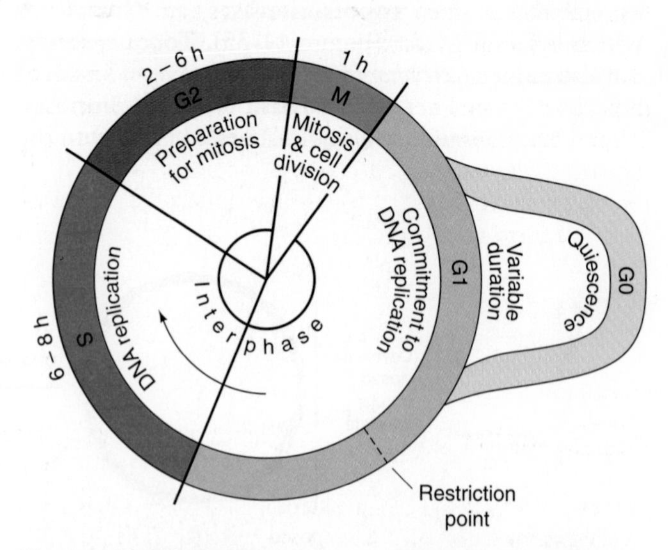

Figure 11-39 Growth cycle of a eukaryotic cell, including a quiescent phase G0. Variation in cell-cycle time is due largely to variation in the duration of the G1 phase. (Modified from A. Kornberg and T. A. Baker, *DNA Replication,* 2d ed. Copyright 1992 by W. H. Freeman and Company.)

Table 11-2. Length of Cell-Cycle Periods in Eukaryotes

Organism	Cell period M	G1	S	G2	Total doubling time
Human (h)	1	8	10	5	~24
Plant (h)	1	8	12	8	~29
Yeast (min)	20	25	40	35	~120

SOURCE: A. Kornberg and T. A. Baker, *DNA Replication*, 2d. ed. W. H. Freeman and Company, 1992.

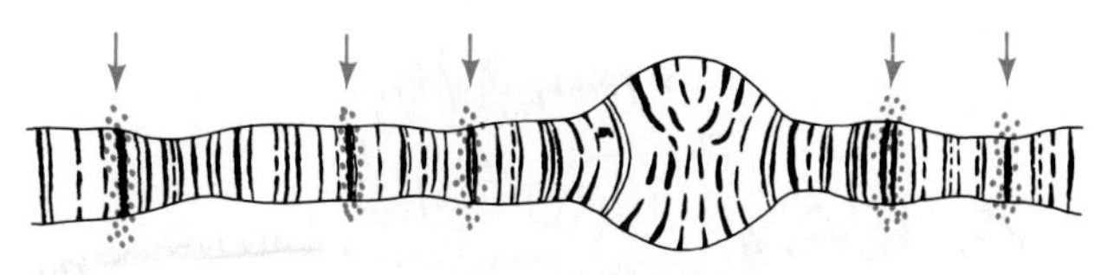

Figure 11-41 Replication pattern in a *Drosophila* polytene chromosome revealed by autoradiography. Several points of replication are seen within a single chromosome, as indicated by the arrows.

Table 11-2 compares the lengths of cell-cycle periods in several eukaryotes.

Origins of Replication

In eukaryotes, replication proceeds from multiple points of origin. This can be demonstrated by a procedure in which a eukaryotic cell is briefly exposed to [^{3}H]-thymidine, in a step called a **pulse** exposure, and then is provided an excess of "cold" (unlabeled) thymidine, in a step called the **chase;** the DNA is then extracted, and autoradiographs are made. Figure 11-40 shows the results of such a procedure, with what appear to be distinct, simultaneously replicating regions along the DNA molecule. Replication appears to begin at several different sites on these eukaryotic chromosomes. Similarly, a pulse-and-chase study of DNA replication in polytene (giant) chromosomes of *Drosophila* by autoradiography reveals many replication regions within single chromosome arms (Figure 11-41). As yet there is no firm proof that these regions are indeed different starting points on a single DNA molecule. However, recent experiments in yeast indicate the existence of approximately 400 replication origins distributed among the 17 yeast chromosomes. The exact structure of the eukaryotic chromosome still remains one of the most exciting problems in genetics (see Chapter 16).

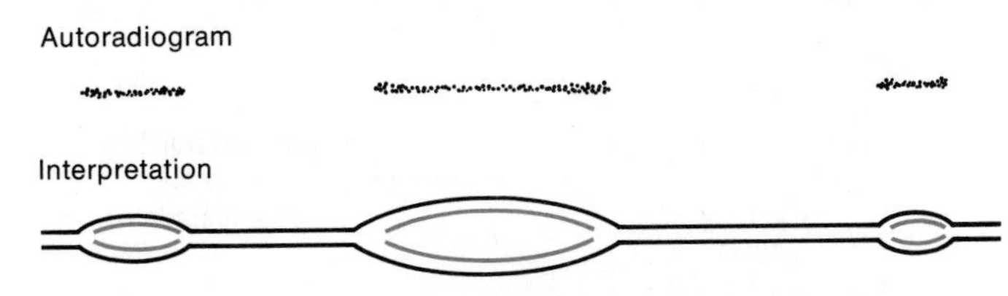

Figure 11-40 A replication pattern in DNA revealed by autoradiography. A cell is briefly exposed to [^{3}H]-thymidine (pulse) and then provided with an excess of nonradioactive ("cold") thymidine (chase). DNA is spread on a slide and autoradiographed. In the interpretation shown here, there are several initiation points for replication within one double helix of DNA.

Eukaryotic DNA Polymerases

There are at least four DNA polymerases, α, β, γ, and δ, in higher eukaryotes. Polymerase α and δ in the nucleus have roles similar to pol II in *E. coli.* Polymerase β is involved in DNA repair and gap filling. The γ polymerase is found in mitochondria and appears to be involved in replication of mitochondrial DNA.

DNA and the Gene

We have now learned that DNA is the genetic material and consists of a linear sequence of nucleotide pairs. The obvious conclusion is that the allele maps represent a genetic equivalent of the nucleotide-pair sequences in DNA. We can validate this assumption if we can show that *the genetic maps are congruent with DNA maps.* This step was first taken using some elegant genetic and biochemical manipulations.

The DNA of the λ phage turns out to be a linear stretch of DNA that can be circularized because each 5′ end of the two strands has an extra terminal extension of 12 bases that is complementary to the other 5′ end (Figure 11-42). Because they are complementary, these ends can pair to join the DNA into a circle; they are known as *cohesive,* or *sticky,* ends.

When a linear object such as a DNA molecule is subjected to shear stress (say, by pipetting or stirring), the mechanics of the stress causes breaks to occur, principally in the middle of the molecule. When DNA from the λ phage is sheared in half, the two halves happen to have different G–C:A–T ratios, which means, as we saw earlier, that their buoyant densities differ and they can be separated by centrifugation in CsCl. When the two half-molecules are separated, their genetic content can be assayed through their ability to recombine with normal λ DNA carrying known mutations (Figure 11-43). First, bacteria are simultaneously infected with mutant DNA and transformed with one or the other of the DNA half-molecules. If the DNA fragment contains the genes for which the injected DNA is mutant, crossovers will occur

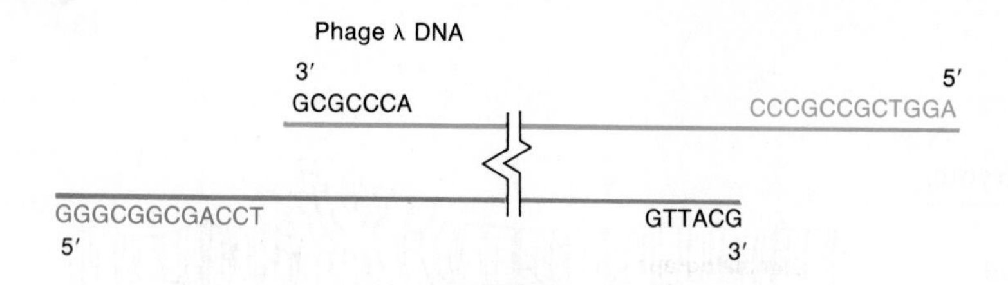

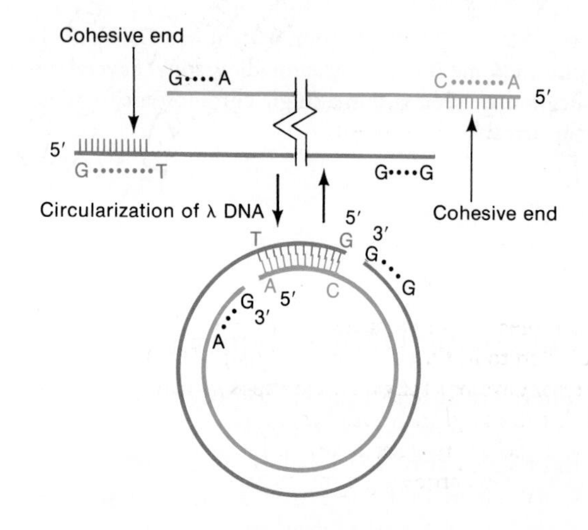

Figure 11-42 Circularization of DNA. The λ DNA is linear in the phage, but once in a host cell, it circularizes as a prelude to insertion or replication. Circularization is achieved by joining the complementary ("sticky") single-stranded ends. (From A. Kornberg, *DNA Synthesis.* Copyright 1974 by W. H. Freeman and Company.)

between the fragment and the mutant DNA and the resultant recombination can be detected by observing plaque-forming phage. If no recombination is detected, the half-molecule does not contain the loci for the mutant genes. A given half-molecule, or DNA fragment of any size, can be tested against a number of mutant markers to see if it recombines with them and thus contains their loci. This process of incorporation of DNA fragments into normal phage DNA and subsequent "rescue" of the affected genes upon lysis is called a **marker-rescue experiment.** *Such experiments demonstrate that one particular half of the DNA molecule carries the information of one particular half of the linkage map.*

Subsequent experiments have demonstrated that the order of genes on a section of DNA matches the order on linkage maps. Dale Kaiser and his associates separated one of the DNA halves carrying a cohesive end and attached the other half to chromatographic material over which DNA could be passed (Figure 11-44). The single-stranded ends dangled free like fish hooks. When the other half of the DNA was sheared into smaller and smaller molecules and then passed over the sticky ends, the complementary sequences stuck. These stuck pieces could be detached easily by raising the temperature to break the hydrogen bonds in the sticky ends. In this way, a series of fractions (of varying size) of one end of the λ DNA were separated and tested for content by marker rescue. The progressive elimination of markers indicated their respective positions along the DNA half-molecule

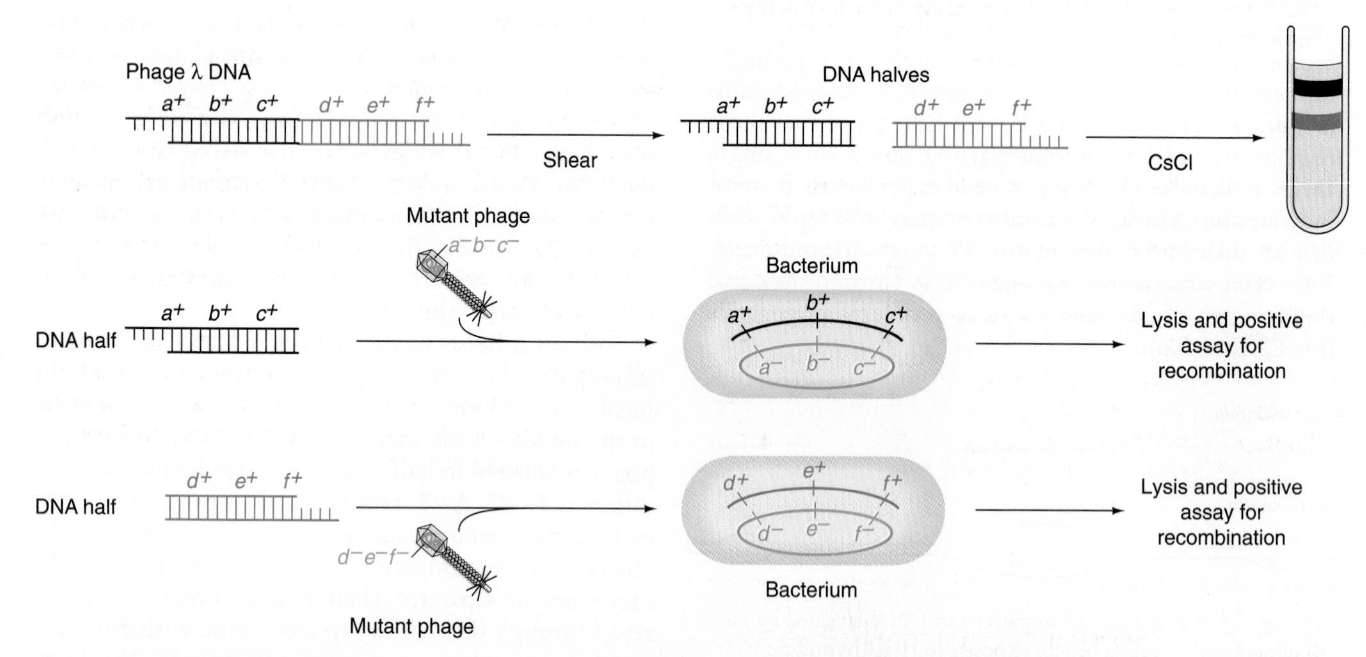

Figure 11-43 When λ DNA is broken, it forms roughly equal halves that happen to have different buoyant densities. Each half can be "rescued" by a multiply mutant phage during simultaneous transformation and infection. From these experiments, it has been shown that only genes from one specific half of the genetic map could be rescued from one specific half of the DNA. (Note that the bacterial chromosome has been omitted here to simplify the diagram.)

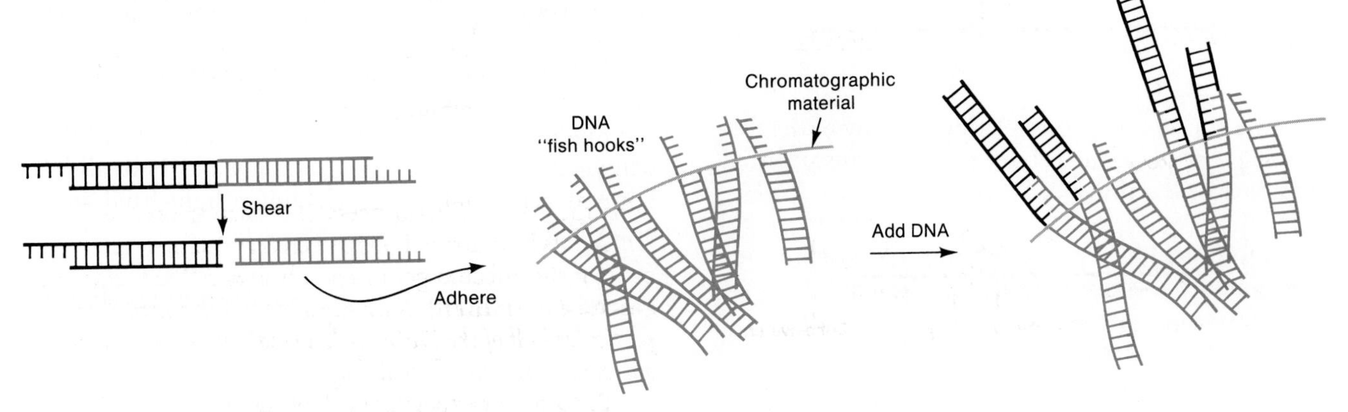

Figure 11-44 One specific sticky end of a DNA molecule can be used as a "fish hook" for the other sticky end plus any genes that may be attached to it. One half of the λ phage DNA is attached to chromatographic material so that the cohesive or sticky ends are exposed. Any DNA passing over this chromatographic material that has single-stranded base sequences complementary to the "fish hook" cohesive ends will stick. Any other DNA will pass through. Later, the "stuck" DNA can be removed from the chromatographic material by heating it to break the hydrogen bonds.

and thus revealed their sequence. Kaiser and his associates showed that an unambiguous arrangement of genes on the DNA can be determined and that this sequence is completely congruent with the genetic map (Figure 11-45).

Message We are now justified in concluding that the sequence of bases in the DNA is indeed congruent with the gene map.

Summary

Experimental work on the molecular nature of hereditary material has demonstrated conclusively that DNA (not protein, RNA, or some other substance) is indeed the genetic material. Using data supplied by others, Watson and Crick created a double helical model with two DNA strands, wound around each other, running in antiparallel fashion. Specificity of binding the two strands together is based on the fit of adenine (A) to thymine (T) and guanine (G) to cytosine (C). The former pair is held by two hydrogen bonds; the latter, by three.

The Watson-Crick model shows how DNA can be replicated in an orderly fashion — a prime requirement for genetic material. Replication is accomplished semiconservatively in both prokaryotes and eukaryotes. One double helix is replicated into two identical helices, each with identical linear orders of nucleotides; each of the two new double helices is composed of one old and one newly polymerized strand of DNA.

Replication is achieved with the aid of several enzymes, including DNA polymerase, gyrase, and helicase. Replication starts at special regions of the DNA called *origins of replication* and proceeds down the DNA in both directions. Since DNA polymerase acts only in a $5' \rightarrow 3'$ direction, one of the newly synthesized strands at each replication fork must be synthesized in short segments and then joined using the enzyme ligase. DNA polymerization cannot begin without a short primer, which is also synthesized with special enzymes.

The marker-rescue technique has demonstrated that the sequence of genes on a chromosome corresponds exactly to the linear sequence of DNA and has shown convincingly that the genetic and chemical entities are one and the same thing.

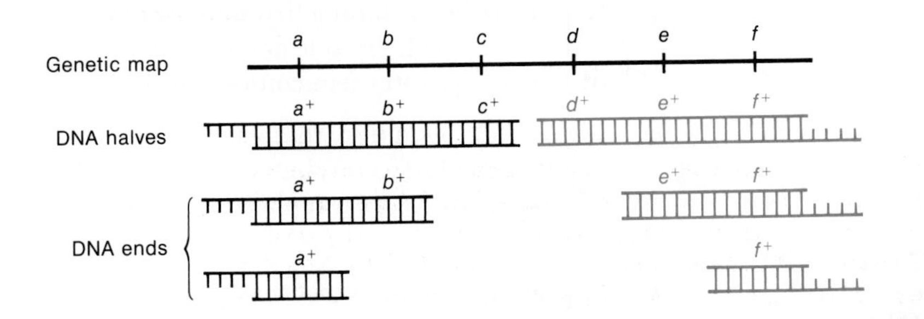

Figure 11-45 As the λ DNA is sheared into progressively smaller and smaller molecules, genes are lost to the fish hooks (Figure 11-44) in the same order as they appear on the genetic map.

Concept Map

Draw a concept map interrelating as many of the following terms as possible. Note that the terms are listed in no particular order.

DNA double helix / nucleotides / hydrogen bonding / semiconservative / 5′ / replication / chromatid / mitosis / DNA polymerase / meiosis / S phase / gene / 3′

Chapter Integration Problem

We discussed mitosis and meiosis in Chapter 3. Considering what we have covered in this chapter concerning DNA replication, draw a graph showing DNA content against time in a cell that undergoes mitosis and then meiosis. Assume a diploid cell.

Solution

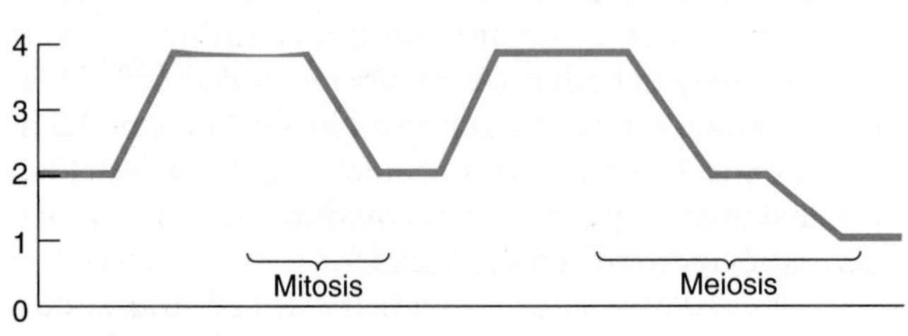

(Solution from Diane K. Lavett)

Solved Problems

1. If the GC content of a DNA molecule is 56 percent, what are the percentages of the four bases (A, T, G, and C) in this molecule?

Solution

If the GC content is 56 percent, then since G = C, the content of G is 28 percent and the content of C is 28 percent. The content of AT is $100 - 56 = 44$ percent. Since A = T, the content of A is 22 percent and the content of T is 22 percent.

2. Describe the expected pattern of bands in a CsCl gradient for *conservative* replication in the Meselson-Stahl experiment. Draw a diagram.

Solution

Refer to Figure 11-18 for an additional explanation. In conservative replication, if bacteria are grown in the presence of ^{15}N and then shifted to ^{14}N, one DNA molecule will be all ^{15}N after the first generation and the other molecule will be all ^{14}N, resulting in one heavy band and one light band in the gradient. After the second generation, the ^{15}N DNA will yield one molecule with all ^{15}N and one molecule with all ^{14}N, whereas the ^{14}N DNA will yield only ^{14}N DNA. Thus, only all ^{14}N or all ^{15}N DNA is generated, again, yielding a light band and a heavy band:

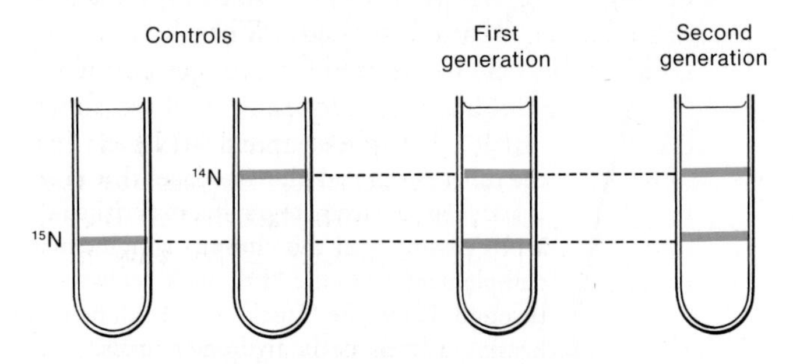

Problems

1. If thymine makes up 15 percent of the bases in a specific DNA molecule, what percentage of the bases is cytosine?

2. If the GC content of a DNA molecule is 48 percent, what are the percentages of the four bases (A, T, G, and C) in this molecule?

3. *E. coli* chromosomes in which every nitrogen atom is labeled (that is, every nitrogen atom is the heavy isotope ^{15}N instead of the normal isotope ^{14}N) are allowed to replicate in an environment in which all the nitrogen is ^{14}N. Using a solid line to represent a heavy polynucleotide chain and a dashed line for a light chain, sketch the following:

   a. The heavy parental chromosome and the products of the first replication after transfer to a ^{14}N medium, assuming that the chromosome is one DNA double helix and that replication is semiconservative.

   b. Repeat (a), but assuming that replication is conservative.

   c. Repeat (a), but assuming that the chromosome is in fact two side-by-side double helices, each of which replicates semiconservatively.

   d. Repeat (c), assuming that each side-by-side double helix replicates conservatively and that the overall *chromosome* replication is semiconservative.

   e. Repeat (d), assuming that the overall chromosome replication is conservative.

f. If the daughter chromosomes from the first division in ^{14}N are spun in a cesium chloride (CsCl) density gradient and a single band is obtained, which of possibilities (a) to (e) can be ruled out? Reconsider the Meselson-Stahl experiment: What does it *prove?*

4. R. Okazaki found that the immediate products of DNA replication in *E. coli* include single-stranded DNA fragments approximately 1000 nucleotides in length after the newly synthesized DNA is extracted and denatured (melted). When he allowed DNA replication to proceed for a longer period of time, he found a lower frequency of these short fragments and long single-stranded DNA chains after extraction and denaturation. Explain how this result might be related to the fact that all known DNA polymerases synthesize DNA only in a $5' \rightarrow 3'$ direction.

5. When plant and animal cells are given pulses of [^{3}H]-thymidine at different times during the cell cycle, heterochromatic regions on the chromosomes are invariably shown to be "late replicating." Can you suggest what, if any, biological significance this observation might have?

6. On the planet of Rama, the DNA is of six nucleotide types: A, B, C, D, E, and F. A and B are called marzines, C and D are orsines, and E and F are pirines. The following rules are valid in all Raman DNAs:

$$\text{Total marzines} = \text{total orsines} = \text{total pirines}$$

$$A = C = E$$

$$B = D = F$$

a. Prepare a model for the structure of Raman DNA.

b. On Rama, mitosis produces three daughter cells. Bearing this fact in mind, propose a replication pattern for your DNA model.

c. Consider the process of meiosis on Rama. What comments or conclusions can you suggest?

7. If you extract the DNA of the coliphage ϕX174, you will find that its composition is 25 percent A, 33 percent T, 24 percent G, and 18 percent C. Does this make sense in terms of Chargaff's rules? How would you interpret this result? How might such a phage replicate its DNA?

8. The temperature at which a DNA sample denatures can be used to estimate the proportion of its nucleotide pairs that are G–C. What would be the basis for this determination, and what would a high denaturation temperature for a DNA sample indicate?

9. In 1960, Paul Doty and Julius Marmur observed that when DNA is heated to 100°C, all the hydrogen bonds between the complementary strands are destroyed and the DNA becomes single-stranded (see the figure shown below). If the solution is cooled slowly, some double-stranded DNA is formed that is biologically normal (for example, it may have transforming ability). Presumably, this **reannealing,** or **renaturation,** process occurs when two single strands happen to collide in such a way that the complementing base sequences can align and reconstitute the original double helix (see the figure). This reannealing is very specific and precise, making it a powerful tool because stretches of complementary base sequences in *different* DNAs

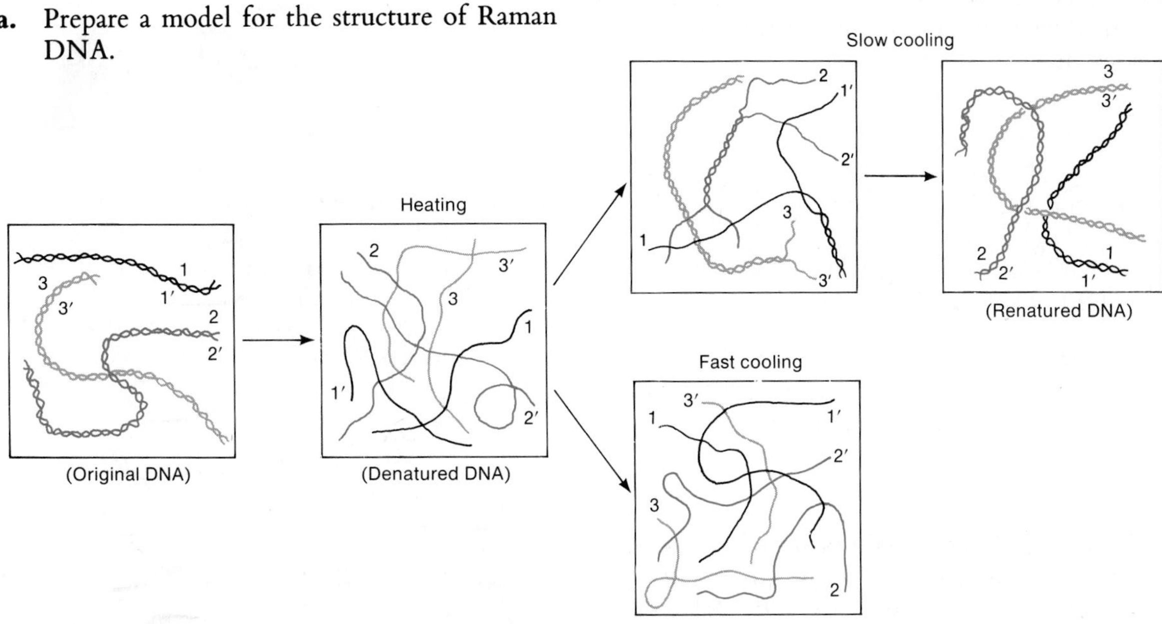

also will anneal after melting and mixing. Thus, the efficiency of annealing provides a measure of similarity between two different DNAs.

Now suppose that you extract DNA from a small virus, denature it, and allow it to reanneal with DNA taken from other strains that carry either a deletion, an inversion, or a duplication. What would you expect to see on inspection with an electron microscope?

10. DNA extracted from a mammal is heat-denatured and then slowly cooled to allow reannealing. The following graph shows the results obtained. There are two "shoulders" in the curve. The first shoulder indicates the presence of a very rapidly annealing part of the DNA—so rapid, in fact, that it occurs before strand interactions take place.

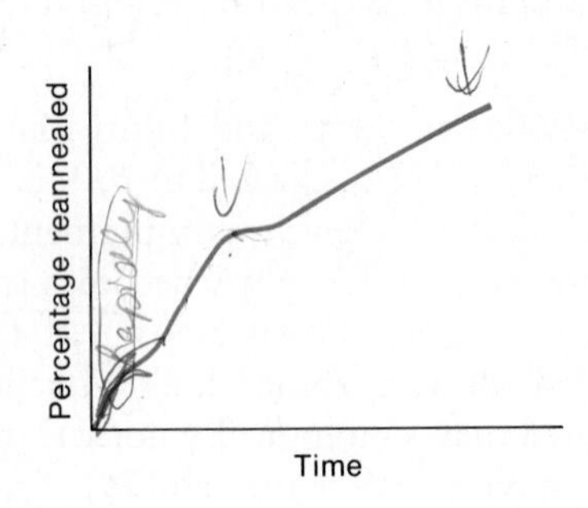

a. What could this part of the DNA be?

b. The second shoulder is a rapidly reannealing part as well. What does this evidence suggest?

11. Design tests to determine the physical relationship between highly repetitive and unique DNA sequences in chromosomes. (HINT: it is possible to vary the size of DNA molecules by the amount of shearing they are subjected to.)

12. Viruses are known to cause cancer in mice. You have a pure preparation of virus DNA, a pure preparation of DNA from the chromosomes of mouse cancer cells, and pure DNA from the chromosomes of normal mouse cells. Viral DNA will specifically anneal with cancer cell DNA, but not with normal-cell DNA. Explore the possible genetic significance of this observation, its significance at the molecular level, and its medical significance.

13. Ruth Kavenaugh and Bruno Zimm have devised a technique to measure the maximal length of the longest DNA molecules in solution. They studied DNA samples from the three *Drosophila* karyotypes shown in the accompanying figure. They found the longest molecules in karyotypes (a) and (b) to be of similar length and about twice the length of the longest molecule in (c). Interpret these results.

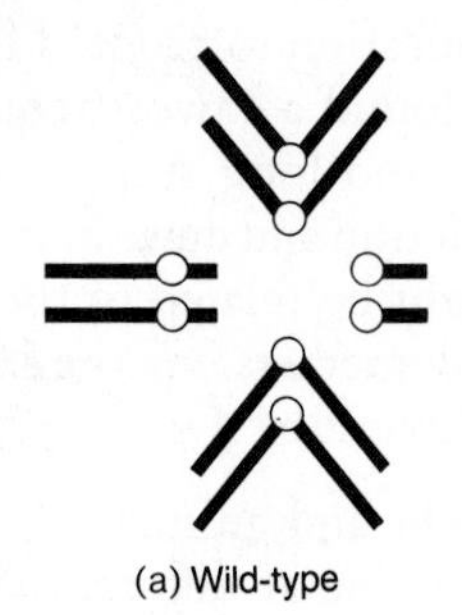

(a) Wild-type

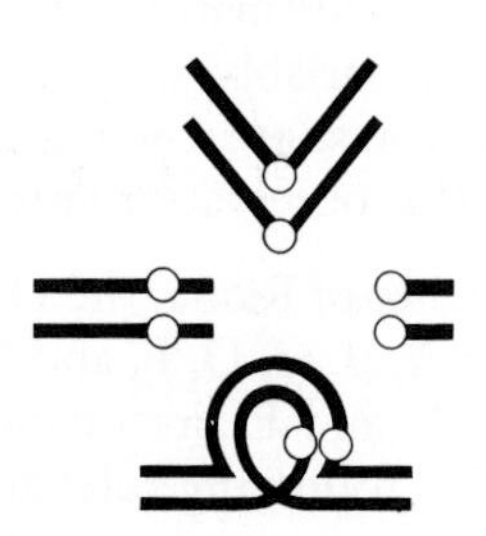

(b) Pericentric inversion

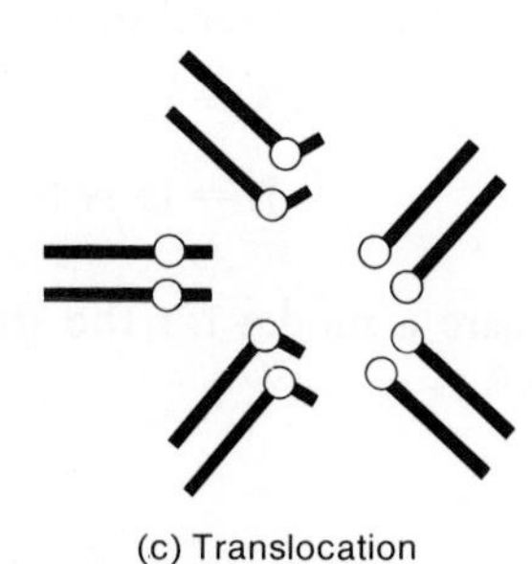

(c) Translocation

14. In the harlequin chromosome technique, you allow *three* rounds of replication in bromodeoxyuridine and then stain the chromosomes. What result do you expect to obtain?

12 The Nature of the Gene

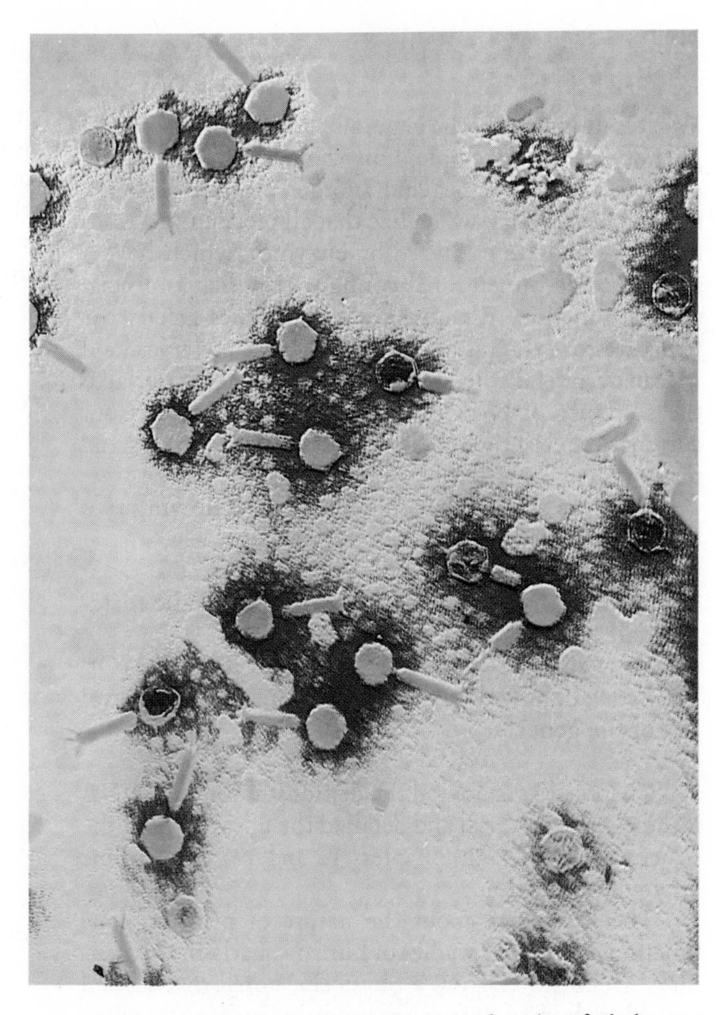

False-color transmission electron micrograph of unidentified phage; magnification, 58,800×. (CNRI/Science Photo Library/Photo Researchers)

KEY CONCEPTS

The one-gene–one-enzyme hypothesis postulates that genes control the structure of proteins.

▪

Studies on hemoglobin demonstrated that a gene mutation can be tied to an altered amino acid sequence in a protein.

▪

The linear sequence of nucleotides in a gene determines the linear sequence of amino acids in a protein.

▪

Fine structural analysis of the *rII* genes in phage T4 showed that the gene consists of a linear array of subelements that can mutate and recombine with one another; these subelements were later correlated with nucleotide pairs.

▪

A gene can be defined as a unit of function by a complementation test.

How Genes Work

What is the nature of the gene, and how do genes control phenotypes? For example, how can one allele of a gene produce a wrinkled pea and another produce a round, smooth pea? Today we realize that all biochemical reactions of a cell are catalyzed by enzymes, which have a specific three-dimensional configuration that is crucial to their function. We now know that genes specify the structures of proteins, some of which are enzymes, and we can even relate the structure of the genetic material to the structure of proteins. Table 12-1 summarizes our current model of the relationship between genotype and phenotype.

How did we arrive at this point? By following two lines of inquiry:

1. What is the physical structure of the genetic material?
2. How does the genetic material exert its effect? (Or, in detail, how does the structure operate at the level of the gene?)

Chapter 11 recounted the demonstration that DNA is the genetic material and detailed the unraveling of the structure of DNA. In Chapters 12 and 13, we examine how genes function.

The first clues about the nature of primary gene function came from studies of humans. Early in this century, Archibald Garrod, a physician, noted that several hereditary human defects are produced by recessive mutations. Some of these defects can be traced directly to metabolic defects that affect the basic body chemistry—an observation that has led to the suggestion of "inborn errors" in metabolism. We know now, for example, that phenylketonuria, which is caused by an autosomal recessive allele, results from an inability to convert phenylalanine into tyrosine. Consequently, phenylalanine accumulates and is spontaneously converted into a toxic compound, phenylpyruvic acid. In a different example, the inability to convert tyrosine into the pigment melanin produces an albino. Garrod's observations focused attention on metabolic control by genes.

Table 12-1. Model of the Relationship Between Genotype and Phenotype

1. The characteristic features of an organism are determined by the phenotype of its parts, which are in turn determined by the phenotype of the component cells.
2. The phenotype of a cell is determined by its internal chemistry, which is controlled by enzymes that catalyze its metabolic reactions.
3. The function of an enzyme depends on its specific three-dimensional structure, which in turn depends on the specific linear sequence of amino acids in the enzyme, since all enzymes are proteins.
4. The enzymes present in a cell, and structural proteins as well, are determined by the genotype of the cell.
5. Genes specify the linear sequence of amino acids in polypeptides and hence in proteins; thus, genes determine phenotypes.

The One-Gene–One-Enzyme Hypothesis

Clarification of the actual function of genes came from research in the 1940s on *Neurospora* by George Beadle and Edward Tatum, who later received a Nobel prize for their work. Before we describe their actual experiments, let's jump ahead and examine some aspects of **biosynthetic pathways,** based on our current understanding. We now know that molecules are synthesized as a series of steps, each one catalyzed by an enzyme. For instance, a biosynthetic pathway might have four steps, where 1 is the starting material and 5 is the final product:

$$1 \xrightarrow{A} 2 \xrightarrow{B} 3 \xrightarrow{C} 4 \xrightarrow{D} 5$$

Each step is catalyzed by an enzyme: A, B, C, or D. In turn, each enzyme is specified by a particular gene. We might say that gene *a* specifies enzyme A, gene *b* specifies enzyme B, and so on. Therefore, if we inactivate the gene responsible for an enzyme, we eliminate one required step and the pathway is interrupted.

In the following diagram enzyme B is eliminated due to a mutation in gene *b*:

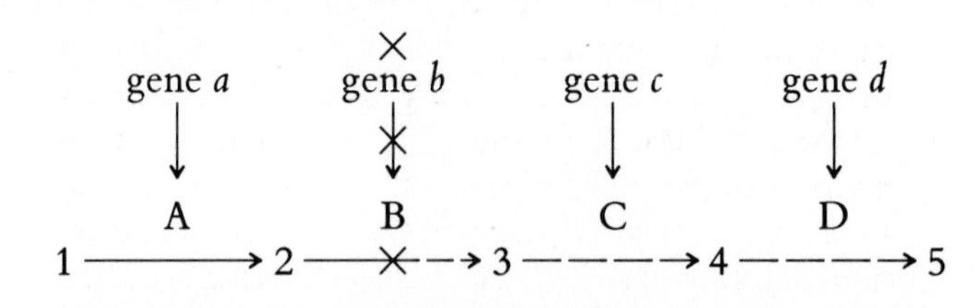

Now the cell cannot carry out the reaction that converts compound 2 to compound 3. It is blocked at compound 2, and cannot go further. But what happens if we add back different intermediate compounds? Suppose that, for example, we feed the cell compound 3 or 4. Can the cell then synthesize the final product, 5? Yes, with either compound 3 or 4 it can, because subsequent steps are not blocked. What if we add more of compound 1? No, adding compound 1 will not allow the synthesis of product 5 because one of the subsequent steps is blocked.

Let's test our understanding of the concept of biosynthetic pathways by looking at a sample problem that illustrates this principle:

Let's assume we've isolated five mutants, 1 to 5, that cannot synthesize compound G for growth. We know the five compounds, A to E, that are required in the biosynthetic pathway, but we do not know the order in which they are synthesized by the wild-type cell. We have tested each compound for its ability to support the growth of each mutant, with the following results:

		Compound tested (+ = growth; − = no growth)					
		A	*B*	*C*	*D*	*E*	*G*
Mutant	1	−	−	−	+	−	+
	2	−	+	−	+	−	+
	3	−	−	−	−	−	+
	4	−	+	+	+	−	+
	5	+	+	+	+	−	+

a. What is the order of compounds A to E in the pathway? How do we approach this type of problem? First, let's work out the underlying principle, and then draw a diagram. The main points are these: A mutation blocks a biosynthetic pathway by canceling out one enzyme needed in the pathway. The mutant therefore lacks one compound needed in the pathway and thus cannot make any of the compounds that come after the blocked step. If we add a compound that is normally synthesized before the block, it will not help matters: the mutant will still be unable to synthesize the rest of the compounds in the pathway. However, the mutant does have the enzymes to make these later compounds; thus, if we add either the blocked compound or any compound that comes after the block, the mutant will grow. So in our testing of mutants with blocks at different steps in the pathway, the compounds that occur latest in the pathway will support the growth of the most mutants, and compounds that occur earliest in the pathway will allow the growth of the fewest mutants. A look at our table shows us that the compound supporting the fewest mutants is compound E, with no mutants supported, followed by A (one mutant), then by C, B, D, and G, the final product. Now we can construct our diagram.

$$E \longrightarrow A \longrightarrow C \longrightarrow B \longrightarrow D \longrightarrow G$$

b. At which point in the pathway is each mutant blocked? Clearly, a mutant that is blocked between E and A cannot be supported by E but can be supported by all the other compounds. Thus, we see that mutant 5 must be blocked in the E–A conversion. We also see that mutant 4 cannot be supported by E or A, so it must be blocked in the A–C conversion. By similar logic, we obtain the order 5–4–2–1–3, which we can insert into the diagram as follows:

$$E \xrightarrow{5} A \xrightarrow{4} C \xrightarrow{2} B \xrightarrow{1} D \xrightarrow{3} G$$

Now we can comprehend how Beadle and Tatum first worked out this experiment, using one particular biosynthetic pathway of *Neurospora.*

The Experiments of Beadle and Tatum

Beadle and Tatum analyzed mutants of *Neurospora,* a fungus with a haploid genome. They first irradiated *Neurospora* to produce mutations and then tested cultures from ascospores for interesting mutant phenotypes. They detected numerous **auxotrophs**—strains that cannot grow on a minimal medium unless the medium is supplemented with one or more specific nutrients. In each case, the mutation that generated the auxotrophic requirement was inherited as a single-gene mutation: each gave a 1:1 ratio when crossed with a wild type (recall that *Neurospora* is haploid). Figure 12-1 depicts the procedure Beadle and Tatum used.

One set of mutant strains required arginine to grow on a minimal medium. These strains provided the focus for much of Beadle and Tatum's further work. First, they found that the mutations mapped into three different locations on separate chromosomes, even though the same supplement (arginine) satisfied the growth requirement for each mutant. Let's call the three loci the *arg-1*, *arg-2*, and *arg-3* genes. Beadle and Tatum discovered that the auxotrophs for each of the three loci differed in their response to the chemical compounds ornithine and citrulline, which are related to arginine (Figure 12-2). The *arg-1* mutants grew when supplied with either ornithine or citrulline or arginine in addition to the minimal medium. The *arg-2* mutants grew on either arginine or citrulline but not on ornithine. The *arg-3* mutants grew only when arginine was supplied. We can see this more easily by looking at Table 12-2 (see page 337).

It was already known that cellular enzymes often interconvert related compounds such as these. Based on the properties of the *arg* mutants, Beadle and Tatum and their colleagues proposed a biochemical model for such conversions in *Neurospora:*

$$\text{precursor} \xrightarrow{\text{enzyme X}} \text{ornithine} \xrightarrow{\text{enzyme Y}} \text{citrulline} \xrightarrow{\text{enzyme Z}} \text{arginine}$$

Note how this relationship easily explains the three classes of mutants shown in Table 12-2. The *arg-1* mutants have a defective enzyme X, so they are unable to convert the precursor into ornithine as the first step in producing arginine. However, they have normal enzymes Y and Z, and so the *arg-1* mutants are able to produce arginine if supplied with either ornithine or citrulline. The *arg-2* mutants lack enzyme Y, and the *arg-3* mutants lack enzyme Z. Thus, a mutation at a particular gene is assumed to interfere with the production of a single enzyme. The defective enzyme, then, creates a

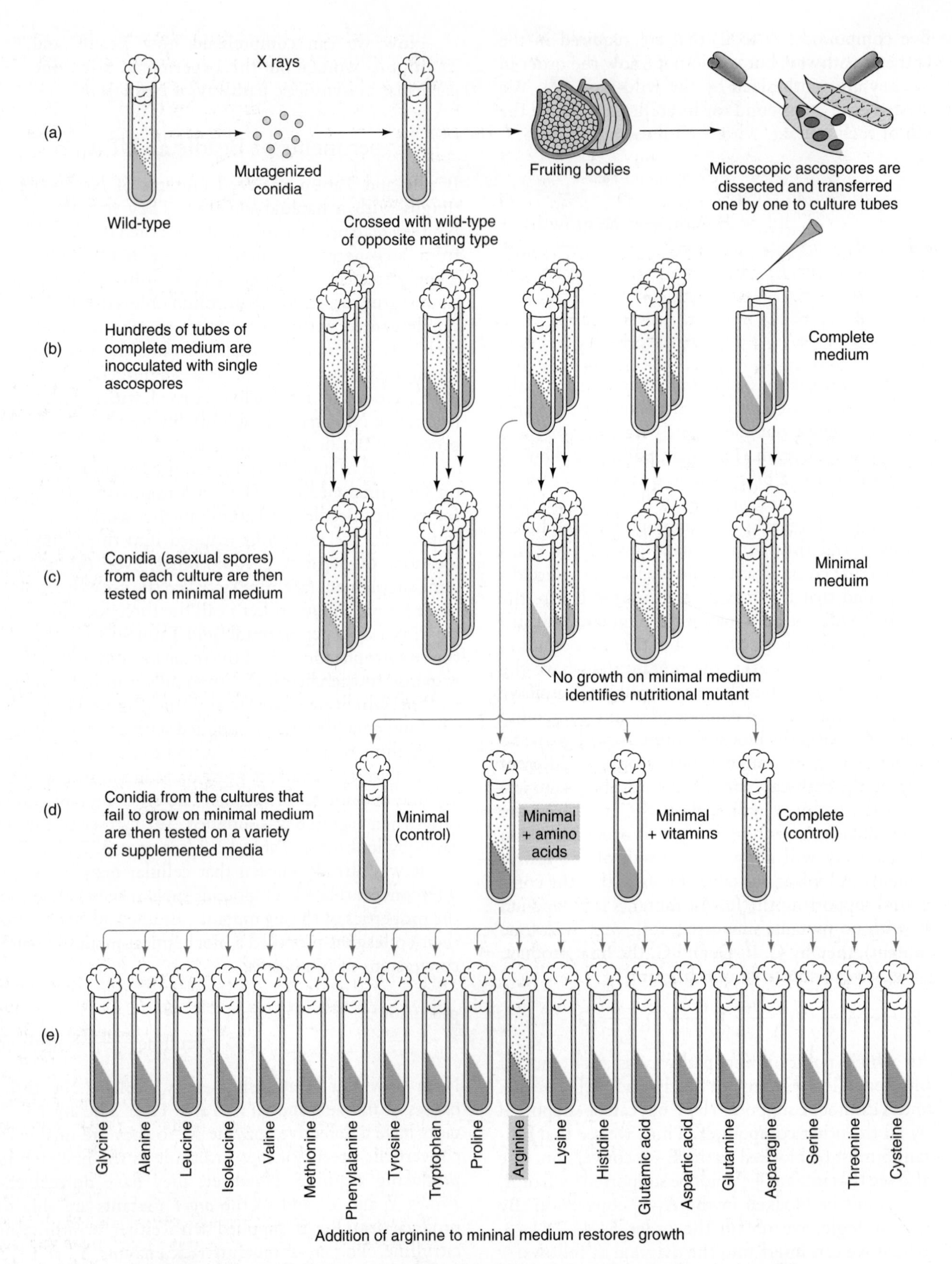
(a)
X rays
Mutagenized conidia
Wild-type
Crossed with wild-type of opposite mating type
Fruiting bodies
Microscopic ascospores are dissected and transferred one by one to culture tubes
(b)
Hundreds of tubes of complete medium are inocculated with single ascospores
Complete medium
(c)
Conidia (asexual spores) from each culture are then tested on minimal medium
Minimal meduim
No growth on minimal medium identifies nutritional mutant
(d)
Conidia from the cultures that fail to grow on minimal medium are then tested on a variety of supplemented media
Minimal (control)
Minimal + amino acids
Minimal + vitamins
Complete (control)
(e)
Glycine
Alanine
Leucine
Isoleucine
Valine
Methionine
Phenylalanine
Tyrosine
Tryptophan
Proline
Arginine
Lysine
Histidine
Glutamic acid
Aspartic acid
Glutamine
Asparagine
Serine
Threonine
Cysteine
Addition of arginine to mininal medium restores growth

Figure 12-1 The procedure used by Beadle and Tatum. See Chapter 5 for a review of the *Neurospora* life cycle and genetics. (Little, Brown & Co.)

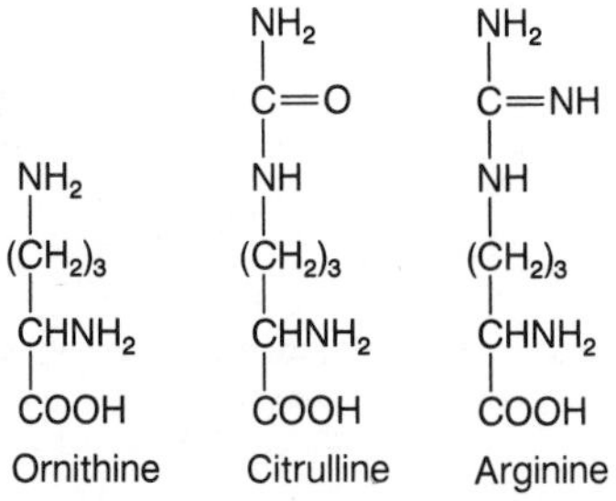

	NH_2	NH_2
	|	|
	C=O	C=NH
	|	|
NH_2	NH	NH
|	|	|
$(CH_2)_3$	$(CH_2)_3$	$(CH_2)_3$
|	|	|
$CHNH_2$	$CHNH_2$	$CHNH_2$
|	|	|
COOH	COOH	COOH
Ornithine	Citrulline	Arginine

Figure 12-2 Chemical structures of arginine and the related compounds citrulline and ornithine. In the work of Beadle and Tatum, different *arg* auxotrophic mutants of *Neurospora* were found to grow when the medium was supplemented with citrulline, or with ornithine, as an alternative to arginine.

block in some biosynthetic pathway. The block can be circumvented by supplying to the cells any compound that normally comes after the block in the pathway.

We can now diagram a more complete biochemical model:

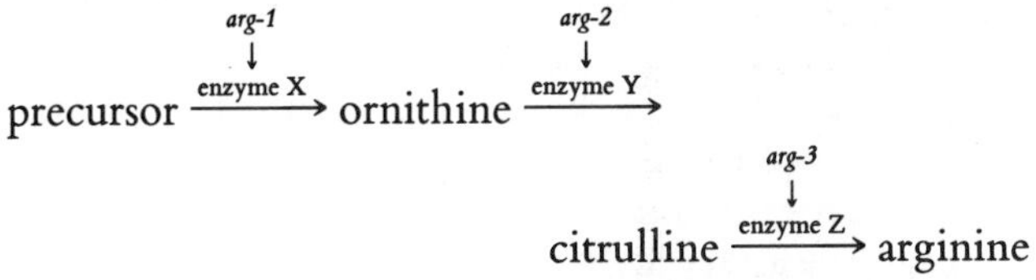

$$\text{precursor} \xrightarrow[\text{enzyme X}]{arg\text{-}1\ \downarrow} \text{ornithine} \xrightarrow[\text{enzyme Y}]{arg\text{-}2\ \downarrow} \text{citrulline} \xrightarrow[\text{enzyme Z}]{arg\text{-}3\ \downarrow} \text{arginine}$$

Note that this entire model was inferred from the properties of the mutant classes detected through genetic analysis. Only later were the existence of the biosynthetic pathway and the presence of defective enzymes demonstrated through independent biochemical evidence.

This model, which has become known as the **one-gene–one-enzyme hypothesis,** provided the first exciting insight into the functions of genes: genes somehow were responsible for the function of enzymes, and each gene apparently controlled one specific enzyme.

Table 12-2. Growth of *arg* Mutants in Response to Supplements

Mutant	Supplement		
	Ornithine	Citrulline	Arginine
	(+ = growth; − = no growth)		
arg-1	+	+	+
arg-2	−	+	+
arg-3	−	−	+

Other researchers obtained similar results for other biosynthetic pathways, and the hypothesis soon achieved general acceptance. It was subsequently refined, as we shall see later in this chapter. Nevertheless, the one-gene–one-enzyme hypothesis became one of the great unifying concepts in biology, because it provided a bridge that brought together the concepts and research techniques of genetics and biochemistry.

Message Genes control biochemical reactions by controlling the production of enzymes.

We should pause to ponder the significance of this discovery. Let's summarize what it established:

1. Biochemical reactions in vivo (in the living cell) occur as a series of discrete, stepwise reactions.
2. Each reaction is specifically catalyzed by a single enzyme.
3. Each enzyme is specified by a single gene.

Gene–Protein Relationships

The one-gene–one-enzyme hypothesis was an impressive step forward in our understanding of gene function, but just *how* do genes control the functioning of enzymes? Enzymes are proteins, and thus we must review the basic facts of protein structure in order to follow the next step in the study of gene function.

Protein Structure

In simple terms, a **protein** is a macromolecule composed of **amino acids** attached end to end in a linear string. The general formula for an amino acid is $H_2N{-}CHR{-}COOH$, in which the side chain, or R (reactive) group, can be anything from a hydrogen atom (as in the amino acid glycine) to a complex ring (as in the amino acid tryptophan). There are 20 common amino acids in living organisms (Table 12-3), each having a different R group. Amino acids are linked together in proteins by covalent (chemical) bonds called **peptide bonds.** A peptide bond is formed through a condensation reaction that involves the removal of a water molecule (Figure 12-3).

Several amino acids linked together by peptide bonds form a molecule called a **polypeptide;** the proteins found in living organisms are large polypeptides. For instance, the α-chain of human hemoglobin contains

Table 12-3. The 20 Amino Acids Common in Living Organisms

Amino acid	Abbreviation 3-Letter	1-Letter	Amino acid	Abbreviation 3-Letter	1-Letter
Alanine	Ala	A	Leucine	Leu	L
Arginine	Arg	R	Lysine	Lys	K
Asparagine	Asn	N	Methionine	Met	M
Aspartic acid	Asp	D	Phenylalanine	Phe	F
Cysteine	Cys	C	Proline	Pro	P
Glutamine	Gln	Q	Serine	Ser	S
Glutamic acid	Glu	E	Threonine	Thr	T
Glycine	Gly	G	Tryptophan	Trp	W
Histidine	His	H	Tyrosine	Tyr	Y
Isoleucine	Ile	I	Valine	Val	V

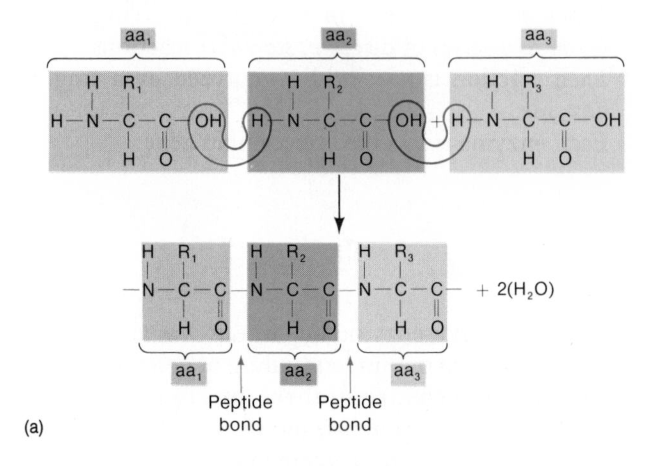

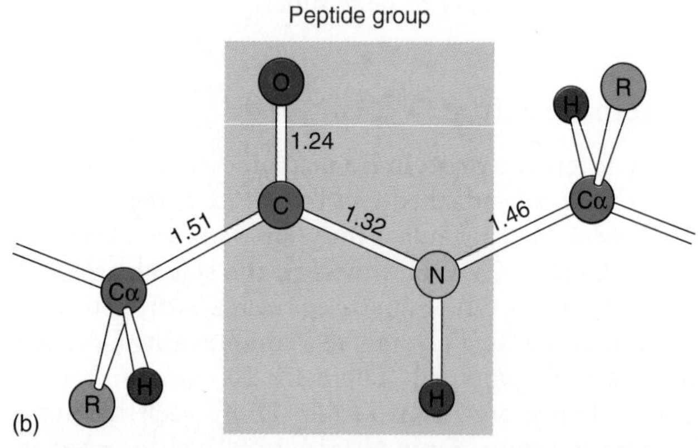

Figure 12-3 The peptide bond. (a) A polypeptide is formed by the removal of water between amino acids to form peptide bonds. Each aa indicates an amino acid. R_1, R_2, and R_3 represent R groups (side chains) that differentiate the amino acids. R can be anything from a hydrogen atom (as in glycine) to a complex ring (as in tryptophan). (b) The peptide group is a rigid planar unit with the R groups projecting out from the C—N backbone. Standard bond distances (in Angstroms) are shown. (Part b from L. Stryer, *Biochemistry,* 3d ed. Copyright 1983 by W. H. Freeman and Company.)

141 amino acids, and some proteins consist of more than 1000 amino acids.

Proteins have a complex structure that is traditionally thought of as having four levels. The linear sequence of the amino acids in a polypeptide chain is called the **primary structure** of the protein. Figure 12-4 shows the linear sequence of tryptophan synthetase (an enzyme) and beef insulin (a hormonal protein).

The **secondary structure** of a protein refers to the interrelationships of amino acids that are close together in the linear sequence. This spatial arrangement often results from the fact that polypeptides can bend into regularly repeating (periodic) structures, generated by hydrogen bonds between the CO and NH groups of different residues. Two of the basic periodic structures are the α-helix (Figure 12-5) and the β-pleated sheet (Figure 12-6).

A protein also has a three-dimensional architecture, termed the **tertiary structure,** which is generated by electrostatic, hydrogen, and Van der Waals bonds that form between the various amino acid R groups, causing the protein chain to fold back on itself. In many cases amino acids that are far apart in the linear sequence are brought close together in the tertiary structure. Often, two or more folded structures will bind together to form a **quaternary structure;** this structure is **multimeric** because it is composed of several separate polypeptide chains, or monomers.

Figure 12-7 depicts the four levels of protein structure. In Figure 12-7c we can see the tertiary structure of myoglobin. Note how the α-helix is folded back on itself to generate the three-dimensional shape of the protein. Figure 12-7d shows the combining of four subunits (two α-chains and two β-chains) to form the quaternary structure of hemoglobin. Figures 12-8 and 12-9 show the

10 20
Met-Glu-Arg-Tyr-Glu-Ser-Leu-Phe-Ala-Gln-Leu-Lys-Glu-Arg-Lys-Glu-Gly-Ala-Phe-Val-

30 40
Pro-Phe-Val-Thr-Leu-Gly-Asp-Pro-Gly-Ile-Glu-Gln-Ser-Leu-Lys-Ile-Ile-Asp-Thr-Leu-

50 60
Ile-Glu-Ala-Gly-Ala-Asp-Ala-Leu-Glu-Leu-Gly-Ile-Pro-Phe-Ser-Asp-Pro-Leu-Ala-Asp-

70 80
Gly-Pro-Thr-Ile-Gln-Asn-Ala-Thr-Leu-Arg-Ala-Phe-Ala-Ala-Gly-Val-Thr-Pro-Ala-Gln-

90 100
Cys-Phe-Glu-Met-Leu-Ala-Leu-Ile-Arg-Gln-Lys-His-Pro-Thr-Ile-Pro-Ile-Gly-Leu-Leu-

110 120
Met-Tyr-Ala-Asn-Leu-Val-Phe-Asn-Lys-Gly-Ile-Asp-Glu-Phe-Tyr-Ala-Gln-Cys-Glu-Lys-

130 140
Val-Gly-Val-Asp-Ser-Val-Leu-Val-Ala-Asp-Val-Pro-Val-Gln-Glu-Ser-Ala-Pro-Phe-Arg-

150 160
Gln-Ala-Ala-Leu-Arg-His-Asn-Val-Ala-Pro-Ile-Phe-Ile-Cys-Pro-Pro-Asn-Ala-Asp-Asp-

170 180
Asp-Leu-Leu-Arg-Gln-Ile-Ala-Ser-Tyr-Gly-Arg-Gly-Tyr-Thr-Tyr-Leu-Leu-Ser-Arg-Ala-

190 200
Gly-Val-Thr-Gly-Ala-Glu-Asn-Arg-Ala-Ala-Leu-Pro-Leu-Asn-His-Leu-Val-Ala-Lys-Leu-

210 220
Lys-Glu-Tyr-Asn-Ala-Ala-Pro-Pro-Leu-Gln-Gly-Phe-Gly-Ile-Ser-Ala-Pro-Asp-Gln-Val-

230 240
Lys-Ala-Ala-Ile-Asp-Ala-Gly-Ala-Ala-Gly-Ala-Ile-Ser-Gly-Ser-Ala-Ile-Val-Lys-Ile-

250 260
Ile-Glu-Gln-His-Asn-Ile-Glu-Pro-Glu-Lys-Met-Leu-Ala-Ala-Leu-Lys-Val-Phe-Val-Gln-

268
Pro-Met-Lys-Ala-Ala-Thr-Arg-Ser

(a)

A chain
Gly-Ile-Val-Glu-Gln-Cys-Cys-Ala-Ser-Val-Cys-Ser-Leu-Tyr-Gln-Leu-Glu-Asn-Tyr-Cys-Asn
5 10 15 21

S—S (between Cys 6 and Cys 11); S–S bridges from A chain Cys 7 and Cys 20 to B chain Cys 7 and Cys 19

B chain
Phe-Val-Asn-Gln-His-Leu-Cys-Gly-Ser-His-Leu-Val-Glu-Ala-Leu-Tyr-Leu-Val-Cys-Gly-Glu-
5 10 15 20
Arg-Gly-Phe-Phe-Tyr-Thr-Pro-Lys-Ala
25 30

(b)

Figure 12-4 Linear sequences of two proteins. (a) The *E. coli* tryptophan synthetase A protein, 268 amino acids long. (b) Bovine insulin protein. Note that the amino acid cysteine can form unique "sulfur bridges," because it contains sulfur.

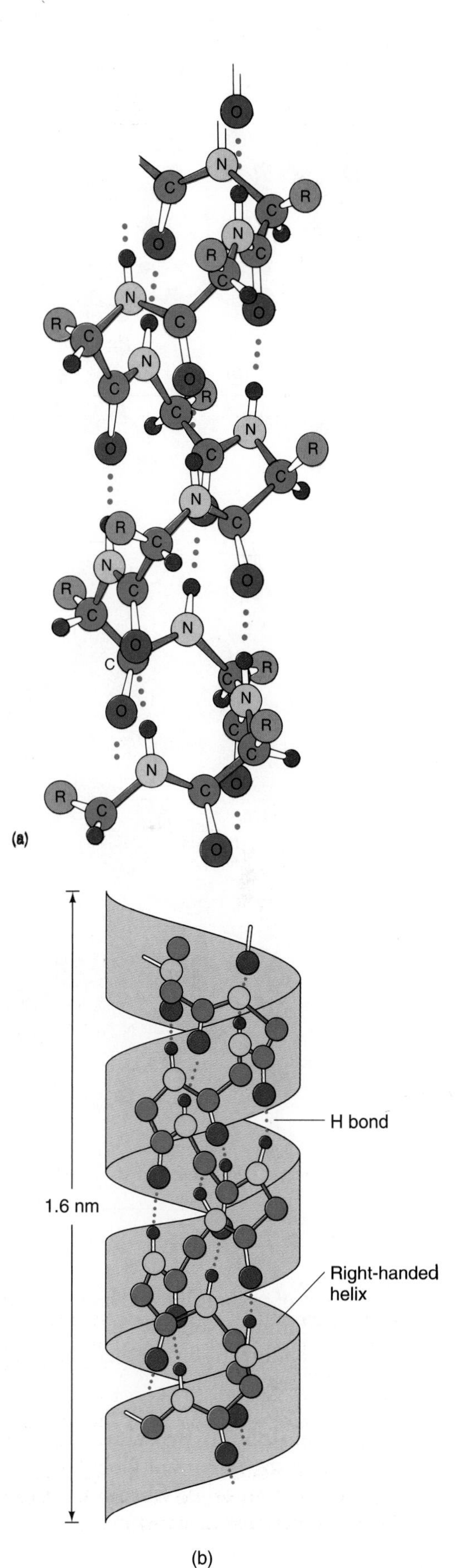

Figure 12-5 The α-helix, a common basis of secondary protein structure. (a) The spiral backbone of the protein can be seen as the red-colored lines. Each R is a specific side chain on one amino acid. The red dots are weak hydrogen bonds that bond the CO group of residue n to the NH group of residue $n + 4$, thereby stabilizing the helical shape. (Reprinted from Linus Pauling, *The Nature of the Chemical Bond.* Copyright 1939 and 1940 by Cornell University. Third edition copyright 1960 by Cornell University. Used by permission of Cornell University Press.) (b) A different perspective of the α-helix; the R groups have been omitted for simplicity.

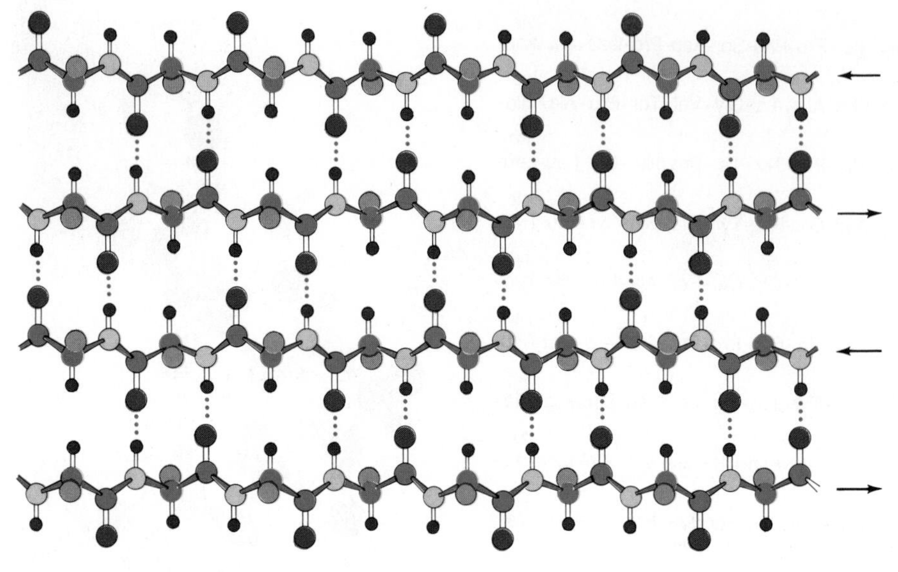

(a)

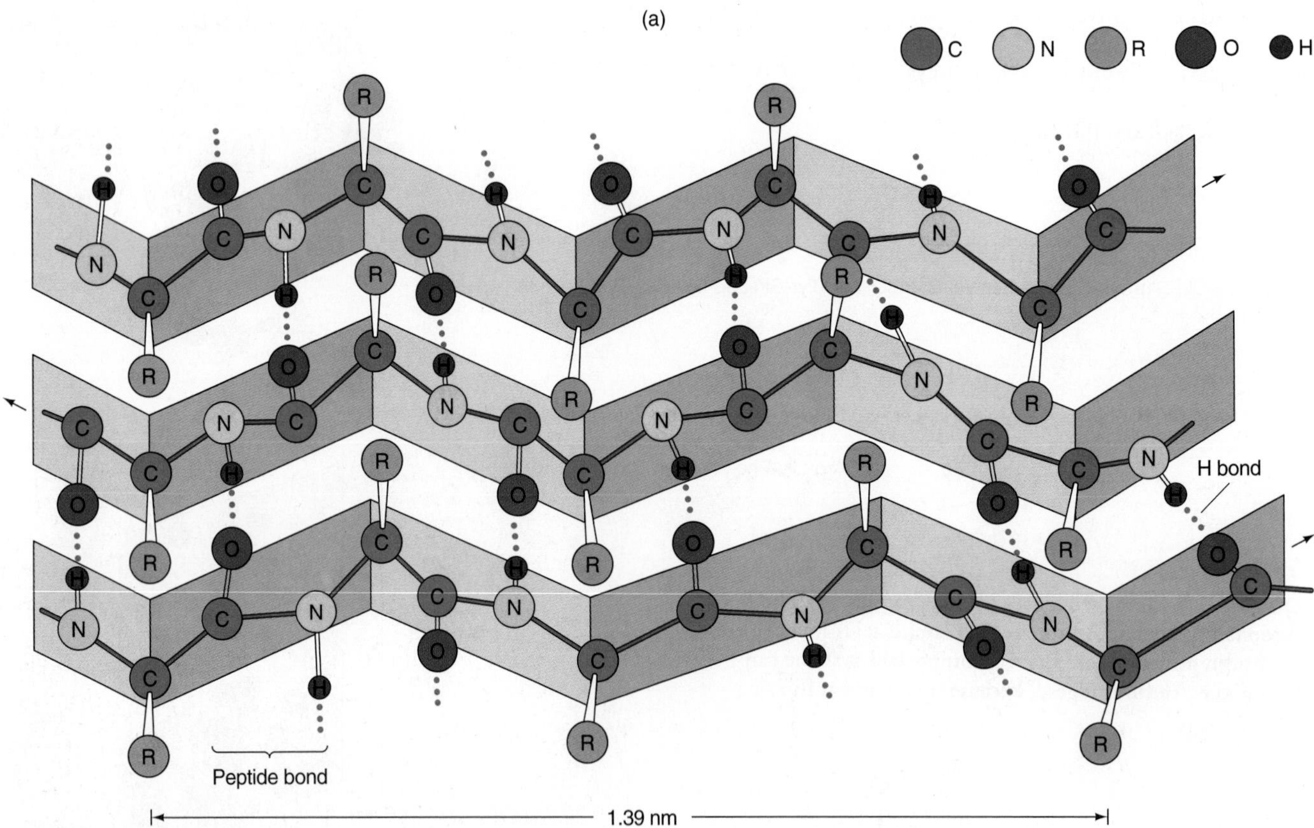

(b)

Figure 12-6 Two views of the antiparallel β-pleated sheet, another common form of secondary protein structure. Adjacent strands run in opposite directions. Hydrogen bonds between NH and CO groups of adjacent strands stabilize the structure. The side chains (R) are above and below the plane of the sheet. (Part a from L. Stryer, *Biochemistry,* 3d ed. Copyright 1983 by W. H. Freeman and Company.)

structures of myoglobin and hemoglobin in more detail. The combining of subunits to form a multimeric enzyme can be seen directly in the electron microscope in some cases (Figure 12-10).

Many proteins are compact structures; such proteins are called **globular proteins.** Enzymes and antibodies are among the important globular proteins. Other, unfolded proteins, called **fibrous proteins,** are important components of such structures as hair and muscle.

Message The linear sequence of a protein folds up to yield a unique three-dimensional configuration. This configuration creates specific sites to which substrates bind and at which catalytic reactions occur. The three-dimensional structure of a protein, which is crucial for its function, is determined solely by the primary structure (linear sequence) of amino acids. Therefore, genes can control enzyme function by controlling the primary structure of proteins.

(a) Primary structure

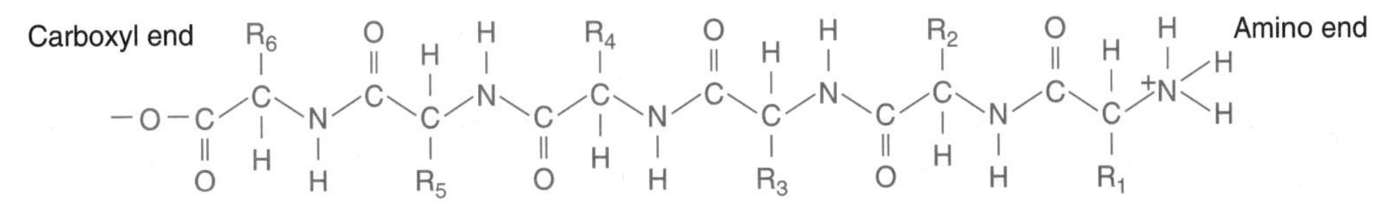

(b) Secondary structure

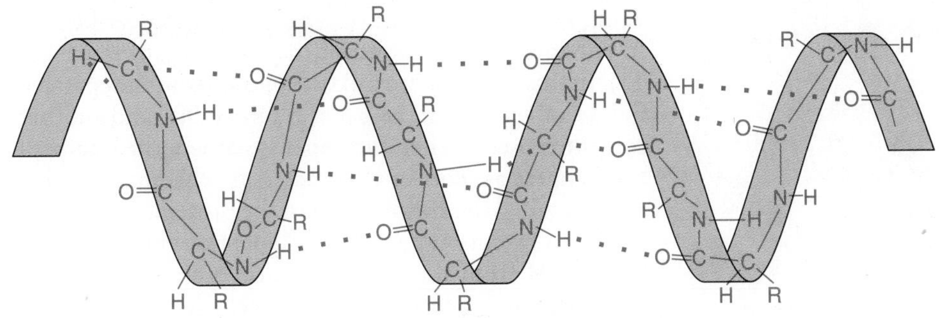

Hydrogen bonds between amino acids at different locations in polypeptide chain

(c) Tertiary structure

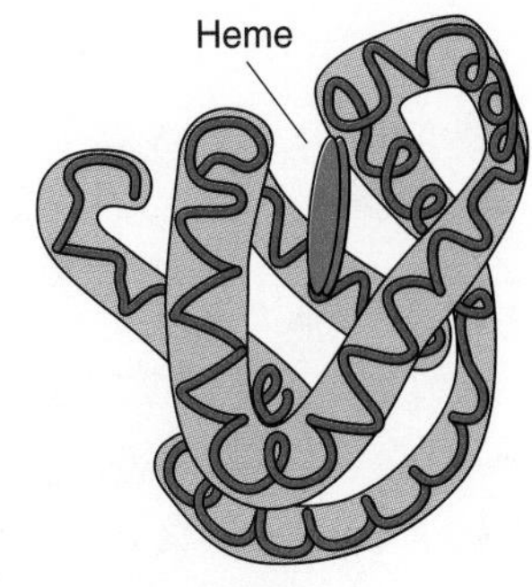

β polypeptide

(d) Quarternary structure

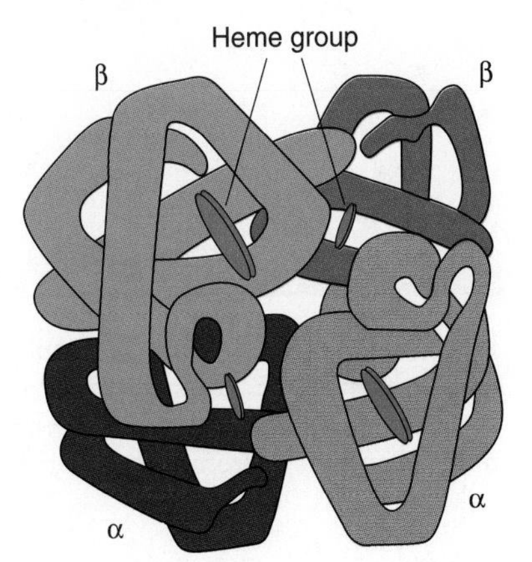

Figure 12-7 Different levels of protein structure. (a) Primary structure. (b) Secondary structure. The polypeptide shown in part a is drawn into an α-helix by hydrogen bonds. (c) Tertiary structure: the three-dimensional structure of myoglobin. (d) Quaternary structure: the arrangement of two α subunits and two β subunits to form the complete quaternary structure of hemoglobin.

Figure 12-8 Folded tertiary structure of myoglobin, an oxygen-storage protein. Each dot represents an amino acid. The heme group, a cofactor that facilitates the binding of oxygen, is shown in blue. (From L. Stryer, *Biochemistry,* 2d ed. Copyright 1981 by L. Stryer. Based on R. E. Dickerson, *The Proteins,* 2d ed., vol. 2. Edited by H. Neurath. Copyright 1964 by Academic Press.)

Figure 12-9 A model of the hemoglobin molecule (shown as contoured layers to provide a visualization of the three-dimensional shape). The different shadings indicate the different polypeptide chains, two α and two β subunits, that combine to form the quaternary structure of the protein (see Figure 12-7d). The disks are heme groups—complex structures containing iron. (After M. F. Perutz, "The Hemoglobin Molecule." Copyright 1964 by Scientific American, Inc. All rights reserved.)

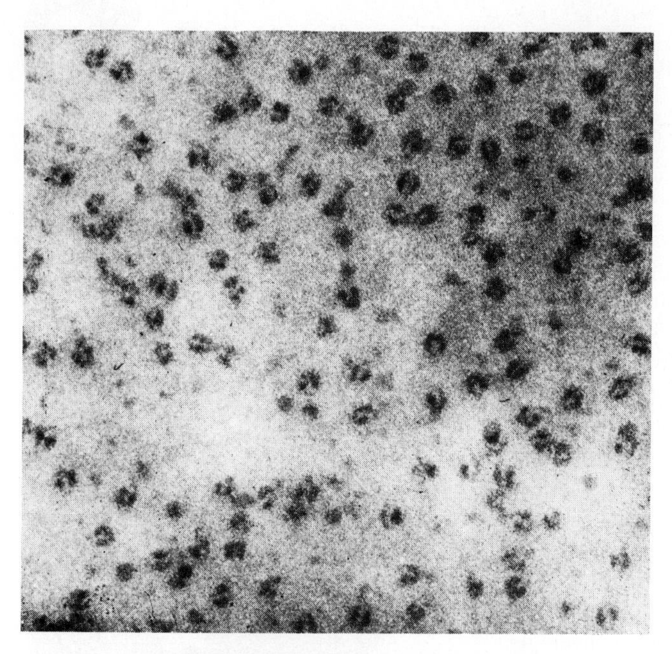

Figure 12-10 Electron micrograph of the enzyme aspartate transcarbamylase. Each small "glob" is an enzyme molecule. Note the quaternary structure: the enzyme is composed of subunits. (Photograph from Jack D. Griffith.)

Determining Protein Sequence

If we purify a particular protein, we find that we can specify a particular ratio of the various amino acids that make up that specific protein. But the protein is not formed by a random hookup of fixed amounts of the various amino acids; each protein has a unique, characteristic sequence. For a small polypeptide, the amino acid sequence can be determined by clipping off one amino acid at a time and identifying it. However, large polypeptides cannot be readily "sequenced" in this way.

Frederick Sanger worked out a brilliant method for deducing the sequence of large polypeptides. There are several different **proteolytic enzymes**—enzymes that can break peptide bonds only between specific amino acids in proteins. Proteolytic enzymes can break a large protein into a number of smaller fragments, which can then be separated according to their migration speeds in a solvent on chromatographic paper. Because different fragments will move at different speeds in various solvents, **two-dimensional chromatography** can be used to enhance the separation of the fragments (Figure 12-11). In this technique, a mixture of fragments is separated in one solvent; then the paper is turned 90° and another solvent is used. When the paper is stained, the polypeptides appear as spots in a characteristic chromatographic pattern called the **fingerprint** of the protein. Each of the spots can be cut out, and the polypeptide fragments can be washed from the paper. Because each spot contains only small polypeptides, their amino acid sequences can easily be determined.

Using different proteolytic enzymes to cleave the protein at different points, we can repeat the experiment to obtain other sets of fragments. The fragments from the different treatments overlap, because the breaks are made in different places with each treatment. The problem of solving the overall sequence then becomes one of fitting together the small-fragment sequences—almost like solving a tricky jigsaw or crossword puzzle (Figure 12-12).

Using this elegant technique, Sanger confirmed that the sequence of amino acids (as well as the amounts of the various amino acids) is specific to a particular protein. In other words, the amino acid sequence is what makes insulin insulin.

Relationship Between Gene Mutations and Altered Proteins

We now know that the change of just one amino acid is sometimes enough to alter protein function. This was first shown in 1957 by Vernon Ingram, who studied the globular protein hemoglobin—the molecule that transports oxygen in red blood cells. As we saw in Figures 12-7d and 12-9, hemoglobin is made up of four polypeptide chains: two identical α-chains, each containing 141

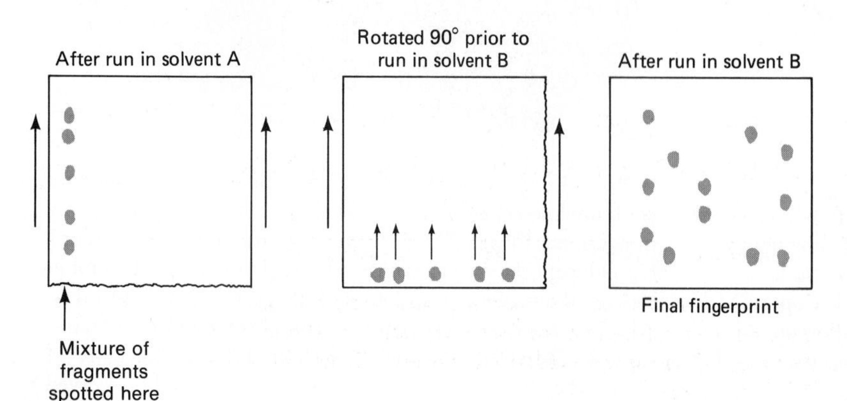

Figure 12-11 Two-dimensional chromatographic fingerprinting of a polypeptide fragment mixture. A protein is digested by a proteolytic enzyme into fragments that are only a few amino acids long. A piece of chromatographic filter paper is then spotted with this mixture and dipped into solvent A. As solvent A ascends the paper, some of the fragments become separated. The paper is then turned 90° and further resolution of the fragments is obtained as solvent B ascends.

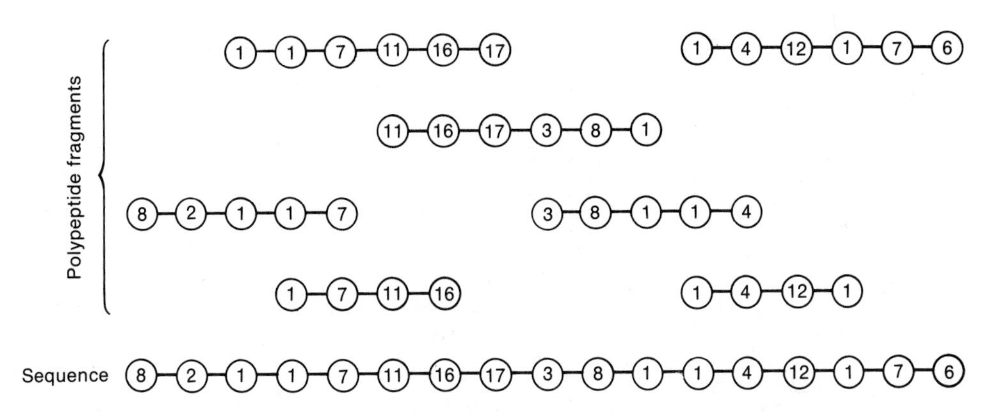

Figure 12-12 Alignment of polypeptide fragments to reconstruct an entire amino acid sequence. Different proteolytic enzymes can be used on the same protein to form different fingerprints, as shown here. The amino acid sequence of each fragment can be determined rather easily, and due to the overlap of amino acid sequences from different fingerprints, the entire amino acid sequence of the original protein can be determined. Using this procedure, it took Sanger about six years to determine the sequence of the insulin molecule, a relatively small protein.

amino acids, and two identical β-chains, each containing 146 amino acids.

Ingram compared hemoglobin A (HbA), the hemoglobin from normal adults, with hemoglobin S (HbS), the protein from people homozygous for the mutant gene that causes sickle-cell anemia, the disease in which red blood cells take on a sickle-cell shape (see Figure 4-2). Using Sanger's technique, Ingram found that the fingerprint of HbS differs from that of HbA in only one spot. Sequencing that spot from the two kinds of hemoglobin, Ingram found that only one amino acid in the fragment differs in the two kinds. Apparently, of all the amino acids known to make up a hemoglobin molecule, a substitution of valine for glutamic acid at just one point, position 6 in the β-chain, is all that is needed to produce the defective hemoglobin (Figure 12-13). Unless pa-

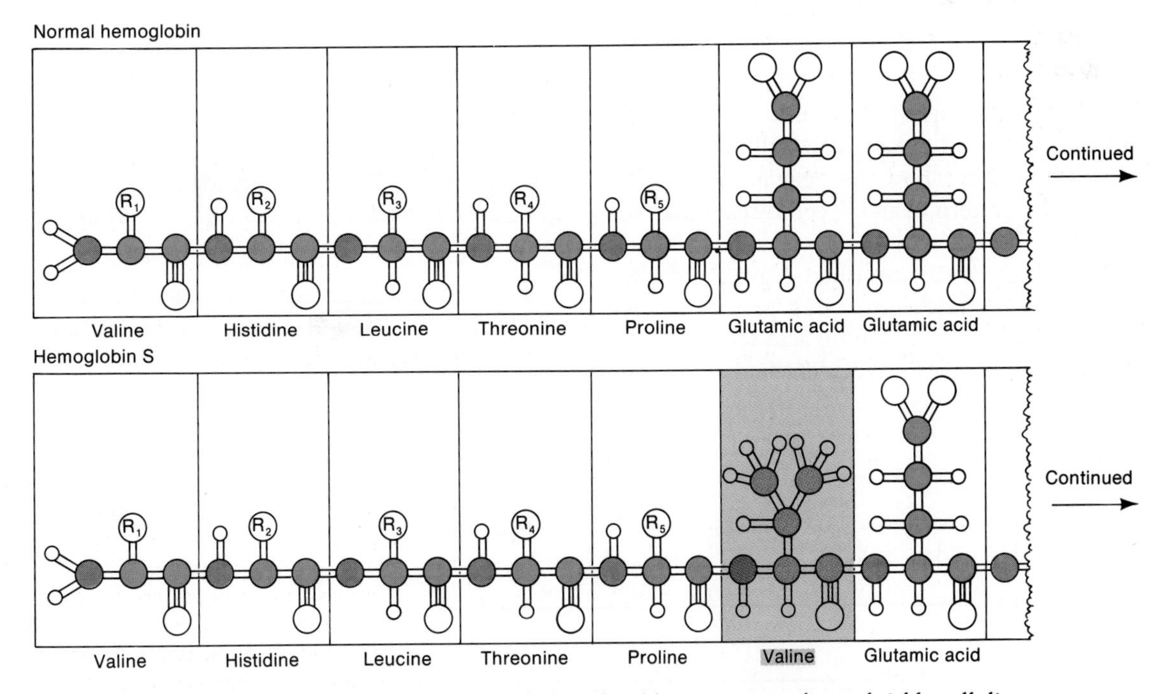

Figure 12-13 The difference at the molecular level between normalcy and sickle-cell disease. Shown are only the first seven amino acids; all the rest not shown are identical. (From Anthony Cerami and Charles M. Peterson, "Cyanate and Sickle-Cell Disease." Copyright 1975 by Scientific American, Inc. All rights reserved.)

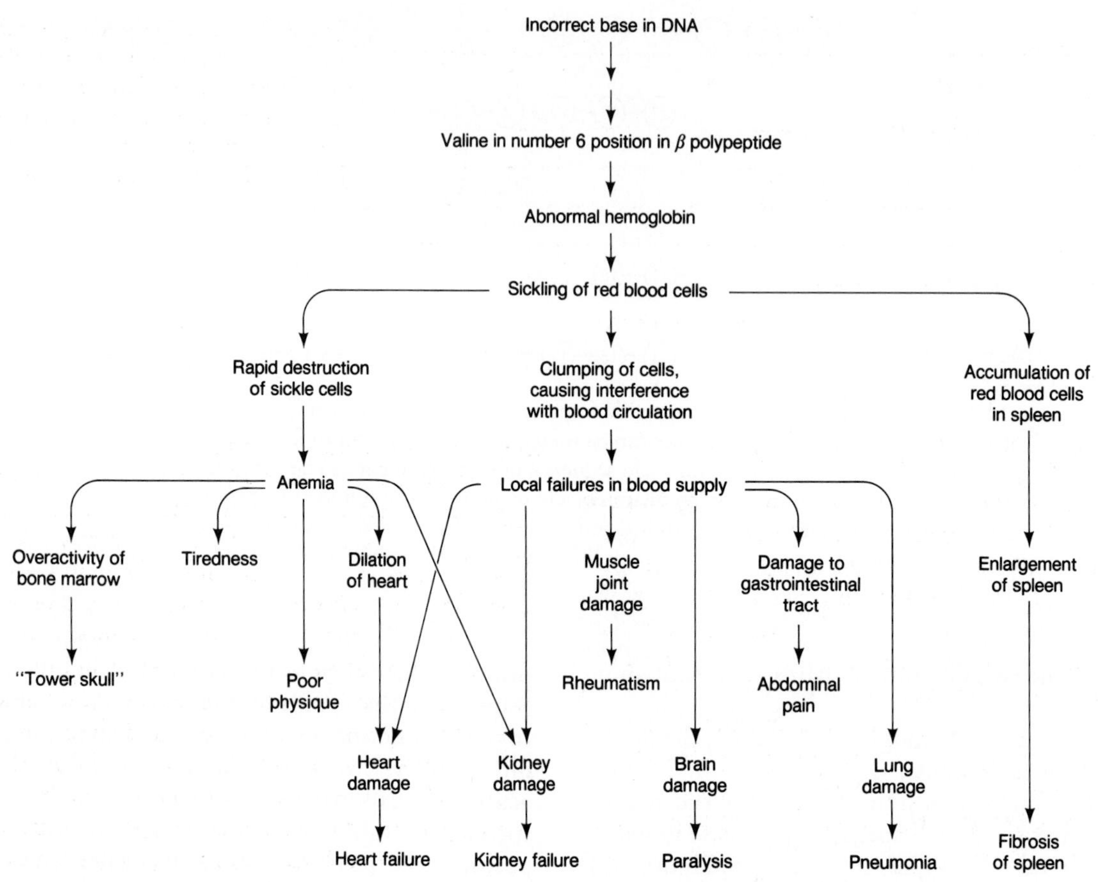

Figure 12-14 The compounded consequences of one amino acid substitution in hemoglobin to produce sickle-cell anemia.

tients with HbS receive medical attention, this single error in one amino acid in one protein will hasten their death. Figure 12-14 shows how this gene mutation ultimately leads to the pattern of sickle cell disease that we previously saw in Figure 1-9.

Notice what Ingram accomplished. A gene mutation that had been well established through genetic studies was connected with an altered amino acid sequence in a protein! Subsequent studies have identified numerous changes in hemoglobin, and each one is the consequence of a single amino acid difference. (Figure 12-15 shows a few examples.) We can conclude that one mutation in a gene corresponds to a change of one amino acid in the sequence of a protein.

Message Genes determine the specific primary sequences of amino acids in specific proteins.

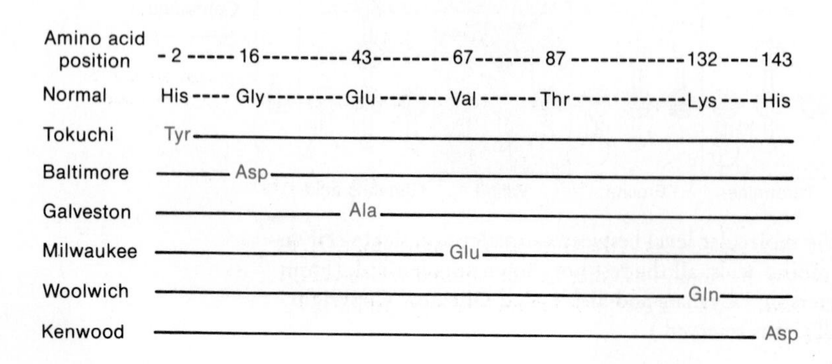

Figure 12-15 Some single amino acid substitutions found in human hemoglobin. Amino acids are normal at all residue positions except those indicated. Each type of change causes disease. (Names indicate areas in which cases were first identified.)

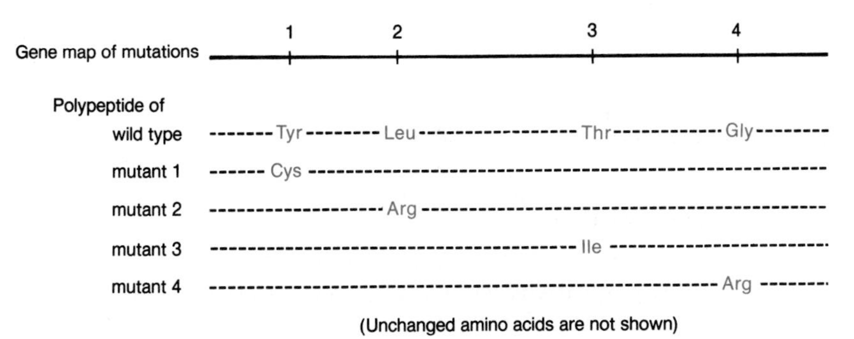

Figure 12-16 Simplified representation of the colinearity of gene mutations. The genetic map of point mutations (determined by recombinational analysis) corresponds linearly to the changed amino acids in the different mutants (determined by fingerprint analysis).

Colinearity of Gene and Protein

Once the structure of DNA was determined by Watson and Crick, it became apparent that the structure of proteins must be encoded in the linear sequence of nucleotides in the DNA. (We'll see in Chapter 13 how this genetic code was deciphered.) Following Ingram's demonstration that one mutation alters one amino acid in a protein, a relationship was sought between the linear sequence of mutant sites in a gene and the linear sequence of amino acids in a protein. (It is possible to map mutational sites within a gene due to studies on genetic fine structure, to be described in detail later in this chapter.)

Charles Yanofsky probed the relationship between altered genes and altered proteins by studying the enzyme tryptophan synthetase in *E. coli.* This enzyme catalyzes the conversion of indole glycerol phosphate into tryptophan. Two genes, *trpA* and *trpB*, control the enzyme. Each gene controls a separate polypeptide; after the A and B polypeptides are produced, they combine to form the active enzyme (a multimeric protein). Yanofsky analyzed mutations in the *trpA* gene which resulted in alterations of the tryptophan synthetase A subunit. He ordered the mutations by P1 transduction (see p. 289) to produce a detailed gene map, and he also determined the amino acid sequence of each respective altered tryptophan synthetase. His results were similar to Ingram's for hemoglobin: each mutant had a defective polypeptide associated with a specific amino acid substitution at a specific point. However, Yanofsky was able to show an exciting correlation that Ingram was not able to observe due to the limitations of his system. He found an exact match between the sequence of the mutational sites in the gene map of the *trpA* gene and the location of the corresponding altered amino acids in the A polypeptide chain. The farther apart two mutational sites were in map units, the more amino acids there were between the corresponding substitutions in the polypeptide (Figure 12-16). Thus, Yanofsky demonstrated **colinearity**—the correspondence between the linear sequence of the gene and that of the polypeptide. Figure 12-17 shows the complete set of data.

Message The linear sequence of nucleotides in a gene determines the linear sequence of amino acids in a protein.

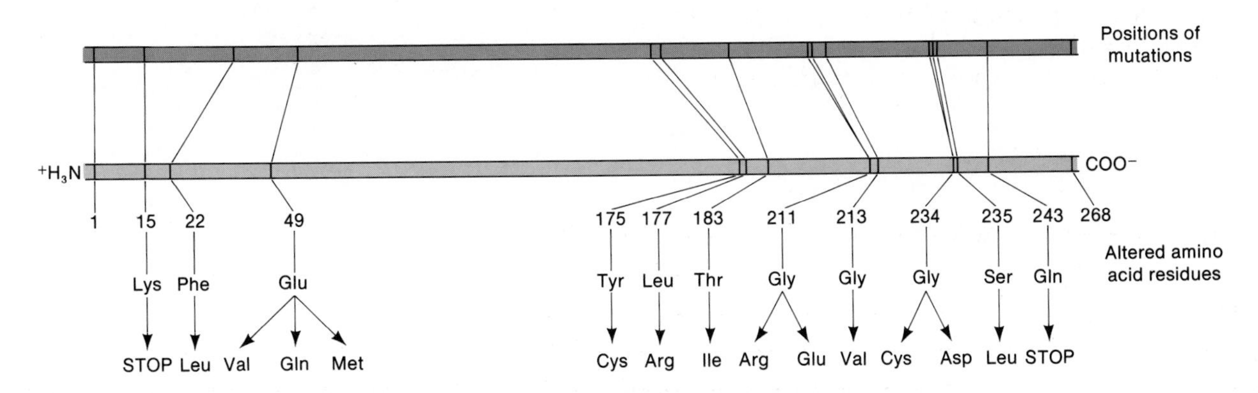

Figure 12-17 Actual colinearity shown in the A protein of tryptophan synthetase from *E. coli.* There is a linear correlation between the mutational sites and the altered amino acid residues. (Based on C. Yanofsky, "Gene Structure and Protein Structure." Copyright 1967 by Scientific American. All rights reserved.)

Enzyme Function

How can a single amino acid substitution, such as that in sickle-cell hemoglobin (Figure 12-13), have such a profound effect on protein function and the phenotype of an organism? Take enzymes, for example. Enzymes are known to do their job of catalysis by physically grappling with their substrate molecules, twisting or bending the molecules to make or break chemical bonds. Figure 12-18 shows the gastric digestion enzyme carboxypeptidase in its relaxed position and after grappling with its substrate molecule, glycyltyrosine. The substrate molecule fits into a notch in the enzyme structure; this notch is called the **active site.**

Figure 12-19 diagrams the general concept. [Note that there are two basic types of reactions performed by enzymes: (1) the breakdown of a substrate into simpler products, and (2) as shown in Figure 12-19, the synthesis of a complex product from one or more simpler substrates.]

Much of the globular structure of an enzyme is nonreactive material that simply supports the active site. So we might expect that amino acid substitutions throughout most of the structure would have little effect, whereas very specific amino acids would be required for the part of the enzyme molecule that gives the precise shape to the active site. Hence, the possibility arises that a functional enzyme does not always require a unique amino acid sequence for the *entire* polypeptide. This has proved to be the case: in a number of systems, numerous positions in a polypeptide can be filled by several alternative amino acids and enzyme function is retained. But at certain other positions in the polypeptide, only the wild-type amino acid will preserve activity; in all likelihood, these amino acids form critical parts of the active sites.

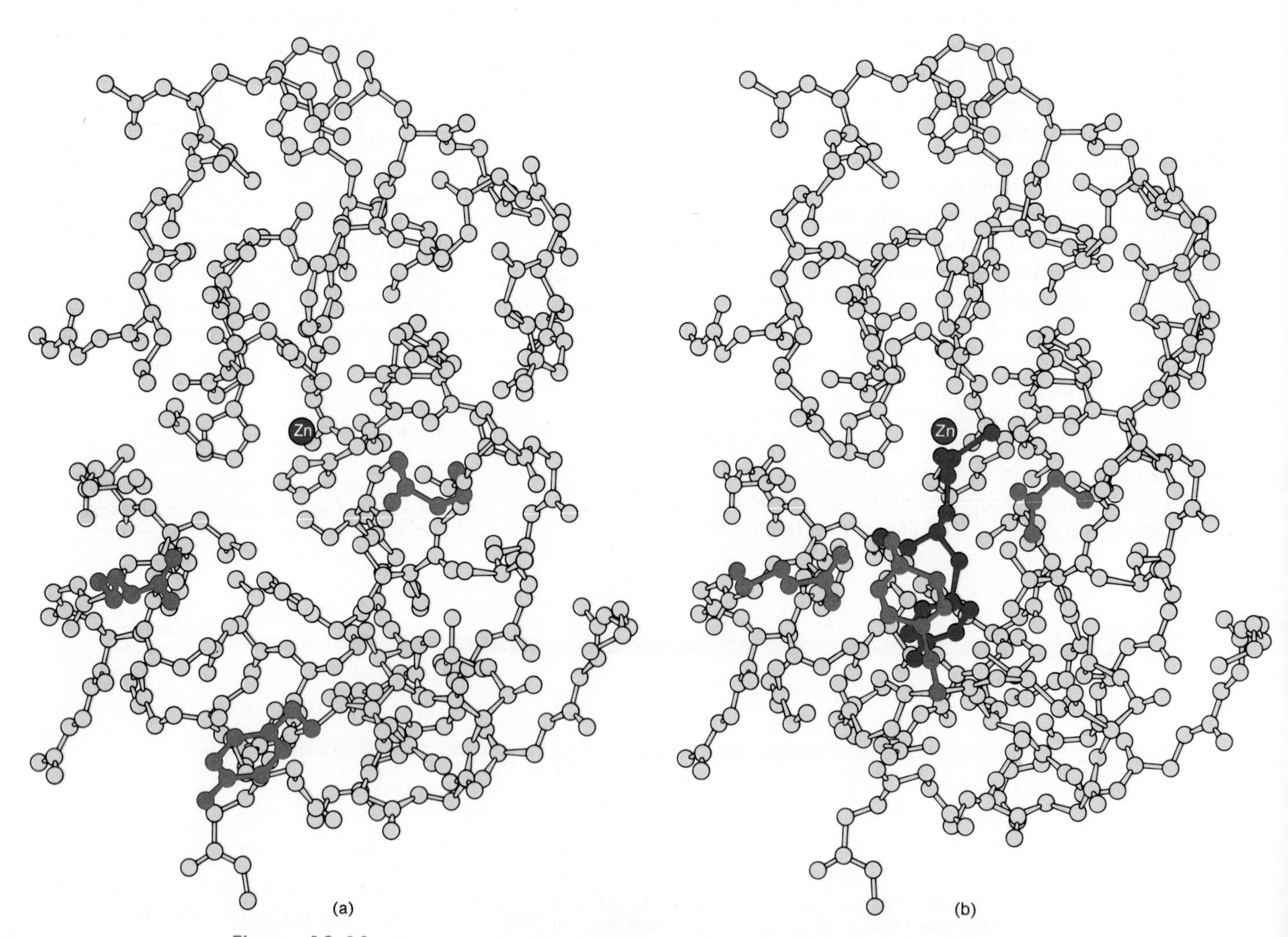

Figure 12-18 The active site of the digestive enzyme carboxypeptidase. (a) The enzyme without substrate. (b) The enzyme with its substrate (purple) in position. Three crucial amino acids (red) have changed positions to move closer to the substrate. Carboxypeptidase carves up proteins in the diet. (From W. N. Lipscomb, *Proceedings of the Robert A. Welch Foundation Conferences on Chemical Research* 15, 1971, 140–41.)

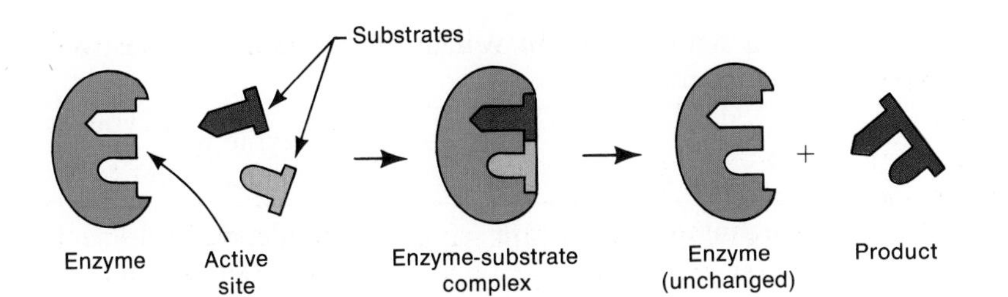

Figure 12-19 Schematic representation of the action of a hypothetical enzyme in putting two substrate molecules together. The "lock-and-key" fit of the substrate into the enzyme's active site is very important in this model.

Some of these critical amino acids in carboxypeptidase are indicated in Figure 12-18 in red.

> Message Protein architecture is the key to gene function. A gene mutation typically results in a substitution of a different amino acid into the polypeptide sequence of a protein. The new amino acid may have chemical properties that are incompatible with the proper protein architecture at that particular position; in such a case, the mutation will lead to a nonfunctional protein.

Genes and Cellular Metabolism: Genetic Diseases

When we think of enzyme activity in terms of cellular metabolism, we realize that the inactivation of one or more enzymes can have staggering consequences. Most of us have been amazed by the charts on laboratory walls showing the myriad interlocking, branched, and circular pathways along which the cell's chemical intermediates are shunted like parts on an assembly line. Bonds are broken, molecules cleaved, molecules united, groups added or removed, and so on. The key fact is that almost every step, represented by an arrow on the metabolic chart, is controlled (mediated) by an enzyme, and each of these enzymes is produced under the direction of a gene that specifies its function. Change one critical gene, and the entire assembly line can break down.

Humans provide some startling examples. The list in Table 12-4 gives some representative examples and suggests the magnitude of genetic involvement in human disease. Figure 12-20 shows a corner of the human metabolic map to illustrate how a set of diseases, some of them common and familiar to us, can stem from the blockage of adjacent steps in biosynthetic pathways.

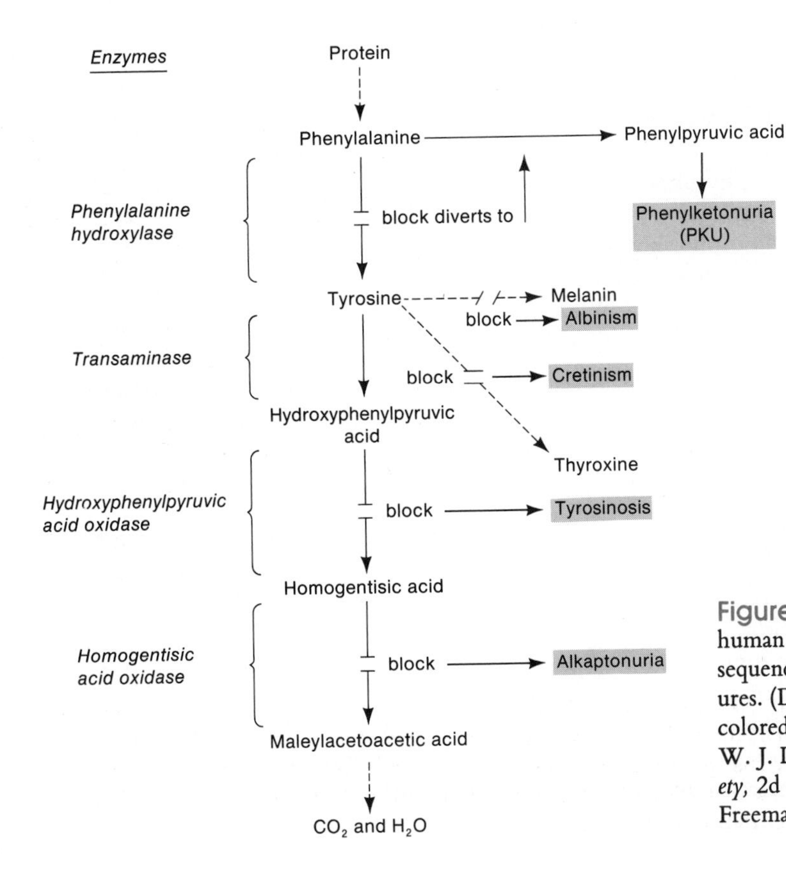

Figure 12-20 One small part of the human metabolic map, showing the consequences of various specific enzyme failures. (Disease phenotypes are shown in colored boxes.) (After I. M. Lerner and W. J. Libby, *Heredity, Evolution, and Society*, 2d ed. Copyright 1976 by W. H. Freeman and Company.)

Table 12-4. Representative Examples of Enzymopathies: Inherited Disorders in Which Altered Activity (Usually Deficiency) of a Specific Enzyme Has Been Demonstrated in Humans

Condition	Enzyme with deficient activity[1]	Condition	Enzyme with deficient activity[1]
Acatalasia	Catalase	Granulomatous disease	Reduced nicotinamide-adenine dinucleotide phosphate (NADPH) oxidase
Acid phosphatase deficiency	Acid phosphatase	Hydroxyprolinemia	Hydroxyproline oxidase
Albinism	Tyrosinase	Hyperlysinemia	Lysine-ketoglutarate reductase
Aldosterone deficiency	18-Hydroxydehydrogenase	Hypophosphatasia	Alkaline phosphatase
Alkaptonuria	Homogentisic acid oxidase	Immunodeficiency disease	Adenosine deaminase
Angiokeratoma, diffuse (Fabry disease)	Ceramide trihexosidase		Uridine monophosphate kinase
Apnea, drug-induced	Pseudocholinesterase	Krabbe disease	Galactosylceramide β-galactosidase
Argininemia	Arginase	Leigh necrotizing encephalomyelopathy	Pyruvate carboxylase
Argininosuccinic aciduria	Argininosuccinase	Maple-sugar urine disease	Keto acid decarboxylase
Ataxia, intermittent	Pyruvate decarboxylase	Niemann-Pick disease	Sphingomyelinase
Citrullinemia	Argininosuccinic acid synthetase	Ornithinemia	Ornithine ketoacid aminotransferase
Crigler-Najjar syndrome	Glucuronyl transferase	Pentosuria	Xylitol dehydrogenase (L-xylulose reductase)
Cystathioninuria	Cystathionase	Phenylketonuria	Phenylalanine hydroxylase
Ehlers-Danlos syndrome, type V	Lysyl oxidase	Refsum disease	Phytanic acid oxidase
Farber lipogranulomatosis	Ceramidase	Richner-Hanhart syndrome	Tyrosine aminotransferase
Galactosemia	Galactose 1-phosphate uridyl transferase	Sandhoff disease (GM_2 gangliosidosis, type II)	Hexosaminidase A, B
Gangliosidosis, GM_1; generalized, type I, or infantile form	β-Galactosidase A, B, C	Tay-Sachs disease	Hexosaminidase A
Gangliosidosis, GM_1; type II, or juvenile form	β-Galactosidase B, C	Wolman disease	Acid lipase
Gaucher disease	Glucocerebrosidase	Xeroderma pigmentosum	Ultraviolet-specific endonuclease
Gout	Hypoxanthine-guanine phosphoribosyltransferase		
	Phosphoribosyl pyrophosphate (PRPP) synthetase (increased activity)		

SOURCE: Victor A. McKusick, *Mendelian Inheritance in Man*, 4th ed. Copyright 1975 by Johns Hopkins University Press.

[1] The form of gout due to increased activity of PRPP is the only disorder listed that is due to *increased* enzymatic activity.

Genetic Observations Explained by Enzyme Structure

When the significance of the gene control of cellular chemistry became clear, a lot of other things fell into place. Many genetic observations now could be explained and tied together in a single conceptual model. Thus, armed with the understanding of the gene–protein relationship and how enzymes function, we can now re-examine some of the genetic findings discussed in earlier chapters and look at them in terms of the biochemistry involved.

Temperature-Sensitive Alleles

Recall that some mutants appear to be wild-type at normal temperatures but can be detected as mutants at high or low temperatures (see page 181). We now know that such mutations result from the substitution of an amino

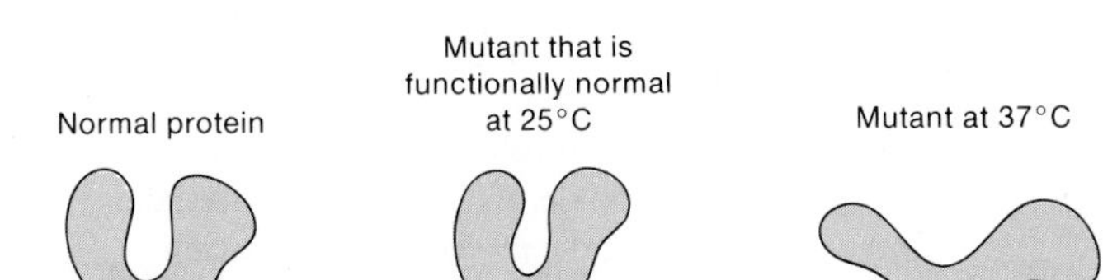

Figure 12-21 Schematic representation of protein conformational distortion, which is probably the basis for temperature sensitivity in certain mutants. An amino acid substitution that has no significant effect at normal (permissive) temperatures may cause significant distortion at abnormal (restrictive) temperatures.

acid which produces a protein that is functional at normal temperatures, called **permissive** temperatures, but distorted and nonfunctional at high or low temperatures, called **restrictive** temperatures (Figure 12-21).

As we have seen, conditional mutations such as temperature-sensitive mutations can be very useful to geneticists. Stocks of the mutant culture can easily be maintained under permissive conditions, and the mutant phenotype can be studied intensively under restrictive conditions. Such mutants can be very handy in the genetic dissection of biological systems. For example, with a temperature-sensitive allele we can shift to a restrictive temperature at various times during development in order to determine the time at which a gene is active.

Genetic Ratios

Most "classical" (Mendelian) gene-interaction ratios can be explained readily by the one-gene–one-enzyme concept. For example, recall the 9:7 F_2 dihybrid ratio for pea-flower pigment (page 100):

P   *AA bb* (white) × *aa BB* (white)

↓

F_1 all *Aa Bb* (purple)

↓

F_2

9	*A–B–*	(purple)
3	*A–bb*	(white)
3	*aaB–*	(white)
1	*aabb*	(white)

We can easily explain this result if we imagine a biosynthetic pathway that leads ultimately to a purple petal pigment in which there are two colorless (white) precursors:

$$\text{white precursor 1} \xrightarrow[\text{enzyme A}]{A \text{ allele}} \text{white precursor 2} \xrightarrow[\text{enzyme B}]{B \text{ allele}} \text{purple pigment}$$

(Can you work out models to explain such ratios as 9:3:4 and 13:3?)

Dominance and Recessiveness

The meaning of dominance and recessiveness also becomes a little clearer in light of the biochemical model. In most cases, dominance represents the presence of enzyme function, whereas recessiveness represents the lack of enzyme function. A heterozygote has one dominant allele that can produce the functional enzyme:

$$\text{allele } a \longrightarrow \text{nonfunctional enzyme (does nothing)}$$

$$\text{precursor X} \xrightarrow[\text{enzyme A}]{\text{allele } A} \text{product Y}$$

If phenotype Y is due to the presence of product Y and phenotype X is due to the absence of product Y, then it is clear that the heterozygote will show phenotype Y and allele *A* is dominant over allele *a*.

However, this is not the only possible model. We can build in the concept of a threshold. Suppose that phenotype Y is produced only when the concentration of product Y exceeds some threshold level. Further suppose that the homozygote *AA* produces more enzyme—and hence more product—than the heterozygote *Aa*. In this case, the phenotype of the heterozygote will depend on the relationship between the threshold and the amount of product Y that is produced by the heterozygote. In Figure 12-22a the heterozygote A_1A_2 does produce enough product Y to exceed the threshold, so the heterozygote has phenotype Y and therefore A_2 is dominant over A_1. However, the situation could be like the one shown for the alleles B_1 and B_2, where the heterozygote B_1B_2 does not exceed the threshold. In this case (which is less common), B_1 is dominant over B_2 and the dominant phenotype is the one involving a lack of product Y.

Figure 12-22b illustrates a situation in which no threshold exists. The heterozygote has an intermediate phenotype—exactly the situation observed in cases of incomplete dominance.

What determines whether a threshold exists and whether a gene will behave like *A*, *B*, or *C*? Probably many interacting factors are involved: other genes, the chemical nature of the product, and (last but not least) the effect of the environment on that particular cell type. Finally, note that Figure 12-22 shows a simple linear relationship between enzyme concentration and the number of active alleles—but this need not be the case, as we'll see in Chapter 17.

Let's go back to the *arg* mutants in *Neurospora*, studied by Beadle and Tatum. Their recessiveness can be demon-

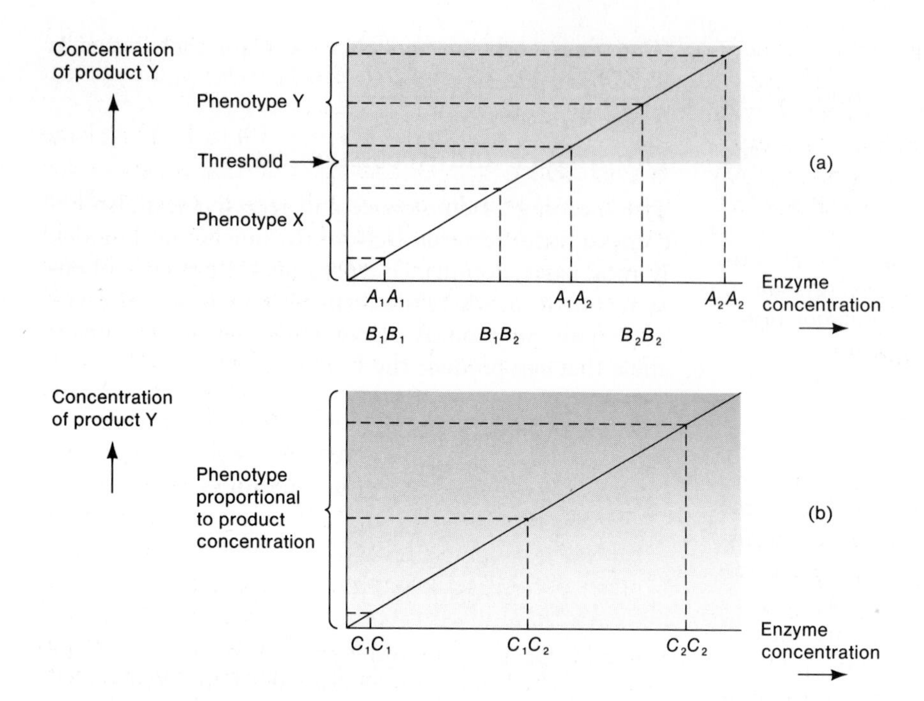

Figure 12-22 Hypothetical curves relating enzyme concentration to the amount of product. Two basic situations are possible. In (a) there is a threshold level of the product above which a sharply contrasting phenotype, Y, is observed. Depending on where the product levels in the three possible genotypes occur in relation to this threshold, the heterozygote will show either phenotype X (for example, B_1B_2) or Y (for example, A_1A_2). If allele 2 is the active one, you can see that the active allele (A_2) can be dominant, which is normally the case. Less frequently, the inactive allele (B_1) can be dominant. In (b) there is no threshold, and the heterozygote has an intermediate phenotype. This second situation explains incomplete dominance.

strated by making heterokaryons between two *arg* mutants. If an *arg-1* mutant is placed on a minimal medium with an *arg-2* mutant, the two cell types fuse and form a **heterokaryon** composed of both nuclear types in a common cytoplasm (Figure 12-23). Recall that *arg-1* and *arg-2* each lack a different enzyme needed in the pathway to arginine. As a heterokaryon, the two mutants can complement one another, so that their combined abilities will produce arginine in the shared cytoplasm.

How does this show recessiveness in the case of the *arg* mutants? The heterokaryon is, in effect, heterozygous at the *arg-1* and *arg-2* loci, and the fact that the heterokaryon can produce arginine shows that the wild-type genes at these two loci are dominant; hence, the mutant genes *arg-1* and *arg-2* are recessive.

Genetic Fine Structure

Until the beginning of this chapter, our genetic and cytological analysis led us to regard the chromosome as a linear (one-dimensional) array of genes, strung rather like beads on an unfastened necklace. Indeed, this model is sometimes called the **bead theory.** According to the

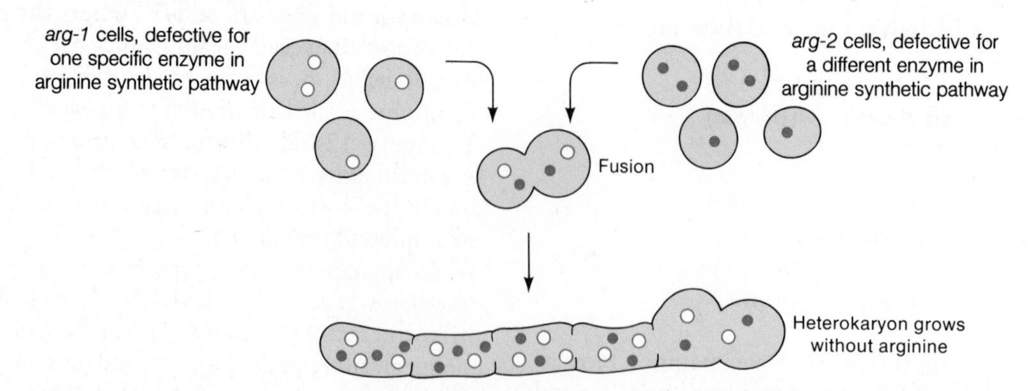

Figure 12-23 Formation of a heterokaryon of *Neurospora,* demonstrating both complementation and recessiveness. Vegetative cells of this normally haploid fungus can fuse, allowing the nuclei from the two strains to intermingle within the same cytoplasm. If each strain is blocked at a different point in a metabolic pathway, as are *arg-1* and *arg-2* mutants, all functions are present in the heterokaryon and the *Neurospora* will grow; in other words, complementation takes place. The growth must be due to the presence of the wild-type gene; hence, the mutant genes *arg-1* and *arg-2* must be recessive.

bead theory, the existence of a gene as a unit of inheritance is recognized through its mutant alleles. All of these alleles affect a single phenotypic character, all map to one chromosome locus, all give mutant phenotypes when paired, and all show Mendelian ratios when intercrossed. Several tenets of the bead theory are worth emphasizing:

1. The gene is viewed as a fundamental unit of *structure,* indivisible by crossing-over. Crossing-over occurs between genes (the beads in this model) but never within them.
2. The gene is viewed as the fundamental unit of *change,* or mutation. It changes in toto from one allelic form to another; there are no smaller components within it that can change.
3. The gene is viewed as the fundamental unit of *function* (although the precise function of the gene is not specified in this model). Parts of a gene, if they exist, cannot function.

Yet, how can we reconcile the fact that the gene consists of a series of nucleotides with the view that the gene is the smallest unit of mutation and recombination? Seymour Benzer's work in the 1950s showed that the bead theory was not correct. Benzer was able to use a genetic system in which extremely small levels of recombination could be detected. He demonstrated that whereas a gene can be defined as a unit of function, a gene can be subdivided into a linear array of sites that are mutable and that can be recombined. The smallest units of mutation and recombination are now known to be correlated with single nucleotide pairs.

Fine-Structure Analysis of the Gene

As we shall see, Benzer's classic analysis of the fine structure of the gene deals with the following material:

1. The life cycle of the bacteriophage.
2. Plaque morphology and the *rII* system of phage T4.
3. The concept of "selection" in genetic crosses with bacteriophages.
4. Deletion mapping.
5. Destruction of the bead theory.
6. Complementation and the difference between complementation and recombination.

The Life Cycle of the Bacteriophage

Benzer chose the bacteriophage genome as the system for his high-resolution genetic studies. Therefore, Benzer's work is much easier to understand if we first review the material covered on pages 282 to 285. Let's summarize some of the essential points. Recall that bacteriophages, or phages, are viruses that attack bacteria. (The bacterial species used most frequently for these studies is *E. coli.*) Like other viruses, a phage usually consists of a protein coat containing the viral DNA, and tail fibers that allow the virus to attach to the bacterial cell wall. The phage nucleic acid is injected into the cell; this programs the synthesis of proteins that stimulate viral replication and the production of new viral DNA and new coat and tail proteins. After the new phage particles assemble, the bacterial cell is lysed and the particles are released into the surrounding medium. If the infected cell is on a lawn of bacteria immobilized on the agar surface of a petri plate, the new phage progeny will infect neighboring cells, and their progeny in turn will infect other neighboring cells until a small clearing, visible to the naked eye, is produced. This clearing is termed a **plaque** (see Figure 10-21). Different phages make different types of plaques. Phages also differ in the bacterial strains that they infect —a property termed the **host range.** Initially, phages were characterized according to their host range and given numbers in the "T" series. Thus, we have phages T1, T2, T3, T4, T5, and so on. Many experiments relating to the nature of the gene and the genetic code have been carried out with phage T4 (Figure 12-24).

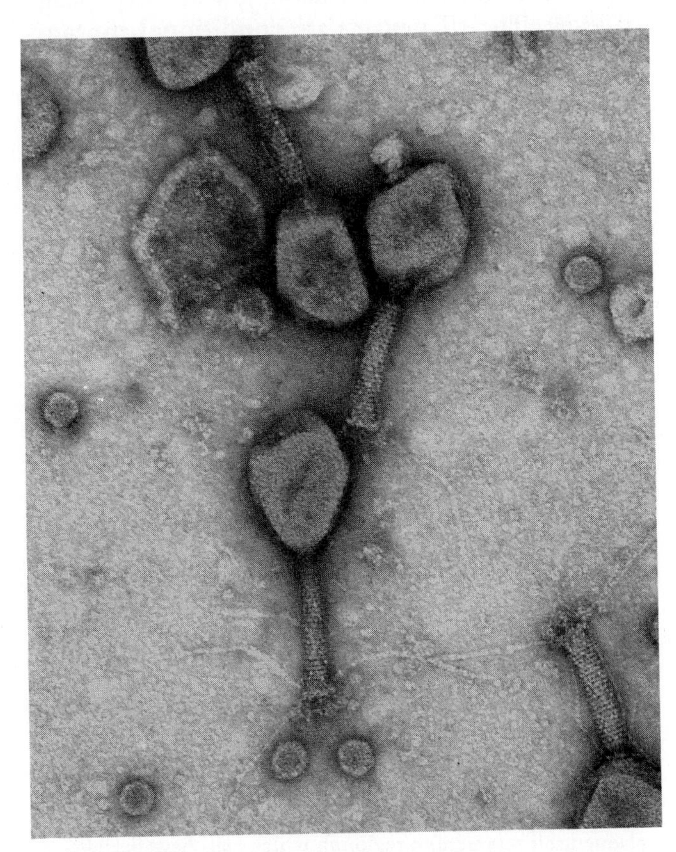

Figure 12-24 Enlargement of the *E. coli* phage T4 showing details of structure: Note head, tail, and tail fibers. This was the phage used by Benzer in his experiments on the nature of the *rII* (rapid lysis) gene. (Photograph from Jack D. Griffith.)

The *rII* System

Benzer sought genetic markers that could be used with bacteriophages. The size and shape of a plaque, he realized, were heritable traits of the virus, and so he began a genetic analysis of plaque morphology. One type of mutant T4 phage produced larger, ragged plaques that were easy to distinguish from wild-type plaques (Figure 12-25). The large plaque size resulted from rapid lysis of the bacteria, so the mutants were termed ***r*** (for rapid lysis) **mutants.** Benzer analyzed the *r* mutants genetically, and mapped the mutations responsible for the *r* phenotype into two loci: *rI* and *rII*. He then studied the *rII* mutants intensively.

One extraordinary property of *rII* mutants made all of Benzer's work possible: *rII* mutants have a different host range than that of wild-type phages. Two related but different strains of *E. coli,* termed B and K(λ), can be used as different hosts for phage T4. Both bacterial strains can distinguish *rII* mutants from wild-type phages. *E. coli* B allows both to grow, but plaques of different sizes result: wild-type phages produce small plaques, and *rII* mutants produce large plaques. *E. coli* K, an abbreviation for *E. coli* K(λ), does not permit the growth of *rII* mutants, but it does allow wild-type phages to grow. The *rII* mutants are **conditional mutants,** namely mutants that can grow under one set of conditions but not another. *E. coli* B is said to be **permissive** for *rII* mutants, because it allows phage growth, whereas *E. coli* K is said to be **nonpermissive** for *rII* mutants, because it does not allow phage growth. Figure 12-26 and Table 12-5 show the growth characteristics and plaque morphology of these phages on each host strain.

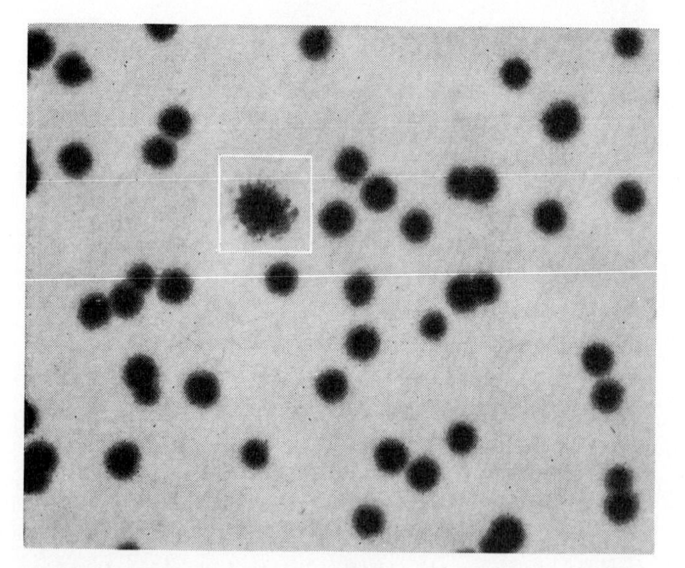

Figure 12-25 A spontaneous mutational event is disclosed by the one mottled plaque *(in the square)* among dozens of normal plaques produced when wild-type phage T4 is plated on a layer of colon bacilli of strain B. Each plaque contains some 10 million progeny descended from a single phage particle. The plaque itself represents a region in which cells have been destroyed. Mutants found in abnormal plaques provide the raw material for genetic mapping. (From S. Benzer, "The Fine Structure of the Gene." Copyright 1962 by Scientific American, Inc. All rights reserved.)

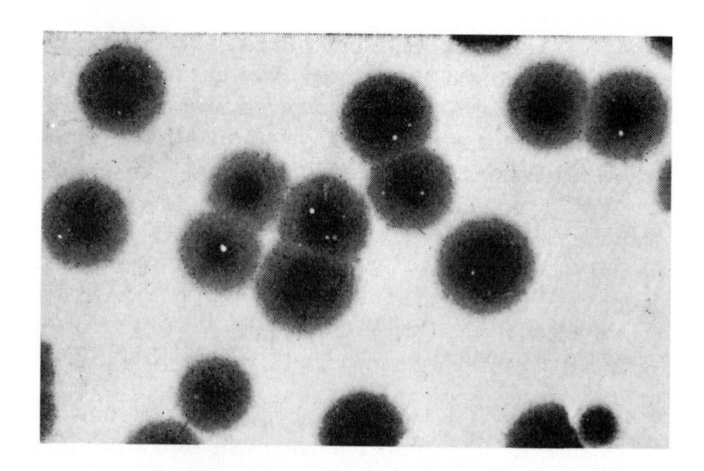

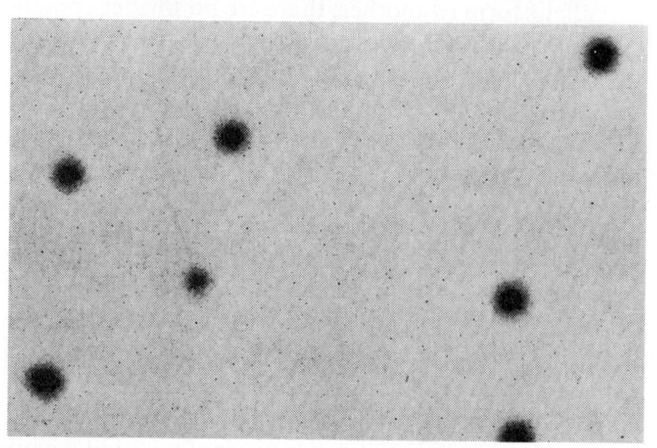

Figure 12-26 Duplicate replatings of a mixed phage population, obtained from a mottled plaque like the one shown in Figure 12-25, give contrasting results, depending on the host. Replated on *E. coli* of strain B *(top)*, *rII* mutants produce large plaques. If the same mixed population is plated on strain K *(bottom)*, only the wild-type phages produce plaques. (From S. Benzer, "The Fine Structure of the Gene." Copyright 1962 by Scientific American, Inc. All rights reserved.)

Table 12-5. Plaque Phenotypes Produced by Different Combinations of *E. coli* and Phage Strains

	E. coli strain	
T4 phage strain	B	K
rII	Large, round	No plaques
rII$^+$	Small, ragged	Small, ragged

Selection in Genetic Crosses of Bacteriophages

Benzer crossed various *rII* mutants of the T4 phage and obtained recombination frequencies, which he then used to map mutations within the *rII* gene region. How does one carry out crosses with different phages? In order to do this, phages of two different types are used to infect the same bacterial cell, as described on page 284. If the ratio of phages to bacteria is high enough, then virtually every bacterium will be infected with at least one phage of each type. Once inside the cell, the DNA molecules from the two phages have an opportunity to recombine with one another, generating recombinant phages, which are recovered among the progeny.

Let's see how this works. Suppose we wish to cross two *rII* mutants and recover wild-type recombinants. Because wild-type and *rII* mutants make plaques that can be distinguished from one another, we could cross two different *rII* mutants in *E. coli* B and examine the progeny on *E. coli* B (Figure 12-27, photograph at top right), hoping to find small wild-type plaques among the large parental *rII* plaques. If the recombination frequency is high enough to yield 2 to 3 percent or more wild-type

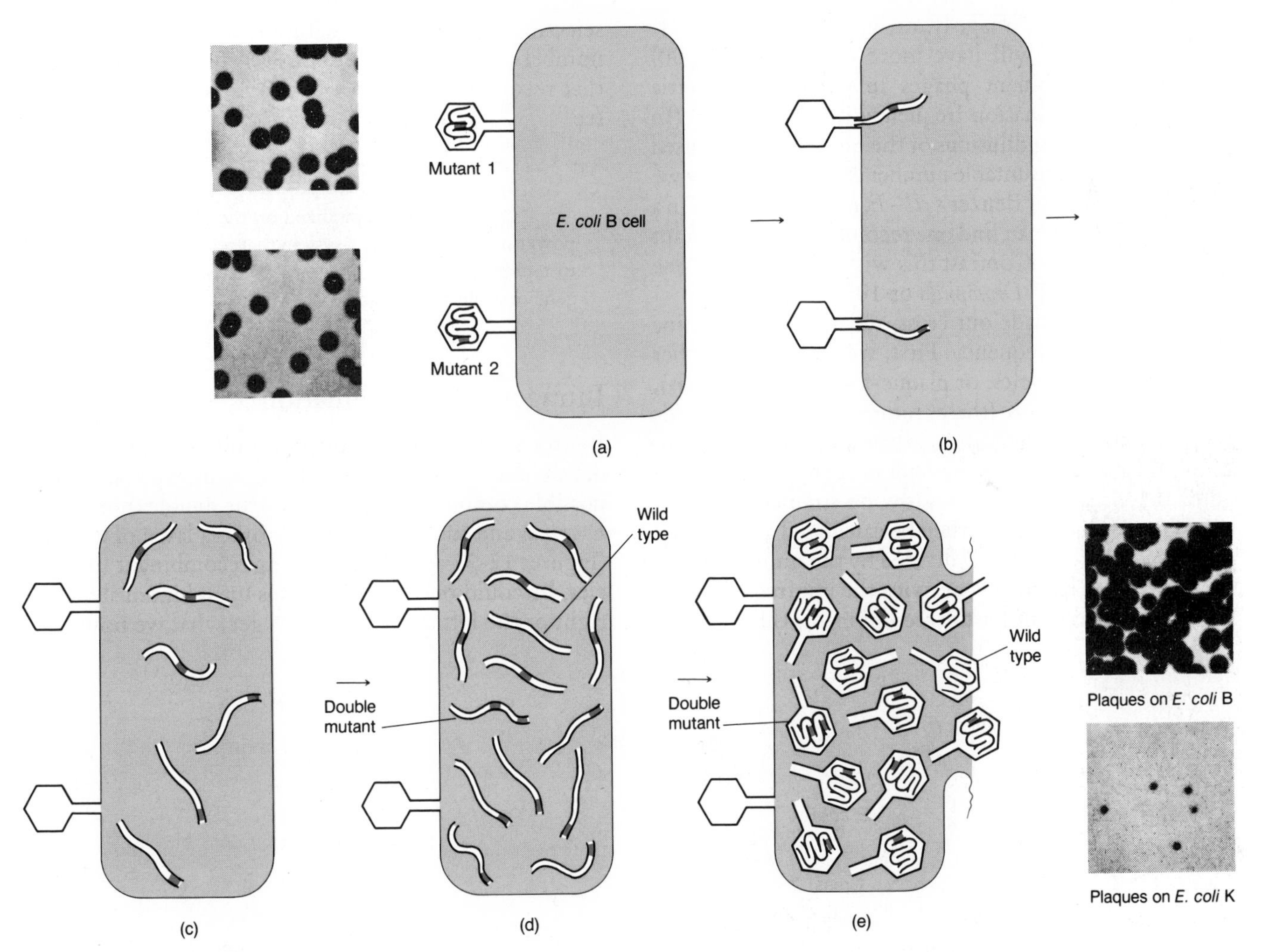

Figure 12-27 The process of recombination permits parts of the DNA of two different phage mutants to be reassembled in a new DNA molecule that may contain both mutations or neither of them. Mutants obtained from two different cultures are introduced into a broth of *E. coli* strain B. Crossing occurs when DNA from each mutant type infects a single bacillus. Most of the DNA replicas are of one type or the other, but occasionally recombination will produce either a double mutant or a wild-type recombinant containing neither mutation. When the progeny of the cross are plated on strain B, all grow successfully, producing many plaques. Plated on strain K, only the wild-type recombinants are able to grow. A single wild-type recombinant can be detected among as many as 100 million progeny. (Modified from S. Benzer, "The Fine Structure of the Gene." Copyright 1962 by Scientific American, Inc. All rights reserved.)

plaques, then this method would suffice. However, for recombination that is less frequent than 1 percent, a lot of work would be involved in generating a map of numerous *rII* mutations.

Instead of plating the progeny phages from the cross on *E. coli* B, however, we could plate the progeny on *E. coli* K (Figure 12-27, bottom photograph at right), so that only the wild-type recombinant phages could grow. Even if the recombination frequency is very low (say, 0.01 percent), we could easily detect the recombinant wild-type phages. Why? Because a typical phage lysate (the phage mixture released after lysis of the bacteria) from such an infection (whether it involves a cross or not) contains in excess of 10^9 phages per milliliter (ml). If we mix 0.1 ml of such a phage lysate with 0.1 ml of *E. coli* K bacteria, then we will have more than 10^5 (100,000) wild-type recombinant phages infecting the bacteria when the recombination frequency is 0.01 percent. (In practice, increasing dilutions of the phage lysate are used until one yields a countable number of plaques.) Now we can see the power of Benzer's *rII*–*E. coli* B/K system. In a single milliliter, it can find one recombinant or revertant per 10^9 organisms. Contrast this with trying to find one recombinant in 10^9 *Drosophila* or 10^9 mice!

Once we've made our cross, we need to determine the recombinant frequency. First, we count the number of active virus particles, or plaque-forming units **(pfu),** that grew on *E. coli* K (these, remember, are only wild-type recombinant phages), and the number that grew on *E. coli* B (these represent the total progeny phages, since all of the virus particles can grow on strain B). The recombinant frequency can then be calculated as twice the number of pfu on *E. coli* K divided by the number of pfu on *E. coli* B. Why do we use twice the pfu frequency for *E. coli* K? To account for the recombinants that are double mutants and which we cannot detect; such mutants should be present at the same frequency as the wild-type recombinants. Figures 12-27 and 12-28 provide diagrammatic representations of these principles.

Finally, in any cross of this type, we need to plate each parental lysate on *E. coli* K to see how many revertants to wild-type there were in the population. **Back (reverse) mutations** occur at some very low but real frequency. It is important to monitor this frequency and to compare it with our calculated frequency of recombination in order to be sure that recombination—not back reversion of the parental types—has occurred.

In summary, Benzer's use of the *rII* system and two different bacterial hosts provided him with a method for selecting for rare events without having to screen large numbers of plaques. This allowed him to demonstrate that recombination occurred within a gene, as we shall see.

> Message Benzer capitalized on the fantastic resolving power made possible by a system that selects for rare events in rapidly multiplying phages; this allowed him to map a gene in molecular detail.

Intragenic Recombination

Benzer started with an initial sample of eight independently derived *rII* mutant strains, crossing them in all possible combinations of pairs by the double infection of *E. coli* B and subsequent plating onto a lawn of *E. coli* K (Figures 12-27 and 12-28). Using recombinant frequencies, he could map the mutations unambiguously to the right or the left of each other to get what we now call a

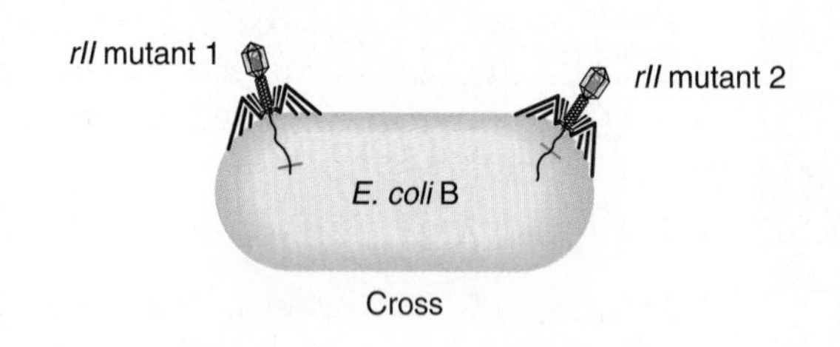

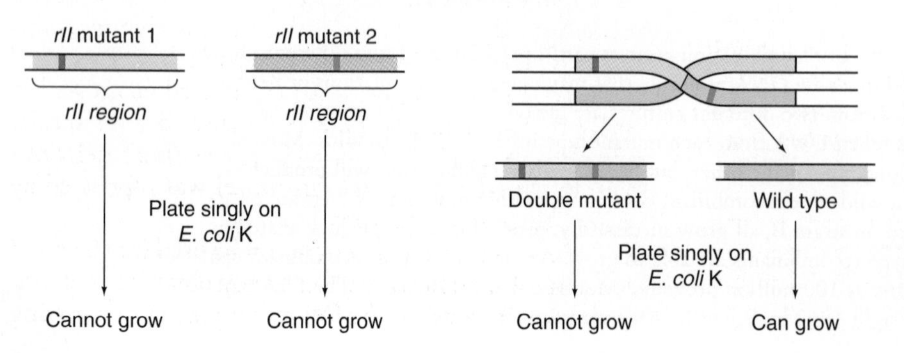

Figure 12-28 Selection of intragenic recombinants at the *rII* locus of phage T4. Mutants within the gene *rII* cannot grow on *E. coli* K. When two different phages carrying different alleles of *rII* infect the same bacterial cell, some progeny phages can grow on *E. coli* K; in other words, some progeny have become rII^+. This result indicates that recombination has occurred *within* a single gene and not just between genes.

gene map (in this case, the map units are the frequency of rII^+ plaques):

rII^1 rII^7 rII^4 rII^5 rII^2 rII^8 rII^3 rII^6

Recombination within a gene, called **intragenic recombination,** seems to be the rule rather than the exception. It can virtually always be found at any locus if a suitable selection system is available to detect recombinants. In other words, a mutant allele can be pictured as a length of genetic material (the gene) that contains a damaged or non-wild-type part — a **mutational site** — somewhere. This partial damage is what causes the non-wild-type phenotype. Different alleles produce different phenotypic effects because they involve damage to different sites within the wild-type allele.

Thus, an allele a^1 can be represented as

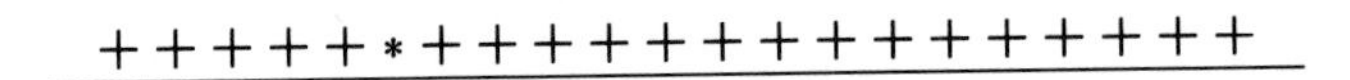

where the asterisk (*) is the mutant site within an otherwise normal gene (denoted by the sites marked +). A cross between a^1 and another mutant allele, a^2, can be represented as

+ + + + + * + + + + + + + + + + + + + + +

×

+ + + + + + + + + + + + + + * + + + + + +

and it is easy to see how the normal, wild-type gene

+ +

could be generated by a simple crossover anywhere between the two mutational sites.

Benzer showed that, contrary to the classical view, genes were not indivisible but could be subdivided by recombination. Extending his analysis to hundreds of *rII* alleles, Benzer found that the minimal recombinant frequency in a cross between a pair of different mutant alleles was 0.01 percent, even though his analytical system was capable of detecting recombinant frequencies as low as 0.0001 percent if they occurred. This led to the idea that genes were composed of small units, and that recombination could occur between but not within these units. We now know that the single nucleotide pair is the smallest unit of recombination.

> Message A gene is composed of subunits that are not divisible by recombination. Each subunit consists of a single nucleotide pair.

Recombination within genes permits us to construct detailed gene maps, such as the map for the *rII* region shown in Figure 12-29. Making crosses between various mutations provides us with the relative frequencies of intragenic recombinants, and these reveal the order and relative positions of the mutational sites within a gene. It should be noted that recombination *within* genes is the same as recombination *between* genes, except that the scale is different.

Mutational Sites

Using Deletions in Mapping Mutational Sites

Benzer extended his fine-structure analysis to the properties of mutational sites. A useful tool for these studies is **deletion mapping.** Deletions are mutations that result from the elimination of segments of DNA. Deletions can be intercrossed and mapped just like point mutations. The deleted region is represented by a bar. If no wild-type recombinants are produced in a cross between different deletions, then the bars are shown as overlapping. A typical deletion map might be as follows:

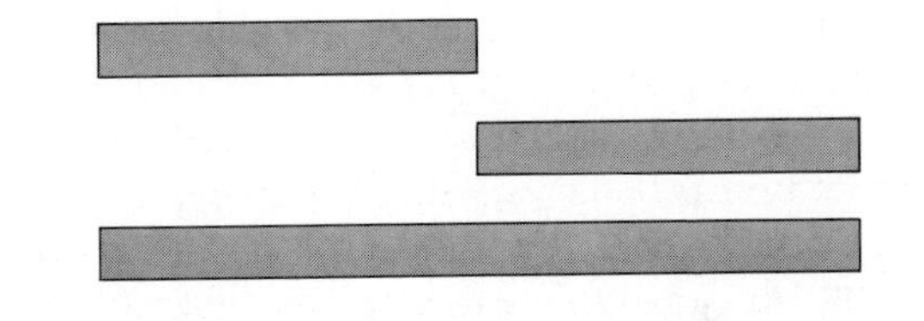

Such deletion maps are useful in delineating regions of the gene to which new point mutations can be assigned. Mutations cannot be converted into wild types in recombination tests against deletion mutations when the DNA corresponding to the wild-type region for that particular mutation is no longer present. Therefore, mutations can be ordered against a set of deletions by rapid tests that do not depend on extensive quantitative measurements.

Deletion Mapping of the *rII* Region

By using deletion mutants, Benzer could rapidly locate new mutational sites in his *rII* gene map. He had found some special *rII* mutants that would not give recombinants when crossed with any of several other types of *rII* mutants, but would give recombinants when crossed to still other *rII* mutants. Benzer realized that his special mutants behaved as if they contained short deletions within the *rII* region, and this model was supported by their lack of reversion to wild types.

Mutants carrying deletions can be used for the rapid location of mutational sites in newly obtained mutants. For example, consider the following gene map, showing

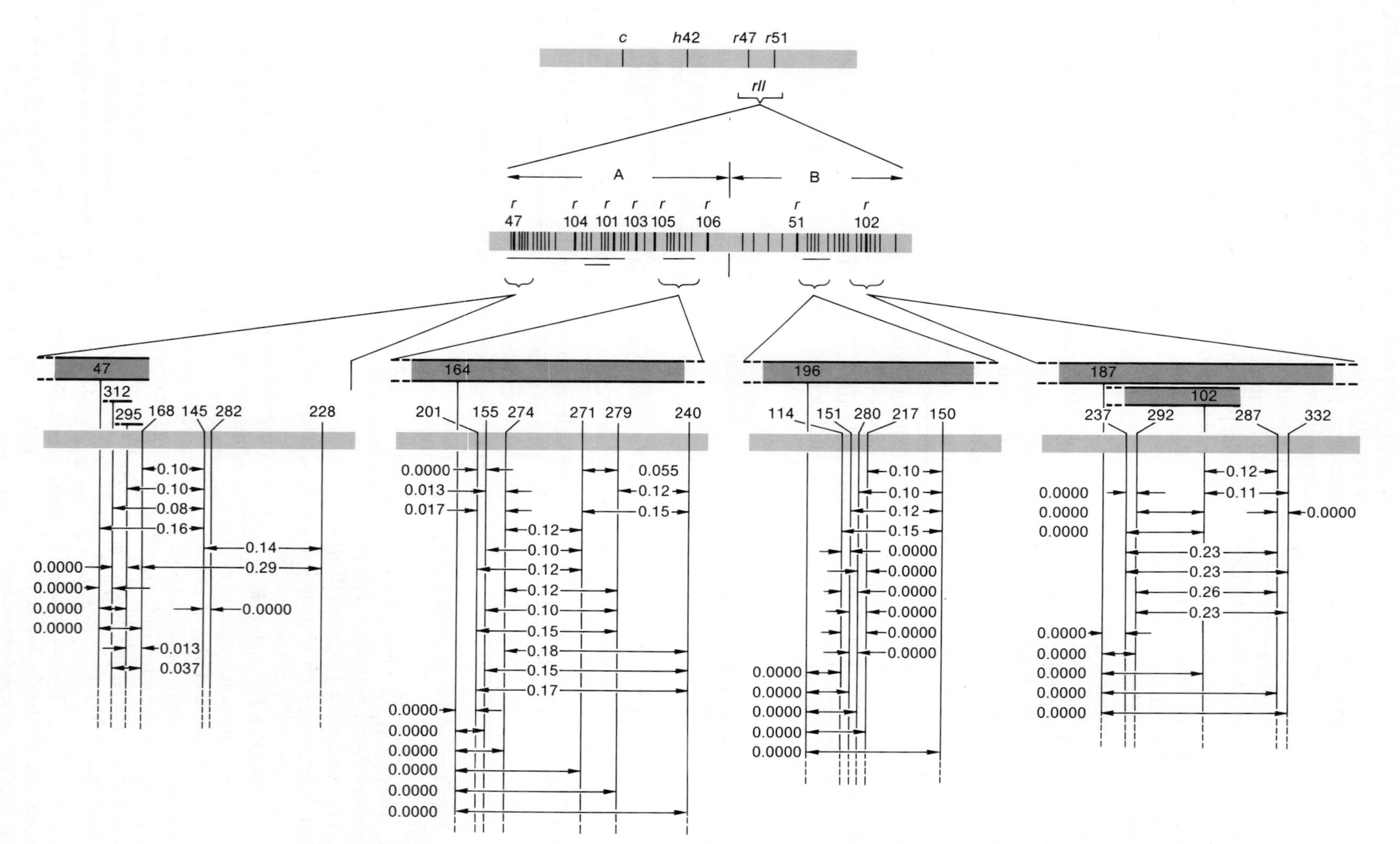

Figure 12-29 Detailed recombination map of the *rII* region of the phage T4 chromosome. The map unit is the percentage of rII^+ recombinants in crosses between the *rII* mutants. Typical regions are progressively enlarged. Numbers on the map represent mutational sites; numbers 47, 64, 196, 187, and 102 represent deletions. Note that the two portions of the *rII* region, A and B, are two different functional units of the region, as described later in the text. (After S. Benzer, *Proceedings of the National Academy of Sciences USA* 41, 1955, 344. From G. S. Stent and R. Calendar, *Molecular Genetics,* 2d ed. Copyright 1978 by W. H. Freeman and Company.)

12 identifiable mutational sites:

1 2 3 4 5 6 7 8 9 10 11 12

Let us suppose that one special mutant, D_1, fails to give rII^+ recombinants when crossed with mutants carrying altered sites *1, 2, 3, 4, 5, 6, 7,* or *8*; therefore, D_1 behaves as if it has a deletion of sites *1* to *8*:

9 10 11 12

Another special mutant, D_2, fails to give rII^+ recombinants when crossed with *5, 6, 7, 8, 9, 10, 11,* or *12*; therefore, D_2 behaves as it if involves a deletion of sites *5* to *12*:

1 2 3 4

These overlapping deletions now define three areas of the gene. Let's call them i, ii, and iii:

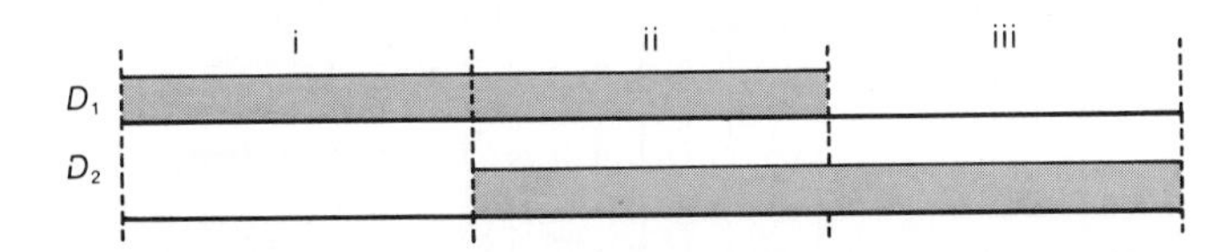

A new mutant that gives rII^+ recombinants when crossed with D_1 but not when crossed with D_2 must have its mutational site in area iii. One that gives rII^+ recombinants when crossed with D_2 but not with D_1 must have its mutational site in area i. A new mutant that does not give rII^+ recombinants with either D_1 or D_2 must have its mutational site in area ii. For example, assume that a mutant in area iii is crossed with D_1. We would envision the cross schematically as follows, where the mutant site is shown with a red bar:

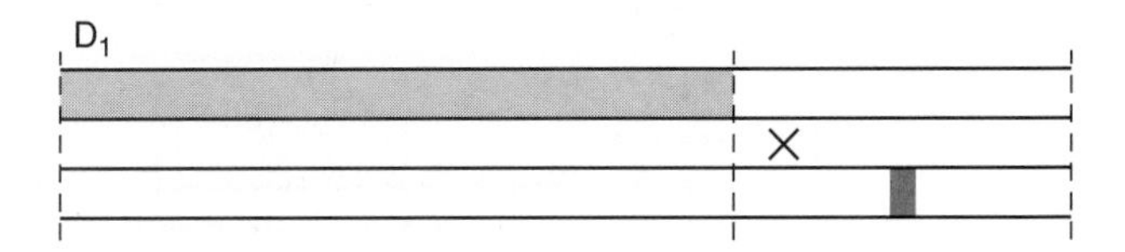

and draw the actual pairing involved as follows:

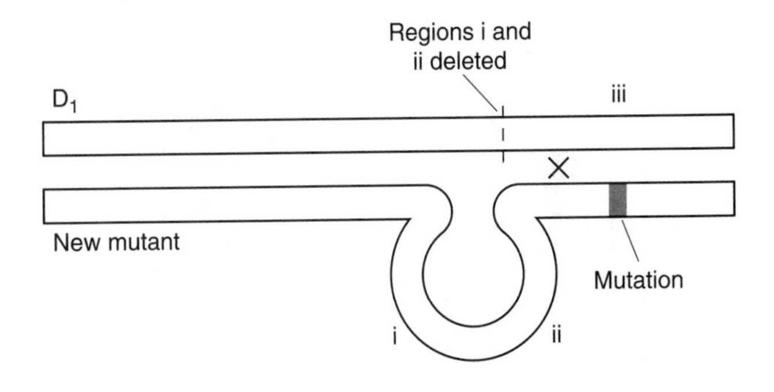

The more deletions there are in the tester set, the more areas can be uniquely designated and the more rapidly new mutational sites can be located (Figure 12-30). Once assigned to a region, a mutation can be mapped against other alleles in the same region to obtain an accurate position. Figure 12-31 shows the complexity of Benzer's actual map.

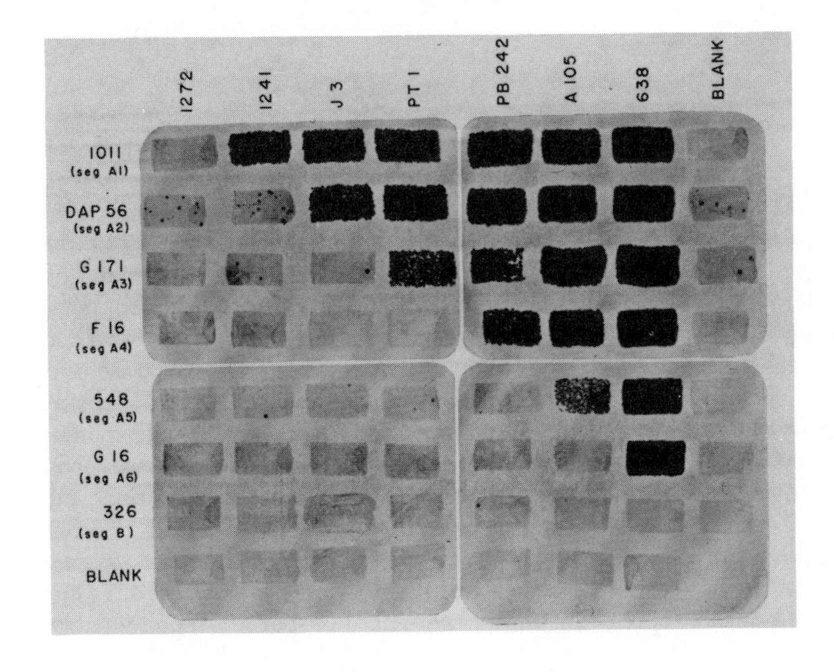

Figure 12-30 Crosses for mapping *rII* mutations. The photograph is a composite of four plates. Each row shows a given mutant tested against the reference deletions of Figure 12-32. The results show each of these mutations to be located in a different segment. Plaques appearing in the blanks are due to revertants present in the mutant stock. (Figures 12-31, 12-33, and 12-34 from S. Benzer, *Proceedings of the National Academy of Sciences USA* 47, 1961, 403–416.)

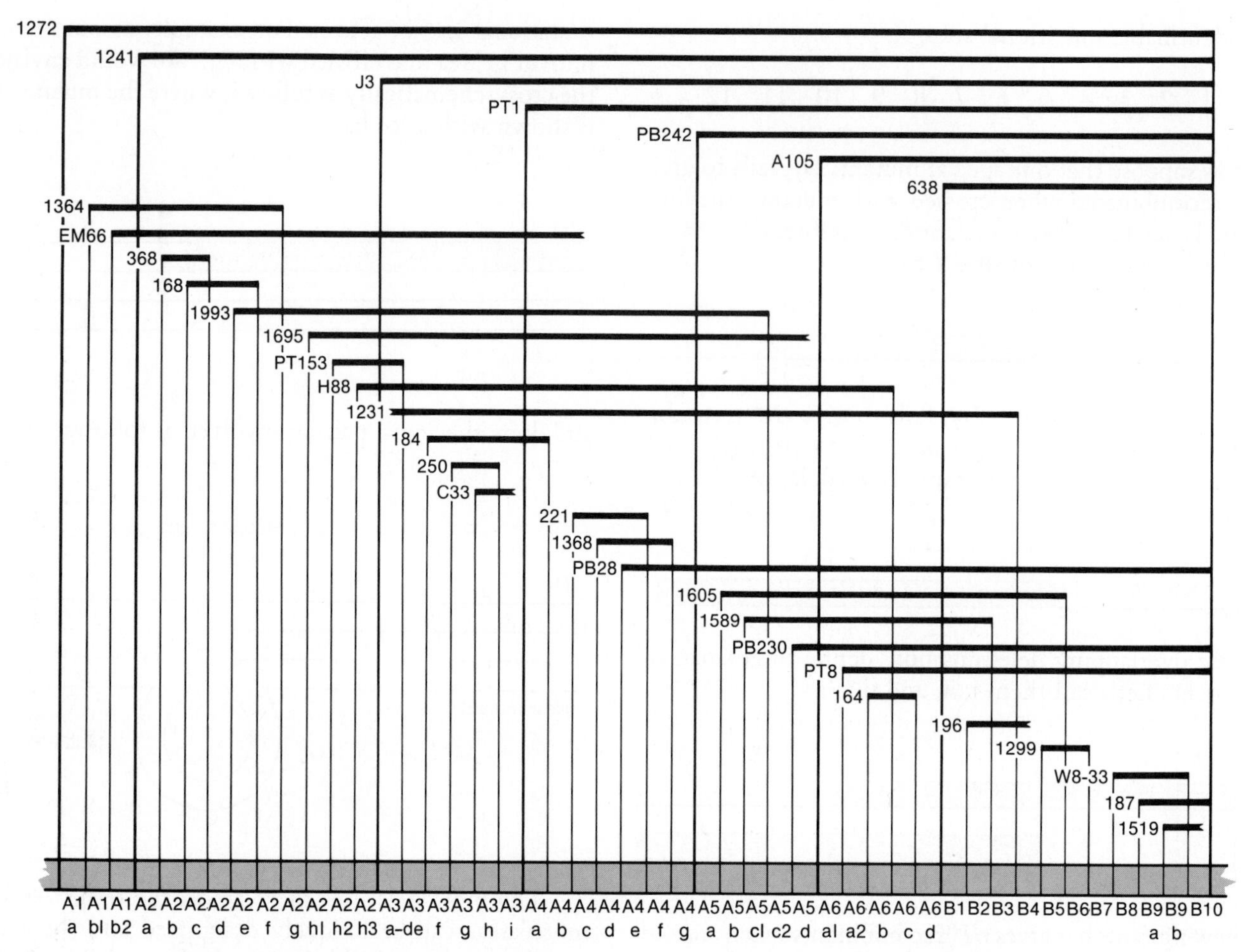

Figure 12-31 Detailed deletion map of the *rII* gene. Each deletion *(horizontal black bar)* has an identification number. Along the bottom are the arbitrary identification numbers of the regions defined by the deletions. Note that some deletions extend out of the *rII* gene. (From G. S. Stent and R. Calendar, *Molecular Genetics,* 2d ed. Copyright 1978 by W. H. Freeman and Company. After S. Benzer, *Proceedings of the National Academy of Sciences USA* 47, 1961, 403–416.)

Analysis of Mutational Sites

The use of deletions enabled Benzer to define the **topology** of the gene—the manner in which the parts are interconnected. His genetic experiments showed the gene to consist of a linear array of mutable subelements. Benzer's next step was to examine the **topography** of the gene—differences in the properties of the subelements. Operationally, he determined this by asking whether all of the subelements or sites were equally mutable, or whether mutations were prevalent at some sites and rare at others. For this study, it was essential to work with the smallest mutable subelements possible. Instead of multisite mutations (deletions) that exhibited no reversion, Benzer used revertible mutations, since they probably represented small alterations, or point mutations. Also, he discarded mutants with high reversion rates, because high reversion interferes with recombination tests.

Benzer used his deletion mutants to rapidly map the set of point mutations. He first localized each mutation into short deletion segments and then crossed all of the point mutations within a segment against one another; any two revertible mutations that failed to recombine with one another represented mutations at the same site.

Figure 12-32 shows the distribution of 1612 spontaneous mutations in the *rII* locus. In Benzer's own words, "That the distribution is not random leaps to the eye." This extraordinary nonrandom distribution demonstrates that all sites are not equally mutable. Benzer termed sites that are more mutable than other sites **hot spots.** The most prominent hot spot was represented by

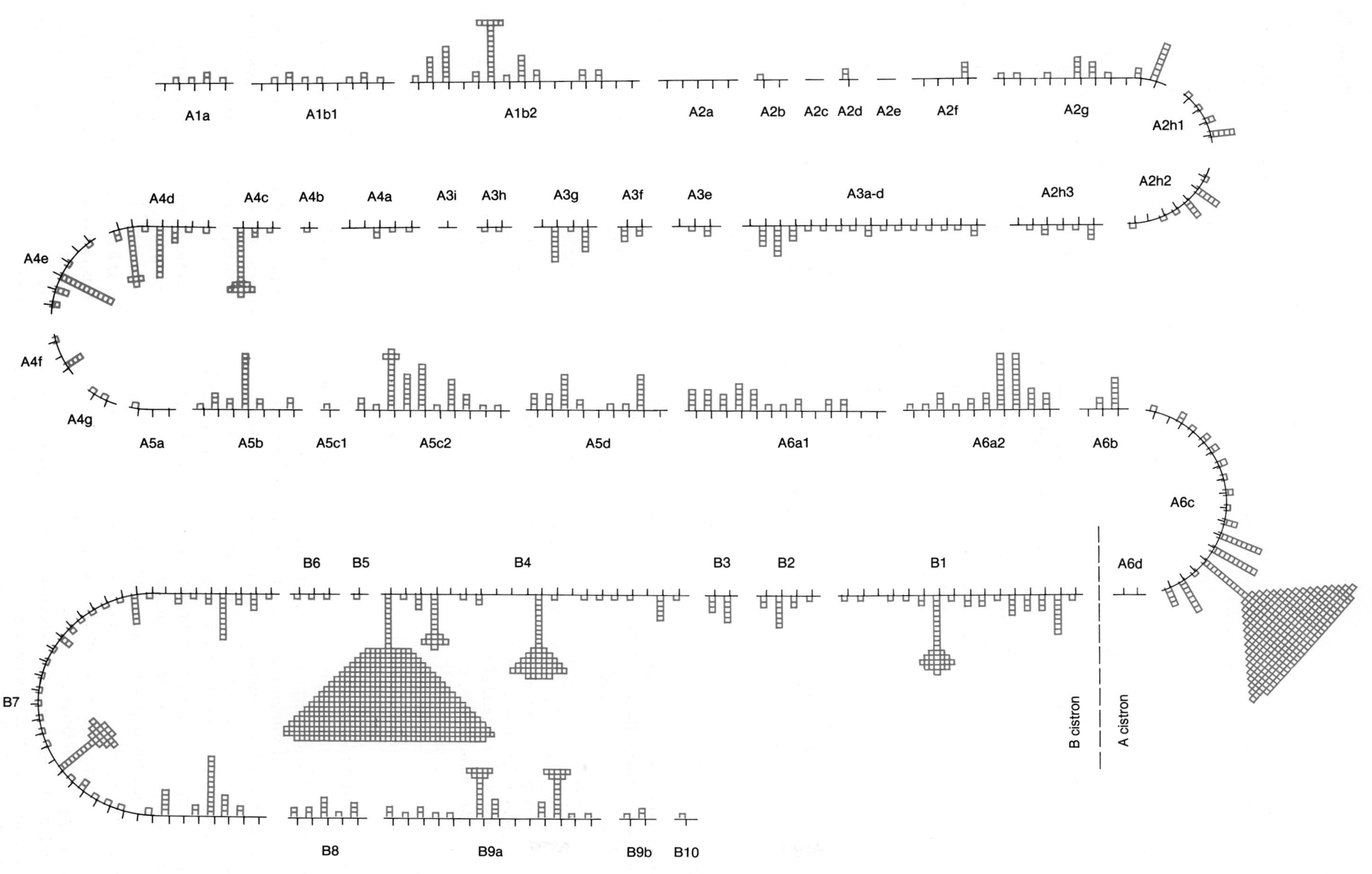

Figure 12-32 The distribution of 1612 spontaneous mutations in the *rII* locus. Each occurrence of an independent mutation is depicted by a square. When mutations are identical, the squares are drawn on top of one another. Although many positions are represented by only a single square, others have a large number of squares; these sites are called "hot spots." The two subdivisions of the *rII* locus, *A* and *B*, are indicated in this diagram.

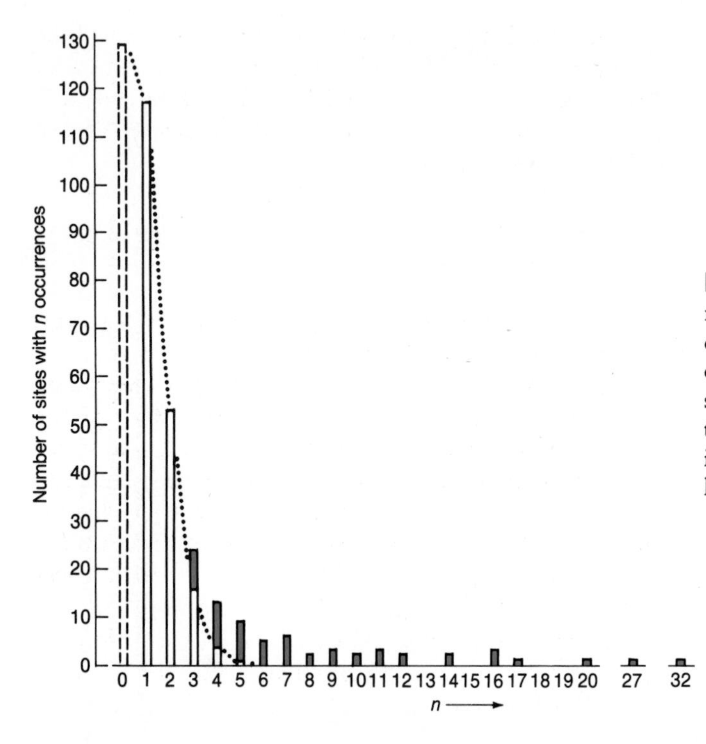

Figure 12-33 Distribution of occurrences of spontaneous mutations at various sites. The dotted curve indicates a Poisson distribution fitted to the number of sites having one and two occurrences. It predicts a minimum estimate for the number of sites of comparable mutability that have zero occurrences due to chance (dashed column at $n = 0$). Blue bars indicate the minimum numbers of sites that have mutation rates significantly higher than the one- and two-occurrence class.

over 500 repeated occurrences in the collection of 1612 mutations. By examining a Poisson distribution calculated to fit the number of sites having only one or two occurrences (Figure 12-33), Benzer could show that at least 60 sites were truly more mutable than those with only one or two occurrences. Mutations were not observed at all (by chance) in at least 129 sites in this collection, though testing with mutagens later proved these sites to be as mutable as those represented by one or two occurrences. When Benzer extended the analysis to include mutagen-induced mutations, results were similar to the spontaneous mutation crosses: There were also hot spots, although often different ones than the spontaneous hot spots, among the mutations generated by mutagenic agents (see Chapter 17).

The size of a mutational site was also of interest. The physical size of the *rII* region can be used to calculate the approximate number of base pairs in the region. From this figure, the number of mutational sites was determined to be approximately one-fifth of the number of nucleotide pairs. In other words, the smallest mutable site was five nucleotide pairs or less. The deciphering of the genetic code (Chapter 13), together with the demonstration by Ingram, Yanofsky, and others (described earlier in this chapter) that single amino acid substitutions resulted from single mutations, allowed Benzer to conclude that a mutation could result from the alteration of a single nucleotide pair. (The direct sequencing of DNA, described in Chapter 15, has since confirmed these conclusions in many examples.)

Message The gene can be divided into a linear array of mutable subelements that correspond to individual nucleotide pairs.

Destruction of the Bead Theory

Let's look again at the bead theory in light of Benzer's work. With the aid of deletion mapping, Benzer was able to map an extraordinary number of mutations in the *rII* locus against each other. His experiments have shown that mutations in the same gene can indeed recombine with one another. This result contradicts the bead theory of classical genetics, which held that recombination could occur between genes, but not within genes.

Benzer's analysis of the fine structure of the gene demonstrated that each gene consists of a linear array of subelements, and that these sites within a gene can be altered by mutation and can undergo recombination. This finding also contradicts the bead theory, one tenet of which implies that only the gene as a whole is mutable, not parts of the gene.

Subsequent work by several investigators identified each genetic site as a base pair in double-stranded DNA. Therefore, Benzer's contribution bridged the gulf between classical genetics and the knowledge of the chemical structure of DNA revealed by Watson and Crick. According to the bead theory, the Watson-Crick structure made no sense. However, Benzer's demonstration that genes do indeed have fine structures that can be

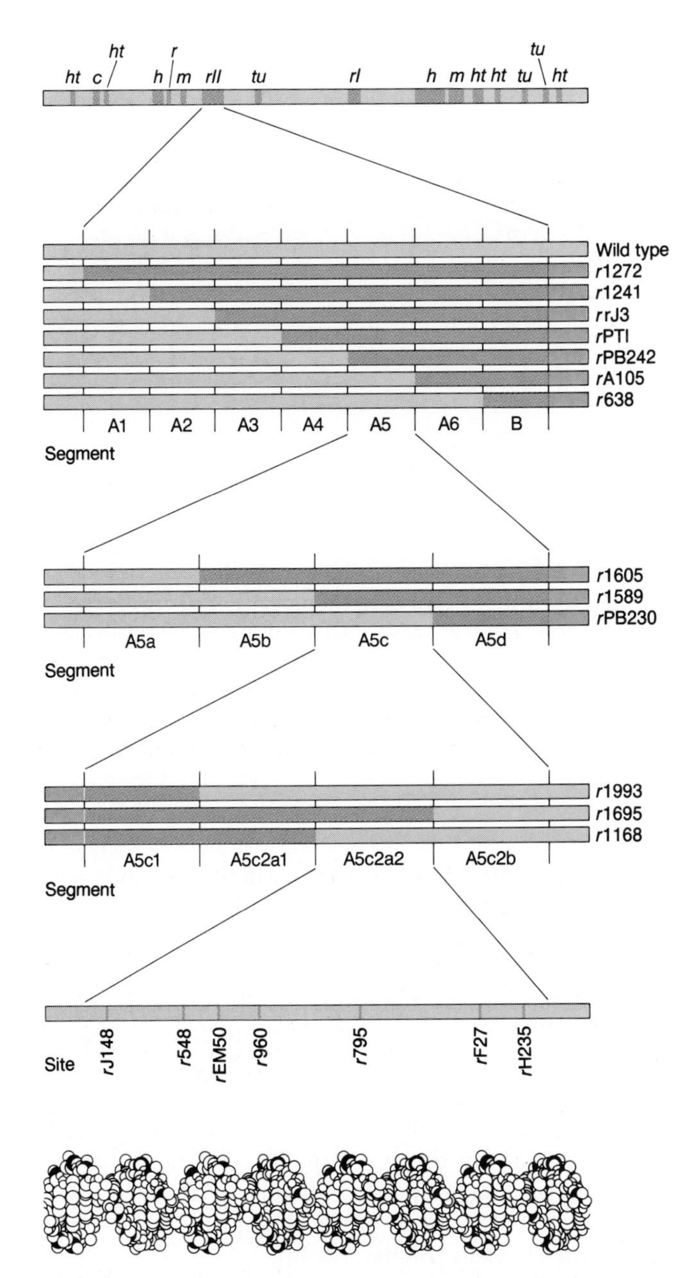

Figure 12-34 Fine-structure analysis of the *rII* locus. This mapping technique localizes the position of a given mutation in progressively smaller segments of the DNA molecule contained in phage T4. The *rII* region represents only a few percent of the entire molecule *(top)*. The mapping is done by crossing an unknown mutant with reference mutants having deletions (color) of known extent in the *rII* region. Each site represents the smallest mutable unit in the DNA molecule, a single base pair. The molecular segment *(extreme bottom)*, estimated to be roughly in proper scale, contains a total of about 40 base pairs. (From S. Benzer, "The Fine Structure of the Gene." Copyright 1962 by Scientific American, Inc. All rights reserved.)

revealed solely by genetic analysis allowed a fusion of the two disciplines and helped to launch the modern era of molecular genetics. Figure 12-34 illustrates the fine-structure analysis of the *rII* locus and its correspondence with the DNA structure.

Complementation

In another part of his studies, Benzer carried out a series of experiments designed to define the gene in terms of function. Benzer worked out the concept of **complementation.** Although complementation often appears to be a foreign concept, so-called complementation for normally haploid organisms simply represents a situation that we have been dealing with routinely in the genetic analysis of diploids. Complementation is often confused with recombination: we will consider the differences between them a little later.

Before we describe Benzer's complementation experiments with *rII* mutants, let's first review complementation in diploids.

Complementation in Diploids

Suppose that we are studying a phenotype in a diploid cell that requires the active product of two genes. As an example, let's use the situation for flower color in peas that was detailed in Chapter 4, in which purple flower color is dependent on two loci: *A* and *B*. Let's say that mutation in *A* or *B* leads to the loss of purple color and results in white flowers, and that the wild-type allele is dominant in each case. Thus, *A* is dominant to *a* and *B* is dominant to *b*. The phenotypes resulting from each combination of genes are as follows:

Genotype	Color
A A B B	purple
A a B B	purple
A A B b	purple
A a B b	purple
a a B B	white
a a B b	white
a a b b	white
A A b b	white
A a b b	white

What happens when we cross an *A A b b* plant against an *a a B B* plant? Both of these plants are white. (Can you see why? Recall that *both A* and *B* are required for purple flowers.) However, the cross will yield offspring that are *A a B b*, and, since both *A* and *B* are dominant to *a* and *b*, respectively, all progeny from this cross will produce purple flowers.

Let's see why this is so in terms of cellular biosynthetic pathways.

Because each parent is homozygous at each locus, we know that all gametes from one parent are *A b* and that all gametes from the other parent are *a B*. The zygote resulting from the union of these two gametes, and all cells descending from this zygote, will complete the two-step cellular pathway that allows the production of purple flower pigment, because each of the two relevant chromosomes supplies one of the functional genes. We can diagram this as follows, where (1) is the starting compound in the cellular pathway, (2) is an intermediate compound, (3) is the final product for purple pigment, and A and B are the enzymes produced by genes *A* and *B*:

Chromosome set 1 ├── *A* ──┤ ··· ├── *b* ──┤

$$\text{Pathway} \quad (1) \xrightarrow{\text{enzyme A}} (2) \xrightarrow{\text{enzyme B}} (3)$$

Chromosome set 2 ├── *a* ──┤ ··· ├── *B* ──┤

Note that one chromosome set (from one parent) supplies enzyme A but not enzyme B and that the second chromosome set (from the other parent) supplies enzyme B but not enzyme A. Since only one "good" copy of the relevant gene is required to produce each enzyme in this pathway, the two partially defective chromosome sets compensate for their respective deficiencies; that is, they help one another out, or **complement** one another.

Now let's look at complementation in haploids.

Complementation in Bacteriophage T4

Benzer wanted to find out whether the entire *rII* region of phage T4 acts as a single functional unit, or whether it is made up of subunits that function independently. Therefore, he tested the mutations that he had mapped in the *rII* region to see whether various pairwise combinations of the mutations would restore the wild-type phenotype. In other words, Benzer looked for complementation in *E. coli* host cells that were temporarily "diploid" for the T4 chromosome. To do this, he carried out a mixed infection with different *rII* mutants (Figure 12-35). His criterion for the wild-type phenotype was the ability to lyse *E. coli* K hosts. (Recall that *rII* mutants cannot do this but that wild-type phages can.)

Complementation tests like the one Benzer conducted are carried out in one cycle of infection; they do not involve the multiple cycles of reinfection required for plaque formation. Samples of the two phages to be tested are spread over a strip of host bacteria on a section of a petri plate at a high ratio of phages to bacteria in order to ensure that essentially every bacterium is infected with both phages. After a period of incubation, the growth or the absence of growth of the bacteria in the strip indicates whether or not the bacteria have lysed as a result of the phage infection.

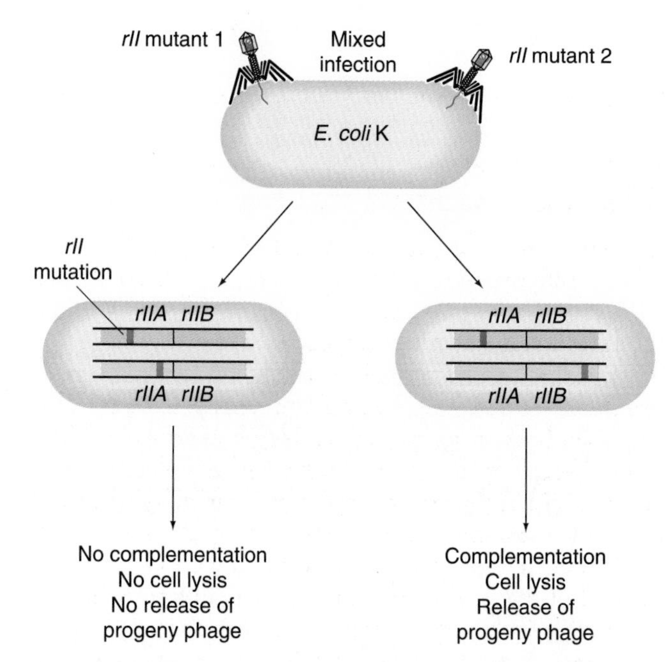

Figure 12-35 Complementation test: A schematic view of *rII* complementation. Two different mutants of *rII* are used to simultaneously infect *E. coli* K (mixed infection). Normally, an *rII* mutant cannot lyse *E. coli* K or generate progeny phages. However, if the two different mutants can complement one another, then lysis and phage growth will result. If the two *rII* mutants cannot complement one another, then no lysis or phage growth will result.

Pairwise tests of many different mutants allowed Benzer to separate the mutations into two groups, labeled A and B. All mutations in the A group complemented those in the B group, whereas no mutations in the A group complemented any other mutations in the A group and no mutations in the B group complemented any other mutations in the B group. Benzer found that all mutations in group A mapped in one half of the *rII* locus and that all mutations in group B mapped in the other half of the *rII* locus (Figure 12-36).

The following diagram depicts a model for the *rII* locus based on Benzer's results:

Genes (phage 1) ├── $rIIA^+$ ──┼── $rIIB^-$ ──┤

$$\text{Pathway} \quad (1) \xrightarrow{\text{enzyme A}} (2) \xrightarrow{\text{enzyme B}} \text{lysis}$$

Genes (phage 2) ├── $rIIA^-$ ──┼── $rIIB^+$ ──┤

In our model, we assume that two genes, termed *rIIA* and *rIIB*, are involved in lysis. We can envision two steps, (1) and (2), each controlled by an enzyme specified by one of

Figure 12-36 Gene map of *rII*.

the wild-type *rII* genes. Thus, $rIIA^+$ and $rIIB^+$ would specify enzymes A and B, respectively, while $rIIA^-$ and $rIIB^-$ are two different *rII* mutants with defects in either the *rII* A or *rII* B region, respectively.

Because each phage chromosome can program the synthesis of one of the required functions for the cellular pathway, lysis can occur, which is the wild-type phenotype. Therefore, the two phages can complement one another. Notice how this situation is virtually identical to the example given previously for purple flower color in plants. The only difference is that the diploid cells in the case of the plant persist and remain diploid, whereas the double-infection experiment in T4 phage creates a temporary diploid-like cell for the phage chromosome. During the existence of this "diploid," complementation can occur. We should also note how similar complementation with T4 phage in *E. coli* is to complementation in *Neurospora* heterokaryons, described on page 350.

Recombination and Complementation

It is important to understand the difference between recombination and complementation. Recombination represents the creation of new combinations of genes through the physical breakage and rejoining of chromosomes. The progeny from a cross in which recombination has occurred have *new genotypes* that are different from the parental genotypes. Complementation, on the other hand, does not involve any change in the genotypes of individual chromosomes; rather, it represents the *mixing of gene products.* Complementation occurs during the time that two chromosomes are in the same cell and can each supply a function. Afterward, each respective chromosome remains unaltered. In the case of *rII* mutants, complementation occurs when two different phage chromosomes, with mutations in different *rII* genes, are in the same host cell. However, progeny that result from this complementation carry only the parental genotypes. Figure 12-37 diagrams some of the differences between these two basic processes. The arrangement shown in Figure 12-37a will allow phage growth and lysis (complementation) without a rearrangement of phage chromosomes, whereas the mutant pair shown in Figure 12-37c will not. Recombination can occur in each situation, as shown in parts b and d of the figure.

How can we distinguish operationally between complementation and recombination? To distinguish them definitively, we could conduct an additional genetic analysis: in the case of *rII* mutants, we could analyze the genotypes of the progeny phages. To do this, we first perform a mixed infection of *E. coli* K with the two phages we are testing. We then plate the progeny from this mixed infection on both *E. coli* B and *E. coli* K (Figure 12-38). All phages can grow on B, but only *rII* recombinants can form plaques on K.

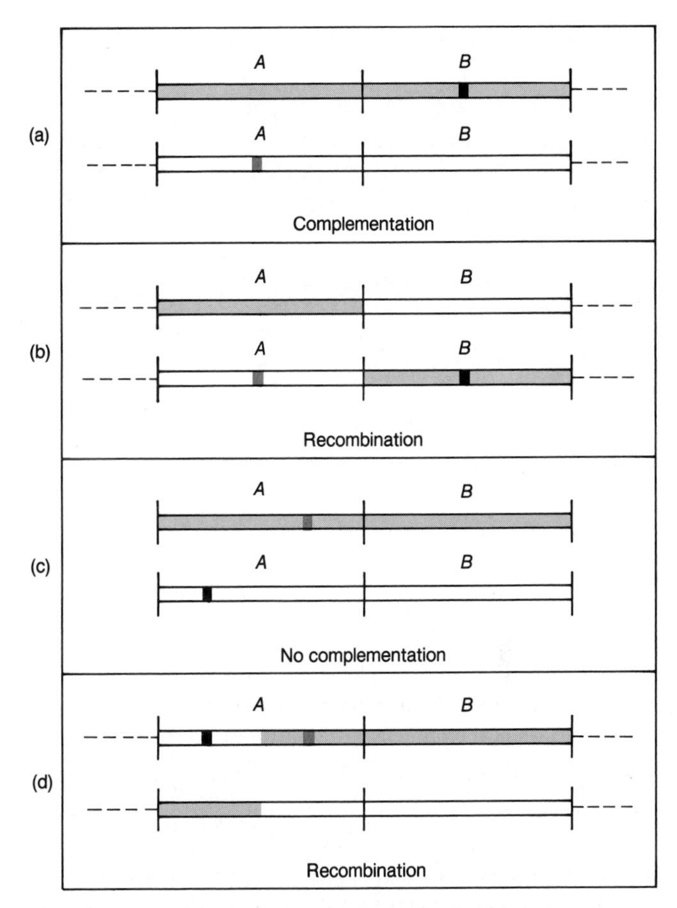

Figure 12-37 A comparison of the genetic consequences of complementation and recombination. (a) The mutant pairs can complement one another, since wild-type gene *A* and gene *B* products can mix in the cytoplasm. (b) If recombination occurs, then a rearrangement of the genomes will take place. (c) The mutant pairs cannot complement one another because neither mutant can contribute a wild-type gene *A* product. (d) A relatively rare recombination event could result in a wild-type phage.

Two important details allow our test to work. First, in the case of *rII* mutants, the incidence of recombination is never more than a few percent (see Figure 12-29), and it is rarely that high. Therefore, lysis will occur in only a very small fraction of the infected cells and will not interfere with the interpretation of the test. Second, in our second step, plating the progeny phages, the ratio of phages to bacteria is kept low, so that each bacterium is infected initially by no more than one phage; complementation cannot result in plaque formation in this situation.

Let's look at our tests results. If the progeny phages from the mixed infection of *E. coli* K result from comple-

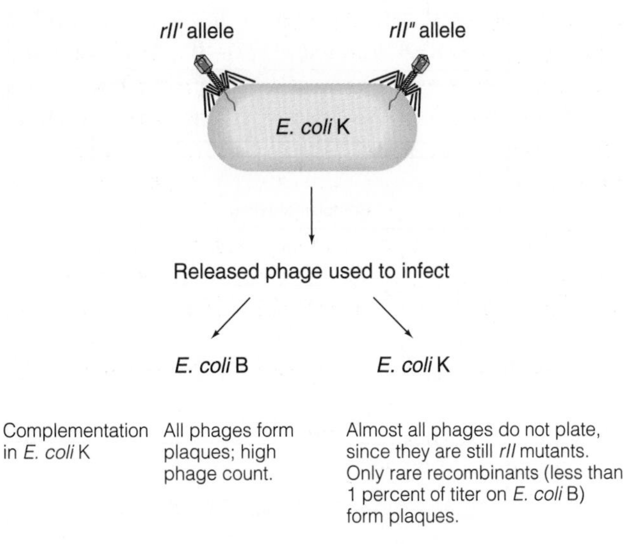

rII′ allele
rII″ allele
E. coli K
Released phage used to infect
E. coli B
E. coli K
Complementation in E. coli K
All phages form plaques; high phage count.
Almost all phages do not plate, since they are still rII mutants. Only rare recombinants (less than 1 percent of titer on E. coli B) form plaques.
Recombination in E. coli K
All phages form plaques; very low phage count.
Approximately 50 percent of all phages that plate on E. coli B are wild-type recombinants and will form plaques on E. coli K.

Figure 12-38 Analysis of phages resulting from the mixed infection of *E. coli* K. If complementation occurs between two *rII* mutants, then the progeny phages that are released will still be principally *rII* mutants and will fail to plate on *E. coli* K in single-infection experiments. If recombination but no complementation occurs, then there will be a sharply reduced yield of progeny phages due to the rarity of recombination. However, the few resulting phages will consist of approximately 50 percent wild-type recombinants and will form plaques on *E. coli* K in single-infection experiments.

mentation, then virtually all of the phages will still be *rII* mutants and will not plate on K. However, if recombination is required for phage growth and lysis during the mixed infection, then the analysis of the very low titer of progeny phages will show that about one-half are rII^+ recombinants and will plate on K (Figure 12-38).

The Cistron

Mutations that fail to complement one another must affect the same functional unit; mutations that do complement one another must affect different functional units. Benzer called this unit of genetic function a **cistron.**

The cistron gets its name from the ***cis-trans* test,** which is the term Benzer used to describe a complete complementation test involving mutational sites on the same chromosome *(cis)* and on opposite chromosomes *(trans)*, as shown in Figure 12-39. The *cis* portion of the test is really a control; the *trans* portion is the actual test for complementation. In summary, the *cis-trans* (or complementation) test is performed to determine whether two mutational sites are located within the same functional unit or in different functional units.

> Message **A cistron** is a genetic region within which there is normally no complementation between mutations. The cistron is equivalent to the gene.

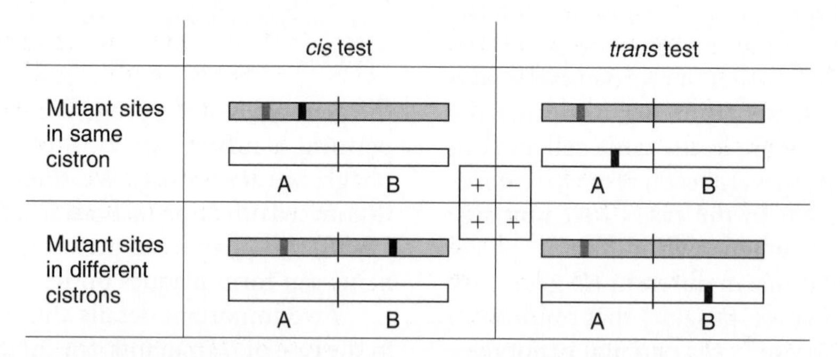

cis test
trans test
Mutant sites in same cistron
Mutant sites in different cistrons
A
B
+
−
+
+

Figure 12-39 The *cis-trans* test. This test uses complementation to show whether two mutational sites are located within the same functional unit (cistron) in a gene or in different functional units: Mutations that lie within the same cistron cannot complement one another. In the *trans* step, an *E. coli* K cell is infected with two different *rII* mutants, so that the mutations are on different chromosomes. If progeny phages do not grow, complementation has not occurred and the mutants are in the same functional unit, or cistron *(upper right box)*. If progeny phages do grow, complementation has occurred and the mutants are in different cistrons *(lower right box)*. In the *cis* test, used as a control, an *E. coli* K cell is infected with a T4 double mutant and a wild-type T4, so that both mutations are on a single chromosome *(upper and lower left boxes)*. A + represents a successful complementation; a − indicates no complementation.

Complementation and the Concept of the Gene

Of the three tenets defining a gene according to the classical (bead) theory, the one aspect that has held up and seems most essential is the gene as a unit of function. We now see that the gene is equivalent to the cistron, which we can consider to be a functional unit that can be defined experimentally by a *cis-trans* complementation test. There are occasional exceptions to this operational definition expressed by the *cis-trans* test. For example, some genes encode polypeptides that form multimeric proteins; in rare cases these display **intragenic (intracistronic) complementation,** in which individual monomers of the protein are inactive but they combine to form the active, functional multimer. Also, some mutations affect the expression of more than one gene, and complementation would fail to occur with mutations in these genes. However, these exceptions do not upset the basic concept of the gene as a unit of function.

We now know that a cistron is a region of the genetic material that codes for one polypeptide chain. Therefore, the one-gene–one-enzyme hypothesis could be referred to more precisely as the **one-cistron–one-polypeptide hypothesis,** thereby also emphasizing that cistrons, or genes, can code for proteins other than enzymes.

Summary

The work of Beadle and Tatum in the 1940s showed that one gene codes for one protein, often an enzyme. When an enzyme fails to function normally due to a mutation, a variant phenotype results. These variant phenotypes are often the basis of genetic diseases in any organism, including humans. In order to understand how abnormal enzymes can cause phenotypic change, we need to understand the structure of proteins. Composed of a specific linear sequence of amino acids connected through peptide bonds, the proteins assume specific three-dimensional shapes as a result of the interaction of the 20 amino acids that in different combinations constitute the polypeptide chain. Different areas of this folded chain are sites for the attachment and interaction of substrates. Furthermore, many functional enzymes and other proteins are built by combining several polypeptide chains into a multimeric form.

Specific amino acid changes can be detected in a protein by the technique of fingerprinting and amino acid sequencing. Work of this type by Sanger, Ingram, Yanofsky, and others has demonstrated colinearity between a mutational site on the genetic map and the position of an altered amino acid in a protein.

Further work by Benzer and others illustrated that the gene could be dissected into smaller and smaller pieces. A structural unit of mutation and one of recombination were identified and equated with a single nucleotide pair. A cistron is defined at the phenotypic level as a genetic region within which there is no complementation between mutations. This is the unit that codes for the structure of a single functional polypeptide. Thus, the gene is equivalent to the cistron.

In the early days of genetics, genes were represented as indivisible beads on a chain. We have now arrived at a far different picture of the gene. Multiple intragenic mutational sites exist, and recombination may occur anywhere within a gene. In addition, a closer connection between genotype and phenotype was realized when it was established that one cistron is responsible for the synthesis of one polypeptide.

Concept Map

Draw a concept map interrelating as many of the following terms as possible. Note that the terms are listed in no particular order.

allele / mutation / biosynthetic pathway / enzyme / catalysis / auxotroph / dominant / recessive / complementation / cistron / deletion / peptide bond / polypeptide / amino acid sequence / recombination

Chapter Integration Problem

We saw in Chapter 4 that the wild-type hemoglobin allele, Hb^A, displayed dominance to Hb^S (the allele of sickle-cell anemia) with regard to anemia, but codominance with regard to the electrophoretic properties of hemoglobin. Explain the different types of dominance in terms of protein sequence, based on the knowledge of protein structure learned in Chapter 12.

Solution

The protein hemoglobin has two α and two β polypeptide chains, closely fitted together in an elaborate quaternary structure which holds the heme groups that carry oxygen to the tissues. The primary structure of a protein — the linear sequence of its component amino acids — determines the protein's ultimate tertiary and quaternary structures, and hence its activity. Because of the change from glutamic acid to valine at position 6 in the β-chain, the hemoglobin of sickle-cell anemia (HbS) is changed structurally, so that it carries less oxygen than normal adult hemoglobin (HbA). Heterogenotes ($Hb^A\ Hb^S$) still maintain sufficient oxygen-carrying capacity to avoid severe symptoms of anemia; hence, Hb^A shows dominance to Hb^S with regard to anemia. The substitution of valine for glutamic acid also changes the electric charge of the molecule, so that HbS moves more slowly than HbA on gel electrophoresis (see Figure 4-4). Thus, the two hemoglobin molecules (HbA and HbS) show codominance with respect to electrophoretic properties, because they both can be resolved and detected by this technique.

Solved Problems

1. Various pairs of *rII* mutants of phage T4 are tested in *E. coli* in both the *cis* and *trans* positions, and the "burst size" (the average number of phage particles produced per bacterium) for each test pair is compared. Results for six different *r* mutants—*rM*, *rN*, *rO*, *rP*, and *rS*—are as follows (+ indicates the wild-type allele).

Cis genotype	Burst size	*Trans* genotype	Burst size
rM rN/++	245	*rM* +/+ *rN*	250
rO rP/++	256	*rO* +/+ *rP*	268
rR rS/++	248	*rR* +/+ *rS*	242
rM rO/++	270	*rM* +/+ *rO*	0
rM rP/++	255	*rM* +/+ *rP*	255
rM rR/++	264	*rM* +/+ *rR*	0
rM rS/++	240	*rM* +/+ *rS*	240
rN rO/++	257	*rN* +/+ *rO*	268
rN rP/++	250	*rN* +/+ *rP*	0
rN rR/++	245	*rN* +/+ *rR*	255
rN rS/++	259	*rN* +/+ *rS*	0
rP rR/++	260	*rP* +/+ *rR*	245
rP rS/++	253	*rP* +/+ *rS*	0

If we assign the mutation *rO* to the A cistron, what are the locations of the other five mutations with respect to the A and B cistrons? (NOTE: This problem is similar to Problem 14 in the following problem set.)

Solution

In solving problems like this, we first look for the underlying principle and then attempt to draw a diagram to help us work out the solution. The key principle is that in the *trans* position, mutations in the same cistron will not complement one another and thus will yield no progeny phage (no burst), whereas mutations in different cistrons will complement and thus will yield progeny phages of normal burst size. Since *rO* is in the A cistron, we start with the initial diagram

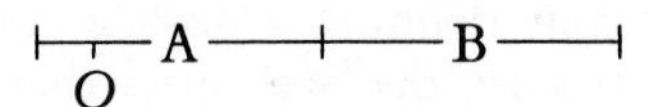

We can now look at the test results in the problem and assign each mutation to a cistron based on whether or not complementation occurs in a mixed infection with *rO* mutants. From the table, it is evident that *N* and *P* complement *O* and must be in a different cistron (the B cistron), whereas *M* does not complement *O* and must be in the same cistron (the A cistron). This leaves us with *R* and *S*, which can be assigned to the A and B cistrons, respectively, based on their complementation tests with *M*, *N*, and *P*. Therefore, we have the arrangement

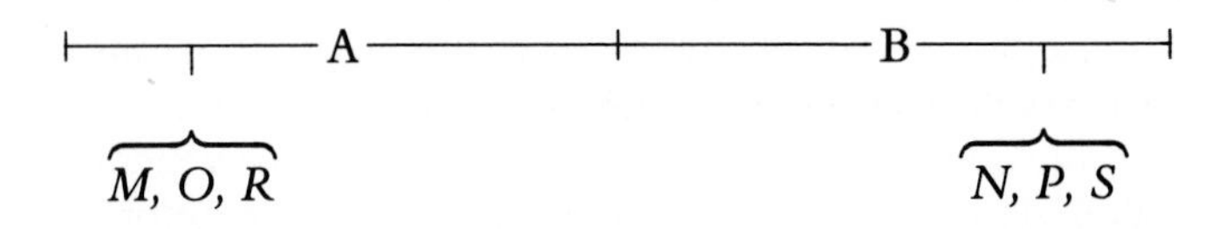

We can check this assignment by examining the other crosses to find out if the results are consistent with our answer.

2. The following deletion map shows four deletions (1 to 4) involving the *rIIA* cistron of phage T4:

1 ————————
2 ——————————
3 ————————————
4 ——————————————

Five point mutations (*a* to *e*) are tested against these four deletion mutants for their ability (+) or inability (−) to give wild-type (r^+) recombinants; the results are

	a	*b*	*c*	*d*	*e*
1	+	+	+	+	+
2	+	+	+	−	−
3	+	−	+	−	−
4	−	−	+	−	−

What is the order of the point mutations?

Solution

The key principle here is that point mutations can recombine with deletions that *do not* extend past the mutation, but they cannot recombine to yield wild-type phages with deletions that *do* extend past the mutation.

For the test results given in the problem, any mutation that recombines with deletion 1 must be to the right of the deletion, any mutation that recombines with deletion 2 must be to the right of deletion 2, and so on. Let's look at point mutation *a*. It recombines with deletions 1, 2, and 3 but not with deletion 4. Therefore, it is to the right of deletions 1, 2, and 3 but not to the right of deletion 4. We can therefore place point mutation *a* in the interval between deletions 3 and 4. Point Mutation *b* recombines with deletions 1 and 2 and must be to the right of them. It does not recombine with deletions 3 and 4, so it is in the interval between deletion 2 and 3. Point mutation *c* recombines with all the deletions and is to the right of all deletions, even deletion 4. Finally, both point mutations *d* and *e* recombine only with deletion 1 and must therefore be in the interval between deletions 1 and 2. The solution we have just derived can be summarized as follows:

```
1 ______________ e,d
2 _________________ b
3 ______________________ a
4 __________________________ c
```

Problem 18 in the following problem set is similar to this sample problem. See if you can apply the reasoning set forth here when you solve Problem 18.

Problems

1. A common weed, Saint-John's-wort, is toxic to albino animals. It also causes blisters on animals that have white areas of fur. Suggest a possible genetic basis for this reaction.

2. In humans, the disease galactosemia causes mental retardation at an early age because lactose in milk cannot be broken down, and this failure affects brain function. How would you provide a secondary cure for galactosemia? Would you expect this phenotype to be dominant or recessive?

3. Amniocentesis is a technique in which a hypodermic needle is inserted through the abdominal wall of a pregnant woman and into the amnion, the sac that surrounds the developing embryo, in order to withdraw a small amount of amniotic fluid. This fluid contains cells that come from the embryo (not from the woman). The cells can be cultured; they will divide and grow to form a population of cells on which enzyme analyses and karyotypic analyses can be performed. Of what use would this technique be to a genetic counselor? Name at least three specific conditions under which amniocentesis might be useful (NOTE: This technique involves a small but real risk to the health of both the woman and the embryo; take this fact into account in your answer.)

4. Table 12-6 shows the ranges of enzymatic activity (in units we need not worry about) observed for enzymes involved in two recessive metabolic diseases in humans. Similar information is available for many metabolic genetic diseases.

a. Of what use is such information to a genetic counselor?

b. Indicate any possible sources of ambiguity in interpreting studies of an individual patient.

c. Reevaluate the concept of dominance in the light of such data.

5. Two albinos marry and have a normal child. How is this possible? Suggest at least two ways. (This question appeared first in Chapter 4. Reconsider it now in light of biosynthetic pathways.)

6. In humans, PKU (phenylketonuria) is a disease caused by an enzyme inefficiency at step A in the following simplified reaction sequence, and AKU (alkaptonuria) is due to an enzyme inefficiency in one of the steps summarized as step B here:

$$\text{phenylalanine} \xrightarrow{A} \text{tyrosine} \xrightarrow{B} CO_2 + H_2O$$

A person with PKU marries a person with AKU. What phenotypes do you expect for their children?

Table 12-6.

		Range of enzyme activity		
Disease	Enzyme involved	Patients	Parents of patients	Normal individuals
Acatalasia	Catalase	0	1.2–2.7	4.3–6.2
Galactosemia	Gal-1-P uridyl transferase	0–6	9–30	25–40

(**a**) All normal; (**b**) all having PKU only; (**c**) all having AKU only; (**d**) all having both PKU and AKU; (**e**) some having AKU and some having PKU.

7. Three independently isolated tryptophan-requiring strains of yeast are called trpB, trpD, and trpE. Cell suspensions of each are streaked on a plate supplemented with just enough tryptophan to permit weak growth for a trp^- strain. The streaks are arranged in a triangular pattern so that they do not touch one another. Luxuriant growth is noted at both ends of the trpE streak and at one end of the trpD streak.

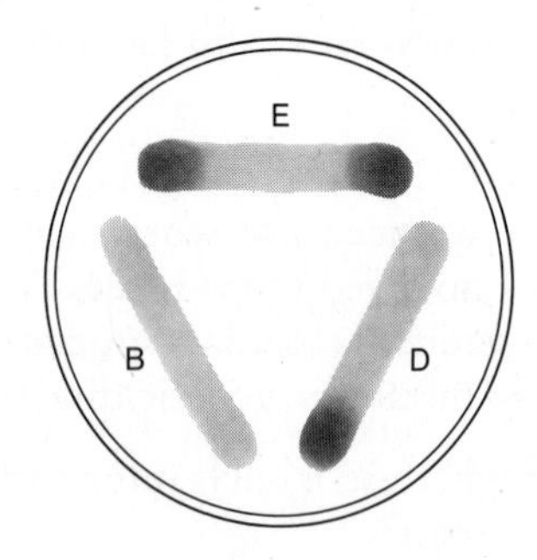

***a.** Do you think complementation is involved?

b. Briefly explain the patterns of luxuriant growth.

c. In what order in the tryptophan-synthesizing pathway are the enzymatic steps defective in trpB, trpD, and trpE?

d. Why was it necessary to add a small amount of tryptophan to the medium in order to demonstrate such a growth pattern?

8. In *Drosophila* pupae, certain structures called imaginal disks can be detected as thickenings of the skin; after metamorphosis, these imaginal disks develop into specific organs of the adult fly. George Beadle and Boris Ephrussi devised a means of transplanting eye imaginal disks from one larva into another larval host. When the host metamorphoses into an adult, the transplant can be found as a colored eye located in its abdomen. The researchers took two strains of flies that were phenotypically identical in terms of bright scarlet eyes: one due to the sex-linked mutant vermilion (v); the other due to cinnabar (cn) on chromosome 2. If v disks are transplanted into v hosts or cn disks into cn hosts, then the transplants develop as mutant scarlet eyes. Transplanted cn or v disks in wild-type hosts develop wild-type eye colors. A cn disk in a v host develops a mutant eye color, but a v disk in a cn host develops wild-type eye color. Explain these results, and outline the experiments you would propose to test your explanation.

9. In *Drosophila,* the autosomal recessive *bw* causes a dark brown eye and the unlinked autosomal recessive *st* causes a bright scarlet eye. A homozygote for both genes has a white eye. Thus, we have the following corespondences between genotypes and phenotypes:

$$st^+st^+\ bw^+bw^+ = \text{red eye (wild-type)}$$
$$st^+st^+\ bw\,bw = \text{brown eye}$$
$$st\,st\ bw^+bw^+ = \text{scarlet eye}$$
$$st\,st\ bw\,bw = \text{white eye}$$

Construct a hypothetical biosynthetic pathway showing how the gene products interact and why the different mutant combinations have different phenotypes.

10. Several mutants are isolated, all of which require compound G for growth. The compounds (A to E) in the biosynthetic pathway to G are known, but not their order in the pathway. Each compound is tested for its ability to support the growth of each mutant (1 to 5). In the following table, + indicates growth and − indicates no growth:

	Compound tested					
	A	B	C	D	E	G
Mutant 1	−	−	−	+	−	+
2	−	+	−	+	−	+
3	−	−	−	−	−	+
4	−	+	+	+	−	+
5	+	+	+	+	−	+

a. What is the order of compounds A to E in the pathway?

b. At which point in the pathway is each mutant blocked?

c. Would a heterokaryon composed of double mutants 1,3 and 2,4 grow on a minimal medium? 1,3 and 3,4? 1,2 and 2,4 and 1,4?

11. In *Neurospora* (a haploid), assume that two genes participate in the synthesis of valine. Their mutant alleles are *val-1* and *val-2*, and their wild-type alleles are $val\text{-}1^+$ and $val\text{-}2^+$. The two genes are linked on the same chromosome, and a crossover occurs between them, on average, in one of every two meioses.

a. In what proportion of meioses are there no crossovers between the genes?

b. Use the map function to determine the recombinant frequency between these two genes.

c. Progeny from the cross *val-1 val-2*$^+$ × *val-1*$^+$ *val-2* are plated on a medium containing no valine. What proportion of the progeny will grow?

d. The *val-1 val-2*$^+$ strains accumulate intermediate compound B, and the *val-1*$^+$ *val-2* strains accumulate intermediate compound A. The *val-1 val-2*$^+$ strains grow on valine or A, but the *val-1*$^+$ *val-2* strains grow only on valine and not on B. Show the pathway order of A and B in relation to valine, and indicate which gene controls each conversion.

12. In a certain plant, the flower petals are normally purple. Two recessive mutations arise in separate plants and are found to be on different chromosomes. Mutation 1 (m_1) gives blue petals when homozygous ($m_1 m_1$). Mutation 2 (m_2) gives red petals when homozygous ($m_2 m_2$). Biochemists working on the synthesis of flower pigments in this species have already described the following pathway:

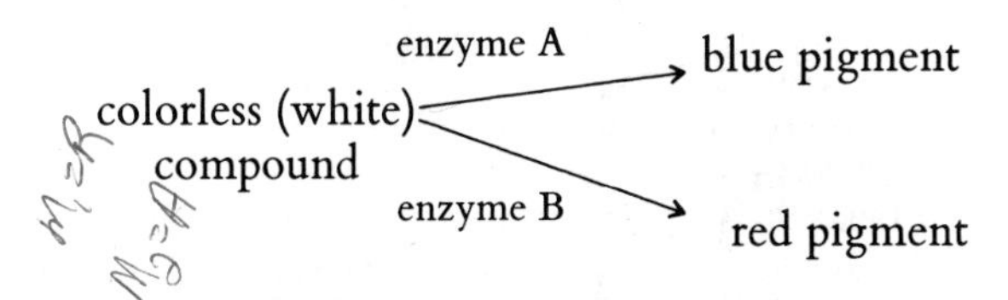

a. Which mutant would you expect to be deficient in enzyme A activity?

b. A plant has the genotype $M_1 m_1 M_2 m_2$. What would you expect its phenotype to be?

c. If the plant in (**b**) is selfed, what colors of progeny would you expect, and in what proportions?

d. Why are these mutants recessive?

13. In sweet peas, the synthesis of purple anthocyanin pigment in the petals is controlled by two genes, *B* and *D*. The pathway is

white intermediate —(gene *B* enzyme)→ blue intermediate —(gene *D* enzyme)→ anthocyanin (purple)

a. What color petals would you expect in a pure-breeding plant unable to catalyze the first reaction?

b. What color petals would you expect in a pure-breeding plant unable to catalyze the second reaction?

c. If the plants in (**a**) and (**b**) are crossed, what color petals will the F_1 plants have?

d. What ratio of purple:blue:white plants would you expect in the F_2?

14. Various pairs of *rII* mutants of phage T4 are tested in *E. coli* in both the *cis* and *trans* positions. Comparisons are made of the "burst size" (the average number of phage particles produced per bacterium). Results for six different *r* mutants—*rU*, *rV*, *rW*, *rX*, *rY*, and *rZ*—are as follows:

Cis genotype	Burst size	*Trans* genotype	Burst size
rU rV/+ +	250	*rU* +/+ *rV*	258
rW rX/+ +	255	*rW* +/+ *rX*	252
rY rZ/+ +	245	*rY* +/+ *rZ*	0
rU rW/+ +	260	*rU* +/+ *rW*	250
rU rX/+ +	270	*rU* +/+ *rX*	0
rU rY/+ +	253	*rU* +/+ *rY*	0
rU rZ/+ +	250	*rU* +/+ *rZ*	0
rV rW/+ +	270	*rV* +/+ *rW*	0
rV rX/+ +	263	*rV* +/+ *rX*	270
rV rY/+ +	240	*rV* +/+ *rY*	250
rV rZ/+ +	274	*rV* +/+ *rZ*	260
rW rY/+ +	260	*rW* +/+ *rY*	240
rW rZ/+ +	250	*rW* +/+ *rZ*	255

If we assign *rV* to the A cistron, what are the locations of the other five *rII* mutations with respect to the A and B cistrons?

15. There is evidence that occasionally during meiosis either one or both homologous centromeres will divide and segregate precociously at the first division rather than at the second division (as is the normal situation). In *Neurospora*, *pan2* alleles produce a pale ascospore, aborted ascospores are completely colorless, and normal ascospores are black. In a cross between two complementing alleles *pan2x* × *pan2y*, what ratios of black:pale:colorless would you expect in asci resulting from the precocious division of (**a**) one centromere? (**b**) both centromeres? (Assume that *pan2* is near the centromere.)

Table 12-7.

	1	2	3	4	5	6	7	8	9	10	11	12	13	14
1	−	+	+	+	−	+	+	−	−	+	+	+	+	−
2	+	−	−	−	+	+	+	+	+	+	+	−	+	−
3	+	−	−	−	+	+	+	+	+	+	+	−	+	−
4	+	−	−	−	+	+	+	+	+	+	+	−	+	−
5	−	+	+	+	−	+	+	−	−	+	+	+	+	−
6	+	+	+	+	+	−	−	+	+	−	−	+	−	−
7	+	+	+	+	+	−	−	+	+	−	−	+	−	−
8	−	+	+	+	−	+	+	−	−	+	+	+	+	−
9	−	+	+	+	−	+	+	−	−	+	+	+	+	−
10	+	+	+	+	+	−	−	+	+	−	−	+	−	−
11	+	+	+	+	+	−	−	+	+	−	−	+	−	−
12	+	−	−	−	+	+	+	+	+	+	+	−	+	−
13	+	+	+	+	+	−	−	+	+	−	−	+	−	−
14	−	−	−	−	−	−	−	−	−	−	−	−	−	−

16. *Protozoon mirabilis* is a hypothetical single-celled haploid green alga that orients to light by means of a red eye-spot. By selecting cells that do not move toward the light, 14 white-eye-spot mutants (eye^-) are isolated after mutation. It is possible to fuse haploid cells to make diploid individuals. The 14 eye^- mutants are paired in all combinations, and the color of the eye-spot is scored in each. Table 12-7 shows the results, where + indicates a red eye-spot and − indicates a white eye-spot.

a. Mutant 14 obviously is different from the rest. Why might this be?

b. Excluding mutant 14, how many complementation groups are there, and which mutants are in which group?

c. Three crosses are made, with the following results:

Mutants Crossed	Number of Progeny eye^+	eye^-	Total
1 × 2	31	89	120
2 × 6	5	113	118
1 × 14	0	97	97

Explain these genetic ratios with symbols.

d. How many genetic loci are involved altogether, and which of the 14 mutants are at each locus?

e. What is the linkage arrangement of the loci? (Draw a map.)

17. You have the following map of the *rII* locus:

A cistron B cistron
rX *rY* *rZ* *rD* *rE* *rF*

You detect a new mutation, *rW*, and you find that it does not complement any of the mutants in the A or B cistron. You find that wild-type recombinants are obtained in crosses with *rX*, *rY*, *rE*, and *rF* but not with *rZ* or *rD*. Suggest possible explanations for these results. Describe tests you would use to choose between the explanations.

18. The following map shows four deletions (1 to 4) involving the *rIIA* cistron of phage T4:

1
2
3
4

Five point mutations (*a* to *e*) in *rIIA* are tested against these four deletion mutants for their ability (+) or inability (−) to give r^+ recombinants, with the following results:

	a	*b*	*c*	*d*	*e*
1	+	+	−	+	+
2	+	+	−	−	−
3	−	−	+	−	+
4	+	−	+	+	+

a. What is the order of the point mutants?

b. Another strain of T4 has a point mutation in the *rIIB* cistron. This strain is mixed in turn with each of the *rIIA* deletion mutants, and the mixtures are used to infect *E. coli* K at a multiplicity of infection great enough that each host cell will be infected by at least one *rIIA* and one *rIIB* mutant. A normal plaque is formed with deletions 1, 2, and 3, but no plaque forms with deletion 4. Given that the B cistron is to the right of A, explain the behavior of deletion 4. Does your explanation affect your answer to (a)?

19. In a phage, a set of deletions is intercrossed in pairwise combinations. The following results are obtained (a + indicates that wild-type recombinants are obtained from that cross):

	1	2	3	4	5
1	−	+	−	+	−
2	+	−	+	+	−
3	−	+	−	−	−
4	+	+	−	−	+
5	−	−	−	+	−

a. Construct a deletion map from this table.

b. The first geneticists to do a deletion-mapping analysis in the mythical schmoo-phage SH4 (which lyses schmoos) came up with this unique set of data:

	1	2	3	4
1	−	−	+	−
2	−	−	−	+
3	+	−	−	−
4	−	+	−	−

Show why this is a unique result by drawing the only deletion map that is compatible with this table. (Don't let your mind be shackled by conventional expectations.)

20. In a haploid eukaryote, four alleles of the *cys-2* gene are obtained. Each allele requires cysteine, and all alleles map to the same locus. The four strains bearing these mutant alleles are crossed to wild types to obtain a set of eight cultures representing the four mutant alleles in association with each mating type. Then the mutant alleles are intercrossed in all pairwise combinations. The haploid meiotic products from each cross are plated on a medium containing no cysteine. In some crosses, *cys*$^+$ prototrophs are observed at low frequencies. The results follow, where the numbers represent the frequencies of *cys*$^+$ colonies per 10^4 meiotic products plated:

		Mating type A′			
		1	2	3	4
Mating type A″	1	0	14	2	20
	2	14	0	12	6
	3	2	12	0	18
	4	20	6	18	0

a. Draw a map of the four mutant sites within the *cys-2* gene. Provide a measurement of the relative intersite distances.

b. Do you see any evidence that mutation might be involved in the production of the prototrophs?

***21.** In *Neurospora,* there is a gene controlling the production of adenine, and mutants in this gene are called *ad-3* mutants. The *his-2* locus is 2 m.u. to the left, and the *nic-2* locus is 3 m.u. to the right of the *ad-3* locus (*his-2* controls histidine, and *nic-2* controls nicotinamide). Thus, the genetic map is

his-2 —— 2 m.u. —— *ad-3* —— 3 m.u. —— *nic-2*

Three different *ad-3* auxotrophs are detected: *ad-3*^{a}, *ad-3*^{b}, and *ad-3*^{c}. (Use *a*, *b*, and *c* as their labels.) The following crosses are made:

Cross 1: *his-2*$^+$ *a nic-2*$^+$ × *his-2 b nic-2*

Cross 2: *his-2*$^+$ *a nic-2* × *his-2 c nic-2*$^+$

Cross 3: *his-2 b nic-2* × *his-2*$^+$ *c nic-2*$^+$

The ascospores are then plated on a minimal medium containing histidine and nicotinamide, and *ad-3*$^+$ prototrophs are picked up. Table 12-8 shows the results obtained. What is the map order of the *ad-3* mutants and the genetic distance between them?

Table 12-8.

Genotype of *ad-3*$^+$ recombinants	Number of *ad-3*$^+$ spores picked up		
	Cross 1	Cross 2	Cross 3
his-2 nic-2	0	6	0
his-2$^+$ *nic-2*$^+$	0	0	0
his-2 nic-2$^+$	15	0	5
his-2$^+$ *nic-2*	0	0	0
Total ascospores scored	41,236	38,421	43,600

***22.** The *Notch* locus of *Drosophila* is assumed to be a single cistron. William Welshons has recovered two classes of *Notch* mutants. Class I mutants show a dominant effect on the wings, bristles, and eyes, and the mutation acts as a recessive lethal. Class II mutations are all nonlethal, and all have recessive mutant phenotypes affecting eyes, bristles, or wings. Heterozygotes for a class I and a class II mutation are viable, but they exhibit the dominant phenotype of the class I mutant and the recessive phenotype of the class II mutant. Construct an explanation, and describe ways to test your model.

23. In a hypothetical diploid organism, squareness of cells is due to a threshold effect: more than 50 units of "square factor" per cell will produce a square phenotype and less than 50 units will produce a round phenotype. Allele s^f is a functional gene that causes the synthesis of the square factor. Each s^f allele contributes 40 units of square factor; thus, s^fs^f homozygotes have 80 units and are phenotypically square. A mutant allele (s^n) arises; it is nonfunctional, contributing no square factor at all.

a. Which allele will show dominance, s^f or s^n?

b. Are functional alleles necessarily always dominant? Explain your answer.

c. In a system such as this one, how might a specific allele become changed in evolution so that its phenotype shows recessive inheritance at generation 0 and dominant inheritance at a later generation?

24. Some genes in humans are known to have a multiple (or "pleiotropic") effect on a phenotype. Does this constitute an invalidation of the one-gene–one-enzyme hypothesis? Explain.

***25.** Consider the following three biochemical sequences in *Neurospora* (genes controlling the catalyzing enzymes are indicated for some reactions):

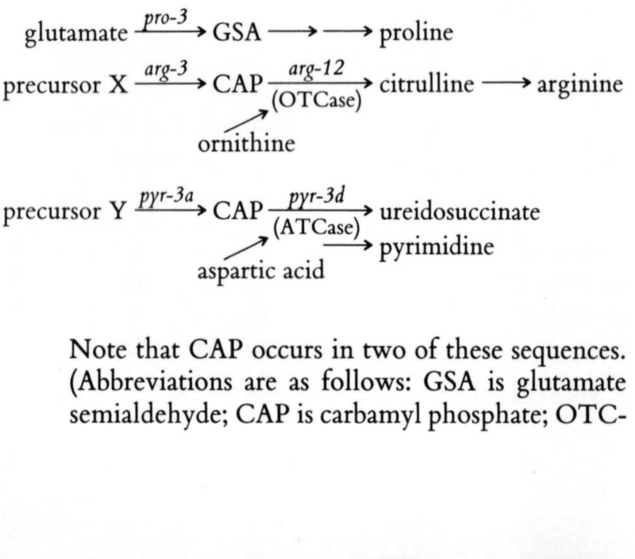

$$\text{glutamate} \xrightarrow{pro\text{-}3} \text{GSA} \longrightarrow \longrightarrow \text{proline}$$

$$\text{precursor X} \xrightarrow{arg\text{-}3} \text{CAP} \xrightarrow[\text{(OTCase)}]{arg\text{-}12} \text{citrulline} \longrightarrow \text{arginine}$$

ornithine ↗ (enters at the OTCase step)

$$\text{precursor Y} \xrightarrow{pyr\text{-}3a} \text{CAP} \xrightarrow[\text{(ATCase)}]{pyr\text{-}3d} \text{ureidosuccinate} \longrightarrow \text{pyrimidine}$$

aspartic acid ↗ (enters at the ATCase step)

Note that CAP occurs in two of these sequences. (Abbreviations are as follows: GSA is glutamate semialdehyde; CAP is carbamyl phosphate; OTCase is ornithine transcarbamoylase; ATCase is aspartate carbamoyl transferase.)

a. The *arg-3* mutants require arginine, and the *pyr-3a* mutants require pyrimidine. Because CAP is found in *both* sequences, what does this observation suggest?

b. The gene *pyr-3* seems to control two consecutive enzymic conversions. How does this fact fit with your answer to (**a**)? What can you suggest about the probable structure of the *pyr-3* enzyme?

c. The *arg-12* mutants partially suppress *pyr-3a* mutants, and the *pyr-3d* mutants partially suppress *arg-3* mutants. Is this consistent with your answer to (**a**)?

d. All *pro-3* mutants require proline, but an *arg-12 pro-3* genotype does not require proline; in this case the label added as ornithine ends up in GSA and proline. The conversion of ornithine to GSA is catalyzed by the enzyme ornithine transaminase (OTA). Why do you think *pro-3* single mutants do not have access to ornithine?

e. In rats, enzyme blocks corresponding to *pyr-3a* do *not* lead to pyrimidine requirement. Compare rats and *Neurospora* in this regard.

***26.** Explain how Benzer was able to calculate the number of sites with zero occurrences, as depicted in Figure 12-34.

27. In *Collinsia parviflora,* the petal color is normally purple. Four recessive mutations are induced, each of which produces white petals. The four pure-breeding lines are then intercrossed in the following combinations, with the results indicated:

Mutant crosses	F_1	F_2
1 × 2	all purple	$\frac{1}{2}$ purple, $\frac{1}{2}$ white
1 × 3	all purple	$\frac{9}{16}$ purple, $\frac{7}{16}$ white
1 × 4	all white	all white

a. Explain all these results clearly, using diagrams wherever possible.

b. What F_1 and F_2 do you predict from crosses of 2 × 3 and 2 × 4?

***28.** A biologist is interested in the genetic control of leucine synthesis in the haploid filamentous fungus *Aspergillus.* He treats spores with mutagen and obtains five point mutations (*a* to *e*), all of which are leucine auxotrophs. He first makes heterokaryons between them to check on their functional relationships. He determines the following results, where + indicates that the heterokaryon grew, and − that the heterokaryon did not grow, on a medium lacking leucine:

	a	*b*	*c*	*d*	*e*
a	−	+	+	+	−
b		−	+	+	+
c			−	+	+
d				−	+
e					−

The biologist then intercrosses the mutations in all possible combinations. From each cross, he tests 500 ascospore progeny by inoculating them onto a medium lacking leucine. The results follow (the numbers represent the number of leucine prototrophs in the 500 progeny):

	a	*b*	*c*	*d*	*e*
a	0	125	128	126	0
b		0	124	2	125
c			0	124	127
d				0	123
e					0

Explain both sets of data genetically. (Note that the two leucine prototrophs from the *b–d* cross were found *not* to be due to reversion.)

29. In *Drosophila,* the eye phenotype "star" is caused by recessive mutations (*s*) mapping to one location on the second chromosome. This region is flanked to the left by the *A* locus (allele *A* or *a*) and to the right by the *B* locus (allele *B* or *b*):

A/*a* — *star* — *B*/*b*

Six independently induced *star* mutations are each made homozygous, with both *AA BB* and *aa bb* constitutions, and the six are intercrossed to study complementation at the *star* locus. The results follow, where + indicates wild eye and *s* indicates star eye, both being phenotypes of the F_1.

	AA ss BB					
aa ss bb	1	2	3	4	5	6
1	*s*	+	*s*	*s*	+	[+]
2		*s*	+	[+]	*s*	+
3			*s*	*s*	+	+
4				*s*	+	+
5					*s*	+
6						*s*

a. How many cistrons are at the *star* location, and which mutational sites are in each cistron?

b. The heterozygotes in brackets are allowed to produce gametes. In both cases, $star^+$ recombinant gametes are identified. The gametes are tested for the flanking marker conformation, which is *a B* in both the 1 × 6 heterozygote and the 2 × 4 heterozygote. Order the cistrons in relation to the *A a* and *B b* loci.

13

DNA Function

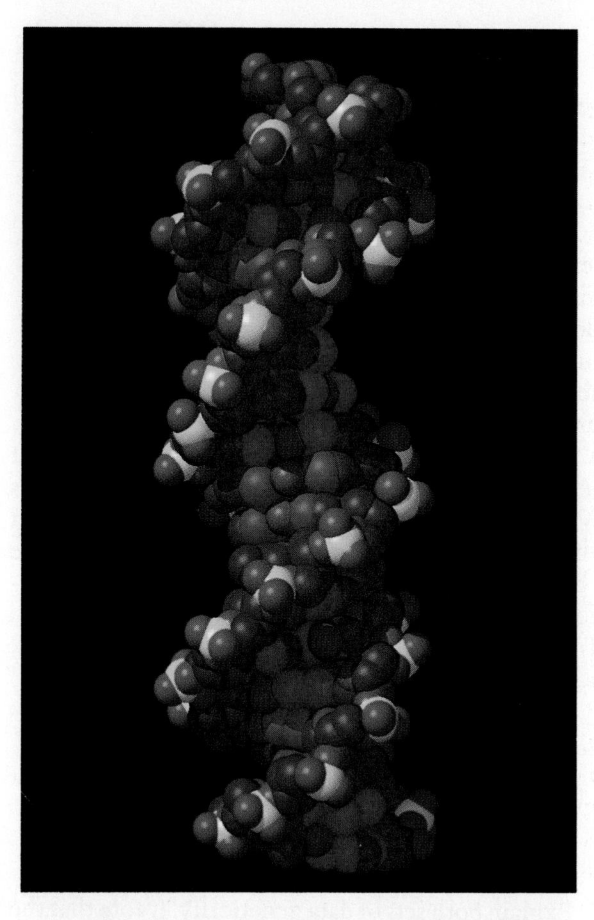

Computer model of DNA. (J. Newdol, Computer Graphics Laboratory, University of California, San Francisco. Copyright Regents, University of California.

KEY CONCEPTS

DNA is transcribed into an mRNA molecule, which is then translated during protein synthesis.

Translation requires transfer RNAs and ribosomes.

The genetic code is a nonoverlapping triplet code.

Special sequences signal the initiation and termination of both transcription and translation.

In eukaryotes, the initial RNA transcript is processed in several ways to generate the final mRNA.

Many eukaryotic genes contain segments of DNA, termed introns, that interrupt the normal gene coding sequence.

The primary eukaryotic transcript is spliced in one of a variety of ways to remove the RNA encoded by the intron and to yield the final mRNA.

Overview

The genetic information embodied in DNA can either be copied into more DNA during replication or be translated into protein. These are the processes of **information transfer** that constitute DNA function. We considered replication in some detail in Chapter 11. Let's now explore the way in which genetic information is turned into protein. Only a part of this story has been revealed through purely genetic analysis; mutants have been useful, of course, but only as tools to facilitate the biochemistry. However, the kind of reasoning that molecular biologists employ to investigate this aspect of DNA function illustrates an analytical approach also characteristic of work in genetics.

We shall see in this chapter that one of the first clues as to how DNA directs the synthesis of proteins came from bacteriophages, when it was shown that gene expression resulted in the synthesis of RNA molecules from a DNA template, a process termed **transcription.** RNA is synthesized from only one strand of a double-stranded DNA helix. This transcription is catalyzed by an enzyme, **RNA polymerase,** and follows rules similar to those of replication. RNA has similarities to DNA, although ribose is the sugar used in RNA, and uracil replaces thymine. Extraction of RNA from a cell yields several basic varieties: ribosomal, transfer, and messenger RNA, and small amounts of other small RNAs. The three sizes of **ribosomal RNA (rRNA)** combine with an array of specific proteins to form **ribosomes,** which are the machines used for protein synthesis. The ribosome-mediated synthesis of a polypeptide from a messenger RNA molecule is termed **translation. Transfer RNA (tRNA)** comprises a group of rather small RNA molecules, each with specificity for a particular amino acid; they carry the amino acids to the ribosome, where they can be attached to a growing polypeptide.

Messenger RNA (mRNA) molecules, which are produced on the DNA template, contain the information that is translated into proteins. The sequence of bases in mRNA determines the sequence of amino acids. We shall see how this is accomplished and how this **genetic code** was deciphered. Also, we shall see how certain sequences signal the initiation and termination of protein synthesis. In eukaryotes, many mRNA molecules are processed before translation, adding additional features to the mechanism of information transfer.

Transcription

Early investigators had good reason for thinking that information is not transferred directly from DNA to protein. DNA is found in the nucleus (of a eukaryotic cell), whereas protein is known to be synthesized in the cytoplasm. Another nucleic acid, ribonucleic acid (RNA), is necessary as an intermediary.

Early Experiments Suggesting RNA Intermediate

If cells are fed radioactive RNA precursors, then the labeled RNA shows up first of all in the nucleus, indicating that the RNA is synthesized there. In a **pulse-chase** experiment, a brief pulse of labeled RNA precursors is given. These are incorporated into RNA molecules. The cells are then transferred to medium with unlabeled RNA precursors. This "chases" the label out of the RNA, since as the RNA breaks down, only the unlabeled precursors are used to synthesize new RNA molecules. The pulse-chase protocol allows one to track a population of RNA molecules, synthesized almost simultaneously, over time. In samples taken after the chase, the labeled RNA is found in the cytoplasm (Figure 13-1). Apparently, the RNA is synthesized in the nucleus and then moves into the cytoplasm, where proteins are synthesized. Thus, it is a good candidate as an information-transfer intermediary between DNA and protein.

In 1957, Elliot Volkin and Lawrence Astrachan made a significant observation. They found that one of the most striking molecular changes when *E. coli* is infected with the phage T2 is a rapid burst of RNA synthesis. Furthermore, this phage-induced RNA "turns over" rapidly, as shown in the following experiment. The infected bacteria are first pulsed with radioactive uracil (a specific precursor of RNA); the bacteria are then chased with cold uracil. The RNA recovered shortly after the pulse is labeled, but that recovered somewhat longer after the chase is unlabeled, indicating that the RNA has a very short lifetime. Finally, when the nucleotide contents of

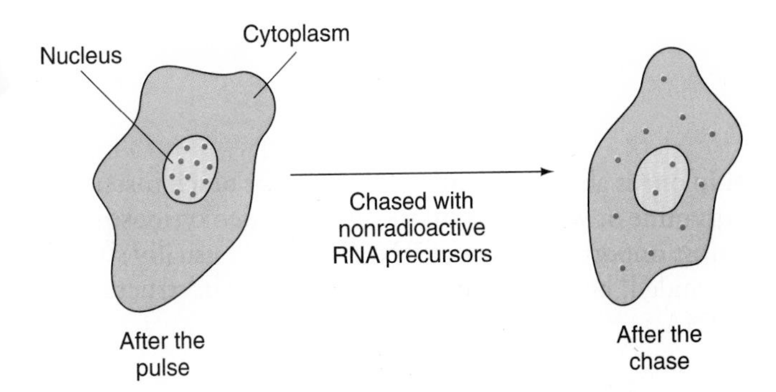

Figure 13-1 RNA synthesized during one short time period is labeled by feeding the cell a brief "pulse" of radioactive RNA precursors, followed by a "chase" of nonradioactive precursors. In an autoradiograph, the labeled RNA appears as dark grains. Apparently, the RNA is synthesized in the nucleus and then moves out into the cytoplasm.

E. coli and T2 DNA are compared with the nucleotide content of the phage-induced RNA, the RNA is found to be very similar to the phage DNA.

The tentative conclusion from the two experiments described above is that RNA is synthesized from DNA and that it is somehow used to synthesize protein. We can now outline three stages of information transfer (Figure 13-2): *replication* (the synthesis of DNA), *transcription* (the synthesis of an RNA copy of a portion of the DNA), and *translation* (the synthesis of a polypeptide directed by the RNA sequence).

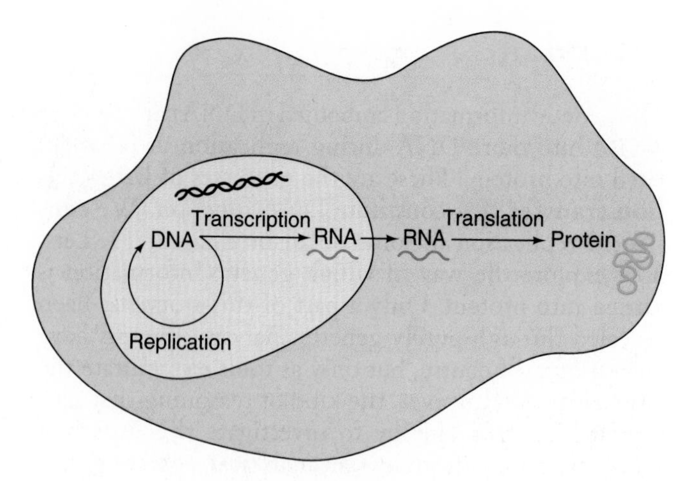

Figure 13-2 The three processes of information transfer: replication, transcription, and translation.

Properties of RNA

Although RNA is a long-chain macromolecule of nucleic acid (as is DNA), it has very different properties. First, RNA is usually single-stranded, not a double helix. Second, RNA has ribose sugar, rather than deoxyribose, in its nucleotides (hence, its name):

—CH_2 O Base; H H H H; OH OH — Ribose

—CH_2 O Base; H H H H; OH H — Deoxyribose

Third, RNA has the pyrimidine base **uracil** (abbreviated *U*) instead of thymine. However, uracil does form hydrogen bonds with adenine just as thymine does:

Uracil

No one is absolutely sure why RNA has uracil instead of thymine or why it has ribose instead of deoxyribose. The most important aspect of RNA is that it is usually single-stranded, but otherwise it is very similar in structure to DNA.

Complementarity and Asymmetry in RNA Synthesis

The similarity of RNA to DNA suggests that transcription may be based on the complementarity of bases, which is also the key to DNA replication. A transcription enzyme, RNA polymerase, could carry out transcription from a DNA template strand in a fashion quite similar to replication (Figure 13-3).

In fact, this model of transcription is confirmed cytologically (Figure 13-4). The fact that RNA can be synthesized with DNA acting as a template is also demonstrated by synthesis in vitro of RNA from nucleotides in the presence of DNA, using an extractable RNA polymerase. Whatever source of DNA is used, the RNA synthesized has an (A + U)/(G + C) ratio similar to the (A + T)/(G + C) ratio of the DNA (Table 13-1). This experiment does not indicate whether the RNA is synthesized from both DNA strands or from just one, but it does indicate that the linear frequency of the A–T pairs (in comparison with the G–C pairs) in the DNA is precisely mirrored in the relative abundance of (A + U) in the RNA. (These points are difficult to grasp without drawing some diagrams; Problem 2 at the end of this chapter provides some opportunities to clarify these notions.)

To test the complementarity of DNA with RNA, investigators can apply the specificity and precision of

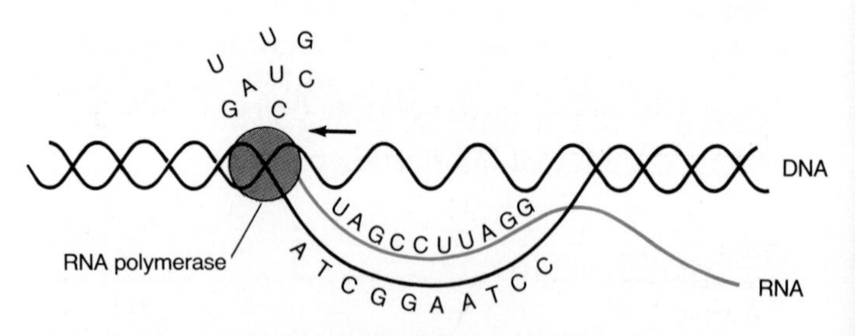

Figure 13-3 Synthesis of RNA on a single-stranded DNA template using free nucleotides. The process is catalyzed by RNA polymerase. Uracil (U) pairs with adenine (A).

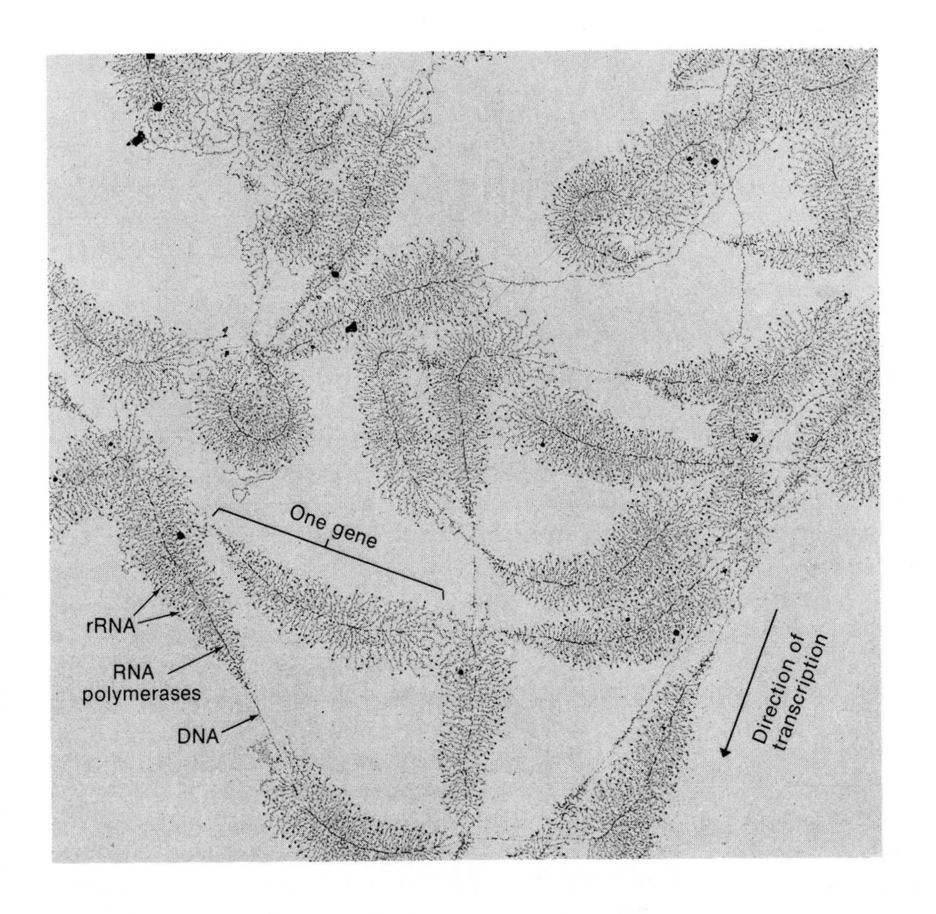

Figure 13-4 Tandemly repeated ribosomal RNA (rRNA) genes being transcribed in the nucleolus of *Triturus viridiscens* (an amphibian). (rRNA is a component of the ribosome, a cellular organelle.) Along each gene, many RNA polymerase molecules are attached and transcribing in one direction. The growing RNA molecules appear as threads extending out from the DNA backbone. The shorter RNA molecules are nearer the beginning of transcription; the longer ones have almost been completed. Hence, the "Christmas tree" appearance. (Photograph from O. L. Miller, Jr., and Barbara A. Hamkalo.)

nucleic acid hybridization. DNA can be denatured and mixed with the RNA formed from it. On slow cooling, some of the RNA strands anneal with complementary DNA to form a DNA-RNA hybrid. The DNA-RNA hybrid differs in density from the DNA-DNA duplex, so its presence can be detected by ultracentrifugation in cesium chloride. Nucleic acids will anneal in this way only if there are stretches of base-sequence complementarity, so the experiment does prove that the RNA transcript is complementary in base sequence to the parent DNA.

Can we determine whether RNA is synthesized from both or only one of the DNA strands? It seems reasonable that only one strand is used, because transcription of RNA from both strands would produce two complementary RNA strands from the same stretch of DNA, and these strands presumably would produce two different kinds of protein (with different amino acid sequences). In fact, a great deal of chemical evidence confirms that transcription usually takes place on only one of the DNA strands (although not necessarily the same strand throughout the entire chromosome).

Table 13-1. Nucleotide Ratios in Various DNAs and in Their Transcripts (in vitro)

DNA source	$\frac{(A+T)}{(G+C)}$ of DNA	$\frac{(A+U)}{(G+C)}$ of RNA
T2 phage	1.84	1.86
Cow	1.35	1.40
Micrococcus (bacterium)	0.39	0.49

The hybridization experiment can be extended to explore this problem. If the two strands of DNA have distinctly different purine : pyrimidine ratios, they can be purified separately because they have different densities in cesium chloride (CsCl). The RNA made from a stretch of DNA can be purified and annealed separately to each of the strands to see whether it is complementary to only one. J. Marmur and his colleagues were able to separate the strands of DNA from the *B. subtilis* phage SP8. They denatured the DNA, cooled it rapidly to prevent reannealing of the strands, and then separated the strands in CsCl. They showed that the SP8 RNA hybridizes to only one of the two strands, proving that transcription is **symmetrical**—that it occurs only on one DNA strand.

Although, for each gene, RNA is transcribed from only one of the DNA strands, the same DNA strand is not necessarily transcribed throughout the entire chromosome or through all stages of the life cycle. The RNA

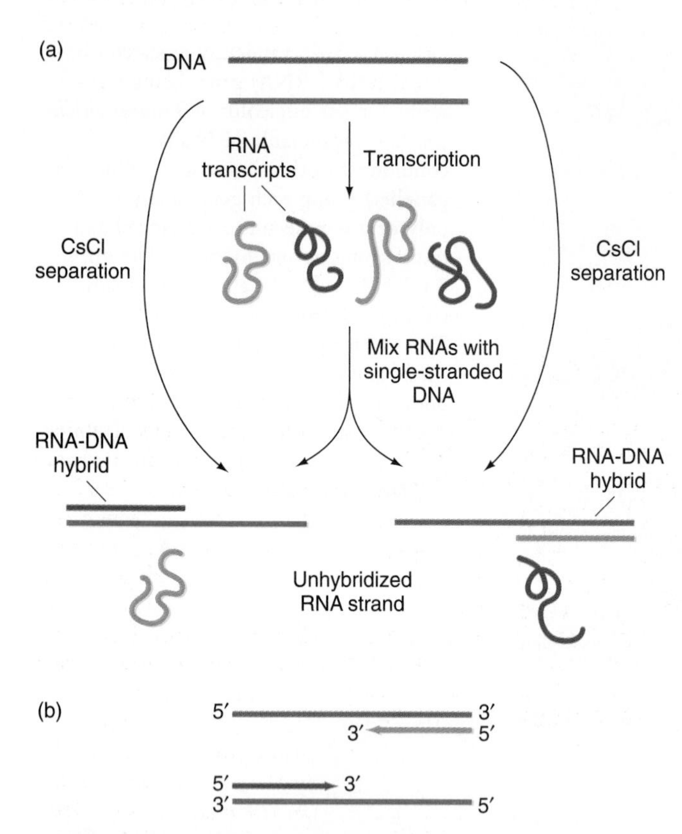

Figure 13-5 (a) DNA–RNA hybridization demonstrates that each RNA transcript is complementary to only one strand of the parent DNA. In this example, each of the two DNA strands is transcribed, but transcription is asymmetrical—only one strand is transcribed at any particular location. (b) A map of this hypothetical genome, showing the direction of transcription for the two transcripts from opposite DNA strands.

produced at different stages in the cycle of a phage hybridizes to different segments of the chromosome, showing the different genes that are activated at each stage (Figure 13-5). In λ phage, each of the two DNA strands is partially transcribed at a different stage. In phage T7, however, the same strand is transcribed for both early-acting and late-acting genes. Figure 13-6 shows a sequence of RNA made from the DNA template strand. The DNA template strand for a given mRNA is termed the **sense strand.** The complementary DNA strand is called the **antisense strand.** Note that the mRNA has the same sequence as the antisense strand.

RNA Polymerase

In most prokaryotes, a single RNA polymerase species transcribes all types of RNA. Figure 13-7 shows the structure of RNA polymerase from *E. coli.* We can see that the enzyme consists of four different subunit types. The beta (β) subunit has a molecular weight of 150,000 daltons, beta prime (β') 160,000, alpha (α) 40,000, and sigma (σ) 70,000. The σ subunit can dissociate from the rest of the complex, leaving the **core enzyme.** The complete enzyme with σ is termed the **RNA polymerase holoenzyme** and is necessary for correct initiation of transcription, whereas the core enzyme can continue transcription after initiation.

Next, let's look at the three distinct stages of transcription: **initiation, elongation,** and **termination.**

Initiation

The regions of the DNA that signal initiation of transcription in prokaryotes are termed **promoters.** (We consider their role in gene control in Chapter 17). Figure 13-8 shows the promoter sequences from 12 different transcription initiation points on the *E. coli* genome. The bases are aligned according to homologies, or similar base sequences, that appear just before the first base transcribed (designated the "initiation site" in Figure 13-8).

Note in Figure 13-8 that two regions of partial homology appear in virtually each case. These regions have been termed the -35 and -10 regions because of their locations relative to the transcription initiation point. At the bottom of Figure 13-8 an ideal, or concensus, sequence of a promoter is given. Physical experiments have confirmed that RNA polymerase makes contact with these two regions when binding to the DNA. The enzyme then unwinds DNA and begins the synthesis of an RNA molecule.

The dissociative subunit of RNA polymerase, the σ factor, allows RNA polymerase to recognize and bind specifically to promoter regions. First, the holoenzyme searches for a promoter (Figure 13-9a) and initially binds loosely to it, recognizing the -35 and -10 regions. The resulting structure is termed a *closed promoter complex* (see Figure 13-9b). Then, the enzyme binds more tightly, unwinding bases near the -10 region. When the bound

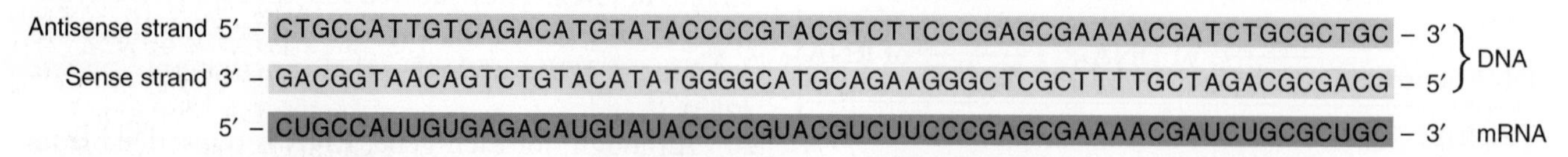

Figure 13-6 The mRNA sequence is complementary to the DNA template strand from which it is synthesized. The sequence shown here is from the gene for the enzyme β-galactosidase, which is involved in lactose metabolism.

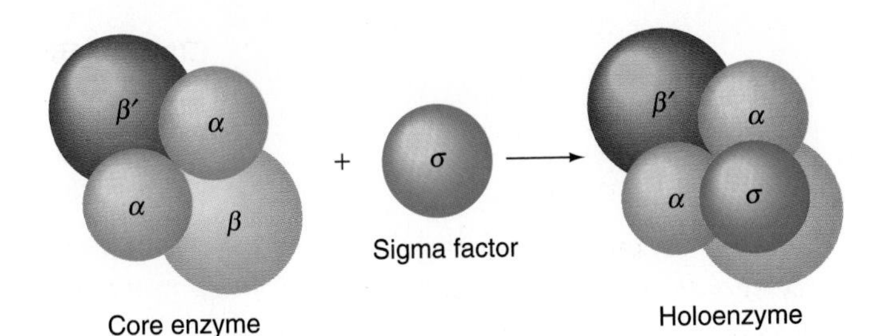

Figure 13-7 The structure of RNA polymerase. The core enzyme contains two α polypeptides, one β polypeptide, and one β′ polypeptide. The addition of the σ subunit allows initiation at promoter sites. (Promoters are discussed in the next section.)

Gene(s)	Sequence (−35 region … −10 region … Initiation site (+1))
lac	ACCCCAGGCTTTACACTTTATGCTTCCGGCTCGTATGTTGTGTGGAATTGTGAGCGG
*lac*1	CCATCGAATGGCGCAAAACCTTTCGCGGTATGGCATGATAGCGCCCGGAAGAGAGTC
ga/P2	ATTTATTCCATGTCACACTTTTCGCATCTTTGTTATGCTATGGTTATTTCATACCAT
*ara*B, A, D	GGATCCTACCTGACGCTTTTTATCGCAACTCTCTACTGTTTCTCCATACCCGTTTTT
*ara*C	GCCGTGATTATAGACACTTTTGTTACGCGTTTTTGTCATGGCTTTGGTCCCGCTTTG
trp	AAATGAGCTGTTGACAATTAATCATCGAACTAGTTAACTAGTACGCAAGTTCACGTA
*bio*A	TTCCAAAACGTGTTTTTTGTTGTTAATTCGGTGTAGACTTGTAAACCTAAATCTTTT
*bio*B	CATAATCGACTTGTAAACCAAATTGAAAAGATTTAGGTTTACAAGTCTACACCGAAT
*t*RNA^{Tyr}	CAACGTAACACTTTACAGCGGCGCGTCATTTGATATGATGCGCCCCGCTTCCCGATA
*rrn*D1	CAAAAAAATACTTGTGCAAAAAATTGGGATCCCTATAATGCGCCTCCGTTGAGACGA
*rrn*E1	CAATTTTTCTATTGCGGCCTGCGGAGAACTCCCTATAATGCGCCTCCATCGACACGG
*rrn*A2	AAAATAAATGCTTGACTCTGTAGCGGGAAGGCGTATTATGCACACCCCGCGCCGCTG

−35 region −10 region

General plan: T G T T G A C A-----11–15 bp-----T A T A A T-----5–8 bp----- Initiation site

Figure 13-8 Promoter sites have regions of similar sequences, as indicated by the colored region in the 12 different promoter sequences in *E. coli*. The gene (or genes) governed by each promoter sequence is indicated on the left. Numbering is given in terms of the number of bases before (−) or after (+) the RNA synthesis initiation point. (After J. Darnell, H. Lodish, and D. Baltimore, *Molecular Cell Biology*. Copyright 1986 by W. H. Freeman and Company.)

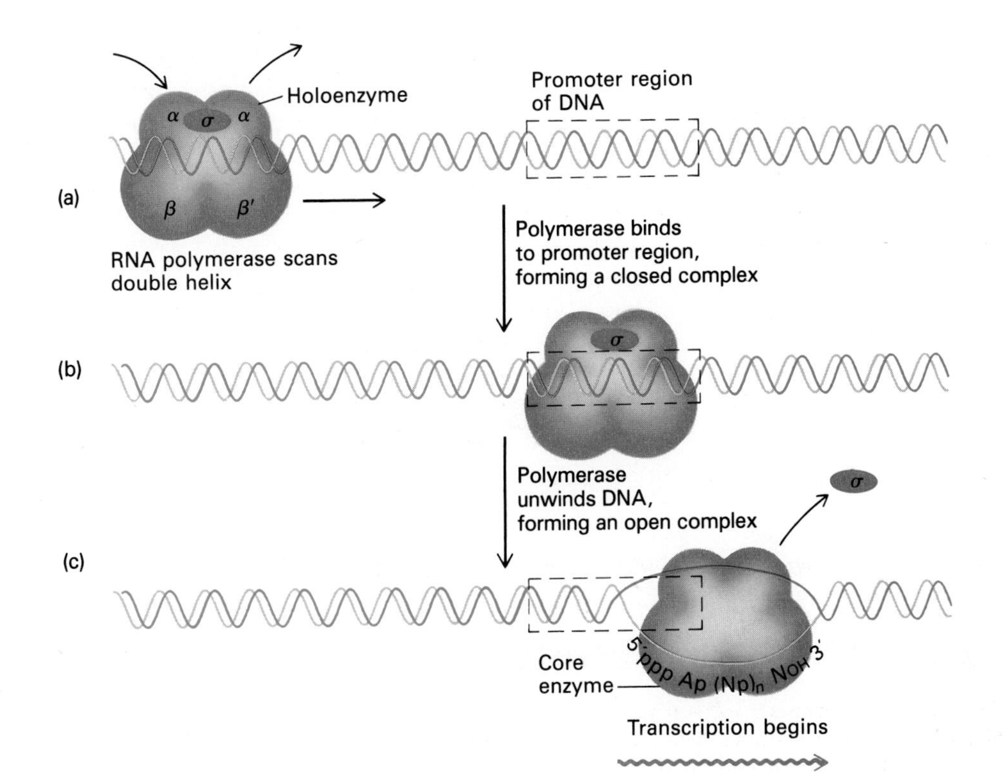

Figure 13-9 Initiation of transcription. (a) RNA polymerase searches for a promoter site. (b) It recognizes a promoter site and binds tightly, forming a closed complex. (c) The holoenzyme unwinds a short stretch of DNA, forming an open complex. Transcription begins, and the σ factor is released. The RNA transcript is shown, beginning with adenosine triphosphate (pppA), carrying on through an indeterminate number of nucleotides [$(Np)_n$], and ending with NOH. (Modified from J. Darnell, H. Lodish, and D. Baltimore, *Molecular Cell Biology, 2d ed.* Copyright 1990 by Scientific American Books, Inc.)

polymerase causes this local denaturation of the DNA duplex, it is said to form an *open promoter complex* (Figure 13-9c). This initiation step, the formation of an open complex, requires the sigma factor.

Elongation

Shortly after initiating transcription, the sigma factor dissociates from the RNA polymerase. The RNA is always synthesized in the 5′ → 3′ direction (Figures 13-10 and 13-11), with nucleoside triphosphates (NTPs) acting as substrates for the enzyme. The equation below represents the addition of each ribonucleotide.

$$\text{NTP} + (\text{NMP})_n \xrightarrow[\text{RNA polymerase}]{\text{DNA, } Mg^{2+}} (\text{NMP})_{n+1} + \text{PP}_i$$

The energy for the reaction is derived from splitting the high-energy triphosphate into the monophosphate and releasing the inorganic diphosphates (PP_i), as shown in Figure 13-10. Figure 13-11 gives a physical picture of elongation. Note how a "transcription bubble" must be maintained, since the transcription takes place on a double-stranded template. The bubble must move along the DNA duplex during elongation. Certain sequences may cause stalling or pausing, which becomes critical for termination of transcription (see below).

Termination

RNA polymerase also recognizes signals for chain termination, which involves the release of the nascent RNA and enzyme from the template. There are two major mechanisms for termination in *E. coli.*

In the first the termination is direct. The terminator sequences contain about 40 bp, ending in a GC-rich stretch that is followed by a run of six or more A's on the template strand. The corresponding GC sequences on the RNA are so arranged that the transcript in this region is able to form complementary bonds *with itself,* as can be seen in Figure 13-12. The resulting double-stranded RNA section is called a **hairpin loop.** It is followed by

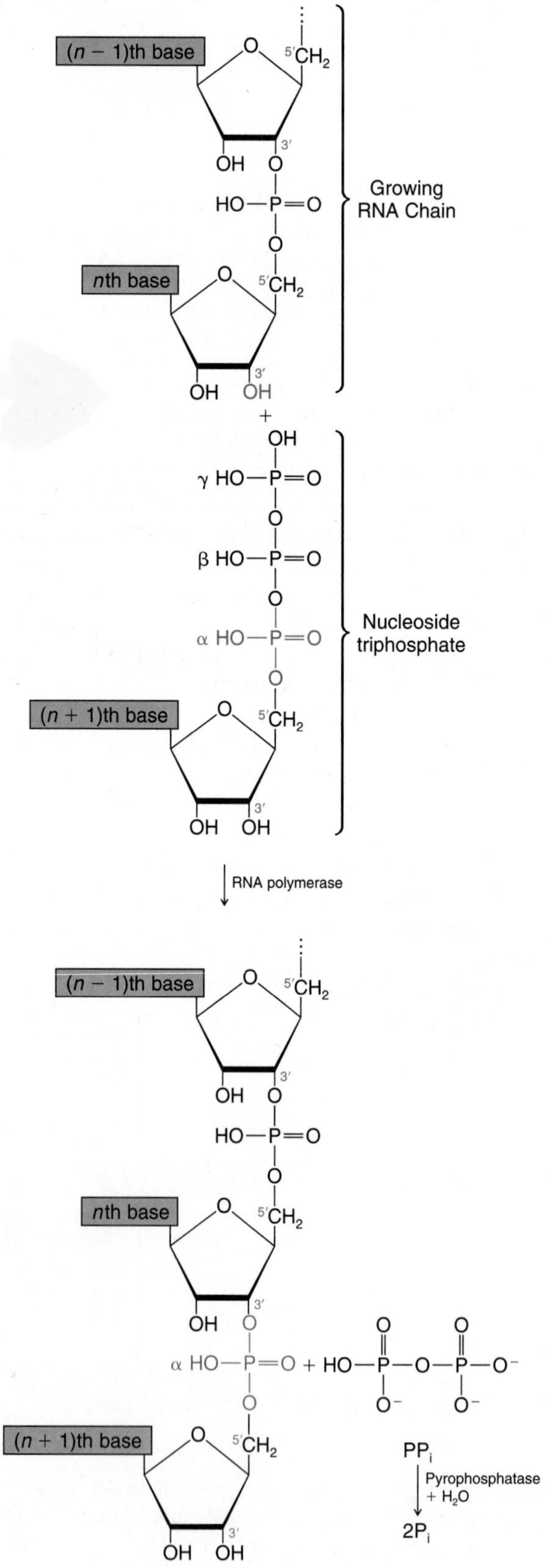

(n − 1)th base
5′CH2
3′
OH O
HO—P=O
O
nth base
5′CH2
3′
OH OH
Growing RNA Chain
+
OH
γ HO—P=O
O
β HO—P=O
O
α HO—P=O
O
(n + 1)th base
5′CH2
3′
OH OH
Nucleoside triphosphate
RNA polymerase
(n − 1)th base
5′CH2
3′
OH O
HO—P=O
O
nth base
5′CH2
3′
OH O
α HO—P=O + HO—P—O—P—O−
O
(n + 1)th base
5′CH2
3′
OH OH
PPi
Pyrophosphatase + H2O
2Pi

Figure 13-10 The sequential addition of nucleotides occurs one at a time in the 5′ to 3′ direction. The chain grows by the formation of a bond between the 3′ hydroxyl end of the growing strand and a nucleotide triphosphate, releasing one pyrophosphate ion (PP_i). This results in the net addition of one phosphate, which is incorporated into the backbone of the new strand. DNA grows by reaction with deoxyribonucleoside triphosphates, and RNA grows by reaction with ribonucleoside triphosphates. (From J. Darnell, H. Lodish, and D. Baltimore, *Molecular Cell Biology,* 2d ed. Copyright 1990 by Scientific American Books, Inc.)

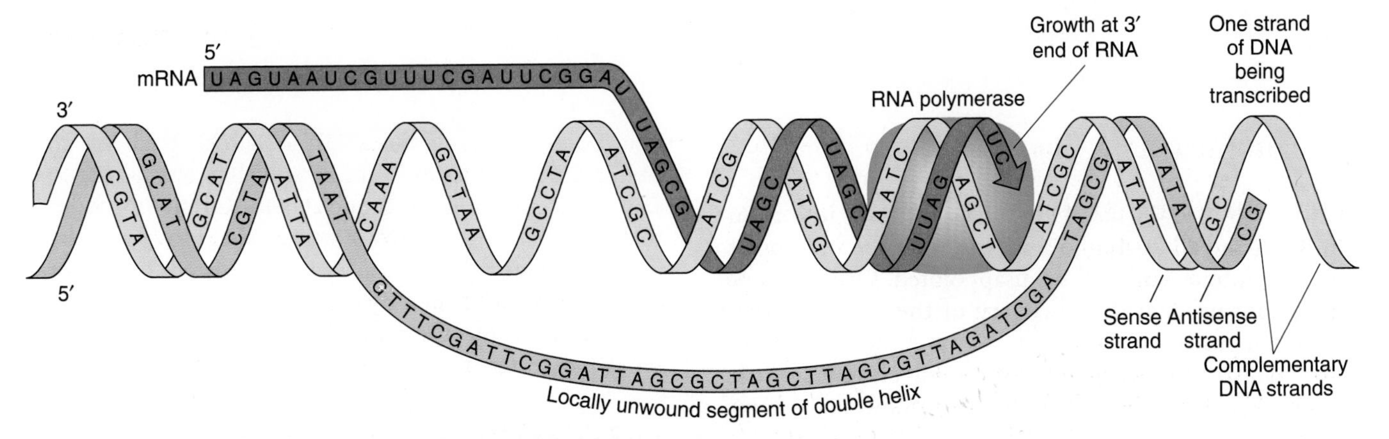

Figure 13-11 Transcription by RNA polymerase. An RNA strand is synthesized in the 5′ → 3′ direction from a locally single-stranded region of DNA. (After E. J. Gardner, M. J. Simmons, and D. P. Snustad, *Principles of Genetics,* 8th ed. Copyright 1991 by John Wiley and Sons, Inc.)

the terminal run of U's that correspond to the A residues on the DNA template. The hairpin loop and section of U residues appear to serve as a signal for the release of RNA polymerase and termination of transcription.

In the second type, the help of an additional protein factor, termed **rho,** is required in order to recognize the termination signals. mRNAs with rho-dependent termination signals do not have the string of U residues at the end of the RNA, and usually do not have hairpin loops. A model for rho-dependent termination is shown in Figure 13-13. Rho is a hexamer consisting of six identical subunits, and uses the hydrolysis of ATP to ADP and P_i to drive the termination reaction. The first step in termination is the binding of rho to a specific site on the RNA termed *rut* (Figure 13-13a and b). After binding, rho pulls the RNA off the RNA polymerase, probably by translocating along the mRNA, as depicted in Figure 13-13b and c. The *rut* sites are located just upstream from (that is, 5′ from) sequences at which the RNA polymerase tends to pause.

The efficiency of both mechanisms of termination is influenced by surrounding sequences and other protein factors, as well.

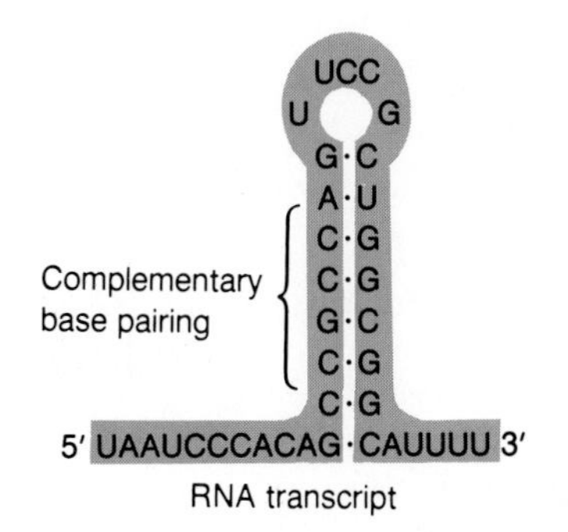

Figure 13-12 The structure of a termination site for RNA polymerase in bacteria. The hairpin structure forms by complementary base pairing *within* the RNA strand.

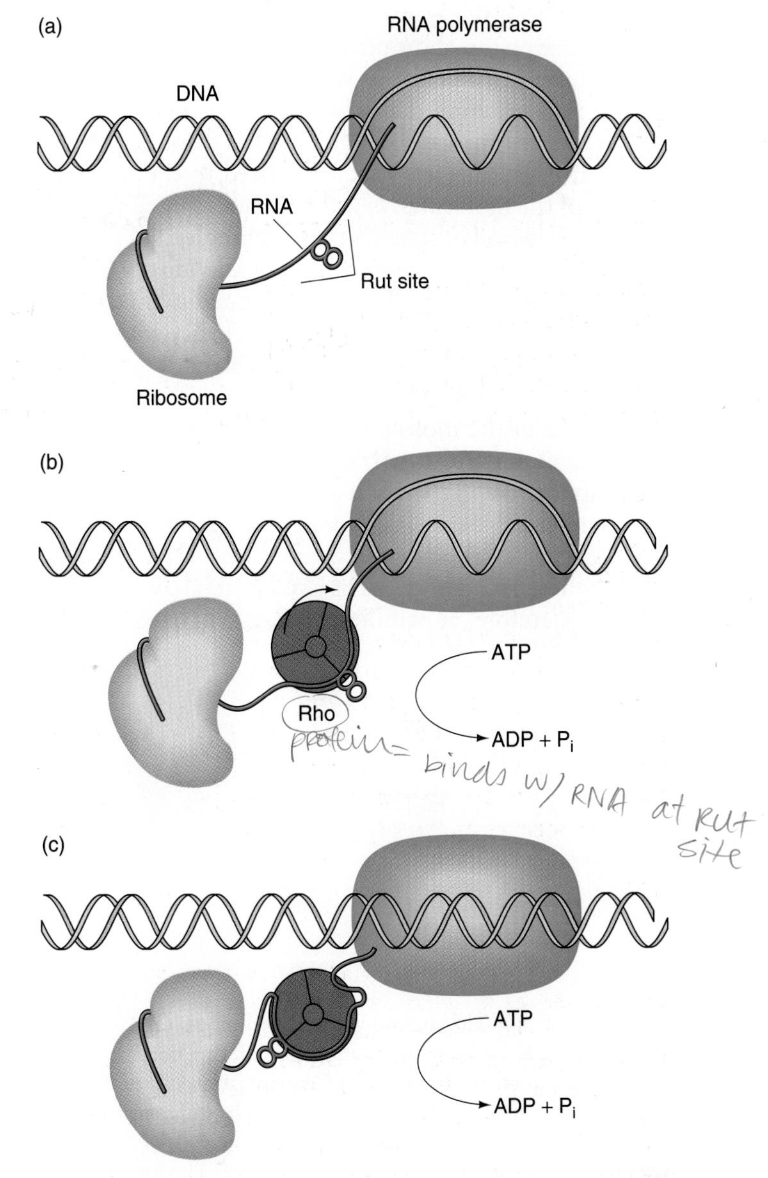

Figure 13-13 A model for rho action on a nascent cotranslated mRNA. (Adapted from J. P. Richardson, *Cell* 64, 1991, 1047–1049.)

Translation

Although mRNA directs protein synthesis, if you mix mRNA and all 20 amino acids in a test tube and hope to make protein, you will be disappointed. Other components are needed; the discovery of the nature of these components has provided the key to understanding the mechanism of translation. A simple but elegant centrifugation technique helped reveal these additional components.

Application of Sucrose Gradients

The **sucrose density gradient** is created in a test tube by layering successively lower concentrations of sucrose solution, one on top of the other. The material to be studied is carefully placed on top. When the solution is centrifuged in a machine that allows the test tube to swivel freely, the sedimenting material travels through the gradient at different rates that are related to the sizes and shapes of the molecules. Large molecules migrate farther in a given period of time than smaller molecules do. The separated molecules can be collected individually by capturing sequential drops from a small opening in the bottom of the tube (Figure 13-14). The velocity with which a fraction moves the fixed distance to the tube bottom indicates its sedimentation (S) value, which is a measure of the size of the molecules in the fraction.

It is important to note the difference between a sucrose gradient and the CsCl gradient that we considered in Chapter 11. In a CsCl gradient, the molecules being studied have a density somewhere in between the lowest and highest concentrations of CsCl generated in the gradient. Therefore, at equilibrium they will band at a specific point on the gradient. In a sucrose gradient, the molecules being studied are denser than any of the sucrose concentrations used, and at equilibrium they would form a pellet in the bottom of the tube. However, they migrate toward the bottom at varying speeds depending on their size and shape. By comparing the different positions of each molecule in the gradient at a particular time, we can determine the relative sizes of the molecules.

Different Classes of RNA Molecules

Using the separatory powers of the sucrose-gradient technique, it is possible to identify several macromolecules and macromolecular aggregates in a typical protein-synthesizing system. The main components are transfer RNA (tRNA), ribosomes, and messenger RNA (mRNA). Transfer RNA is a class of small (4S) RNA molecules of rather similar type and function. In fact, complete nucleotide sequences have been determined for a large number of tRNA molecules from different organisms.

Ribosomes, which are the sites of protein synthesis, are cellular organelles composed of very complex aggregations of ribosomal proteins and ribosomal RNA (rRNA) components. In *E. coli,* for instance, at least three separate RNA molecules can be distinguished by size in ribosomes: 23S, 16S, and 5S. We do not know precisely how rRNA is bound up with the protein components or how each component functions. Under the electron microscope, a ribosome appears globular. On chemical treatment, it splits into two main globular entities (50S and 30S in *E. coli*). We shall consider the structure of tRNA and rRNA molecules in detail later.

Both rRNA and tRNA molecules will form RNA-DNA hybrids in vitro, indicating that they are tran-

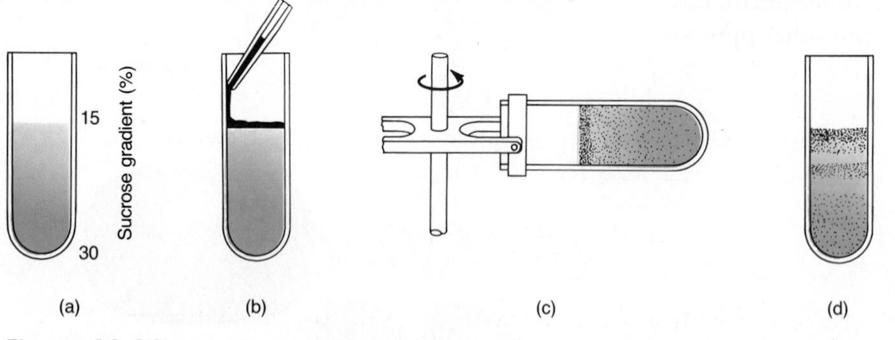

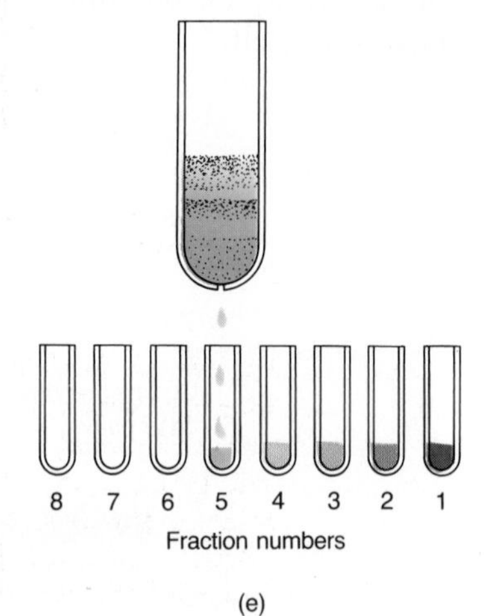

Figure 13-14 The sucrose-gradient technique. (a) A sucrose density gradient is created in a centrifuge tube by layering solutions of differing densities. (b) The sample to be tested is placed on top of the gradient. (c) Centrifugation causes the various components (fractions) of the sample to sediment differentially. (d) The different fractions appear as bands in the centrifuged gradient. (e) The different bands can be collected separately by collecting samples from the bottom of the tube at fixed time intervals. The S value for the fraction is based on its position in the gradient, which is related to the time at which it drips from the bottom of the tube. (From A. Rich, "Polyribosomes." Copyright 1963 by Scientific American, Inc. All rights reserved.)

Table 13-2. RNA Molecules in *E. coli*

Type	Percentage of cell RNA	Sedimentation coefficient, S	Molecular weight	Number of nucleotides
Ribosomal RNA (rRNA)	80	23	1.2×10^6	3700
		16	0.55×10^6	1700
		5	3.6×10^4	1700
Transfer RNA (tRNA)	15	4	2.5×10^4	75
Messenger RNA (mRNA)	5		Heterogeneous	Varies

SOURCE: Adapted from L. Stryer, *Biochemistry*, 2d ed. Copyright 1981 by W. H. Freeman and Company.

scribed from the DNA. Table 13-2 summarizes the main types of RNA that can be found in a typical protein-synthesizing system.

Other components are needed to make protein synthesis work in vitro. These include several enzymes, several protein "factors," and a chemical source of energy. The energy donor is needed because an orderly structure is being created out of a mess of components—a process that requires energy because the system must lose entropy (a measure of disorder or randomness). We will examine the process of protein synthesis after we first consider the genetic code.

The Genetic Code

If genes are segments of DNA and if DNA is just a string of nucleotide pairs, then how does the sequence of nucleotide pairs dictate the sequence of amino acids in proteins? The analogy to a code springs to mind at once. The cracking of the genetic code is the story told in this section. The experimentation was sophisticated and swift, and it did not take long for the code to be deciphered once its existence was strongly indicated.

Simple logic tells us that if nucleotide pairs are the *letters* in a code, then a combination of letters could form words representing different amino acids. We must ask how the code is read. Is it overlapping or nonoverlapping? Then we must ask how many letters in the mRNA make up a word, or **codon,** and which specific codon or codons represent each specific amino acid.

Overlapping versus Nonoverlapping Codes

Figure 13-15 shows the difference between an overlapping and a nonoverlapping code. In the example, a three-letter, or **triplet,** code is shown. For the nonoverlapping code, consecutive amino acids are specified by consecu-

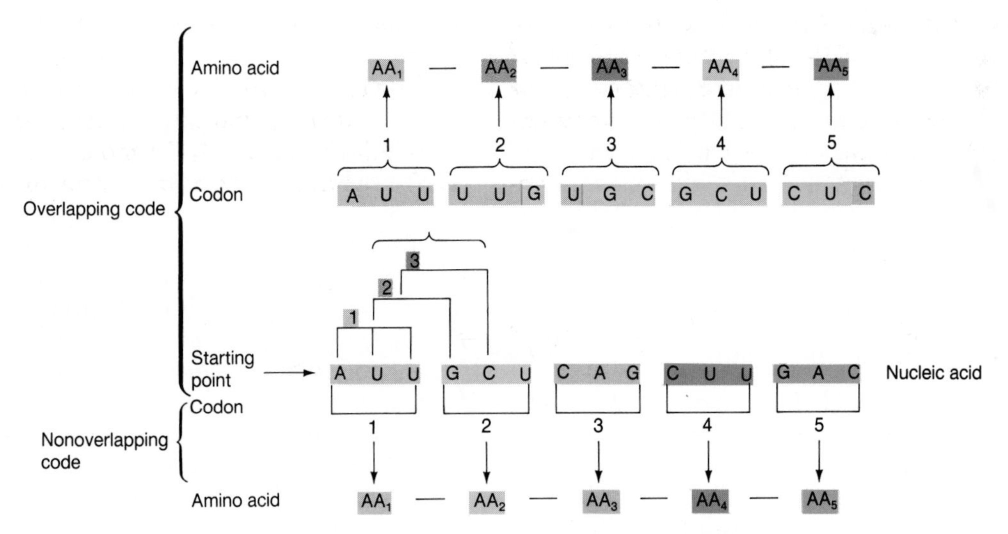

Figure 13-15 The difference between an overlapping and a nonoverlapping code. The case illustrated is for a code with three letters (a triplet code). An overlapping code uses codons that employ some of the same nucleotides as other codons for the translation of a single protein, as shown in the top of the diagram (for the mRNA sequence shown at the bottom of the diagram). In a nonoverlapping code, a protein is translated by reading codons that do not share any of the same nucleotides. Note that the designation of an amino acid as "AA_2," "AA_3," etc., in the nonoverlapping model does not mean it is necessarily the same amino acid as its numerical counterpart in the overlapping model. This is because the triplets making up the respective codons for the two AA_3's, for example, are different and would more than likely code for different amino acids. Identical amino acid numbers between the two models merely indicate the same amino acid position on the protein chain.

tive code words (codons), as shown in the bottom of the figure. For an overlapping code, consecutive amino acids are encoded in the mRNA by codons that share some consecutive bases; for example, the last two bases of one codon may also be the first two bases of the next codon. Overlapping codons are shown in the upper portion of the figure. Thus, for the sequence AUUGCUCAG in a nonoverlapping code, the first three amino acids are encoded by the three triplets AUU, GCU, and CAG, respectively. However, for an overlapping code, the first three amino acids are encoded by the triplets AUU, UUG, and UGC if the overlap is two bases, as shown in the figure.

By 1961, it was already clear that the genetic code was nonoverlapping. The analysis of mutationally altered proteins — in particular, the nitrous acid–generated mutants of tobacco mosaic virus — showed that only a single amino acid changes at one time in one region of the protein. This is predicted by a nonoverlapping code. As you can see from Figure 13-15 an overlapping code predicts that a single base change will alter as many as three amino acids at adjacent positions in the protein.

It should be noted that while the use of an overlapping *code* was ruled out by the analysis of single proteins, nothing precluded the use of alternative reading frames to encode amino acids in two different proteins. In the example here, one protein might be encoded by the series of codons that reads AUU, GCU, CAG, CUU, and so on. A second protein might be encoded by codons that are shifted over by one base and therefore read UUG, CUC, AGC, UUG, and so on. This is an example of storing the information encoding two different proteins in two different reading frames, while still using a genetic code that is read in a nonoverlapping manner during the translation of a *specific* protein. Some examples of such shifts in reading frame have been found.

Number of Letters in the Code

Reading an mRNA molecule from one particular end, only one of four different bases, A, U, G, or C, can be found at each position. Thus, if the words are one letter long, only four words are possible. This cannot be the genetic code, because we must have a word for each of the 20 amino acids commonly found in cellular proteins. If the words are two letters long, then $4^2 = 16$ words are possible; for example, AU, CU, or CC. This vocabulary still is not large enough.

If the words are three letters long, then $4^3 = 64$ words are possible; for example, AUU, GCG, or UGC. This code provides more than enough words to describe the amino acids. We can conclude that the code word must consist of at least three nucleotide pairs. However, if all words are "triplets," then we have a considerable excess of possible words over the 20 needed to name the common amino acids.

Use of Suppressors to Demonstrate a Triplet Code

Convincing proof that a codon is, in fact, three letters long (and no more than three) came from beautiful genetic experiments first reported in 1961 by Francis Crick, Sidney Brenner, and their co-workers, who used mutants in the *rII* locus of T4 phage. Mutations causing the rII phenotype (see Chapter 12) were induced using a chemical called *proflavin,* which was thought to act by the addition or deletion of single nucleotide pairs in DNA. (This assumption is based on experimental evidence not presented here.) The following examples illustrate the action of proflavin on double-stranded DNA.

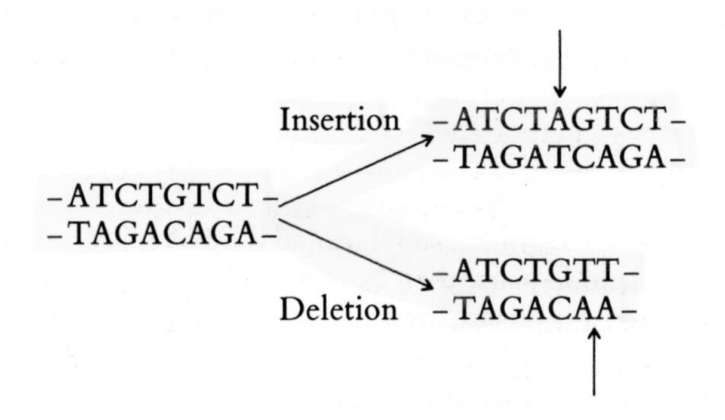

Then, starting with one particular proflavin-induced mutation called FCO, Crick and his colleagues again used proflavin to induce "reversions" (reversals of the mutation) that were detected by their wild-type plaques on *E. coli* strain K(λ). Genetic analysis of these plaques revealed that the "revertants" were not identical true wild types, thereby suggesting that the back mutation was not an exact reversal of the original forward mutation. In fact, the reversion was found to be caused by the presence of a *second mutation* at a different site from — but in the same cistron as — that of FCO; this second mutation "suppressed" mutant expression of the original FCO. A **suppressor mutation** counteracts or suppresses the effects of another mutation. Properties of suppressor mutations include the following:

1. A suppressor mutation is at a different site from that of the mutation it counteracts. Therefore, the original mutation can be recovered by genetic crosses between the wild type and the revertant, since the revertant carries both the suppressor and the original mutation.
2. A suppressor mutation may be within the same gene as the mutation it suppresses (internal suppressor), as

in the example just given, or it may be in a different gene (external suppressor).
3. Different suppressors may exert their effects in different ways. For instance, some suppressors act at the level of transcription or translation; others alter the physiology of the cell.

The suppressor mutation could be separated from the original forward mutation by recombination, and as we have seen, when this was done, the suppressor was shown to be an *rII* mutation itself (Figure 13-16).

How can we explain these results? If we assume that reading is polarized—that is, if the cistron is read from one end only—then the original proflavin-induced addition or deletion could be mutant because it interrupts a normal reading mechanism that establishes the group of bases to be read as words. For example, if each three bases on the resulting mRNA make a word, then the "reading frame" might be established by taking the first three bases from the end as the first word, the next three as the second word, and so on. In that case, a proflavin-induced addition or deletion of a single pair on the DNA would shift the reading frame on the mRNA from that corresponding point on, causing all following words to be misread. Such a **frameshift mutation** could reduce most of the genetic message to gibberish. However, the proper reading frame could be restored by a compensatory insertion or deletion somewhere else, leaving only a short stretch of gibberish between the two. Consider the following example in which three-letter English words are used to represent the codons:

THE FAT CAT ATE THE BIG RAT

Delete C: THE FAT↑ATA TET HEB IGR AT

Insert A: THE FAT ATA↑ ATE THE BIG RAT

The insertion suppresses the effect of the deletion by restoring most of the sense of the sentence. By itself, however, the insertion also disrupts the sentence:

THE FAT CAT AAT ETH EBI GRA T

If we assume that the FCO mutant is caused by an addition, then the second (suppressor) mutant would have to be a deletion because, as we have seen, this would restore the reading frame of the resulting message (a second insertion would not correct the frame). In the following diagrams, we use a hypothetical nucleotide chain to represent RNA for simplicity. We also assume that the code words are three letters long and are read in one direction (left to right in our diagrams).

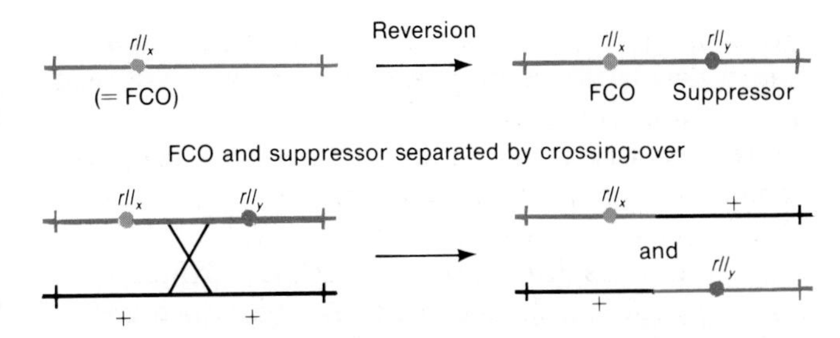

Figure 13-16 The suppressor of an initial *rII* mutation is shown to be an *rII* mutation itself after separation by crossing-over. The original mutant, FCO, was induced by proflavin. Later, when the FCO strain was treated with proflavin again, a revertant was found, which on first appearance seemed to be wild-type. However, it was found that a second mutation within the *rII* region had been induced, and the double mutant rII_x rII_y was shown not to be quite identical with the original wild type.

1. Wild-type message

CAU CAU CAU CAU CAU

2. rII_a message: distal words changed (x) by frameshift mutation (words marked ✓ are unaffected)

Addition ↓

CAU ACA UCA UCA UCA U

✓ x x x x

3. rII_a rII_b message: few words wrong, but reading frame restored for later words

Deletion ↓

CAU ACA UCU CAU CAU

✓ x x ✓ ✓

The few wrong words in the suppressed genotype could account for the fact that the revertants (suppressed phenotypes) Crick and his associates recovered did not look exactly like the true wild types phenotypically.

We have assumed here that the original frameshift mutation was an addition, but the explanation works just as well if we assume that the original FCO mutation is a deletion and the suppressor is an addition. If the FCO is defined as plus, then suppressor mutations are automatically minus. Experiments have confirmed that a plus cannot suppress a plus and a minus cannot suppress a minus. In other words, two mutations of the same sign never act as suppressors of each other. However, very interestingly, combinations of *three* pluses or *three* minuses have been shown to act together to restore a wild-type phenotype.

This observation provided the first experimental confirmation that a word in the genetic code consists of three successive nucleotide pairs, or a triplet. This is because three additions or three deletions within a gene automatically restore the reading frame in the mRNA if the words are triplets. For example,

```
          Deletions
          ↓  ↓        ↓
CAU   CAU   CAU   CAU   CAU   CAU   CAU
CAU   ACA   UAU   CAU   CAU   CAU
 ✓     x     x     ✓     ✓     ✓
```

Proof that the genetic deductions about proflavin were correct came from an analysis of proflavin-induced mutations in a gene with a protein product that could be analyzed. George Streisinger worked with the gene that controls the enzyme lysozyme, which has a known amino acid sequence. He induced a mutation in the gene with proflavin and selected for proflavin-induced revertants, which were shown genetically to be double mutants (with mutations of opposite sign). When the protein of the double mutant was analyzed, a stretch of different amino acids lay between two wild-type ends, just as predicted:

Wild type
– Thr – Lys – Ser – Pro – Ser – Leu – Asn – Ala –

Revertant type
– Thr – Lys – Val – His – His – Leu – Met – Ala –

Degeneracy of the Genetic Code

Crick's work also suggested that the genetic code is **degenerate.** That expression is not a moral indictment! It simply means that each of the 64 triplets must have some meaning within the code, so that at least some amino acids must be specified by two or more different triplets. If only 20 triplets are used (with the other 44 being nonsense, in that they do not code for any amino acid), then most frameshift mutations can be expected to produce nonsense words, which presumably would stop the protein-building process. If this were the case, then the suppression of frameshift mutations would rarely, if ever, work. However, if all triplets specify some amino acid, then the changed words would simply result in the insertion of incorrect amino acids into the protein. Thus, Crick reasoned that many or all amino acids must have several different names in the base-pair code; this hypothesis was later confirmed biochemically.

Review

Let's summarize what the work up to this point demonstrates about the genetic code:

1. The code is nonoverlapping.
2. Three bases code for an amino acid. These triplets are termed *codons.*
3. The code is read from a fixed starting point and continues to the end of the coding sequence. We know this because a single frameshift mutation anywhere in the coding sequence alters the codon alignment for the rest of the sequence.
4. The code is degenerate in that some amino acids are specified by more than one codon.

Cracking the Code

The deciphering of the genetic code — determining the amino acid specified by each triplet — is one of the most exciting genetic breakthroughs of the past two decades. Once the necessary experimental techniques became available, the genetic code was broken in a rush.

The first breakthrough was the discovery of how to make synthetic mRNA. If the nucleotides of RNA are mixed with a special enzyme (polynucleotide phosphorylase), a single-stranded RNA is formed in the reaction. No DNA is needed for this synthesis, and so the nucleotides are incorporated at random. The ability to synthesize mRNA offered the exciting prospect of creating specific RNA sequences and then seeing what kinds of amino acids were incorporated when acting as mRNA. The first synthetic messenger obtained, poly(U), was made by reacting only uracil nucleotides with the RNA-synthesizing enzyme, producing – UUUU –. In 1961, Marshall Nirenberg and Heinrich Mathaei mixed poly(U) with the protein-synthesizing machinery of *E. coli* (ribosomes, aminoacyl-tRNAs, a source of chemical energy, several enzymes, and a few other things) in vitro and *observed the formation of a protein!* Of course, the main excitement centered on the question of the amino acid sequence of this protein. It proved to be polyphenylalanine — a string of phenylalanine molecules attached to form a polypeptide. Thus, the triplet UUU must code for phenylalanine:

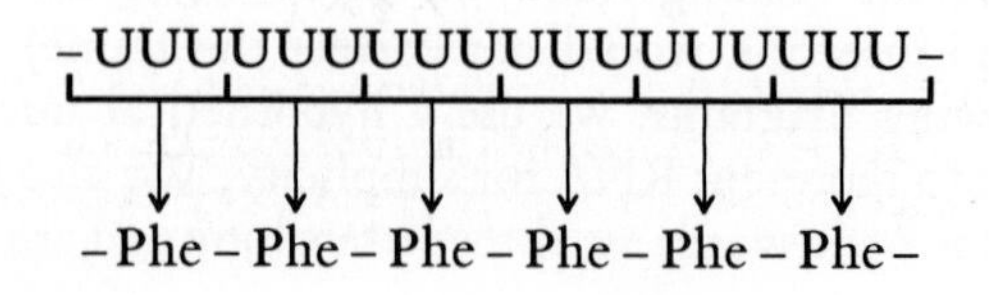

This type of analysis was extended by mixing nucleotides in a known fixed proportion when making synthetic mRNA. In one experiment, the nucleotides uracil and guanine were mixed in a ratio of 3 : 1. When nucleotides are incorporated at random into synthetic mRNA, the relative frequency at which each triplet will appear in the sequence can be calculated on the basis of the relative proportion of the various nucleotides present (Table 13-3). In the table, note that UUU is used as the baseline frequency against which the other frequencies are measured in determining their respective ratios. For example, UUG, with a probability of $p(\text{UUG}) = 9/64$, would be expected only one-third as often as UUU, with its probability of $p(\text{UUU}) = 27/64$. Stated alternatively, $p(\text{UUG})/p(\text{UUU}) = 9/27 = 1/3 = 0.33$, which is the ratio for UUG given in Table 13-3.

If these codons each code for a different amino acid (that is, are not redundant), we would expect the amino acids generated by this particular mix of guanine and uracil to occur in ratios similar to those of the various codons. Although there is some redundancy among these codons, the ratios of the amino acids actually obtained from this mix of bases (see Table 13-4) are indeed quite similar to the ratios seen for the codon frequencies in Table 13-3. (In Table 13-4, phenylalanine is used as the baseline in determining ratios.)

From this evidence, we can deduce that codons consisting of one guanine and two uracils (G + 2U) code for valine, leucine, and cysteine, although we cannot distinguish the specific sequence for each of these amino acids. Similarly, one uracil and two guanines (U + 2G) must code for tryptophan, glycine, and perhaps one other. It looks as though the Watson-Crick model is correct in predicting the importance of the precise sequence (not just the ratios of bases). Many provisional assignments (such as those just outlined for G and U) were soon obtained, primarily by groups working with Nirenberg or with Severo Ochoa.

Table 13-3. Expected Frequencies of Various Codons in Synthetic mRNA Composed of $\frac{3}{4}$ Uracil and $\frac{1}{4}$ Guanine

Codon	Probability	Ratio*
UUU	$p(\text{UUU}) = \frac{3}{4} \times \frac{3}{4} \times \frac{3}{4} = \frac{27}{64}$	1.00
UUG	$p(\text{UUG}) = \frac{3}{4} \times \frac{3}{4} \times \frac{1}{4} = \frac{9}{64}$	0.33
UGU	$p(\text{UGU}) = \frac{3}{4} \times \frac{1}{4} \times \frac{3}{4} = \frac{9}{64}$	0.33
GUU	$p(\text{GUU}) = \frac{1}{4} \times \frac{3}{4} \times \frac{3}{4} = \frac{9}{64}$	0.33
UGG	$p(\text{UGG}) = \frac{3}{4} \times \frac{1}{4} \times \frac{1}{4} = \frac{3}{64}$	0.11
GGU	$p(\text{GGU}) = \frac{1}{4} \times \frac{1}{4} \times \frac{3}{4} = \frac{3}{64}$	0.11
GUG	$p(\text{GUG}) = \frac{1}{4} \times \frac{3}{4} \times \frac{1}{4} = \frac{3}{64}$	0.11
GGG	$p(\text{GGG}) = \frac{1}{4} \times \frac{1}{4} \times \frac{1}{4} = \frac{1}{64}$	0.03

* UUU is used as the baseline frequency against which the frequencies of the other codons are measured in establishing the respective ratios. For example, the ratio for UUG is derived from $p(\text{UUG})/p(\text{UUU}) =$ 0.33.

Table 13-4. Observed Frequencies of Various Amino Acids in Protein Translated from mRNA Composed of $\frac{3}{4}$ Uracil and $\frac{1}{4}$ Guanine

Amino acid	Ratio*
Phenylalanine	1.00
Leucine	0.37
Valine	0.36
Cysteine	0.35
Tryptophan	0.14
Glycine	0.12

* Phenylalanine is used as the baseline concentration against which the concentrations of other amino acids are measured in deriving their respective ratios. Note the correlations with the ratios in Table 13-3.

Before we consider other code words, we will examine tRNA molecules, which further explain the link between the mRNA codon and amino acid recognition.

tRNA Recognition of the Codon

Is it the tRNA or the amino acid itself that recognizes the portion of the mRNA, or codon, that codes for a specific amino acid? A very convincing experiment has answered this question. In the experiment, cysteinyl-tRNA (tRNA^{Cys}, the tRNA specific for cysteine) charged with cysteine was treated with nickel hydride, which converted the cysteine (while still bound to tRNA^{Cys}) into another amino acid, alanine, without affecting the tRNA:

$$\text{Cysteine—tRNA}^{\text{Cys}} \xrightarrow{\text{nickel hydride}} \text{alanine—tRNA}^{\text{Cys}}$$

Protein synthesized with this hybrid species had alanine wherever we would expect cysteine. Thus, the experiment demonstrated that the amino acids are illiterate; they are inserted at the proper position because the tRNA "adapters" recognize the mRNA codons and insert their attached amino acids appropriately. We would expect, then, to find some site on the tRNA that recognizes the mRNA codon by complementary base pairing.

Figure 13-17a shows several functional sites of the tRNA molecule. The site that recognizes an mRNA codon is called the **anticodon;** its bases are complementary and antiparallel to the bases of the codon. Another operationally identifiable site is the amino acid attachment site. The other arms probably assist in binding the tRNA to the ribosome. Figure 13-17b shows a specific tRNA (yeast alanine tRNA). The "flattened" cloverleafs shown in these diagrams are not the normal conformation of tRNA molecules; tRNA normally exists as an

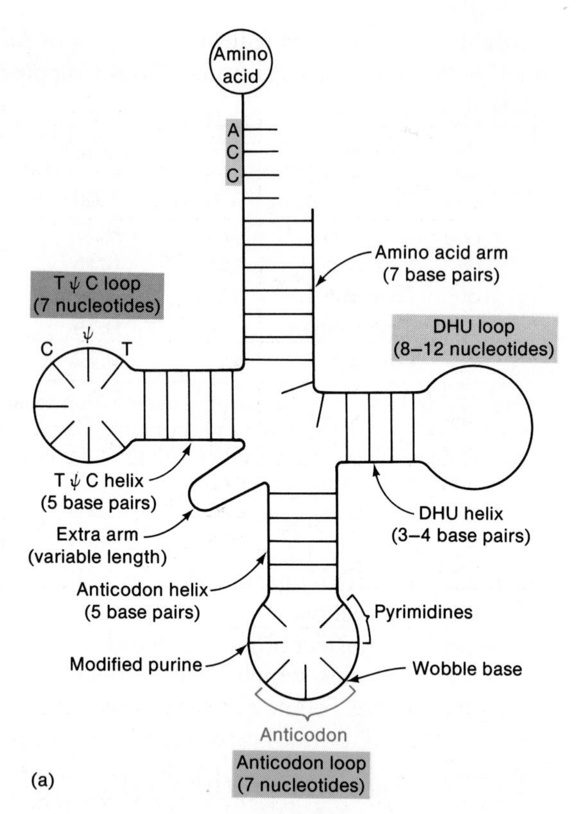

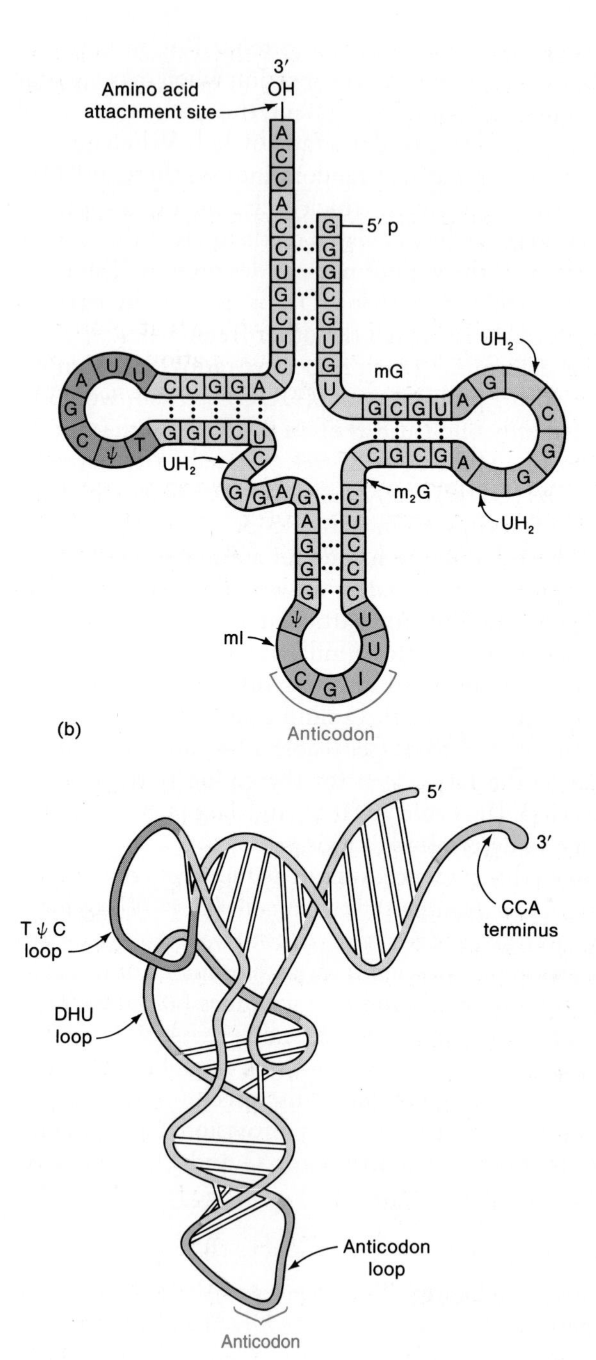

Figure 13-17 The structure of transfer RNA. (a) The functional areas of a generalized tRNA molecule. (b) The specific sequence of yeast alanine tRNA. Arrows indicate several kinds of rare modified bases. (c) The actual three-dimensional structure of yeast phenylalanine tRNA. The symbols ψ, mG, m_2G, mI, and DHU (or UH_2) are abbreviations for modified bases pseudouridine, methylguanosine, dimethylguanosine, methylinosine, and dihydrouridine, respectively. [Part (a) from S. Arnott, "The Structure of Transfer RNA," *Progress in Biophysics and Molecular Biology* 22, 1971, 186; parts (b) and (c) from L. Stryer, *Biochemistry,* 2d ed. Copyright 1981 by W. H. Freeman and Company; part (c) based on a drawing by Dr. Sung-Hou Kim.]

L-shaped folded cloverleaf, as shown in Figure 13-17c. These diagrams are supported by very sophisticated chemical analysis of tRNA nucleotide sequences and by X-ray crystallographic data on the overall shape of the molecule. Although tRNA molecules share many structural similarities, each has a unique three-dimensional shape that allows recognition by the correct synthetase, which catalyzes the joining of a tRNA with its specific amino acid to form an aminoacyl-tRNA. (Synthetases will be discussed under "Protein Synthesis.") The specificity of charging the tRNAs is crucial to the integrity of protein synthesis.

Where does tRNA come from? If radioactive tRNA is put into a cell nucleus in which the DNA has been partially denatured by heating, the radioactivity appears (by autoradiography) in localized regions of the chromosomes. These regions probably reflect the location of tRNA genes; they are regions of DNA that produce tRNA rather than mRNA, which produces a protein. The labeled tRNA hybridizes to these sites because of the complementarity of base sequences between the tRNA and its parent gene. A similar situation holds for rRNA. Thus, we see that even the one-gene–one-polypeptide idea is not completely valid. Some genes do not code for protein; rather, they specify RNA components of the translational apparatus.

Message Some genes code for proteins; other genes specify RNA (tRNA or rRNA) as their final product.

How does tRNA get its fancy shape? It probably folds up spontaneously into a conformation that produces maximal stability. Transfer RNA contains many "odd" or modified bases (such as pseudouracil, ψ) in its nucleotides; these play a direct role in folding and also have been implicated in other tRNA functions. You may have noticed some unusual base pairing within the loops of the tRNA in Figure 13-17b; G is hydrogen-bonded to U (instead of C). This apparent mismatching will be discussed in the next section.

First letter	Second letter: U	Second letter: C	Second letter: A	Second letter: G	Third letter
U	UUU Phe	UCU Ser	UAU Tyr	UGU Cys	U
	UUC Phe	UCC Ser	UAC Tyr	UGC Cys	C
	UUA Leu	UCA Ser	UAA Stop	UGA Stop	A
	UUG Leu	UCG Ser	UAG Stop	UGG Trp	G
C	CUU Leu	CCU Pro	CAU His	CGU Arg	U
	CUC Leu	CCC Pro	CAC His	CGC Arg	C
	CUA Leu	CCA Pro	CAA Gln	CGA Arg	A
	CUG Leu	CCG Pro	CAG Gln	CGG Arg	G
A	AUU Ile	ACU Thr	AAU Asn	AGU Ser	U
	AUC Ile	ACC Thr	AAC Asn	AGC Ser	C
	AUA Ile	ACA Thr	AAA Lys	AGA Arg	A
	AUG Met	ACG Thr	AAG Lys	AGG Arg	G
G	GUU Val	GCU Ala	GAU Asp	GGU Gly	U
	GUC Val	GCC Ala	GAC Asp	GGC Gly	C
	GUA Val	GCA Ala	GAA Glu	GGA Gly	A
	GUG Val	GCG Ala	GAG Glu	GGG Gly	G

Figure 13-18 The genetic code.

New Code Words

Specific code words were finally deciphered through two kinds of experiments. The first involved making "mini mRNAs," each only three nucleotides in length. These, of course, are too short to promote translation into protein, but they do stimulate the binding of aminoacyl-tRNAs (aa-tRNA's) to ribosomes in a kind of abortive attempt at translation. It is possible to make a specific mini mRNA and determine *which* aminoacyl-tRNA it will bind to ribosomes.

For example, the G + 2U problem discussed previously can be resolved by using the following mini mRNAs:

GUU—stimulates binding of valyl-tRNA

UUG—stimulates binding of leucyl-tRNA

UGU—stimulates binding of cysteinyl-tRNA

Analogous mini RNAs provided a virtually complete cracking of all the $4^3 = 64$ possible codons.

The second kind of experiment that was useful in cracking the genetic code involved the use of *repeating copolymers.* For instance, the copolymer designated $(AGA)_n$, which is a long sequence of AGAAGAAGAAGAAGA, was used to stimulate polypeptide synthesis in vitro. From the sequence of the resulting polypeptides and the possible triplets that could occur in the respective RNA copolymer, many code words could be verified. (This kind of experiment is detailed in Problem 10 at the end of this chapter. In solving it, you can put yourself in the place of H. Gobind Khorana, who received a Nobel prize for directing the experiments.)

Figure 13-18 gives the genetic code dictionary of 64 words. Inspect this figure carefully, and ponder the miracle of molecular genetics. Such an inspection should reveal several points that require further explanation.

Multiple Codons for a Single Amino Acid

As we saw in our discussion of degeneracy, the number of codons for a single amino acid varies, ranging from one (tryptophan = UGG) to as many as six (serine = UCU or UCC or UCA or UCG or AGU or AGC). Why? The answer is complex but not difficult; it can be divided into two parts:

1. Certain amino acids can be brought to the ribosome by several *alternative* tRNA types (species) having different anticodons, whereas certain other amino acids are brought to the ribosome by only one tRNA.
2. Certain tRNA species can bring their specific amino acids in response to several codons, not just one, through a loose kind of base pairing at one end of the codon and anticodon. This sloppy pairing is called **wobble.**

Message The degree of degeneracy for a given amino acid is determined by the number of codons for that amino acid that have only one tRNA each, plus the number of codons for the amino acid that share a tRNA through wobble.

We had better consider wobble first, and it will lead us into a discussion of the various species of tRNA. Wobble is caused by the third nucleotide of an anticodon (at

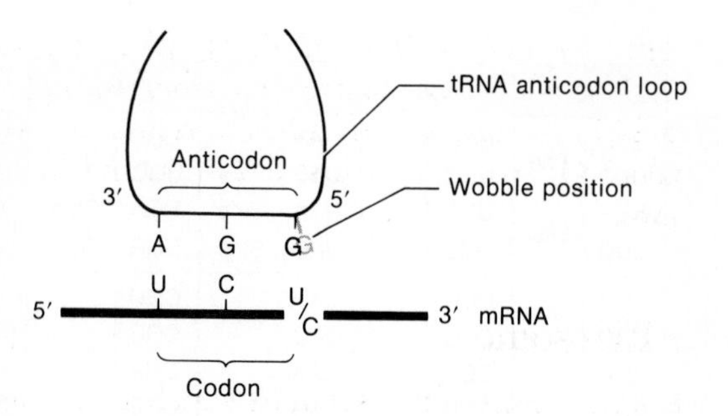

Figure 13-19 In the third site (5′ end) of the anticodon, G can take either of two wobble positions, thus being able to pair with either U or C. This means that a single tRNA species carrying an amino acid (in this case, serine) can recognize two codons—UCU and UCC—in the mRNA.

the 5′ end) that is not quite aligned (Figure 13-19). This out-of-line nucleotide sometimes can form hydrogen bonds not only with its normal complementary nucleotide in the third position of the codon but also with a different nucleotide in that position. Crick established certain "wobble rules" that dictate which nucleotides can and cannot form new hydrogen-bonded associations through wobble (Table 13-5). In the table, I (inosine) is one of the rare bases found in tRNA, often in the anticodon.

Figure 13-19 shows the possible codons that one tRNA serine species can recognize. As the wobble rules indicate, G can pair with U or with C. Table 13-6 lists all the codons for serine and shows how different tRNAs can service these codons. This is a good example of the effects of wobble on the genetic code.

Sometimes there can be an additional tRNA species that we represent as $tRNA^{Ser_4}$; it has an anticodon identical with any of the three anticodons shown in Table 13-6, but it differs in its nucleotide sequence elsewhere in the tRNA molecule. These four tRNAs are called **isoaccepting tRNAs** because they accept the same amino acid, but they are probably all transcribed from different tRNA genes.

Table 13.5. Codon-Anticodon Pairings Allowed by the Wobble Rules

5′ end of anticodon	3′ end of codon
G	U or C
C	G only
A	U only
U	A or G
I	U, C, or A

Table 13-6. Different tRNAs That Can Service Codons for Serine

Codon	tRNA	Anticodon
UCU UCC	$tRNA^{Ser_1}$	AGG + wobble
UCA UCG	$tRNA^{Ser_2}$	AGU + wobble
AGU AGC	$tRNA^{Ser_3}$	UCG + wobble

Stop Codons

The second point you may have noticed in Figure 13-18 is that some codons do not specify an amino acid at all. These codons are labeled as **stop** or **termination codons.** They can be regarded as similar to periods or commas punctuating the message encoded in the DNA.

One of the first indications of the existence of stop codons came in 1965 from Brenner's work with the T4 phage. Brenner analyzed certain mutations ($m_1 - m_6$) in a single cistron that controls the head protein of the phage. These mutants had two things in common. First, the head protein of each mutant was a shorter polypeptide chain than that of the wild type. Second, the presence of a suppressor mutation (*su*) in the host chromosome would cause the phage to develop a head protein of normal (wild-type) chain length despite the presence of the *m* mutation (Figure 13-20).

Brenner examined the ends of the shortened proteins and compared them with wild-type protein, recording for each mutant the next amino acid that *would* have been inserted to continue the wild-type chain. These amino acids for the six mutations were glutamine, lysine, glutamic acid, tyrosine, tryptophan, and serine. There is no immediately obvious pattern to these results, but Brenner brilliantly deduced that certain codons for each of these

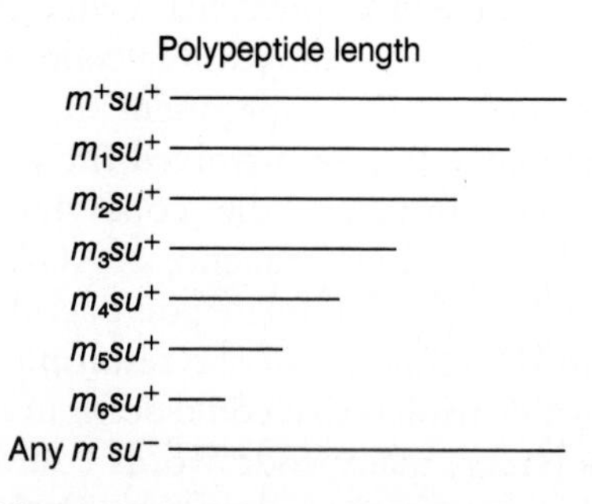

Figure 13-20 Polypeptide chain lengths of phage T4 head protein in wild type *(top)* and various amber mutants (*m*). An amber suppressor (*su*) leads to phenotypic development of the wild-type chain.

amino acids are similar in that each of them can mutate to the codon UAG by a single change in a DNA nucleotide pair. He therefore postulated that UAG is a stop (or termination) codon — a signal to the translation mechanism that the protein is now complete.

UAG was the first stop codon deciphered; it is called the **amber codon.** Mutants that are defective owing to the presence of an abnormal amber codon are called *amber mutants,* and their suppressors are *amber suppressors.* UGA, the **opal codon,** and UAA, the **ochre codon,** are also stop codons and also have suppressors. Stop codons often are called **nonsense codons** because they designate no amino acid. Not surprisingly, stop codons do not act as mini mRNAs in binding aa-tRNA to ribosomes in vitro. We will discuss stop codons and suppressors further after we consider the process of protein synthesis.

Protein Synthesis

We can regard **protein synthesis** as a chemical reaction, and we shall take this approach at first. Then we shall take a three-dimensional look at the physical interactions of the major components.

In protein synthesis as a chemical reaction:

1. Each amino acid (aa) is attached to a tRNA molecule specific to that amino acid by a high-energy bond derived from ATP. The process is catalyzed by a specific enzyme called a **synthetase** (the tRNA is said to be "charged" when the amino acid is attached):

$$\text{aa}_1 + \text{tRNA}_1 + \text{ATP} \xrightarrow{\text{synthetase}_1} \text{aa}_1\text{—tRNA}_1 + \text{ADP}$$

   There is a separate synthetase for each amino acid.

2. The energy of the charged tRNA is converted into a peptide bond linking the amino acid to another one on the ribosome:

$$\text{aa}_1\text{—tRNA}_1 + \text{aa}_2\text{—tRNA}_2 \xrightarrow[\text{on a ribosome}]{\text{peptidyl transferase}} \underbrace{\text{aa}_1\text{—aa}_2}_{\text{Small polypeptide}}\text{—tRNA}_2 + \text{tRNA}_1 \text{ (released)}$$

3. New amino acids are linked by means of a peptide bond to the growing chain:

$$\text{aa}_3\text{—tRNA}_3 + \text{aa}_1\text{—aa}_2\text{—tRNA}_2 \longrightarrow \underbrace{\text{aa}_1\text{—aa}_2\text{—aa}_3}_{\text{Larger polypeptide}}\text{—tRNA}_3 + \text{tRNA}_2 \text{ (released)}$$

4. This process continues until aa_n (the final amino acid) is added. Of course, the whole thing works only in the presence of mRNA, ribosomes, several additional protein factors, enzymes, and inorganic ions.

The Ribosome

Ribosomes consist of two subunits which in prokaryotes sediment as 50S and 30S particles and which associate to form a 70S particle, as seen in Figure 13-21a. The eukaryotic counterparts are 60S and 40S for the large and small subunits, and 80S for the complete ribosome (Figure 13-21b). Ribosomes contain specific sites that enable them to bind to the mRNA, the tRNAs, and specific protein factors required for protein synthesis. Let's look at a general picture of protein synthesis on the ribosome, and then examine each of the steps in the process in more detail.

Figure 13-22 shows a polypeptide being synthesized on the ribosome. The mRNA binds to the 30S subunit. The tRNAs bind to two sites on the ribosome. These sites overlap the subunits. The **A site** is the entry site for an aminoacyl-tRNA (a tRNA carrying a single amino acid). The peptidyl-tRNA carrying the growing polypeptide chain binds at the **P site.** Each new amino acid is added by the transfer of the growing chain to the new aminoacyl-tRNA, forming a new peptide bond. The deacylated tRNA is then released from the P site, and the ribosome moves one codon farther along the message, transferring the new peptidyl-tRNA to the P site, and leaving the A site vacant for the next incoming aminoacyl-tRNA.

We can separate the process of protein synthesis into three distinct steps. **Initiation, elongation,** and **termination.** Let's examine each of these steps in detail, using prokaryotes as an example.

Initiation

Three Steps of Initiation. In addition to mRNA, ribosomes, and specific tRNA molecules, initiation requires the participation of several factors, termed **initiation factors IF1, IF2,** and **IF3.** In *E. coli* and in most other prokaryotic organisms, the first amino acid in any newly synthesized polypeptide is *N*-formylmethionine. It is inserted not by tRNA^{Met}, however, but by an **initiator tRNA** called $\text{tRNA}^{\text{fMet}}$. This initiator tRNA has the normal methionine anticodon but inserts *N*-formylmethionine rather than methionine (Figure 13-23). In *E. coli,* AUG and GUG, and on rare occasions UUG, serve as initiation codons. When one of these triplets occurs in the initiation position, it is recognized by *N*-formylMet-tRNA and methionine appears as the first

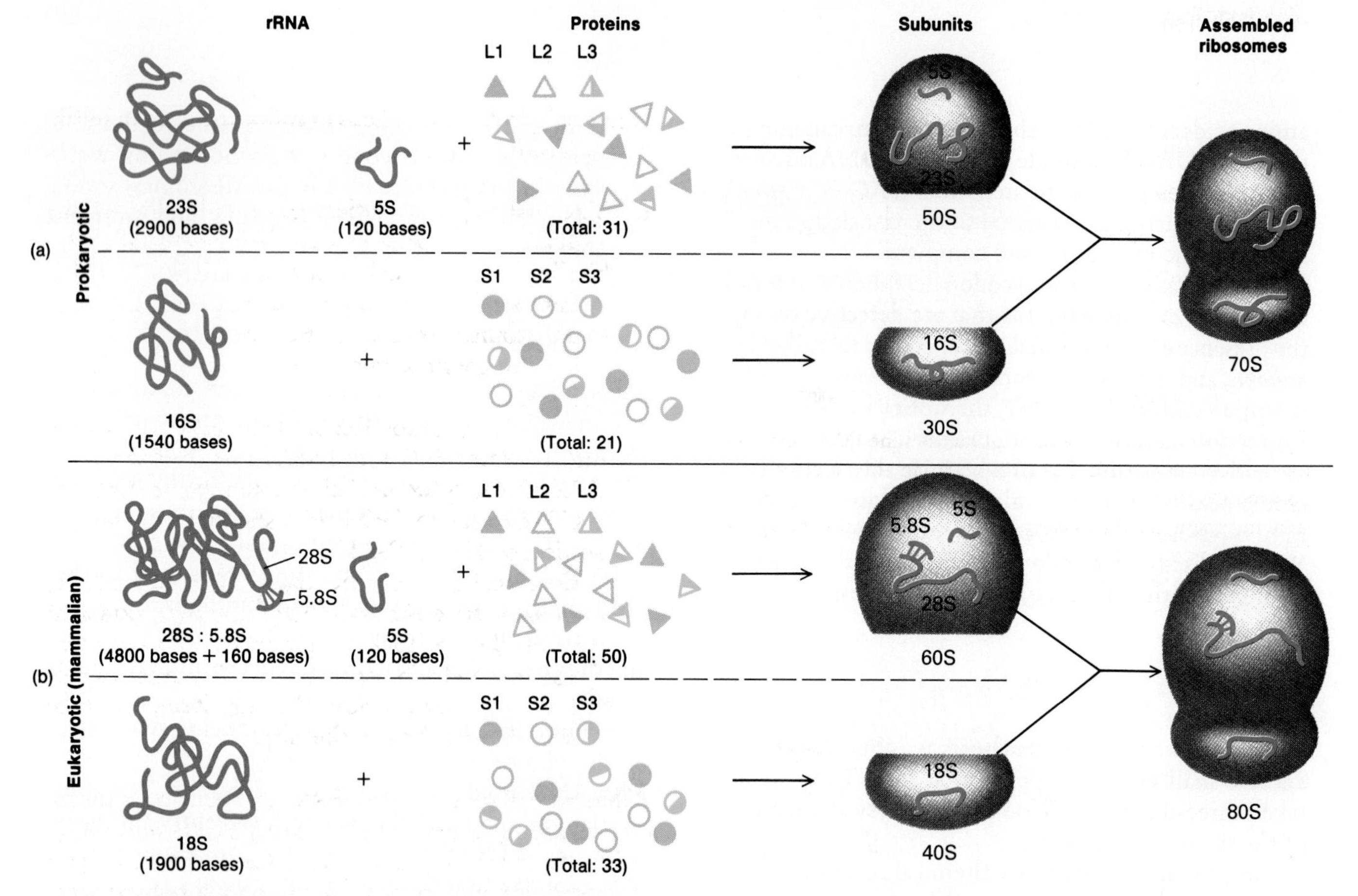

Figure 13-21 Ribosomes contain a large and a small subunit. Each subunit contains both rRNA of varying lengths and a set of proteins (designated by different colors). There are two principal rRNA molecules in all ribosomes. Ribosomes from prokaryotes also contain one 120-base-long rRNA which sediments at 5S, whereas eukaryotic ribosomes have two small rRNAs: a 5S RNA molecule similar to the prokaryotic 5S, and a 5.8S molecule 160 bases long. The large subunit proteins are named L1, L2, etc., and the small subunit proteins S1, S2, etc. (From J. Darnell, H. Lodish, and D. Baltimore, *Molecular Cell Biology,* 2d ed. Copyright 1990 by Scientific American Books, Inc.)

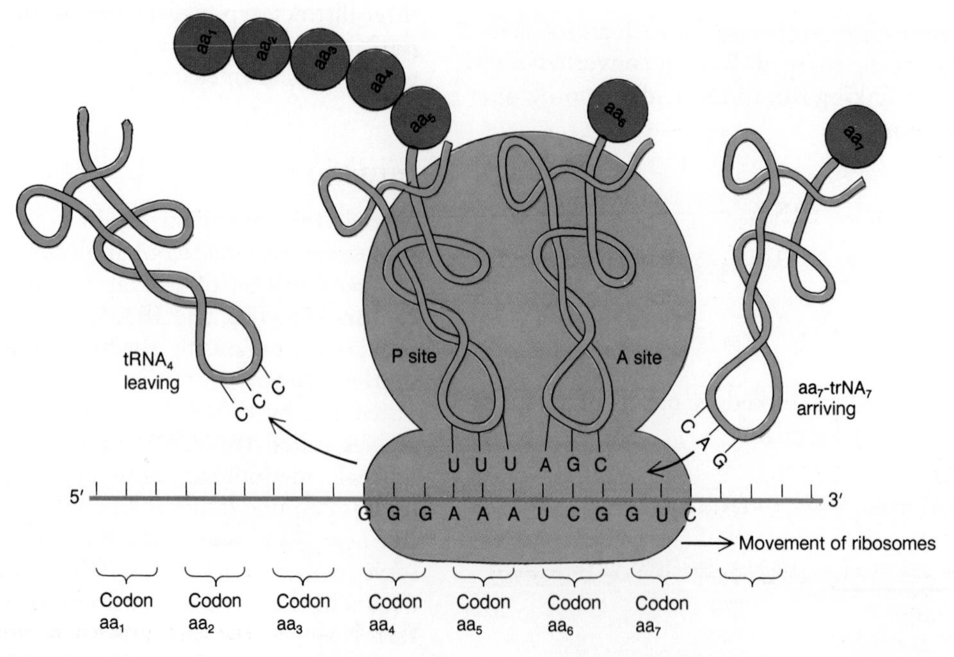

Figure 13-22 The addition of a single amino acid to the growing polypeptide chain during translation of mRNA.

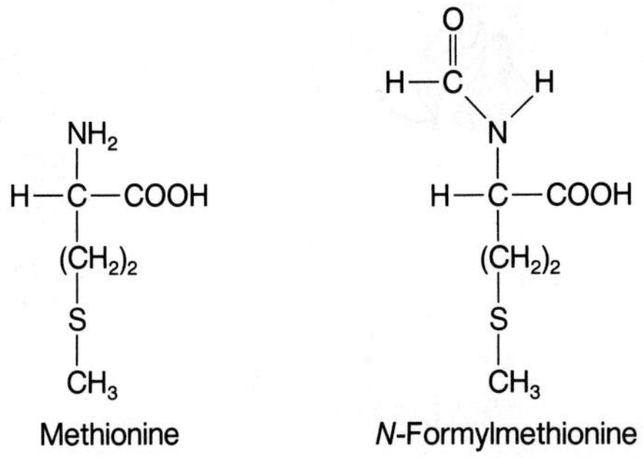

NH$_2$ | H—C—COOH | (CH$_2$)$_2$ | S | CH$_3$
Methionine

O=C—H | N—H | H—C—COOH | (CH$_2$)$_2$ | S | CH$_3$
N-Formylmethionine

Figure 13-23 The structures of methionine (Met) and *N*-formylmethionine (fMet). A tRNA bearing fMet can initiate a polypeptide chain in prokaryotes but cannot be inserted in a growing chain; a tRNA bearing Met can be inserted in a growing chain but will not initiate a new chain. Both these tRNAs bear the same anticodon complementing the codon AUG.

amino acid in the chain. Let's examine the steps in initiation in detail.

1. The first step in initiation involves binding of the mRNA to the 30S subunit (Figure 13-24). The binding is stimulated by the initiation factor IF3. When not engaged in protein synthesis, the ribosomal subunits exist in the free form; they assemble into complete ribosomes as a result of the initiation process.
2. The initiation factor IF2 binds to GTP and to the initiator fMet-tRNA, and stimulates the binding of fMet-tRNA to the initiation complex, leading the fMET-tRNA into the P site, as shown in the middle portion of Figure 13-24.
3. A ribosomal protein splits the GTP bound to IF2, helping to drive the assembly of the two ribosomal subunits (Figure 13-24, bottom). At this stage the factors IF2 and IF3 are released. (The exact role of IF1 is not completely clear, although it seems to take part in the recycling of the ribosome.)

Ribosome Binding Sites. How are the correct initiation codons selected from the many AUG and GUG codons in a mRNA molecule? John Shine and Lynn Dalgarno first noticed that true initiation codons were preceded by sequences that paired well with the 3′ end of 16S rRNA. Figure 13-25 shows some of these sequences.

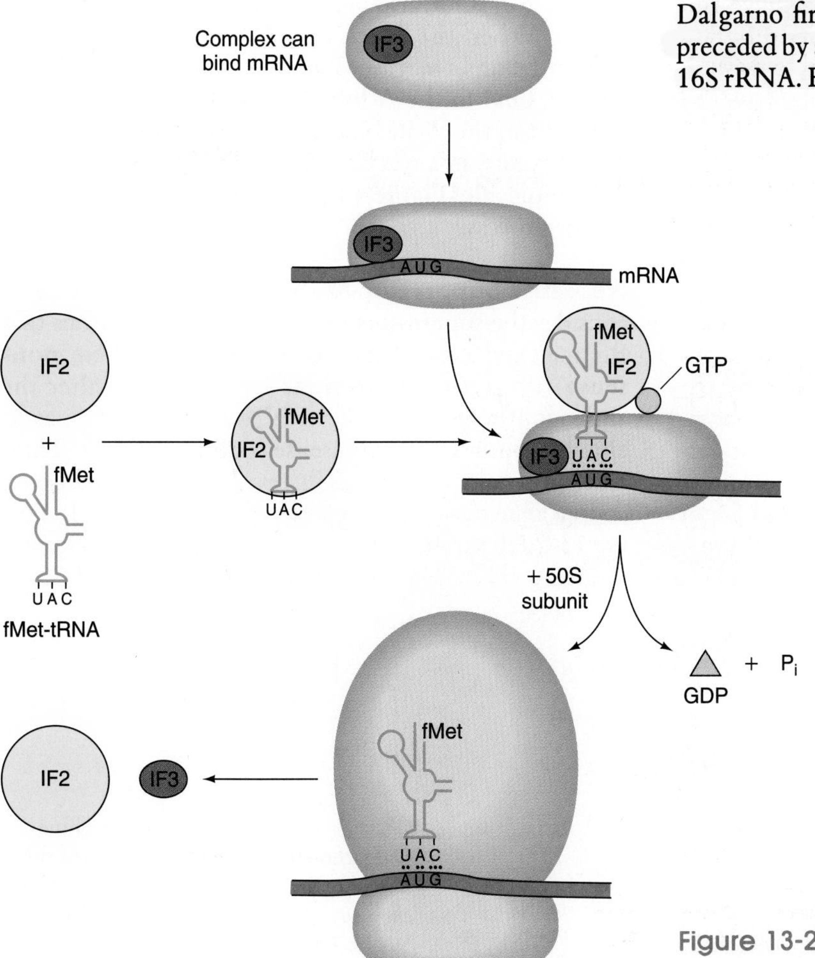

Figure 13-24 The steps involved in initiation of translation (see text).

AGCACGAGGGGAAAUCUGAUGGAACGCUAC	*E. coli trpA*
UUUGGAUGGAGUGAAACGAUGGCGAUUGCA	*E. coli araB*
GGUAACCAGGUAACAACCAUGCGAGUGUUG	*E. coli thrA*
CAAUUCAGGGUGGUGAAUGUGAAACCAGUA	*E. coli lacI*
AAUCUUGGAGGCUUUUUUAUGGUUCGUUCU	ϕX174 phage A protein
UAACUAAGGAUGAAAUGCAUGUCUAAGACA	Qβ phage replicase
UCCUAGGAGGUUUGACCUAUGCGAGCUUUU	R17 phage A protein
AUGUACUAAGGAGGUUGUAUGGAACAACGC	λ phage *cro*

Pairs with 16S rRNA — Pairs with initiator tRNA

Figure 13-25 Ribosomal binding site sequences in *E. coli* and its bacteriophages share certain common features, which are shown in the colored regions. The initiation codon (color) is separated by several bases from a short sequence (color) that is complementary to the 3′ end of 16S rRNA. (After L. Stryer, *Biochemistry,* 3d ed. Copyright 1988 by W. H. Freeman and Company.)

There is a short but variable separation between the Shine-Delgarno sequence and the initiation codon. Figure 13-26 depicts the base pairing between idealized mRNA and the 16S rRNA that results in ribosome-mRNA complexes leading to protein initiation in the presence of fMet-tRNA.

Elongation

Figure 13-27 details the steps in elongation, which are aided by three protein factors, EF-Tu, EF-Ts, and EF-G. The steps are as follows:

1. The elongation factor EF-Tu mediates the entry of aminoacyl-tRNAs into the A site. To achieve this EF-Tu first binds to GTP. This activated EF-Tu–GTP complex binds to the tRNA. Next, hydrolysis of the GTP of the complex to GDP helps drive the binding of the aminoacyl-tRNA to the A site, at which point the EF-Tu is released (Figure 13-27a), leaving the new tRNA in the A site (Figure 13-27b).
2. The elongation factor EF-Ts mediates the release of EF-Tu–GDP from the ribosome and also the regeneration of EF-Tu–GTP.
3. In the "translocation" step, the polypeptide chain on the peptidyl-tRNA is transferred to the aminoacyl-tRNA on the A site in a reaction catalyzed by the enzyme peptidyltransferase (Figure 13-27c). The ribosome then translocates by moving one codon farther along the mRNA, going in the 5′ → 3′ direction. This step is mediated by the elongation factor EF-G (Figure 13-27d) and is driven by splitting a GTP to GDP. This action releases the uncharged tRNA from the P site and transfers the newly formed peptidyl-tRNA from the A site to the P site (Figure 13-27e).

Termination

Release Factors. In our previous discussion of the genetic code, we described the three chain-termination codons UAG, UAA, and UGA. Interestingly, these three triplets are not recognized by a tRNA, but instead by protein factors, termed **release factors,** which are abbreviated **RF1** and **RF2.** RF1 recognizes the triplets UAA and UAG, and RF2 the triplets UAA and UGA. A third factor, **RF3,** also helps to catalyze the chain termination. When the peptidyl-tRNA is in the P site, the release factors, in response to the chain-terminating codons, bind to the A site. The polypeptide is then released from the P site, and the ribosomes dissociate into two subunits, in a reaction driven by the hydrolysis of a GTP molecule. Figure 13-28 provides a schematic view of this process.

Nonsense Suppressor Mutations. It is interesting to consider the suppressors of the nonsense mutations that Brenner and coworkers defined. Many of these **nonsense suppressor mutations** are known to alter the anticodon loop of specific tRNAs in such a way as to allow recognition of a nonsense codon in mRNA. Thus, an amino acid is inserted in response to the nonsense codon, and translation continues past that triplet. In Figure 13-29 the amber mutation replaces a wild-type codon with the chain-terminating nonsense codon UAG. By

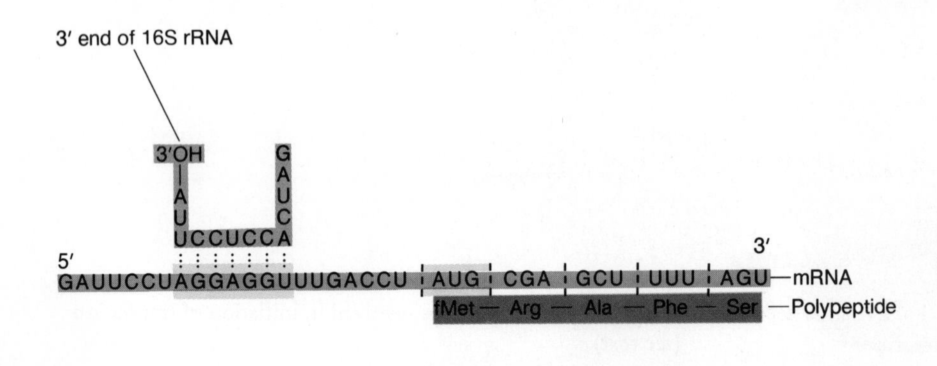

Figure 13-26 Binding of the Shine-Dalgarno sequence on an mRNA to the 3′ end of 16S rRNA. (After L. Stryer, *Biochemistry,* 3d ed. Copyright 1988 by W. H. Freeman and Company.)

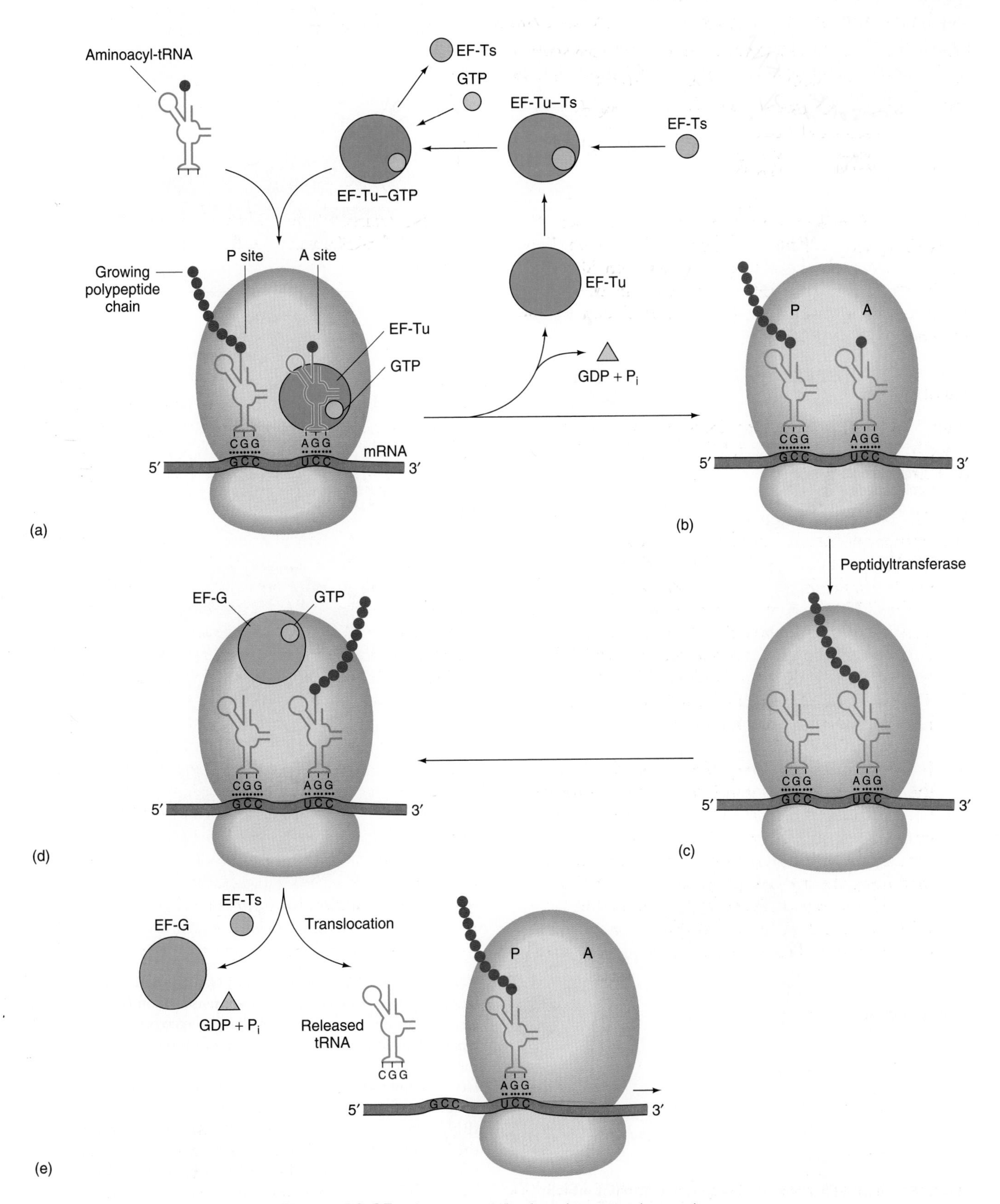

Figure 13-27 The steps involved in elongation (see text).

itself, the UAG would result in prematurely cutting off the protein at the corresponding position. The suppressor mutation in this case produces a $tRNA^{Tyr}$ with an anticodon that recognizes the mutant UAG stop codon. The suppressed mutant thus contains tyrosine at that position in the protein.

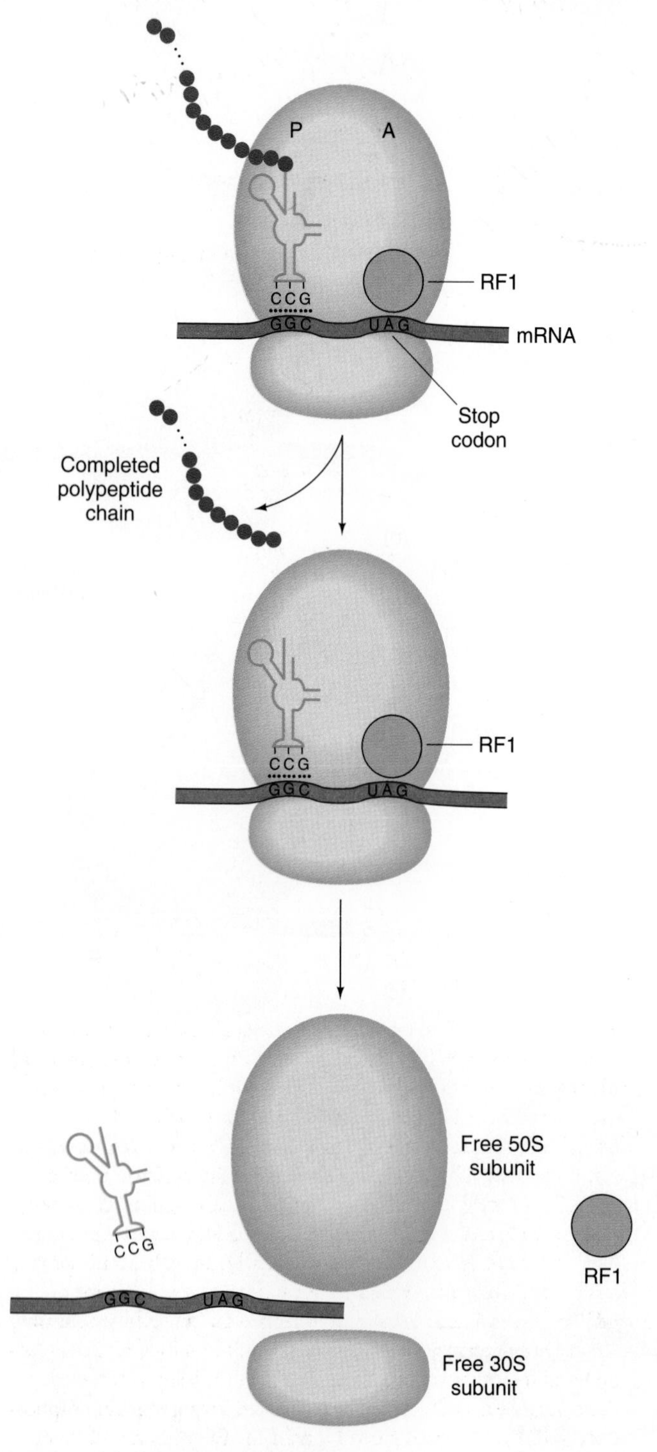

Figure 13-28 The steps leading to termination of protein synthesis (see text).

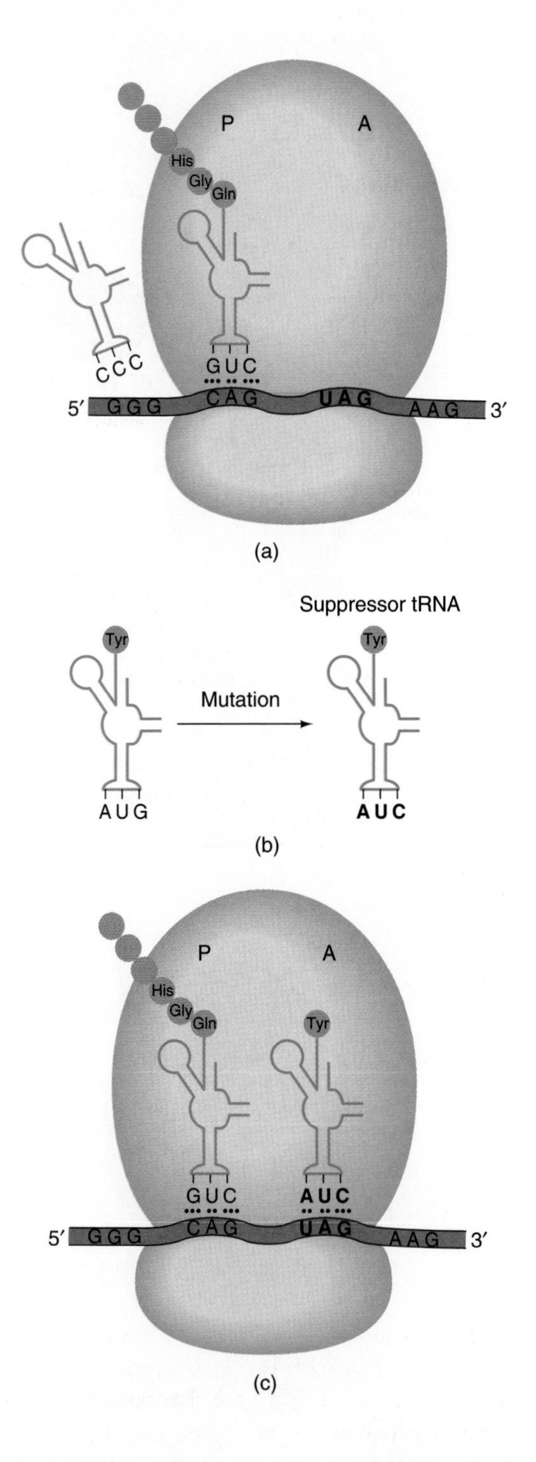

Figure 13-29 (a) Termination of translation. Here the translation apparatus cannot go past a nonsense codon (UAG in this case), because there is no tRNA that can recognize the UAG triplet. This leads to the termination of protein synthesis and to the subsequent release of the polypeptide fragment. The release factors are not shown here. (b) The molecular consequences of a mutation that alters the anticodon of a tyrosine tRNA. This tRNA can now read the UAG codon. (c) The suppression of the UAG codon by the altered tRNA, which now permits chain elongation. (Adapted from James D. Watson, John Tooze, and David T. Kurtz, *Recombinant DNA: A Short Course.* Copyright 1983 by W. H. Freeman and Company.)

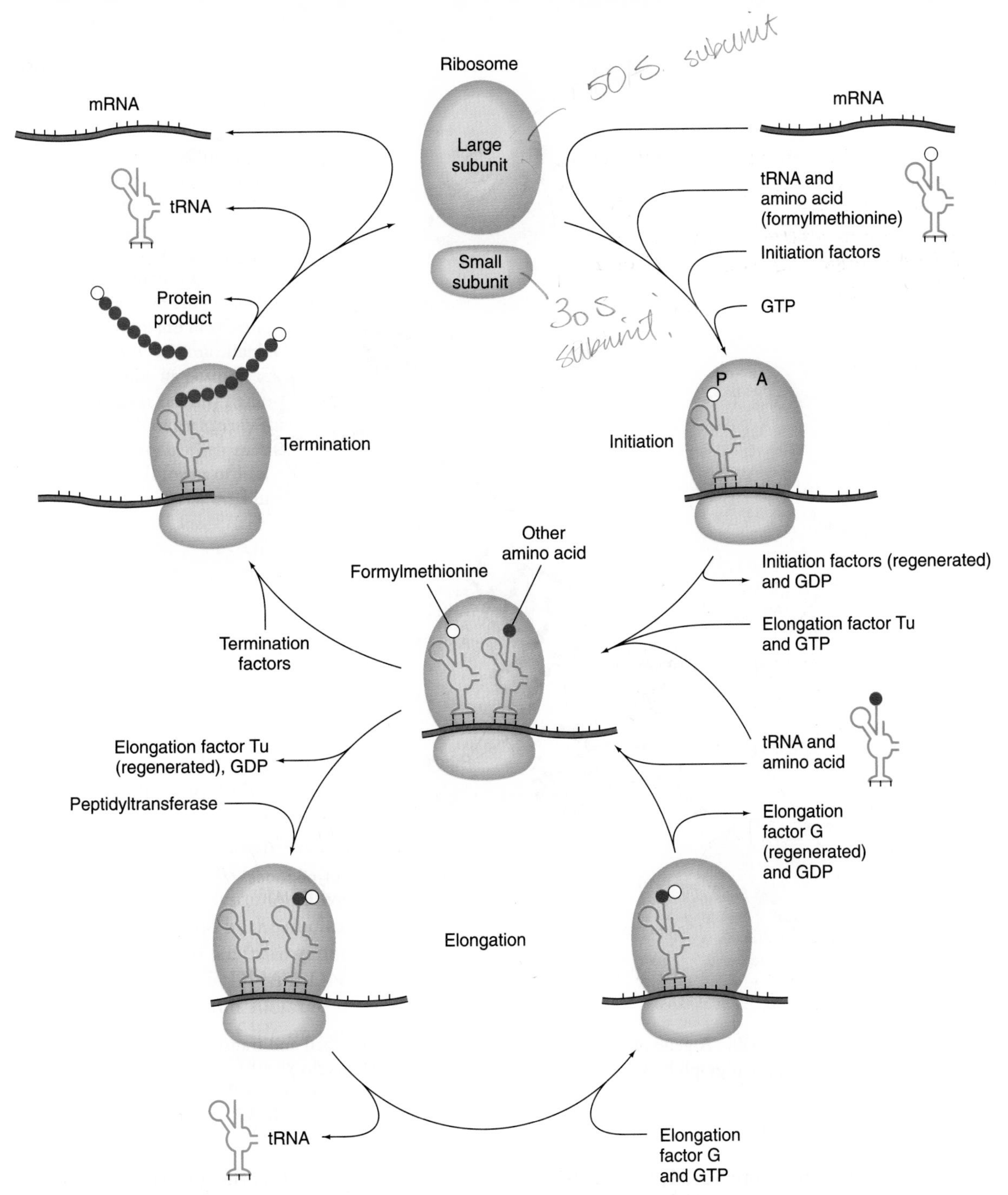

Figure 13-30 The transactions of the ribosome. At initiation, the ribosome recognizes the starting point in a segment of mRNA and binds a molecule of tRNA bearing a single amino acid. In all bacterial proteins, this first amino acid is *N*-formylmethionine. In elongation, a second amino acid is linked to the first one. The ribosome then shifts its position on the mRNA molecule, and the elongation cycle is repeated. When the stop codon is reached, the chain of amino acids folds spontaneously to form a protein. Subsequently, the ribosome splits into its two subunits, which rejoin before a new segment of mRNA is translated. Protein synthesis is facilitated by a number of catalytic proteins (initiation, elongation, and termination factors) and by guanosine triphosphate (GTP), a small molecule that releases energy when it is converted into guanosine diphosphate (GDP). (Adapted from Donald M. Engelman and Peter B. Moore, "Neutron-Scattering Studies of the Ribosome." Copyright 1976 by Scientific American, Inc. All rights reserved.)

What happens to normal termination signals at the ends of proteins in the presence of a suppressor? Many of the natural termination signals consist of two chain-termination signals in a row. Nonsense suppressors are sufficiently inefficient in translating through chain-terminating triplets that the probability of suppression at two codons in a row is small. Consequently, very few protein copies are produced that carry many extraneous amino acids resulting from translation beyond the natural stop codon.

Overview of Protein Synthesis

Figure 13-30 provides a summary of the steps in protein synthesis that we have covered in this section. A direct visualization of protein synthesis can be seen in the elec-

the fact that this is the only workable biochemical option in the earth environment (biochemical predestination)? Whatever the answer, the wonderful uniformity of the molecular basis of life is firmly established. Minor variations do exist, but they do not detract from the central uniformity of the mechanism we have described.

Message The processes of information storage, replication, transcription, and translation are fundamentally similar in all living systems. In demonstrating this fact, molecular genetics has provided a powerful unifying force in biology. We now know some of the tricks that life uses to achieve persistent order in a randomizing universe.

Eukaryotic RNA

Several aspects of RNA synthesis and processing in eukaryotes are distinctly different from their counterparts in prokaryotes.

RNA Synthesis

Whereas a single RNA polymerase species synthesizes all RNAs in prokaryotes, there are three different RNA polymerases in eukaryotic systems:

1. RNA polymerase I synthesizes rRNA.
2. RNA polymerase II synthesizes mRNA. The mRNA molecules are **monocistronic,** whereas many mRNAs are **polycistronic** (code for several cistrons) in prokaryotes.
3. RNA polymerase III synthesizes tRNAs and also small nuclear and cellular RNA molecules.

RNA Processing

The primary RNA transcript produced in the nucleus usually is processed in several ways before its transport to the cytoplasm, where it is used to program the translation machinery (Figure 13-33). Figure 13-34 depicts these processing events in detail. First a **cap** consisting of a 7-methylguanosine residue linked to the 5′ end of the transcript by a triphosphate bond is added during transcription. Then stretches of adenosine residues are added at the 3′ ends. These **poly(A) tails** are 150 to 200 residues long. Following these modifications, a crucial **splicing** step removes internal portions of the RNA transcript. The uncovering of this process, and the corresponding realization that genes are "split," with coding regions interrupted by "intervening sequences," constitutes one of the most important discoveries in molecular genetics in the last decade.

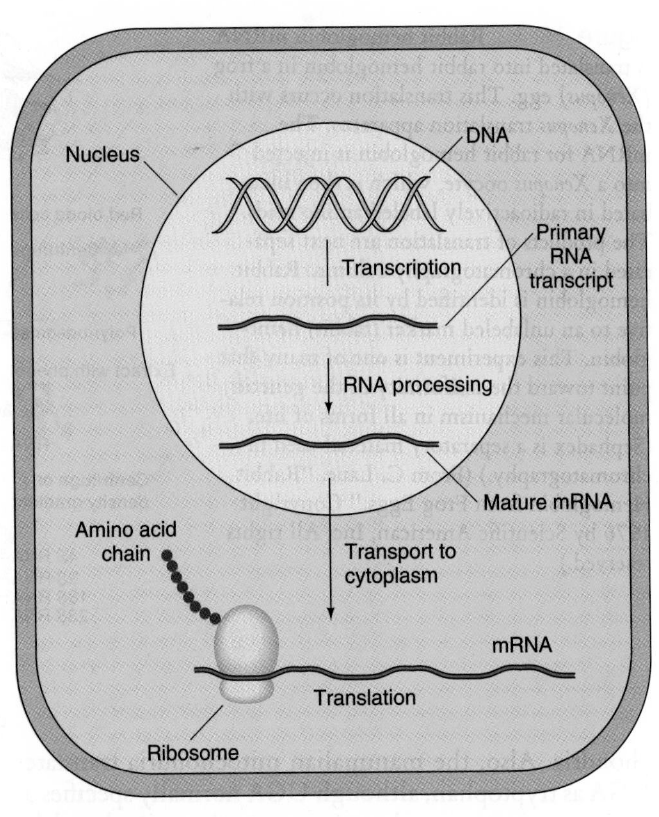

Figure 13-33 Gene expression in eukaryotes. The mRNA is processed in the nucleus before transport to the cytoplasm. (From J. E. Darnell, Jr., "The Processing of RNA." Copyright 1983 by Scientific American, Inc. All rights reserved.)

Split Genes

Studies of mammalian viral DNA transcripts first suggested a lack of correspondence between the genetic map and specific mRNA molecules. As recombinant DNA techniques (see Chapter 14) facilitated the physical analysis of eukaryotic genes, it became apparent that *primary* RNA transcripts were being shortened by the elimination of internal segments before transport into the cytoplasm. In most higher eukaryotes studied, this was found to be true not only for mRNA but also for rRNA—and even for tRNA in some cases.

Figure 13-35 shows the organization of the gene for chicken ovalbumin, a 386-amino acid polypeptide. The DNA segments that code for the structure of the protein are interrupted by intervening sequences, termed **introns.** In Figure 13-35, these segments are designated with the letters A to G. The primary transcript is processed by a series of "splicing" reactions, much in the same way that a tape-recorded message can be cut and pasted back together. Splicing removes the introns and brings together the coding regions, termed **exons,** to form an mRNA, which now consists of a sequence that is completely colinear with the ovalbumin protein. The

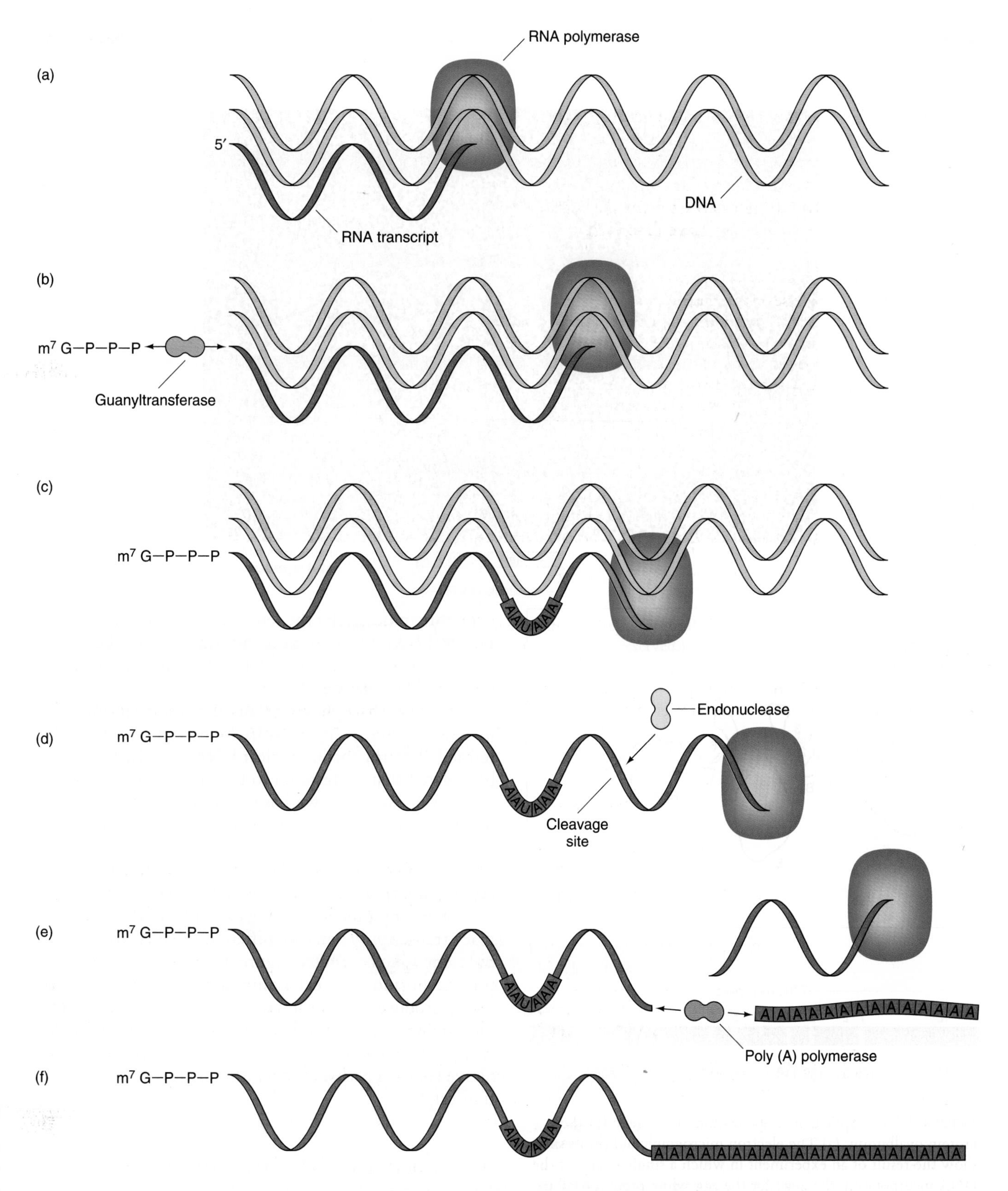

Figure 13-34 Processing of primary mRNA. (a) Transcription is mediated by RNA polymerase. (b) Early in transcription an enzyme, guanyltransferase, adds 7-methylguanosine (m^7Gppp) to the 5′ end of the mRNA. (c) The sequence AAUAAA, near the 3′ end, helps signal a cleavage event (d) by an endonuclease approximately 20 bp farther downstream. (e) An enzyme, poly(A) polymerase, then adds a poly(A) tail, made up of 150 to 200 adenosine residues, to the site of this cleavage at the 3′ end. (f) This yields the complete primary mRNA. (From J. E. Darnell, Jr., "The Processing of RNA." Copyright 1983 by Scientific American, Inc. All rights reserved.)

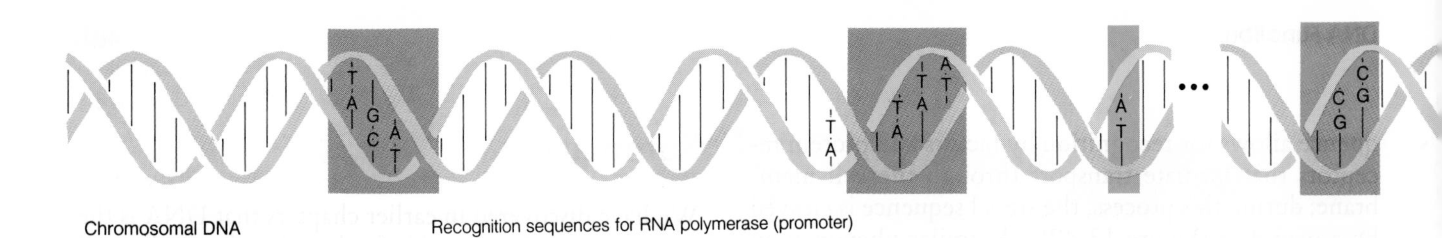

(a)

Initiation site

Coding strand

TGTTGACA TATAAT A GAGG

ACAACTGT ATATTA T CTCC

Template strand

(b)

Ribosomal binding site

mRNA A GAGG

Translation

Protein chain

(c)

Figure 13-44 Genetic information is stored in the double helix of DNA. (a) Each strand of the helix is a chain of nucleotides, each comprising a deoxyribose sugar and a phosphate group, which form the strand's backbone, as well as one of four bases: adenine (A), guanine (G), thymine (T), or cytosine (C). The information is encoded in the sequence of the bases along a strand. The complementarity of the bases (A always pairs with T and G with C) is the basis of the replication of DNA from generation to generation and of its expression (shown here for bacterial DNA) as protein. Transcription is regulated, in part, by the promoter site upstream from the initiation site. (b) and (c) Expression begins with the transcription of the DNA base sequence into a strand of mRNA, which corresponds to the coding strand of the DNA except for the fact that uracil (U) replaces thymine. Transcription into RNA and translation into protein are regulated by special sequences in the DNA and the RNA, respectively. The transcribing enzyme, RNA polymerase, binds to the promoter region before the transcription-initiation site; beyond the end of the structural gene, a termination region causes the polymerase to cease transcription. mRNA is translated on cellular organelles called ribosomes; each triplet of bases (codon) encodes a particular amino acid and specifies its incorporation into the growing protein chain. A ribosome binding site on the RNA allows translation to begin at a "start" codon, which is always AUG for the amino acid methionine (Met). Translation proceeds until a "stop" codon is reached (UAA is one of three possibilities), which signals the end of translation and the detachment of the completed protein chain from the ribosome.

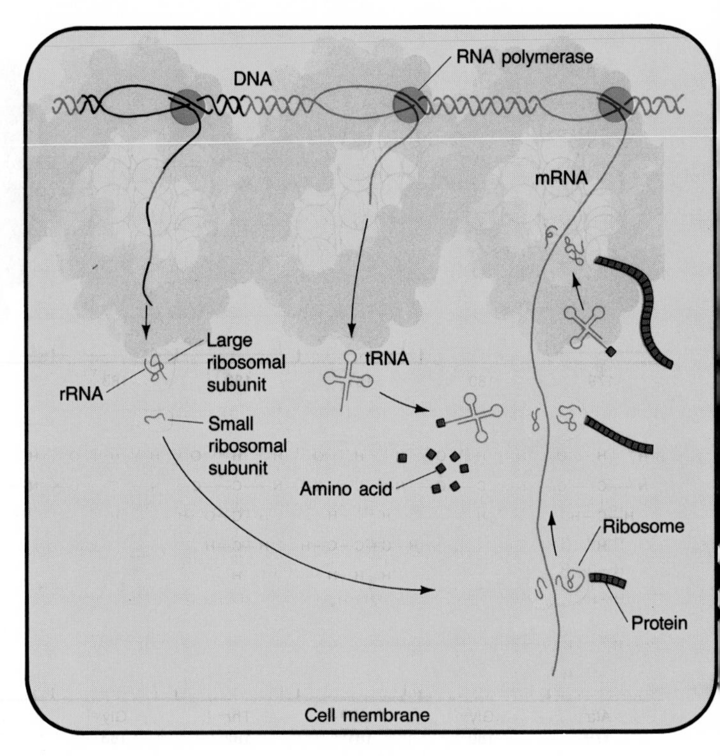

(a) Prokaryote

Figure 13-45 Protein synthesis involves the same three types of RNA in prokaryotic (a) and eukaryotic (b) cells, but with an important difference: in eukaryotes, the protein-coding sequences of DNA (exons) are often separated by intervening sequences (introns) that must be excised from a primary transcript to make mRNA. In both kinds of cell, tRNA, rRNA, and mRNA are made by the transcription of one strand of the DNA double helix. Three different polymerase enzymes catalyze these reactions in eukaryotes, whereas in prokaryotes there is only a single type of polymerase. In both kinds of cell, the tRNA and rRNA primary transcripts must be processed. The ends of the tRNA transcript are cut, and the molecule assumes a looped structure. A single rRNA transcript is cut in several places to form two major types of rRNA, which are then bound to protein molecules to form ribosomal subunits. In prokaryotes, which have no nucleus, the mRNA transcript is generally not processed; ribosomes and tRNAs carrying amino acids begin translating the mRNA into a sequence of amino acids (a protein) as it is being made. In eukaryotes, the nuclear envelope probably facilitates the removal of introns and the splicing of exons from the primary mRNA transcript by protecting it from immediate translation. The mRNA is "read" only after it (like tRNA and the ribosomal subunits) has exited the nucleus through pores in the envelope. Only in the cytoplasm are the mature mRNA, tRNA, and ribosomes united. (From J. E. Darnell, Jr., "RNA.")

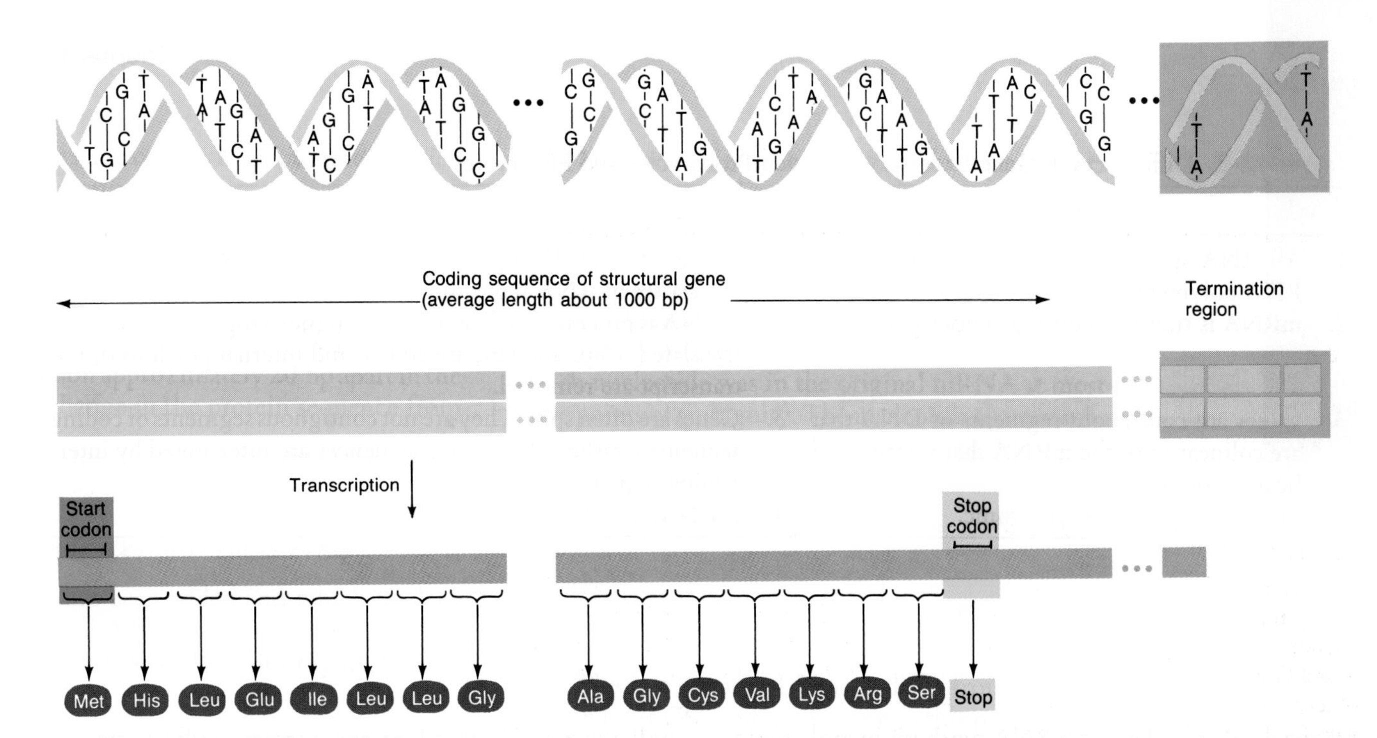

Messenger RNA (mRNA) molecules are of many sizes and base sequences. These are the molecules that contain information for the structure of proteins. The sequence of codons in mRNA determines the sequence of amino acids that will constitute a polypeptide. Each codon is specific for one amino acid, but several different codons may code for the same amino acid; that is, there is redundancy in the genetic code. In addition, there are three codons for which there are no tRNAs; these stop codons terminate the process of translation. Figure 13-44 summarizes the way in which genetic information is turned into protein.

RNA is processed in eukaryotes before transport to the cytoplasm. Caps and tails are added, and internal portions of the primary transcript are removed. Many genes are therefore "split" in eukaryotes, and the coding segments of a gene are not colinear with the processed mRNA. Figure 13-45 and Table 13-8 summarize some

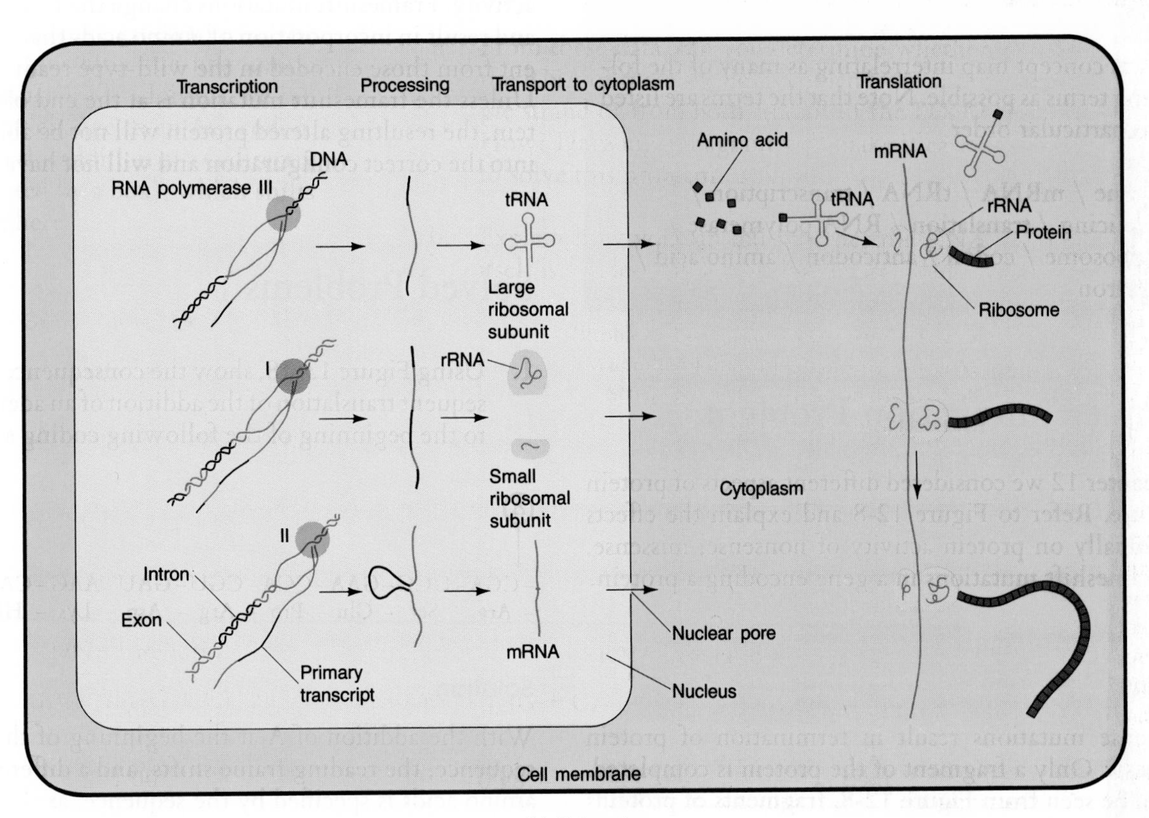

(b) Eukaryote

g. A child dies.

h. The oxygen-transport capacity of the body is severely impaired.

i. The tRNA anticodon that lines up is one of a type that brings an unsuitable amino acid.

j. Nucleotide-pair substitution occurs in the DNA of the gene for hemoglobin.

17. The comparison of physical distances with map distances is facilitated by our knowledge of the genetic code. The code permits us to "translate" numbers of amino acids in a protein into numbers of nucleotides in a corresponding stretch of DNA.

a. Consider two mutant forms of a particular enzyme in *Neurospora.* The two proteins are known to differ at only two positions in their polypeptide chains. The two positions are separated by 40 other amino acids. Crosses of the two mutants regularly give a recombinant frequency of 1.0×10^{-5} for the two mutations. Approximately how many nucleotides in *Neurospora* lie between a pair of markers that give a recombinant frequency of 10^{-5}?

b. Consider two other mutant forms of the enzyme. Each has the normal (wild-type) number of amino acids, but these enzymes differ from each other and from the wild-type enzyme at amino acid 68 in the polypeptide chain. The wild-type protein has arginine at position 68, whereas one mutant has glutamine and the other has serine. When mutants 1 and 2 are crossed, six kinds of offspring differing in the amino acid at position 68 are observed. Two of these are parental types in which glutamine or serine is present at position 68, and one has arginine like the wild-type. Three novel types—having lysine, asparagine, or histidine—make up the remainder. The table below shows the frequencies of these classes in the progeny of the cross. Deduce the nucleotide sequences at codon 68 for each of the six types. What is the frequency of recombinants for crosses in which the markers are at adjacent positions in the DNA?

Amino acid at position 68	Frequency among offspring of the cross
Gln	about 0.50
Ser	about 0.50
Arg	4.0×10^{-7}
His	2.0×10^{-7}
Asn	1.0×10^{-11}
Lys	2.0×10^{-7}

(Problem 17 is from Franklin W. Stahl, *The Mechanics of Inheritance,* 2d ed., p. 192. Copyright 1969, Prentice-Hall, Inc., Englewood Cliffs, New Jersey. Reprinted by permission.)

18. An induced cell mutant is isolated from a hamster tissue culture because of its resistance to α-amanitin (a poison derived from a fungus). Electrophoresis shows that the mutant has an altered RNA polymerase; *just one* electrophoretic band is in a position different from that of the wild-type polymerase. The cells are presumed to be diploid. What does this experiment tell you about ways in which to detect recessive mutants in such cells?

19. A double-stranded DNA molecule with the sequence shown below produces, in vivo, a polypeptide that is five amino acids long.

a. Which strand of DNA is transcribed, and in which direction?

b. Label the 5′ and the 3′ ends of each strand.

c. If an inversion occurs between the second and third triplets from the left and right ends, respectively, and the same strand of DNA is transcribed, how long will the resultant polypeptide be?

d. Assume that the original molecule is intact and that transcription occurs on the bottom strand from left to right. Give the base sequence, and label the 5′ and 3′ ends of the anticodon that inserts the *fourth* amino acid into the nascent polypeptide. What is this amino acid?

TAC ATG ATC ATT TCA CGG AAT TTC TAG CAT GTA
ATG TAC TAG TAA AGT GCC TTA AAG ATC GTA CAT

14

Recombinant DNA

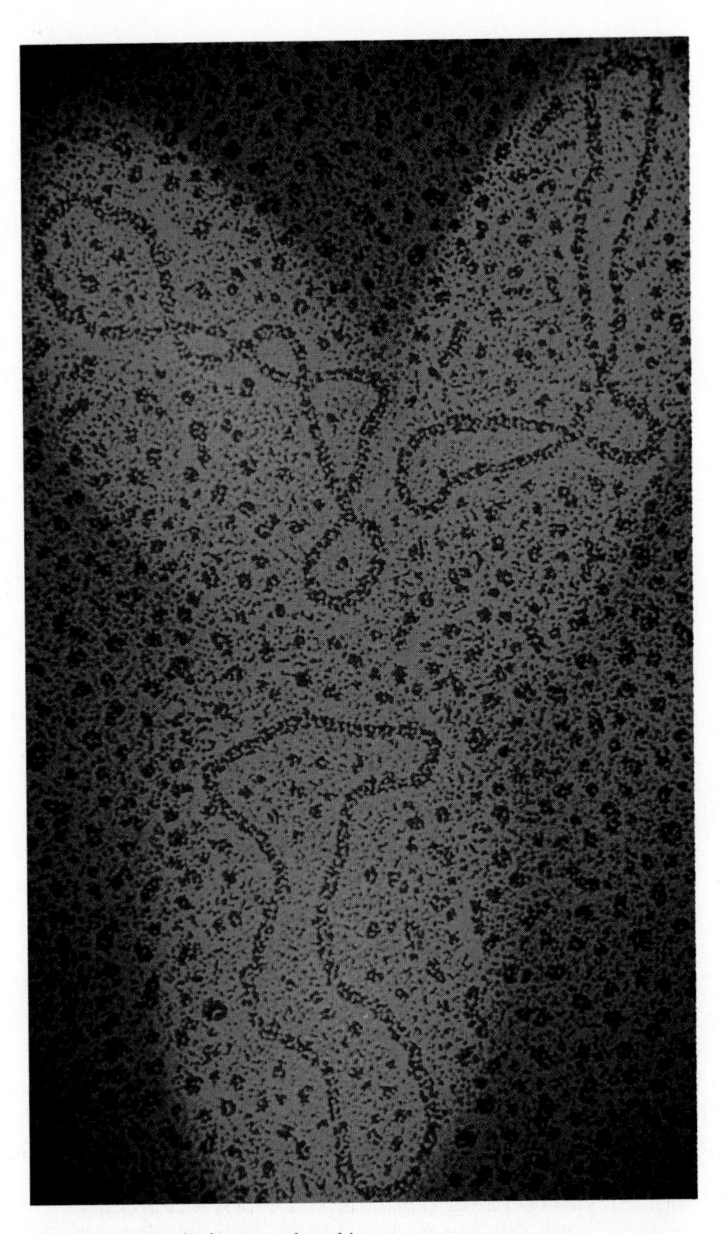

Plasmid DNA. (Schering-Plough)

KEY CONCEPTS

Restriction enzymes cleave DNA at specific sequences, facilitating the joining of nonhomologous DNAs from different sources.

When DNA fragments are inserted into small elements capable of replication, the segment of DNA can be selectively amplified in certain hosts.

It is now possible to rapidly determine the nucleotide sequence of segments of DNA.

The ability to create and analyze recombinant DNA molecules has generated a new era in molecular genetics. Until now, the complexity of eukaryotic genomes has made a detailed analysis of gene regulation extremely difficult. However, we can now isolate distinct fragments of DNA from any organism and recombine them into a smaller genome that is much easier to analyze. These **recombinant DNA molecules** can be made by joining nonhomologous DNAs from virtually any source, including DNAs from widely differing species. Usually, the vehicle, or **vector,** into which the DNA fragment has been incorporated is capable of selective replication at some stage, thus allowing amplification of the specific fragment of DNA. The increased number of copies of the DNA segment under study greatly enhances detailed molecular analysis. Bacteria are frequently the host organisms for the maintenance and amplification of recombinant DNA molecules derived from bacteria or higher organisms.

Together with methods for forming recombinant DNA molecules, great technical advances have been made in the analysis of fragments of genomes, including the sequence determination of long stretches of DNA and synthesis of fragments of DNA with any desired sequence. In this chapter we shall review these methods, and in the following chapter describe their application to studying gene control, analyzing chromosome organization in higher cells, engineering plants and certain vertebrates to carry desirable traits, and detecting genetic diseases in humans.

Restriction Enzymes

The key breakthrough in recombinant DNA technology was the use of restriction enzymes, which make sequence-specific cuts in DNA. As we shall see, restriction enzymes generate fragments of DNA with sequence-specific ends that can be spliced into small, self-replicating vector molecules and reintroduced into different hosts.

Discovery of Restriction Enzymes

The initial work that led to the discovery of enzymes that cleave DNA at specific sequences involved the phenomena of host modification and restriction, studied by Werner Arber and his colleagues in the 1960s. Certain *E. coli* hosts degrade, or restrict, infecting phage DNA, unless the DNA is modified by the addition of a methyl group (a process called **methylation**) to adenine or cytosine at specific sequences in the DNA. The degradation is caused by specific nucleases, or **restriction enzymes.**

Specificity of Restriction Enzymes

The next step in understanding restriction phenomena did not come until 1970, when Hamilton Smith studied the restriction enzyme from the bacterium *Haemophilus influenzae.* This enzyme, called *Hind*II, cuts T7 phage DNA into 40 specific fragments. What do the cleavage sites have in common? To find out, Smith took the mixture of fragments and marked the 5′ ends of the cleavage sites by attaching the radioisotope ^{32}P. He then used hydrolyzing enzymes to break the labeled fragments into still smaller pieces. He was able to separate the labeled end fragments in small segments only a few nucleotides long, which because of their small size he was then able to sequence.

Smith found that the label was always attached to an adenine or a guanine (the purines). Fragments with labeled adenine were always A-A or A-A-C; fragments with labeled guanine were always G-A or G-A-C. Smith suggested that such fragments could be produced if the enzyme cleaves only at the points indicated by the arrows in the following specific sequence (where "Py" represents a pyrimidine and "Pu" represents a purine):

```
              ↓
5′  -G-T-Py-Pu-A-C-  3′
           •
3′  -C-A-Pu-Py-T-G-  5′
           ↑
```

Note the rotational symmetry of this sequence: a rotation of 180° around the dot in the center leaves the sequence unchanged. Verify that cleavage at the arrows will produce just the 5′-end fragments that Smith identified: Pu-A- and Pu-A-C-. *Hind*II apparently is very specific in identifying the sequence at which it will cleave. Furthermore, Smith showed that the host-induced methylation that confers protection from *Hind*II activity occurs at this same cleavage site: the adenines one base away from the cleavage site are methylated (m):

```
                 m
5′  -G-T-Py-Pu-A-C-  3′
           •
3′  -C-A-Pu-Py-T-G-  5′
       m
```

Studies of other restriction enzymes have yielded similar results. For example, the enzyme *Eco*RI is produced by a gene on an R plasmid in *E. coli. Eco*RI cleaves the circular DNA of SV40 (a small mammalian virus) at only one site:

```
       ↓ m
5′  -G-A-A-T-T-C-  3′
          •
3′  -C-T-T-A-A-G-  5′
           m ↑
```

Again, the sequence is rotationally symmetrical, but in this case the DNA ring is converted to a linear molecule. Furthermore, unlike *Hind*II, which cuts straight across the two strands and thus forms *blunt* ends, *Eco*RI cuts the strands in a staggered pattern, leaving single-stranded tails at the sites of cleavage (see Figure 14-1). Because these tails can hydrogen-bond with complementary base sequences from other, similarly formed, tails, they are called **sticky,** or **cohesive, ends.** As we shall see, this "stickiness" permits splicing together of nonhomologous DNA fragments—an invaluable characteristic in recombinant DNA studies. Once again, methylation of adenines near the cleavage site can provide protection against the cleaving action (the methylated bases are indicated in the sequence just displayed, although of course, either methylation or cleavage would occur at such a sequence, not both).

Dozens of restriction enzymes with different sequence specificities have now been identified, some of which are shown in Table 14-1 (page 418). Restriction enzymes are powerful tools for analysis, because they can locate specific base sequences and cut the DNA in a very specific way.

Message Restriction enzymes make sequence-specific cuts. They produce a heterogeneous population of fragments with identical ends.

Restriction Maps

The restriction enzyme target sites can be used as markers for DNA. The DNA from a specific source is subjected to successive digestion by different restriction enzymes. When the fragments are separated electrophoretically on polyacrylamide gels, the "map" of the restriction sites can be deduced. Figure 14-2 shows bands resulting from stained DNA fragments, generated by restriction enzyme digestion, on a gel after electrophoresis. Detailed maps, such as the one shown in Figure 14-3, can be deduced by different methods, for example, by following the pattern of fragments produced as a partial digestion proceeds to completion. Restriction maps are highly specific for each DNA molecule under study.

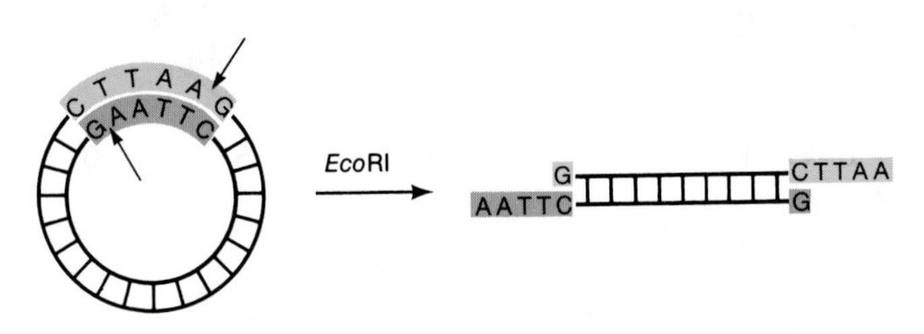

Figure 14-1 Cleavage of SV40 DNA by the restriction enzyme *Eco*RI produces a linear DNA with sticky ends.

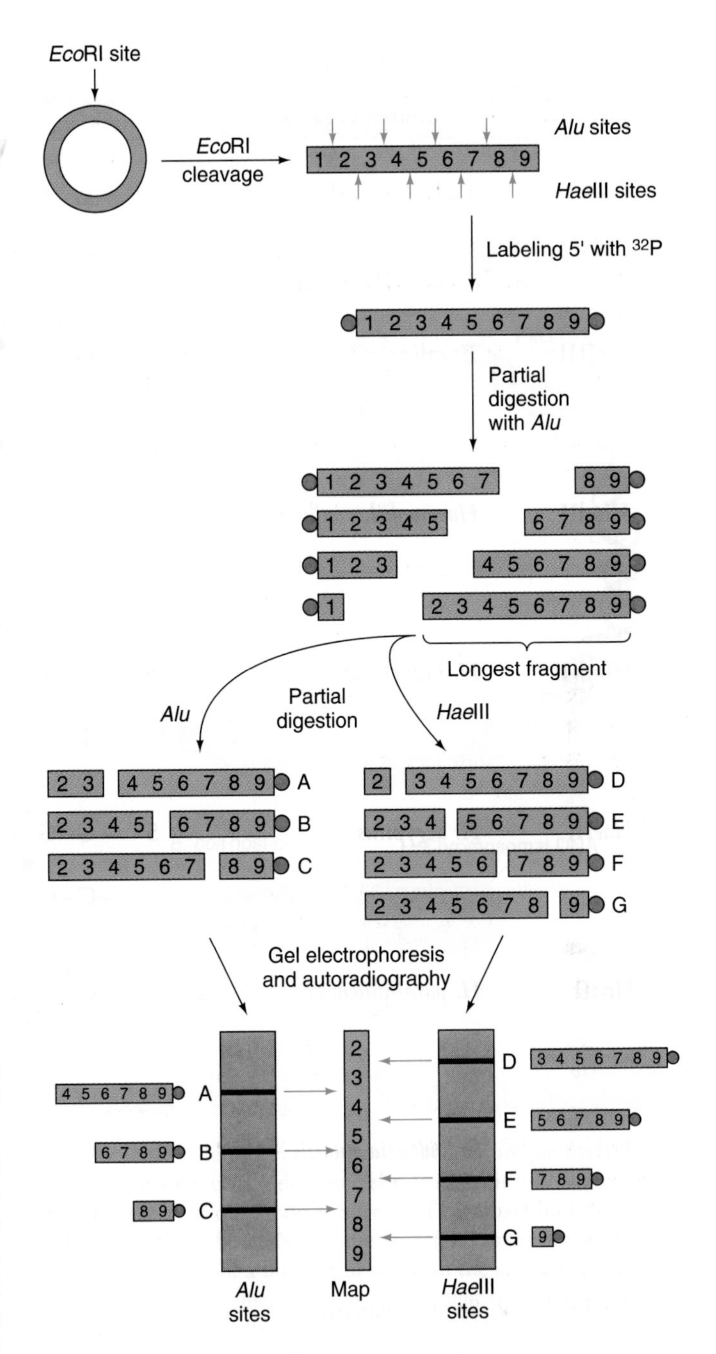

Figure 14-2 Identification of the sequence of restriction-enzyme sites in a circular DNA molecule. In this example, the ring of DNA with a single *Eco*RI site is opened into a linear molecule by cleavage with *Eco*RI. The molecule is labeled on its 5′ ends with ^{32}P and then partially digested with *Alu* (restriction enzyme from *Arthrobacter lutens,* which could just as easily be *Hae*III, the restriction enzyme from *Haemophilus aegyptius*) to produce fragments of varying lengths. Each fragment now carries the radioactive label at only one end. The longest fragment is selected (after testing on a gel to compare lengths) and is divided into two samples. One sample is partially digested with *Alu,* and the other is partially digested with *Hae*III. The new sets of fragments are separated on gels. The location of the labeled fragment can be determined by autoradiography of the gels. The sequence of restriction sites can now be read in order as the two gels are compared.

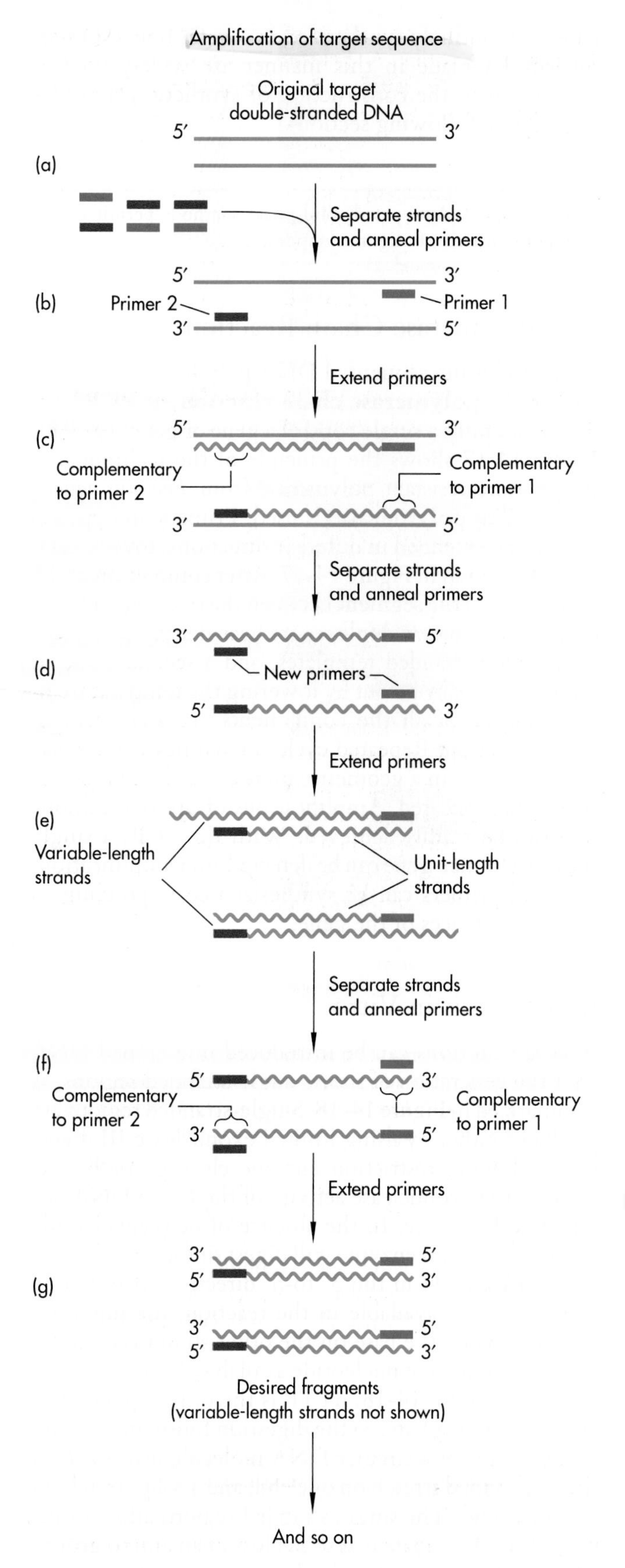

Figure 14-17 The polymerase chain reaction. (a) Double-stranded DNA containing the target sequence. (b) Two primers are chosen or are created that have sequences complementing primer binding sites at the 3′ ends of the target gene on the two strands. The strands are separated by heating, allowing the two primers to anneal to the primer binding sites. Together, the primers thus flank the targeted sequence. (c) *Taq* polymerase then synthesizes the first set of complementary strands in the reaction. These first two strands are of varying length, since they do not have a common stop signal. They extend beyond the ends of the target sequence as delineated by the primer binding sites. (d) The two duplexes are heated again, exposing four binding sites. (For simplicity only the two new strands are shown.) The two primers again bind to their respective strands at the 3′ ends of the target region. (e) *Taq* polymerase again synthesizes two complementary strands. Although the template strands at this stage are variable in length, the two strands just synthesized from them are precisely the length of the target sequence desired. This is because each new strand begins at the primer binding site, *at one end of the target sequence,* and proceeds until it runs out of template, *at the other end of the sequence.* (f) Each new strand now begins with one primer sequence and ends with the primer binding sequence for the other primer. Following strand separation, the primers again anneal and the strands are extended to the length of the target sequence. [The variable-length strands from part (c) are also producing target-length strands.] (g) The process can be repeated indefinitely, each time creating two double-stranded DNA molecules identical with the target sequence. (From J. D. Watson, M. Gilman, J. Witkowski, and M. Zoller, *Recombinant DNA,* 2d ed. Copyright 1992 by Scientific American Books.)

the plasmid is recircularized and the gap is filled by polymerase I and ligase.

Alternatively, the incorporation of base analogs can be employed, which is done by creating short, single-stranded gaps within a duplex (without cleaving both strands). The gap is then repaired using a modified form of DNA polymerase I (see Chapter 10), together with three nucleotide triphosphates and *N*-4-hydroxycytosine in place of T (actually, dTTP). Both tautomeric forms of *N*-4-hydroxycytosine (keto and enol) are present (Figure 14-19), allowing pairing with either G or A. The analog is incorporated in place of T but can also pair with G. Therefore, this in vitro method results in T → C transitions.

Less specific changes can be effected at single-stranded gaps by carrying out the polymerization reaction in the absence of one of the four deoxynucleotide triphosphates. At a low rate in vitro, polymerases will add nucleotides at random from across the base complementary to the missing base.

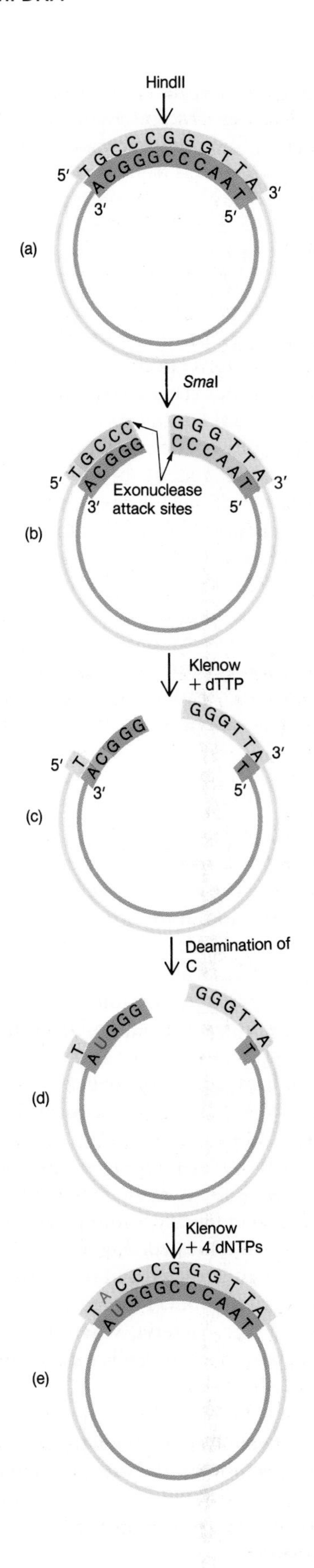

Figure 14-18 Creation of a substitution mutant through the deamination of cytosine. A restriction enzyme site—in this case, *Sma*I (a, b)—is treated with a modified form of DNA polymerase I, termed the Klenow fragment, in the presence of dTTP. (c) The Klenow fragment, acting as an exonuclease, performs a 3′ → 5′ digestion of single strands until a T residue is encountered. (d) The deamination of C residues on exposed single strands is effected by the addition of bisulfite. The molecule is now repaired by the addition of all four dNTPs and the Klenow fragment. (e) The result in this case is the alteration of a G–C base pair to an A–U base pair.

Additional methods for introducing mutations in vitro into cloned DNA segments are covered in the following chapter.

Recombinant DNA and Social Responsibility

Recombinant DNA techniques have revolutionized biology with their revelations about gene and chromosome organization, and they promise enormous potential benefits for humanity. However, as these techniques began to be exploited in the early 1970s, some scientists (and some other people) began to express concern about the possible hazards of manipulating gene segments. For example, SV40 is a mammalian virus known to cause cancer in monkeys, and *E. coli* is a bacterium that normally lives in the human digestive tract. When DNA from SV40 is

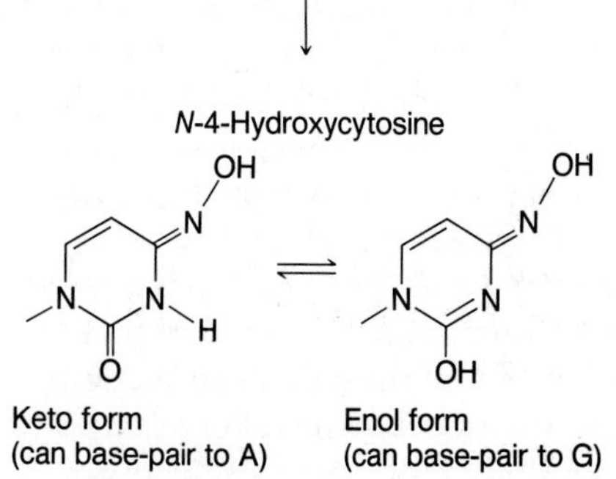

Cytosine

Hydroxylamine

N-4-Hydroxycytosine

Keto form
(can base-pair to A)

Enol form
(can base-pair to G)

Figure 14-19 The formation of *N*-4-hydroxycytosine with hydroxylamine.

inserted into *E. coli,* is it possible that a carcinogenic bacterium might be produced, escape, and thrive as a parasite in humans? Others have wondered whether the combination of genes from eukaryotes with prokaryotic cells might generate new types of pathogenic organisms against which humans would have no natural defenses. Of course, the question of whether such fears are reasonable could be settled definitely only by carrying out the experiments to explore the possible results.

In an unprecedented step, 11 eminent molecular biologists published a letter in 1974 pointing out some of their concerns about the potential biohazards of work with recombinant DNA. They called for the development of guidelines to regulate such research. They asked scientists to observe a moratorium on certain kinds of experiments deemed particularly hazardous (cloning genes for toxins or cancer-causing agents). For the first time in history, a group of scientists publicly declared certain areas of scientific inquiry to be "off-limits" and called for possible restrictions on such research.

The call for a moratorium attracted widespread public notice and caused many people to conclude that recombinant DNA research is dangerous. Under considerable public pressure, the National Institutes of Health (NIH) in the United States set out to establish categories of biohazards and guidelines for conducting experiments in each category. Eventually (on June 23, 1976), the NIH announced categories for experiments based on their potential hazards and defined four categories of physical conditions to contain the experiments. These conditions range from the P1 requirements of standard sterile techniques and commonsense precautions to the P4 facilities exercising the most extreme precautions against the escape of any organisms. Standards were set as well for biological restrictions that would minimize the chances of survival if escape occurred. For example, special strains of genetically enfeebled *E. coli* were constructed to use as recipients of recombinant DNA. These bacteria cannot survive except under special laboratory conditions.

Over the decade of the 1970s, evidence accumulated to indicate that the biohazards of recombinant DNA research were not so serious as some had feared. Meanwhile, increasing impatience built up in the scientific community about the delays in scientific progress and the regulation of scientific research by those with no training in the field. The NIH guidelines have now been relaxed significantly, and the use of recombinant DNA techniques has become routine laboratory practice.

Nonetheless, the turmoil about recombinant DNA did raise important social issues. For example, what is the social responsibility of scientists who are developing powerful new technologies? Should the people who are doing the experiments be the ones who set the guidelines? At what point should the public have an input? Who should be legally liable for any accidental damage that results from scientific research? Should limits be placed on the freedom of scientists to design and conduct research projects? Should a scientist attempt to foresee the possible adverse effects from the future use of discoveries and refuse to advance knowledge in certain directions that might have unfortunate applications?

Other kinds of questions are raised by the controversy over recombinant DNA. Can we predict with confidence the properties of an organism that has been modified by inserting DNA from a totally unrelated source? Can there be deleterious effects that will not be detected until large populations have been exposed for years (as was the case with oral contraceptives)? In the long run, will increasing sophistication in DNA manipulation inevitably lead to the genetic manipulation of human beings? If so, who will decide the conditions?

Although much of the worry about dangers of recombinant DNA research has been laid to rest, the issue has served to raise far more profound questions about the relationship between science and society. These questions have not been answered satisfactorily. They are likely to persist and become even more important in the coming years, as we confront the issues raised by prenatal diagnosis, forensic applications, and the possibility of gene therapy.

Summary

Restriction enzymes recognize specific nucleotide sequences and cleave the DNA molecule at such sites; they provide a powerful tool for fragmenting DNA in a controlled fashion. Coupled with electrophoretic gels that permit the separation of strands varying in length (by as little as a single base), restriction enzymes have made genetic engineering simple. Large DNA molecules can be cut into small fragments; the fragments can be separated, and their base sequences can be determined. Such a study determined the entire base sequence for the DNA of the phages ϕX174 and λ.

Recombinant DNA is produced by linking DNA fragments that have sticky ends. Two molecules are prepared with complementary single-stranded ends, and these ends are then annealed. A gene from a eukaryote can be isolated or constructed and then inserted into the DNA of a bacterial plasmid. This recombinant DNA can then be inserted into bacteria, where the recombinant plasmid can persist as a self-replicating cytoplasmic entity.

Recombinant DNA technology provides powerful insights into the structure and regulation of genes. It also promises a way to produce modified organisms that will greatly benefit humankind. Figure 14-20 provides a stunning example of how organisms can be modified by genetic engineering. However, like any powerful new technique, genetic engineering also involves potential hazards for society that must be assessed carefully.

Actinorhodin

Medermycin

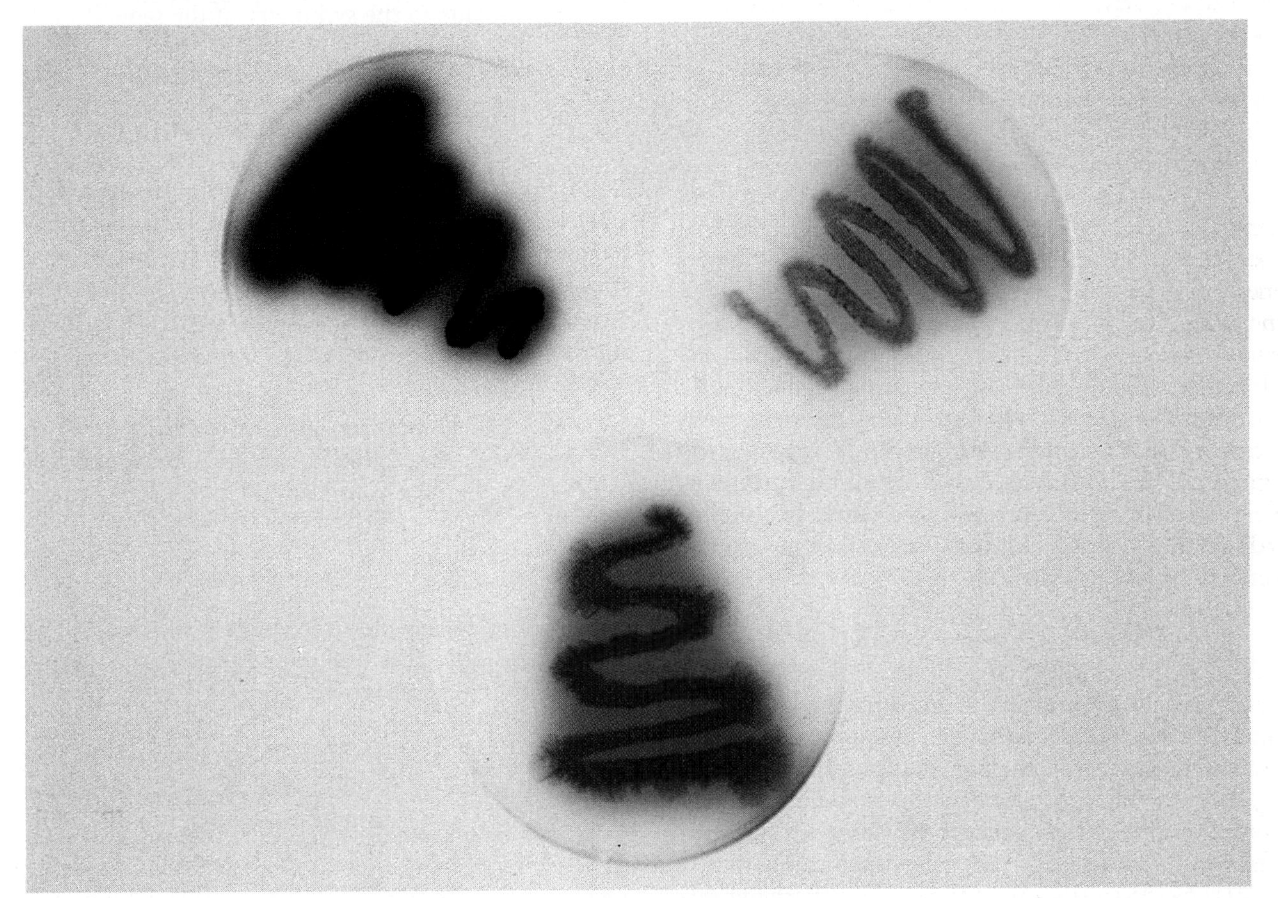

Mederrhodin A

Figure 14-20 Genetically engineered bacteria. This figure shows how new antibiotics can be synthesized by genetically engineered bacteria. Different species of *Streptomyces* produce related but different antibiotics. One species produces actinorhodin, which is blue at alkaline pH. A second species produces medermycin, which is brown. When the genes from the second species were cloned and transformed on a plasmid into the first species, some of the transformed bacteria had acquired the capacity to synthesize a new antibiotic, mederrhodin A, which gives a reddish-purple color. The structures of the three compounds are shown here, together with the streaks of the bacteria synthesizing each of the three compounds. The new antibiotic, mederrhodin A, is similar to medermycin but carries a hydroxyl group at the 6 position. The cloned segment probably contains a gene that encodes a hydrolyase that is expressed when transformed into a different species. (Mervyn J. Bibb, John Innes Institute, United Kingdom.)

Concept Map

Draw a concept map interrelating as many of the following terms as possible. Note that the terms are listed in no particular order.

restriction enzyme / recombinant DNA / cloning / chromosome walking / transformation / in vitro mutagenesis / vector / complementation / sequencing / genetic dissection

Chapter Integration Problem

In Chapter 13 we studied the structure of tRNA molecules. Suppose that you want to clone the gene from organism X that encodes a certain tRNA. You possess the purified tRNA and also an *E. coli* plasmid that contains a single *Eco*RI cutting site and also confers resistance to ampicillin (Amp^R). How can you clone the gene of interest?

Solution

You can use the tRNA itself to probe for the DNA containing the gene. One method is to digest the DNA from organism X with *Eco*RI and then to insert this DNA into the plasmid, which you also have cut with *Eco*RI. Amp^R colonies resulting from a transformation with the relegated plasmid containing the inserts then can be tested against the probe, using applications of Southern hybridization. You can examine those colonies hybridizing with the probe further to verify how much of the tRNA sequence they contain. Alternatively, you can run *Eco*RI-digested DNA from organism X on a gel and then identify the correct fragment by probing with the tRNA. This fragment can be cut out of the gel and used as a source of highly enriched DNA to clone into the plasmid cut with *Eco*RI.

Solved Problems

1. The restriction enzyme *Hin*dIII cuts DNA at the sequence AAGCTT, and the restriction enzyme *Hpa*II cuts DNA at the sequence CCGG. On average, how frequently will each enzyme cut double-stranded DNA? (In other words, what is the average spacing between restriction sites?)

Solution

We need only to consider one strand of DNA, because both sequences will be present on the opposite strand at the same site due to the symmetry of the sequences:

5′-AAGCTT-3′ and 5′-CCGG-3′
3′-TTCGAA-5′ 3′-GGCC-5′

The frequency of the six-base-long *Hin*dIII sequence is $1/4^6 = \frac{1}{4096}$, since there are four possibilities at each of the six positions. Therefore, the average spacing between *Hin*dIII sites is approximately 4 kb. For *Hpa*II, the frequency of the four-base-long sequence is $1/4^4$, or $\frac{1}{256}$. The average spacing between *Hpa*II sites is approximately 0.25 kb.

2. From Table 14-2 determine whether the *Eco*RI enzyme or the *Sma*I enzyme would be more useful for cloning. Explain your answer.

Solution

Both restriction enzymes recognize a six-base-pair sequence, so both would be expected to have approximately the same number of recognition sites per genome. The major difference between the two is that *Eco*RI leaves staggered ends while *Sma*I leaves blunt ends. Staggered ends are much easier to manipulate during cloning because of the base-pairing capacity inherent in them.
(Solution from Diane K. Lavett)

Problems

1. The restriction enzyme *Eco*RI cuts DNA at the sequence GTTAAC, and the enzyme *Hae*III cuts DNA at the sequence GGCC. On average, how frequently will each enzyme cut double-stranded DNA? (In other words, what is the average spacing between restriction sites?)

2. The bacteriophage ϕX174 has in its head a single strand of DNA as its genetic material. On infection of a bacterial cell, the phage forms a complementary strand on the infective strand to yield a double-stranded replicative form (RF). Design an experiment using ϕX174 to determine whether or not transcription occurs on both strands of the RF double helix.

3. You have a purified DNA molecule, and you wish to map restriction-enzyme sites along its length. After digestion with *Eco*RI, you obtain four fragments: 1, 2, 3, and 4. After digestion of each of these fragments with *Hin*dII, you find that fragment 3 yields two subfragments (3_1 and 3_2) and that fragment 2 yields three (2_1, 2_2, and 2_3). After digestion of the entire DNA molecule with *Hin*dII, you recover four pieces: A, B, C, and D. When these pieces are treated with *Eco*RI, piece D yields fragments 1 and 3_1, A yields 3_2 and 2_1, and B yields 2_3 and 4. The C piece is identical with 2_2. Draw a restriction map of this DNA.

4. After treating *Drosophila* DNA with a restriction enzyme, the fragments are attached to plasmids and selected as clones in *E. coli.* Using this "shotgun" technique, David Hogness has recovered every DNA sequence of *Drosophila* in a cloned line.

a. How would you identify the clone that contains DNA from a particular chromosomal region of interest to you?

b. How would you identify a clone coding for a specific tRNA?

5. You have isolated and cloned a segment of DNA that is known to be a unique sequence in the genome. It maps near the tip of the X chromosome and is about 10 kb in length. You label the 5′ ends with ^{32}P and cleave the molecule with *Eco*RI. You obtain two fragments: one is 8.5 kb long; the other is 1.5 kb. You separate the 8.5 kb fragment into two fractions, partially digesting one with *Hae*III and the other with *Hin*dII. You then separate each sample on an agarose gel. You obtain the following results by autoradiography:

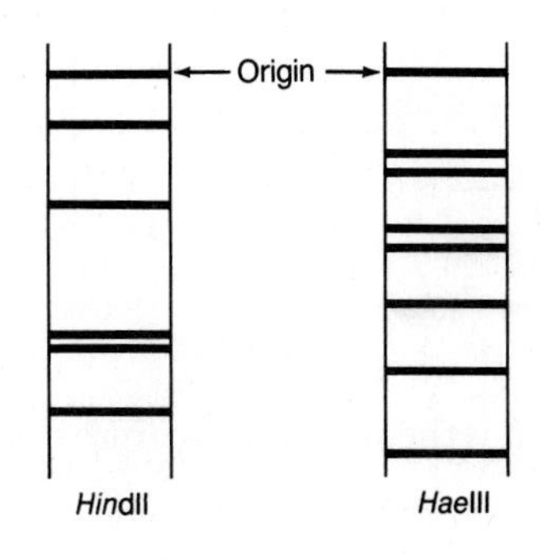

Draw a restriction-enzyme map of the complete 10-kb molecule.

6. As shown at the top of the autoradiographs in the accompanying figure, the Maxam-Gilbert technique has been used for sequencing two different DNA fragments from the *a* mating-type locus in yeast. The bases above each gel are the ones attacked by the reagent. What are the base sequences of these two DNA fragments?

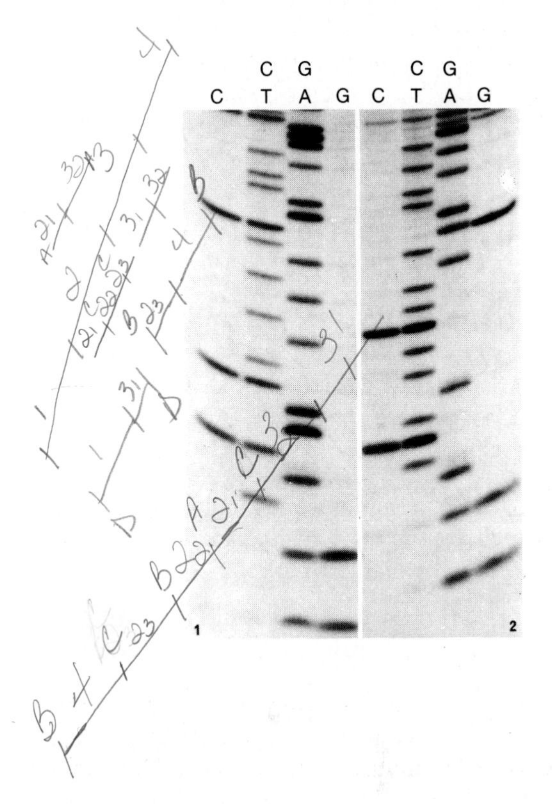

7. Design an experiment to allow the purification of DNA sequences in the Y chromosome.

8. Calculate the average distances (in nucleotide pairs) between the restriction sites in organism X for the following restriction enzymes:

*Alu*I	5′ AGCT 3′
	3′ TCGA 5′
*Eco*RI	5′ GAATTC 3′
	3′ CTTAGG 5′
*Acy*I	5′ G Pu CG Py C 3′
	3′ C Py GC Pu G 5′

(NOTE: Py = any pyrimidine; Pu = any purine.)

9. a. A fragment of mouse DNA with *Eco*RI sticky ends carries the gene *M*. This DNA fragment, which is 8 kb long, is inserted into the bacterial plasmid pBR322 at the *Eco*RI site. The recombinant plasmid is cleaved with three different restriction enzymes. The patterns of ethidium bromide fragments, following electrophoresis on agarose gels, are shown in this diagram:

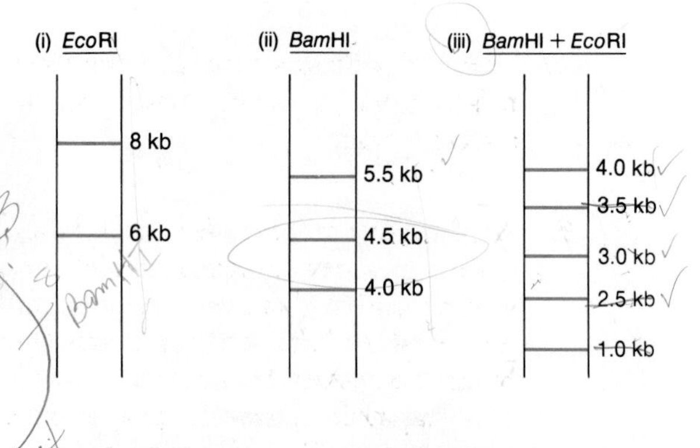

A Southern blot is prepared from gel (iii). Which fragments will hybridize to a probe (^{32}P) of pBR plasmid DNA?

b. Gene *X* is carried on a plasmid consisting of 5300 nucleotide pairs (5300 bp). Cleavage of the plasmid with the restriction enzyme *Bam*HI gives fragments 1, 2, and 3, as indicated in the diagram shown below (*B* = *Bam*HI restriction site). Tandem copies of gene *X* are contained within a single *Bam*HI fragment. If gene *X* encodes a protein X of 400 amino acids, indicate the approximate positions and orientations of the gene-*X* copies.

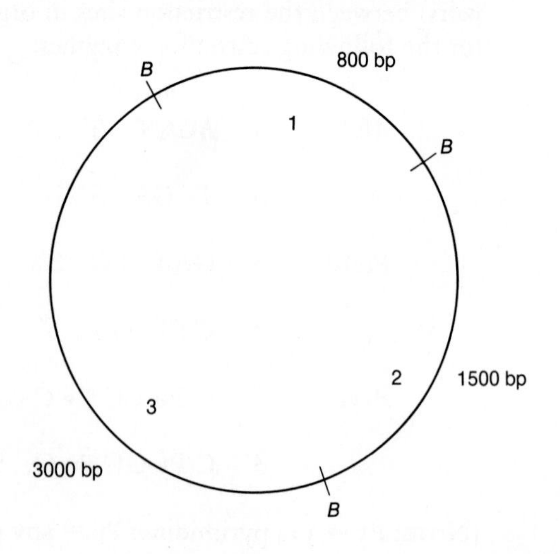

(Problem 9 courtesy of Joan McPherson.)

10. A linear fragment of DNA is cleaved with the individual restriction enzymes *Hind*III and *Sma*I and then with a combination of the two enzymes. The fragments obtained are

*Hind*III	2.5 kb, 5.0 kb
*Sma*I	2.0 kb, 5.5 kb
*Hind*III and *Sma*I	2.5 kb, 3.0 kb, 2.0 kb

a. Draw the restriction map.

b. The mixture of fragments produced by the combined enzymes is cleaved with the enzyme *Eco*RI, resulting in the loss of the 3-kb fragment (band stained with ethidium bromide on an agarose gel) and the appearance of a band stained with ethidium bromide representing a 1.5-kb fragment. Mark the *Eco*RI cleavage site on the restriction map.

(Problem 10 courtesy of Joan McPherson.)

11. A viral DNA fragment carrying a specific gene *V* is transfected (introduced into the cell by transformation) into a muscle-cell culture. Following incubation with ^{32}P-labeled ribonucleotides, the virus-encoded RNA product is isolated at two timed intervals. The radiolabeled viral RNA is treated as follows. First, it is hybridized to a specific cDNA previously constructed from viral-gene-*V* mature mRNA. Second, the hybrid is treated with RNase. Finally, the hybrid is denatured and electrophoresed on a gel, which is then subjected to autoradiography. The following results suggest that the pathologic nature of the virus is time-related (the number of nucleotides is indicated on the bands observed):

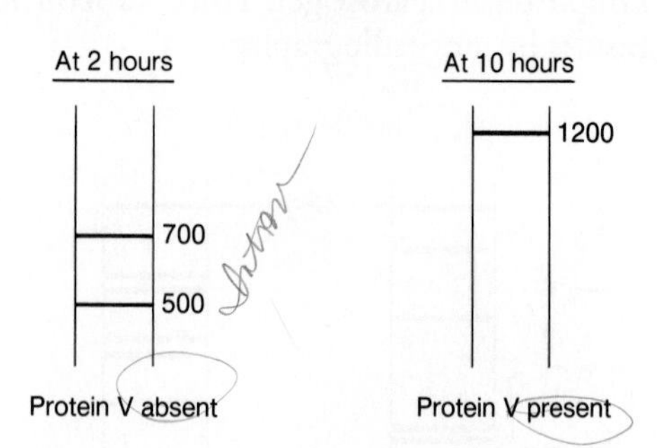

a. What is the size of the mature mRNA for gene *V*?

b. Draw a diagram of each hybrid and indicate what the illustrated bands represent.

c. Why is protein V not produced until after 2 hours?

(Problem 11 courtesy of Joan McPherson.)

12. The gene for β-tubulin has been cloned from *Neurospora* and is available. List a step-by-step procedure for cloning the same gene from the related fungus *Podospora,* using as the cloning vector the pBR *E. coli* plasmid shown here, where Kan = kanamycin and Tet = tetracycline:

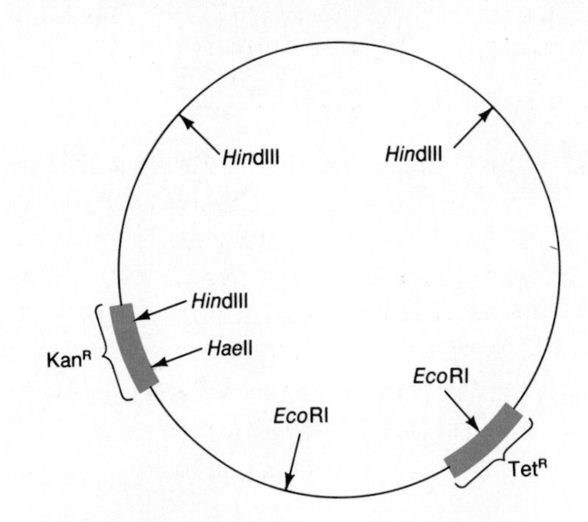

13. A linear phage chromosome is labeled at both ends with ^{32}P and digested with restriction enzymes. *Eco*RI produces fragments of sizes 2.9, 4.5, 6.2, 7.4, and 8.0 kb. An autoradiogram developed from a Southern blot of this digest shows radioactivity associated with the 6.2 and 8.0 fragments. *Bam*HI cleaves the same molecule into fragments of sizes 6.0, 10.1, and 12.9; the label is found to be associated with the 6.0 and 10.1 fragments. When *Eco*RI and *Bam*HI are used together, fragments of sizes 1.0, 2.0, 2.9, 3.5, 6.0, 6.2, and 7.4 kb are produced.

a. Draw a restriction-enzyme target site map of this molecule, showing relative positions and distances apart.

b. A radioactive probe made from a cloned phage gene *X* is added to Southern blots of single-enzyme digests that have used nonradioactive phage DNA. The autoradiograms show hybridization associated with the 4.5, 10.1, and 12.9 fragments. Draw in the approximate location of gene *X* on the restriction map.

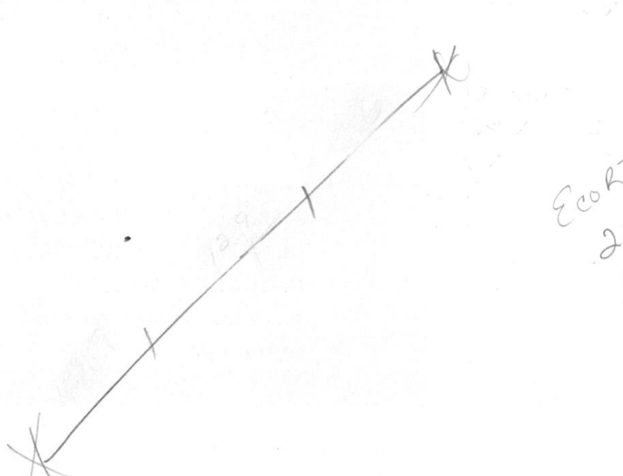

15

Applications of Recombinant DNA

Transgenic tobacco plant expressing the luciferase gene from a firefly. (Keith Wood, Promega, Madison, Wisconsin)

KEY CONCEPTS

Genes can be altered at specific sites by in vitro mutagenesis techniques.

Recombinant DNA technology permits the production of human proteins in bacteria.

Recombinant DNA allows the generation of transgenic animals and plants.

Genetic diseases can be detected at an early stage.

In the previous chapter we examined methods for creating and analyzing recombinant DNA molecules. In the following section we will see how these methods allow us to alter genes at specific sites, and to use these altered genes to change the characteristics of bacteria, plants, and certain vertebrates. Recombinant DNA methodology opens up new frontiers in the detection and understanding of human genetic diseases, as we shall see in the following sections.

Applications of Recombinant DNA Technology Using Prokaryotes

The essence of recombinant DNA technology is that fragments of DNA from any organism can be inserted into a bacterially based vector molecule. These vectors are amplified in bacteria to provide large amounts of DNA; then the insert can be studied in a variety of ways. Of course, the insert can be from prokaryotic or eukaryotic organisms, and indeed many eukaryotic genes have been intensively studied in this way at the structural and functional levels. Much has been learned about the organization of eukaryotic genes at the structural level by sequencing fragments cloned in bacteria. Much has also been learned at the functional level: eukaryotic genes are sometimes found to be expressed in bacteria in their intact state; in other instances, the eukaryotic sequences have to be manipulated by removing introns or adding bacterial regulatory signals in order to function in a bacterial cell. Furthermore, as we will see below, directed mutagenesis enables many functional aspects of eukaryotic genes to be dissected in bacterial settings.

For the first time in history, geneticists can program the alteration of the genetic material. The late 1970s and the early 1980s saw the development of techniques that permit the introduction of point mutations, deletions, and insertions into segments of cloned DNA, and alterations of specific base pairs. By directing specific mutations into predetermined segments of DNA, we make possible many experiments. Two methods for programming specific alterations of genes are **site-directed mutagenesis** and **gene synthesis.**

Site-Directed Mutagenesis

We have already described methods involving specific or random base substitutions generated in vitro in cloned DNA. However, they all involve short regions surrounding a favorable restriction site. How can we create mutations at specific places that do not happen to be so favorably situated? There is a powerful method for creating specific mutations that circumvents this limitation. This technique employs synthetic oligonucleotides, which are short DNA segments (usually 15 to 25 bases in length) constructed in vitro (page 431).

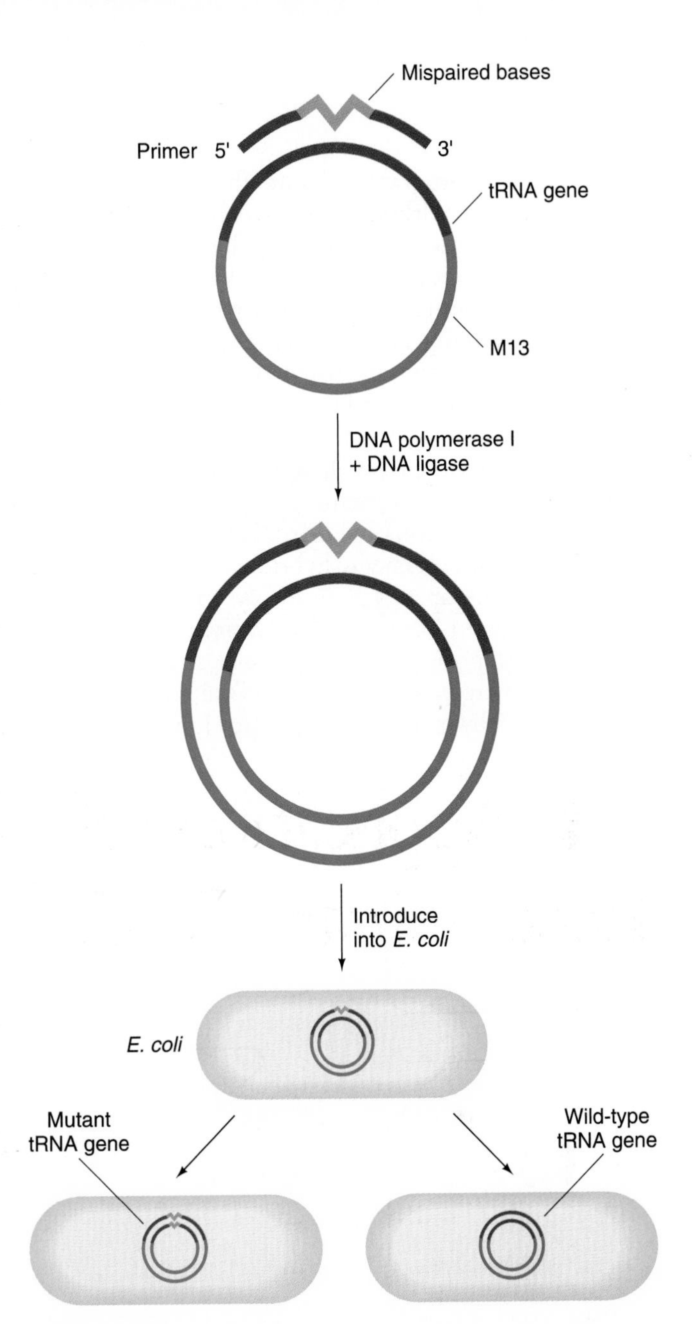

Figure 15-1 Creation of a substitution mutant by use of a synthetic oligonucleotide. The 12-to 15-base oligonucleotide is constructed so that it is complementary to a region of a DNA strand, but with one or two mismatches. When mixed with a clone of the complementary strand, the oligonucleotide will anneal to it even though the match is not exact, so long as the hybridization conditions are not stringent and the mismatches are in the middle of the oligonucleotide segment. The segment then serves as a primer for DNA polymerase I, which synthesizes the remainder of the complementary strand. When the resulting double-stranded molecule is introduced into *E. coli,* the molecule replicates to re-create either the mutated sequence or the original wild-type sequence.

As a first step, the gene of interest is cloned into a single-stranded phage vector, such as the phage M13. The synthetic oligonucleotide serves as a primer for the in vitro synthesis of the complementary strand of the M13 vector (Figure 15-1). Any desired specific base change can be programmed into the sequence of the synthetic primer. Although there will be a mispaired base when the synthetic oligonucleotide hybridizes with the complementary sequence on the M13 vector, one or two mismatched bases can be tolerated when hybridization occurs at a low temperature and a high salt concentration. After DNA synthesis is mediated by DNA polymerase in vitro, the M13 is replicated in *E. coli,* in which case many of the resulting phages will be the desired mutant.

It should be noted that the synthetic oligonucleotide can serve as a labeled probe to distinguish wild-type from mutant phages by using hybridization. The mismatched base will allow the primer to hybridize with both types of phage at low temperature, but only with the complementary mutant phage at high temperature.

Synthetic primers containing mispaired bases have been used to create an amber (UAG) suppressor from a human tRNA by converting the anticodon to a sequence that recognizes the UAG triplet instead of the AAA triplet (lysine codon), as depicted in Figure 15-2. The altered tRNA now inserts lysine in response to the UAG codon.

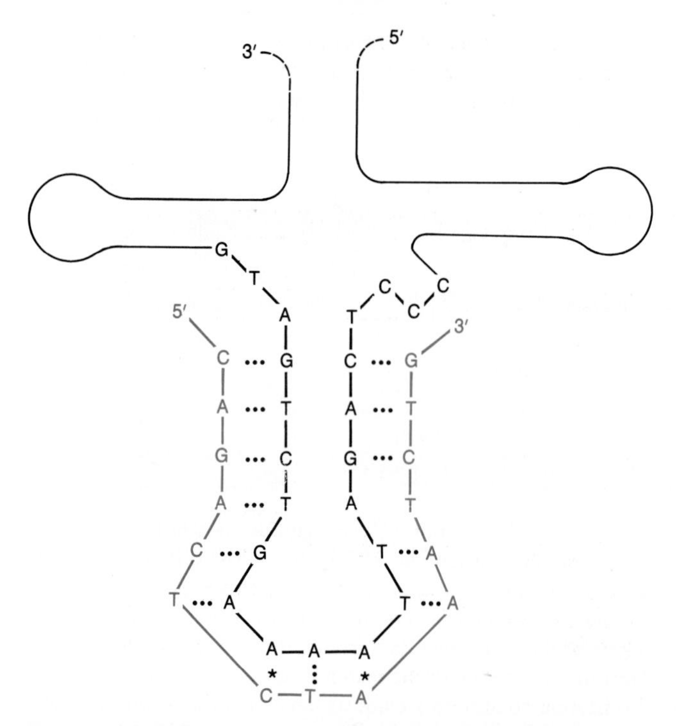

Figure 15-2 The portion of the DNA sense strand (black) for the gene coding for the lysine tRNA, showing the sequences transcribing the anticodon region. (The DNA strand is shown in the shape of the corresponding tRNA it transcribes for illustrative purposes; it does not actually assume this shape.) Base-paired with this is a synthetic oligonucleotide (red) with two mismatched base pairs (*). This can serve as a primer in the formation of a complementary strand as described in Figure 15-1. When the sense strand of this antisense mutant is produced, it codes for a (3′)AUC(5′) anticodon, which recognizes the (5′)UAG(3′) stop codon of the mRNA. Consequently, the tRNA incorporates a lysine at the stop codon, thus serving as an amber (UAG) suppressor.

Applications of Gene Synthesis

The Khorana method (page 431) was applied by Herbert Boyer's group to make the gene coding for a small human growth-regulating hormone, somatostatin. The hormone is a short polypeptide with the sequence shown in Figure 15-3. Boyer's group synthesized the gene using overlapping fragments. Then they added a triplet specifying methionine and an *Eco*RI cleavage site on the amino end; on the other end, they placed two consecutive stop triplets and a *Bam*HI site (Figure 15-4). The entire gene was inserted into a plasmid carrying the bacterial gene *β*-galactosidase, within which there is an *Eco*RI site. The other end of the somatostatin gene hybridized with a *Bam*HI site elsewhere in the plasmid. The *E. coli* selected for their possession of the plasmid were found to produce a protein chimera containing part of *β*-galactosidase fused to somatostatin via a methionine residue. Methionine is cleaved by cyanogen bromide, so the active hormone could be liberated by such treatment (Figure 15-5).

A good example of gene synthesis utilizing synthetic oligonucleotides produced by automated solid-phase synthesis was provided by John Abelson and Jeffrey Miller and their coworkers in 1986 when they constructed artificial tRNA genes in vitro. Specifically, they constructed the phenylalanine tRNA gene, with two changes in the anticodon, so that it would read the UAG (amber) codon and become an amber suppressor tRNA. The gene can be synthesized by combining six oligonucleotides containing the two alterations. These oligonucleotides are synthesized individually, with short overlaps to allow annealing in vitro. As in the preceding example, specific restriction-enzyme cutting sites are present at each end of the duplex that is generated. Figure 15-6 portrays the steps in the synthesis. After annealing, ligase is added and the gene is inserted into an expression vector that contains a promoter. Because a nonsense suppressor results, the ability to suppress a nonsense mutation in the *lacZ* gene, which encodes *β*-galactosidase, is

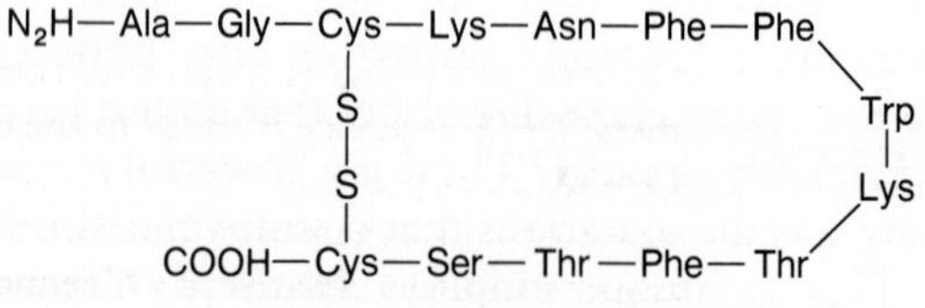

N_2H—Ala—Gly—Cys—Lys—Asn—Phe—Phe
S
S
COOH—Cys—Ser—Thr—Phe—Thr
Trp
Lys

Figure 15-3 The amino acid sequence of the hormone somatostatin.

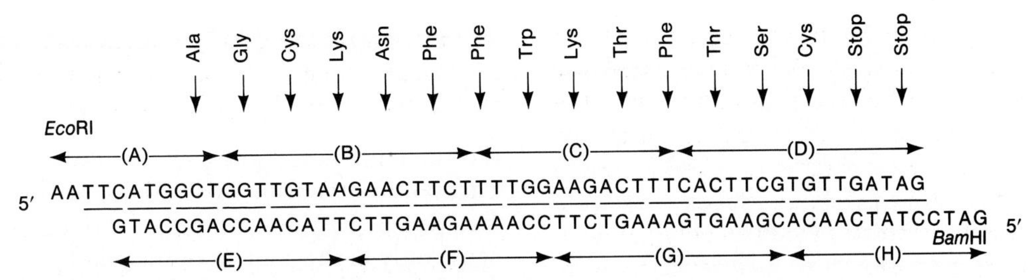

Figure 15-4 The overlapping complementary sequences (indicated by the letters in parentheses) synthesized to produce the somatostatin gene. A triplet specifying methionine is added to the 5′ end of the somatostatin coding region (inferred from the amino acid sequence); Adjacent to this, an *Eco*RI restriction sequence is added. A *Bam*HI restriction sequence is added at the other end of the "artificial" gene. (From K. Itakura et al., "Expression in *Escherichia coli* of a Chemically Synthesized Gene for the Hormone Somatostatin." *Science* 198, 1977, 1056–1063. Copyright 1977 by the American Association for the Advancement of Science.)

monitored in order to verify the presence of an active suppressor tRNA. The entire experiment can be completed in several days, in contrast to the procedure originally utilized by Khorana to construct a tRNA molecule.

Such techniques, in addition to direct cloning, already have been used to produce recombinant plasmids bearing DNA sequences for human insulin, growth hormone, interferon, and blood-clotting factors. Commercially profitable quantities of such human proteins can be obtained from bacterial cultures.

Message Recombinant DNA methodology permits us to program any desired change into the genetic content of microorganisms.

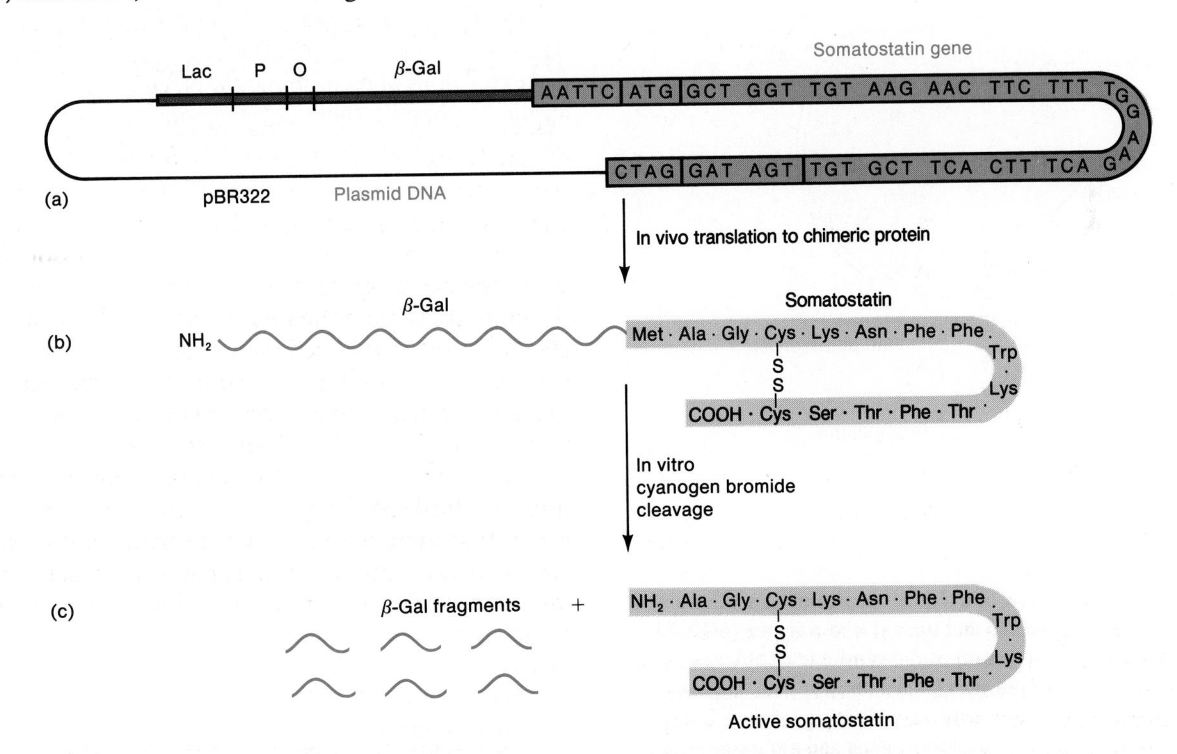

Figure 15-5 The production of somatostatin by *E. coli.* (a) The plasmid carrying the synthetic DNA sequence is added to the bacterial cell. (β-Gal = β-galactosidase gene; Lac, P, and O = regulatory sequences for β-Gal.) (b) A chimeric polypeptide is produced. (c) The desired somatostatin is liberated by treatment with cyanogen bromide. (From K. Itakura et al., "Expression in *Escherichia coli* of a Chemically Synthesized Gene of the Hormone Somatostatin." *Science* 198, 1977, 1056–1063. Copyright 1977 by the American Association for the Advancement of Science.)

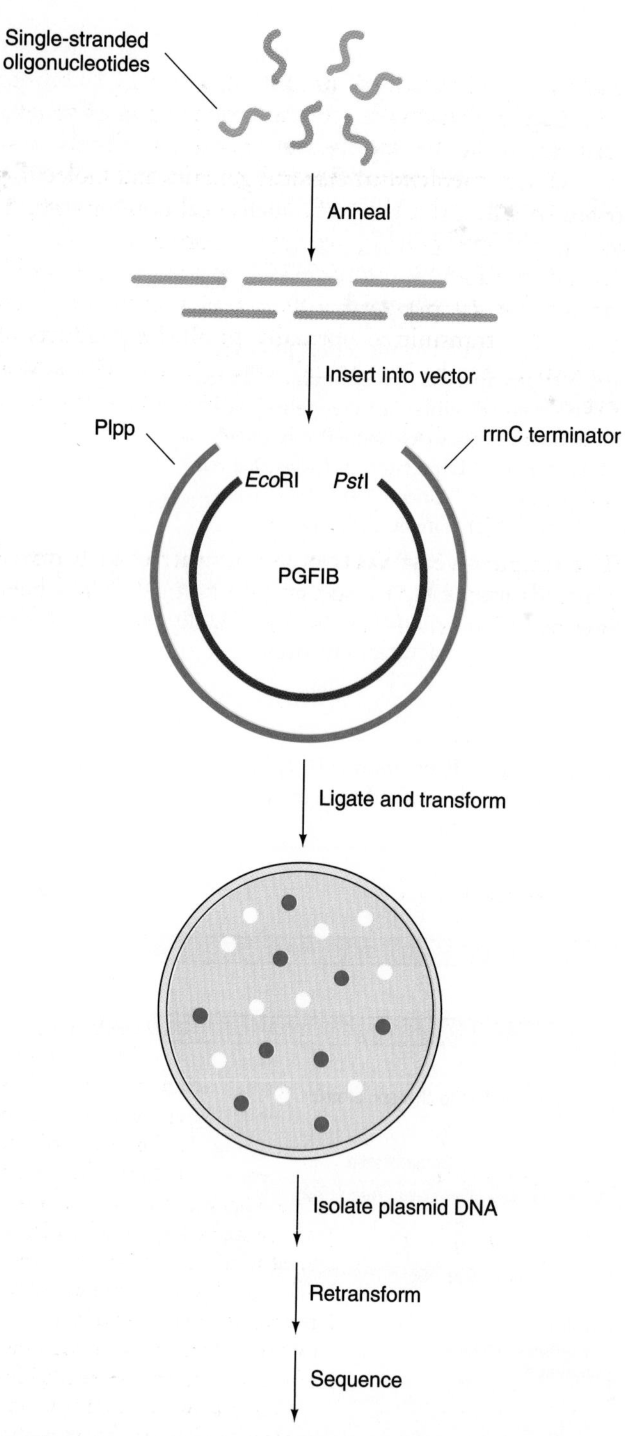

Figure 15-6 Synthesis of a tRNA suppressor gene. Six oligonucleotides are annealed and ligated into a vector (pGFIB) constructed for the expression of the synthetic tRNA gene. The vector contains restriction sites (*Eco*RI and *Pst*I) for the insertion of the synthesized segment, as well as a promoter (Plpp) derived from the lipopolysaccharide gene and a transcription terminator (rrnC) derived from ribosomal RNA genes. The plasmid is used to transform cells that carry an amber mutation in the *lacZ* gene. These cells cannot synthesize β-galactosidase owing to the chain-terminating amber mutation. However, transformants that synthesize an active suppressor tRNA will be able to produce β-galactosidase (indicated by the blue colonies). These transformants are then chosen for further analysis.

Recombinant DNA Technology in Eukaryotes: An Overview

To the eukaryote biologist, the ultimate interest is the eukaryotic organism itself. Many of the structural and functional questions to be asked about a gene really make sense only in the context of the whole organism. Therefore, it would be of considerable significance and interest to be able to manipulate eukaryotic DNA in the convenience of a bacterial vector and then return the DNA back to the eukaryotic organism from which it came. Furthermore, there is simply no way to detect the expression of some eukaryotic genes in bacterial clones, and therefore the clones in a gene bank need to be tested back in a eukaryotic cell, defective for the gene in question, in order to find the expression of the clone of interest. As we shall see, these kinds of reintroductions into eukaryotic cells are now possible, even routine, in several eukaryotes.

In addition to the preceding considerations, the genomes of eukaryotes are much larger and much more complex than, say, those of bacteria. For example, in contrast to the *E. coli* genome of approximately 4000 kilobases (kb), the simple eukaryote *Neurospora crassa* has a haploid genome size of 27,000 kb; in humans, the equivalent figure is 3,000,000 kb. This means that special extensions of recombinant DNA technology must be applied to handle these large genomes, and we will explore some of these later in the chapter.

How to Get DNA into Eukaryotic Cells

It is perhaps a surprise that bacterial cells will admit DNA through their cell membranes under the appropriate physiological conditions. However, this process now constitutes the basis for bacterial transformation. It is perhaps an even bigger surprise that eukaryotic cells will also take up DNA out of a solution across the cell membrane. Once again, the conditions have to be right. In addition, for DNA to be taken up in plants and fungi, a cell wall must be eroded away to form naked protoplasts. Special enzymes are used to break down the polysaccharides that constitute cell walls; in fungi, one of the enzymes used is extracted from the digestive glands of snails.

Alternatively, DNA can be injected into a cell under the microscope using a microsyringe and a micromanipulator. This permits extra DNA to be inserted into the egg cells of animals from *Drosophila* to humans. There are bizarre techniques, too; one that deserves mention is a specially adapted gun that shoots DNA-coated microscopic tungsten projectiles into a plant cell through the cell wall. The point is, in conclusion, that there are now some standard and some bizarre ways of introducing

DNA into the cells of eukaryotic organisms. Of course, once the DNA is inside the cell, it faces numerous possible fates depending on the nature of the DNA itself, its vector sequences, and the idiosyncracies of the organism in question. We will follow examples from several different organisms to illustrate some of the processes involved. A eukaryotic organism that develops from a cell into which new DNA has been introduced is called a **transgenic organism.** We will next examine the production of several different representative transgenic organisms and some of their unique features.

Transgenic Yeast

The yeast *Saccharomyces cerevisiae* has become the *E. coli* of the eukaryotes. One of the main reasons is that the classical genetics of yeast is extremely well developed, and the availability of thousands of mutants affecting hundreds of different phenotypes provides a rich source of genetic markers in the development of yeast as a molecular system. Today the blend of classical genetics and molecular biology is indeed a powerful analytical combination in any organism. In yeast, another important advantage is the occurrence of a circular 6.3 kb natural yeast plasmid, named the **2μ plasmid** after its circumference. This plasmid is transmitted normally to all the products of meiosis either by cell-to-cell contact or in the sexual cycle. The 2μ plasmid forms the basis for several specially engineered, sophisticated yeast vectors.

Yeast Vectors

The simplest **yeast vectors** are derivatives of bacterial plasmids into which a section of yeast DNA has been inserted (Figure 15-7a). When transformed into yeast

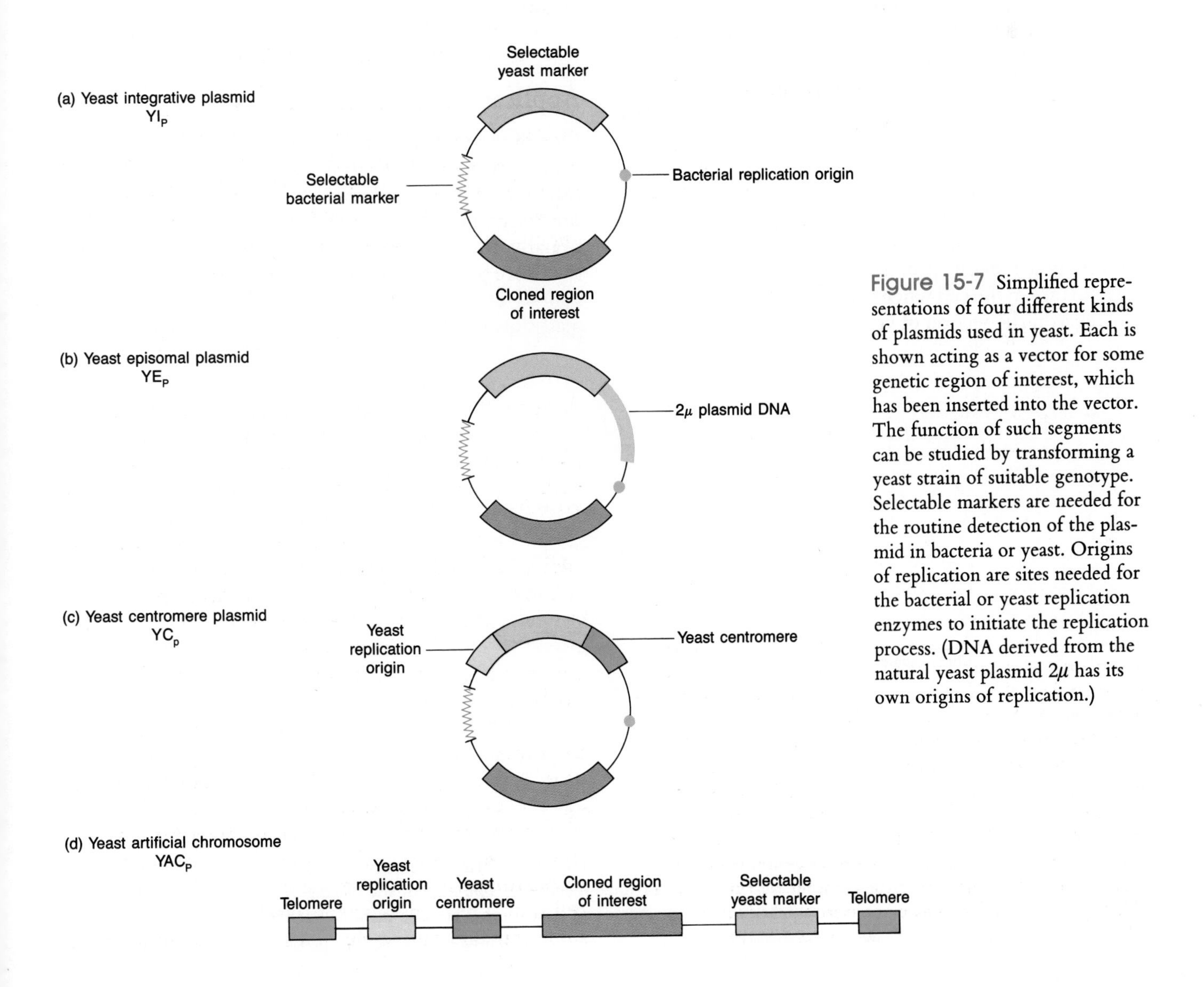

Figure 15-7 Simplified representations of four different kinds of plasmids used in yeast. Each is shown acting as a vector for some genetic region of interest, which has been inserted into the vector. The function of such segments can be studied by transforming a yeast strain of suitable genotype. Selectable markers are needed for the routine detection of the plasmid in bacteria or yeast. Origins of replication are sites needed for the bacterial or yeast replication enzymes to initiate the replication process. (DNA derived from the natural yeast plasmid 2μ has its own origins of replication.)

cells, these plasmids can be inserted into yeast chromosomes by homologous recombination, involving a single or a double crossover (Figure 15-8). Either the entire plasmid is inserted, or the targeted allele is replaced by the allele on the plasmid. Such integrations can be detected by plating cells on a medium that selects for the allele on the plasmid. Because bacterial plasmids do not replicate in yeast, integration is the only way to generate a stable transgenic phenotype. One of the uses of this type of vector is to select bacterial clones that contain a yeast gene that cannot be detected or selected in bacterial cells: successful transformation of a yeast mutant tells the investigator that the clone has the equivalent wild-type gene.

If the 2μ plasmid is used as the basic vector and other bacterial and yeast segments are spliced into it (Figure 15-7b), then a construct is obtained that has several useful properties. First, the 2μ segment confers the ability to replicate autonomously in the yeast cell, and insertion is not necessary for a stable transgenic phenotype. Second, genes can be introduced into yeast, and their effects can be studied in that organism; then the plasmid can be recovered and put back into *E. coli,* provided that a bacterial replication origin and a selectable bacterial marker are on the plasmid. Such **shuttle vectors** are very useful in the routine cloning and manipulation of yeast genes.

With any autonomously replicating plasmid, there is the possibility that a daughter cell will not inherit a copy because the partitioning of copies to daughter cells is essentially a random process depending on where the plasmids are in the cell when the new cell wall is formed. However, if the section of yeast DNA containing a centromere is added to the plasmid (Figure 15-7c), then the nuclear spindle that ensures the proper segregation of chromosomes will treat the plasmid in somewhat the same way and partition it to daughter cells more efficiently at cell division. The addition of a centromere is one step toward the formation of an artificial chromosome. A further step has been made by linearizing a plasmid containing a centromere and adding the DNA from yeast telomeres to the ends (Figure 15-7d). If this construct contains yeast replication origins **(autonomous replication sequences, ARS),** then it behaves in many ways like a small yeast chromosome at mitosis and meiosis. For example, when two haploid cells—one bearing a *trp*$^-$ *ura*$^+$ artificial chromosome and another bearing a *trp*$^+$ *ura*$^-$ artificial chromosome are brought together to form a diploid, some tetrads will show the clean segregations expected if these two elements are behaving as regular chromosomes. In other words, two ascospores will show *trp*$^-$ *ura*$^+$ and the other two will show *trp*$^+$ *ura*$^-$ genotypes.

Applications of Yeast Vectors

One of the great assets of genetic analysis is its incisiveness. Through the analysis of specific blocks in gene function, normal biological processes can be dissected precisely and conclusive inferences can be drawn. Traditionally, the experimenter had to make use of mutations that were produced essentially at random. In vitro mutagenesis allows the production of changes at specific places in specific genes. Such an in vitro–mutated gene

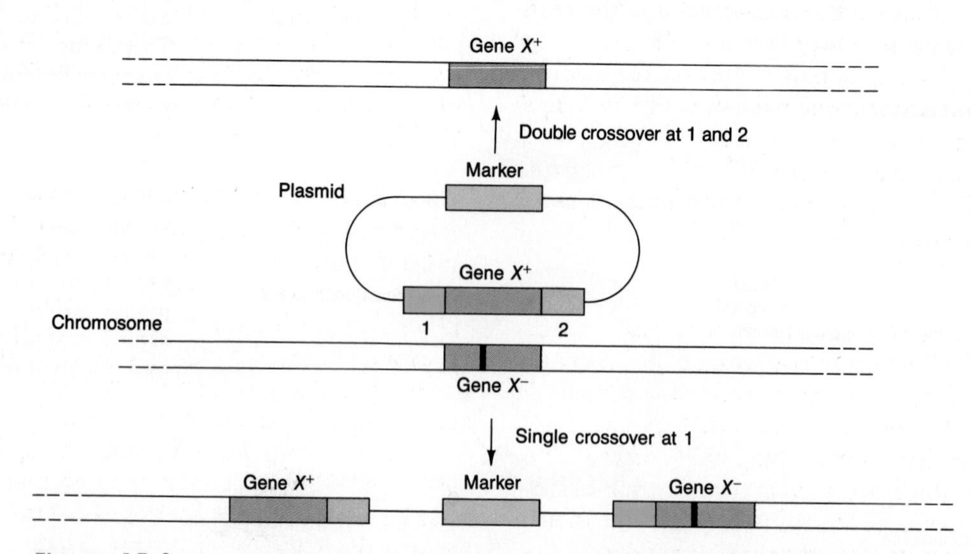

Figure 15-8 Two ways in which a recipient yeast strain bearing a defective X^- can be transformed by a plasmid bearing an active allele (gene X^+). The mutant site of gene X^- is represented as a vertical black bar. Single crossovers at position 2 are also possible but are not shown.

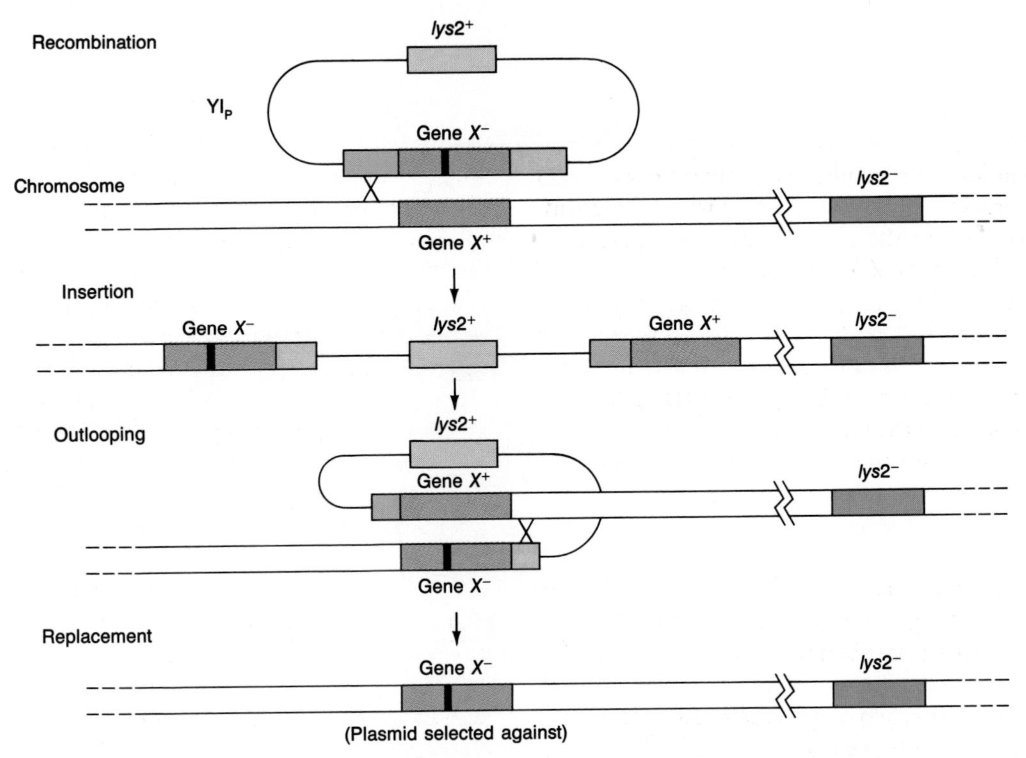

Figure 15-9 A two-step method for replacing an active gene X^+ with a deliberately engineered mutant allele (X^-) for the purpose of observing its effects on phenotype. The *lys2*$^+$ allele is a yeast marker that can be selected *for* (through the prototrophy that it confers) or *against* (by plating on the chemical α-aminoadipate).

then is inserted back into the organism to replace the resident wild-type gene, and the effects are observed. The yeast integrative plasmid provides a model of how this can be achieved in a eukaryote. A mutated gene and its flanking regions, carried on an integrating plasmid, provide a region of homology in which a crossover can occur; the entire plasmid then is inserted into the wild-type yeast chromosome at the proper locus (Figure 15-9). Owing to the presence of two copies of the gene in question (one mutated and one wild-type) on one chromosome, pairing can occur by looping, and a crossover at a different site can excise the plasmid—this time bearing the wild-type allele and leaving the mutant allele in the normal chromosomal locus.

Selection for Loss of Plasmid. In some cases, we can select for the elimination of the plasmid sequence. For example, if the plasmid selection marker is *lys2*$^+$ and the recipient yeast chromosome bears *lys2*$^-$ (Figure 15-9), then insertion produces a strain with *lys2*$^+$ and *lys2*$^-$. The chemical α-aminoadipate permits the growth only of *lys2*$^-$ strains, so strains that have lost the insert can then be selected by plating on this chemical. About half these strains retain the original wild-type gene of interest (gene X^+); the remainder retain the plasmid-borne mutant allele X^-. In other cases, the mutant phenotype of the mutant gene X^- can be selected directly by appropriate platings.

Note that YI_p plasmids carry two yeast elements: the gene under investigation X, and the selectable marker. Both can undergo homologous recombination with their respective chromosomal loci. The specificity of insertion can be targeted more efficiently by making a restriction cut in the plasmid at the gene in question (gene X). Since the ends are recombinogenic (recombine preferentially), they direct the plasmid's entry to that specific site (Figure 15-10).

Gene Inactivation. Sometimes, all that is desired is to specifically inactivate a gene of interest. A way of achieving such gene disruptions in one step is actually to insert another gene with a selectable function into the middle of a wild-type allele of the gene of interest, carried in a plasmid. A linear derivative of such a construct will then insert specifically at the wild-type locus, automatically disrupting it by virtue of the selectable gene inside it.

Thus, if the gene of interest is *lys2* (Figure 15-11), a *his3*$^+$ gene might be chosen to be inserted into *lys2*$^+$ carried in a plasmid. A linear form of this would be used to transform a strain of genotype *lys2*$^+$ *his3*$^-$, and *his3*$^+$ transformants would be selected. Such transformants are found also to be *lys2*$^-$ by virtue of the replacement of *lys2*$^+$ by the disrupted *lys2*$^+$ allele from the plasmid.

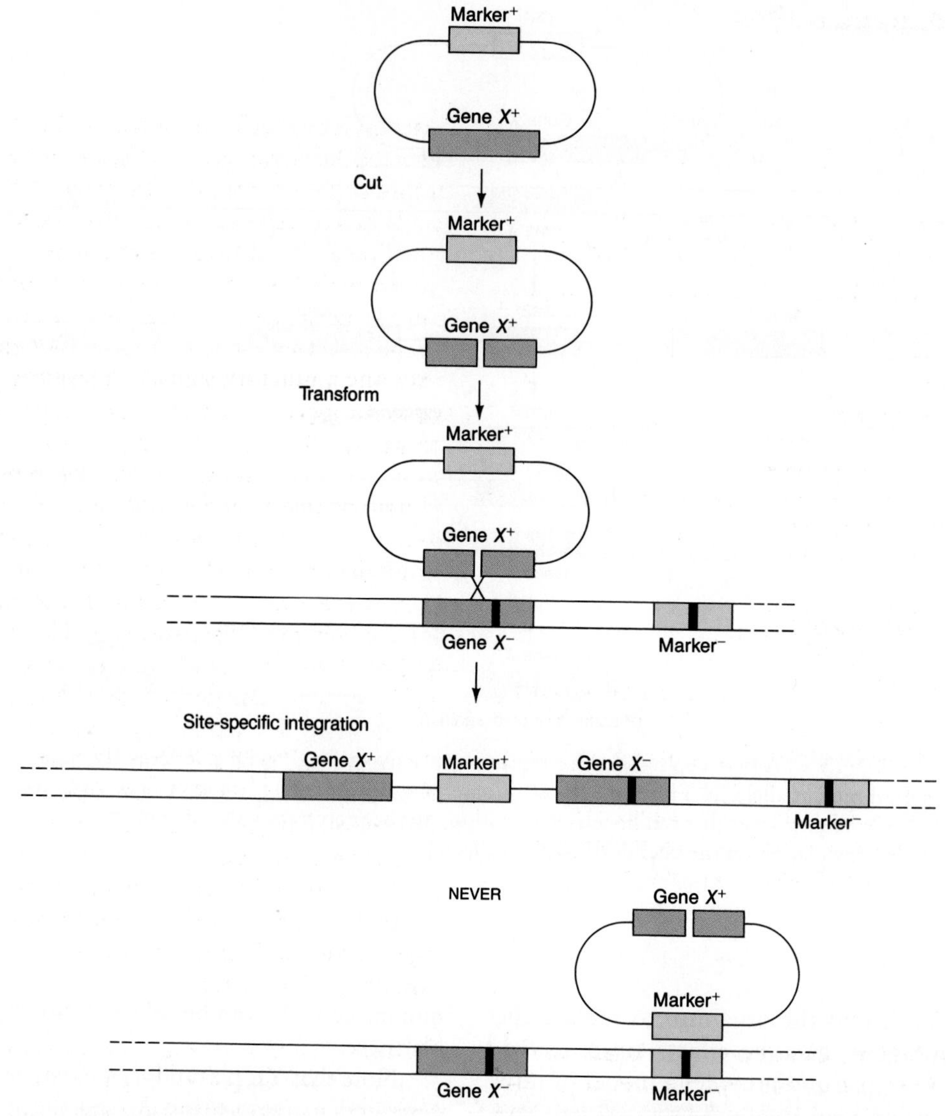

Figure 15-10 Integrative yeast plasmids often bear more than one region of homology to yeast chromosomes; the site specificity of integration can be increased by cutting the desired region, thereby producing recombinogenic ends. Because these free ends are preferable sites for recombination, the alternative alignment—of the homologous markers—does not occur *(bottom)*.

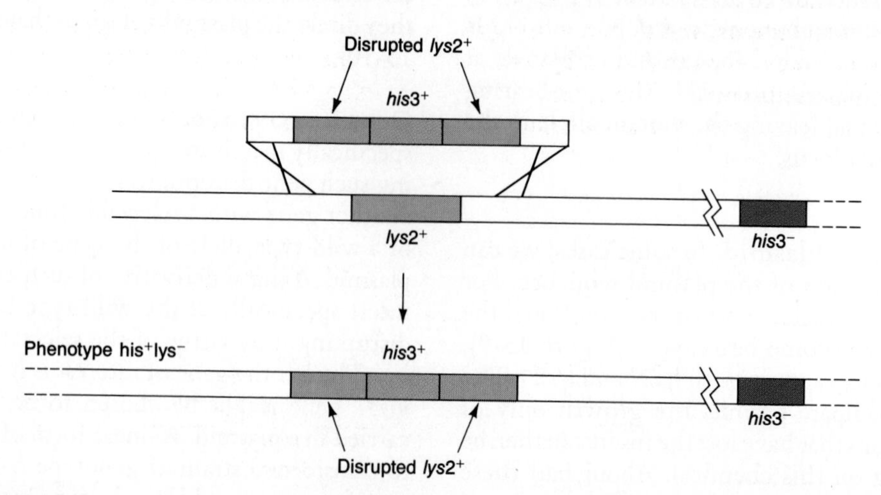

Figure 15-11 A single-step method for replacing a wild-type gene with a disrupted (inactivated) version carried on a linear DNA fragment for the purpose of specifically knocking out gene function. Integrants are selected as *his*$^+$ in this case.

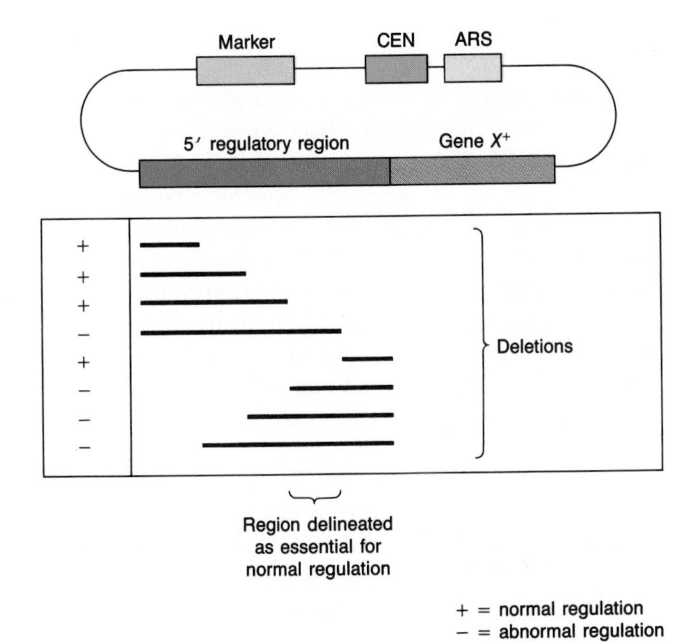

Figure 15-12 The regulation of the yeast gene X^+ can be studied by manipulating its regulatory region through deletion analysis in vitro and then transforming the constructs into a yeast strain bearing a defective allele X^-. (CEN = centromere sequence; ARS = autonomously replicating sequence).

Studying Regulation. Centromere plasmids can be used to study the regulatory elements upstream (5′) of a gene (Figure 15-12). The relevant coding region and its upstream region can be spliced into a plasmid, which can be selected by a separate yeast marker such as *URA3*. The upstream region can be manipulated by inducing a series of deletions, which are achieved by cutting the DNA, using a special exonuclease to chew away the DNA in one direction to different extents, and then rejoining it. The experimental objective is then to determine which of these deletions still permits normal functioning of the gene. Proper function is assayed by transforming the plasmid into a recipient in which the chromosome locus carries a defective mutant allele and then monitoring for return of gene function in the recipient. The results generally define a specific region that is necessary for normal function and regulation of the gene.

In such regulatory studies, it is often more convenient to use a gene whose product is easy to assay, instead of the structural gene of interest. A gene that is used to study the regulatory signals of another gene is termed a **reporter gene.** Therefore, if the regulation of gene *X* is of interest, the upstream regulatory regions of gene *X* are spliced to the reporter gene. The reporter gene has a phenotype that is easier to monitor than that of gene *X*, so the normal regulatory signals of gene *X* are expressed through the reporter. A gene that has been extensively used as a reporter is the bacterial *lacZ* gene, which codes for the enzyme β-galactosidase. This enzyme normally breaks down lactose, but it can also break down an analog of lactose, *X-gal* (5-bromo-4-chloro-indolyl-β,D-galactoside), very efficiently to yield 5-bromo-4-chloro-indigo, which is bright blue. The blue color is expressed as a blue yeast colony whenever it is active. Generally, the fusions are constructed in such a way that the upstream regulatory regions of gene *X* plus a few codons from the structural gene *X* are fused in the correct reading frame to the region coding for the enzymatically active portion of β-galactosidase. These constructs can be transformed in a nonintegrative vector.

Retrieval. Autonomously replicating vectors can also be used as retrieval agents. If a particularly interesting phenotype is produced by a specific mutant allele, it is useful to be able to retrieve that allele easily and examine its structure and function. This can be achieved by transformation with a gapped, centromeric plasmid that bears a deleted form of the gene of interest. The gap is repaired using information from the in situ mutant locus. The mutant sequence in the targeted locus is thus introduced into the plasmid. The repaired plasmid is then simply reisolated from the strain and examined at the molecular level (Figure 15-13).

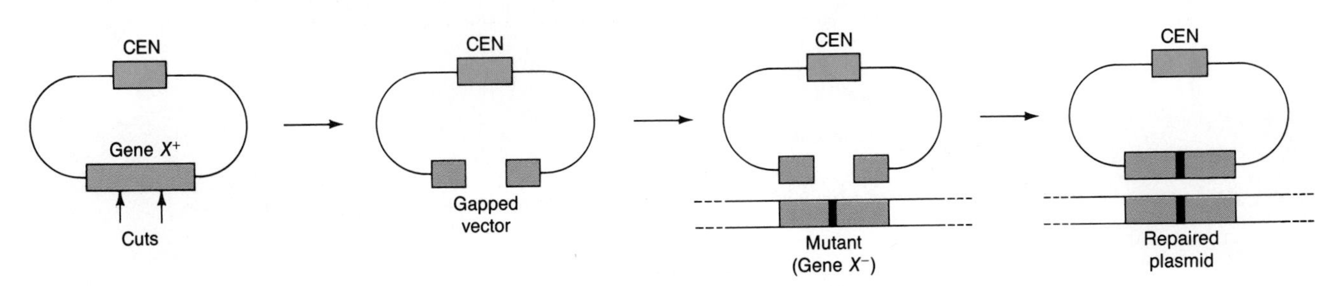

Figure 15-13 A gapped yeast centromeric plasmid is repaired by DNA copied from the homologous chromosome locus. This provides a convenient way of retrieving a mutant sequence of particular interest.

Future Applications. Yeast artificial chromosomes hold great promise as cloning vectors for large sections of mammalian (especially human) DNA. Consider that, for example, the size of the region coding the VIII blood-clotting factor in humans is known to span about 190 kb and that the gene for Duchenne muscular dystrophy spans probably more than 1000 kb! Furthermore, the large size of mammalian genomes in general means that banks or libraries of bacterial vectors are huge. Yeast artificial chromosomes, on the other hand, can carry much longer inserts and are potentially very useful in this regard.

The yeast system is by far the most sophisticated at present. However, the same techniques are being applied to other organisms—especially ones with well-defined genetic systems, such as filamentous fungi like *Neurospora* and *Aspergillus.*

Transgenic Plants

Because of their immense economic significance, plants have long been the subject of genetic analysis aimed at developing improved varieties. The advent of recombinant DNA technology has introduced a new dimension to this effort because the genome modifications made possible by this technology are almost limitless. No longer is breeding confined to selecting variants within the species. DNA can now be introduced from other species of plants, animals, or even bacteria!

The Ti Plasmid

The only vectors routinely used to produce transgenic plants are derived from a soil bacterium called *Agrobacterium tumefaciens.* This bacterium causes what is known as *crown gall disease,* in which the infected plant produces uncontrolled growths (tumors, or galls), normally at the base (or crown) of the plant. The key to tumor production is a large (200-kb) circular DNA plasmid—the **Ti** (*t*umor-*i*nducing) **plasmid.** When the bacterium infects a plant cell, a part of the Ti plasmid—a region called the **T-DNA**—is transferred and inserted, apparently more or less at random, into the genome of the host plant (Figure 15-14). The functions required for this transfer are outside the T-DNA on the Ti plasmid. The T-DNA itself carries several interesting functions, including the production of the tumor and the synthesis of compounds called *opines.* Opines are actually synthesized by the host plant under the direction of the T-DNA. The bacterium then uses the opines for its own purposes, calling on opine-utilizing genes on the Ti plasmid outside the T-DNA. Two important opines are nopaline and octopine; two separate Ti plasmids produce them. The structure of Ti is shown in Figure 15-15.

Using the Ti Plasmid as a Vector

The natural behavior of the Ti plasmid appears to make it well-suited for the role of a plant vector. If the DNA of interest could be spliced into the T-DNA, then it seems

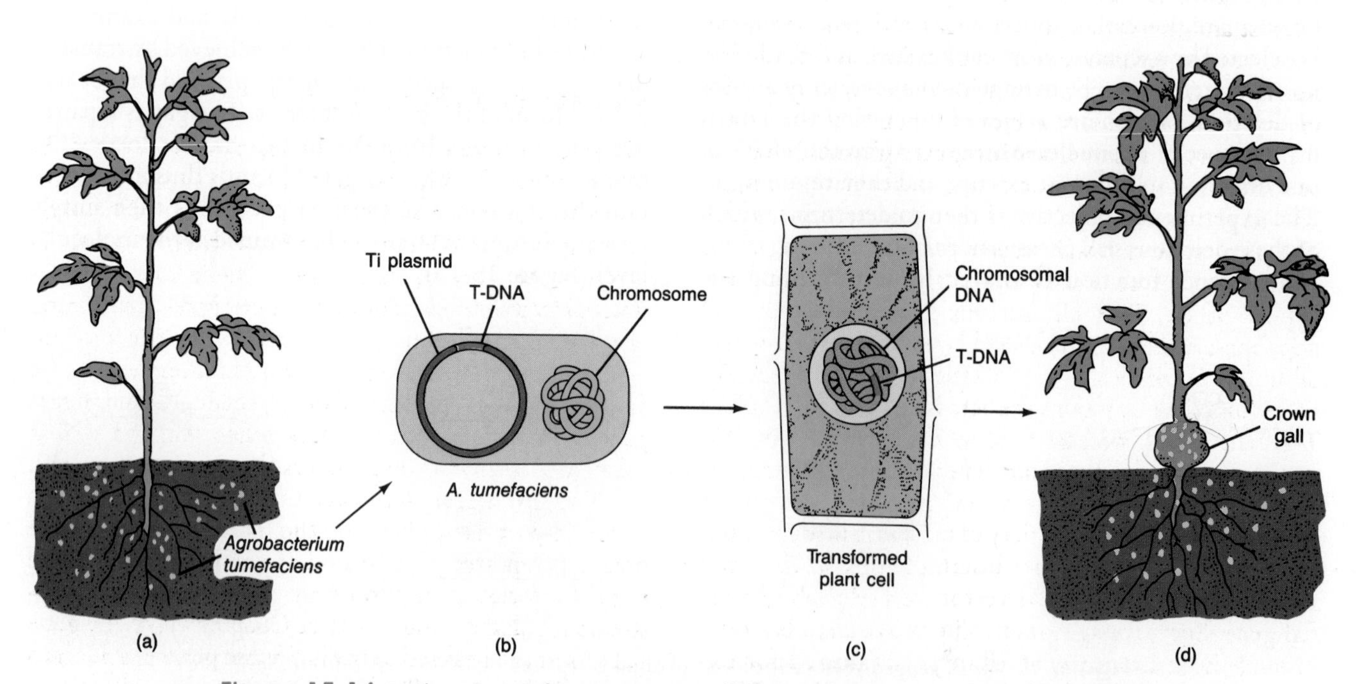

Figure 15-14 In the process of causing crown gall disease, the bacterium *Agrobacterium tumefaciens* inserts a portion of its Ti plasmid—a region called *T-DNA*—into a chromosome of the host plant.

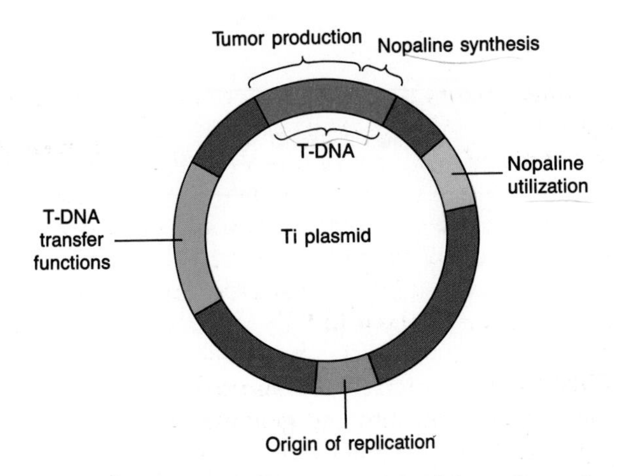

Figure 15-15 Simplified representation of the major regions of the Ti plasmid of *A. tumefaciens.* The T-DNA, when inserted into the chromosomal DNA of the host plant, directs the synthesis of nopaline, which is then utilized by the bacterium for its own purposes. T-DNA also directs the plant cell to divide in an uncontrolled manner, producing a tumor.

likely that the whole package would be inserted in a stable state into a plant chromosome. This system has indeed been made to work essentially in this way, but with some necessary modifications. Let's follow through a typical protocol.

Ti plasmids are too large to permit easy manipulation and cannot readily be made smaller, since they contain few unique restriction sites. Consequently, a smaller, intermediate vector that *can* be spliced receives the insert of interest and the various other genes and gene segments necessary for recombination, replication, and antibiotic resistance. Once engineered with the desired gene elements, this intermediate vector can then be inserted into the Ti plasmid, forming a **cointegrate plasmid** that can be introduced into a plant cell by transformation. Figure 15-16a shows one method of creating the cointegrate. The Ti plasmid that will receive the intermediate vector is first attenuated; that is, it has the entire right-hand region of its T-DNA, including tumor genes and nopaline-synthesis genes deleted, rendering it incapable of tumor formation—a "nuisance" aspect of the T-DNA function. It retains the left-hand border (L) of its T-DNA, which will be used as the crossover site for incorporation of the intermediate vector. The intermediate vector has had a convenient cloning segment spliced in, containing a variety of unique restriction sites. The gene of interest has been inserted at this site in Figure 15-16. Also spliced into the vector are a selectable bacterial gene (Spc^R) for spectinomycin resistance; a bacterial kanamycin-resistance gene (Kan^R), engineered for expression in plants; and two segments of T-DNA. One segment carries the nopaline-synthesis gene *(nos)*, plus the right-hand T-DNA border sequence (R). The second T-DNA segment comes from near the left-hand border and provides a section for recombination with a homologous portion of region L, which was retained in the disarmed Ti plasmid. After the intermediate vectors are introduced to *Agrobacterium* cells containing the disarmed Ti plasmids (by conjugation with *E. coli*), plasmid recombinants (cointegrates) can be selected by plating on spectinomycin. The selected bacterial colonies will contain only the Ti plasmid, since the intermediate vector is incapable of replication in *Agrobacterium.*

As Figure 15-16b shows, following spectinomycin selection for the cointegrates, bacteria containing the recombinant double or "cointegrant" plasmid are then used to infect cut segments of plant tissue, such as punched out leaf disks. If bacterial infection of plant cells occurs, anything between the left and right T-DNA border sequences can be inserted into the plant chromosomes. If the leaf disks are placed on a medium containing kanamycin, the only plant cells that will go through cell division are those that have acquired the Kan^R gene from T-DNA transfer. The growth of such cells results in a clump, or **callus,** which is an indication that transformation has occurred. These calluses can be induced to form shoots and roots and then be transferred to soil, where they develop into transgenic plants (Figure 15-16b). Often only one T-DNA insert is detectable in such plants, where it segregates at meiosis like a regular Mendelian allele (Figure 15-17). The insert can be detected by a T-DNA probe in a Southern hybridization, or by the detection of the chemical nopaline in the transgenic tissue.

Expression of Cloned DNA

What about expressing the DNA cloned into the T-DNA? This can, of course, be any DNA the investigator wants to test in the plant being used. One particularly striking foreign DNA that has been inserted using T-DNA is the gene for the enzyme luciferase, which is isolated from fireflies. The enzyme catalyzes the reaction of a chemical called *luciferin* with ATP; in this process, light is emitted, which explains why fireflies glow in the dark. A transgenic tobacco plant expressing the luciferase gene will also glow in the dark when watered with a solution of luciferin (see photograph at chapter opening).

This might seem like a playful experiment, but it has a very important application: the luciferase gene can be used as a reporter gene to study various aspects of gene regulation during development. For example, the upstream regulatory sequences (see Chapter 17) of any gene of interest can be fused to the luciferase gene and put into a plant via T-DNA. Then the luciferase gene will follow the same developmental pattern as the normally regu-

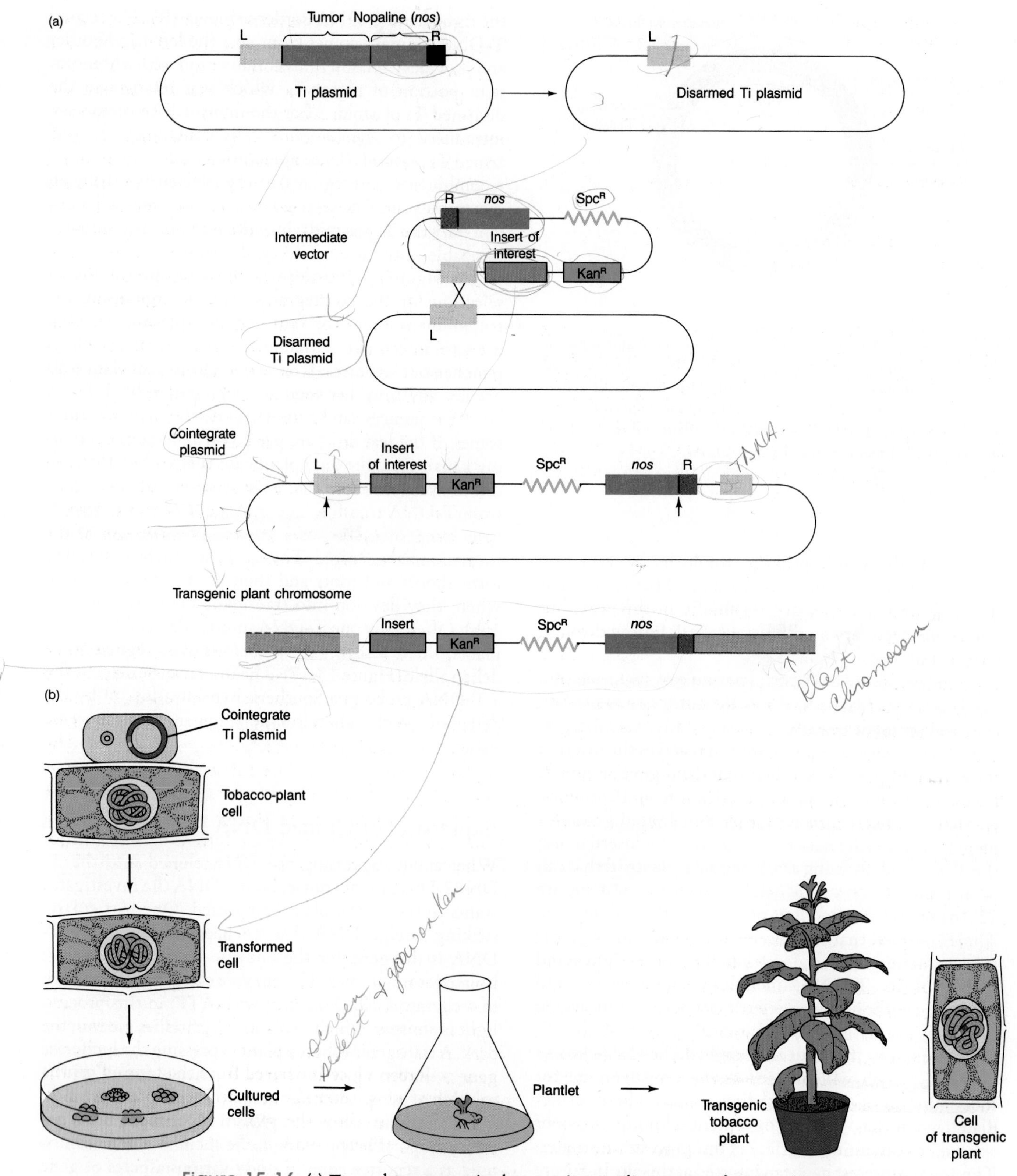

Figure 15-16 (a) To produce transgenic plants, an intermediate vector of manageable size is used to clone the segment of interest. In the method shown here, the intermediate vector is then recombined with an attenuated ("disarmed") Ti plasmid to generate a cointegrate structure bearing the insert of interest and a selectable plant kanamycin-resistance marker between the T-DNA borders, which are all the T-DNA that is necessary to promote insertion. (b) The generation of a transgenic plant through the growth of a cell transformed by T-DNA.

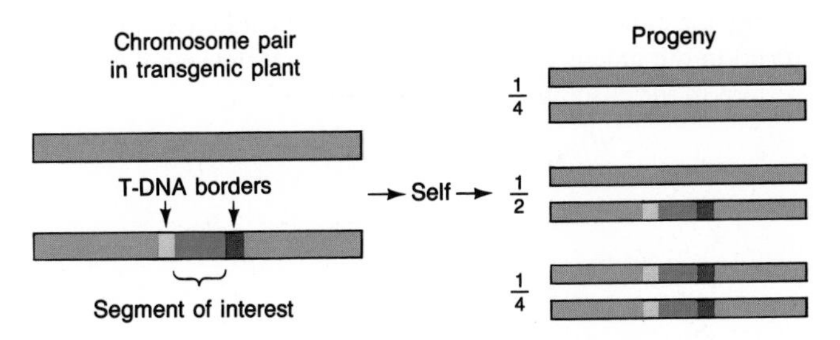

Figure 15-17 T-DNA and any DNA contained within it are inserted into a plant chromosome in the transgenic plant and then transmitted in a Mendelian pattern of inheritance.

lated gene does, but the luciferase gene will announce its activity prominently by glowing at various times or in various tissues, depending on the regulatory sequence. An agriculturally important example of inserting foreign DNA via T-DNA is a bacterial gene for resistance to the herbicide glyphosate. This gene confers resistance to the transgenic plant, enabling it to withstand the field application of glyphosate as a weed killer.

Transgenic Animals

There are several ways of producing transgenic animals. Two major examples are covered elsewhere in this book. The first is the production of transgenic *Drosophila* by the injection of plasmid vectors containing P elements into the fly embryo (page 595). The second major example is the production of transgenic mammals by injecting special plasmid vectors into a fertilized egg. In both cases, the extra DNA can find its way into the germ line cells, is then passed on to the progeny desired from these cells, and behaves from then on rather like a regular nuclear gene.

Gene Therapy

Gene therapy is an important experimental development in the use of transgenic animals (and other organisms too). Here the functions absent in a defective gene of the host are provided via the vector and ultimately expressed in the transgenic animal. The technique has been used in microbes routinely, of course, but is of great relevance in the case of humans in that it offers the hope of correcting hereditary diseases. Gene therapists have solved several hereditary problems in mammals other than humans.

One example is the correction of a growth-hormone deficiency in mice. The recessive mutation *little* (*lit*) results in dwarf mice. Even though the mouse's growth-hormone gene is present and apparently normal, no messenger RNA (mRNA) is produced.

As the initial step in correcting this deficiency, homozygous *lit lit* eggs are injected with about 5000 copies of a 5-kilobase linear DNA fragment that contains the rat growth-hormone structural gene *(RGH)* fused to a regulator-promoter sequence from a mouse metallothionine gene (*MP*). The normal job of metallothionine is to detoxify heavy metals, so that the regulatory sequence is responsive to the presence of heavy metals in the animal. The eggs are then implanted into pseudo-pregnant mice, and the baby mice are raised. About 1 percent of these babies turn out to be transgenic, showing increased size when heavy metals are administered during development. A representative transgenic mouse is then crossed to a *lit lit* female; the ensuing pedigree is shown in Figure 15-18. We can see that mice two to three times the weight of their *lit lit* relatives (Figure 15-19) are produced down through the generations, with the transgenic rat growth-hormone gene acting as a dominant marker, always heterozygous in this pedigree.

This kind of technology in mammals, *Drosophila,* and plants is not so controlled as it is in fungi such as yeast. The site of insertion of the introduced DNA in higher eukaryotes can be highly variable, and the DNA is generally not found at the homologous locus. Hence, gene therapy provides not a genuine correction of the original problem but a masking of it.

Regulation

Transgenic *Drosophila* provide us with another example of the use of the bacterial *lacZ* gene as a reporter in the study of gene regulation during development. The *lacZ* gene is fused to the upstream regulatory region of a *Drosophila* heat shock gene, which is normally activated by high temperatures. This construct is then used to generate transgenic flies. Following heat shock, the flies are killed and bathed in X-gal (page 449). The resulting pattern of blue tissues provides information on the major sites of action of the heat shock gene (Figure 15-20).

Screening for Genetic Diseases

Recessive mutations that follow Mendelian inheritance are responsible for over 500 genetic diseases. Homozygous individuals resulting from marriages involving two carriers of the same recessive trait will be affected by the disease. Screening cells derived from the fetus offers the possibility of predicting genetic defects at an early enough stage to allow the option of abortion to prevent the birth of afflicted individuals. The enzymes or proteins that are altered or missing in a number of genetic diseases are known (refer to the list of "inborn errors of metabolism," Table 12-4). To detect such genetic defects, fetal cells are taken from the amniotic fluid, separated from other components, and cultured to allow the analysis of chromosomes, proteins, and enzymic reac-

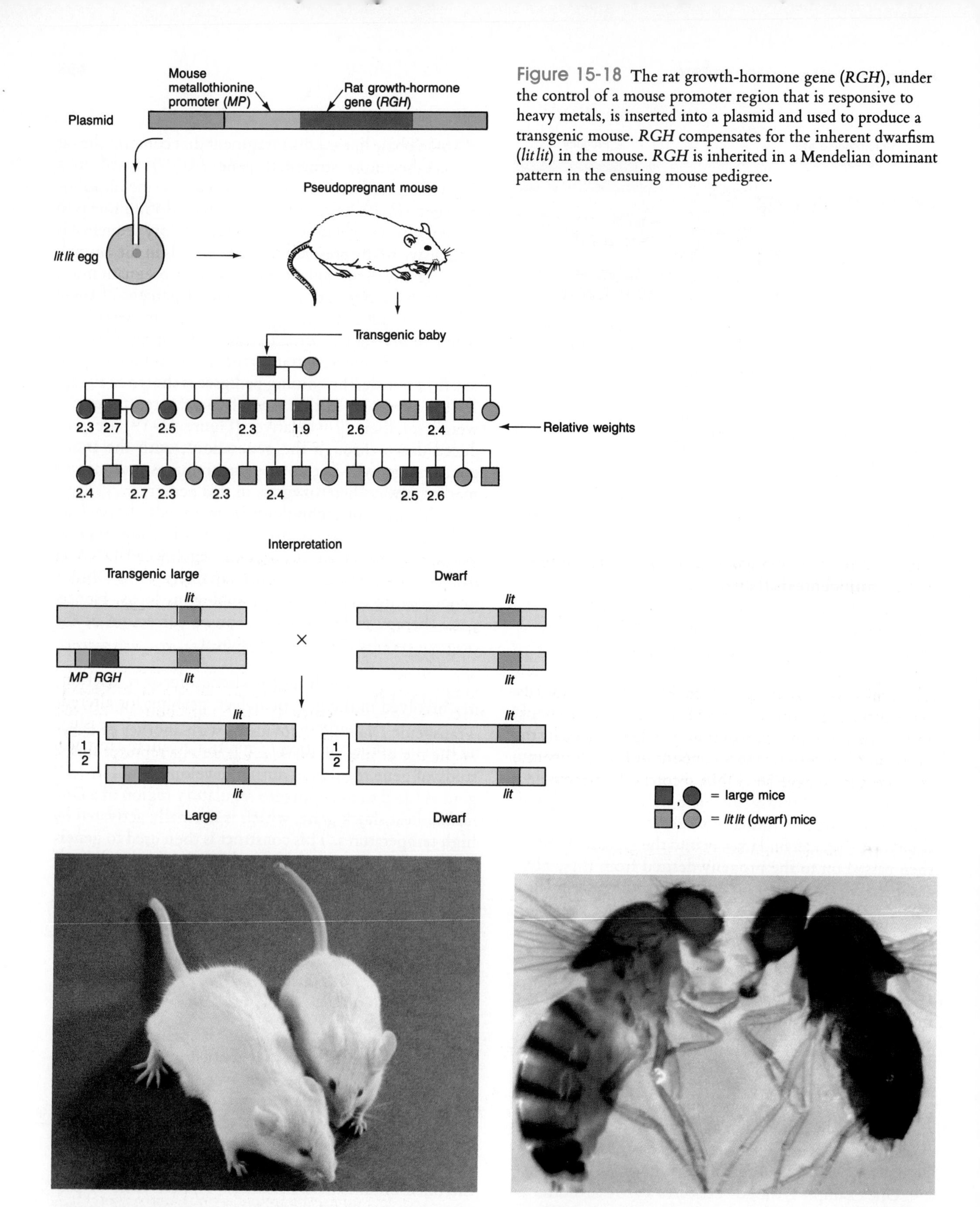

Figure 15-18 The rat growth-hormone gene (*RGH*), under the control of a mouse promoter region that is responsive to heavy metals, is inserted into a plasmid and used to produce a transgenic mouse. *RGH* compensates for the inherent dwarfism (*lit lit*) in the mouse. *RGH* is inherited in a Mendelian dominant pattern in the ensuing mouse pedigree.

Figure 15-19 Transgenic mouse. The mice are siblings, but the mouse on the left was derived from an egg transformed by injection with a new gene composed of the mouse metallothionein promoter fused to the rat growth-hormone structural gene. (This mouse weighs 44 g and its untreated sibling 29 g.) The new gene is passed on to progeny, in a Mendelian manner, and so is proven to be chromosomally integrated. (R. L. Brinster.)

Figure 15-20 Transgenic *Drosophila* expressing a bacterial β-galactosidase gene. *Drosophila* was transformed with a construct consisting of the *E. coli lac* Z gene driven by a *Drosophila* heat-shock promoter. The resulting flies were heat-shocked, killed immediately, and stained for β-galactosidase activity, detected by production of a blue pigment. Transformed fly is at right, normal fly at left. (John Lis.)

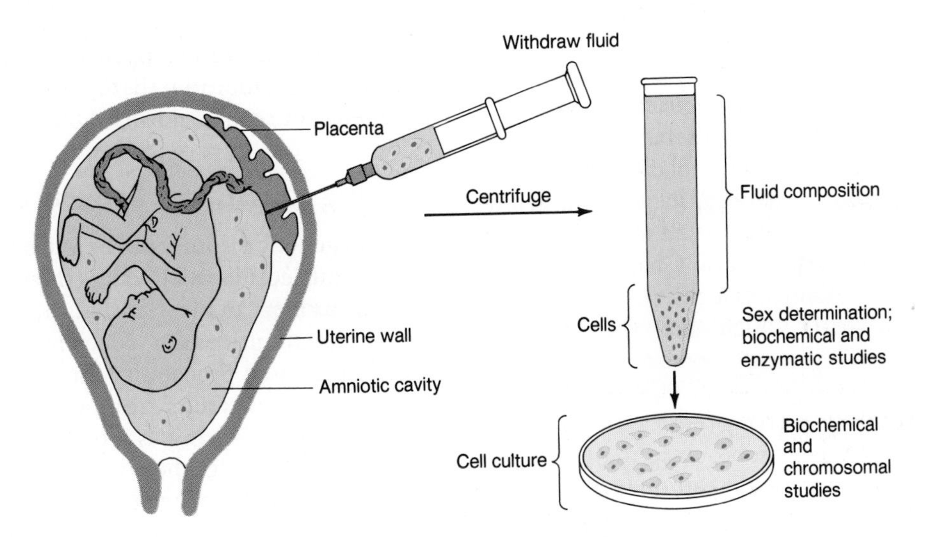

Figure 15-21 Amniocentesis.

tions, and other biochemical properties. This process, termed **amniocentesis** (Figure 15-21), can already pinpoint a series of known disorders. Table 15-1 lists examples of genetic diseases that can be detected by amniocentesis. Relying on physiological properties or on the presence or absence of enzymic activity in cultured fetal cells limits the screening procedure to those disorders that affect characters or proteins expressed in the cultured cells. The use of recombinant DNA greatly increases our ability to screen for genetic diseases, however, since we can analyze the DNA directly. In principle, if we could clone out the gene being tested and compare its sequence with that of a cloned normal gene, we could determine whether suspected defects were present. Of course, this would be a very laborious procedure, so shortcuts have to be devised to allow more rapid screening. Three useful techniques that have been used for this purpose involve searching for alterations of restriction sites involved in the genetic defect, probing for altered sequences with synthetic oligonucleotides, and measuring linkage of the mutation in question to altered restriction sites.

Table 15-1 Some Common Genetic Diseases

Inborn Errors of Metabolism	Approximate Incidence Among Live Births
1. Cystic fibrosis	1/1600 Caucasians
2. Duchenne muscular dystrophy	1/3000 boys (X-linked)
3. Gaucher's disease (defective glucocerebrosidase)	1/2500 Ashkenazi Jews; 1/75,000 others
4. Tay-Sachs disease (defective hexosaminidase A)	1/3500 Ashkenazi Jews; 1/35,000 others
5. Essential pentosuria (a benign condition)	1/2000 Ashkenazi Jews; 1/50,000 others
6. Classic hemophilia (defective clotting factor VIII)	1/10,000 boys (X-linked)
7. Phenylketonuria (defective phenylalanine hydroxylase)	1/5000 Celtic Irish; 1/15,000 others
8. Cystinuria (mutated gene unknown)	1/15,000
9. Metachromatic leukodystrophy (defective arylsulfatase A)	1/40,000
10. Galactosemia (defective galactose 1-phosphate uridyl transferase)	1/40,000
Hemoglobinopathies	**Approximate Incidence among live births**
1. Sickle-cell anemia (defective β-globin chain)	1/400 U.S. blacks. In some West African populations, the frequency of heterozygotes is 40%.
2. β-thalassemia (defective β-globin chain)	1/400 among some Mediterranean populations

NOTE: Although the vast majority of the over 500 recognized recessive genetic diseases are extremely rare, in combination they represent an enormous burden of human suffering. As is consistent with Mendelian mutations, the incidence of some of these diseases is much higher in certain racial groups than in others.

SOURCE: J. D. Watson, J. Tooze, and D. T. Kurtz, *Recombinant DNA: A Short Course.* Copyright 1983 by W. H. Freeman and Company.

Alterations of Restriction Sites

Sickle-cell anemia is an example of a genetic disease that is caused by a well-characterized alteration. Affecting approximately 0.25 percent of U.S. blacks, the disease results from an altered hemoglobin, in which a valine residue is substituted for a glutamic acid residue at position 6 in the β-globin chain (see also Chapter 12). The GAG → GTG change eliminates a cleavage site for the restriction enzyme *Mst*II, which cuts the sequence CCTNAGG (where N represents any of the four bases). The change from CCTGAGG to CCTGTGG can thus be recognized by Southern blotting (Figure 14-10) using labeled β-globin cDNA as a probe, since the DNA derived from individuals with sickle-cell disease will lack one fragment contained in the DNA from normal individuals, and in addition, there will be a large (uncleaved) fragment not seen in normal DNA (Figure 15-22).

Type of Hb	Amino acid sequence Nucleotide sequence
A	-Pro-Glu-Glu- -CCT-GAG-GAG- *Mst*II
S	-Pro-Val-Glu -CCT-GTG-GAG-

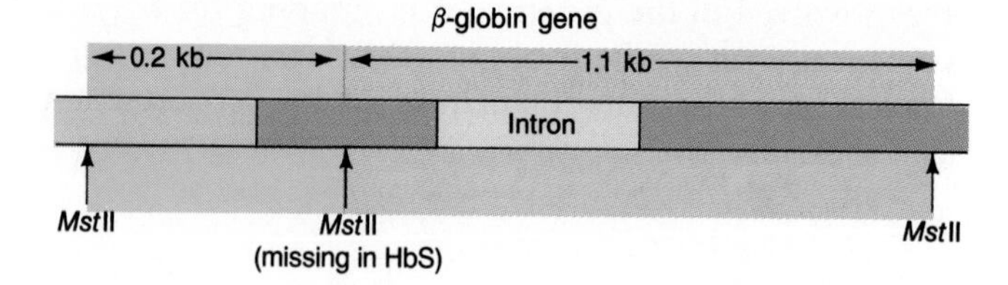

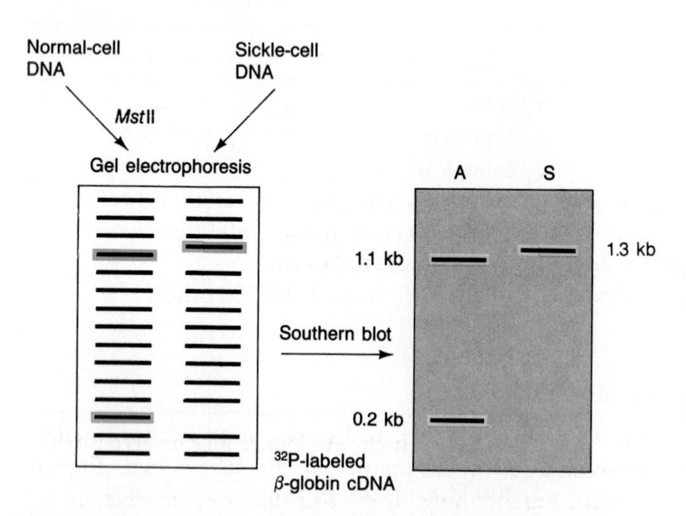

Figure 15-22 Detection of the sickle-cell globin gene by Southern blotting. The base charge (A → T) that causes sickle-cell anemia destroys an *Mst*II site that is present in the normal β-globin gene. This difference can be detected by Southern blotting. (Modified from J. D. Watson, J. Tooze, and D. T. Kurtz, *Recombinant DNA: A Short Course.* Copyright 1983 by W. H. Freeman and Company.)

Probing for Altered Sequences

When a genetic disorder can be attributed to a change in a specific nucleotide in all cases, then synthetic oligonucleotide probes can identify that change. The best example is alpha$_1$-antitrypsin deficiency, which leads to a greatly increased probability for developing pulmonary emphysema and results from a single base change at a known position. Using as a probe a synthetic oligonucleotide that contains the wild-type sequence in the relevant region of the gene, Southern blot analysis can be employed to determine whether the DNA contains the wild-type or the mutant sequence. At higher temperatures, a complementary sequence will hybridize, whereas a sequence containing even a single mismatched base will not.

Linkage to Altered Restriction Sites

What if a genetic defect itself does not alter a restriction site? Sometimes linkage to a restriction-site alteration can be measured. This strategy is derived from the observation that if a specific genetic region is cloned and sequenced and this sequence is compared with equivalent homologous regions in other individuals, a small percentage of nucleotide differences is seen. One obvious reason for this is the redundancy of the genetic code. But whatever the reason, these differences sometimes create or destroy restriction-enzyme target sequences. In fact, in eukaryotic DNA in general, it is quite easy to find such differences, using a protocol such as the following.

When used as a probe, most cloned segments of, say, human DNA are capable of detecting restriction-site variation. So let's begin with a randomly chosen cloned fragment and use it in Southern hybridizations against restriction-enzyme-digested DNA preparations from a sample of people. If we are unlucky, we will see the same pattern on all autoradiograms: either one band or more, depending on whether the specific restriction enzyme used happened to cut into the region spanned by the probe. But eventually it is likely that we will find a variant—a fragment pattern unique from the others that represents a restriction site with a different location (Figure 15-23).

Such variations in restriction-enzyme sites are almost always neutral; they do not represent coding or other differences that result in phenotypic changes at the cellular level. However, they do result in detectable differences at the molecular level, and they can be used as a marker. Such markers are sorely needed in mapping and manipulating large genomes. The coexistence in a popu-

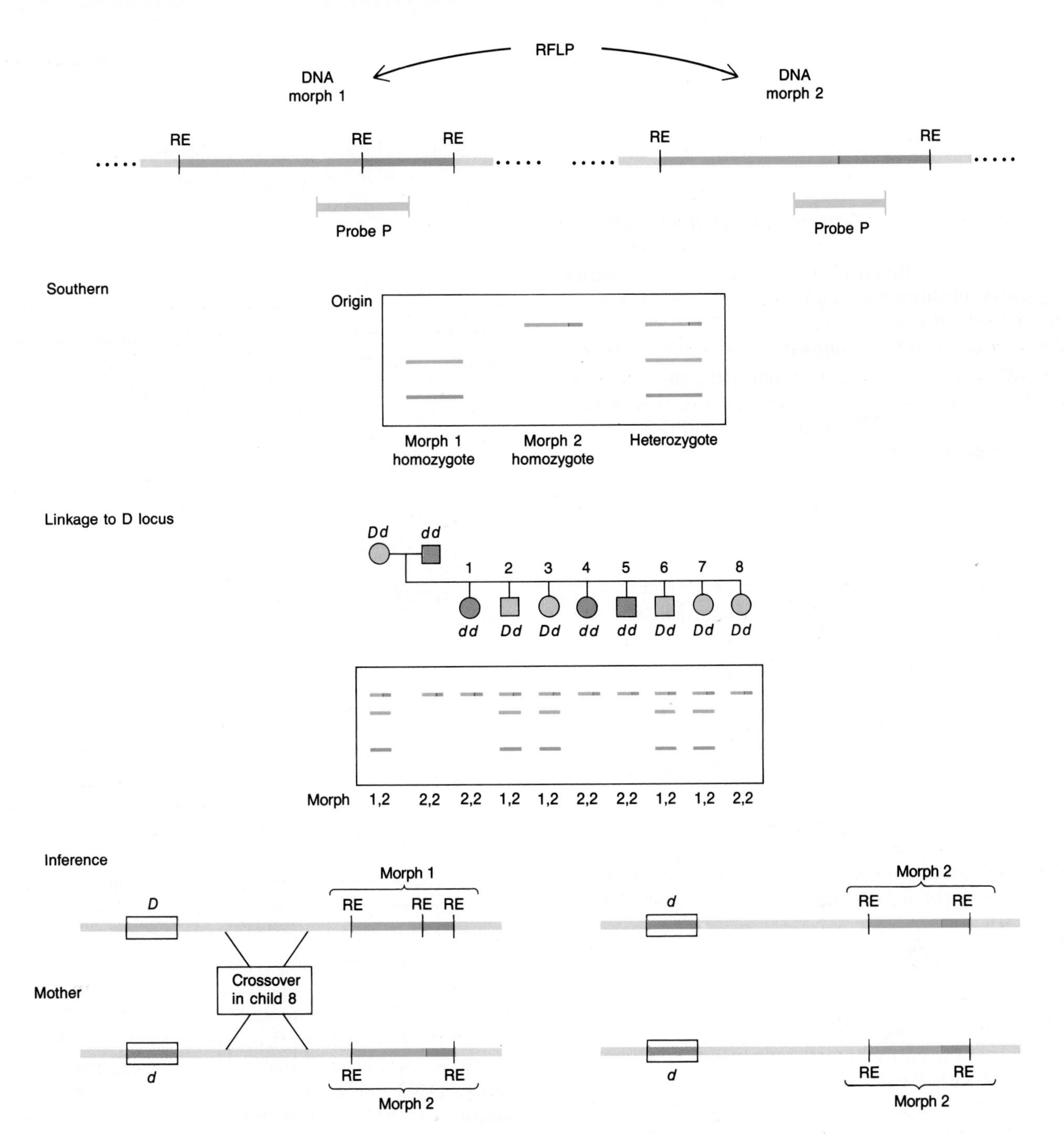

Figure 15-23 The detection and inheritance of a restriction fragment length polymorphism (RFLP). A probe P detects two DNA "morphs" when the DNA is cut by a certain restriction enzyme (RE). The pedigree of the dominant disease phenotype D shows linkage of the *D* locus to the *RFLP* locus; only child 8 is recombinant.

lation of two or more phenotypes, generally attributable to the alleles of one gene, is called **polymorphism** (Greek: "many forms"). At the molecular level, the phenotypes are the restriction fragments of varying size. The coexistence of two or more restriction fragment patterns, revealed by hybridization to a particular probe, is called **restriction fragment length polymorphism (RFLP).** In the most useful RFLPs, none of the variants are particularly rare; we can then examine pedigrees for other phenotypes, such as diseases, and look for linkage between the disease locus and the RFLP loci (Figure 15-23). Furthermore, the RFLP loci can be mapped in relation to each other.

Another example of this method is diagrammed in Figure 15-24, where *Hpa*I-site polymorphism is very closely linked to the sickle-cell β-globin gene. Examining linkage to restriction-site changes is simply doing on the DNA level the same type of analysis that has already

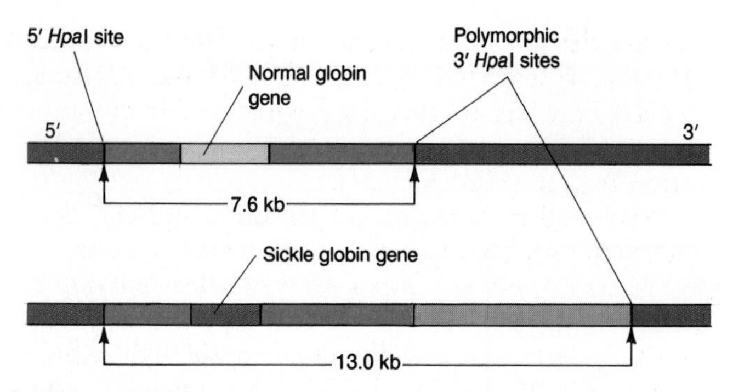

Figure 15-24 *Hpa*I-site polymorphism is diagnostic for the sickle *β*-globin gene in humans.

been applied to fully developed organisms: namely, using an identifiable character to give additional information about the genotype of an individual. For instance, colorblindness is used to yield information about the state of the locus governing the disease hemophilia on the X chromosome (see Problem 32 in Chapter 5). Because the *Cb* (colorblindness) locus is closely linked to the *Hb* (hemophilia) locus, the probability of the presence of the *Hb* alleles in a colorblind individual can be determined, provided the genotypes of the parents are known. The example given in Figure 15-24 employs the same principle. However, instead of determining colorblindness, we are determining the restriction-enzyme cleavage pattern of the DNA.

RFLPs as Map Reference Points

Using RFLPs as map reference points has been so successful that the entire human gene map is now liberally sprinkled with RFLP loci (for an example, see Figure 15-25). These loci are immensely helpful landmarks to use as a guide around the genome. As an example, if a geneticist is interested in cloning and studying a gene that causes a human disease, finding a linked RFLP can be a starting point. If the RFLP is less than about one map unit (1 m.u. = ~1000 kb) away, then the DNA identified by the RFLP probe can be used as a starting point for a chromosome walk (page 427) that ends with a clone of the disease-causing gene.

As we have seen, in some cases, a mutational event that causes a mutant phenotype, such as a disease, either creates or destroys a restriction site. Then the RFLP becomes useful as a diagnostic test for the presence of the disease allele in heterozygotes or individuals in which the disease has not yet manifested itself. For example, a certain mutation in a hemoglobin gene might produce a recessive allele that can cause anemia. This same nucleotide change creates a new restriction site, say, for the enzyme *Hin*dIII. Therefore, a probe that spans that site will pick up two fragments from the mutant allele and one larger fragment from the wild-type allele. A heterozygous carrier for this disease will show both the two smaller fragments and the larger fragment; that individual can be appropriately counseled concerning having children. Note also that such "direct hits," where the origin of the RFLP is actually in the gene of interest, provide a way of directly locating and sequencing the gene. This is because the probe provides immediate access to this genetic region.

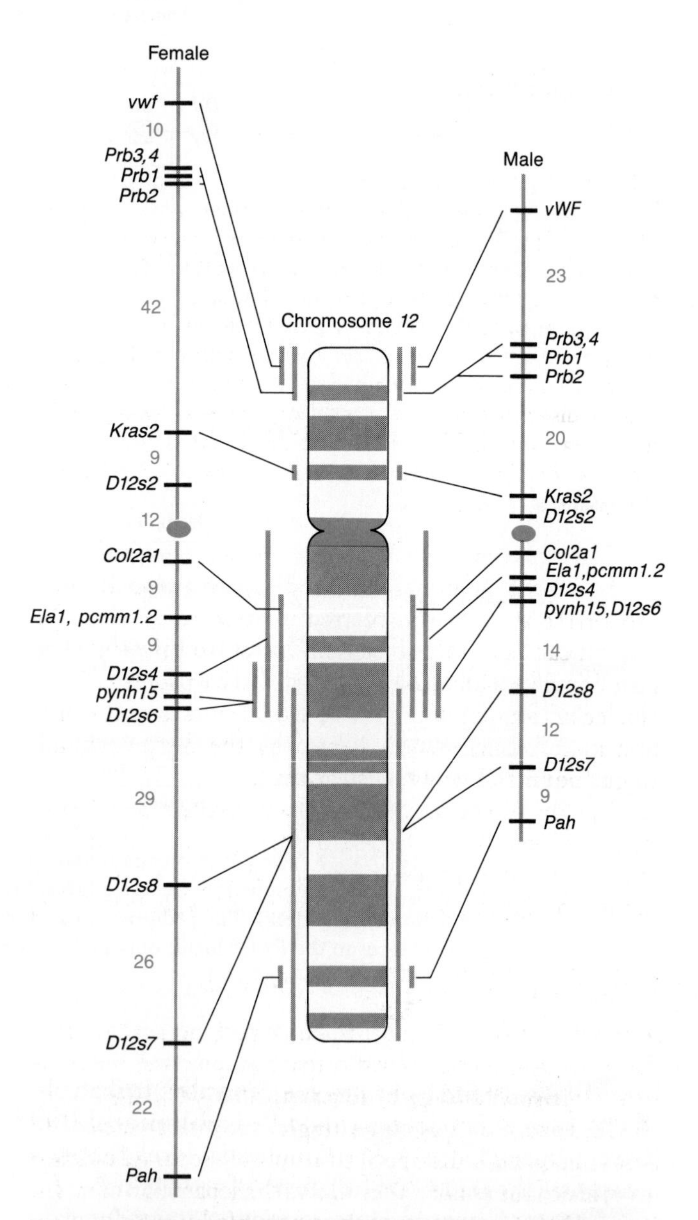

Figure 15-25 The human chromosome 12, showing the location of RFLP marker loci that have been detected in various pedigrees. Recombinant frequencies are shown for meiosis in men and in women; note that crossing-over appears to be more frequent in women.

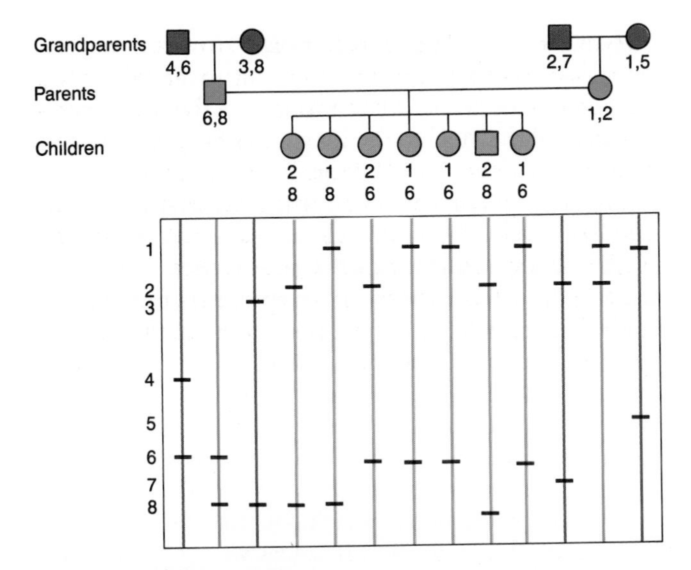

Figure 15-26 A pedigree illustrating the inheritance of alleles of the RFLP locus *VNTR* (*v*ariable *n*umber *t*andem *r*epeat). In this polymorphism a specific base sequence is repeated a variable number of times between two restriction sites. Thus, it is not the restriction sites per se that vary, but rather the number of repeats between the restriction sites. When these sections are cut by the restriction enzyme, the resulting fragment lengths depend on the number of repeats, and the fragments will migrate in distinctive patterns on an electrophoretic gel, as seen in the autoradiograph presented here below the family tree. This family shows eight alleles, that is, eight fragments of different length at this locus.

Once the gene has been located and sequenced, a protein sequence can be inferred and a short stretch of the protein can be synthesized. Antibodies to this synthetic protein will usually also bind to the native protein. Thus, labeled antibodies will reveal the distribution of the protein in cells and tissues affected by the disease, which might point the way to a treatment.

We have considered simple two-allele restriction fragment length dimorphism as examples of RFLPs. However, some probes pick up multiple alleles. The family in Figure 15-26 illustrates such a situation.

Pulsed Field Gel Electrophoresis

A technique that is useful in eukaryote mapping, especially in conjunction with restriction enzyme technology, is **pulsed field gel electrophoresis (PFGE).** In PFGE, instead of applying a single, uniform electric field across a gel, pulses are applied from two separate fields at an angle of about 90°. This allows the separation of much larger DNA fragments than is possible by regular electrophoresis. Apparently, the two fields allow the larger molecules to "snake" more efficiently through the gel, and fragments hundreds of kilobases long will move at a rate proportional to their size. The PFGE technique is so effective that the intact chromosomes of yeast and other fungi can be separated on a gel. Thus, for the first time, cloned genes can be associated with specific chromosomes without any need for meiotic or mitotic recombination analysis (Figure 15-27).

Mammalian chromosomes are too large to be separated in this way, but special restriction enzymes come to the rescue. These enzymes—the so-called *rare cutters,* such as *Not*I, *Pvu*I or *Mlu*I—cleave mammalian DNA into fragments of several hundred to 2000 kb, which then can be separated by PFGE. The human genome measures about 3 million kb, and there are 23 unique chromosomes in the haploid set. Thus, the average chromosome measures about 100,000 kb. The fragments generated by the rare-cutting enzymes are manageable and significant portions of a chromosome. Standard restriction mapping techniques can be applied to these large fragments, and long-range maps can be produced. If probes of known chromosome location are shown to hybridize to specific PFGE fragments, then a chromosomal link-up can ultimately be achieved.

Identifying Disease Genes

The General Approach

The genetic approach to a biological system is first to obtain mutants in some aspect of that system, then to use these mutants to isolate the gene and perhaps its encoded protein. Thus, genetic dissection generally begins with a pretty good idea of what the gene function is. **Reverse genetics** is a novel approach, made possible by molecular techniques, that works essentially in the other direction. Reverse genetics begins with a piece of DNA or a gene of unknown function and then deduces function. Often the

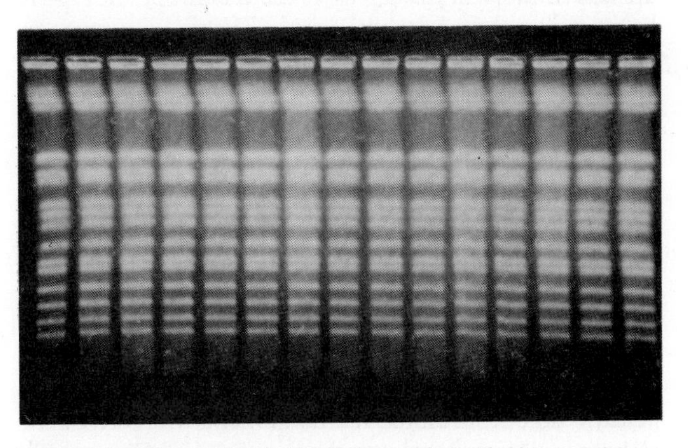

Figure 15-27 Pulsed field gel electrophoresis of uncut chromosomal DNA from different strains of yeast. Sixteen bands are resolved. Since there are 22 chromosomes in yeast, some have obviously comigrated in the gel. (All lanes were loaded identically.) (Source: Bio-Rad Laboratories.)

sequence with unknown function is a cloned piece of DNA that is interesting for some reason. For example, if a particular developmental stage of an organism is being studied, mRNA can be collected from that stage and cDNA made from it. These cDNAs should represent the genes that are active at this developmental stage. A specific cDNA can be cloned, and parts of it changed by directed mutagenesis (see Figures 15-1 and 15-2). Then by transformation the mutated DNA fragment can be introduced into cells of the organism in question. Once inside the cell, transforming DNA can be targeted to the proper locus for the gene, where it replaces the in situ wild-type segment by homologous recombination. The mutant phenotype is then assessed, and from this some idea of the function of the normal DNA sequence is deduced.

Sometimes the process begins with a *protein* of unknown function. From the amino acid sequence of the protein, an appropriate section is chosen from which a DNA coding sequence can be inferred using the code dictionary. From this an oligonucleotide is made that is used as a probe to recover the gene from genomic DNA. Once again, directed mutagenesis is used to create a mutant phenotype from which an idea of the role of the protein can be deduced.

The Approach Used in Isolating Human Disease Genes

Reverse genetics has become one of the standard methods of finding and deducing the function of hereditary disease genes in humans. Although the functions of many human disease genes are well understood (for example the so-called inborn errors of metabolism), many are still unknown. In such cases the mutant phenotype is known, but the underlying cause of this phenotype is not. Good examples are Duchenne's muscular dystrophy and cystic fibrosis; no primary defect for these conditions was known, but reverse genetic methods have now led to the successful isolation of both these genes, and a better understanding of the causes of these disorders.

As an example, we will follow the methods used to identify the genomic sequence of the cystic fibrosis (CF) gene. The search for this important gene, for which 1 in 20 people are carriers of the mutant CF allele, was a tour de force involving the international collaboration of several research teams. The search was particularly difficult because no chromosome rearrangements were available to help in narrowing down the location (see Chapter 8). The analysis provided a model for other searches for genes of unknown function.

The major symptoms of cystic fibrosis are well known, most notably serious digestive and respiratory problems and extremely salty sweat. However, no primary biochemical defect was known up to the time of the discovery of the gene. The technique of RFLP mapping had located the gene to the long arm of chromosome 7, between 7q22 and 7q31.1. The CF gene was thought to be inside this region, flanked by the markers *met* (a protooncogene, see Chapter 23) on one end and a molecular marker D788 at the other end. But between these markers lay 1.5 centiMorgans (map units) of DNA, the equivalent of 1.5 million nucleotide pairs. Additional markers within the region were obtained by using as probes DNA sequences from a chromosome 7–specific library. This library had been derived from a process called *flow sorting,* in which chromosomes are passed under an electronic device that sorts them by size.

However, the two key techniques that were used to traverse the huge genetic distances were chromosome walking and a related technique called **chromosome jumping.** This latter technique provides a way of jumping across potentially unclonable areas of DNA, and also generates spaced-out landmarks along the sequence that can be used as initiation points for bidirectional chromosomal walking. Chromosome jumping is illustrated in Figure 15-28.

In chromosome jumping, large fragments are created by partial restriction cleavage of high-molecular-weight DNA believed to contain the gene of interest. Each DNA fragment is then circularized, thus bringing the beginning and end of a given fragment together. This juncture is cut out and cloned into a phage vector, which, together with the junction segments formed from the other large fragments, makes up a "jumping library." A probe from the beginning of the stretch of DNA under investigation can be used to screen the jumping library to find the clone that contains the beginning sequence. This segment will be joined to a DNA sequence from farther along the chromosome. The distance between these two sections on the chromosome corresponds to the length of the fragment that brought the sections together during circularization. The second sequence in the junction fragment can be identified by another probe, which can be used either to start a chromosome *walk* back toward the first probe or to serve as the beginning probe for a second chromosome *jump.* Figure 15-28a shows the formation of the junction fragment containing the sequence complementary to the starting probe; Figure 15-28b shows a series of jumps toward the gene of interest.

A restriction map of the overall region was obtained with rare-cutting restriction enzymes, and the restriction sites used to position and orient the sequences obtained from jumping and walking. Once enough sequencing had been done to cover representative parts of the overall region, then the hunt for genes which might be the CF gene began. Genes were sought by several techniques. First, it was known that genes in humans generally are

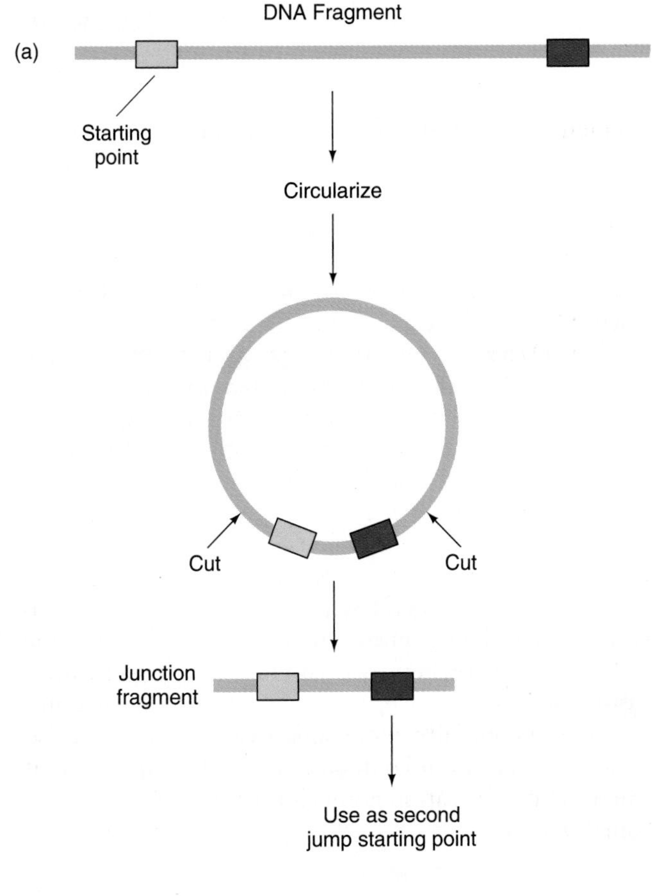

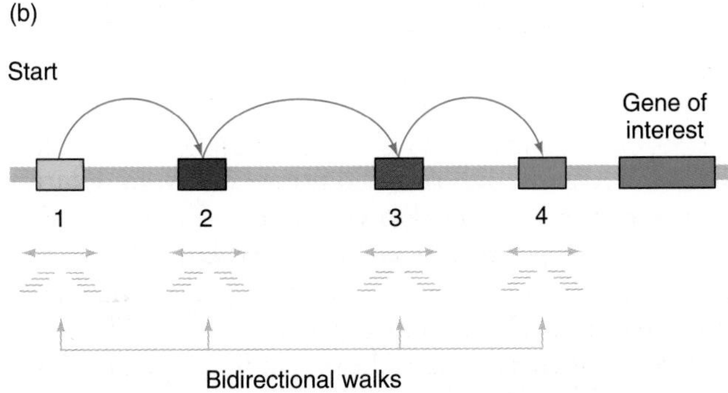

Figure 15-28 The method of chromosome jumping. (See the text for details.)

preceded at the 5′ end by clusters of cytosines and guanines, called **CpG islands,** and several of these were found. Second, it was reasoned that a gene would show homology to the DNA of other animals, because of evolutionary conservation, so candidate sequences were used to probe what were called "zoo blots" of genomic DNA from a range of animals. Third, genes should have appropriate start and stop signals. Fourth, genes should be transcribed, and transcripts should be found.

Ultimately, a strong candidate gene was found spanning 250 kilobase-pairs of the region. From cultured sweat gland cells, cDNA was prepared, and a 6500-nucleotide cDNA homologous to the candidate gene was detected. Upon sequencing this cDNA in normal and CF patients, it was found that the cDNA of the patients showed a deletion of three base pairs, eliminating a phenylalanine from the protein. Thus the CF gene had been found. From its sequence, an amino acid sequence was inferred, and from this the three dimensional structure of the protein was predicted. This protein showed structural similarities to ion-transport proteins in other systems, suggesting that a transport defect is the primary cause of CF. When used to transform CF cell lines, the wild-type gene restored normal function—the final confirmation that the isolated sequence was in fact the CF gene.

Summary

Methods are now available to manipulate eukaryotic DNA in bacterial vectors, to change specific base pairs in a gene, and then to reintroduce the DNA into the eukaryotic organism from which it came. The resulting transgenic organisms (yeast, plants, and certain vertebrates) have unique features and are invaluable for studying many biological processes. Recombinant DNA methodology has wide applications in the study of human genetic diseases. Probing for altered restriction fragment sizes allows the early detection of diseases such as sickle-cell anemia. Such techniques as RFLP mapping, chromosome walking, and chromosome jumping have helped to locate genes for genetic diseases such as cystic fibrosis.

Concept Map

Draw a concept map interrelating as many of the following terms as possible. Note that the terms are listed in no particular order.

recombinant DNA / reverse genetics / gene therapy / chromosome jumping / RFLP / transgenic / genetic screening

Chapter Integration Problem

In Chapter 2 we learned how pedigrees can be used to trace a family's genetic history, and in Chapter 14 we examined Southern blots. We can incorporate these ideas into the concepts discussed in this chapter to help find information about human genetic diseases. Huntington's disease (HD) is a lethal neurodegenerative disorder that exhibits autosomal-dominant inheritance. Because the onset of symptoms is usually not until the third, fourth,

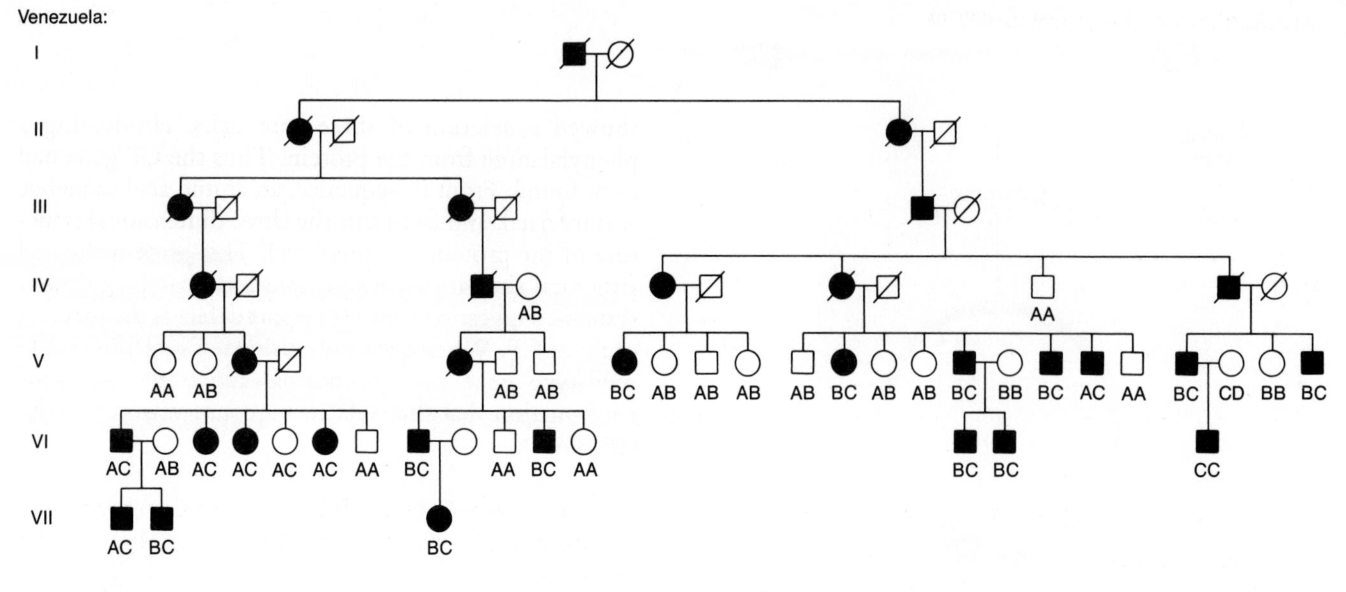

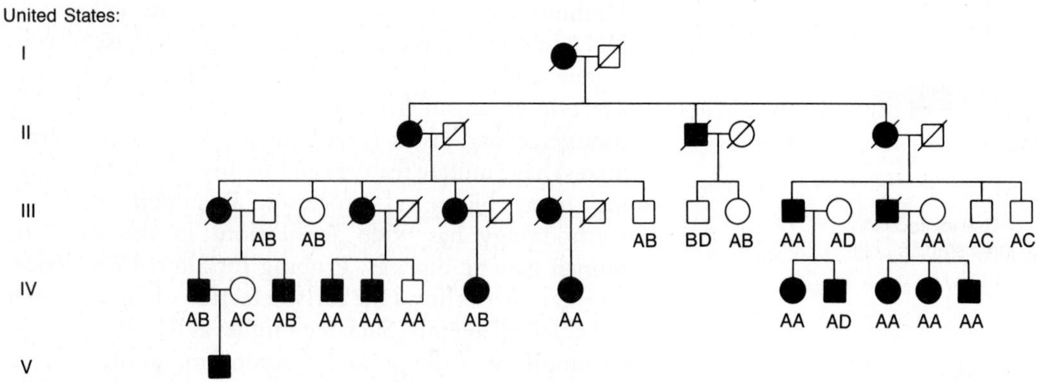

or fifth decade of life, patients with HD usually have already had their children, and some of them inherit the disease. There has been little hope of a reliable pre-onset diagnosis until recently, when a team of scientists searched for and found a cloned probe (called *G8*) that revealed a DNA polymorphism (actually a tetramorphism) relevant to HD. The probe and its four hybridizing DNA types are shown here; the vertical lines represent *Hin*dIII cutting sites:

Extent of homology of G8 probe

17.5	3.7	1.2	2.3	8.4	DNA A
17.5	4.9		2.3	8.4	DNA B
15.0	3.7	1.2	2.3	8.4	DNA C
15.0	4.9		2.3	8.4	DNA D

a. Draw the Southern blots expected from the cells of people who are homozygous (AA, BB, CC, and DD) and all who are heterozygous (AB, AC, and so on). Are they all different?

b. What do the DNA differences result from in terms of restriction sites? Do you think they are probably trivial or potentially adaptive? Explain.

c. When human-mouse cell lines were studied, the G8 probe bound only to DNA containing human chromosome 4. What does this tell you?

d. Two families showing HD—one from Venezuela, and one from the United States—are checked to determine their G8 hybridizing DNA type. The results are shown in the pedigree at the top of this page, where solid black symbols indicate HD and slashes indicate family members who were dead in 1983. What linkage associations do you see, and what do they tell you?

e. If a 20-year-old member of the Venezuelan family needs genetic counseling, what test would you devise and what advice would you give for each outcome? Repeat this for the U.S. family.

f. How might these data be helpful in finding the primary defect of HD?

g. Could these results be useful in counseling other HD families? Explain.

h. Are there any exceptional individuals in the pedigrees? If so, account for them.

Solution

a.

	AA	*BB*	*CC*	*DD*	*AB*	*AC*	*AD*	*BC*	*BD*	*CD*
17.5	—	—			—	—	—	—	—	
15.0			—	—		—	—	—	—	—
8.4	—	—	—	—	—	—	—	—	—	—
4.9		—		—	—		—	—	—	—
3.7	—		—		—	—	—	—		—
2.3	—	—	—	—	—	—	—	—	—	—
1.2	—		—		—	—	—	—		—

AD and *BC* are identical. The rest are different.

b. The differences in restriction sites come from differences in DNA sequence. There is no evidence on which to base a judgment of either trivial or potentially adaptive differences.

c. The sequence that gave rise to the G8 probe is located on chromosome 4.

d. For each family, construct a 2×2 table for each polymorphism. Do not include people who marry into the family. Calculate χ^2. This is done below for the relevant polymorphism in each family.

Venezuela

	Disease	No disease	Total
C present	19	1	20
C absent	0	15	15
Total	19	16	35

$$\chi^2 = \frac{(19-10.6)^2}{10.6} + \frac{(1-9.4)^2}{9.4} + \frac{(0-8.4)^2}{8.4} + \frac{(15-7.6)^2}{7.6}$$

$$= 6.7 + 7.5 + 8.4 + 7.2 = 29.8,$$

highly significant

United States

	Disease	No disease	Total
A present	13	7	20
A absent	0	1	1
Total	13	8	21

$$\chi^2 = \frac{(13-12.4)^2}{12.4} + \frac{(7-7.6)^2}{7.6} + \frac{(0-12.3)^2}{12.3} + \frac{(1-0.4)^2}{0.4}$$

$$= 0.029 + 0.047 + 12.3 + 0.9 = 13.276,$$

highly significant

Huntington's disease is linked with *C* in the family from Venezuela and with *A* in the family from the United States.

e. In each case, test for the relevant polymorphism by digesting with *Hin*dIII and probing with G8.

In the family from Venezuela, there is one crossover individual (VI, 5) among the 20 that carry the *C* polymorphism. Therefore, the G8 probe is $100\%(1/20) = 5$ m.u. from the Huntington's disease gene. If the person tests positive for the *C* polymorphism, there is a 95 percent chance of having the gene for Huntington's disease and a 5 percent chance of not having it. If the person tests negative for the *C* polymorphism, there is a 5 percent chance that he has the Huntington's disease gene and a 95 percent chance that he does not have the gene.

The situation in the family from the United States is unclear in comparison with the situation in the family from Venezuela. In the American family, the *A* polymorphism is present in three people with no family history of Huntington's disease who married into the family in question. The lower χ^2 value in this family, as compared with the one from Venezuela, is a reflection of these individuals. Their presence makes it impossible to identify crossover individuals unambiguously. It can be assumed that the G8 probe is identifying a polymorphism that is approximately 5 m.u. from the Huntington locus, just as it did in the family from Venezuela. However, the conclusions from testing of the individual in question would vary with the polymorphism genotype of his affected parent.

For instance, if the affected parent were *AA* and the individual in question were *A*–, the chance of the person's having inherited the Huntington's disease gene would be 50 percent. If, however, the affected parent were *AD* and the individual in question were *A*–, the chance of having inherited the Huntington's disease gene would be 95 percent, unless the unaffected parent also carried *A*. In that case, the risk would be 50 percent.

f. The G8 probe can be used to identify the region in which the Huntington's disease gene is located. The locus can be isolated by means of chromosome walking. The gene can be transcribed and translated, and the protein product can be identified.

g. Once the protein product of the Huntington's disease gene is identified, members of other families can be tested for the protein directly.

h. One exceptional person was identified already: Venezuela VI, 5, who is a crossover between the polymorphism and the Huntington's disease gene. In the U.S. pedigree there is no individual who is an obligate crossover.

(Solution from Diane K. Lavett)

Solved Problems

1. DNA studies are performed on a large family that shows a certain autosomal-dominant disease of late onset (approximately 40 years of age). A DNA sample from each family member is digested with the restriction enzyme *Taq*I and run on an electrophoretic gel. A Southern blot is then performed, using a radioactive probe consisting of a portion of human DNA cloned in a bacterial plasmid. The autoradiogram is shown below, aligned with the family pedigree. Affected members are shown in black.

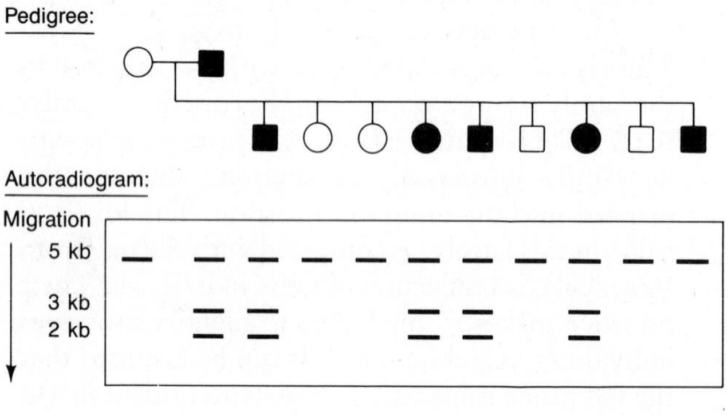

Pedigree:

Autoradiogram:

Migration

5 kb

3 kb

2 kb

a. Analyze fully the relationship between the DNA variation, the probe DNA, and the gene for the disease. Draw the relevant chromosome regions.

b. How do you explain the last son?

c. Of what use would these results be in counseling people from this family who subsequently married?

Solution

a. All individuals have the 5-kb band, indicating that the band sequence is not involved with the gene in question. All affected individuals, except the last son, have the 3-kb and 2-kb bands. These two bands are not seen in unaffected individuals. The suggestion is that the two bands are close to or part of the gene in question.

b. The last affected son indicates that the two bands are not part of the gene in question. He represents a crossover between the gene in question and the two bands.

c. The 2-kb and 3-kb bands are closely linked to the dominant allele. Their presence in an individual would indicate a high risk of developing the disorder, while their absence would indicate a low risk of developing the disorder. Exact risk cannot be stated until the map units between the two bands and the gene in question are determined. Although a rough estimate of map units can be made from the pedigree, the sample size is too small to make the estimate reliable.

(Solution from Diane K. Lavett)

2. A yeast plasmid carrying the yeast $leu2^+$ gene is used to transform nonrevertible haploid $leu2^-$ yeast cells. Several leu^+-transformed colonies appear on a leucineless medium. Thus, $leu2^+$ DNA presumably has entered the recipient cells, but now you have to decide what has happened to it inside these cells. Crosses of transformants to $leu2^-$ testers reveal that there are three types of transformants, A, B, and C, reflecting three different fates of the $leu2^+$ in the transformation. The results are

Type A × $leu2^-$ ⟶ $\frac{1}{2}$ leu^-
$\frac{1}{2}$ leu^+, × standard $leu2^+$
⟶ $\frac{3}{4}$ leu^+
$\frac{1}{4}$ leu^-

Type B × $leu2^-$ ⟶ $\frac{1}{2}$ leu^-
$\frac{1}{2}$ leu^+, × standard $leu2^+$
⟶ 100% leu^+
0% leu^-

Type C × $leu2^-$ ⟶ 100% leu^+

What three different fates of the $leu2^+$ DNA do these results suggest? Be sure to explain *all* the results according to your hypotheses. Use diagrams if possible.

Solution

If the yeast plasmid remains unintegrated, then it replicates independently of the chromosomes. During meiosis, the daughter plasmids would be distributed to the daughter cells, resulting in 100 percent transformation. This was observed in type C.

If one copy of the plasmid is inserted, when crossed

with a $leu2^-$ line, the resulting offspring would have a ratio of 1 leu^+: 1 leu^-. This is seen in type A and type B.

When the resulting leu^+ cells are crossed with standard $leu2^+$ lines, the data from type A cells suggest that the inserted gene is segregating independently of the standard $leu2^+$ gene, and data from type B cells suggest that the inserted gene is located in the same locus as the standard $leu2^+$ allele. Therefore, the gene in type A cells did not insert at the *leu2* site, and the gene in type B cells did.

(Solution from Diane K. Lavett)

Problems

1. Transgenic tobacco plants were obtained in which the vector Ti plasmid was designed to insert the gene of interest plus an adjacent kanamycin resistance gene. The inheritance of chromosomal insertion was followed by testing progeny for kanamycin resistance. Two plants typified the results obtained generally. When plant 1 was backcrossed to wild-type tobacco, 50 percent of the progeny were kanamycin-resistant and 50 percent were sensitive. When plant 2 was backcrossed to the wild type, 75 percent of the progeny were kanamycin-resistant, and 25 percent were sensitive. What must have been the difference between the two transgenic plants? What would you predict about the situation regarding the gene of interest?

2. In *Neurospora,* which has seven chromosomes, the following chromosomal rearrangements were obtained in different strains.

   a. A paracentric inversion of chromosome 1 (the largest chromosome)

   b. A pericentric inversion of chromosome 1.

   c. A reciprocal translocation in which about half of chromosome 1 was exchanged with about half of chromosome 7 (the smallest chromosome).

   d. A unidirectional insertional translocation, in which a part of one chromosome was inserted into another.

   e. A disomic ($n + 1$)

   f. A monosomic ($2n - 1$)

   g. A tandem duplication of a large part of chromosome 1.

   From all these strains and a normal wild type, DNA was isolated carefully to avoid mechanical breakage, and the samples were subjected to pulsed field gel electrophoresis. Predict the bands you would expect to see in each case.

3. In a bacterial vector you have cloned a plant gene that codes for a photosynthesis protein. Now you wish to find out if this gene is active in roots and other nonphotosynthetic tissue. How would you go about this? Describe the experimental details as well as you can.

4. In *Neurospora* you have two chromosome 5 probes that detect RFLP loci approximately 20 m.u. apart. In Southern blots of a *Pst*I digest of strain 1, probe A picks up two fragments of 1 and 2 kb, and probe B picks up two fragments of 4 and 1.5 kb. In Southern blots of strain 2, probe A detects one band of 3 kb, and probe B detects one band of 5.5 kb. Draw the Southern banding patterns you would see in *Pst*I digests of individual ascospore cultures from asci, using both probes simultaneously. Be sure to state how many different ascus types you expect, and be sure to account for the occurrence of crossovers.

5. A cystic fibrosis mutation in a certain pedigree involves a single nucleotide pair change. This change destroys an *Eco*RI restriction site normally found in this position. How would you use this information in counseling individuals in this family about their likelihood of being carriers? State the precise experiments needed. Assume that you detect that a woman in this family is a carrier, and it transpires she is married to an unrelated man who is also a heterozygote for cystic fibrosis, but in his case it is a different mutation in the same gene. How would you counsel this couple about the risks of a child's having cystic fibrosis?

6. In yeast, you have cloned a piece of wild-type DNA that you have sequenced, and it clearly contains a gene, but you do not know what gene it is. Therefore, to investigate further, you would like to find out its mutant phenotype. How would you use the cloned wild-type gene to do this? Show your experimental steps clearly. (HINT: think about molecular gene disruption methods.)

7. How would you use pulsed field gel electrophoresis to find out what chromosome a cloned gene is on?

8. Bacterial glucuronidase converts a colorless substance called *X-gluc* into a bright-blue indigo pigment. The gene for glucuronidase also works in plants if given a plant promoter region. How would you use this gene as a reporter gene to find out in which tissues a plant gene you have just cloned is normally active? (Assume X-gluc is easily taken up by the plant tissues.)

9. In mouse *Hin*dIII restriction digests, a certain probe picks a simple RFLP consisting of two alternative alleles of 1.7 kb and 3.8 kb. A mouse heterozygous for a dominant allele for bent tail and also heterozygous for the above RFLP is mated to a wild-type mouse that shows only the 3.8-kb fragment. Forty percent of the bent-tail progeny are homozygous for the 3.8-kb allele, and sixty percent are heterozygous for the 3.8- and the 1.7-kb forms.

a. Is the bent-tail locus linked to the RFLP locus? Draw the parental and progeny chromosomes to illustrate your answer.

b. What RFLP types do you predict among the wild-type offspring, and in what proportions?

10. Genes *A* and *B*, which map on yeast chromosome 4, are used as genetic markers in a study of two different haploid populations of yeast. The two populations express different allelic forms of the genes: in population 1, gene *A* gives *A1* and gene *B* gives *B1*; in population 2, gene *A* gives *A2* and gene *B* gives *B2*. These alleles are distinguished by the *Hin*dIII restriction map of the DNA in the region of the genes:

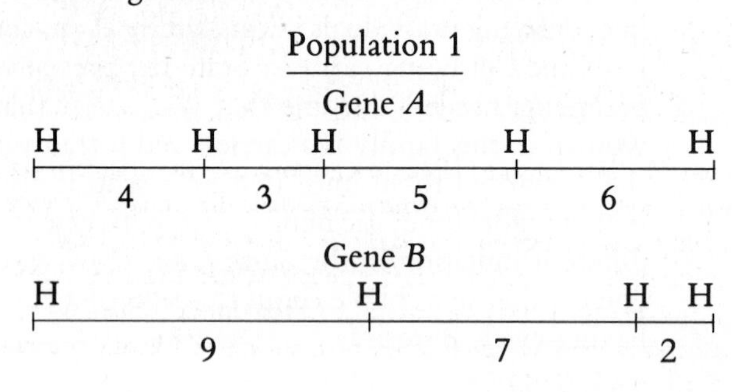

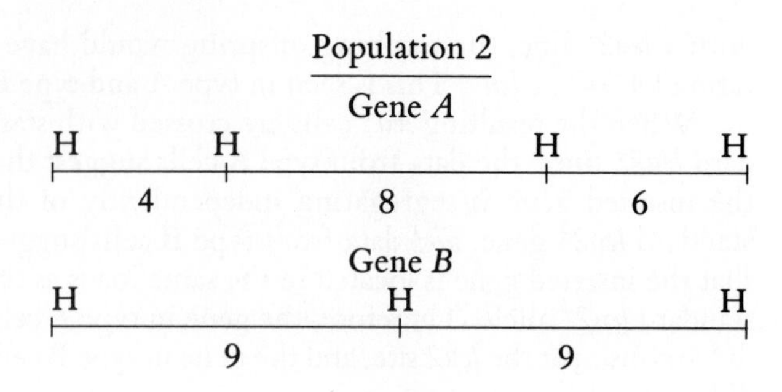

The tetrad products are examined, and the DNA fragments corresponding to genes *A* and *B*, respectively, are given for each type:

Spore type	Frequency	DNA (*Hin*dIII fragments) Gene *A*	Gene *B*
1	15%	4, 3, 5, 6	9, 9
2	15%	4, 8, 6	9, 7, 2
3	35%	4, 3, 5, 6	9, 7, 2
4	35%	4, 8, 6	9, 9

a. What are the allelic forms of the tetrad products?

b. How did they arise?

c. Draw the appropriate linkage map.

(Problem 10 courtesy of Joan McPherson.)

16

The Structure and Function of Eukaryotic Chromosomes

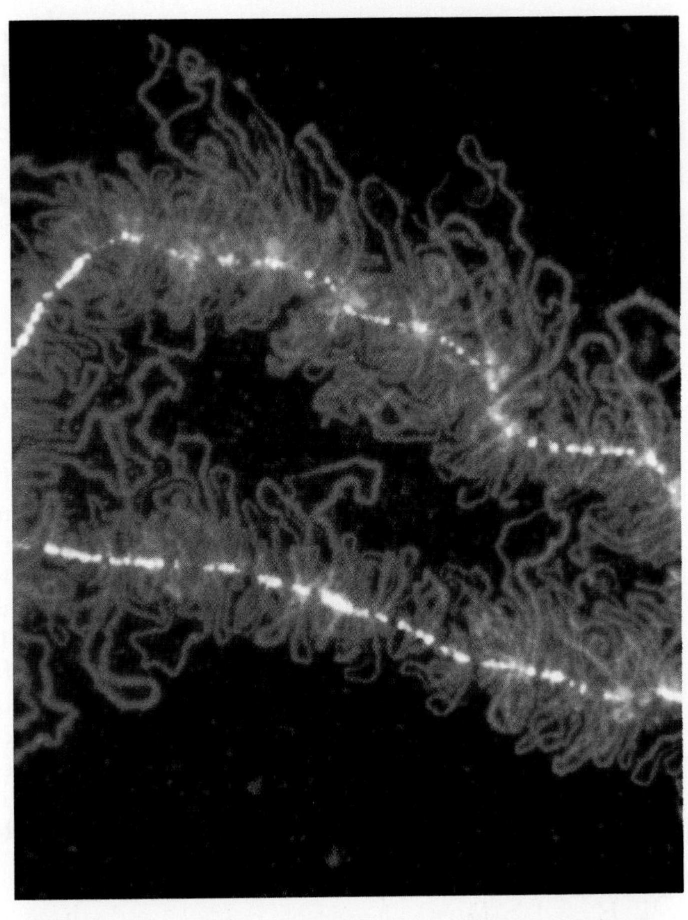

Lampbrush chromosomes. The chromosomes of some animals take on this lampbrush appearance during meiotic diplotene in females. The lampbrush structure is thought to reflect the underlying organization of all chromosomes, a central scaffold (here stained brightly), and projecting lateral loops (stained red) formed by a folded continuous strand of DNA with associated histone proteins. (M. Roth and J. Gall.)

KEY CONCEPTS

A chromosome contains only one long DNA molecule.

▪

DNA winds itself around protein spools and the spooled unit then coils, loops, and supercoils, forming a chromosome.

▪

A large proportion of active and inactive eukaryotic DNA is present in multiple copies.

▪

Centromeres have specialized molecular sequences that attach to spindles and telomeres have specialized sequences that permit complete replication of chromosome ends.

▪

Replication and transcription both occur with the protein spools in place.

One cell of the prokaryote *E. coli* contains about 1.3 mm of DNA (about 4,200 kb). In stark contrast, a human cell contains about 2 m of DNA (6,000,000 kb). The human body consists of approximately 10^{13} cells, and therefore contains about 2 times 10^{13} m of DNA. Some idea of the extreme length of this DNA can be obtained from comparing it with the distance from earth to the sun, 1.5 times 10^{11} m: the DNA in your body could stretch to the sun and back about 50 times. This peculiar fact makes the point that the DNA of eukaryotes is obviously efficiently packed. In fact, the packing occurs at the level of the nucleus, where the 2 m of DNA in a human cell is packed into 46 chromosomes, all in a nucleus 0.006 mm in diameter. In this chapter we have to translate what we have learned about the structure and function of eukaryotic genes into the "real world" of the nucleus. Instead of envisioning replication and transcription machinery moving along the airy-looking straight lines that we have used to represent genes in previous chapters, we must now come to grips with the fact that these processes take place in what must be very much like the inside of a densely wound ball of wool.

First, we will discuss aspects of the architecture of the eukaryotic genome, examining its topology at the molecular level. Many of the chromosomal landmarks that we have viewed at the level of light microscopy, such as bands, centromeres, and nucleolar organizers, need to be reexamined in higher resolution analyses. Once this picture is developed, we will examine the problems of replication and transcription in densely packed DNA. The analytical procedures that are used in this subject area are a powerful combination of genetics, molecular biology, and light and electron microscopy.

One DNA Molecule per Chromosome

If eukaryotic cells are broken, and the contents of their nuclei are examined under the electron microscope, the chromosomes appear as masses of spaghetti-like fibers with diameters of about 30 nm. Some examples are shown in the electron micrograph in Figure 16-1. In the 1960s, Ernest DuPraw studied such chromosomes carefully and showed that there are no ends protruding from the fibrillar mass. This suggests that each chromosome is one, long fine fiber folded up in some way. If the fiber somehow corresponds to a DNA molecule, then we arrive at the idea that each chromosome is one, densely folded, DNA molecule.

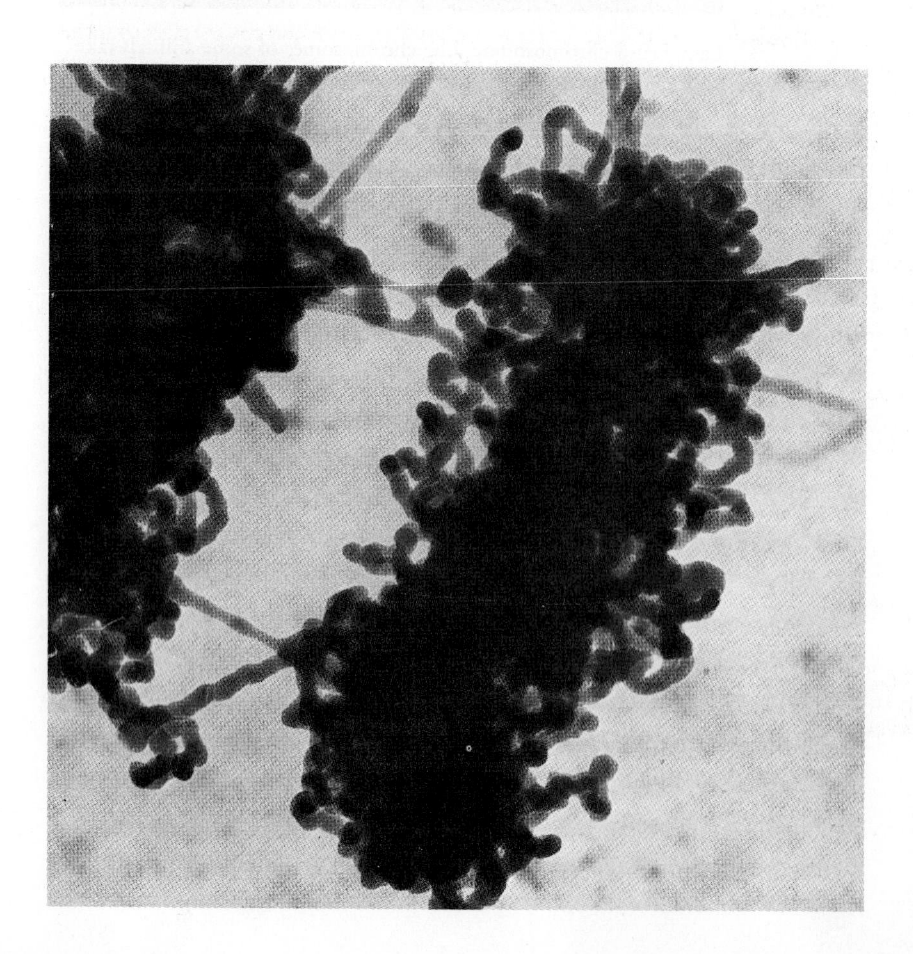

Figure 16-1 Electron micrograph of metaphase chromosomes from a honeybee. The chromosomes each appear to be composed of one continuous fiber 30 nm wide. (From E. J. DuPraw, *Cell and Molecular Biology.* Copyright 1968 by Academic Press.

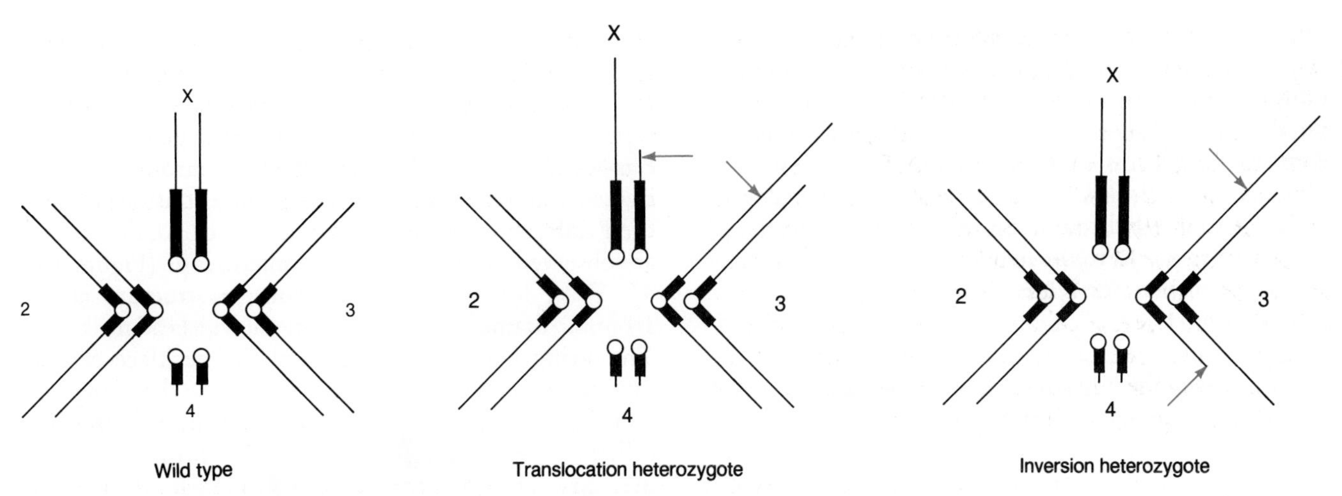

Figure 16-2 Genomes of *Drosophila* used to study DNA lengths in chromosomes. The chromosomes of three females are shown, one a normal wild type, one heterozygous for an X–3 translocation, and one heterozygous for a pericentric inversion of chromosome 3. Arrows mark breakpoints. Only the translocation produced a DNA molecule that was significantly longer.

In 1973, Ruth Kavenoff and Bruno Zimm performed experiments that showed this was most likely the case. They studied *Drosophila* DNA by using a viscoelastic recoil technique, which measures the size of DNA molecules in solution by measuring their elastic recoil properties. Put simply, the procedure is analogous to stretching out a coiled spring and measuring how long it takes to return to its fully coiled state. DNA is stretched by spinning a paddle in the solution and is then allowed to recoil into its relaxed state. The recoil time is known to be proportional to the size of the largest molecules present. In their study of *Drosophila melanogaster,* which has four pairs of chromosomes, Kavenoff and Zimm obtained a value of 41 times 10^9 daltons for the largest DNA molecule in the wild-type genome. Then they studied two chromosomal rearrangements (Figure 16-2). One, a translocation involving chromosome 3, resulted in a chromosome that was one-third larger than the largest wild-type chromosome. Their viscoelastic measurement in this case was 58 times 10^9 daltons, also about one-third larger. The other rearrangement, a pericentric inversion, altered the arm ratio for chromosome 3, but not its size. The viscoelastic measurement was 42 times 10^9 daltons, not significantly different from that of the wild type. In other words, making a chromosome bigger made the DNA of that chromosome bigger by exactly the same proportion, and rearranging the chromosome but keeping the same length had no effect on DNA size. It looked like the chromosome was indeed one strand of DNA, continuous from one end through the centromere, to the other end. Kavenoff and Zimm also were able to piece together electron micrographs of DNA molecules about 1.5 cm long, each presumably corresponding to a *Drosophila* chromosome (Figure 16-3).

Today, geneticists can demonstrate directly that certain chromosomes contain single DNA molecules by using pulse-field gel electrophoresis, a technique for separating very long DNA molecules by size (Chapter 15). If the DNA of an organism with relatively small chromosomes, such as *Neurospora,* is run for long periods of time in this apparatus, then the number of bands that separate on the gel is equal to the number of chromo-

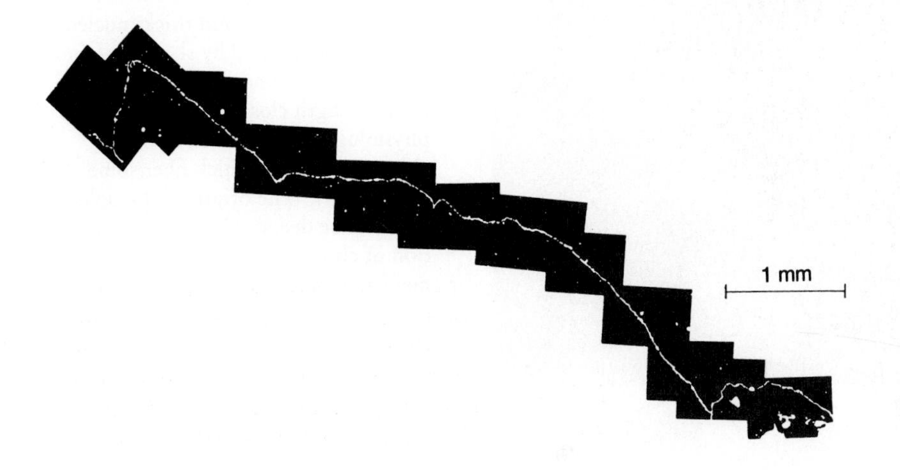

Figure 16-3 Composite electron micrograph of a DNA molecule from *Drosophila.* The overall length is 1.5 cm and is thought to correspond to one chromosome (From R. Kavenoff, L. C. Klotz, and B. H. Zimm. *Cold Spring Harbor Symp. Quant. Biol.,* 38, 1974, 4.)

somes (seven in the case of *Neurospora*). If each chromosome contained more than one DNA molecule, then the number of bands would be expected to be greater than the number of chromosomes. Such separations cannot be done for organisms with large chromosomes (such as humans and *Drosophila*) because the DNA molecules are too large to move through the gel, but, nevertheless, all the evidence points to the general principle that a chromosome contains one DNA molecule.

> Message Each eukaryotic chromosome contains a single, long, folded DNA molecule.

The Role of Histone Proteins in Packaging DNA

We have seen that because the length of a chromosomal DNA molecule is much greater than the length of a chromosome, there must be an efficient packaging system. What are the mechanisms that pack DNA into chromosomes? How is the very long DNA thread converted into the relatively thick dense rod that is a chromosome? The overall mixture of material that chromosomes are composed of is given the general name **chromatin.** It is DNA and protein. If chromatin is extracted and treated with differing concentrations of salt, then different degrees of compaction, or condensation, are observed under the electron microscope (Figure 16-4). With low salt concentrations, a structure about 10 nm in diameter is seen that resembles a bead necklace. The string between the beads of the necklace can be digested away with the enzyme DNase, so the string can be inferred to be DNA. The beads on the necklace are called **nucleosomes,** and these can be shown to consist of special chromosomal proteins, called **histones,** and DNA. Histone structure is remarkably conserved across the gamut of eukaryotic organisms, and nucleosomes are always found to contain an octamer of two units each of histones H2A, H2B, H3, and H4. The DNA is wrapped twice around the octamer as shown in Figure 16-5a. When salt concentrations are higher, the nucleosome bead necklace gradually assumes a coiled form called a **solenoid** (Figure 16-5b). This solenoid produced in

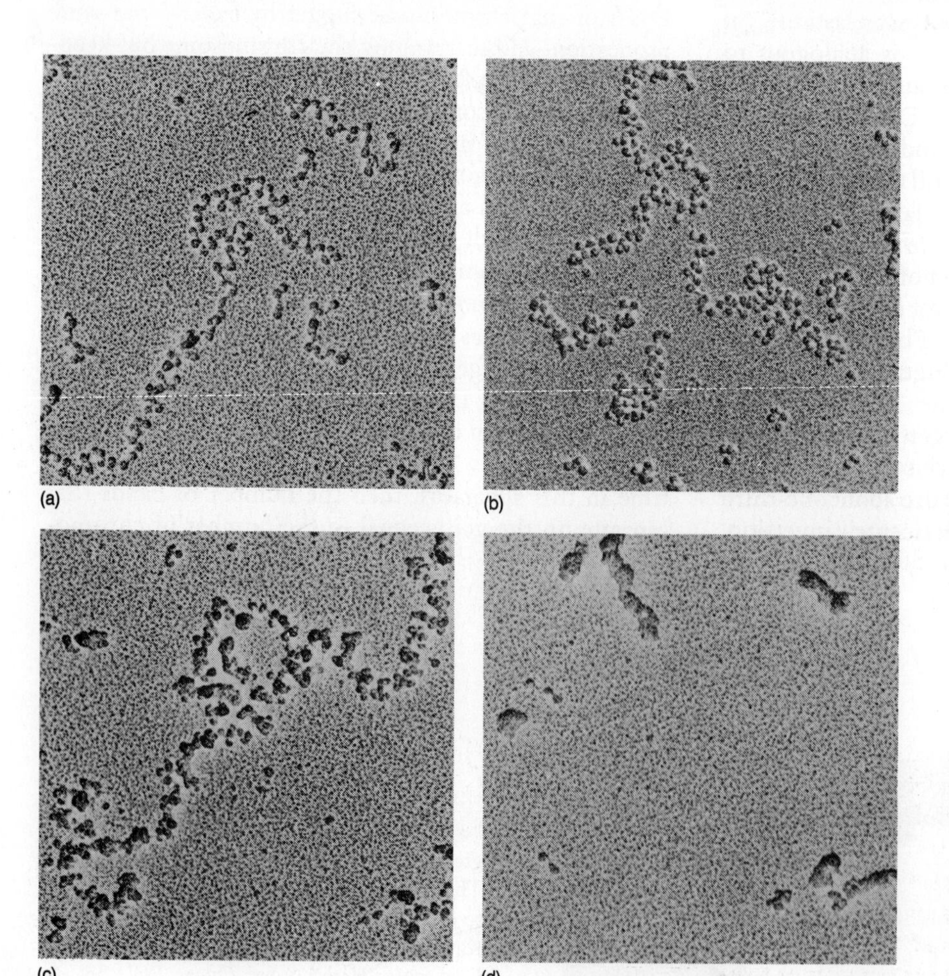

Figure 16-4 Condensation of chromatin with increasing salt concentration is demonstrated in electron micrographs made by Fritz Thoma and Theo Koller. At a very low salt concentration, as in (a), chromatin forms a loose fiber about 10 nm thick; nucleosomes are connected by short stretches of DNA. At a concentration with an ionic strength closer to that of normal physiological conditions, as in (d), chromatin forms a thick fiber some 30 nm thick. The origin of this solenoid can be deduced by an examination of chromatin at increasing intermediate ionic strengths, as in (b) and (c). It arises from a shallow coiling of the nucleosome filament. The chromatin is enlarged here about 80,000 diameters.

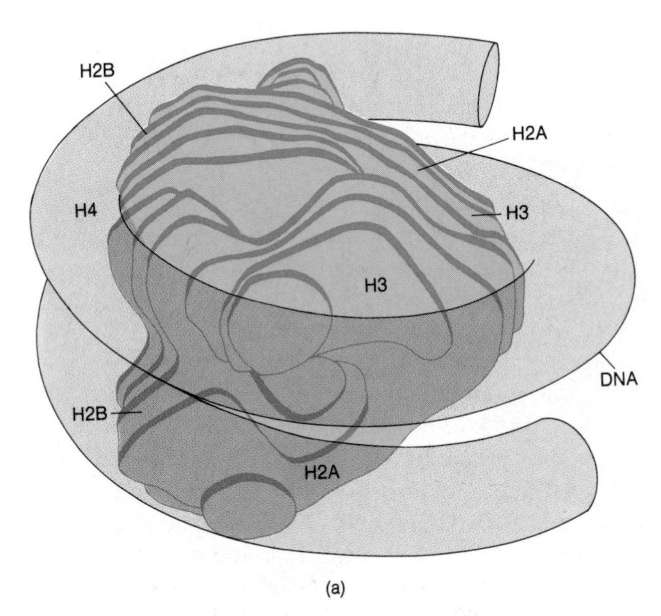

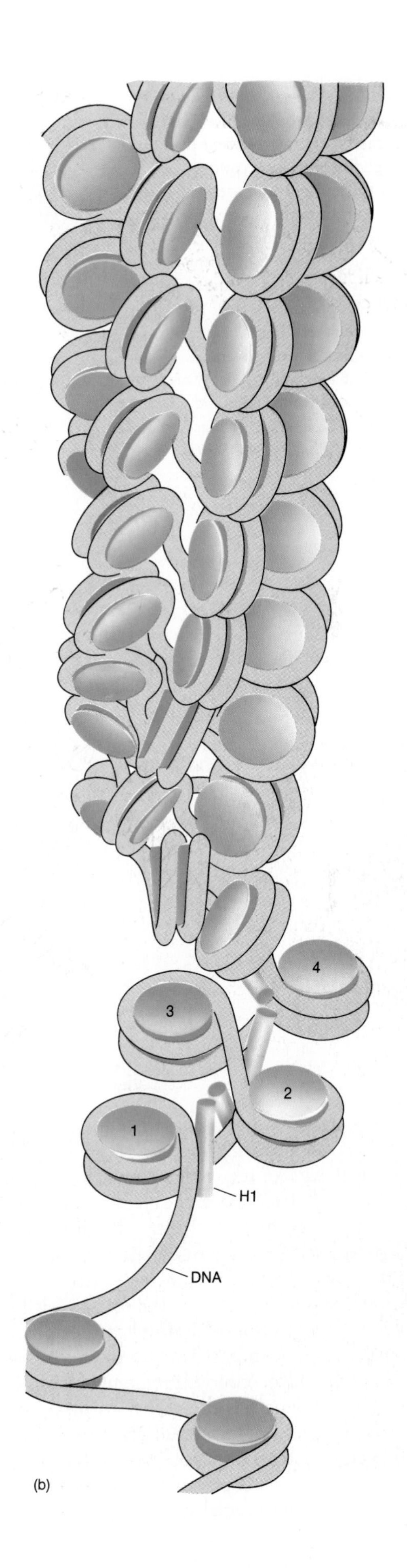

Figure 16-5 (a) Model of a nucleosome. The DNA is shown wrapped twice around a histone octamer. (b) Model of the 30-nm solenoid; histone octamers are shown as disks. The additional histone H1 is shown running down the center of the coil, probably acting as a stabilizer. With increasing salt concentrations, the zigzag pattern of nucleosomes 1, 2, 3, 4 closes up to form a solenoid with six nucleosomes per turn. (From R. Kornberg and A. Klug, "The Nucleosome." *Scientific American,* 1981.)

vitro is 30 nm in diameter, and probably corresponds to the in vivo spaghetti-like structures we first encountered in Figure 16-1. The solenoid is thought to be stabilized by another histone, H1, that runs down the center of the structure, as the figure shows.

We see then that to achieve its first level of packaging, DNA winds onto histones, which act somewhat like spools. Further coiling results in the solenoid conformation. However, it takes one more level of packaging to convert the solenoids into the three-dimensional structure we call the chromosome.

Higher Order Coiling

Many cytogenetic studies show that when viewed under the microscope chromosomes are visibly coiled, and Figure 16-6 shows a good example from the nucleus of a protozoan. Whereas the diameter of the solenoids is 30 nm, the diameter of these coils is the same as the diameter of the chromosome during cell division, often about 700 nm. What produces these supercoils? One clue comes from observing mitotic metaphase chromosomes from

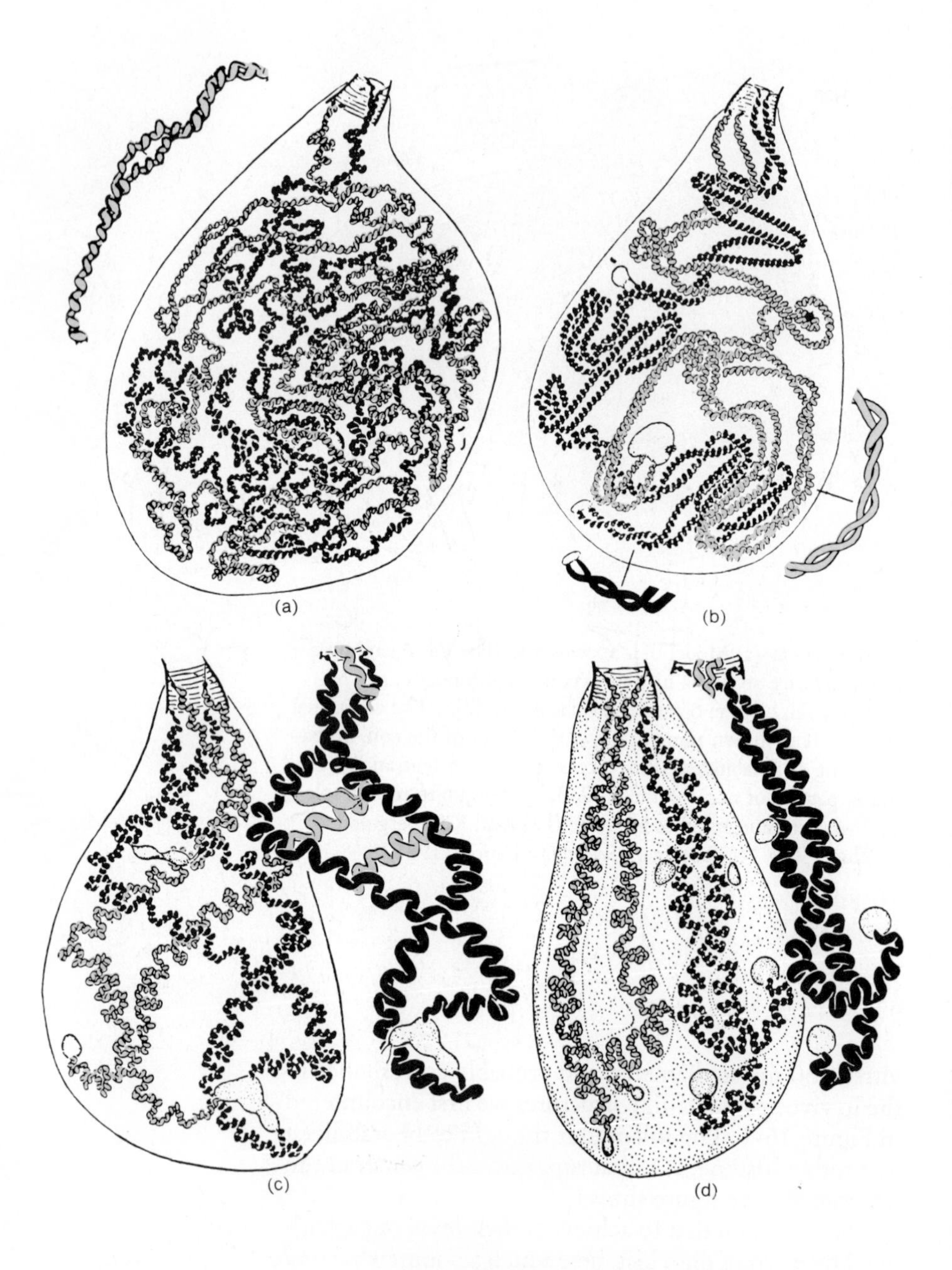

Figure 16-6 Drawings of chromosomes in meiotic prophase in a protozoan, demonstrating different degrees of coiling and supercoiling visible with the light microscope. Two large chromosomes are shown: one colored and the other black; (a) to (d) is a progression. (a) Coiling is seen though duplication becomes apparent. (b) Duplication is well advanced. (c) Supercoiling is beginning. (d) Supercoiling is well advanced. (From L. R. Cleveland, "The Whole Life Cycle of Chromosomes and Their Coiling Systems," *Transactions of the American Philosophical Society* 39, 1949, 1.)

which the histone proteins have been removed chemically. After such treatment, the chromosomes have a densely staining central core called the **scaffold,** as shown in Figure 16-7 and in the electron micrograph on the opening page of this chapter. Projecting laterally from this scaffold are loops of DNA. At high magnifications, it is clear from electron micrographs that each DNA loop begins and ends at the scaffold. It has been discovered that the central scaffold in metaphase chromosomes is largely composed of the enzyme topoisomerase II (see page 323). You will remember from Chapter 11 that this enzyme has the ability to pass a strand of DNA through another cut strand. Evidently, this central scaffold manipulates the vast skein of DNA during replication, preventing many possible problems of unwinding DNA strands at this crucial stage. In any case, it is well established that there is a scaffold in eukaryotic chromosomes, and it seems to be a major organizing device for these chromosomes.

Now to return to the question of how the supercoiling of the chromosome is produced. The best evidence suggests that the solenoids arrange in loops emanating from the central scaffold matrix, which itself is in the form of a spiral. We see the general idea in Figure 16-8, which shows a representation of loosely coiled interphase chromosomes and the more tightly coiled metaphase chromosomes. How do the loops attach to the scaffold? There appear to be special regions called **scaffold attachment regions,** or **SARs.** The evidence for these is as follows. When histoneless chromatin is treated with restriction enzymes, the DNA loops are cut off the scaffold, but special regions remain attached to the scaf-

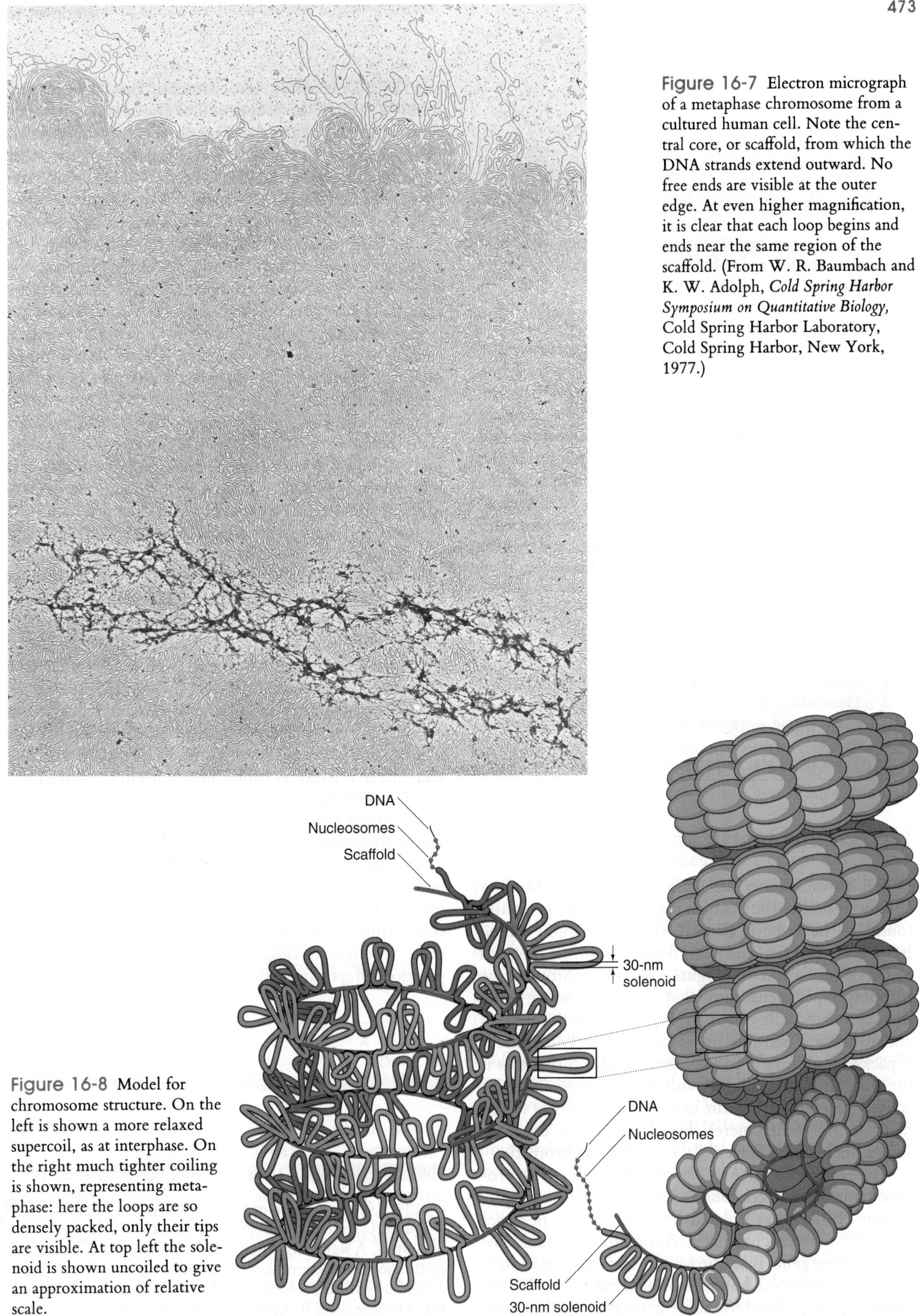

Figure 16-7 Electron micrograph of a metaphase chromosome from a cultured human cell. Note the central core, or scaffold, from which the DNA strands extend outward. No free ends are visible at the outer edge. At even higher magnification, it is clear that each loop begins and ends near the same region of the scaffold. (From W. R. Baumbach and K. W. Adolph, *Cold Spring Harbor Symposium on Quantitative Biology,* Cold Spring Harbor Laboratory, Cold Spring Harbor, New York, 1977.)

Figure 16-8 Model for chromosome structure. On the left is shown a more relaxed supercoil, as at interphase. On the right much tighter coiling is shown, representing metaphase: here the loops are so densely packed, only their tips are visible. At top left the solenoid is shown uncoiled to give an approximation of relative scale.

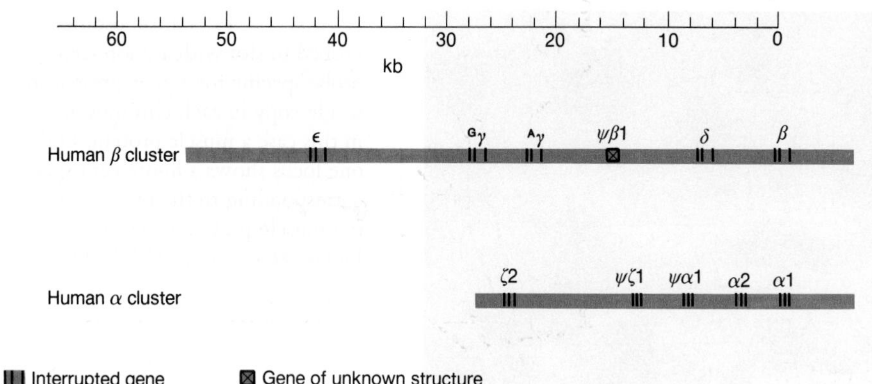

Figure 16-18 The human α-like and β-like globin gene families are each organized into a single cluster that includes functional genes and pseudogenes; the latter are denoted here by ψ (psi). (After B. Lewin, *Genes.* John Wiley, 1983.)

Functional Repetitive Sequences

Dispersed Gene Families. Several types of proteins are coded by families of homologous genes spread throughout the genome. Such families may comprise only a few genes or very many, as some examples illustrate: actins, 5 to 30; keratins more than 20; myosin heavy chain, 5 to 10; tubulins, 3 to 15; insect eggshell proteins, 50; globins, up to 5; immunoglobin variable region, 500; ovalbumin, 3; and histones, 100 to 1000. The exact DNA sequences of the genes within a family may diverge as have the genes for human hemoglobins that we studied on page 217, and the different homologous genes may come to have slightly different functions. Some genes within families have become nonfunctional giving rise to untranscribed **pseudogenes,** as illustrated in Figure 16-18.

Tandem Gene Family Arrays. Cells need large amounts of the products of some genes, and families of these genes have evolved as tandem arrays. A good example is seen in the nucleolar organizer (NO), which was easily observed in cytological preparations of nuclei long before its function was understood; it was easily observed because it does not stain with normal chromatin stains. The role of the NO has been revealed by a variety of genetic and molecular studies. In the 1960s, it was suggested that NOs might be tandem arrays of genes that code for rRNA. Ferruccio Ritossa and Sol Spiegelman tested this hypothesis by constructing *Drosophila melanogaster* strains having different numbers of NOs per cell. In this species, the NOs are located in the heterochromatin of the X chromosome and in the short arm of the Y chromosome. (In *Drosophila,* the X chromosome is always depicted with the centromere on the right. We assume this orientation of the chromosome in discussing positions on it.) Several mutant strains of *Drosophila* exist with inversions of the X chromosome in which the left breakpoint is near the *scute* locus and the right breakpoint is in the heterochromatic region; these inversion mutants are named for the recessive scute phenotype they confer (a reduced number of bristles). The inversion scute-8 has

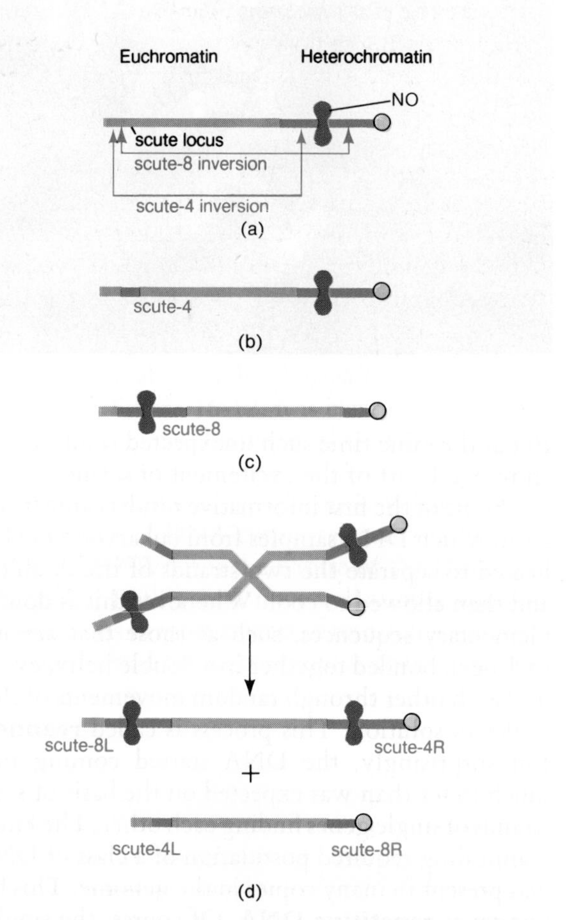

Figure 16-19 The scute-4 and scute-8 inversions. (a) Breakpoints for the two inversions on the *Drosophila* X chromosome. (b) The result of the scute-4 inversion. (c) The result of the scute-8 inversion. (d) A crossover between scute-4 and scute-8 in a heterozygous female yields one crossover product with two nucleolar organizers and another product with no nucleolar organizer. These products are identified as scute-8L scute-4R and scute-4L scute-8R, respectively. The notation scute-8L scute-4R indicates that the left part of the chromosome is derived from the scute-8 chromosome and the right part is derived from the scute-4 chromosome.

Table 16-1. Chromosomal Complements with Various Numbers of NOs per Cell

Chromosomes	Number of NOs Per Cell
scute-4L scute-8R / Y♂	1
scute-4L scute-8R / wild-type / X♀	1
wild-type X / Y♂	2
wild-type / wild-type / X / X♀	2
scute-8L scute-4R / wild-type / X♀	3
scute-8L scute-4R / scute-8L scute-4R	4

its right breakpoint between the NO and the centromere; the inversion scute-4 has its right breakpoint on the distal side of NO (Figure 16-19). From females heterozygous for scute-4 and scute-8, crossover products can be recovered that carry either two or zero NOs. Different numbers of NOs per cell then can be obtained by appropriate genetic combinations (Table 16-1).

By annealing radioactive ribosomal RNA to known amounts of DNA, Ritossa and Spiegelman measured the amount of DNA homologous to rRNA. They found a linear relationship between the number of NOs per cell and the amount of 18S and 28S rRNA that hybridized (Figure 16-20). This result demonstrates that the rRNA loci are located in the NOs. Subsequently, hybridization in situ has confirmed the NO location of the DNA corresponding to rRNA. Furthermore, the sizes of the 18S and 28S rRNA's are now known, as are the percentages of the total DNA that hybridize to them. Thus it was estimated that each *Drosophila* X chromosome has about 200 genes for rRNA. Obviously, such redundancy is one way of ensuring a large amount of rRNA per cell.

Message The nucleolar organizer, which is cytologically distinct, is a tandem array of genes that code for ribosomal RNA.

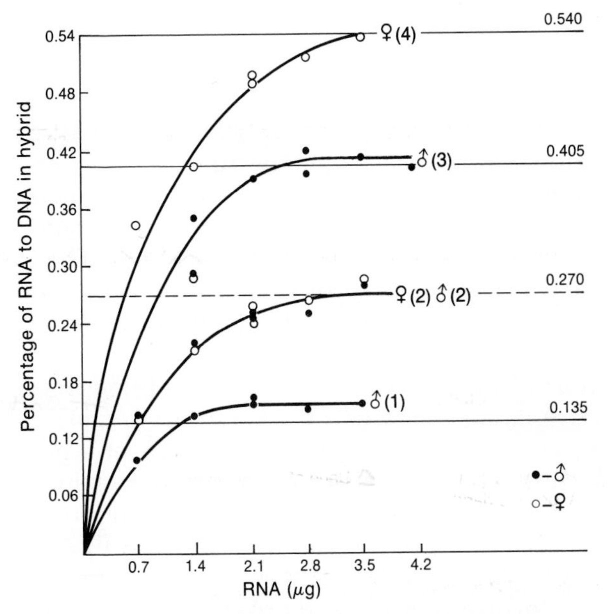

Figure 16-20 The amount of rRNA that hybridizes to a constant amount of DNA. The plateau is reached when all DNA complementary to the rRNA is hybridized. Each curve is obtained using DNA samples isolated from individuals with a particular number of the nucleolus organizer (NO) regions per cell (in parentheses). (From F. M. Ritossa and S. Spiegelman, *Proceedings of the National Academy of Sciences USA* 53, 1965, 737.)

We now know that the NOs on the *Drosophila* X and Y chromosomes contain 250 and 150 tandem copies of rRNA genes. One human NO has about 250 copies. In Figure 13-4 (page 377) we encountered an electron micrograph of the transcription of rRNA tandem arrays in an amphibian.

Another example of a tandem array is the genes for tRNA. In humans there are about 50 chromosomal sites corresponding to the different tRNA types, and at each site there are between 10 to 100 copies.

Finally, the histone genes are arranged in tandem arrays in some species, and examples are shown in Figure 16-21. For histone arrays, as for the other tandemly re-

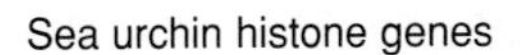

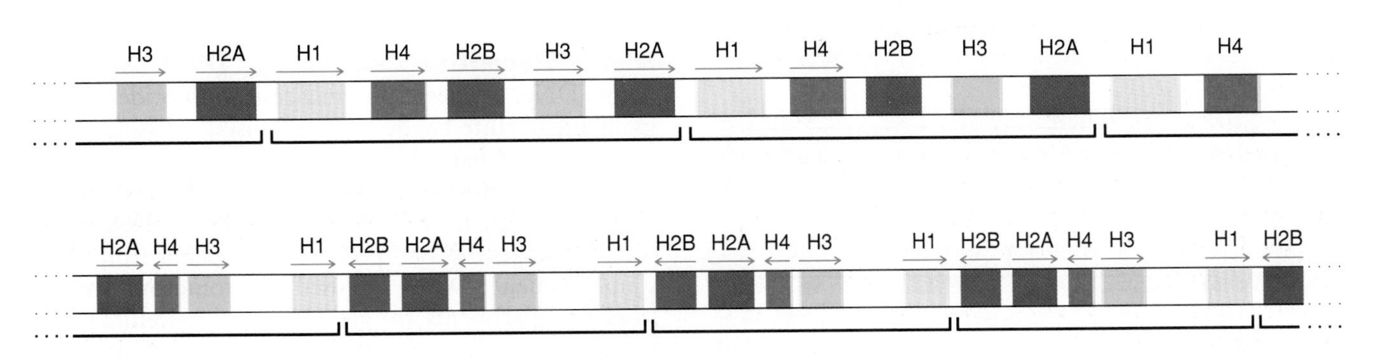

Figure 16-21 Tandem repeats of histone genes in sea urchin and fruit fly. Only a small fraction of the repeats is shown. Arrows indicate direction of transcription.

peated arrays, sequencing analysis has shown that the multiple copies are identical. Because one might expect that mutation would lead to some differences at noncrucial sites, it seems there is some mechanism for maintaining constancy across the members of the array. The most likely mechanism is some kind of gene conversion, a process that will be discussed in Chapter 19.

Noncoding Functional Sequences. Telomeres, the tips of chromosomes, have tandem arrays of simple DNA sequences that do not code for an RNA or a protein product, but nevertheless have a definite function. For example in the ciliate *Tetrahymena,* there is repetition of the sequence TTGGGG, and in humans it is TTAGGG. The telomeric repeats are there to solve a functional problem that is inherent in the replication of linear DNA molecules. Figure 16-22 shows the problem: for the leading strand the polynucleotide addition can always extend to the end because it is automatically primed from behind. However, at the tip the lagging strand reaches a point where its system of RNA priming cannot work, and an unpolymerized section remains. This would result in chromosome shortening. However, an enzyme called telomerase adds the simple repeat units to the ends. It seems likely that these additional units bend back on

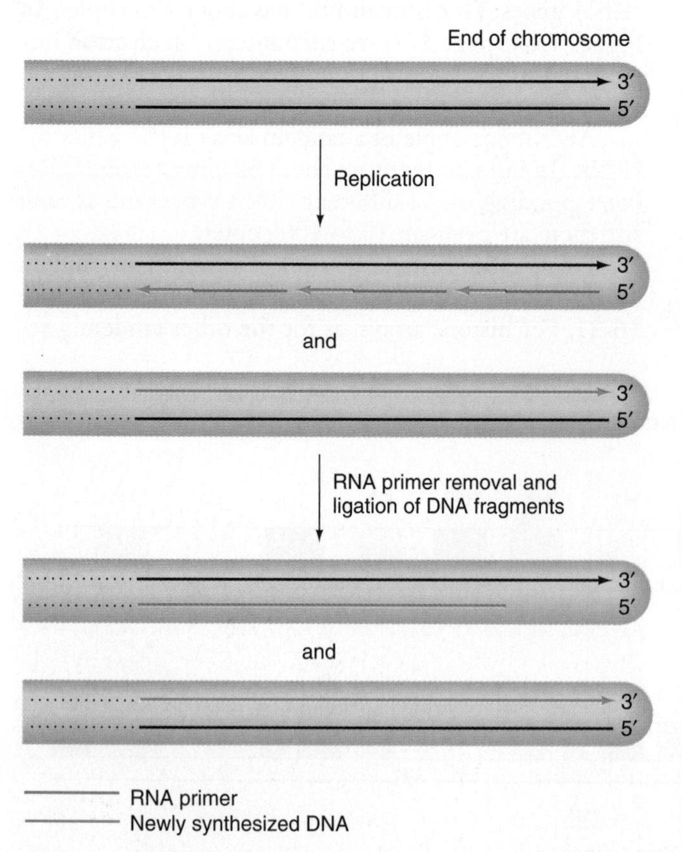

Figure 16-22 The replication problem at chromosome ends. There is no way of priming the last section of the lagging strand, and a shortened chromosome would result.

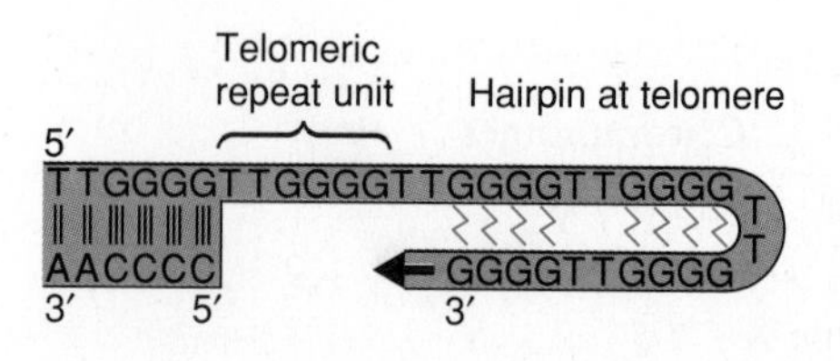

Figure 16-23 Model for the way that telomeric repeats solve the replication problem at chromosome ends. By folding itself into a hairpin secured by an unusual type of hydrogen bonding between Gs, the DNA gains a priming site.

themselves in a hairpin structure made possible by an unusual type of hydrogen bonding of Gs and Gs (Figure 16-23). This provides a 3′ end for filling in the remaining section. The addition of extra telomeric repeats is under careful genetic control, as demonstrated by the fact that yeast mutants have been isolated that are defective in this control. It is interesting to note that through recombinant DNA technology it has been possible to transfer telomeres from other organisms onto yeast artificial chromosomes. Generally they work! However any additional units that are added to the ends in vivo are yeast units: apparently the telomerase does not use the ends of the DNA as a template, and its specificity resides within itself. Figure 16-24 demonstrates the positions of the telomeric DNA through in situ hybridization.

Sequences with No Known Function

For some repetitive DNA, no function is known. This category contains DNA sequences that generally have many more copies in the genome than the functional sequences discussed above. The combined size of this class is surprising; for example, it has been calculated that about 20 percent of the human genome consists of nonfunctional repetitive sequences of one kind or another. We will consider three types.

Highly Repetitive Centromeric DNA. After genomic DNA has been spun in a cesium chloride density gradient, satellite bands often are visible, distinct from the main DNA band. Upon isolation, such satellite DNA is found to consist of multiple tandem repeats of short nucleotide sequences, stretching up to hundreds of kilobases in length. When probes are prepared from such simple sequence DNA, and used in chromosomal in situ labeling experiments, the great bulk of the satellite DNA is found to reside in the heterochromatic regions flanking the centromeres. There can be either one or several basic units, but usually they are less than 10 bases long. For example, in *Drosophila melanogaster,* the sequence AATAACATAG is found in tandem arrays around all

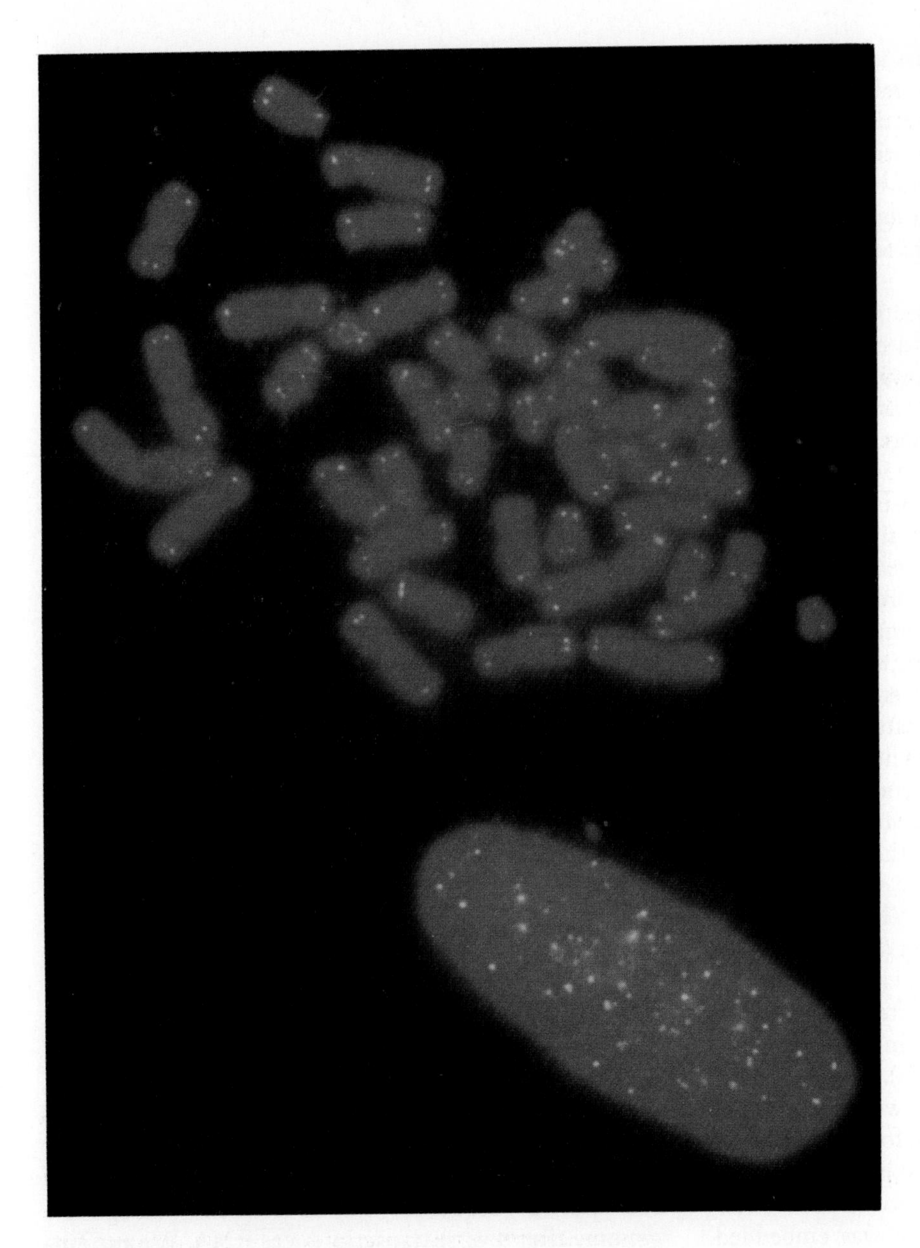

Figure 16-24 Chromosomes probed in situ with a telomere-specific DNA probe that has been coupled to a substance that can fluoresce yellow under the microscope. Each sister chromatid binds the probe at both ends. An unbroken nucleus is shown at the bottom of the photograph. (Robert Moyzis)

centromeres. Similarly, in the guinea pig, the shorter sequence CCCTAA is arrayed flanking the centromeres. In situ labeling of a mouse satellite DNA is shown in Figure 16-25.

Because the centromeric repeats are a nonrepresentative sample of the genomic DNA, the G + C content can be significantly different from the rest of the DNA.

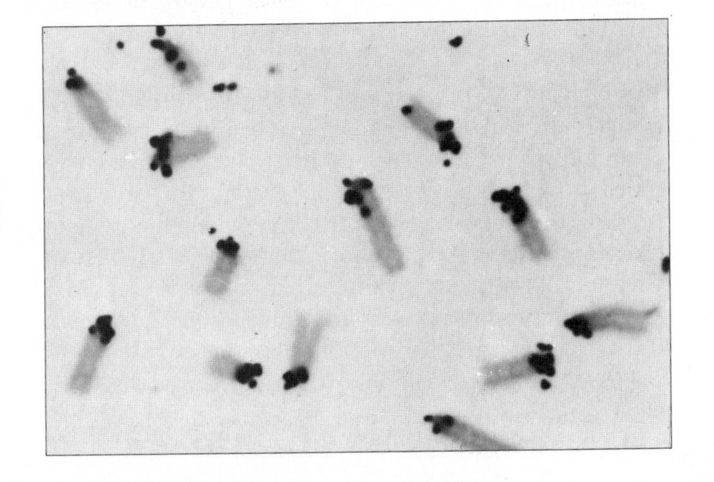

Figure 16-25 Autoradiographic localization of simple-sequence mouse DNA to the centromeres. A radioactive probe for simple-sequence DNA was added to chromosomes whose DNA had been denatured. (Note that all mouse chromosomes have their centromeres at one end.) (From M. L. Pardue and J. G. Gall. *Science* 168, 1970, 1356.)

This is why the DNA forms a separate band in a cesium chloride gradient. There is no demonstrable function for centromeric repetitive DNA, neither is there any understanding of its relation to heterochromatin or to genes in the heterochromatin. Some organisms have staggering amounts of this DNA; for example as much as 50 percent of kangaroo DNA can be centromeric satellite DNA.

VNTRs. A special class of tandem repeats shows variable number at different loci and in different individuals. These are called **VNTRs, or variable number tandem repeats.** The VNTR loci in humans are 1 to 5 kb sequences consisting of variable numbers of a repeating unit 15 to 100 nucleotides long. If a VNTR probe is available, and the total genomic DNA is cut with a restriction enzyme that has no target sites within the VNTR arrays, then a Southern blot reveals a large number of different-sized fragments that are bound by the probe. Because of the variability in the number of tandem repeats from individual to individual, the set of fragments that shows up on the Southern autoradiogram is highly individualistic. In fact these patterns are called **DNA fingerprints** (Figure 16-26). They are used extensively in forensic medicine. In a criminal investigation, blood or semen can be used to prepare DNA fingerprints, and the patterns compared with those of a suspect. Minute samples of tissue such as follicle cells clinging to single hairs can be used by amplifying VNTR sequences through the polymerase chain reaction (PCR).

One of the first probes that detected human VNTRs was obtained in 1985 by Alec Jeffries from a quadruple repeat of a 33-bp sequence found within the first intron of the myoglobin gene (Figure 16-27a). Once some of the VNTRs in the DNA fingerprint were cloned and sequenced, it was found that the reason for hybridization of the probe to the VNTRs was a 13-bp core sequence that was common to all the repeats and to the probe. The longer sequences into which the core was embedded were not necessarily similar.

DNA fingerprints can be made in many different species of organisms, so VNTRs must be common and the technique has great value in testing genetic individuality.

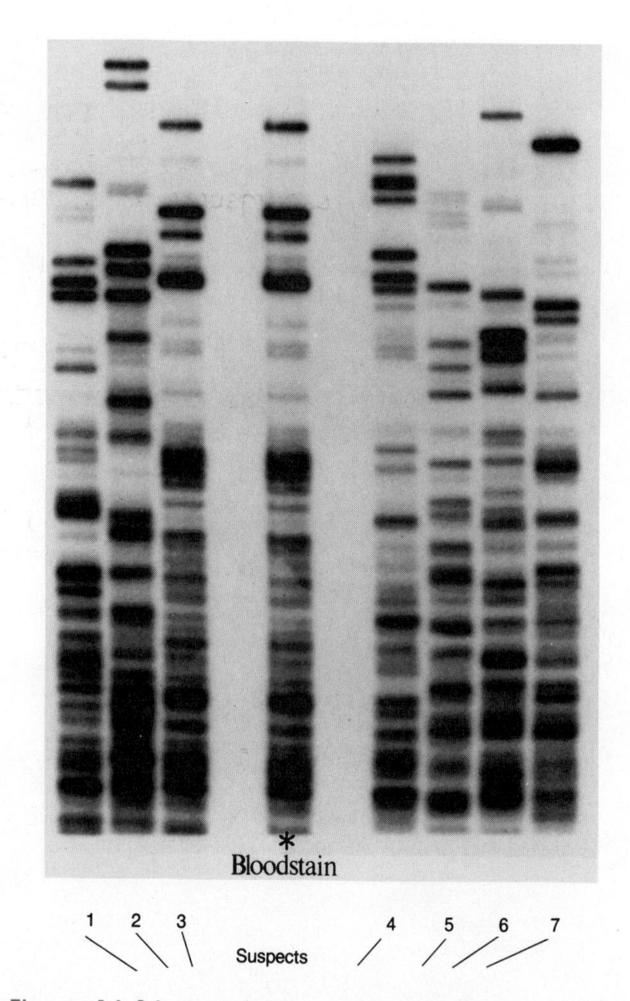

Figure 16-26 DNA fingerprints from a bloodstain at the scene of a crime and from the blood of seven suspects. (Cellmark Diagnostics, Germantown, MD)

Transposed Sequences. In Chapter 20 we will discuss a special class of DNA elements that can move from one chromosomal position to another. These elements are called **transposons.** Transposons account for large proportions of human and other genomes. The first type in this class is genuine transposons that move as DNA. In Chapter 20 we will encounter examples such as the Ac/Ds transposons in corn and P transposons in *Drosophila.* Many genomes show multiple copies of such elements, or truncated versions of them, dispersed throughout the genome.

The second type of transposed repetitive sequence is **retroposons,** sequences that have spread through the genome after reverse transcription of RNA. We encountered the enzyme reverse transcriptase in Chapter 14. The enzyme copies RNA into DNA. Subsequently, these DNA copies can find their way back into the chromosomes at numerous positions. A good example is the human *Alu* sequence, so named because it often contains an *Alu* restriction enzyme target site. The human genome contains hundreds of thousands of whole and partial *Alu* sequences, scattered between genes and within introns, and making up about 5 percent of human DNA. The full *Alu* sequence is about 200 nucleotides long and bears remarkable resemblance to 7SL RNA, an RNA that is part of a complex involved in secretion of newly synthesized polypeptides through the endoplasmic reticulum. Presumably the *Alu* sequences have originated as reverse transcripts of these RNA molecules. Short interspersed repeats such as *Alu* sequences are collectively called **SINEs (for short interspersed elements).**

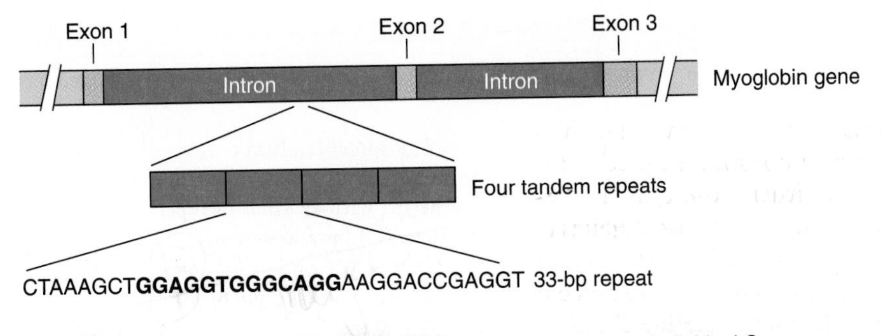

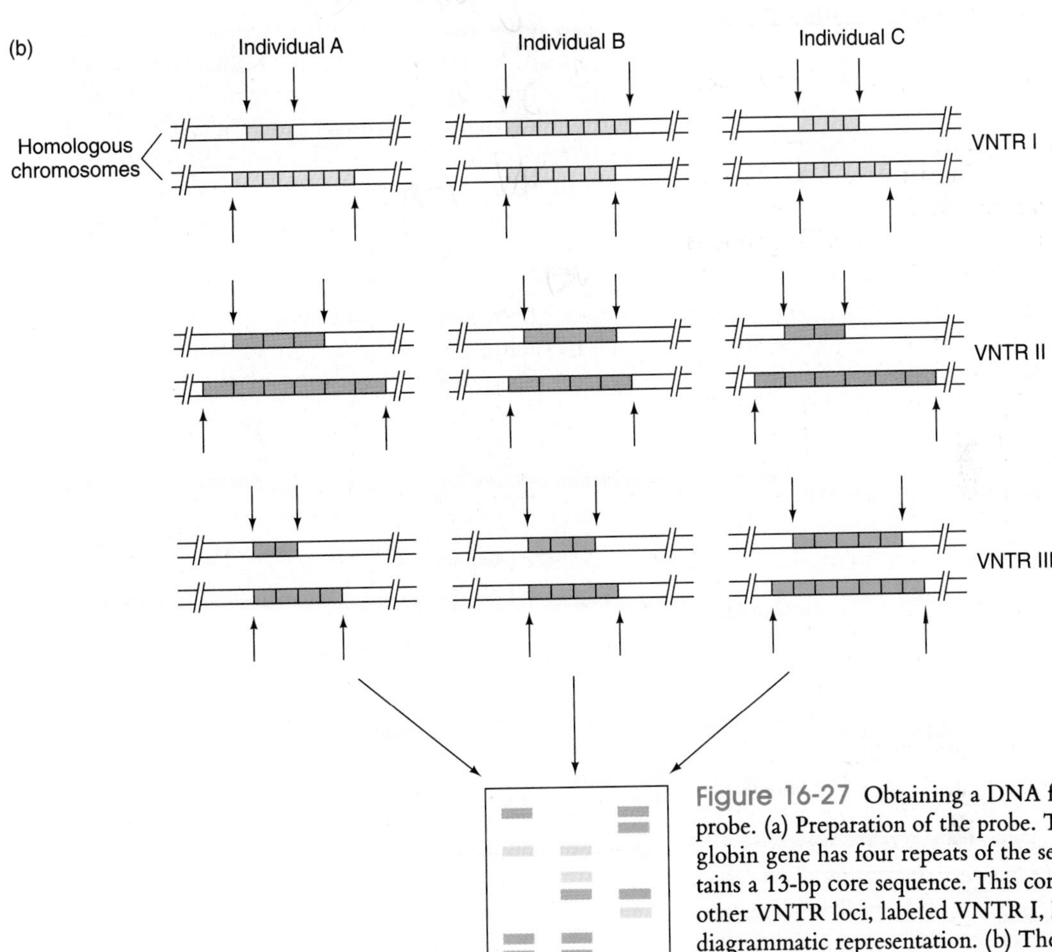

Figure 16-27 Obtaining a DNA fingerprint using a VNTR probe. (a) Preparation of the probe. The first intron of the myoglobin gene has four repeats of the sequence shown, which contains a 13-bp core sequence. This core sequence is found at other VNTR loci, labeled VNTR I, II, and III in this simple diagrammatic representation. (b) The number of repeats at the three VNTR loci with the core sequence. The Southern blot has been probed with the 33-bp repeat in (a), and shows the DNA fingerprints of three individuals. (From J. D. Watson, M. Gilman, J. Witkowski, and M. Zoller. *Recombinant DNA,* 2d ed., Copyright 1992, by Scientific American Books.)

Other examples of this class of moderately repetitive elements are the many scattered pseudogenes that have clearly been created by the reverse transcription process because they do not contain the introns that are found in the original functional gene.

The third type of transposed element that we shall consider is one that shows sequence homology to retroviruses, RNA viruses that replicate through a DNA stage, which can integrate into host chromosomes. Examples of such elements are the copia elements of *Drosophila,* (5 kb sequences present at about 50 copies per genome), the Ty elements of yeast (6 kb elements with about 30 full copies per genome and about 100 solo copies of their direct repeats called delta sequences), and the **LINEs (long interspersed elements)** of mammals (1 to 5 kb elements present in 20,000 to 40,000 copies per human genome). Elements in this class have open reading frames that potentially code for enzymes used in transposition.

Spacer DNA

The final category of DNA is spacer DNA. This is basically what is left after all the recognizable units have been

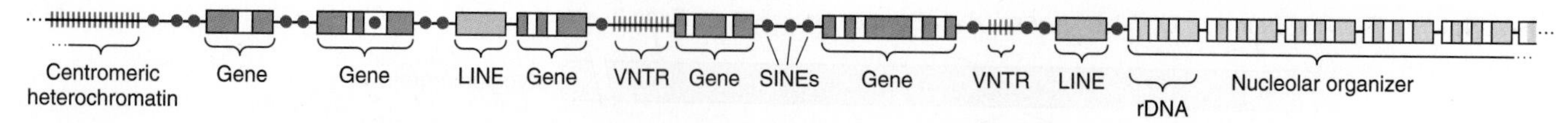

Figure 16-28 General depiction of a eukaryotic chromosomal landscape. A small region of a chromosome is shown that happens to have five protein-coding genes, one end of a nucleolar organizer, and one end of centromeric heterochromatin. Various kinds of repetitive DNAs are shown. (Each chromosome would normally have several thousand genes.)

identified. Needless to say, little is known about spacer DNA. Possibly its only function is to space.

> **Message** Single copy genes are embedded in a complex array of tandem and dispersed types of repetitive DNA most of which have no known function.

Figure 16-28 shows a stylized diagram of the overall organization of a hypothetical eukaryotic chromosome, summarizing much of the above discussion. However, we can now study the architecture of a real eukaryotic chromosome because one complete chromosome of yeast (chromosome 3) has been fully sequenced. This *tour de force* was carried out through a collaboration of 35 European research laboratories. The arrangement of the genes and other elements is shown in Figure 16-29. The entire chromosome is 315 kb long. Analysis of the complete sequence showed that there are 182 open reading frames (shown in dark blue). Of these supposed 182 protein-coding genes, only 34 had been discovered by classical genetic techniques, and these are shown in yellow in the figure. The possible functions of the unknown open

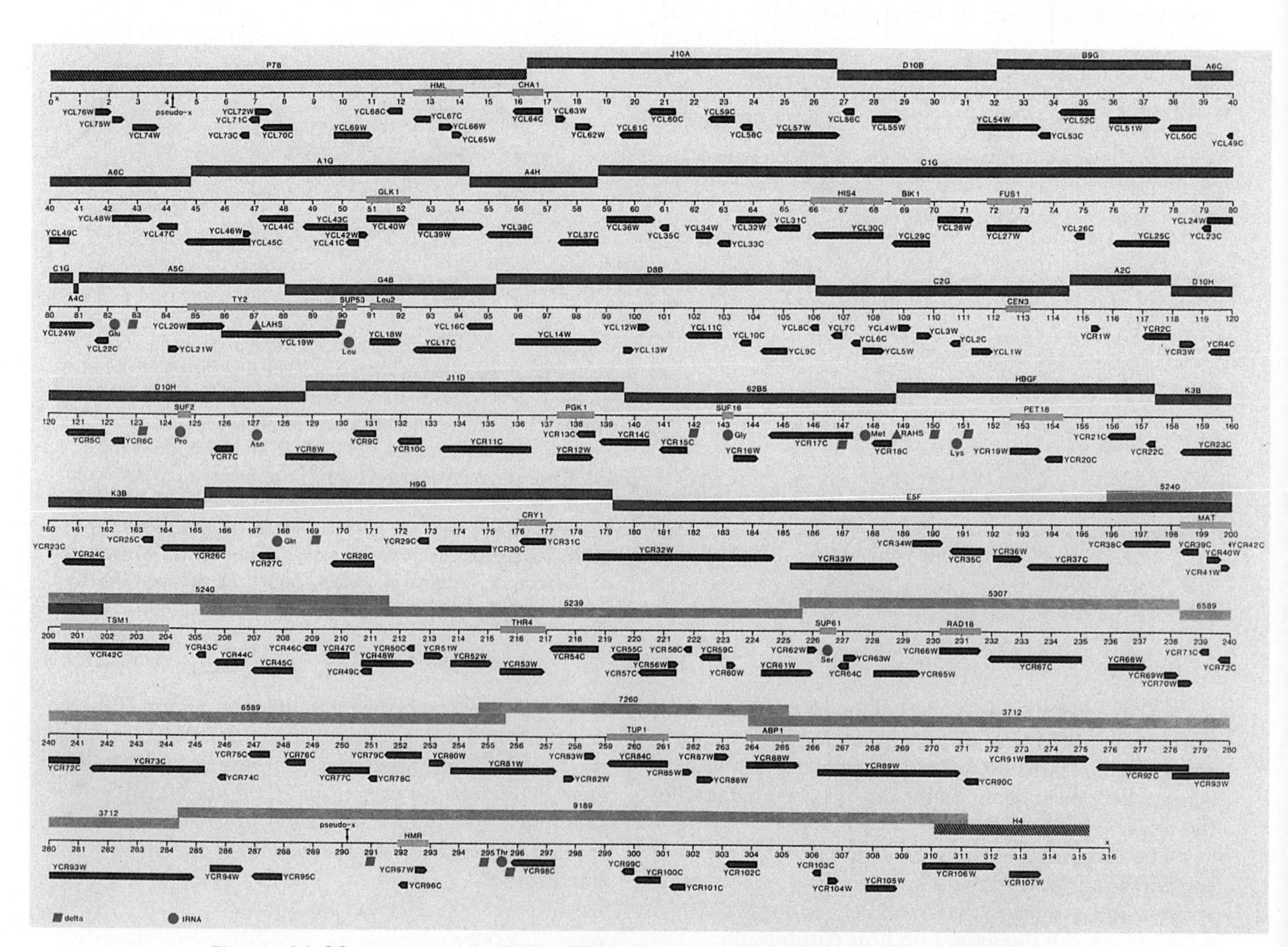

Figure 16-29 The genetic landscape of chromosome 3 in yeast, determined by sequencing the entire chromosome. Genes previously detected from mutant phenotypes are shown in yellow. Open reading frames, which likely are protein-coding genes, were detected by sequence analysis, and are shown in dark blue. The clones used in sequencing are shown above the line in light blue. Delta sequences are derived from the transposon Ty. (S. G. Oliver et al., *Nature* 357, 1992, 38–46.)

reading frames is being explored by comparison with other DNA sequences in the computer data bank, and by looking for mutant phenotypes that might be caused by gene disruptions. The delta sequences are the solitary terminal repeats mentioned earlier, derived from Ty transposons. Note that yeast has virtually no interspersed repetitive DNA, so the genes are seen to be relatively close together, separated by short spacer sequences.

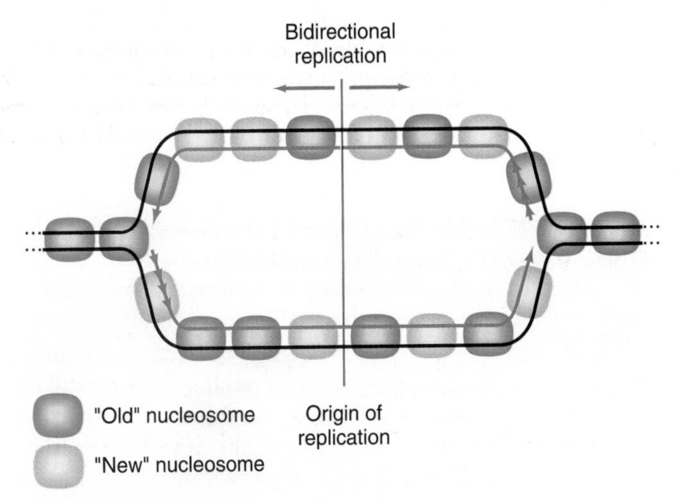

Figure 16-30 The dispersive model for the association of newly formed nucleosomes with the daughter DNA molecules at the replication fork.

Replication and Transcription of Chromatin

We have seen that the central scaffold of the chromosome contains topoisomerase II, which must surely play a role in replication, but the details of how polymerases and associated factors gain access to the center of the chromatin is not clear. There are many other crucial questions that cannot be answered fully. A major one is how the replication machinery can pass through the nucleosomes. Nucleosomes appear not to leave the DNA in this process, but may disassemble in situ. There is still controversy about the arrangement of "old" and "new" nucleosomes on the two daughter DNA molecules, but in balance most of the evidence favours the dispersive model, in which each daughter molecule gets some old and some new nucleosomes (Figure 16-30).

Similar problems exist for understanding transcription. If RNA polymerase is drawn to the same scale as the solenoid structure, it is seen to be approximately the same size (Figure 16-31). Therefore it seems unlikely that the enzyme could do its job without a great deal of loosening of the DNA–histone complex. Indeed there is evidence that when a gene is being transcribed in interphase it becomes sensitive to DNase attack as would be predicted if there was relaxing of the nucleosome spools to expose unwound DNA. This effect can be demonstrated in chick cells that are synthesizing globin proteins. If the nuclei of such cells are exposed to DNase, then subsequent restriction enzyme digestion following the removal of histones reveals no intact globin genes. However, in cells where the gene was not being expressed, a clear globin gene fragment is obtained at the end of the procedure. One possible conformation of the loosened arrays of histones is shown in Figure 16-32.

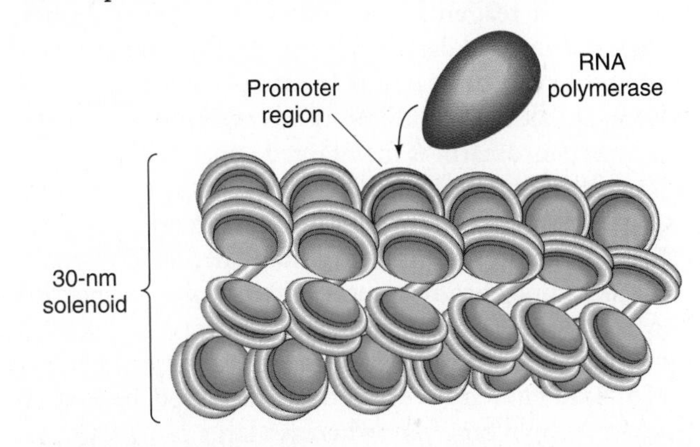

Figure 16-31 The relative sizes of RNA polymerase and the 30-nm solenoid. Transcription seems impossible from DNA so tightly wound on nucleosomes.

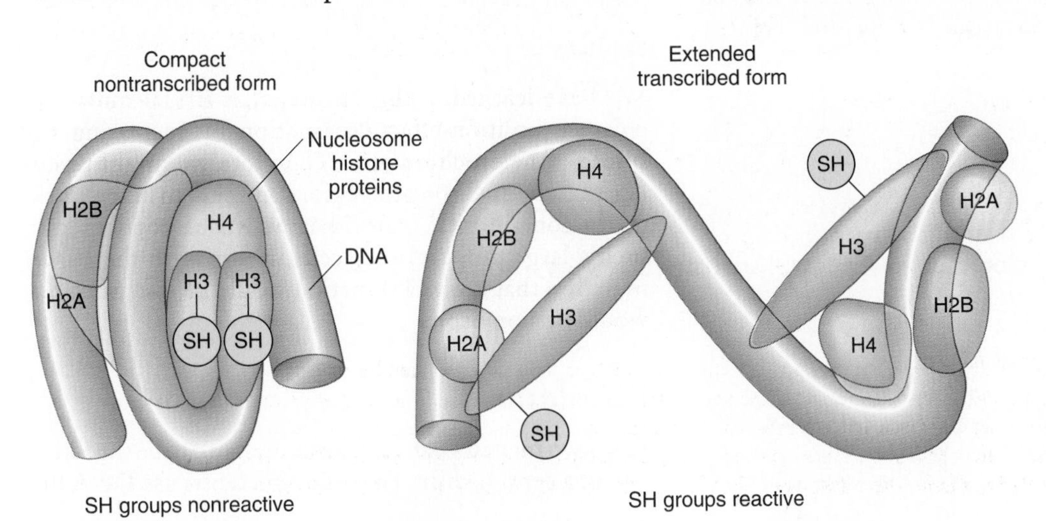

Figure 16-32 Compact form of nucleosomes, and a proposed unfolded transcriptional form. (After C. P. Pryor et al. *Cell* 34, 1983, 1033.)

Message In chromatin the transcription and replication processes act on DNA with nucleosomes still bound.

Summary

Each eukaryotic chromosome represents one DNA molecule continuous from one end through the centromere to the other end. This DNA is efficiently packed to achieve the chromosome shape. The first level of organization is winding of the DNA onto octamers of histone proteins to form nucleosomes. The nucleosomal strand is coiled into a solenoid structure. The solenoid is thrown into lateral loops fastened to a central scaffold at specific scaffold attachment regions. This whole structure is arranged in large supercoils. The compactness of the supercoils at metaphase makes it easy for chromosomes to segregate and reveals a banding pattern when stained with Giemsa reagent. The light bands contain housekeeping genes, and dark bands contain the genes that are tissue-specific in their action. In interphase the supercoils are less compact. The difference between euchromatin and heterochromatin is maintained genetically, and this system is being investigated by analysis of mutations that alter position-effect variegation. Centromeres have specialized structures to permit spindle fiber attachment.

Eukaryotic chromosomes contain a large proportion of repetitive DNA. Some is in tandem arrays and some is interspersed throughout the genome. Some repetitive DNA, both tandemly repeated and dispersed, consists of genes of known function, whereas a large proportion has no known function. Centromeric heterochromatin contains highly repetitive short tandem repeats. An important type of dispersed repeats is transposable elements of several types. Telomeres have specialized repetitive structures to permit DNA replication to proceed to the end of the chromosome.

Replication and transcription both take place through nucleosomes, but the nucleosomal strands partially unwind and histones might partially disassemble. Nucleosome replication is dispersive, with new and old nucleosomes dispersed on both leading and lagging strands.

Concept Map

Draw a concept map interrelating as many of the following terms as possible. Note that the terms are listed in no particular order.

histones / chromosome / transcription / position-effect variegation / DNA / nucleosomes / bands / loops / scaffold / heterochromatin

Chapter Integration Problem

A yellow bodied *Drosophila* stock (*yy*) is transformed with a vector containing a fully functional wild-type allele y^+, which codes for brown body color. Two transgenic lines are obtained: one has a fully brown body, and the other has some brown sectors and some yellow sectors. Give a possible explanation for the difference between the two transgenic flies, and give a test of your idea.

Solution

Evidently the *y* allele is recessive, otherwise no brown tissue would be expected in the transformed individuals (Chapter 2). Therefore the transgenic flies, which, as we saw in Chapter 13, probably contain two *y* alleles and one y^+ allele, are expected to be brown if the y^+ allele is fully dominant over two recessive *y* alleles. We saw in the chapter on aneuploidy that this kind of dominance is often encountered (Chapter 9). Indeed it is found in one line in the present experiment. The other line shows variegation, and this should give us a clue to the location of the transgene. If the transgene is inserted into a chromosome next to heterochromatin, it could show position-effect variegation. No other mechanism for variegation (for example, somatic crossing-over, X chromosome inactivation) makes sense in this case. The position of the transgene in both lines can be mapped using standard linkage analysis, as covered in Chapter 5. We predict the transgene to be at a locus next to heterochromatin in the yellow and brown line and in the other line we expect it to be in euchromatin, possibly at the true yellow locus, but more likely at a new site.

Solved Problem

Investigators have found that *Drosophila E(var)* mutations are rarer than *Su(var)* mutations. Why do you think this is so?

Solution

We have learned in this chapter that *E(var)* mutations enhance position-effect variegation by increasing the spread of heterochromatin. Therefore we might postulate that the euchromatic regions adjacent to heterochromatin contain vital genes that would be condensed in such a large proportion of cells in flies with the *E(var)* mutation that the vital function would be lost. Such *flies* would be nonviable.

Problems

1. Two men in a paternity dispute claim to be the father of a certain child. Forensic geneticists use DNA fin-

gerprints to settle the case, and the relevant ones are in the figure below (M is the mother, C the child, and F1 and F2 are the two men who claim parentage):

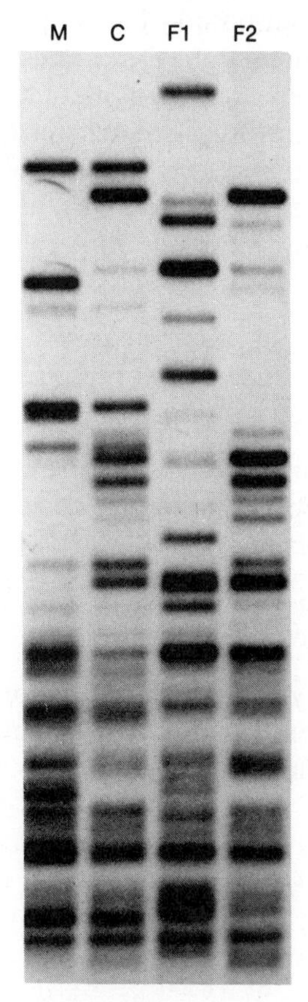

a. Who is the father? Explain.

b. Attribute as many bands as possible to one or the other parent.

c. Are there any bands that can not be attributed to a parent?

d. If the child did show a band that neither parent possessed, how might this be explained? (Such situations do sometimes occur.)

e. Explain what a band actually represents, using diagrams.

(DNA fingerprint from Cellmark Diagnostics, Germantown, MD)

2. Suppose that an essential wild-type allele in *Drosophila* is translocated next to heterochromatin on one homolog, and that the other homolog carries the gene in its normal chromosomal position, but the allele is recessive and lethal, Predict the organismal phenotype in **(a)** a *Su(var)* mutant, **(b)** an *E(var)* mutant, and **(c)** a normal background. Assume first that the gene product is diffusible throughout the body. Then make the alternative assumption that it is used only in cells where it is synthesized.

3. Predict the in situ chromosomal hybridization pattern in humans using probes consisting of

   a. a single copy gene

   b. a fragment of telomere sequence

   c. rDNA

   d. a SINE

   e. satellite DNA

4. Assume that a SINE occurs on average every 50 kb, and you cut DNA with a restriction enzyme that cuts on average every 100 kb. You denature the DNA with heat, then cool it slowly to promote reannealing. Predict what you might see with the electron microscope.

5. Forensic geneticists prepared DNA fingerprints from the blood of a rape victim, from semen taken from her body, and from samples obtained from three suspects. The resultant Southern blots are shown below:

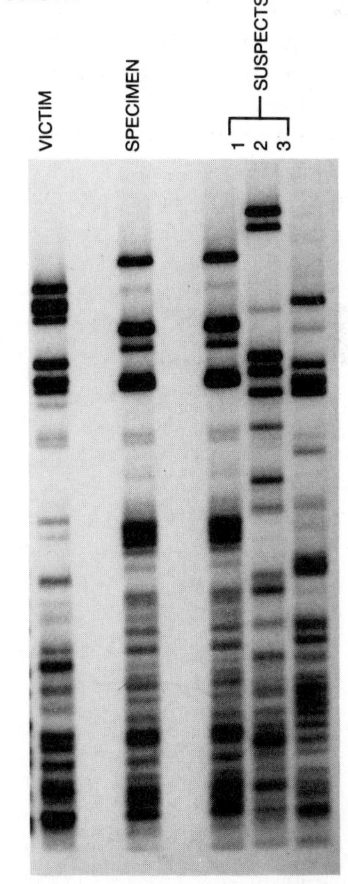

a. Is there a clear case of guilt of one of the suspects?

b. Can any of the suspects be absolved of guilt? Explain your answers.

(DNA fingerprint from Cellmark Diagnostics, Germantown, MD)

6. A VNTR probe revealed simple DNA fingerprints in the family shown below:

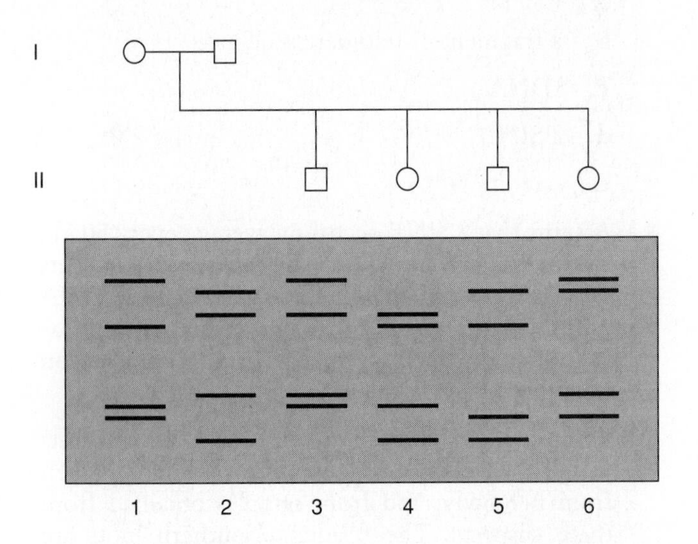

From these data, what can you say about the number of VNTR loci, their linkage, and the number and size of the alleles?

7. Transposons can be used as mutagens. They inactivate genes by inserting into them or near them. In one experiment in yeast, single copies of a transposon were added transgenically to haploid cells containing no transposon. The single copies were allowed to integrate at random throughout the genome. Then the strains that resulted from this procedure were examined to determine their phenotype. It was found that 70 percent of these had an abnormal phenotype that could be attributed to the transposon insertion, but 30 percent appeared fully wild type.

   a. How would you make sure that the strains with the wild-type phenotype actually do have a transposon insertion?

   b. How would you make sure that the abnormal phenotypes were in fact caused by the transposon insertion?

   c. How do you explain the 30 percent that are unaffected phenotypically?

   d. Discuss your answer in relation to the gene arrangement revealed by the complete sequence of chromosome 3 (Figure 16-29).

17

Control of Gene Expression

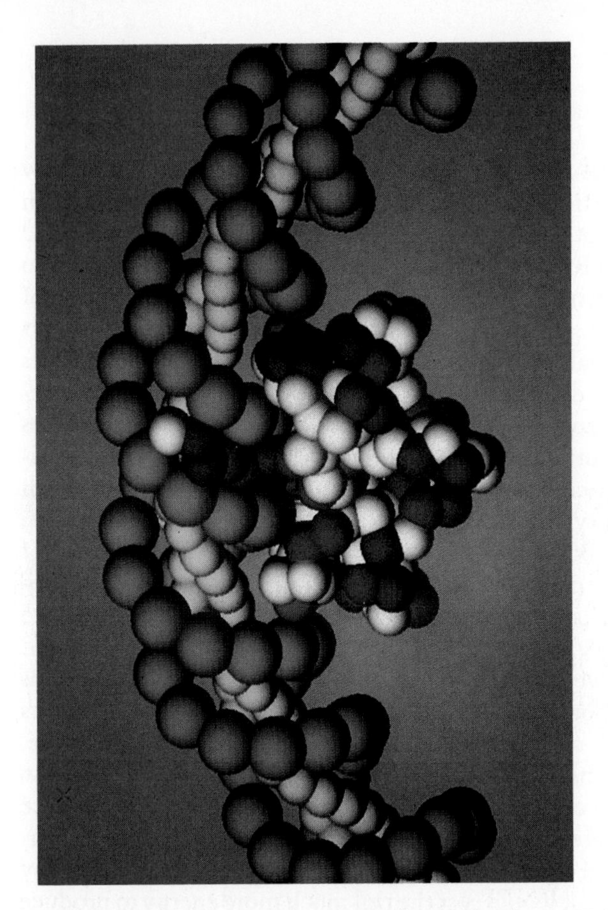

Computer model of repressor and operator. (Brian Matthews, University of Oregon)

KEY CONCEPTS

Gene regulation is most often mediated by proteins that react to environmental signals by raising or lowering the transcription rates of specific genes.

▪

Negative control in prokaryotes is exemplified by the *lac* system, in which a repressor protein blocks transcription by binding to a site on the DNA termed the *operator.*

▪

Positive control in prokaryotes occurs when protein factors are required to activate transcription.

▪

Many regulatory proteins have common structural features.

▪

Transcriptional control of eukaryotes is also mediated by trans-acting protein factors, which bind to specific regulatory sequences.

▪

Additional regulatory sites on the DNA, termed *enhancers,* modulate gene expression in eukaryotes by interacting with specific regulatory proteins.

Until now, we have discussed what genes are and how genetic change occurs and is inherited. Also, we saw in Chapter 13 how genes are expressed by being transcribed into RNA molecules, many of which are translated into proteins. But how does the cell *regulate* the expression of all its genes? We can see that control of gene expression is crucial to an organism. In higher cells, specific cell types have differentiated to the point that they are highly specialized. An eye cell in humans synthesizes the proteins important for eye color but does not produce the detoxification enzymes that are synthesized in liver cells. Each cell type has arranged to express only some of its genes.

Bacteria also have a need to regulate the expression of their genes. Enzymes involved in sugar metabolism provide an example. Metabolic enzymes are required to break down different carbon sources to yield energy. However, there are many different types of compounds that bacteria could use as carbon sources, including sugars such as lactose, glucose, maltose, rhamnose, raffinose, melibiose, galactose, and xylose. Several enzymes allow each of these compounds to enter the cell and to catalyze different steps in sugar breakdown. If a cell were to synthesize simultaneously all the enzymes it might possibly need, it would cost the cell much more energy to produce the enzymes than it could ever derive from breaking down any of the prospective carbon sources. Therefore, the cell has devised mechanisms to repress all the genes encoding enzymes that are not needed and to activate those genes at a time when the enzymes are needed. Clearly, to do this, two requirements must be met:

1. A method must be found for turning on or off each specific gene or group of genes.
2. The cell must be able to recognize situations in which it should activate or repress a specific gene or group of genes.

Let's review the current model for gene control and then discuss its development.

Basic Control Circuits

The first system we will focus on is concerned with lactose metabolism in *Escherichia coli.* We have now learned a lot about how this system works. Figure 17-1 shows a physical model for the control of the lactose enzymes.

The metabolism of lactose requires two enzymes: a permease to transport lactose into the cell and β-galactosidase to cleave the lactose molecule to yield glucose and galactose. Permease and β-galactosidase are encoded by two contiguous genes, Z and Y, respectively. A third gene, the A gene, encodes an additional enzyme, termed *transacetylase,* but this enzyme is not required for lactose metabolism, and we will not concentrate on it for now. All three genes are transcribed into a single or **polycistronic** messenger RNA (mRNA) molecule.

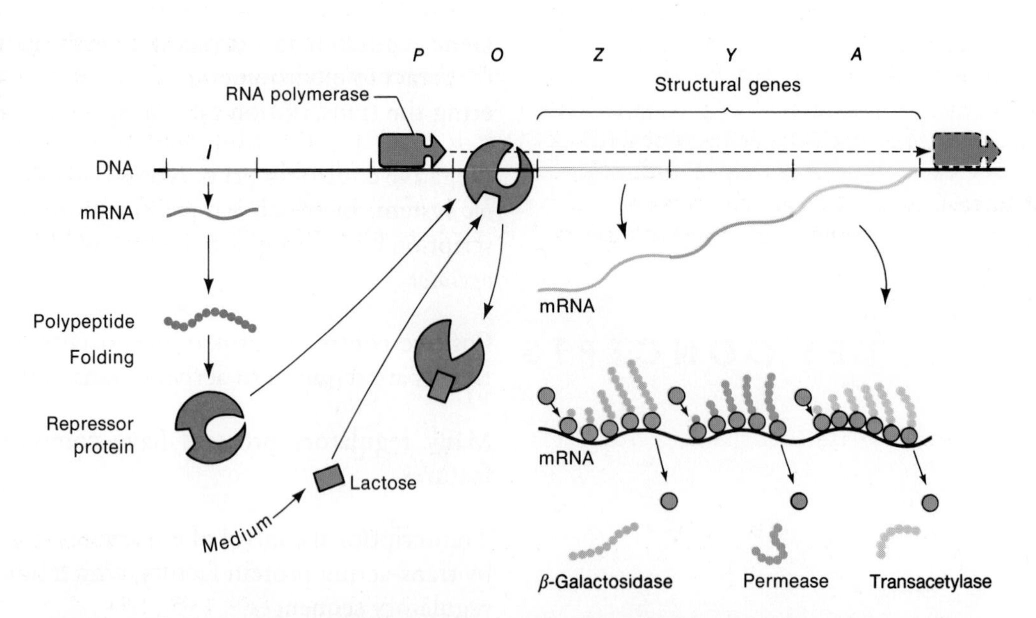

Figure 17-1 Regulation of the *lactose* operon. The I gene continually makes repressor. The repressor binds to the O (operator) region, blocking the RNA polymerase bound to P from transcribing the adjacent structural genes. When lactose is present, it binds to the repressor and changes its shape so that the repressor no longer binds to O. The RNA polymerase is then able to transcribe the Z, Y, and A structural genes, so the three enzymes are produced.

Thus, it can be seen that by regulating the production of this mRNA, the regulation of the synthesis of all three enzymes can be coordinated. A fourth gene, the *I* gene, which maps near but not directly adjacent to the *Z*, *Y*, and *A* genes, encodes a **repressor** protein, so named because it can block the expression of the *Z*, *Y*, and *A* genes. The repressor binds to a region of DNA near the beginning of the *Z* gene and near the point at which transcription of the polycistronic mRNA begins. The site on the DNA to which the repressor binds is termed the **operator.** One necessary property of the repressor is that it be able to recognize a specific short sequence of DNA—namely, a specific operator. This ensures that the repressor will bind only to the site on the DNA near the genes that it is controlling and not to other random sites all over the chromosome. By binding to the operator, the repressor prevents the initiation of transcription by RNA polymerase. Normally, RNA polymerase binds to specific regions of the DNA at the beginning of genes or groups of genes, termed **promoters** (see Chapter 13), so that it can initiate transcription at the proper starting points. The *POZYA* segments shown in Figure 17-1 constitutes an **operon,** which is a genetic unit of coordinate expression.

The *lac* repressor is a molecule with two recognition sites—one that can recognize the specific operator sequence for the *lac* operon and another that can recognize lactose and certain analogs of lactose. When the repressor binds to lactose derivatives, it undergoes a conformational change; this slight alteration in shape changes the operator binding site so that the repressor loses affinity for the operator. Thus, in response to binding lactose derivatives, the repressor falls off the DNA. This satisfies the second requirement for such a control system—the ability to recognize conditions under which it is worthwhile to activate expression of the *lac* genes. The relief of repression for systems such as *lac* is termed **induction;** derivatives of lactose that inactivate the repressor and lead to expression of the *lac* genes are termed **inducers.**

Other bacterial systems operate by using protein "activator" molecules, which must bind to DNA as a prerequisite of transcription. Still additional mechanisms of control require proteins that allow the continuation of transcription in response to intracellular signals. Before we examine some of these control circuits in detail, let's review the classic work that initially described bacterial control systems, for these studies are landmarks in the use of genetic analysis.

Discovery of the *lac* System: Negative Control

The first major breakthrough in understanding gene control came in the 1950s with the detailed genetic analysis, by François Jacob and Jacques Monod, of the enzymes concerned with lactose metabolism in *E. coli* and of λ phage immunity. Jacob and Monod used the lactose metabolism system of *E. coli* (see Figure 17-2) to attack the problem of enzyme induction (originally termed *adaptation*)—that is, the appearance of a specific enzyme only in the presence of its substrates. This phenomenon had been observed in bacteria for many years. How could a cell possibly "know" precisely which enzymes to synthesize? How could a particular substrate induce the appearance of a specific enzyme?

For the *lac* system, such an induction phenomenon could be illustrated when, in the presence of certain galactosides termed *inducers,* cells produced over 1000 times more of the enzyme β-galactosidase, which cleaves β-galactosides, than they produced when grown in the absence of such sugars. What role did the inducer play in the induction phenomenon? One idea was that the inducer was simply activating a pre-β-galactosidase intermediate that had accumulated in the cell. However, when Jacob and Monod followed the fate of radioactively labeled amino acids added to growing cells either before or after the addition of an inducer, they could

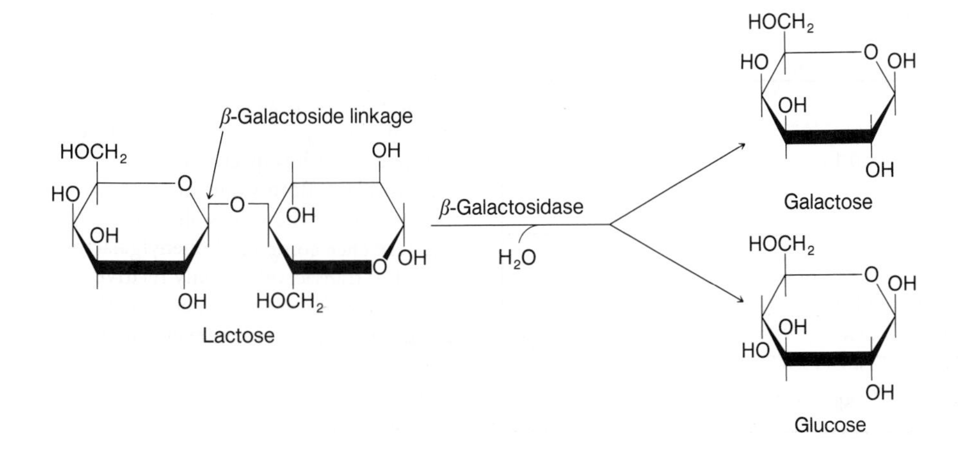

Figure 17-2 The metabolism of lactose. The enzyme β-galactosidase catalyzes a reaction in which water is added to the β-galactoside linkage to break lactose into separate molecules of galactose and glucose. The enzyme lactose permease is required to transport lactose into the cell.

Table 17-1. Synthesis of β-Galactosidase and Permease in Haploid and Heterozygous Diploid Strains

Strain	Genotype	β-Galactosidase		Permease	
		Noninduced	Induced	Noninduced	Induced
1	$I^+ Z^+ Y^+$	−	+	−	+
2	$I^- Z^+ Y^+$	+	+	+	+
3	$I^+ Z^- Y^+ / FI^- Z^+ Y^+$	−	+	−	+
4	$I^- Z^- Y^+ / FI^+ Z^+ Y^-$	−	+	−	+
5	$I^- Z^- Y^+ / FI^- Z^+ Y^+$	+	+	+	+
6	$\nabla(I, Z, Y) / FI^- Z^+ Y^+$	+	+	+	+

NOTE: Bacteria were grown in glycerol as a carbon source and induced by IPTG. The presence of the maximal level of the enzyme is indicated by "+"; the absence or very low level of an enzyme is indicated by "−"; "∇" indicates deletion.

diploid for the *lac* genes. This partial diploid state persists for several hours, during which time measurements of gene activity can be carried out. In their experiment, the F^- recipients were $I^- Z^-$. They could not synthesize active β-galactosidase. Shortly after introduction of an $I^+ Z^+$ chromosome segment from the donor Hfr, the recipient cells gained the capacity to synthesize β-galactosidase. As can be seen from Figure 17-6, however, after about an hour, the synthesis of β-galactosidase stopped. If an inducer such as IPTG was added at this point, then the synthesis of β-galactosidase continued. In other words, after about an hour, β-galactosidase synthesis became inducible. The simplest interpretation of these data was that the *I* gene product was absent in the cytoplasm of the $I^- Z^-$ recipient, so that when the $I^+ Z^+$ chromosome was introduced, β-galactosidase could begin to be synthesized at the maximal rate. However, the *I* gene product could also be synthesized, and after an hour the concentration of I product in the cytoplasm was sufficient to be able to block, or "repress," β-galactosidase synthesis. The addition of IPTG inactivated the *I* gene product, allowing induction.

The discovery of F′ factors (see Chapter 10) carrying the *lac* region, allowed the construction of stable partial diploids, and facilitated more direct complementation tests. Recall the complementation tests performed by Seymour Benzer (Chapter 12), in which complementation occurring in the trans position implies the action of a diffusable product. Tests with I^+ and I^- genes showed that I^+ is dominant over I^- in the trans position, strengthening the conclusion that the *I* gene product acted through the cytoplasm as a repressor. Table 17-1 shows the effect of various combinations of mutations, in the induced and noninduced state, on the production of β-galactosidase and permease.

A piece of evidence in support of the repressor model was the characterization of I^s mutations. Although mapping within the *I* gene, these mutations prevented induction of the *lac* enzymes by lactose or by the synthetic inducer IPTG (Figure 17-4). Moreover, they were dominant in trans to both an I^+ and an I^- allele (Table 17-2). The I^s mutation eliminates response to an inducer, presumably by altering the stereospecific binding site and destroying inducer binding. Therefore, even in the presence of IPTG, these molecules can still block *lac* enzyme synthesis. This would also explain their dominance, since the I^s repressor would be active, even in the presence of the wild-type repressor that was inactivated by the inducer. The I^s mutations clearly pointed to a direct interaction between the *I* gene product and the inducer.

Message The I^- mutation affects the *DNA binding region* of the repressor, thus preventing binding and allowing transcription, even in the *absence* of inducer; the I^s mutation affects the *inducer binding region* of the repressor, thus repressing transcription, even in the *presence* of inducer.

The Operator and the Operon

The specificity of the repressor, which results in turning off *lac* enzyme synthesis, suggests a stereospecific complex with an element that Jacob and Monod termed the

Table 17-2. Synthesis of β-Galactosidase and Permease by the Wild-Type and by Strains Carrying Different Alleles of the *I* Gene

Genotype	Inducer	β-Galactosidase	Permease
$I^+ Z^+ Y^+$	None	−	−
	IPTG	+	+
$I^s Z^+ Y^+$	None	−	−
	IPTG	−	−
$I^s Z^+ Y^+ / FI^+$	None	−	−
	IPTG	−	−
$I^s Z^+ Y^+ / FI^-$	None	−	−
	IPTG	−	−

NOTE: Bacteria were grown in glycerol with and without the inducer IPTG. Presence of the indicated enzyme is represented by "+"; absence or low levels, by "−."

Table 17-3. Synthesis of β-Galactosidase and Permease by Haploid and Heterozygous Diploid Operator Mutants

Genotype	β-Galactosidase		Permease	
	Noninduced	Induced	Noninduced	Induced
$O^+ Z^+ Y^+$	−	+	−	+
$O^+ Z^+ Y^+ / FO^+ Z^- Y^+$	−	+	−	+
$O^c Z^+ Y^+$	+	+	+	+
$O^+ Z^+ Y^- / FO^c Z^+ Y^+$	+	+	+	+
$O^+ Z^+ Y^+ / FO^c Z^- Y^+$	−	+	+	+
$O^+ Z^- Y^+ / FO^c Z^+ Y^-$	+	+	−	+
$I^s O^+ Z^+ Y^+ / FI^+ O^c Z^+ Y^+$	+	+	+	+

NOTE: Bacteria were grown in glycerol with and without the inducer IPTG. The presence and absence of enzyme are indicated by + and −, respectively.

operator. The operator was postulated to be a region of DNA near the beginning of the set of genes it controlled. The researchers sought mutations in the operator that would allow synthesis of the *lac* enzymes even in the presence of an active repressor. These mutations should be dominant in the cis position. Whereas trans dominance reflects a diffusible product, cis dominance reflects the action of an element that affects only the genes directly adjacent to it. No diffusible product is altered by the mutation. By selecting for constitutivity (unrepressed synthesis) in cells with two copies of the *lac* region, to eliminate the effects of single I^- mutations, such mutations were detected and labeled O^c, for **operator constitutive.** As Table 17-3 indicates, strains carrying these mutations are capable of synthesizing maximal amounts of enzyme in the presence of IPTG, but can also synthesize 10 to 20 percent of these levels in the absence of an inducer. The O^c mutations are indeed dominant in the cis position, as shown in Table 17-3. Mapping experiments have pinpointed the operator locus between *I* and *Z*.

As we have seen, the *OZYA* segment constitutes a genetic unit of coordinate expression that Jacob and Monod termed the *operon.* Figure 17-7 depicts a simplified operon model for the *lac* system. The *lac* operon is said to be under the **negative control** of the *lac* repressor, since the repressor normally blocks expression of the *lac* enzymes in the absence of an inducer.

Let's review the model in Figure 17-7, as postulated by Jacob and Monod. The *Z* and *Y* genes code for the structure of two enzymes required for the metabolism of the sugar lactose, β-galactosidase and permease, respectively. The *A* gene codes for transacetylase. All three genes are linked together on the chromosome. Their transcription into a polycistronic (single) mRNA provides the basis for coordinate control at the level of mRNA synthesis. The synthesis of the polycistronic *lac* mRNA can be blocked by the action of a repressor protein molecule, which binds to an operator region near the start point for transcription. The repressor is the product of the *I* gene. Therefore, mutations in the *I* gene that prevent the synthesis of a functional repressor result in

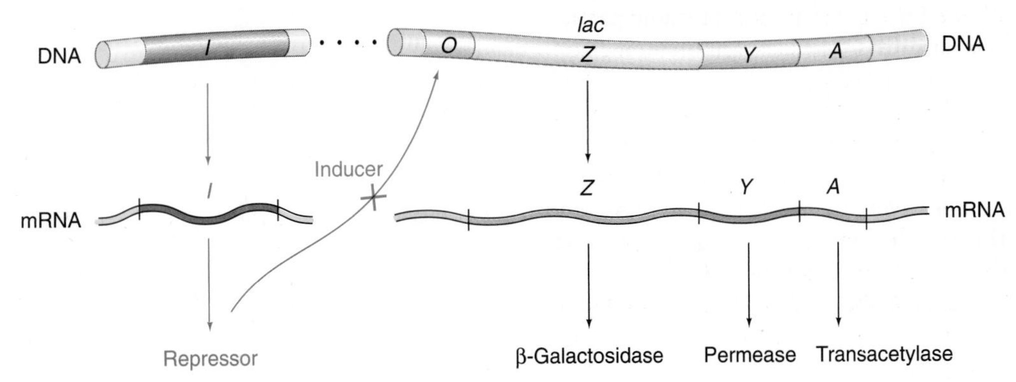

Figure 17-7 A simplified *lac* operon model. The three genes *Z*, *Y*, and *A* are coordinately expressed. The product of the *I* gene, the repressor, blocks the expression of the *Z*, *Y*, and *A* genes by interacting with the operator (O). The inducer can inactivate the repressor, thereby preventing interaction with the operator. When this happens, the operon is fully expressed. Mutations in *I* or O can also result in expression of the three *lac* enzymes, even in the absence of an inducer.

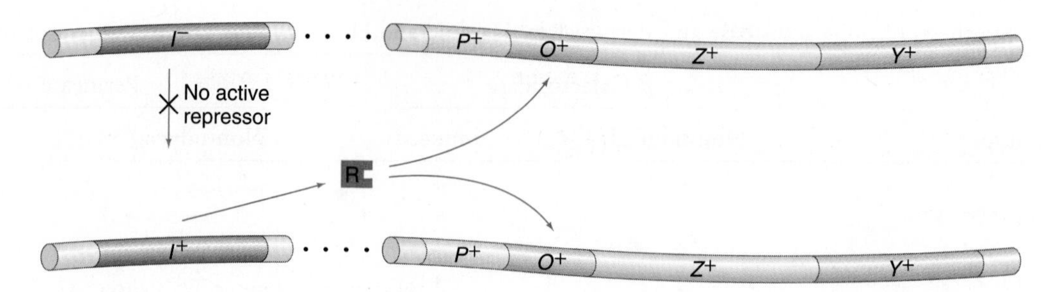

Figure 17-8 I^- mutations are recessive to wild-type. Although no active repressor is synthesized from the I^- gene, the wild-type (I^+) gene provides a functional repressor that binds to both operators in a diploid cell and blocks *lac* operon expression (in the absence of an inducer).

unrepressed, or constitutive, synthesis of the *lac* enzymes. Repression can also be overcome by certain galactosides, termed *inducers,* which inactivate the repressor by binding to it and altering its affinity for the operator. In this manner, the inducer can pull the repressor off the DNA.

We can now understand the properties of some of the diploids used for complementation tests, in light of the operon model. Figure 17-8 shows how I^- mutations are recessive to wild-type, because one functional I gene is all that is needed to produce a repressor that can bind to both operators in a diploid. On the other hand, Figure 17-9 shows a diploid cell carrying one copy of a wild-type I gene and one copy of an I^s gene. The I^s mutation alters the inducer binding site so that repressor no longer binds to inducer, although it still recognizes the operator. These diploid cells will be Lac⁻, because the altered repressor will always bind to the operator, even in the presence of an inducer, and block synthesis of the *lac* enzymes.

Operator mutations (O^c) are cis-dominant, because they are dominant only for genes directly linked to them on the same chromosome, as diagrammed in Figure 17-10. Thus, if an altered operator is in the same cell with a second chromosome that contains a wild-type operator, the repressor will recognize the wild-type operator and repress the genes linked to it, but will not recognize the altered operator. Therefore, the genes linked to the altered operator are expressed even in the absence of inducer.

Allostery

The *lac* repressor is a protein with two different binding sites. One site recognizes the inducer molecule; the second site recognizes the *lac* operator sequence on the DNA. Interaction of the repressor with the inducer lowers the affinity of the repressor for the operator. This change in affinity for operator, in response to the binding of an inducer at a distant site, is mediated by a conformational change in the repressor protein. In one conformation, the repressor binds the operator well; in a second conformation, it does not. Proteins that function this way are termed **allosteric** proteins. Allosteric transitions, the change from one conformation to another, occur in many different proteins.

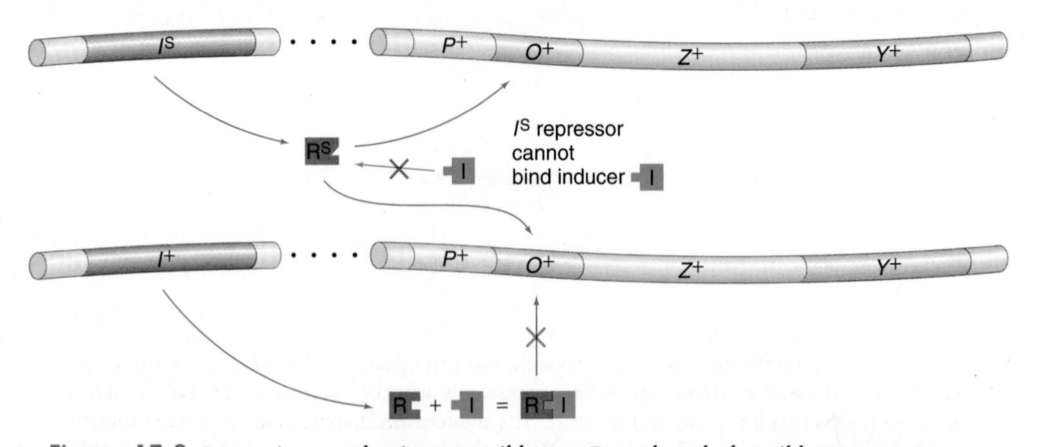

Figure 17-9 I^s mutations are dominant to wild-type. Even though the wild-type repressor is inactivated by an inducer, the I^s repressor is not and can bind to both operators in a diploid cell. Therefore, no enzyme is produced in an I^s/I^+ diploid, even in the presence of an inducer.

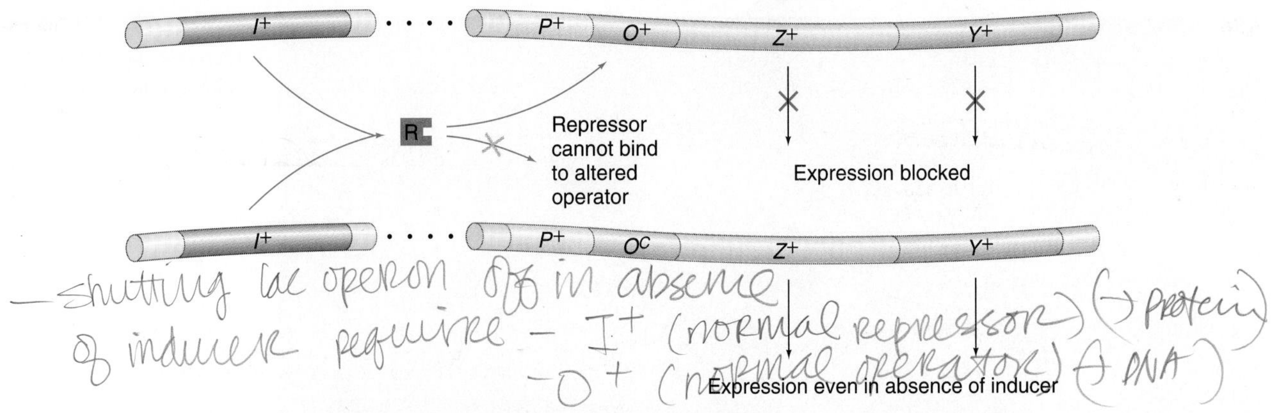

Figure 17-10 O^c mutations are dominant in the cis position. Because a repressor cannot bind to O^c operators, genes linked to an O^c operator are expressed even in the absence of an inducer. However, genes linked to an O^+ operator are still subject to repression.

The *lac* Promoter

Genetic experiments have suggested that an element essential for *lac* transcription is located between *I* and *O* in the operon model for the *lac* system. This element, termed the *promoter* (***P***), is postulated to serve as an initiation site for transcription. Promoter mutations affect the transcription of all the genes in the operon in a similar manner. Promoter mutations are cis-dominant, as would be expected for a site on the DNA that serves as a recognition element for transcription initiation, since each promoter governs transcription only for those genes in the operon adjacent to it on the *same* DNA molecule. As outlined in Chapter 13, in vitro experiments have demonstrated that RNA polymerase binds to the promoter region and that repressor binding to the operator can block RNA polymerase from binding to the promoter. Mutant analysis, physical experiments, and comparisons with other promoters have identified two binding regions for RNA polymerase in a typical prokaryotic promoter. Figure 17-11 summarizes this body of information (see also Figure 13-8).

Message The *lac* operon is a cluster of structural genes that specify enzymes involved in lactose metabolism. These genes are controlled by the coordinated actions of cis-dominant promoter and operator regions. The activity of these regions is, in turn, determined by a repressor molecule specified by a separate regulator gene. Figure 17-1 integrates all this information into a single picture.

Characterization of the *lac* Repressor and the *lac* Operator

Several genetic experiments have argued strongly that the repressor is a protein—instead of, for instance, an RNA molecule—the most compelling of which was the discovery of suppressible nonsense mutations in the *I* gene, since the resulting nonsense codons exert their effect by signaling termination of the polypeptide chain during translation. The decisive experiment, however, was provided by Walter Gilbert and Benno Müller-Hill, who in 1966 isolated and purified the repressor by monitoring the binding of the radioactively labeled inducer

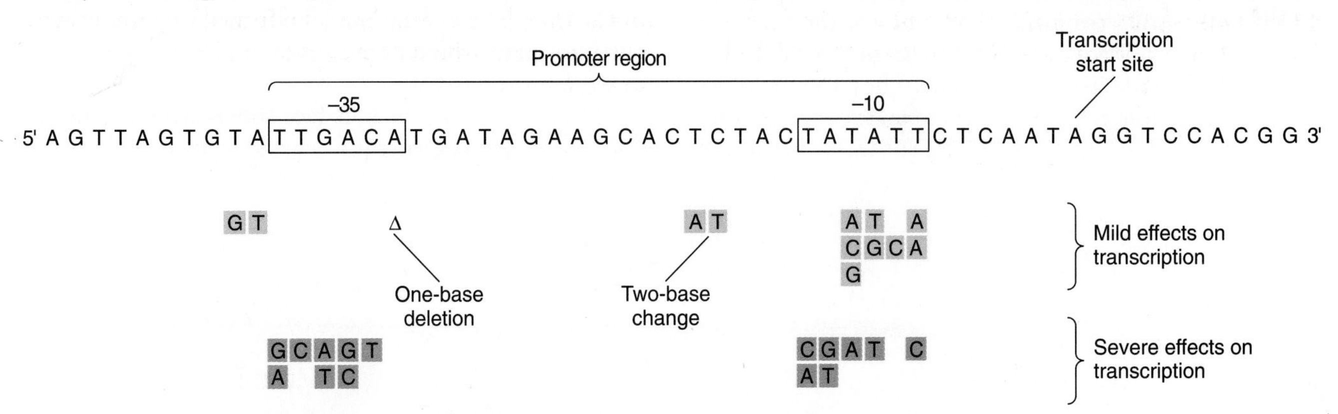

Figure 17-11 Specific DNA sequences are important for efficient transcription of *E. coli* genes by RNA polymerase. The boxed sequences at approximately 35 and 10 nucleotides before the transcription start site are highly conserved in all *E. coli* promoters. Mutations in these regions have mild (gold) and severe (orange) effects on transcription. The mutations may be changes of single nucleotides or pairs of nucleotides, or a deletion (Δ) may occur. (From J. D. Watson, M. Gilman, J. Witkowski, and M. Zoller. *Recombinant DNA*, 2nd ed. Copyright 1992 by Scientific American Books.)

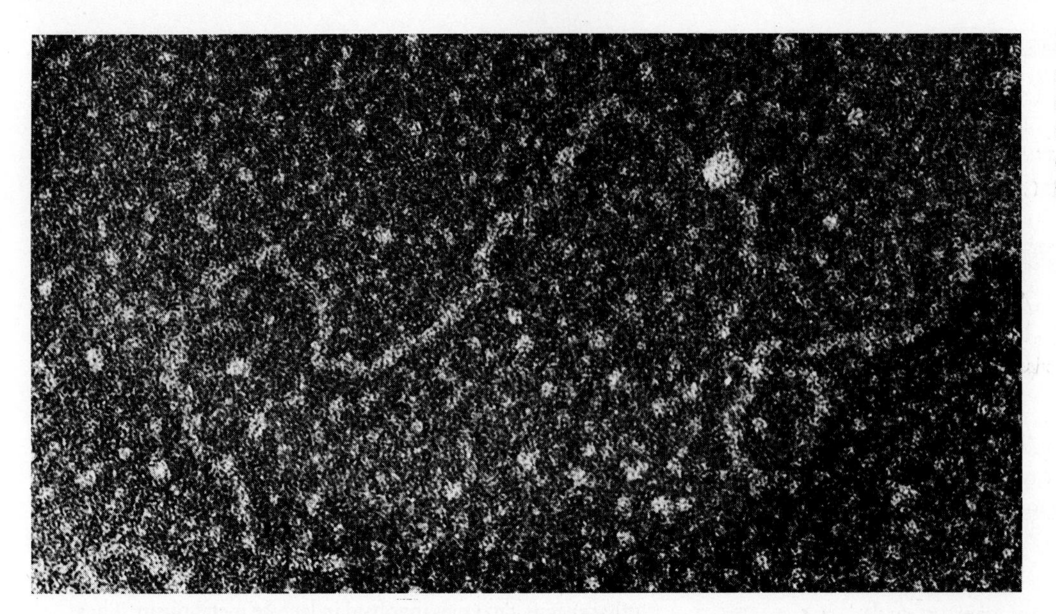

Figure 17-12 The Lac repressor (large whitish sphere) bound to DNA at the operator region of the operon. (Photograph courtesy of Jack D. Griffith.)

IPTG. They demonstrated that the repressor is a protein consisting of four identical subunits, each with a molecular weight of approximately 38,000 daltons. Each molecule contains four IPTG-binding sites. (A more detailed description of the repressor is given later in the chapter.) In vitro, repressor binds to DNA containing the operator (see Figure 17-12) and comes off the DNA in the presence of IPTG. Gilbert and his coworkers have shown that the repressor can protect specific bases in the operator from chemical reagents. These experiments provide crucial proofs of the mechanism of repressor action formulated by Jacob and Monod.

Gilbert used the enzyme DNase to break apart the DNA bound to the repressor. He was able to recover short DNA strands that had been shielded from the enzyme activity by the repressor molecule and that presumably represented the operator sequence. This sequence was determined, and each operator mutation was shown to involve a change in the sequence (Figure 17-13). These results confirm the identity of the operator locus as a specific sequence of 17 to 25 nucleotides situated just before the structural *Z* gene. They also show the incredible specificity of repressor-operator recognition, which is disrupted by a single base substitution. When the sequence of bases in the *lac* mRNA (transcribed from the *lac* operon) was determined, the first 21 bases on the 5′ initiation end proved to be complementary to the operator sequence Gilbert had determined.

Catabolite Repression of the *lac* Operon: Positive Control

There is an additional control system superimposed on the repressor-operator system. This system exists because cells have specific enzymes that favor glucose uptake and metabolism. If both lactose *and* glucose are present, synthesis of β-galactosidase is not induced until all the glucose has been utilized. Thus, the cell conserves its metabolic machinery (that, for example, induces the *lac* enzymes) by utilizing any existing glucose before going through the steps of creating new machinery to exploit the lactose. The operon model outlined previously will not account for the suppression of induction by glucose, so we must modify it.

Studies indicate that in fact some catabolic breakdown product of glucose (no exact identity is yet known)

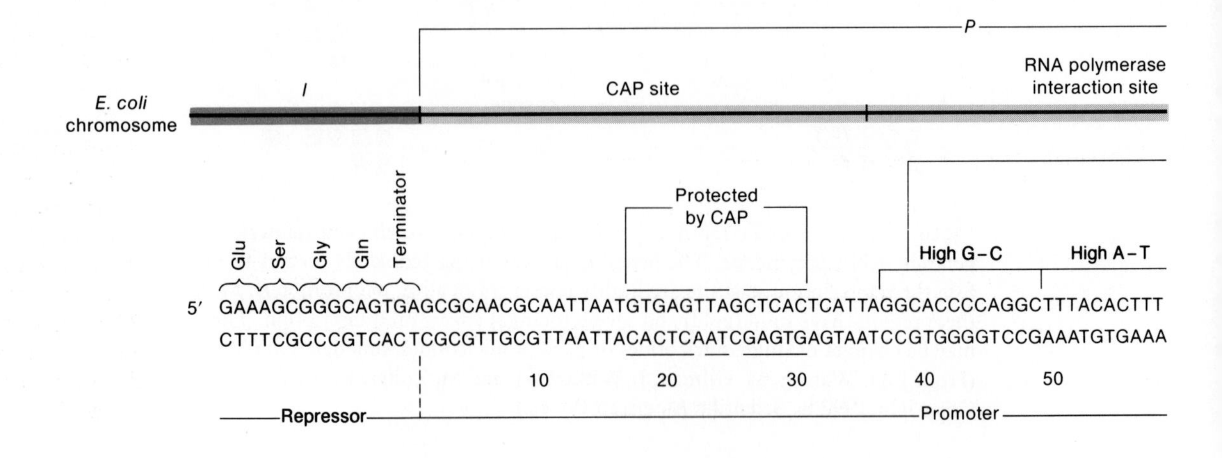

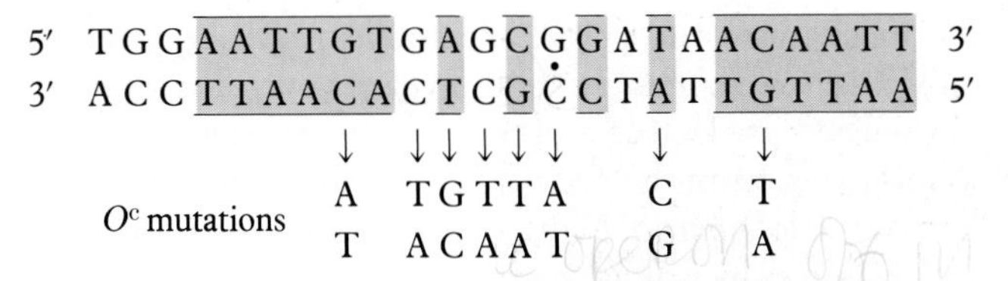

Figure 17-13 The DNA base sequence of the lactose operator and the base changes associated with eight O^c mutations. Regions of twofold rotational symmetry are indicated by color and by a dot at their axis of symmetry. (From W. Gilbert, A. Maxam, and A. Mirzabekov, in N. O. Kjeldgaard and O. Maløe, eds., *Control of Ribosome Synthesis.* Academic Press, 1976. Used by permission of Munksgaard International Publishers, Ltd., Copenhagen.)

prevents activation of the *lac* operon by lactose, so this effect was originally called **catabolite repression.** The effect of the glucose catabolite is exerted on an important cellular constituent called *cyclic adenosine monophosphate (cAMP)* in a way we will see shortly.

When glucose is present in high concentrations, the cAMP concentration is low; as the glucose concentration decreases, the concentration of cAMP increases correspondingly. The high concentration of cAMP is necessary for activation of the *lac* operon. Mutants that cannot convert ATP to cAMP cannot be induced to produce β-galactosidase, because the concentration of cAMP is not great enough to activate the *lac* operon. In addition, there are other mutants that do make cAMP but cannot activate the *lac* enzymes, because they lack yet another protein, called *CAP (catabolite activator protein)*, made by the *crp* gene. The CAP protein forms a complex with cAMP, and it is this complex that activates the *lac* operon (Figure 17-14).

How does catabolite repression fit into our model for the structure and regulation of the *lac* operon? Recall the technique that Gilbert used to identify the operator base sequence. In a similar experiment, the CAP-cAMP complex was added to DNA, and the DNA then was subjected to digestion by the enzyme DNase. The surviving strands are presumably those shielded from digestion by an attached CAP-cAMP complex; the sequence of these strands (Figure 17-15) clearly is different from the operator sequence (Figure 17-13), but it also has a rotational twofold symmetry.

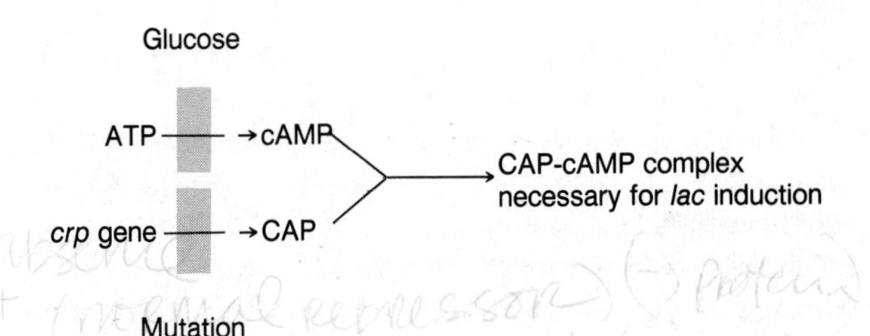

Figure 17-14 Catabolite control of the *lac* operon. The operon is inducible by lactose to the maximal levels when cAMP and CAP form a complex. The *lac* operon cannot be expressed at full levels if formation of cAMP is blocked by excess glucose or if formation of CAP is blocked by mutation of the *crp* gene. (CAP = catabolic activator protein; cAMP = cyclic adenosine monophosphate; *crp* = structural gene responsible for synthesizing CAP.)

The entire *lac* operon can be inserted into λ phage in such a way that the initiation of transcription is prompted by the phage gene adjacent to *lac*. In this case, the transcribed product carries a complementary copy of the base sequence from the *lac* control regions (sequences not transcribed in the *lac* mRNA). We already know the amino acid sequences for the repressor and β-galactosidase, so these sequences can be identified, and the remaining sequences can be assigned to the control regions (Figure 17-16). We can also fit the known repressor, CAP-cAMP, and RNA polymerase binding sites

GTGAGTTAGCTCAC
CACTCAATCGAGTG

Figure 17-15 The DNA base sequence to which the CAP-cAMP complex binds. Regions of twofold rotational symmetry are indicated by the colored boxes and by a dot at their axis of symmetry.

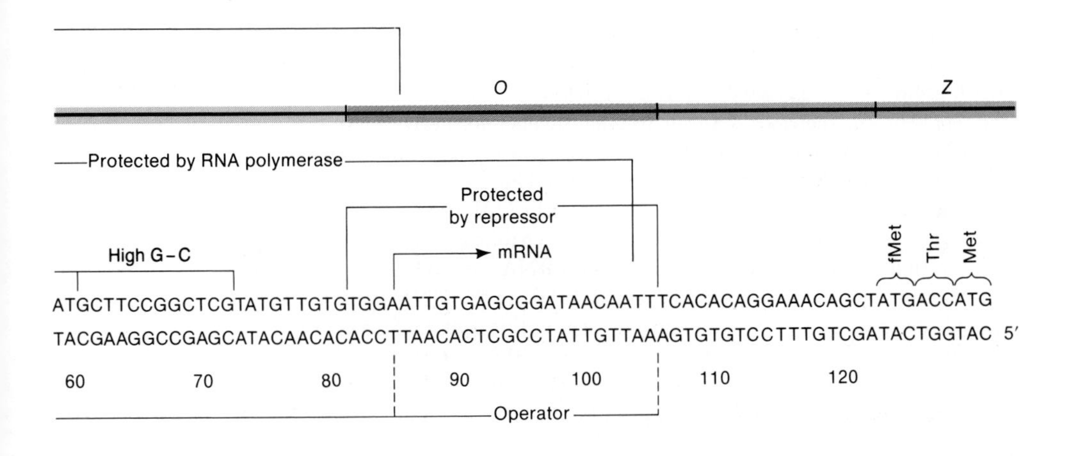

Figure 17-16 The base sequence and the genetic boundaries of the control region of the *lac* operon, with partial sequences for the structural genes. (After R. C. Dickson, J. Abelson, W. M. Barnes, and W. S. Reznikoff, "Genetic Regulation: The *Lac* Control Region." *Science* 187, 1975, 27. Copyright 1975 by the American Association for the Advancement of Science.)

into the detailed model, as shown in Figures 17-16 and 17-17.

Glucose control is accomplished because a glucose-breakdown product inhibits formation of the CAP-cAMP complex required for the RNA polymerase to attach at the *lac* promoter site. Even when there is a shortage of glucose catabolites and CAP-cAMP forms, the mechanism for lactose metabolism will be created only if lactose is present. This level of control is accomplished because lactose must bind to the repressor protein to remove it from the operator site and permit transcription of the *lac* operon. Thus, the cell conserves its energy and resources by producing the lactose-metabolizing enzymes only when they are both needed and useful.

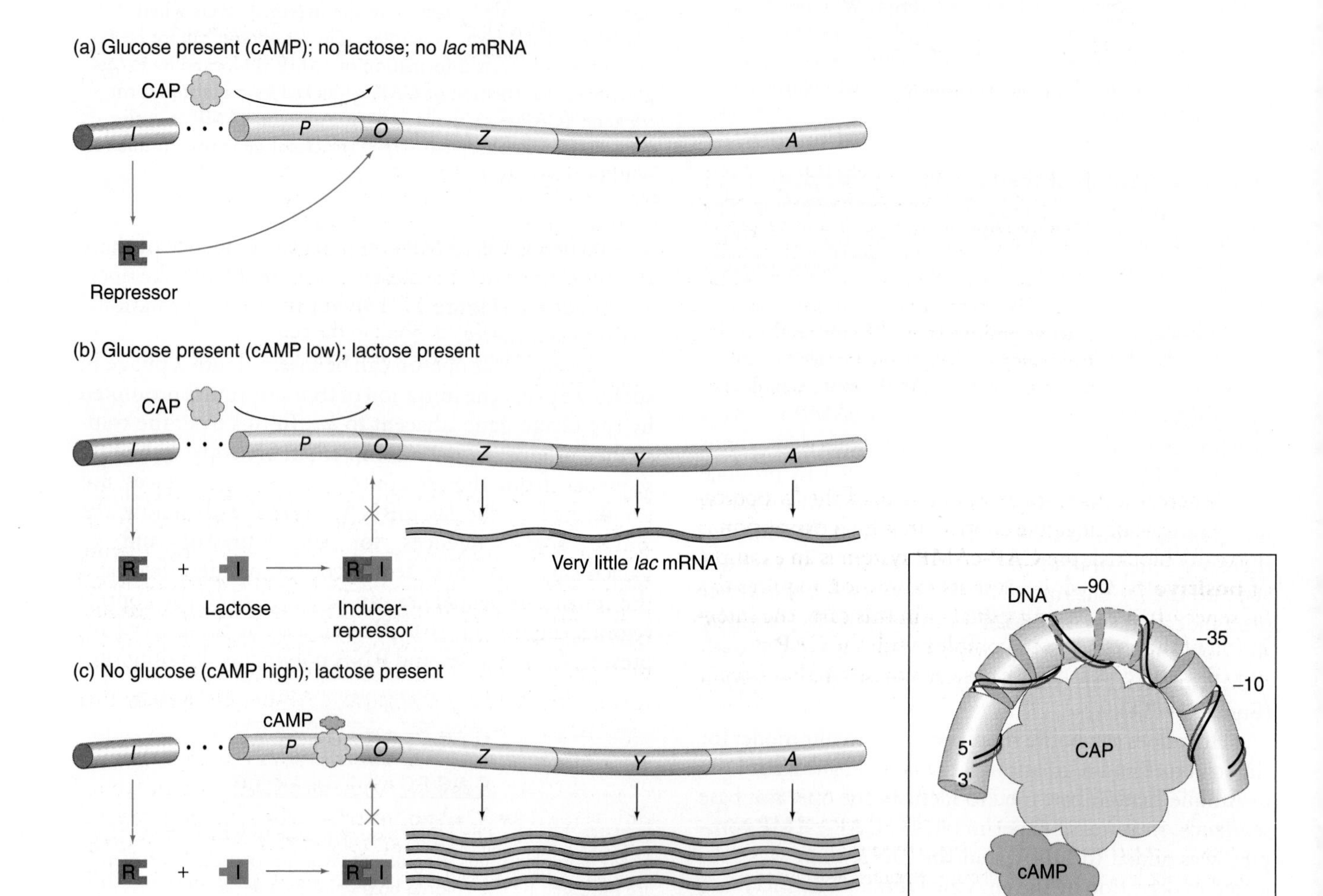

Figure 17-17 Negative and positive control of the *lac* operon by the *lac* repressor and catabolite activator protein (CAP), respectively. (a) In the absence of lactose to serve as an inducer, the *lac* repressor is able to bind the operator; regardless of the levels of cAMP and presence of CAP, mRNA production is repressed. (b) With lactose (or its metabolite *allolactose*) present to bind the repressor, the repressor is unable to bind the operator; however, only small amounts of mRNA are produced because the presence of glucose keeps the levels of cAMP low, and thus the cAMP-CAP complex does not form and bind the promoter. (c) With the repressor inactivated by lactose and with high levels of cAMP present (owing to the absence of glucose), the cAMP binds the CAP, activating it and enabling it to bind the promoter; the *lac* operon is thus activated, and large amounts of mRNA are produced. *Inset:* When CAP binds the promoter, it creates a bend greater than 90° in the DNA. Apparently, RNA polymerase binds more effectively when the promoter is in this bent configuration. [Redrawn from B. Gartenberg and D. M. Crothers, *Nature* 333, 1988, 824. (See H-N, Lie-Johnson et al., *Cell* 47, 1986, 995.) Adapted from J. Darnell, H. Lodish, and D. Baltimore, *Molecular Cell Biology,* 2d ed. Copyright 1990 by Scientific American Books, Inc.]

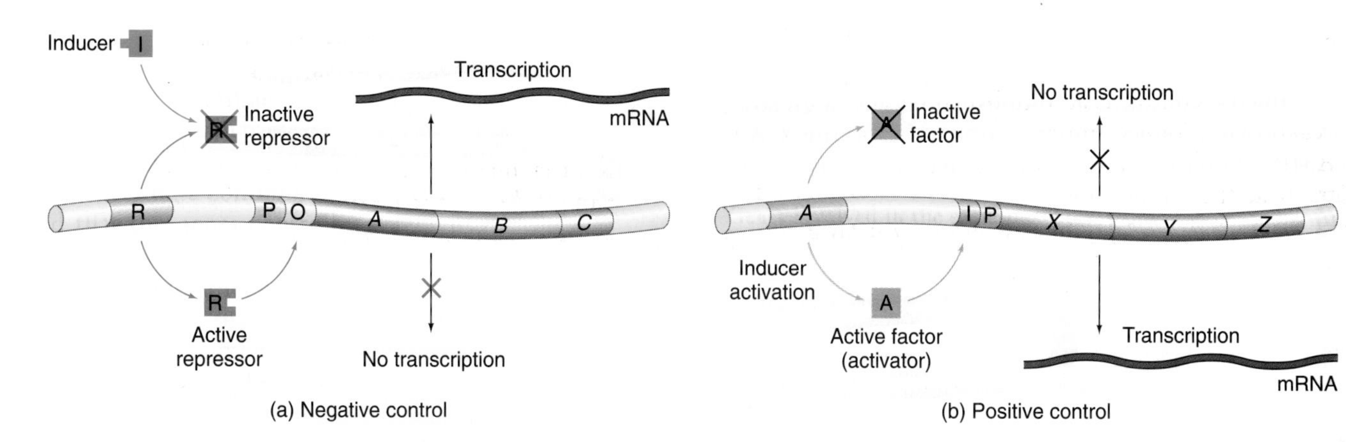

Figure 17-18 Comparison of positive and negative control. The basic aspects of negative and positive control are depicted. (a) In negative control, an active repressor (encoded by the *R* gene in the example shown here) blocks gene expression of the *A,B,C* operon by binding to an operator site (*O*). An inactive repressor allows gene expression. The repressor can be inactivated either by an inducer or by mutation. (b) In positive control, an active factor is required for gene expression, as shown for the *X,Y,Z* operon here. Small molecules can convert an inactive factor into an active one, as in the case of cyclic AMP and the CAP protein. An inactive positive control factor results in no gene expression. The activator binds to the control region of the operon, termed *I* in this case. (The positions of both *O* and *I* with respect to the promoter, *P*, in the two examples are arbitrarily drawn.)

Whereas inducer-repressor control of the *lac* operon is an example of negative control, in which expression is normally blocked, the CAP-cAMP system is an example of **positive control,** because its expression requires the presence of an activating signal — in this case, the interaction of the CAP-cAMP complex with the CAP region. Figure 17-18 distinguishes between these two basic types of control systems.

> Message The *lac* operon has an added level of control so that the operon remains inactive in the presence of glucose even if lactose is also present. High concentrations of glucose catabolites produce low concentrations of cyclic adenosine monophosphate (cAMP), which must form a complex with CAP to permit the induction of the *lac* operon.

By using different combinations of controlling elements, bacteria have evolved numerous strategies for regulating gene expression. Some examples follow.

Dual Positive and Negative Control: The Arabinose Operon

The metabolism of the sugar arabinose is catalyzed by three enzymes encoded by the *araB*, *araA*, and *araD* genes. Figure 17-19 depicts this operon. Expression is activated at the **initiator** region, *araI*. Within this region, the product of the *araC* gene, when bound to arabinose, can activate transcription, perhaps by directly affecting RNA polymerase binding in the *araI* region, which contains the promoter for the *araB*, *araA*, and *araD* genes. This represents positive control, since the product of the regulatory gene *(araC)* must be active in order for the operon to be expressed. An additional positive control is mediated by the same CAP-cAMP system that regulates *lac* expression.

In the presence of arabinose, both binding of the C product to the initiator region and CAP protein are required to allow RNA polymerase to bind to the promoter for the *araB*, *araA*, and *araD* genes (Figure 17-20a). In the absence of arabinose, the *araC* product assumes a different conformation and actually represses the *ara* operon by binding both to *araI* and to an operator region, *araO*, thereby forming a loop (Figure 17-20b) that prevents transcription. Thus, the *araC* protein has two conformations that promote two opposing functions at two alternative binding sites. The conformation is dependent on whether the inducer, arabinose, is bound to the protein.

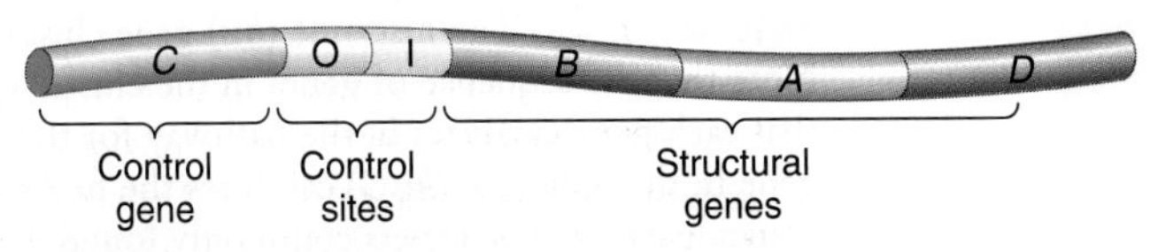

Figure 17-19 Map of the *ara* region. The *BAD* genes together with the *I* and *O* sites constitute the *ara* operon.

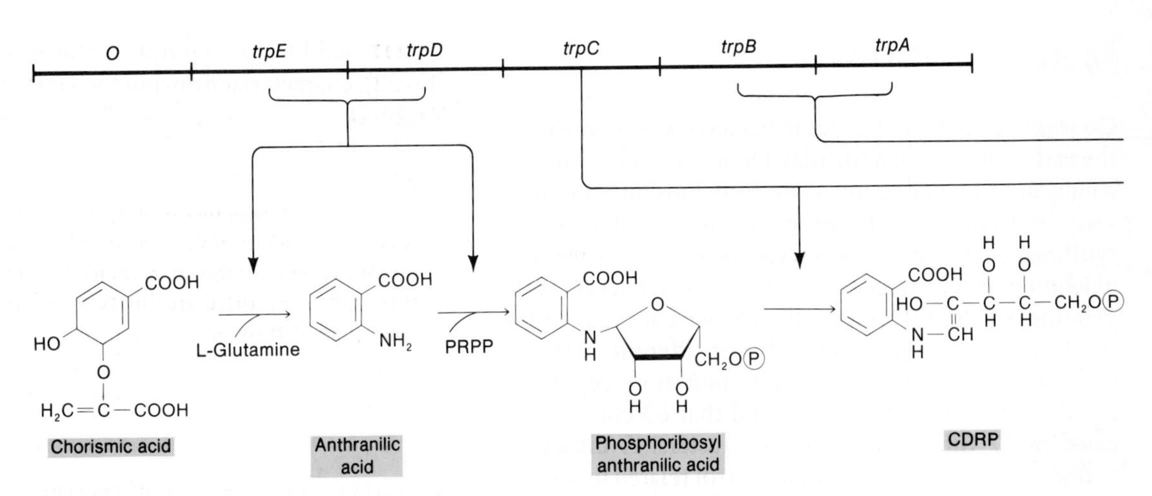

Figure 17-22 The genetic sequence of cistrons in the *trp* operon of *E. coli* and the sequence of reactions catalyzed by the enzyme products of the *trp* structural genes. The products of genes *trpD* and *trpE* form a complex that catalyzes specific steps, as do the products of genes *trpB* and *trpA*. Tryptophan synthetase is a tetrameric enzyme formed by the products of *trpB* and *trpA*. It catalyzes a two-step process leading to the formation of tryptophan. (PRPP = phosphoribosyl pyrophosphate.) (After S. Tanemura and R. H. Bauerle, *Genetics* 95, 1980, 545–559.)

fied for a cluster of genes controlling enzymes in the pathway for tryptophan production. Synthesis of tryptophan is shut off when there is an excess of tryptophan in the medium. Jacob and Monod suggested that the cluster of five *trp* cistrons in *E. coli* forms another operon, differing from the *lac* operon in that the tryptophan repressor will bind to the *trp* operator only when it *is* bound to tryptophan (Figure 17-22). (Recall that the *lac* repressor binds to the operator *except* when it is bound to lactose.) A second control pathway also modulates tryptophan biosynthesis at the level of enzyme activity. This is termed **feedback inhibition.** Here, the first enzyme in the pathway, encoded by the *trpE* and *trpD* genes, is inhibited by tryptophan itself.

As with the *lac* operon, further analysis of the *trp* operon has revealed yet another level of control superimposed on the basic repressor-operator mechanism. In studying mutant strains (carrying a mutation in *trpR*, the repressor locus) that continue to produce *trp* mRNA in the presence of tryptophan, Yanofsky found that removal of tryptophan from the medium leads to almost a tenfold increase in *trp* mRNA production in these strains, even though the Trp repressor was inactive, and thus could not account for the increase through normal derepression of the operator because of low tryptophan levels. Furthermore, Yanofsky identified the region responsible for this increase by isolating a totally constitutive mutant strain that produces *trp* mRNA at this tenfold maximal level, even in the presence of tryptophan, and he showed that this mutation has a deletion located between the operator and the *trpE* cistron (see the map in Figure 17-22).

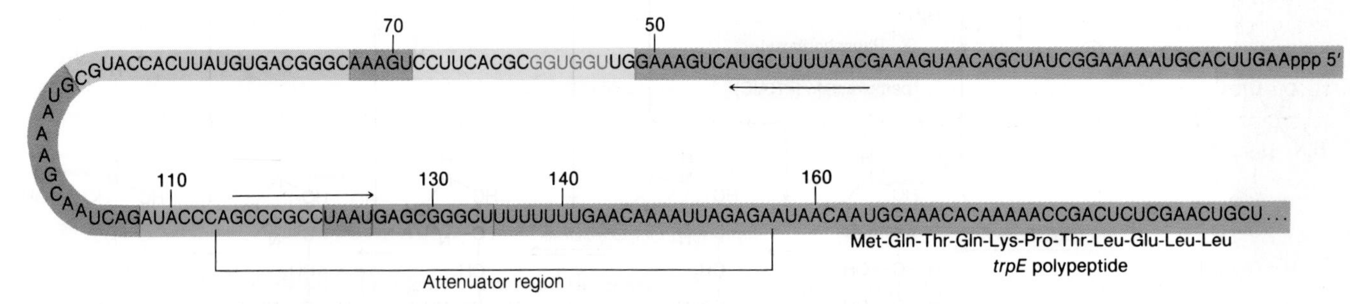

Figure 17-23 The leader sequence, showing the attenuator segment of the *trp* operon, along with the beginning of the *trpE* structural sequence (showing the amino acid sequence of the *trpE* polypeptide). For reference purposes, the color scheme for the various segments is repeated in Figures 17-25 and 17-26. (Modified from G. S. Stent and R. Calendar, *Molecular Genetics,* 2d ed. Copyright 1978 by W. H. Freeman and Company. Based on unpublished data provided by Charles Yanofsky.)

Indole-3-glycerol phosphate

Indole

L-Serine

L-Tryptophan

Yanofsky was able to isolate the polycistronic *trp* operon mRNA. On sequencing it, he found a long sequence, termed the **leader sequence,** of approximately 160 bases at the 5′ end before the first triplet in the *trpE* gene. The deletion mutant that always produces *trp* mRNA at maximal levels has a deletion extending from base 130 to base 160 (Figure 17-23). Yanofsky called the element inactivated by the deletion the **attenuator,** because its presence apparently leads to a reduction in the rate of mRNA transcription when tryptophan is present. Figure 17-24 shows the position of these elements in the *trp* operon. But what is the role of the leader sequence in bases 1 to 130? A surprising observation provides the key to solving this problem.

While studying mRNAs transcribed by the *trp* operon (using *trpR*$^-$ mutants), Yanofsky discovered that even in the presence of high levels of tryptophan (which should cause the attenuator region to reduce the rate of transcription tenfold), the first 141 bases of the leader sequence were always transcribed at the maximal rate, though the full-length mRNA occurred, as expected, at levels only one-tenth as great. Another way of stating this is that, even in the presence of high concentrations of tryptophan, the first 141 bases are transcribed in maximal numbers, but because of some attenuating mechanism in this region only 1 in 10 of the mRNAs can be transcribed farther (to completion). This suggests that the end of the attenuator acts as a mRNA chain terminator that, *in the presence of tryptophan,* halts transcription of 9 out of 10 mRNAs. In the absence of tryptophan, this attenuator is somehow deactivated, and every mRNA goes to completion—hence, the tenfold increase. In those *trpR*$^-$ mutants in which the attenuator is also deleted, there is no block to extension of the mRNA, so transcription is carried through in every case (that is, the maximal rate of production occurs), regardless of whether tryptophan is present or not.

What causes the interference with termination at the attenuator in the absence of tryptophan? Figure 17-25 presents a model based on alternative secondary structures formed by the mRNA in the leader region. The model proposes that one of the two conformations favors transcription termination and that the other favors elongation. Translation of part of the leader sequence would promote the conformation that favors termination.

It is known that a portion of the leader sequence near the beginning is in fact translated and yields a short peptide of 14 amino acids. There are two tryptophan codons in the translated stretch of the leader mRNA (Figure 17-26). When excess tryptophan is present, there is a sufficient supply of Trp-tRNA to allow efficient translation through the relevant portion of the leader mRNA. The mRNA passes through the ribosome at a sufficiently fast rate that segment 2 of the leader section is drawn into the ribosome before it can form a loop-stem structure with segment 3, as shown in Figure 17-25b. Segment 3 is thus able to form a transcription-termination stem-loop with segment 4. In conditions of low tryptophan levels, however, translation is slowed in segment 1 at the Trp codons by the relative unavailability of Trp-tRNA. As Figure 17-25c shows, this allows the stem-loop between segments 2 and 3 to form, thus preventing the formation of the transcription-terminating loop between segments 3 and 4. Consequently, under conditions of low tryptophan, transcription is not stopped by the attenuator. In this manner, an additional tenfold range of tryptophan biosynthetic enzymes is superimposed on the normal range that is achieved by repressor-operator interaction.

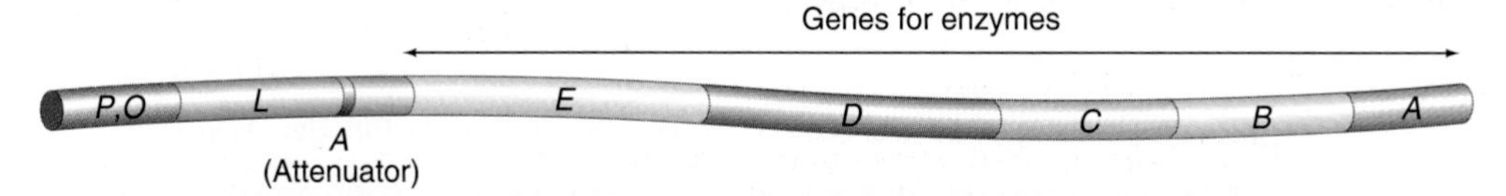

Figure 17-24 Diagram of the *trp* operon showing the promoter (*P*), operator (*O*), and attenuator (*A*) control sites and the genes for the leader sequence (*L*) and enzymes of the tryptophan pathway (*E, D, C, B,* and *A*). (Adapted from L. Stryer, *Biochemistry,* 3d ed. Copyright 1988 by W. H. Freeman and Company.)

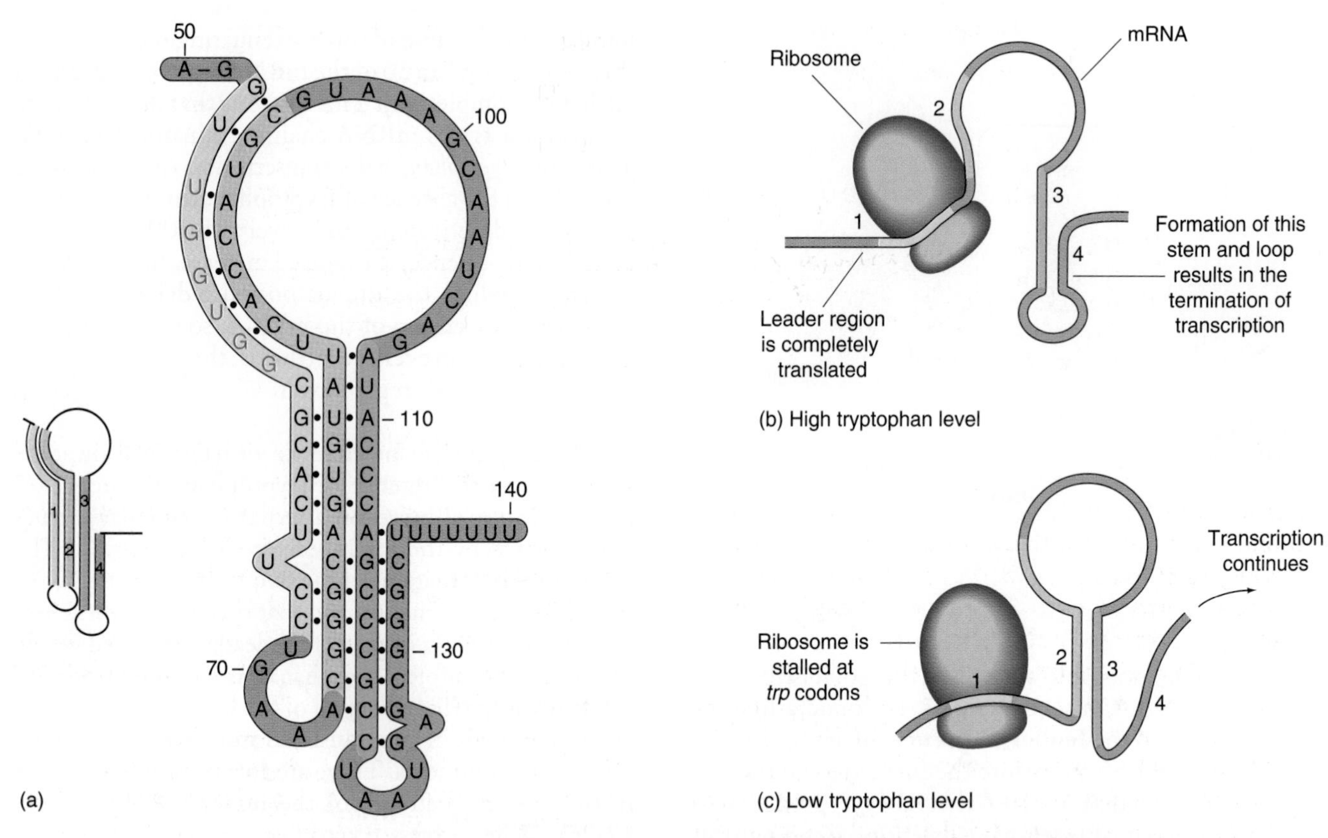

Figure 17-25 Model for attenuation in the *trp* operon. (a) Proposed secondary structures in *E. coli* terminated *trp* leader RNA. Four regions can base-pair to form three stem-and-loop structures. When tryptophan is abundant, segment 1 of the *trp* mRNA is fully translated. Segment 2 enters the ribosome (although it is not translated), which enables segments 3 and 4 to base-pair. This base-paired region somehow signals RNA polymerase to terminate transcription. In contrast, when tryptophan is scarce (c), the ribosome is stalled at the codons of segment 1. Segment 2 interacts with 3 instead of being drawn into the ribosome, and so segments 3 and 4 cannot pair. Consequently, transcription continues. (After D. L. Oxender, G. Zurawski, and C. Yanofsky, *Proc. Natl. Acad. Sci.* 76, 1979, 5524.)

The analysis of numerous point mutations in the *trp* leader sequence that favor or disfavor the respective secondary structures lends strong support to Yanofsky's attenuation model.

Several operons for enzymes in biosynthetic pathways have attentuation controls similar to the one described for tryptophan (Figure 17-27). For instance, the leader region of the *his* operon, which encodes the enzymes of the histidine biosynthetic pathway, contains a translated region with seven consecutive histidine codons. Mutations at outside loci that result in lowering levels of normal charged *his* tRNA produce partially constitutive levels of the enzymes encoded by the *his* operon.

Message The *trp* operon is regulated by a negative repressor-operator control system that represses the synthesis of tryptophan enzymes when tryptophan is present in the medium. A second level of control involves an attenuator region where termination of transcription is induced by the presence of tryptophan.

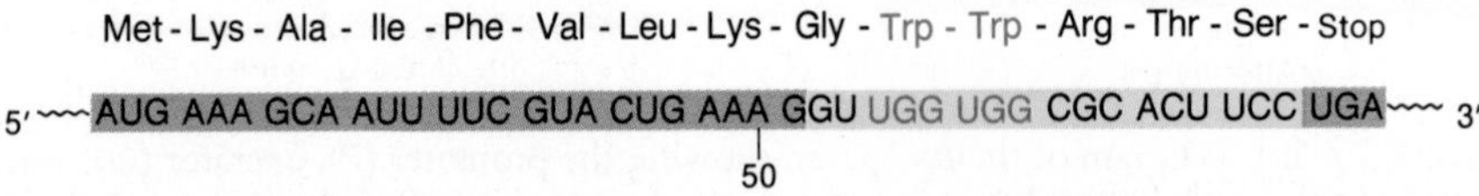

Met - Lys - Ala - Ile - Phe - Val - Leu - Lys - Gly - Trp - Trp - Arg - Thr - Ser - Stop

5′ AUG AAA GCA AUU UUC GUA CUG AAA GGU UGG UGG CGC ACU UCC UGA 3′

50

Figure 17-26 The translated portion of the *trp* leader region, shown with the corresponding sequence of the leader mRNA. Translation of the leader sequence ends at the stop codon.

(a)
Met - Lys - His - Ile - Pro - Phe - Phe - Phe - Ala - Phe - Phe - Phe - Thr - Phe - Pro - Stop
5′ AUG AAA CAC AUA CCG UUU UUC UUC GCA UUC UUU UUU ACC UUC CCC UGA 3′

(b)
Met - Thr - Arg - Val - Gln - Phe - Lys - His - His - His - His - His - His - His - Pro - Asp -
5′ AUG ACA CGC GUU CAA UUU AAA CAC CAC CAU CAU CAC CAU CAU CCU GAC 3′

Figure 17-27 Amino acid sequence of the leader peptide and base sequence of the corresponding portion of mRNA from (a) the phenylalanine operon and (b) the histidine operon. Note that 7 of the 15 residues in the phenylalanine operon leader are phenylalanine and that 7 consecutive residues from the histidine operon leader peptide are histidine. (From L. Stryer, *Biochemistry.* Copyright 1981 by W. H. Freeman and Company.)

The λ Phage: A Complex of Operons

At the time they proposed the operon model, Jacob and Monod suggested that the genetic activity of temperate phages might be controlled by a system analogous to the *lac* operon. In the lysogenic state, the prophage genome is inactive (repressed). In the lytic phase, the phage genes for reproduction are active (induced). Since Jacob and Monod proposed the idea of operon control for phages, the genetic system of the λ phage has become one of the better understood control systems. This phage does indeed have an operon-type system controlling its two functional states. By now, you should not be surprised to learn that this system has proved to be more complex than initially suggested.

Allan Campbell induced and mapped many conditionally lethal mutations in the λ phage (Figure 17-28), providing clear evidence for the clustering of genes with related functions. Furthermore, mutations in the *N*, *O*, and *P* genes prevent most of the genome from being expressed after phage infection, with only those loci lying between *N* and *O* being active. We shall soon see the significance of this observation.

When normal bacteria are infected by wild-type λ phage particles, two possible sequences may follow: (1) a phage may be integrated into the bacterial chromosome as an inert prophage (thus lysogenizing the bacterial cell) or (2) the phage may produce the necessary enzymes to guide the production of the products needed for phage maturation and cell lysis. When wild-type phage particles are placed on a lawn of sensitive bacteria, clearings (plaques) appear where bacterial cells are infected and lysed, but these plaques are turbid because lysogenized bacteria (which are resistant to phage superinfection) grow within the plaques.

Mutant phages that form clear plaques can be selected as a source of phages that are unable to lysogenize cells. Such *clear* (*c*) mutants prove to be analogous to *I* and *O* mutants in *E. coli.* For example, conditional mutants for a site called *cI* are unable to establish a lysogenic state under restrictive conditions but also fail to induce lysis after superinfecting a cell that has been lysogenized by a wild-type prophage. Apparently, the *cI* mutation produces a defective repressor in the phage control system.

Virulent mutant λ phage have have been isolated that do not lysogenize cells but that do grow and enter the lytic cycle after superinfecting a lysogenized cell. Thus, these mutant phage are insensitive to the λ repressor. Genetic mapping has revealed *two* operators, designated O_L and O_R, located on the left and right of *cI*, respectively. When the λ phage infects a non-immune host, RNA polymerase begins transcription from the two pro-

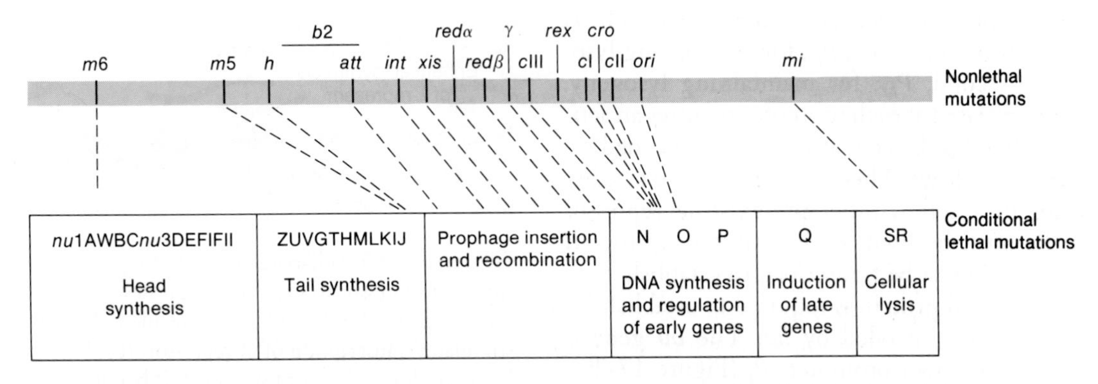

Figure 17-28 The genetic map of the λ phage. The positions of nonlethal and conditionally lethal mutations are indicated, and the characteristic clusters of genes with related functions are shown. (From A. Campbell, *The Episomes.* Harper & Row, 1969.)

somehow triggered by damaged DNA that blocks replication, a protease activity of the protein encoded by *recA* is activated. This results in the cleavage of the LexA repressor, allowing high levels of synthesis of the SOS functions. UV light and other agents that damage DNA induce the SOS functions. Interestingly, the RecA protease also cleaves the λ repressor (the product of the λcI gene), as well as the repressors of several other phages related to λ. This is the physical basis for the classical UV-light induction of λ lysogens.

Structure of Regulatory Proteins

Protein-sequence analyses and structural comparisons indicate that a number of DNA-binding regulatory proteins share important features. Many consist of a DNA-binding domain, located at one end of the protein, which protrudes from the main "core" of the protein. In certain cases, the core protein contains the inducer-binding site. This arrangement (Figure 17-32) holds for the Lac, λ cI, and λ Cro repressors, as well as for the CAP protein.

It has been postulated that protruding α helices fit into the major groove of the DNA. A model of the *lac* repressor suggests how such an operator-repressor complex might look (Figure 17-33). Here two α helices from the repressor protein interact with the two consecutive major grooves of the DNA of the operator site. The helices are connected by a turn in the protein secondary structure. This **helix-turn-helix motif** (Figure 17-34) is common to many regulatory proteins.

Striking partial-sequence homologies of the DNA binding domains of several regulatory proteins suggest that other such proteins bind to their own operator sites in a similar fashion. It is hoped that the determination of the three-dimensional structure of different repressor-operator complexes by X-ray crystallography will allow the elucidation of the rules for DNA sequence recognition by regulatory proteins.

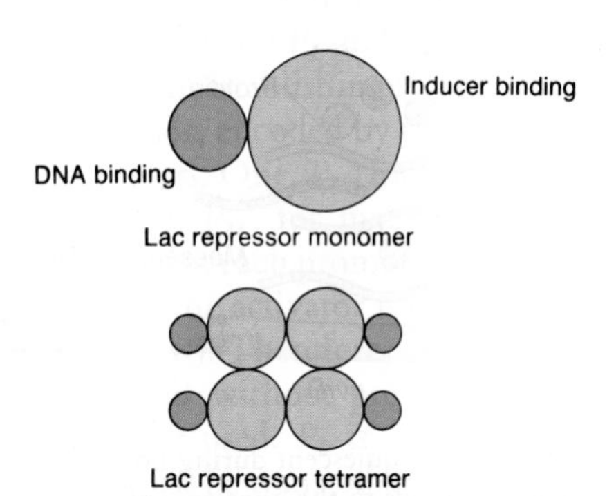

Figure 17-32 Schematic diagram showing the arrangement of domains in the Lac repressor. All mutations affecting DNA and operator binding result in alterations in the amino-terminal end of the protein, whereas mutants defective in inducer binding or aggregation result in alterations in the remaining portion of the protein.

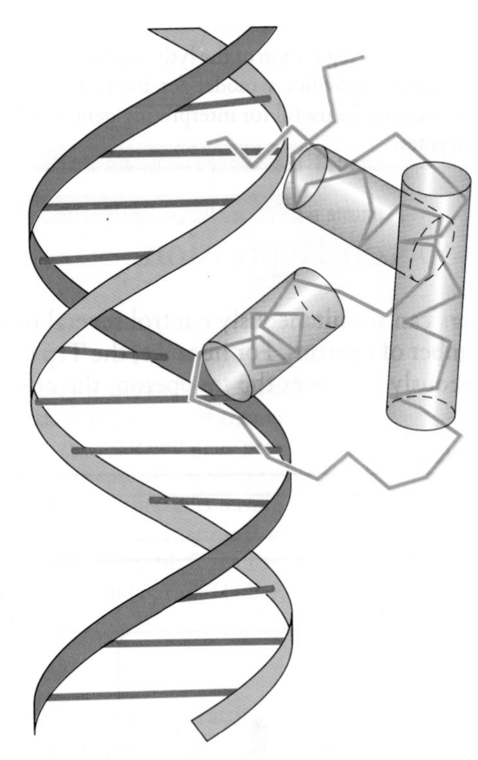

Figure 17-33 A model for the complex between the *lac* repressor and the *lac* operator derived from two-dimensional nuclear magnetic resonance studies. The amino terminus of the *lac* repressor is shown; cylinders represent α helices. (Adapted from R. Boelens et al. *Journal of Molecular Biology* 193, 1987, 213–216.)

Transcription: Gene Regulation in Eukaryotes — An Overview

There are several levels of control of gene expression in eukaryotes. We have already seen in Chapter 16 that chromosome structure can profoundly influence gene activity. Processing of the primary gene transcript (Chapter 13) provides an example of control at a post-transcriptional level. For instance, in the case of the calcitonin peptide in the thyroid and the CGRP peptide in the hypothalmus (Figure 13-37), alternative splicing pathways in two different organs generate different translation products from the same primary transcript.

What about control of the rate of transcription? What is known about the control signals for eukaryotic genes and about their specific regulation at the level of transcription? There are some similarities between tran-

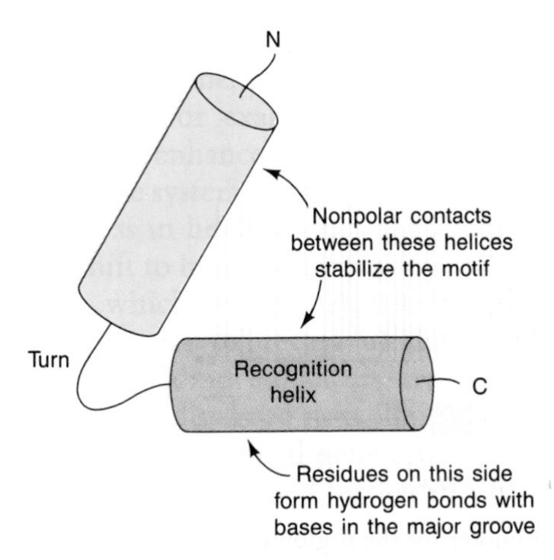

Figure 17-34 Helix-turn-helix motif of DNA binding proteins. Each monomer of these dimeric proteins contains a helix-turn-helix; the two units are separated by 34Å—the pitch of the DNA helix. (From L. Stryer, *Biochemistry,* 3d ed. Copyright 1988 by W. H. Freeman and Company.)

scriptional control circuits in prokaryotes and eukaryotes. In bacterial systems, positive and negative regulators serve as *trans*-acting factors that operate on *cis*-dominant regulatory sequences, such as operators, initiators, and binding sites within promoters. *Trans*-acting factors and *cis*-acting regulatory sequences have also been characterized in diverse eukaryotic systems. Moreover, there is a remarkable conservation of control mechanisms from one eukaryotic species to another.

Cis Control of Transcription

As mentioned in Chapter 13, there are three different RNA (ribonucleic acid) polymerases in eukaryotic systems. All mRNA molecules are synthesized by RNA polymerase II. In order to achieve maximal rates of transcription, RNA polymerase II requires two basic *cis*-acting control sequences: promoters, and additional sequence elements termed **enhancers** in mammalian cells and **upstream activating sequences (UASs)** in yeast. Each of these elements is recognized by *trans*-acting factors, which serve as positive control elements. Properly functioning promoters are required for transcription initiation; enhancers serve to maximize the rate of transcription for promoters. Enhancers and promoters were originally distinguished by the distance from the point of initiation at which they operate. Promoters function near the initiation site, whereas enhancers can usually function at great distances from this point. More recent work has shown that the internal structure of these elements is more similar than originally imagined.

Promoters

In the same manner that a comparison of promoter sequences in bacteria (Figure 13-8) led to an understanding of important elements in prokaryotic promoters, a compilation of DNA sequences upstream of mRNA start sites in eukaryotes has revealed conserved sequence elements. Figure 17-35 gives a schematic view of sequence elements that have been identified as part of eukaryotic promoters. The **TATA** box is involved in directing RNA polymerase to begin transcribing approximately 30 base pairs (bp) downstream in mammalian systems and 60 to 120 bp downstream in yeast. The TATA box works most efficiently together with two other upstream sequences located approximately 40 bp and 110 bp, respectively, from the start of transcription. The **CCAAT** box serves as one of these sequences, and a GC-rich segment often serves as the other sequence (Figure 17-35). Site-directed mutational analysis has demonstrated that altering bases in the TATA box or in the upstream sequences lowers in vivo transcription rates. In one version of this experiment by Richard Myers, Kit Tilly, and Tom Maniatis (Figure 17-36), a "saturation mutagenesis" in vivo alters, in series, each of the bases in the promoter for the β-globin gene. The effect on transcription rates is measured for changes at each base in the promoter sequence. Base changes outside the TATA box and the upstream sequences have no effect on levels of transcription, whereas an alteration in either of these elements severely lowers transcription rates.

Unlike prokaryotic promoters, eukaryotic promoters do not provide sufficient recognition signals for RNA polymerase to initiate transcription in vivo. The TATA box and the upstream sequences must each be recognized

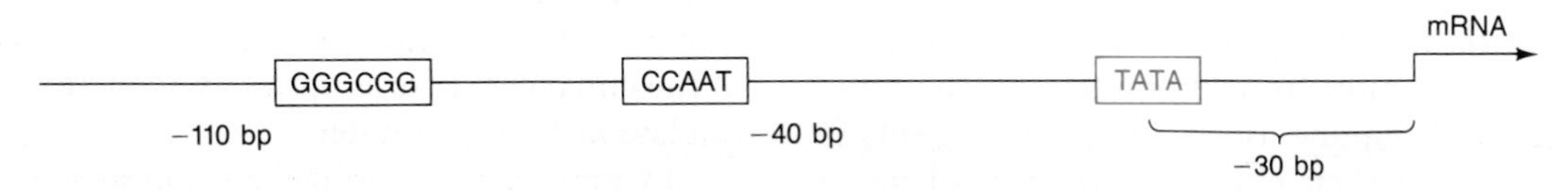

Figure 17-35 Promoter in higher eukaryotes. The TATA box is located approximately 30 base pairs (bp) from the mRNA start site. Usually, two or more upstream elements are found 40 to 110 bp upstream of the mRNA start site. The CCAAT box and the GC box are shown here. Other upstream elements include the sequences GCCACACCC and ATGCAAAT.

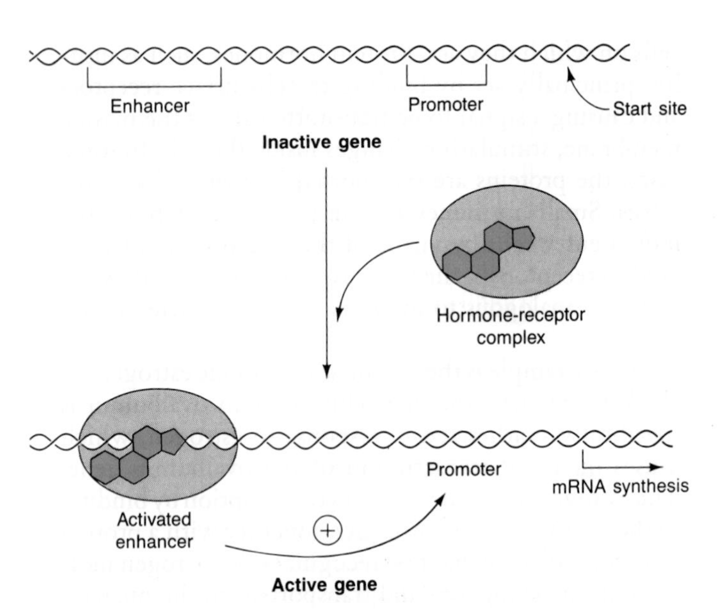

Figure 17-39 The action of steroid hormones at enhancer sequences. A glucocorticoid hormone (blue) binds to a soluble receptor protein. This complex, in turn, binds to enhancer sequences, such as the one found in the DNA of the mouse mammary tumor virus (MMTV), and enables them to stimulate the transcription of hormone-responsive genes. (From L. Stryer, *Biochemistry,* 3d ed. Copyright 1988 by W. H. Freeman and Company.)

Trans-Acting Transcription Factors

A large number of transcription factors have now been identified in eukaryotic cells. Some of these bind to enhancer regions, as is the case for the steroid-receptor complex just discussed. Others bind to *cis*-acting elements in the promoter. *Trans*-acting factors are required to allow RNA polymerase II to initiate transcription, and together with RNA polymerase they form a preinitiation complex, as pictured in Figure 17-40. Several proteins that recognize the CCAAT box have been identified from mammalian cells. These factors can distinguish between subunits of CCAAT elements. GC boxes are recognized by additional factors. For instance, the Spl protein recognizes the upstream GC boxes in the SV40 promoter (Figure 17-38). Factors that bind to TATA box sequences have been identified in *Drosophila* and yeast.

Two examples of *trans*-acting factors in yeast that operate on the enhancer-like upstream activating sequences (UASs) are the GCN4 and the GAL4 proteins. GCN4 activates the transcription of many genes involved in amino acid biosynthetic pathways. In response to amino acid starvation, GCN4 protein levels rise and, in turn, increase the levels of the amino acid biosynthetic genes. The UASs recognized by GCN4 contain the principal recognition sequence element ATGACTCAT.

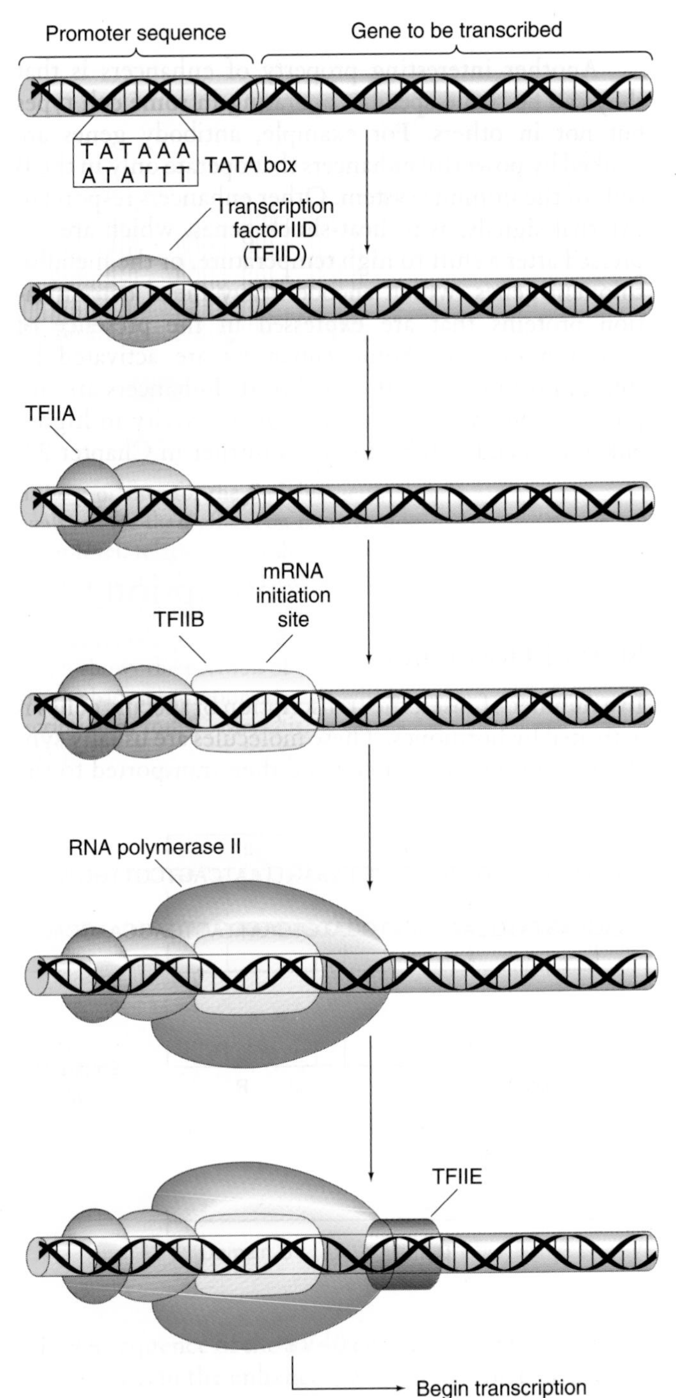

Figure 17-40 Formation of a transcription complex. Interactions among the TATA box and four transcription factors (TFIIA, B, D, and E) lead to the formation, with RNA polymerase II, of a preinitiation complex (PIC) required for transcription. (From W. K. Purves, G. L. Orians, and H. C. Heller, *Life, the Science of Biology,* 3d ed. Copyright 1992 by Sinauer Associates, Inc.)

GAL4 is involved in activating UASs for the promoters of genes involved in galactose metabolism. GAL4 binds to four sites with similar 17-bp sequences within these promoters.

Figure 17-41 Structural models for DNA binding domains. (a) The helix-turn-helix motif; (b) the zinc finger; (c) the leucine zipper; (d) the helix-loop-helix motif. (From J. D. Watson, M. Gilman, J. Witkowski, and M. Zoller, *Recombinant DNA,* 2d ed. Copyright 1992 by Scientific American Books.)

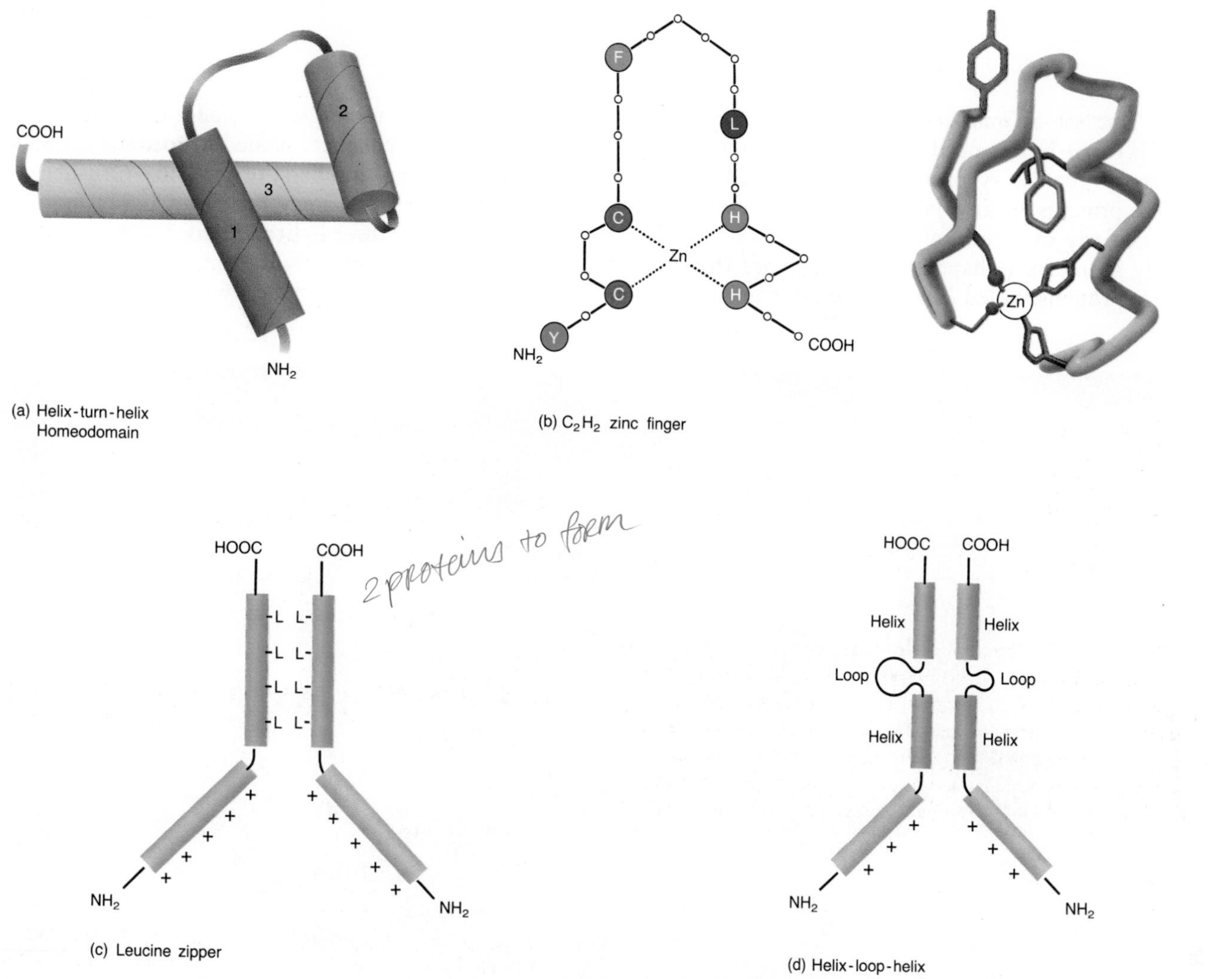

Domain Structure of Transcription Factors

Several interesting aspects of *trans*-acting proteins have emerged from recent studies. It is clear that many of these proteins have two separate domains: one of which recognizes and binds to *cis*-acting DNA sequences, and the second of which activates transcription. This can be seen in the GAL4 protein, which consists of 881 amino acids. A fragment of the protein containing only the first 147 amino acids can bind to DNA at specific UAS sites but cannot activate transcription. However, fragments containing later sequences in the protein cannot bind to DNA but can activate transcription when attached to fragments that bind DNA. Also, when the 144 carboxyl-terminal residues of GAL4 are fused to the DNA-binding segment of the human estrogen receptor, this hybrid protein can activate transcription at the binding site normally recognized by the estrogen receptor. This experiment shows not only that several different proteins have similar domain structures but also that the mechanisms of activation are similar in diverse systems. In a related experiment, a hybrid was constructed between two different steroid hormone–receptor molecules: the glucocorticoid receptor and the progesterone receptor. Here the DNA-binding region of the glucocorticoid receptor was substituted for the progesterone-receptor binding region, but the activating portion of the progesterone receptor was left intact. The fused protein responded by binding to glucocorticoid regulatory sequences and stimulating the transcription of the *cis*-linked genes—but only in response to progesterone.

Structural Motifs Found in Transcription Factors

DNA binding domains of eukaryotic transcription factors often fall into one of several structural families, as shown in Figure 17-41.

One set of *trans*-acting regulatory proteins has a **helix-turn-helix** motif similar to that found in many bacterial regulatory proteins (Figure 17-34). Many of these proteins are involved in *Drosophila* development, particularly in determining the spatial arrangement of body segments (Chapter 22). The genes that encode these proteins, called *homeotic proteins,* contain a similar sequence in the region that encodes the DNA binding domain. This sequence, named a **homeotic box,** encodes the recognition helix of the helix-turn-helix segment for these regulatory proteins. As Figure 17-41a shows, there are three helices. Helix 3 is the recognition helix and makes important contacts with the DNA. Helices 1 and 2 make contacts with other proteins.

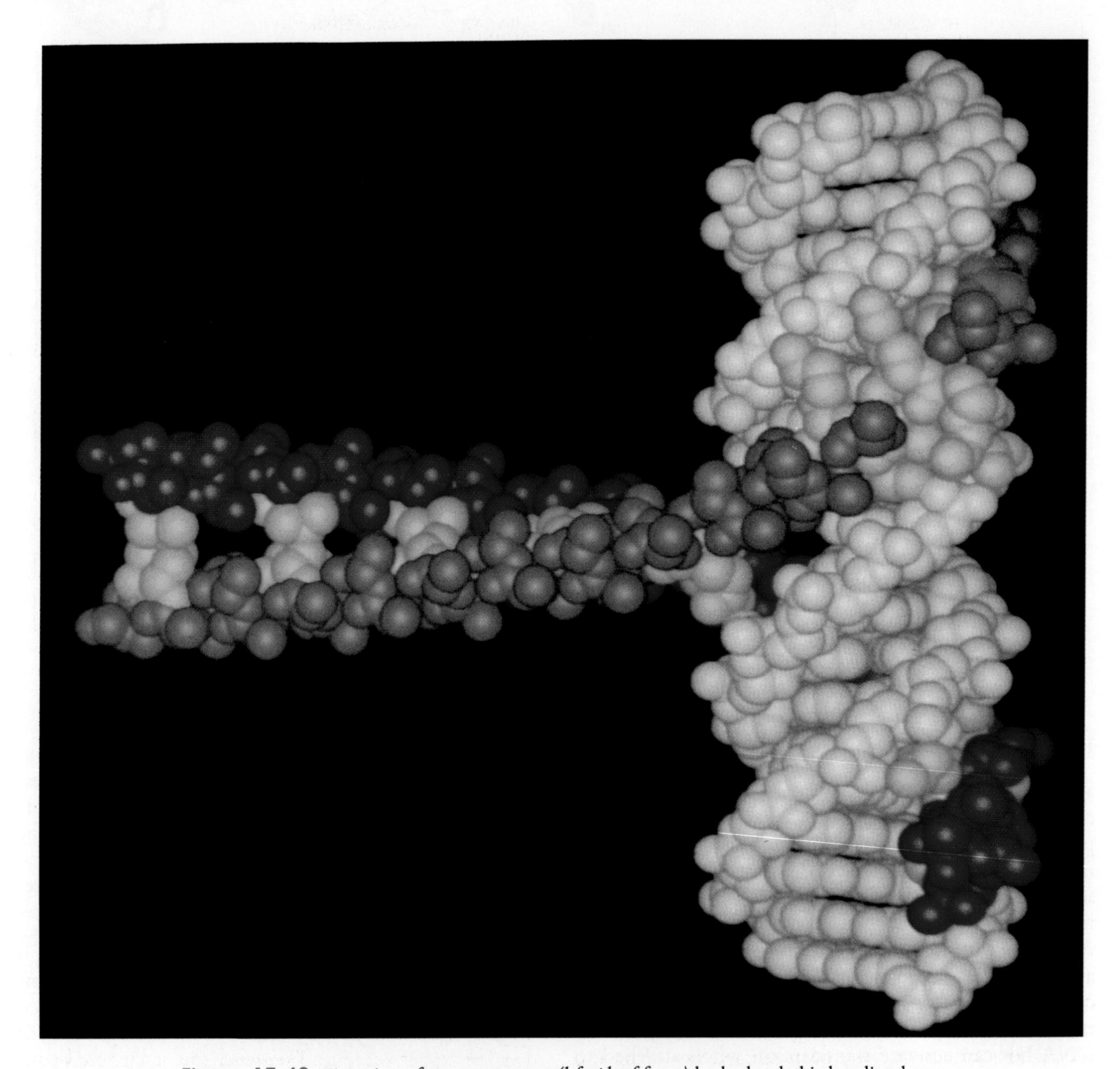

Figure 17-42 Zippering of two monomers *(left side of figure)* by hydrophobic bonding between their leucine residues (yellow spheres) can facilitate binding to the DNA molecule (white) and, thus, activation and silencing of genes. The yellow spheres in contact with the DNA represent not leucine but positively charged amino acids that may strengthen the binding of the protein to the DNA. The green cluster of atoms represents asparagine, which may help the DNA-binding segments, located in the major groove, to wrap around the DNA. (From S. L. McKnight, "Molecular Zippers in Gene Regulation," *Scientific American,* April 1991, pp. 53–64. Copyright 1991 by Scientific American, Inc. All rights reserved.)

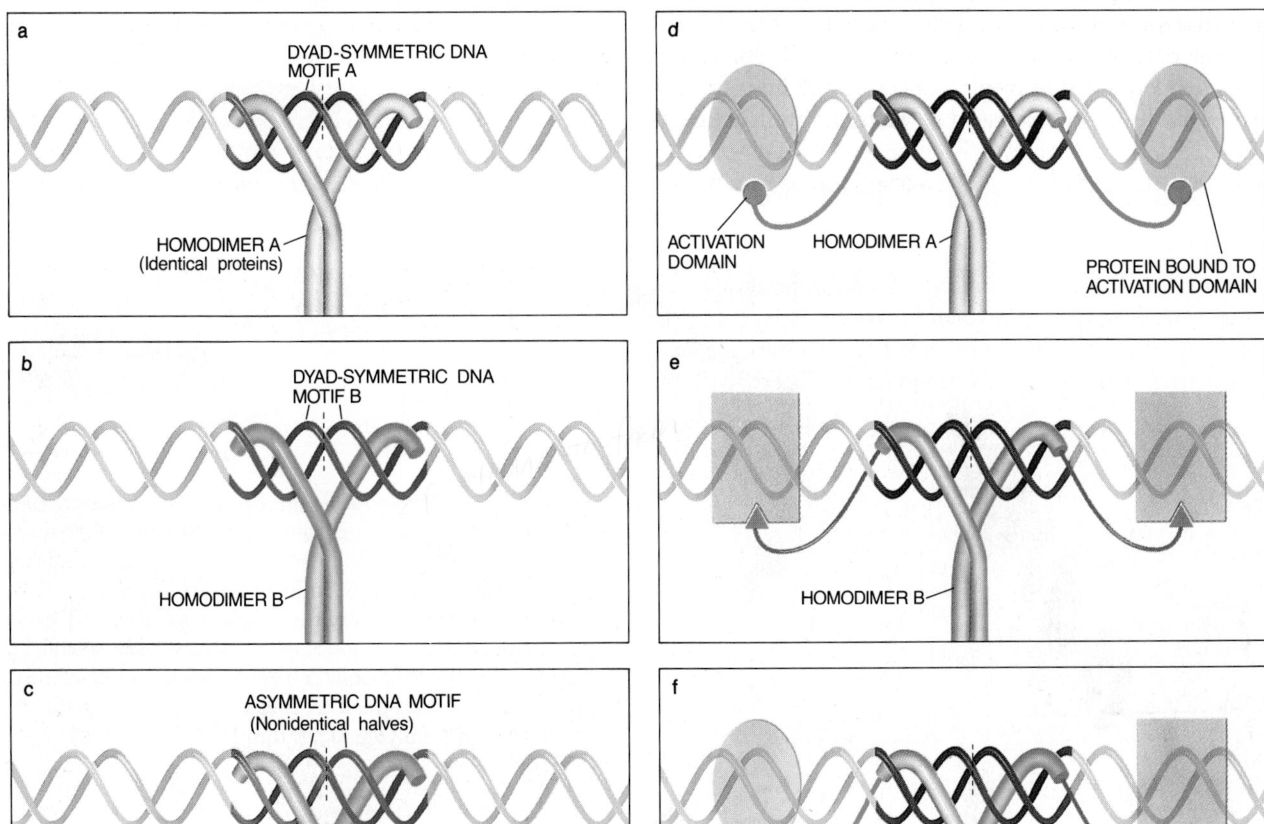

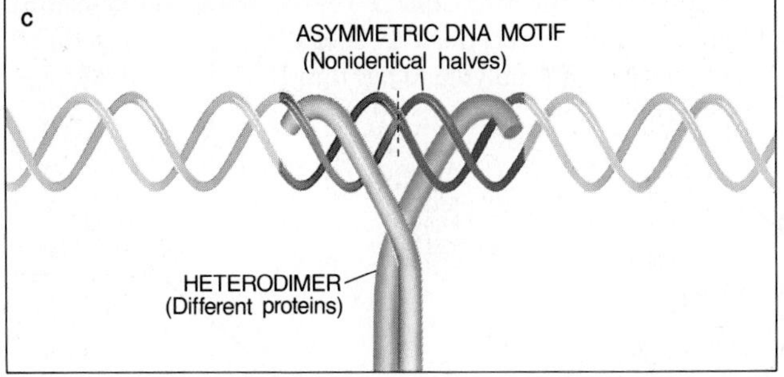

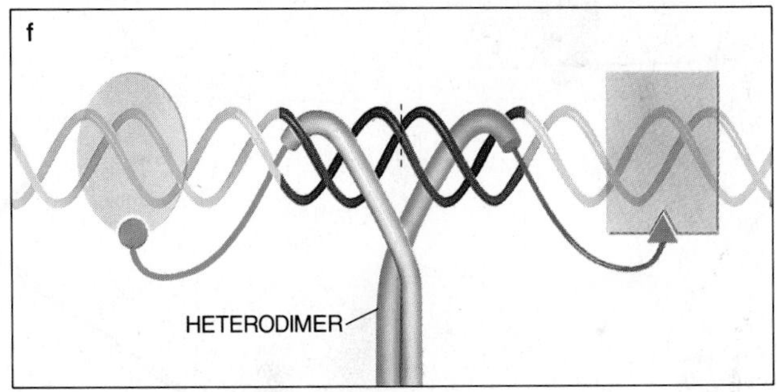

Figure 17-43 Zippered pairs of heterodimers have two possible advantages for an organism. (1) The subunits of homodimers tend to recognize only dyad-symmetric (twinlike) DNA sequences (panels a and b). (c) Heterodimers, the subunits of which may bind to different sequences, free the DNA from such constraints of symmetry and thus increase the number of sequence motifs available for gene regulation. (2) On the other hand, the increased diversity afforded by heterodimers could arise at the protein level. The subunits of homodimers (panels d and e) each bind identical activating proteins. (f) Heterodimers, however, can bind two different proteins, thus increasing the adaptive strategies for turning genes on and off. (From S. L. McKnight, "Molecular Zippers in Gene Regulation," *Scientific American,* April 1991, pp. 16–42. Copyright 1991 by Scientific American, Inc. All rights reserved.)

A second common structural motif arises from a cysteine- and histidine-rich region in a DNA binding domain that complexes zinc. The protrusions in the protein structure resemble fingers (Figure 17-41b); for this reason, the structures have been termed **zinc fingers.** Zinc fingers have been found in steroid receptor proteins and among transcription factors for many mRNA promoters.

Leucine zipper proteins can form dimers through a hydrophobic interface formed by leucine residues spaced 7 amino acids apart, as shown in Figure 17-41c. Flanking the leucine zipper is a DNA binding domain containing many lysine and arginine residues. Dimer formation is required for DNA binding. Figure 17-42 shows a leucine zipper protein binding to DNA. A number of protooncogenes (see Chapter 23), such as *c-jun* and *c-fos*, encode leucine zipper proteins. **Heterodimers,** zippered pairs of nonidentical proteins, also can form, and offer benefits to the organism by increasing the number of usable DNA sequences and the variety of protein combinations that can be used to turn proteins on and off, as illustrated by Figure 17-43. For instance, the heterodimers of c-Jun and c-Fos proteins bind DNA much more strongly than **homodimers** of c-Fos or c-Jun alone.

Helix-loop-helix proteins (Figure 17-41d) also form dimers, but not via a leucine zipper. The dimer interface consists of two helices linked by a loop. Some protooncogenes (for example, *c-myc*) encode helix-loop-helix proteins, as do genes involved in differentiation.

Mechanism of Action of Transcription Factors

Transcription factors that bind to DNA can be shown to increase transcription in vivo and in vitro. Figure 17-44 shows one type of experiment used to demonstrate increased transcription in vivo. Here, two different plas-

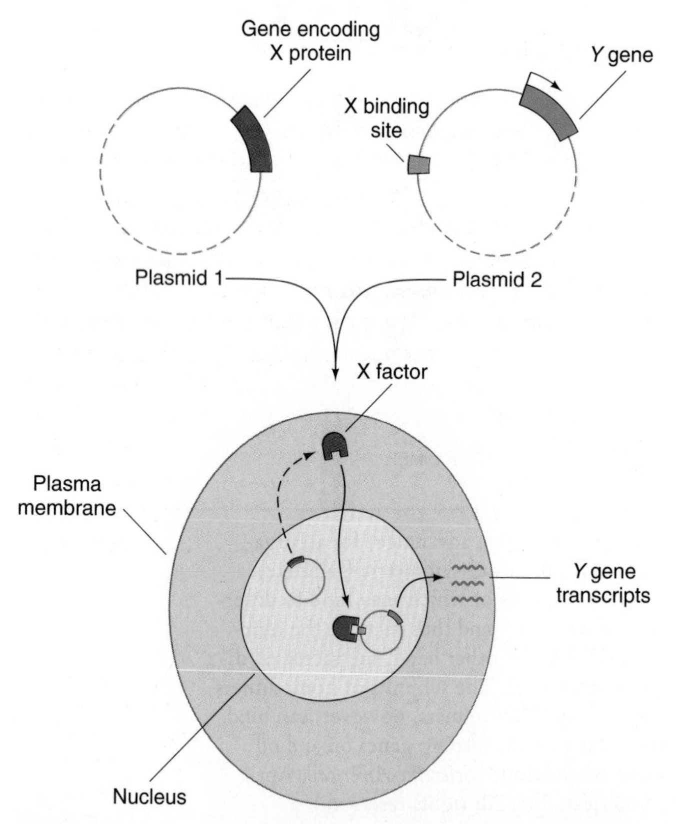

Figure 17-44 An in vivo test for transcription factor activity. A host cell receives two engineered plasmids. One contains a gene encoding protein X, the putative transcription factor being investigated, in a form which will allow expression of protein X at high levels. The second plasmid contains a test gene *Y* with a binding site for X. The host cell lacks both gene *Y* and the gene for protein X. Following introduction of the plasmids, the host cell is tested for the presence of transcripts for gene *Y*. If gene *Y* mRNA levels increase in the presence of X, the protein is considered a positively acting regulator; if levels go down, it is viewed as negatively acting. The use of plasmids that produce a mutated or rearranged factor can elucidate important domains in the protein. (From J. Darnell, H. Lodish, and D. Baltimore, *Molecular Cell Biology*, 2d ed. Copyright 1990 by Scientific American Books.)

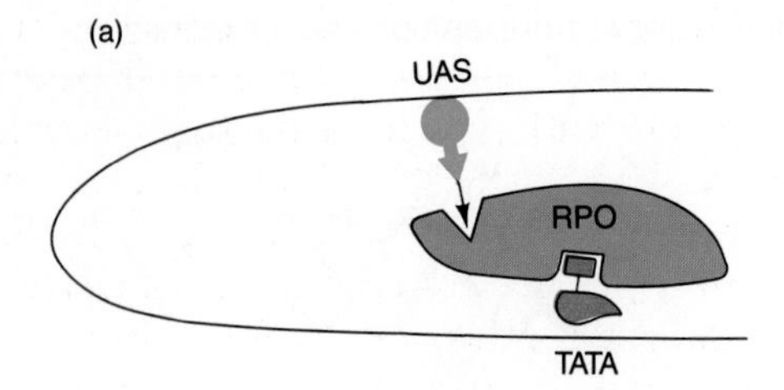

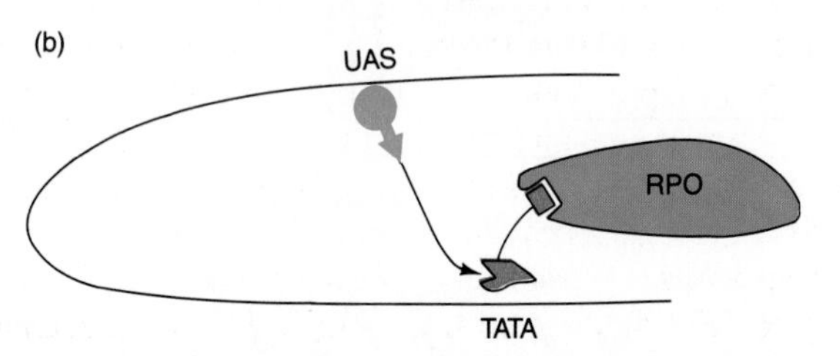

Figure 17-45 Two possible mechanisms of transcriptional activation in eukaryotes at enhancer (or UAS) sites. (a) An activator bound at a UAS (solid blue circle) and a TATA box factor (red) bound at a TATA box both contact RNA polymerase (RPO). Activation regions of the activator and the TATA box factor are drawn as a triangle and a rectangle, respectively. The DNA between the UAS and TATA box has been looped. (b) The activator contacts the TATA box factor, which in turn contacts the RNA polymerase. (From L. Guarente, *Cell* 52, 1988, 303–305.)

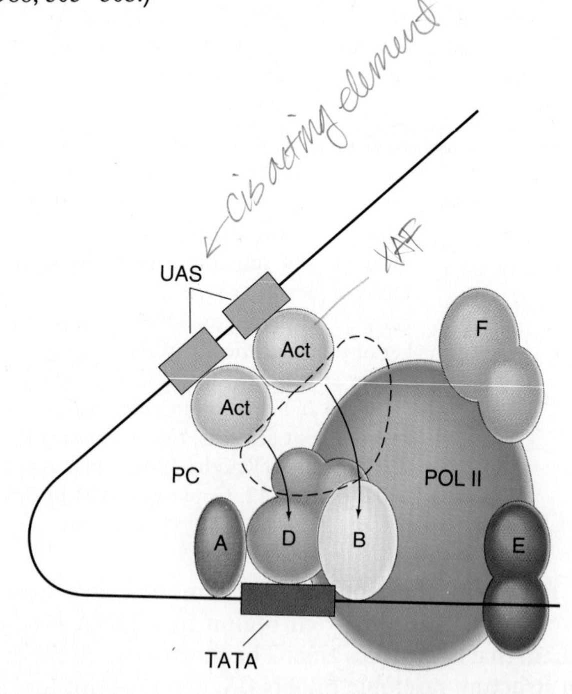

Figure 17-46 Active activator-potentiated preinitiation complex. Loop formation allowing activators (Act) to enhance transcription by interacting with transcription factors. The positioned activators are indicated by the area circumscribed by the dashed line, which indicates the possibility that additional protein factors may bind to TFIIB and TFIID. (From R. G. Roeder, "The Complexities of Eukaryotic Transcription Initiation: Regulation of Preinitiation Complex Assembly," TIBS 16, 1991, 402–408. Elsevier Science Publishers, U.K.)

mids are used. One plasmid encodes the *trans*-acting factor X, and the second contains the gene *Y* to be transcribed, together with the *cis*-controlling elements recognized by X. After introducing both plasmids into the same cell, transcripts from gene *Y* are measured. In vitro systems utilize purified protein factors and measure RNA polymerase transcription from DNA containing a gene with binding sites for the transcription factor.

Most models for the action of *trans*-acting factors at a distance from the genes they activate involve some type of DNA looping reminiscent of the model proposed for the bacterial *ara* operon (Figure 17-20). Figure 17-45 outlines two variations of such a model for upstream activating sequences, and Figure 17-46 details a model for activation of the preinitiation complex (PIC) (recall Figure 17-40). In this model multiply bound activators (Act) interact with transcription factors D and B (TFIID and TFIIB), stabilizing their binding to the complex. These interactions may require additional cofactors in some systems.

5-Methylcytosine Regulation

Genes which are transcriptionally inactive contain 5-methylcytosine (5-methyl-C) residues (Figure 17-47), usually as part of p5-methyl-CpG sequences, whereas genes which are active are free of 5-methyl-C. The methyl groups are added after replication of the DNA susceptible to methylation. Several percent of the cytosines in DNA from mammals are methylated. Although no direct experiments have proved that methylation controls transcription, the correlation between methylation and inactive transcription, together with the inherited tissue-specific pattern of methylation, points to a key role. Experiments with a cytosine analog, 5-azacytodine, which cannot be methylated at the C-5 position, reveal that the addition of 5-azacytodine can activate expression of certain previously unexpressed genes.

Post-Transcriptional Control

Differential Processing

We saw in Chapter 13 how many eukaryotic transcripts are processed. Poly(A) tails are added, and the primary transcript is spliced to yield the final mRNA. Figure 13-37 details how the same transcript can yield two different mRNAs, depending on which of several different splicing pathways is utilized. Thus, from the same transcript one splicing pathway results in the production of the calcitonin peptide in the thyroid, whereas a second splicing pathway ultimately generates the CGRP peptide in the hypothalamus. Differential poly(A) addition can also result in different mRNAs from the same primary transcript. Table 17-4 provides some interesting examples of differential processing of pre-mRNA.

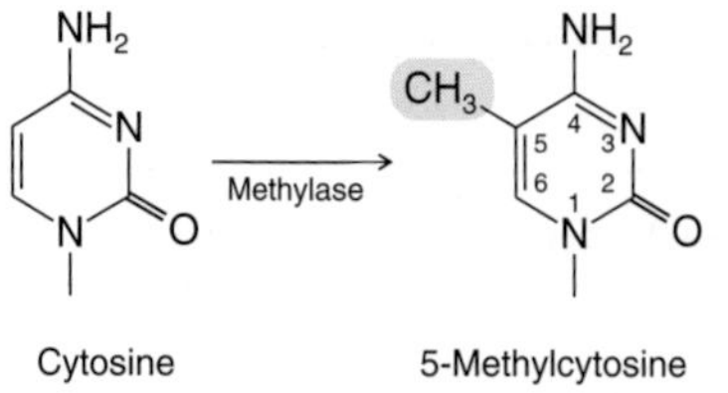

Figure 17-47 Cytosine can be converted to 5-methylcytosine by methylases.

Control of mRNA Degradation

Clearly the concentration of a given mRNA in the cytoplasm depends not only on its rate of synthesis, but also on its stability. By controlling the rate of mRNA turnover, cells have an additional opportunity to fine-tune mRNA levels. Poly(A) tails help to stabilize mRNAs. Without poly(A) tails, messengers are rapidly degraded. Other factors affecting mRNA stability in specific cases include hormones, low-molecular-weight compounds, and specific sequences at the untranslated 3′ end of the mRNA. For example, one class of mRNAs with short half-lives contain repeats of sequences of the form AUUUA at their 3′ ends, as shown in Figure 17-48a. Transplanting one of these sequences into the more stable β-globin gene reduces the half-life of the resulting mRNA from more than 10 hours to less than 2 hours, as seen in Figure 17-48b.

Message Eukaryotic genes are activated by two *cis*-acting control sequences—promoters and enhancers—which are recognized by *trans*-acting proteins. The *trans*-acting proteins allow the RNA polymerse to initiate transcription and to achieve maximal rates of transcription.

Control of Ubiquitous Molecules in Eukaryotic Cells

Several classes of molecules are found in abundance in every eukaryotic cell: histones, the translational apparatus, membrane components, and so on. Several strategies have evolved to maintain a sufficient amount of these products in the eukaryotic cell. These strategies include continuous and repeated transcription throughout the cell cycle, repetition of genes, and extrachromosomal amplification of specific gene sequences.

Table 17-4. Examples of Differential Processing of Pre-mRNA

Gene Products	Product Differences	Basis of Differential Control	
		Poly(A) Choice	Splicing Variation
Mammals:			
Calcitonin and CGRP (rat)	Two proteins, calcitonin (Ca^{2+}-regulating hormone) and calcitonin gene-related peptide (CGRP), which probably functions in taste, are formed preferentially in thyroid and brain, respectively.	+	+
Kininogens (cow)	Two prehormones which both yield bradykinin, which controls blood pressure, differ in their protease susceptibility.	+	
Immunoglobulin heavy chains (mouse)	These differ at their carboxyl ends in different antibody molecules.	+	
Troponin (rat)	Two different muscle proteins appear in different rat skeletal muscles.		+
Myosin light chains (rat)*	Two proteins, 140 and 180 amino acids long, respectively, appear in different "fast twitch" muscles.		+
Preprotachykinin (cow)	Two prehormones that release substance P, a neuropeptide that affects smooth muscle, are formed; the larger prehormone also releases substance K, a neuropeptide of unknown function.		+
Fibronectin (rat)	The structural protein has at least six forms.		+
Drosophila:			
Tropomyosin	Two muscle proteins differ between embryos and adults.		+
Myosin heavy chains	Two muscle proteins differ between embryos and adults; a third form is found in pupae and adults.	+	
Ultrabithorax	Different gene products that control thoracic segment development are found in early embryos and larvae.	+	
Glycinamide ribotide transformylase	This enzyme is encoded by a long mRNA; a shorter mRNA that shares several exons encodes a second polypeptide.	+	

* The rat myosin light-chain proteins are processed from two different overlapping primary transcripts with different 5′ ends. The resulting mRNAs also contain different exons within the region of overlap, plus four exons that are similar. An almost-identical situation exists in chicken myosin genes.

SOURCE: J. Darnell, H. Lodish, and D. Baltimore, *Molecular Cell Biology*, 2d. ed. Copyright 1990 by Scientific American Books, Inc.

Genetic Redundancy

Genes coding for rRNA and tRNA have been extensively analyzed because the availability of purified gene products provides a probe with which to recover complementary DNA. Simple DNA-RNA hybridization shows that each *Xenopus* chromosome carrying a nucleolus organizer (NO) has 450 copies of the DNA coding for 18S and 28S rRNA. In contrast, there are 20,000 copies in each nucleus of the genes coding for 5S rRNA, and these genes are not located in the NO region. Donald Brown and his collaborators during the 1970s analyzed the NO DNA extensively and showed that it carries tandem duplications of a large (40S) transcription unit

(a) 3' Sequences of unstable mRNAs

GMCSF	UAAUAUUUAUAUAUUUAUAUUUUUAAAAUAUUUAUUUAUUUAUUUAUUUAA
IFNβ	UUUUGAAAUUUUUAUUAAAUUAUGAGUUAUUUUUAUUUAUUUAAAUUUUAUUUU
IL-1	UUAUUUUUUAAUUAUUAUUUAUAUAUGUAUUUAUAAAUAUAUUUAAGAUAAU
TNF	AUUAUUUAUUAUUUAUUUAUUAUUUAUUUAUUUA
c-Fos	GUUUUUAAUUUAUUUAUUAAGAUGGAUUCUCAGAUAUUUAUAUUUUUAUUUU
c-Myc	UAAUUUUUUUUAUUUAAGUACAUUUUGCUUUUUAAAGUUGAUUUUUUUCU

(b) Activity of $A(U)_nA$ repeats

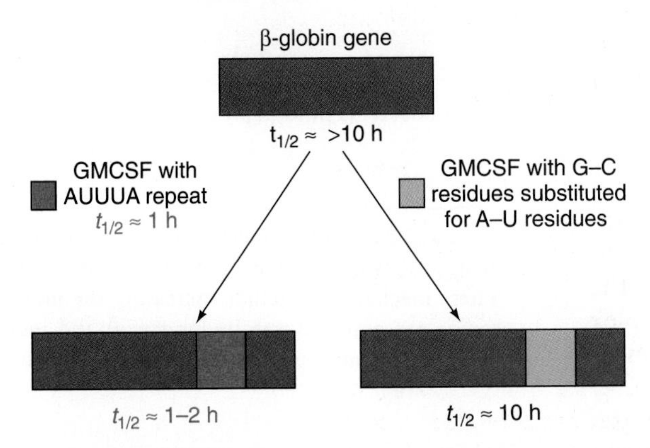

Figure 17-48 Effect of a 3′ untranslated sequence on mRNA stability. (a) Sequences of 3′ untranslated regions of mRNAs encoding human granulocyte-monocyte–stimulating factor (GMCSF), fibroblast interferon *β* (IFN*β*), interleukin 1 (IL-1), tumor necrosis factor (TNF), and two protooncogene proteins (c-Fos and c-Myc). Characteristic $A(U)_nA$ sequences are shown in red; some of these are overlapping. Numerous other protooncogene and cell growth factor mRNAs contain these sequences. (b) Effect of $A(U)_nA$ sequences on mRNA half-life. Cultured cells were transfected with recombinant *β*-globin DNA containing a 62-bp segment of GMCSF DNA encoding either the $A(U)_nA$-rich segment shown in part a or an altered segment in which many A–U residues were replaced with G–C residues. The destabilizing effect of the $A(U)_nA$ sequences was clearly demonstrated by the decrease in half-life ($t_{1/2}$) of the corresponding mRNAs. (See G. Shaw and R. Kamen, *Cell* 46, 1986, 659. From J. Darnell, H. Lodish, and D. Baltimore, *Molecular Cell Biology,* 2d ed. Copyright 1990 by Scientific American Books.)

separated from its neighbors by nontranscribed spacers. The initial transcript is then processed to release the smaller rRNAs finally found in the ribosome.

Max Birnstiel and his associates have shown that the five histone genes in sea urchins also are found as a tightly linked unit (see Figure 16-21). There are several hundred copies of a repeating unit consisting of the genes in the sequence *H4–H2B–H3–H2A–H1*. Transcription begins at *H4* and ends at *H1*; each gene is transcribed as a single message. As we have seen in earlier chapters, tandem duplications can pair asymmetrically between homologous chromosomes or within the same chromosome; exchanges then increase or reduce the number of copies. It remains to be seen how (or whether) the number of tandem repeats is kept constant in a particular species.

Gene Amplification

Specific gene amplification has been studied extensively in the case of the rRNA genes in amphibian oocytes. As noted, each chromosome of *Xenopus* carries about 450 copies of the DNA coding for 18S and 28S rRNA. However, the oocyte contains up to 1000 times this number of copies of these genes. The oocyte is a very specialized cell

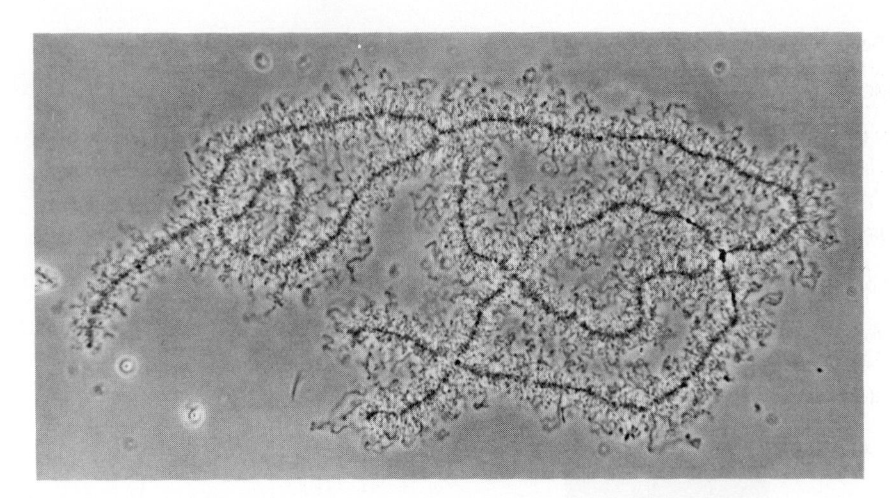

Figure 17-49 The lampbrush chromosomes of an amphibian oocyte. (Photograph courtesy of J. Gall.)

that is loaded with the nutritive material needed to maintain the embryo until the tadpole stage without any ingestion of food. The cytoplasm also is prepared with the translational apparatus needed to carry out the complex program of differentiation in the many cells of the developing embryo.

During oogenesis, the maturing oocyte increases greatly in size as material is poured in from adjacent nurse cells. (Amazingly, so many ribosomes are present in the egg at fertilization that *an an* homozygotes carrying no genes for 18S and 28S rRNA will develop and differentiate to the twitching tadpole stage before death.) The oocyte nucleus also contributes material as the cell proceeds to the first meiotic prophase, where further development is arrested. The amphibian meiotic chromosomes are enormous, with numerous lateral loops representing regions of intense genetic activity (Figure 17-49). In the oocyte nucleus, there are hundreds of extrachromosomal nucleoli of varying size. Each nucleolus contains a different-sized ring of rDNA that has been replicated and released from the chromosome. The steps involved in this process of DNA amplification are completely unknown. The DNA rings actively produce rRNA that is assembled into ribosomes, which are stored in the nucleolus until their release during meiosis.

Other examples of genetic amplification are known. Specific puffs of the polytene chromosomes in dipterans are found to produce excess DNA rather than mRNA. Also, George Rudkin showed in the early 1960s that the polytene chromosomes from *Drosophila* salivary glands themselves may represent specific amplifications of euchromatic DNA about a thousandfold, whereas the proximal heterochromatic regions may be replicated only a few times. Thus, in addition to the mechanism for replicating chromosomal DNA normally for cell division, there exists a means for amplifying specific loci and not others.

Message Cellular mechanisms exist to ensure an adequate supply of the gene products that are vital to all cells. These mechanisms include increasing the number of gene copies per chromosome (redundancy) and increasing the number of gene copies in the cell (amplification).

Summary

The operon model explains how prokaryotic genes are controlled through a mechanism that coordinates the activity of a number of related genes. In negative control, the initiation of transcription is controlled at the operator by a repressor with binding affinities to the operator that may be altered by inducer molecules. Inactivation of the repressor — the negative control element — is required for active transcription. In positive control, transcription initiation requires the activation of a factor. Sometimes one control system is superimposed on another. For instance, superimposed on the repressor-operator system for the *lac* operon is the cAMP-CAP positive control system. Modulation of transcription by an attenuator sequence supplements the repressor-operator control in the *trp* operon. In phage λ, multiple binding sites for repressors, as well as repressors of repressors, add additional levels of control.

A major problem in transcriptional control in eukaryotes is understanding how thousands of promoters can be regulated to yield desired levels of mRNA. It is now clear that promoters are governed both by the number and type of *cis*-acting control elements and by the action of regulatory proteins that recognize these elements. Also modulating the levels of transcription are enhancers — *cis*-acting sequences that, when recognized by functional regulatory proteins, can greatly increase the activity of promoters. In some aspects, transcriptional control in eukaryotes is similar to that found in prokaryotes: namely, *trans*-acting factors recognize *cis*-acting (or *cis*-dominant) sites in both cases. The mechanism of action may even be similar in many cases, as evidenced by the looping out of DNA in the *ara* system in bacteria and by enhancer action in eukaryotes. Also, structural motifs of some regulatory proteins share a common thread as in the case of the helix-turn-helix motif in bacterial regulatory proteins and in the homeotic proteins in *Drosophila.* However, other structural motifs appear to be unique to classes in eukaryotic regulatory proteins, as exemplified by the zinc finger and leucine zipper motifs for many *trans*-acting factors. Eukaryotic mRNA levels are controlled in several different ways, including synthesis and degradation.

Concept Map

Draw a concept map interrelating as many of the following terms as possible. Note that the terms are listed in no particular order.

environment / promoter / operator / gene / operon / RNA polymerase / mRNA / enhancer / *trans*-acting factors / regulation

Chapter Integration Problem

In Chapter 10 we learned how Hfr strains transfer a segment of the chromosome into a recipient strain, and in this chapter we saw how Jacob, Pardee, and Monod initially exploited this phenomenon to create temporary partial diploids. Explain what happens to *lacZ* expression when an Hfr transfers $I^+ Z^+$ genes into a recipient which is $I^- Z^-$, under conditions which prevent protein synthesis in the donor strain. How would this change if we use a recipient that is $I^+ Z^-$ and transfer the same $I^+ Z^+$ genes from the Hfr? Draw a diagram showing the results of these two experiments.

Solution

As shown in Figure 17-6, the introduction of $I^+ Z^+$ genes into the cytoplasm of the cells containing no repressor results in an initial burst of β-galactosidase synthesis, which lasts until enough repressor can build up in the cytoplasm to block further β-galactosidase synthesis (see the diagram that follows). In the case of a recipient that is $I^+ Z^-$, there is already repressor in the cytoplasm, so there is no initial burst of β-galactosidase synthesis:

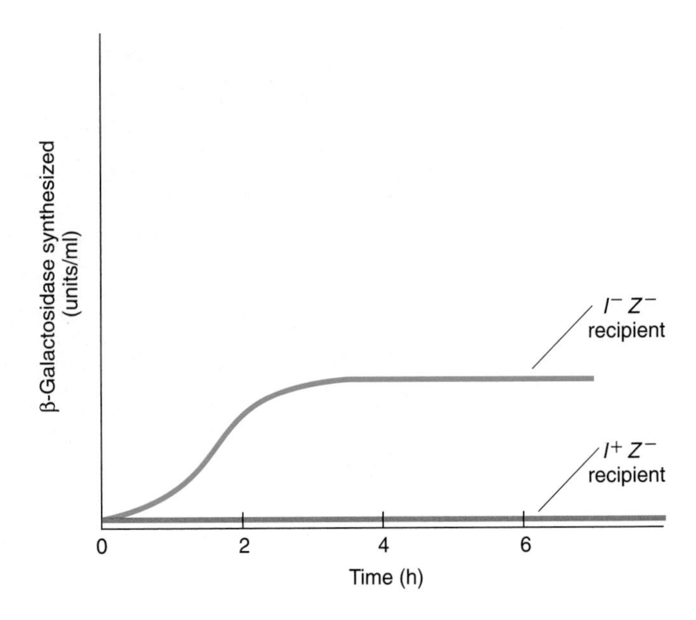

Solved Problems

This set of four solved problems, which are similar to Problem 4 at the end of this chapter, are designed to test understanding of the operon model. Here we are given several diploids and are asked to determine whether *Z* and *Y* gene products are made in the presence or absence of an inducer. Use a table similar to the one in Problem 4 as a basis for your answers.

1. $\dfrac{I^- P^- O^c Z^+ Y^+}{I^+ P^+ O^+ Z^- Y^-}$

Solution

One way to approach these problems is first to consider each chromosome separately and then to construct a diagram. The following figure diagrams this diploid:

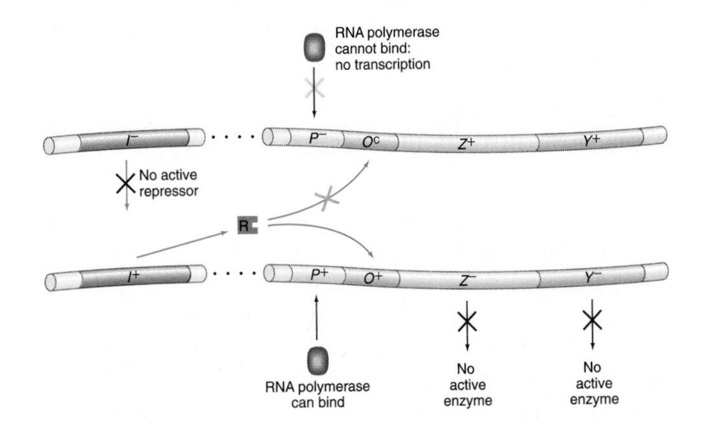

The first chromosome is P^-, so no *lac* enzyme can be synthesized off it. The second chromosome (O^+) can be transcribed, and this transcription is repressible. However, the structural genes linked to the good promoter are defective; thus, no active Z product or Y product can be produced. The symbols to add to your table are "$----$."

2. $\dfrac{I^+P^-O^+Z^+Y^+}{I^-P^+O^+Z^+Y^-}$

Solution

The first chromosome is P^-, so no enzyme can be synthesized off it:

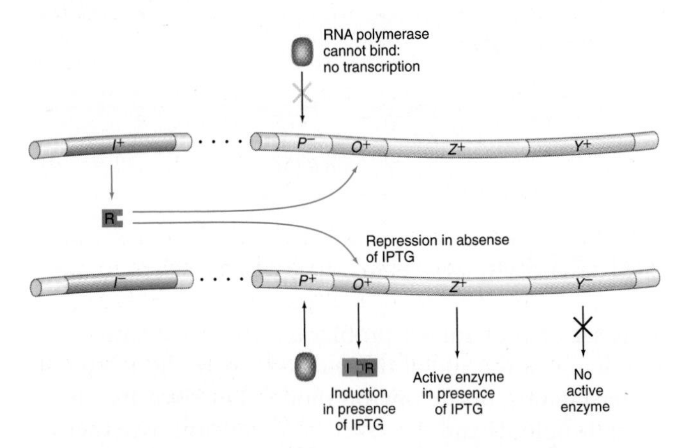

The second chromosome is O^+, so transcription will be repressed by the repressor supplied from the first chromosome, which can act *in trans* through the cytoplasm. However, only the Z gene from this chromosome is intact. Therefore, in the absence of an inducer, no enzyme will be made; in the presence of an inducer, only the Z gene product β-galactosidase will be produced. The symbols to add to the table are "$-+--$."

3. $\dfrac{I^+P^+O^cZ^-Y^+}{I^+P^-O^+Z^+Y^-}$

Solution

Because the second chromosome is P^-, we need only consider the first chromosome:

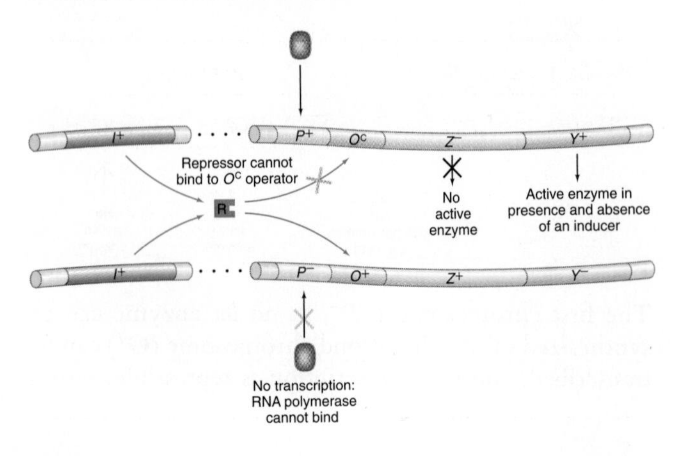

This chromosome is O^c, so that enzyme is made in the absence of an inducer, although only the Y gene is active. The entries in the table should be "$--++$."

4. $\dfrac{I^sP^+O^+Z^+Y^-}{I^-P^+O^cZ^-Y^+}$

Solution

In the presence of an I^s repressor, all wild-type operators will be shut off, both with and without an inducer:

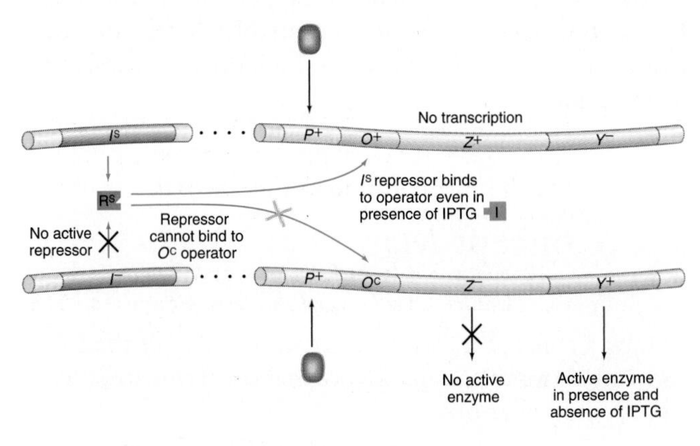

Therefore, the first chromosome will be unable to produce any enzyme. However, the second chromosome has an altered (O^c) operator and can produce enzyme in both the absence and presence of an inducer. Only the Y gene is active on this chromosome, so the entries in the table should be "$--++$."

Problems

1. Explain why I^- mutations in the *lac* system are normally recessive to I^+ mutations and why I^+ mutations are recessive to I^s mutations.

2. What do we mean when we say that O^c mutations in the *lac* system are *cis*-dominant?

3. The genes shown in Table 17-5 are from the *lac* operon system of *E. coli.* The symbols *a*, *b*, and *c* represent the repressor (I) gene, the operator (O) region, and the structural gene (Z) for β-galactosidase, although not necessarily in that order. Furthermore, the order in which the symbols are written in the genotypes is not necessarily the actual sequence in the *lac* operon.

a. State which symbol (*a*, *b*, or *c*) represents each of the *lac* genes I, O, and Z.

b. In Table 17-5, a superscript minus sign on a gene symbol merely indicates a mutant, but you know there are some mutant behaviors in

Table 17-5.

Genotype	Activity (+) or Inactivity (−) of *z* Gene: Inducer Absent	Inducer Present
$a^- b^+ c^+$	+	+
$a^+ b^+ c^-$	+	+
$a^+ b^- c^-$	−	−
$a^+ b^- c^+ / a^- b^+ c^-$	+	+
$a^+ b^+ c^+ / a^- b^- c^-$	−	+
$a^+ b^+ c^- / a^- b^- c^+$	−	+
$a^- b^+ c^+ / a^+ b^- c^-$	+	+

this system that are given special mutant designations. Use the conventional gene symbols for the *lac* operon to designate each genotype in the table.

(Problem 3 is from J. Kuspira and G. W. Walker, *Genetics: Questions and Problems.* Copyright 1973 by McGraw-Hill.)

4. The map of the *lac* operon is

I POZY

The promoter (*P*) region is the start site of transcription through the binding of the RNA polymerase molecule before actual mRNA production. Mutationally altered promoters (P^-) apparently cannot bind the RNA polymerase molecule. Certain predictions can be made about the effect of P^- mutations. Use your predictions and your knowledge of the lactose system to complete Table 17-6. Insert a "+" where an enzyme is produced and a "−" where no enzyme is produced.

5. In a haploid eukaryotic organism, you are studying two enzymes that perform sequential conversions of a nutrient A supplied in the medium:

$$A \xrightarrow[E_1]{} B \xrightarrow[E_2]{} C$$

Treatment of cells with mutagen produces three different mutant types with respect to these functions. Mutants of type 1 show no E_1 function; all type 1 mutations map to a single locus on linkage group II. Mutations of type 2 show no E_2 function; all type 2 mutations map to a single locus on linkage group VIII. Mutants of type 3 show no E_1 or E_2 function; all type 3 mutants map to a single locus on linkage group I.

a. Compare this system with the *lac* operon of *E. coli,* pointing out the similarities and the differences. (Be sure to account for each mutant type at the molecular level.)

b. If you were to intensify the mutant hunt, would you expect to find any other mutant types on the basis of your model? Explain.

6. In *Neurospora,* all mutants affecting the enzymes carbamyl phosphate synthetase and aspartate transcarbamylase map at the *pyr-3* locus. If you induce *pyr-3* mutations by ICR-170 (a chemical mutagen), you find that either both enzyme functions are lacking or only the transcarbamylase function is lacking; in no case is the synthetase activity lacking when the transcarbamylase activity is present. (ICR-170 is assumed to induce frameshifts.) Interpret these results in terms of a possible operon.

7. In 1972, Suzanne Bourgeois and Alan Jobe showed that a derivative of lactose, allolactose, is the true natural inducer of the *lac* operon, rather than lactose itself. Lactose is converted to allolactose by the enzyme β-galactosidase. How does this result explain the early finding that many Z^- mutations,

Table 17-6.

Part	Genotype	β-Galactosidase: No Lactose	β-Galactosidase: Lactose	Permease: No Lactose	Permease: Lactose
Example	$I^+ P^+ O^+ Z^+ Y^+ / I^+ P^+ O^+ Z^+ Y^+$	−	+	−	+
(a)	$I^- P^+ O^c Z^+ Y^- / I^+ P^+ O^+ Z^- Y^+$				
(b)	$I^+ P^- O^c Z^- Y^+ / I^- P^+ O^c Z^+ Y^-$				
(c)	$I^s P^+ O^+ Z^+ Y^- / I^+ P^+ O^+ Z^- Y^+$				
(d)	$I^s P^+ O^+ Z^+ Y^+ / I^- P^+ O^+ Z^+ Y^+$				
(e)	$I^- P^+ O^c Z^+ Y^- / I^- P^+ O^+ Z^- Y^+$				
(f)	$I^- P^- O^+ Z^+ Y^+ / I^- P^+ O^c Z^+ Y^-$				
(g)	$I^+ P^+ O^+ Z^- Y^+ / I^- P^+ O^+ Z^+ Y^-$				

which are not polar, still do not allow the induction of *lac* permease and transacetylase by lactose?

8. Certain *lacI* mutations eliminate operator binding by the *lac* repressor but do not affect the aggregation of subunits to make a tetramer, the active form of the repressor. These mutations are partially dominant to wild-type. Can you explain the partially I^- phenotype of the $I^- I^+$ heterodiploids?

9. Explain the fundamental differences between negative control and positive control.

10. Mutants that are *lacY*$^-$ retain the capacity to synthesize β-galactosidase. However, even though the *lacI* gene is still intact, β-galactosidase can no longer be induced by adding lactose to the medium. How can you explain this?

11. What analogies can you draw between transcriptional *trans*-acting factors that activate gene expression in eukaryotes and prokaryotes? Give an example.

12. Compare the arrangement of *cis*-acting sites in the control regions of eukaryotes and prokaryotes.

13. Explain how models for bacterial operons such as *ara* relate to eukaryotic *trans*-acting proteins and their mechanism of action.

14. It is now known that the Lac repressor of *E. coli* has a leucine zipper at the carboxyl-terminal region of the protein which is required to allow dimers to associate into tetramers. It is also known that a weak operator sequence, O_2, exists early in the *Z* gene, to which the repressor can also bind, in addition to the normal O_1. The elements are as shown here:

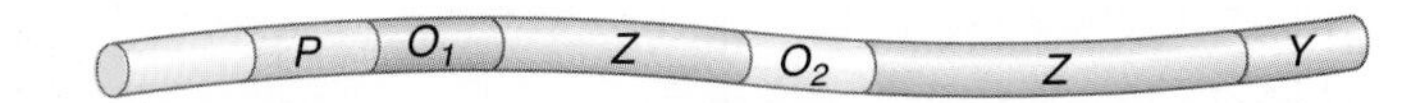

Using Figures 17-20 and 17-32 for reference, devise a model which explains why mutations in *lacI* that eliminate the leucine zipper reduce the ability of the repressor to block *lac* operon transcription completely. Draw a diagram.

15. One interesting mutation in *lacI* results in repressors with 100-fold increased binding to both operator and nonoperator DNA. These repressors display a "reverse" induction curve, allowing β-galactosidase synthesis in the *absence* of inducer (IPTG), but partially repressing β-galactosidase expression in the *presence* of IPTG. How can you explain this? Note that when IPTG binds repressor, it does not completely destroy operator affinity, but rather it reduces affinity 1,000-fold. Also, as cells divide and new operators are generated by the synthesis of daughter strands, repressor must find the new operators by searching along the DNA, rapidly binding to and dissociating from nonoperator sequences.

18

Mechanisms of Genetic Change I: Gene Mutation

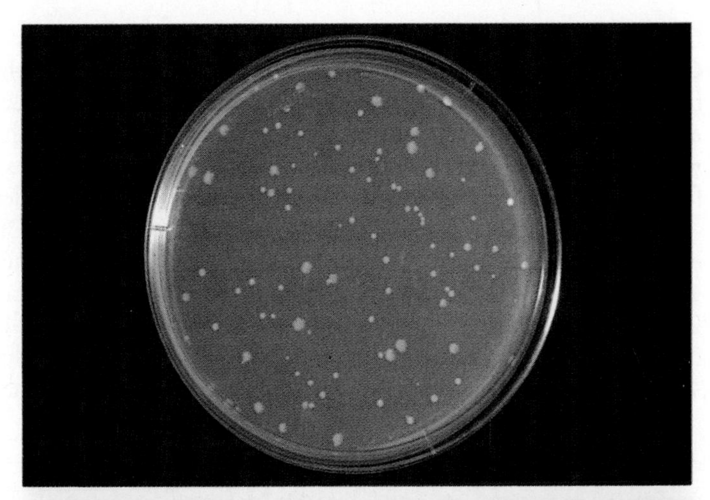

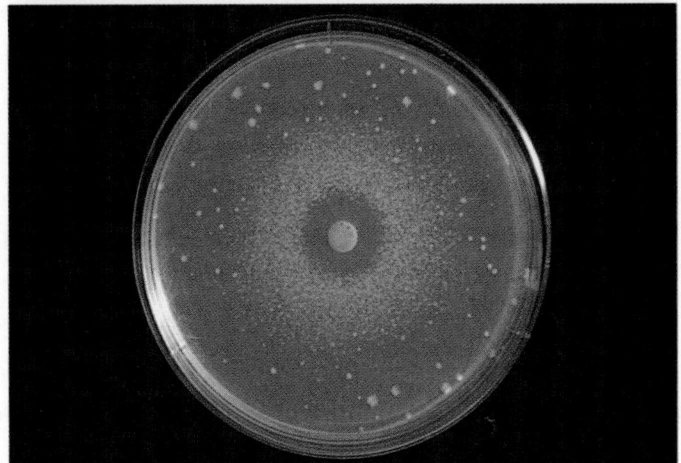

Salmonella test (Ames test) for mutagens. (*Top*) Petri plate containing about 10^9 bacteria that cannot synthesize histidine (the small number of visible colonies are spontaneous revertants). (*Bottom*) Plate containing a disc with a mutagen, which produces a large number of revertants that can synthesize histidine; revertants appear as a ring of colonies around the disc. (Kristien Mortelmans, SRI International, Menlo Park, California)

KEY CONCEPTS

Mutations can occur spontaneously due to several different mechanisms, including errors of DNA replication and spontaneous damage to the DNA.

Mutagens are agents that increase the frequency of mutagenesis, usually by altering the DNA.

Potentially mutagenic and carcinogenic compounds can be detected easily by mutagenesis tests with bacterial systems.

Biological repair systems eliminate many potentially mutagenic alterations in the DNA.

Cells lacking certain repair systems have higher than normal mutation rates.

Genetic change can result from a number of processes. Consider an individual organism that represents a variant from an established control population. What mechanism could have produced the change that created this variant individual? The following mechanisms provide four possibilities:

1. Gene mutation
2. Recombination
3. Transposable genetic elements
4. Chromosomal rearrangements

In previous chapters, we have considered genetic changes merely as useful and interesting phenomena. Let's now examine, at the molecular levels, the processes that lead to each type of genetic change. This chapter explores the mechanisms of gene mutation; Chapter 19 describes the events leading to recombination; and Chapter 20 introduces the concept of transposable genetic elements, which can bring about genetic change by moving from one chromosome location to another. Many chromosomal rearrangements are large deletions, duplications, or inversions that can result from recombination and from transposable elements.

Table 18-1. Summary of Changes at the Molecular Level in Gene Mutations

Type of Mutation	Result and Example(s)
Forward mutations	
Single-nucleotide-pair (base-pair) substitutions	
At DNA level	
Transition	Purine replaced by a different purine, or pyrimidine replaced by a different pyrimidine: AT ⟶ GC GC ⟶ AT CG ⟶ TA TA ⟶ CG
Transversion	Purine replaced by a pyrimidine, or pyrimidine replaced by a purine: AT ⟶ CG AT ⟶ TA GC ⟶ TA GC ⟶ CG TA ⟶ GC TA ⟶ AT CG ⟶ AT CG ⟶ GC
At protein level	
Silent mutation	Triplet codes for same amino acid: AGG ⟶ CGG both code for Arg
Neutral mutation	Triplet codes for different but functionally equivalent amino acid: AAA ⟶ AGA changing basic Lys to basic Arg (at many positions, will not alter protein function)
Missense mutation	Triplet codes for a different and nonfunctional amino acid.
Nonsense mutation	Triplet codes for chain termination: CAG ⟶ UAG changing from a codon for Gln to an amber termination codon
Single-nucleotide-pair addition or deletion: frameshift mutation	Any addition or deletion of base pairs that is not a multiple of 3 results in a frameshift in DNA segments that code for proteins.
Addition or deletion of several to many nucleotide pairs	Any addition or deletion of base pairs that is not a multiple of 3 results in a frameshift in DNA segments that code for proteins.
Reverse mutations	
Exact reversion	AAA (Lys) wild-type $\xrightarrow{\text{forward}}$ GAA (Glu) mutant $\xrightarrow{\text{reverse}}$ AAA (Lys) wild-type

The Molecular Basis of Gene Mutations

Gene mutations can arise spontaneously or they can be induced. **Spontaneous mutations** are naturally occurring mutations and arise in all cells. **Induced mutations** are produced when an organism is exposed to a mutagenic agent, or mutagen; such mutations typically occur at much higher frequencies than spontaneous mutations do.

To understand the mechanisms of gene mutation requires analysis at the level of DNA and protein molecules. Preceding chapters have described models for DNA and protein structure and have discussed the nature of mutations that alter these structures. Table 18-1 draws together this information to provide an explanation of the nature of gene mutation at the molecular level.

Technical advances in the mid-1970s ushered in an exciting new era in molecular genetics, permitting the first direct determination of the sequence of large segments of DNA and also of the sequence changes resulting from mutations. This has greatly increased our understanding of the pathways that lead to mutagenesis and ha even helped to unravel the mysteries of mutational ho spots — genetic sites with a penchant for mutating

Table 18-1. *(Continued)*

Type of Mutation	Result and Example(s)
Equivalent reversion	UCC (Ser) $\xrightarrow{\text{forward}}$ UGC (Cys) $\xrightarrow{\text{reverse}}$ AGC (Ser) wild-type — mutant — wild-type CGC (Arg, basic) $\xrightarrow{\text{forward}}$ CCC (Pro, not basic) $\xrightarrow{\text{reverse}}$ CAC (His, basic) wild-type — mutant — pseudo-wild-type
Intragenic suppressor mutations	
Frameshift of opposite sign at second site within gene	CAT CAT CAT CAT CAT CAT (+) (−) ↓ ↓ CAT XCA TAT CAT CAT CAT ✓ × × ✓ ✓ ✓
Second-site missense mutation	Still not fully understood at the level of protein function; explained in terms of a second distortion that restores a more or less wild-type protein conformation after a primary distortion.
Extragenic suppressor mutations	
Nonsense suppressors	A gene (for example, for tyrosine tRNA) undergoes a mutational event in its anticodon region that enables it to recognize and align with a mutant nonsense codon (say, amber UAG) to insert an amino acid (here, tyrosine) and permit completion of the translation.
Missense suppressors	A heterogeneous set of mutations with molecular mechanisms that are not fully understood. One missense suppressor in *E. coli* is an abnormal tRNA that carries glycine but inserts it in response to arginine codons. Although all wild-type arginine codons are mistranslated, the observed mutations are not lethal, probably due to the low efficiency of abnormal substitution.
Frameshift suppressors	Very few examples have been found; in one, a four-nucleotide anticodon in a single tRNA can read a four-letter codon caused by a single-nucleotide-pair insertion.
Physiological suppressors	A defect in one chemical pathway is circumvented by another mutation (for example, one that opens up another chemical pathway to the same result, or one that permits more efficient transport of a compound produced in small quantities due to the original mutation). Thus, these mutations act as one form of missense suppressors (a very heterogeneous group).

Much work until now on the molecular basis of mutation has been carried out in single-celled bacteria and their viruses. However, many mutations leading to inherited diseases in humans have been analyzed recently. We'll review some of the recent findings of all of these studies. We shall also consider biological repair mechanisms, since repair systems play a key role in mutagenesis, operating to lower the final observed mutation rates. For example, in *Escherichia coli,* with all repair systems functioning, base substitutions occur at rates of 10^{-10} to 10^{-9} per base pair per cell per generation.

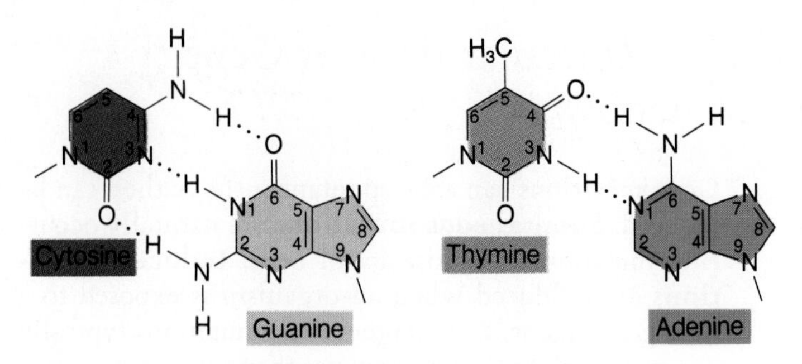

Figure 18-1 Base pairs in DNA. The normal Watson-Crick base pairs are shown.

Spontaneous Mutations

Spontaneous mutations arise from a variety of sources, including errors in DNA replication, spontaneous lesions, and transposable genetic elements. The first two are discussed below; the third is examined in Chapter 20.

Errors in DNA Replication

An error in DNA replication can occur when an illegitimate nucleotide pair (say, A – C) forms during DNA synthesis, leading to a base substitution.

Each of the bases in DNA can appear in one of several forms, called **tautomers,** which are isomers that differ in the positions of their atoms and in the bonds between the atoms. The forms are in equilibrium. The **keto** form of each base is normally present in DNA (Figure 18-1), whereas the **imino** or **enol** forms of the bases are rare. The ability of the wrong tautomer of one of the standard bases to mispair and cause a mutation during DNA replication was first noted by Watson and Crick when they formulated their model for the structure of DNA (Chapter 11). Figure 18-2 demonstrates some possible mispairs resulting from changes of one tautomer to another, termed **tautomeric shifts.**

Mispairs can also occur when one of the bases becomes **ionized.** This type of mispair may occur more frequently than mispairs involving imino and enol forms of bases.

Transitions. All of the mispairs described above lead to **transition mutations,** in which a purine is substituted for a purine or a pyrimidine for a pyrimidine (Figure 18-3). The bacterial DNA polymerase III (Chapter 11) has an editing capacity that recognizes such mismatches and excises them, thus greatly reducing the observed mutations. Another repair system (described later in this chapter) corrects many of the mismatched bases that escape correction by the polymerase editing function.

Transversions. In **transversion mutations,** a pyrimidine is substituted for a purine, or vice versa. Transversions cannot be generated by the mismatches depicted in

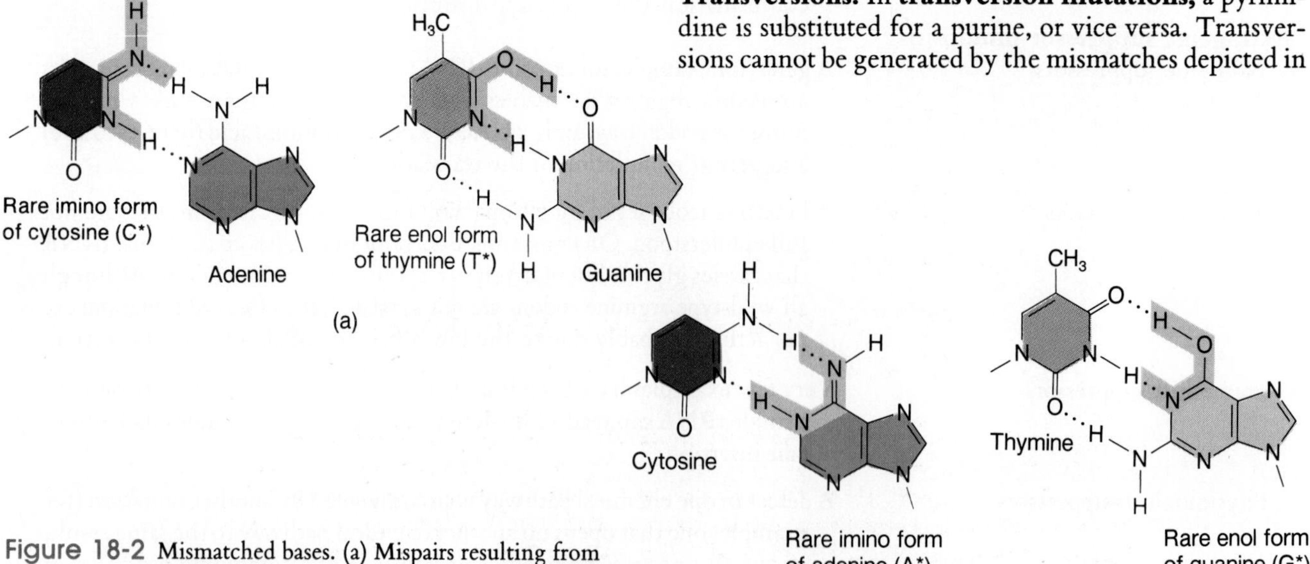

Figure 18-2 Mismatched bases. (a) Mispairs resulting from rare tautomeric forms of the pyrimidines; (b) mispairs resulting from rare tautomeric forms of the purines.

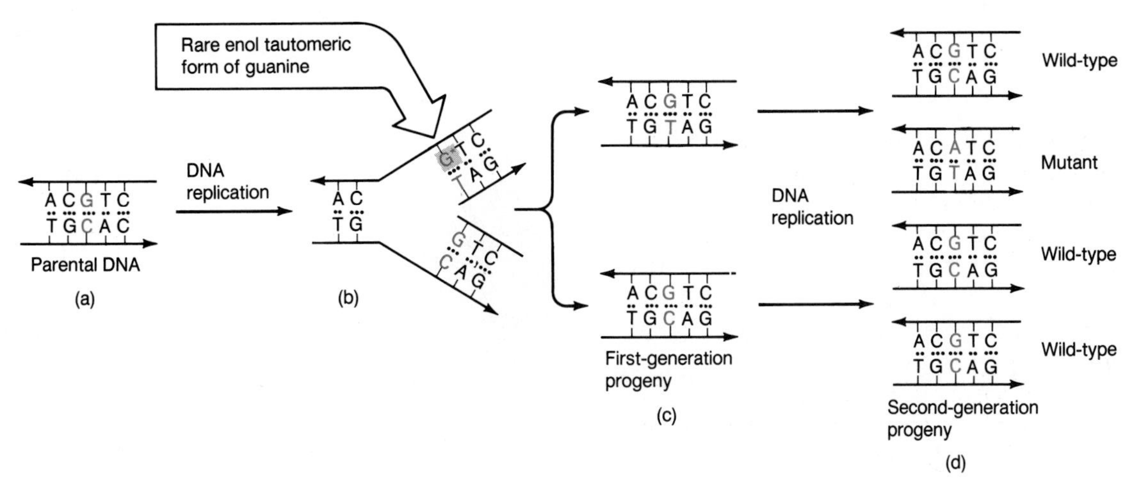

Figure 18-3 Mutation via tautomeric shifts in the bases of DNA. (a) In the example diagrammed, a guanine undergoes a tautomeric shift to its rare enol form (G*) at the time of replication. (b) In its enol form, it pairs with thymine. (c) and (d) During the next replication, the guanine shifts back to its more stable keto form. The thymine incorporated opposite the enol form of guanine, seen in (b), directs the incorporation of adenine during the subsequent replication, shown in (c) and (d). The net result is a GC → AT mutation. If a guanine undergoes a tautomeric shift from the common keto form to the rare enol form at the time of incorporation (as a nucleoside triphosphate, rather than in the template strand diagrammed here), it will be incorporated opposite thymine in the template strand and cause an AT → GC mutation. (From E. J. Gardner and D. P. Snustad, *Principles of Genetics,* 5th ed. Copyright 1984 by John Wiley & Sons, New York.)

Figure 18-2. With bases in the DNA in the normal orientation, creation of a transversion by a replication error would require, at some point during replication, mispairing of a purine with a purine or a pyrimidine with a pyrimidine. Although the dimensions of the DNA double helix render such mispairs energetically unfavorable, we now know from X-ray diffraction studies that G–A pairs, as well as other purine–purine pairs, can form.

Frameshift Mutations. Replication errors can also lead to **frameshift mutations.** Recall from Chapter 13 that such mutations result in greatly altered proteins.

In the mid-1960s, George Streisinger and his coworkers deduced the nucleotide sequence surrounding different sites of frameshift mutations in the lysozyme gene of phage T4. They found that these mutations often occurred at repeated sequences and formulated a model to account for frameshifts during DNA synthesis. In the Streisinger model (Figure 18-4), frameshifts arise when loops in single-stranded regions are stabilized by the "slipped mispairing" of repeated sequences.

With the advent of DNA sequencing in the mid-1970s, such models could be tested directly. Recall from Chapter 12 that in 1961 Seymour Benzer had demon-

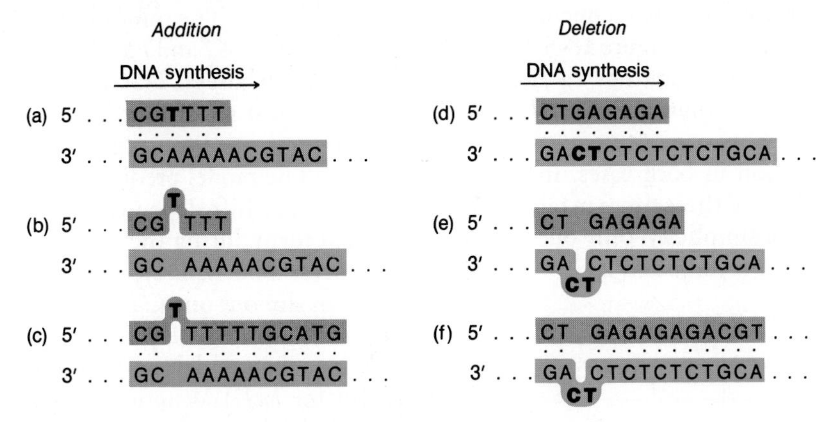

Figure 18-4 A simplified version of the Streisinger model for frameshift formation. (a) to (c) During DNA synthesis, the newly synthesized strand slips, looping out one or several bases. This loop is stabilized by the pairing afforded by the repetitive-sequence unit (the A bases, in this case). An addition of one base pair, A–T, will result at the next round of replication in this example (d) to (f). If instead of the newly synthesized strand, the template strand slips, then a deletion results. Here the repeating unit is a CT dinucleotide. After slippage, a deletion of two base pairs (C–G and T–A) would result at the next round of replication.

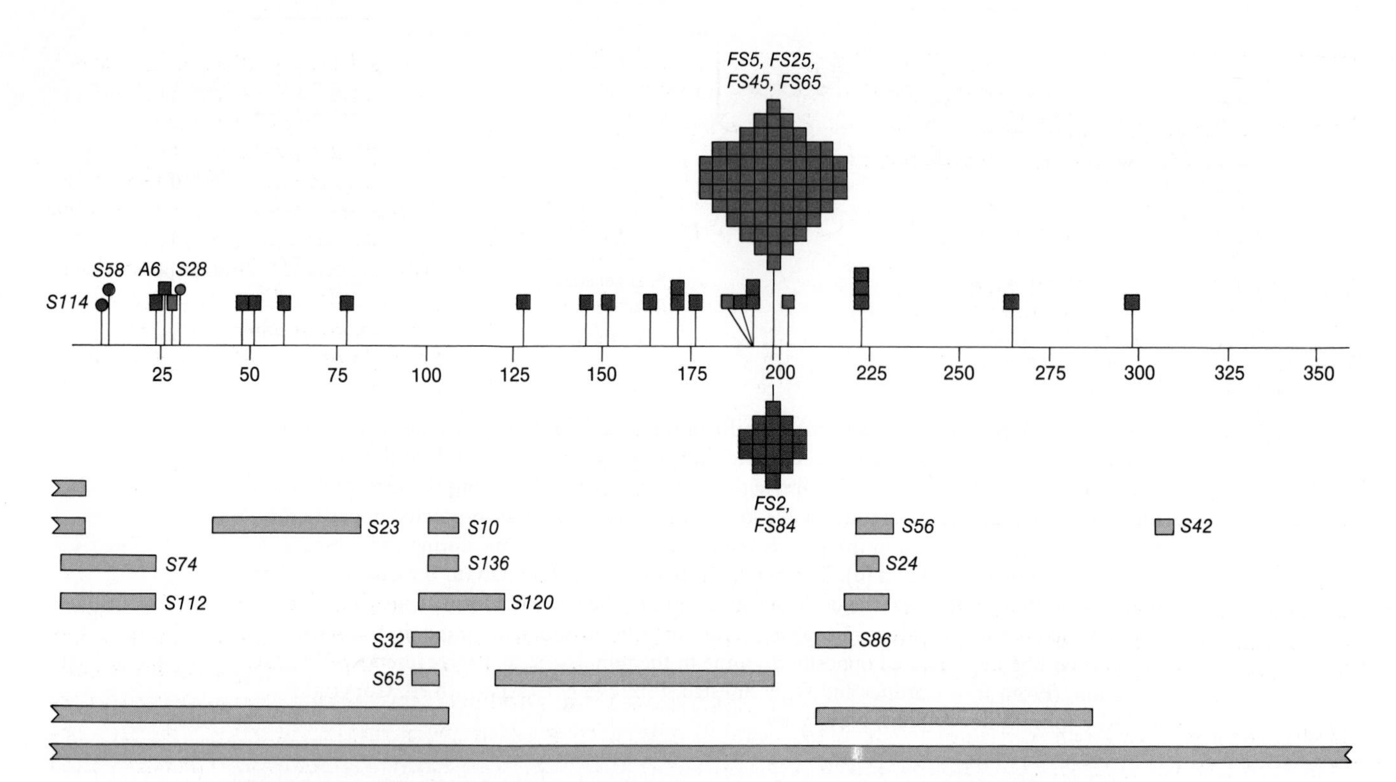

Figure 18-5 The distribution of 140 spontaneous mutations in *lacI*. Each occurrence of a point mutation is indicated by a box. Red boxes designate fast-reverting mutations. Deletions (gold) are represented below. The *I* map is given in terms of the amino-acid number in the corresponding *I*-encoded *lac* repressor. Allele numbers refer to mutations that have been analyzed at the DNA sequence level. The mutations *S114* and *S58* (circles) result from the insertion of transposable elements (see Chapter 19). *S28* (red circle) is a duplication of 88 base pairs. (From P. J. Farabaugh, U. Schmeissner, M. Hofer, and J. H. Miller, *Journal of Molecular Biology* 126, 1978, 847.)

strated the existence of **hot spots,** or sites in a gene that are much more mutable than other sites. In the 1970s, Jeffrey Miller and his coworkers examined mutational hot spots in the *lacI* gene of *E. coli.* The *lacI* work showed that certain hot spots result from repeated sequences, just as predicted by the Streisinger model. Figure 18-5 depicts the distribution of spontaneous mutations in the *lacI* gene. Compare this with the distribution in the *rII* genes of T4 seen by Benzer. Note how one or two mutational sites dominate the distribution in both cases. In *lacI*, a four-base-pair sequence repeated three times in tandem is the cause of the hot spots (for simplicity, only one strand of the double strand of DNA is indicated):

−GTCTGG CTGG CTGG CTGG C

FS5, FS25, FS45, FS65 ↑↓

wild-type 5′−GTCTGG CTGG CTGG C−3′

FS2, FS84 ↓

GTCTGG CTGG C

The major hot spot, represented here by the mutations *FS5*, *FS25*, *FS45*, and *FS65*, results from the addition of one extra set of the four bases CTGG to one strand of the DNA. This hot spot reverts at a high rate, losing the extra set of four bases. The minor hot spot, represented here by the mutations *FS2* and *FS84*, results from the loss of one set of the four bases CTGG. This mutant does not readily regain the lost set of four base pairs.

How can the Streisinger model explain these observations? The model predicts that the frequency of a particular frameshift depends on the number of base pairs that can form during the slipped mispairing of repeated sequences. The wild-type sequence shown for the *lacI* gene can slip out one CTGG sequence and stabilize this by forming nine base pairs. (Can you work this out by applying the model in Figure 18-4 to the sequence shown for *lacI*?) Whether a deletion or an addition is generated depends on whether the slippage occurs on the template or on the newly synthesized strand, respectively. In a similar fashion, the addition mutant can slip out one CTGG sequence and stabilize this with 13 base pairs (verify this for the *FS5* sequence shown for *lacI*), which explains the rapid reversion of mutations such as *FS5*. However, there are only five base pairs available to

S74, S112 75 bases

CAATTCAGG GTGGTGAA TGTGAAACC------CGC GTGGTGAA CCAGG

Site (no. of b.p.)	Sequence repeat	No. of bases deleted	Occurrences	
20 to 95	GTGGTGAA	75	2	*S74, S112*
146 to 269	GCGGCGAT	123	1	*S23*
331 to 351	AAGCGGCG	20	2	*S10, S136*
316 to 338	GTCGA	22	2	*S32, S65*
694 to 707	CA	13	1	*S24*
694 to 719	CA	25	1	*S56*
943 to 956	G	13	1	*S42*
322 to 393	None	71	1	*S120*
658 to 685	None	27	1	*S86*

Figure 18-6 Deletions in *lacI*. Deletions occurring in *S74* and *S112* are shown at the top of the figure. As indicated by the red bars, one of the sequence repeats and all of the intervening DNA is deleted, leaving one copy of the repeated sequence. All mutations were analyzed by direct DNA sequence determination. (From P. J. Farabaugh, U. Schmeissner, M. Hofer, and J. H. Miller, *Journal of Molecular Biology* 126, 1978, 847.)

stabilize a slipped-out CTGG in the deletion mutant, accounting for the infrequent reversion of mutations such as *FS2* in the sequence shown for *lacI*.

Deletions and Duplications. Large **deletions** (more than a few base pairs) represent a sizable fraction of spontaneous mutations, as you can visualize from Figure 18-5. Deletions of up to several thousand base pairs in size have been studied extensively at the DNA sequence level. The majority, although not all, of the deletions occur at repeated sequences. Figure 18-6 shows the results for the first 12 deletions analyzed at the DNA sequence level, presented by Miller and his coworkers in 1978. Further studies have shown that hot spots for deletions involve the longest sequences that are repeated. **Duplications** of segments of DNA have been observed in many organisms. Like deletions, they often occur at sequence repeats.

How do deletions and duplications form? Several mechanisms could account for this. Deletions may be generated as replication errors. For example, an extension of the Streisinger model of slipped mispairing (Figure 18-4) could explain why deletions predominate at short repeated sequences. Alternatively, deletions and duplications could be generated by recombinational mechanisms (to be described in Chapter 19) employed by one or a number of cellular enzyme systems that recognize sequence repeats.

Spontaneous Lesions

In addition to replication errors, **spontaneous lesions,** naturally occurring damage to the DNA, can also generate mutations. Two of the most frequent spontaneous lesions result from depurination and deamination.

Depurination, the more common of the two, involves the interruption of the glycosidic bond between the base and deoxyribose and the subsequent loss of a guanine or an adenine residue from the DNA (Figure 18-7). A mammalian cell spontaneously loses about 10,000 purines from its DNA during a 20-hour cell-generation period at 37°C. If these lesions were to persist, they would result in significant genetic damage because, during replication, the resulting **apurinic sites** cannot specify a base complementary to the original purine. However, as we see later in the chapter, efficient repair systems remove apurinic sites. Under certain conditions (to be described later), a base can be inserted across from an apurinic site; this will frequently result in a mutation.

The **deamination** of cytosine yields uracil (Figure 18-8a). Unrepaired uracil residues will pair with adenine during replication, resulting in the conversion of a G–C pair to an A–T pair (a **GC → AT transition**). One of the repair enzymes in the cell, uracil-DNA glycosylase, recognizes uracil residues in the DNA and excises them, leaving a gap that is subsequently filled in (a process to be described later in the chapter). An exciting discovery in 1978 revealed that the specificity of this repair enzyme is the cause of one type of mutational hot spot! DNA sequence analysis of GC → AT transition hot spots in the *lacI* gene has shown that 5-methylcytosine residues occur at the position of each hot spot. (Certain bases in prokaryotes and eukaryotes are methylated; see page 416.) Some of the data from this *lacI* study are shown in Figure 18-9. The height of each bar on the graph represents the frequency of mutations at each of a number of sites. It can be seen that the position of 5-methylcytosine residues correlates nicely with the most mutable sites. This is because the deamination of 5-methylcytosine (Figure 18-8b) generates thymine (5-methyluracil), which is not recognized by the enzyme uracil-DNA glycosylase and thus is not repaired. Therefore, C → T transitions generated by deamination are found more frequently at 5-methylcytosine sites, which occur both in bacteria and in higher cells.

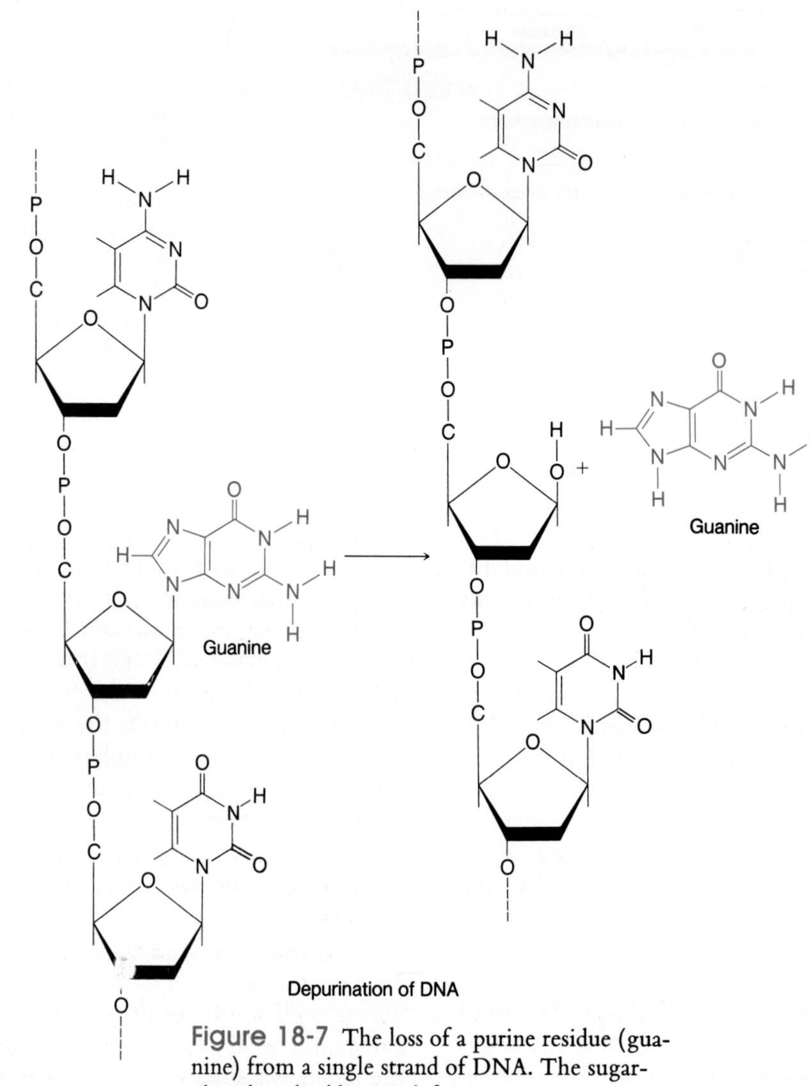

Depurination of DNA

Figure 18-7 The loss of a purine residue (guanine) from a single strand of DNA. The sugar-phosphate backbone is left intact.

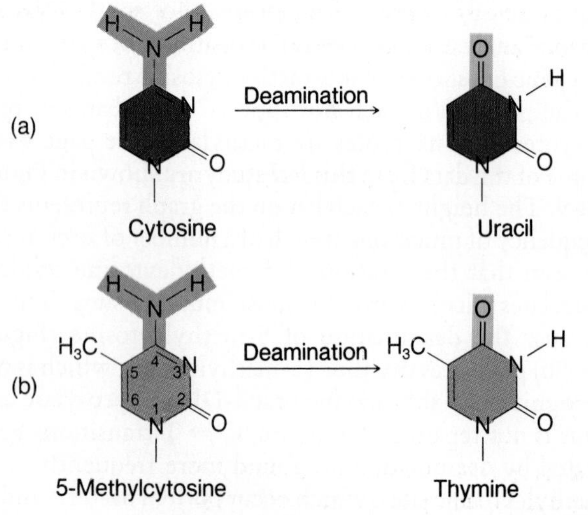

Figure 18-8 Deamination of (a) cytosine and (b) 5-methylcytosine.

Oxidatively damaged bases represent a third type of spontaneous lesion implicated in mutagenesis. Active oxygen species, such as superoxide radicals ($O_2^- \cdot$), hydrogen peroxide (H_2O_2), and hydroxyl radicals (OH), are produced as byproducts of normal aerobic metabolism. These can cause oxidative damage to DNA, which results in mutation and which has been implicated in a number of human diseases. Figure 18-10 shows two products of oxidative damage. The 8-oxo-7,8-dihydrodeoxyguanine (8-oxodG, or "GO") product frequently mispairs with A, resulting in a high level of G → T transversions.

Message Spontaneous mutations can be generated by different processes. Replication errors and spontaneous lesions generate most of the base-substitution and frameshift mutations. Replication errors may also cause some deletions that occur in the absence of mutagenic treatment.

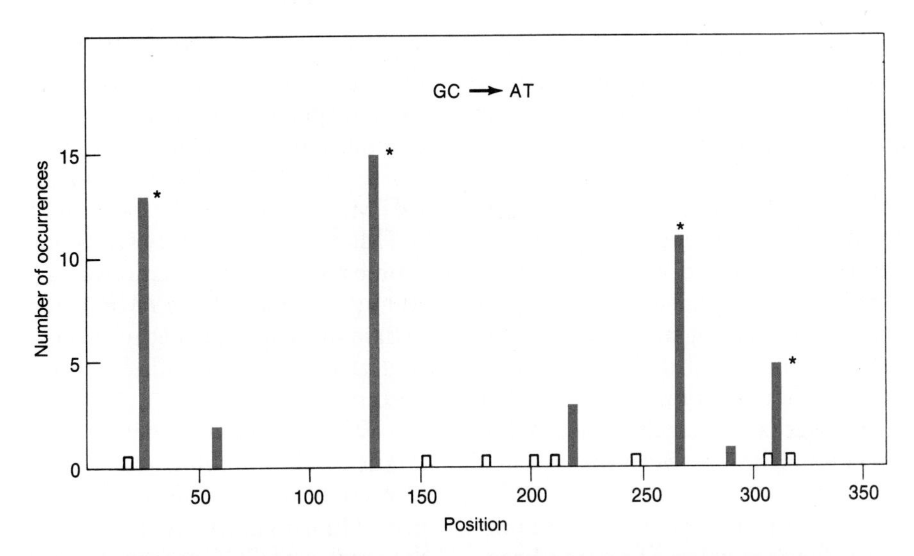

Figure 18-9 5-Methylcytosine hot spots in *E. coli.* Nonsense mutations occurring at 15 different sites in *lacI* were scored. All result from the GC → AT transition. The asterisks (*) mark the position of 5-methylcytosines. Open bars depict sites at which the GC → AT change could be detected but at which no mutations occurred in this particular collection. It can be seen that 5-methylcytosine residues are hot spots for the GC → AT transition. Of 50 independently occurring mutations, 44 were at the four 5-methylcytosine sites and only six were at the 11 unmethylated cytosines. (From C. Coulondre et al., *Nature* 274, 1978, 775.)

Thymine glycol

8-Oxo-7,8-dihydrodeoxyguanine (8-oxodG)

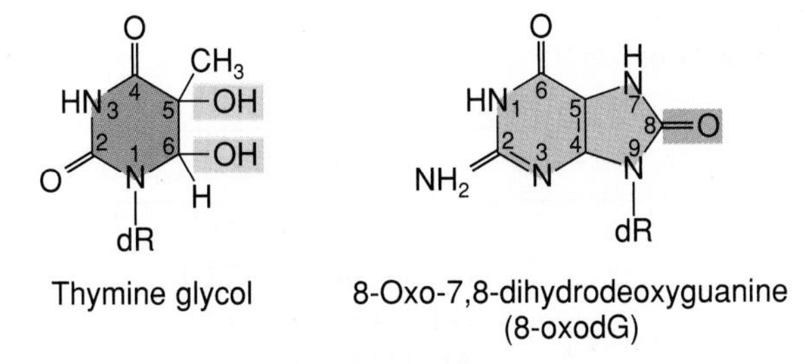

Figure 18-10 DNA damage products formed after attack by oxygen radicals. dR = deoxyribose.

Spontaneous Mutations and Human Diseases

DNA sequence analysis has begun to reveal the mutations responsible for a number of human hereditary diseases. The previously discussed studies of bacterial mutations allow us to suggest mechanisms that cause these human disorders.

A number of these disorders are due to **deletions or duplications** involving repeated sequences. For example, mitochondrial encephalomyopathies are a group of disorders affecting the central nervous system or the muscles. They are characterized by dysfunction of oxidative phosphorylation (a function of the mitochondria) and by changes in mitochondrial structure. These disorders have been shown to result from deletions that occur between repeated sequences. Figure 18-11 depicts one of

WT

···· ACCT ACCTCCCTCACCA AAGC ······ ~ 5000 bp ······ TTCA ACCTCCCTCACCA TTGG ·····

Deletion of ~ 5000 bp

KS

······ ACC**A** ACCTCCCTCACCA TTGG ·····

Figure 18-11 Sequences of wild-type (WT) mitochondrial DNA, and deleted DNA (KS) from a patient with Kearns-Sayre/chronic external opthalmoplegia plus syndrome. The 13-base boxed sequence is identical in both WT and KS, and serves as a breakpoint for the DNA deletion. A single base (**bold** type) is altered in KS, aside from the deleted segment.

these deletions. Note how similar it is in form to the spontaneous *E. coli* deletions shown in Figure 18-6. A second example involves Fabry disease, this inborn error of glycosphingolipid catabolism results from mutations in the X-linked gene encoding the enzyme α-galactosidase A. Many of these mutations are gene rearrangements, resulting from either deletions or duplications between short direct repeats. Table 18-2 shows the short repeats at the breakpoints of rearrangements leading to Fabry disease, and also some rearrangements in the globin genes, resulting in anemias and thalassemias, that have been analyzed. All of these deletions occurred either by a slipped mispairing mechanism, such as pictured in Figure 18-4, or by recombination between the repeated sequences.

A common mechanism that is responsible for a number of genetic diseases is the **expansion of a three-base-pair repeat,** as in the case of fragile X syndrome (Figure 18-12). This syndrome is the most common form of inherited mental retardation, occurring in close to 1 of 1500 males and 1 of 2500 females. It is evidenced cytologically by a fragile site in the X chromosome that results in a break in vitro.

The inheritance of fragile X syndrome is unusual in that 20 percent of the males with a fragile X chromosome are phenotypically normal but transmit the affected chromosome to their daughters, who also appear normal. These males are said to be "normally transmitting males" (NTMs). However, the sons of the daughters of the NTMs frequently display symptoms. The fragile X syndrome results from mutations in a $(CGG)_n$ repeat in the coding sequence of the *FMR-1* gene. Patients with the disease show specific methylation, induced by the mutation, at a nearby CpG cluster, resulting in reduced *FMR-1* expression.

Why do symptoms develop in some persons with a fragile X chromosome and not in others? The answer seems to lie in the number of CGG repeats in the *FMR-1* gene. Humans normally show a considerable variation in the number of CGG repeats in the *FMR-1* gene, ranging from 6 to 54, with 29 repeats representing the most frequent allele. (The variation of CGG repeats produces a corresponding variation in the number of arginine residues [CGG is an arginine codon] in the *FMR-1*–encoded protein.) Both NTMs and their daughters have a much larger number of repeats, ranging from 50 to 200. These

Table 18–2. Breakpoint Sequences in Mammalian Germinal Rearrangements Involving Short Direct Repeats

Rearrangement	5′ Breakpoint	3′ Breakpoint
Hb Leiden	TCCT **GA** GGAG	AGGA **GA** AGTC
Hb Lyon	GGGC **AA** GGTG	GGTG **AA** CGTG
Hb Freiburg	TGAA **GT** TGGT	GTTG **GT** GGTG
Hb Niteroi	GGTT **CTTTG** AGTC	AGTC **CTTTG** GGGA
Hb Gun Hill	AGTG **AGCTGCA** CTGT	GACA **AGCTGCA** CGTG
Hb Tochigi	TATG **GG** CAAC	CTAA **GG** TGAA
Hb St. Antoine	TGAT **GGC** CTGG	GCCT **GGC** TCAC
Hb Coventry	TAAT **GCCC** TGGC	CCTG **GCCC** ACAA
$\gamma\delta\beta$-Thal 1	TCCC **AG** CACT	GAAA **AG** TCTG
Hb BK	GGTA **TCT** GGAG	AATT **TCT** ATTA
Indian HPFH	CGCG **CCACT** GCAC	ATCC **CCACT** ATAT
Dutch β^0-Thal	AACC **AAATTT** GCAC	GAGA **AAATTT** TTGC
Turkish β-Thal	GTCT **ACCC** TTGG	TTGG **ACCC** AGAG
$-(\alpha)^{20.5}$	CCTA **GGC** AACA	TAAG **GGC** CACG
RB 1	AGCT **TTTATAC** TTGA	TGAA **TTTAAAC** ATAA
Pro-α2(I)	TTTC **TTTC** TAAG	GTGG **TTTC** CCTG
Fabry Family A	GAAC **CCA** GAAC	AGCT **CCA** CCTC
Fabry Family B	CATC **AAG** GGTA	TTAA **AAG** ACTT
Fabry Family D	AGCT **TAGACA** GGTA	AATG **TAGATA** AAGA
Fabry Family E	CAGC **AGAACT** GGGG	GGAA **AGAACT** TTGA
Fabry Family J	TTTA **AC** CAGG	TAAT **AC** ATTT
	GAAA **AT** TTTA	ATGT **AT** AGGC

NOTE: For each gene rearrangement, the 5′ and 3′ breakpoint regions are shown. Spaces have been inserted between the direct repeats (in **bold**) and the four adjacent 5′ and 3′ nucleotides which may be involved in the recombinational event. CTGG/CCAG tetranucleotides and CTG/CAG trinucleotides are indicated by highlighting. Hb = hemoglobin; Thal = thalassemia.
SOURCE: R. Kornreich, D.F. Bishop, and R.J. Desnick, *Journal of Biological Chemistry* 2, 1990, 9319.

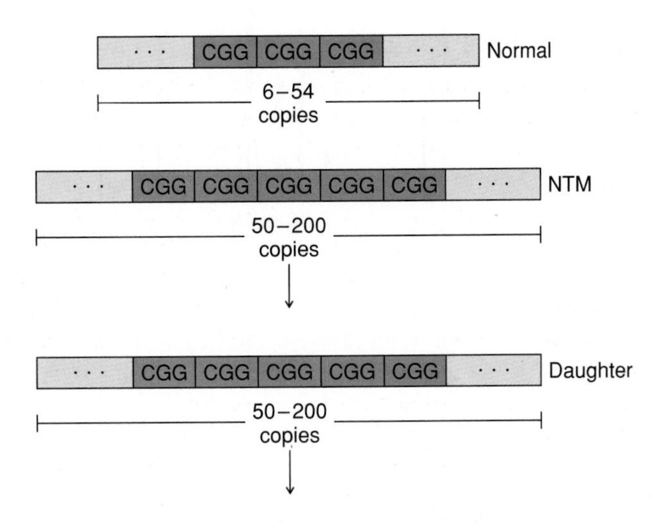

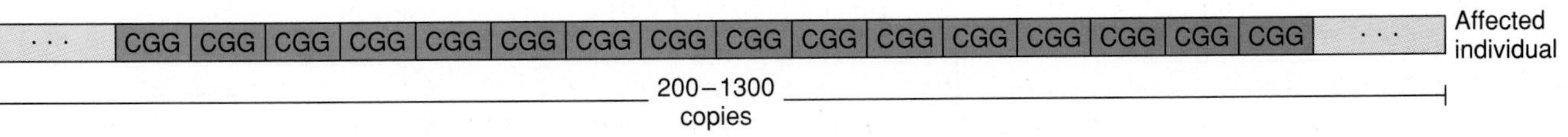

Figure 18-12 Expansion of the CGG triplet in the *FMR-1* gene seen in the fragile X syndrome. Normal individuals have from 6 to 54 copies of the CGG repeat, while individuals from susceptible families display an increase (premutation) in the number of repeats: Normally transmitting males (NTMs) and their daughters are phenotypically normal but display 50 to 200 copies of the CGG triplet; the number of repeats expands to some 200 to 1300 in individuals showing full symptoms of the disease.

increased repeats have been termed **premutations.** All premutation alleles are unstable. The males and females with symptoms of the disease, as well as many carrier females, have additional insertions of DNA, suggesting repeat numbers of 200 to 1300. It has been shown that the frequency of expansion increases with the size of the DNA insertion (and thus, presumably, with the number of repeats). Apparently, the number of repeats in the premutation alleles found in NTMs and their daughters is above a certain threshold, and thus is much more likely to expand to a full mutation than is the case for normal individuals.

The proposed mechanism for these repeats is a slipped mispairing during DNA synthesis, just as shown previously (see diagram on page 533) for the *lacI* hot spot involving a one-step expansion of the four-base-pair sequence CTGG. However, the extraordinarily high frequency of mutation at the three-base-pair repeats in the fragile X syndrome suggests that in human cells, after a threshold level of about 50 repeats, the replication machinery cannot faithfully replicate the correct sequence, and large variations in repeat numbers result.

A second inherited disease, X-linked spinal and bulbar muscular atrophy (known as Kennedy's disease), also results from the amplification of a three-base-pair repeat, in this case a repeat of the CAG triplet. Kennedy's disease, which is characterized by progressive muscle weakness and atrophy, results from mutations in the gene that codes for the androgen receptor. Normal individuals have an average of 21 CAG repeats in this gene, while affected patients have repeats ranging from 40 to 52.

Myotonic dystrophy, the most common form of adult muscular dystrophy, is yet another example of sequence expansion causing a human disease. Susceptible families display an increase in severity of the disease in successive generations; this is caused by the progressive amplification of a CTG triplet at the 3′ end of a transcript. Normal individuals possess, on average, five copies of the CTG repeat; mildly affected individuals have approximately 50 copies, and severely affected people have more than 1,000 repeats of the CTG triplet.

Induced Mutations

Mutational Specificity

Induced mutations have played an essential role in genetic analysis. When we observe the distribution of mutations induced by different mutagens, we see a distinct specificity that is characteristic of each mutagen. Such **mutational specificity** was first noted in the *rII* system

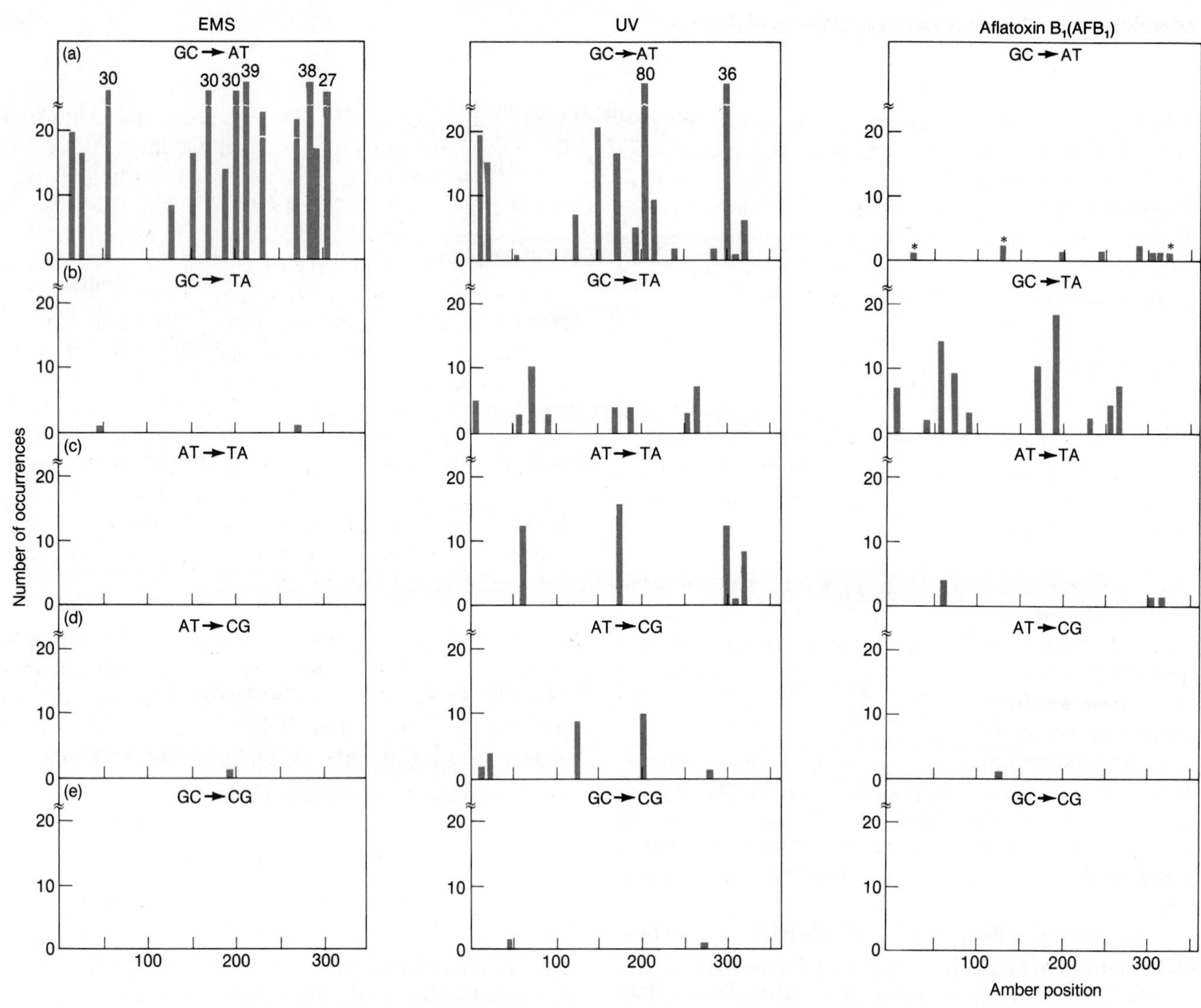

Figure 18-13 Specificity of mutagens. The distribution of mutations among 36 sites in the *lacI* gene is shown for three mutagens: EMS, UV light, and aflatoxin B_1. The height of each bar represents the number of occurrences of mutations at the respective site. Some hot spots are shown off-scale, with the number of occurrences indicated directly above the respective peak. For instance, in the UV-generated collection, one site resulting from a GC → AT transition is represented by 80 occurrences. Each mutational site represented in the figure generates an amber (UAG) codon in the corresponding mRNA. The mutations are arranged according to the type of base substitution involved. Asterisks mark the position of 5-methylcytosines. (Redrawn from C. Coulondre and J. H. Miller, *Journal of Molecular Biology* 117, 1977, 577; and P. L. Foster et al., *Proceedings of the National Academy of Sciences USA* 80, 1983, 2695.)

by Benzer in 1961. Specificity arises from a given mutagen's "preference" both for a certain *type* of mutation (for example, GC → AT transitions) and for certain mutational *sites* (hot spots). Now that we can determine DNA sequences, we are able to examine mutational specificity at the molecular level. Figure 18-13 shows the mutational specificity in *lacI* of three mutagens described later: ethyl methanesulfonate (EMS), ultraviolet (UV) light, and aflatoxin B_1 (AFB_1). The graphs show the distribution of base-substitution mutations that create chain-terminating UAG codons. Figure 18-13 is similar to Figure 12-32, which shows the distribution of mutations in *rII*, except that the specific sequence changes are known for each *lacI* site, allowing the graphs to be broken down into each category of substitution.

Figure 18-13 reveals the two components of mutational specificity. First, each mutagen shown favors a specific category of substitution. For example, EMS and UV favor GC → AT transitions, whereas AFB_1 favors GC → TA transversions. This preference is related to the different mechanisms of mutagenesis. Second, even within the same category, there are large differences in mutation rate. This can be seen best with UV light for the GC → AT changes. Some aspect of the surrounding

DNA sequence must cause these differences. In some cases, the cause of mutational hot spots can be determined by DNA sequence studies, as previously described for 5-methylcytosine residues and for certain frameshift sites (Figures 18-5 and 18-9). In many examples of mutagen-induced hot spots, however, the precise reason for the high mutability of specific sites is still unknown.

Mechanisms of Mutagenesis

Mutagens induce mutations by at least three different mechanisms. They can either replace a base in the DNA, alter a base so that it specifically mispairs with another base, or damage a base so that it can no longer pair with any base under normal conditions.

Incorporation of Base Analogs. Some chemical compounds are sufficiently similar to the normal nitrogen bases of DNA that they occasionally are incorporated into DNA in place of normal bases; such compounds are called **base analogs.** Once in place, these analogs have pairing properties unlike those of the normal bases; thus, they can produce mutations by causing incorrect nucleotides to be inserted opposite them during replication. The original base analog exists in only a single strand, but it can cause a nucleotide-pair substitution that is replicated in all DNA copies descended from the original strand.

For example, **5-bromouracil (5-BU)** is an analog of thymine that has bromine at the C-5 position in place of the CH_3 group found in thymine. This change does not involve the atoms that take part in hydrogen bonding during base-pairing, but the presence of the bromine significantly alters the distribution of electrons in the base. The normal structure (the keto form) of 5-BU pairs with adenine, as shown in Figure 18-14a. 5-BU can frequently change to either the enol form or an ionized form; the latter pairs in vivo with guanine (Figure 18-14b). Thus, the nature of the pair formed during replication will depend on the form of 5-BU at the moment of pairing (Figure 18-15). 5-BU causes transitions almost exclusively, as predicted in Figures 18-14 and 18-15.

Another analog widely used in research is **2-aminopurine (2-AP),** which is an analog of adenine that can pair with thymine but can also mispair with cytosine when protonated, as shown in Figure 18-16. Therefore, when 2-AP is incorporated into DNA by pairing with thymine, it can generate AT → GC transitions by mispairing with cytosine during subsequent replications. Or, if 2-AP is incorporated by mispairing with cytosine, then GC → AT transitions will result when it pairs with thymine. Genetic studies have shown that 2-AP, like 5-BU, is very specific for transitions.

Specific Mispairing. Some mutagens are not incorporated into the DNA but instead alter a base, causing specific mispairing. Certain **alkylating agents,** such as **ethyl methanesulfonate (EMS)** and the widely used **nitrosoguanidine (NG),** operate via this pathway:

$$C_2H_5-O-\overset{\overset{\displaystyle O}{\|}}{\underset{\underset{\displaystyle O}{\|}}{S}}-H_3C$$

EMS

$$O{=}N-N(CH_3)-C(=N-H)-N(H)-NO_2$$

NG

Although such agents add alkyl groups (an ethyl group in the case of EMS and a methyl group in the case

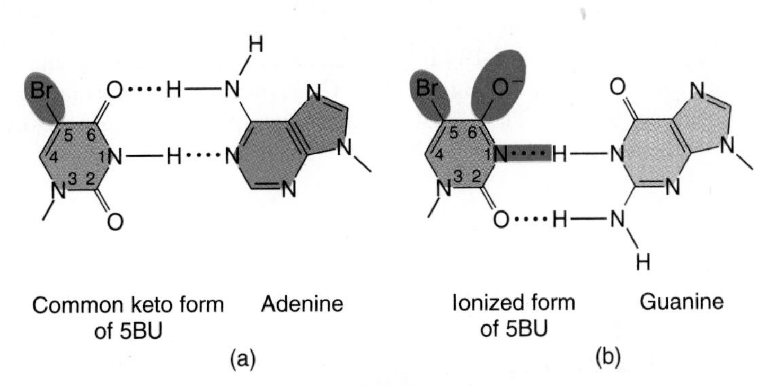

Figure 18-14 Alternative pairing possibilities for 5-bromouracil (5-BU). 5-BU is an analog of thymine that can be mistakenly incorporated into DNA as a base. It has a bromine atom in place of the methyl group. (a) In its normal keto state, 5-BU mimics the pairing behavior of the thymine it replaces, pairing with adenine. (b) The presence of the bromine atom, however, causes a relatively frequent redistribution of electrons, so that 5-BU can spend part of its existence in the rare ionized form. In this state, it pairs with guanine, mimicking the behavior of cytosine and thus inducing mutations during replication.

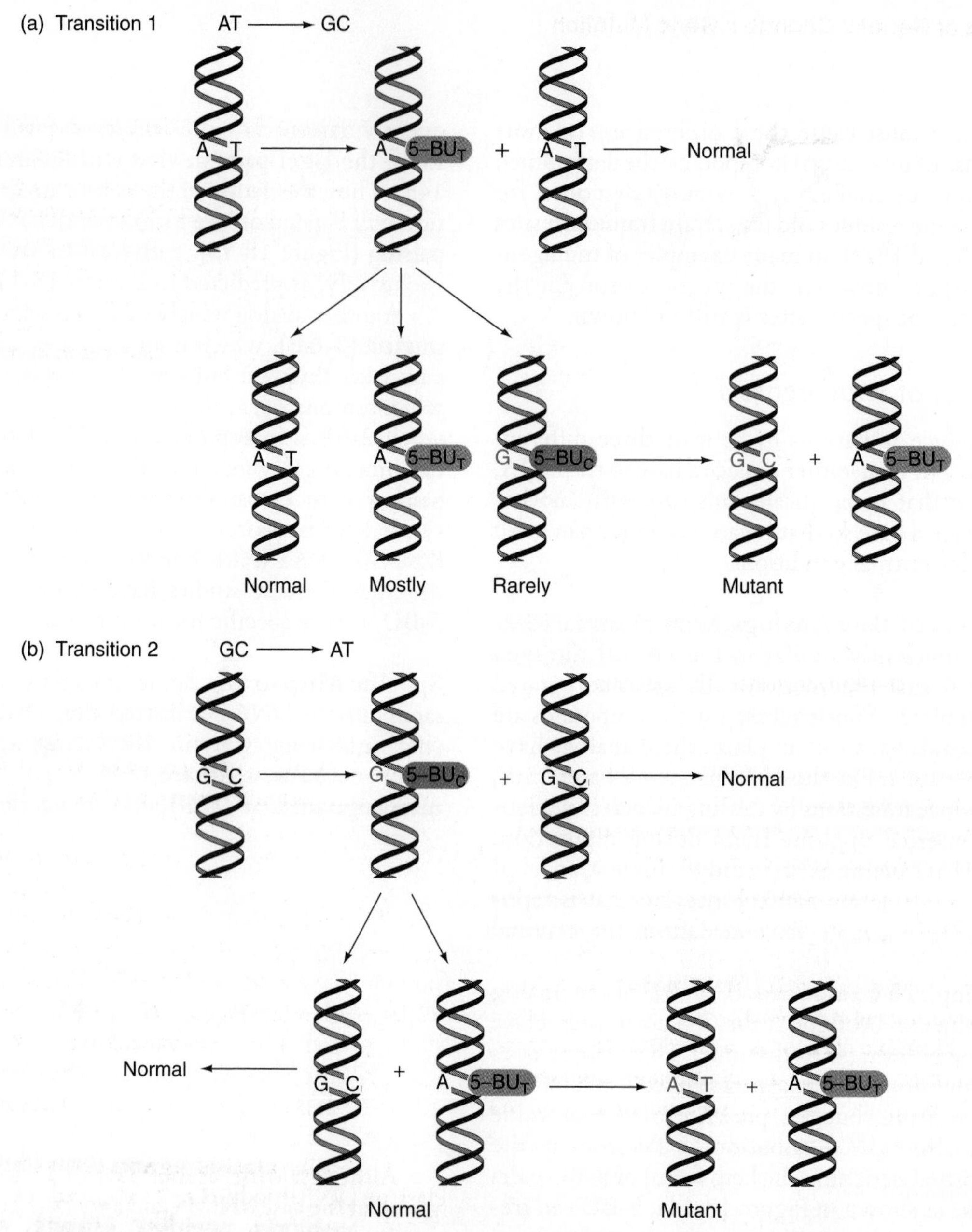

Figure 18-15 The mechanism of 5-BU mutagenesis. 5-BU causes mutations when it is incorporated in one form and then shifts to another form. (a) In its normal keto state, 5-BU pairs like thymine (5-BU_T). Thus, 5-BU is incorporated across from adenine and subsequently mispairs with guanine, resulting in AT → GC transitions. (b) In its ionized form, 5-BU pairs like cytosine (5-BU_C). Thus, 5-BU is misincorporated across from guanine and subsequently pairs with adenine, resulting in GC → AT transitions.

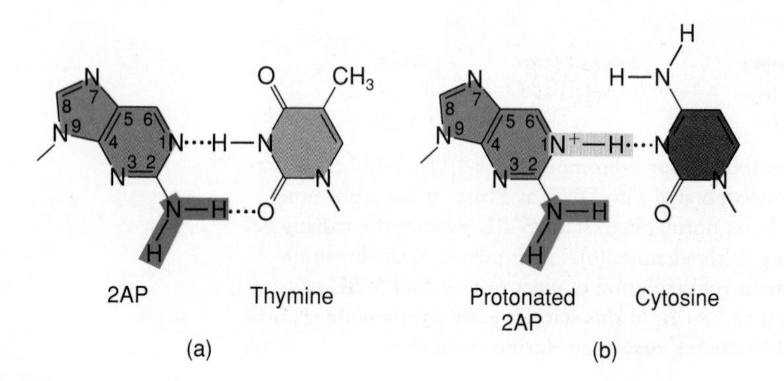

Figure 18-16 Alternative pairing possibilities for 2-aminopurine (2-AP), an analog of adenine. Normally, 2-AP pairs with thymine (a), but in its protonated state it can pair with cytosine (b).

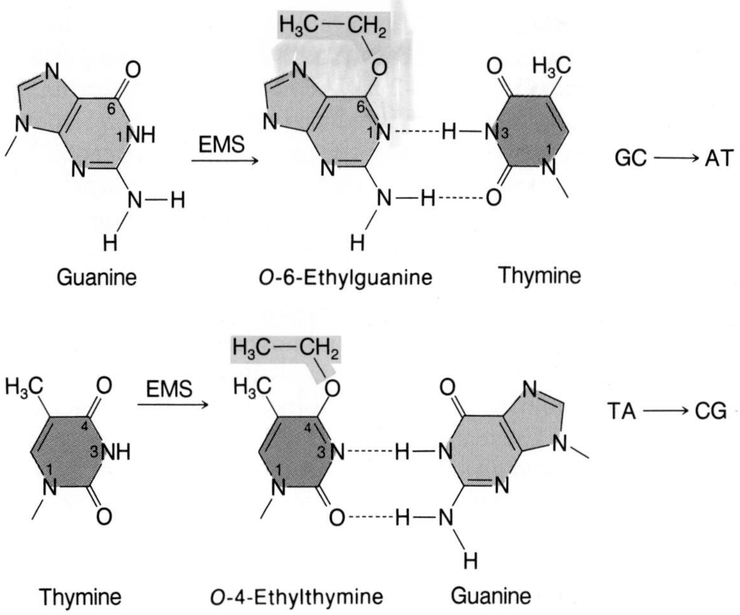

Figure 18-17 Alkylation-induced specific mispairing. The alkylation (in this case, EMS-generated ethylation) of the *O*-6 position of guanine, and the *O*-4 position of thymine, can lead to direct mispairing with thymine and guanine, respectively, as shown here. In bacteria, where mutations have been analyzed in great detail, the principal mutations detected are GC → AT transitions, indicating that the *O*-6 alkylation of guanine is most relevant to mutagenesis.

of NG) to many positions on all four bases, mutagenicity is best correlated with an addition to the oxygen at the 6 position of guanine to create an *O*-6-alkylguanine. This leads to direct mispairing with thymine, as shown in Figure 18-17, and would result in GC → AT transitions at the next round of replication. As expected, determinations of mutagenic specificity for EMS and NG show a strong preference for GC → AT transitions (see the data for EMS shown in Figure 18-13).

Hydroxylamine (HA) is a specific inducer of GC → AT transitions, particularly in phage and *Neurospora;* the effects of HA are less specific in *E. coli.* Its structure is

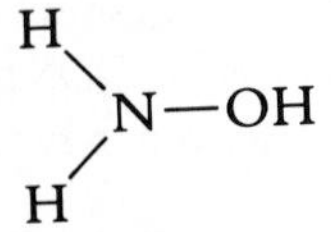

$$\begin{matrix} H \\ & \diagdown \\ & & N{-}OH \\ & \diagup \\ H \end{matrix}$$

The relative specificity of HA is very probably due to the fact that it preferentially hydroxylates the amino nitrogen at cytosine C-4, creating *N*-4-hydroxycytosine, which can pair like thymine (Figure 18-18). *N*-4-Hydroxycytosine prepared in vitro has the same ability to pair like thymine and cause mutations (see page 432), which strongly supports the proposed mechanism.

Other examples of mutagens that alter bases specifically are those that deaminate cytosine to uracil (see Figure 18-8). For instance, **bisulfite ions** convert cytosine to uracil, as does **nitrous acid (NA).** The uracil residue, if unrepaired, will pair with adenine instead of guanine, generating a C → T transition. Nitrous acid also deaminates adenine to generate hypoxanthine, which can form A-C mispairs (Figure 18-19).

The **intercalating agents** form another important class of DNA modifiers. This group of compounds includes **proflavin, acridine orange,** and a class of chemicals termed **ICR compounds** (Figure 18-20a). These agents are planar molecules, which mimic base pairs and are able to slip themselves in **(intercalate)** between the stacked nitrogen bases at the core of the

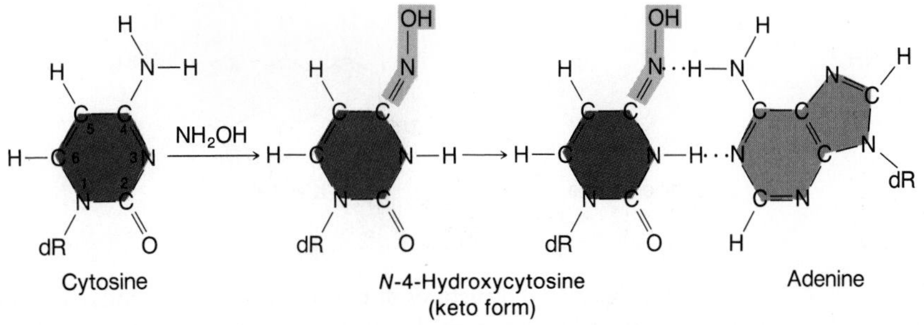

Figure 18-18 A possible explanation for the GC → AT specificity of hydroxylamine (HA) in some organisms. Cytosine is modified to *N*-4-hydroxycytosine, which pairs like thymine, resulting in a GC → AT transition. dR = deoxyribose.

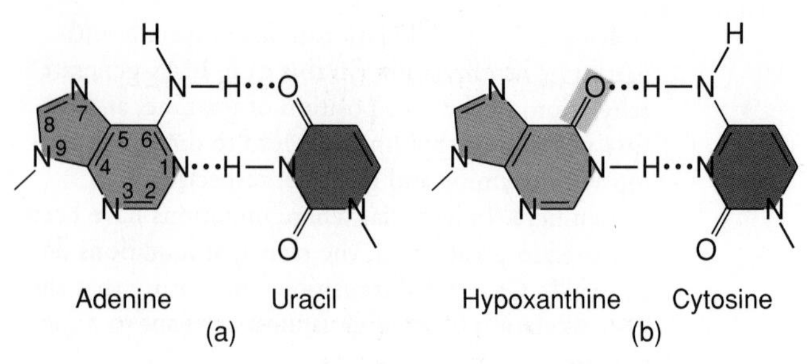

Figure 18-19 Nitrous acid (NA) mutagenesis. (a) NA deaminates cytosine to form uracil, which bonds like thymine. (b) NA deaminates adenine to form hypoxanthine, which bonds like guanine. These altered bonding patterns can lead to mutations. For example, AT may become GC, or GC may become AT. (Modified from E. Freese, in *Structure and Function of Genetic Elements,* Brookhaven Symposia in Biology No. 12, Brookhaven National Laboratory, Upton, N.Y., 1959.)

DNA double helix (Figure 18-20b). In this intercalated position, the agent can cause single-nucleotide-pair insertions or deletions. Intercalating agents may also stack between bases in single-stranded DNA; in so doing, they may stabilize bases that are looped out during frameshift formation, as depicted in the Streisinger model (Figure 18-4). A model for the actions of intercalating agents is shown in Figure 18-21.

Loss of Specific Pairing. A large number of mutagens damage one or more bases so that specific pairing is no longer possible. The result is a replication block, because DNA synthesis will not proceed past a base that cannot specify its complementary base by hydrogen bonding. This is for a good reason: The insertion of bases across from noncoding lesions would lead to frequent mutations.

The **bypass** of such replication blocks by inserting bases requires the activation of a special system, the **SOS system** (Figure 18-22). The name SOS comes from the idea that this system is induced as an emergency response to prevent cell death in the presence of significant DNA damage. SOS induction is a last resort, allowing the cell to trade a certain level of mutagenesis for ultimate survival. (This "induction" is really the activation of gene expression, described more fully in Chapter 17.)

Exactly how the SOS bypass system functions is not clear, although in *E. coli* it is known to be dependent on at least three genes, *recA* (which is also involved in general recombination, as described in Chapter 19), *umuC,* and *umuD.* Current models for SOS bypass suggest that the *umuC* and *umuD* proteins combine with the polymerase III DNA replication complex to loosen its otherwise strict specificity and permit replication past noncoding lesions.

Mutagens are dependent on the SOS system for their action when they cause the loss of specific pairing, thereby generating noncoding lesions and blocking replication. The category of SOS-dependent mutagens is important, since it includes most carcinogens, such as ultraviolet (UV) light and aflatoxin B_1 (discussed below) and benzo(a)pyrene (see Figure 18-26).

How is the SOS system involved in the recovery of mutants induced by certain mutagens? Does the SOS system lower the fidelity of DNA replication so much (to permit the bypass of noncoding lesions) that many replication errors occur, even for undamaged DNA? If this hypothesis were correct, most mutations generated by different SOS-dependent mutagens would be similar, rather than specific to each mutagen. Most mutations would result from the action of the SOS system itself on undamaged DNA. The mutagen, then, would play the indirect role of inducing the SOS system. Studies of mutational specificity, however, have shown that this is not

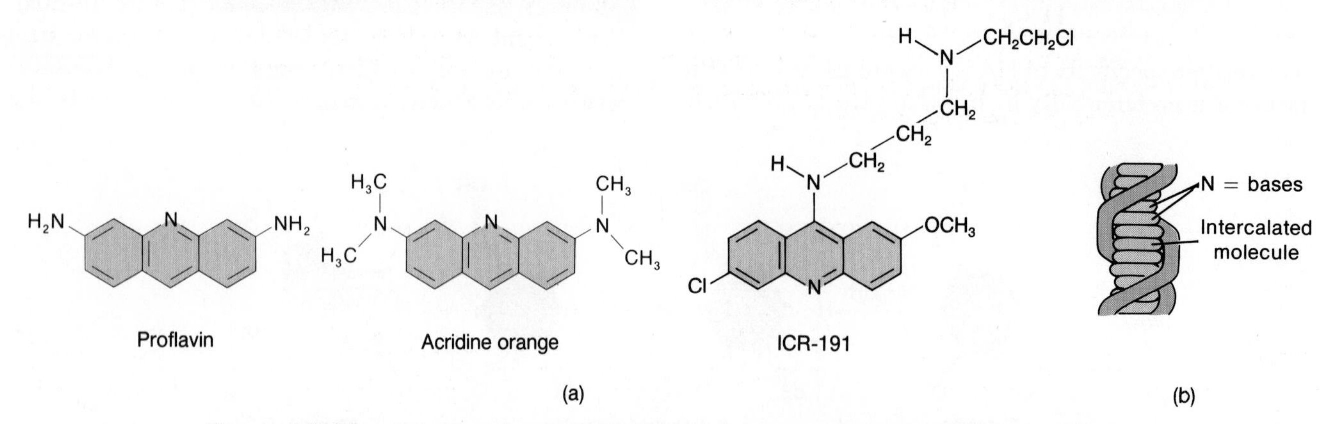

Figure 18-20 Intercalating agents. (a) Structures of the common agents proflavin, acridine orange, and ICR-191. (b) An intercalating agent slips between the nitrogenous bases stacked at the center of the DNA molecule. This occurrence can lead to single-nucleotide-pair insertions and deletions. (From L. S. Lerman, *Proceedings of the National Academy of Sciences USA* 49, 1963, 94.)

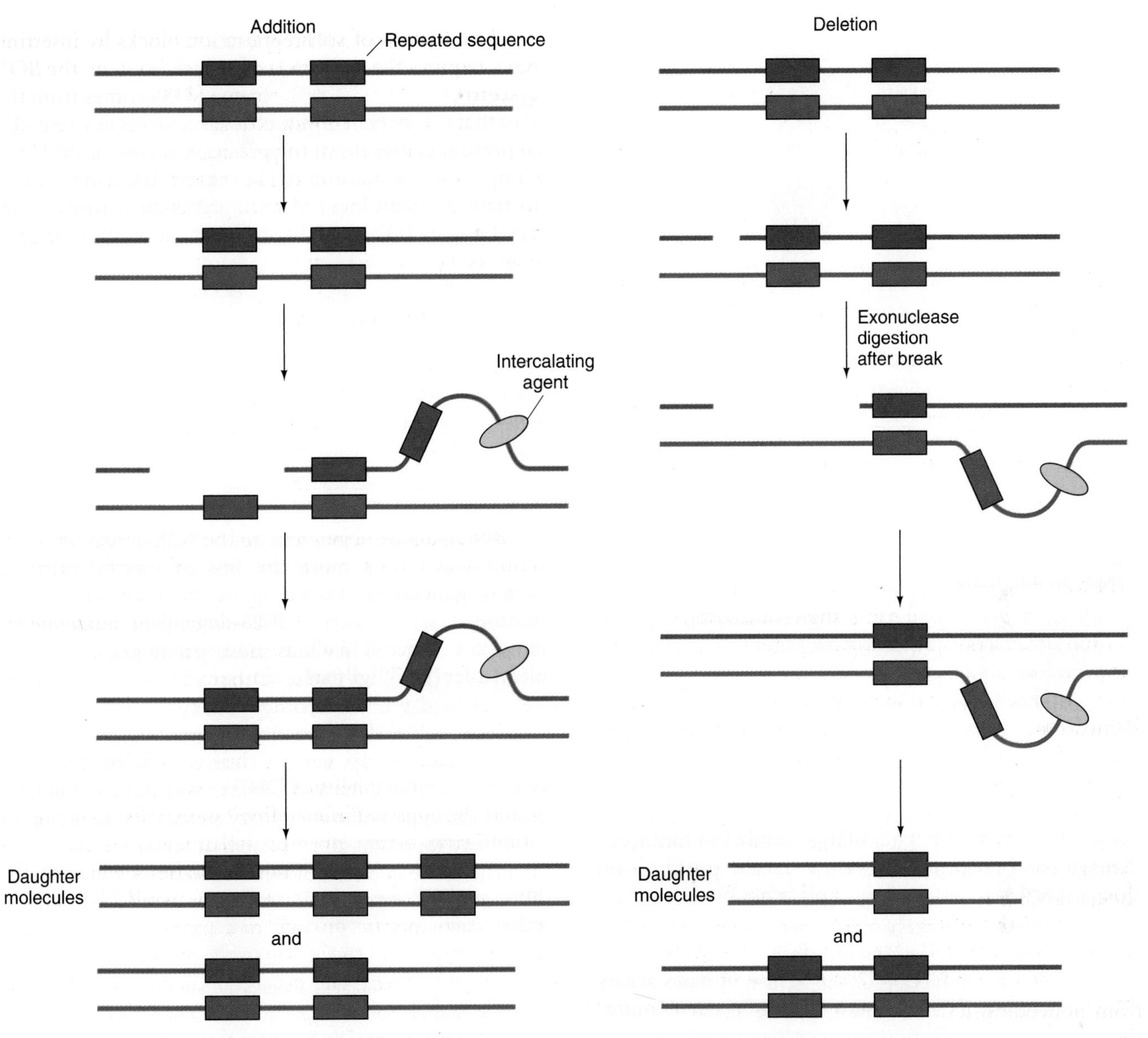

Figure 18-21 Model showing how intercalating agents can cause short deletions or insertions. Here we assume that the agents are active only during DNA processing (repair or recombination), when they insert into a single-stranded loop and stabilize it. (See also Figures 18-4 and 18-20.) For an addition to occur, the loop can form only if there is a short, repeat length of complementary sequence.

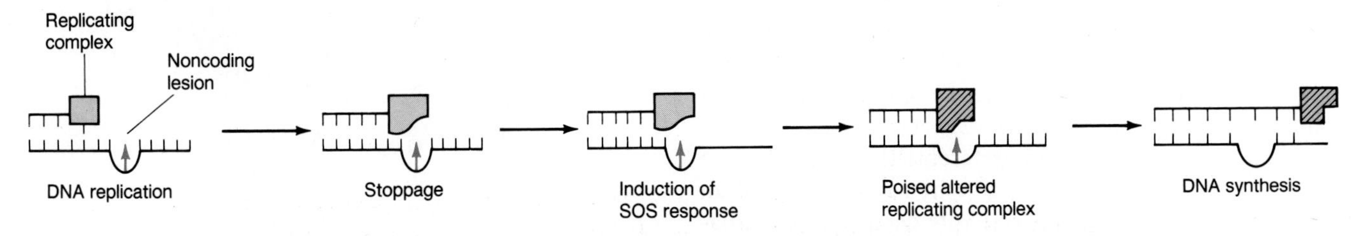

Figure 18-22 The SOS bypass system. This highly schematic diagram represents the stoppage of DNA replication in response to a noncoding lesion. The SOS system relieves this blockage, perhaps by altering the replicating complex.

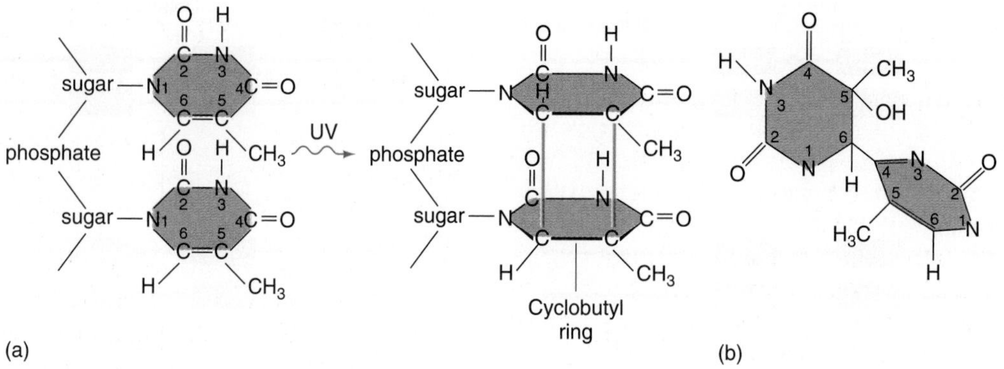

Figure 18-23 UV photoproducts. (a) Cyclobutane-pyrimidine photodimer: here, adjacent thymine residues are joined by UV irradiation to produce a thymine dimer. (b) The 6-4 photoproduct: here, the C-6 of a thymine molecule is linked by UV to the C-4 of an adjacent thymine.

the case. Instead, a series of different SOS-dependent mutagens have markedly different specificities. You can see this by comparing the specificities of UV light and aflatoxin B_1 in Figure 18-13. Each mutagen induces a unique distribution of mutations. Therefore, the mutations must be generated in response to specific damaged base pairs. The type of lesion differs in many cases. Some of the most widely studied lesions include UV photoproducts, apurinic sites, and bulky chemical additions on specific bases.

Ultraviolet (UV) light generates a number of photoproducts in DNA. Two different lesions that occur at adjacent pyrimidine residues—the cyclobutane-pyrimidine photodimer and the 6-4 photoproduct (Figure 18-23)—have been most strongly correlated with mutagenesis. These lesions interfere with normal base-pairing; hence, induction of the SOS system is required for mutagenesis. The insertion of bases across from UV photoproducts leads most frequently to transition mutations, but other base substitutions (transversions) and frameshifts are also stimulated by UV light, as are duplications and deletions. The mutagenic specificity of UV light is illustrated in Figure 18-13.

Aflatoxin B_1 (AFB_1) is a powerful carcinogen. It generates **apurinic sites** following the formation of an addition product at the N-7 position of guanine (Figure 18-24). Studies with apurinic sites generated in vitro have demonstrated a requirement for the SOS system and have also shown that SOS bypass of these sites leads to the preferential insertion of an adenine across from an apurinic site. This predicts that agents that cause depurination at guanine residues should preferentially induce GC → TA transversions. Can you see why the insertion of an adenine across from an apurinic site derived from a guanine would generate this substitution at the next round of replication? Figure 18-13 shows the genetic analysis of many base substitutions induced by AFB_1. You can verify that most of the substitutions are indeed GC → TA transversions.

AFB_1 is a member of a class of chemical carcinogens that represent **bulky addition products** when they bind covalently to DNA. Other examples include the diol epoxides of **benzo(a)pyrene,** a compound produced by internal combustion engines. For many different compounds it is not yet clear which DNA addition products play the principal role in mutagenesis. In some cases, the mutagenic specificity suggests that depurination may represent an intermediate step in mutagenesis; in others, the question of which mechanism is operating is completely open.

Message Mutagens induce mutations by a variety of mechanisms. Some mutagens mimic normal bases and are incorporated into DNA, where they can mispair. Others damage bases and either cause specific mispairing or destroy pairing by causing nonrecognition of bases. In the latter case, a bypass system, the SOS system, must be induced in order to allow replication past the lesion.

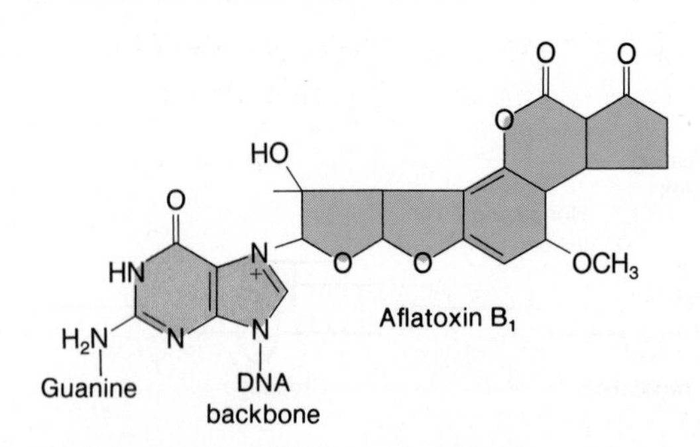

Figure 18-24 The binding of metabolically activated aflatoxin B_1 to DNA.

Reversion Analysis

In this chapter, we have considered the different pathways leading to mutagenesis and have observed that mutagenic processes are often very specific. We can now begin to see how testing for the reversion of a mutation can tell us something about the nature of the mutation or the action of a mutagen. For example, if a mutation cannot be reverted by action of the mutagen that induced it, then the mutagen must have some relatively specific unidirectional action. In the case of a mutation induced by hydroxylamine (HA), for instance, it would be reasonable to expect that the original mutation is GC → AT, which of course cannot be reverted by another specific GC → AT event. Similarly, mutations that can be reverted by proflavin are in all likelihood frameshift mutations; thus mutations induced by nitrous acid (NA) which are transitions, should not be revertible by proflavin.

Transversions cannot be induced by the agents mentioned above, but they are known definitely to be common among spontaneous mutations, as shown by studies of DNA and protein sequencing. Thus, in the reversion test, if a mutation reverts spontaneously but does not revert in response to a transition mutagen or a frameshift mutagen, then, by elimination, it is probably a transversion.

Table 18-3 summarizes some reversion expectations based on simple assumptions. The system outlined in Table 18-3 merely illustrates the kinds of inferences possible from reversion analysis. Recall that mutagen specificities depend on the organism, the genotype, the gene studied, and the action of biological repair systems. Note also that the kinds of logic employed in the reversion test rely heavily on the assumption that the reversion events are not due to suppressors or transposable elements; either of these would make inference from reversion more difficult.

The Relationship Between Mutagens and Carcinogens

There is increasing awareness of a correlation between mutagenicity and carcinogenicity. One study showed that 157 of 175 known carcinogens (approximately 90 percent) are also mutagens. The **somatic mutation theory** of cancer holds that these agents cause cancer by inducing the mutation of somatic cells. Thus mutagenesis is of great relevance to our society. We are faced not only with the genetic time bomb of germinal mutation — with its potential for increasing inherited disease over the long term — but also with the somatic genetic disease of cancer with its overwhelming immediacy.

Induced Mutations and Human Cancer

Understanding the specificity of mutagens in bacteria has led to the direct implication of certain environmental mutagens in the causation of human cancers. Ultraviolet (UV) light and aflatoxin B_1 (AFB_1) have long been suspected as resulting in skin cancer and liver cancer, respectively. Now, DNA sequence analysis of mutations in a human cancer gene has provided direct evidence of their involvement. The gene in question is termed ***p53***, and is one of a number of **tumor-suppressor genes** — genes which encode proteins that suppress tumor formation. (We will learn more about these genes in Chapter 23). A sizable proportion of human cancer patients have mutated tumor-suppressor genes.

Liver cancer is prevalent in southern Africa and East Asia, and a high exposure to AFB_1 in these regions has been correlated with the high incidence of liver cancer. When *p53* mutations in cancer patients were analyzed, G → T transversions, the signature of AFB_1-induced mutations, were found in liver cancer patients from South Africa and East Asia, but not in patients from these

Table 18-3. Different Types of Point Mutations Theoretically Distinguishable by Their Reversion Behaviors in Response to a Battery of Specific Mutagens

Mutation	Reversion Mutagen: NA	Reversion Mutagen: HA or EMS	Reversion Mutagen: Proflavin	Spontaneous Reversion
Transition (GC → AT)	+	−	−	+
Transition (AT → GC)	+	+	−	+
Transversion	−	−	−	+
Frameshift	−	−	+	+

NOTE: + indicates a measurable rate of reversion due to a given mutagen. NA = nitrous acid; HA = hydroxylamine; EMS = ethyl methanesulfonate.

regions with lung, colon, or breast cancer. On the other hand, *p53* mutations in liver cancer patients from areas of low AFB_1 exposure did not result from G → T transversions. These findings, together with the results from the mutagenic specificity studies of AFB_1 (see Figure 18-13), allow us to conclude that AFB_1-induced mutations are a prime cause of liver cancer in South Africa and East Asia.

Sequencing *p53* mutations has also strengthened the link between UV and human skin cancers. The majority of invasive human squamous cell carcinomas analyzed so far have *p53* mutations, all of them involving mutations at dipyrimidine sites, most of which are C → T substitutions. This is the profile of UV-induced mutations. In addition, several tumors have *p53* mutations resulting from a CC → TT double base change, which is found only among UV-induced mutations.

The modern environment exposes each individual to a wide variety of chemicals in drugs, cosmetics, food preservatives, pesticides, compounds used in industry, pollutants, and so on. Many of these compounds have been shown to be carcinogenic and mutagenic. Examples include the food preservative AF-2, the food fumigant ethylene dibromide, the antischistosome drug hycanthone, several hair-dye additives, and the industrial compound vinyl chloride; all are potent, and some have subsequently been subjected to government control. However, hundreds of new chemicals and products appear on the market each week. How can such vast numbers of new agents be tested for carcinogenicity before much of the population has been exposed to them?

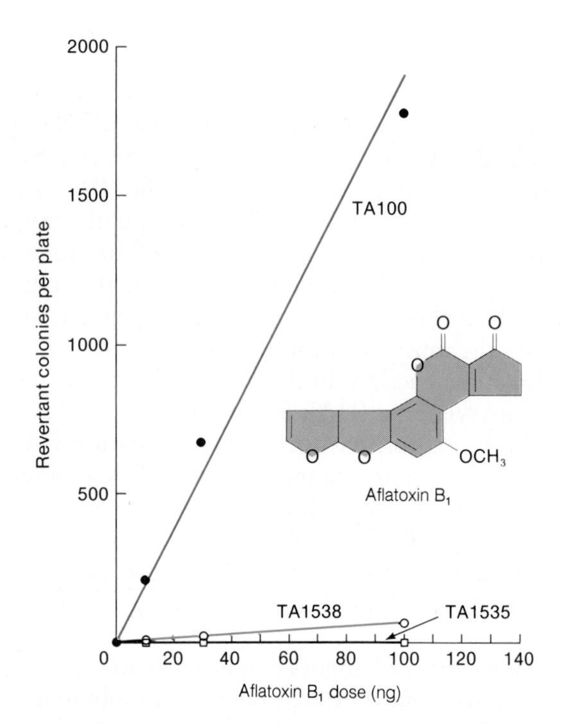

Figure 18-25 Ames test results showing the mutagenicity of aflatoxin B_1, which is also a potent carcinogen. TA100, TA1538, and TA1535 are strains of *Salmonella* bearing different *his* auxotrophic mutations. The TA100 strain is highly sensitive to reversion through base-pair substitution. The TA1535 and TA1538 strains are sensitive to reversion through frameshift mutation. The test results show that aflatoxin B_1 is a potent mutagen that causes base-pair substitutions but not frameshifts. (From J. McCann and B. N. Ames, in *Advances in Modern Toxicology,* Vol. 5. Edited by W. G. Flamm and M. A. Mehlman. Copyright by Hemisphere Publishing Corp., Washington, DC.)

The Ames Test

Many test systems have been devised to screen for carcinogenicity. These tests are time-consuming, typically involving laborious research with small mammals. More rapid tests do exist that make use of microbes (such as fungi or bacteria) and test for mutagenicity rather than carcinogenicity. The most widely used test was developed in the 1970s by Bruce Ames, who worked with *Salmonella typhimurium.* This **Ames test** uses two auxotrophic histidine mutations, which revert by different molecular mechanisms (Figure 18-25). Further properties were genetically engineered into these strains to make them suitable for mutagen detection. First, they carry a mutation that inactivates the excision-repair system (described later). Second, they carry a mutation that eliminates the protective lipopolysaccharide coating of wild-type *Salmonella,* so that any escaping bacteria will be unable to survive in such natural (but chemically hostile) environments as sewers or intestines.

Bacteria are evolutionarily a long way removed from humans. Can the results of a test on bacteria have any real significance in detecting chemicals that are dangerous for humans? First, we have seen that the genetic and chemical nature of DNA is identical in all organisms, so that a compound acting as a mutagen in one organism is likely to have some mutagenic effects in other organisms. Second, Ames devised a way to simulate the human metabolism in the bacterial system. In mammals, much of the important processing of ingested chemicals occurs in the liver, where externally derived compounds normally are detoxified or broken down. In some cases, the action of liver enzymes can create a toxic or mutagenic compound from a substance that was not originally dangerous (Figure 18-26). Ames incorporated mammalian liver enzymes in his bacterial test system, using rat livers for this purpose. Figure 18-27 outlines the procedure used in the Ames test.

The Ames test has detected potential carcinogens among a wide variety of heterogeneous types of chemicals. Of course, chemicals detected by this test can be regarded not only as potential carcinogens (sources of somatic mutations) but also as possible causes of muta-

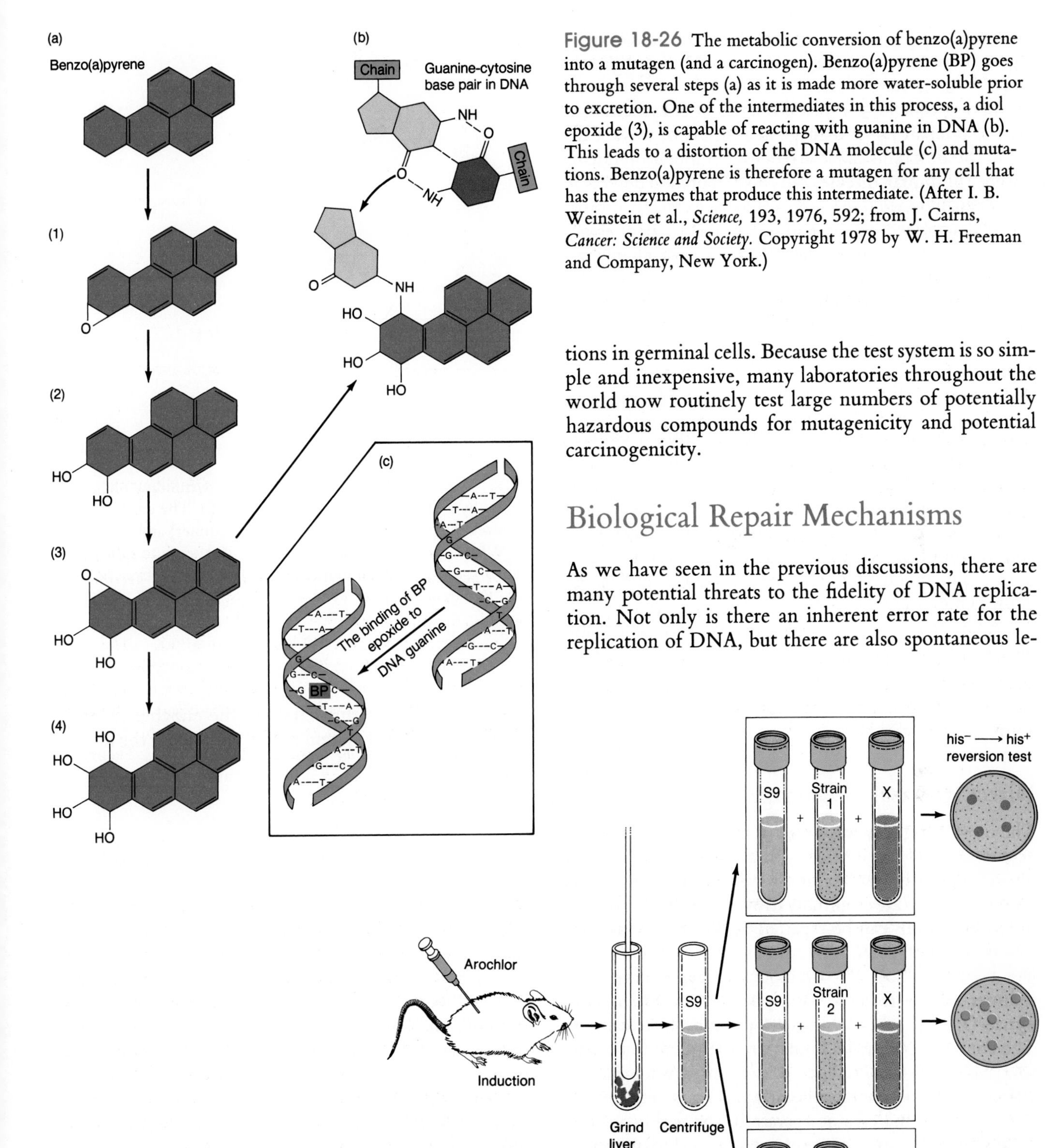

Figure 18-26 The metabolic conversion of benzo(a)pyrene into a mutagen (and a carcinogen). Benzo(a)pyrene (BP) goes through several steps (a) as it is made more water-soluble prior to excretion. One of the intermediates in this process, a diol epoxide (3), is capable of reacting with guanine in DNA (b). This leads to a distortion of the DNA molecule (c) and mutations. Benzo(a)pyrene is therefore a mutagen for any cell that has the enzymes that produce this intermediate. (After I. B. Weinstein et al., *Science*, 193, 1976, 592; from J. Cairns, *Cancer: Science and Society*. Copyright 1978 by W. H. Freeman and Company, New York.)

tions in germinal cells. Because the test system is so simple and inexpensive, many laboratories throughout the world now routinely test large numbers of potentially hazardous compounds for mutagenicity and potential carcinogenicity.

Biological Repair Mechanisms

As we have seen in the previous discussions, there are many potential threats to the fidelity of DNA replication. Not only is there an inherent error rate for the replication of DNA, but there are also spontaneous le-

Figure 18-27 Summary of the procedure used for the Ames test. First rat liver enzymes are mobilized by injecting the animals with Arochlor. (Enzymes from the liver are used because they carry out the metabolic processes of detoxifying and toxifying body chemicals.) The rat liver is then homogenized, and the supernatant of solubilized liver enzymes (S9) is added to a suspension of auxotrophic bacteria in a solution of the potential carcinogen (X). This mixture is plated on a medium containing no histidine, and revertants of mutant strains 1 and 2 are looked for. A control experiment containing no potential carcinogen is always run simultaneously. The presence of revertants indicates that the chemical is a mutagen and possibly a carcinogen as well.

sions that can provoke additional errors. Moreover, mutagens in the environment can damage DNA and greatly increase the mutation rate.

Living cells have evolved a series of enzymatic systems that repair DNA damage in a variety of ways. Failure of these systems can lead to a higher mutation rate. A number of human diseases can be attributed to defects in DNA repair, as we'll see later. Let's first examine some of the characterized repair pathways, and then consider how the cell integrates these systems into an overall strategy for repair.

We can divide repair pathways into several categories.

Avoidance of Errors Before They Happen

Some enzymatic systems neutralize potentially damaging compounds before they even react with DNA. One example of such a system involves the detoxification of superoxide radicals produced during oxidative damage to DNA: the enzyme **superoxide dismutase** catalyzes the conversion of the superoxide radicals to hydrogen peroxide, and the enzyme **catalase,** in turn, converts the hydrogen peroxide to water.

Another error avoidance pathway depends on the protein product of the *mutT* gene: This enzyme prevents the incorporation of 8-oxodG (see Figure 18-10) into DNA by hydrolyzing the triphosphate of 8-oxodG back to the monophosphate.

Direct Reversal of Damage

The most straightforward way to repair a lesion, once it occurs, is to reverse it directly, thereby regenerating the normal base. Reversal is not always possible, since some types of damage are essentially irreversible. In a few cases, however, lesions can be repaired in this way. One case involves a mutagenic photodimer caused by UV light (see Figure 18-23). The cyclobutane–pyrimidine photodimer can be repaired by a **photoreactivating enzyme (PRE),** whereas the 6-4 photoproduct cannot. The enzyme operates by binding to the photodimer and splitting it, in the presence of certain wavelengths of visible light, to generate the original bases (Figure 18-28). This enzyme cannot operate in the dark, so other repair pathways are required to remove UV damage.

Alkyltransferases are also enzymes involved in the direct reversal of lesions. They remove certain alkyl groups that have been added to the 0 to 6 positions of guanine (Figure 18-17) by such agents as NG and EMS. The methyltransferase from *E. coli* has been well studied. This enzyme transfers the methyl group from *O*-6-methylguanine to a cysteine residue on the protein (Figure 18-29). When this happens, the enzyme is inactivated, so this repair system can be saturated if the level of alkylation is high enough.

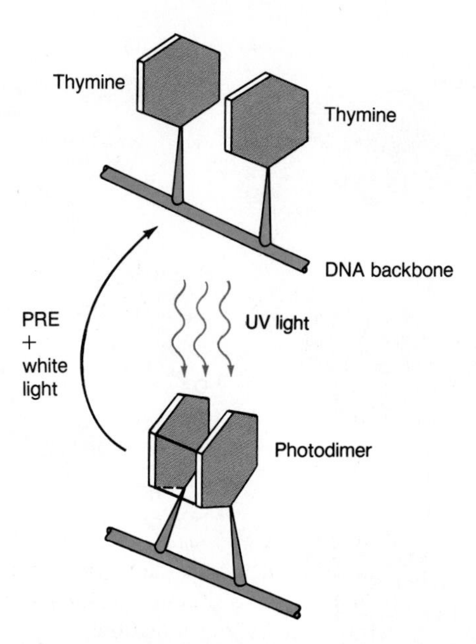

Figure 18-28 Repair of a UV-induced pyrimidine photodimer by a photoreactivating enzyme (PRE). The enzyme recognizes the photodimer (here, a thymine dimer) and binds to it. When light is present, the PRE uses its energy to split the dimer into the original monomers. (After J. D Watson, *Molecular Biology of the Gene,* 3d ed. Copyright 1976 by W. A. Benjamin).

Excision–Repair Pathways

General Excision Repair. There are several pathways for excising altered bases, together with a stretch of neighboring bases, and then repairing the gap by DNA synthesis. One general pathway is encoded in *E. coli* by three genes termed *uvrA*, *uvrB*, and *uvrC*. This system recognizes any lesion that creates a significant distortion of the DNA double helix. An **endonuclease** called uvrABC nuclease makes a cut several base pairs away on either side of the damaged base, and a 12-base-long segment of single-stranded DNA is removed. The short gap is then filled in by **repair synthesis** (mediated by DNA polymerase I) and is sealed by DNA ligase (Figure 18-30).

Many types of lesions are removed by means of this general excision–repair process, including the principal UV photoproducts (Figure 18-23) as well as the bulky chemical additions resulting from the binding of compounds such as AFB_1 and epoxides of benzo(a)pyrene. This is a crucial repair pathway in humans also. A rare inherited disorder, xeroderma pigmentosum (see Table

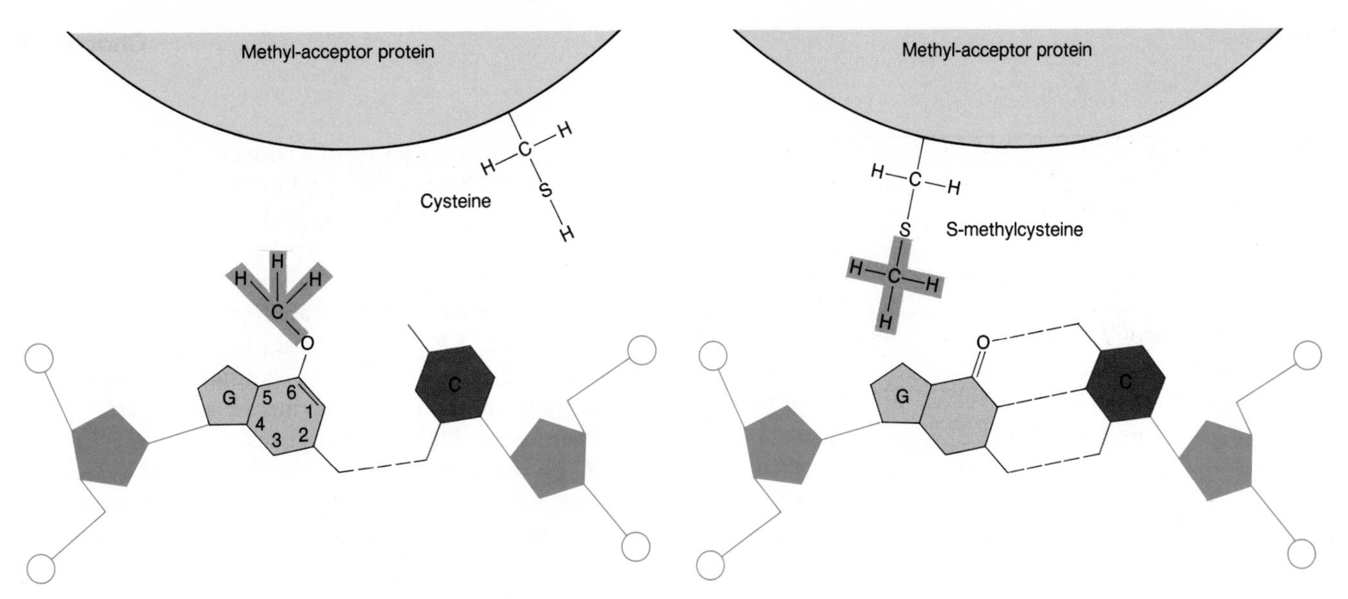

Figure 18-29 Direct reversal of DNA damage by an alkyltransferase. Methylation of a guanine residue by nitrosoguanidine (NG) is repaired by this novel process. The NG adds a methyl group (CH_3) at various sites in the DNA, including an oxygen atom at position 6 of guanine *(left)*. This disrupts the hydrogen bonding of guanine to a cytosine. The repair is accomplished by a methyl-acceptor protein, one of the enzymes known as alkyltransferases. A cysteine residue on the protein acts as the methyl acceptor: it binds the CH_3 group, thereby restoring the guanine to its original state *(right)*. (From P. Howard-Flanders, "Inducible Repair of DNA." Copyright 1981 by Scientific American, Inc. All rights reserved.)

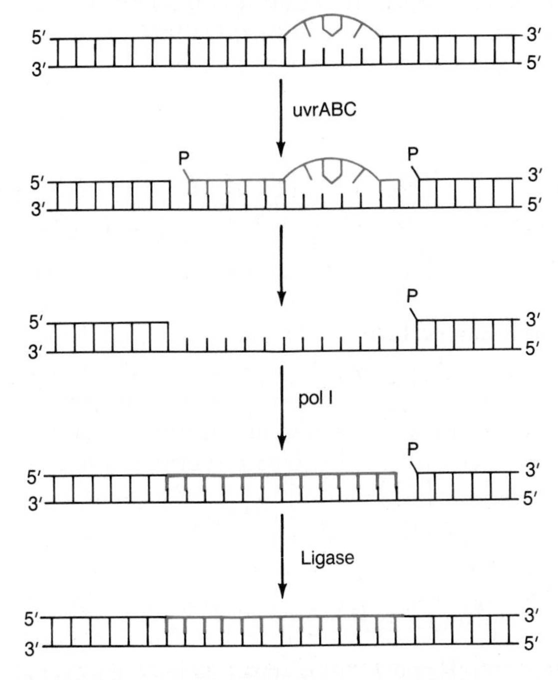

Figure 18-30 A model for excision repair of a DNA-distorting lesion in *E. coli.* Here uvrABC nuclease hydrolyzes the eighth phosphodiester bond 5′ and the fourth phosphodiester bond 3′ to a pyrimidine photodimer, producing a 12-nucleotide-long single-stranded DNA fragment and 3′-OH and 5′-phosphoryl termini. The oligonucleotide carrying the damage is removed, and the resulting gap is filled by DNA polymerase I (pol I) and sealed by DNA ligase. This diagram shows the removal of a 12-nucleotide fragment. The uvrABC nuclease also generates 13-base-long oligonucleotides. (From A. Sancar and W. D. Rupp, *Cell,* 33, 1983, 249.)

18-5), results from a deficiency in one of the excision–repair enzymes. Most people with this disorder die of skin cancer before they reach the age of 30, so we can see how vital a role the excision–repair pathway plays.

Specific Excision Pathways. Certain lesions are too subtle to cause a distortion large enough to be recognized by the *uvrABC*-encoded general excision–repair system. Thus, additional excision pathways are necessary.

AP Endonuclease Repair Pathway. All cells have endonucleases that attack the sites left after the spontaneous loss of single purine or pyrimidine residues. For convenience, the apurinic and apyrimidinic sites are termed **AP sites** because they are biochemically equivalent (see Figure 18-7). The **AP endonucleases** are vital to the cell, since, as noted earlier, spontaneous depurination is a relatively frequent event. These enzymes introduce chain breaks by cleaving the phosphodiester bonds at AP sites. This initiates an excision–repair process mediated by three further enzymes—an exonuclease, DNA polymerase I, and DNA ligase (Figure 18-31). (The last two enzymatic steps are identical to the corresponding steps in the excision–repair process shown in Figure 18-30).

Due to the efficiency of the AP endonuclease repair pathway, it can be the final step of other repair pathways. Thus, if damaged base pairs can be excised, leaving an AP site, the AP endonucleases can complete the restoration to the wild type. This is what happens in the DNA glycosylase repair pathway.

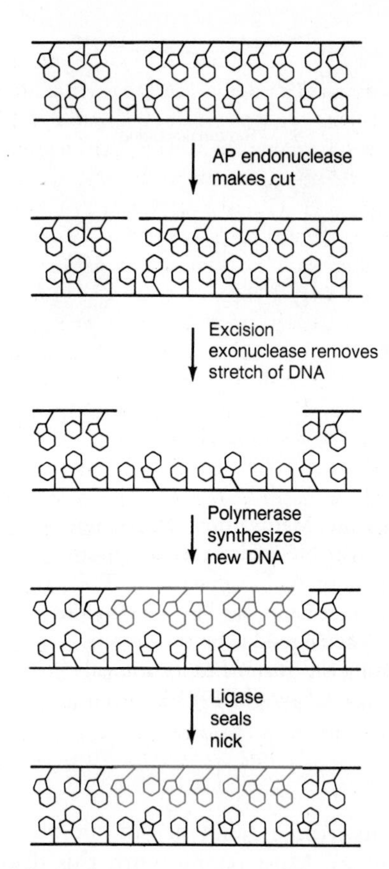

Figure 18-31 Repair of AP (apurinic or apyrimidinic) sites. AP endonucleases recognize AP sites and cut the phosphodiester bond. A stretch of DNA is removed by an exonuclease, and the resulting gap is filled in by DNA polymerase I and DNA ligase. (After B. Lewin, *Genes*. Copyright 1983 by John Wiley.)

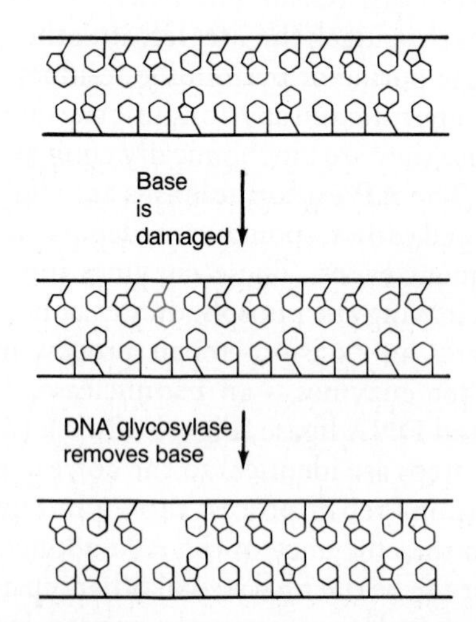

Figure 18-32 Action of DNA glycosylases. Glycosylases remove altered bases and leave an AP site. The AP site is subsequently excised by the AP endonucleases diagrammed in Figure 18-31. (After B. Lewin, *Genes*. Copyright 1983 by John Wiley.)

DNA Glycosylase Repair Pathway. **DNA glycosylases** do not cleave phosphodiester bonds but instead cleave *N*-glycosidic (base–sugar) bonds, liberating the altered base and generating an AP site (Figure 18-32). The AP site resulting from glycosylase action is then repaired by the AP endonuclease repair pathway (Figure 18-31). Thus, AP endonucleases cleave the phosphodiester bond, and exonuclease excision is followed by repair mediated by DNA polymerase and ligase.

Numerous DNA glycosylases exist. One, uracil-DNA glycosylase, removes uracil from DNA. Uracil residues, which result from the spontaneous deamination of cytosine (Figure 18-8), can lead to a C → T transition if unrepaired. It is possible that the natural pairing partner of adenine in DNA is thymine (5-methyluracil), rather than uracil, in order to allow the recognition and excision of these uracil residues. If uracil were a normal constituent of DNA, such repair would not be possible.

There is also a glycosylase that recognizes and excises hypoxanthine, the deamination product of adenine. Other glycosylases remove alkylated bases (such as 3-methyladenine, 3-methylguanine, and 7-methylguanine), ring-opened purines, oxidatively damaged bases, and, in some organisms, UV photodimers. New glycosylases are still being discovered.

The GO System. Two glycosylases, the products of the *mutM* and *mutY* genes, work together to prevent mutations arising from the 8-oxodG, or "GO," (Figure 18-10) lesion in DNA. Together with the product of the *mutT* gene mentioned above, these glycosylases form the GO system. When GO lesions are generated in DNA by spontaneous oxidative damage, a glycosylase encoded by *mutM* removes the lesion (Figure 18-33). Still, some GO lesions persist and mispair with adenine. A second glycosylase, the product of the *mutY* gene, removes the adenine from this specific mispair, leading to restoration of the correct cytosine by repair synthesis (mediated by DNA polymerase I) and allowing subsequent removal of the GO lesion by the *mutM* product.

Postreplication Repair

Mismatch Repair. Some repair pathways are capable of recognizing errors even after DNA replication has already occurred. One such system, termed the **mismatch repair system,** can detect mismatches that occur during DNA replication. Suppose you were to design an enzyme system that could repair replication errors.

What would this system have to be able to do? At least three things:

1. Recognize mismatched base pairs.
2. Determine which base in the mismatch is the incorrect one.

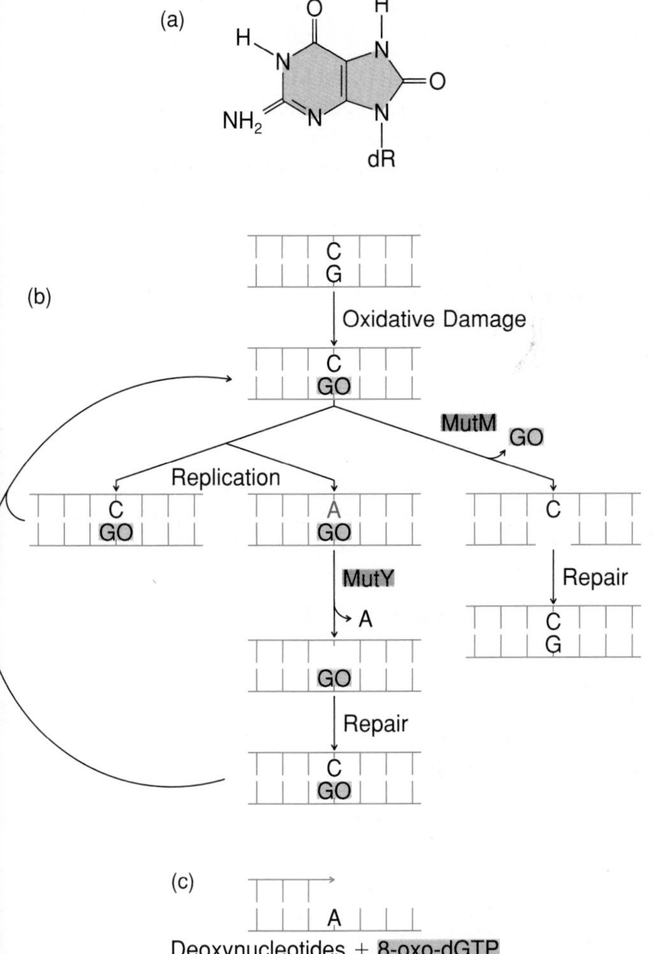

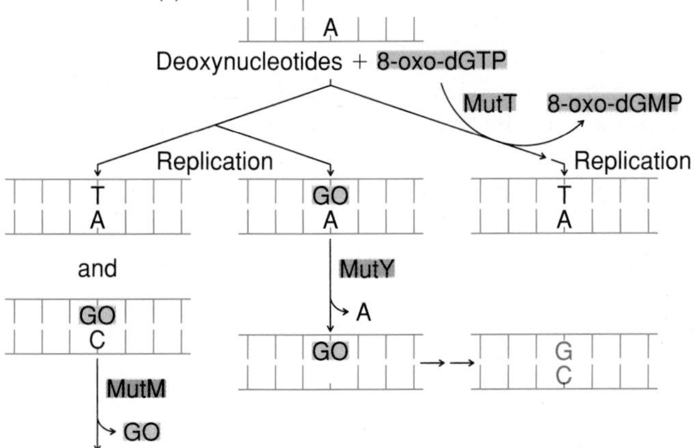

Figure 18-33 The GO system. (a) 7,8-dihydro-8-oxo-guanine (8-hydroxyguanine), GO. This is the structure of the predominant tautomeric form of the GO lesion. (b) The GO lesions, generated by oxidative damage, can be removed by the *mutM* gene product (MutM protein) and subsequent repair can restore the original G–C base pair. Translesion synthesis by replicative DNA polymerases is frequently inaccurate, leading to the misincorporation of A opposite the GO lesion. The *mutY* glycosylase (MutY) removes the misincorporated adenine from the A–GO mispairs. Repair polymerases are much less error prone during translesion synthesis and can lead to a C–GO pair, a substrate for MutM. (c) Oxidative damage can also generate 8-oxo-dGTP from dGTP in the deoxynucleotide pool. MutT is active on 8-oxo-dGTP and hydrolyzes it to 8-oxo-dGMP, effectively removing the triphosphate from the deoxynucleotide pool. If 8-oxo-dGTP were not removed, inaccurate replication could result in the misincorporation of 8-oxo-dGTP opposite template A residues, leading to A–GO mispairs. MutY could be involved in the mutation process because it is active on the A–GO substrate and would remove the template A, leading to the AT → CG transversions that are characteristic of a *mutT* strain. The 8-oxo-dGTP could also be incorporated opposite template cytosines, resulting in a damaged C–GO pair that could be corrected by MutM. (From M. Michaels and J. H. Miller, *Journal of Bacteriology* 174, 1992, 6321.)

3. Excise the incorrect base and carry out repair synthesis.

The second point is the crucial property of such a system. Unless it is capable of discriminating between the correct and the incorrect bases, the mismatch repair system could not determine which base to excise. If, for example, a G–T mismatch occurs as a replication error, how can the system determine whether G or T is incorrect? Both are normal bases in DNA. But replication errors produce mismatches on the newly synthesized strand, so it is the base on this strand that must be recognized and excised.

To distinguish the old, template strand from the newly synthesized strand, the mismatch repair system takes advantage of the normal delay in the postreplication methylation of the sequence

5′—G—A—T—C—3′
3′—C—T—A—G—5′

The methylating enzyme is **adenine methylase,** which creates 6-methyladenine on each strand. However, it takes the adenine methylase several minutes to recognize and modify the newly synthesized GATC stretches. During that interval, the mismatch repair system can operate because it can now distinguish the old strand from the new one by the methylation pattern. Methylating the 6-position of adenine does not affect base-pairing, and it provides a convenient tag that can be detected by other enzyme systems. Figure 18-34 shows the replication fork during mismatch correction. Note that only the old strand is methylated at GATC sequences right after replication.

Once the mismatched site has been identified, the mismatch repair system corrects the error. Figure 18-35 depicts two models proposed to explain how the mismatch repair system carries out the correction.

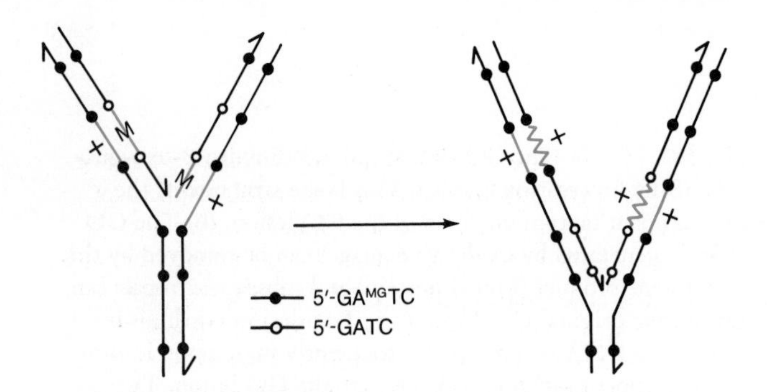

Figure 18-34 Mismatch correction during replication. When mismatched base pairs (+/M) occur as replicational errors (mutations M), the incorrect bases must be on the newly synthesized strands. The new strands can be recognized for a brief period while the GATC sequences close to the replication fork remain unmethylated (–o–). The incorrect bases can then be efficiently repaired by excision and resynthesis (–∧∧∧–). The mismatch repair system conserves the correct DNA sequence because the repair is strand-directed. (A^{MG} = adenine with a methyl group at the 6 position.) (From F. Bourguignon-Van Horen et al., *Biochimie,* 64, 1982, 559.)

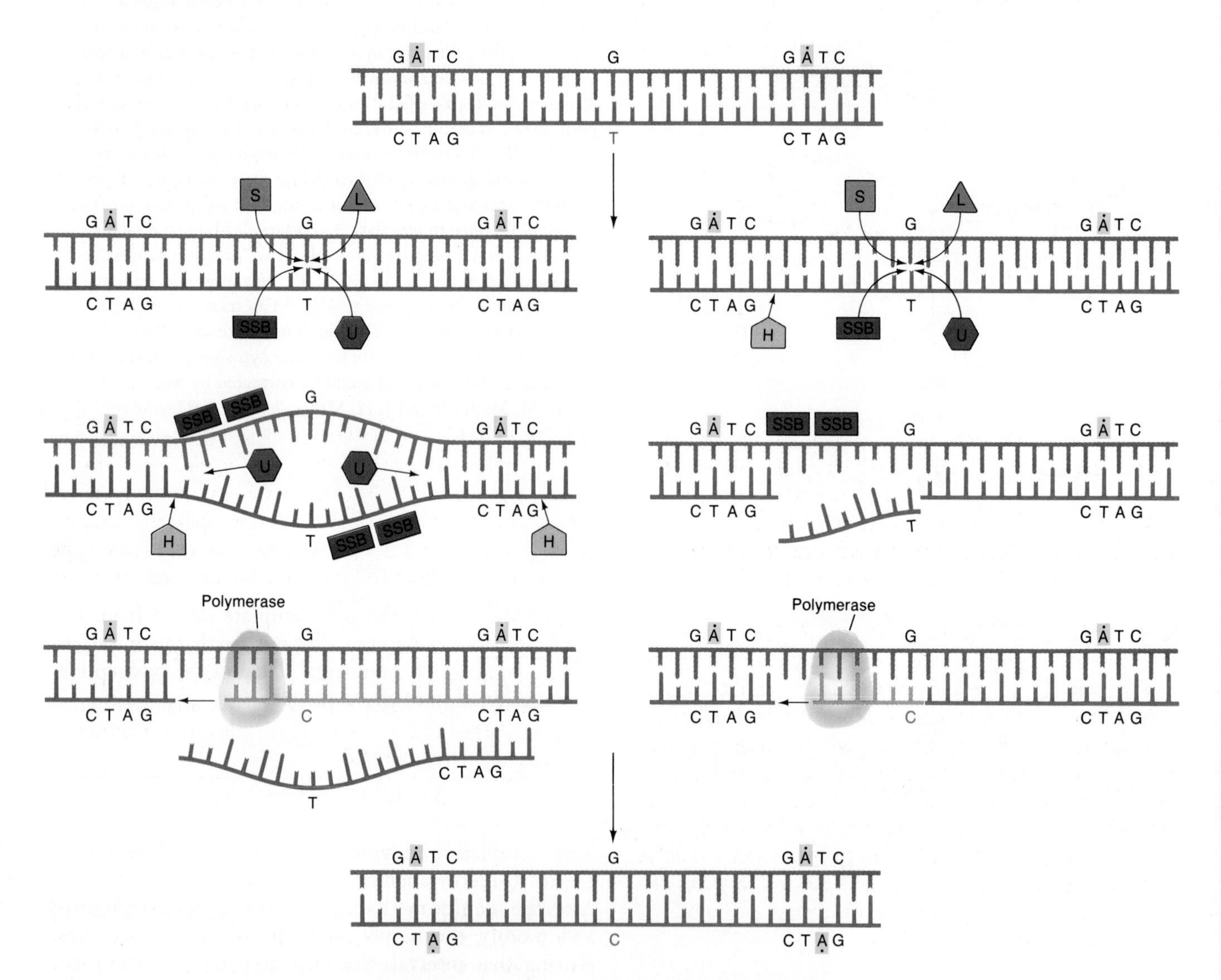

Figure 18-35 Two proposed models to explain how mismatch repair corrects errors after a DNA strand has been synthesized. In both models, proteins called MutL (L) and MutS (S) interact with the mismatch site (G–T), and a protein called MutH (H) cleaves the newly synthesized strand. The repair apparatus distinguishes the parental strand from the new one by means of the methyl group (black dots) within the parental GATC sequences. The strands surrounding the mismatch are separated with the help of a protein called MutU (U) and are stabilized by a single-stranded binding protein (SSB). The main difference between the two models has to do with where the strand containing the incorrect nucleotide is cleaved: at two flanking GATC sequences *(left)* or at one GATC sequence and the mismatch itself *(right)*. In either case, polymerases synthesize a new segment in place of the excised one, and the corrected strand is eventually methylated like the parental copy. (From M. Radman and R. Wagner "The High Fidelity of DNA Duplication." Copyright 1988 by Scientific American, Inc. All rights reserved.)

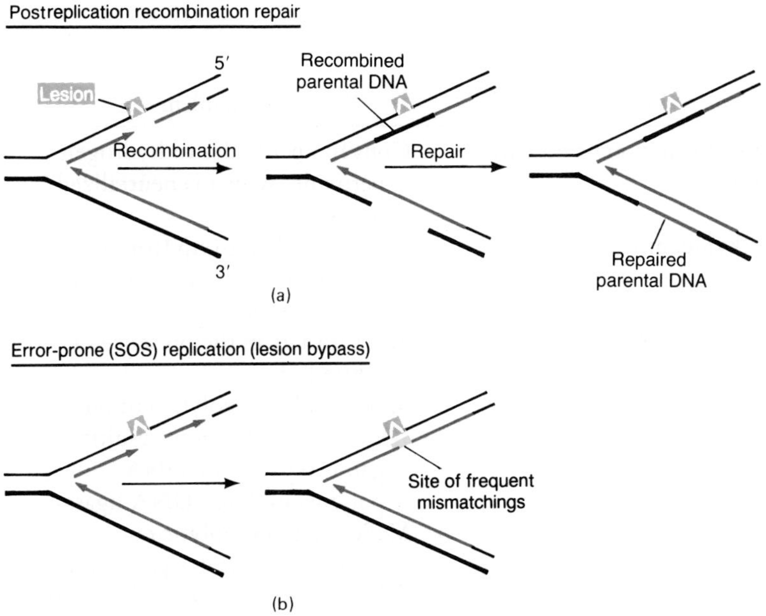

Postreplication recombination repair

Error-prone (SOS) replication (lesion bypass)

Figure 18-36 Schemes for postreplication repair. (a) In recombinational repair, replication jumps across a blocking lesion, leaving a gap in the new strand. A *recA*-directed protein then fills in the gap, using a piece from the opposite parental strand (because of DNA complementarity, this filler will supply the correct bases for the gap). Finally, the *recA* protein repairs the gap in the parental strand. (b) In SOS bypass, when replication reaches a blocking lesion, the SOS system inserts the necessary number of bases (often incorrect ones) directly across from the lesion and replication continues without a gap. Note that with either pathway the original blocking lesion is still there and must be repaired by some other repair pathway. (Adapted from A. Kornberg and T. Baker, *DNA Replication,* 2nd ed. Copyright 1992 by W. H. Freeman and Co., New York.)

A second mismatch repair activity has been discovered in bacteria. An **adenine glycosylase** enzyme encoded by the *mutY* gene recognizes and removes G–A mispairs, as well as 8-oxodG–A mispairs (see the GO system above). This creates an AP site, which is then repaired by the AP endonuclease pathway.

Recombinational Repair. The *recA* gene, which is involved in SOS bypass (Figure 18-22), is also involved in postreplication repair. Here the DNA replication system stalls at a UV photodimer or other blocking lesions and then restarts past the block, leaving a single-stranded gap. In recombinational repair, this gap is patched by DNA cut from the sister molecule (Figure 18-36a). This process seems to lead to few errors. SOS bypass, by contrast, is highly mutagenic, as described earlier. Here the replication system continues past the lesion (Figures 18-22 and 18-36b), accepting noncomplementary nucleotides for new strand synthesis.

Strategy for Repair

We can now assess the overall repair-system strategy used by the cell. The many different repair systems available to the cell are summarized in Table 18-4. It would be convenient if enzymes could be used to directly reverse each specific lesion or excise particular altered bases. However, it would cost too much energy to encode and synthesize a different repair enzyme for each possible type of cell damage. Therefore, a general excision repair system is used to remove any type of damaged base that causes a recognizable distortion in the double helix. When lesions are too subtle to cause such a distortion, specific excision systems, glycosylases, or removal systems are designed. To eliminate replication errors, a postreplication mismatch repair system operates; finally, postreplication recombinational systems eliminate gaps across from blocking lesions that have escaped the other repair systems.

> **Message** Repair enzymes play a crucial role in reducing genetic damage in living cells. The cell has many different repair pathways at its command to eliminate potentially mutagenic errors.

Mutators

As the preceding description of repair processes indicates, normal cells are programmed for error avoidance. The repair processes are so efficient that the observed base-substitution rate is as low as 10^{-10} to 10^{-9} per base pair per cell per generation in *E. coli.* However, mutant strains with increased spontaneous mutation rates have been detected. Such strains are termed **mutators.** In many cases, the mutator phenotype is due to a defective repair system.

In *E. coli,* the mutator loci *mutH, mutL, mutU,* and *mutS* affect components of the postreplication mismatch repair system (see Figure 18-35), as does the *dam* locus, which specifies the enzyme *d*eoxy*a*denosine *m*ethylase. Strains that are *dam*⁻ cannot methylate adenines at GATC sequences (see Figure 18-34), and so the mis-

c. In the reversion experiment for mutant 5, a particularly interesting prototrophic derivative is obtained. When this type is crossed to a standard wild-type strain, the progeny consists of 90 percent prototrophs and 10 percent auxotrophs. Provide a full explanation for these results, including a precise reason for the frequencies observed.

13. You are using nitrous acid to "revert" mutant *nic-2* alleles in *Neurospora.* You treat cells, plate them on a medium without nicotinamide, and look for prototrophic colonies. You obtain the following results for two mutant alleles. Explain these results at the molecular level, and indicate how you would test your hypotheses.

a. With *nic-2* allele 1, you obtain no prototrophs at all.

b. With *nic-2* allele 2, you obtain three prototrophic colonies, and you cross each separately with a wild-type strain. From the cross prototroph A × wild-type, you obtain 100 progeny, all of which are prototrophic. From the cross prototroph B × wild-type, you obtain 100 progeny, of which 78 are prototrophic and 22 are nicotinamide-requiring. From the cross prototroph C × wild-type, you obtain 1000 progeny, of which 996 are prototrophic and 4 are nicotinamide-requiring.

14. Devise imaginative screening procedures for detecting the following:

a. Nerve mutants in *Drosophila.*

b. Mutants lacking flagella in a haploid unicellular alga.

c. Supercolossal-sized mutants in bacteria.

d. Mutants that overproduce the black compound melanine in normally white haploid fungus cultures.

e. Individual humans (in large populations) whose eyes polarize incoming light.

f. Negatively phototrophic *Drosophila* or unicellular algae.

g. UV-sensitive mutants in haploid yeast.

19 Mechanisms of Genetic Change II: Recombination

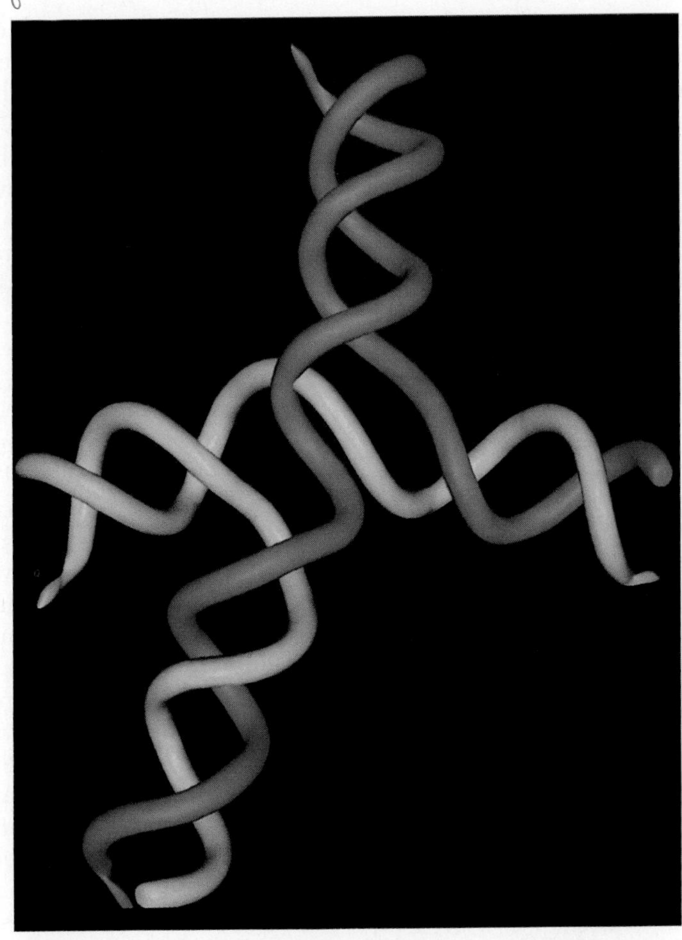

Computer model of exchange site. (Julie Newdol, Computer Graphics Laboratory, University of California, San Francisco. Copyright by Regents, University of California.)

KEY CONCEPTS

Recombination occurs at regions of homology between chromosomes through the breakage and reunion of DNA molecules.

▪

Models for recombination, such as the Holliday model, involve the creation of a heteroduplex branch, or cross bridge, that can migrate and the subsequent splicing of the intermediate structure to yield different types of recombinant DNA molecules.

▪

The Holliday model can be applied to explain genetic crosses.

▪

Many of the enzymes involved in recombination in bacteria have been identified.

▪

Specific recombination systems catalyze recombination events only between certain sequences.

A normal crossover (a reciprocal interchromosomal recombination event) is an extraordinary process. Somehow the genetic material from one parental chromosome and the genetic material from the other parental chromosome are "cut up and pasted together" during each meiosis, and this is done with complete reciprocity. In other words, neither chromosome gains or loses any genes in the process. In fact, it is probably correct to say that neither chromosome gains or loses even one nucleotide in the exchange. How is this remarkable precision attained? Many interesting phenomena provide important clues about the nature of the answer.

General Homologous Recombination

Some recombination events occur only at specific sequences. We consider these special events toward the end of the chapter. Here we focus on the process that results in recombination at any large region of homology between chromosomes.

The Breakage and Reunion of DNA Molecules

Throughout our analysis of linkage, we have implicitly assumed that crossing-over occurs by some process of breakage and reunion of chromatids. The evidence against the copy-choice hypothesis (Chapter 5) provides good *indirect* evidence in favor of breakage and reunion. Furthermore, there is good genetic and cytological evidence that crossing-over occurs during the prophase stage of meiosis, rather than during interphase when chromosomal DNA is replicating. A small amount of DNA synthesis does occur during prophase, but certainly chromosome replication is not associated with crossing-over. One of the first direct proofs that chromosomes (albeit viral chromosomes) can break and rejoin came from experiments on λ phage done in 1961 by Matthew Meselson and Jean Weigle.

Meselson and Weigle simultaneously infected *E. coli* with two strains of λ. One strain, which had the genetic markers *c* and *mi* at one end of the chromosome, was "heavy" because the phages were produced from cells grown in heavy isotopes of carbon (^{13}C) and nitrogen (^{15}N). The other strain was c^+ mi^+ for the markers and had "light" DNA because it was harvested from cells grown on the normal light isotopes ^{12}C and ^{14}N. The two DNAs (chromosomes) can be represented as shown in Figure 19-1a. The multiply infected cells were then incubated in a light medium until they lysed.

The progeny phages released from the cells were spun in a cesium chloride density gradient. A wide band was obtained, indicating that the viral DNAs ranged in density from the heavy parental value to the light parental value, with a great many intermediate densities (Figure 19-1b). Interestingly, some recombinant phages were recovered with density values very close to the heavy parental value. They were of genotype *c* mi^+, and they must have arisen through an exchange event between the two markers (Figure 19-1c). The heavy density of the chromosome would be expected because only the small tip of the chromosome carrying the mi^+ allele would come from the light parental chromosome. When

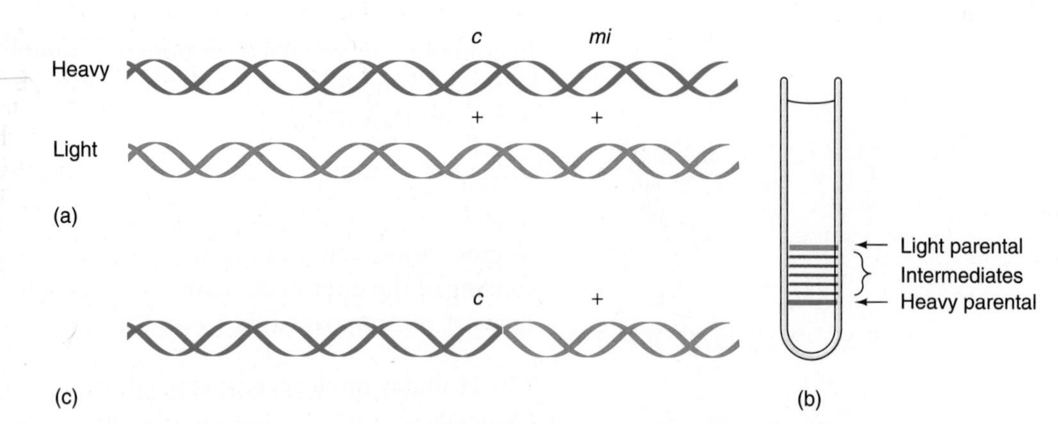

Figure 19-1 Evidence for chromosome breakage and reunion in λ phages. (a) The chromosomes of the two λ strains used to multiply infect *E. coli.* (b) Bands produced when progeny phages are spun in a cesium chloride density gradient. The fact that intermediate densities are obtained indicate a range of chromosome compositions with partly light and partly heavy components. (c) The chromosome of the heavy *c* mi^+ progeny resulting from crossover between the two markers. The density of this crossover product confirms that the crossover involved a physical breakage and reunion of the DNA.

heavy $c^+ mi^+$ phages were crossed with light $c\ mi$, the heavy recombinants were found to be $c^+ mi$, and the light recombinants were found to be $c\ mi^+$, as expected. These results can be explained in only one way: the recombination event must have occurred through the physical breakage and reunion of DNA. Of course, we have to be careful about extrapolating from viral to eukaryotic chromosomes. However, this evidence shows that the breakage and reunion of DNA strands is a chemical possibility.

Chiasmata: The Crossover Points

In Chapter 5, we made the simple assumption that chiasmata are the actual sites of crossovers. Mapping analysis indirectly supports this idea: since an average of one crossover per meiosis produces 50 genetic map units, there should be correlation between the size of the genetic map of a chromosome and the observed mean number of chiasmata per meiosis. This correlation has been made in well-mapped organisms.

However, the harlequin chromosome-staining technique (see Chapter 11) has made it possible to test the idea directly. In 1978, C. Tease and G. H. Jones prepared harlequin chromosomes in meioses of the locust. Remember that the harlequin technique produces sister chromatids: one dark, and the other light. When a crossover occurs, it can involve two dark, two light, or dark and light nonsister chromatids as shown in Figure 11-21b (page 318). This last situation is crucial because mixed (part dark and part light) crossover chromatids are produced. Tease and Jones found that the dark-light transition occurs right at the chiasma—proving beyond reasonable doubt that these are the crossover sites and settling a question that had been unresolved since the early part of the century (Figure 19-2).

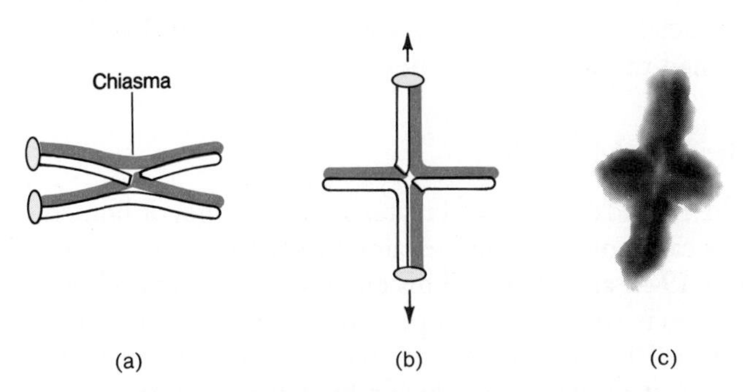

Figure 19-2 Crossing-over between dark- and light-stained nonsister chromatids in a meiosis in the locust. (a) Representation of the chiasma. (b) The best stage for observing is when the centromeres have pulled apart slightly, forming a cross-shaped structure with the chiasma at the center. (c) Photograph of the stage shown in (b). (Photo courtesy of C. Tease and G. H. Jones, *Chromosoma* 69, 1978, 163-178.)

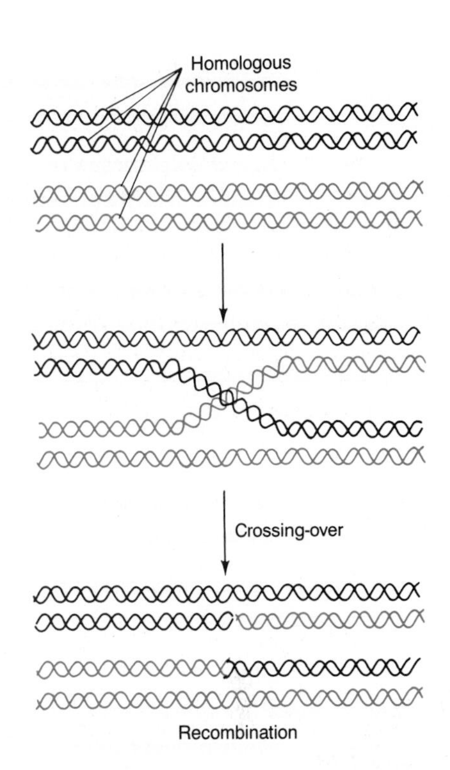

Figure 19-3 The molecular event of recombination may be schematically represented by two double-stranded molecules breaking and rejoining. (From J. Darnell, H. Lodish, and D. Baltimore, *Molecular Cell Biology*. Copyright 1986 by Scientific American Books.)

The Holliday Model

Much of the available information on the mechanism of interchromosomal recombination, especially at the chemical level, has come from studies of bacteria and phages. In addition, we shall consider a different kind of information based on the genetic analysis of eukaryotes. This is a study of what can be called the *genetics of genetics!* Several clues about recombination mechanisms have emerged in studies of eukaryotes, resulting in the construction and refinement of models of recombination. Let's examine our current model of recombination and see how it explains experimental results from different genetic systems. Then we can consider some of the enzymes involved in the recombination process.

Any model for recombination must explain the basic sequence of events depicted in Figure 19-3. The most plausible recombination model was originally formulated by Robin Holliday and has subsequently been modified by others, in particular Matthew Meselson and Charles Radding. The **Holliday model** is typical of the various recombination models that have been proposed for both eukaryotes and prokaryotes. The concept of hybrid, or **heteroduplex,** DNA put forth in this model provides a useful way of expressing the present state of

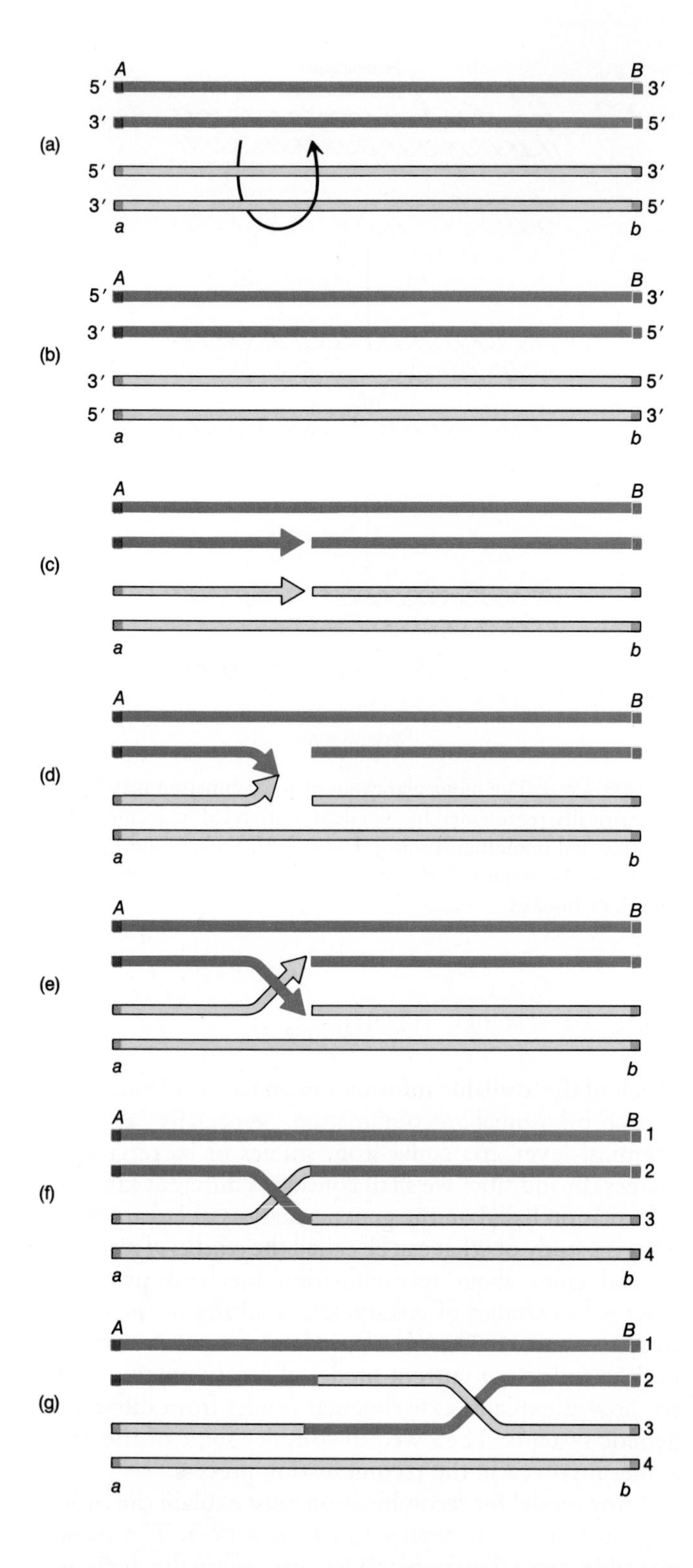

Figure 19-4 A prototype mechanism for genetic recombination. (a) Two homologous double helices are shown. Each pair represents a chromatid, and the two pairs represent two nonsister chromatids. (b) The helices are rotated around the horizontal axis (see arrow) and aligned so that the bottom strand of the first helix has the same polarity as the top strand of the second helix. (c) Two parallel or two antiparallel strands are cut. (d) The free ends leave the complementary strands to which they had been hydrogen-bonded. (e) The free ends become associated with the complementary strands in the homologous double helix. (f) Ligation creates partially heteroduplex double helices. This is the Holliday structure. (g) Migration of the branch point occurs by continued strand transfer by the two polynucleotide chains involved in the crossover. (From H. Potter and D. Dressler, *Cold Spring Harbor Symposium on Quantitative Biology* 43, 1979, 970. Cold Spring Harbor Laboratory, Cold Spring Harbor, N.Y.)

knowledge about crossing-over. Figures 19-4, 19-5, and 19-6 depict the Holliday model in its entirety. In some of the figures pertinent to this discussion of recombination, the DNA strands have been numbered so that they can be followed from one figure to another. The model is based on the creation of a branch, or **cross bridge,** its migration along the two heteroduplex strands, termed **branch migration,** and the subsequent splicing of the intermediate structure in one of two ways to yield different types of recombinant molecules. Let's work through the Holliday model in Figure 19-4.

Enzymatic Cleavage and the Creation of Heteroduplex DNA

Looking at 19-5a, we can see that two homologous double helices are aligned, although note that in 19-4b they have been rotated so that the bottom strand of the first helix has the same polarity as the top strand of the second helix (3′ → 5′, in this case). Then a nuclease cleaves the two strands that have the same polarity (Figure 19-4c). The free ends leave their original complementary strands (Figure 19-4d) and undergo hydrogen bonding with the complementary strands in the homologous double helix (Figure 19-4e). Ligation produces the structure shown in Figure 19-4f. This partially heteroduplex double helix is a crucial intermediate in recombination. It has been termed the **Holliday structure.**

Branch Migration

The Holliday structure creates a cross bridge, or branch, that can move, or migrate, along the heteroduplex (Figures 19-4f and 19-4g). This phenomenon of branch migration is a distinctive property of the Holliday structure. Figure 19-5 portrays a more realistic view of this structure as it might appear during branch migration.

Resolution of Holliday Structure

Figure 19-6 demonstrates, in schematic form, how the Holliday structure can be converted to the recombinant structures with which we are familiar. In 19-6a, we can

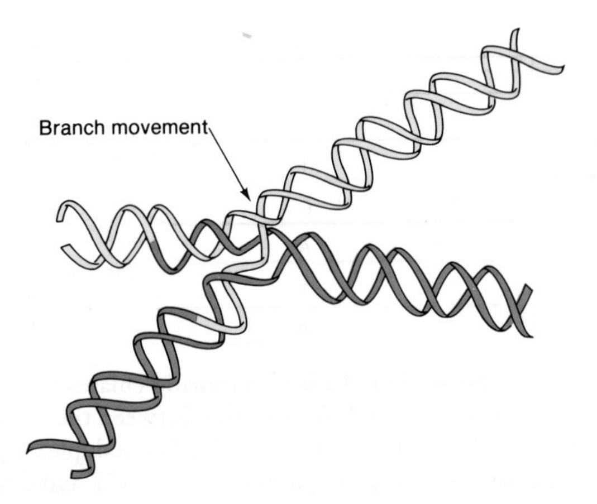

Figure 19-5 Branch migration, the movement of the crossover point between DNA complexes. (After T. Broker, *Journal of Molecular Biology* 81, 1973, 1; from J. D. Watson et al., *Molecular Biology of the Gene,* 4th ed. Copyright 1987 by Benjamin Cummings.)

see the structure that we arrived at in Figure 19-4g drawn out in an extended form. Compare 19-4g and 19-6a until you are convinced that these two structures are indeed equivalent. If we rotate the bottom portion of this structure, as shown in Figure 19-6b, we can generate the form depicted in Figure 19-6c. This last form can be converted back to two unconnected double helices by enzymatically cleaving only two strands. As indicated in 19-6c, cleavage can occur in either of two ways, each of which generates a different product (Figure 19-6d). These cleaved structures can be viewed more simply (Figure 19-6e). Repair synthesis produces the final recombinant molecules (Figure 19-6f). Note the two different types of recombinants.

Application of the Holliday Model to Genetic Crosses

Let's examine how the model shown in Figures 19-4, 19-5, and 19-6 relate to a genetic cross. We can set up a hypothetical cross of + × *m*, in which the + site corresponds to a G–C nucleotide pair and the *m* site corre-

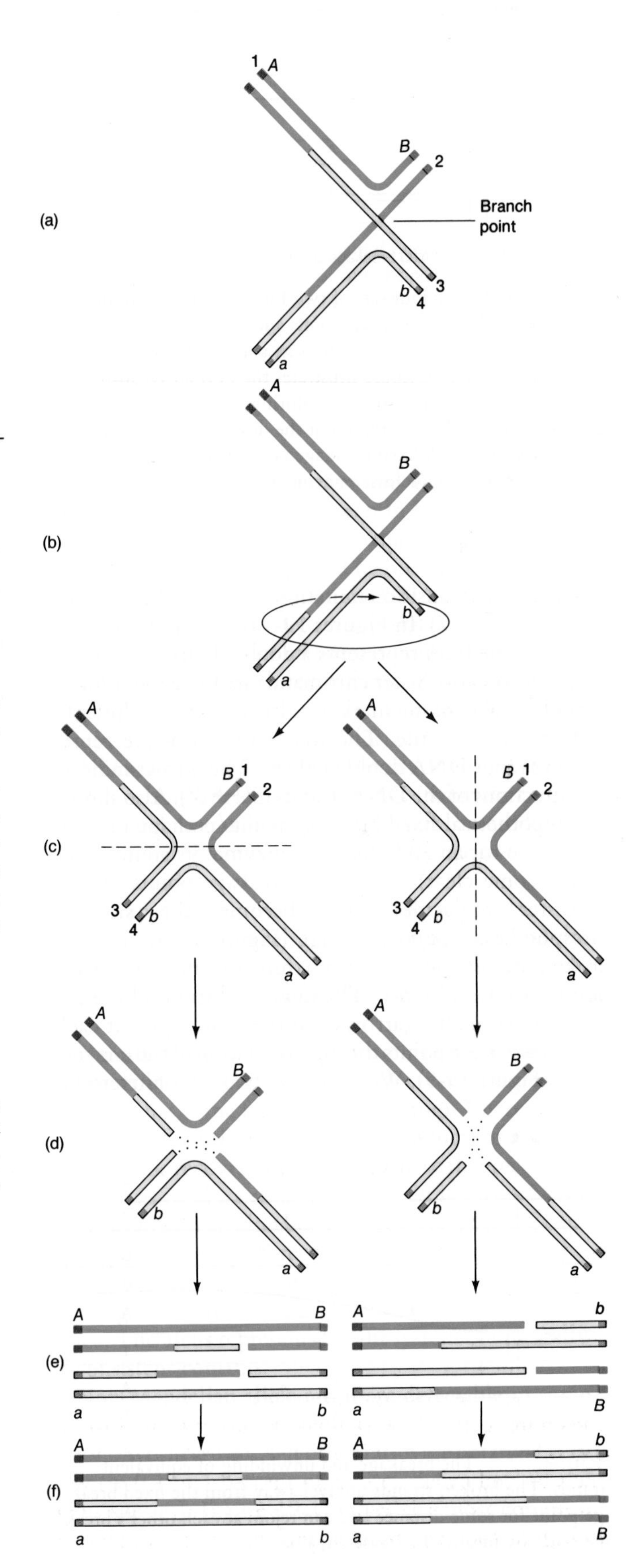

Figure 19-6 (a) The Holliday structure shown in an extended form. (b) The rotation of the structure shown in (a) can yield the form depicted in (c). Resolution of the structure shown in (c) can proceed in two ways, depending on the points of enzymatic cleavage, yielding the structures shown in (d). The dotted lines show which segments will rejoin to form recombinant strands for each particular cleavage scheme. The strands are shown linearly in (e) and can be repaired to the forms shown in (f). (From H. Potter and D. Dressler, *Cold Spring Harbor Symposium on Quantitative Biology* 43, 1970, 970. Cold Spring Harbor Laboratory, Cold Spring Harbor, N.Y.)

pair insertion or deletion mutations show gene conversions of the 6:2 or 2:6 type and only rarely of the 5:3, 3:5, or 3:1:1:3 type. Base-pair transition mutations show gene conversion of the 3:5, 5:3, or 3:1:1:3 type, and only rarely of the 6:2 or 2:6 type.

a. In terms of the hybrid DNA model, propose an explanation for these observations.

b. Leblon and Rossignol also have shown that there are far fewer 6:2 than 2:6 conversions for insertions and far more 6:2 than 2:6 conversions for deletions (where the ratios are $+:m$). Explain these results in terms of heteroduplex DNA. (You might also think about the excision of thymine photodimers.)

c. Finally, the researchers have shown that when a frame-shift mutation is combined in a meiosis with a transition mutation at the same locus in a cis configuration, the asci showing *joint* conversion are all 6:2 or 2:6 for *both* sites (that is, the frame-shift conversion pattern seems to have "imposed its will" on the transition site). Propose an explanation for this.

9. At the *grey* locus in the Ascomycete fungus *Sordaria*, the cross $+ \times g_1$ is made. In this cross, heteroduplex DNA sometimes extends across the site of heterozygosity, and two heteroduplex DNA molecules are formed (as discussed in this chapter). However, correction of heteroduplex DNA is not 100 percent efficient. In fact, 30 percent of all heteroduplex DNA is not corrected at all, whereas 50 percent is corrected to + and 20 percent is corrected to g_1. What proportion of aberrant-ratio asci will be **(a)** 6:2? **(b)** 2:6? **(c)** 3:1:1:3? **(d)** 5:3? **(e)** 3:5?

10. Noreen Murray crossed α and β, two alleles of the *me-2* locus in *Neurospora*. Included in the cross were two markers, *trp* and *pan*, which each flank *me-2* at a distance of 5 m.u. The ascospores were plated onto a medium containing tryptophan and pantothenate but no methionine. The methionine prototrophs that grew were isolated and scored for the flanking markers, yielding the results shown in Table 19-3.

Table 19-3.

Cross	Genotype of *me-2*$^+$ prototrophs			
	trp +	+ *pan*	*trp pan*	+ +
trp α + × + β *pan*	26	59	16	56
trp β + × + α *pan*	84	23	87	15

Interpret these results in light of the models presented in this chapter. Be sure to account for the asymmetries in the classes.

11. In *Neurospora*, the cross $A\,x \times a\,y$ is made, in which *x* and *y* are alleles of the *his-1* locus and *A* and *a* are mating-type alleles. The recombinant frequency between the *his-1* alleles is measured by the prototroph frequency when ascospores are plated on a medium lacking histidine; the recombinant frequency is measured as 10^{-5}. Progeny of parental genotype are backcrossed to the parents, with the following results. All $a\,y$ progeny backcrossed to the $A\,x$ parent show prototroph frequencies of 10^{-5}. When $A\,x$ progeny are backcrossed to the $a\,y$ parent, two prototroph frequencies are obtained: one-half of the crosses show 10^{-5}, but the other half show the much higher frequency of 10^{-2}. Propose an explanation for these results, and describe a research program to test your hypothesis. (NOTE: intragenic recombination is a *meiotic* function that occurs in a diploid cell. Thus, even though this is a haploid organism, dominance and recessiveness could be involved in this question.)

20

Mechanisms of Genetic Change III: Transposable Genetic Elements

Jumping gene in snapdragon *(Antirrhinum majus)*. (Heinz Saedler, Max-Planck-Institut)

KEY CONCEPTS

A series of genetic elements can occasionally move or transpose from one position on the chromosome to another position on the same chromosome or on a different chromosome.

In bacteria, insertion sequences, transposons, and phage *mu* are examples of transposable genetic elements.

Transposable elements can mediate chromosomal rearrangements.

In higher cells, transposable elements have been extensively characterized in yeast, *Drosophila,* and maize and in mammalian systems.

In eukaryotes, some transposable elements utilize an RNA intermediate during transposition, whereas in prokaryotes transposition occurs exclusively at the DNA level.

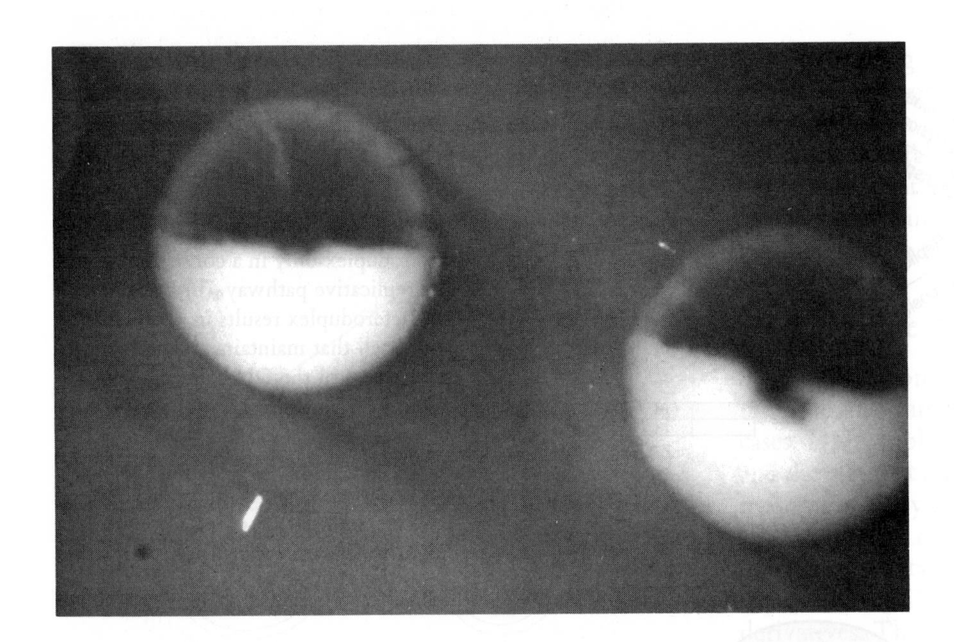

Figure 20-21 Colonies resulting from the experiment illustrated in Figures 20-19 and 20-20. A dye is used that stains Z^+ individuals blue. One half of this colony is Z^+ (dark area), and the other is Z^- (white). (Photo courtesy of N. Kleckner.)

Molecular Consequences of Transposition

The molecular consequences of transposition reveal an additional piece of evidence concerning the mechanism of transposition: on integration into a new target site, transposable elements generate a repeated sequence of the target DNA in both replicative and conservative transposition. Figure 20-22 depicts the integration of IS1 into a gene. In the example shown, the integration event results in the repetition of a 9-bp target sequence. Analysis of many integration events reveals that the repeated sequence does not result from reciprocal site-specific recombination (as is the case in λ phage integration; see page 572); rather, it is generated during the process of integration itself. The number of base pairs is a characteristic of each element. In bacteria, 9-bp and 5-bp repeats are most common.

The preceding observations have been incorporated into somewhat complicated models of transposition.

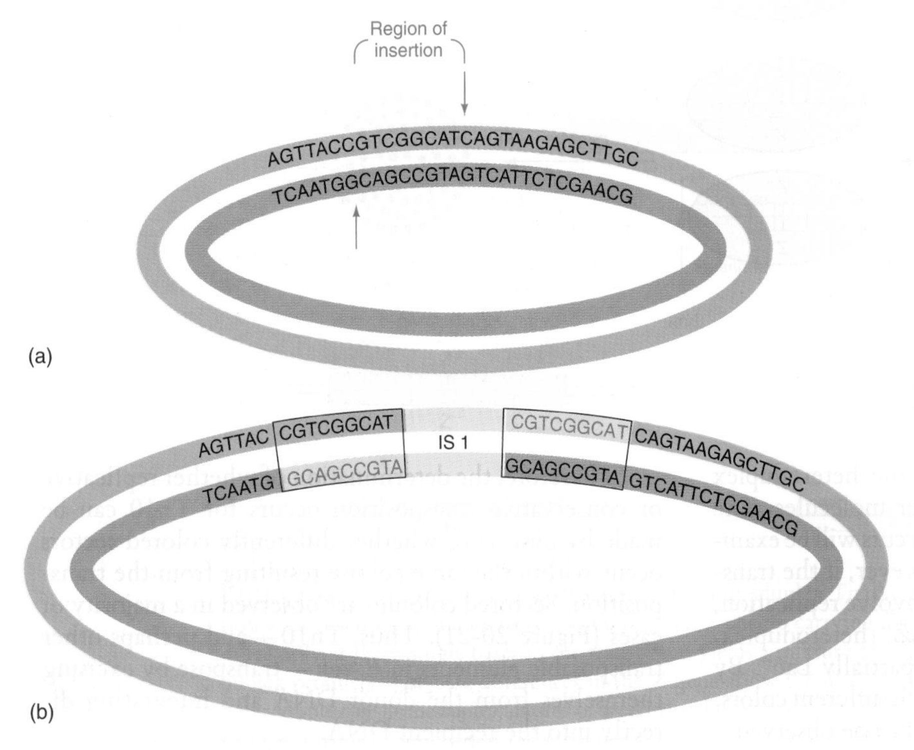

Figure 20-22 Duplication of a short sequence of nucleotides in the recipient DNA is associated with the insertion of a transposable element; the two copies bracket the inserted element. Here the duplication that attends the insertion of IS1 is illustrated in a way that indicates how the duplication may come about. IS1 insertion causes a nine-nucleotide duplication. If the two strands of the recipient DNA are cleaved (arrow) at staggered sites that are nine nucleotides apart, as shown in (a), followed by insertion of IS1 between the resulting single-stranded ends, then the subsequent filling in of single strands on each side of the newly inserted element, indicated by red letters in (b), with the right complementary nucleotides could account for the duplicated sequences (boxes). (From S. M. Cohen and J. A. Shapiro, "Transposable Genetic Elements." Copyright 1980 by Scientific American, Inc. All rights reserved.)

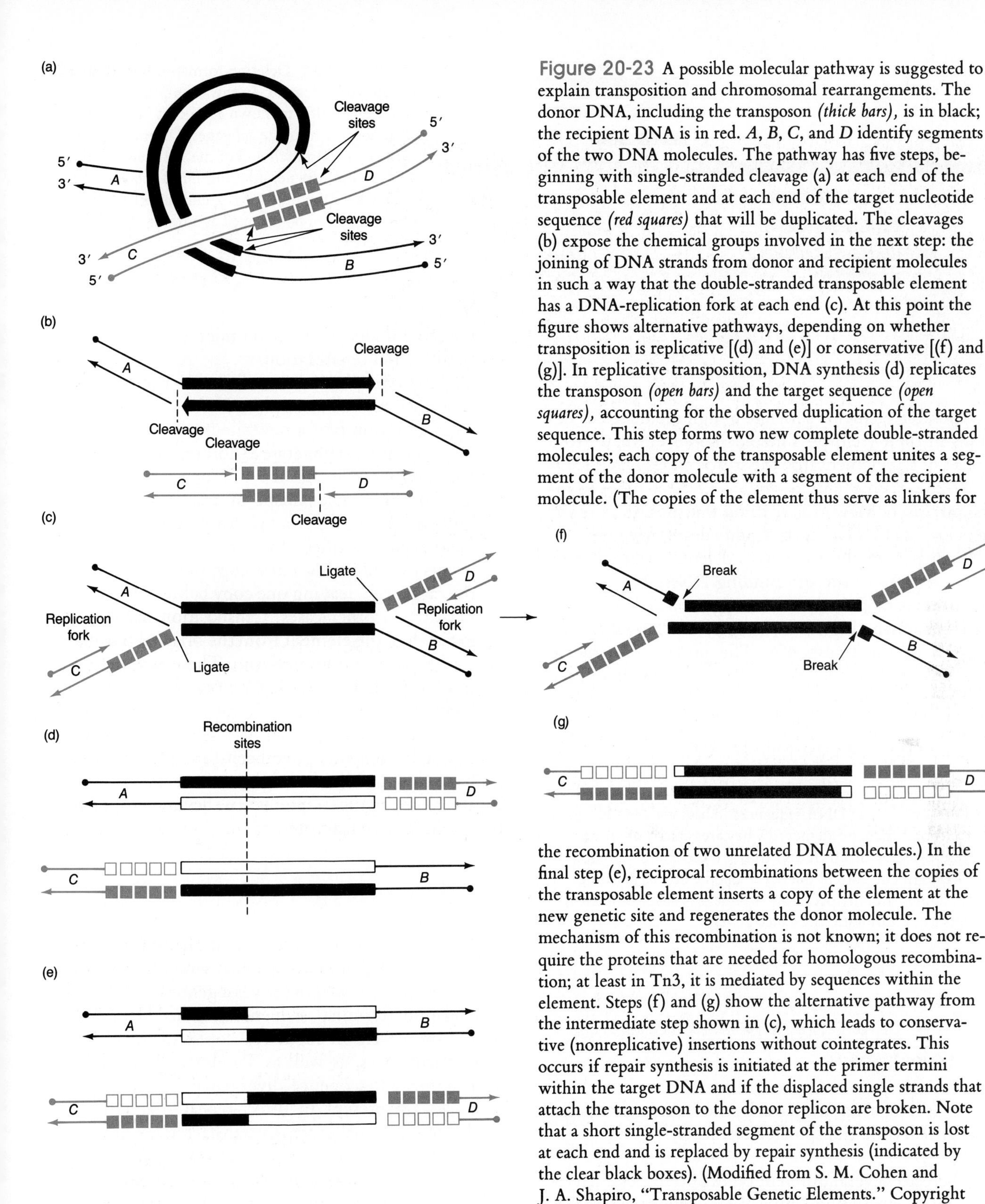

Figure 20-23 A possible molecular pathway is suggested to explain transposition and chromosomal rearrangements. The donor DNA, including the transposon *(thick bars)*, is in black; the recipient DNA is in red. *A*, *B*, *C*, and *D* identify segments of the two DNA molecules. The pathway has five steps, beginning with single-stranded cleavage (a) at each end of the transposable element and at each end of the target nucleotide sequence *(red squares)* that will be duplicated. The cleavages (b) expose the chemical groups involved in the next step: the joining of DNA strands from donor and recipient molecules in such a way that the double-stranded transposable element has a DNA-replication fork at each end (c). At this point the figure shows alternative pathways, depending on whether transposition is replicative [(d) and (e)] or conservative [(f) and (g)]. In replicative transposition, DNA synthesis (d) replicates the transposon *(open bars)* and the target sequence *(open squares)*, accounting for the observed duplication of the target sequence. This step forms two new complete double-stranded molecules; each copy of the transposable element unites a segment of the donor molecule with a segment of the recipient molecule. (The copies of the element thus serve as linkers for the recombination of two unrelated DNA molecules.) In the final step (e), reciprocal recombinations between the copies of the transposable element inserts a copy of the element at the new genetic site and regenerates the donor molecule. The mechanism of this recombination is not known; it does not require the proteins that are needed for homologous recombination; at least in Tn3, it is mediated by sequences within the element. Steps (f) and (g) show the alternative pathway from the intermediate step shown in (c), which leads to conservative (nonreplicative) insertions without cointegrates. This occurs if repair synthesis is initiated at the primer termini within the target DNA and if the displaced single strands that attach the transposon to the donor replicon are broken. Note that a short single-stranded segment of the transposon is lost at each end and is replaced by repair synthesis (indicated by the clear black boxes). (Modified from S. M. Cohen and J. A. Shapiro, "Transposable Genetic Elements." Copyright 1980 by Scientific American, Inc. All rights reserved.)

Most models postulate that staggered cleavages are made at the target site and at the ends of the transposable element by a transposase enzyme that is encoded by the element. One end of the transposable element is then attached by a single strand to each protruding end of the staggered cut. Subsequent steps depend on which mode of transposition occurs (replicative or conservative). Figure 20-23 depicts this model of transposition.

Message In prokaryotes, transposition occurs by at least two different pathways. Some transposable elements can replicate a copy of the element into a target site, leaving one copy behind at the original site. In other cases, transposition involves the direct excision of the element and its reinsertion into a new site.

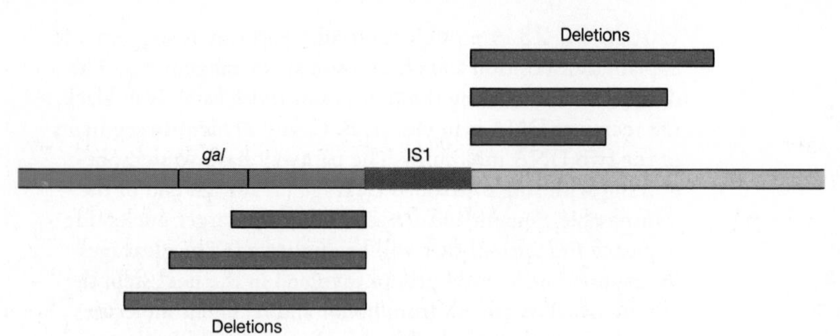

Figure 20-24 Deletion formation mediated by a transposable element. In this example, the transposable element IS1 is shown at a point in the *E. coli* chromosome near the *gal* genes. Deletions can be generated from each end of the IS1 element, extending into the neighboring DNA sequences. In cases where the deletions extend into the *gal* regions, they can be detected as a result of the Gal^- phenotype.

Rearrangements Mediated by Transposable Elements

Transposable elements generate a high incidence of deletions in their vicinity. These deletions emanate from one end of the element into the surrounding DNA (Figure 20-24). Such events, as well as element-induced inversions, can be viewed as aberrant transposition events. Transposons also give rise to readily detectable deletions in which part of the element is deleted together with varying lengths of the surrounding DNA. This process of **imprecise excision** is now recognized as deletions or inversions emanating from the internal ends of the IR segments of the transposon. The process of **precise excision**—the loss of the transposable element and the restoration of the gene that was disrupted by the insertion—also occurs, although at very low rates compared with the frequencies of the events just described.

Message Some DNA sequences in bacteria and phages act as mobile genetic elements. They are capable of joining different pieces of DNA and thus are capable of splicing DNA fragments into or out of the middle of a DNA molecule. Some naturally occurring mobile or transposable elements carry antibiotic-resistance genes.

Review of Transposable Elements in Prokaryotes

Let's examine what we have learned up to this point about prokaryotic transposable elements:

1. There are several different types of transposable elements, including insertion sequences (IS1, IS2, . . . , and so on), transposons (Tn1, Tn2, . . . , and so on), and phage mu.
2. Two copies of a transposable element can act in concert to transpose the DNA segments in between them. Some of the antibiotic-resistance-conferring transposons are obviously formed in this manner, with two insertion sequences flanking the genes for antibiotic resistance.
3. Most of the transposable elements have recognizable inverted repeat (IR) structures, some of which can be observed under the electron microscope after denaturation and renaturation.
4. Transposable elements are found in bacterial chromosomes, as well as in plasmids.
5. After insertion into a new site on the DNA, transposable elements generate a short repeated sequence, commonly consisting of 9 or 5 bp.
6. The detailed mechanism of transposition is not known, but two different pathways for transposition have been identified. In some cases, transposition occurs by replicating a new copy of the element into the target site, leaving one copy behind at the original site. In other cases, transposition involves the excision of the element from the original site and its reintegration into a new site. These two modes of transposition are called *replicative* and *conservative,* respectively.

Transposable elements have been found in eukaryotes and have close similarities to those observed in bacteria. The remainder of the chapter will be devoted to transposable elements in eukaryotes.

Ty Elements in Yeast

Figure 20-25 shows the structure of one of the **Ty elements** in yeast: the Ty1 sequence, which is present in approximately 35 copies in the yeast genome. The 38-bp-long termini (terminal sequences), called δ *(delta) sequences,* are present in about 100 copies of the genome. Yeast δ sequences, as well as Ty elements as a whole, show significant sequence divergence. The terminal δ sequences are present in direct-repeat orientation, in contrast to transposable elements in bacteria, which carry inverted repeat (IR) sequences. However, like prokaryotic transposons, Ty elements generate a repeated sequence of target DNA (in this case, 5 bp) during transpo-

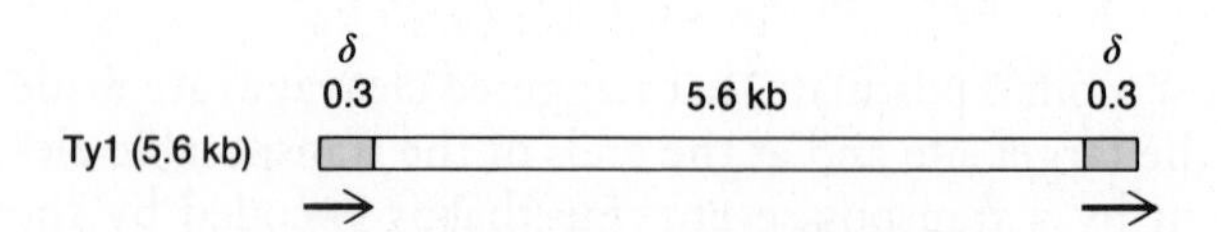

Figure 20-25 The structure of a yeast transposable element. The Tyl sequence occurs approximately 35 times in the yeast genome. It contains two copies of delta (δ) sequence in direct orientation at each end. Delta occurs approximately 100 times in the yeast genome.

sition. Also, Ty elements cause mutations by insertion into different genes in the yeast chromosome. It is now known that Ty elements transpose through an RNA intermediate (see the following section on retroviruses).

Transposable Elements in *Drosophila*

It is now estimated that many spontaneous mutations and chromosomal rearrangements in *Drosophila* are caused by transposable elements. As much as 10 percent of the *Drosophila* chromosome may be composed of families of dispersed, repetitive DNA sequences that move as discrete elements. Three types of transposable elements have been characterized: the *copia*-like elements, the fold-back (FB) elements, and the P elements. Their structures are summarized in Figure 20-26.

copia-like Elements

The **copia-like elements** comprise at least seven families, ranging in size from 5 to 8.5 kb. Members of each family appear at 10–100 positions in the *Drosophila* chromosome. Each member carries a long, direct terminal repeat and a short, imperfect inverted repeat (Figure 20-26) and is structurally similar to a yeast Ty element (Figure 20-25). Also, *copia*-like elements cause a duplication of a characteristic number of base pairs of *Drosophila* DNA on insertion. Certain classic *Drosophila* mutations result from the insertion of *copia*-like and other elements. For example, the white-apricot (w^a) mutation for eye color is caused by the insertion of an element from the *copia* family (from which these elements derive their name) into the white locus. Some *copia*-like families have surprising properties. For instance, all the insertion mutations detected so far that result from the *gypsy* family of *copia*-like elements are suppressible by a specific allele at one particular outside locus. That is, the phenotypes resulting from the *gypsy* insertions are affected by unlinked genes. The mechanism of the effect is unknown.

FB Elements

The **FB elements** range in size from a few hundred to a few thousand base pairs. These elements have sequence homologies, but different elements have sequence differences also. Each carries long inverted repeats at its termini (see Figure 20-26). Sometimes the entire element consists of inverted repeats, but a central sequence separates the inverted repeats in other elements. In either

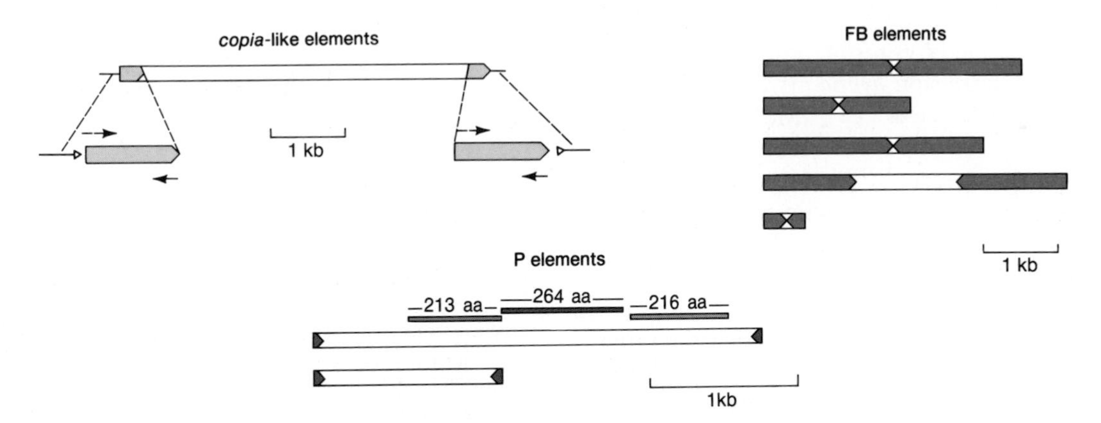

Figure 20-26 Summary of the structures of three classes of *Drosophila* transposable elements. The *copia*-like elements carry long direct terminal repeats. Each repeat makes up about 5 percent of the length of the element. These repeats are shown on an expanded scale below the element to illustrate the presence of short, imperfect inverted repeats at the ends of each long direct repeat (→) and the presence of a few base pairs of duplicate target sequence (▷) flanking the element after insertion. Such duplication of the target sequence is a virtually universal feature of transposable element insertion. The different genomic copies of the family elements are very similar in structure to one another.

The FB elements make up a family of heterogeneous but cross-homologous sequences, ranging in size from a few hundred base pairs to several kilobases. Each FB element carries long terminal inverted repeats. In some cases, the entire element consists of these inverted repeats. In other cases, a central sequence is located between the inverted repeats. The inverted repeat sequences themselves are internally repetitious, having a substructure made up primarily of 31-bp tandem repeats. The number of these 31-bp tandem repeats can differ not only between FB elements but also between the termini of a single FB element.

The P elements have a very different structure from that of the *copia*-like and FB elements. P elements carry perfect terminal inverted repeats of 31 bp. A fraction of the P elements (about one-third in the one strain examined) are very similar in sequence to one another and are 2.9 kb in length. The remainder of the P elements are more heterogeneous, but all appear to have structures that are consistent with their having been derived from the 2.9-kb element by one or more internal deletions. DNA sequence analysis of the 2.9-kb element reveals three long, open translational reading frames, which are indicated. (From G. Robin, in *Mobile Genetic Elements*, J. A. Shapiro, ed., pp. 329–361. Copyright 1983 by Academic Press.)

case, FB elements can literally fold back on themselves owing to the inverted repeats (hence, their name). Several unstable mutations in *Drosophila* have been shown to be caused by the insertion of FB elements. Mutations can result either from the interruption of a gene-coding sequence by FB-element insertion or from the effects on gene expression due to FB-element insertion in or near a control region. The properties of some of the FB-insertion mutations suggest that FB elements can excise themselves from the genome and promote chromosomal rearrangements at high frequencies.

P Elements

Of all the transposable elements in *Drosophila,* the most intriguing and useful to the geneticist are the **P elements.** These elements were discovered as a result of studying **hybrid dysgenesis**—a phenomenon that occurs when females from laboratory strains of *Drosophila melanogaster* are mated to males derived from natural populations. In such crosses, the laboratory stocks are said to possess an **M cytotype** (cell type), and the natural stocks are said to possess a **P cytotype.** In a cross of M♀ × P♂, the progeny show a range of surprising phenotypes that are manifested in the germ line, including sterility, a high mutation rate, and a high frequency of chromosomal aberation and nondisjunction. These hybrid progeny are termed **dysgenic,** or biologically deficient (hence, the expression *hybrid dysgenesis*). Interestingly, the reciprocal cross P♀ × M♂ produces no dysgenic offspring. An important observation is that a large portion of the dysgenically induced mutations are unstable—that is, they revert to wild-type or to other mutant alleles at very high frequencies. This instability is generally restricted to the germ line of individuals possessing an M cytotype.

These findings led to the hypothesis that the mutations are caused by the insertion of foreign DNA into specific genes, thereby rendering them inactive. According to this hypothesis, reversion usually would result from the spontaneous excision of these inserted sequences. This hypothesis has been critically tested by isolating dysgenically derived unstable mutants at the eye-color locus *white.* A plasmid constructed to carry the white locus was used as a probe to recover the dysgenesis-mutated *white* genes. (This type of experiment is explained in Chapter 14). The majority of the mutations were found to be caused by the insertion of a genetic element into the middle of the *white*$^+$ gene. The element, called the *P element,* was found to be present in 30 to 50 copies per genome in P strains but to be completely absent in M strains. The P elements vary in size, ranging from 0.5 to 2.9 kb in length (this size difference reflects partially deleted elements derived from a single complete P element), but there is always a 31-bp perfect inverted repeat at their ends. There can be as many as three open reading frames in the central area of the P element, suggesting that the largest elements have the coding potential for three protein products.

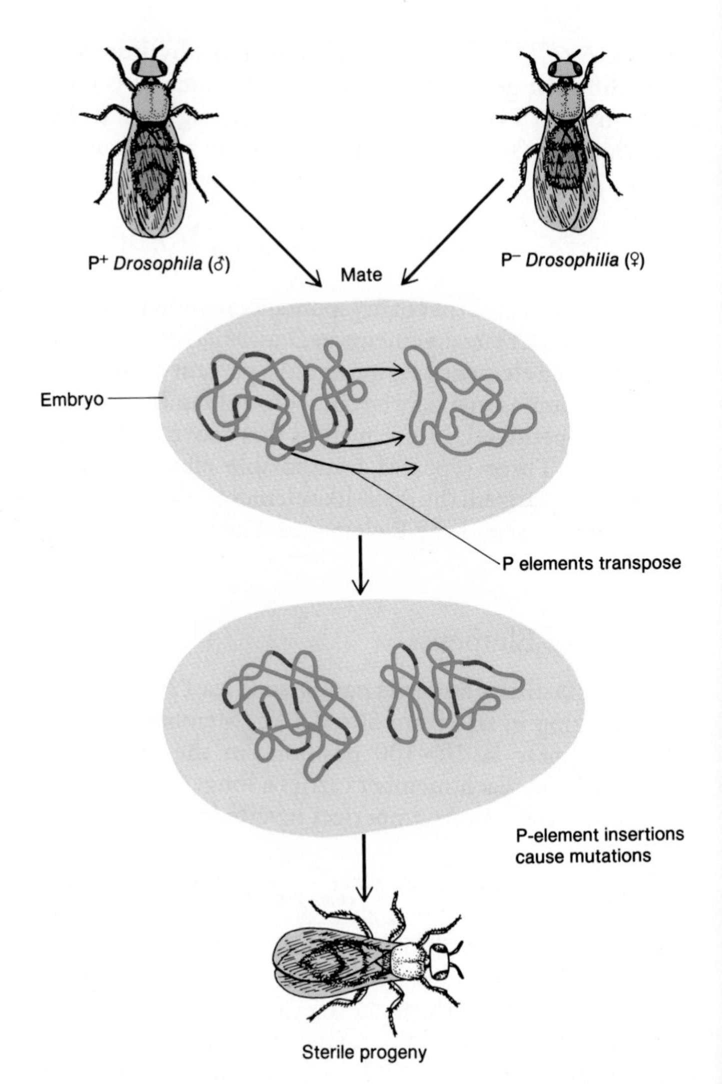

Figure 20-27 The phenomenon known as hybrid dysgenesis results from the mobilization of DNA sequences called *P* elements in *Drosophila* embryos. When a sperm from a P-carrying strain fertilizes an egg from a non-P-carrying strain, the P elements transpose throughout the genome, usually disrupting vital genes. (After J. D. Watson, J. Tooze, and D. T. Kurtz, *Recombinant DNA: A Short Course.* Copyright 1983 by W. H. Freeman and Company.)

The current explanation of hybrid dysgenesis is based on the proposal that P elements encode both the transposase product and P-repressor products. According to this model, which is depicted in Figure 20-27, the transposase is responsible for mobilization of the P elements, whereas the repressor prevents transposase production, thereby blocking transposition of the element.

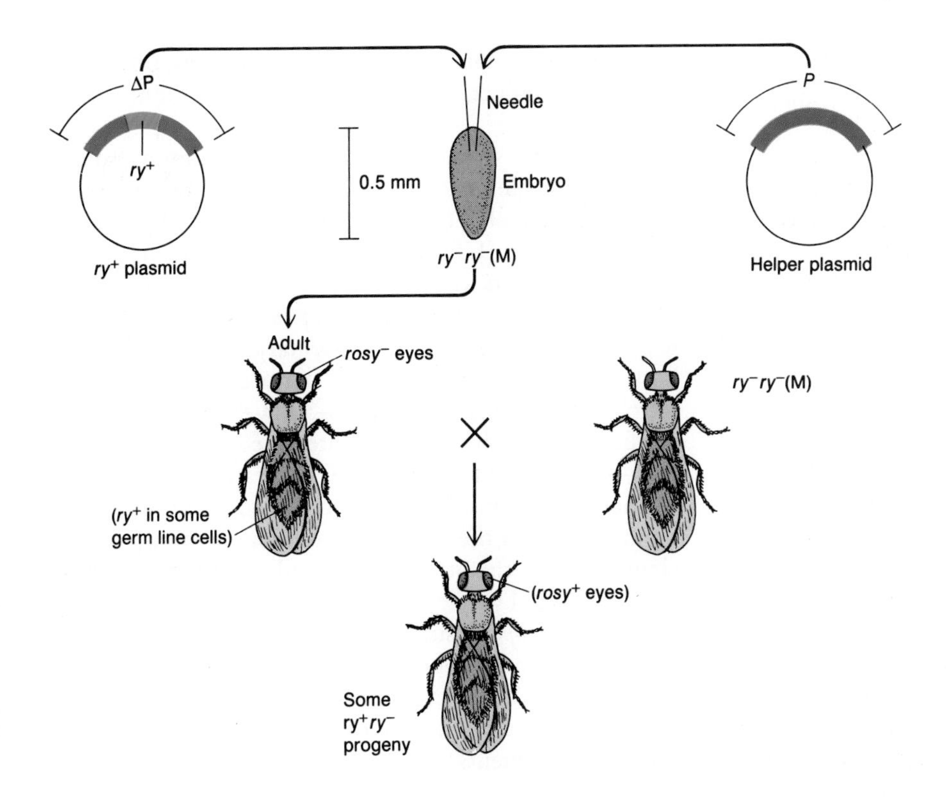

Figure 20-28 P element–mediated gene transfer in *Drosophila.* The $rosy^+$ (ry^+) eye-color gene is inserted into a deleted P element (ΔP) carried on a bacterial plasmid. At the same time, a helper plasmid bearing an intact P element is used. Both are injected into a ry^- embryo, where ry^+ transposes with the ΔP element into the chromosomes of the germ line cells.

(We consider repressor proteins in general in Chapter 17.) In the P cytotype, the high copy number of P elements leads to the abundant production of repressor, so that the P elements are immobilized. For some reason, most laboratory strains have no P elements; consequently, there is no repressor in the cytoplasm. In hybrids from the cross M♀ × P♂, the P elements are in a repressor-free environment and can transpose throughout the genome, causing a variety of damage expressed as the various manifestations of hybrid dysgenesis. On the other hand, as we noted earlier, P♀ × M♂ crosses do not result in dysgenesis. This is because P cytotype females already have high levels of P repressor.

Quite apart from their interest as a genetic phenomenon, the P elements have become major tools of the modern *Drosophila* geneticist. Two main analytical techniques are possible.

Using P Elements to Locate Genes. P elements are used to isolate any *Drosophila* gene of interest. First, the investigator simply looks for mutants of that gene in progeny of dysgenic crosses. Then a vector is constructed in which a P element is inserted. This vector is used as a probe to identify and isolate DNA segments containing P elements (see Chapter 14 for more details of these types of experiments); in a subset of these segments, P elements are found inserted into the gene of interest. These genes can then be cloned (Chapter 14) and studied.

Using P Elements to Insert Genes. The second major analytical technique stems from the discovery by Gerald Rubin and Allan Spradling that P element DNA can be used as an effective vehicle for transferring donor genes into the germ line of a recipient fly. Rubin and Spradling devised the following experimental procedure (Figure 20-28). The recipient genotype is homozygous for the *rosy* (ry^-) mutation, which confers a characteristic eye color and is of M cytotype. From this strain, embryos are collected at the completion of about nine nuclear divisions. At this stage, the embryo is one multinucleate cell, and the nuclei destined to form the germ cells are clustered at one end. (P elements mobilize only in germ line cells.) Two types of DNA are injected into embryos of this type. The first is a bacterial plasmid carrying a deleted P element into which the ry^+ gene has been spliced. This deleted element is not able to transpose owing to the deletion, so a helper plasmid bearing a complete element is also injected. Flies developing from these embryos are phenotypically still *rosy* mutants, but their offspring contain a large proportion of ry^+ individuals. These ry^+ descendants show Mendelian inheritance of the newly acquired ry^+ gene, suggesting that it is located on a chromosome. This has been confirmed by in situ hybridization, which shows that the ry^+ gene, together with the deleted P element, has been inserted into one of several distinct chromosome locations. None appears exactly at the normal locus of the *rosy* gene. These new ry^+ genes are found to be inherited in a stable fashion.

Message P elements in *Drosophila* are a type of transposon that causes hybrid dysgenesis. They are very useful in two ways to the genetic analyst. First, they can be used through transposon mutagenesis to recover selectively any gene with a recognizable mutant phenotype. Marking a specific gene by transposon mutagenesis is termed **transposon tagging.** Second, P elements in *Drosophila* can be used as efficient vehicles for the transfer of specific genes to given recipient genotypes.

Retroviruses

Retroviruses are single-stranded RNA animal viruses that employ a double-stranded DNA intermediate for replication. The RNA is copied into DNA by the enzyme **reverse transcriptase.** The life cycle of a typical retrovirus is shown in Figure 20-29. Some retroviruses, such as mouse mammary tumor virus (MMTV) and Rous sarcoma virus (RSV), are responsible for the induction of cancerous tumors. When integrated into host chromosomes as double-stranded DNA, these retroviruses are termed **proviruses.** Proviruses, like the mu phage in bacteria, can be considered transposable elements, since they can, in effect, transpose from one location to another.

Retroviruses have structural similarities to some transposable elements from bacteria and other organisms. In particular, the ends of the proviruses have long terminal repeats (LTRs) reminiscent of the sequences of the Ty1 elements in yeast and the long terminal repeats of the *copia*-like elements in *Drosophila.* Also, integration results in the duplication of a short target sequence in the host chromosome. For example, in the case of the mouse retrovirus shown in Figure 20-30, a 4-bp sequence is duplicated on each side of the integrated provirus. (Note how the staggered cuts and the duplication of the short segment of the host DNA are similar to those shown in Figure 20-22 for the integration of bacterial transposable elements.) On the other hand, similarities between retroviruses and Ty1 elements in yeast and *copia*-like elements in *Drosophila* suggest that the Ty1 and *copia*-like elements might also be integrated forms of retrovirus-like elements. This was nicely confirmed in 1985 by Jef Boeke and Gerald Fink and their co-workers in the following experiments.

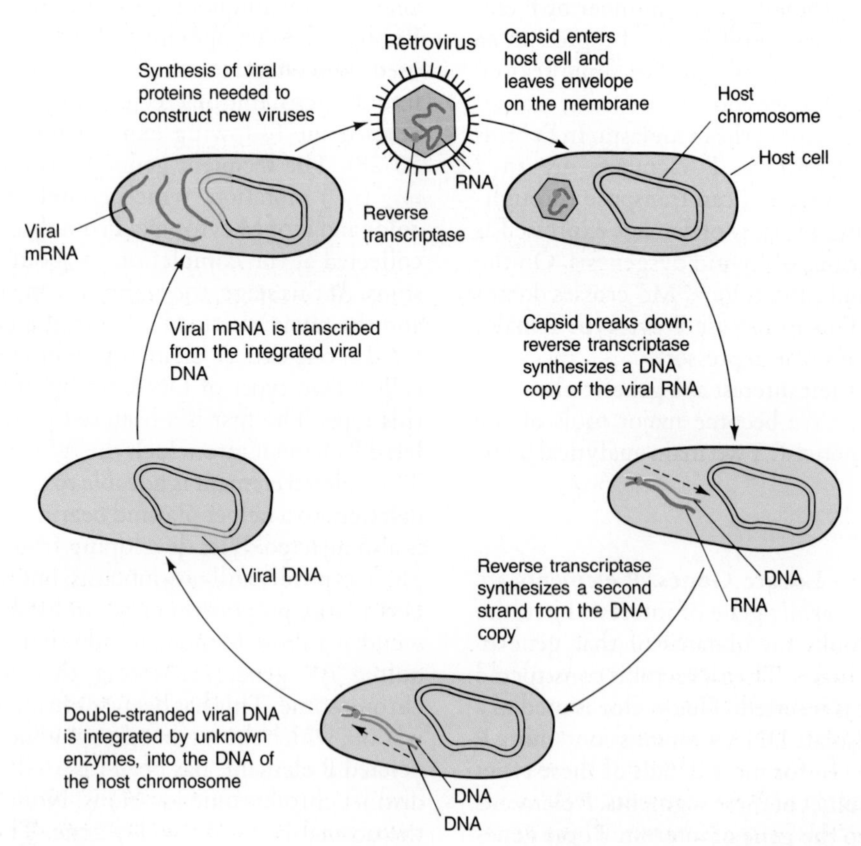

Figure 20-29 The life cycle of a retrovirus. Viral RNA is shown in red; DNA, in blue.

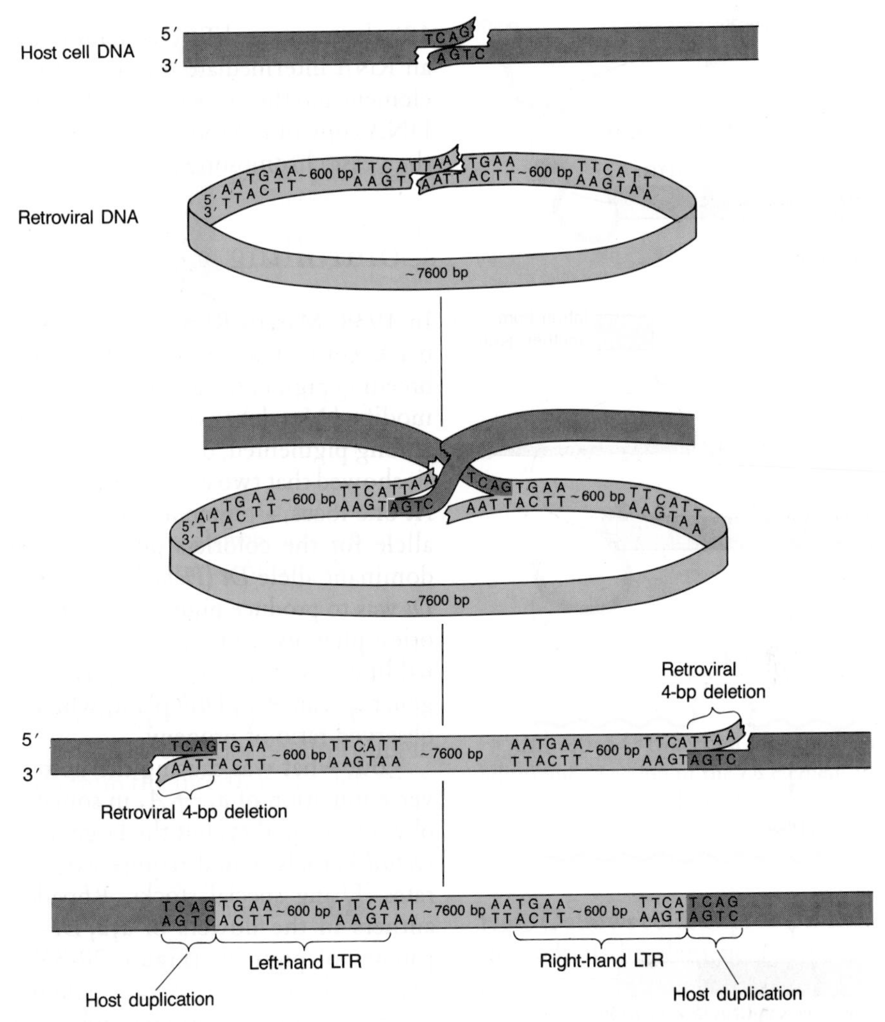

Figure 20-30 The integration of a circular retrovirus DNA molecule with two tandem LTRs. The circular DNA is shown integrating into host cell DNA at a site at which the sequence TCAG happens to occur. The integration occurs by staggered cleavages of both the cellular DNA and the viral DNA. The illustrated case is a mouse retrovirus that makes a 4-bp repeat of cellular DNA at the site of integration; for other retroviruses, the repeat is 4 to 6 bp long. Also note that there is a 4-bp double deletion of retroviral DNA. (From C. Shoemaker and D. Baltimore, *Proceedings of the National Academy of Sciences* 77, 1980, 3932. After J. Darnell *et al., Molecular Cell Biology.* Copyright 1986 by Scientific American Books.

Transposition via an RNA Intermediate

Figure 20-31 diagrams the experimental design used by Boeke and Fink and their colleagues to alter a yeast Ty element, cloned on a plasmid. First, a promoter was inserted near the end of the element that could be activated by the addition of galactose to the medium. The use of a galactose-sensitive promoter allows the manipulation of the expression of Ty RNA. In the presence of galactose, more transcription of Ty RNA occurs. Second, an intron (page 400) from another yeast gene was introduced into the Ty transposon coding region.

The addition of galactose greatly increases the frequency of transposition of the altered Ty element. This suggests the involvement of RNA, because galactose-stimulated transcription begins at the galactose-sensitive promoter and continues through the element (Figure 20-31). The key experimental result, however, is the fate of the transposed Ty DNA. When the researchers examined the Ty DNA resulting from transpositions, they found that the intron had been removed! Because introns are excised only during RNA processing (see Chapter

wild-type fungus in a number of ways: it is slow growing, it shows maternal inheritance of slow growth, and it has abnormal amounts of cytochromes. Cytochromes are mitochondrial electron-transport proteins necessary for the proper oxidation of foodstuffs to generate energy in the form of ATP. Like most organisms, wild-type *Neurospora* has three main types of cytochrome: *a*, *b*, and *c*. In *poky*, however, there is no cytochrome *a* or *b*, and there is an excess of cytochrome *c*.

How can maternal inheritance be demonstrated in a haploid organism? It is possible to cross some fungi in such a way that one parent contributes the bulk to the cytoplasm to the progeny; this cytoplasm-contributing parent is called the female parent, even though no true sex is involved. Mitchell demonstrated maternal inheritance for the poky phenotype in the following crosses:

poky ♀ × wild-type ♂ ⟶ all poky
wild-type ♀ × poky ♂ ⟶ all wild type

However, in such crosses, any nuclear genes that differed between the parental strains were observed to segregate in the normal Mendelian manner and to produce 1:1 ratios in the progeny (Figure 21-6). All *poky* progeny behaved like the original *poky* strain, transmitting the *poky* phenotype down through many generations when crossed as females. Because *poky* does not behave like a strain arising from a nuclear mutation, the mutation responsible has been termed an **extranuclear mutation** or a **cytoplasmic mutation.**

But where in the cytoplasm is the mutation carried? The important clue is that several aspects of the mutant phenotype seem to involve mitochondria. For instance, the cells' slow growth suggests a lack of ATP energy, which is normally produced by mitochondria. Also, there are abnormal amounts of cytochromes in the mutants, and cytochromes are known to be located in the mitochondrial membranes. These indications led geneticists to conclude that mitochondria were involved and inspired researchers to design several interesting experiments designed to investigate mitochondrial autonomy.

David Luck labeled mitochondria with radioactive choline, a membrane component, and then followed their division autoradiographically in an unlabeled medium. He found that even after several doublings of mass, the radiation was distributed evenly among the mitochondria. Luck concluded that mitochondria propagate by division of previously existing mitochondria. If mitochondria were synthesized de novo (anew), some of the resulting population of mitochondria would be expected to be unlabeled and the original mitochondria should remain heavily labeled.

In 1965, a research group led by Edward Tatum extracted purified mitochondria from a mutant called

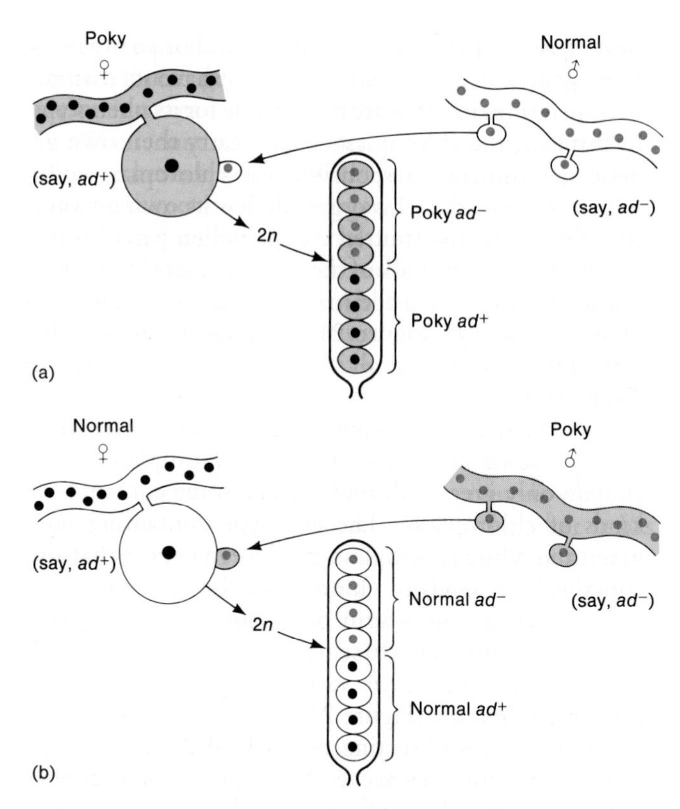

Figure 21-6 Explanation of the different results from reciprocal crosses of poky and normal *Neurospora.* The parent contributing most of the cytoplasm to the progeny cells is called female. Color shading represents cytoplasm with the poky determinants. The nuclear locus with the alleles ad^+ and ad^- is used to illustrate the segregation of the nuclear genes in the expected 1:1 Mendelian ratio.

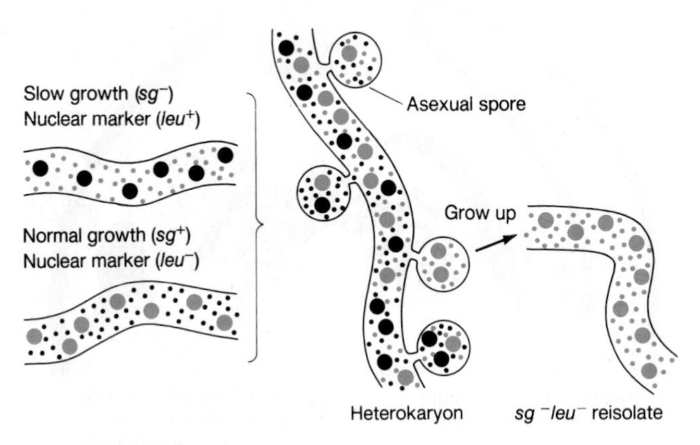

Figure 21-7 The heterokaryon test is used to detect extranuclear inheritance in filamentous (threadlike) fungi. A strain with a possible extranuclear mutation (here, sg^-, causing slow growth) is combined with a strain having a nuclear mutation (leu^-) to form a heterokaryon. Cultures having the phenotype caused by leu^- can be derived from the heterokaryon. If some of these cultures also are sg^- in phenotype, then *sg* is very likely to be an extranuclear gene, borne in an organelle. Since no nuclear recombination normally occurs in a heterokaryon, the sg^- phenotype must have been acquired by cytoplasmic contact.

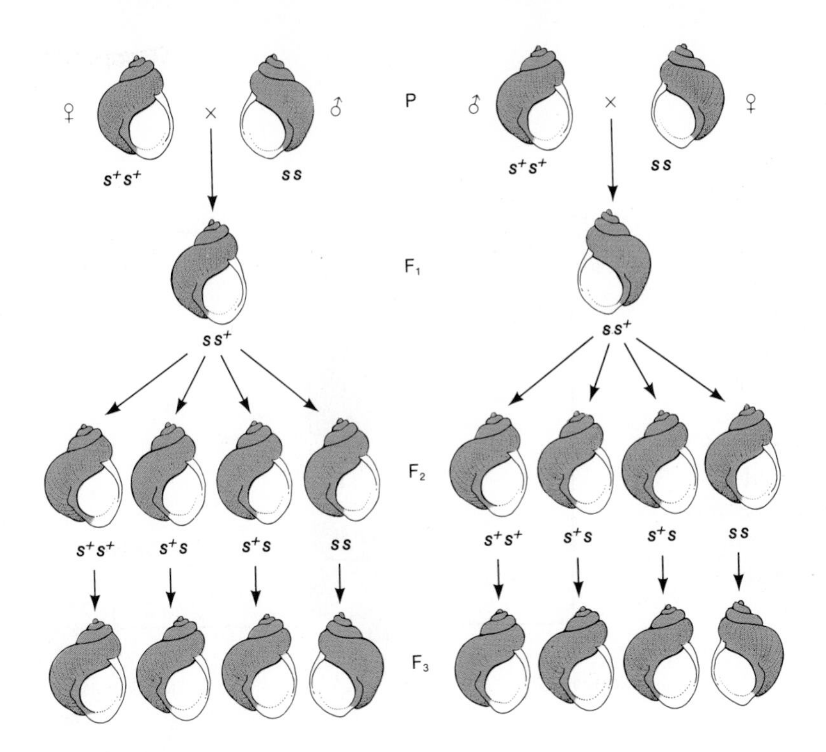

Figure 21-8 The inheritance of dextral (s^+) and sinistral (s) alleles of the nuclear gene for shell coiling in a species of water snail. The direction of coiling is determined by the nuclear genotype of the mother rather than the genotype of the individual itself. This explanation accounts for the initial difference between reciprocal crosses as well as the phenotypes of later generations. No organelle inheritance is involved (such a hypothesis would not explain the phenotypes of later generations).

abnormal, which is similar to poky. With an ultrafine needle and syringe, the experimenters injected the mitochondria into wild-type recipient cells, using appropriate controls. The recipient cells were subcultured, and after several of these subcultures, the *abnormal* phenotype appeared! In transferring the mitochondria, the experimenters had transferred the hereditary determinants of the *abnormal* phenotype. Presumably, then, the extranuclear genes involved in this phenotype are located in the mitochondria. These inferences gained credibility with the discovery of DNA in the mitochondria of *Neurospora* and other species (but more of that story later).

The Heterokaryon Test

Maternal inheritance is one criterion for recognizing organelle-based inheritance. Another diagnostic test has been applied in filamentous fungi such as *Neurospora* and *Aspergillus.* In principle, this test could be applied to other systems involving heterokaryons. A heterokaryon is made between the putative extranuclear mutant (say, a slow-growth mutant) and a strain carrying a known nuclear mutation. If the phenotype of the nuclear mutation can be recovered from the heterokaryon in combination with the phenotype of the slow-growth mutant being tested, a good hypothesis is that the growth mutant arose from an organelle-based mutation. This is because generally no diploidy occurs in a heterokaryon, so that there is no genetic exchange between nuclei. The slow-growth phenotype must have been transferred solely by cytoplasmic contact. The segregation of the pure extranuclear type from the mixed cytoplasm of the heterokaryon presumably involves the CSAR process (Figure 21-7).

Message Extranuclear inheritance can be recognized by uniparental (usually maternal) transmission through a cross or by transmission via cytoplasmic contact.

Shell Coiling in Snails: A Red Herring

Does maternal inheritance always indicate extranuclear inheritance? Usually, but not always. It is possible for a maternal-inheritance pattern of reciprocal crosses to be generated by nuclear genes. In 1923, Alfred Sturtevant (whom you will remember from his studies on crossing-over in *Drosophila*) found a good example in the water snail *Limnaea.* Sturtevant analyzed the results of crosses between snails that differed in the direction of their shell coiling. On looking into the opening of the shells, an observer can see that some snails coil to the right (dextral coiling) and that others coil to the left (sinistral coiling). All the F_1 progeny of the cross dextral ♀ × sinistral ♂ were dextral, but all the F_1 progeny of the cross sinistral ♀ × dextral ♂ were sinistral. Thus far, the situation seems similar to the one observed with *poky* or with chloroplast inheritance. However, the F_2 progenies in both pedigrees were all dextral!

The F_3 generation in this study revealed that the inheritance of coiling direction involves nuclear genes rather than extranuclear genes. (The F_3 is produced by individually selfing the F_2 snails; this is possible because snails are hermaphroditic.) Sturtevant found that three-fourths of the F_3 snails were dextral and one-fourth were sinistral (Figure 21-8). This ratio reveals a Mendelian segregation in the F_2 generation. Apparently, dextral

(s^+) is dominant to sinistral (s), but Sturtevant concluded that, strangely enough, the shell-coiling phenotype of any individual animal is determined by the genotype (not the phenotype) of its mother! We now know that this happens because the genotype of the mother's body determines the initial cleavage pattern of the developing embryo. Note that the segregation ratios shown in Figure 21-8 would never appear in the phenotypes of true organelle genes. This example shows that one generation of crosses is not enough to provide conclusive evidence that maternal inheritance is due to organelle-based inheritance. The term **maternal effect** can be used to describe the results in cases like the shell-coiling example to distinguish them from organelle-based inheritance.

Extranuclear Genes in *Chlamydomonas*

If an investigator could somehow design an ideal experimental organism with a simple life cycle, the result would be something like the unicellular freshwater alga *Chlamydomonas* (Figure 21-9). Like fungi, algae rarely have different sexes, but they do have mating types. In many algal and fungal species, two mating types are determined by alleles at one locus. A cross can occur only if the parents are of different mating types. The mating types are physically identical but physiologically different. Such species are called **heterothallic** (literally, "different bodied"). In *Chlamydomonas,* the mating-type alleles are called mt^+ and mt^- (in *Neurospora* they are A and a; in yeast, a and α). Figure 21-10 diagrams the *Chlamydomonas* life cycle. This organism has played a central role in research on organelle genetics. As you might guess, tetrad analysis (see Chapter 6) is possible and in fact is routinely performed.

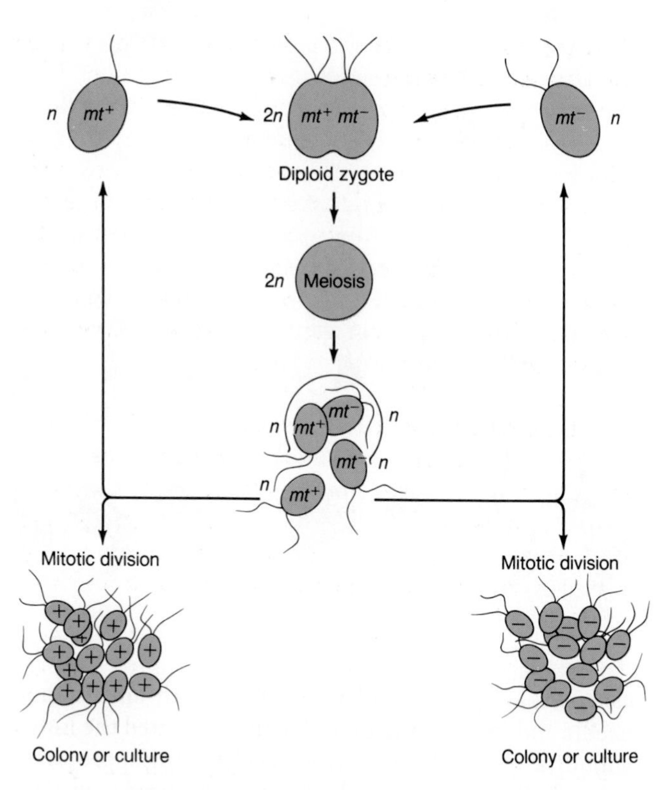

Figure 21-10 The life cycle of *Chlamydomonas,* a unicellular green alga. All diploid zygotes are heterozygous for the mating-type alleles mt^+ and mt^- because only algae differing in these alleles can mate.

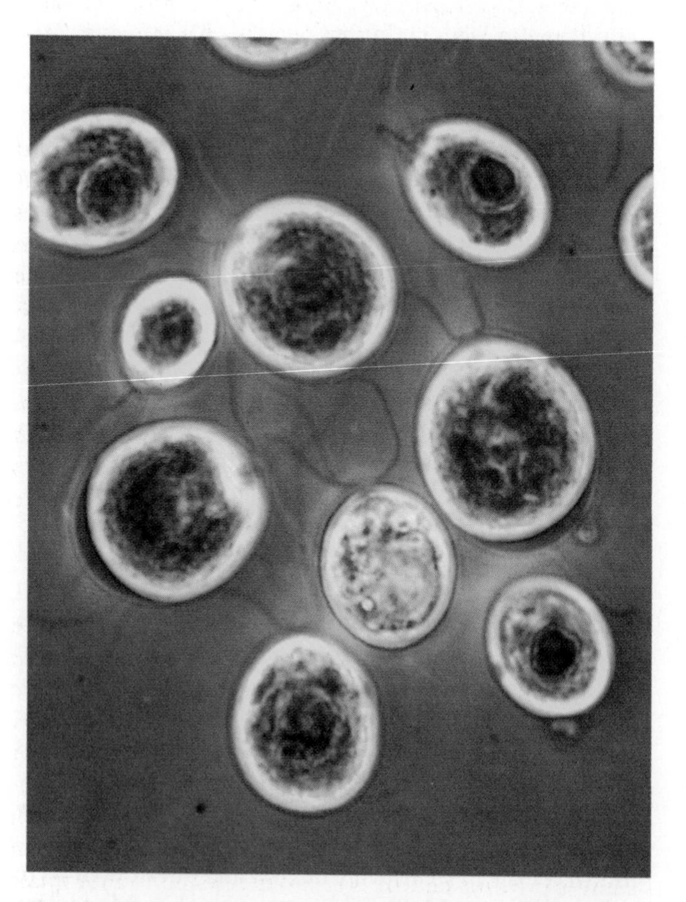

Figure 21-9 Living cells of the unicellular alga *Chlamydomonas reinhardtii.* Note the pair of flagella and the large single chloroplast. (M. I. Walker/Science Source/Photo Researchers)

In 1954, Ruth Sager isolated a streptomycin-sensitive mutant of *Chlamydomonas* with a peculiar inheritance pattern. In the following crosses, *sm-r* and *sm-s* indicate streptomycin resistance and sensitivity, respectively, and *mt* is the mating type gene discussed earlier:

$$sm\text{-}r\ mt^+ \times sm\text{-}s\ mt^- \longrightarrow \text{progeny all } sm\text{-}r$$

$$sm\text{-}s\ mt^+ \times sm\text{-}r\ mt^- \longrightarrow \text{progeny all } sm\text{-}s$$

Here we see a difference in reciprocal crosses; all progeny cells show the streptomycin phenotype of the mt^+ parent.

Like the maternal inheritance phenomenon, this is a case of **uniparental inheritance.** In fact, Sager referred to the mt^+ mating type as the female, using this analogy. However, there is no observable physical distinction between the mating types, as there would be if true sex were involved, nor is there a difference in the contribution of cytoplasm as seen in *Neurospora.* In these crosses, the conventional nuclear marker genes (such as *mt* itself) all behave normally and give 1 : 1 progeny ratios. For example, half of the progeny of the cross *sm-r* mt^+ × *sm-s* mt^- are *sm-r* mt^+ and half are *sm-r* mt^-.

To determine the sm-s or sm-r phenotypes of progeny, the cells are plated onto a medium containing streptomycin. While performing these experiments, Sager observed that the drug streptomycin itself acts as a mutagen. If a pure strain of *sm-s* cells is plated onto streptomycin, quite a significant proportion of streptomycin-resistant colonies will appear. In fact, treatment with streptomycin will produce a whole crop of different kinds of mutant phenotypes, all of which show uniparental inheritance and hence presumably are phenotypes of extranuclear mutations. Furthermore, streptomycin treatment does not produce nuclear mutants. Most of the mutants produced by streptomycin treatment show either resistance to one of several different drugs or defective photosynthesis. The photosynthetic mutants are unable to make use of CO_2 from the air as a source of carbon, so they survive only if soluble carbon (in some form such as acetate) is added to their medium. Table 21-2 lists several different extranuclear mutations produced by the streptomycin treatment.

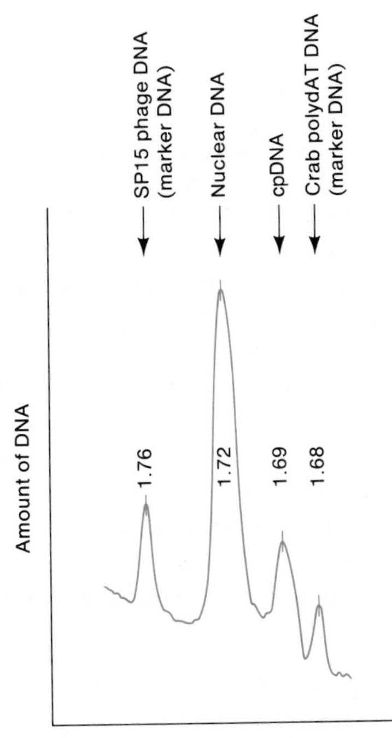

Figure 21-11 The chloroplast DNA (cpDNA) of *Chlamydomonas* can be detected in a CsCl density gradient as a band of DNA distinct from the nuclear DNA. In this particular experiment, two other DNA types were sedimented along with the *Chlamydomonas* DNA to act as known reference points in the gradient. The numbers represent buoyant densities (g/cm³) of the various DNA types. (After Ruth Sager, "Genetic Analysis of Chloroplast DNA in Chlamydomonas," in *Cytoplasmic Genes and Organelles,* Academic Press.)

Table 21-2. Some *Chlamydomonas* Mutations Showing Uniparental Inheritance

Gene	Mutant phenotype
ac1 – ac4	Requires acetate
tm1 and *tm3 – tm9*	Cannot grow at 35°C
tm2	Conditional: grows at 35°C only in presence of streptomycin
ti1 – ti5	Forms tiny colonies on all media
ery1	Resistant to erythromycin at concentration of 50 μg/ml
kan1	Resistant to kanamycin at concentration of 100 μg/ml
spc1	Resistant to spectinomycin at concentration of 50 μg/ml
spi1 – spi5	Resistant to spiramycin at concentration of 100 μg/ml
ole1 – ole3	Resistant to oleandomycin at concentration of 50 μg/ml
car1	Resistant to carbamycin at concentration of 50 μg/ml
ele1	Resistant to eleosine at concentration of 50 μg/ml
ery3 and *ery11*	Resistant to erythromycin, carbamycin, oleandomycin, and spiramycin (at concentrations listed above for individual resistances)
sm2 and *sm5*	Resistant to streptomycin at concentration of 500 μg/ml
sm3	Resistant to streptomycin at concentration of 50 μg/ml
sm4	Requires no streptomycin for survival

NOTE: All these mutants were produced by treatment with streptomycin except for the *ti* mutants, which were produced by treatment with nitrosoguanidine.

SOURCE: R. Sager, *Cytoplasmic Genes and Organelles,* Academic Press.

These experiments reveal the existence of a mysterious "uniparental genome" in *Chlamydomonas* — that is, a group of genes that all show uniparental transmission in crosses. Where is this genome located? What is the mechanism of the uniparental transmission? When there seems to be no physical difference between mating types, why are these genes transmitted only by the mt^+ parent?

Evidence arose to suggest that the uniparental genome in this case is in fact chloroplast DNA (cpDNA). About 15 percent of the DNA of a *Chlamydomonas* cell in rapidly growing culture is found in the single chloroplast of the cell. This DNA forms a band in a cesium chloride (CsCl) gradient that is distinct from the band of nuclear DNA (Figure 21-11). Furthermore, the precise position

Table 21-3. Buoyant Densities of cpDNA in Progeny from Various *Chlamydomonas* Crosses

Cross	Buoyant density of zygote cpDNA
^{14}N $mt^+ \times$ ^{14}N mt^-	1.69
^{15}N $mt^+ \times$ ^{15}N mt^-	1.70
^{15}N $mt^+ \times$ ^{14}N mt^-	1.70
^{14}N $mt^+ \times$ ^{15}N mt^-	1.69

SOURCE: Ruth Sager, *Cytoplasmic Genes and Organelles*, Academic Press.

of the cpDNA band can be altered by adding the heavy isotope of nitrogen, ^{15}N, to the growth medium. Using this technique, the cpDNA of the two parents in a cross can be labeled differently: one light (^{14}N), and one heavy (^{15}N). The buoyant densities of the cpDNA from these parental cells are 1.69 and 1.70, respectively. Although this difference seems small, it provides appreciably different band positions in a CsCl gradient. With differently labeled parents, the zygote DNA can be examined to see how it compares with the parents. You can see from the results in Table 21-3 that the cpDNA of the mt^- parent is in fact lost, inactivated, or destroyed in some way. This loss of cpDNA from the mt^- parent, of course, parallels the loss of uniparental genes (such as the *sm* genes) borne by the mt^- parent.

Other experiments have been performed using similar logic. For example, crosses can be made between *Chlamydomonas* strains having cpDNAs with distinctly different restriction-enzyme digest patterns on electrophoretic gels. Again, the specific cpDNA digest pattern passed on to the progeny is that of the mt^+ parent only.

Message The cpDNA of *Chlamydomonas* is inherited uniparentally. Because the behavior of this DNA parallels the behavior of the uniparentally transmitted genes, it can be inferred that these genes are located in the cpDNA.

Mapping Chloroplast Genes in *Chlamydomonas*

We have seen (Table 21-2) that *Chlamydomonas* has a large number of uniparentally inherited genes. Is there linkage among these genes? Are they all in one linkage group (on one chromosome), or are they arranged in several linkage groups? Or does each gene assort independently as if it were on its own separate piece of DNA? Of course, the way to approach this question in the true tradition of classical genetics is to perform a recombination analysis. But here we run into a problem. Both sets of parental DNA must be present if there is to be an opportunity for recombination, and we have seen that the cpDNA of the mt^- parent is eliminated in the zygote cell. Luckily, there is a way out of this dilemma.

In crosses of mt^+ *sm-r* $\times$ mt^- *sm-s*, about 0.1 percent of the progeny zygotes are found to contain both *sm-r* and *sm-s*. The presence of both alleles can be inferred from the fact that the products of meiosis of such cells show both sm-r and sm-s phenotypes among their number. Segregation of the two alleles presumably arises from a CSAR process in the zygote or at a later stage of cell division. Such zygotes are called **biparental zygotes,** for obvious reasons, and their genetic condition is described as a **cytohet** ("*cyto*plasmically *het*erozygous") or a **heteroplasmon.** It appears as though the inactivation of the mt^- parent's cpDNA fails in these rare zygotes. However, whether this is what happens or not, the cytohets provide just the opportunity we need to study recombination: these cells contain *both* sets of parental cpDNA. The rarity of cytohet zygotes does pose a problem for research, but their frequency can be increased by treating the mt^+ parent with ultraviolet light before mating. After such treatment, 40 to 100 percent of the progeny zygotes are cytohets. (Note in passing that this observation implies that the m^+ cell plays a normal role in actively eliminating the mt^- cell's cpDNA. The ultraviolet irradiation of the mt^+ cell must inactivate such a function.)

Message In *Chlamydomonas,* biparental zygotes (or cytohets) must be the starting point for all studies on the segregation and recombination of chloroplast genes.

A good map of the cpDNA has been obtained using the cytohets. One of the most profitable ways of studying linkage relationships is through the cosegregation patterns of two separate extrachromosomal genes in the same cross. For example, in a cross of the type mt^+ *sm-s* ac^+ $\times$ mt^- *sm-r* ac^-, biparental zygotes are first obtained. Cytoplasmic segregation can be detected in the two daughter cells arising from each product of meiosis, or in the daughter cells arising from the subsequent few mitotic divisions. It is reasonable to assume that if the two genes are linked closely, then when one gene undergoes cytoplasmic segregation, the other probably will too. More distantly linked genes should show proportionately lower frequencies of cosegregation.

Sager was able to quantify these procedures by using standard cell populations. She obtained a consistent additive linkage map of the genes in the "uniparental genome." However, the map has one major inconsistency, which can be resolved only by assigning the shape of a circle to the cpDNA molecule. The map resulting from this and other mapping techniques is shown in Figure 21-12. Circular cpDNA has now been demonstrated directly in many plant species. It is worth recalling the

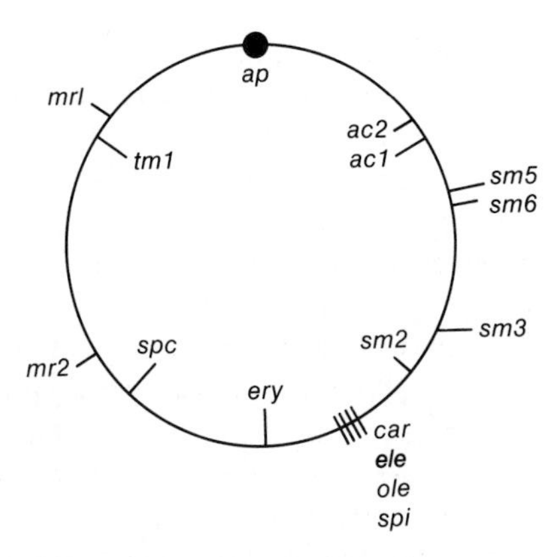

Figure 21-12 Circular map of *Chlamydomonas* cpDNA derived from genetic analysis. The map must be circular to accommodate all the linkage data. (After Ruth Sager, "Genetic Analysis of Chloroplast DNA in Chlamydomonas," in *Cytoplasmic Genes and Organelles,* Academic Press.)

history of classical genetic analysis. The genetic maps for nuclear chromosomes also were derived before their physical counterparts were demonstrated to exist. A great deal more detail is now available on the cpDNA maps in several organisms; we shall take another look at such maps after we have considered the extranuclear genome of yeast.

Mitochondrial Genes in Yeast

Possibly the greatest success story in the clarification of extranuclear inheritance has been the development of the current view of the mitochondrial genome in bakers' yeast *(Saccharomyces cerevisiae)*. This achievement provides a classic example of the power of genetic analysis in combination with modern molecular techniques. Let's start with the contributions of genetic analysis.

Three kinds of mutants have been of particular importance: the *petite, ant*^{R}, and *mit*$^-$ mutants. In the 1940s, Boris Ephrussi and his colleagues first described some curious mutants in yeast. The wild-type cells of yeast form relatively large colonies on the surface of a solidified culture medium. Among these large, or "grande," colonies an occasional small, or "petite," colony is found. When isolated, the **petite mutants** prove to be of three types on the basis of their inheritance patterns. The first type is called **segregational petites** because on crossing to a grande strain, half the ascospores give rise to grande colonies and the other half give rise to petite colonies. This 1 : 1 Mendelian segregation obviously indicates that the petite phenotype is due to a nuclear mutation in these cases. The second type is the **neutral petites,** which in crosses to a grande strain give ascospores that all grow into grande colonies — a clear case of uniparental inheritance. The third type of petite mutants is the **suppressive petites,** which give some ascospores that grow into grande colonies and some that grow into petite colonies. The ratio of grandes to petites is variable but strain-specific: some suppressive petites give exclusively petite offspring in such crosses. Thus the suppressive petites obviously show a non-Mendelian inheritance pattern, and some show uniparental inheritance.

In a yeast cross, the two parental cells fuse and apparently contribute equally to the cytoplasm of the resulting diploid cell (Figure 21-13). Furthermore, the inheritance of the neutral and suppressive petites is independent of mating type. In this sense, then, yeast is clearly quite different from *Chlamydomonas.* Nevertheless, since their inheritance is obviously extranuclear, the neutral and suppressive types have become known as the **cytoplasmic petites.**

Several properties of cytoplasmic petites point to the involvement of mitochondria in their phenotype:

1. In cytoplasmic petites, the mitochondrial electron-transport chain is defective. Because this chain is responsible for ATP synthesis, petites must rely on the less efficient process of fermentation to provide

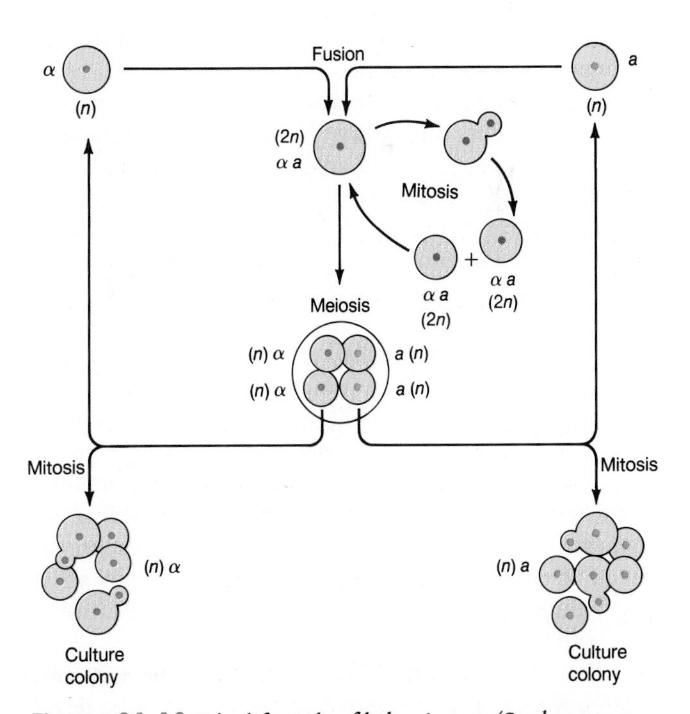

Figure 21-13 The life cycle of bakers' yeast *(Saccharomyces cerevisiae)*. The nuclear alleles *a* and *α* determine mating type. Cell fusion between haploid *a* and *α* cells produces a diploid cell. Normally, the cell then goes through a diploid mitotic cycle (budding). However, the cell can be induced (by plating on a special medium) to undergo meiosis (sporulation), producing haploid products. Note that budding involves the formation of a small growth on the side of the parent cell; this bud eventually enlarges and separates to become one of the daughter cells.

their ATP. If they are placed in a medium containing a nonfermentable energy source such as glycerol, they will not grow.

2. These petites show no mitochondrial protein synthesis. Mitochondria normally possess their own unique protein-synthesizing apparatus, consisting of a unique set of tRNA molecules and unique ribosomes, all of which are quite different from those operating outside the mitochondrion in the **cytosol,** or nonorganellar phase of the cytoplasm.
3. Mitochondria in all organisms have their own unique mtDNA that, as well as being smaller in amount, is quite different from nuclear DNA. The **mitochondrial DNA (mtDNA)** of petites is radically different from that of wild types. Neutral petites totally lack mtDNA, whereas suppressive petites show altered base ratios compared with the grandes from which they spring.

The second major class of yeast mutants, the ***ant*^R^ mutants,** were initially recognized at the phenotypic level by their resistance to antibiotics supplied in the medium. For example, strains have been obtained that are resistant to chloramphenicol (cap^R), erythromycin (ery^R), spiromycin (spi^R), paramomycin (par^R), and oligomycin (oli^R). These mutations each show a non-Mendelian inheritance pattern similar to that in the suppressive petites; that is, in a cross such as $ery^R \times ery^S$, a strain-specific non-Mendelian ratio is seen among the random ascospore progeny. However, examination of specific meiosis by tetrad analysis reveals several now-familiar phenomena. Let's trace the cross through its sequential stages (Figure 21-14). When the parental cells fuse, the fusion product, as well as being diploid, is a cytohet. The diploid cells can then be allowed to bud mitotically. During this mitotic division, because of CSAR, the daughter cells become either ery^R or ery^S. Therefore, when meiosis is induced by shifting the cells onto a special medium, all of the haploid products of any single meiosis are identical with respect to erythromycin sensitivity or resistance. At the level of single meioses, then, we see uniparental inheritance at work. (Note that the nuclear mating-type alleles *a* and α always segregate in a 1 : 1 ratio.) Similar results have been shown by all the ant^R mutations, pointing once again to their location in an "extrachromosomal genome."

The third important class of mutants, the ***mit*^−^ mutants,** were last to be discovered and required the development of special selective techniques. These mutants are similar to petite mutants in that they exhibit small colony size and abnormal electron-transport-chain functions, but they differ in that they have normal protein synthesis and are able to revert. In a way, mit^- mutants are like point-mutation petites. The inheritance of mit^- mutants is comparable to the patterns shown by ant^R types; that is, they show cytoplasmic segregation and also uniparental inheritance at meiosis.

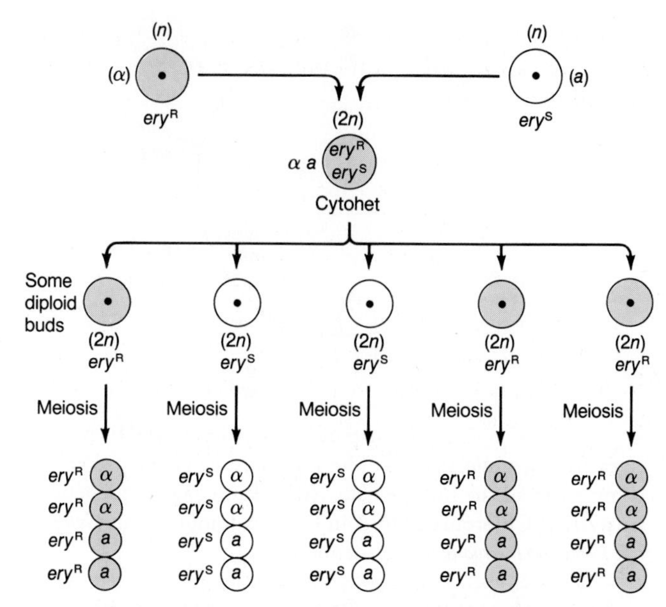

Figure 21-14 The special inheritance pattern shown by certain drug-resistant phenotypes in yeast. When diploid buds are induced to sporulate (undergo meiosis), the products of each meiosis show uniparental inheritance. (Only a representative sample of the diploid buds is shown.) Note that the nuclear genes, represented here by the mating-type alleles *a* and α, segregate in a strictly Mendelian pattern. The alleles ery^R and ery^S determine erythromycin resistance and sensitivity, respectively.

Mapping the Mitochondrial Genome in Yeast

The demonstration of genetic determinants constituting an extranuclear genome immediately raises the question of the physical interrelationship of these determinants. Are they all linked together on one mitochondrial "chromosome," or are they located on separate structural units? Mapping the yeast mitochondrial genome has proceeded using many different approaches. A few representative analytical methods follow.

Recombination Mapping. We can set out to look for recombinants by using a cross between parents differing in two extranuclear gene pairs (a kind of "dihybrid cross"). For example, we can carry out the cross $ery^R\ spi^R \times ery^S\ spi^S$, allow the resulting diploid cell to bud through several cell generations, and then induce the resulting cells to sporulate. We can then identify the genotype of each bud cell by observing the phenotype common to all its ascospores.

Four genotypes can result from such a cross (Figure 21-15) and all are in fact observed. An early cross yielded the following results:

ery^R spi^R	63 tetrads
ery^S spi^S	48 tetrads
ery^S spi^R	7 tetrads
ery^R spi^S	1 tetrad

The genotypes ery^S spi^R and ery^R spi^S, of course, represent recombinants that have been produced by CSAR during the formation of diploid buds. (Note that these recombinants cannot have been produced by meiosis because the products of any given meiosis are identical with respect to the drug-resistance phenotypes.)

Message In yeast, there is cytoplasmic segregation and recombination during bud formation in diploid cytohets. The segregation and recombination may be detected directly in cultures of the diploid buds or may be detected by observing the products of meiosis that result when the buds are induced to sporulate.

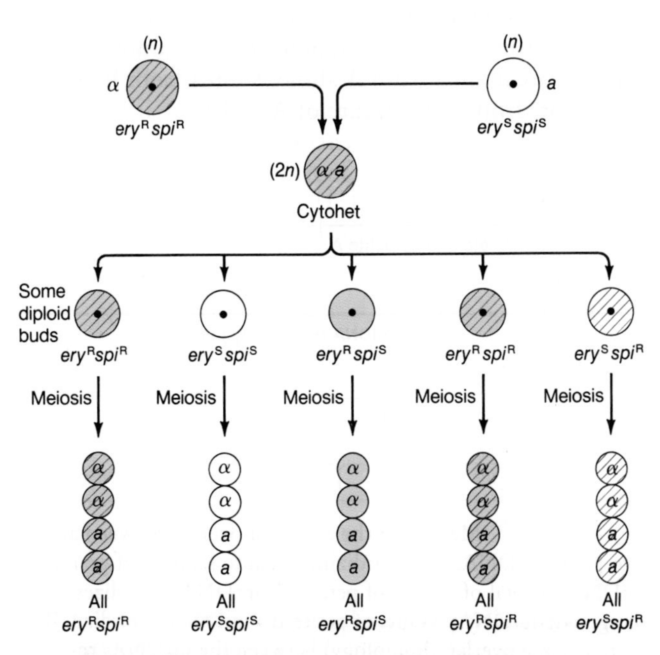

Figure 21-15 The study of inheritance in a cross between yeast cells differing with respect to two different drug-resistance alleles (ery^R = erythromycin resistance; spi^R = spiramycin resistance). Each diploid bud can be classified as parental or recombinant on the basis of these results. Note that the identity of the extranuclear genotype for all four products of meiosis confirms that CSAR must occur during the production of the diploid buds.

We seem to be just a short step away from the development of a complete map of all the drug-resistance genes in yeast. Unfortunately, the technique of recombination mapping proved to be of only limited usefulness. For one thing, recombination involving mitochondria was shown to be a population phenomenon, similar to phage recombination. This is to say, so many rounds of recombination are occurring that most genes appeared to be unlinked. Linkage is detectable only for genes that are very close together. Furthermore, the process of recombination has been shown to be strongly influenced by a specific genetic factor, ω (omega), that is present in some mitochondrial genomes and not in others. The most useful mapping developments came from less conventional analyses, examples of which are described in the following subsections.

Mapping by Petite Analysis. The petite mutations, the drug-resistance (ant^R) mutations, and the mit^- mutations are apparently inherited on the mitochondrial genome of yeast. Some very effective techniques for mapping that genome were developed through the combined study of these classes of mutants. Most of these approaches are based on the discovery that petites represent deletions of the mtDNA. This fact opens up a new and different kind of genetic analysis that has been combined with new techniques of DNA manipulation to produce a rather complete genetic map of the mtDNA.

The pivotal observation came in studies where drug-resistant grande strains (such as ery^R) were used as the starting material for the induction of petite mutants. We have seen that petite mutants form spontaneously, but they can also be induced at high frequencies by the use of various specific mutagens, notably ethidium bromide. It is of interest to determine the drug resistance of petite cultures obtained from grande ery^R strains. Are they still ery^R, or has the petite mutation caused them to become ery^S? The answer cannot be determined directly because drug resistance cannot be tested in petite cells.

However, genetic trickery comes to the rescue. The induced petites are combined with grande ery^S cells (which, of course, lack resistance to erythromycin) to form diploids, and the diploids are allowed to form diploid buds. Depending on the type of petite used, varying amounts of petite and grande diploid buds are formed. It is the grande diploid buds that are of interest here because they *can* be tested for drug resistance. If the original petites retain the drug resistance, then some of these diploid grande cells may be expected to have acquired that resistance through the recombination of CSAR. In fact, for some petites, the derived grande diploid buds do prove to be of two phenotypes: ery^S (derived from the grande ery^S haploid) and ery^R (which must be derived from the petite haploid). This result indicates that only

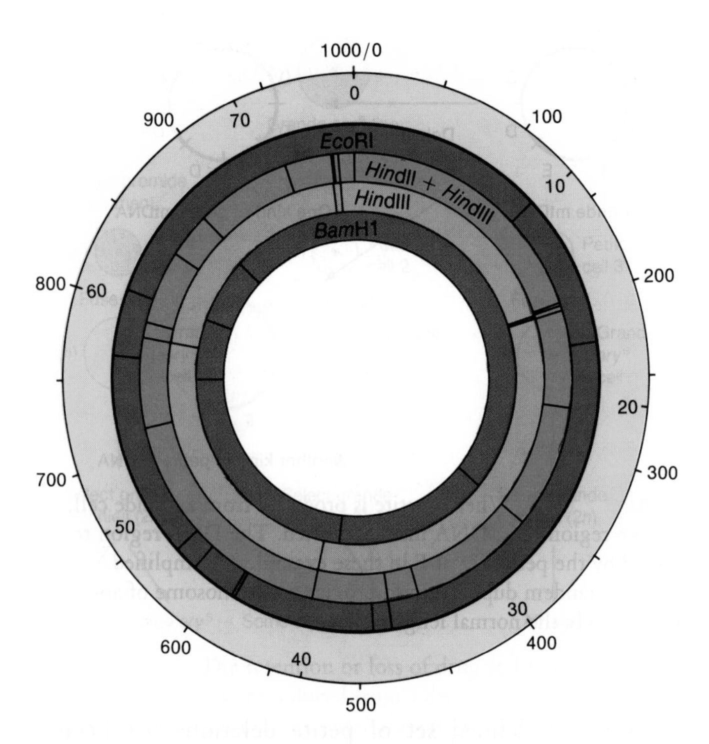

Figure 21-20 Map showing the target sites of various restriction enzymes on the yeast mtDNA. An arbitrary scale from 0 to 1000 is indicated on the outside of the outer circle; a scale indicating thousands of DNA base pairs is indicated on the inside of the outer circle. The inner four bands show the fragments produced by treatment with particular restriction enzymes: *Eco*RI, *Hin*dII + *Hin*dIII, *Hin*dIII, and *Bam*HI. The position for the zero point of this map is arbitrary. (After J. P. M. Sanders et al., "The Organization of Genes in Yeast Mitochondrial DNA, III." *Molecular and General Genetics* 157, 1977.)

Southern hybridization (see page 425) to a restriction-enzyme digest. The band that "lights up" on autoradiography represents the approximate location of the RNA. Further similar experiments using different restriction enzymes will narrow the region of hybridization down to a precise locus on the mtDNA.

What about the ant^R and mit^- genes? One approach is to correlate the retained restriction fragments in a number of different petites with, say, the antibiotic-resistance markers also retained in those strains. In this way, specific markers may be associated with specific regions of the mtDNA, as defined by the restriction-enzyme target sites. Another approach is to take a petite that retains only a single drug-resistance gene — say, ery^R. The mtDNA is extracted from the petite and made radioactive for use as a probe. Its hybridization with the various restriction fragments can be tested, thus locating the portion of the mtDNA that is retained by this particular petite — and hence the locus of the *ery* gene. For example, if the mtDNA from the ery^R petite hybridizes with fragment X from one restriction enzyme and with fragment Y from another enzyme, then the *ery* locus must be located within the overlap region of these two fragments.

An Overview of the Mitochondrial Genome

Many specific areas of the yeast map have been subjected to more detailed dissection, such as sequencing and intron analysis. The present state of the yeast mtDNA map, based on the types of techniques we have considered, is shown in Figure 21-21, together with the human mtDNA map for comparison. The various genes in yeast discussed in this chapter are shown together with their protein products. We can see that cap^R, ery^R, and spi^R are in fact alterations of the large mitochondrial rRNA genes, and that par^R is associated with changes in the small rRNA gene. Other ant^R mutations such as oli^R are associated with alterations in various subunits of the enzyme ATPase. On the other hand, mit^- mutations are in fact lesions in several subunits of cytochrome oxidase (I–III) or in the cytochrome *b* gene. In addition, several other genes are indicated, such as those for the mitochondrial tRNAs and a gene for a ribosome-associated protein.

The map contains some surprises too. Most prominent are the introns in several genes. Subunit I of cytochrome oxidase contains nine introns! The discovery of introns in the mitochondrial genes is particularly surprising because they are relatively rare in yeast nuclear genes. Another set of surprises was the **unassigned reading frames (URFs).** These are sequences that have correct initiation codons and are uninterrupted by stop codons. These "genes in search of a function" are the subject of current research. Some URF's within introns appear to specify proteins important in the splicing out of the introns themselves at the RNA level. Notice that the human mtDNA is by comparison much smaller and more compact. There seems to be much less spacer DNA between the genes.

The overall view of the mitochondrial genome shows it to have two main functions: it codes for some proteins that are actually in or associated with the electron-transport chain, and it codes for some proteins, all the tRNAs, and both rRNAs necessary for mitochondrial protein synthesis. But it is striking that the remaining necessary components for both of these functions are encoded by nuclear genes with mRNA that is translated on cytosolic ribosomes followed by transport of the products to the mitochondrion (Figure 21-22). Why this peculiar division of labor exists between nuclear and mitochondrial DNA is not known. Another curiosity is that some specific subunits are encoded by mtDNA in one organism but by nuclear DNA in another. Evidently, an

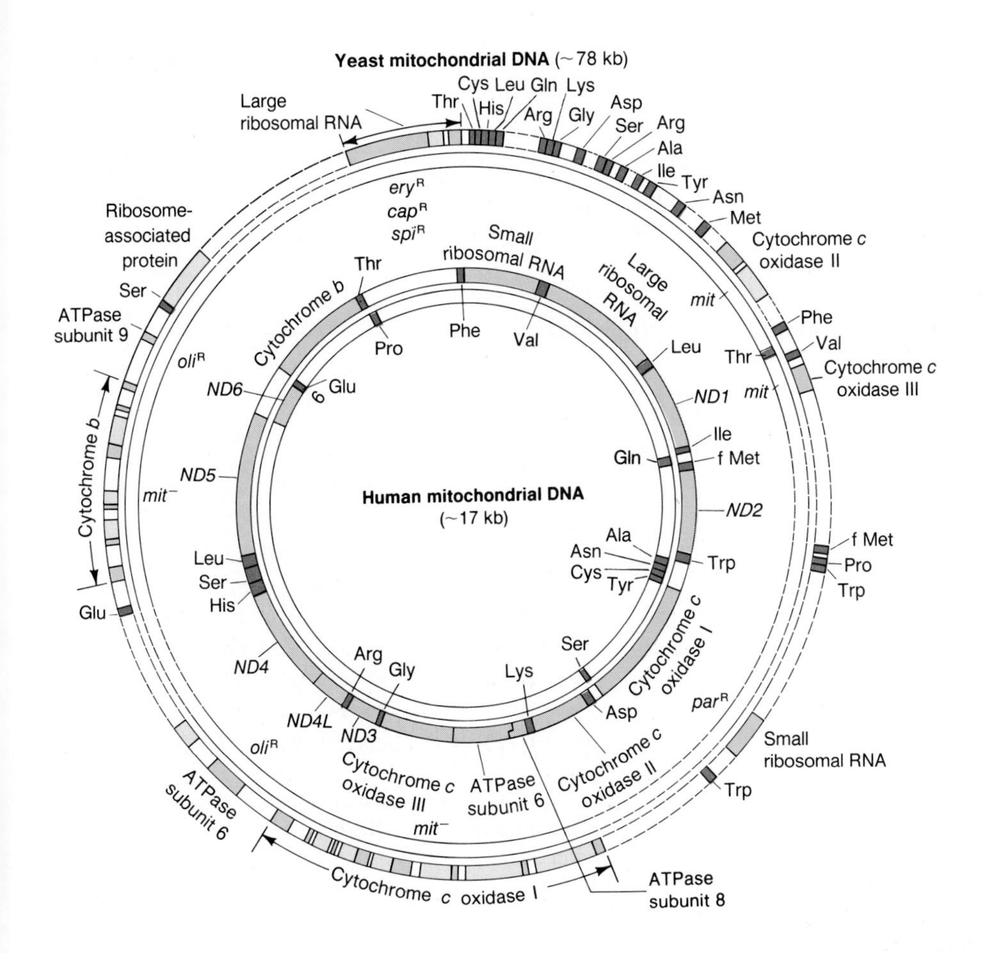

Figure 21-21 Maps of yeast and human mtDNA's. Each map is shown as two concentric circles corresponding to the two strands of the DNA helix. The human map has been produced exclusively by physical techniques. The yeast map has been produced by a combination of genetic and physical techniques, as discussed in the text. Note that the mutants used in the yeast genetic analysis are shown opposite their corresponding structural genes. Blue = exons and uninterrupted genes, red = tRNA genes, and yellow = URF's (unassigned reading frames). tRNA genes are shown by their amino-acid abbreviations; ND genes code for subunits of NADH dehydrogenase. (Note that the human map is not drawn to the same scale as the yeast map.)

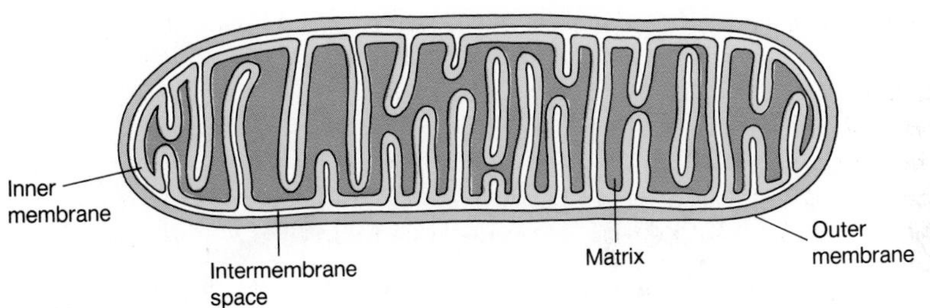

Mitochondrial DNA
Blocked by acridines, ethidium bromide
Messenger RNA
Blocked by chloramphenicol, tetracycline, macrolides
Proteins
Nuclear DNA
Blocked by α amanitin
Messenger RNA
Blocked by cycloheximide
Proteins
Cytochrome c oxidase
ATPase complex
Cytochrome b
Ubiquinol-cytochrome c reductase complex
Matrix
Cytoplasm
Inner membrane
Intermembrane space
Outer membrane

Figure 21-22 Cooperation of yeast mtDNA and nuclear DNA in coding for the protein components of the inner mitochondrial membrane. (a) Overview of mitochondrial structure. (b) Details of membrane constitution. Three functional proteins, cytochrome, C oxidase, ATPase and cytochrome *b*, show the cooperation. The mtDNA supplies six proteins, three subunits of cytochrome *c* oxidase, two subunits of ATPase, and cytochrome *b*. Part of the dissection of this complex scheme was through the use of drugs that block specific reactions as shown.

evolutionary transposition of information has occurred between these organelles. There is also evidence of transposition between mitochondria and chloroplasts. Furthermore, inactive pseudogenes are detectable in the nucleus, showing homology with mitochondrial genes.

Genes for 25 yeast and 22 human mitochondrial tRNAs are shown on the maps in Figure 21-21. These tRNAs carry out all the translation that occurs in mitochondria. These are far fewer than the minimum of 32 required to translate nucleus-derived mRNA. The economy is achieved by a "more wobbly" wobble pairing (see page 389) of tRNA anticodons. The tRNA specificities in human mtDNA are shown in Figure 21-23. Notice that the codon assignments are in some cases different from the nuclear code. It is also known that there is variation between the mitochondria of different species. Hence, the genetic code is obviously not universal, as had been supposed for many years.

First letter	Second letter: U	Second letter: C	Second letter: A	Second letter: G	Third letter
U	Phe	Ser	Tyr	Cys	U
	Phe	Ser	Tyr	Cys	C
	Leu	Ser	*Stop*	*(Stop)* Trp	A
	Leu	Ser	*Stop*	Trp	G
C	Leu	Pro	His	Arg	U
	Leu	Pro	His	Arg	C
	Leu	Pro	Gln	Arg	A
	Leu	Pro	Gln	Arg	G
A	Ile (Met)	Thr	Asn	Ser	U
	Ile	Thr	Asn	Ser	C
	(Ile) Met	Thr	Lys	(Arg) *Stop*	A
	Met	Thr	Lys	(Arg) *Stop*	G
G	Val	Ala	Asp	Gly	U
	Val	Ala	Asp	Gly	C
	Val	Ala	Glu	Gly	A
	Val	Ala	Glu	Gly	G

Figure 21-23 The genetic code of the human mitochondrion. The functions of the 22 tRNA types are shown by the 22 non-*stop* codon boxes.

Message Mitochondrial DNA has genes for mitochondrial translation components (mainly rRNAs and tRNAs) and for some subunits of the proteins associated with mitochondrial ATP production. Less understood regions include the introns, unassigned reading frames, and spacer DNA.

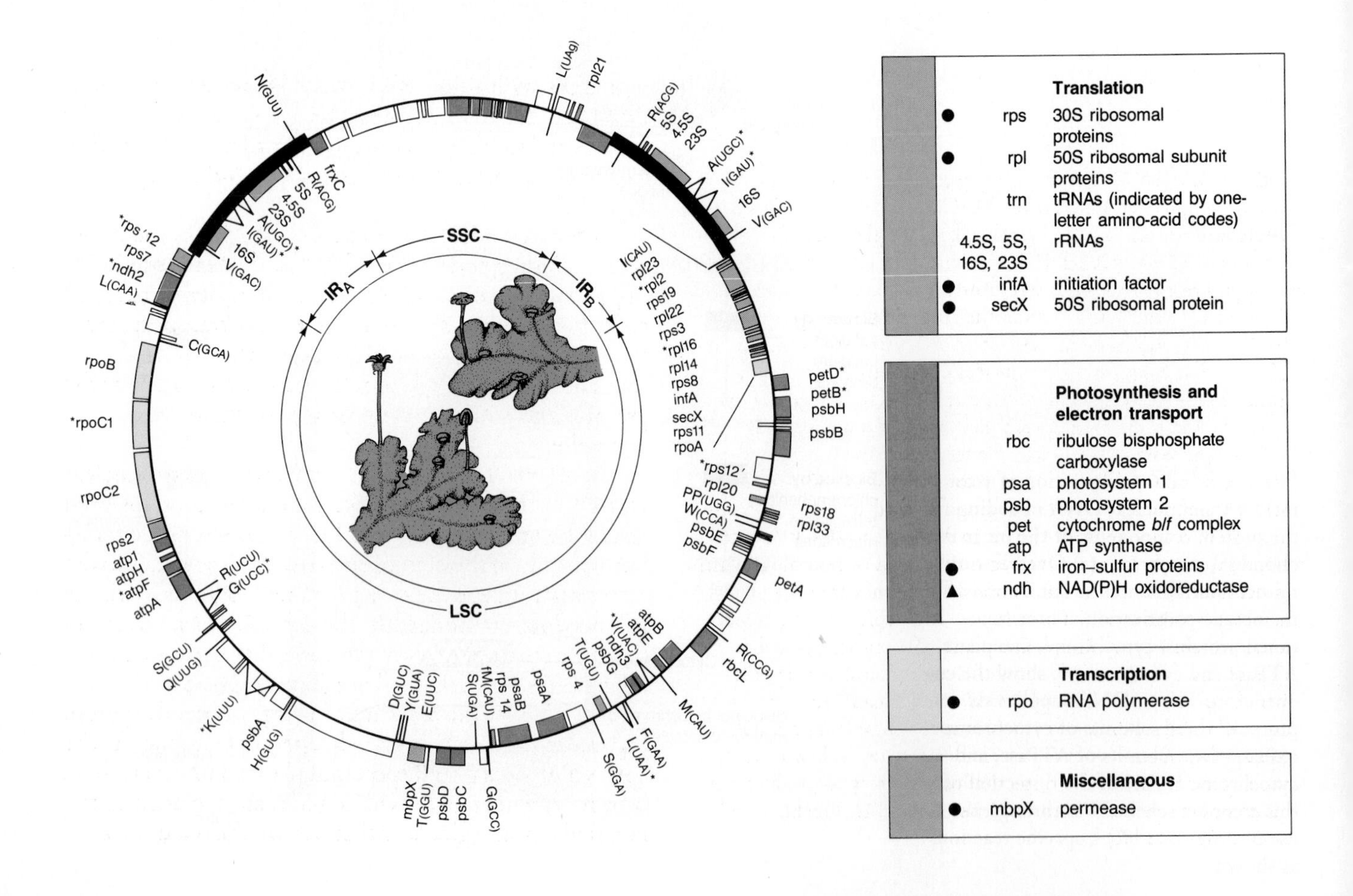

An Overview of the Chloroplast Genome

Although many of the concepts behind chloroplast inheritance have been derived from the kind of genetic studies we have followed in *Mirabilis* and *Chlamydomonas,* the current perspective on the overall organization of cpDNA has come mainly from molecular studies. In fact, the cpDNA of several species has been completely sequenced. As an example Figure 21-24 shows the organization of the cpDNA from the liverwort *Marchantia polymorpha.*

Typically, cpDNA molecules range from 120 to 200 kb in different plant species. In *Marchantia,* the molecular size is 121 kb. Figure 21-24 shows the functions of most of the genes in the *Marchantia* cpDNA. However, the sequencing of 121,000 nucleotides does not automatically tell us what genes are present. The presence of genes is inferred by the detection of **open reading frames (ORFs)** — long sequences that begin with a start codon but are uninterrupted by stop codons except at the termini. Determining what these specific genes code for is more difficult; although most ORFs have been assigned, some of the ORFs are still unassigned reading frames (URFs). Most of the protein-encoding genes have been assigned by comparing the predicted amino acid sequences with those of known chloroplast, mitochondrial, and bacterial proteins. Other genes, such as the rRNAs and tRNAs, have been assigned by hybridization experiments, using known genes from other organisms as probes.

Figure 21-25 Mutations in nuclear genes can also result in white leaves. Here a green *Cc* corn plant, heterozygous for a recessive allele causing albino leaves, was selfed and the progeny were $\frac{1}{4}$ *cc* and fully albino. Because they cannot photosynthesize, these albinos die as soon as the food reserve in the seed (laid down by the maternal plant) is exhausted. Such mutations can be in the synthesis of chlorophyll itself or in one of the other nuclear proteins that interact with the chloroplast-encoded proteins to produce a functional photosynthetic light reaction. (Anthony Griffiths)

Figure 21-24 The chloroplast genome of the liverwort *Marchantia polymorpha.* IR_A and IR_B, LSC, and SSC on the inner circle indicate the inverted repeats, large single-copy and small single-copy regions, respectively. Genes shown inside the map are transcribed clockwise, and those outside are transcribed anticlockwise. Genes for rRNAs in the IR regions are represented by 16S, 23S, 4.5S and 5S, respectively. Genes for tRNAs are indicated by the one-letter amino acid code with the unmodified anticodon. Protein genes identified are indicated by gene symbols, and the remaining open boxes represent unidentified ORFs, approximately to scale. Genes containing introns are marked with asterisks. The boxes to the right of the gene map summarize the functions of the genes identified to date; groups with related functions are shown in different shades. ORFs are shown in color. Genes identified on the basis of homology with genes from bacterial or mitochondrial genomes are marked ● and ▲, respectively. The central drawing depicts a male *(above)* and a female *(below) Marchantia* plant. The antheridia and archegonia are elevated on specialized stalks above the thallus, which contains the chloroplasts. *Marchantia* can also reproduce asexually: disks of green tissue (gemmae) grow from the bottom of cup-shaped structures on the thallus surface. When mature, the gemmae separate from the thallus and grow to produce new gametophyte plants. (From K. Umesono and H. Ozeki, *Trends in Genetics* 3, 1987.)

There are about 136 genes in the *Marchantia* molecule, including four kinds of rRNA, 31 kinds of tRNA, and about 90 protein genes. Of the 90 protein genes, 20 code for photosynthesis and electron-transport functions. Genes coding for translation functions take up about one-half of the chloroplast genome and include the proteins and RNA types necessary for translation in the organelle.

Like mtDNA, cpDNA cooperates with nuclear DNA to provide subunits for functional proteins used inside the organelle. The nuclear components are translated outside in the cytosol and then transported into the chloroplast, where they are assembled together with the components synthesized in the organelle. Mutations in the nuclear components of photosynthetic complexes are inherited as Mendelian alleles (Figure 21-25).

Also, notice the presence of a large inverted repeat in Figure 21-24. Such inverted repeats are found in the cpDNA of virtually all species of plants. However, there is some variation as to which genes are included in the inverted repeat region and therefore in the relative size of that region. One of the mysteries of the inverted repeat is that the duplicates are always identical within a species, yet to date no mechanism is known that ensures this complete identity.

Mitochondrial Diseases in Humans

Several human diseases are known that are caused by mutations in the mitochondrial DNA. These mutations are all in the mitochondrial genes for protein subunits of the mitochondrial electron transport chain. The mutations result in defects of mitochondrial ATP production. Humans that carry these mutations have **mitochondrial cytopathies,** diseases associated with various combinations of defects of the brain, heart, muscle, kidney, and liver.

The mutations themselves can be point mutations or deletions or duplications of the mtDNA, as shown in Figure 21-26. As expected, such mitochondrial cyto-

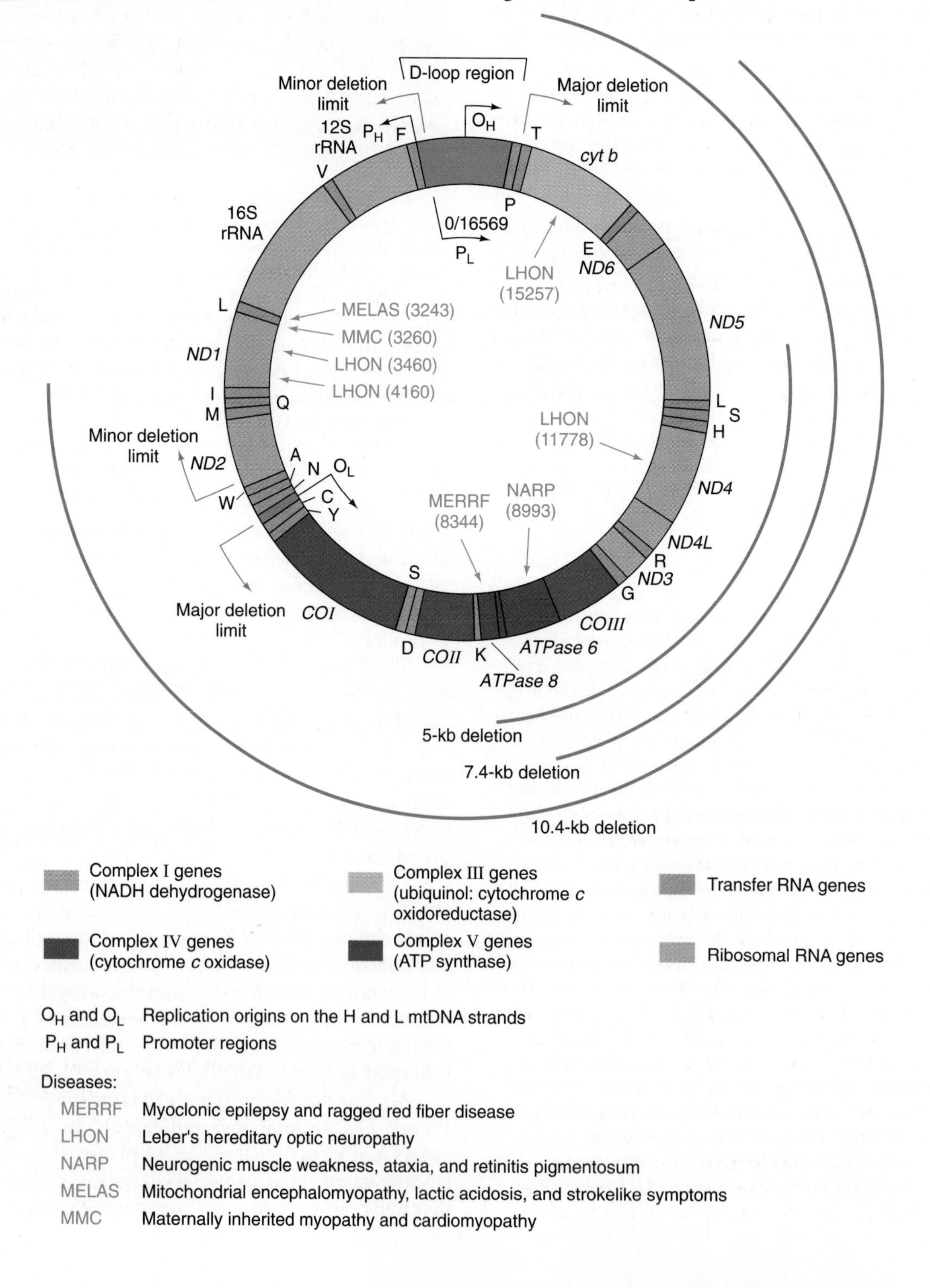

pathies show the inheritance features characteristic of mitochondrial mutations. First, they are passed on from one generation to the next only through maternal parents. Second, they show cytoplasmic segregation when present in heteroplasmic mixtures. Hence, a person with a mixture of normal and abnormal mitochondria can have cells with widely differing ratios of these mitochondria because of cytoplasmic segregation. Similarly, a woman with mild expression of a mitochondrial disease can produce children with disease expression ranging from none to severe, again because of cytoplasmic segregation.

Incidentally, mutations in the nucleus-encoded subunits of the mitochondrial electron transport chain (Figure 21-22) produce a similar spectrum of disease phenotypes. However, these mutations can be distinguished in pedigrees because they show the Mendelian inheritance typical of nuclear genes.

Extragenomic Plasmids in Eukaryotes

We saw in Chapter 10 that plasmids are common in bacteria. However, plasmids are also regularly, although less commonly, encountered in eukaryotes. Because plasmids are generally not associated with nuclear chromosomes, they can show a kind of non-Mendelian inheritance pattern similar in some ways to that of organelle DNA.

Most eukaryotic plasmids are silent at the phenotypic level and can only be detected using molecular techniques. The best known of these is the "two-micron circle" (2μ) plasmid of yeast. This circular plasmid is probably located in the nucleoplasm. If a haploid strain containing the plasmid is mated with another haploid strain lacking the plasmid, the asexual or sexual descendants all tend to have the plasmid. Although mysterious biologically, the 2μ plasmid has assumed a prominent role in the molecular biology of yeast because it has been genetically engineered to act as a gene vector for that organism (see Chapter 15). When nuclear DNA is spliced into the 2μ vector and used to transform a recipient cell type, the insert is carried to the nucleus, where it may become incorporated into the chromosomes via a variety of recombination mechanisms.

However, most eukaryotic plasmids are mitochondrial. An interesting and well-studied type of mitochondrial plasmid is associated with cytoplasmic male sterility in corn. Male sterility in plants is of great importance in agriculture. Plants with the male-sterile trait produce no functional pollen. In the case of cytoplasmic male sterility, when male sterile plants are crossed as female parents using normal fertile plants as males, all of the progeny prove to be male-sterile and the inheritance pattern is clearly maternal. The best-studied type of cytoplasmic male sterility is in corn where it is associated with two linear plasmids, S1 and S2, that are located within the mitochondria in addition to the mtDNA. The precise way in which the S1 and S2 plasmids are involved with male sterility is not yet fully understood, although much is known about the molecular properties of these plasmids. One intriguing property is that S1 and S2 can recombine with the mtDNA.

Male sterility in corn and other crop plants is used to facilitate the production of hybrid seed issuing from a cross between genetically different lines; such seeds usually result in larger, more vigorous plants. The big problem is how to prevent self-pollination, which interferes with the production of hybrid seed. Male sterility is useful here because selfing is impossible since there is no functional pollen. One breeding scheme is illustrated in Figure 21-27.

Our final example of eukaryotic plasmids is a type that determines senescence (aging) in the fungus *Neurospora*. Most *Neurospora* strains do not senesce and, given a large enough supply of medium, will keep growing forever. However, certain populations contain senescent individuals. One such population is from Hawaii, where the senescent strains are called "kalilo," a Hawaiian word meaning "at death's door." Kalilo strains will grow only a certain distance through the culture medium before they die. However, crossing these strains before they die reveals that the ability to die is inherited in a strict maternal fashion. It has been found that the senescence determinant is a linear 9-kb mitochondrial plasmid called kalDNA. This plasmid can exist autonomously—and apparently innocuously—in the cytoplasm, but the onset of death is preceded by the physical insertion of kalDNA into the mtDNA. At death, most mtDNA molecules contain the full-length kalDNA insert (Figure 21-28). Presumably, the strain dies because the kalDNA interferes with mitochondrial function. It can therefore be seen that kalDNA behaves like a molecular parasite.

Notice that the plasmids are associated with a definite phenotype in the examples from corn and *Neurospora,* whereas the yeast $2\text{-}\mu$ plasmid has no detectable phenotypic manifestation. The existence and evolution-

Figure 21-26 Locations of the major mitochondrial DNA mutations that result in human disease. Point mutations are shown by nucleotide number on the mtDNA map. Most of the deletions found in CEOP (chronic external ophthalmoplegia plus), KSS (Kearns-Sayre syndrome), and Pearson's syndrome occur within the major deletion limits indicated; others are found within the region called the minor deletion limit. Some examples of these deletions are shown as red arcs. Gene symbols are explained in Figure 21-24. (From D. C. Wallace, *Science* 256, 629, 1992.)

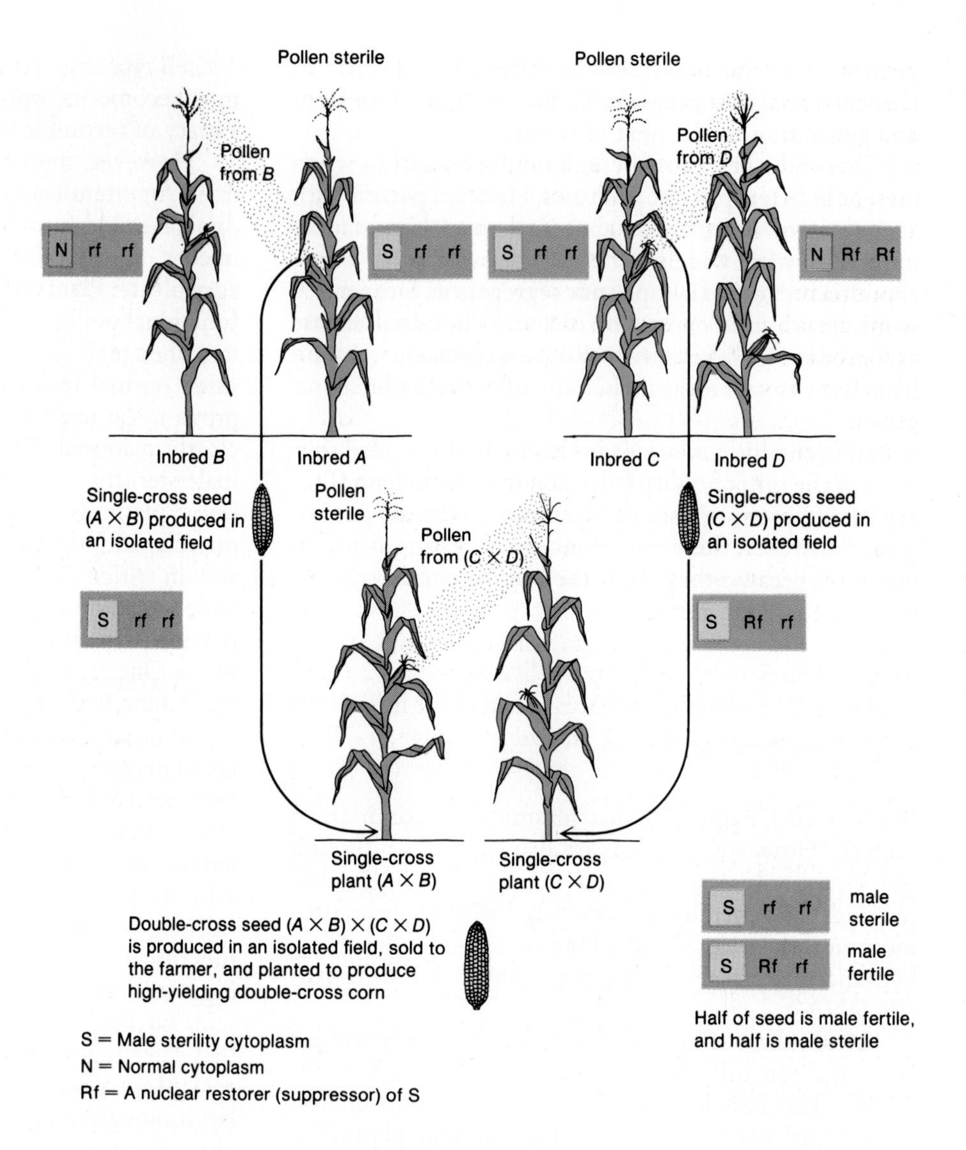

Figure 21-27 The use of cytoplasmic male sterility to facilitate the production of hybrid corn. In this scheme, the hybrid corn is generated from four pure parental lines: *A*, *B*, *C*, and *D*. Such hybrids are called double-cross hybrids. At each step, appropriate combinations of cytoplasmic genes and nuclear restorer genes ensure that the female parents will not self and that male parents will have fertile pollen. (From J. Janick et al., *Plant Science.* Copyright 1974 by W. H. Freeman and Company.)

ary position of eukaryotic plasmids is still a mystery, but the study of the structure and behavior of such elements will undoubtedly provide important clues about the fundamental properties of the genetic material.

> Message Extragenomic eukaryotic plasmids are inherited in a non-Mendelian manner that often mimics the inheritance patterns of organelle DNA.

How Many Copies?

For the genetic and the biochemical approaches to the study of organelle genomes that we have examined, it does not matter much how many copies of organelle

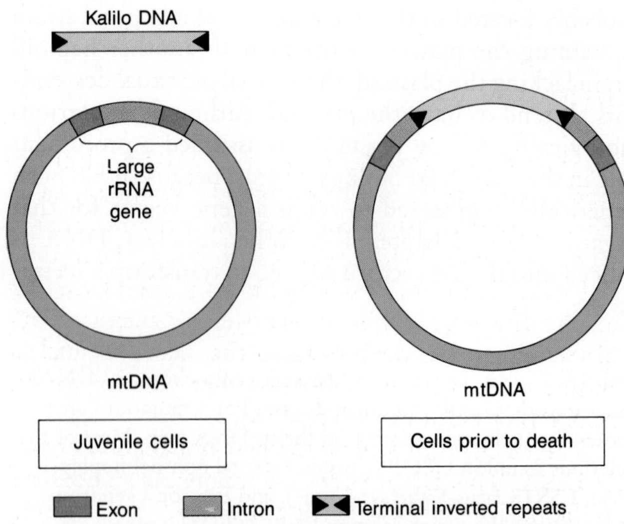

Figure 21-28 Senescence is *Neurospora* is associated with the insertion of a linear plasmid called kalilo DNA into the mtDNA. Here insertion is shown in the intron of a gene for rRNA, but other insertion sites are known.

genome are present per cell. However, it is a question that occurs to most people interested in organelle inheritance.

The number of copies turns out to vary among species. More surprisingly, it also can vary within a single species. The leaf cells of the garden beet have about 40 chloroplasts per cell. The chloroplasts themselves contain specific areas that stain heavily with DNA stains; these areas are called **nucleoids,** and they are a feature commonly found in many organelles. Each beet chloroplast contains 4 to 18 nucleoids, and each nucleoid can contain from 4 to 8 cpDNA molecules. Thus, cells of a single beet leaf can contain as many as $40 \times 18 \times 8 = 5760$ copies of the chloroplast genome! Although *Chlamydomonas* has only one chloroplast per cell, the chloroplast contains 500 to 1500 cpDNA molecules, commonly observed to be packed in nucleoids.

What about mitochondria? A "typical" haploid yeast cell can contain 1 to 45 mitochondria, each having 10 to 30 nucleoids, with four or five molecules in each nucleoid. The mitochondrial nucleoids of the unicellular organism *Euglena gracilis* are shown in Figure 21-29.

How does this genome duplication relate to the CSAR process? How many genomes are present at the beginning of CSAR? Do all copies of the genome actually become involved in CSAR? Does CSAR occur when organelles fuse, and must the nucleoids fuse also? How does CSAR segregate the many copies of the genome scattered through the cell? Few answers to such questions are available at present. The situation is rather like the one faced by the early geneticists, who knew about genes and linkage groups but knew nothing about their relation to the process of meiosis.

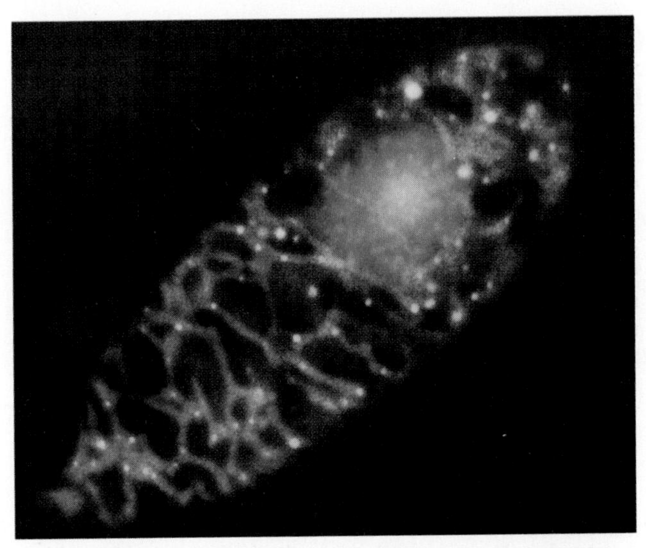

Figure 21-29 Fluorescent staining of a cell of *Euglena gracilis.* With the dyes used, the nucleus appears red because of the fluorescence of large amounts of nuclear DNA. The mitochondria fluoresce green, and within mitochondria the concentrations of mtDNA (nucleoids) fluoresce yellow. (From Y. Huyashi and K. Veda, *Journal of Cell Science* 93, 565, 1989.)

Summary

The extranuclear genome supplements the nuclear genome of eukaryotic organisms which we examined in earlier chapters. The existence of the extranuclear genome is recognized at the genetic level chiefly by uniparental transmission of the relevant mutant phenotypes. At the cellular level, a combination of genetic and biochemical techniques has demonstrated that the extranuclear genome is organelle DNA—either mitochondrial (mtDNA) or chloroplast (cpDNA).

The mtDNA of yeast is the best-understood organelle DNA. This DNA codes for unique mitochondrial translation components, and it also codes for some components of the respiratory enzymes found in the mitochondrial membranes. Mutations in the translational-component genes typically produce drug-resistant phenotypes; mutations in the respiratory-enzyme genes typically produce phenotypes involving respiratory insufficiency.

Chloroplast DNA is larger and more complex than mtDNA. Mutations in cpDNA typically produce photosynthetic defects or drug resistance.

Organelle genes have been fully investigated in only a few organisms, but these studies have produced general analytical techniques that have widespread application in the study of extranuclear genomes in other organisms.

Eukaryotic plasmids are extragenomic elements that are also inherited in a non-Mendelian manner. Most are mitochondrial in location.

Concept Map

Draw a concept map interrelating as many of the following terms as possible. Note that the terms are listed in no particular order.

maternal inheritance / cpDNA / cytoplasmic segregation / mtDNA / variegation / heterokaryon test / maternal effect / cytoplasm / Mendelian ratios

Chapter Integration Problem

1. The accompanying human pedigree concerns a rare visual abnormality in which the person affected loses central vision while retaining peripheral vision.

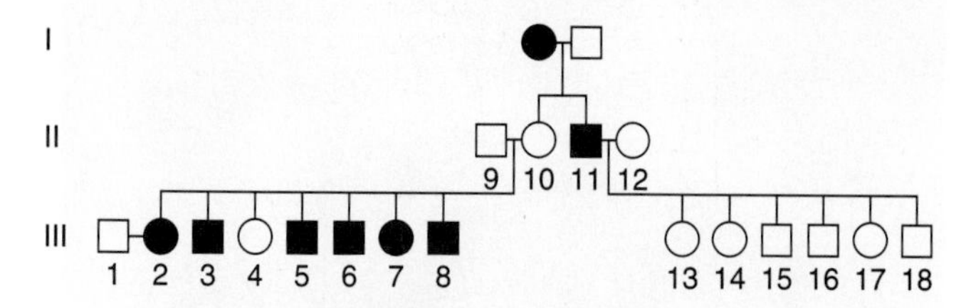

a. What inheritance pattern is shown? Can it be explained by nuclear inheritance? Mitochondrial inheritance?

Molecular geneticists studied the mitochondrial DNA of the 18 members of generations II and III. A restriction fragment 212 base pairs long of the mtDNA from each person was digested with another restriction enzyme, *Sfa*N1, with the following results.

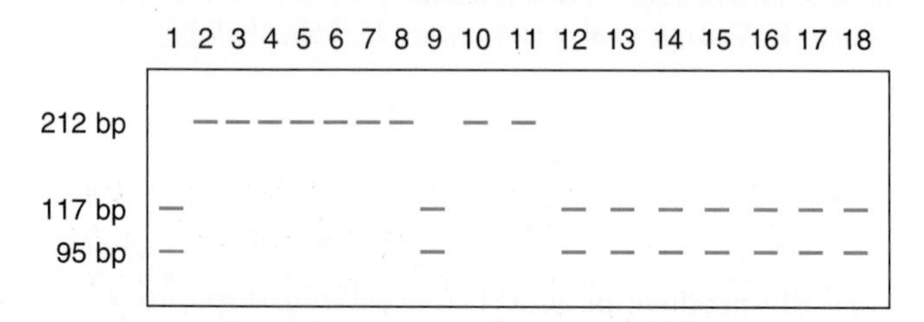

b. What inheritance pattern is shown by these restriction fragments?

c. How does the restriction pattern inheritance relate to the inheritance of the disease?

d. How can you explain individuals 4 and 10?

e. What is the likely nature of the mutation?

f. How would this analysis be useful in counseling this family?

Solution

a. Based on the pedigree alone, it is possible, but unlikely, that the disease is caused by a dominant nuclear allele. But we would have to invoke lack of penetrance in individual 10, who would have to carry the allele because it is passed on to her children. In addition, we have to explain the ratios in generation III. The matings 9×10 and 11×12 would have to be $Aa \times aa$; and the phenotypic ratio of affected to normal then expected among the children in each family is 1:1. So overall this is not an attractive model to explain the results.

The results can also be explained by maternal inheritance of the disease. Individuals 4 and 10, however, require special explanation. Once again, we can invoke incomplete penetrance. But alternatively, we could invoke cytoplasmic segregation; we have learned that cells can be mixtures of normal and abnormal cytoplasmic determinants (here, mitochondria), and cytoplasmic segregation can skew the ratio from cell to cell. The mother in the first generation would have to be a heteroplasmon and by chance pass along predominantly normal mitochondria to her daughter (10) who does not express the disease. Then, by a skewing in the other direction, 10 would pass along mainly abnormal mitochondria to six out of her seven children.

b. The restriction patterns clearly show maternal inheritance. This is expected because we are dealing with mtDNA.

c. There is obviously a close correlation between the presence of the large 212-bp fragment and the disease. If this same correlation were to be found in other similar pedigrees, one could formulate a model in which the mutation that causes the disease simultaneously causes the loss of a *Sfa*N1 restriction site.

*Sfa*N1 site

Normal mtDNA: 117 | 95

Mutant mtDNA: 212

d. The possibility that 4 and 10 are heteroplasmons is now less attractive, because if there were mixtures, we would expect to see that the restriction-enzyme patterns of some persons in the family have all three bands—95, 117, and 212—but none were seen. Therefore the most likely explanation is the incomplete penetrance of a mitochondrial disease.

e. According to the model, the most likely kind of mutation would be a nucleotide pair substitution, because if the mutation is at the *Sfa*N1 site, no nucleotides are lost or gained, since $117 + 95 = 212$.

f. If the model is upheld by other studies, appearance of the 212-bp fragment after *Sfa*N1 digestion would be a diagnostic marker for the mutation. All women with this marker could pass the disease on to their children, whereas men with the marker could not transmit the disease.

Note that in answering this question we have combined concepts of Mendelian inheritance, cytoplasmic inheritance, mutation, and DNA restriction analysis. The question is based on patterns shown in a true pedigree for the disease Leber's hereditary optic neuropathy (LHON), which is believed to be mitochondrially based.

Solved Problems

1. In a strain of *Chlamydomonas* that carries the mt^+ allele, a temperature-sensitive mutation arises that renders cells unable to grow at higher temperatures. This mutant strain is crossed to a wild-type stock, and all the progeny of both mating types are temperature-sensitive. What can you conclude about the mutation?

Solution

We are told that the mutation arose in a mt^+ stock. Therefore, the cross must have been

$$mt^+\ ts \times mt^-\ ts^+$$

and the progeny must have been $mt^+\ ts$ and $mt^-\ ts$. This is a clear-cut case of uniparental inheritance from the mt^+ parent to all the progeny. In *Chlamydomonas,* this type of inheritance pattern is diagnostic of genes in chloroplast DNA, so the mutation must have occurred in the chloroplast DNA.

2. Due to evolutionary conservation, organelle DNA shows homology across a wide range of organisms. Consequently, DNA probes derived from one organism often hybridize with the DNA of other species. Two probes derived from the cpDNA and mtDNA of a fir tree are hybridized to a Southern blot of the restriction digests of the cpDNA and mtDNA of two pine trees, R and S, that had been used as parents in a cross. The autoradiograms follow (numbers are in kb):

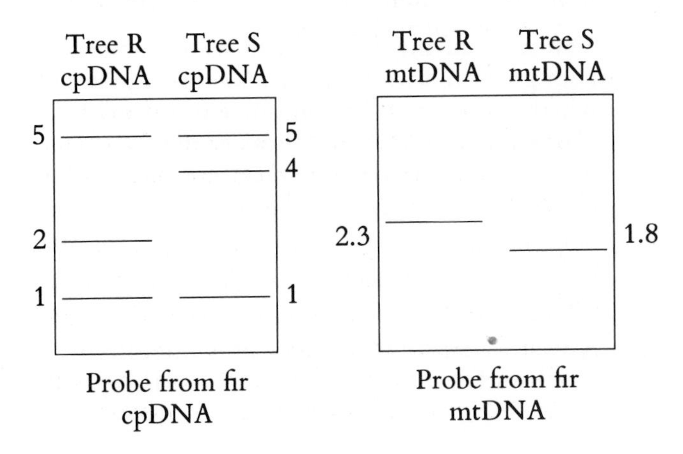

Probe from fir cpDNA

Probe from fir mtDNA

The cross R ♀ × S ♂ is made, and 20 progeny are isolated. They are all identical in regard to their hybridization to the two probes. The autoradiogram for each progeny is

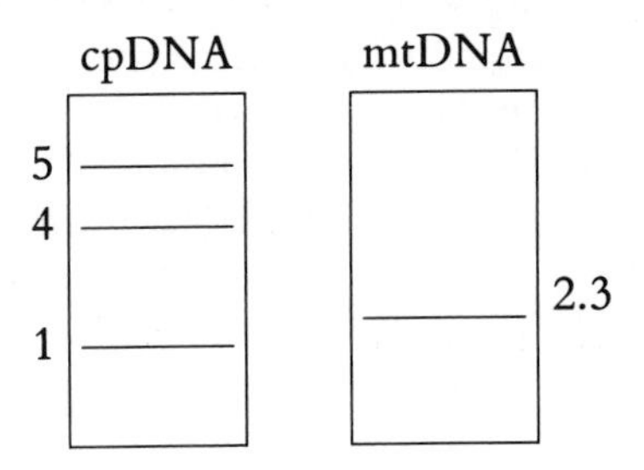

a. Explain the probe hybridization of parents and progeny.

b. Explain the progeny results. Compare and contrast them with the results in this chapter.

c. What do you predict from the cross S ♀ × R ♂?

NOTE: This question is based on results shown in several conifer species.

Solution

a. For both cpDNA and mtDNA, the total amount of DNA hybridized by the probe is different in plants R and S. Hence, we can represent the DNA something like this (other fragment arrangements are possible):

R cpDNA: 5 | 1 | 2

S cpDNA: 5 | 1 | 4

Probe

R mtDNA: 2.3

S mtDNA: 1.8

Probe

Thus, the probes reveal a (presumably neutral) restriction-fragment-length polymorphism of both the cpDNA and the mtDNA. These are useful organelle markers in the cross.

b. We can see that all of the progeny have inherited their mtDNA from the maternal parent R, because all show the same R mtDNA fragment hybridized by the probe. This is what we might have predicted, based on the predominantly maternal inheritance encountered in this chapter. However, cpDNA is apparently inherited exclusively paternally, because all progeny show the 5/4/1 pattern of the paternal plant S. This is surprising, but it is the only explanation of the data. In fact, all of the gymnosperms studied so far show paternal inheritance of cpDNA. This cause is unknown, but the phenomenon contrasts with that in angiosperms.

c. From this cross, we can predict that all progeny will show the paternal 5/2/1 pattern for cpDNA and the maternal 1.8-kb band for mtDNA.

Problems

1. How do the nuclear and organelle genomes cooperate at the protein level?
2. Name and describe two tests for cytoplasmic inheritance.
3. What is the basis for the green-white color variegation in the leaves of *Mirabilis?* If the cross is made

   variegated ♀ × green ♂

   what progeny types can be predicted? What about the reciprocal cross?
4. In *Neurospora* the mutant *stp* exhibits erratic stop–start growth. The mutant site is known to be in the mitochondrial DNA. If a *stp* strain is used as the female parent in a cross to a normal strain acting as the male, what type of progeny can be expected? What about the progeny from the reciprocal cross?
5. If a yeast cell carrying an antibiotic-resistance mutation in its mtDNA is crossed to a normal cell and tetrads are produced, what ascus types can you expect with respect to resistance?
6. A new antibiotic mutation (ant^R) is discovered in a certain yeast. Cells of genotype ant^R are treated with ethidium bromide, and petite colonies are obtained. Some of these petites prove to have lost the ant^R determinant.

   a. What can you conclude about the location of the ant^R gene?

   b. Why didn't all of the petites lose the ant^R gene?
7. Two corn plants are studied. One is resistant (R) and the other is susceptible (S) to a certain pathogenic fungus. The following crosses are made, with the results shown:

   S ♀ × R ♂ ⟶ progeny all S

   R ♀ × S ♂ ⟶ progeny all R

   What can you conclude about the location of the genetic determinants of R and S?
8. In *Chlamydomonas,* a certain probe picks up a restriction-fragment-length polymorphism in cpDNA. There are two morphs, as follows:

   Morph 1: two bands, sizes 2 and 3 kb

   Morph 2: two bands, sizes 3 and 5 kb

   If the following crosses are made

   mt^+ morph 1 × mt^- morph 2

   mt^+ morph 2 × mt^- morph 1

   what progeny types can be predicted from these crosses? Be sure to draw the DNA morphs with their restriction sites. Also draw a sketch of the autoradiogram.
9. In yeast, the following cross is dihybrid for two mitochondrial antibiotic genes:

   $MATa\ oli^R\ cap^R \times MAT\alpha\ oli^S\ cap^S$

   ($MATa$ and $MAT\alpha$ are the mating-type alleles in yeast.) What types of tetrads can be predicted from this cross?
10. In the genus *Antirrhinum,* a yellowish leaf phenotype called prazinizans (pr) is inherited as follows:

    normal ♀ × pr ♂ ⟶ 41,203 normal + 13 variegated

    pr ♀ × normal ♂ ⟶ 42,235 pr + 8 variegated

    Explain these results based on a hypothesis involving cytoplasmic inheritance. (Explain both the majority *and* the minority classes of progeny.)
11. You are studying a plant with tissue comprising both green and white sectors. You wish to decide whether this phenomenon is due to (1) a chloroplast mutation of the type discussed in this chapter, or (2) a dominant nuclear mutation that inhibits chlorophyll production and is present only in certain tissue layers of the plant as a mosaic. Outline the experimental approach you would use to resolve this problem.
12. A dwarf variant of tomato appears in a research line. The dwarf is crossed as female to normal plants,

and all of the F_1 progeny are dwarfs. These F_1 individuals are selfed, and the F_2 progeny are all normal. Each of the F_2 individuals is selfed, and the resulting F_3 generation is $\frac{3}{4}$ normal and $\frac{1}{4}$ dwarf. Can these results be explained by (1) cytoplasmic inheritance? (2) cytoplasmic inheritance plus nuclear suppressor gene(s)? (3) maternal effect on the zygotes? Explain your answers.

13. Assume that diploid plant A has a cytoplasm genetically different from that of plant B. To study nuclear-cytoplasmic relations, you wish to obtain a plant with the cytoplasm of plant A and the nuclear genome predominantly of plant B. How would you go about producing such a plant?

14. Two species of *Epilobium* (fireweed) are intercrossed reciprocally as follows:

♀ *E. luteum* × ♂ *E. hirsutum* ⟶ all very tall

♀ *E. hirsutum* × ♂ *E. luteum* ⟶ all very short

The progeny from the first cross are backcrossed as females to *E. hirsutum* for 24 successive generations. At the end of this crossing program, the progeny still are all tall, like the initial hybrids.

a. Interpret the reciprocal crosses.

b. Explain why the program of backcrosses was performed.

15. One form of male sterility in corn is maternally transmitted. Plants of a male-sterile line crossed with normal pollen give male-sterile plants. In addition, some lines of corn are known to carry a dominant nuclear restorer gene (*Rf*) that restores pollen fertility in male-sterile lines.

a. Research shows that the introduction of restorer genes into male-sterile lines does not alter or affect the maintenance of the cytoplasmic factors for male sterility. What kind of research results would lead to such a conclusion?

b. A male-sterile plant is crossed with pollen from a plant homozygous for gene *Rf*. What is the genotype of the F_1? The phenotype?

c. The F_1 plants from part b are used as females in a testcross with pollen from a normal plant (*rf rf*). What would be the result of this testcross? Give genotypes and phenotypes, and designate the kind of cytoplasm.

d. The restorer gene already described can be called *Rf-1*. Another dominant restorer, *Rf-2*, has been found. *Rf-1* and *Rf-2* are located on different chromosomes. Either or both of the restorer alleles will give pollen fertility. Using a male-sterile plant as a tester, what would be the result of a cross where the male parent was:

(i) Heterozygous at both restorer loci?

(ii) Homozygous dominant at one restorer locus and homozygous recessive at the other?

(iii) Heterozygous at one restorer locus and homozygous-recessive at the other?

(iv) Heterozygous at one restorer locus and homozygous-dominant at the other?

16. Treatment with streptomycin induces the formation of streptomycin-resistant mutant cells in *Chlamydomonas*. In the course of subsequent mitotic divisions, some of the daughter cells produced from some of these mutant cells show the normal phenotype. Suggest a possible explanation of this phenomenon.

17. Cosegregation mapping is performed in *Chlamydomonas* on four chloroplast markers: *m1*, *m2*, *m3*, and *m4*. The markers are considered pairwise in heterozygous condition, and cosegregation frequencies are obtained as follows:

	m1	*m2*	*m3*	*m4*
m1	—	29.0	18.0	18.4
m2		—	10.9	26.2
m3			—	8.8
m4				—

(For example, *m1* and *m2* cosegregate in 29 percent of the cell divisions followed.) Draw a rough genetic map based on these results.

18. In *Aspergillus*, a "red" mycelium arises in a haploid strain. You make a heterokaryon with a nonred haploid that requires *para*-aminobenzoic acid (PABA). From this heterokaryon you obtain some PABA-requiring progeny cultures that are red, along with several other phenotypes. What does this information tell you about the gene determining the red phenotype?

19. On page 611 an experiment is described in which abnormal mitochondria are injected into normal *Neurospora*. The text mentions that "appropriate controls" are used. What controls would you use?

20. Adrian Srb crossed two closely related species, *Neurospora crassa* and *N. sitophila*. In the progeny of some of these crosses, a phenotype called aconidial (ac) appeared that involves a lack of conidia (asexual spores). The observed inheritance was

♀ *N. sitophila* × ♂ *N. crassa* ⟶ $\frac{1}{2}$ ac, $\frac{1}{2}$ normal

♀ *N. crassa* × ♂ *N. sitophila* ⟶ all normal

a. What is the explanation of this result? Explain all components of your model with symbols.

b. From which parent(s) did the genetic determinants for the ac phenotype originate?

c. Why were neither of the parental types ac?

21. Several crosses involving poky or nonpoky strains A, B, C, D, and E were made in *Neurospora*. Explain the results of the following crosses, and assign genetic symbols for each of the strains involved. (Note that poky strain D behaves just like poky strain A in all crosses.)

	Cross	Progeny
a.	nonpoky B ♀ × poky A ♂	all nonpoky
b.	nonpoky C ♀ × poky A ♂	all nonpoky
c.	poky A ♀ × nonpoky B ♂	all poky
d.	poky A ♀ × nonpoky C ♂	$\frac{1}{2}$ poky, all identical (e.g., D); $\frac{1}{2}$ nonpoky, all identical (e.g., E)
e.	nonpoky E ♀ × nonpoky C ♂	all nonpoky
f.	nonpoky E ♀ × nonpoky B ♂	$\frac{1}{2}$ poky; $\frac{1}{2}$ nonpoky

22. In yeast, an antiobiotic-resistance haploid strain ant^R arises spontaneously. It is combined with a normal ant^S strain of opposite mating type to form a diploid culture that is then allowed to go through meiosis. Three tetrads are isolated:

Tetrad 1	Tetrad 2	Tetrad 3
α ant^R	α ant^R	a ant^S
α ant^R	a ant^R	a ant^S
a ant^R	a ant^R	α ant^S
a ant^R	α ant^R	α ant^S

a. Interpret these results.

b. Explain the origin of each ascus.

c. If an ant^R grande strain were used to generate petites, would you expect some of the petites to be ant^S? Explain your answer.

23. In yeast, two haploid strains are obtained that are both defective in their cytochromes; the mutants are designated *cyt1* and *cyt2*. The following crosses are made:

$cyt1^- \times cyt1^+$

$cyt2^- \times cyt2^+$

One tetrad is isolated from each cross:

$cyt1^-$	$cyt2^-$
$cyt1^-$	$cyt2^-$
$cyt1^+$	$cyt2^-$
$cyt1^+$	$cyt2^-$

a. From these tetrad patterns, explain the differences in the two underlying mutations.

b. What other ascus types could be expected from each cross?

c. How might the two genes involved here interact at the functional level?

24. In a marker-retention analysis in yeast, a multiply resistant strain apt^R bar^R cob^R is used to induce 500 petites. The table shows, for different pairs of markers, the number of petites in which only one marker of the pair is lost.

Genes	Petites in which first marker is lost	Petites in which second marker is lost
apt bar	87	120
apt cob	27	18
bar cob	48	69

(For example, 87 petites were apt^S bar^R, and 120 were apt^R bar^S.) Use these results to draw a rough map of these mitochondrial genes.

25. A grande yeast culture of genotype cap^R ery^R-oli^R par^R is used to obtain petites by treatment with ethidium bromide. The petites are tested for (1) their drug resistance, and (2) the ability of their mtDNA to hybridize with various specific mtRNA types. In all, 12 petites are tested, and Table 21-4 shows the results. Use this information to plot a

Table 21-4.

Petite culture	Drug resistance (R) or sensitivity (S)				Ability (+ or −) of petite mtDNA to hybridize with mtRNAs						
	cap	ery	oli	par	$rRNA_{large}$	$rRNA_{small}$	$tRNA_1$	$tRNA_2$	$tRNA_3$	$tRNA_4$	$tRNA_5$
1	R	S	S	S	+	−	−	−	−	−	−
2	S	S	S	S	−	−	−	−	−	+	−
3	S	S	S	S	−	−	−	+	−	−	+
4	S	S	R	S	−	−	+	−	−	−	−
5	S	R	S	S	+	+	−	−	−	−	−
6	R	S	S	S	−	−	−	−	+	−	−
7	S	S	R	R	−	−	−	+	−	−	−
8	S	S	S	S	−	+	−	−	−	+	+
9	S	R	S	S	+	−	−	−	−	−	−
10	S	S	S	S	−	−	+	−	+	−	−
11	R	S	R	R	−	+	+	+	+	+	+
12	S	R	S	S	+	+	−	−	−	−	−

map of the mtDNA, showing the sequence of the 11 genetic loci involved in these phenotypes. Be sure to state your assumptions and draw a complete map.

26. The mtDNA is compared from two haploid strains of baker's yeast. Strain 1 (mating type α) is from North America, and strain 2 (mating type *a*) is from Europe. A single restriction enzyme is used to fragment the DNAs, and the fragments are separated on an electrophoretic gel. The sample from strain 1 produces two bands, corresponding to one very large and one very small fragment. Strain 2 also produces two bands, but they are of more intermediate sizes. If a standard diploid budding analysis is performed, what results do you expect to observe in the resulting cells and in the tetrads derived from them? In other words, what kinds of restriction-fragment patterns do you expect?

27. In yeast, some strains are found to have in their cytoplasm circular DNA molecules that are 2 micrometers (μm) in circumference. In some of these strains, this 2-μm DNA has a single *Eco*RI restriction site; in other strains, there are two such sites. A strain with one site is mated to a strain with two sites. All of the resulting diploid buds are found to contain both kinds of 2-μm DNA.

a. Is the 2-μm DNA inherited in the same fashion as mtDNA?

b. If you used radioactive 2-μm DNA as a probe, predict the results of hybridizing it to a Southern blot of *Eco*RI-treated DNA of ascospores from these diploid cells.

28. Circular mitochondrial DNA is cut with two restriction enzymes A and B, with the following results:

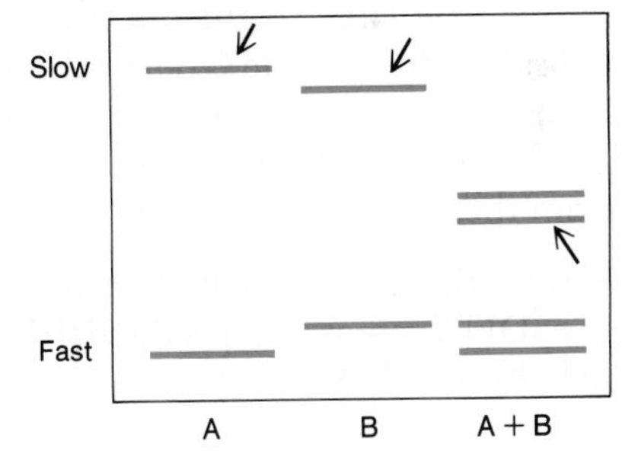

(The arrows indicate the bands that bound a radioactive mt rRNA-derived cDNA probe in a Southern blot.) Draw a rough map of the positions of the restriction site(s) of A and B, and also show approximately where the mt rRNA gene is located.

29. You are interested in the mitochondrial genome of a fungal species for which genetic analysis is very difficult but from which mtDNA can be extracted easily. How would you go about finding the positions of the major mitochondrially encoded genes in this species? (Assume some evolutionary conservation for such genes.)

22

Developmental Genetics: Cell Fate and Pattern Formation

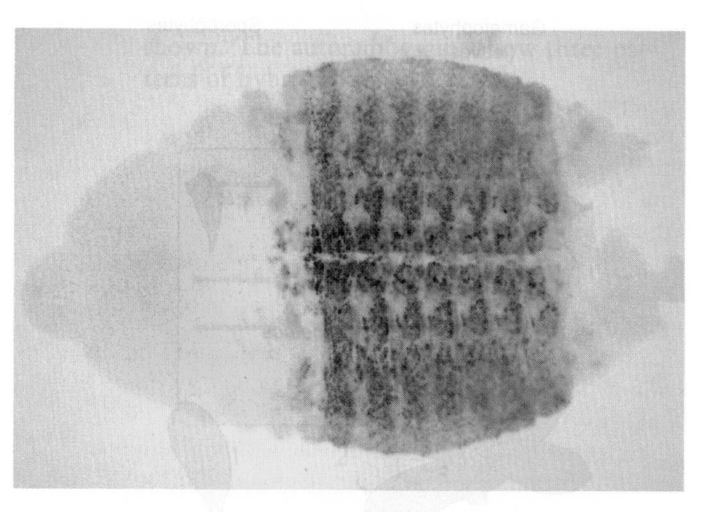

A "pelt" of a *Drosophila* embryo double stained for two proteins encoded by homeotic genes of the bithorax gene complex (BX-C). In blue, the expression of the *Ultrabithorax (Ubx)* gene is shown, while in brown, *abdominal-A (abd-A)* expression is visualized. The anterior edge of the *Ubx* expression domain is anterior to that of *abd-A*. The homeotic gene products determine the identity of the segments. (W. Bender, Harvard Medical School)

KEY CONCEPTS

A programmed set of instructions in the genome of higher organisms leads to the establishment of developmental fates.

The zygote is totipotent, giving rise to every adult cell type; as development proceeds, successive decisions restrict each cell lineage to its particular fate.

Gradients of maternally derived regulatory proteins establish polarity along the major body axes of the egg; these proteins control the local transcriptional activation of genes encoding master regulatory proteins in the zygote.

Many proteins that act as master regulators of early development are transcription factors; others are components of pathways that mediate signaling between cells.

Some fate decisions are made autonomously by individual cells; many fate decisions require communication and collaboration between cells.

The same basic set of genes identified in *Drosophila* and the regulatory proteins they encode are conserved in mammals and appear to govern major developmental events in many — perhaps in all — higher animals.

In all higher organisms, life begins with a single cell, the newly fertilized egg. It reaches maturity with thousands, millions, or even trillions of cells combined into a complex organism with many integrated organ systems. The goal of developmental biology is to unravel the fascinating and mysterious processes that achieve the transfiguration of egg into adult.

The processes underlying the formation of the general body plan of an organism must be encoded in the genome. The fact that species faithfully transmit their basic characteristics from parent to offspring is a reflection of this principle. Polar bears beget polar bears, koalas beget koalas. How do the processes underlying **pattern formation,** the construction of complex form, operate reliably to execute the developmental program? The question reflects one of the major aims of developmental genetics — to understand pattern formation at the molecular level through a combined approach that uses the tools of genetics, embryology, and molecular biology.

During development, cells adopt specific **fates,** that is, the capacity to **differentiate** into particular kinds of cells. The process of commitment to a particular fate, **determination,** is a gradual one. A cell does not go directly from being totally uncommitted **(totipotent)** to being earmarked for a single fate. Rather, as zygotic cell division proceeds, periodic decisions are made in each cell lineage to specify more exactly the fates of the daughter cells. Some of these fate decisions are made autonomously by the dividing cell. Other decisions are made by committee — that is, the fate of a cell becomes dependent upon input from neighboring cells.

Figure 22-1 Identical quadruplets. (Erika Stone)

In this chapter, we will consider several examples of the genetic dissection of cell fate decisions. If this chapter were written 10 years ago, we would have to have been satisfied with a phenomenological description of mutant phenotypes and with microsurgical manipulations of embryos. Not so today. The combination of genetics and recombinant DNA techniques has revolutionized our understanding of development. We are now in a position to identify the protein products contributing to these developmental events and to build a general framework of the molecular mechanisms involved. Perhaps the most exciting development is that recombinant DNA tools have allowed researchers to fish out related genes from very different organisms. From such studies, we are coming to realize that the same basic set of regulatory proteins — including certain classes of transcription factors and of molecules that pass information between cells — govern major developmental events in many — maybe in all — higher animals.

Cell Fate: When Do the Cells and Nuclei of Higher Organisms Lose Their Totipotency?

By definition, the newly fertilized, single-celled zygote is **totipotent:** it gives rise to every cell type in the adult. At what point do cells begin to become irreversibly restricted in their developmental potentials? One way to address this question is by testing, at various stages, whether a cell or the nucleus of a cell is still capable of supporting complete development from egg to adult. Note that there is not just one answer — and note that the answers differ from species to species, as will be seen from the following examples.

There are cases of multiple births in which "identical" siblings are derived from a single fertilized human egg (Figure 22-1). Thus, we can conclude that genetic information in people is faithfully reproduced through at least the first three cleavages after fertilization (two cleavages produce only four cells and identical quintuplets have been recorded). But it is known that early embryonic development is controlled primarily by maternal genotype. What of the nuclei in cells resulting from later divisions? Do they remain totipotent?

Many highly differentiated organisms can regenerate new organs and tissues. As examples, a starfish can regrow a lost "arm," a reptile can re-form a lost tail, and the human body can repair a damaged liver. However, such regeneration is possible only in certain tissues. Regeneration of a complete organism from a single somatic cell is not observed among animals in nature.

Figure 22-14 Outline of a mating scheme for recovering maternal-effect lethal mutations in *Drosophila melanogaster.* In this example, mutations mapping to chromosome 2 are sought. All indicated marker mutations map to chromosome 2. Flies simultaneously homozygous for the recessive eye color mutations *cn* (cinnabar) and *bw* (brown) have white eyes instead of the normal red. *CyO* is a balancer chromosome, which carries the dominant *Cy* (curly wings) mutation and has multiple inversions that prevent recombination between *CyO* and the structurally normal *cn bw* chromosomal homolog. Thus, the *cn bw* chromosome is always passed on as an intact genetic unit in this mating scheme. Males in the initial generation (G0) are treated with a mutagen. An asterisk (*) indicates mutagenized chromosome. Sons (G1 males) are recovered and mated in individual vials to females as shown. For each vial of G1 offspring, an individual vial must be set up for the G2 and G3 matings. Finally, in the G3, females homozygous for a single mutagenized *cn bw* chromosome are recovered and can be tested for maternal-effect lethality. If they show this lethality, the mutant chromosome can be recovered from among their siblings and placed into a stock. Embryos from maternal-effect lethal mutants are examined for A/P or D/V defects.

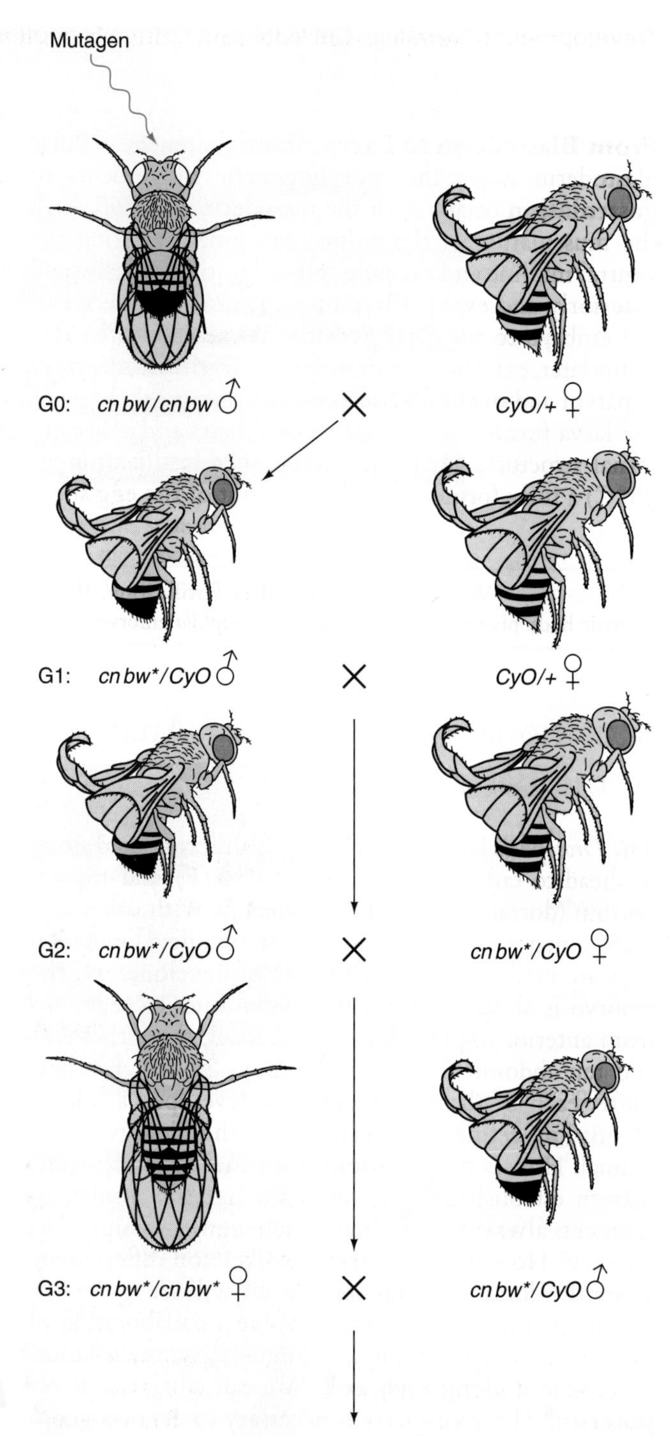

ing the genome for mutations that alter the A/P or D/V patterns of the larva. These mutageneses fall into two general classes: screen for mutations that are recessive maternal-effect lethal (Figure 22-14) and for mutations that are homozygous lethal (Figure 22-15). Homozygotes for **recessive maternal-effect lethal mutations** are viable themselves, but homozygous mutant females produce aberrant and inviable offspring.

These mutational screens take several generations to execute. Because "balancer" chromosomes (chromosomes containing multiple inversions and a dominant morphological marker mutation; see Chapter 8 for a discussion of inversions and their effect on crossing over) can suppress all recombinations between themselves and a structurally normal homologous chromosome, it is possible to follow an entire mutagenized chromosome at once. Thus, in separate experiments, large numbers of recessive maternal-effect lethal and homozygous lethal mutations were produced. Those lethals in which death was embryonic were studied further; the exoskeletons of dying animals were examined, and those with abnormal A/P or D/V patterns were checked for allelism by recombination and complementation testing.

The Molecular Analysis of Developmental Mutations. Recombinant DNA techniques like those described in Chapters 14 and 15 have permitted the cloning of many of these mutated genes. The cloned genes have then been analyzed in many ways.

Transcript localization analyses make use of histochemical techniques. First, complementary DNAs (cDNAs) for the genes are sequenced; from each cDNA sequence, the sequence of the polypeptide it encodes can be inferred. The locations of these gene products in the embryo can then be identified in one of two ways.

1. The gene transcript is localized by a technique that makes use of **RNA in situ hybridization.** The embryo is made permeable so that it takes up labeled denatured (single-stranded) cDNA. The single-stranded cDNA specifically hybridizes to complementary messenger RNAs (mRNAs). By labeling the cDNA with an enzyme that can create a colored product from an uncolored substrate, the position of the hybridized cDNA in the embryo can be detected

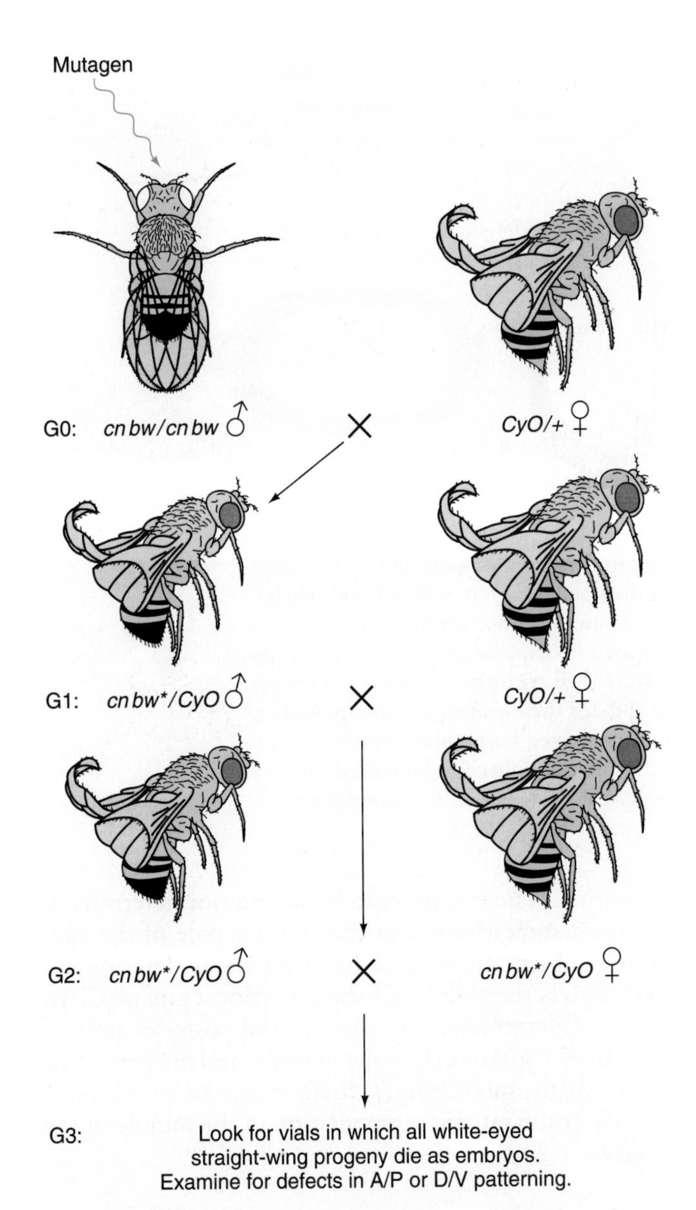

Figure 22-15 Outline of a mating scheme for recovering homozygous lethal mutations in *Drosophila melanogaster.* As in Figure 22-14, mutations on chromosome 2 are sought. All mutations are as described in Figure 22-14. The logic of the scheme is that if no white-eyed, straight-winged flies are recovered in G3, then a recessive lethal mutation on chromosome 2 must have arisen. Recessive lethal mutants are visually screened for those producing dead embryos, and then the exoskeletons of the dead animals are examined to determine if a mutation affecting A/P or D/V pattern formation is present.

histochemically. This identifies where the gene transcript is located at a particular time in development.

2. The protein product of the gene is localized by **immunohistochemistry.** The technique is similar except that the tag to identify the position of the protein is the antibody made against the polypeptide. (Determining the polypeptide sequence from the gene sequence allows one to make purified antibodies that will bind tightly and specifically to that polypeptide alone.) The antibody is attached to an enzyme, and a histochemical stain is then used to locate the position of the protein to which the antibody–enzyme complex binds.

Another analytic method makes use of **in vitro mutagenesis techniques.** P elements are used for germ-line transformation in *Drosophila* (see Chapter 20). The cloned genes affecting early developmental processes are mutated in a test tube and put back in the fly. The mutated genes are then analyzed to see how the mutation alters their function.

Message We only need to know the chromosomal location of a gene in order to clone it. Recombinant DNA techniques allow us to physically purify genes, their transcripts, and their protein products and to study the spatial and temporal patterns of mRNA or protein localization in the embryo.

The Nature of Localized Cytoplasmic Determinants

We will see that cell fate is determined during development by the selective local activation of particular master regulatory proteins. If the initial egg were truly homogeneous, then it is difficult to imagine how the specific master regulatory proteins could be activated in the correct parts of the early zygote. Indeed, in several instances, embryologists have shown that the egg cytoplasm is *not* homogeneous. Rather, specific regions of the egg contain particular molecules—**localized determinants.** These determinants provide the initial geographic cues—asymmetries—that create regional identities within the egg and that set up polarities (like longitudes and latitudes on the surface of the earth) along what will become the primary body axes.

The **polar granules** of *Drosophila* represent a classic example of such a localized determinant. In animals, one of the first developmental decisions is to distinguish the germ line (the cells that will form gametes) from the soma (all the nonreproductive tissues). In a series of elegant experiments Karl Illmensee and Anthony Mahowald demonstrated that these granules are associated with germ-line determination. We can conduct an experiment to demonstrate this.

First, we subject the posterior pole of a newly fertilized *Drosophila* egg to ultraviolet irradiation. This causes loss of the germ line; no pole cells form in the embryo and the resulting adults have no gonads. Using a glass

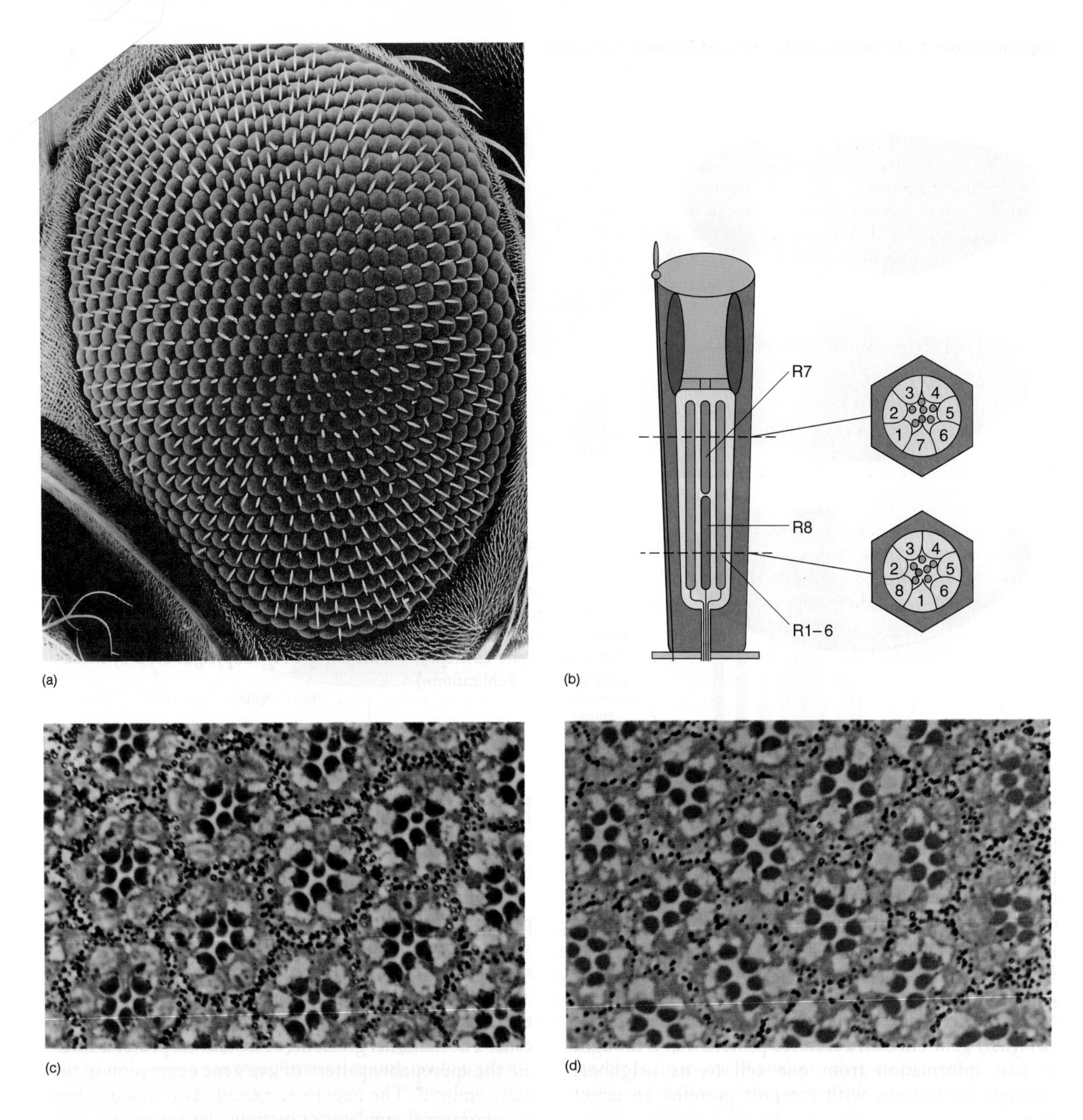

Figure 22-22 The structure of the *Drosophila* compound eye. (a) Scanning electron micrograph showing the left eye of a fly, with its many individual facets in a hexagonal array. (b) Diagrams showing the tubular structure of an individual facet and cross-section cuttings through the middle of the tube. (c) Histological cross section at the upper mid-tube level of a wild-type fly, showing the positions of pigmentation in photoreceptors R1 through R7 (photoreceptor R8 is located beneath R7). (d) A comparable cross section through a *sev* mutant fly. Note that one of the photoreceptors (R7) is missing from every facet. (From Ernst Hafen and Konrad Basler, *Development,* Supplement 1, 1991, 123.)

D/V axis. The maternal D/V determinant is the *dl* protein. As with *bcd*, the *dl* protein is a transcription factor and is present in a gradient. This concentration of *active dl* protein controls the local expression of certain regulatory genes in broad regions along the D/V axis. Interactions of the D/V proteins produced by these genes then leads to a cascade of regulatory events among zygotic genes at the blastoderm stage and during gastrulation that ultimately subdivides the D/V axis into a series of fine lines of cell fate.

Message Studies of the molecular mechanisms producing the *Drosophila* body plan reinforce the view that cell fate is established more and more exactly through many gradual steps. These studies also demonstrate the power of combining genetic and molecular approaches to attack complex biological phenomena.

Communication Between Cells and the Establishment of Cell Fate

While the preceding section may seem to imply that the process of determination, involving a cascade of transcription factor activation, is autonomous to each cell, in fact this is probably not true. There are hints that interactions between cells, such as those involving the secreted *wg* protein mentioned above, may underlie many events in the establishment of cell fate decisions. This makes sense, since it is certainly important that biological systems should be able to compensate for occasional perturbations, such as the death of a group of cells in one part of the embryo. If all aspects of fate allocation and determination were truly autonomous, such compensation would probably be impossible. In the following sections, we'll see two concrete examples of the importance of intercellular interactions.

The *Drosophila* Visual System

One glimpse into the mechanisms underlying developmental choices comes from systems where we can follow the fate of a specific cell. One of the most spectacular of such systems is the compound eye of *Drosophila* (Figure 22-22). The adult eye is a hexagonal array of about 800 individual facets, with each facet composed of an identical arrangement of 20 cells. Eight of these cells are photoreceptors, the light-sensing cells of the eye. Six of the photoreceptors (R1 to R6) are arranged in an outer ring and two are central, with the nucleus of one (R7) sitting above the other (R8).

The cells of the facet are not related to one another by lineage. Donald Ready and Seymour Benzer demonstrated this in the following way. Using mitotic crossing-over (see Chapter 6), a clone of cells homozygous for the *w* (white eye) mutation can be generated in a background of cells that are w^+ (the wild-type red eye). Expression of the *white* gene product is **autonomous,** that is, the color (red or white) of each cell directly reflects its genotype at the *white* locus. If the cells of each facet were related by descent from a common ancestral cell, then you would expect to see certain kinds of mixed facets but not others. Let's suppose for example that a hypothetical ancestral facet cell divided to give rise to one precursor for R1, R2, R7, and R8, and to a second precursor for R3, R4, R5, and R6. You would then expect to see, for example, facets where R7 and R8 were *w* while R1 to R6 were w^+, as well as facets in which R3 and R4 were *w* while the other 6 photoreceptors were w^+. But you would not expect facets in which R1 and R4 were *w* while the others were w^+. In reality, however, all possible combinations of mosaic facets are observed, indicating no fixed lineage relationship between the cells. Rather, cells are recruited during late larval development because they are near one another at a critical point in the development of the eye (Figure 22-23).

Message As demonstrated by the analysis of mosaic facets, cell fates in the *Drosophila* visual system do not arise by lineage relationships, but rather are determined by the positions of cells at critical developmental times.

The photoreceptor cells of the *Drosophila* eye become recruited progressively into the developing facet. It is possible to screen for mutants that lack the R7 photoreceptor cell by using various techniques: one procedure is to test the ability of mutagenized flies to see in ultraviolet light, an ability that depends upon the presence of the R7 cell; another is to shine a bright enough light through a fly's head to see directly if R7 is there.

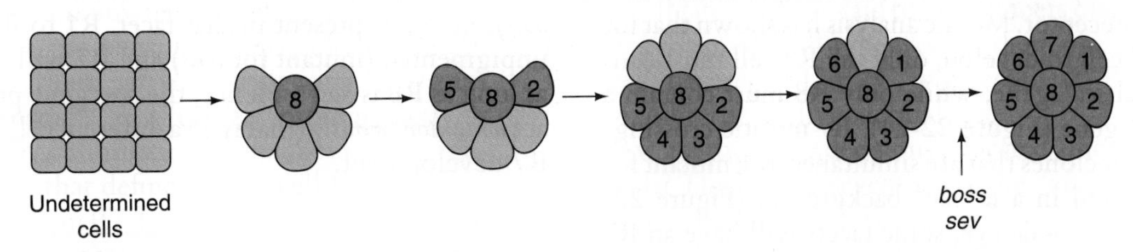

Figure 22-23 The sequential recruitment of cells into the developing facets of the *Drosophila* eye. R8 is recruited first, then R2 and R5, and so on, as depicted in the diagram. R7 is the last photoreceptor to be recruited, and the wild-type functions of the *sev* and *boss* genes are required for R7 formation (see Figure 22-24). (Modified from Ernst Hafen and Konrad Basler, *Development,* Supplement 1, 1991, 123.)

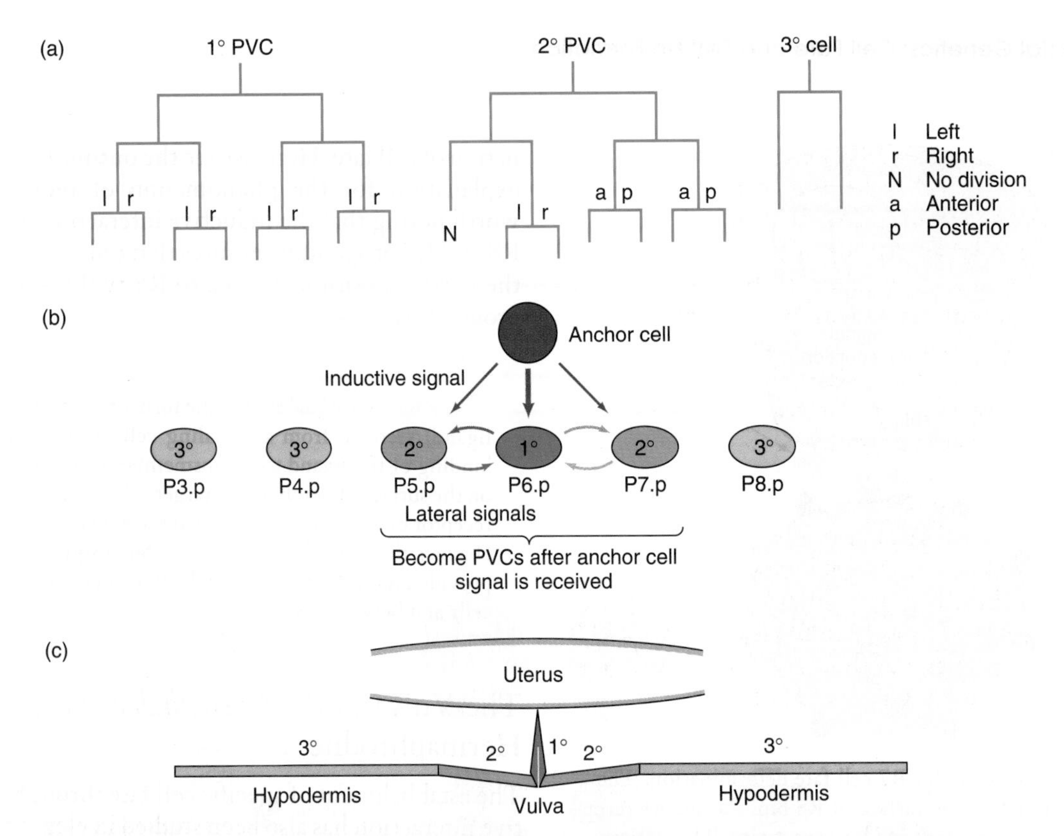

Figure 22-26 A model for the signalling that sets up primary (1°), secondary (2°), and tertiary (3°) fates among Pn.p cells in *C. elegans.* (a) The cell lineages characteristic of 1°, 2°, and 3° cells. (b) The anchor cell sends out an inductive signal that converts cells from a 3° to a higher fate. The cell receiving the strongest signal becomes the 1° PVC cell and sends out a signal that helps convert its two immediate neighbors to the 2° PVC fate. (c) Only the descendants of the 1° and 2° PVCs form the vulva. The descendants of the 3° cell are integrated into the hypodermis (skin) of the worm. (Part a from R. Horvitz and P. Sternberg, *Nature* 351, 1991, 357. Parts b and c from Iva Greenwald, *Trends in Genetics* 7, 1991, 366.)

cause the 3° fate appears in the absence of instructions from the anchor cell, the 3° fate can be regarded as a **ground state** or **default state.**

Paul Sternberg and Robert Horvitz studied vulval development in mutant genotypes in which the position of the anchor cell relative to the row of Pn.p cells was shifted from its normal location nearest to P6.p. For example, in a worm in which the anchor cell was nearest to P4.p, P4.p became the 1° PVC while P3.p and P5.p became the two 2° PVCs. From many other observations like this one, Sternberg and Horvitz concluded that the Pn.p cell nearest to the anchor cell becomes the 1° PVC and that its neighbors, the next closest two Pn.p cells become the two 2° PVCs. Thus, it appears that the anchor cell induces three of the six cells to divert from their 3° fate to the 1° and 2° fates.

What mechanism can account for these results? We can postulate that an inducing substance is released by the anchor cell. As this hypothetical substance diffuses away from the anchor cell and through the embryo, a gradient of concentration occurs. So the closest hypodermis cell experiences the highest concentration and therefore becomes a 1° cell, cells exposed to the next highest concentration become 2° cells, and cells exposed to little or no inducer become 3° cells.

How can this hypothesis be tested? We should be able to predict the phenotypes of mutations defective in the signaling process. In the absence of signaling or in the absence of signal reception, all Pn.p cells should remain in the 3° default state, producing worms lacking a vulva (vulvaless). Such vulvaless mutations are easy to identify, since the hundreds of eggs of the self-fertilizing hermaphrodite hatch inside the uterus of the vulvaless mother, producing a so-called "bag of worms" phenotype. Of particular interest are two genes identified by vulvaless mutations: *lin-3* and *let-23*. (*lin* stands for *lineage defective*—some of the lineages in the worm show abnormal division and fate patterns in such mutants; *let* stands for *lethal*—the phenotype for the homozygote with these mutations.) In *lin-3* and *let-23* mutants, all Pn.p cells exhibit 3° fates.

Molecular analysis of these genes confirms that the PVC fate decisions depend upon intercellular signaling. As with the R7 fate decision in the *Drosophila* eye, in the *C. elegans* vulva the inductive signal from the anchor cell to the Pn.p cells involves a **ligand–receptor interaction.** The anchor-cell signal is the product of the *lin-3* gene and is a secreted protein that is similar to a vertebrate growth factor called EGF (epidermal growth factor). All of the Pn.p cells express a receptor for this signal, very likely the product of the *let-23* gene. This gene encodes a transmembrane receptor that is similar to the vertebrate EGF receptor and contains a protein tyrosine kinase activity in its cytoplasmic domain. As with the *sevenless* receptor in *Drosophila,* the binding of the *lin-3* ligand to the *let-23* receptor initiates a signal transduction pathway in the Pn.p cells that causes them to adopt 1° or 2° PVC fates. How are the 1° and 2° fates distinguished? The Pn.p cell nearest the anchor cell adopts the 1° PVC fate, presumably because it receives the strongest inductive signal from the anchor cell. The 2° PVCs are prevented from also adopting the 1° fate, possibly because they receive a **lateral inhibition signal** from the 1° PVC (see Figure 22-26b). This signal may also involve a ligand–receptor interaction, with the *lin-12* gene encoding the receptor. (In *lin-12* mutants, all three PVCs acquire a 1° fate and none exhibits a 2° fate.) The ligand for the *lin-12* receptor is not yet known.

Message Many aspects of cell fate involve communication between neighboring cells.

Intercellular Signaling Systems

In cell fate decisions, communication between cells can be **inductive,** leading to the acquisition of a new fate, or it can be **inhibitory,** suppressing the ability of a cell to adopt the same fate as that of its neighbors. Presumably, these signaling systems ultimately modulate the activity of key transcription factors whose role is to determine the activity of the structural genes that give each cell its differentiated characteristics, although in the examples we've looked at here, the transcription factors have not yet been identified. By means of intercellular communication, an organism can ensure that all appropriate fates are assigned within a developing embryo (or tissue or organ). Undoubtedly, such intercellular communication systems underlie may fate choices in higher organisms.

Homeotic Mutations and the Establishment of Segment Identity

At the turn of the century, William Bateson collected examples of monstrosities in which one body part was turned into another; he called this phenomenon **homeosis.** Early in the history of *Drosophila* genetics, researchers uncovered mutations that regularly produced certain homeotic transformations. To cite two extreme examples, **bithorax mutations** can cause the entire third thoracic segment (T3) to be transformed into a second thoracic segment (T2), giving rise to flies with four wings instead of the normal two (Figure 22-27), and dominant **Antennapedia *(Antp)* mutations** transform

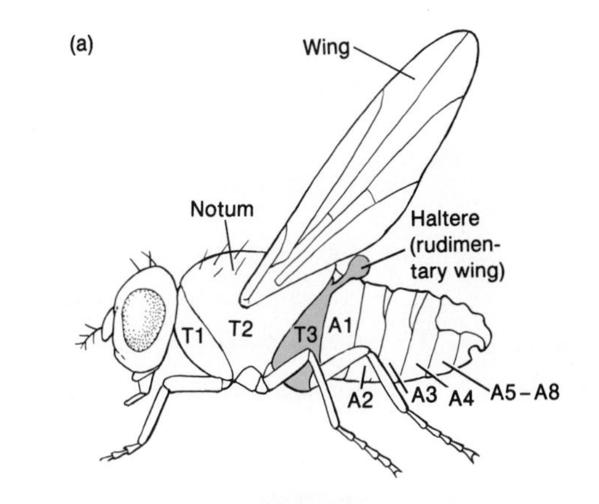

Figure 22-27 The homeotic transformation of the third thoracic segment (T3) of *Drosophila* into an extra second thoracic segment (T2). (a) Diagram showing the normal thoracic and abdominal segments; note the rudimentary wing structure normally derived from T3. Most of the thorax of the fly, including the wings and the notum, comes from T2. (b) A wild-type fly with one copy of T2 and one of T3. (c) A triple bithorax mutant homozygote completely transforms T3 into a second copy of T2. Note the second notum and second pair of wings (T2 structures) and the absence of the halteres (T3 structures). (From E. B. Lewis, *Nature* 276, 1978, 565. Photographs courtesy of E. B. Lewis. Reprinted by permission of *Nature.* Copyright 1978 by Macmillan Journals Ltd.)

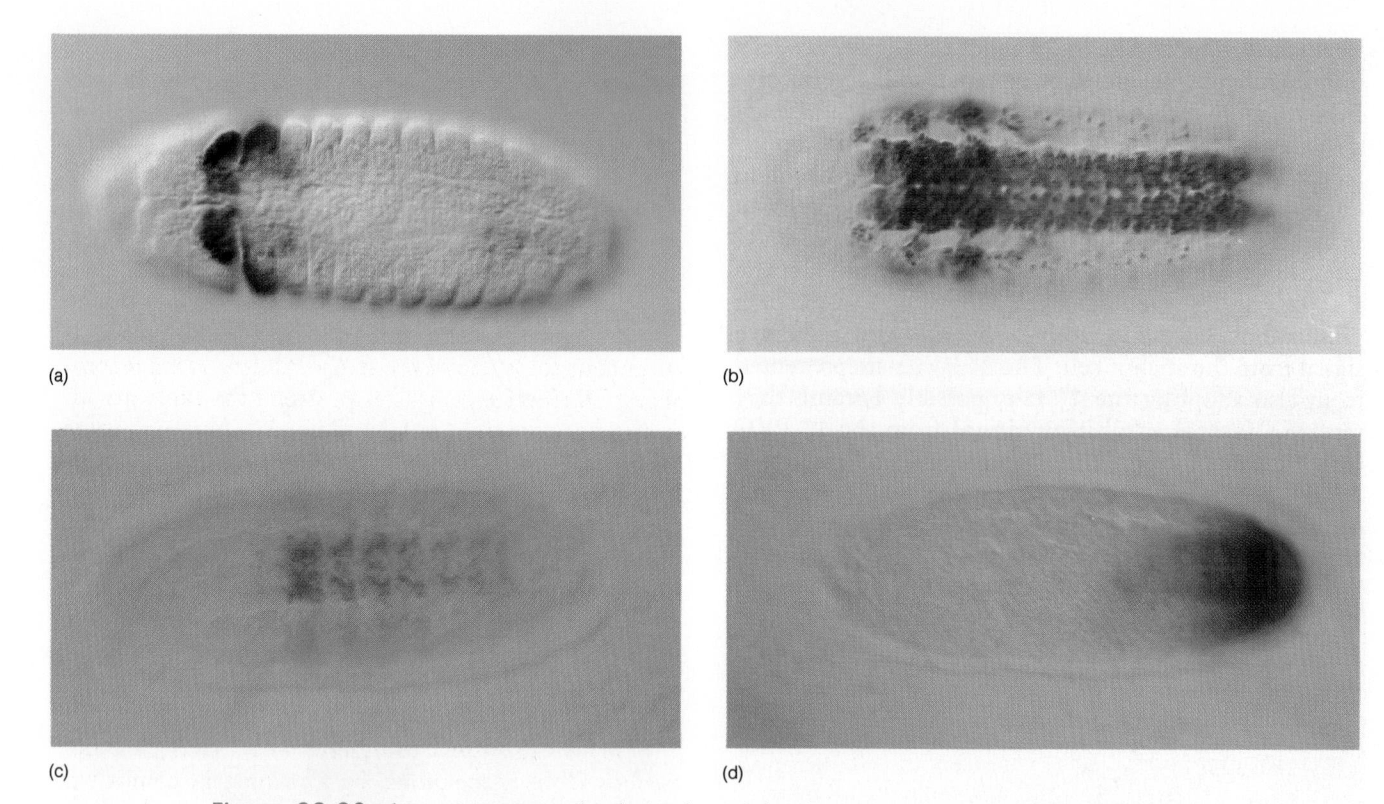

Figure 22-28 Photomicrographs of embryos that exhibit the protein expression patterns of several homeotic genes that confer segment identity in *Drosophila.* (a) *Scr*; (b) *Antp*; (c) *Ubx*; (d) *Abd-B*. Note that the anterior boundary of homeotic gene expression is ordered from *Scr* (most anterior) to *Antp*, *Ubx*, and *Abd-B* (most posterior). (Parts a and b from T. Kaufman, Indiana University. Parts c and d from S. Celniker and E. B. Lewis, California Institute of Technology.)

antenna into leg (see the photograph on the first page of Chapter 23).

The primary homeotic genes map in two clusters on chromosome 3 in *Drosophila* and play major roles in the development of segment identity (Figure 22-28). The **bithorax gene complex (BX-C),** studied in depth by Ed Lewis and colleagues, contains related genes controlling the segmental identity of the third thoracic segment and all of the abdominal segments, not only in the adult but in the embryo as well. The **Antennapedia gene complex (ANT-C),** similarly characterized in Thomas Kaufman's laboratory, contains a cluster of genes controlling the identity of the head and thoracic segments.

Once Bill McGinnis, Michael Levine, Walter Gehring, and Matthew Scott cloned some of the homeotic genes, they realized that all homeotic genes share a very similar 180-base-pair DNA sequence, the **homeobox.** The homeobox encodes a 60-amino-acid polypeptide segment—the **homeodomain.** The term "homeodomain" is used, even though we now know that it is a more general motif, also present in many proteins that are *not* encoded by homeotic genes. The homeodomain, which is a portion of the protein product of each homeotic gene, binds to DNA in a sequence-specific manner and activates or represses the transcription of specific target genes. As shown by Walter Gehring, Matt Scott, Claude Desplan, and others, the homeodomain sequence in large part confers the specificity of action of the different homeotic gene products.

ANT-C and BX-C mutations can affect the entire body plan of the developing *Drosophila* larva. For example, deleting the entire BX-C locus produces larvae in which every segment from T3 on back adopts a T2 segmental phenotype (Figure 22-29).

How do the ANT-C and BX-C genes normally get turned on in the right places? As we discussed earlier in this chapter, they appear to be regulated by the gap genes; moreover, the homeotic genes are able to regulate one another. This multiple regulation produces a stable pattern of gene expression in which, for example, only the *Abd-B* gene is expressed in the most posterior abdominal segments, *Ubx* in the most anterior abdominal segments, *Antp* in the thoracic region, and *Scr* in the head (see Figure 22-28). *(Abd-B* stands for *Abdominal-B, Ubx* for *Ultrabithorax,* and *Scr* for *Sex combs reduced.)* Typically, only one of the homeotic genes remains stably expressed in a given cell.

The homeotic genes, once established in their given cells and properly activated, are able to maintain their own expression at least until adulthood, and perhaps for

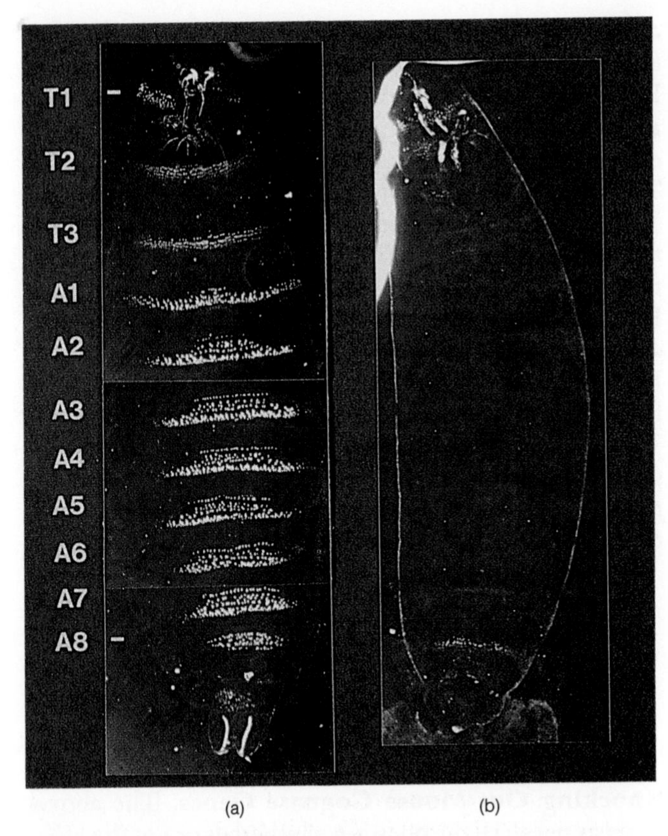

Figure 22-29 Photomicrographs of the exoskeleton patterns of (a) a wild-type *Drosophila* larva and (b) a larval homozygote lacking the entire bithorax gene complex (BX-C⁻). Note that the faint denticle bands characteristic of T2 are now repeated down the entire BX-C⁻ embryo. (S. Celniker and E. B. Lewis, California Institute of Technology)

the lifetime of the fly. This contrasts with most of the maternal, gap, and pair-rule genes, which are only expressed transiently during segmentation. The continuous expression of the homeotics is necessary to maintain proper segmental identity. Mitotic crossing-over reveals, for example, that normal Ubx^{+} gene function is necessary to maintain T3 in its proper state of determination. When homozygous *Ubx* mutant clones are induced, even late in development, these clones express a mutant bithorax phenotype, in which the homozygous *Ubx* T3 cell is transformed to a T2 phenotype.

The homeotic genes thus serve not only as molecules necessary to initiate a particular level of segmental determination, but also as memory molecules for maintaining the determined state.

Message Once cell identity is established, it must be remembered. The properties of the homeotic genes and their proteins provide a model for the mechanisms underling this maintenance of the determined state.

Applying the Fly and Worm Lessons to Other Organisms

Mouse Molecular Genetics

How universal are the developmental principles uncovered in *Drosophila* and in *C. elegans?* Until recently, the type of genetic analysis possible in *Drosophila* and *Caenorhabditis* has not been feasible in most other organisms, at least not without a huge investment to develop comparable genetic tools. However, in the last few years, recombinant DNA technology has provided the tools for addressing the generality of the *Drosophila* findings. Some of the most spectacular advances have come from studying early mouse development.

Finding Cognates of *Drosophila* Developmental Genes. Once the homeobox was discovered, Bill McGinnis, Michael Levine, and Walter Gehring used DNA hybridization techniques to see if similar sequences exist in other animals. They performed "zoo blots" (Southern blots of restriction-enzyme-digested DNA from many organisms), using radioactive *Drosophila* homeobox DNA as the probe. By adjusting the conditions of hybridization, they discovered homologous homeobox sequences in many different animals, including humans and mice. Some of these mammalian homeobox genes are very similar in sequence to the *Drosophila* genes. Perhaps the most striking case is the similarity between the clusters of homeobox genes in mammals called the **Hox complexes** and the insect ANT-C and BX-C homeotic gene clusters, now collectively called the **HOM-C (homeotic gene complex)** (Figure 22-30).

The ANT-C and BX-C clusters, which are far apart on chromosome 3 of *Drosophila melanogaster,* are tightly linked in more primitive insects such as the flour beetle, *Tribolium castaneum.* The tight linkage of the HOM-C cluster is considered the general case. Moreover, the genes of the HOM-C cluster are arranged on the chromosome in an order that is colinear with their spatial pattern of expression: The genes at the 3′ end of the complex are transcribed near the anterior end of the embryo; going toward the 5′ end, the genes are transcribed progressively more posteriorly (see Figure 22-28).

The mouse Hox clusters (and the human) show similar features, although there are four Hox clusters, each located on a different chromosome. These clusters are **"paralogous,"** meaning that the structure (order of genes) in each cluster is very similar, as if the entire cluster had been quadruplicated during vertebrae evolution. The genes near the 3′ end of each Hox cluster are quite similar not only to each other but also to one of the insect HOM-C 3′ genes. Similar relationships hold throughout the clusters. Finally, and notably, the Hox genes are expressed in a segmental fashion in the devel-

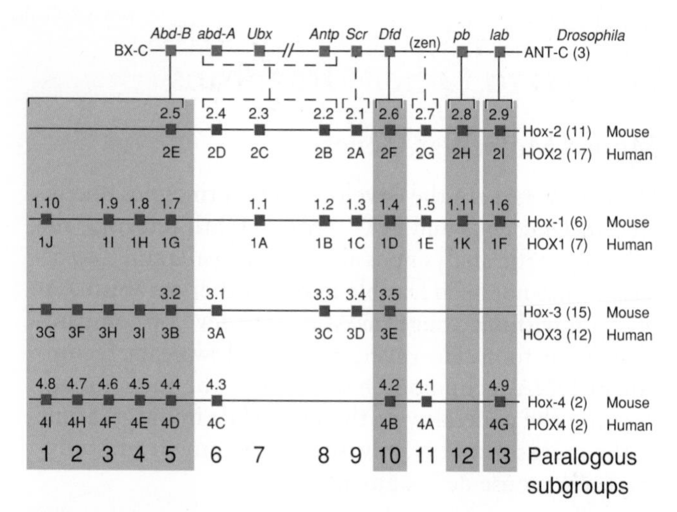

Figure 22-30 Chromosome maps comparing the *Drosophila* HOM-C with the mouse Hox and human HOX complexes. Note that mammals have four complexes, each on a different chromosome. Each square indicates a different homeobox gene (encoding a homeodomain protein); purple squares indicate the mammalian genes and red squares indicate the fly genes. Geneticists use different nomenclature for mouse and human genes, and the designations of the equivalent genes are shown above and below the squares. The shaded regions and solid brackets indicate genes with established molecular evidence of equivalence between fly and mammals. The dashed brackets indicate genes whose equivalence between flies and mammals is less firmly established. The chromosomes on which the clusters are located are shown in parentheses. All of the genes in a cluster are transcribed in the same direction on the chromosome. (From R. Krumlauf, *BioEssays* 14, 1992, 245.)

oping segmental somites and central nervous system of the mouse (and presumably the human) embryo! Each Hox gene is expressed in a continuous block beginning at a specific anterior limit and running posteriorly to the end of the developing vertebral column (Figure 22-31). The anterior limit differs for different Hox genes. Within each Hox cluster, the 3′-most genes have the most anterior limits. These limits proceed more and more posteriorly for the more 5′ Hox genes. Thus, the Hox gene clusters appear to be arranged and expressed in an order that is strikingly similar to that of the insect HOM-C genes. How can such disparate organisms—fly, mouse, human—have such similar gene sequences? The simplest interpretation is that the Hox and HOM-C genes are the vertebrate and insect descendants of a homeobox gene cluster present in a common ancestor some 600 million years ago!

> Message From the HOM-C/Hox example and other examples uncovered in the last few years, it is clear that many genes with important developmental roles are highly conserved during animal evolution.

Knocking Out Mouse Cognate Genes. The above analysis sidesteps the question of whether or not the Hox genes and the HOM-C genes perform a similar developmental role. Some very clever technology is allowing this question to be addressed. Techniques for mixing different cell types in cultured mouse embryos and reim-

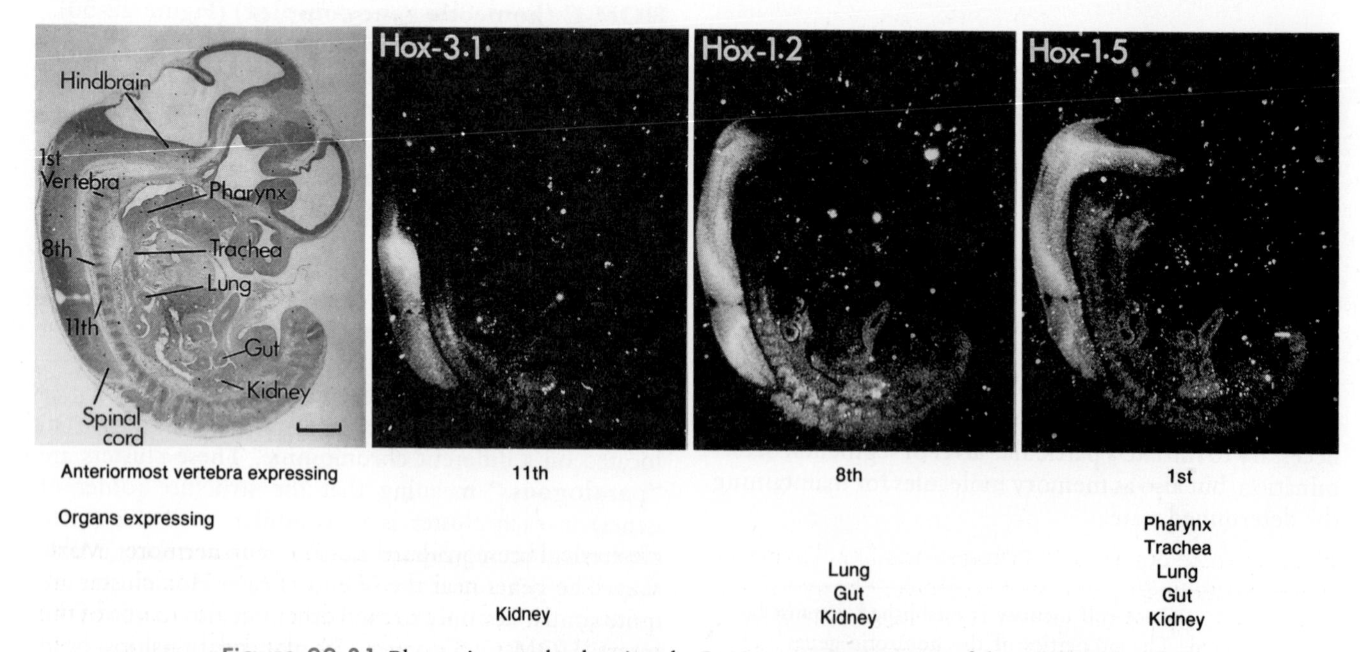

Figure 22-31 Photomicrographs showing the RNA expression patterns of three mouse Hox genes in the vertebral column of a sectioned 12.5-day-old mouse embryo. Note that the anterior limit of each of the expression patterns is different. (From S. J. Gaunt and P. B. Singh, *Trends in Genetics* 6, 1990, 208.)

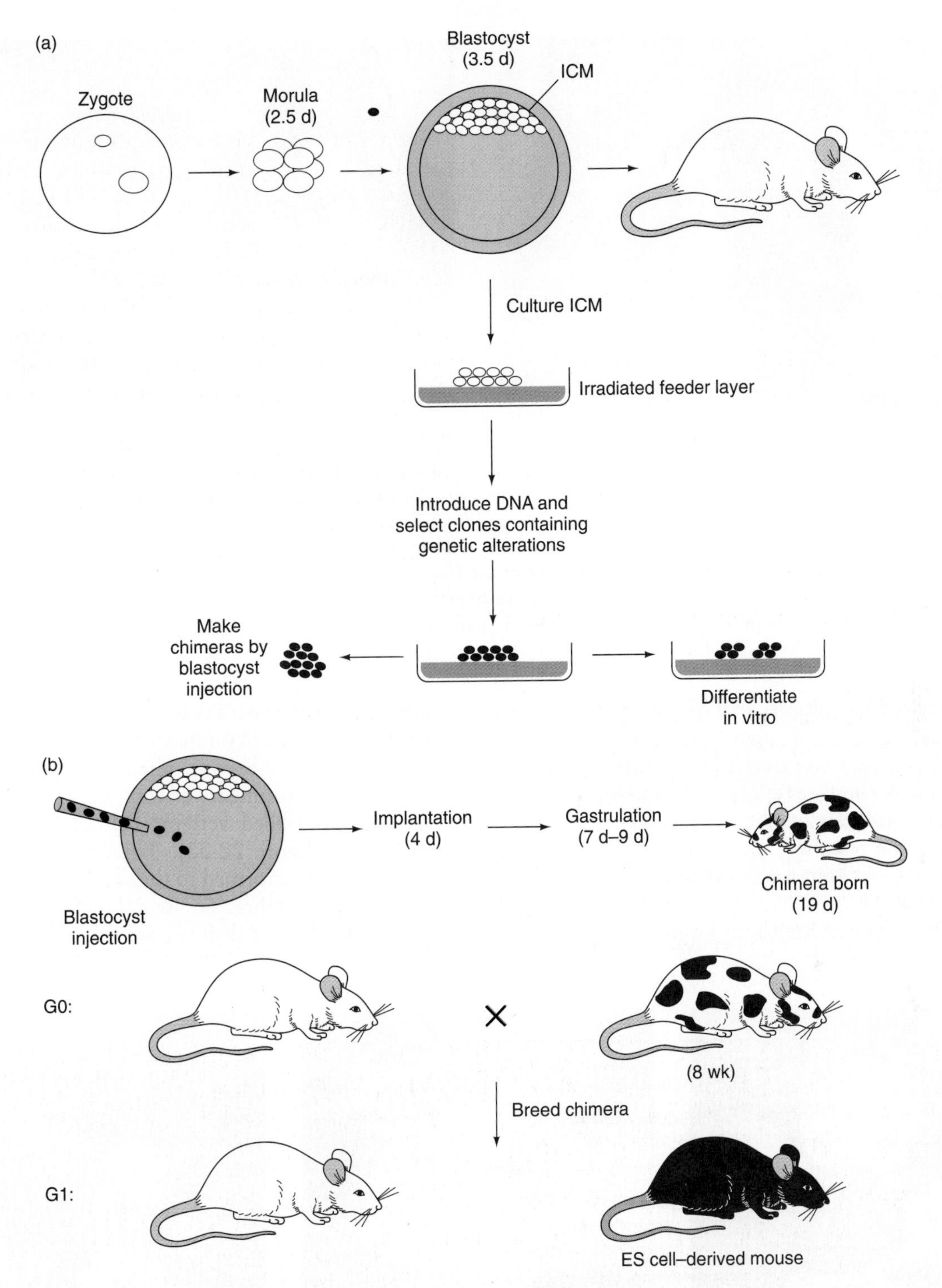

Figure 22-32 The production of embryonic stem (ES) cell lines and ES chimeric mice. (a) The process of establishing ES cell lines from cultures of the inner cell mass (ICM) of developing mouse embryos. (b) The ES cell injection process that leads to incorporation of the mutated ES cells into the germ-line and soma of a recipient embryo, producing ES chimeric mice and, ultimately, ES cell–derived mice. (From A. L. Joyner, *BioEssays* 13, 1991, 649.)

planting these embryos in surrogate mothers have been in use for about 25 years. More recently, Elizabeth Robertson and Allen Bradley have developed techniques to grow mouse germ-line precursors (embryonic stem cells, or ES cells) under conditions in which they can successfully be injected into the inner cell mass of host embryos and successfully repopulate the germ line of the resulting chimeric mouse (Figure 22-32). Furthermore, Mario Capecchi developed recombinant DNA techniques allowing individual cultured ES cells to be injected with altered, defective versions of standard mouse genes, which can replace their normal counterparts by homologous recombination that exploits DNA repair enzymes present in the ES cells. Hence, researchers can now pro-

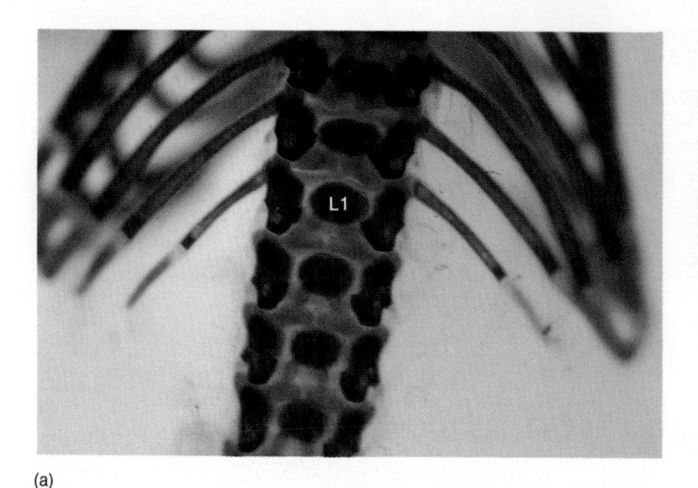

(a) (b)

Figure 22-33 The phenotype of a homeotic mutant mouse. Mice homozygous for a "targeted knockout" of the *Hox-3.1* gene were generated by using cultured ES cells, as shown in Figure 22-32. (a) A close-up of the thoracic and lumbar vertebrae of a homozygous *Hox-3.1*$^-$ mouse. Note the ribs coming from L1, the first lumbar vertebra. L1 in wild-type mice has no ribs. (b) An unexpected second phenotype of the *Hox-3.1* knockout. Note that the homozygous mutant mouse on the right has clenched fingers, while the wild-type mouse on the left has normal fingers. (From H. Le Mouellic, Y. Lallemand, and P. Brûlet, *Cell* 69, 1992, 251.)

duce a **"targeted knockout"** of a specific gene. ES cells containing a knockout are grown into a colony and injected into host embryos which are in turn implanted into surrogate mothers carrying genetic markers (such as coat color) that differentiate them from both the host and the ES cells. The resulting adult mice are bred and F1 individuals heterozygous for the knockout mutation are easily identified. Such F1 mice are then interbred to observe the phenotype of knockout homozygotes.

Applying Knockout Technology to the Hox Gene Cluster. One recently reported knockout of *Hox-3.1* causes a striking embryonic phenotype: the production of ribs on the first lumbar vertebra, L1, which ordinarily is the first nonribbed vertebra behind those vertebrae bearing ribs (Figure 22-33). Thus, the L1 vertebra is homeotically transformed to the segmental identity of a more anterior vertebra. To use geneticists' jargon, *Hox-3.1*$^-$ has caused a fate shift toward anterior. Clearly, this

(a) (b) (c)

Figure 22-34 Homeotic flower mutants in *Arabidopsis thaliana.* (a) Wild-type, with four petals, six stamens, and an ovary. (b) The *agamous-3* mutant, with extra petals and sepals, and no stamens or ovary. (c) The *apetala3* mutant; the outer two whorls have been removed to show the transformation of stamens into carpels. (From Eliot M. Myerowitz, J. L. Bowman, L. L. Brockman, G. N. Drews, T. Jack, L. E. Sieburth, and D. Weigel, *Development,* Supplement 1, 1991, 157. Photographs courtesy of J. L. Bowman.)

Hox gene seems to control segmental fate in a manner extremely similar to the HOM-C genes, since, for example, the absence of the *Ubx* gene also causes a fate shift toward anterior. All the work to date suggests that the Hox clusters not only have strong evolutionary parallels to the HOM-C genes, but they may have considerable functional parallels as well. A strong message is emerging that developmental strategies in animals are quite ancient, and that a mammal and a fly are put together using the same regulatory devices.

Homeotic Genes and Plant Development

Clearly, the emphasis in this chapter has been on principles underlying animal development. We should not assume that the same governing principles will emerge in plant development. The life strategy of plants and animals is quite different, reflecting their ancient evolutionary divergence. As examples, while animals separate germ line from soma early in development, plants make this decision quite late, and cell migration is essentially nonexistent in plant development because of the constraints imposed by having cell walls.

While there is considerable information on developmental genetics in several plants, including maize, tobacco, and snapdragon, a small weed, *Arabidopsis thaliana,* has several features that are particularly handy for the geneticist: small genome size (about 10^8 base pairs of DNA per haploid genome, far smaller than other plants), small chromosome number ($n = 5$), short generation time, and small plant size. Embryonic lethal mutations in *Arabidopsis* are in the process of being generated and characterized. Elliot Myerowitz and colleagues have identified homeotic genes affecting flower development in *Arabidopsis* (Figure 22-34), and these are under active genetic and molecular investigation. These studies have already identified a new class of transcription factors present in flowering plants. We can anticipate that in the near future, information from this system will rival that of the best genetic systems in animals.

Summary

The details of how animal development proceeds will undoubtedly differ from species to species. However, the examples that we have looked at here do portray general themes.

Early in life, each cell is totipotent: it has the potential to differentiate into a number of cell types. But in later stages of development, the type of cell it can become is increasingly restricted. In other words, determination — the commitment of a cell to a specific fate — is successively restricted to ever-narrowing possibilities.

The egg must quickly establish polarity — the orientations of its body axes. Flies create polarity through maternally contributed localized determinants. In frogs, the site of fertilization serves as a reference point for asymmetries in the initial cleavages (Figure 22-35). Regardless of its origins, polarity is achieved.

Polarity creates positional information, and this is used to control the differential expression of regulatory genes that establish cell fates. Many of these regulatory genes encode transcription factors that determine the constellation of genes expressed in a given cell. Others (for example, those we examined in the *Drosophila* eye and the *C. elegans* vulva) encode components of signaling systems that allow groups of cells to allocate cell fate in a cooperative manner.

Many of the regulatory genes of the fly have cognates in mammals; these mammalian cognates appear to serve important (and, in the case we examined, analogous) developmental roles.

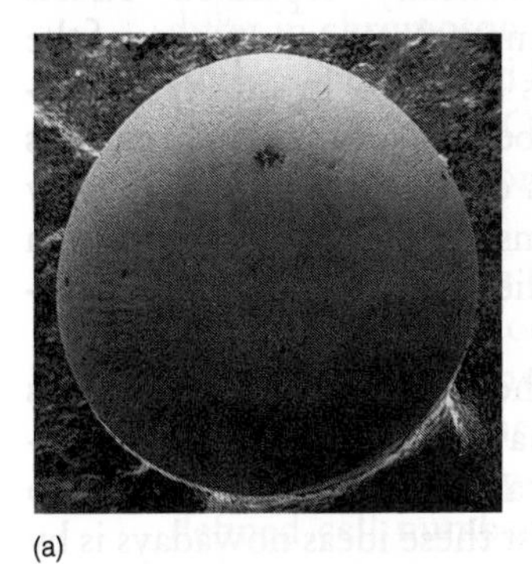
(a)

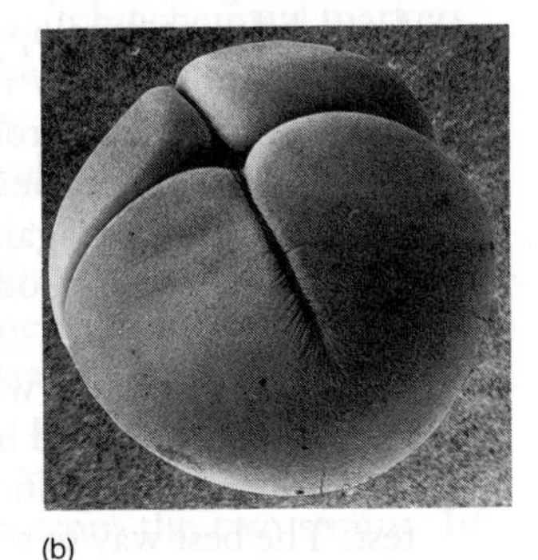
(b)

(c)

(d)

Figure 22-35 Early cleavages of an amphibian egg. It can be seen that asymmetries soon develop due to the presence of yolk at the lower (vegetal) pole of the zygote cell. (a) Undivided egg. (b) Second cleavage. (c) Eight-cell stage. (d) 16-cell stage. (Photographs courtesy of Lloyd M. Beidler, Florida State University.)

At the 3′ end of each variable gene is a segment that contains a 7-bp sequence and a 9-bp sequence, separated by a 12-bp spacer. The 7- and 9-bp elements always have the same base sequences, but the 12-bp sequence appears to be random. 5′ to every joining region is almost the same structure, with the same 7- and 9-bp sequences, except that they are in opposite orientation and the spacer is 23 bp long.

During antibody gene assembly, recombination always occurs between a segment with a 12-bp spacer and one with a 23-bp spacer. Thus, we can envision that recombination involves forming a heteroduplex between the regions immediately adjacent to the variable and joining regions based on the complementarity of the 7-bp and 9-bp sequences. Two proteins, encoded by *RAG-1* and *RAG-2* (for *r*ecombination-*a*ctivating *g*enes 1 and 2), are thought to bind to these sequences and somehow mediate these site-specific recombination events.

Is this sort of programmed DNA rearrangement limited to the immune system? Perhaps not. Interestingly, the *RAG-1* gene is also expressed in the brain, suggesting that some neural functions may involve site-specific DNA rearrangements as well.

Transcriptional Regulation by Tissue-Specific Enhancers

Transcriptional regulation of gene activity is probably the most commonly encountered level of control. Recall from Chapters 13 and 17 that promoters and enhancers are DNA sequences that play a role in transcription. **Promoters** are the binding sites for RNA polymerase to start transcription; **enhancers** maximize a promoter's rate of transcription. Promoters and enhancers are ***cis*-acting:** they are on the DNA strand being transcribed.

One of the most striking observations in eukaryotic gene regulation was the identification of **tissue-specific enhancers**—*cis*-acting regulatory sequences that can confer the activation of a promoter, and hence the expression of a gene, on specific cell types. Antibody genes, for example, are expressed only in B cells (and their plasma cell descendants); this is because antibody genes have B-cell-specific enhancers.

Properties of Tissue-Specific Enhancers

In some genes, regulation can be controlled by simple sets of enhancers. For example, in *Drosophila,* vitellogenins are large egg-yolk proteins made in the female adult's ovary and fat body (an organ that is essentially the fly's liver) and transported into the developing oocyte. Two distinct enhancers regulate a vitellogenin gene, one driving expression in the ovaries and the other in the fat body (Figure 23-7).

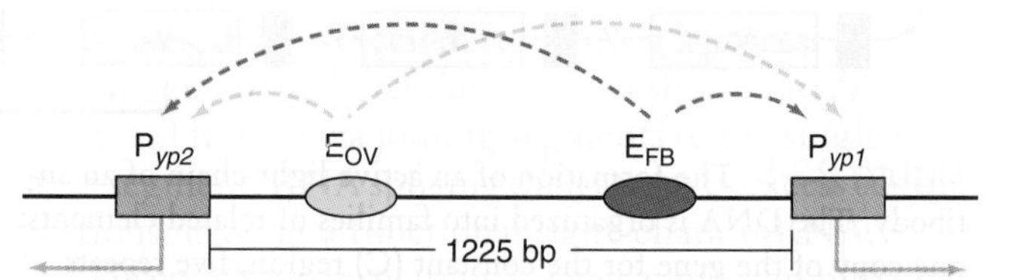

Figure 23-7 The enhancer elements of yolk protein (vitellogenin) genes in *Drosophila melanogaster.* Two yolk protein genes (*yp1* and *yp2*) encode very similar proteins and are transcribed in opposite directions from nearby promoters. (P_{yp1} and P_{yp2}). (Their transcription start sites are 1225 bp apart.) One enhancer (E_{OV}) drives expression in the follicle cells of the ovary and another (E_{FB}) drives expression in the fat body. Each enhancer causes tissue-specific expression of both *yp1* and *yp2*, indicating that enhancers can act on the nearest promoters in both directions along the DNA.

By contrast, the array of enhancers for a gene can be quite complex, reflecting similarly complex patterns of gene expression that must be controlled. The *dpp* (*de*ca*p*enta*p*legic) gene in *Drosophila,* for example, contains a number of enhancers interspersed along a 50-kb interval of DNA (one kb, or kilobase, equals 1000 bases). Each of these enhancers regulates the expression of *dpp* in a different site in the developing animal (Figure 23-8).

In prokaryotes, *cis*-regulatory elements reside in the immediate neighborhood of the promoter. By contrast, as we have just seen, tissue-specific enhancers in eukaryotes can be quite far from the promoter, acting from a distance of 30 kb or more! Further, these enhancers are typically orientation-independent: they can function whether their base sequence is in the same or opposite orientation relative to the orientation of transcription.

The existence of enhancers that act at great distances helps to explain several puzzling features of higher eukaryotes. One reason why genes in higher eukaryotes are so much bigger than those in lower organisms is that they typically contain many enhancers. Furthermore, the tissue-specific regulation of a gene may be quite complex, requiring the action of numerous enhancers, while DNA is essentially linear (that is, one-dimensional). Therefore, enhancers may have to be spaced over considerable distances from the promoters they regulate. In some cases, only a subset of enhancers regulating a gene are upstream (5′ to the promoter), while other enhancers lie within introns and downstream (on the 3′ side) of the transcribed region, as in the case of the *dpp* gene (see Figure 23-8). Apparently mechanisms have evolved to achieve remote control of gene expression.

Using Reporter Genes To Find Enhancers

How can enhancers be studied and exploited? The typical way that they are dissected is by use of a **"reporter-gene" construct** (Figure 23-9). To make the construct,

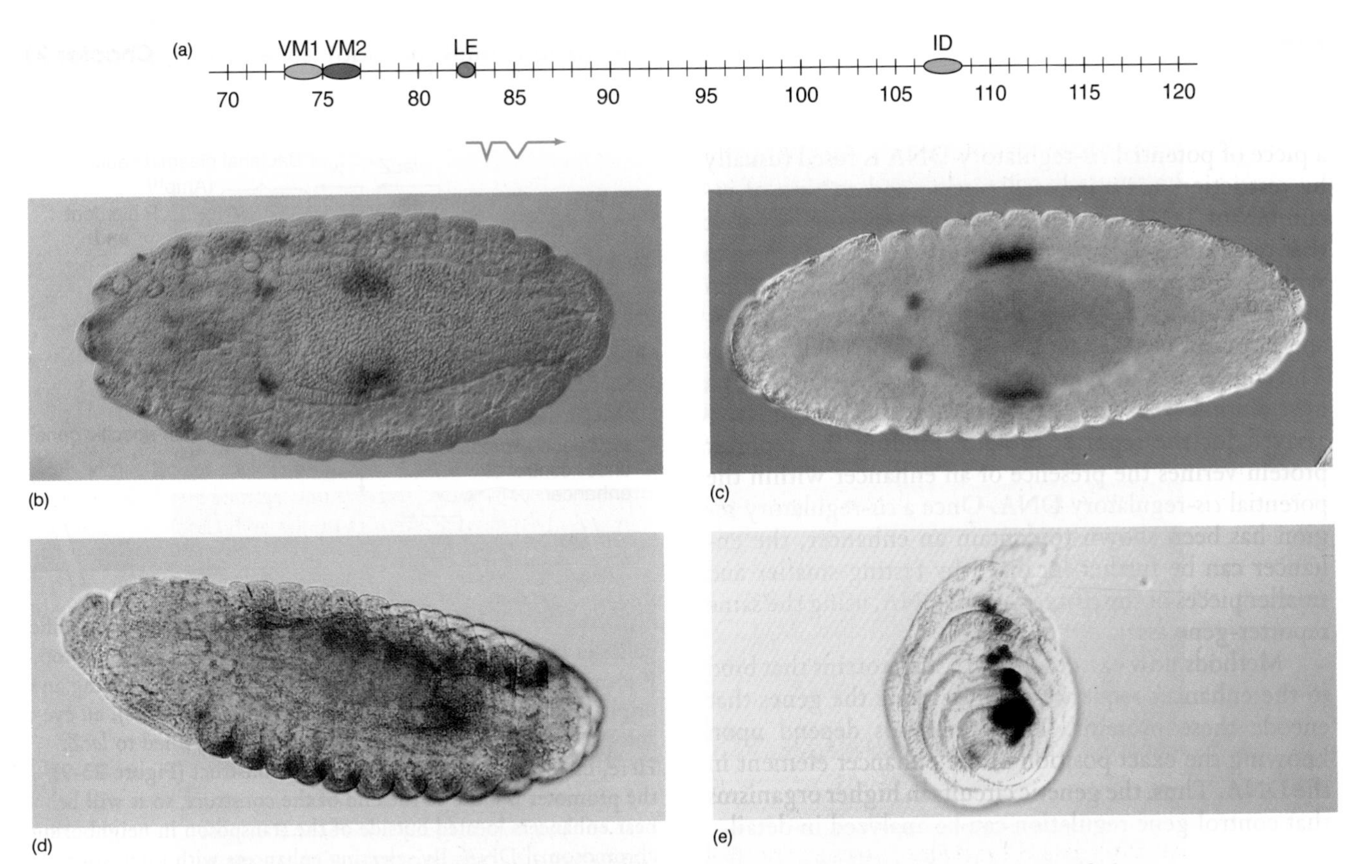

Figure 23-8 Some examples of complex tissue-specific regulation of a gene. (a) A simplified genetic and molecular map of the *dpp* gene of *Drosophila.* Units on the map are in kb. The basic transcription unit of the gene is shown below the map coordinate line. The abbreviations above the line point to the sites of a few of the many tissue-specific enhancers that regulate this transcript. The sites have been mapped by means of mutations in the gene and by means of reporter-gene constructs (see Figure 23-9). (b) Reporter-gene assay for expression of the *dpp* gene in two portions of the embryonic visceral mesoderm, the precursor of the gut musculature. VM1 = enhancer driving anterior *v*isceral *m*esoderm expression (left-hand block of blue staining in figure); VM2 = enhancer driving posterior visceral mesoderm expression. In (b), (d) and (e), the blue staining is due to a histochemical assay for *E. coli* β-galactosidase activity (the protein encoded by the *lacZ* reporter gene). (c) RNA in situ hybridization assay of *dpp* expression in the embryonic visceral mesoderm. Note that the blue reporter-gene expression pattern in (b) is the same as the brown *dpp* RNA expression pattern shown here, confirming the reliability of the reporter-gene assay. (d) Reporter-gene expression driven by a different enhancer (LE) of *dpp* in the *l*ateral *e*ctoderm of an embryo. (e) Reporter-gene expression driven by ID, one of many enhancer elements driving *i*maginal *d*isk expression of *dpp.* (An imaginal disk is a flat circle of cells in the larva that gives rise to one of the adult appendages.) A blue sector of *dpp* reporter-gene expression in a leg imaginal disk is shown. (Parts b and c courtesy of D. Hursh, part d courtesy of R. W. Padgett, and part e courtesy of R. Blackman and M. Sanicola).

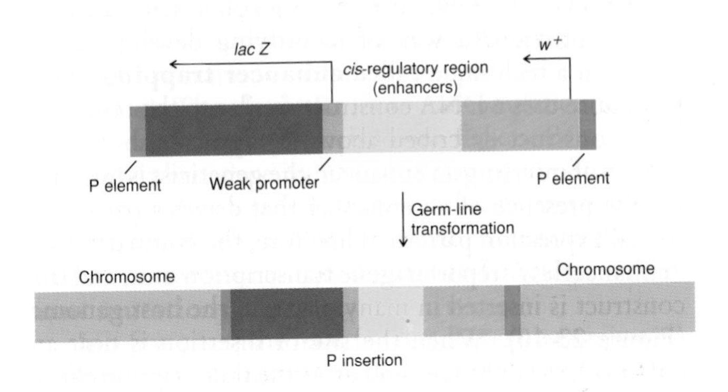

Figure 23-9 Use of a reporter gene construct in *Drosophila* to identify enhancers. The top line represents a portion of a plasmid, bracketed by P element ends so that the material in between can be inserted into the genome by P element transformation (see Chapter 20). The transformants can be identified because of the white gene (w^+) marker in the construct. A region of DNA thought to contain one or more enhancers is inserted immediately adjacent to a "weak" promoter, that is, a promoter that by itself cannot initiate transcription. The promoter is joined to the *lacZ* structural gene (the reporter gene), which encodes β-galactosidase. If there are any enhancers in the construct, they will induce tissue-specific expression of *lacZ.* The embryos and imaginal disks in Figure 23-8b, d, and e are examples of expression from such reporter-gene constructs.

ratio of 1.0); thus *Sxl* is turned on in females but remains off in males.

What happens when *Sxl* is on? The high level of transcription-factor activity turns on an **early *Sxl* promoter** that makes some active *Sxl* protein. The X:A transcription factor appears not to be present after the blastoderm stage, and so the early *Sxl* promoter is never activated again. However, there is a **late *Sxl* promoter** that is expressed in both sexes for the rest of development. The *Sxl* protein is an RNA-binding protein that can alter the splicing of the nascent *Sxl* transcript coming from this late promoter. The female-specific splicing pattern of *Sxl* produces an mRNA that can make active *Sxl* protein identical to that produced from the early promoter mRNA. Thus, by **autoregulation** controlled at the level of RNA splicing, *Sxl* activity is maintained throughout development in individuals with an X:A ratio of 1.0.

When the X:A ratio is 0.5, the *Sxl* switch is set in the "off" position. The early promoter is not activated and no active *Sxl* protein is made. In the absence of active *Sxl* protein, the "default" splicing pattern for the transcript made from the late promoter takes effect, producing an mRNA that contains a stop codon shortly after the translation-initiation codon. The small protein produced from this male-specific spliced mRNA has no biological activity. Hence, *Sxl* remains off throughout development in individuals with an X:A ratio of 0.5.

Role of *tra* in Sex Determination. The female-specific RNA-binding activity of *Sxl* not only controls its own splicing, it also initiates the female differentiation pathway by altering the splicing of the *tra* gene product (see Figure 23-18). As with *Sxl*, the female-specific *tra* mRNA encodes an active RNA-binding protein while the male-specific mRNA encodes an inactive product. However, the *tra* protein does not autoregulate itself; rather, it directly or indirectly controls the sex-specific splicing of the final gene in the pathway, *dsx*. Both the female-specific and the male-specific spliced versions of the nascent *dsx* transcript have biological activity, serving as repressors of male-specific or female-specific transcription, respectively. While a *dsx* protein represses transcription of sex-specific genes of one sex, the genes specific to the other sex remain active, giving the individual its sexual phenotype.

Message In *Drosophila,* the state of sexual differentiation is established and maintained through regulation of transcription and RNA splicing, respectively.

Autonomy of Sexual Determination. In *Drosophila,* studies have shown that the sexual phenotype is autonomous; that is, every cell appears to determine its own secondary sexual phenotype autonomously. Furthermore, the sexual phenotype of a cell is decided by the time of blastoderm formation, early in embryogenesis. The autonomy of sex differentiation in *Drosophila* is easily seen in an X-chromosome mosaic, or **gynandromorph** (see Chapter 9, Figure 9-25). Occasionally, an X chromosome is lost during an early mitotic division of the embryo. When this occurs, the embryo becomes a mosaic of XX AA cells and X0 AA cells. If the XX AA cells end up in an area of the fly epidermis that is sexually dimorphic, they show a female phenotype; conversely, X0 AA cells in such an area are male.

Sex Determination in Mammals

Originally, because mammals and flies had a similar sex chromosome dimorphism, it was assumed that the X:A ratio model would apply to mammals as well. However, once sex-chromosome aneuploids were recognized, it was clear that the basis of mammalian sex determination and differentiation was completely different (Table 23-1). In humans, XXY individuals are phenotypically male, with a syndrome of moderate abnormalities (Klinefelter syndrome; see Figure 9-20). By comparison, XXY flies are phenotypically normal females. X0 humans have a number of abnormalities (Turner syndrome; see Figure 9-17), including short stature, mental retardation, and just traces of gonads, but they are clearly female in morphology. By comparison, X0 flies are males. In mammals, the sex determination mechanism is based on the presence or absence of a Y chromosome. Without a Y, the individual develops as a female; with it, as a male.

Another difference between flies and mammals is that sex in mammals is nonautonomous. Individuals mosaic for XX and XY tissues arise occasionally; they typically have a generalized appearance characteristic of one or the other sex, but are not cell-by-cell mosaics in phenotype, as the *Drosophila* gynandromorphs are.

Table 23-1. The Effect of Sex Chromosome Dosage on Somatic Sexual Phenotype in Diploid *Drosophila* and Humans

Sex chromosome constitution	Somatic sexual phenotype	
	Drosophila	Humans
Euploidy		
XX	♀	♀
XY	♂	♂
Aneuploidy		
XXY	♀	♂
X0	♂	♀

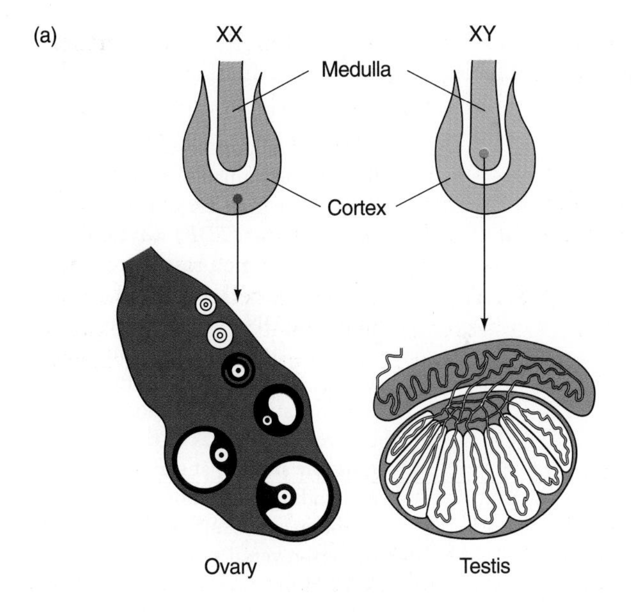

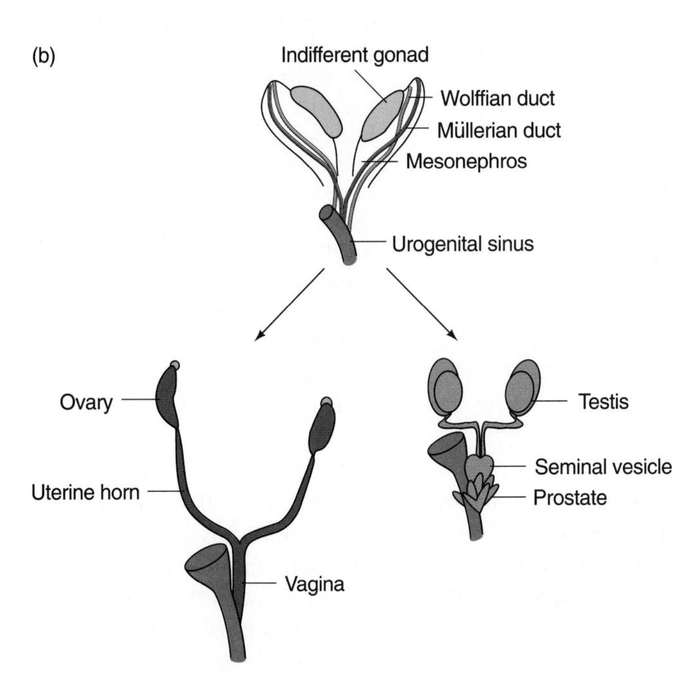

Figure 23-20 Development of the mammalian urogenital system. (a) The embryonic genital ridge consists of a medulla surrounded by a cortex. Female germ cells migrate into the cortex, and become organized into an ovary. Male germ cells migrate into the medulla and become organized into a testis. (b) In the initial urogenital organization at the indifferent gonad stage, precursors of both male (Wolffian) and female (Müllerian) ducts are present. If a testis is present, it secretes two hormones, testosterone and a polypeptide hormone called Müllerian-inhibiting substance (MIS, or anti-Müllerian hormone). This causes the Müllerian ducts to regress and the Wolffian ducts to develop into the male reproductive ducts. If an ovary is present, testosterone and MIS are absent and the opposite happens: the Wolffian ducts regress and the Müllerian ducts develop into the female reproductive ducts. (From Ursula Mittwoch, *Genetics of Sex Differentiation.* Copyright 1973 by Academic Press.)

Mammalian Reproductive Development. The observation of nonautonomy in mammalian sex determination can be understood in view of the biology of the reproductive system: secondary sexual phenotypes are driven by the presence or absence of the testes.

The gonad forms within the first two months of human gestation. Primordial germ cells migrate into the genital ridge, which sits atop the rudimentary kidney. The chromosomal sex of the germ cells determines whether they will migrate superficially or deeply into the gonadal ridge and whether they will organize into a testis or an ovary (Figure 23-20). If they form a testis, the Leydig cells of the testis secrete testosterone, an androgenic (male-determining) steroid hormone. This hormone binds to androgen receptors; the androgen–receptor complex binds to androgen-responsive enhancer elements, leading to the activation of male-specific gene expression. In chromosomally female individuals, no Leydig cells form in the gonad, no testosterone is produced, androgen receptor is not activated, and individuals continue along a female pathway of development. Hence, it is the presence or absence of a testis that determines the sexual phenotype, through the endocrine release of testosterone. Indeed, in XY individuals lacking the androgen receptor, development proceeds along a completely female pathway even though these individuals have testes. This is a genetic syndrome called the testicular feminization (*Tfm*) syndrome (see Figure 3-24, page 69).

Role of the Y Chromosome in Sex Determination. What initiates the sex determination pathway? Molecular genetic analysis has focused on identifying the locus on the Y chromosome that drives testis formation. This gene has been called ***testis-determining factor on the Y chromosome*** (*TDF* in humans, *Tdy* in mice). *TDF/Tdy* has very likely been identified, through the mapping and characterization by Robin Lovell-Badge and Peter Goodfellow of a genetic syndrome common to mice and humans that almost certainly affects this factor. This syndrome is called **sex reversal.** Sex-reversed XX individuals are phenotypic males and have been shown to carry a fragment of the Y chromosome in their genome. In general, these Y chromosome duplications arise by an illegitimate recombination between the X and Y chromosomes that fuses a piece of the Y chromosome to a tip of one of the X chromosomes. The portion of the Y chromosome that includes these duplications has been cloned in a "chromosome walk" (see pages 527–528 and Figure 14-12). Lovell-Badge and Goodfellow have identified from this region a transcript that is expressed in the appropriate location of the developing kidney capsule (Figure 23-21).

The gene encoding this transcript was named the ***sex reversal on Y*** gene (*SRY* in humans, *Sry* in mice), since it

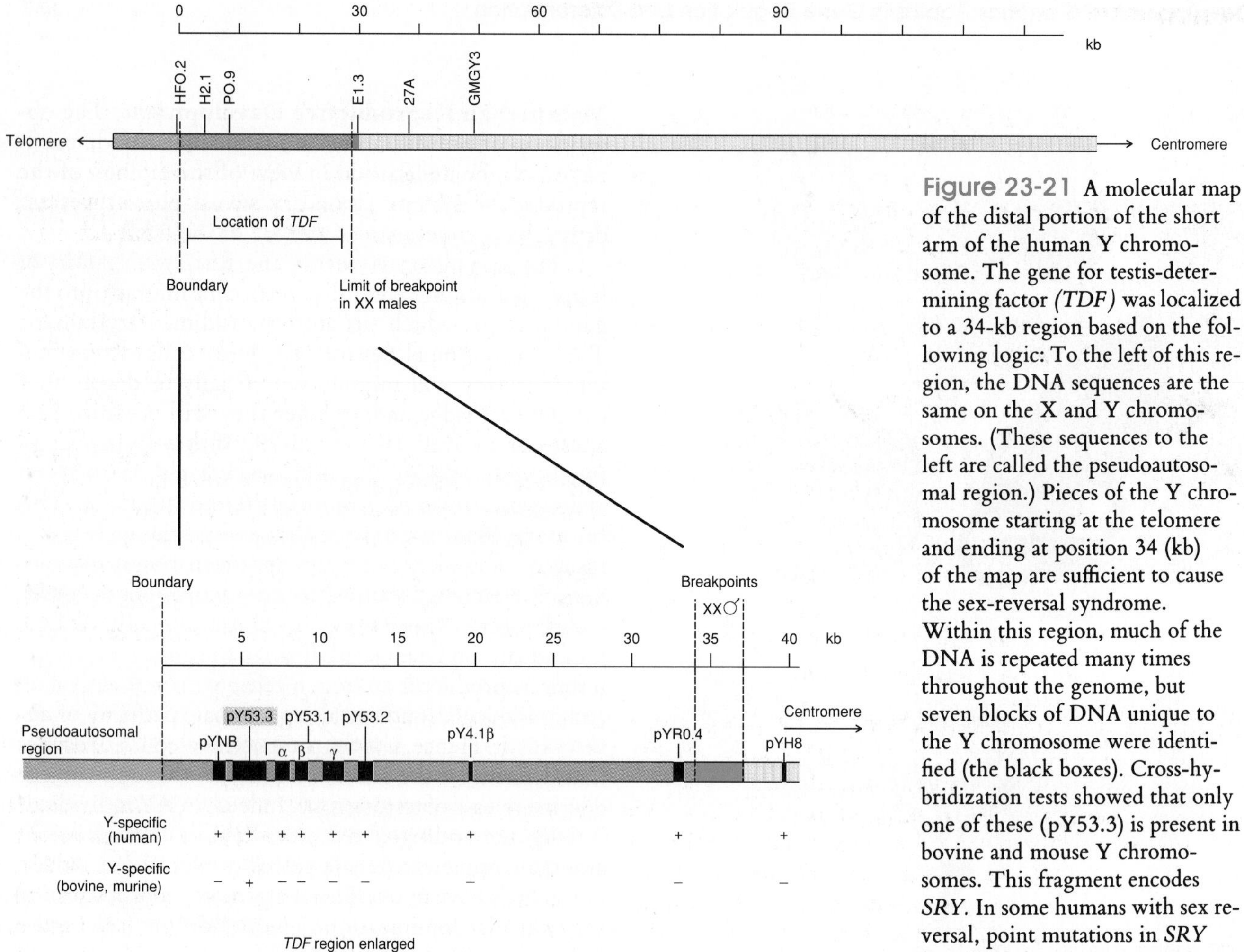

Figure 23-21 A molecular map of the distal portion of the short arm of the human Y chromosome. The gene for testis-determining factor *(TDF)* was localized to a 34-kb region based on the following logic: To the left of this region, the DNA sequences are the same on the X and Y chromosomes. (These sequences to the left are called the pseudoautosomal region.) Pieces of the Y chromosome starting at the telomere and ending at position 34 (kb) of the map are sufficient to cause the sex-reversal syndrome. Within this region, much of the DNA is repeated many times throughout the genome, but seven blocks of DNA unique to the Y chromosome were identified (the black boxes). Cross-hybridization tests showed that only one of these (pY53.3) is present in bovine and mouse Y chromosomes. This fragment encodes *SRY*. In some humans with sex reversal, point mutations in *SRY* have been found. (From A. H. Sinclair et al., *Nature* 346, 1990, 240.)

was identified on the basis of the sex reversal syndrome, but it is very likely to be the same gene as *TDF/Tdy*. Lovell-Badge and Goodfellow used a **transgene** to provide spectacular evidence in support of this identity (Figure 23-22): A cloned 14-kb genomic fragment of the mouse Y chromosome, including the *Sry* gene, was inserted into the mouse genome by germ-line transformation. An XX offspring containing this inserted *Sry* DNA (the transgene) was completely male in external and internal phenotype and, as predicted, possessed the somatic tissues of the testis—including the Leydig cells that make testosterone! (It should be noted, however, that this mouse was sterile. The sterility is probably a consequence of having two X chromosomes in a male germ cell, since XXY male mice are similarly sterile. Thus, a single genetic unit was directly shown to alter the mammalian sexual phenotype profoundly, completely consistent with the role of *SRY/Sry* as the testis-determining factor.

How does *SRY/Sry* contribute to sex determination? While not certain, it is likely that the *SRY/Sry* protein is a transcription factor. Several pieces of evidence point to this function. First, *SRY/Sry* protein has a polypeptide sequence, called an HMG (for high-mobility-group) domain, that is found in several other transcription factors. Second, some sex-reversed XY morphological females have point mutations in the HMG domain of *SRY/Sry;* while recombinant *SRY/Sry* protein made from normal males is a strong DNA-binding protein, recombinant protein made from the *SRY/Sry* genes of these sex-reversed individuals shows reduced or no DNA-binding activity. With the *SRY/Sry* protein sequence in hand, many avenues for answering the age-old questions about the biological basis of sexual phenotype can be pursued.

Message Mammalian sex differentiation is established by a path quite different from that in *Drosophila.* In mammals, a gene on the Y chromosome causes germ cells to organize the gonad into a testis rather than an ovary. Sex is determined nonautonomously and is based on the presence or absence of androgens produced by the testis.

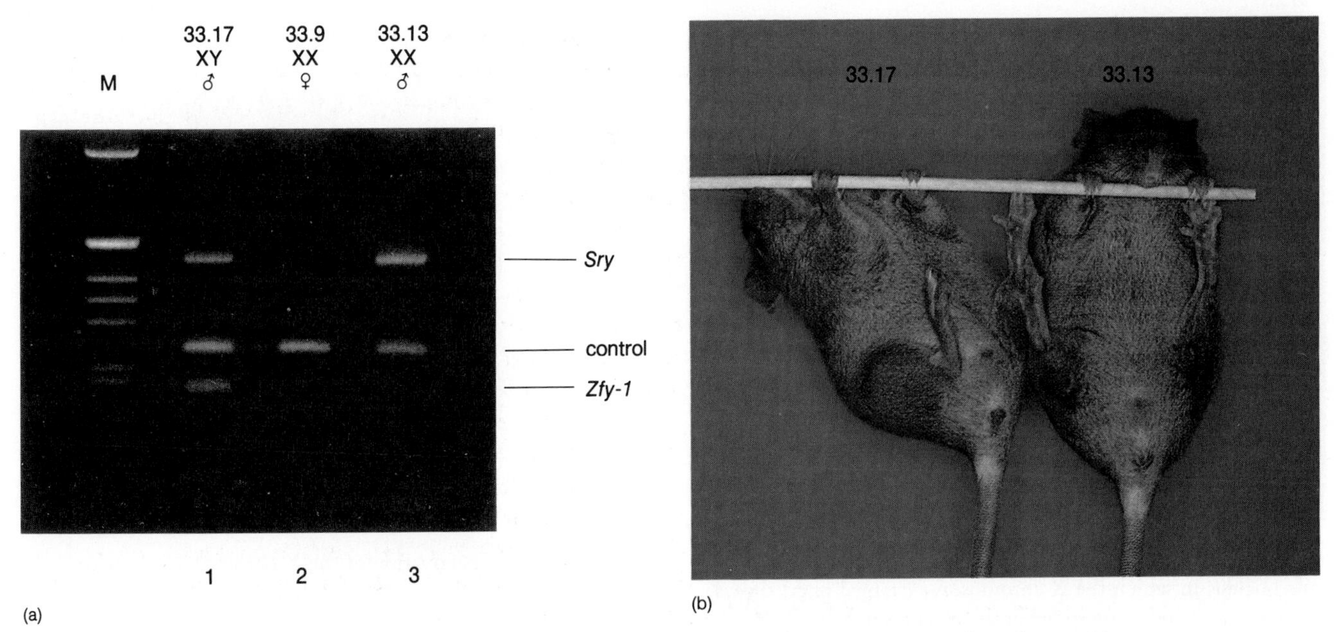

Figure 23-22 A transgenic mouse that proves *Sry* can cause the sex-reversal syndrome. (a) Amplification of genomic DNA by the polymerase chain reaction (PCR; see page 531) shows that mouse 33.13 lacks a DNA marker for the presence of a Y chromosome (*Zfy-1*), but that it does contain the *Sry* transgene. M, marker bands. (b) The external genitalia of sex-reversed transgenic mouse 33.13 are indistinguishable from those of a normal male sib (33.17). (From P. Koopman, J. Gubbay, N. Vivian, P. Goodfellow, and R. Lovell-Badge, *Nature* 351, 1991, 117.)

Dosage Compensation

Before leaving the topic of sex determination and differentiation, we should consider one important consequence of having a sexually dimorphic chromosome pair, namely aneuploidy. Monosomy for an autosome produces many deleterious effects (as discussed in Chapter 9). Why doesn't monosomy for the X chromosome do likewise? One possible answer might be that X-linked genes are all involved in sex determination. However, this is not true: Most genes on X chromosomes in both flies and humans appear to encode proteins contributing to the standard array of bodily processes—intermediary metabolism, cell structure, gene regulation, etc.—just like the genes on the autosomes. Rather, a special mechanism equalizes the gene activity of most X-linked genes in the two sexes. This phenomenon is called **dosage compensation.** As with sex determination and differentiation, dosage compensation is accomplished quite differently in flies and mammals.

In *Drosophila,* dosage compensation is accomplished by **hyperactivation of X chromosome genes** in males. That is, each X-linked gene in the male is transcribed at the same rate as the combined rate of the two copies of that gene in the female. The *Sxl* gene controls the sex-specificity of dosage compensation: In females, although *Sxl* protein activity is required to *activate* the sex-differentiation genes, *Sxl* activity also *prevents* hyperactivation of the female's X chromosomes. There is a series of male-specific lethal mutations in which homozygous females are viable but homozygous males die during development. In all mutants, there is a failure in X chromosome hyperactivation. These male-specific lethal genes are thought to be repressed by *Sxl* protein activity, so that they function only when *Sxl* protein is inactive. (Given that *Sxl* controls splicing of *tra* in female sex differentiation, we can speculate that the repression of the male-specific lethal genes will also prove to be at the level of regulated RNA splicing). One of these genes, *mle (male-lethal),* has been cloned and antibodies directed against its protein product have been made. Interestingly, immunohistochemistry shows that the protein is found specifically on the X chromosome in the salivary-gland polytene chromosomes of male larvae (Figure 23-23), whereas it is absent from chromosomes in females. One possibility is that male X-hyperactivation occurs by affecting general chromatin condensation, so that X-linked genes in males are more accessible to transcription factors.

As we described earlier (Chapter 3), in mammals, dosage compensation is accomplished by a very different route, namely by **X-inactivation,** thereby maintaining only a single active X chromosome, regardless of sex (Figure 23-24). The monosomic X in an XY individual remains active, whereas one of the two X chromosomes

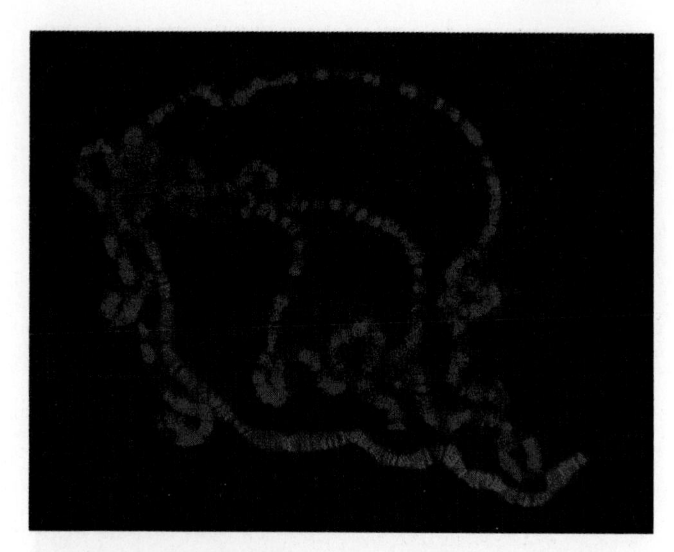

Figure 23-23 A male *Drosophila* larval salivary gland polytene nucleus in which the X chromosome (stained pink) shows the presence of *mle* protein, whereas the autosomes (counterstained blue) lack *mle* protein. An antibody directed against *mle* protein was used to stain the nucleus by immunohistochemical techniques. (From M. Kuroda, M. Kernan, R. Kreber, B. Ganetzky, and B. Baker, *Cell* 66, 1991, 95; cover of September 6, 1991, issue.)

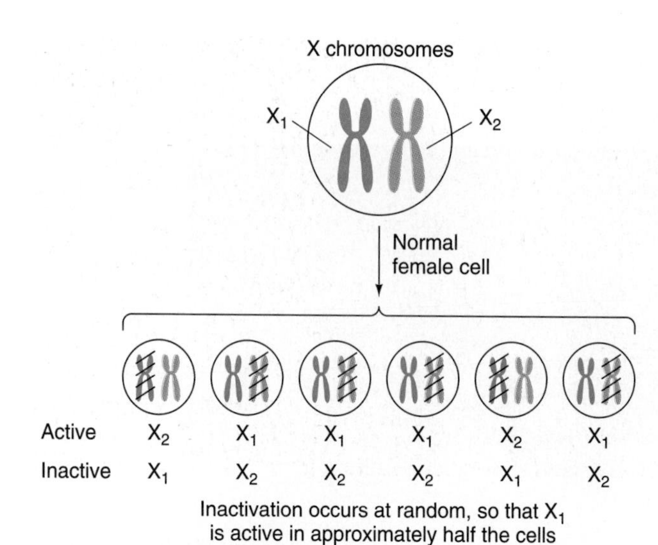

Figure 23-24 Inactivation of the X chromosome in mammals. X-inactivation occurs randomly in normal individuals. (Modified from J. D. Watson, M. Gilman, J. Witkowski, and M. Zoller, *Recombinant DNA,* 2d ed. Copyright 1992 by Scientific American Books.)

in an XX individual is inactivated and condensed into a highly compact **Barr body** (see Figure 3-27, page 70). X-inactivation is not a sex-specific event, as XXY males and XX females both have one active X and one Barr body. X-inactivation occurs early in development and is random; that is, in XX individuals, the maternally derived X chromosome is inactivated in some cells and the paternally derived one in other cells (Figure 23-24). The random mosaic hair pigmentation pattern in calico cats (see Figure 3-29, page 71) is due to X-inactivation; calico cats are virtually all females.

A single locus acts as a master inactivation switch for the entire X chromosome; its location has been mapped by using translocations between the X chromosome and an autosome. What is the nature of this molecular on/off switch? At present, it is unknown, but an intriguing candidate gene, called *XIST (X-inactive specific transcript),* maps in the immediate vicinity of the master inactivation switch. Contrary to other X-chromosome transcripts, the transcript from *XIST* is apparently only expressed from the inactive X in a cell. Molecular studies of *XIST* may provide important insights into the process of X-inactivation.

Message Various mechanisms have evolved in species with heteromorphic sex chromosomes to counteract the potential gene-dosage problems caused by the difference in X chromosome number in males and females.

Cancer as a Developmental Genetic Disease

In genetics, we study the abnormal (mutations) in order to make inferences about how wild-type genes function and how normal processes operate. Similarly, by studying abnormal cell proliferation (cancer), we can hope to gain insights into the mechanisms underlying normal growth control. In addition, we can hope to learn how to improve our diagnosis, treatment, and control of this major group of diseases.

How Cancer Cells Differ from Normal Cells

Malignant tumors or cancers, are clonal. They are aggregates of cells, all derived from an initial aberrant founder cell that, although surrounded by normal tissue, is no longer integrated into that environment. Cancer cells often differ from their normal neighbors by a host of specific phenotypic changes, such as rapid division, invasion of new cellular territories, high metabolic rate, and altered shape. For example, when cells from normal epithelial cell sheets are grown in cell culture, they divide until they form a continuous monolayer. Then, they somehow recognize that they have formed an epithelial sheet and stop dividing. In contrast, malignant cells derived from epithelial tissue continue to proliferate, piling up on one another (Figure 23-25). Clearly, the factors regulating normal cell differentiation have been altered. What, then, is the underlying cause of cancer? Many

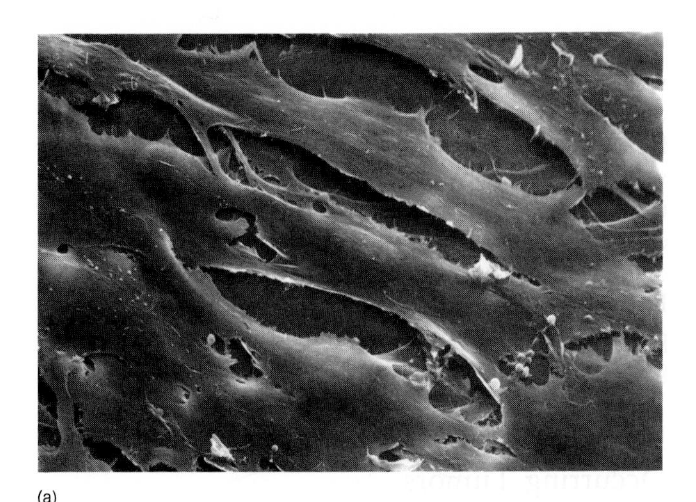
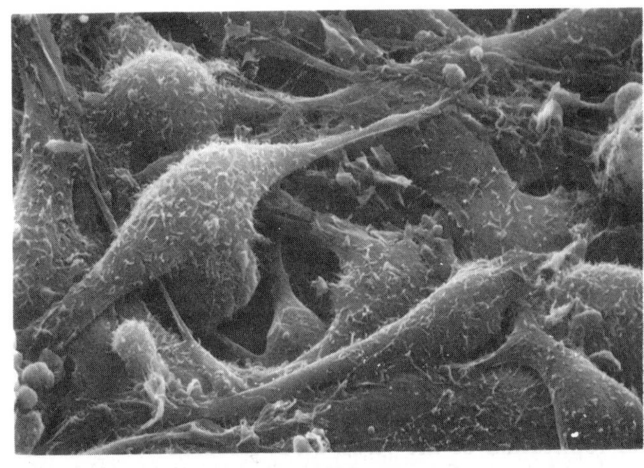

(a) (b)

Figure 23-25 Scanning electron micrographs of (a) normal cells and (b) cells transformed with Rous sarcoma virus. (a) A normal cell line called 3T3. Note the organized monolayer structure of the cells. (b) A transformed derivative of 3T3. Note how the cells are rounder and pile up on one another. (From J. Darnell, H. Lodish, and D. Baltimore, *Molecular Cell Biology,* 2d ed. Copyright 1990 by Scientific American Books.)

different cell types can be converted to a malignant state. Is there a common theme to the ontogeny of these different types of cancer, or do they each arise in a quite different way? It is becoming clear that we can think about cancer in a general way: as occurring by the production of multiple mutations in a single cell that cause it to proliferate out of control.

Evidence for the Genetic Origin of Cancers

As briefly discussed in Chapter 7 (pages 195–198), several lines of evidence have pointed to a genetic origin for the transformation of cells from the benign to the cancerous state. For many years, it has been known that most carcinogenic agents (chemicals and radiation) are also mutagenic. There are occasional instances in which certain cancers are inherited as highly penetrant single Mendelian factors; an example is familial retinoblastoma (see Figure 7-24). Perhaps reflecting the more general case are less penetrant susceptibility alleles that increase the probability of developing a particular type of cancer. Such alleles have been recombinationally and sometimes molecularly mapped in the human genome, using RFLP mapping or related techniques (see Chapter 15). **Oncogenes,** genes that are responsible for causing cancers in animals, have been isolated from **tumor viruses,** viruses that can transform normal cells in certain animals into tumor-forming cells. Such oncogenes can also be isolated from tumor cells, using cell-culture assays that can distinguish between some types of benign and malignant cells. These studies also indicate that, in most cases, a tumor does not arise as a single genetic event, but rather involves a **multiple-hit** process, in which two or more mutations must arise within a single cell for it to become cancerous. In the following sections, we will consider further the genetic origin of cancers, and the nature of the gene products that are mutated in oncogenes.

Message Many, perhaps all, tumors arise through a series of sequential mutational events that lead to a state of uncontrolled proliferation.

Tumor Viruses, Oncogenes, and Proto-oncogenes

Back in 1910, Peyton Rous showed evidence for a "filterable agent" (the term then for a virus) that induced a tumor called a sarcoma in chickens. Since then, many other viruses have been shown to cause cancers as well. Typically, the cancers that are caused by a given virus arise only in a specific cell type.

Virally induced tumors take months or years to appear. A speedier system is needed in order to study tumorigenesis. This is achieved by the technique of **cell-culture transformation:** normal cells in culture are infected with appropriate tumor viruses, thereby transforming the normal, slowly proliferating cells that cease growth when confluent (having formed monolayers) into malignant, rapidly proliferating cells that will continue to grow after confluence (Figure 23-25). Such cell-culture assay systems for cellular transformation have been invaluable in isolating and studying oncogenes.

Both DNA and RNA tumor viruses have been identified. The Rous sarcoma virus, an RNA tumor virus, is a

7. Tumors are clonal, in that they derive from a single cell. Some of the evidence for this statement was based on the phenomenon of dosage compensation in humans. Explain how dosage compensation might have helped to reveal the clonal nature of tumors.

8. You are studying a mouse gene that is expressed in the kidneys of male mice. You have previously cloned this gene. Now you wish to identify the segments of DNA that control the tissue-specific and sex-specific expression of this gene. Describe an experimental approach that would allow you to do so.

9. In *Drosophila,* homozygotes for mutations in the *tra* gene transform XX individuals into phenotypic males with regard to somatic secondary sexual characteristics.

   a. Such XX;*tra* homozygotes do not hyperactivate their X chromosomes. Explain the significance of this observation in terms of the pathway of sex determination and differentiation.

   b. The gonad in *Drosophila* forms from somatic mesoderm tissue and germ-line cells. XX;*tra* homozygotes are sterile and have rudimentary gonads. You suspect that the reason for this sterility is that the somatic tissue is sexually transformed to male by *tra* but the germ-line cells remain female. Design an experiment that would test this prediction.

10. XYY humans are fertile males. XXX humans are fertile females. What do these observations tell you about the mechanisms of sex determination and dosage compensation?

11. Humans that are mosaics of XX and XY tissue occasionally occur. They generally exhibit a uniform sexual phenotype. Some individuals are phenotypically female, others male. Explain these observations in terms of the mechanism of sex determination in mammals.

12. Can a recessive oncogene be identified in the cell transformation assay used to identify the *ras* or *myc* oncogenes? Why or why not?

13. There are dominant mutations of the *Sxl* gene, called Sxl^M alleles, that are male-lethal, but viable in females. Reversions of these alleles can be readily induced with mutagen treatment. These reversions are male-viable, but prove to be recessive female-lethal alleles of *Sxl*, called Sxl^f alleles. Provide a possible explanation for these observations, keeping in mind that *Sxl* ordinarily is dispensable for male development.

24

Quantitative Genetics

Quantitative variation in flower color, flower diameter, and number of flower parts in the composite flowers of *Gaillardia pulchella.* (J. Heywood, *Journal of Heredity,* May/June 1986)

KEY CONCEPTS

In natural populations, variation in most characters takes the form of a continuous phenotypic range rather than discrete phenotypic classes. In other words, the variation is quantitative, not qualitative.

▪

Mendelian genetic analysis is extremely difficult to apply to such continuous phenotypic distributions, so statistical techniques are employed instead.

▪

A major task of quantitative genetics is to determine the ways in which genes interact with the environment to contribute to the formation of a given quantitative trait distribution.

▪

The genetic variation underlying a continuous character distribution can be the result of segregation at a single genetic locus or at numerous interacting loci which produce cumulative effects on the phenotype.

▪

The estimated ratio of genetic to environmental variation is *not* a measure of the relative contribution of genes and environment to phenotype.

▪

Estimates of genetic and environmental variance are specific to the single population and the particular set of environments in which the estimates were made.

Ultimately, the goal of genetics is the analysis of the genotype of organisms. But the genotype can be identified—and therefore studied—only through its phenotypic effect. We recognize two genotypes as different from each other because the phenotypes of their carriers are different. Basic genetic experiments, then, depend on the existence of a simple relationship between genotype and phenotype. That is why studies of DNA sequences are so important, because we can read off the genotype directly from this most basic of all phenotypes. In general, we hope to find a uniquely distinguishable phenotype for each genotype and only a simple genotype for each phenotype. At worst, when one allele is completely dominant, it may be necessary to perform a simple genetic cross to distinguish the heterozygote from the homozygote. Where possible, geneticists avoid studying genes that have only partial penetrance and incomplete expressivity (page 103) because of the difficulty of making genetic inferences from such traits. Imagine how difficult (if not impossible) it would have been for Benzer to study the fine structure of the gene in phage, if the only effect of the rII mutants had been to be able to grow on *E. coli* K almost as well as wild-type phage could. (See page 352.) For the most part, then, the study of genetics presented in the previous chapters has been the study of allelic substitutions that cause *qualitative* differences in phenotype.

However, the actual variation among organisms is usually quantitative, not qualitative. Wheat plants in a cultivated field or wild asters at the side of the road are not neatly sorted into categories of "tall" and "short," any more than humans are neatly sorted into categories of "black" and "white." Height, weight, shape, color, metabolic activity, reproductive rate, and behavior are characteristics that vary more or less continuously over a range (Figure 24-1). Even when the character is intrinsically countable (such as eye facet or bristle number in *Drosophila*), the number of distinguishable classes may be so large that the variation is nearly continuous. If we consider extreme individuals—say, a corn plant 8 feet tall and another one 3 feet tall—a cross between them will not reproduce a Mendelian result. Such a corn cross will produce plants about six feet tall, with some clear variation among siblings. The F_2 from selfing the F_1 will not fall into two or three discrete height classes in ratios of 3:1 or 1:2:1. Instead, the F_2 will be continuously distributed in height from one parental extreme to the other. This behavior of crosses is not an exception, but is the rule, for most characters in most species. Mendel obtained his simple results because he worked with horticultural varieties of the garden pea that differed from each other by single allelic differences that had drastic phenotypic effects. Had Mendel conducted his experiments on the natural variation of the weeds in his garden, instead of abnormal pea varieties, he would never have discovered Mendel's laws. In general, size, shape, color, physiological activity, and behavior do not assort in a simple way in crosses.

The fact that most phenotypic characters vary continuously does not mean that their variation is the result of some genetic mechanisms different from the Mendelian genes we have been dealing with. The continuity of phenotype is a result of two phenomena. First, each genotype does not have a single phenotypic expression but a norm of reaction (see Chapter 1) that covers a wide phenotypic range. As a result, the phenotypic differences

Figure 24-1 Quantitative inheritance of bract color in Indian paintbrush *(Castilleja hispida)*. The left photograph shows the extremes of the color range, and the right one shows examples from throughout the phenotypic range.

between genotypic classes become blurred, and we are not able to assign a particular phenotype unambiguously to a particular genotype. Second, many segregating loci may have alleles that make a difference to the phenotype being observed. Suppose, for example, that five equally important loci affect the number of flowers that will develop in an annual plant and that each locus has two alleles (call them + and −). For simplicity, also suppose that there is no dominance and that a + allele adds one flower whereas a − allele adds nothing. Thus, there are $3^5 = 243$ different possible genotypes [three possible genotypes (++,+−, and −−) at each of five loci], ranging from

$$\frac{+\,+\,+\,+\,+}{+\,+\,+\,+\,+}$$

through

$$\frac{+\,+\,+\,+\,+}{-\,-\,-\,-\,-} \quad \text{to} \quad \frac{-\,-\,-\,-\,-}{-\,-\,-\,-\,-}$$

but there are only 11 phenotypic classes (10, 9, 8, . . . ,0) because many of the genotypes will have the same numbers of + and − alleles. For example, although there is only one genotype with 10 + alleles and therefore an average phenotypic value of 10, there are 51 different genotypes with 5 + alleles and 5 − alleles, for example,

$$\frac{+\,+\,+\,+\,-}{+\,-\,-\,-\,-} \quad \text{and} \quad \frac{+\,+\,-\,+\,-}{+\,+\,-\,-\,-}.$$

Thus, many different genotypes may have the same average phenotype. At the same time, because of environmental variation, two individuals of the same genotype may not have the same phenotype. This lack of a one-to-one correspondence between genotype and phenotype obscures the underlying Mendelian mechanism. If we cannot study the behavior of the Mendelian factors controlling such traits directly, then what can we learn about their genetics?

Using current experimental techniques, geneticists can answer the following questions about the genetics of a continuously varying character in a population (say, height in a human population). These questions constitute the study of **quantitative genetics**—the study of the genetics of continuously varying characters:

1. Is the observed variation in the character influenced *at all* by genetic variation? Are there alleles segregating in the population that produce some differential effect on the character, or is all the variation simply the result of environmental variation and developmental noise (pages 13–15)?
2. If there is genetic variation, what are the norms of reaction of the various genotypes?
3. How important is genetic variation as a source of total phenotypic variation? Are the norms of reaction and the environments such that nearly all the variation is a consequence of environmental difference and developmental instabilities, or does genetic variation predominate?
4. Do many loci (or only a few) vary with respect to the character? How are they distributed over the genome?
5. How do the different loci interact with each other to influence the character? Is there dominance or epistasis (interaction among genes at different loci)?
6. Is there any nonnuclear inheritance (for example, any maternal effect)?

In the end, the purpose of asking these questions is to be able to predict what kinds of offspring will be produced by crosses of different phenotypes.

The precision with which these questions can be framed and answered varies greatly. In experimental organisms on the one hand, it is relatively simple to determine whether there is any genetic influence at all, but extremely laborious experiments are required to localize the genes (even approximately). In humans, on the other hand, it is extremely difficult to answer even the question of the presence of genetic influence for most traits because it is almost impossible to separate environmental from genetic effects in an organism that cannot be manipulated experimentally. As a consequence, we know a relatively large amount about the genetics of bristle number in *Drosophila* but essentially nothing about the genetics of complex human traits, except that a few (such as skin color) clearly are influenced by genes whereas others (such as the specific language spoken) clearly are not. It is the purpose of this chapter to develop the basic statistical and genetic concepts needed to answer these questions and to provide some examples of the applications of these concepts to particular characters in particular species.

Some Basic Statistical Notions

In order to consider the answers to these questions about the most common kinds of genetic variation, we must first examine a number of statistical tools that are essential in the study of quantitative genetics.

Distributions

The outcome of a cross for a Mendelian character can be described in terms of the proportions of the offspring that fall into several distinct phenotypic classes or often simply in terms of the presence or absence of a class. For

Figure 24-8 shows a scatter diagram of points for two variables, y and x, together with a straight line expressing the general linear trend of y with increasing x. This line, called the **regression line of y on x,** has been positioned so that the deviations of the points from the line are as small as possible. Specifically, if Δy is the distance of any point from the line in the y direction, then the line has been chosen so that

$$\Sigma(\Delta y)^2 = \text{a minimum}$$

Any other straight line passed through the points on the scatter diagram will have a larger total squared deviation of the points from it.

Obviously, we cannot find this **least-squares line** by trial and error. It turns out, however, that if the slope b of the line is calculated by

$$b = \frac{\text{cov } xy}{s_x^2}$$

and if a is then calculated from

$$a = \bar{y} - b\bar{x}$$

so that the line passes through the point $\bar{x},\bar{y}$, then these values of b and a will yield the least-squares prediction equation.

Note that the prediction equation cannot predict y exactly for a given x, because there is scatter around the line. The equation predicts the *average* y for a given x, if large samples are taken.

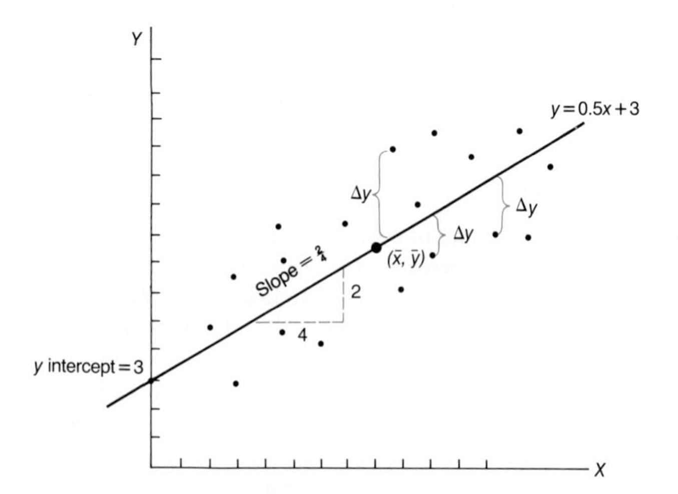

Figure 24-8 A scatter diagram showing the relation between two variables, x and y, with the regression line of y on x. This line, with a slope of $\frac{2}{4}$ minimizes the squares of the deviations (Δy).

Samples and Populations

The preceding sections have described the distributions and some statistics of particular assemblages of individuals that have been collected in some experiments or sets of observations. For some purposes, however, we are not really interested in the particular 100 undergraduates or 27 snakes that have been measured. Instead, we are interested in a wider world of phenomena, of which those particular individuals are representative. Thus, we might want to know the average height *in general* of undergraduates or the average seed weight *in general* of plants of the species *Crinum longifolium.* That is, we are interested in the characteristics of a **universe,** of which our small collection of observations is only a **sample.** The characteristics of any particular sample are, of course, not identical with those of the universe but vary from sample to sample. Two samples of undergraduates drawn from the universe of all undergraduates will not have exactly the same mean ($\bar{x}$) or variance (s^2), nor will these values for a sample typically be exactly equal to the mean and variance of the universe. We distinguish between the statistics of a sample and the values in the universe by using Roman letters, such as $\bar{x}$, s^2, and r, for sample values and Greek letters μ (mean), σ^2 (variance), and ρ (correlation) for the values in the universe.

The sample mean $\bar{x}$ is an approximation of the true mean μ of the universe. It is a statistical *estimate* of that true mean. So, too, s^2, s, and r are estimates of σ^2, σ, and ρ in the universe from which respective samples have been taken.

If we are interested in a particular collection of individuals not for its own sake but as a way of obtaining information about a universe, then we want the sample statistics such as $\bar{x}$, s^2, and r to be good estimates of the true values of μ, σ^2, and ρ, and so on. There are many criteria of what a "good" estimate is, but the one that seems clearly desirable is that if we take a very large number of samples, then the average value of the estimate over these samples should be the true value in the universe; that is, the estimate should be **unbiased.** It turns out that $\bar{x}$ is indeed an unbiased estimate of μ. If a very large number of samples are taken and $\bar{x}$ is calculated in each one, then the average of these $\bar{x}$ values will be μ. Unfortunately, s^2, as we have defined it, is not an unbiased estimate of σ^2. (It tends to be a little too small, so that the average of many s^2 values is less than σ^2.) The amount of bias is precisely related to the size N of each sample, and it can be shown that $[N/(N-1)]s^2$ is an unbiased estimate of σ^2. Thus, whenever we are interested in the variance of a set of measurements — not as a characteristic of the particular collection but as an estimate of a universe which the sample represents — then the appropriate quantity to use is $[N/(N-1)]s^2$ rather than s^2 itself. Note that this new quantity is equivalent to dividing the sum of squared deviations by $N-1$ instead of N in the first place, so that

$$\left(\frac{N}{N-1}\right)s^2 = \left(\frac{N}{N-1}\right)\frac{1}{N}\sum(x_i-\bar{x})^2$$

$$=\frac{1}{N-1}\sum(x_i-\bar{x})^2$$

All these considerations about bias also apply to the sample covariance. In the formula for the correlation coefficient (page 709), however, the factor $N/(N-1)$ would appear in both the numerator and the denominator and therefore cancel out, so we can ignore it for the purposes of computation.

Genotypes and Phenotypic Distribution

Using the concepts of distribution, mean, and variance, we can understand the difference between quantitative and Mendelian genetic traits. Suppose that a population of plants contains three genotypes, each of which has some differential effect on growth rate. Further, assume that there is some environmental variation from plant to plant because of inhomogeneity in the soil in which the population is growing and that there is some developmental noise (see page 15). For each genotype, there will be a separate distribution of phenotypes with a mean and a standard deviation that depend on the genotype and the set of environments. Suppose that these distributions look like the three height distributions in Figure 24-9a. Finally, assume that the population consists of a mixture of the three genotypes but in the unequal proportions 1:2:3 (*aa*:*Aa*:*AA*). Then the phenotypic distribution of individuals in the population as a whole will look like the black line in Figure 24-9b, which is the result of summing the three underlying separate genotypic distributions, weighted by their frequencies in the population. The mean of this total distribution is the average of the three genotypic means, again weighted by the frequencies of the genotypes in the population. The variance of the total distribution is produced partly by the environmental variation within each genotype and partly by the slightly different means of the three genotypes.

Two features of the total distribution are noteworthy. First, there is only a single mode. Despite the existence of three separate genotypic distributions underlying it, the population distribution as a whole does not reveal the separate modes. Second, any individual whose height lies between the two arrows could have come from any one of the three genotypes, because they overlap so much. The result is that we cannot carry out any simple Mendelian analysis to determine the genotype of an individual organism. For example, suppose that the three genotypes are the two homozygotes and the heterozygote for a pair of alleles at a locus. Let *aa* be the short homozygote and *AA* be the tall one, with the heterozygote being of intermediate height. Because there is so much overlap of the phenotypic distributions, we cannot know to which genotype a given individual belongs. Conversely, if we cross a homozygote *aa* and a heterozygote *Aa*, the offspring will not fall into two discrete classes in a 1:1 ratio but will cover almost the entire range of phenotypes smoothly. Thus, we cannot know that the cross is in fact *aa* × *Aa* and not *aa* × *AA* or *Aa* × *Aa*.

Suppose we grew the hypothetical plants in Figure 24-9 in an environment that exaggerated the differences between genotypes, for example, by doubling the growth rate of all genotypes. At the same time, we were very careful to provide all plants with exactly the same environment. Then, the phenotypic variance of each separate genotype would be reduced because all the plants are grown under identical conditions; at the same time the differences between genotypes would be exaggerated by the more rapid growth. The result (Figure 24-10) would be a separation of the population as a whole into three nonoverlapping phenotypic distributions, each characteristic of one genotype. We could now carry out a perfectly conventional Mendelian analysis of plant

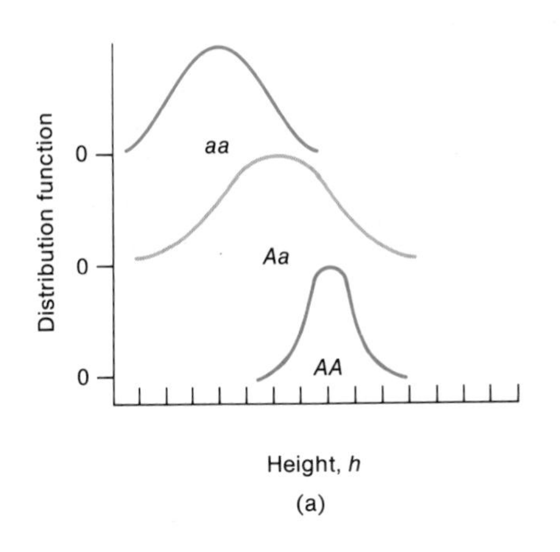

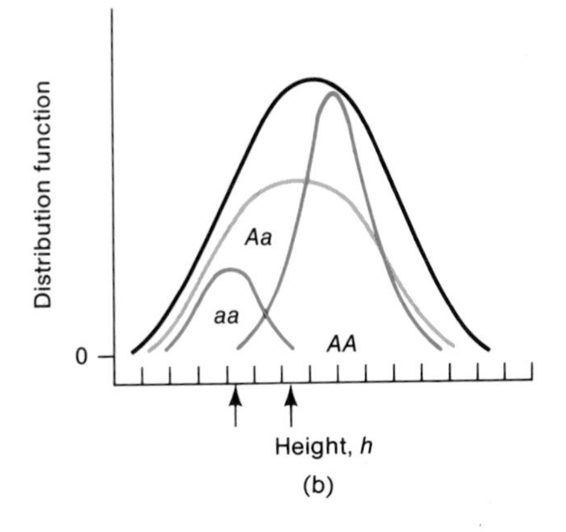

Figure 24-9 (a) Phenotypic distributions of three genotypes. (b) A population phenotypic distribution results from mixing individuals of the three genotypes in a proportion 1:2:3 (*aa*:*Aa*:*AA*).

gous genotypes found in natural populations. Only easily inbred species such as corn or *Drosophila* and clonal species such as strawberries have been studied to any degree. The outcomes of such studies resemble Figure 24-15. No genotype is consistently above or below other genotypes; instead, there are small differences among genotypes, and the direction of these differences is not consistent over a wide range of environments.

These facts have two important consequences. First, the selection of "superior" genotypes in domesticated animals and cultivated plants will result in very specifically adapted varieties that may not show their superior properties in other environments. To some extent, this problem is overcome by deliberately testing genotypes in a range of environments (for example, over several years and in several locations). It would be even better, however, if plant breeders could test their selections in a variety of controlled environments in which different environmental factors could be separately manipulated. The consequences of actual plant-breeding practices can be seen in Figure 24-16, where the yields of two varieties of corn are shown as a function of different farm environments. Variety 1 is an older variety of hybrid corn; variety 2 is a later "improved" hybrid. These performances are compared at a low planting density, which prevailed when variety 1 was developed, and at a high planting density characteristic of farming practice when hybrid 2 was selected. At the high density, the new variety is clearly superior to the old variety in all environments (Figure 24-16a). At the low density, however, the situation is quite different. First, note that the new variety is less sensitive to environment than the older hybrid, as evidenced by its flatter norm of reaction. Second, the new "improved" variety is actually poorer under the best farm conditions. Third, the yield improvement of the new variety is not apparent under the low densities characteristic of earlier agricultural practice.

The second consequence of the nature of reaction norms is that even if it should turn out that there is genetic variation for various mental and emotional traits in the human species, which is by no means clear, this variation is unlikely to favor one genotype over another across a range of environments. We must beware of hypothetical norms of reaction for human cognitive traits that show one genotype unconditionally superior to another. Even putting aside all questions of moral and political judgment, there is simply no basis for describing different human genotypes as "better" or "worse" on any scale, unless the investigator is able to make a very exact specification of environment.

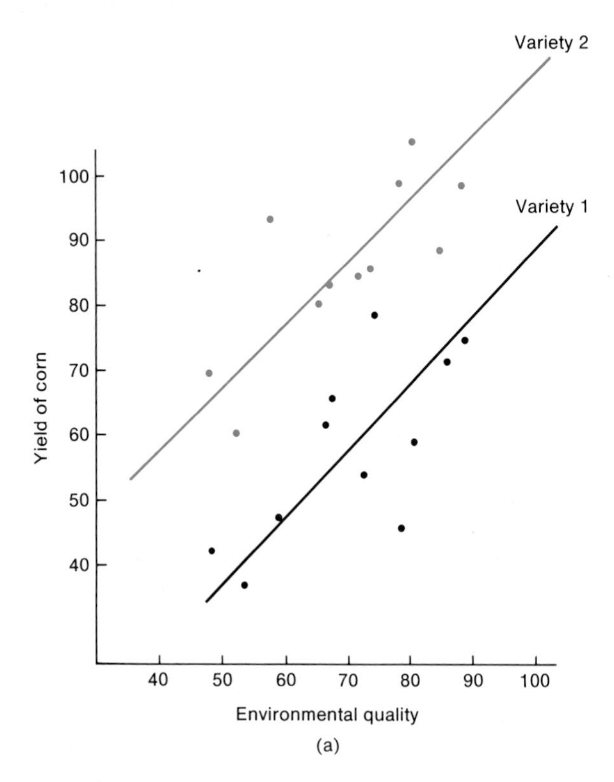

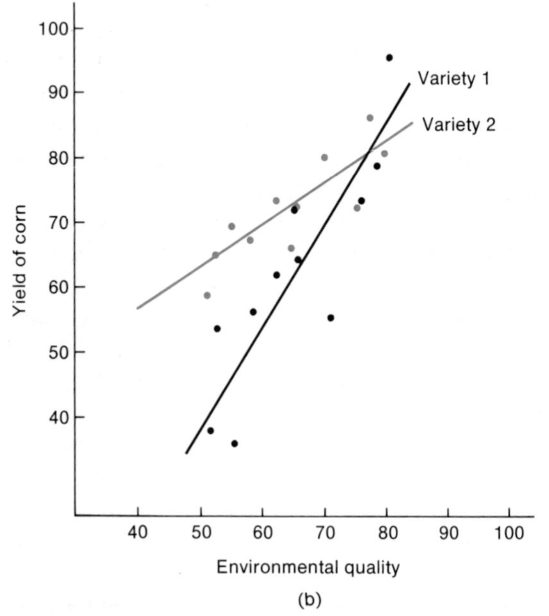

Figure 24-16 Yields of grain of two varieties of corn in different environments: (a) at a high planting density; (b) at a low planting density. (Data courtesy of W. A. Russell, *Proceedings of the 29th Annual Corn and Sorghum Research Conference,* 1974.)

Message Norm-of-reaction studies show only small differences among natural genotypes, and these differences are not consistent over a wide range of environments. Thus, "superior" genotypes in domesticated animals and cultivated plants may be superior only in certain environments. If it should turn out that humans exhibit genetic variation for various mental and emotional traits, this variation is unlikely to favor one genotype over another across a range of environments.

Quantifying Heritability

If a trait is shown to have some heritability in a population, then it is possible to quantify the degree of heritability. In Figure 24-9, we see that the variation among phenotypes in a population arises from two sources. First, there are average differences between the genotypes; second, each genotype exhibits phenotypic variance because of environmental variation. The total phenotypic variance of the population (s_p^2) can then be broken into two portions: the variance among genotypic means (s_g^2), and the remaining variance (s_e^2). The former is called the **genetic variance,** and the latter is called the **environmental variance;** however, as we shall see, these names are quite misleading. Moreover, the breakdown of the phenotypic variance into the sum of environmental and genetic variance leaves out the possibility of some covariance between genotype and environment. If the phenotype is the sum of a genetic and an environmental effect, $P = G + E$, then as we showed on page 711,

$$s_p^2 = s_g^2 + s_e^2 + 2 \text{ cov } ge$$

If genotypes are not distributed randomly across environments, there will be some covariance between genotype and environmental values and the covariance will be hidden in the genetic and environmental variances. For example, suppose it were true (we do not know) that there are genes that influence musical ability. Parents with such genes might themselves be musicians, who would create a more musical environment for their children, who would then have both the genes and the environment promoting musical performance. The result would be an increase in the phenotypic variances of musical ability and an erroneous estimate of genetic and environmental variances.

The degree of heritability can be defined as the portion of the total variance that is due to genetic variance:

$$H^2 = \frac{s_g^2}{s_p^2} = \frac{s_g^2}{s_g^2 + s_e^2}$$

H^2, so defined, is called the **broad heritability** of the character.

It must be stressed that this measure of "genetic influence" tells us what portion of the population's *variation* in phenotype can be assigned to *variation* in genotype. It does not tell us what portions of an *individual's* phenotype can be ascribed to its heredity and to its environment. This latter distinction is not a reasonable one. An individual's phenotype is a consequence of the interaction between its genes and its sequence of environments. It clearly would be silly to say that you owe 60 inches of your height to genes and 10 inches to environment. All measures of the "importance" of genes are framed in terms of the proportion of variance ascribable to their variation. This approach is a special application of the more general technique of the **analysis of variance** for apportioning relative weight to contributing causes. The method was, in fact, invented originally to deal with experiments in which different environmental and genetic factors were influencing the growth of plants. (For a sophisticated but accessible treatment of the analysis of variance written for biologists, see R. Sokal and J. Rohlf, *Biometry,* 2d ed., W. H. Freeman and Company, 1980.)

Methods of Estimating H^2

Genetic variance and heritability can be estimated in several ways. Most directly, we can obtain an estimate of s_e^2 by making a number of homozygous lines from the population, crossing them in pairs to reconstitute individual heterozygotes, and measuring the phenotypic variance *within* each heterozygous genotype. Because there is no genetic variance within a genotypic class, these variances will (when averaged) provide an estimate of s_e^2. This value then can be subtracted from the value of s_p^2 in the original population to give s_g^2. Using this method, any covariance between genotype and environment in the original population will be hidden in the estimate of genetic variance and inflate it.

Other estimates of genetic variance can be obtained by considering the genetic similarities between relatives. Using simple Mendelian principles, we can see that half the genes of full siblings will (on average) be identical. For identification purposes, we can label the alleles at a locus carried by the parents differently, so that they are, say, A_1A_2 and A_3A_4. Now the older sibling has a probability of $\frac{1}{2}$ of getting A_1 from its father, as does the younger sibling, so the two siblings have a chance of $\frac{1}{2} \times \frac{1}{2} = \frac{1}{4}$ of both carrying A_1. On the other hand, they might both have received an A_2 from their father, so again, they have a probability of $\frac{1}{4}$ of carrying a gene in common that they inherited from their father. Thus, the chance is $\frac{1}{4} + \frac{1}{4} = \frac{1}{2}$ that both siblings will carry an A_1 or that both siblings will carry an A_2. The other half of the time, one sibling will inherit an A_1 and the other will inherit an A_2. So, as far as paternally inherited genes are concerned, full siblings have a 50 percent chance of carrying the same allele. But the same reasoning applies to their maternally inherited gene. Averaging over their paternally and maternally inherited genes, half the genes of full siblings are identical between them. Their **genetic correlation,** which is equal to the chance that they carry the same allele, is $\frac{1}{2}$.

If we apply this reasoning to half-siblings, say, with a common father but with different mothers, we get a different result. Again, the two siblings have a 50 percent chance of inheriting an identical gene from their father, for a less active gene product or one with no activity at all and if one unit of gene product is sufficient to allow full physiological activity of the organism, then we would expect complete dominance of one allele over the other, as Mendel observed for flower color in peas. If, on the other hand, physiological activity is proportional to the amount of active gene product, we would expect the heterozygote phenotype to be exactly intermediate between the homozygotes (show no dominance).

For many quantitative traits, however, neither of these simple cases is the rule. In general, heterozygotes are not exactly intermediate between the two homozygotes but are closer to one or the other (show partial dominance). The complexity of biochemical and developmental pathways is such that the phenotype of a heterozygote may not be intermediate between the two homozygotes, even though there is an equal mixture of the primary products of the two alleles in the heterozygote. Indeed, in some cases, the heterozygote phenotype may lie outside the phenotypic range of the homozygotes altogether—a feature termed **overdominance.** For example, newborn babies who are intermediate in size have a higher chance of survival than very large or very small newborns. Thus, if survival were the phenotype of interest, heterozygotes for genes influencing growth rate would show overdominance.

Suppose that two alleles, *a* and *A*, segregate at a locus influencing height. In the environments encountered by the population, the mean phenotypes (heights) and frequencies of the three genotypes might be:

	aa	*Aa*	*AA*
Phenotype	10	18	20
Frequency	0.36	0.48	0.16

There is genetic variance in the population; the phenotypic means of the three genotypic classes are different. Some of the variance arises because there is an average effect on phenotype of substituting an allele *A* for an allele *a*; that is, the average height of all individuals with *A* alleles is greater than that of all individuals with *a* alleles. By defining the average effect of an allele as the average phenotype of all individuals that carry it, we necessarily make the average effect of the allele depend on the frequencies of the genotypes.

The average effect is calculated by simply counting the *a* and *A* alleles and multiplying them by the heights of the individuals in which they appear. Thus, 0.36 of all the individuals are homozygous *aa*, each *aa* individual has two *a* alleles, and the average height of *aa* individuals is 10 cm. Heterozygotes make up 0.48 of the population, each has only one *a* allele, and the average phenotypic measurement of *Aa* individuals is 18 cm. The total "number" of *a* alleles is 2(0.36) + 1(0.48). Thus, the average effect of all the *a* alleles is

$$\bar{a} = \text{average effect of } a = \frac{2(0.36)(10) + 1(0.48)(18)}{2(0.36) + 1(0.48)}$$

$$= 13.20 \text{ cm}$$

and, by a similar argument

$$\bar{A} = \text{average effect of } A = \frac{2(0.16)(20) + 1(0.48)(18)}{2(0.16) + 1(0.48)}$$

$$= 18.80 \text{ cm}$$

This average difference in effect between *A* and *a* alleles of 5.60 cm accounts for some of the variance in phenotype—but not for all of it. The heterozygote is not exactly intermediate between the homozygotes; there is some dominance.

We would like to separate that part of the variation among genotypes that is the result of substituting *a* alleles for *A* alleles, the so-called **additive effect** of the alleles, from the variation caused by dominance. The reason is that the effect of selective breeding depends on the additive variation and not on the variation caused by dominance. Thus, for purposes of plant and animal breeding or for making predictions about evolution by natural selection, we must determine the additive variation. An extreme example will illustrate the principle. Suppose that there is overdominance and that the phenotypic means and frequencies of three genotypes are:

	AA	*Aa*	*aa*
Phenotype	10	12	10
Frequency	0.25	0.50	0.25

It is apparent (and a calculation like the preceding one will confirm) that there is no average difference between the *a* and *A* alleles, because each has an effect of 11 units. So there is no *additive* variation although there is obviously variation in phenotype among the genotypes. The largest individuals are heterozygotes. If a breeder attempts to increase height in this population by selective breeding, mating these heterozygotes together will simply reconstitute the original population. Selection will be totally ineffective. This illustrates the general law that the effect of selection depends on the *additive* genetic variation and not on genetic variation in general.

We partition the total genetic variance in a population into **additive genetic variation** s_a^2, the variance that arises because there is an average difference between the carriers of *a* alleles and the carriers of *A* alleles, and a component called the **dominance variance** s_d^2, which results from the fact that heterozygotes are not exactly intermediate between the monozygotes. Thus

$$s_g^2 = s_a^2 + s_d^2$$

The components of variance in the first example, where $aa = 10$, $Aa = 18$, and $AA = 20$, can be calculated using the definitions of mean and variance developed earlier in this chapter. Remembering that a mean is the sum of the values of a variable, each weighted by the frequency with which that value occurs (see page 707), we can calculate the mean phenotype to be

$$\bar{x} = \Sigma f_i x_i = (0.36)(10) + (0.48)(18) + (0.16)(20)$$
$$= 15.44 \text{ cm}$$

The total genetic variance that arises from the variation among the mean phenotypes of the three genotypes is

$$s_g^2 = \Sigma f_i(x_i - \bar{x})^2 = (0.36)(10 - 15.44)^2 + (0.48)(18 - 15.44)^2 + (0.16)(20 - 15.44)^2$$
$$= 17.13 \text{ cm}^2$$

The frequency of allele a is (by counting alleles)

$$f_a = \frac{2(aa) + 1(Aa)}{2}$$
$$= \frac{2(0.36) + 1(0.48)}{2} = 0.60$$

and the frequency of the A allele is

$$f_A = \frac{2(AA) + 1(Aa)}{2}$$
$$= \frac{2(0.16) + 1(0.48)}{2} = 0.40$$

The variance of allelic means is then

$$s^2 = f_a(\bar{a} - \bar{x})^2 + f_A(\bar{A} - \bar{x})^2$$
$$= (.60)(13.20 - 15.44)^2 + (.40)(18.80 - 15.44)^2$$
$$= 7.525 \text{ cm}^2$$

But we want the variance among diploid individuals that results from the allelic effects, and every diploid individual carries two alleles, so

$$s_a^2 = (2)(7.525) = 15.05 \text{ cm}^2$$

and

$$s_d^2 = s_g^2 - s_a^2 = 17.13 - 15.05 = 2.08 \text{ cm}^2$$

The total phenotypic variance can now be written as

$$s_p^2 = s_g^2 + s_e^2 = s_a^2 + s_d^2 + s_e^2$$

We define a new kind of heritability, the **heritability in the narrow sense (h^2),** as

$$h^2 = \frac{s_a^2}{s_p^2} = \frac{s_a^2}{s_a^2 + s_d^2 + s_e^2}$$

It is this heritability, not to be confused with H^2, that is useful in determining whether a program of selective breeding will succeed in changing the population. The greater the h^2 is, the greater the difference is between selected parents and the population as a whole that will be preserved in the offspring of the selected parents.

Message The effect of selection depends on the amount of *additive* genetic variance and not on the genetic variance in general. Therefore, the narrow heritability h^2, not the broad heritability H^2, is relevant for a prediction of response to selection.

What has been described as the "dominance" variance is really more complicated. It is all the genetic variation that cannot be explained by the average effect of substituting A for a. If there is more than one locus affecting the character, then any epistatic interactions between loci will appear as variance not associated with the average effect of substituting alleles at the A locus. In principle, we can separate this **interaction variance (s_i^2)** from the dominance variance s_d^2. In practice, however, this cannot be done with any semblance of accuracy, so all the nonadditive variance appears as "dominance" variance.

Estimating Genetic Variance Components

Genetic components of variance can be estimated from covariance between relatives, but the derivation of these estimates is beyond the scope of an elementary text.

There is, however, another way to estimate h^2 that provides an insight into its real meaning. If we plot the phenotypes of the offspring against the average phenotypes of their two parents (the midparent value), we may observe a relationship like the one illustrated in Figure 24-20. The regression line will pass through the mean of all the parents and the mean of all the offspring, which will be equal to each other because no change has occurred in the population between generations. Moreover, taller parents have taller children and shorter parents have shorter children, so that the slope of the line is positive. But the slope is not unity; very short parents have children who are somewhat taller and very tall parents have children who are somewhat shorter than they themselves are. This slope of less than unity for the regression line arises because heritability is less than perfect. If the phenotype were additively inherited with

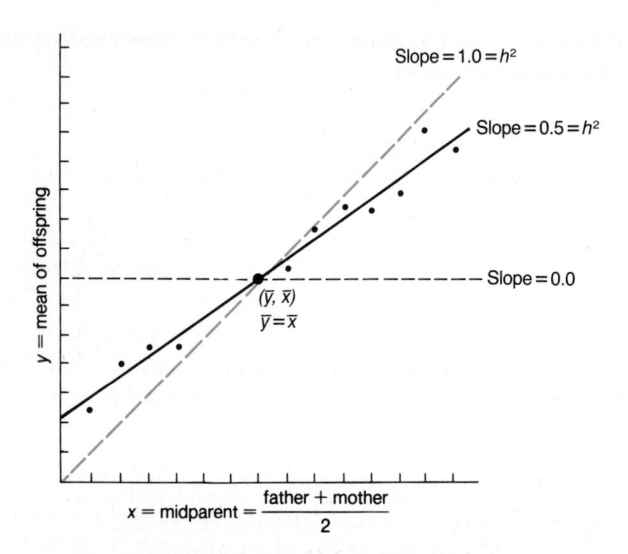

Figure 24-20 The regression (black line) of offspring measurements (y) on midparents (x) for a trait with narrow heritability (h^2) of 0.5. The blue line shows the regression slope if the trait were perfectly heritable.

complete fidelity, then the height of the offspring would be identical with the midparent value and the slope of the line would be 1. On the other hand, if the offspring had no heritable similarity to their parents, all parents would have offspring of the same average height and the slope of the line would be 0. This suggests that the slope of the regression line of the offspring value on the midparent value is an estimate of additive heritability. In fact, the relationship is precise.

The fact that the slope equals the additive heritability now allows us to use h^2 to predict the effects of artificial selection. Suppose that we select parents for the next generation who are on the average 2 units above the general mean of the population from which they were chosen. If $h^2 = 0.5$, then the offspring who form the next, selected generation will lie $0.5(2.0) = 1.0$ unit above the mean of the present population, since the regression coefficient predicts how much increase in y will result from a unit increase in x. We can define the **selection differential** as the difference between the selected parents and the unselected mean, and the **selection response** as the difference between their offspring and the previous generation. Then

$$\text{Selection response} = h^2 \times \text{selection differential}$$

or

$$h^2 = \frac{\text{selection response}}{\text{selection differential}}$$

The second expression provides us with yet another way to estimate h^2: by selecting for one generation and comparing the response with the selection differential. Usually this is carried out for several generations, and the average response is used.

Remember that any estimate of h^2, just as for H^2, depends on the assumption of no greater environmental correlation between closer relatives. Moreover, h^2 in one population in one set of environments will not be the same as h^2 in a different population at a different time. Figure 24-21 shows the range of heritabilities reported in various studies for a number of traits in chickens. The very small ranges are generally close to zero. For most traits for which a substantial heritability has been reported in some population, there are big differences from study to study.

Partitioning Environmental Variance

Environmental variance, like genetic variance, can also be further subdivided. In particular, developmental noise is usually confounded with environmental variance, but when a character can be measured on the left and right sides of an organism or over repeated body segments, it is possible to separate noise from environment. Table 24-3 shows the complete partitioning of variation for two

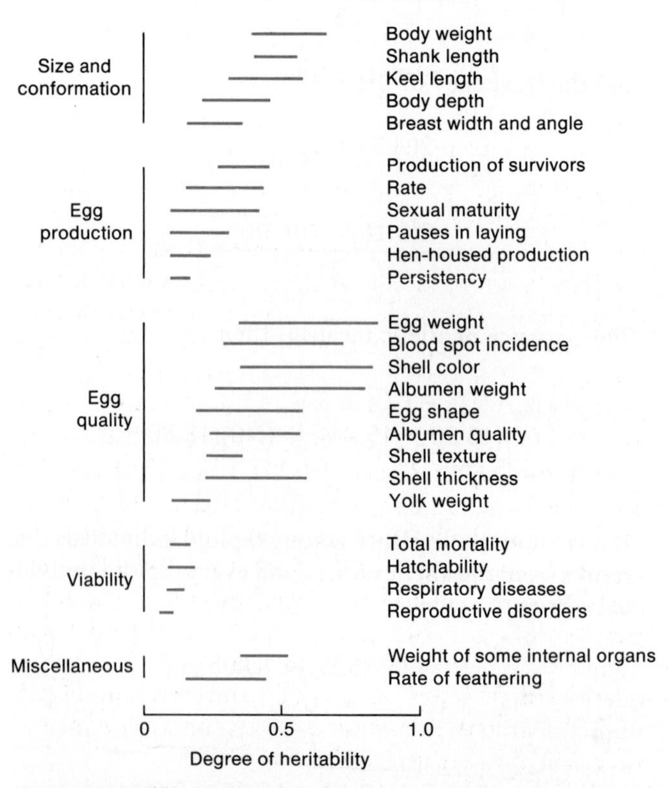

Figure 24-21 Ranges of heritabilities (h^2) reported for a variety of characters in chickens. (From I. M. Lenner and W. J. Libby, *Heredity, Evolution and Society.* Copyright 1976 by W. H. Freeman and Company.)

Table 24-3. Partition of Total Phenotypic Variance for Two Characters in Population of *Drosophila melanogaster*

Source of Variation		Percentage of Variance: Number of Abdominal Bristles	Percentage of Variance: Ovary Size
Additive genetic	s_a^2 } s_g^2	52	30
Dominance + epistatic variance	s_d^2 } s_g^2	9	40
Environmental variance	s_e^2 } s_e^2	1	3
Developmental noise	s_n^2 } s_e^2	38	27
Total	s_p^2	100	100

SOURCE: D. Falconer, *Quantitative Genetics.* Longman Group Limited. Copyright 1981.

characters in a population of *Drosophila melanogaster* raised under standard laboratory conditions. For each character, there is a substantial h^2 ($h^2 = s_a^2/s_p^2$), so we might expect selective breeding to increase or decrease bristle number and ovary size. Furthermore, nearly all the nongenetic variation is due to developmental noise. These values, however, are a consequence of the relatively rigorously controlled environment of the laboratory. Presumably s_e^2 in nature would be considerably larger, with a consequent diminution in the relative sizes of h^2 and of s_a^2. As always, such studies of variation are applicable only to a particular population in a given distribution of environments.

The Use of h^2 in Breeding

Even though h^2 is a number that applies only to a particular population and a given set of environments, it is still of great practical importance to breeders (Figure 24-22). A poultry geneticist interested in increasing, say, growth rate, is not concerned with the genetic variance over all possible flocks and all environmental distributions. Given a particular flock (or a choice between a few particular flocks) under the environmental conditions approximating present husbandry practice, the question becomes can a selection scheme be devised to increase growth rate and, if so, how fast? If one flock has a lot of genetic variance and another only a little, the breeder will choose the former to carry out selection. If the heritability in the chosen flock is very high, then the mean of the population will respond quickly to the selection imposed, because most of the superiority of the selected parents will appear in the offspring. The higher h^2 is, the higher the parent-offspring correlation is. If, on the other hand, h^2 is low, then only a small fraction of the increased growth rate of the selected parents will be reflected in the next generation.

If h^2 is very low, some alternative scheme of selection or husbandry may be needed. In this case, H^2 together with h^2 can be of use to the breeder. Suppose that h^2 and H^2 are both low. This means that there is a lot of environmental variance compared with genetic variance. Some scheme of reducing s_e^2 must be used. One method is to change the husbandry conditions so that environmental variance is lowered. Another is to use **family selection.** Rather than choosing the best individuals, the

Figure 24-22 Quantitative genetic theory has been extensively applied to poultry breeding. (Larry Lefever/Grant Heilman Photography, Inc.)

10. Using the concepts of norms of reaction, environmental distribution, genotypic distribution, and phenotypic distribution, try to restate the following statement in more exact terms: "80 percent of the difference in IQ performance between the two groups is genetic." What would it mean to talk about the heritability of a difference between two groups?

11. Describe an experimental protocol involving studies of relatives that could estimate the broad heritability of alcoholism. Remember that you must make an adequate observational definition of the trait itself!

12. A line selected for high bristle number in *Drosophila* has a mean of 25 sternopleural bristles, whereas a low-selected line has a mean of only 2. Marker stocks involving the two large autosomes II and III are used to create stocks with various mixtures of chromosomes from the high (h) and low (l) lines. The mean number of bristles for each chromosomal combination is as follows:

$\frac{h\ h}{h\ h}$ 25.1	$\frac{h\ h}{l\ h}$ 22.2	$\frac{l\ h}{l\ h}$ 19.0
$\frac{h\ h}{h\ l}$ 23.0	$\frac{h\ h}{l\ l}$ 19.9	$\frac{l\ h}{l\ l}$ 14.7
$\frac{h\ l}{h\ l}$ 11.8	$\frac{h\ l}{l\ l}$ 9.1	$\frac{l\ l}{l\ l}$ 2.3

What conclusions can you reach about the distribution of genetic factors and their actions from these data?

13. Suppose that number of eye facets is measured in a population of *Drosophila* under various temperature conditions. Further suppose that it is possible to estimate total genetic variance s_g^2 as well as the phenotypic distribution. Finally, suppose that there are only two genotypes in the population. Draw pairs of norms of reaction that would lead to the following results:

a. An increase in mean temperature decreases the phenotypic variance.

b. An increase in mean temperature increases H^2.

c. An increase in mean temperature increases s_g^2 but decreases H^2.

d. An increase in temperature *variance* changes a unimodal into a bimodal phenotypic distribution (one norm of reaction is sufficient here).

14. Francis Galton compared the heights of male undergraduates with the heights of their fathers, with the results shown in the following graph.

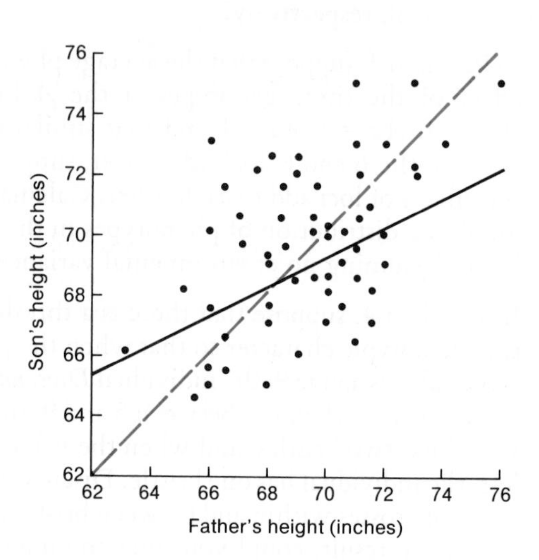

The average height of all fathers is the same as the average height of all sons, but the individual height classes are not equal across generations. The very tallest fathers had somewhat shorter sons, whereas the very short fathers had somewhat taller sons. As a result, the best line that can be drawn through the points on the scatter diagram has a slope of about 0.67 *(solid line)* rather than 1.00 *(dashed line)*. Galton used the term regression to describe this tendency for the phenotype of the sons to be closer than the phenotype of their fathers to the population mean.

a. Propose an explanation for this regression.

b. How are regression and heritability related here?

(Graph after W. F. Bodmer and L. L. Cavalli-Sforza, *Genetics, Evolution, and Man.* Copyright 1976 by W. H. Freeman and Company.)

25 Population Genetics

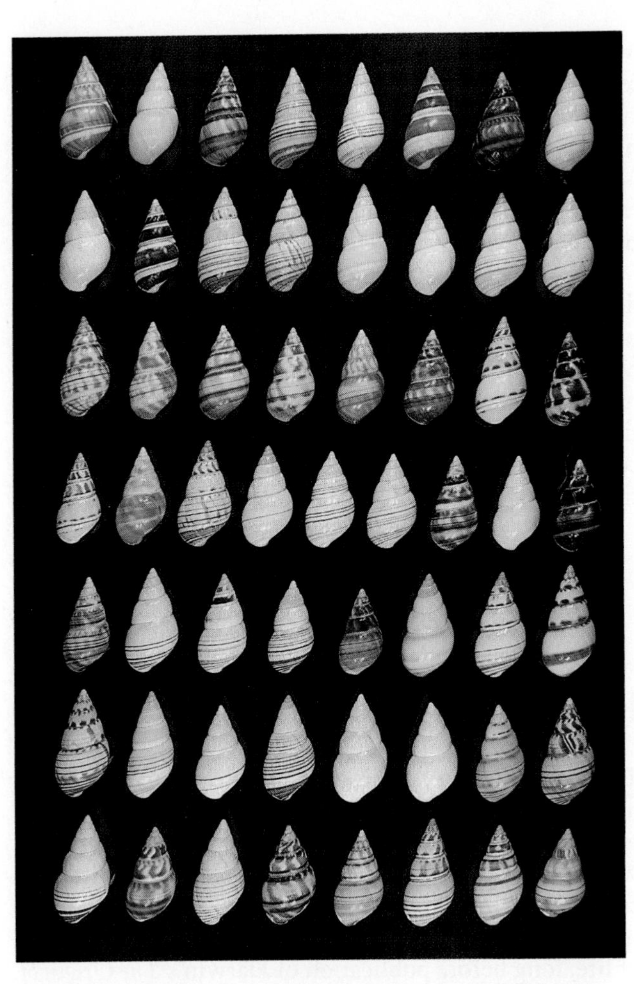

Shell color polymorphism in *Liguus fascitus.* (From David Hillis, *Journal of Heredity,* July–August 1991.)

KEY CONCEPTS

The goal of population genetics is to understand the genetic composition of a population and the forces that determine and change that composition.

▪

In any species, a great deal of genetic variation within and between populations arises from the existence of various alleles at different gene loci.

▪

A fundamental measurement in population genetics is the frequency at which the alleles can occur at any gene locus of interest.

▪

The frequency of a given allele in a population can be changed by recurrent mutation, selection, or migration, or by random sampling effects.

▪

In an idealized population, in which no forces of change are acting, a randomly interbreeding population would show constant genotypic frequencies for a given locus.

Mendel's investigations of heredity — indeed all the interest in heredity in the nineteenth century — arose from two related problems: how to breed improved crops and how to understand the nature and origin of species. What is common to these problems (and differentiates them from the problems of transmission and gene action) is that they are concerned with *populations* rather than with *individuals*. Studies of gene replication, protein synthesis, development, and chromosome movement focus on processes that go on within the cells of individual organisms. But the transformation of a species, either in the natural course of evolution or by the deliberate intervention of human beings, is a change in the properties of a collectivity — of an entire population or a set of populations.

Message Population genetics relates the heritable changes in populations of organisms to the underlying individual processes of inheritance and development.

Darwin's Revolution

The modern theory of evolution is so completely identified with the name of Charles Darwin (1809 – 1882) that many people think that the concept of organic evolution was first proposed by Darwin, but that is certainly not the case. Most scholars had abandoned the notion of fixed species, unchanged since their origin in a grand creation of life, long before publication of Darwin's *The Origin of Species* in 1859. By that time, most biologists agreed that new species arise through some process of evolution from older species; the problem was to explain *how* this evolution could occur.

Darwin's theory of the mechanism of evolution begins with the variation that exists among organisms within a species. Individuals of one generation are qualitatively different from one another. Evolution of the species as a whole results from the differential rates of survival and reproduction of the various types, so that the relative frequencies of the types change over time. Evolution, in this view, is a sorting process. For Darwin, evolution of the group resulted from the differential survival and reproduction of individual variants *already existing* in the group — variants arising in a way unrelated to the environment.

Message Darwin proposed a new explanation to account for the accepted phenomenon of evolution. He argued that the population of a given species at a given time includes individuals of varying characteristics. The population of the next generation will contain a higher frequency of those types that most successfully survive and reproduce under the existing environmental conditions. Thus, the frequencies of various types within the species will change over time.

There is an obvious similarity between the process of evolution as Darwin described it and the process by which the plant or animal breeder improves a domestic stock. The plant breeder selects the highest-yielding plants from the current population and (as far as possible) uses them as the parents of the next generation. If the characteristics causing the higher yield are heritable, then the next generation should produce a higher yield. It was no accident that Darwin chose the term **natural selection** to describe his model of evolution through differential rates of reproduction of different variants in the population. As a model for this evolutionary process, he had in mind the selection that breeders exercise on successive generations of domestic plants and animals.

We can summarize Darwin's theory of evolution through natural selection in three principles:

1. **Principle of variation.** Among individuals within any population, there is variation in morphology, physiology, and behavior.
2. **Principle of heredity.** Offspring resemble their parents more than they resemble unrelated individuals.
3. **Principle of selection.** Some forms are more successful at surviving and reproducing than other forms in a given environment.

Clearly, a selective process can produce change in the population composition only if there are some variations to select among. If all individuals are identical, no amount of differential reproduction of individuals can affect the composition of the population. Furthermore, the variation must be in some part heritable if differential reproduction is to alter the population's genetic composition. If large animals within a population have more offspring than small ones but their offspring are no larger on average than those of small animals, then no change in population composition can occur from one generation to another. Finally, if all variant types leave, on average, the same number of offspring, then we can expect the population to remain unchanged.

Message Darwin's principles of variation, heredity, and selection must hold true if there is to be evolution by a variational mechanism.

Variation and Its Modulation

Population genetics is the translation of Darwin's three principles into precise genetic terms. As such, it deals with the description of genetic variation in populations and with the experimental and theoretical determination of how that variation changes in time and space.

Message Population genetics is the study of inherited variation and its modulation in time and space.

Observations of Variation

Population genetics necessarily deals with genotypic variation, but by definition, only phenotypic variation can be observed. The relation between phenotype and genotype varies in simplicity from character to character. At one extreme, the phenotype may be the observed DNA sequence of a stretch of the genome. In this case, the distinction between genotype and phenotype disappears, and we can say that we are, in fact, directly observing the genotype. At the other extreme lie the bulk of characters of interest to plant and animal breeders and to most evolutionists — the variations in yield, growth rate, body shape, metabolic ratio, and behavior that constitute the obvious differences between varieties and species. These characters have a very complex relation to genotype, and we must use the methods introduced in Chapter 24 to say anything at all about the genotypes. But as we have seen in Chapter 24, it is not possible to make very precise statements about the genotypic variation underlying quantitative characters. For that reason, most of the study of experimental population genetics has concentrated on characters with simple relations to the genotype, much like the characters studied by Mendel. A favorite object of study for human population geneticists, for example, has been the various human blood groups. The qualitatively distinct phenotypes of a given blood group — say, the MN group — are coded for by alternative alleles at a single locus, and the phenotypes are insensitive to environmental variations.

The study of variation, then, consists of two stages. The first is a description of the phenotypic variation. The second is a translation of these phenotypes into genetic terms and the redescription of the variation genetically. If there is a perfect one-to-one correspondence between genotype and phenotype, then these two steps merge into one, as in the case of the MN blood group. If the relation is more complex — for example, as the result of dominance, so that heterozygotes resemble homozygotes, it may be necessary to carry out experimental crosses or to observe pedigrees to translate phenotypes into genotypes. This is the case for the human ABO blood group (see page 89).

The simplest description of Mendelian variation is the frequency distribution of genotypes in a population. Table 25-1 shows the frequency distribution of the three genotypes at the MN blood-group locus in several human populations. Note that there is variation both within and between populations. More typically, instead of the frequencies of the diploid genotypes, the frequencies of the alternative alleles are used. If f_{AA}, f_{Aa}, and f_{aa} are the proportions of the three genotypes at a locus with two alleles, then the frequencies $p(A)$ and $q(a)$ of the alleles are obtained by counting alleles. Since each homozygote AA consists only of A alleles and only half the alleles of each heterozygote Aa are type A, the total frequency (p) of A alleles in the population is

Table 25-1. Frequencies of Genotypes for Alleles at MN Blood-Group Locus in Various Human Populations

	Genotype			Allele frequencies	
Population	*MM*	*MN*	*NN*	*p(M)*	*q(N)*
Eskimo	0.835	0.156	0.009	0.913	0.087
Australian aborigine	0.024	0.304	0.672	0.176	0.824
Egyptian	0.278	0.489	0.233	0.523	0.477
German	0.297	0.507	0.196	0.550	0.450
Chinese	0.332	0.486	0.182	0.575	0.425
Nigerian	0.301	0.495	0.204	0.548	0.452

SOURCE: W. C. Boyd, *Genetics and the Races of Man*. D. C. Heath, 1950.

$$p = f_{AA} + \tfrac{1}{2}f_{Aa} = \text{frequency of } A$$

Similarly, the frequency q of a alleles is given by

$$q = f_{aa} + \tfrac{1}{2}f_{Aa} = \text{frequency of } a$$
$$p + q = f_{AA} + f_{aa} + f_{Aa} = 1.00$$

If there are multiple alleles, then the frequency for each allele is simply the frequency of its homozygote plus half the sum of the frequencies for all the heterozygotes in which it appears. Table 25-1 shows the values of p and q for each of the MN blood-group populations.

As an extension of p, which represents the **gene frequency** or **allele frequency,** we can describe variation at more than one locus simultaneously in terms of the **gametic frequencies.** Locus S (the secretor factor) is closely linked to the MN locus in humans. Table 25-2 shows the gametic frequencies of the four gametic types (MS, Ms, NS, and Ns) in various populations. The gametic frequency is obtained by summing up all the contributions of the different heterozygotes and homozygotes to the total pool of gametes. For example, the frequency of the MS gamete is given by

$$\begin{aligned} g(MS) &= \text{frequency of } MS/MS \\ &\quad + \tfrac{1}{2}\text{ frequency of } MS/NS + \tfrac{1}{2}\text{ frequency } MS/Ms \\ &\quad + \tfrac{1}{2}\text{ frequency of } MS/Ns \end{aligned}$$

Note that the last term in the sum involves the frequency of the double heterozygote MS/Ns. There is no contribution from the other double heterozygote Ms/NS, because it produces no MS gametes. If there were recombination, the Ms/NS heterozygote would produce MS gametes at a rate proportional to the recombination frac-

Table 25-2. Frequencies of Gametic Types for MNS System in Various Human Populations

Population	Gametic type: *MS*	*Ms*	*NS*	*Ns*	Heterozygosity (*H*): From gametes	From alleles
Ainu	0.024	0.381	0.247	0.348	0.672	0.438
Ugandan	0.134	0.357	0.071	0.438	0.658	0.412
Pakistan	0.177	0.405	0.127	0.291	0.704	0.455
English	0.247	0.283	0.080	0.390	0.700	0.469
Navaho	0.185	0.702	0.062	0.051	0.467	0.286

SOURCE: A. E. Mourant, *The Distribution of the Human Blood Groups.* Blackwell Scientific Pub., 1954.

tion. Thus, to give a gametic frequency description of a population at more than one locus simultaneously, we need to be able to distinguish coupling from repulsion double heterozygotes and to know the recombination fraction between the genes. For the human MNS system, there is essentially no recombination, and the different types of heterozygotes can be distinguished by pedigree analysis.

A *measure* of genetic variation (as opposed to its *description* by gene frequencies) is the amount of **heterozygosity** at a locus in a population, which is given by the total frequency of heterozygotes at a locus. If one allele is in very high frequency and all others are near zero, then there will be very little heterozygosity because, by necessity, most individuals will be homozygous for the common allele. We expect heterozygosity to be greatest when there are many alleles at a locus, all at equal frequency. In Table 25-1, the heterozygosity is simply equal to the frequency of the MN genotype in each population. When more than one locus is considered, there are two possible ways of calculating heterozygosity. First, we can average the frequency of heterozygotes at each locus separately. Alternatively, we can take the gametic frequencies, as in Table 25-2, and calculate the proportion of all individuals who carry two different gametic forms. The results of both calculations are given in Table 25-2. (See the discussion of Hardy-Weinberg equilibrium on page 750 for the calculation of heterozygosity.)

Simple Mendelian variation can be observed within and between populations of any species at various levels of phenotype, from external morphology down to the amino acid sequence of enzymes and other proteins. Indeed, with the new methods of DNA sequencing, variations in DNA sequence (such as third-position variants that are not differentially coded in amino acid sequences and even variations in nontranslated intervening sequences) have been observed. Every species of organism ever examined has revealed considerable genetic variation, or **polymorphism,** reflected at one or more levels of phenotype, either within populations or between populations, or both. Genetic variation that might be the basis for evolutionary change is ubiquitous. The tasks for population geneticists are to describe that ubiquitous variation quantitatively in terms that allow evolutionary predictions and to build a theory of evolutionary change that can use these observations in prediction.

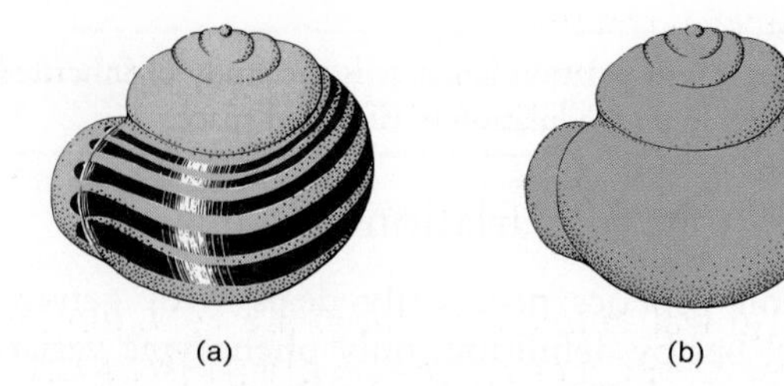

Figure 25-1 Shell patterns of the snail *Cepaea nemoralis:* (a) banded yellow; (b) unbanded pink.

It is quite impossible in this text to provide an adequate picture of the immense richness of genetic variation that exists in species. We can consider only a few examples of the different kinds of Mendelian variation to gain a superficial sense of the genetic diversity within species. Each of these examples can be multiplied many times over in other species and with other traits.

Morphological Variation. The shell of the land snail *Cepaea nemoralis* may be pink or yellow, depending on two alleles at a single locus, with pink dominant to yellow. Also, the shell may be banded or unbanded (Figure 25-1) as a result of segregation at a second linked locus, with unbanded dominant to banded. Table 25-3 shows the variation of these two loci in several European colonies of the snail. The populations also show polymorphism for the number of bands and the height of the shells, but these characters have complex genetic bases.

Examples of naturally occurring morphological variation within plant species are *Plectritis* (see Figure 1-8); *Collinsia* (blue-eyed Mary, page 46), and clover (see Figure 4-5).

Chromosomal Polymorphism. Although the karyotype is often regarded as a distinctive characteristic of a

Table 25-3. Frequencies of Snails *(Cepaea nemoralis)* with Different Shell Colors and Banding Patterns in Three French Populations

Population	Yellow: Banded	Yellow: Unbanded	Pink: Banded	Pink: Unbanded
Guyancourt	0.440	0.040	0.337	0.183
Lonchez	0.196	0.145	0.564	0.095
Peyresourde	0.175	0.662	0.100	0.062

SOURCE: Maxime Lamotte, *Bulletin Biologique de France et Belgique,* supplement 35, 1951.

Table 25-4. Frequencies of Plants with Supernumerary Chromosomes and of Translocation Heterozygotes in a Population of *Clarkia elegans* from California

No supernumeraries or translocations	Supernumeraries	Translocations	Both translocations and supernumeraries
0.560	0.265	0.133	0.042

SOURCE: H. Lewis, *Evolution* 5, 1951, 142–157.

species, in fact, numerous species are polymorphic for chromosome number and morphology. Extra chromosomes (supernumeraries), reciprocal translocations, and inversions segregate in many populations of plants, insects, and even mammals.

Table 25-4 gives the frequencies of supernumerary chromosomes and translocation heterozygotes in a population of the plant *Clarkia elegans* from California. The "typical" species karyotype would be hard to identify.

Immunological Polymorphism. A number of loci in vertebrates code for antigenic specificities such as the ABO blood types. Over 40 different specificities on human red cells are known, and several hundred are known in cattle. Another major polymorphism in humans is the HLA system of cellular antigens, which are implicated in tissue graft compatibility (Chapter 17). Table 25-5 gives the allelic frequencies for the ABO blood-group locus in some very different human populations. The polymorphism for the HLA system is vastly greater. There appear to be two main loci, each with five distinguishable alleles. Thus, there are $5^2 = 25$ different possible gametic types, making 25 different homozygous forms and $(25)(24)/2 = 300$ different heterozygotes. All genotypes are not phenotypically distinguishable, however, so only 121 phenotypic classes can be seen. L. L. Cavalli-Sforza and W. F. Bodmer report that, in a sample of only 100 Europeans, 53 of the 121 possible phenotypes were actually observed!

Protein Polymorphism. In recent years, studies of genetic polymorphism have been carried down to the level of the polypeptides coded by the structural genes themselves. If there is a nonredundant codon change in a structural gene (say, GGU to GAU), this will result in an amino acid substitution in the polypeptide produced at translation (in this case, glycine to aspartic acid). If a specific protein could be purified and sequenced from separate individuals, then it would be possible to detect genetic variation in a population at this level. In practice, this is tedious for large organisms and impossible for small ones unless a large mass of protein can be produced from a homozygous line.

There is, however, a practical substitute for sequencing that makes use of the change in the physical properties of a protein when an amino acid is substituted. Five amino acids (glutamic acid, aspartic acid, arginine, lysine, and histidine) have ionizable side chains that give a protein a characteristic net charge, depending on the pH of the surrounding medium. Amino acid substitutions may directly replace one of these charged amino acids, or a noncharged substitution near one of them in the polypeptide chain may affect the degree of ionization of the charged amino acid, or a substitution at the joining between two α helices may cause a slight shift in the three-dimensional packing of the folded polypeptide. In all these cases, the net charge on the polypeptide will be altered.

To detect the change in net charge, protein can be subjected to the method of gel electrophoresis. Figure 25-2 shows the outcome of such an electrophoretic sepa-

Table 25-5. Frequencies of the Alleles I^A, I^B, and i at the ABO Blood-Group Locus in Various Human Populations

Population	I^A	I^B	i
Eskimo	0.333	0.026	0.641
Sioux	0.035	0.010	0.955
Belgian	0.257	0.058	0.684
Japanese	0.279	0.172	0.549
Pygmy	0.227	0.219	0.554

SOURCE: W. C. Boyd, *Genetics and the Races of Man.* D. C. Heath, 1950.

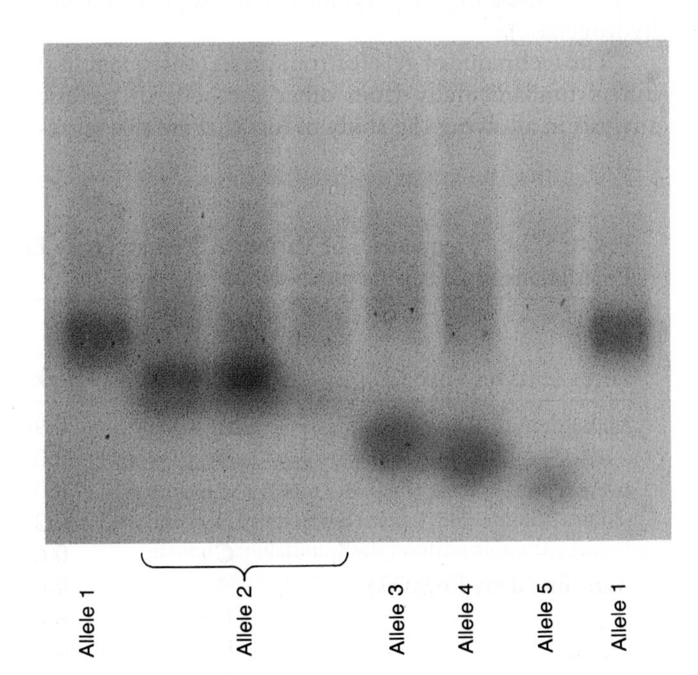

Figure 25-2 Electrophoretic gel showing homozygotes for five different alleles at the *esterase-5* locus in *Drosophila pseudoobscura.* Repeated samples of the same allele are identical, but there are repeatable differences between alleles.

The Effect of Sexual Reproduction on Variation

The evolutionary theorists of the nineteenth century encountered a fundamental difficulty in dealing with Darwin's theory of evolution through natural selection. The possibility of continued evolution by natural selection is limited by the amount of genetic variation. But biologists of the nineteenth century, including Darwin, believed in one form or another of **blending inheritance,** a model postulating that the characteristics of each offspring are some intermediate mixture of the parental characters. Such a model of inheritance has fatal implications for a theory of evolution that depends on variation.

Suppose that some trait (say, height) has a distribution in the population and that individuals mate more or less at random. If intermediate individuals mated with each other, they would produce only intermediate offspring according to a blending model. The mating of a tall with a short individual also would produce only intermediate offspring. Only the mating of tall with tall individuals and short with short individuals would preserve extreme types. The net result of all matings would be an increase in intermediate types and a decrease in extreme types. The variance of the distribution would shrink, simply as a result of sexual reproduction. In fact, it can be shown that the variance is *cut in half* in each generation, so that the population would be essentially uniformly intermediate in height before very many generations had passed. There then would be no variation on which natural selection could operate. This was a very serious problem for the early Darwinists; it made it necessary for Darwin to assume that new variation is generated at a very rapid rate by the inheritance of characters acquired by individuals during their lifetimes.

The rediscovery of Mendelism changed this picture completely. Because of the discrete nature of the Mendelian genes and the segregation of alleles at meiosis, a cross of intermediate with intermediate individuals does *not* result in all intermediate offspring. On the contrary, extreme types (homozygotes) segregate out of the cross. To see the consequence of Mendelian inheritance for genetic variation, consider a population in which males and females mate with each other at random with respect to some gene locus A; that is, individuals do not choose their mates preferentially with respect to the partial genotype at the locus. Such random mating is equivalent to mixing all the sperm and all the eggs in the population together and then matching randomly drawn sperm with randomly drawn eggs.

If the frequency of allele A is p in both the sperm and the eggs and the frequency of allele a is $q = 1 - p$, then the consequences of random unions of sperm and eggs are shown in Figure 25-9. The probability that both the sperm and the egg will carry A is $p \times p = p^2$, so this will

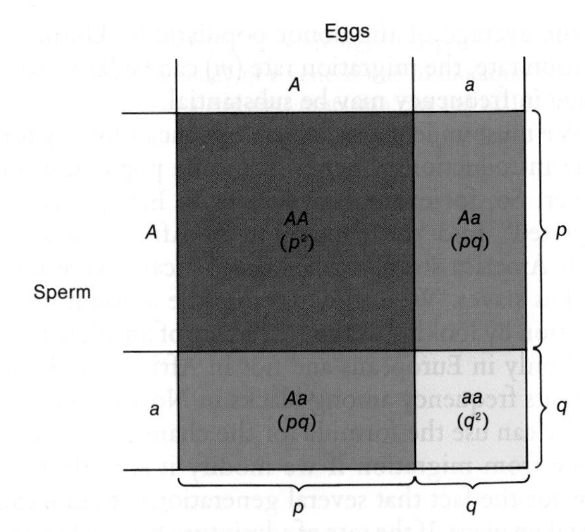

Figure 25-9 The Hardy-Weinberg equilibrium frequencies that result from random mating. The frequencies of A and a among both eggs and sperm are p and q ($=1 - p$), respectively. The total frequencies of the zygote genotypes are p^2 for AA, $2pq$ for Aa, and q^2 for aa. The frequency of the allele A in the zygotes is the frequency of AA plus half the frequency of Aa, or $p^2 + pq = p(p + q) = p$.

be the frequency of AA homozygotes in the next generation. In like manner, the chance of heterozygotes Aa will be $(p \times q) + (q \times p) = 2pq$, and the chance of homozygotes aa will be $q \times q = q^2$. The three genotypes, after a generation of random mating, will be in the frequencies p^2:$2pq$:q^2. As the figure shows, the allelic frequency of A has not changed and is still p. Therefore, in the second generation, the frequencies of the three genotypes will again be p^2:$2pq$:q^2, and so on, forever.

Message Mendelian segregation has the property that random mating results in an equilibrium distribution of genotypes after only one generation, so that genetic variation is maintained.

The equilibrium distribution

$$\begin{array}{ccc} AA & Aa & aa \\ p^2 & 2pq & q^2 \end{array}$$

is called the **Hardy-Weinberg equilibrium** after those who independently discovered it. (A third independent discovery was made by the Russian geneticist Sergei Tschetverikov.)

The Hardy-Weinberg equilibrium means that sexual reproduction does not cause a constant reduction in genetic variation in each generation; on the contrary, the amount of variation remains constant generation after generation, in the absence of other disturbing forces. The equilibrium is the direct consequence of the segregation of alleles at meiosis in heterozygotes.

Numerically, the equilibrium shows that irrespective of the particular mixture of genotypes in the parental generation, the genotypic distribution after one round of mating is completely specified by the allelic frequency p. For example, consider three hypothetical populations, all having the same frequency of $A (p = 0.3)$:

	AA	Aa	aa
I	0.3	0.0	0.7
II	0.2	0.2	0.6
III	0.1	0.4	0.5

After one generation of random mating, each of the three populations will have the same genotypic frequencies:

AA	Aa	aa
$(0.3)^2 = 0.09$	$2(0.3)(0.7) = 0.42$	$(0.7)^2 = 0.49$

and they will remain so indefinitely.

One consequence of the Hardy-Weinberg proportions is that rare alleles are virtually never in homozygous condition. An allele with a frequency of 0.001 occurs in homozygotes at a frequency of only one in a million; most copies of such rare alleles are found in heterozygotes. In general, since two copies of an allele are in homozygotes but only one copy of that allele is in each heterozygote, the relative frequency of the allele in heterozygotes (as opposed to homozygotes) is

$$\frac{2pq}{2q^2} = \frac{p}{q}$$

which for $q = 0.001$ is a ratio of 999:1. The general relation between homozygote and heterozygote frequencies as a function of allele frequencies is shown in Figure 25-10.

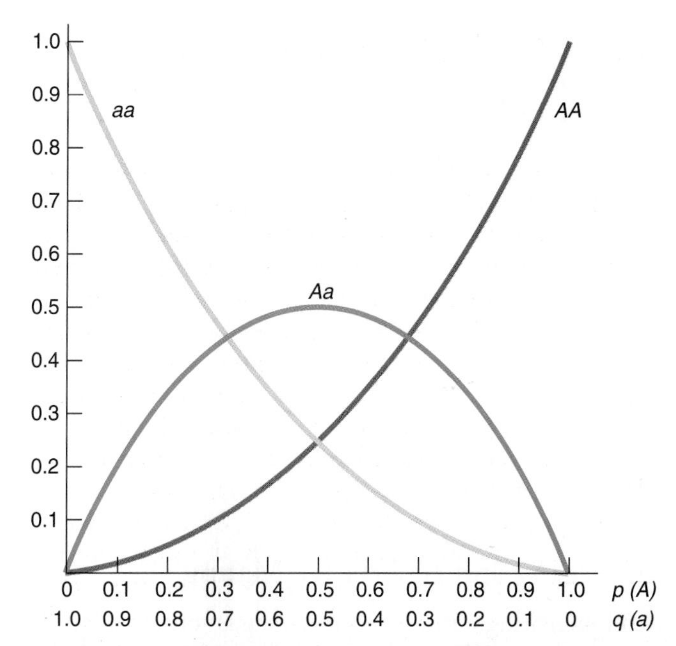

Figure 25-10 Curves showing the proportions of homozygotes AA (blue line), homozygotes aa (yellow line), and heterozygotes Aa (green line) in populations of different allelic frequencies if the populations are in Hardy-Weinberg equilibrium.

In our derivation of the equilibrium, we assumed that the allelic frequency p is the same in sperm and eggs. The Hardy-Weinberg equilibrium theorem does not apply to sex-linked genes if males and females start with unequal gene frequencies (see Problem 7 at the end of the chapter).

The Hardy-Weinberg equilibrium was derived on the assumption of "random mating," but we must carefully distinguish two meanings of that process. First, we may mean that individuals do not choose their mates on the basis of some heritable character. Human beings are random-mating with respect to blood groups in this first sense, because they generally do not know the blood type of their prospective mates, and even if they did, it is unlikely that blood type would be used as a criterion for choice. In the first sense, random mating will occur with respect to genes that have no effect on appearance, behavior, smell, or other characteristics that directly influence mate choice.

There is a second sense of random mating that is relevant when there is any subdivision of a species into subgroups. If there is genetic differentiation between subgroups so that the frequencies of alleles differ from group to group and if individuals tend to mate within their own subgroup **(endogamy),** then with respect to the species as a whole, mating is not at random and frequencies of genotypes will depart more or less from Hardy-Weinberg frequencies. In this sense, human beings are not random-mating, because ethnic and racial groups differ from each other in gene frequencies and people show high rates of endogamy, not only within major races but also within local ethnic groups. Spaniards and Russians differ in their ABO blood group frequencies, Spaniards marry Spaniards and Russians marry Russians, so there is unintentional nonrandom mating with respect to ABO blood groups. Table 25-11 shows random mating in the first sense and nonrandom mating in the second sense for the MN blood group. Within Eskimos, Egyptian, Chinese, and Australian subpopulations, females do not choose their mates by MN type, and, thus, Hardy-Weinberg equilibrium exists *within* the subpopulations. But Egyptians do not mate with Eskimos or Australian aborigines, so the nonrandom associations in the human species *as a whole* result in large differences in genotype frequencies and departure from Hardy-Weinberg equilibrium.

Table 25-11. Comparison between Observed Frequencies of Genotypes for the MN Blood-Group Locus and the Frequencies Expected from Random Mating

	Observed			Expected		
Population	*MM*	*MN*	*NN*	*MM*	*MN*	*NN*
Eskimo	0.835	0.156	0.009	0.834	0.159	0.008
Egyptian	0.278	0.489	0.233	0.274	0.499	0.228
Chinese	0.332	0.486	0.182	0.331	0.488	0.181
Australian aborigine	0.024	0.304	0.672	0.031	0.290	0.679

NOTE: The expected frequencies are computed according to the Hardy-Weinberg equilibrium, using the values of p and q computed from the observed frequencies.

Inbreeding and Assortative Mating

Random mating with respect to a locus is common, but it is not universal. Two kinds of deviation from random mating must be distinguished. First, individuals may mate with each other nonrandomly either because of their degree of common ancestry or their degree of genetic relationship. If mating between relatives occurs more commonly than would occur by pure chance, then the population is **inbreeding.** If mating between relatives is less common than would occur by chance, then the population is said to be undergoing **enforced outbreeding,** or **negative inbreeding.**

Second, individuals may tend to choose each other as mates, not because of their degree of genetic relationship but because of their degree of resemblance to each other at some locus. Bias toward mating of like with like is called **positive assortative mating.** Mating with unlike partners is called **negative assortative mating.** Assortative mating is never complete.

Inbreeding levels in natural populations are a consequence of geographical distribution, of the mechanism of reproduction, and of behavioral characteristics. If close relatives occupy adjacent areas, then simple proximity may result in inbreeding. The seeds of many plants, for example, fall very close to the parental source and the pollen is not widely spread, so a high frequency of sib mating occurs. Some plants (such as corn) can be self-pollinated as well as cross-pollinated, so that wind pollination results in some very close inbreeding. Yet other plants, like the peanut, are obligatorily selfed. Many small mammals (such as house mice) live and mate in restricted family groups that persist generation after generation. Humans, on the other hand, generally have complex mating taboos and proscriptions that reduce inbreeding.

Assortative mating for some traits is common. In humans, there is a positive assortative mating bias for skin color and height, for example. An important difference between assortative mating and inbreeding is that the former is specific to a trait whereas the latter applies to the entire genome. Individuals may mate assortatively with respect to height but at random with respect to blood group. Cousins, on the other hand, resemble each other genetically on the average to the same degree at all loci.

For both positive assortative mating and inbreeding, the consequence to population structure is the same: there is an increase in homozygosity above the level predicted by the Hardy-Weinberg equilibrium. If two individuals are related, they have at least one common ancestor. Thus, there is some chance that an allele carried by one of them and an allele carried by the other are both descended from the identical DNA molecule. The result is that there is an extra chance of **homozygosity by descent,** to be added to the chance of homozygosity ($p^2 + q^2$) that arises from the random mating of unrelated individuals. The probability of homozygosity by descent is called the **inbreeding coefficient (*F*).** Figure 25-11 illustrates the calculation of the probability of homozygosity by descent. Individuals I and II are full sibs because they share both parents. We label each allele in the parents uniquely to keep track of them. Individuals I and II mate to produce individual III. If individual I is A_1A_3 and the gamete that it contributes to III contains the allele A_1, then we would like to calculate the probability that the gamete produced by II is also A_1. The chance is $\frac{1}{2}$ that II will receive A_1 from its father, and if it does, the chance is $\frac{1}{2}$ that II will pass A_1 on to the gamete in question. Thus, the probability that III will receive an A_1 from II is $\frac{1}{2} \times \frac{1}{2} = \frac{1}{4}$, and this is the chance that III—the product of a full-sib mating—will be homozygous by descent.

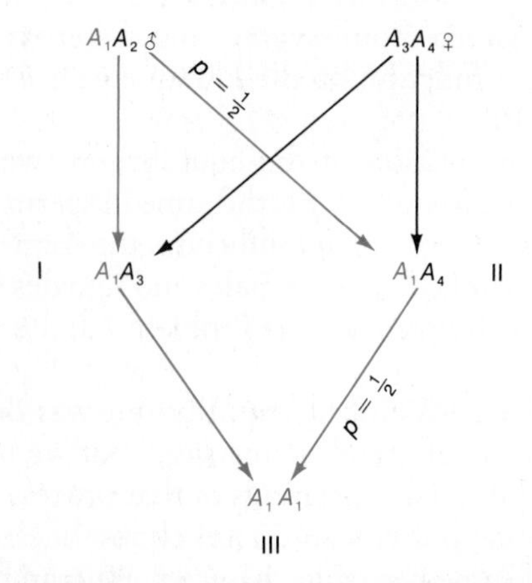

Figure 25-11 Calculation of homozygosity by descent for an offspring (III) of a brother-sister (I-II) mating. The probability that II will receive A_1 from its father is $\frac{1}{2}$; if it does, the probability that II will pass A_1 on to the generation producing III is $\frac{1}{2}$. Thus, the probability that III will receive an A_1 from II is $\frac{1}{2} \times \frac{1}{2} = \frac{1}{4}$.

Such close inbreeding can have deleterious consequences. Let's consider a rare deleterious allele a that, when homozygous, causes a metabolic disorder. If the frequency of the allele in the population is p, then the probability that a random couple will produce a homozygous offspring is only p^2 (from the Hardy-Weinberg equilibrium). Thus, if p is, say, $\frac{1}{1000}$, the frequency of homozygotes will be 1 in 1,000,000. Now suppose that the couple are brother and sister. If one of their common parents is a heterozygote for the disease, they may both receive it and may both pass it on to the offspring they produce. The probability of a homozygous aa offspring is

probability one or the other grandparent is Aa

× probability a is passed to male sib

× probability a is passed to female sib

× probability of a homozygous aa offspring from $Aa \times Aa$

$$= (2pq + 2pq) \times \tfrac{1}{2} \times \tfrac{1}{2} \times \tfrac{1}{4}$$

$$= \frac{pq}{4}$$

We assume that the chance that both grandparents are Aa is negligible. If p is very small, then q is nearly 1.0 and the chance of an affected offspring is close to $p/4$. For $p = \frac{1}{1000}$, there is 1 chance in 4000 of an affected child, compared to the one-in-a-million chance from a random mating. In general, for full sibs, the ratio of risks will be

$$\frac{p/4}{p^2} = \frac{1}{4p}$$

so the rarer the gene, the worse the *relative* risk of a defective offspring from inbreeding. For more distant relatives the chance of homozygosity by descent is, of course, less but still substantial. For first cousins, for example, the relative risk is $1/16p$ compared with random mating.

The population consequences of inbreeding depend on its intensity and form. Next we consider some examples. In experimental genetics (especially in plant and animal breeding), generation after generation of systematic selfing, full-sib, parent-offspring, or some other form of mating between relatives may be used to increase homozygosity. Such systematic inbreeding between close relatives eventually leads to complete homozygosity of the population, but at different rates. Referring to Table 24-2, we can see the amount of heterozygosity left within lines after various numbers of generations of inbreeding. Which allele is fixed within a line is a matter of chance. If, in the original population from which the inbred lines are taken, allele A has frequency p and allele a has frequency $q = 1 - p$, then a proportion p of the homozygous lines established by inbreeding will be homozygous AA and a proportion q of the lines will be aa. What inbreeding does is take the genetic variation present *within* the original population and convert it into variation *between* homozygous inbred lines sampled from the population (Figure 25-12).

In a natural population, there will be some fraction of mating between relatives (or even selfing if that is possible) because of spatial proximity. However, there is no continuity of inbreeding within any specific family. If some proportion of wind-pollinated plants are selfed in a particular generation, these are not necessarily the progeny of selfed plants in the previous generation; they are distributed at random over selfed and outcrossed progeny. A consequence of such random inbreeding is that there is an equilibrium frequency of homozygotes and heterozygotes similar to the Hardy-Weinberg equilibrium, but with more homozygotes. Thus, genetic variation is still preserved, in contrast to the result of systematic experimental inbreeding.

Suppose that a population is founded by some small number of individuals who mate at random to produce the next generation. Also assume that no further immigration into the population ever occurs again. (For example, the rabbits now in Australia probably have descended from a single introduction of a few animals in the nineteenth century.) In later generations, then, everyone is related to everyone else, because their family trees have common ancestors here and there in their

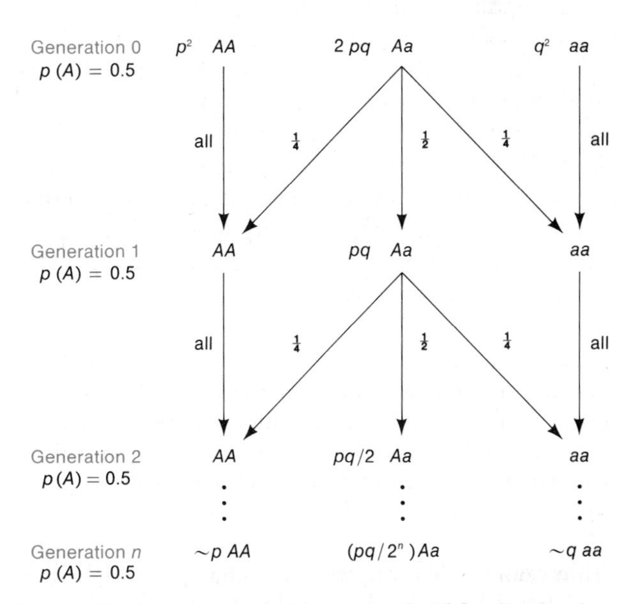

Figure 25-12 Repeated generations of self-fertilization (or inbreeding) will eventually split a heterozygous population into a series of completely homozygous lines. The frequency of AA lines among the homozygous lines will be equal to the frequency of allele A in the original heterozygous population.

Figure 25-14 (a) A blue jay eating a monarch butterfly, which (b) induces vomiting in the jay. Because of this experience, the jay later will refuse to eat a viceroy butterfly that is similar in appearance to the monarch, although jays that have never tried monarchs will eat the viceroys with no ill effects. (Photographs courtesy of Lincoln Brower.)

Furthermore, the environment is not a fixed situation that is experienced passively by the organism. The environment of an organism is defined by the activities of the organism itself. Dry grass is part of the environment of a junco, so juncos that are more efficient at gathering it may waste less energy in nest building and thus have a higher reproductive fitness. But dry grass is part of a junco's environment *because juncos gather it to make nests.* The rocks among which the grass grows are not part of the junco's environment, although the rocks are physically present there. But the rocks are part of the environment of thrushes; these birds use the rocks to break open snails. Moreover, the environment that is defined by the life activities of an organism evolves as a result of those activities. The structure of the soil that is in part determinative of the kinds of plants that will grow is altered by the growth of those very plants. Environment is both the cause and the result of the evolution of organisms. As primitive plants evolved photosynthesis, they changed the earth's atmosphere from one that had had essentially no free oxygen and a high concentration of carbon dioxide to the atmosphere that we know today, which contains 21 percent oxygen and only 0.03 percent carbon dioxide. Plants that evolve today must do so in an environment created by the evolution of their own ancestors.

Darwinian or reproductive fitness is not to be confused with "physical fitness" in the everyday sense of the term, although they may be related. No matter how strong, healthy, and mentally alert the possessor of a genotype may be, that genotype has a fitness of zero if, for some reason, the possessor is sterile. Thus, such statements as "the unfit are outreproducing the fit so the species may become extinct" are meaningless. The fitness of a genotype is a consequence of all the phenotypic effects of the genes involved. Thus, an allele that doubles the fecundity of its carriers but at the same time reduces the average lifetime of its possessors by 10 percent will be more fit than its alternatives, despite its life-shortening property. The most common example is parental care. An adult bird that expends a great deal of its energy gathering food for its young will have a lower probability of survival than one that keeps all the food for itself. But a totally selfish bird will leave no offspring because its young cannot fend for themselves. As a consequence, parental care is favored by natural selection.

Two Forms of the Struggle for Existence

Darwin saw the "struggle for existence" as having two quite different forms, with different consequences for fitness. In one form, the organism "struggles" with the environment directly. Darwin's example was the plant that is struggling for water at the edge of a desert. The fitness of a genotype in such a case does not depend on whether it is frequent or rare in the population, because fitness is not mediated through the interactions of individuals but is a direct consequence of the individual's physical relationship to the external environment. Fitness is then **frequency-independent.**

The other form of struggle is between organisms competing for a resource in short supply or otherwise interacting so that their relative abundances determine fitness. An example is mimicry in butterflies. Some species of butterflies (such as the brightly colored orange and black monarchs) are distasteful to birds, which learn, after a few trials, to avoid attacking them (Figure 25-14).

It is then advantageous for a palatable species (such as the viceroy butterfly) to evolve to look like the distasteful one, because birds will avoid the tasty mimics as well as the distasteful models. But as the frequency of the mimics increases, birds will increasingly have the experience that butterflies with this morphology are, in fact, good to eat. They will no longer avoid them, and the mimics will lose their fitness advantage. These are examples of **frequency-dependent fitness.**

For reasons of mathematical convenience, most models of natural selection are based on frequency-independent fitness. In actual fact, however, a very large number of selective processes (perhaps most) are frequency-dependent. The kinetics of the evolutionary process depend on the exact form of frequency dependence, and, for that reason alone, it is difficult to make any generalizations. The result of *positive* frequency dependence (such as competing predators, where fitness increases with increasing frequency) is quite different from the case of *negative* frequency dependence (such as the butterfly mimics, where fitness of a genotype declines with increasing frequency). For the sake of simplicity and to illustrate the main qualitative features of selection, we deal only with models of frequency-independent selection in this chapter, but convenience should not be confused with reality.

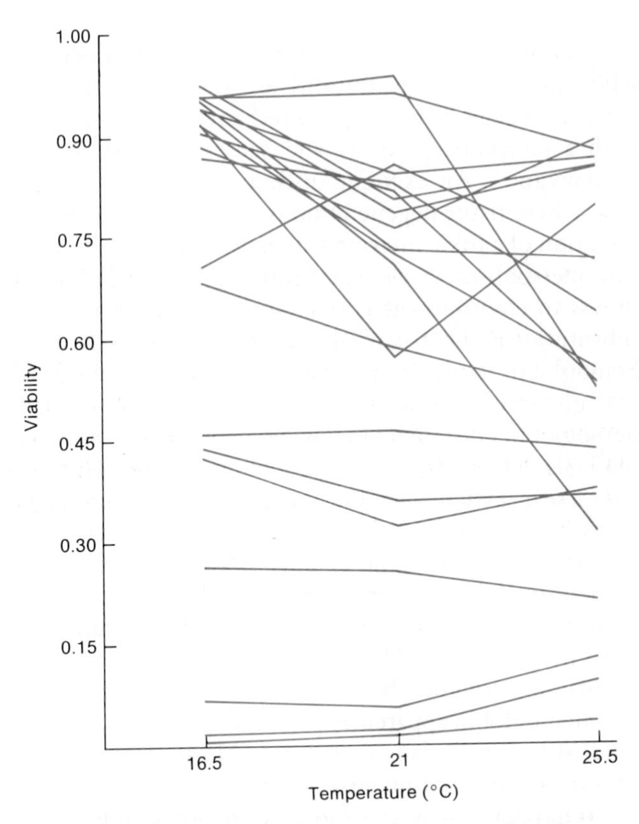

Figure 25-15 Viabilities of various chromosomal homozygotes of *Drosophila pseudoobscura* at three different temperatures.

Measuring Fitness Differences

For the most part, the differential fitness of different genotypes can be most easily measured when the genotypes differ at many loci. In very few cases (except for laboratory mutants, horticultural varieties, and major metabolic disorders) does the effect of an allelic substitution at a single locus make enough difference to the phenotype to be reflected in measurable fitness differences. Figure 25-15 shows the probability of survival from egg to adult—that is, the **viability**—of a number of second-chromosome homozygotes of *Drosophila pseudoobscura* at three different temperatures. As is generally the case, the fitness (in this case, a component of the total fitness, viability) is different in different environments. A few homozygotes are lethal or nearly so at all three temperatures, whereas a few have consistently high viability. Most genotypes, however, are not consistent in viability between temperatures, and no genotype is unconditionally the most fit at all temperatures. The fitness of these chromosomal homozygotes was not measured in competition with each other; all are measured against a common standard, so we do not know whether they are frequency-dependent. An example of frequency-dependent fitness is shown in the estimates for inversion homozygotes and heterozygotes of *Drosophila pseudoobscura* in Table 25-12.

Table 25-12. Comparison of Fitnesses for Inversion Homozygotes and Heterozygotes in Laboratory Populations of *Drosophila pseudoobscura* when Measured in Different Competitive Combinations

	Homozygotes			Heterozygotes		
Experiment	*ST / ST*	*AR / AR*	*CH / CH*	*ST / AR*	*ST / CH*	*AR / CH*
ST and *AR* alone	0.8	0.5	—	1.0	—	—
ST and *CH* alone	0.8	—	0.4	—	1.0	—
AR and *CH* alone	—	0.86	0.48	—	—	1.0
ST, *AR*, and, *CH* together	0.83	0.15	0.36	1.0	0.77	0.62

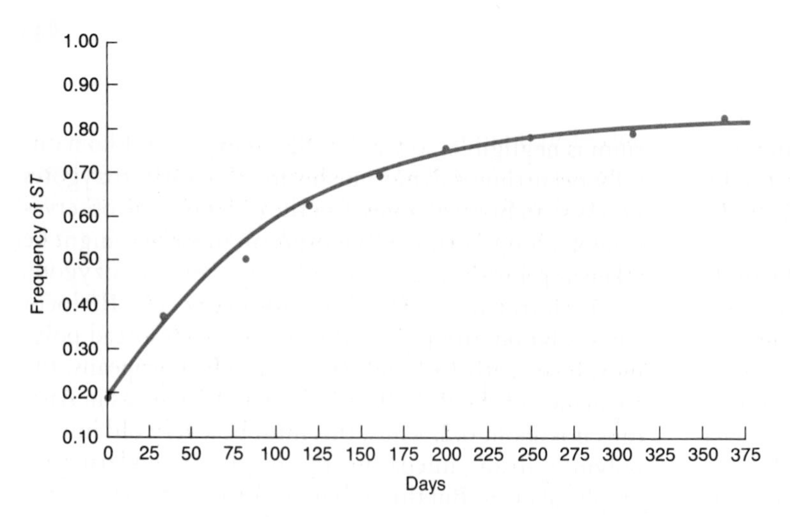

Figure 25-18 Changes in the frequency of the inversion Standard (*ST*) in competition with Chiricahua (*CH*) in a laboratory population *Drosophila pseudoobscura.* The points show the actual frequencies in successive generations. The solid line shows the theoretical course of change, if the fitnesses of the genotypes were as given in the text.

Multiple Adaptive Peaks

We must avoid taking an overly simplified view of the consequences of selection. At the level of the gene—or even at the level of the partial phenotype—the outcome of selection for a trait in a given environment is not unique. Selection to alter a trait (say, to increase size) may be successful in a number of ways. In 1952, F. Robertson and E. Reeve successfully selected to change wing size in *Drosophila* in two different populations. However, in one case the *number* of cells in the wing changed, whereas in the other case the *size* of the wing cells changed. Two different genotypes had been selected, both causing a change in wing size. The initial state of the population at the outset of selection determined which of these selections occurred.

The way in which the same selection can lead to different outcomes can most easily be illustrated by a simple hypothetical case. Suppose that the variation of two loci (there will usually be many more) influences a character and that (in a particular environment) intermediate phenotypes have the highest fitness. (For example, newborn babies have a higher chance of surviving birth if they are neither too big nor too small.) If the alleles act in a simple way in influencing the phenotype, then the three genetic constitutions $AaBb$, $AAbb$, and $aaBB$ will produce a high fitness because they will all be intermediate in phenotype. On the other hand, very low fitness will characterize the double homozygotes $AABB$ and $aabb$. What will the result of selection be? We can predict the result by using the mean fitness $\overline{W}$ of a population. As previously discussed, selection acts in most simple cases to increase $\overline{W}$. Therefore, if we calculate $\overline{W}$ for every possible combination of gene frequencies at the two loci, we can determine which combinations yield high values of $\overline{W}$. Then we should be able to predict the course of selection by following a curve of increasing $\overline{W}$.

The surface of mean fitness for all possible combinations of allelic frequency is called an **adaptive surface,** or an **adaptive landscape** (Figure 25-19). The figure is like a topographic map. The frequency of allele A at one locus is plotted on one axis, and the frequency of allele B

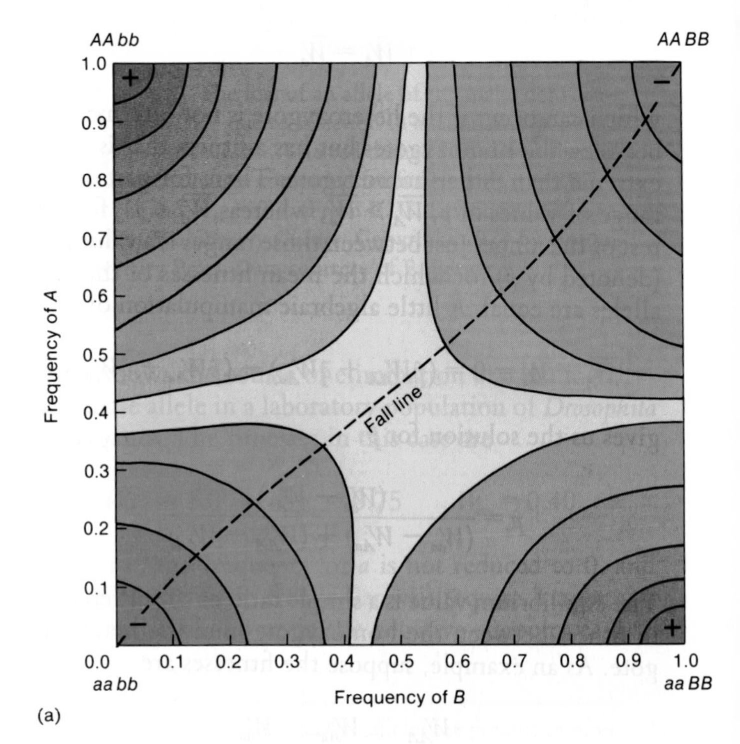

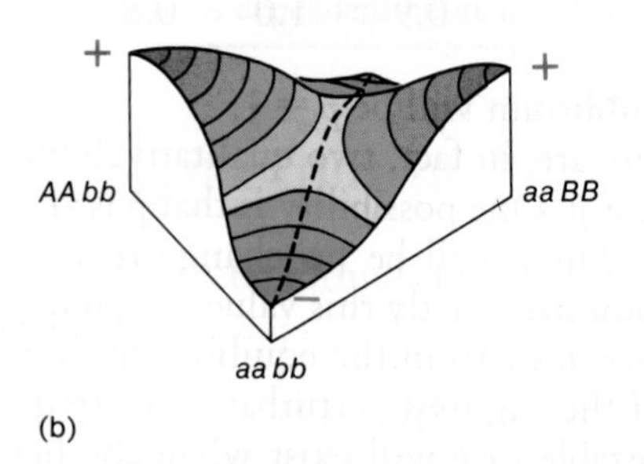

Figure 25-19 An adaptive landscape with two adaptive peaks (red), two adaptive valleys (blue), and a topographic saddle in the center of the landscape. The topographic lines are lines of equal mean fitness. If the genetic composition of a population always changes in such a way as to move the population "uphill" in the landscape, then the final composition will depend on where the population began with respect to the fall (dashed) line. (a) Topographic map of the adaptive landscape. (b) A perspective sketch of the surface shown in the map.

at the other locus is plotted on the other axis. The height above the plane (represented by topographic lines) is the value of $\overline{W}$ that the population would have for a particular combination of frequencies of A and B. According to the rule of increasing fitness, selection should carry the population from a low-fitness "valley" to a high-fitness "peak." However, Figure 25-19 shows that there are two adaptive peaks, corresponding to a fixed population of $AA\,bb$ and a fixed population of $aa\,BB$, with an adaptive valley between them. Which peak the population will ascend—and therefore what its final genetic composition will be—depends on whether the initial genetic composition of the population is on one side or the other of the dashed "fall line" shown in the figure.

Message Under identical conditions of natural selection, two populations may arrive at two different genetic compositions as a direct result of natural selection.

The existence of multiple adaptive peaks for a selective process means that some differences between species are the result of history and not of environmental differences. For example, African rhinoceroses have two horns, and Indian rhinoceroses have one (Figure 25-20). We need not invent a special story to explain why it is better to have two horns on the African plains and one in India. It is much more plausible that the trait of having horns was selected, but that two long, slender horns and one short, stout horn are simply alternative adaptive features, and that historical accident differentiated the species. Explanations of adaptations by natural selection do not require that every difference between species be differentially adaptive.

It is important to note that nothing in the theory of selection requires that the different adaptive peaks be of the same height. The kinetics of selection is such that $\overline{W}$ increases, not that it necessarily reaches the highest possible peak in the field of gene frequencies. Suppose, for example, that a population is near the peak $AA\,bb$ in Figure 25-19 and that this peak is lower than the $aa\,BB$ peak. Selection alone cannot carry the population to $aa\,BB$, because that would require a temporary decrease in $\overline{W}$ as the population descended the $AA\,bb$ slope, crossed the saddle, and ascended the other slope. Thus, the force of selection is myopic. It drives the population to a *local* maximum of $\overline{W}$ in the field of gene frequencies—not to a *global* one.

Artificial Selection

In contrast to the difficulties of finding simple, well-behaved cases in nature that exemplify the simple formulas of natural selection, there is a vast record of the effectiveness of artificial selection in changing populations phenotypically. These changes have been produced by laboratory selection experiments and by selection of

(a)

(b)

Figure 25-20 Differences in horn morphology in two geographically separated species of rhinoceros: (a) the African rhinoceros; (b) the Indian rhinoceros. (Part a from Anthony Bannister/NHPA; part b from K. Ghani/NHPA.)

Chapter 4

Bodmer, W. F., and L. L. Cavalli-Sforza. 1976. *Genetics, Evolution, and Man.* New York: W. H. Freeman and Co. A very readable, well-illustrated book, including a clear account of HLA genetics.

Griffiths, A. J. F., and F. R. Ganders. 1984. *Wildflower Genetics.* Vancouver: Flight Press. A field guide to plant variation in natural populations and its genetic basis, including examples relevant to this chapter.

Griffiths, A. J. F., and J. McPherson. 1989. *One Hundred + Principles of Genetics.* New York: W. H. Freeman and Co. A short book that distills genetics down to its basic key concepts; useful for rapid review of the whole subject.

Hutt, W. B. 1979. *Genetics for Dog Breeders.* New York, W. H. Freeman and Co. A short book that will make genetics immediately relevant to all dog owners and breeders.

Searle, A. G. 1968. *Comparative Genetics of Coat Color in Mammals.* New York: Academic Press. A classic treatment of the subject with many examples relevant to this and other chapters.

Silvers, W. K. 1979. *The Coat Colors of Mice.* New York: Springer-Verlag. A standard handbook on the subject, including many examples of gene interaction.

Wright, M., and S. Walters, eds. 1981. *The Book of the Cat.* New York: Summit Books. A fascinating book that has an excellent chapter on gene interaction in determining the coat colors of domestic cats.

Chapter 5

O'Brien, S. J., ed. 1984. *Genetic Maps.* Cold Spring Harbor Press. A compendium of the detailed maps of 80 well-analyzed organisms.

Peters, J. A., ed. 1959. *Classic Papers in Genetics.* Englewood Cliffs, N.J.: Prentice-Hall. A collection of important papers in the history of genetics.

White, R., et al. 1985. "Construction of Linkage Maps with DNA Markers for Human Chromosomes." *Nature* 313: 101–104. An extension of the techniques of this chapter to DNA markers.

Chapter 6

Finchman, J. R. S., P. R. Day, and A. Radford. 1979. *Fungal Genetics,* 3d ed. London: Blackwell. A large, standard technical work. Good for tetrad analysis.

Kemp, R. 1970. *Cell Division and Heredity.* London: Edward Arnold. A short, clear introduction to genetics. Good for map functions and tetrad analysis.

Murray, A. W., and J. W. Szostak. 1983. "Construction of Artificial Chromosomes in Yeast." *Nature* 305: 189–193. The first creation of new chromosomes by splicing together known telomere, centromere, replicator, and then DNA frgaments by recombinant DNA technology. Includes tetrad analysis of markers on the new chromosomes.

Puck, T. T., and F-T. Kao. 1982. "Somatic Cell Genetics and Its Application to Medicine." *Annual Review of Genetics* 16: 225–272. A technical but readable review.

Ruddle, F. H., and R. S. Kucherlapati. 1974. "Hybrid Cells and Human Genes." *Scientific American* (July). A popular account of the use of cell hybridization in mapping human genes.

Stahl, F. W. 1969. *The Mechanics of Inheritance,* 2d ed. Englewood Cliffs, N.J.: Prentice-Hall. A short introduction to genetics, including some advanced material presented with a novel approach.

Chapter 7

Induced Mutations—A Tool in Plant Research. 1981. Vienna: International Atomic Energy Agency. A collection of papers by eminent workers in agricultural genetics illustrating the practical uses of mutations in plant breeding.

Lawrence, C. W. 1971. *Cellular Radiobiology.* London: Edward Arnold. A short standard text.

Lindsley, D. L., and E. H. Grell. 1972. *Genetic Variations of* Drosophila melanogaster. Washington, D.C.: Carnegie Institute of Washington. A reference book for fruitfly researchers, containing all the thousands of known *Drosophila* mutations. Fascinating browsing for anyone interested in genetics generally and in the major contributions made by *Drosophila* research.

Neuffer, M. G., L. Jones, and M. S. Zuber. 1968. *The Mutants of Maize.* Madison: Crop Science Society of America. A color catalog of the many, and often bizarre, mutants used by corn geneticists.

Schull, W. J., et al. 1981. "Genetic Effect of the Atomic Bombs: A Reappraisal." *Science* 213: 1220–1227. A summary of all the indicators of potential genetic effects of the Hiroshima and Nagasaki explosions, concluding that "In no instance is there a statistically significant effect of parental exposure; but for all indicators the observed effect is in the direction suggested by the hypothesis that genetic damage resulted from the exposure."

deSerres, F. J., and A. Hollaender, eds. 1982. *Chemical Mutagens: Principles and Methods for Their Detection,* Vol. 7. New York: Plenum Press. One of a set of useful volumes on this important class of mutagens that is particularly relevant to human mutation.

Chapters 8 and 9

Dellarco, V. L., P. E. Voytek, and A. Hollaender. 1985. *Aneuploidy: Etiology and Mechanisms.* New York: Plenum. A collection of research summaries on aneuploidy in humans and experimental organisms.

Epstein, C. J., et al. 1983. "Recent Developments in Prenatal Diagnosis of Genetic Diseases and Birth Defects." *Annual Review of Genetics* 17: 49–83. Includes amniocentesis.

Feldman, M. G., and E. R. Sears. 1981. "The Wild Gene Resources of Wheat." *Scientific American* (January). A general discussion of the genomes of wheat and its relatives, and how new genes can be introduced.

Friedmann, T. 1971. "Prenatal Diagnosis of Genetic Disease." *Scientific American* (November). An early article on amniocentesis and its uses.

Fuchs, F. 1980. "Genetic Amniocentesis." *Scientific American* (August).

German, J., ed. 1974. *Chromosomes and Cancer.* New York: Wiley. A large technical work, but readable, describing the relation of chromosome changes and cancer.

deGrouchy, J., and C. Turleau. 1984. *Clinical Atlas of Human Chromosomes.* New York: Wiley. A systematic examination of all the human chromosomes and the aberrations associated with them.

Hassold, T. J., and P. A. Jacobs, 1984. "Trisomy in Man." *Annual Review of Genetics* 18: 69–98. A comprehensive summary of trisomy, including a discussion of the maternal age effect.

Hulse, J. H., and D. Spurgeon. 1974. "Triticale." *Scientific American* (August). An account of the development and possible benefits of this wheat-rye amphidiploid.

Lawrence, W. J. C. 1968. *Plant Breeding.* London: Edward Arnold (*Studies in Biology,* No. 12). A short introduction to the subject.

Mangelsdof, P. C. 1986. "The Origins of Corn." *Scientific American* (August).

Maniatis, T. E., et al. 1980. "The Molecular Genetics of Human Hemoglobins." *Annual Review of Genetics* 14: 145–178. A useful summary, which could be profitably read at this point in the course or after reading the material on molecular genetics.

Patterson, D. 1987. "The Causes of Down Syndrome." *Scientific American* (August).

Shepherd, J. F. 1982. "The Regeneration of Potato Plants from Protoplasts." *Scientific American* (May). A review by one of the leaders in this field.

Swanson, C. P., T. Mertz, and W. J. Young. 1967. *Cytogenetics.* Englewood Cliffs, N.J.: Prentice-Hall.

Chapter 10

Adelberg, E. A. 1966. *Papers on Bacterial Genetics.* Boston: Little, Brown.

Brock, T. D. 1990. *The Emergence of Bacterial Genetics.* Cold Spring Harbor, N.Y.: Cold Spring Harbor Laboratory Press. A complete and detailed treatise of the origins of bacterial genetics, with particular emphasis on the early history of the field.

Hayes, W. 1968. *The Genetics of Bacteria and Their Viruses,* 2d ed. New York: Wiley. The standard and classic text, written by a pioneer in the subject.

Lewin, B. 1977. *Gene Expression,* Vol. 1: *Bacterial Genomes.* New York: Wiley. An excellent set of volumes, all of which are relevant to various sections of this text.

Lewin, B. 1977. *Gene Expression,* Vol. 3: *Plasmids and Phages.* New York: Wiley.

Stent, G. S., and R. Calendar. 1978. *Molecular Genetics,* 2d ed. New York: W. H. Freeman and Co. A lucidly written account of the development of our present understanding of the subject, based mainly on experiments in bacteria and phage.

Chapter 11

Dickerson, R. E. 1983. "The DNA Helix and How It is Read." *Scientific American* (December). An article with some beautiful color models of DNA structures.

Kornberg, A., and T. Baker. 1992. *DNA Replication,* 2d ed. New York: W. H. Freeman and Co.

Wang, J. C. 1982. "DNA Topoisomerases." *Scientific American* (July). Diagrams different topological forms of DNA.

Watson, J. D. 1968. *The Double Helix.* New York: Atheneum. An enjoyable personal account of Watson and Crick's discovery, including the human dramas involved.

Chapter 12

Benzer, S. 1962. "The Fine Structure of the Gene." *Scientific American* (January). A popular version of the author's pioneer experiments.

Felsenfeld, G. 1985. "DNA." *Scientific American* (October).

Radman, M., and R. Wagner. 1988. "The High Fidelity of DNA Duplication." *Scientific American* (August).

Watson, J. D., et al. 1987. *The Molecular Biology of the Gene,* 4th ed. Menlo Park, Calif.: Benjamin/Cummings. A superb development of the subject, written in a highly readable style and well illustrated.

Yanofsky, C. 1967. "Gene Structure and Protein Structure." *Scientific American* (May). This article gives the details of colinearity at the molecular level.

Chapter 13

Crick, F. H. C. 1962. "The Genetic Code." *Scientific American* (October). This article and the following one are popular accounts of code-cracking experiments.

Crick, F. H. C. 1966. "The Genetic Code: III." *Scientific American* (October).

Darnell, J. E., Jr. 1985. "RNA." *Scientific American* (October).

Doolittle, R. F. 1985. "Proteins." *Scientific American* (October).

Lake, J. A. 1981. "The Ribosome." *Scientific American* (August). Three-dimensional model of the ribosome.

Lane, C. 1976. "Rabbit Haemoglobin from Frog Eggs." *Scientific American* (August). This article describes experiments illustrating the universality of the genetic system.

Lawn, R. M., and G. A. Vehar. 1986. "The Molecular Genetics of Hemophilia." *Scientific American* (March).

Miller, O. L. 1973. "The Visualization of Genes in Action." *Scientific American* (March). A discussion of electron microscopy of transcription and translation.

Moore, P. B. 1976. "Neutron-Scattering Studies of the Ribosome." *Scientific American* (October). This article gives the details of ribosome substructure.

Nirenberg, M. W. 1963. "The Genetic Code: II." *Scientific American* (March). Another account of early code-cracking experiments.

Radman, M., and R. Wagner. 1988. "The High Fidelity of DNA Duplication." *Scientific American* (August).

Rich, A., and S. H. Kim. 1978. "The Three-Dimensional Structure of Transfer RNA." *Scientific American* (January). A presentation of the experimental evidence behind the structure described in this chapter.

Weinberg, R. A. 1985. "The Molecules of Life." *Scientific American* (October).

Chapter 14

Britten, R. J., and D. Kohne. 1968. "Repeated Sequences in DNA." *Science* 161: 529–540. One of the important summaries of the theoretical basis for distinguishing DNAs by renaturation.

Broda, P. 1979. *Plasmids.* New York: W. H. Freeman and Co. One of the few technical books on the subject.

antimorph A mutant expressing some agent that antagonizes a normal gene product.
antiparallel A term used to describe the opposite orientations of the two strands of a DNA double helix; the 5′ end of one strand aligns with the 3′ end of the other strand.
AP sites Apurinic or apyrimidinic sites resulting from the loss of a purine or pyrimidine residue from the DNA.
Arg Arginine (an amino acid).
ascospore A sexual spore from certain fungus species in which spores are found in a sac called an ascus.
ascus In fungi, a sac that encloses a tetrad or an octad of ascospores.
asexual spores *See* **spore.**
Asn Asparagine (an amino acid).
Asp Aspartate (an amino acid).
ATP (adenosine triphosphate) The "energy molecule" of cells, synthesized mainly in mitochondria and chloroplasts; energy from the breakdown of ATP drives many important reactions in the cell.
attached X A pair of *Drosophila* X chromosomes joined at one end and inherited as a single unit.
attenuator A region adjacent to the structural genes of the *trp* operon; this region acts in the presence of tryptophan to reduce the rate of transcription from the structural genes.
autonomous controlling element A controlling element that apparently has both regulator and receptor functions combined in the single unit, which enters a gene and causes an unstable mutation.
autonomous replication sequence (ARS) A segment of a DNA molecule necessary for the initiation of its replication; generally a site recognized and bound by the proteins of the replication system.
autopolyploid A polyploid formed from the doubling of a single genome.
autoradiography A process in which radioactive materials are incorporated into cell structures, which are then placed next to a film or photographic emulsion, thus forming a pattern on the film corresponding to the location of the radioactive compounds within the cell.
autoregulation The control of the transcription of a gene by its own gene product.
autosome Any chromosome that is not a sex chromosome.
auxotroph A strain of microorganisms that will proliferate only when the medium is supplemented with some specific substance not required by wild-type organisms.

Back mutation *See* **reversion.**
bacteriophage (phage) A virus that infects bacteria.
balanced polymorphism Stable genetic polymorphism maintained by natural selection.
Balbiani ring A large chromosome puff.
Barr body A densely staining mass that represents an inactivated X chromosome.
base analog A chemical whose molecular structure mimics that of a DNA base; because of the mimicry, the analog may act as a mutagen.
bead theory The disproved hypothesis that genes are arranged on the chromosome like beads on a necklace, indivisible into smaller units of mutation and recombination.
bimodal distribution A statistical distribution having two modes.
binary fission The process in which a parent cell splits into two daughter cells of approximately equal size.
biparental zygote A *Chlamydomonas* zygote that contains cpDNA from both parents; such cells generally are rare.
blastoderm In an insect embryo, the layer of cells that completely surrounds an internal mass of yolk.
blastula An early developmental stage of lower vertebrate embryos, in which the embryo consists of a single layer of cells surrounding the central yolk.
blending inheritance A discredited model of inheritance suggesting that the characteristics of an individual result from the smooth blending of fluidlike influences from its parents.
brachydactyly A human phenotype of unusually short digits, generally inherited as an autosomal dominant.
branch migration The process by which a single "invading" DNA strand extends its partial pairing with its complementary strand as it displaces the resident strand.
bridging cross A cross made to transfer alleles between two sexually isolated species by first transferring the alleles to an intermediate species that is sexually compatible with both.
broad heritability (H^2) The proportion of total phenotypic variance at the population level that is contributed by genetic variance.
bud A daughter cell formed by mitosis in yeast; one daughter cell retains the cell wall of the parent, and the other (the bud) forms a new cell wall.
buoyant density A measure of the tendency of a substance to float in some other substance; large molecules are distinguished by their differing buoyant densities in some standard fluid. Measured by density-gradient ultracentrifugation.
Burkitt lymphoma A cancer of the lymphatic system manifested by tumors in the jaw, often associated with a translocation bringing a specific oncogene next to a novel regulatory element.

C Cytosine, or cytidine.
callus An undifferentiated clone of plant cells.
cAMP (cyclic adenosine monophosphate) A molecule that plays a key role in the regulation of various processes within the cell.
canalized character A character whose phenotype is kept within narrow boundaries even in the presence of disturbing environments or mutations.
cancer A syndrome that involves the uncontrolled and abnormal division of eukaryotic cells.
CAP (catabolite activator protein) A protein whose presence is necessary for the activation of the *lac* operon.
carbon source A nutrient (such as sugar) that provides carbon "skeletons" needed in the organism's synthesis of organic molecules.
carcinogen A substance that causes cancer.
carrier An individual who possesses a mutant allele but does not express it in the phenotype because of a dominant allelic partner; thus, an individual of genotype *Aa* is a carrier of *a* if there is complete dominance of *A* over *a*.

cassette model A model to explain mating-type interconversion in yeast. Information for both *a* and α mating types is assumed to be present as silent "cassettes"; a copy of either type of cassette may be transposed to the mating-type locus, where it is "played" (transcribed).

catabolite activator protein *See* **CAP.**

catabolite repression The inactivation of an operon caused by the presence of large amounts of the metabolic end product of the operon.

cation A positively charged ion (such as K^+).

cDNA *See* **complementary DNA.**

cell autonomous A genetic trait in multicellular organisms in which only genotypically mutant cells exhibit the mutant phenotype. Conversely, a *nonautonomous* trait is one in which genotypically mutant cells cause other cells (regardless of their genotype) to exhibit a mutant phenotype.

cell division The process by which two cells are formed from one.

cell fate The ultimate differentiated state to which a cell has become committed.

cell lineage A pedigree of cells related through asexual division.

cellular blastoderm The stage of blastoderm in insects after the nuclei have each been packaged in an individual cellular membrane.

centimorgan (cM) *See* **map unit.**

central dogma The hypothesis that information flows only from DNA to RNA to protein; although some exceptions are now known, the rule is generally valid.

centromere A kinetochore; the constricted region of a nuclear chromosome, to which the spindle fibers attach during division.

character Some attribute of individuals within a species for which various heritable differences can be defined.

character difference Alternative forms of the same attribute within a species.

chase *See* **pulse-chase experiment.**

chiasma (plural, **chiasmata**) A cross-shaped structure commonly observed between nonsister chromatids during meiosis; the site of crossing-over.

chimera *See* **mosaic.**

chi-square (χ^2) test A statistical test used to determine the probability of obtaining the observed results by chance, under a specific hypothesis.

chloroplast A chlorophyll-containing organelle in plants that is the site of photosynthesis.

chromatid One of the two side-by-side replicas produced by chromosome division.

chromatid conversion A type of gene conversion that is inferred from the existence of identical sister-spore pairs in a fungal octad that shows a non-Mendelian allele ratio.

chromatid inference A situation in which the occurrence of a crossover between any two nonsister chromatids can be shown to affect the probability of those chromatids being involved in other crossovers in the same meiosis.

chromatin The substance of chromosomes; now known to include DNA, chromosomal proteins, and chromosomal RNA.

chromocenter The point at which the polytene chromosomes appear to be attached together.

chromomere A small beadlike structure visible on a chromosome during prophase of meiosis and mitosis.

chromosome A linear end-to-end arrangement of genes and other DNA, sometimes with associated protein and RNA.

chromosome aberration Any type of change in the chromosome structure or number.

chromosome loss Failure of a chromosome to become incorporated into a daughter nucleus at cell division.

chromosome map *See* **linkage map.**

chromosome mutation Any type of change in the chromosome structure or number.

chromosome puff A swelling at a site along the length of a polytene chromosome; the site of active transcription.

chromosome rearrangement A chromosome mutation involving new juxtapositions of chromosome parts.

chromosome set The group of different chromosomes that carries the basic set of genetic information for a particular species.

chromosome theory of inheritance The unifying theory stating that inheritance patterns may be generally explained by assuming that genes are located in specific sites on chromosomes.

chromosome walking A method for the dissection of large portions of DNA, in which a cloned portion of DNA, usually eukaryotic, is used to screen recombinant DNA clones from the same genome bank for other clones containing neighboring sequences.

***cis* conformation** In a heterozygote involving two mutant sites within a gene or within a gene cluster, the arrangement $a_1 a_2/++$.

***cis* dominance** The ability of a gene to affect genes next to it on the same chromosome.

***cis-trans* test** A test to determine whether two mutant sites of a gene are in the same functional unit or gene.

cistron Originally defined as a functional genetic unit within which two mutations cannot complement. Now equated with the term gene, as the region of DNA that encodes a single polypeptide (or functional RNA molecule such as tRNA or rRNA).

clone (1) A group of genetically identical cells or individuals derived by asexual division from a common ancestor. (2) *(colloquial)* An individual formed by some asexual process so that it is genetically identical to its "parent." (3) *See* **DNA clone.**

cM (centimorgan) *See* **map unit.**

code dictionary A listing of the 64 possible codons and their translational meanings (the corresponding amino acids).

codominance The situation in which a heterozygote shows the phenotypic effects of both alleles equally.

codon A section of DNA (three nucleotide pairs in length) that codes for a single amino acid.

coefficient of coincidence The ratio of the observed number of double recombinants to the expected number.

cohesive ends Ends of DNA which are cut in a staggered pattern, and then can hydrogen-bond with complementary base sequences from other similarly formed ends.

cointegrate The product of the fusion of two circular elements to form a single, larger circle.

colinearity The correspondence between the location of a mutant site within a gene and the location of an amino

non-Mendelian ratio An unusual ratio of progeny phenotypes that does not reflect the simple operation of Mendel's laws; for example, mutant : wild ratios of 3 : 5, 5 : 3, 6 : 2, or 2 : 6 in tetrads indicate that gene conversion has occurred.

nonparental ditype (NPD) A tetrad type containing two different genotypes, both of which are recombinant.

nonsense codon A codon for which no normal tRNA molecule exists; the presence of a nonsense codon causes termination of translation (ending of the polypeptide chain). The three nonsense codons are called amber, ocher, and opal.

nonsense mutation A mutation that alters a gene so as to produce a nonsense codon.

nonsense suppressor A mutation that produces an altered tRNA that will insert an amino acid during translation in response to a nonsense codon.

norm of reaction The pattern of phenotypes produced by a given genotype under different environmental conditions.

Northern blot Transfer of electrophoretically separated RNA molecules from a gel onto an absorbent sheet, which is then immersed in a labeled probe that will bind to the RNA of interest.

NPD *See* **nonparental ditype.**

nu body *See* **nucleosome.**

nuclease An enzyme that can degrade DNA by breaking its phosphodiester bonds.

nucleoid A DNA mass within a chloroplast or mitochondrion.

nucleolar organizer A region (or regions) of the chromosome set physically associated with the nucleolus and containing rRNA genes.

nucleolus An organelle found in the nucleus, containing rRNA and amplified multiple copies of the genes coding for rRNA.

nucleoside A nitrogen base bound to a sugar molecule.

nucleosome A nu body; the basic unit of eukaryotic chromosome structure; a ball of eight histone molecules wrapped about by two coils of DNA.

nucleotide A molecule composed of a nitrogen base, a sugar, and a phosphate group; the basic building block of nucleic acids.

nucleotide pair A pair of nucleotides (one in each strand of DNA) that are joined by hydrogen bonds.

nucleotide-pair substitution The replacement of a specific nucleotide pair by a different pair; often mutagenic.

null allele An allele whose effect is either an absence of normal gene product at the molecular level or an absence of normal function at the phenotypic level.

nullisomic A cell or individual with one chromosomal type missing, with a chromosome number such as $n - 1$ or $2n - 2$.

nurse cells The sister cells of the oocyte in insects. The nurse cells produce the bulk of the cytoplasmic contents of the mature oocyte.

ocher codon The codon UAA, a nonsense codon.

octad An ascus containing eight ascospores, produced in species in which the tetrad normally undergoes a postmeiotic mitotic division.

oncogene A gene that contributes to the production of a cancer. Oncogenes are generally mutated forms of normal cellular genes.

opal codon The codon UGA, a nonsense codon.

open reading frame *See* **ORF.**

operator A DNA region at one end of an operon that acts as the binding site for repressor protein.

operon A set of adjacent structural genes whose mRNA is synthesized in one piece, plus the adjacent regulatory signals that affect transcription of the structural genes.

ORF (open reading frame) A section of a sequenced piece of DNA that begins with a start codon and ends with a stop codon; it is presumed to be the coding sequence of a gene.

organogenesis The production of organ systems during animal embryogenesis.

organelle A subcellular structure having a specialized function — for example, the mitochondrion, the chloroplast, or the spindle apparatus.

origin of replication The point of specific sequence at which DNA replication is initiated.

overdominance A phenotypic relation in which the phenotypic expression of the heterozygote is greater than that of either homozygote.

P element A *Drosophila* transposable element that has been used as a tool for insertional mutagenesis and for germline transformation.

paracentric inversion An inversion not involving the centromere.

parental ditype (PD) A tetrad type containing two different genotypes, both of which are parental.

parologous genes Two genes or clusters of genes at different chromosomal locations in the same organism that have structural similarities indicating that they derived from a common ancestral gene.

parthenogenesis The production of offspring by a female with no genetic contribution from a male.

partial diploid *See* **merozygote.**

particulate inheritance The model proposing that genetic information is transmitted from one generation to the next in discrete units ("particles"), so that the character of the offspring is not a smooth blend of essences from the parents (*compare* **blending inheritance**).

pathogen An organism that causes disease in another organism.

patroclinous inheritance Inheritance in which all offspring have the nucleus-based phenotype of the father.

pattern formation The developmental processes by which the complex shape and structure of higher organisms occurs.

PD *See* **parental ditype.**

pedigree A "family tree," drawn with standard genetic symbols, showing inheritance patterns for specific phenotypic characters.

penetrance The proportion of individuals with a specific genotype who manifest that genotype at the phenotype level.

peptide *See* **amino acid.**

peptide bond A bond joining two amino acids.

pericentric inversion An inversion that involves the centromere.

permissive conditions Those environmental conditions under which a conditional mutant shows the wild-type phenotype.

petite A yeast mutation producing small colonies and altered mitochondrial functions. In cytoplasmic petites (neutral and suppressive petites), the mutation is a deletion in mitochondrial DNA; in segregational petites, the mutation occurs in nuclear DNA.

phage *See* **bacteriophage.**

Phe Phenylalanine (an amino acid).

phenocopy An environmentally induced phenotype that resembles the phenotype produced by a mutation.

phenotype (1) The form taken by some character (or group of characters) in a specific individual. (2) The detectable outward manifestations of a specific genotype.

phenotypic sex determination Sex determination by nongenetic means.

phenylketonuria (PKU) A human metabolic disease caused by a mutation in a gene coding for a phenylalanine-processing enzyme, which leads to mental retardation if not treated; inherited as an autosomal recessive phenotype.

Philadelphia chromosome A translocation between the long arms of chromosomes 9 and 22, often found in the white blood cells of patients with chronic myeloid leukemia.

phosphodiester bond A bond between a sugar group and a phosphate group; such bonds form the sugar-phosphate backbone of DNA.

piebald A mammalian phenotype in which patches of skin are unpigmented because of lack of melanocytes; generally inherited as an autosomal dominant.

pilus (plural, pili) A conjugation tube; a hollow hairlike appendage of a donor *E. coli* cell that acts as a bridge for transmission of donor DNA to the recipient cell during conjugation.

plant breeding The application of genetic analysis to development of plant lines better suited for human purposes.

plaque A clear area on a bacterial lawn, left by lysis of the bacteria through progressive infections by a phage and its descendants.

plasmid Autonomously replicating extrachromosomal DNA molecule.

plate (1) A flat dish used to culture microbes. (2) To spread cells over the surface of solid medium in a plate.

pleiotropic mutation A mutation that has effects on several different characters.

point mutation A mutation that can be mapped to one specific locus.

Poisson distribution A mathematical expression giving the probability of observing various numbers of a particular event in a sample when the mean probability of an event on any one trial is very small.

poky A slow-growing mitochondrial mutant in *Neurospora.*

polar gene conversion A gradient of conversion frequency along the length of a gene.

polar granules Cytoplasmic granules localized at the posterior end of a Drosophila oocyte and early embryo. These granules are associated with the germ-line and posterior determinants.

polar mutation A mutation that affects the transcription or translation of the part of the gene or operon on only one side of the mutant site—for example, nonsense mutations, frame-shift mutations, and IS-induced mutations.

poly(A) tail A string of adenine nucleotides added to mRNA after transcription.

polyacrylamide A material used to make electrophoretic gels for separation of mixtures of macromolecules.

polycistronic mRNA An mRNA that codes for more than one protein.

polydactyly More than five fingers and/or toes. Inherited as an autosomal dominant phenotype.

polygenes *See* **multiple-factor hypothesis.**

polymerase chain reaction (PCR) A method for amplifying specific DNA segments which exploits certain features of DNA replication.

polymorphism The occurrence in a population (or among populations) of several phenotypic forms associated with alleles of one gene or homologs of one chromosome.

polypeptide A chain of linked amino acids; a protein.

polyploid A cell having three or more chromosome sets, or an organism composed of such cells.

polysaccharide A biological polymer composed of sugar subunits—for example, starch or cellulose.

polytene chromosome A giant chromosome produced by an endomitotic process in which the multiple DNA sets remain bound in a haploid number of chromosomes.

position effect Used to describe a situation where the phenotypic influence of a gene is altered by changes in the position of the gene within the genome.

position-effect variegation Variegation caused by the inactivation of a gene in some cells through its abnormal juxtaposition with heterochromatin.

positive assortative mating A situation in which like phenotypes mate more commonly than expected by chance.

positive control Regulation mediated by a protein that is required for the activation of a transcription unit.

primary structure of a protein The sequence of amino acids in the polypeptide chain.

Pro Proline (an amino acid).

probe Defined nucleic acid segment that can be used to identify specific DNA molecules bearing the complementary sequence, usually through autoradiography.

product of meiosis One of the (usually four) cells formed by the two meiotic divisions.

product rule The probability of two independent events occurring simultaneously is the product of the individual probabilities.

proflavin A mutagen that tends to produce frame-shift mutations.

prokaryote An organism composed of a prokaryotic cell, such as bacteria and blue-green algae.

prokaryotic cell A cell having no nuclear membrane and hence no separate nucleus.

promoter A regulator region a short distance from the 5′ end of a gene that acts as the binding site for RNA polymerase.

prophage A phage "chromosome" inserted as part of the linear structure of the DNA chromosome of a bacterium.

prophase The early stage of nuclear division during which chromosomes condense and become visible.

propositus In a human pedigree, the individual who first came to the attention of the geneticist.

proto-oncogene A gene that, when mutated or otherwise affected, becomes an oncogene.

protoplast A plant cell whose wall has been removed.

prototroph A strain of organisms that will proliferate on minimal medium (*compare* **auxotroph**).

provirus A virus "chromosome" integrated into the DNA of the host cell.

pseudodominance The sudden appearance of a recessive phenotype in a pedigree, due to deletion of a masking dominant gene.

pseudogene An inactive gene derived from an ancestral active gene.

puff *See* **chromosome puff.**

pulse-chase experiment An experiment in which cells are grown in radioactive medium for a brief period (the pulse) and then transferred to nonradioactive medium for a longer period (the chase).

pulsed-field gel electrophoresis An electrophoretic technique in which the gel is subjected to electrical fields alternating between different angles, allowing very large DNA fragments to "snake" through the gel, and hence permitting efficient separation of mixtures of such large fragments.

Punnett square A grid used as a graphic representation of the progeny zygotes resulting from different gamete fusions in a specific cross.

pure-breeding line or strain A group of identical individuals that always produce offspring of the same phenotype when intercrossed.

purine A type of nitrogen base; the purine bases in DNA are adenine and guanine.

pyrimidine A type of nitrogen base; the pyrimidine bases in DNA are cytosine and thymine.

quantitative variation The existence of a range of phenotypes for a specific character, differing by degree rather than by distinct qualitative differences.

quaternary structure of a protein The multimeric constitution of the protein.

R plasmid A plasmid containing one or several transposons that bear resistance genes.

random genetic drift Changes in allele frequency that result because the genes appearing in offspring do not represent a perfectly representative sampling of the parental genes.

random mating Mating between individuals where the choice of a partner is not influenced by the genotypes (with respect to specific genes under study).

reading frame The codon sequence that is determined by reading nucleotides in groups of three from some specific start codon.

realized heritability The ratio of the single-generation progress of selection to the selection differential of the parents.

reannealing Spontaneous realignment of two single DNA strands to re-form a DNA double helix that had been denatured.

receptor element A controlling element that can insert into a gene (making it a mutant) and can also excise (thus making the mutation unstable); both of these functions are nonautonomous, being under the influence of the regulator element.

recessive allele An allele whose phenotypic effect is not expressed in a heterozygote.

recessive phenotype The phenotype of a homozygote for the recessive allele; the parental phenotype that is not expressed in a heterozygote.

reciprocal crosses A pair of crosses of the type genotype A ♀ × genotype B ♂ and genotype B ♀ × genotype A ♂.

reciprocal translocation A translocation in which part of one chromosome is exchanged with a part of a separate nonhomologous chromosome.

recombinant An individual or cell with a genotype produced by recombination.

recombinant DNA A novel DNA sequence formed by the combination of two nonhomologous DNA molecules.

recombinant frequency (RF) The proportion (or percentage) of recombinant cells or individuals.

recombination (1) In general, any process in a diploid or partially diploid cell that generates new gene or chromosomal combinations not found in that cell or in its progenitors. (2) At meiosis, the process that generates a haploid product of meiosis whose genotype is different from either of the two haploid genotypes that constituted the meiotic diploid.

recombinational repair The repair of a DNA lesion through a process, similar to recombination, that uses recombination enzymes.

recon A region of a gene within which there can be no crossing-over; now known to be a nucleotide pair.

reduction division A nuclear division that produces daughter nuclei each having one-half as many centromeres as the parental nucleus.

redundant DNA *See* **repetitive DNA.**

regression A term coined by Galton for the tendency of the quantitative traits of offspring to be closer to the population mean than are their parents' traits. It arises from dominance, gene interaction, and nongenetic influences on traits.

regression coefficient The slope of the straight line that most closely relates two correlated variables.

regulator element *See* **receptor element.**

regulatory genes Genes that are involved in turning on or off the transcription of structural genes.

repetitive DNA Redundant DNA; DNA sequences that are present in many copies per chromosome set.

replication DNA synthesis.

replication fork The point at which the two strands of DNA are separated to allow replication of each strand.

replicon A chromosomal region under the influence of one adjacent replication-initiation locus.

reporter gene A gene whose phenotypic expression is easy to monitor; used to study promoter activity at different times or developmental stages, in recombinant DNA constructs in which the reporter is attached to a promoter region of particular interest.

repressor protein A molecule that binds to the operator and prevents transcription of an operon.

repulsion conformation Two linked heterozygous gene pairs in the arrangement *A b / a B*.

resolving power The ability of an experimental technique to distinguish between two genetic conditions (typically discussed when one condition is rare and of particular interest).

restriction enzyme An endonuclease that will recognize specific target nucleotide sequences in DNA and break the DNA chain at those points; a variety of these enzymes are known, and they are extensively used in genetic engineering.

restrictive conditions Environmental conditions under which a conditional mutant shows the mutant phenotype.

retinoblastoma A cancer of the human retina; predisposition to retinoblastoma is inherited as an autosomal dominant.

retroposon A transposón that was created by reverse transcription of an RNA molecule.

retrovirus An RNA virus that replicates by first being converted into double-stranded DNA.

reverse genetics The experimental procedure that begins with a cloned segment of DNA, or a protein sequence, and uses this (through directed mutagenesis) to introduce programmed mutations back into the genome in order to investigate function.

reverse transcriptase An enzyme that catalyzes the synthesis of a DNA strand from an RNA template.

reversion The production of a wild-type gene from a mutant gene.

RF *See* **recombinant frequency.**

RFLP mapping A technique in which DNA restriction fragment length polymorphisms are used as reference loci for mapping in relation to known genes or other RFLP loci.

rho A protein factor required to recognize certain transcription termination signals in *E. coli.*

ribonucleic acid *See* **RNA.**

ribosomal RNA *See* **rRNA.**

ribosome A complex organelle that catalyzes translation of messenger RNA into an amino acid sequence. Composed of proteins plus rRNA.

ribozymes RNAs with enzymatic activities, for instance, the self-splicing RNA molecules in *Tetrahymena.*

RNA (ribonucleic acid) A single-stranded nucleic acid similar to DNA but having ribose sugar rather than deoxyribose sugar and uracil rather than thymine as one of the bases.

RNA in situ hybridization A technique that is used to identify the spatial pattern of expression of a particular transcript (usually an mRNA). In this technique, the DNA probe is labeled, either radioactively or by chemically attaching an enzyme that can convert a substrate to a visible dye. A tissue or organism is soaked in a solution of single-stranded labeled DNA under conditions that allow the DNA to hybridize to complementary RNA sequences in the cells; unhybridized DNA is then removed. Radioactive probe is detected by autoradiography. Enzyme-labeled probe is detected by soaking the tissue in the substrate; the dye develops in sites where the transcript of interest was expressed.

RNA polymerase An enzyme that catalyzes the synthesis of an RNA strand from a DNA template.

rRNA (ribosomal RNA) A class of RNA molecules, coded in the nucleolar organizer, that have an integral (but poorly understood) role in ribosome structure and function.

S (Svedberg unit) A unit of sedimentation velocity, commonly used to describe molecular units of various sizes (because sedimentation velocity is related to size).

SARs Scaffold attachment regions; the positions along DNA where it is anchored to the central scaffold of the chromosome.

satellite A terminal section of a chromosome, separated from the main body of the chromosome by a narrow constriction.

satellite chromosomes Chromosomes that seem to be additions to the normal genome.

satellite DNA DNA that forms a separate band in a density gradient because of its different nucleotide composition.

saturation mutagenesis Induction and recovery of large numbers of mutations in one area of the genome, or in one function, in the hope of identifying all the genes in that area, or affecting that function.

scaffold The central framework of a chromosome to which the DNA solenoid is attached as loops; composed largely of topoisomerase.

SCE *See* **sister-chromatid exchange.**

secondary sexual characteristics The sex-associated phenotypes of somatic tissues in sexually dimorphic animals.

secondary structure of a protein A spiral or zigzag arrangement of the polypeptide chain.

second-division segregation pattern A pattern of ascospore genotypes for a gene pair showing that the two alleles separate into different nuclei only at the second meiotic division, as a result of a crossover between that gene pair and its centromere; can only be detected in a linear ascus.

second-site mutation The second mutation of a double mutation within a gene; in many cases, the second-site mutation suppresses the first mutation, so that the double mutant has the wild-type phenotype.

sector An area of tissue whose phenotype is detectably different from the surrounding tissue phenotype.

sedimentation The sinking of a molecule under the opposing forces of gravitation and buoyancy.

segmentation The process by which the correct number of segments are established in a developing segmented animal.

segregation (1) Cytologically, the separation of homologous structures. (2) Genetically, the production of two separate phenotypes, corresponding to two alleles of a gene, either in different individuals (meiotic segregation) or in different tissues (mitotic segregation).

segregational petite A petite that in a cross with wild type produces $\frac{1}{2}$ petite and $\frac{1}{2}$ wild-type progeny; caused by a nuclear mutation.
selection coefficient *(s)* The proportional excess or deficiency of fitness of one genotype in relation to another genotype.
selection differential The difference between the mean of a population and the mean of the individuals selected to be parents of the next generation.
selection progress The difference between the mean of a population and the mean of the offspring in the next generation born to selected parents.
selective neutrality A situation in which different alleles of a certain gene confer equal fitness.
selective system An experimental technique that enhances the recovery of specific (usually rare) genotypes.
self To fertilize eggs with sperms from the same individual.
self-assembly The ability of certain multimeric biological structures to assemble from their component parts through random movements of the molecules and formation of weak chemical bonds between surfaces with complementary shapes.
semiconservative replication The established model of DNA replication in which each double-stranded molecule is composed of one parental strand and one newly polymerized strand.
semisterility (half sterility) The phenotype of individuals heterozygotic for certain types of chromosome aberration; expressed as a reduced number of viable gametes and hence reduced fertility.
Ser Serine (an amino acid).
sex chromosome A chromosome whose presence or absence is correlated with the sex of the bearer; a chromosome that plays a role in sex determination.
sex determination The genetic or environmental process by which the sex of an individual is established.
sex linkage The location of a gene on a sex chromosome.
sex reversal A syndrome known in humans and mice in which chromosomally XX individuals develop as males. In some cases, sex reversal is now known to be due to the translocation of the testis-determining region of the Y chromosome to the tip of the X chromosome in such individuals.
sexduction Sexual transmission of donor *E. coli* chromosomal genes on the fertility factor.
sexual spore *See* **spore.**
shotgun technique Cloning a large number of different DNA fragments as a prelude to selecting one particular clone type for intensive study.
shuttle vector A vector (e.g. a plasmid) constructed in such a way that it can replicate in at least two different host species, allowing a DNA segment to be tested or manipulated in several cellular settings.
sickle-cell anemia Potentially lethal human disease caused by a mutation in a gene coding for the oxygen-transporting molecule hemoglobin. The altered molecule causes red blood cells to be sickle shaped. Inherited as an autosomal recessive.
signal sequence The N-terminal sequence of a secreted protein, which is required for transport through the cell membrane.
signal transduction cascade A series of sequential events, such as protein phosphorylations, that pass a signal received by a transmembrane receptor through a series of intermediate molecules until final regulatory molecules, such as transcription factors, are modified in response to the signal.
silent mutation Mutation in which the function of the protein product of the gene is unaltered.
SINE Short interspersed element. A type of small repetitive DNA sequence found throughout a eukaryotic genome.
sister-chromatid exchange (SCE) An event similar to crossing-over that can occur between sister chromatids at mitosis or at meiosis; detected in harlequin chromosomes.
site-specific recombination Recombination occurring between two specific sequences that need not be homologous; mediated by a specific recombination system.
S-9 mix A liver-derived supernatant used in the Ames test to activate or inactivate mutagens.
solenoid structure The supercoiled arrangement of DNA in eukaryotic nuclear chromosomes produced by coiling the continuous string of nucleosomes.
somatic cell A cell that is not destined to become a gamete; a "body cell," whose genes will not be passed on to future generations.
somatic mutation A mutation occurring in a somatic cell.
somatic-cell genetics Asexual genetics, involving study of somatic mutation, assortment, and crossing-over, and of cell fusion.
somatostatin A human growth hormone.
SOS repair The error-prone process whereby gross structural DNA damage is circumvented by allowing replication to proceed past the damage through imprecise polymerization.
Southern blot Transfer of electrophoretically separated fragments of DNA from the gel to an absorbent sheet such as paper. This sheet is then immersed in a solution containing a labeled probe that will bind to a fragment of interest.
spacer DNA DNA found between genes; its function is unknown.
specialized (restricted) transduction The situation in which a particular phage will transduce only specific regions of the bacterial chromosome.
specific-locus test A system for detecting recessive mutations in diploids. Normal individuals treated with mutagen are mated to testers that are homozygous for the recessive alleles at a number of specific loci; the progeny are then screened for recessive phenotypes.
spindle The set of microtubular fibers that appear to move eukaryotic chromosomes during division.
splicing The reaction that removes introns and joins together exons in RNA.
spontaneous mutation A mutation occurring in the absence of mutagens, usually due to errors in the normal functioning of cellular enzymes.
spore (1) In plants and fungi, sexual spores are the haploid cells produced by meiosis. (2) In fungi, asexual spores are

somatic cells that are cast off to act either as gametes or as the initial cells for new haploid individuals.

sporophyte The diploid sexual-spore-producing generation in the life cycle of plants—that is, the stage in which meiosis occurs.

stacking The packing of the flattish nitrogen bases at the center of the DNA double helix.

staggered cuts The cleavage of two opposite strands of duplex DNA at points near one another.

standard deviation The square root of the variance.

statistic A computed quantity characteristic of a population, such as the mean.

statistical distribution The array of frequencies of different quantitative or qualitative classes in a population.

steroid receptor A family of related proteins that act as transcription factors when bound to their cognate hormones. Not all members of this family actually bind to steroids; the name derives from the first family member that was discovered, which was indeed a steroid hormone receptor.

strain A pure-breeding lineage, usually of haploid organisms, bacteria, or viruses.

structural gene A gene encoding the amino acid sequence of a protein.

subvital gene A gene that causes the death of some proportion (but not all) of the individuals that express it.

sum rule The probability that one or the other of two mutually exclusive events will occur is the sum of their individual probabilities.

supercoil A closed double-stranded DNA molecule that is twisted on itself.

superinfection Phage infection of a cell that already harbors a prophage.

supersuppressor A mutation that can suppress a variety of other mutations; typically a nonsense suppressor.

suppressive petite A petite that in a cross with wild type produces progeny of which variable non-Mendelian proportions are petite.

suppressor mutation A mutation that counteracts the effects of another mutation. A suppressor maps at a different site than the mutation it counteracts, either within the same gene or at a more distant locus. Different suppressors act in different ways.

synapsis Close pairing of homologs at meiosis.

synaptonemal complex A complex structure that unites homologs during the prophase of meiosis.

syncytial blastoderm In insects, the stage of blastoderm preceding the formation of cell membranes around the individual nuclei of the early embryo.

syncytium A single cell with many nuclei.

T (1) Thymine, or thymidine. (2) *See* **tetratype.**

tandem duplication Adjacent identical chromosome segments.

targeted gene knockout The introduction of a null mutation in a gene by a designed alteration in a cloned DNA sequence that is then introduced into the genome through homologous recombination and replacement of the normal allele.

tautomeric shift The spontaneous isomerization of a nitrogen base to an alternative hydrogen-bonding condition, possibly resulting in a mutation.

T-DNA A portion of the Ti plasmid that is inserted into the genome of the host plant cell.

telocentric chromosome A chromosome having the centromere at one end.

telomere The tip (or end) of a chromosome.

telophase The late stage of nuclear division when daughter nuclei re-form.

temperate phage A phage that can become a prophage.

temperature-sensitive mutation A conditional mutation that produces the mutant phenotype in one temperature range and the wild-type phenotype in another temperature range.

template A molecular "mold" that shapes the structure or sequence of another molecule; for example, the nucleotide sequence of DNA acts as a template to control the nucleotide sequence of RNA during transcription.

teratogen An agent that interferes with normal development.

terminal redundancy In phage, a linear DNA molecule with single-stranded ends that are longer than is necessary to close the DNA circle.

tertiary structure of a protein The folding or coiling of the secondary structure to form a globular molecule.

testcross A cross of an individual of unknown genotype or a heterozygote (or a multiple heterozygote) to a tester individual.

tester An individual homozygous for one or more recessive alleles; used in a testcross.

testicular feminization syndrome A human condition, caused by a mutation in a gene coding for androgen receptors, in which XY males develop into women; inherited as an X-linked recessive.

tetrad (1) Four homologous chromatids in a bundle in the first meiotic prophase and metaphase. (2) The four haploid product cells from a single meiosis.

tetrad analysis The use of tetrads (definition 2) to study the behavior of chromosomes and genes during meiosis.

tetraparental mouse A mouse that develops from an embryo created by the experimental fusion of two separate blastulas.

tetraploid A cell having four chromosome sets; an organism composed of such cells.

tetratype (T) A tetrad type containing four different genotypes, two parental and two recombinant.

Thr Threonine (an amino acid).

three-point testcross A testcross involving one parent with three heterozygous gene pairs.

thymidine The nucleoside having thymine as its base.

thymine A pyrimidine base that pairs with adenine.

thymine dimer A pair of chemically bonded adjacent thymine bases in DNA; the cellular processes that repair this lesion often make errors that create mutations.

Ti plasmid A circular plasmid of *Agrobacterium tumifaciens* that enables the bacterium to infect plant cells and produce a tumor (crown gall tumor).

totipotency The ability of a cell to proceed through all the stages of development and thus produce a normal adult.

***trans* conformation** In a heterozygote involving two mutant sites within a gene or gene cluster, the arrangement $a_1 +/+ a_2$.
transcription The synthesis of RNA using a DNA template.
transcription factor A protein that binds to a *cis*-regulatory element (e.g., an enhancer) and thereby, directly or indirectly, affects the initiation of transcription.
transduction The movement of genes from a bacterial donor to a bacterial recipient using a phage as the vector.
transfection The process by which exogenous DNA in solution is introduced into cultured cells.
transmembrane receptor A protein that spans the plasma membrane of a cell, with the extracellular portion of the protein having the ability to bind to a ligand and the intracellular portion having an activity (such as a protein kinase) that can be induced upon ligand binding.
transfer RNA *See* **tRNA.**
transformation (1) The directed modification of a genome by the external application of DNA from a cell of different genotype. (2) Conversion of normal higher eukaryotic cells in tissue culture to a cancer-like state of uncontrolled division.
transgenic organism One whose genome has been modified by externally applied new DNA.
transient diploid The stage of the life cycle of predominantly haploid fungi (and algae) during which meiosis occurs.
transition A type of nucleotide-pair substitution involving the replacement of a purine with another purine, or of a pyrimidine with another pyrimidine—for example, GC→AT.
translation The ribosome-mediated production of a polypeptide whose amino acid sequence is derived from the codon sequence of an mRNA molecule.
translocation The relocation of a chromosomal segment in a different position in the genome.
transmission genetics The study of the mechanisms involved in the passage of a gene from one generation to the next.
transposable genetic element A general term for any genetic unit that can insert into a chromosome, exit, and relocate; includes insertion sequences, transposons, some phages, and controlling elements.
transposition *See* **translocation.**
transposon A mobile piece of DNA that is flanked by terminal repeat sequences and typically bears genes coding for transposition functions.
transversion A type of nucleotide-pair substitution involving the replacement of a purine with a pyrimidine, or vice versa—for example, GC→TA.
triplet The three nucleotide pairs that compose a codon.
triploid A cell having three chromosome sets, or an organism composed of such cells.
trisomic Basically a diploid with an extra chromosome of one type, producing a chromosome number of the form $2n + 1$.
tritium A radioactive isotope of hydrogen.
tRNA (transfer RNA) A class of small RNA molecules that bear specific amino acids to the ribosome during translation; the amino acid is inserted into the growing polypeptide chain when the anticodon of the tRNA pairs with a codon on the mRNA being translated.
Trp Tryptophan (an amino acid).
true-breeding line or strain *See* **pure-breeding line or strain.**
truncation selection A breeding technique in which individuals in whom quantitative expression of a phenotype is above or below a certain value (the truncation point) are selected as parents for the next generation.
tumor-suppressor gene A gene encoding a protein that suppresses tumor formation.
tumor virus A virus that is capable of inducing a cancer.
Turner syndrome An abnormal human female phenotype produced by the presence of only one X chromosome (XO).
twin spot A pair of mutant sectors within wild-type tissue, produced by a mitotic crossover in an individual of appropriate heterozygous genotype.
2μ (2 micron) plasmid A naturally occurring extragenomic circular DNA molecule found in some yeast cells, with a circumference of 2μ. Engineered to form the basis for several types of gene vectors in yeast.
Tyr Tyrosine (an amino acid).

U Uracil, or uridine.
underdominance A phenotypic relation in which the phenotypic expression of the heterozygote is less than that of either homozygote.
unequal crossover A crossover between homologs that are not perfectly aligned.
uniparental inheritance The transmission of certain phenotypes from one parental type to all the progeny; such inheritance is generally produced by organelle genes.
unstable mutation A mutation that has a high frequency of reversion; a mutation caused by the insertion of a controlling element, whose subsequent exit produces a reversion.
uracil A pyrimidine base that appears in RNA in place of thymine found in DNA.
URF Unassigned reading frame. An open reading frame (ORF) whose function has not yet been determined.
uridine The nucleoside having uracil as its base.

Val Valine (an amino acid).
variable A property that may have different values in various cases.
variable region A region in an immunoglobin molecule that shows many sequence differences between antibodies of different specificities; the part of the antibody that binds to the antigen.
variance A measure of the variation around the central class of a distribution; the average squared deviation of the observations from their mean value.
variant An individual organism that is recognizably different from an arbitrary standard type in that species.
variate A specific numerical value of a variable.
variation The differences among parents and their offspring or among individuals in a population.
variegation The occurrence within a tissue of sectors with differing phenotypes.

vector In cloning, the plasmid or phage chromosome used to carry the cloned DNA segment.

viability The probability that a fertilized egg will survive and develop into an adult organism.

virulent phage A phage that cannot become a prophage; infection by such a phage always leads to lysis of the host cell.

VNTR (variable number tandem repeat) A chromosomal locus at which a particular repetitive sequence is present in different numbers in different individuals or in the two different homologs in one diploid individual.

wild type The genotype or phenotype that is found in nature or in the standard laboratory stock for a given organism.

wobble The ability of certain bases at the third position of an anticodon in tRNA to form hydrogen bonds in various ways, causing alignment with several possible codons.

X:A ratio The ratio between the X chromosome and the number of sets of autosomes.

X chromosome inactivation In female mammalian embryos, the early random inactivation of the genes on one of the X chromosomes, leading to mosaicism for functions coded by heterozygous X-linked genes (*see* **dosage compensation** and **Barr body**).

X hyperactivation In *Drosophila,* the process by which the structural genes of the male X chromosome are transcribed at the same rate as the two X chromosomes of the female combined.

X linkage The inheritance pattern of genes found on the X chromosome but not on the Y.

X-and-Y linkage The inheritance pattern of genes found on both the X and Y chromosomes (rare).

X-ray crystallography A technique for deducing molecular structure by aiming a beam of X rays at a crystal of the test compound and measuring the scatter of rays.

Y linkage The inheritance pattern of genes found on the Y chromosome but not on the X (rare).

zygote The cell formed by the fusion of an egg and a sperm; the unique diploid cell that will divide mitotically to create a differentiated diploid organism.

zygotic induction The sudden release of a lysogenic phage from an Hfr chromosome when the prophage enters the F^- cell, and the subsequent lysis of the recipient cell.

Answers to Selected Problems

Chapter 2

2. To determine whether the *Drosophila* is AA or Aa, a testcross should be done. By definition, this means using a fly that is aa. If the original fly is AA, all progeny will have the A phenotype. If the original fly is Aa, one half the progeny will have the A phenotype and one half will have the aa phenotype.

3. The parents must be $Bb \times Bb$. Their black progeny must be BB and Bb in a 1 : 2 ratio, and their white progeny must be bb.

5. **a.** $(\frac{1}{6})^3$
b. $(\frac{1}{6})^3$
c. $(\frac{1}{6})^3$
d. $(\frac{5}{6})^3$
e. $3(\frac{1}{6})^3$
f. $2(\frac{1}{6})^3$
g. $(\frac{1}{6})^2$
h. $\frac{5}{9}$

6. **a.** 1. 0.0054
2. 0.0396
3. 0.0006
4. 0.0594
5. 0.4162
6. 0.9946

b. 11

8. Recessive

10. $\frac{5}{8}$

13. The results suggest that winged ($A-$) is dominant to wingless (aa).

		Number of Progeny Plants	
Pollination	Genotypes	Winged	Wingless
Winged (selfed)	$AA \times AA$	91	1*
Winged (selfed)	$Aa \times Aa$	90	30
Wingless (selfed)	$aa \times aa$	4*	80
Winged × wingless	$AA \times aa$	161	0
Winged × wingless	$Aa \times aa$	29	31
Winged × wingless	$AA \times aa$	46	0
Winged × winged	$AA \times A-$	44	0
Winged × winged	$AA \times A-$	24	0

The five unusual plants are most likely due either to human error in classification or to contamination. Alternatively, they could result from environmental effects on development. For example, too little water may have prevented the seed pods from becoming winged even though they are genetically winged.

14. **a.** **Pedigree 1:** recessive
Pedigree 2: dominant
Pedigree 3: dominant
Pedigree 4: recessive

b. **Genotypes of pedigree 1:**
Generation 1: AA, aa
Generation 2: Aa, Aa, Aa, $A-$, $A-$, Aa
Generation 3: Aa, Aa
Generation 4: aa

Genotypes of pedigree 2:
Generation 1: Aa, aa, Aa, aa
Generation 2: aa, aa, Aa, Aa, aa, aa, Aa, Aa, aa
Generation 3: aa, aa, aa, aa, aa, $A-$, $A-$, $A-$, Aa, aa
Generation 4: aa, aa, aa

Genotypes of pedigree 3:
Generation 1: Aa, aa
Generation 2: Aa, aa, aa, Aa
Generation 3: aa, Aa, aa, aa, Aa, aa
Generation 4: aa, Aa, Aa, Aa, aa, aa

Genotypes of pedigree 4:
Generation 1: aa, $A-$, Aa, Aa
Generation 2: Aa, Aa, Aa, aa, $A-$, aa, $A-$, $A-$, $A-$, $A-$, $A-$
Generation 3: Aa, aa, Aa, Aa, aa, Aa

16. **a.** recessive

b. $\frac{1}{8}$

19. recessive

22. 1 $Aann$ dwarf : 1 $aann$ normal : 1 $aaNn$ neurofibromatosis : 1 $AaNn$ dwarf, neurofibromatosis

23. **a.** $CcSs \times CcSs$
b. $CCSs \times CCss$
c. $CcSS \times ccSS$
d. $ccSs \times ccSs$
e. $Ccss \times Ccss$
f. $CCSs \times CCSs$
g. $CcSs \times Ccss$

25. **a.** 1. $\frac{9}{128}$
2. $\frac{9}{128}$
3. $\frac{9}{64}$
4. $\frac{55}{64}$

b. 1. $\frac{1}{32}$
2. $\frac{1}{32}$
3. $\frac{1}{16}$
4. $\frac{15}{16}$

Chapter 3

2. Both will be $AaBbCc$.

3. 1 $ad^+ a$, white; 1 $ad^- a$, purple; 1 $ad^+ \alpha$, white; 1 $ad^- \alpha$, purple

5.

	Mitosis	Meiosis
fern	sporophyte gametophyte	prothallus
moss	sporophyte gametophyte	archeogonia antheridia
flowering plant	sporophyte gametophyte	flowers
pine tree	sporophyte gametophyte	pinecones
mushroom	sporophyte gametophyte	hyphae
frog	somatic cells	gonads
butterfly	somatic cells	gonads
snail	somatic cells	gonads

7. **e.** Chromosome pairing

8. $\frac{1}{8}$

9. $(\frac{1}{2})^{n-1}$

12. X-linked dominant

13. **e.** 0

14. The data suggest Y-linkage. Another explanation is an autosomal gene that is dominant in males and recessive in females.

16. **a.** X-linked

b. P AY, Aa, AY

F_1 $AY, A-, aY, A-, AY, aY, aY, A-, aY, A-$

18. **a.** Let B = black and b = orange.

Females	Males
$X^BX^B = BB$ = black	$X^BY = BY$ = black
$X^bX^b = bb$ = orange	$X^bY = bY$ = orange
$X^BX^b = Bb$ = tortoise	

b. P X^bX^b (orange) × X^BY (black) or $bb \times BY$

F_1 X^BX^b tortoise-shell female or Bb

X^bY orange male or bY

c. P X^BX^B *(black)* × X^bY (orange) or $BB \times bY$

F_1 X^BX^b tortoise female or Bb

X^BY black male or BY

d. Mother, tortoise-shell; father, black.

e. Mother, tortoise-shell; father, orange.

19. **e.** $\frac{1}{4}$

21. **a.** Autosomal recessive: excluded

b. Autosomal dominant: consistent.

c. X-linked recessive: excluded

d. X-linked dominant: excluded

e. Y-linked: excluded

23. Let B = red, b = brown, S = long, and s = short.

P $bbss \times BBSY$

F_1 $\frac{1}{2}$ $BbSs$ females

$\frac{1}{2}$ $BbsY$ males

F_1 gametes

female: $\frac{1}{4}$ BS : $\frac{1}{4}$ Bs : $\frac{1}{4}$ bS : $\frac{1}{4}$ bs

male: $\frac{1}{4}$ Bs : $\frac{1}{4}$ bs : $\frac{1}{4}$ BY : $\frac{1}{4}$ bY

F_2 females

$\frac{1}{16}$ $BBSs$	red, long
$\frac{1}{16}$ $BBss$	red, short
$\frac{2}{16}$ $BbSs$	red, long
$\frac{2}{16}$ $Bbss$	red, short
$\frac{1}{16}$ $bbSs$	brown, long
$\frac{1}{16}$ $bbss$	brown, short

F_2 males

$\frac{1}{16}$ $BBSY$	red, long
$\frac{1}{16}$ $BBsY$	red, short
$\frac{2}{16}$ $BbSY$	red, long
$\frac{2}{16}$ $BbsY$	red, short
$\frac{1}{16}$ $bbSY$	brown, long
$\frac{1}{16}$ $bbsY$	brown, short

The final phenotypic ratio is

$\frac{3}{8}$ red, long $\frac{1}{8}$ brown, short

$\frac{3}{8}$ red, short $\frac{1}{8}$ brown, long

25. This can most easily be explained by X-inactivation in the woman's cells.

28. Autosomal dominant with expression limited to males

Chapter 4

1.

Genotype	Phenotype
1 *AA*	A
1 AB	AB
1 *AO*	A
1 *BO*	B

4. e. 0 percent

5. a.

Cross	Parents	Progeny	Conclusion
1	*ba* × *ba*	→ 3 *b*–:1 *aa*	black is dominant to albino
2	*bs* × *aa*	→ 1 *ba*:1 *sa*	black is dominant to sepia
3	*ca* × *ca*	→ 3 *c*–:1 aa	cream is dominant to albino
4	*sa* × *ca*	→ 1 *ca*:2 *s*–:1 *aa*	sepia is dominant to albino
5	*bc* × *aa*	→ 1 *ba*:1 *ca*	black is dominant to cream
6	*bs* × *c*–	→ 1 *b*–:1 *s*–	black is dominant to sepia
7	*bs* × *s*–	→ 1 *b*–:1 *s*–	black is dominant to cream
8	*bc* × *sc*	→ 2 *b*–:1 *sc*:1 *cc*	sepia is dominant to cream
9	*sc* × *sc*	→ 3 *s*–:1 *cc*	sepia is dominant to cream
10	*ca* × *aa*	→ 1 *ca*:1 *aa*	cream is dominant to abino

The order of dominance is $b > s > c > a$.

b.

1 *bb* black	1 *bc* black
1 *bs* black	1 *sc* sepia

6.

Parents	Child
a. AB × O	3. B
b. A × O	2. A
c. A × AB	4. AB
d. O × O	1. O

8. The platinum allele is a pleiotropic allele that governs coat color in the heterozygous state and is lethal in the homozygous recessive state.

10. a. Approximately one-half of the males are missing.

b. If the normal-looking female were heterozygous for an X-linked recessive allele that was lethal in either the homozygous or the hemizygous state, then all the female progeny and one half of the male progeny would survive.

c. In order to test this explanation, assume that the original cross was $Aa \times AY$. One half of the females from this cross should be heterozygous and one half should be homozygous. These F_1 females could be crossed individually with normal males and the sex ratio of the progeny could be determined for each cross.

12. a. The ratio of hairy to hairless is 7:2.

b. The ratio of hairy to hairless is 2:1.

13. A dihybrid cross in which $A\!-\!bb$ has the same appearance as $aaB-$.

16. a. Dominant autosomal

b. The pedigree exhibits pleiotropy and variable expressivity. Reduced penetrance is not evident.

c. The pleiotropy indicates that a gene product that exists in a number of different organs is defective. The variable expressivity is due to epistasis by one or more other genes.

18. a. and b. The following allelic dominance relationships can be seen: black > brown > yellow. Arbitrarily assign the following genotypes for homozygotes: B^lB^l = black, B^rB^r = brown, B^yB^y = yellow. A white phenotype is composed of two categories: the double homozygote and one class of the mixed homozygote-heterozygote. Let lack of color be caused by cc. Color will therefore be $C-$.

Cross 1:	P	$B^rB^r CC \times B^yB^y CC$
	F_1	$B^rB^y CC$
	F_2	$3\ B^r\!-CC : 1\ B^yB^y CC$
Cross 2:	P	$B^lB^l CC \times B^rB^r CC$
	F_1	$B^lB^r CC$
	F_2	$3\ B^l\!-CC : 1\ B^rB^r CC$
Cross 3:	P	$B^lB^l CC \times B^yB^y CC$
	F_1	$B^lB^y CC$
	F_2	$3\ B^l\!-CC : 1\ B^yB^y CC$
Cross 4:	P	$B^lB^l cc \times B^yB^y CC$
	F_1	$B^lB^y Cc$
	F_2	$9\ B^l\!-C- : 3\ B^yB^y C- : 3\ B^l\!-cc : 1\ B^yB^y cc$
Cross 5:	P	$B^lB^l cc \times B^rB^r CC$
	F_1	$B^lB^r Cc$
	F_2	$9\ B^l\!-C- : 3\ B^rB^r C- : 3\ B^l\!-cc : 1\ B^rB^r cc$
Cross 6:	P	$B^lB^l CC \times B^yB^y cc$
	F_1	$B^lB^y Cc$
	F_2	$9\ B^l\!-C- : 3\ B^yB^y C- : 3\ B^l\!-cc : 1\ B^yB^y cc$

22. Sun-red is dominant to orange and orange dominant to pink. If c^{sr} = sun-red, c^o = orange, and c^p = pink, then the crosses and the results are

b. P $aa\,BB\,DD\,EE \times AA\,bb\,DD\,EE$ yellow 1 × yellow 2

F_1 $Aa\,Bb\,Dd\,Ee$ red

F_2 9 $A{-}\,B{-}\,DD\,EE$ red

3 $aa\,B{-}\,DD\,EE$ yellow

3 $A{-}\,bb\,DD\,EE$ yellow

1 $aa\,bb\,DD\,EE$ yellow

P $aa\,BB\,DD\,EE \times AA\,BB\,dd\,EE$ yellow 1 × brown

F_1 $Aa\,BB\,Dd\,EE$ red

F_2 9 $A{-}\,BB\,D{-}\,EE$ red

3 $aa\,BB\,D{-}\,EE$ yellow

3 $A{-}\,BB\,dd\,EE$ brown

1 $aa\,BB\,dd\,EE$ yellow

P $aa\,BB\,DD\,EE \times AA\,BB\,DD\,ee$ yellow 1 × orange

F_1 $Aa\,BB\,DD\,Ee$ red

F_2 9 $A{-}\,BB\,DD\,E{-}$ red

3 $aa\,BB\,DD\,E{-}$ yellow

3 $A{-}\,BB\,DD\,ee$ orange

1 $aa\,BB\,DD\,ee$ orange

P $AA\,bb\,DD\,EE \times AA\,BB\,dd\,EE$ yellow 2 × brown

F_1 $AA\,Bb\,Dd\,EE$ red

9 $AA\,B{-}\,D{-}\,EE$ red

3 $AA\,bb\,D{-}\,EE$ yellow

3 $AA\,B{-}\,dd\,EE$ brown

1 $AA\,bb\,dd\,EE$ yellow

P $AA\,bb\,DD\,EE \times AA\,BB\,DD\,ee$ yellow 2 × orange

F_1 $AA\,Bb\,DD\,Ee$ red

9 $AA\,B{-}\,DD\,E{-}$ red

3 $AA\,bb\,DD\,E{-}$ yellow

3 $AA\,B{-}\,DD\,ee$ orange

1 $AA\,bb\,DD\,ee$ yellow

P $AA\,BB\,dd\,EE \times AA\,BB\,DD\,ee$ brown × orange

F_1 $AA\,BB\,Dd\,Ee$ red

F_2 9 $AA\,BB\,D{-}\,E{-}$ red

3 $AA\,BB\,dd\,E{-}$ brown

3 $AA\,BB\,D{-}\,ee$ orange

1 $AA\,BB\,dd\,ee$ orange

c. $\text{yellow}_2 \xrightarrow{} \text{orange} \xrightarrow{} \text{yellow}_1 \xrightarrow{} \text{brown} \xrightarrow{} \text{red}$

(arrows from B, E, A, D acting on the respective steps)

B E A D

41. a. and b. The disorder is governed by an autosomal recessive allele and there are two genes that result in deaf-mutism.

I. $Aa\,BB \times Aa\,BB$, $AA\,Bb \times AA\,Bb$

II. Individuals 1, 3–6: $A{-}\,BB$

Individuals 9, 10, 12–15: $AA\,B{-}$

Individuals 2, 7: $aa\,BB$

Individuals 8, 11: $AA\,bb$

III. All $Aa\,Bb$

Chapter 5

For the following, *CO* is used to designate single recombinants and *DCO* is used to designate double recombinants.

1. 45 percent

4. The two genes are 33.3 map units (m.u.) apart.

7. a. $\frac{1}{4}$

b. $\frac{1}{2}$

c. 45 percent

d. 38 percent

8. a. 4 percent

b. 4 percent

c. 46 percent

d. 8 percent

11. a. $MF\,/\,mf$

b. $mf\,/\,mf$

c. Sex is determined by the male contribution. The two parental gametes are $M\,F$, determining maleness ($M\,F\,/m\,f$), and $m\,f$, determining femaleness ($mf\,/\,mf$). Occasional recombination would yield Mf, determining a hermaphrodite ($Mf\,/\,mf$), and $m\,F$, determining total sterility ($m\,F\,/\,mf$).

d. Recombination in the male yielding Mf

e. Hermaphrodites are rare because the genes are tightly linked.

13. 24 m.u.

15. a. $an\,f^+\,br^+\,/\,an\,f^+\,br^+ \times an^+\,f\,br\,/\,an^+\,f\,br$

b.

an —— *f* —— *br*

16.72 m.u. 4.78 m.u.

c. 0.431

17. Let F = fat, L = long tail, and Fl = flagella.

F —— *L* —— *Fl*

9.3 m.u. 15.3 m.u.

19. 1. *b a c*
2. *b a c*
3. *b a c*
4. *a c b*
5. *a c b*

20.

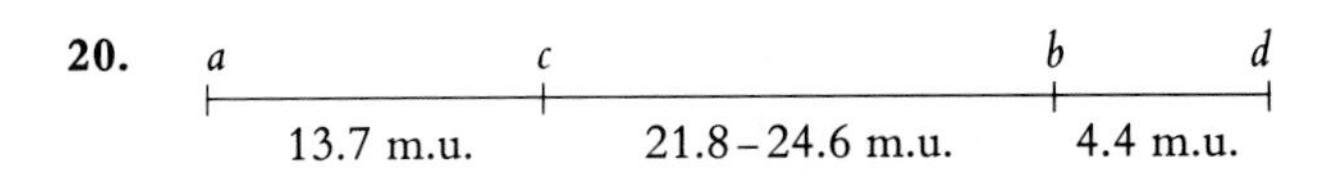

21. a. The hypothesis is that the genes are not linked. Therefore, a 1 : 1 : 1 : 1 ratio is expected.

b. χ^2 is 0.76.

c. The p value is between 0.50 and 0.90.

d. Between 50 percent and 90 percent of the time values this extreme from the prediction would be obtained by chance alone.

e. Accept the initial hypothesis.

f. The two genes are assorting independently. The genotypes of all individuals are

P	$dp^+ dp^+ ee \times dp\,dp\,e^+e^+$
F_1	$dp^+ dp\, e^+ e$
Tester	$dp\,dp\,ee$
Progeny	long ebony $dp^+ dp\, ee$
	long gray $dp^+ dp\, e^+ e$
	short gray $dp\,dp\,e^+ e$
	short ebony $dp\,dp\,ee$

22. a. and **b.**

gal — 32 m.u. — *rad*   *aro* — 20 m.u. — *asp*

25. a. 0.34
b. 0.34
c. 0.015
d. 0.06

27. Assume there is no linkage.
1. 2.1266, nonsignificant. Therefore, the hypothesis of no linkage cannot be rejected.
2. 6.6, nonsignificant. The hypothesis of no linkage cannot be rejected.
3. 66.0, significant. The hypothesis of no linkage must be rejected.
4. 11.60, significant. The hypothesis of no linkage must be rejected.

30. a. Blue sclerotics, autosomal dominant; hemophilia, X-linked recessive.

b. If the individuals in the pedigree are numbered as generations I through IV and the individuals in each generation are numbered clockwise, starting from the top right-hand portion of the pedigree, their genotypes are:

I. *bbHh, BbHY*
II. *BbHY, BbHY, bbHY, BbH–, bbHY, BbHh, BbH–, bbH–*
III. *bbH–, BbH–, bbhY, bbHY, BbHY, BbH–, BbHY, BbhY, BbH–, bbHY, BbH–, bbHY, BbH–, BbHY, BbhY, bbHY, bbHY, bbH–, bbHY, bbHY, BbH–, BbHY, BbhY*
IV. *bbH–, BbH–, BbH–, bbHh, bbHh, bbHY, bbHH, bbHY, bbHh, bbH–, bbH–, bbHY, bbHY, bbH–, bbHY, bbHY, BbHY, bbHY, bbHH, bbHY, bbHY, bbHH, bbH–, bbH–, bbH–, bbHY, bbHY, bbHY, bbHh, BbH–, BbHY, bbHY, BbHY, bbH–*

c. There is no evidence of linkage between these two disorders.

d. The two genes exhibit independent assortment.

e. No individual could be considered intrachromosomally recombinant. However, a number show interchromosomal recombination: all individuals in generation III that have both disorders.

32. Woman X, 0.025; woman Y, 0.225.

33. a. *E* — 40 m.u. — *C* — 20 m.u. — *A* — 30 m.u. — *B* — 10 m.u. — *D*

b. For cross 1: no interference. For cross 2: no interference.

Chapter 6

2. a. *arg-6* al-2 and + +

b. *arg-6* +, *arg-6 al-2*, + +, *and* + *al-2*

c. *arg-6* + and + *al-2*

3. a. 0.135
b. 0.27
c. 0.27

6. This problem is analogous to meiosis in organisms that form linear tetrads. Let red = *R* and blue = *r*. Then meiosis is occurring in an organism that is *Rr* (but there are four "alleles" because the "chromosomes" are at the "two-chromatids-per-chromosome" stage), and the "alleles" are at loci far from the centromere.

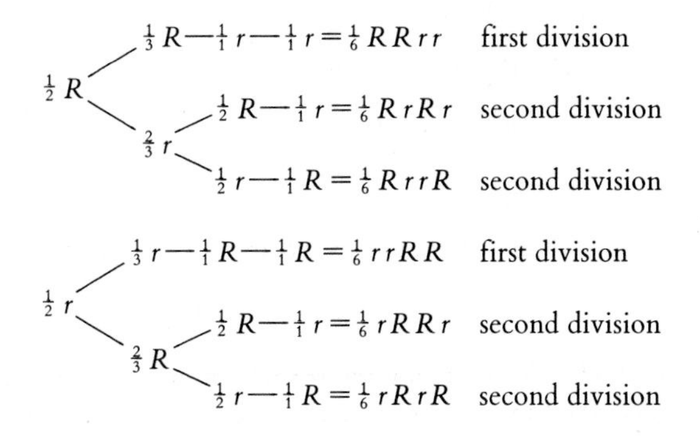

These results indicate one-third first division segregation and two-thirds second division segregation.

7. **a.** 25.5 m.u.

b. No linkage

10. Cross 1:

Cross 2:

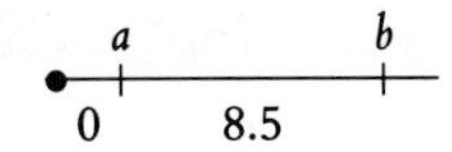

Cross 3:

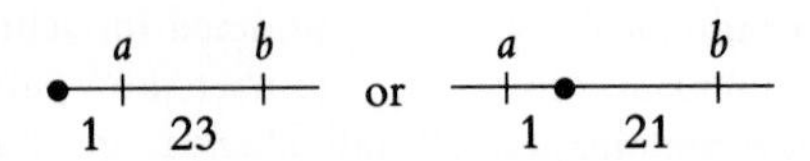

The first diagram is the better interpretation of the data.
Cross 4:

a b
5 11

Cross 5: The genes are considered unlinked to their centromeres in tetrad analysis.
Cross 6:

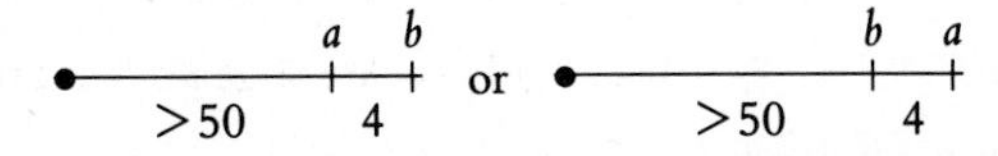

Cross 7:

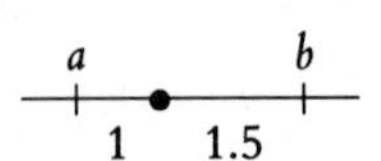

Cross 8: The genes are considered unlinked to their centromeres in tetrad analysis.
Cross 9:

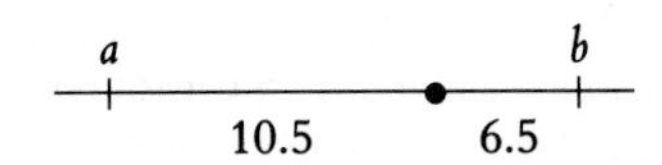

Cross 10:

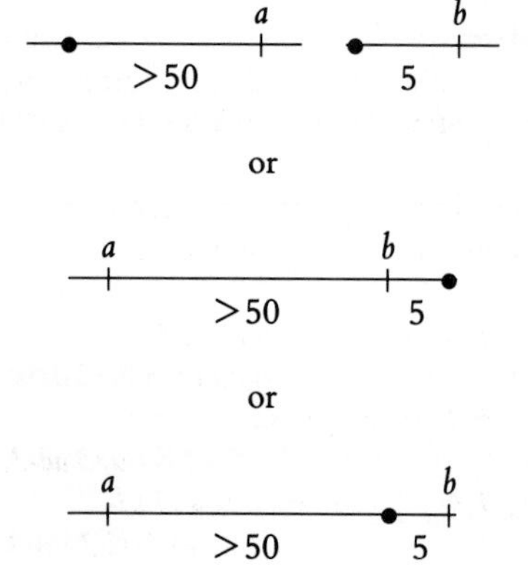

Cross 11:

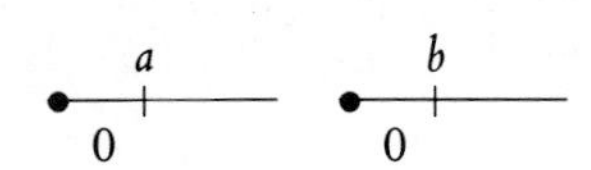

12. Cross 1:

recombinant frequency = 26.5

uncorrected map distance = 26.5 m.u.

corrected map distance = 34.5 m.u.

Cross 2:

recombinant frequency = 19 percent

uncorrected map distance = 19 m.u.

corrected map distance = 29 m.u.

Cross 3:

recombinant frequency = 30 percent

uncorrected map distance = 30 m.u.

corrected map distance = 40 m.u.

13. **a.** The frequency of recombinant asci is 50 percent, which leads to an uncorrected RF of 27 m.u. The corrected RF is 31.5 m.u.

b. Actual 0 crossovers 46 percent
Actual 1 crossovers 38 percent
Actual DCOs 16 percent

c. 35 m.u.

15. *his-4*

24.

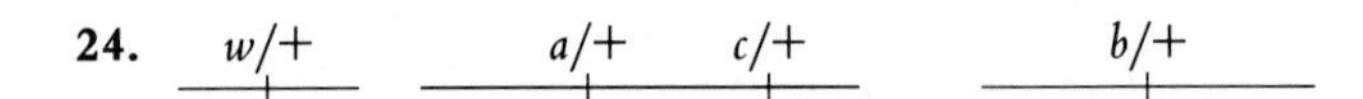

25. centromere – ribo – yellow

26. **a.** All *fpafpa* diploid segregants experienced a crossover between *fpa* and the centromere. The data indicate that *pro* and *paba* are linked to *fpa*. The different frequencies are a measure of the distances involved. Because there are no progeny requiring only *pro*, *pro* is closer to the centromere than *paba*.

b.

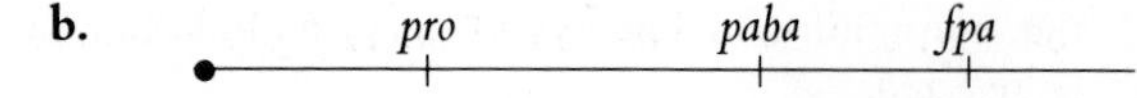

pro-paba: 100%(71)/154 = 71.4 relative m.u.
paba-fpa: 100%(35)/154 = 22.7 relative m.u.
pro-centromere: 100%(9)/154 = 5.8 relative m.u.

c. *pro paba fpa*

27. α: 7
β: 1
γ: 5
δ: 6
ϵ: not on chromosomes 1 through 7.

28. Steroid sulfatase, Xp; phosphoglucomutase-3, 6q; esterase D, 13q; phosphofructokinase, 21; amylase, 1p; galactokinase, 17q.

Chapter 7

1. The petal will now be *Ww*, or blue.

3. Plate the cells on medium lacking proline. Nearly all colonies will come from revertants. The remainder will be second-site suppressors.

6. $\frac{1}{10}^{-6}$ cell divisions

8. Stain pollen grains, which are haploid, from a homozygous *Wx* parent with iodine. Look for red pollen grains, indicating mutations to *wx*, under a microscope.

9. The most straightforward explanation is that a mutation from wild type to black occurred in the germ line of the male wild-type mouse. Thus, he was a gonadal mosaic of wild-type and black germ cells.

11. The mutation rate is 4.25×10^{-5} gametes. You do not have to worry about revertants in this problem because the problem asks for the net mutation frequency to achondroplasia.

12. A shift in the sex ratio is the first indication that a population has sustained lethal genetic damage.

13. **a.** Reddish all over

b. Reddish all over

c. Many small, red spots

d. A few large, red spots

e. Like **c**, but with fewer reddish patches

f. Like **d**, but with fewer reddish patches

g. Some large spots and many small spots

Chapter 8

1. **a.** Deletions lead to a shorter chromosome with missing bands, if banded, an unpaired loop during homologous pairing, and the expression of hemizygous recessive alleles.

b. Duplications lead to a longer chromosome with repeated bands, if banded, and an unpaired loop during homologous pairing; there may be disturbed development.

c. Inversions can be detected by banding and show the typical twisted homologous pairing for heterozygotes; no crossover products are seen for genes within the inversion in the heterozygote.

d. Reciprocal translocations can be detected by banding, show the typical cross structure during homologous pairing, lead to new linkage groups, and show altered linkage relationships. The heterozygote has a high rate of unbalanced gamete production.

2. **a.** 27%

b. 36%

5. A deletion occurred in the left arm of chromosome 2 and leu^+, mating type *a*, and un^+ were lost.

7. The colonies that would not revert most likely had a deletion within the *ad-3B* gene. They could grow with adenine supplementation only.

8.

Allele	Band
b	1
a	2
c	3
e	4
d	5
f	6

9.

Mutant	Defect
1	deletion of at least part of genes *h* and *i*
2	deletion of at least part of genes *k* and *l*
3	deletion of at least part of gene *m*
4	deletion of at least part of genes *k*, *l*, and *m*
5	deletion not within the *h* through *m* genes or a recessive point mutation

10. **a.** An inversion is more likely.

b.

Normal

a b c d e f

Pl sm

Inversion

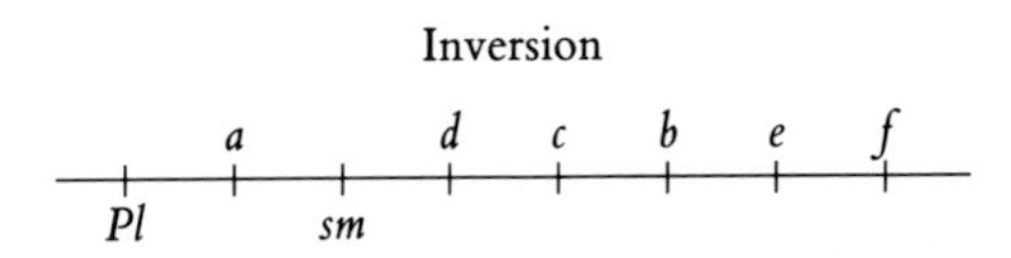

c. The semisterility is the result of crossing-over in the inverted region. All products of crossing-over would have both duplications and deletions.

13. The most likely explanation is that one or both break points were located within essential genes.

14. **a.** + +, + +, + +, + +, *un3 ad-3*, *un3 ad-3*, *un3 ad-3*, *un3 ad-3* 80%

+ +, + +, + *ad-3*, + *ad-3*, *un3* +, *un3* +, *un3 ad-3*, *un3 ad-3* 10%

+ +, + +, + *ad-3*, + *ad-3*, *un3 ad-3*, *un3 ad-3*, *un3* +, *un3* + 10%

b. The aborted spores result from an inversion in the wild type. Crossing-over led to nonviable spores because they were unbalanced. This could be tested by selecting *un3 ad-3* double mutants from the wild type and then crossing them with the ++ inverted strain. The *un3*-to-*ad-3* distance should be altered.

16. a. Okanagan sequence:

d — 20 — *e* — 3 — *a* — 12 — *b* — 2 — *c* — 15 — *f*

Spain sequence:

f — 15 — *c* — 4 — *e* — 3 — *a* — 12 — *b* — 18 — *d*

b.

	a	*b*	*c*	*d*	*e*	*f*
a	0	0	0	0	0	15
b		0	0	0	0	15
c			0	0	0	15
d				0	0	15
e					0	15
f						15

17. a. The aberrant plant is semisterile, which suggests an inversion. Since the *d*–*f* and *y*–*p* frequencies of recombination in the aberrant plant are normal, the inversion must involve *b* through *x*.

b. To obtain recombinant progeny when an inversion is involved, either a double crossover occurred within the inverted region or single crossovers occurred between *f* and the point of inversion, which occurred someplace between *f* and *b*.

19. a. and b.

Class 1: parental

Class 2: parental

Class 3: DCO *y*–*cv* and *B*–*car*

Class 4: reciprocal of class 3

Class 5: DCO *cv*–*v* and *v*–*f*

Class 6: reciprocal of class 5

Class 7: DCO *cv*–*v* and *f*–*car*

Class 8: reciprocal of class 7

Class 9: DCO *v*–*cv* and *v*–*f*

Class 10: reciprocal of class 9

Class 11: This class is identical to the male parent's X chromosome and could not have come from the female parent. Thus, the male sperm must have donated it to the offspring. In *Drosophila,* sex is determined by the ratio of X chromosomes to the number of sets of autosomes. The ratio in males is 1X:2A, where A stands for the autosomes contributed by one parent (the ratio in females is 2X:2A). Thus, this class of males must have arisen from the union of an X-bearing sperm with an egg that was the product of nondisjunction for X and contained only autosomes.

c. Class 11 should have only one sex chromosome, which could be checked cytologically.

26. Cross 1: Independent assortment of 2 genes occurred.

Cross 2: Two genes are linked at 1 m.u. distance. Therefore, a reciprocal translocation took place and both genes were very close to the breakpoint. The black spores resulted from alternate segregation, the white from adjacent segregation.

Cross 3: Half the spores were normal and non-translocated, and half contained both translocated chromosomes.

27. a. 5 percent

b. 45 percent

c. 45 percent

d. 5 percent

e. 20.5 percent

32. a. Breaks in different regions of 17R result in deletion of all genes from the breakpoint.

b. Because there is only 17R from humans, all the human genes expressed must be on 17R.

mouse — 24 m.u. — *c* — 6 m.u. — *b* — 40 m.u. — *a* — 30 m.u. — end

c. The dye could be used to correlate band presence with gene presence.

Chapter 9

1. Klinefelter syndrome: XXY male; Down syndrome: trisomy 21; Turner syndrome: XO female

3. a. 3, 3, 3, 3, 3 3, 3 3, 0, 0

b. 7, 7, 7, 7, 8, 8, 6, 6

5. b.

8. The original parents must have had the following chromosome constitution:

G. hirsutum	26 large, 26 small
G. thurberi	26 small
G. herbaceum	26 large

G. hirsutum is a polyploid derivative of a cross between the two Old World species. This could easily be checked by looking at the chromosomes.

9. a.

$\frac{1}{6}\,FF$ → $\frac{1}{6}\,GG = \frac{1}{36}\,FF\,GG$; $\frac{4}{6}\,Gg = \frac{4}{36}\,FF\,Gg$; $\frac{1}{6}\,gg = \frac{1}{36}\,FF\,gg$

$\frac{4}{6}\,Ff$ → $\frac{1}{6}\,GG = \frac{4}{36}\,Ff\,GG$; $\frac{4}{6}\,Gg = \frac{16}{36}\,Ff\,Gg$; $\frac{1}{6}\,gg = \frac{4}{36}\,Ff\,gg$

$\frac{1}{6}\,ff$ → $\frac{1}{6}\,GG = \frac{1}{36}\,ff\,GG$; $\frac{4}{6}\,Gg = \frac{4}{36}\,ff\,Gg$; $\frac{1}{6}\,gg = \frac{1}{36}\,ff\,gg$

b. $p(FFFfGGgg)$ $\frac{1}{9}$

$p(ffffgggg)$ $\frac{1}{1296}$

10. e. Only achondroplasia and Marfan syndrome are gene disorders rather than disorders characterized by an abnormal chromosome number.

12. a. If most individuals were female, this suggests that the normal allele has been lost (by nondisjunction or deletion) or is nonfunctional (X-inactivation) in the color-blind eye.

b. If most of the individuals were male, this suggests that the male might have two or more cell lines (he is a mosaic, $X^{normal}/Y, X^{cb}/Y$) or that he has two X chromosomes (he has Klinefelter syndrome) and that the same processes as in females could be occurring.

14. One possibility is that the mean age of mothers at birth dropped significantly. Because the older mother is at higher risk for nondisjunction, this would result in the observation. Hospital records could be used to check the age of mothers at birth between 1952 and 1972, as compared with a 20-year period prior to 1952. Another possibility is an increase of amniocentesis among older mothers followed by induced abortion of trisomy 21 fetuses. Because pregnant women 35 and older routinely undergo amniocentesis, the rate for this population may have fallen while the rate for the younger population remained unchanged. This also could be checked through hospital records.

15. a. Loss of one X in the developing fetus after the two-celled stage.

b. Nondisjunction leading to Klinefelter syndrome (XXY) followed by a nondisjunctive event in one cell for the Y chromosome after the two-celled stage, leading to XX and XXYY.

c. Nondisjunction for X at the one-celled stage.

d. Either fused XX and XY zygotes or fertilization of an egg and polar body by one sperm bearing an X and another bearing a Y, followed by fusion.

e. Nondisjunction of X at the two-celled stage or later.

17. a. The extra chromosome must be from the mother. Because the chromosomes are identical, nondisjunction had to have occurred at MII.

b. The extra chromosome must be from the mother. Because the chromosomes are not identical, nondisjunction had to have occurred at MI.

c. The mother correctly contributed one chromosome, but the father did not contribute any chromosome 4. Therefore, nondisjunction occurred in the male during either meiotic division.

d. One cell line lacks a maternal contribution while the other has a double maternal contribution. Because the two lines are complementary, the best explanation is that nondisjunction occurred in the developing embryo during mitosis.

e. Each cell line is normal, indicating that nondisjunction did not occur. The best explanation is that the second polar body was fertilized and was subsequently fused with the developing embryo.

18. a. normal : potato = 3 : 1

b. normal : potato = 1 : 1

20. Radiation could have caused point mutations or induced recombination, but nondisjunction is the more likely explanation.

21. P $a + c + e \times + b + d +$

Selection for $+ + + + +$

Because this rare colony gave rise to both parental types among asexual (haploid) spores, the best explanation is that the rare colony initally contained both marked chromosomes due to nondisjunction. That is, it was disomic. Subsequent mitotic nondisjunction yielded the two parental types, possibly because the disomic was unstable.

23. The two chromosomes are

b_1 b^+ b^+ b_2

If one of the centromeres becomes functionally duplex before meiosis I, the homologous chromosomes will separate randomly. This will result in one daughter cell having one chromosome with one chromatid, and the second daughter cell will have one chromosome with one chromatid and a second chromosome with two chromatids. Meiosis II will lead to a nullisomic (white) and a monosomic (buff) from the first daughter cell and a disomic (black) and a monosomic (buff) from the second daughter cell.

If both centromeres divide prematurely, each daughter cell will get two chromosomes, each with one chromatid. All the ascospores will be black.

If nondisjunction occurred at meiosis I, the result will be one-half black (disomic) and one-half white (nullisomic) spores. Nondisjunction at meiosis II for one of the cells will result in two buff spores (normal meiosis II) and one white (nullisomic) and one buff (two copies of the same gene).

Normal meiosis will yield all buff spores because no crossing-over occurs between the two genes.

Chapter 10

2. $\overline{\text{M Z}}$ X W C

W C N A L

A L B R U

B R $\underline{\text{U M Z}}$

The regions with the bars above or below are identical in sequence. The only possible interpretation is that the donated material was circular, with the F factors inserted at different points in different strains.

3. Strains 2, 3, and 7 are F$^-$; strains 1 and 8 are F$^+$; and strains 4, 5, and 6 are Hfr.

6. a. *arg bio leu*

b. *arg–bio:* 12.76 m.u.
bio–leu: 2.12 m.u.

7. $ade\text{–}Z_2\text{–}Z_1$

8. *pro*–4–5–7–1–6–3–2–8–*ade*

12. The high rate of integration and the preference for the same site originally occupied by the sex factor suggest that the F′ contains some homology with the original site. The source of homology could be a fragment of the sex factor or it could be a chromosomal fragment.

15. a. The two genes are located close together and are cotransformed at a rate of 0.17 percent.

b. Here, when the two genes must be contained on separate pieces of DNA, the rate of cotransformation is much lower, confirming the conclusion in part **a.**

16. a. B

b. A–D–C

17. Interference = −1.5. By definition, the interference is negative.

18. a. *m–r:* 12.8 m.u.
r–tu: 20.8 m.u.
m–tu: 27.2 m.u.

b. *m–r–tu*

c. −.2. A negative value for *I* indicates that the occurrence of one crossover makes a second crossover more likely to occur than it would have been without that first crossover. That is, more double crossovers occur than are expected.

20. The instability of *gal*$^+$ transductants comes from the fact that they are partial diploids and have a tendency to lose a *gal* gene. If the *gal*$^+$ gene is lost, a *gal*$^-$ clone will develop. The stable transductants are not partial diploids and, therefore, do not have a tendency to segregate *gal.* They are not partial diploids because there was an exchange of *gal* alleles rather than an insertion of the second *gal* allele, as with partial diploids. This occurred when λ looped out between the two *gal* alleles and one *gal*$^+$ allele was reconstructed.

23. a. If *trp1* and *trp2* are alleles, then a cross between strains A and B will never result in trp$^+$, unless recombination occurs within the *trp* gene. If they are not allelic, there will be trp$^+$ colonies. Check these colonies for the possibility of recombination.

b. Infect strain C with the Z phage and use the progeny to infect strain B cells (which are immune because they are lysogenic for Z). The strain B cells should be plated onto minimal medium and minimal medium plus cysteine. If the order is *cys–trp2–trp1,* two crossovers will result in *cys*$^+$ *trp2*$^+$ *trp1*$^+$, and the number of colonies on the two media should be approximately the same. If the order is *cys–trp1–trp2,* four crossovers are required for *cys*$^+$ *trp1*$^+$ *trp2*$^+$. Therefore, the number of colonies on medium containing cysteine would be greater than the number of colonies on minimal medium.

24. The order is thus *d–a–e–c.* Notice that *b* is never cotransduced and is therefore distant from this group of genes.

26. a. The colonies are all *cys*$^+$ and either + or − for the other two genes.

b. (a) *cys*$^+$ *leu*$^+$ *thr*$^{+/-}$
(2) *cys*$^+$ *leu*$^{+/-}$ *thr*$^+$
(3) *cys*$^+$ *leu*$^+$ *thr*$^+$

c. Because none grew on minimal medium, no colony was *leu*$^+$ *thr*$^+$. Therefore, medium (1) had *cys*$^+$ *leu*$^+$ *thr*$^-$, and medium (2) had *cys*$^+$ *leu*$^-$ *thr*$^+$. The remaining cultures were *cys*$^+$ *leu*$^-$ *thr*$^-$, and this genotype occurred in 100% − 56% − 5% = 39% of the colonies.

d. *leu* *cys* *thr*

Chapter 11

1. 35 percent

2. G = C = 24 percent, A = T = 26 percent

4. The results suggest that the DNA is replicated in short segments that are subsequently joined by enzymatic action (DNA ligase). Because DNA replication is bidirectional, because there are multiple points along the DNA where replication is initiated, and because DNA polymerase work only in a $5' \rightarrow 3'$ direction, one strand of the DNA is always in the wrong orientation for the enzyme. This requires synthesis in fragments.

6. a. A very plausible model is of a triple helix, which would look like a braid, with each strand interacting by hydrogen bonding to the other two.

b. Replication would have to be terti-conservative. The three strands would separate, and each strand would dictate the synthesis of the other two strands.

c. The reductional division would have to result in three daughter cells, and the equational would have to result in two daughter cells, in either order. Thus, meiosis would yield six gametes.

8. Remember that there are two hydrogen bonds between A and T, while there are three hydrogen bonds between G and C. Denaturation involves the breaking of these bonds, which requires energy. The more bonds that need to be broken, the more energy that must be supplied. Thus the temperature at which a given DNA molecule denatures is a function of its base composition. The higher the temperature of denaturation, the higher the percentage of GC pairs.

10. a. The first shoulder appears before strand interaction takes place, suggesting that the complementary regions are in the same molecule. This is called a palindrome:

ATGCATGGCCA————TGGCCATGCAT
TACGTACCGGT————ACCGGTACGTA

When the strands separate, each strand can base-pair with itself:

ATGCATGGCCA—⌒
TACGTACCGGT—⌣

b. The second shoulder represents sequences that are present in many copies in the genome. Because they are at a higher concentration than unique sequences, they have a higher probability of encountering each other during a given time period.

13. The data suggest that each chromosome is composed of one long, continuous molecule of DNA.

Chapter 12

1. The defective enzyme that results in albinism may not be able to detoxify a chemical component of Saint-John's wort that the wild-type enzyme can detoxify.

4. a. The main use is in detecting carrier parents and in diagnosing a disorder in the fetus.

b. Because the values for normal individuals and carriers overlap for galactosemia, there is ambiguity if a person has 25 to 30 units as a result. That person could be either a carrier or normal.

c. These genes are phenotypically dominant but are incompletely dominant at the molecular level. A minimal level of enzyme activity apparently is enough to ensure normal function and phenotype.

5. One less likely possibility is a germ-line mutation. More likely is that each parent was blocked at a different point in a metabolic pathway. If one were $AA\,bb$ and the other were $aa\,BB$, then the child would be $Aa\,Bb$ and would have sufficient levels of both resulting enzymes to produce pigment.

6. Assuming homozygosity for the normal gene, the mating is $AA\,bb \times aa\,BB$. The children would be normal, $Aa\,Bb$.

9.

$$\text{no pigment (white)} \xrightarrow{\;bw^+ \downarrow\;} \text{scarlet}$$

$$\text{no pigment (white)} \xrightarrow[\;\uparrow st^+\;]{} \text{brown}$$

Scarlet plus brown results in red.

11. a. 60 percent of the meioses will not have a crossover.

b. 20 m.u.

c. 10 percent

d.

$$\rightarrow B \rightarrow A \rightarrow \text{valine}$$
$$\quad\;\uparrow\;\;\;\uparrow$$
$$\quad\;1\;\;\;2$$

13. a. White

b. Blue

c. Purple

d. 9:3:4

16. a. The mutant does not complement any other mutant. The best interpretation is that it is a deletion.

b. 1, 5, 8, and 9; 2, 3, 4, and 12; 6, 7, 10, 11, and 13

c. Mutants 1 and 2 are in different cistrons, so the cross can be written $1 + \times + 2$.

$\frac{1}{4}$ 1 + eye$^-$
$\frac{1}{4}$ 1 2 eye$^-$
$\frac{1}{4}$ + 2 eye$^-$
$\frac{1}{4}$ + + eye$^+$

or three eye$^-$: one eye$^+$

Mutants 2×6 also complement each other. If independent assortment existed, a 3:1 ratio would be observed. Because the ratio is 113:5, there is no independent assortment and the cistrons are linked. Only one of the two recombinant classes can be distinguished: 5 eye$^+$. Because the recombinants should be of equal frequency, the total number of recombinants is 10 out of 118, which leads to 100%(10/118) = 8.47 m.u. distance between the cistrons.

Mutant 14 includes the same cistron (no complementation) as mutant 1.

d. There are three loci, plus mutant 14. Because two of the groups are independently assorting, either mutant 14 is a very large deletion spanning the three loci (and they are therefore on the same chromosome), or it is a separate fourth locus that in some fashion controls the expression of the other three loci, or it is a double mutant with a point mutation within one complementation group (1, 5, 8, 9) and a deletion spanning the two linked complementation groups.

e. Two groups are linked, with 8.5 m.u. between them (2, 3, 4, 12 and 6, 7, 10, 11, 13), and the third group is either on a separate chromosome or more than 50 m.u. from the two other groups.

18. a. $c-e-d-a-b$

b. The suggestion is that deletion 4 spans both cistrons. This does not affect the conclusions from part **a.**

20. a.

```
  2    12    6
+----+-----+----+
1    3     2    4
```

b. no

21.

```
         |<-- 0.072 m.u. -->|
his      a         c        b       nic
--+------+---------+--------+--------+--
         0.031 m.u. 0.023 m.u.
```

24. Pleiotropic effects result from interaction between gene products. As an example, sickle cell anemia results from a point mutation in the globin portion of hemoglobin. The phenotype can include joint damage, brain damage, kidney damage, etc. These effects can be labeled pleiotropic, but because the molecular, cellular, and organismal bases for the damage are understood, the term syndrome is usually used. As another example, a gene product (an enzyme) may act in the branch point of a metabolic pathway. The lack of this gene product would then result in two or more deficiencies, and each endpoint effect would be labeled a pleiotropic effect. A third example would be regulatory proteins (Chapters 17 and 23), which can alter many functions simultaneously.

27. a. Cross 1 × 2: All purple F_1 indicates that two genes are involved. Call the defect in 1 aa and the defect in 2 bb. The cross is

P $aa\,BB \times AA\,bb$

F_1 $Aa\,Bb$

If the two genes assort independently, a 9 : 7 ratio of purple : white would be seen. A 1 : 1 ratio indicates tight linkage. The cross above now needs to be rewritten

P $aB/aB \times Ab/Ab$

F_1 aB/Ab

F_2 1 aB/aB (white) : 2 aB/Ab (purple) : 1 Ab/Ab (white)

Cross 1 × 3: Again, an F_1 of all purple indicates two genes. The 9 : 7 F_2 ratio indicates independent assortment. Therefore, let the cross 3 defect be symbolized by d:

P $aB/aB\,DD \times AB/AB\,dd$

F_1 $aB/AB\,Dd \times aB/AB\,Dd$

F_2 9 $A-BB\,D-$ (purple)

3 $aa\,BB\,D-$ (white)

3 $A-BB\,dd$ (white)

1 $aa\,BB\,dd$ (white)

Cross 1 × 4: All white F_1 and F_2 indicates that the two mutations are in the same gene. The cross is

P $aB/aB\,DD \times aB/aB\,DD$

F_1 same as parents

F_2 same as parents

b. Cross 2 × 3:

P $Ab/Ab\,DD \times AB/AB\,dd$

F_1 $Ab/AB\,Dd$ (purple)

F_2 9 $A-B-D-$ (purple)

3 $Ab/Ab\,D-$ (white)

3 $A-B-dd$ (white)

1 $Ab/Ab\,dd$ (white)

Cross 2 × 4: same as cross 1 × 2

29. a. There are three cistrons:

Cistron 1: mutants 1, 3, and 4

Cistron 2: mutants 2 and 5

Cistron 3: mutant 6

b. The order is A/a–6–(1, 3, 4)–(2, 5)–B/b.

Chapter 13

1. Because RNA can hybridize to both strands, the RNA must be transcribed from both strands. This does not mean, however, that both strands are used as a template within each gene. The expectation is that only one strand is used within a gene but that different genes are transcribed in different directions along the DNA. The most direct test would be to purify a specific RNA, coding for a specific protein, and then hybridize it to the λ genome. Only one strand should hybridize to the purified RNA.

3. A single nucleotide change should result in three adjacent amino acid changes in a protein. One and two adjacent amino acid changes would be expected to be much rarer than the three changes. This is directly opposite of what is observed in proteins.

4. It suggests very little evolutionary change between *E. coli* and humans with regard to the translational apparatus. The code is universal, the ribosomes are interchangeable, the tRNAs are interchangeable, and the enzymes involved are interchangeable.

6. a. Using Figure 13-18, there are eight cases in which knowing the first two nucleotides does not tell you the specific amino acid.

b. If you knew the amino acid, you would not know the first two nucleotides in the cases of Arg, Ser, and Leu (Figure 13-18).

8. a. $\frac{1}{8}$ **c.** $\frac{1}{8}$

b. $\frac{1}{4}$ **d.** $\frac{1}{8}$

9. a. 1 U:5 C: 1 Phe:25 Pro:5 Ser:5 Leu

b. 4 stop: 80 Phe:40 Leu:24 Ile:24 Ser:20 Tyr:6 Pro: 6 Thr:5 Asn:5 His:1 Lys:1 Gln.

c. All amino acids are found in the proportions seen in Figure 13-18.

11. Mutant 1: A simple substitution of Arg for Ser exists, suggesting a nucleotide change. Two codons for Arg are AGA and AGG, and one codon for Ser is AGU. The final U for Ser could have been replaced by either an A or a G.
Mutant 2: The Trp codon (UGG) changed to a stop codon (UGA or UAG).
Mutant 3: Two frame-shift mutations occurred:

5′ GCN CCN (−U)GGA GUG AAA AA(+U or C) UGU/C CAU/C 3′

Mutant 4: An inversion occurred after Trp and before Cys. The DNA original sequence was

3′ CGN GGN ACC TCA CTT TTT ACA/G GTA/G 5′

Therefore, the complementary RNA sequence was

5′ GCN CCN UGG AGU GAA AAA UGU/C CAU/C 3′

The DNA inverted sequence became

3′ CGN GGN ACC AAA AAG TGA ACA/G GTA/G 5′
∧ ∧

Therefore, the complementary RNA sequence was

5′ GCN CCN UGG UUU UUC ACU UGU/C CAU/C 3′
∧ ∧

13. e. With an insertion, the reading frame is disrupted. This will result in a drastically altered protein from the insertion to the end of the protein (which may be much shorter or longer than wild-type because of altered stop signals).

15.

3′ CGT ACC ACT GCA 5′

5′ GCA TGG TGA CGT 3′

5′ GCA UGG UGA CGU 3′

3′ CGU ACC ACU GCA 5′

NH_3-Ala-Trp stop nothing

16. f, d, j, e, c, i, b, h, a, g

18. Cells in long-established culture lines usually are not fully diploid. For reasons that are currently unknown, adaptation to culture frequently results in both karyotypic and gene-dosage changes. This can result in hemizygosity for some genes, which allows for the expression of previously hidden recessive alleles.

Chapter 14

1. GTTAAC occurs, on average, every 4^6 bases. GGCC occurs, on average, every 4^4 bases.

3.

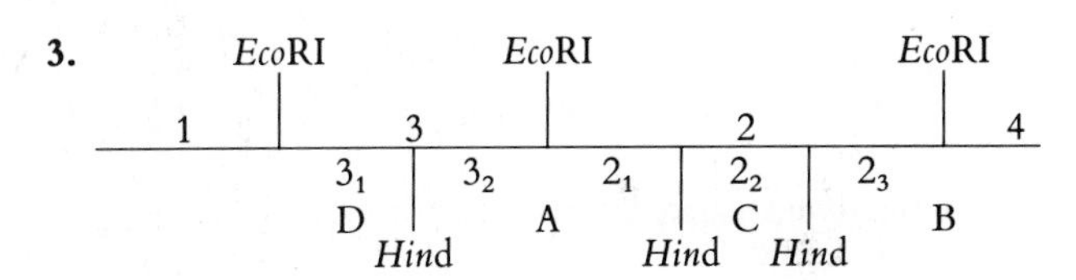

5. Reading from the bottom the sequence is

*Eco*RI-*Hae*-*Hind*-*Hae*-*Hind*-*Hind*-*Hae*-*Hae*-*Hae*-*Hind*-*Hae*-*Hae*-*Hind*

6. Reading from the bottom up,

left column: GGTACAACTATATATCAATTATAAAC

right column: GGATCTATTCTTATGATTATATAG

8. This problem assumes a random distribution of nucleotides.

*Alu*I = every 256 nucleotide pairs

*Eco*RI = every 4096 nucleotide pairs

*Acy*I = every 1024 nucleotide pairs

10. a. 2.5 *Hind* 3.0 *Sma* 2.0

b. 2.5 *Hind* 3.0 *Sma* 2.0
1.5 *Eco*RI 1.5

13. a. and **b.**

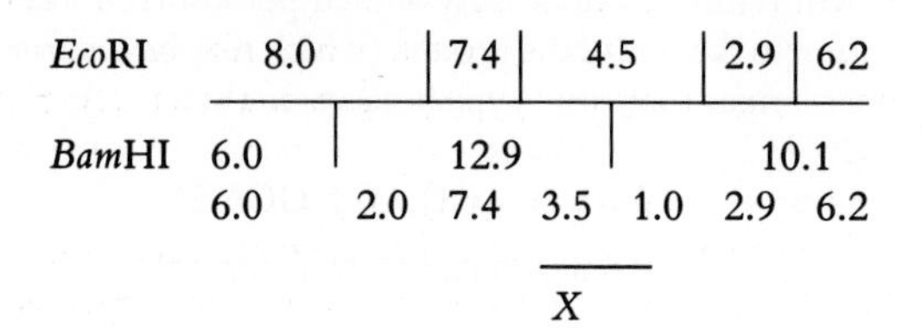

Chapter 15

1. Plant 1 shows a typical heterozygous testcross result when crossed with the wild-type kanamycin-sensitive plant. This indicates that a single copy of the gene integrated. The cross is

P $K^rK^s \times K^sK^s$

F_1 1 K^rK^s resistant

1 K^sK^s sensitive

From this result, it can be concluded that resistance is dominant to sensitivity.

Plant 2 results in a 3 : 1 ratio in a testcross. There are two ways this ratio can be achieved in a testcross: if the plant is actually a tetraploid or if there are two separate integrations that assort independently. Assuming tetraploidy, the cross is

P $K^rK^rK^rK^s \times K^sK^s$

F_1 3 K^rK^s resistant

1 K^sK^s sensitive

Assuming two separate integrations, the cross is

P $K^{r1}K^{s1}\ K^{r2}K^{s2} \times K^{s1}K^{s1}\ K^{s2}K^{s2}$

F_1 $K^{r1}K^{s1}\ K^{r2}K^{s2}$ 25% resistant

$K^{s1}K^{s1}\ K^{r2}K^{s2}$ 25% resistant

$K^{r1}K^{s1}\ K^{s2}K^{s2}$ 25% resistant

$K^{s1}K^{s1}\ K^{s2}K^{s2}$ 25% sensitive

The gene of interest should assort with kanamycin resistance.

3. The first task is to get the bacterial vector containing the gene of interest into the cell type which you wish to test. This can be accomplished by one of several means such as "shooting" the vector into the cell or producing transformed plants.

Assuming that you also have a resistance gene in the vector, you can select for that gene. Once the tissue is expressing the resistance gene, indicating possible integration into the plant chromosome, you can then assay for the protein of interest or for mRNA production.

5. The missing *Eco*RI restriction site can be assayed for all members of the family. DNA isolated from blood (white blood cells) of each individual would be restricted with *Eco*RI, electrophoresed, transferred by Southern blot, and then probed.

If the restriction site is missing within an individual, that person would then know that he carries the allele for cystic fibrosis. Those who have the restriction site could be relieved of worry about carrying the allele.

The two individuals carry two different mutations within the same gene that lead to cystic fibrosis. Because the presence of two copies of a cystic fibrosis gene, whether or not the defect is exactly the same, leads to expression of the disorder, 25 percent of the children from this mating would have cystic fibrosis.

7. After electrophoresis, Southern blot the gel and probe with radioactive copies of the cloned gel.

9. a. Let B = bent tail, b = normal tail, r_1 = 3.8 kb, and r_2 = 1.7 kb. The cross is $Bb\ r_1r_2 \times bb\ r_1r_1$. Among the bent progeny are

40% $Bb\ r_1r_1$

60% $Bb\ r_1r_2$

Eliminating the contribution of the wild type parent, the progeny are

40% $B\ r_1$

60% $B\ r_2$

If independent assortment exists, a 1 : 1 ratio of $r_1 : r_2$ will be seen. But r_1 and r_2 are clearly not assorting independently. Therefore, the two genes are linked with 40 m.u. between them.

b. The original cross is $B\ r_2 / b\ r_1 \times b\ r_1 / b\ r_1$. The wild type progeny for tail conformation are

60% $b\ r_1 / b\ r_1$

40% $b\ r_2 / b\ r_1$

Chapter 16

3. a. One locus on each of the homologous chromosomes would be indicated.

b. Both ends of all chromatids would be indicated.

c. The chromosomal satellite region on all chromosomes containing NORs would be indicated.

d. Many small regions on many chromosomes would be indicated.

e. Many larger regions on many chromosomes would be indicated.

f. Centromeres, telomeres and NORs would be indicated.

4. On average, each DNA fragment contains two SINE sequences, which are palindromic. Each fragment of DNA should be able to fold back on itself and undergo complementary pairing. Alternatively, the fragments may pair with each other.

5. **a.** The pattern of suspect 1 is compatible with him being among the group of individuals who could have committed the rape.

   b. Suspects 2 and 3 could not have committed the rape.

6. To work this problem, number bands in the two parents, then use those numbers to identify bands in their children. If the mother has bands 1–4, marked with an asterisk, and the father has bands 5–8, not marked, the children have the following sets of bands:

Child 1	Child 2	Child 3	Child 4
1*	6	1*	1*
6	2*	5	5
7	3*	4*	7
3*	8	8	4*

Note that four bands are present in each person but that a total of eight different bands exist in this family. Each child receives two bands from each parent. There are, therefore, two loci that are being detected with the probe. The loci are not linked. Four alleles are being detected for each locus.

Chapter 17

2. O^c mutants do not bind the repressor product of the *I* gene, and therefore, the *lac* operon associated with the O^c operator cannot be turned off. Because an operator controls only the genes on the same DNA strand, it is *cis* (on the same strand) and dominant (cannot be turned off).

4.

	β-Galactosidase		Permease	
Part	No lactose	Lactose	No lactose	Lactose
a	+	+	−	+
b	+	+	−	−
c	−	−	−	−
d	−	−	−	−
e	+	+	+	+
f	+	+	−	−
g	−	+	−	+

7. Nonpolar Z^- mutants cannot convert lactose to allolactose, and, thus, the operon is never induced.

9. An operon is turned off by the mediator in negative control, and the mediator must be removed for transcription to occur. An operon is turned on by the mediator in positive control, and the mediator must be added for transcription to occur.

10. The *lacY* gene produces a permease that transports lactose into the cell. A *lacY*$^-$ gene could not transport lactose into the cell, so β-galactosidase will not be induced.

12. The bacterial operon consists of a promotor region that extends approximately 35 bases upstream of the site where transcription is initiated. Within this region is the promoter. Inducers and repressors, both of which are *trans*-acting proteins that bind to the promoter region, regulate transcription of associated cistrons in *cis* only.

    The eukaryotic cistron has the same basic organization. However, the promoter region is somewhat larger. Also, enhancers up to several thousand nucleotides upstream or downstream can influence the rate of transcription. A major difference is that eukaryotes have not been demonstrated to have polycistronic messages.

Chapter 18

2. A nonsense mutation results in a stop codon. Often the truncation of the protein does not eliminate all enzyme activity or change the structure of a structural protein to the point where it is nonfunctional. A frameshift mutation usually results in many amino acid substitutions or, quite frequently, the introduction of a stop codon. When many amino acids in a protein are changed, the enzyme activity usually is eliminated, and any structural protein would be highly altered to the point of being nonfunctional.

5. **a.** Depurination results in the loss of adenine or guanine from the nucleotide. Because the apurinic site cannot specify a complementary base, replication is blocked. Under certain conditions, replication proceeds with a random insertion of a base opposite the apurinic site. In three-fourths of these insertions, a mutation will result.

   b. Deamination of cytosine yields uracil. If left unrepaired, the uracil is paired with adenine during replication, ultimately resulting in a transition mutation.

6. 5-bromouracil is an analog of thymine. It undergoes tautomeric shifts at a higher frequency than thymine and, therefore, is more likely to pair with G than is thymine during replication. At the next replication this will lead to a GC pair rather than the original AT pair.

   Ethylmethanesulfonate is an alkylating agent that produces O-6-ethylguanine. This will pair with thymine, which leads from a GC pair to an AT pair at the next replication.

8. **a.** Mismatch repair occurs when a mismatched nucleotide is inserted during replication. The new, incorrect base is removed and a proper base is inserted. The enzymes involved can distinguish between new and old strands because, in *E. coli,* the old strand is methylated.

   b. Recombination repair occurs when AP sites and UV photodimers block replication (there is a gap in the complementary strand). Recombination occurs with one strand from the sister DNA molecule, which is normal in both strands. This produces two DNA molecules, each with a lesion in one strand. Each lesion is then repaired normally.

   c. Proofreading by DNA polymerase will result in the removal of an incorrect base and the insertion of a correct base.

9. Leaky mutants are mutants with an altered protein product that retains a low level of function. Enzyme activity may, for instance, be reduced rather than abolished.

11. a. Because 5′ UAA 3′ does not contain G or C, a transition to a GC pair in the DNA cannot result in 5′ UAA 3′. 5′ UGA 3′ and 5′ UAG 3′ have the DNA antisense-strand sequence of 3′ ACT 5′ and 3′ ATC 5′. A transition to either of these stop codons occurs from the nonmutant 3′ ATT 5′. A DNA sequence of 3′ ATT 5′, results in an RNA sequence of UAA, itself a stop codon.

b. Yes; an example would be 3′ UGG 5′, which codes for *trp*, to 3′ UAG 5′.

c. No; in the three stop codons the only base that can be acted upon is G (in UAG, for instance). Replacing the G with an A would result in 3′ UAA 5′, a stop codon.

13. To understand these data, recall that half of the progeny should come from the wild-type parent.

a. A lack of revertants suggests either a deletion or an inversion within the gene.

b. Prototroph A: Because 100 percent of the progeny are prototrophic, a reversion at the original mutant site may have occurred.

Prototroph B: Half of the progeny are parental prototrophs, and the remaining prototrophs, 28 percent, are the result of the new mutation. Notice that 28 percent is approximately equal to the 22 percent auxotrophs. The suggestion is that an unlinked suppressor mutation occurred, yielding independent assortment with the *nic* mutant.

Prototroph C: There are 496 "revertant" prototrophs (the other 500 are parental prototrophs) and 4 auxotrophs. This suggests that a suppressor mutation occurred in a site very close ($100\%[4 \times 2]/1000 = 0.8$ m.u.) to the original mutation.

Chapter 19

1. 3, 4, 6

5. First notice that gene conversion is occurring. In the first cross, a_1 converts (1:3). In the second cross, a_3 converts. In the third cross, a_3 converts. Polarity is obviously involved. The results can be explained by the following map, where hybrid DNA enters only from the left.

a_3 ——— a_1 ——— a_2

7. The ratios for a_1 and a_2 are both 3:1. There is no evidence of polarity, which indicates that gene conversion as part of recombination is occurring. The best explanation is that two separate excision-repair events occurred and, in both cases, the repair retained the mutant rather than the wild type.

8. a and **b.** A heteroduplex that contains an unequal number of bases in the two strands has a larger distortion than a simple mismatch. Therefore, the former would be more likely to be repaired. For such a case, both heteroduplex molecules are repaired (leading to 6:2 and 2:6) more often than one (leading to 5:3 or 3:5) or none (leading to 3:1:1:3). The preference in direction (i.e., adding a base rather than subtracting) is analogous to TT dimer repair. In TT dimer repair, the unpaired, bulged nucleotides are treated as correct and the strand with the TT dimer is excised.

A mismatch more often than not escapes repair, leading to a 3:1:1:3 ascus.

Transition mutations would not cause as large a distortion of the helix, and each strand of the heteroduplex should have an equal chance of repair. This would lead to 4:4 (two repairs each in the opposite direction), 5:3 (one repair), 3:1:1:3 (no repairs or two repairs in opposite directions), and, less frequently, 6:2 (two repairs in the same direction).

c. Because excision repair excises the strand opposite the larger buckle (i.e., opposite the frameshift mutation), the *cis* transition mutation will also be retained. The nearby genes are converted because of the length of the excision repair.

9. a. 6:2 = 31.25 percent

b. 2:6 = 5 percent

c. 3:1:1:3 = 11.25 percent

d. 5:3 = 37.5 percent

e. 3:5 = 15 percent

Chapter 20

2. Polar mutations affect the transcription or translation of the part of the gene or operon on only one side of the mutant site, usually described as downstream. Examples are nonsense mutations, frameshift mutations, and IS-induced mutations.

4. R plasmids are the main carriers of drug resistance. They acquire these genes by transposition of drug-resistance genes located between IR (inverted repeat) sequences. Once in a plasmid, the transposon carrying drug resistance can be transferred upon conjugation if it stays in the R plasmid or it can insert into the host chromosome.

6. *P* elements are transposons (genes flanked by inverted repeats, allowing for great mobility). Because they are transposons, they can insert into chromosomes. By inserting specific DNA between the inverted repeats of the *P* elements and injecting the altered transposons into cells, a high frequency of gene transfer will occur.

8. The best explanation is that the mutation is due to an insertion of a transposable element.

10. a. The expression of the tumor is blocked in plant B. This suggests that either plant B can suppress the functioning of the plasmid that causes the tumor, or that plant A provides something to the tissue with the tumor-causing plasmid that plant B does not provide.

b. Tissue carrying the plasmid, when grafted to plant B, appears normal, but the graft produces tumor cells in synthetic medium. This indicates that the plasmid sequences are present and capable of functioning in the right environment. However, the production of normal type A plants from seeds from the graft suggests a permanent loss of the plasmid during meiosis.

Chapter 21

1. Most organelle-encoded polypeptides unite with nucleus-encoded polypeptides to produce active proteins, and these active proteins function in the organelle.

2. Reciprocal crosses reveal cytoplasmic inheritance. Cytoplasmic inheritance also can be demonstrated by doing a series of backcrosses, using hybrid females in each case, so that the nuclear genes of one strain are functioning in cytoplasm from the second strain.

4. a. Stop-start growth

b. Normal growth

6. a. The ant^R gene may be mitochondrial.

b. Some of the petites must have been neutral petites, in which all mitochondrial DNA, including the ant^R gene, was lost. Other petites were suppressive, in which the ant^R gene was retained.

7. The genetic determinants of R and S are cytoplasmic and are showing paternal inheritance.

9. If no crossing-over occurs in the diploid fusion cell, asci will be of two types:

4 oli^R cap^R

4 oli^S cap^S

If crossing-over occurs, the recombinants will be of two types:

4 oli^R cap^S

4 oli^S cap^R

11. If the mutation is in the chloroplast, reciprocal crosses will give different results, while if it is in the nucleus and dominant, reciprocal crosses will give the same results.

13. After the initial hybridization, a series of backcrosses using pollen from B will result in the desired combination of cytoplasm A and nucleus B. With each cross the female contributes all of the cytoplasm and one-half the nuclear contents, while the male contributes one-half the nuclear contents.

15. Let male sterility be symbolized by MS.

a. A line that was homozygous for *Rf* and contained the male-sterility factor would result in fertile males. When this line was crossed, using females, with a line not carrying the restorer gene for two generations, the male-sterility trait would reappear.

b. The F_1 would carry the male-sterility factor and would be heterozygous for the *Rf* gene. Therefore, it would be fertile.

c. The cross is *Rfrf* MS × *rfrf* (no cytoplasmic transmission). The progeny would be $\frac{1}{2}$ *Rfrf* MS (fertile) and $\frac{1}{2}$ *rfrf* MS (sterile).

d.

i.	P	*Rf-1 rf-1 Rf-2 rf-2* × *rf-1 rf-1 rf-2 rf-2* MS	
	F_1	$\frac{1}{4}$ *Rf-1 rf-1 Rf-2 rf-2* MS	fertile
		$\frac{1}{4}$ *Rf-1 rf-1 rf-2 rf-2* MS	fertile
		$\frac{1}{4}$ *rf-1 rf-1 Rf-2 rf-2* MS	fertile
		$\frac{1}{4}$ *rf-1 rf-1 rf-2 rf-2* MS	male sterile
ii.	P	*Rf-1 Rf-1 rf-2 rf-2* × *rf-1 rf-1 rf-2 rf-2* MS	
	F_1	100% *Rf-1 rf-1 rf-2 rf-2* MS	fertile
iii.	P	*Rf-1 rf-1 rf-2 rf-2* × *rf-1 rf-1 rf-2 rf-2* MS	
	F_1	$\frac{1}{2}$ *Rf-1 rf-1 rf-2 rf-2* MS	fertile
		$\frac{1}{2}$ *rf-1 rf-1 rf-2 rf-2* MS	male sterile
iv.	P	*Rf-1 rf-1 Rf-2 Rf-2* × *rf-1 rf-1 rf-2 rf-2* MS	
	F_1	$\frac{1}{2}$ *Rf-1 rf-1 Rf-2 rf-2* MS	fertile
		$\frac{1}{2}$ *rf-1 rf-1 Rf-2 rf-2* MS	fertile

18. The red phenotype in the heterokaryon indicates that the red phenotype is caused by a cytoplasmic organelle allele.

21. Let poky be symbolized by *(c)*. Let the nuclear suppressor of poky be symbolized by *n*. To do these problems, you cannot simply do the crosses in sequence. For instance, the parental genotypes in cross **a** must be written taking cross **c** into consideration.

	Cross	Progeny
a.	*(+)* + × *(c)* +	all *(+)* +
b.	*(+) n* × *(c)* +	$\frac{1}{2}$ *(+) c* : $\frac{1}{2}$ *(+)* +
c.	*(c)* + × *(+)* +	all *(c)* +
d.	*(c)* + × *(+) n*	$\frac{1}{2}$ *(c)* + (= D) : $\frac{1}{2}$ *(c) n* (= E)
e.	*(c) n* × *(+) n*	all *(c) n*
f.	*(c) n* × *(+)* +	$\frac{1}{2}$ *(c) n* : $\frac{1}{2}$ *(c)* +

23. a. The first tetrad shows a 2 : 2 pattern, indicating a nuclear gene, while the second tetrad shows a 0 : 4 pattern, indicating maternal inheritance and a cytoplasmic factor (mitochondrial).

b. The nuclear gene should always show a 1 : 1 segregation pattern. The mitochondrial gene could produce an ascus that was all $cyt2^+$.

c. Both produce proteins involved with the cytochromes. Either the products of the two genes affect different steps in mitochondrial function or they affect the same step if the two proteins interact to form one enzyme.

25. The entire sequence is a circle:

6. a. $p = 0.5249$, $q = 0.4751$

If the population is in equilibrium, the phenotypes should be distributed as follows:

$$L^M L^M = 408$$

$$L^M L^N = 739$$

$$L^N L^N = 334$$

The population is in equilibrium.

b. If mating is random with respect to blood type, then the following frequency of matings should occur:

$$L^M L^M \times L^M L^M = (p^2)(p^2)(741) = 56.25$$

$$L^M L^M \times L^M L^N = (2p^2)(2pq)(741) = 203.6$$

$$L^M L^M \times L^N L^N = (2p^2)(q^2)(741) = 92$$

$$L^M L^N \times L^M L^N = (2pq)(2pq)(741) = 184.28$$

$$L^M L^N \times L^N L^N = (2)(2pq)(q^2)(741) = 166.8$$

$$L^N L^N \times L^N L^N = (q^2)(q^2)(741) = 37.75$$

The mating is random with respect to blood type.

8. a. and **b.**

Population	p	q	Equilibrium?
1	1.0	0.0	yes
2	0.5	0.5	no
3	0.0	1.0	yes
4	0.625	0.375	no
5	0.3775	0.625	no
6	0.5	0.5	yes
7	0.5	0.5	no
8	0.2	0.8	yes
9	0.8	0.2	yes
10	0.993	0.007	yes

c. 0.102

d. $p = 0.56$, $q = 0.44$

10. The frequency of a phenotype in a population is a function of the frequency of alleles that lead to that phenotype in the population. To determine dominance and recessiveness, do standard Mendelian crosses.

12. Dominance is usually wild type because most detectable mutations in enzymes result in lowered or eliminated enzyme function. To be dominant, the heterozygote has approximately the same phenotype as the homozygote dominant. This will be true only when the wild-type allele produces a product and the mutant allele does not.

The chromosomal rearrangements are dominant mutations because so many genes are affected that it is highly unlikely that all of their alleles will be dominant and "cover" for them.

15. Albinos may have been considered lucky and encouraged to breed at very high levels in comparison to nonalbinos. They may also have been encouraged to mate with each other. Alternatively, in the tribes with a very low frequency, albinos may have been considered very unlucky and destroyed at birth or prevented from marriage.

18. a. 0.528

b. 0.75

19. 10

21. 6.5

Index

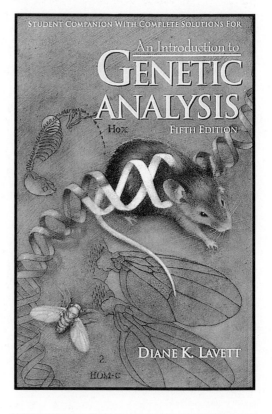